Handbook of Powder Science and Technology

edited by

M. E. Fayed
L. Otten

VAN NOSTRAND REINHOLD COMPANY
NEW YORK CINCINNATI TORONTO LONDON MELBOURNE

Library of Congress Catalog Card Number: 83-6828
ISBM: 0-442-22610-7

Manufactured in the United States of America

Published by Van Nostrand Reinhold Company Inc.
135 West 50th Street
New York, New York 10020

Van Nostrand Reinhold Company Limited
Molly Millars Lane
Wokingham, Berkshire RG11 2PY, England

Van Nostrand Reinhold
480 Latrobe Street
Melbourne, Victoria 3000, Australia

Macmillan of Canada
Division of Gage Publishing Limited
164 Commander Boulevard
Agincourt, Ontario M1S 3C7, Canada

15 14 13 12 11 10 9 8 7 6 5 4 3 2 1

Library of Congress Cataloging in Publication Data

Main entry under title:

Handbook of powder science and technology.

Includes index.
1. Particles. 2. Powders. I. Fayed, M. E.
(Muhammad E.) II. Otten, Lambert.
TP156.P3H35 1983 620′.43 83-6828
ISBN 0-442-22610-7

TO

My Wife Salwa	My Wife Carolyn
and my children	and my children
Marwan and Nora Fayed	Mark and Susan Otten

All of whom have given
up far too much of
their rightful time
that we have used in
preparation of this book.

CONTRIBUTORS

L. G. Austin, Professor, Department of Mineral Engineering, The Pennsylvania State University, University Park, PA. (Ch. 13).

John Bridgwater, Professor, Department of Chemical Engineering, University of Birmingham, Birmingham, England (Ch. 5, Pt. 2).

C. E. Capes, Chairman, Chemical Engineering Division, National Research Council of Canada, Ottawa, Canada (Ch. 7.1, 7.5, 7.6, 7.8).

J. T. Carstensen, Professor of Pharmacy, University of Wisconsin, Madison, Wisconsin, (Ch. 7.3).

Douglas W. Cooper, Associate Professor, Department of Environmental Science and Physiology, School of Public Health, Harvard University, Boston, Mass., (Ch. 19).

Reg. Davies, Senior Consultant, Engineering, E.I. DuPont de Numours & Co., Wilmington, Delaware (Ch. 2).

F. A. L. Dullien, Professor, Department of Chemical Engineering, University of Waterloo, Waterloo, Ontario (Ch. 4).

Norman Epstein, Professor, Department of Chemical Engineering, The University of British Columbia, Vancouver, B.C. (Ch. 11).

A. Fouda, Chemical Engineering Division, National Research Council, Ottawa, Ontario (Ch. 7.1, 7.5, 7.6, 7.8).

A. Goldman, Los Alamos Scientific Laboratory, University of California, Los Alamos, New Mexico (Ch. 1).

John R. Grace, Chairman, Department of Chemical Engineering, The University of British Columbia, Vancouver, B.C. (Ch. 11).

Jacob Katz, Precipitator Technology Inc., Munhall, PA. (Ch. 17).

David Leith, Associate Professor, Environmental Health Engineering, School of Public Health, Harvard University, Boston, Mass. (Ch. 16).

H. D. Lewis, Los Alamos Scientific Laboratory, University of California, Los Alamos, New Mexico (Ch. 1).

R. A. Limons, Homer Research Laboratories, Bethlehem Steel Corp., Bethlehem, PA. (Ch. 7.7).

Thomas P. Meloy, Benedum Professor, College of Mineral and Energy Resources, West Virginia University, Morgantown, West Virginia (Ch. 3).

Wolfgang B. Pietsch, Maschinenfabrik Köppern GmbH & Co. KG, Hattingen/Ruhr, West Germany, (Ch. 7.2, 7.4).

Derek B. Purchas, Consultant Chemical Engineer, Specialist in Filtration and Separation, Pump Green House, Evenlode, Moreton-in-Marsh, Gloucestershire, GL56ONN, Great Britain (Ch. 15).

A. W. Roberts, Professor, Dean of Engineering, Head of the Department of Mechanical Engineering, The University of New Castle, New South Wales, Australia (Ch. 6).

Kunio Shinohara, Professor, Department of Chemical Process Engineering, Hokkaido University, Sapporo, Japan (Ch. 5, Pt. 1).

C. A. Shook, Professor, Department of Chemical Engineering, University of Saskatchewan, Saskatoon, Saskatchewan (Ch. 12).

Keith J. Scott, Chemical Engineering Research Group, Council for Scientific and Industrial Research, Pretoria, South Africa (Ch. 14).

Fred M. Thomson, Senior Consultant, Engineering Service Division, E.I. DuPont de Numours & Co., Wilmington, Delaware (Ch. 9).

S. S. Weidenbaum, Professor, Department of Physical & Ocean Science, United States Coast Guard Academy, New London, Conn. (Ch. 8).

Frederick A. Zenz, Professor, Department of Chemical Engineering, Manhattan College, Riverdale, New York (Ch. 10, Ch. 18).

ACKNOWLEDGMENTS

We with to thank Wendy Mortimer of the University of Waterloo, for her dedication in preparing the index for this text.

Also, we wish to acknowledge the assistance of Larry Hager, Editorial Director of Van Nostrand, for his professional guidance.

Last, but not least, we thank Alberta Gordon, Managing Editor of Van Nostrand Reinhold Company, and her assistant Anne Dempsey, for their professional assistance and cooperation in setting up the format and presentation for this handbook.

FOREWORD

The many pages which follow encompass much of that which is known of the art, science, and technology of the handling, measurement, separation, and use of the constituents of, among many others, such diverse materials as:

- the make-up of muds; face powder, wheat flour, and gun powder; catalyst beds to speed chemical reactions; and corn starch pudding;
- many intricate metallic shapes whose surfaces are formed as accurately as if they were machined to close tolerances;
- the substances which give whiteness or color to most surface coatings, papers, and many other essentials of life; and remove the shine from rayon;
- the components giving solidity and strength to dams, concrete highways, and Limoges china;
- the "dirt" which the wash cloth removes from the skin of the face or body and then is suspended in the wash water along with the particulate grime of the city–coming from, among many causes;
- the black smoke from a furnace or diesel, and the dust coming from the wear, tear and attrition of almost everything to adhere to clothes and then to be removed in the laundry.

These materials appear, and must be worked with, in so many common aspects of every day life and in almost every industry. Particularly the chemical and other process industries are concerned with the utilization and the physical or chemical conversion of powders and larger particles. The chemical engineer must take advantage of their physical parameters–size, density, shape–including angularity or smoothness, and surface characteristics, also in many cases the chemical reactions between the molecules of the solids and/or any surrounding fluid, and always the cost of the pulverulent or granular material and of its processing.

Chemical Engineers tuned their skills in Fluid Dynamics to develop Fluidization, which has become one of the most important uses of powder and particle technology. Over 40 years ago, vapors from petroleum fractions rose to contact catalyst beads; and when the vapors rose fast enough, they suspended the weight of the beads. Similarly, over 30 years ago pyrites were roasted to give SO_2 by being suspended in a rising stream of hot air. In both of these early uses, the chemical changes between the particles and the supporting gas stream were important. The first symposium on Fluidization was part of the Centennial Celebration of the Polytechnic Institute (1954-5) where the first course on Fluidization was given (1959) using as text material that published by Reinhold (*Fluidization*, D. F. Othmer, Editor 1956 and *Fluidization and Fluid Particle Systems*, F. A. Zenz and D. F. Othmer 1960) and by F. A. Zenz in Kirk-Othmer, *Encyclopedia of Chemical Technology* 1957. (Zenz was then a professor at the Polytechnic in the Department of Chemical Engineering of which Othmer was Head).

Looking at this Handbook overall, it is a major compilation of fundamentals and practical experience in the science and technology of powders, and somewhat larger particles. The two editors—seasoned by many years of academic research and teaching, also through their designing and operation of industrial plants—selected twenty-six authors who are outstanding worldwide in their knowledge and first-hand experience in different areas of this very broad field.

This compilation of the combined understanding and experience of this large staff of experts is expected to be used:—first as an introduction to those who have little or no experience in powder technology, including students; and second, as a manual and handbook for technologists, seasoned design engineers, and scientists through its wealth of practical methods for the design of equipment and processes for handling finely divided solids. The book will also serve as a reference for research institutions, colleges, universities, and industrial organizations in such diverse areas as chemical, metallurgical, mining, mechanical and civil engineering, food and pharmaceutical technology, coal and polymer engineering and others.

Here we have served up the theoretical wisdom and technical experience of the science and technology of fine solid particles which has been accumulated and used in countless applications and industry over the ages, principally during the last 50 years. It is predicted that this storehouse of theory and practice will be the guide and mentor for those with interest and work in this field throughout the rest of this century.

Donald F. Othmer
Distinguished Professor
Polytechnic Institute of New York
Brooklyn, New York

CONTENTS

1

Particle Size Analysis: Theory and Statistical Methods*

A. S. Goldman and H. D. Lewis
University of California
Los Alamos, New Mexico

INTRODUCTION

Investigators in the field of particle size measurements are confronted with a multitude of sizing devices. Allen,[1] DallaValle,[2] and Iranis and Calli[3] give insight into the operation and collection of data from such devices. Herdan[4] and Lewis and Goldman (L & G)[5] present procedures for obtaining valid statistical analyses. Herdan's text is a lengthy treatise that gives detailed solutions to problems originating from the fineness of solids and behavior of coarse disperse systems. He also investigates experimental design theory and errors of particle size determination. The L & G effort furnishes general formulas for all forms of particle size data, defines statistical relationships among distributions, suggests standardized notation, and provides rules for mixtures of particulate materials.

It is the purpose of this study to enhance the L & G work by presenting statistical concepts and tests as they are applied to practical examples.

*Based on work performed under the auspices of the U.S. Department of Energy.

1.1. COMPUTATION OF SAMPLE STATISTICS

Types of Measuring Devices

The different instruments available for measuring particle size distributions can be categorized into generic headings including aerosol samplers, sieves, microscopes, sedimentation equipment, classifiers, centrifuges, streaming principle, light scattering, permeametry, and gas adsorption. Although many forms of data are obtained from these varied instruments, they can be categorized and analyzed by general statistical tools. The key to the use of proper computations is to obtain a statistical layout of the data before the analysis.

Statistical Layout

A statistical layout is defined as the arrangement of data in the form necessary for proper statistical analysis. It is desirable to set up the statistical layout as closely as possible to the experimental layout. The experimental layout is directly related to the testing device. For example, the experimental layout from the Coulter counter would consist of volume size intervals and frequencies of size. (See Ref. 6 for a detailed discussion.) If for some reason an experimenter wanted a different measure such as weight fraction, and utilized a Coulter counter, then a statistical layout relating diameter size to weight suggested by Allen[7] or L & G[5] would be appropriate.

In subsequent examples, statistical layouts will be in the same form as experimental layouts. Output from sizing devices will be categorized into two types of statistical layouts will be matched accordingly.

Type I and Type II Data

Many authors confuse the terms *distribution of weights* and *weight distribution.* They erroneously refer to a *size distribution* whose probability measures are determined by weight as a *weight distribution.* For example, in sieve analysis the measure of weight of materials corresponding to intervals of diameter size has been called weight distribution, when in fact it is a distribution of diameters. This nomenclature becomes troublesome when obtaining a measure of frequencies of volume or weight from an instrument such as a Coulter counter which gives a weight distribution, i.e., frequencies of weights in intervals. The correct interpretation of this situation is to refer to sieve data as weight fraction versus size (weight fraction) data, and Coulter counter data as a distribution of volumes or weights. To simplify terminology, data can be classified into two types regardless of the measuring instrument used. These are illustrated in Figure 1.1.

Type I data (Figure 1.1a) is the frequency distribution of some property of the particle such as weight, volume, area, or diameter. Let d represent a linear dimension such as diameter for spherically shaped particles; then the abscissa can be expressed as a function of some power of d. Let N be the total number of particles being measured and n_i the number in the ith interval; the relative frequency is defined as n_i/N, where $\Sigma n_i = N$. Devices for which Type I data is applicable include the microscope (n_i/N vs d_i), Quantimet (n_i/N vs $a_i = C_2 d_i^2$), and Coulter counter (n_i/N vs $v_i = C_3 d_i^3$), where C_2 and C_3 are appropriate constants for a = area and v = volume, respectively.

Type II data (Figure 1.1b) is the moment relationship having ordinate, weight fraction, volume fraction, or relative frequency; and abscissa d. Let A, V, and W denote total area, volume, and weight, respectively, and a_i', v_i', and w_i' respectively denote area, volume, and weight in the ith interval, where $\Sigma a_i' = A$, $\Sigma v_i' = V$, and $\Sigma w_i' = W$. The prime notation is used to distinguish the ordinate measure from the abscissa. Devices for which Type II data is applicable include sieves (w_i'/W vs d_i), Sedibals (w_i'/W vs d_i), or micromerographs (w_i'/W vs d_i).

Type I and Type II data can be combined for moment-related measurements observed on "size," where "size" could be area or weight; i.e. the ordinate of Figure 1.1b and the abscissa of Figure 1.1a are combined. An example can be found when obtaining molecular weight distributions in the field of gel permeation chromatography.[10]

A complete description of Type I and II data can be found in References 5 and 11.

It is important to classify data as to Type I or Type II, then use proper notation and formulas

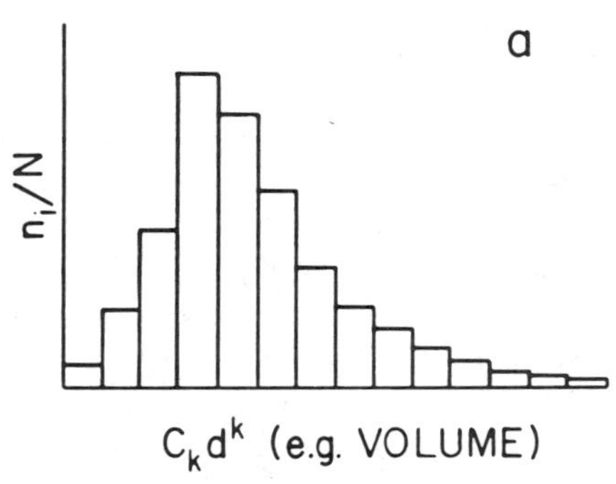

Figure 1.1a. Type I sample data.

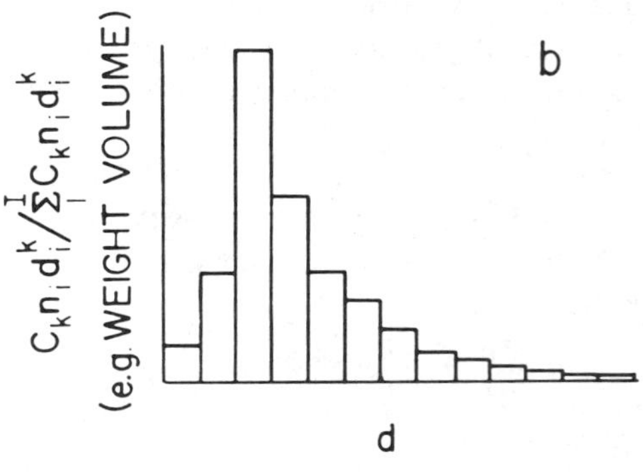

Figure 1.1b. Type II sample data.

available for computing sample statistics (See Refs. 5 and 12). Proper analysis of any data type requires different considerations, depending upon whether the data is *discrete* (a finite number of observations), *grouped* (discrete measurements, with a large sample size [n > 50] assembled into intervals from which a histogram can be drawn), or *continuous* (an "infinite" number of measurements "observed"). Examples are given to illustrate applications of these concepts.

Worked Examples

This section will serve as a guide for obtaining estimates of statistical parameters for any measuring device.

Example 1.1. Type I discrete data (an ungrouped finite number of observations). Assume the experimenter measures approximate diameters of 25 particles from a population of powder.

A. Computing Statistics

Step 1: List the data; d denotes measured diameter in micrometers (microns).
d: 3.14, 2.74, 2.65, 1.22, 1.65, .98, .74, 2.43, 1.69, 1.19, 2.42, 1.73, 1.18, 2.69, .62, 1.58, 1.16, 2.57, 1.01, 1.58, .73, 1.69, 1.96, 2.17, 2.80

Step 2: Sample statistics are computed as follows.
Mean:

$$\bar{d} = \Sigma d_i/N = (3.14 + 2.74 + \cdots + 2.80)/25 = 1.77$$

Variance:

$$s_d^2 = \Sigma(d_i - \bar{d})^2/(\mathrm{N} - 1) = [(3.14 - 1.77)^2 + \cdots + (2.80 - 1.77)^2]/24 = .5550$$

Standard deviation:

$$s_d = \sqrt{s_d^2} = .7450$$

Skewness:

$$g_d = \Sigma(d_i - \bar{d})^3/(\mathrm{N} - 1) = [(3.14 - 1.77)^3 + \cdots + (2.80 - 1.77)^3]/24 = .0662$$

Another measure of asymmetry is:

$$\beta_1 = g_d^2/(s_d^2)^3 = .0256$$

B. Testing for Lognormal Distribution (Wilk-Shapiro Test)

Step 1: List the data by taking $\log_e d$.
$x = \log_e d$ and denotes the log of the measured diameter.
x: 1.14, 1.01, .97, .20, .50, -.02, -.30, .89, .52, .18, .88, .55, .17, .99, -.48, .45, .15, .94, -.01, .46, -.32, .52, .68, .78, 1.03

Step 2: Order the observations and denote the ordered sample $y_1 < y_2 \cdots < y_{25}$. Notation is the same as found in Shapiro and Wilk.[13]
y: -.48, -.32, -.30, -.02, -.01, .15, .17, .18, .2, .45, .46, .5, .52, .52, .55, .68, .78, .88, .89, .94, .97, .99, .1.91, 1.03, 1.14

Step 3: Compute

$$S^2 = \Sigma(y_i - \bar{y})^2 = \Sigma(x_i - \bar{x})^2 = 5.2316$$

Step 4: Compute

$$\begin{aligned} b &= a_n(y_n - y_1) + \cdots + a_{k+2}(y_{k+2} - y_k) \\ &= .4450(1.14 + .48) \\ &\quad + 3.069(1.03 + .32) + \cdots \\ &\quad + .02(.52 - .50) = 2.21682 \end{aligned}$$

where $a_n, a_{n-1}, \cdots a_{k+2}$ values are found in Appendix A1.

Step 5: Compute

$$W = b^2/S^2 = (2.21682)^2/5.2316 = .9393$$

Step 6: The distribution of W is given in Appendix A2. Small values of W are significant, i.e., indicate non-normality. $W = .939$ with n = 25 is slightly greater than the tabular value of .931, significant at the .10 level. Consequently, log normality is doubtful.

Example 1.2. Type I grouped data. Assume the experimenter has an instrument such as a Coulter counter that measures the frequency of volumes of a powder, and a total of 500 counts has been grouped.

A. Computing Sample Statistics

Step 1: Obtain the statistical layout where v_i denotes the volume of particles observed in the ith cell, n_i the frequency, and interval end points are v_i^*, v_{i+1}^*.

VOLUME SIZE INTERVAL (MEASURED IN ANGSTROMS) v_i^*, v_{i+1}^*	v_i = MIDPOINT	n_i = FREQUENCY
150–300	225	10
300–450	375	70
450–600	525	71
600–750	675	114
750–900	825	91
900–1,050	975	64
1,050–1,200	1,125	30
1,200–1,350	1,275	21
1,350–1,500	1,425	13
1,500–1,650	1,575	10
1,650–3,000	2,325	6

Step 2: Sample statistics are computed as follows:

Mean:

$$\bar{v} = \Sigma n_i v_i / \mathrm{N} = (10 \cdot 225 + 70 \cdot 375 + \cdots + 6 \cdot 2{,}325)/500 = 774.9$$

Variance:

$$S_v^2 = \Sigma n_i (v_i - \bar{v})^2/(\mathrm{N} - 1) = [10(225 - 774.9)^2 + 70(375 - 774.9)^2 + \cdots + 6(2{,}325 - 774.9)^2]/499 = 57{,}822{,}791/499 = 115{,}877.33$$

Standard Deviation: $= \sqrt{115{,}877.33} = 340.41$

Skewness:

$$g_v = \Sigma n_i (v_i - \bar{v})^3/(\mathrm{N} - 1) = [10(225 - 774.9)^3 + \cdots + 6(2{,}325 - 774.9)^3]/499 = 2.81182 \times 10^{10}/499 = 56{,}349{,}198$$

$$\beta_1 = (5.6349198 \times 10^7)^2/(1.1587733 \times 10^5)^3 = 20.4070 \times 10^{-1} = 2.0407$$

B. Testing for Log Normality (Chi-square Test)

Step 1: Convert end points to natural logs and set up the statistical layout, where $y = \log_e v$

SIZE INTERVAL (LOG VOLUMES) y_i^*, y_{i+1}^*	y_i = MIDPOINT	n_i = FREQUENCY	f_i = THEORETICAL FREQUENCY	$(n_i - f_i)^2/f_i$
5.01–5.70	5.35	10	12.30	0.43
5.70–6.11	5.90	70	63.45	0.68
6.11–6.40	6.25	71	102.10	9.47
6.40–6.62	6.51	114	100.00	1.96
6.62–6.80	6.71	91	76.55	2.73
6.80–6.96	6.88	64	56.20	1.08
6.96–7.09	7.02	30	32.85	0.25
7.09–7.21	7.15	21	22.50	0.10
7.21–7.31	7.26	13	12.70	0.23
7.31–7.41	7.36	10	8.55	0.25
7.41–8.01	7.71	6	12.56	3.43

Computations in Step 1:

$$\bar{y} = \Sigma n_i y_i / \mathrm{N} = (5.35 \cdot 10) + (5.90 \cdot 70) + \cdots + (7.71 \cdot 6)/500 = 3{,}278.31/500 = 6.56$$

$$s = \sqrt{\Sigma n_i (y_i - \bar{y})^2/499} = \sqrt{95.2053/499} = .4368$$

Values of theoretical frequencies f_i for the normal distribution are calculated as follows:

y_i^*	$(y_i^*-\bar{y})/s$	CUMULATIVE DISTRIBUTION FROM NORMAL TABLES	AREA IN INTERVAL (y_{i-1}^*, y_i^*)	f_i = (AREA IN INTERVAL) × 500
5.01	−3.55	0		
5.70	−1.97	.0246	.0246	12.30
6.11	−1.03	.1515	.1269	63.45
6.40	−0.37	.3557	.2042	102.10
6.62	0.14	.5557	.2000	100.00
6.80	0.55	.7088	.1531	76.55
6.96	0.92	.8212	.1124	56.20
7.09	1.21	.8869	.0657	32.85
7.21	1.49	.9319	.0450	22.50
7.31	1.72	.9573	.0254	12.70
7.14	1.95	.9744	.0171	8.55
8.01	3.32	.9995	.0251	12.56

Step 2: Compute chi-square, compare with critical value, and state conclusion.
Computed:

$$\chi^2 = \Sigma(n_i - f_i)^2/f_i = .43 + .68 + \cdots + 3.43$$
$$= 20.61$$

Tabulated:

$$\chi^2_{.99}(8) = 20.1, \chi^2_{.95}(8) = 15.5$$

(from chi-square table)

The data does not appear to follow the log-normal distribution because the computed value is larger than the tabulated value.

Comment On Sizing Device Output

No sizing device currently exists that either gives a continuous distribution of relative frequencies vs size or gives a discrete distribution of moment weighted material vs size.

The chi-square statistic is *only* applicable with frequency data. It is important not to use this test on weight fraction data. (See Ex. I.3.B).

Example I.3. Type II Grouped Data. Assume the experimenter utilizes sieves to measure the weights of 100 grams of material corresponding to size intervals.

A. Computing Sample Statistics

Step 1: Obtain the statistical layout.

SIZE INTERVAL OF DIAMETER (d_i^*, d_{i+1}^*)	MIDPOINT OF SIZE INTERVAL d_i	WEIGHT OF MATERIAL IN INTERVAL w_i'	WEIGHT FRACTION $w_i'/W = w_i'/\Sigma w_i$
0.80–1.00	0.90	1.55	.0155
1.00–2.00	1.50	7.53	.0753
2.00–3.00	2.50	11.55	.1155
3.00–5.00	4.00	22.68	.2268
5.00–7.00	6.00	13.40	.1340
7.00–9.00	8.00	8.24	.0824
9.00–15.00	12.00	15.46	.1546
15.00–20.00	17.50	5.67	.0567
20.00–25.00	22.50	4.64	.0464
25.00–30.00	27.50	4.12	.0412
30.00–40.00	35.00	3.51	.0351
40.00–50.00	47.50	1.65	.0165

Step 2: Compute sample statistics as follows:
Mean:

$$\bar{d}_3 = \Sigma w_i' d_i / W = (.0155)(.90) + (.0753)(1.50) + \cdots + (.0165)(47.50) = 9.82$$

Variance:

$$s^2_{d_3} = \Sigma w_i'(d_i - \bar{d}_3)^2/W = .0155(.90 - 9.82)^2 + .0753(1.50 - 9.82)^2 + \cdots + .0165(47.5 - 9.82)^2 = 92.64$$

Standard deviation:

$$s_{d_3} = \sqrt{92.64} = 9.63$$

Skewness:

$$g_{d_3} = \Sigma w_i(d_i - \bar{d}_3)^3/W = .0155(.90 - 9.82)^3 + .0753(1.50 - 9.82)^3 + \cdots + .0165(47.5 - 9.82)^3 = 1{,}640.30.$$

$$\beta_1 = (1{,}640.3)^2/(92.64)^3 = 3.3842$$

B. Testing for Log Normality (Subjective Procedure)

There is no test for normality for Type II data. Data may be transformed *from weight to frequencies* but even then a chi-square test cannot be made because of the unknown sample size. A subjective procedure for examining the data includes a plot on log probability paper (see Figure 1.2) and a least squares fit. Least squares programs are quite common and packages such as BIMED[14] are readily available.

Stereology Applications

Methods for finding estimates of parameters (mean, variance, skewness) in three dimensions when two dimensional data are observed are well known.[15,16] If the integral equation model[17] is not correct, it is possible to obtain unreasonable results such as negative variance. However, no other practical model is available at this time. Should this happen, the reader may use techniques given by Goldman, Lewis, and Visscher[18] or Visscher and Goldman.[19] Two examples are presented to illustrate proper procedures.

Example 1.4. The following data was taken from Saltikov.[17] Measurements are taken as frequencies of observed diameters of circles (Type I data) obtained from random sections (slices) through grain beds. The sample mean, variance, and skewness of diameters of the spherical shaped grain are to be computed.

Step 1: List the data.

INTERVAL (OBSERVED DIAM.) d_i^*, d_{i+1}^*	MIDPT. d_i	OBSERVED FREQUENCY f_i
15.85–19.95	17.90	69
19.95–25.12	22.54	138
25.12–31.62	28.37	230
31.62–39.81	35.72	253
39.81–50.12	44.97	161
50.21–63.10	56.61	104

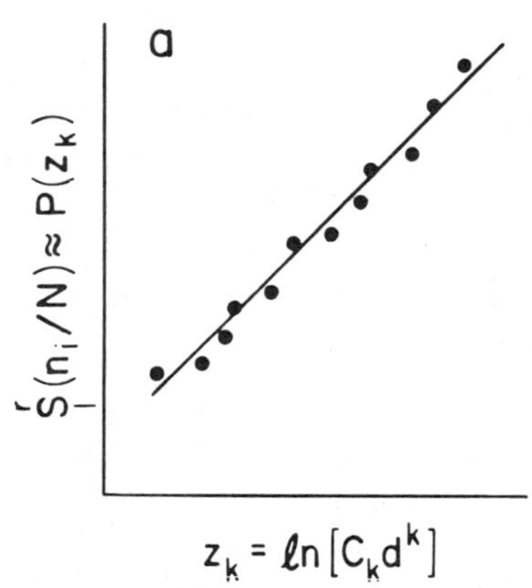

Figure 1.2a. Type I data assuming lognormality. (Plotted on log probability paper.)

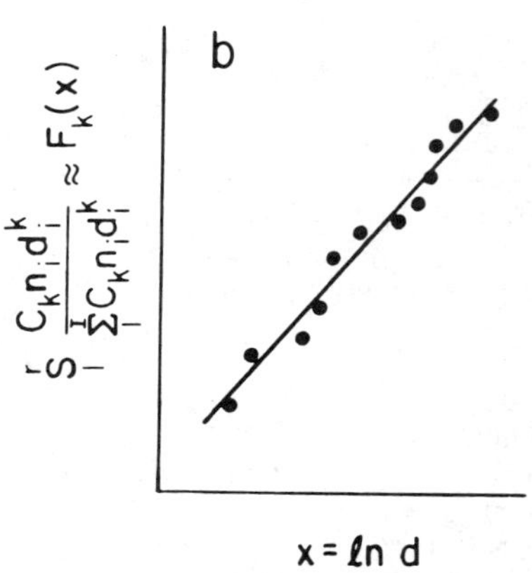

Figure 1.2b. Type II data assuming lognormality. (Plotted on log probability paper.)

Step 2: Compute the harmonic mean H of the observed distribution.

$$H = \frac{\Sigma f_i}{\frac{\Sigma f_i}{d_i}} = \frac{69 + 138 + \cdots + 104}{\frac{69}{17.90} + \cdots + \frac{104}{56.61}} = 31.22$$

(f_i = observed frequency for 2 dimensional data.)

Step 3: Compute the mean of diameters (sizes) of the 3d data:

$$\bar{d} = \Pi/2\, H = \frac{3.1416}{2} \times 31.22 = 49.05$$

Step 4: Compute the first and second moments of the observed diameter distribution, d'_1, d'_2.

$$d'_1 = \frac{\Sigma d_i f_i}{\Sigma f_i} = \frac{69 \times 17.9 + \cdots + 104 \times 56.61}{955} = \frac{33{,}035.49}{955} = 34.59$$

$$d'_2 = \frac{\Sigma d_i^2 f_i}{\Sigma f_i} = \frac{69 \times 17.9^2 + \cdots + 104 \times 56.61^2}{955} = \frac{1{,}259{,}022.2}{955} = 1{,}318.3478$$

Step 5: Compute the second and third moments of the 3 dimensional diameter (size) distribution.

$$m'_2 = \frac{4}{\Pi}\bar{d}\, d'_1 = \frac{4}{3.1416}(49.05)(34.59) = 2{,}160.22$$

$$m'_3 = \tfrac{3}{2}\bar{d}\, d'_2 = \tfrac{3}{2}(49.05)(1{,}318.3478) = 96997.435.$$

Step 6: Compute the variance of diameters (3 dimension).

$$s_{\bar{d}}^2 = m'_2 - \bar{d}^2 = 2{,}160.22 - (49.05)^2 = -245.68.$$

This is a nonsensical answer. Because this method assumes no inaccuracies in either the 2 dimensional sampling assumptions or measurements on the sample, this procedure for estimating the variance of the 3 dimensional distribution gives absurd results.

Example I.5. COMIVE: The recommended procedure, *CO*nstrained *MI*nimum *V*ariance *E*stimation (COMIVE), is used.[18,19] The complex arithmetic forces the use of a computer. The suggested program based on the algorithms given in Ref. 18, 19 and 24 may be obtained from the authors. These methods give the following results for the data of Example I.4.

Step 1:

INTERVAL (OBSERVED DIAMETERS) d_i^*, d_{i+1}^*	MIDPOINT d_i	PLANAR OBSERVED FREQUENCY f_i	DERIVED FREQUENCY OF SPHERICAL DIAMETERS n_i
15.85–19.95	17.90	69	18
19.95–25.12	22.54	138	119
25.12–31.62	28.37	230	271
31.62–39.81	35.72	253	296
39.81–50.12	44.97	161	155
50.12–63.10	56.61	104	97
		955	956

Step 2: Compute mean, variance, and skewness:

$$\text{Mean} = \bar{d} = \frac{\Sigma d_i n_i}{n_i} = \frac{18 \times 17.90 + \cdots + 97 \times 56.61}{956} = 35.28$$

$$\text{Variance} = s_d^2 = \frac{\Sigma d_i^2 n_i - n\bar{d}^2}{\Sigma n_i - 1} = \frac{17.9^2 \times 18 + \cdots + 56.61^2 \times 97 - 956 \times 35.28^2}{955}$$
$$= 100.96.$$

$$\text{Standard deviation} = s_d = \sqrt{100.96} = 10.05$$

$$\text{Skewness} = g_d = \frac{(17.90 - 35.28)^3 18 + \cdots + (56.61 - 35.38)^3 97}{955} = 682.44$$

$$\beta_1 = \frac{682.44^2}{180.96^3} = 0.045$$

Example I.5. It is difficult to find an example that satisfies the demands of the original equation. The following data represent a synthetic example found in Anderssen and Jakeman.[20] The data are contrived to exactly fit the model.

Step 1: List the data:

INTERVAL d_i^*, d_{i+1}^*	MIDPT. d_i	NO. OF CIRCLES f_i
0-.05	.025	7
.05-.10	.075	22
.10-.15	.125	36
.15-.20	.175	47
.20-.25	.225	58
.25-.30	.275	67
.30-.35	.325	74
.35-.40	.375	78
.40-.45	.425	81
.45-.50	.475	81
.50-.55	.525	79
.55-.60	.575	75
.60-.65	.625	70
.65-.70	.675	63
.70-.75	.725	53
.75-.80	.775	43
.80-.85	.825	32
.85-.90	.875	21
.90-.95	.925	11
.95-.100	.975	2

Step 2: Compute the harmonic mean of the circle diameters (observed data):

$$H = \frac{\Sigma f}{\Sigma f/d} = \frac{7 + 22 + \cdots + 2}{\frac{7}{.025} + \frac{22}{.075} + \cdots + \frac{2}{.975}}$$
$$= \frac{1000}{3{,}119.6876} = .321$$

Step 3: Compute the mean of the sphere diameters:

$$\bar{d} = \frac{\Pi}{2} H = \frac{3.14}{2} \times .321 = .504$$

Step 4: Compute the first and second moments of the circle diameter:

$$d_1' = \frac{\Sigma d_i f_i}{f_i} = \frac{471.75}{1{,}000} = .4718$$

$$d_2' = \frac{\Sigma d_i^2 f_i}{f_i} = \frac{267.30}{1{,}000} = .2673$$

Step 5: Compute the second and third moments of the sphere diameter distributions:

$$m_2' = \frac{4}{\Pi} \bar{d}\, d_1' = \frac{4}{\Pi} \times .504 \times .4718 = .3028$$

$$m_3' = \tfrac{3}{2} \bar{d}\, d_2' = \tfrac{3}{2} \times .504 \times .2673 = .2021$$

Step 6: Compute the variance and skewness of diameters:

$$s_d^2 = m_2' - \bar{d}^2 = .3028 - (.504)^2 = .0488$$

$$s_d = \sqrt{s_d^2} = \sqrt{.0488} = .2209$$

$$g_d = m_3' - 3(m_2')\bar{d} + 2(\bar{d})^3$$
$$= .2021 - (3)(.3028)(.504) + 2(.504)^3$$
$$= .0003$$

$$\beta_1 = g_d^2/(s_d^2)^3 = 7.75 \times 10^{-4}$$

Ordinarily, this is a completely worked example; however, the true distribution for this example is known and the parameters are:

$\mu = .50$ (estimate = .504),

$\sigma_d^2 = .05$ (estimate = .0488),

Skewness = 0 (estimators: $g_d = .0003$ and $\beta_1 = .0008$).

The method of moments provides answers that are reassuring. The COMIVE program is used to determine the distribution of sphere diameters as follows:

Comment: The method of moments is not recommended because it is only suitable for contrived data.

1.2. METHODS FOR DATA COMPARISON

Quantitative Relationships between Type I and Type II Data

A quantitative comparison of properties or process behavior of powders necessitates a quantitative comparison of the sample statistics described above in the Worked Examples Section. Two other typical situations involving comparisons of sample data are:

1. Calibration of two or more measuring devices using a standard powder sample. For example, calibration of a Coulter counter, and a micromerograph, using a sample whose distribution had been checked previously be a microscopic count.
2. Comparison of sample data obtained by two or more investigators using two or more measuring devices.

In any situation requiring data comparison, the selection of a comparison technique depends upon the type of measuring device used

INTERVAL	MIDPT.	FREQ. OF SPHERE DIAM.	INTERVAL	MIDPT.	FREQ. OF SPHERE DIAM.
0–.05	.025	8	.50–.55	.525	75
.05–.10	.075	21	.55–.60	.575	73
.10–.15	.125	33	.60–.65	.625	70
.15–.20	.175	43	.65–.70	.675	66
.20–.25	.225	52	.70–.75	.725	60
.25–.30	.275	60	.75–.80	.775	52
.30–.35	.325	66	.80–.85	.825	43
.35–.40	.375	70	.85–.90	.875	33
.40–.45	.425	73	.90–.95	.925	21
.45–.50	.475	75	.95–1.00	.975	6

Estimates of the parameters include $\bar{d} = .4991$, $s_d^2 = .0501$, $g_d = 0$, $\beta_1 = 0$. Results show good agreement indicating the discretized distribution is an excellent approximation to the true distribution and that the program is in working order.

to obtain the data, and whether or not a usable functional model for the data can be assumed. This first discussion concerns comparison of data for which no functional model is recognized, as obtained from the typical "size" measuring devices. Two related methods of

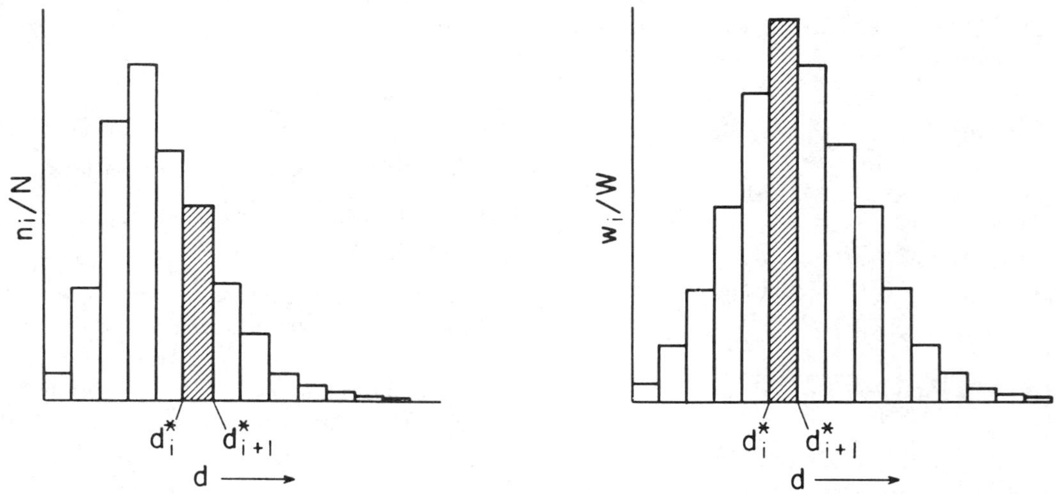

Figure 1.3. Graphical comparison of sample distribution and weight fraction data.

approach in this case are comparison of graphical data and comparison of computed statistics. The second case to be studied will be the lognormal model.

Graphical Data Comparison

Comparison of graphical data involves the following concepts:

1. A comparison of data obtained from two measuring devices demands the transformation of the data from one device into the form of those obtained from the other.
2. The validity of the transformation between moment relationship data and relative frequency data is determined by the total number of particles in the relative frequency sample.
3. The validity of the data comparison depends upon proper selection of interval width, number of "size" intervals, and midpoint value of the transformed data.

These points can be illustrated by use of the moment relationship diagrams and schematic histograms of Figures 1.3 and 1.4. A comparison of Type II data shown in Figure 1.3 would

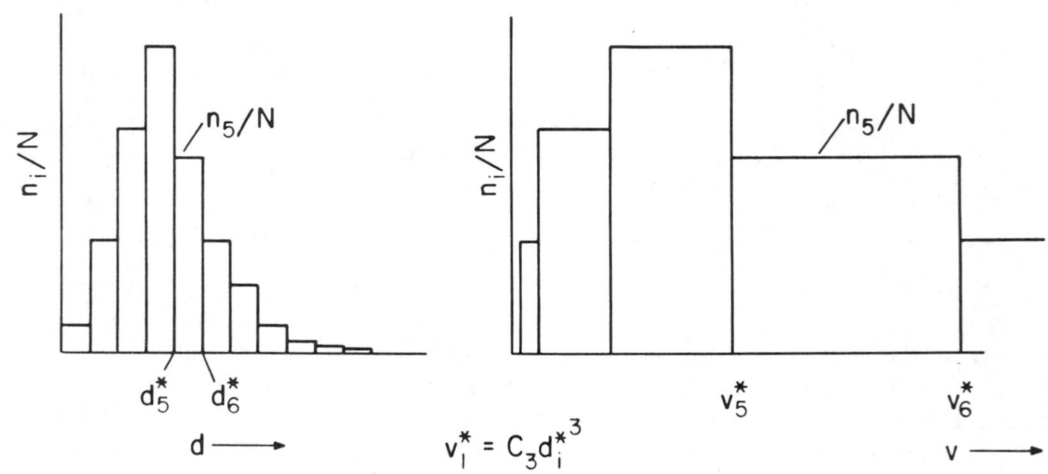

Figure 1.4. Graphical comparison sample distribution of d and volume data.

involve a transformation of the fraction of material w_i'/W in the interval (d_i^*, d_{i+1}^*) to the fraction n_i/N in the same interval, or a transformation of n_i/N to w_i'/W. Results of Monte Carlo experiments[5] have indicated that the proper direction of the transform is from w_i'/W to n_i/N. It requires a total of at least 25,000 particles from a "typical" particle size distribution using n_i/N to obtain reasonable estimates of w_i'/W. To compare weight fraction data with relative frequency data, careful selection of an appropriate "midpoint" value should be made for converting w_i' to n_i. Mellor's midpoint value d_{M_i} is recommended,[21] and is computed

$$d_{M_i} = \{[(d_i^*)^2 + (d_{i+1}^*)^2]\,(d_i^* + d_{i+1}^*)/4\}^{1/3}$$

The number of particles in the ith size interval is

$$n_i = \frac{w_1'}{\rho C_3 d_{M_i}^3}$$

where ρ is the powder density, and C_3 is an appropriate constant related to the shape of a particle, e.g., for spherical particles $C_3 = \Pi/6$. The relationship n_i/N vs d_i or the plot of n_i/N vs the intervals (d_i^*, d_{i+1}^*) is then constructed and compared with the original relative-frequency data.

Data obtained from the automated sedimentation devices, (e.g. the Sedibal* or Sedigraph) pose special problems in that they are obtained as continuous cumulative weight curves. To graphically compare these data with data in the usual form (moment relationship diagrams in general), the cumulative curve should be transformed into the weight fraction diagram. This involves selection of the proper "size" interval width and the proper number of intervals. The following rules of thumb are suggested for interval width selection:

1. The transformed sedimentation data should match in width and number the "size" intervals of the data with which they are being compared.
2. The first rule assumes that the untransformed data have been graphed using the proper number and size interval selection.
3. The proper number of intervals can be taken as approximately 30 for the general powder sample, assuming that 1,000 counts could be made in a microscopic count. The following table indicates the dependence of interval width selection (number of intervals I) on sample size N. This table is just an approximate guide:

The number of "size" intervals for sample size N

N	200	400	600	800	1,000	1,500	2,000
I	12	20	24	27	30	35	39

Coulter counter data and Type I data in general necessitate a somewhat more complicated method of graphical comparison. The basic reason for this is that the abscissa value or "size" is transformed rather than the general fraction representing the amount of material in a given size interval. Consider the comparison of the graphed sample distribution, i.e., histogram, of the dimension d with the histogram of particle volume v as shown schematically in Figure 1.4, where the histograms n_i/N vs d and n_i/N vs v both belong to the general family n_i/N vs $C_k d^k$. The proper comparison involves making a one-to-one correspondence between the intervals (d_i^*, d_{i+1}^*) and (v_i^*, v_{i+1}^*) by the expression

$$v_i^* = C_3 (d_i^*)^3$$

Note that the volume size intervals will not be equal over the size range; however, the relative frequency is the same in corresponding d and v size intervals. In graphically comparing Coulter data with weight fraction, or general moment relationship data, the weight fraction data should be converted to relative frequency data. The comparison is then made by changing the Coulter data to n_i/N vs d as shown in Figure 1.4, observing the rules of thumb and informa-

*The Sedibal, Sedigraph, is an automatic recording balance that measures the cumulative weight of particles settling through a liquid as a function of time.

tion in the above table for interval width selection.

Although the transformation from Coulter counter to weight fraction data would not be proper for N less than approximately 25,000, as indicated by Monte Carlo experiments, many experimenters desire this information.[5] The proper conversion techniques are found in Ref. 6. The main points are summarized below.

1. The end points of the volume size interval are converted to the d size interval by

$$d_i^* = \sqrt[3]{v_i^*/C_3}$$

2. Mellor's equation is used to compute d_{M_i}, and the weight in the interval (d_i^*, d_{i+1}^*) is computed as

$$w_i'/W = n_i\, d_{M_i}^3 / \Sigma n_i\, d_{M_i}^3$$

3. The relationship w_i'/W vs d_i or vs the intervals (d_i^*, d_{i+1}^*), is then plotted.

There appears to be a limited number of sources in the literature concerning transformation of data in the general case.[5, 22, 23] Lognormal transformations are well known and can be found in any text on particle size statistics. A summary of these results will be discussed subsequently. Two examples are given to illustrate how the above formulas can be used as well as a comparison with the work done by Wise.[22]

Example II.1.A. L & G method[5] for converting microscopic data to weight fraction data, where d_i^* is in a geometric progression. Data taken from Wise.[22]

Step 1: List the data:

INTERVAL d_i^*, d_{i+1}^*	FREQUENCY n_i	REL. FREQ. n_i/N	STANDARDIZATION $n_i/N/(d_{i+1}^*-d_i^*)$
$\sqrt{2}$–2	27	.027	.046
2–2$\sqrt{2}$	66	.066	.080
2$\sqrt{2}$–4	142	.142	.121
4–4$\sqrt{2}$	383	.383	.231
4$\sqrt{2}$–8	334	.334	.143
8–8$\sqrt{2}$	34	.034	.010
8$\sqrt{2}$–16	14	.014	.003

Comment: Standardized histogram area must be equal to 1.0.

Step 2: Compute d_{M_i}, $n_i d_{M_i}^3$, and w_i'/W (desired result),

$$d_{M_1} = \{[(\sqrt{2})^2 + (2)^2]\ [(\sqrt{2} + 2)/4]\}^{1/3} = \{6 \times .8536\}^{1/3} = 1.7237$$

.
.
.

$$d_{M_7} = \{[(8\sqrt{2})^2 + (16)^2]\ [8\sqrt{2} + 16)/4]\}^{1/3} = \{384 \times 6.8284\}^{1/3} = 13.7896$$

$$n_1 d_{M_1}^3 = (27)(1.7237)^3 = 138.28 \ldots n_7 d_{M_7}^3 = (14)(13.7896)^3 = 36{,}709.82$$

$$v_1/V = 138.28/(138.28 + \cdots + 36{,}709.82) = .0006$$

.
.
.

$$v_7/V = 36{,}709.82/(138.28 + \cdots + 36.709.82) = .1603$$

Step 3: Summarize results:

d_i^*, d_{i+1}^*	ni	d_{M_i}	$n_i d_{M_i}^3 \times 10^{-3}$	$w_i'/W = v_i'/V$	STANDARDIZATION $w_i'/W/(d_{i+1}^* - d_i^*)$
$\sqrt{2}$-2	27	1.7237	.14	.0006	.001
2-2$\sqrt{2}$	66	2.4377	.96	.0042	.005
2$\sqrt{2}$-4	142	3.4474	5.82	.0254	.022
4-4$\sqrt{2}$	383	4.8754	44.38	.1938	.117
4$\sqrt{2}$-8	334	6.8948	109.47	.4781	.204
8-8$\sqrt{2}$	34	9.7507	31.52	.1376	.042
8$\sqrt{2}$-16	14	13.7896	36.71	.1603	.034
	1,000		229.00	1.0000	

Example II.1.B. Wise's method for converting microscopic data to weight fraction data where d_i^* is in a geometric progression. Wise's method is discussed in Ref. 22. Values of n_i/N are assumed to come from a continuous distribution, and appropriate formulas are derived.

Step 1: Same as Step 1 in Example II.1.A.

Step 2: When d_i falls in a geometric progression, a change of variable makes computations more convenient. The following preliminary results are necessary: Let $d_i^* = 2^{u/2}$. Form the following table:

(a) d_i^*	$\sqrt{2}$	2	2$\sqrt{2}$	4	4$\sqrt{2}$	8	8$\sqrt{2}$	16
(b) u	(u_0=) 1	2	3	4	5	6	7	8 (=u)
(c) n_i/N	.027	.066	.142	.383	.334	.034	.014	
(d) Σn_i/N	0	.027	.093	.235	.618	.952	.989	1.000
(e) $f(d) = 1 - \Sigma n_i/N$	1.00	.973	.907	.765	.382	.048	.014	0
(f) $(d_i^*)^3 \times (e)$	2.8	7.8	20.5	49.0	69.1	24.6	20.3	0

Step 3: Relative proportions by volume are found from

$$\log 2^{1.5} \int_{u_i}^{u} d^3 f(d)\, du + d^3 f(d).$$

The function $d^3 f(d)$ is assumed to be a step function with u_i^* assumed to be a lower end point. For example

$$\int_7^8 d^3 f(d)\, du = \int_7^{7.5} 20.3\, du + \int_{7.5}^{8} 0\, du$$
$$= 10.15.$$

For convenience, the integral is multiplied by 2 and appears as 20.3 below.

A graph of $d^3 f(d)$ (Fig. 1.5) is given in order to assist in preparing a summary table. The summary table is given by (g) through (m).

(g) u_i^*	1	2	3	4	5	6	7	8
(h) $2\int_{u_i^*}^{u} d^3 f(d)\, du$	385.4	374.8	346.5	277.0	158.9	65.2	20.3	0
(i) 3/4 log 2 × (*h*)	200.3	194.9	180.0	144.0	82.6	33.9	10.6	0
(j) $(d_i^*)^3 f(d)$	2.8	7.8	20.5	49.0	69.1	24.6	20.3	0
(k) (*i*) + (*j*)	203.1	202.7	200.5	193.0	151.7	58.5	30.9	0
(l) Relative prop. by volume	.4	2.2	7.5	41.3	93.2	27.6	30.9	
(m) $v_i/V = w_i/W$	.002	.011	.037	.203	.459	.136	.152	

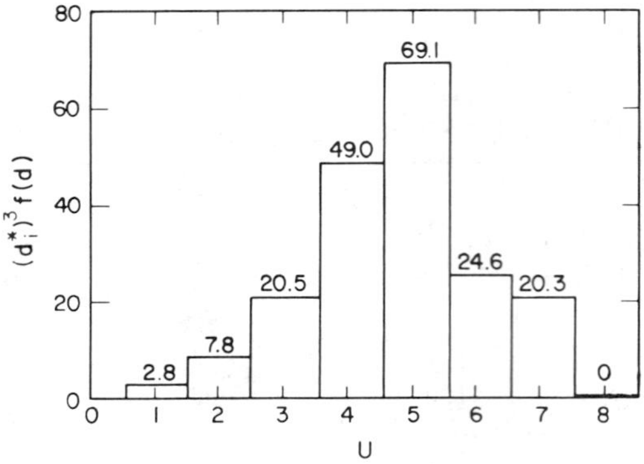

Figure 1.5. A bar chart of $d_i^* f(d)$ vs u.

A comparison of v_i'/V:

d_i^*, d_{i+1}^*	v_i'/V (L & G)	v_i'/V (Wise)
$\sqrt{2}$-2	.001	.002
2-2$\sqrt{2}$	.004	.011
2$\sqrt{2}$-4	.025	.037
4-4$\sqrt{2}$	.194	.203
4$\sqrt{2}$-8	.478	.459
8-8$\sqrt{2}$	.138	.136
8$\sqrt{2}$-16	.160	.152

There is reasonable agreement. The L & G method is recommended because of its simplicity. Wise's method for size intervals not in geometric progression is tedious. It would appear that L & G gives better results.

Even though data from the Coulter counter are Type I (distribution of volumes), many experimenters want to convert the results to Type II (weight fraction). This data conversion is not recommended because results are not accurate even for sample sizes as large as N = 1,000 (see Ref. 6 for more details). There are many ways of doing this transformation.[23] Three methods are compared in the following examples.

Example II.2.A. L & G method[6] for converting Coulter counter data to weight fraction data. Data are taken from Allen (Ref. 7, p. 146).

Step 1: List the data:

INTERVAL v_i^*, v_{i+1}^*	MIDPT. v_i	FREQUENCY n_i	REL. FREQ. n_i/N
.095-.252	.17	6,671	.5048
.252-.796	.52	3,790	.2868
.796-2.25	1.52	1,717	.1299
2.25-5.80	4.03	796	.0602
5.80-17.2	11.50	205.2	.0155
17.2-48.5	32.85	35.5	.0027
48.5-132	90.25	1.0	.0001
		13,215.7	1.0000

Step 2: Use an experimentally determined instrument calibration factor K = 20.6 to transform v to d; i.e., $d = K\sqrt[3]{V}$ (Ref. 7, p. 145). Compute d_{M_i} and the relative proportion to $w_i' = n_i d_{M_i}^3$. Results are tabulated:

d_i^*, d_{i+1}^*	d_{M_i}	n_i	$n_i d_{M_i}^3$	w_i^*/W_A	STANDARDIZATION $w_i'/W_A/(d_{i+1}^* - d_i^*)$
9.4–13	11.2956	6672	9,614,312.	.0828	.023
13–19	16.1853	3790	16,069,448.	.1384	.023
19–27	23.2296	1717	21,522,595.	.1854	.023
27–37	32.2583	796	26,720,048.	.2302	.023
37–53	45.4691	205.5	19,317,924	.1664	.010
53–75	64.6241	35.5	9,581,024	.0825	.004
75–105	90.8257	1.0	749,249	.0065	.0002
		13216.0	103,574,600		

Allen's Computed Weight W_A = 116,079,940[a]

[a] Allen's correction (Ref. 7, p. 145) for total sample weight was used to compute w_i'/W_A in order to compare methods.

Example II.2B. Wise's method for converting Coulter counter data to weight fraction data. The diameter end points turn out to be a geometric progression so that formulas in Example II.1B may be used again.

Step 1: Same as Step 1 in Example II.2A.

Step 2: Let $d_i^* \approx 2^{u/2}$ and make the same computation to obtain d_i^* from v_i^*.

d_i^*, d_{i+1}^*	v_i'/V (L & G)	v_i'/V (Wise)	v_i'/V (Allen)
9.4–13	.083	.095	.080
13–19	.138	.125	.144
19–27	.185	.184	.183
27–37	.230	.245	.227
37–53	.166	.156	.168
53–75	.083	.079	.083
75–105	.007	.008	.004

(a) d_i^*	9.4		13		19		25		37		53		75		105	
(b) u (u_0=)	6.5		7.5		8.5		9.5		10.5		11.5		12.5		13.5(=u)	
(c) n_i/N		.5048		.2868		.1299		.0602		.0155		.0027		.0001		
(d) Σn_i/N	0		.5048		.7916		.9215		.9817		.9972		.9999		1.0000	
(e) $f(d) = 1 - $ (d)	1		.4952		.2084		.0785		.0183		.0028		.0001		0	
(f) $(d_i^*)^3 \times$ (e)	830.6		1,088.0		1,492.4		1,545.1		926.9		416.9		42.2		0	

Step 3: A bar chart of $d^3 f(d)$ vs u (Fig. 1.6) and an example of computing $\int_{u_i}^{u} d^3\, f(d)\, du$ is provided for further assistance.
Computations for two intervals:

$$\int_{12.5}^{13.5} d^3 f(d)\, du = \int_{12.5}^{13.0} 42.2\, du + \int_{13.0}^{13.5} 0$$
$$du = 21.1$$

$$\int_{11.5}^{13.5} d^3 f(d)\, du = \int_{11.5}^{12} 416.9 + \int_{12}^{13} 42.2$$
$$+ \int_{13}^{13.5} 0 = 208.45 + 52.2$$
$$+ 0 = 280.65$$

The summary table is given in (g) thru (n).

There seems to be good agreement, and the L & G method is once again recommended because of its simplicity.

It is important to observe that with a sample size of 1,000, the derived weight fraction distribution is not considered to be valid. Particles of large size have a small probability of being observed; consequently, they contribute little to the size distribution based on frequencies, but contribute an appreciably large amount to the weight fraction distribution. A sample of size 1,000 simply does not include enough large particles for the weight fraction distribution to have its necessary properties. Figure 1.7 which summarizes graphical data comparisons points out the problem. The data come from Example

(g) u_i^* (u_0=)	6.5	7.5	8.5	9.5	10.5	11.5	12.5	13.5
(h) $2\int_{u_i^*}^{u} d^3 f(d)\, du$	11,727.6	9,801	7,291.6	4,317.1	1,845.1	501.3	42.2	0
(i) (3/4) log 2 × (*h*)	6,096.7	5,099.3	3,790.6	2,244.3	959.2	260.6	21.9	0
(j) $(d^*)^3 f(d)$	830.6	1,088.0	1,429.4	1,545.1	0926.9	416.9	42.2	0
(k) (*i*) + (*j*)	6,927.3	6,187.3	5,220.0	3,789.4	1,886.1	677.5	64.1	0
(l) Rel. prop. by vol.	740	967.3	1,430.6	1,903.3	1,208.6	613.4	64.1	

(m) 1.12 × (740 + 967.3 + · · · + 64.1) = 1.12 × 6,927.3 = 7,758.6

(n) $v_i/V = w_i/W$	.095	.125	.185	.245	.156	.079	.008

Finally, a summary of the results of Allen and Marshall (Ref. 23, p. 146), L & G, and Wise is presented:

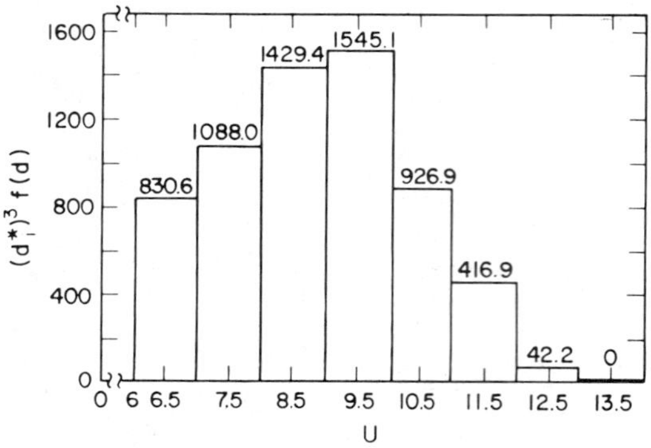

Figure 1.6. A bar chart of $d^3 f(d)$ vs u.

II.1 and the observed (n_i/N) and transformed (w_i'/W) are compared.

Notice that the variance hardly changes but the mean shifts (see Ex. II.3). The lack of a "large enough" sample size is apparent in the weight fraction data. An important criteria for an adequate sample size is for the right tail to asymptotically approach zero as d gets large. This is not the case as shown in Figure 1.7.

Comparison of Sample Statistics

Formulas given on page 20 of Ref. 5 will be used in conjunction with Examples II.1 and II.2 to demonstrate the comparison of sample statistics.

Example II.3. Computing mean, variance, and skewness when microscopic data is transformed to weight fraction.

Step 1: Use Step 1 in Example II.1.A to set up the following statistical layout.

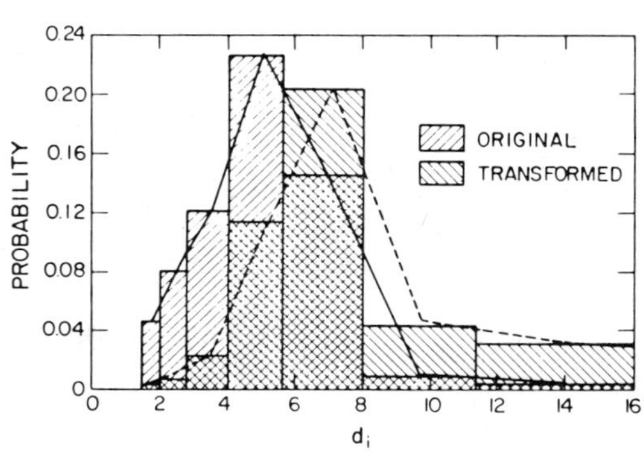

Figure 1.7. A comparison of relative frequency (original) data with weight fraction (transformed).

d_i^*, d_{i+1}^*	MIDPOINT d_i	FREQUENCY n_i
$\sqrt{2}$-2	1.71	27
2-2$\sqrt{2}$	2.41	66
2$\sqrt{2}$-4	3.41	142
4-4$\sqrt{2}$	4.83	383
4$\sqrt{2}$-8	6.83	334
8-8$\sqrt{2}$	9.66	34
8$\sqrt{2}$-16	13.65	14
		1000

Step 2: Compute sample statistics for relative frequency distribution.

$$\bar{d} = \Sigma n_i d_i/\text{N} = (1.7071 \times 27 + \cdots + 13.6568 \times 14)/1{,}000 = 5.34$$

$$s_d^2 = [\Sigma n_i d_i^2/\text{N} - \bar{d}^2]/(\text{N} - 1) = [27 \times 1.7071^2 + \cdots + 14 \times 13.6568^2]/1000 - (5.34)^2]/999 = 3.900$$

$$s_d = 1.97$$

$$g_d = \Sigma n_i(d_i - \bar{d})^3/(\text{N} - 1) = [27(1.71 - 5.34)^3 + \cdots + 14(13.66 - 5.34)^3]/999 = 7.29$$

$$\beta_1 = (g_d)^2/(s_d^2)^3 = 0.90$$

The distribution is nearly symmetrical, with mean = 5.34 and standard deviation = 1.97.

Step 3: Set up the statistical layout for weight fraction data.

d_i^*, d_{i+1}^*	d_{M_i}	w_i'/W
$\sqrt{2}$-2	1.72	.0006
2-2$\sqrt{2}$	2.44	.0042
2$\sqrt{2}$-4	3.45	.0254
4-4$\sqrt{2}$	4.88	.1938
4$\sqrt{2}$-8	6.89	.4781
8-8$\sqrt{2}$	9.75	.1376
8$\sqrt{2}$-16	13.79	.1603

Step 4: Compute sample statistics for weight fraction data.

$$\bar{d}_3 = \Sigma w_i' d_{M_i}/W = (1.72)(.0006) + \cdots + (13.79)(.1603) = 7.89.$$

$$s_{d_3}^2 = \Sigma w_i'(d_{M_i} - \bar{d}_3)^2/W = (1.72 - 7.89)^2 (.0006) + \cdots + (13.79 - 7.89)^2 (.1603) = 8.94.$$

$$s_{d_3} = 2.99$$

$g_{d_3} = \Sigma w_i'(d_{M_i} - \bar{d}_3)^3/W = (1.72 - 7.89)^3$
$(.0006) + \cdots + (13.79 - 7.89)^3$
$(.1603) = 25.00.$

$\beta_1 = (g_d)^2/(s^2_{\bar{d}_3})^3 = .87.$

Lognormal Data Comparison

Detailed information concerning the lognormal distribution may be found in References 5 and 25. Formulas found on pages 21, 22, and 23 of Reference 5 will be used in the following examples to demonstrate proper transformation.

Example II.4. Computing sample statistics of weight fraction data from microscopic data when the lognormal distribution is assumed to be true. An experimenter might want to subjectively decide on his data being lognormal from a plot on log probability paper or if possible use one of the objective tests given in the first section of this chapter.

Step 1: List of data.

INTERVAL d_i^*, d_{i+1}^*	INTERVAL x_i^*, x_{i+1}^*	MIDPT. x_i	FREQ. n_i	THEORETICAL FREQUENCY f_i	CHI-SQUARE $(n_i-f_i)^2/f_i$
$\sqrt{2}$-2	.35-.69	.52	27	10	28.90
2-2$\sqrt{2}$	.69-1.04	.87	66	70	.23
2$\sqrt{2}$-4	1.04-1.39	1.22	142	220	27.65
4-4$\sqrt{2}$	1.39-1.73	1.56	383	340	5.44
4$\sqrt{2}$-8	1.73-2.08	1.91	334	260	21.06
8-8$\sqrt{2}$	2.08-2.43	2.26	34	85	30.60
8$\sqrt{2}$-16	2.43-2.77	2.60	14	14	0

Computations used in determining f_i values:

$\bar{x} = \Sigma n_i x_i/\text{N} = .52 \times 27 + \cdots + 2.6 \times 14$
$= 1.59.$

$s_x^2 = \Sigma n_i(x_i - \bar{x})^2/(\text{N} - 1) = [27(.52 - 1.59)^2$
$+ \cdots + 14(2.60 - 1.59)^2]/999$
$= .1487$

$s_x = .39$

x_{i+1}^*	$(x_{i+1}^* - \bar{x})/s_x$	CU. AREA FROM NORMAL DIST.	THEORETICAL AREA IN x_i^*, x_{i+1}^*	f_i (AREA × N)
.69	−2.31	.01	.01	10
1.04	−1.41	.08	.07	70
1.39	−.51	.30	.22	220
1.73	.36	.64	.34	340
2.08	1.26	.90	.26	260
2.43	2.15	.985	.085	85
2.77	3.03	.999	.014	14

Step 2: Compute chi-square and state conclusion.

$$\chi^2 = 38.90 + \cdots + 30.6 + 0 = 113.88$$

The hypothesis of log normality is rejected, and the formulas given in the preceeding section are said to be "best."

Remark. It is again emphasized that log normality cannot be tested by the chi-square distribution for moment weighted data.

Example II.5: Assume a lognormal distribution fits the data in Example II.4. Formulas from Reference 5 are used to compute sample statistics.

Step 1: List statistics from the log transform distribution (from Ex. II.3).

$$\hat{\mu}_x = \bar{x} = 1.59 \qquad \hat{\sigma}_x^2 = s_x^2 = .1487$$
$$\hat{\sigma}_x = s_x = .39$$

Greek letters imply the lognormal distribution is assumed.

Step 2: Obtain weight fraction statistics.

$$\hat{\mu}_{x3} = \hat{\mu}_x + 3\hat{\sigma}_x^2 = 1.59 + (3)(.1487) = 2.0361$$

$$\hat{\sigma}_{x3}^2 = \hat{\sigma}_x^2 = .1487 \qquad \hat{\sigma}_{x3} = \hat{\sigma}_x = .39$$

$$\begin{aligned}\hat{\mu}_d &= \exp[\hat{\mu}_x + .5\hat{\sigma}_x^2] \\ &= \exp[1.59 + (.5)(.1487)] \\ &= \exp(1.66) = 5.28\end{aligned}$$

$$\begin{aligned}\hat{\mu}_{d3} &= [\exp(3\hat{\sigma}_x^2)\hat{\mu}_d] = [e^{3(.1487)}(5.28)] \\ &= 8.25\end{aligned}$$

$$\begin{aligned}\hat{\sigma}_d^2 &= \exp[2\hat{\mu}_x + 2\hat{\sigma}_x^2] - \exp[2\hat{\mu}_x + \hat{\sigma}_x^2] \\ &= 32.38 - 27.90 = 4.48\end{aligned}$$

$$\hat{\sigma}_d = 2.12$$

$$\begin{aligned}\hat{\sigma}_{d3}^2 &= [e^{6\hat{\sigma}_x^2}\hat{\sigma}_d^2] = e^{(6)(.1487)}\, 4.48 \\ &= 10.93\end{aligned}$$

$$\hat{\sigma}_{d3} = 3.30$$

It is interesting to compare the correct results given in Example II.3 with the improper lognormal transformed results:

	MICROSCOPIC DATA CORRECT (NONPARAMETRIC)	MICROSCOPIC DATA INCORRECT (LOGNORMAL)	WEIGHT FRACTION DATA CORRECT (NONPARAMETRIC)	WEIGHT FRACTION DATA INCORRECT (LOGNORMAL)
Mean	5.34	5.28	7.89	8.25
Variance	3.90	4.48	8.94	10.93
Standard deviation	1.97	2.12	2.99	3.30

This points out the "robustness" of the lognormal assumption. Clearly, the hypothesis was rejected; however, the results are not too far off.

Example II.6. Computing sample statistics of weight fraction data from Coulter counter data assuming a lognormal distribution. For the sake of convenience, assume the data in Example II.2.A follow a lognormal distribution. For this example use the volumetric shape factor instead of the instrument calibration factor k from Ex. II.2.A.

Step 1: Compute sample statistics for mean and variance of the distribution of log v.

$$\begin{aligned}\bar{v} &= \Sigma n_i v_i / N = .17 \times .5048 + .52x\ .2868 \\ &\quad + \cdots + 90.25 \times .0001 = .95\end{aligned}$$

$$\begin{aligned}s_v^2 &= \Sigma n_i (v_i - \bar{v})^2/(N-1) \\ &= [6671 \times (.17 - .95)^2 + \cdots \\ &\quad + 1(90.25 - .95)^2]/999 = 44.18\end{aligned}$$

$\log v_i^*, \log v_{i+1}^*$	$\log v_i$	n_i
-2.35-1.38	-1.86	6,671
-1.38--.23	-.80	3,790
-.23-+.81	.29	1,717
.81-1.76	1.29	796
1.76-2.84	2.30	205.5
2.84-3.88	3.36	35.5
3.88-4.88	4.38	1
		13,216.0

$$\begin{aligned}\hat{\mu}_{z3} &= [-1.86 \times 6{,}671 + \cdots + 4.38 \\ &\quad \times 1]/13{,}216 = -1.01.\end{aligned}$$

$$\begin{aligned}\hat{\sigma}_{z3}^2 &= \{[-.186 - (-1.01)]^2 \times 6{,}671 + \cdots \\ &\quad + [4.38 - (-1.01)]^2 \times 1\}/13{,}215 \\ &= 15{,}057.975/13{,}215 = 1.14\end{aligned}$$

$$\hat{\sigma}_{z3} = 1.07$$

$$C_3 = \frac{\Pi}{6}$$

$$\begin{aligned}\hat{\mu}_d &= (C_3)^{-1/3} \exp\left[\frac{6\hat{\mu}_{z3} + \hat{\sigma}_{z3}^2}{18}\right] \\ &= \left(\frac{\Pi}{6}\right)^{-1/3} \exp\left[\frac{6(-1.01) + 1.14}{18}\right] \\ &= (1.24)(.76) = 0.94\end{aligned}$$

$$\begin{aligned}\hat{\sigma}_d^2 &= \left(\frac{\Pi}{6}\right)^{-2/3} \exp\left[\frac{6\hat{\mu}_{z3} + \hat{\sigma}_{z3}^2}{9}\right] \\ &\quad \exp\left[\left(\frac{\hat{\sigma}_{z3}^2}{9}\right) - 1\right] \\ &= (1.54)(.58)(.14) = .12\end{aligned}$$

$$\hat{\sigma}_x^2 = \hat{\sigma}_{z3}^2/9 = 1.14/9 = .13.$$

Finally,

$$\hat{\mu}_{d_3} = e^{3\hat{\sigma}_x^2}\hat{\mu}_d = e^{.39}\cdot .94 = 1.39$$

$$\hat{\sigma}_{d_3}^2 = [e^{6\hat{\sigma}_x^2}]\,\hat{\sigma}_d^2 = e^{.78}\ (.12) = 0.26$$

Computation of Specific Surface

Specific surface is a sample statistic defined as the ratio of total surface area to weight, $S_w = A/W$, or total surface area to volume, $S_v = A/V$, of the particles in a sample. Methods for computing specific surface are given on pages 24, 25, and 26 in Reference 5. The following examples use sample data from various instruments and include cases where a lognormal distribution is assumed. The lognormal formulas are included in the examples because of the relative ease of computation. It is assumed that the diameters are measured in micrometers.

Example II.7. Computing specific surface from sample statistics. Data are collected from a microscope (Ex. II.1), a Coulter counter (Ex. I.2), and a sieving device (Ex. I.3).

A. Microscope

Step 1: See Step 1 in Example II.1.

Step 2: Use the formula $S_v = (C_2\ \Sigma n_i d_i^2)/(C_3\ \Sigma n_i d_i^3)$ where for spherical particles $C_2 = \Pi$ and $C_3 = \Pi/6$.

$d_i = (d_i^* + d_{i+1}^*)/2 =$ 1.71   2.41   3.41   4.83   6.83   9.66   13.66

$n_i =$ 27   66   142   383   334   34   14

$$S_v\ \text{cm}^{-1} = (\Pi/\Pi/6)[27 \times (1.71)^2 + \cdots + 14 \times (13.66)^2]/[27(1.71)^3 + \cdots + 14(13.66)^3] \times 10^4$$
$$= [6(32{,}414.244)/222{,}594.86] \times 10^4 = .87 \times 10^4$$

$S_w(\text{cm}^2/\text{g}) = S_v/\rho = .87 \times 10^4/\rho$ where ρ = powder density.

B. Coulter Counter

Step 1: See Step 1 in Example I.2.

Step 2: Use the formula

$$S_v = [C_2\ \Sigma n_i d_{L_i}^2/\Sigma n_i v_i]\,10^4$$

where $d_{L_i}^2 = [(d_{i+1}^*)^2 + (d_{i+1}^*)(d_i^*) + (d_i^*)^2]/3$
$d_i^* = (v_i^*/C_3)^{1/3}$

$$C_2 = \Pi \quad \text{and} \quad C_3 = \Pi/6$$

d_i^*	6.59	8.31	9.51	10.46	11.27	11.98	12.61	13.18	1.371	14.20	14.66	17.89
d_{L_i}	7.47	8.92	9.99	10.87	11.62	12.30	12.90	13.45	13.96	14.43	16.30	
n_i	10	70	71	114	91	64	30	21	13	10	6	
v_i	225	375	525	675	825	975	1125	1275	1475	1575	2325	

$$S_v(\text{cm}^{-1}) = \{\Pi[(10)(7.47)^2 + \cdots + 6(16.30)^2]/387{,}450\} \times 10^4$$
$$= [199{,}975/387{,}450]\,10^4$$
$$= .52 \times 10^4$$

C. Sieve (Weight Fraction)

Step 1: See Example I.3, Step 1.

Step 2: Use the formula

$$S_v(\text{cm}^{-1}) = 10^4(C_2/C_3)\ \Sigma(w_i'/w)\,(d_i^2/d_{m_i}^3)$$

where

$$d_{M_i} = \{[(d_i^*)^2 + (d_{i+1}^*)^2](d_i^* + d_{i+1}^*)/4\}^{1/3},\ \text{and}$$

$$(C_2/C_3) = 6.$$

d_i^*	.8	1	2	3	5	7	9	15	20	25	30	40	55
d_i	.9	1.5	2.5	4.0	6.0	8.0	12.0	17.5	22.5	27.5	35.0	47.5	
(w_i'/W)	.02	.08	.12	.23	.13	.08	.15	.06	.05	.04	.03	.02	
d_{M_i}	.9	1.6	2.5	4.1	6.1	8.0	12.2	17.6	22.6	27.6	35.2	47.9	
S_v	.13	.26	.29	.32	.12	.06	.07	.02	.01	.01	.01	.00	

$$S_v(\text{cm}^{-1}) = [13 + .26 + \cdots + .00] \times 10^4 = 1.3 \times 10^4$$

It is sometimes desirable to estimate the specific surface of material in a particular size range for a given sample. The method used is evident in the sieve example. For particles in the size interval $2 \leqslant d_i^* \leqslant 5$, the specific surface would equal .29 + .32 = .61.

Example II.8. Computing specific surface from a lognormal distribution. Data from Example II.7 are assumed to be lognormal, only for illustrative purposes.

A. Microscope

Step 1: See Example II.1, Step 1.

Step 2: See Example II.4, Step 2, computations of sample statistics.

$$\hat{\mu}_x = \bar{x} = 1.59 \quad \text{and} \quad \hat{\sigma}_x^2 = s_x^2 = .1487$$

$$S_v(\text{cm}^{-1}) = (C_2/C_3)[\exp[-(\hat{\mu}_x + 2.5\hat{\sigma}_x^2)]\,10^4 = (\Pi/\Pi/6)\exp[-(1.59 + .37)]\,10^4 = (6)(.14) \times 10^4 = .84 \times 10^4$$

B. Coulter Counter

Step 1: See Example I.2.A, Step 1.

Step 2: See Example I.2.B, Step 1.

$$\hat{\mu}_{z3} = 6.56 \quad \text{and} \quad \hat{\sigma}_{z3}^2 = s^2 = .19$$

$$S_v(\text{cm}^{-1}) = (C_2/C_2^{2/3})\exp[-(6\hat{\mu}_{z3} + 5\hat{\sigma}_{z3}^2)/18]\,10^4 = [\Pi/(\Pi/6)^{2/3}][\exp(-2.24)]\,10^4 = 4.84 \times (.11) \times 10^4 = .51 \times 10^4$$

C. Sieve

Step 1: See Example I.3, Step 1.

Step 2: Find $\hat{\mu}_{x3}$, $\hat{\sigma}_x^2$.

(x_i^*, x_{i+1}^*) INTERVAL $= \log(d_i^*, d_{i+1}^*)$	x_i = MIDPOINT OF SIZE INTERVAL	w_i'/W WEIGHT FRACTION
−.22–0	−.11	.02
0–.69	.35	.08
.69–1.10	.90	.12
1.10–1.61	1.36	.23
1.61–1.95	1.78	.13
1.95–2.20	2.08	.08
2.20–2.71	2.60	.15
2.71–3.00	2.86	.06
3.00–3.22	3.11	.05
3.22–3.40	3.31	.04
3.40–3.69	3.55	.04
3.40–4.01	3.71	.02

$$\hat{\mu}_{x3} = \Sigma x_i(w_i'/W) = -.11 \times .02 + \cdots + 3.71 \times .02 = 1.91$$

$$\hat{\sigma}_x^2 = \Sigma(w_i'/W)(x_i - \bar{x})^2 = (.02)(-.11 - 1.91)^2 + \cdots + .02(3.17 = 1.91)^2 = .9211$$

$$S_v(\text{cm}^{-1}) = (C_2/C_3)\exp[-(\hat{\mu}_{x3} - .5\hat{\sigma}_x^2)]\,10^4 = 6\exp[-(1.45)] \times 10^4 = 1.41 \times 10^4$$

Summarizing results for S_v:

	NONFUNCTIONAL	LOGNORMAL
Microscope	.87 × 10⁴	.84 × 10⁴
Coulter counter	.52 × 10⁴	.51 × 10⁴
Sieve	1.30 × 10⁴	1.41 × 10⁴

The two methods are comparable. The lognormal distribution should be used whenever the hypothesis is acceptable.

The Packing Problem

The packing problem is concerned with the determination of the nature of a randomly distributed array of particles. Analytic attempts at calculation of properties of powders have not been successful. A mathematical model appears hopeless because too many variables are needed to explain the state of the system at any given time. Goldman, Lewis, and Visscher[26] provide a one-dimensional model that solves the "parking" rather than the "packing" problem. L & G[27] have utilized a simplistic approach to provide an answer to the difficult problem of what fractions of two components, a coarse and a fine powder, should be mixed together to maximize packing density.

Example II.9. Computing the binary weight mixing fraction at maximum value of $\hat{\sigma}^2_{x(m)}$, we get variance of a distribution of mixed powders assuming each distribution is approximately lognormal.

Step 1: Use the data listed in Example II.4 and Example II.2(A).

Step 2: From Example II.5: $\hat{\mu}_{x(2)} = 1.59$, $\hat{\sigma}^2_{x(2)} = .1487$ (Ex. II.4 data). Using Example II.2.A, end points are converted to logs and $\hat{\mu}_{x(1)}$ and $\hat{\sigma}^2_{x(1)}$ computed as follows:

x_i^*, x_{i+1}^*	x_i	n_i/N
2.24–2.56	1.74	.505
2.56–2.94	2.75	.287
2.94–3.30	3.12	.130
3.30–3.61	3.46	.060
3.61–3.97	3.79	.020
3.97–4.32	4.15	.003
4.32–4.65	4.49	.000

$$\hat{\mu}_{x(1)} = \bar{x} = (1.74)(.505) + \cdots + (4.15)(.003) = 2.37$$

$$\hat{\sigma}^2_{x(1)} = (1.74 - 2.37)^2\, .505 + \cdots + (4.15 - 2.37)^2\, .003 = .4361$$

Step 3: Compute $\beta'_{(1)}$ and $1 - \beta'_{(1)}$ the binary frequency mixing fraction

$$\beta'_{(1)} = \frac{1}{2} + \frac{\hat{\sigma}^2_{x(1)} - \hat{\sigma}^2_{x(2)}}{2(\hat{\mu}_{x(1)} - \hat{\mu}_{x(2)})^2} = \frac{1}{2} + \frac{.4361 - .1487}{2(2.37 - 1.59)^2} = .74$$

$$1 - (\beta'_{(1)}) = 1 - .74 = .26$$

Step 4: Make final computations of binary weight mixing fractions $\alpha'_{(1)}$ and $1 - \alpha'_{(1)}$.

$$\alpha'_{(1)} = \frac{\beta'_{(1)} \exp\,[3\hat{\mu}_{x(1)} + 4.5\hat{\sigma}^2_{x(1)}]}{\sum_{j=1}^{2} \beta'_{(j)} \exp\,[3\hat{\mu}_{x(j)} + 4.5\hat{\sigma}^2_{x(j)}]} = \frac{6{,}446.8356}{6{,}446.8356 + 230.246} = .97$$

$$\alpha'_{(2)} = .03$$

These results indicate that 97% of powder 1 should be mixed with 3% powder 2 by weight in order to achieve maximum packing density. This result can be used as a guide and additional experiments should yield improved results. Computing the binary weight mixing fraction to maximize the value of $s^2_{d(m)}$ in cases other than the lognormal are unknown.

Qualitative Utilization of Data

Particle size analysis or the measurement of particle size distribution is a common effort in any physical, chemical, or mechanical process. This information and the processing methods are intricate factors that relate directly to material behavior and/or physical properties of the fabricated product. This complex dependency is exemplified by the fact that no analytical procedure exists for predicting the random packing density of spherical particles having a known size distribution. One approach in making such predictions is to compare size data of untested particulates with results of materials whose in-process behavior or as-fabricated properties are known.

Useful comparisons of particulates might be obtained by subjective examination of descriptive quantities such as the median size, total surface area, or tap density of a powder. However, in most processes the characterization of material is strongly dependent upon a dimensionless measure from the size distribution. For example, the chemical activity of two particulate solids having identical composition, specific surface, and particle shape could have different chemical activities if their distributions had unequal coefficients of variation. This would result in different process behavior.

The problem is to determine the acceptability or failure of the size distribution of an untested raw material based upon a comparison with size distribution data of finished products. Particle size measuring devices that can quickly provide sample distribution data such that on-stream analyses are possible lend credence to this approach. There are four different types of comparisons necessary:

1. Replicate distributions of the same sample to determine reproducibility of the "sizing" instrument
2. Distributions from different samples of the same particulate to determine homogeniety of the material
3. Distributions of different materials that have known properties in order to classify powders
4. Distributions of untested and tested product for discrimination purposes

The first two of these comparisons are required in order to justify the final two. In general, there do not exist simple procedures for proper statistical analyses of any of the comparisons. "Sizing" techniques involving counting data require a very large sample size (see above, Graphical Data Comparison) which usually results in rejection of any hypothesis concerning equivalence of distributions. On the other hand, "sizing" techniques that give results in the form of "continuous" data make the analysis difficult because of the unknown sample size and the lack of any feasible statistical procedures.

Examples of applicable methods for comparing particle size data including the Kolmogorov statistic as an approach to classifying untested materials will be given. A more complete study can be found in Reference 28. Difficulties in data comparison arise because many particle sizing instruments display results graphically in the form of hundreds of step functions over time intervals that are small in magnitude. This gives output in the form of a smooth continuous curve instead of discrete sample results. There are no convenient objective statistical tools for handling such data; consequently, a discussion of discretization of data is necessary. An example of output data from six different powders analyzed on a sedigraph device is displayed in Figure 1.8. The problem is to determine what curves are significantly different.

A major obstacle in the statistical analysis of a continuous sample curve is how to discretize the data in order to: (1) obtain good estimates of sample statistics; (2) determine the shape of the distribution; and (3) obtain a satisfactory number of intervals that might be used in lieu of the unknown sample size. An answer to (1) was given above in the section Graphical Data Comparison; however, there is no known solution that will optimize all three conditions simultaneously. Consequently, conditions (2) and (3) must be examined separately.

An approach to condition (3) is given by Kendall and Stuart,[29] (p. 438). If the hypothesized distribution is normal, the Type I error $\alpha = .05$ and the power of the test $1 - \beta = .80$,

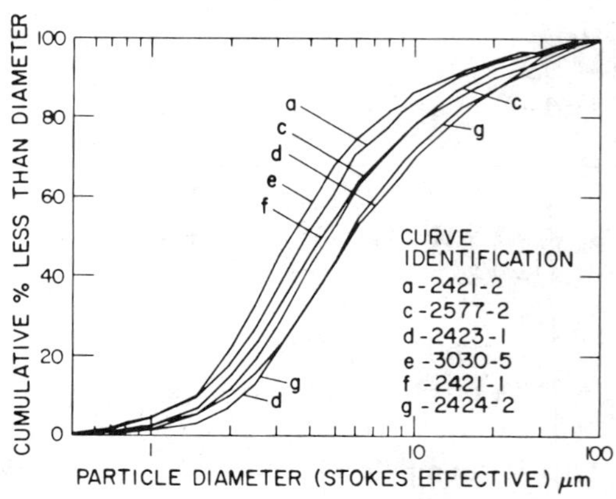

Figure 1.8. Sedigraph results.

Table 1.1. The Number of Intervals Required for Properly Displaying a Distribution

n	30	100	200	300	500	1,000	10,000	100,000
k	5	7	8	9	9	10	13	17

n	1,000,000	2,000,000	100,000,000	500,000,000
k	20	21	27	29

then some values of n (sample size) and k (no. of intervals) are tabulated.

n	200	400	600	800	1,000	1,500	2,000
k	17	20	24	27	30	35	39

The partitions used for estimating statistics will generally not be applicable for determining the shape of the distribution. Irregularities caused by too many intervals give a distorted picture of the distribution. It is usually best to have equal width intervals whose number is determined by $n = 2^k$ where n is the sample size and k is the number of intervals. Table 1.1 can be used as a guide.

Finding a proper sample size is virtually impossible. A sample size of 1,000,000 might be realistic; however, the need for statistical tools is eliminated because any slight differences will reject any hypothesis concerning equal distributions. In practice, the sample represents a very small fraction of the powder being investigated. The number of discretized points may be, in the practical sense, a justifiable estimate of n; however, no really satisfactory answer is available. An example of conditions (1) and (2) follows.

Example II.10. Discretizing a continuous function. The statistical layout of six different powders is depicted in Figure 1.8. Powder 2577-2 (labeled c) will be used in this example for illustrating interval breakdown for both computing sample statistics and data display. A set of eighty points was read from Figure 1.8 and displayed in Table 1.2. Although there are many excellent mathematical smoothing techniques available such as spline fitting, the amount of effort does not justify the small increase in accuracy that might be obtained. All points are considered to be interval end points.

Table 1.2. Eighty Values Read From Figure 1.5

a_i^*	$\frac{w_i'}{w}$	d_i^*	$\frac{w_i'}{w}$	d_i^*	$\frac{w_i'}{W}$	d_i^*	$\frac{w_i'}{w}$
.25	.02	2.00	.15	8.00	.68	18.00	.86
.255	.04	2.25	.17	8.50	.70	18.50	.87
.30	.04	2.00	.21	9.00	.71	19.00	.88
.35	.04	2.75	.24	9.50	.72	19.50	.88
.40	.05	3.00	.28	10.00	.74	20.00	.89
.45	.05	3.25	.32	10.50	.75	21.00	.89
.50	.05	3.50	.35	11.00	.77	22.00	.90
.55	.05	3.75	.38	11.50	.78	23.00	.91
.60	.05	4.00	.40	12.00	.79	24.00	.91
.65	.05	4.25	.44	12.50	.80	25.00	.91
.70	.05	4.50	.46	13.00	.81	26.00	.92
.75	.05	4.75	.48	13.50	.82	27.00	.92
.80	.06	5.00	.50	14.00	.82	28.00	.93
.85	.06	5.25	.52	14.50	.83	29.00	.93
.90	.06	5.50	.55	15.00	.83	30.00	.93
.95	.06	5.75	.58	15.50	.84	35.00	.95
1.00	.07	6.00	.59	16.00	.85	40.00	.96
1.25	.08	6.50	.62	16.50	.85	45.00	.97
1.50	.10	7.00	.65	17.00	.86	50.00	.98
1.75	.12	7.50	.66	17.50	.86	55.00	.99

Because of idiosyncrasies in the Sedigraph, the largest sized particle registers less than 100%; consequently data is usually normalized. For the six powders, 55 micrometers is known to be the largest size and must have a corresponding cumulative weight fraction of 100%. All of the weight fractions in Table 1.2 would have to be adjusted by 100/99 for more nearly correct readings.

These sample statistics were computed using the formulas given in Part 1.1.

Wt. fraction:

$$\bar{d}_3 = 8.64, s_{d_3}^2 = 102.63 \qquad S_w = 5.64 \times 10^4$$

Wt. fraction:

$$(x = \ln d): \bar{x}_3 = 1.60 \qquad s_{x_3}^2 = 1.23;$$

No. fraction:

$$\bar{d} = .30; s_d^2 = .08;$$

No. fraction:

$$(x = \ln d): \bar{x} = 1.28 \qquad s_x^2 = .09.$$

A graph of the 80 interval mesh is shown in Figure 1.9. There are too many intervals to give a reasonable picture of the data. In this case, 80 intervals was an arbitrary choice because the size of n is unknown. Referring to Table 1.1 and initially setting n between 100,000 and 1,000,000 with $k = 18$, a bar graph is depicted in Figure 1.10. It appears to be a judicious choice because of the skewness to the right verifies that for estimating parameters, smaller width intervals should be used in the domain of smaller sized particles and larger width intervals for bigger particles. (These latter choices were used in Figure 1.9). This is in keeping with the premise that probabilities in each interval should be kept as nearly equal as possible. Statistics were computed for the 18 interval break down and are given for comparative purposes.

Wt. fraction:

$$\bar{d}_3 = 8.78; s_{d_3}^2 = 104.69 \qquad S_w = 5.09 \times 10^4;$$

Wt. fraction:

$$(x = \ln d): \bar{x}_3 = 1.61 \qquad s_{x_3}^2 = 1.21;$$

No. fraction:

$$\bar{d} = .40; s_d^2 = .13;$$

No. fraction:

$$(x = \ln d): x = 1.06 \qquad s_x^2 = .12.$$

The sample distribution was fitted by the least squares methods to the lognormal distribution.

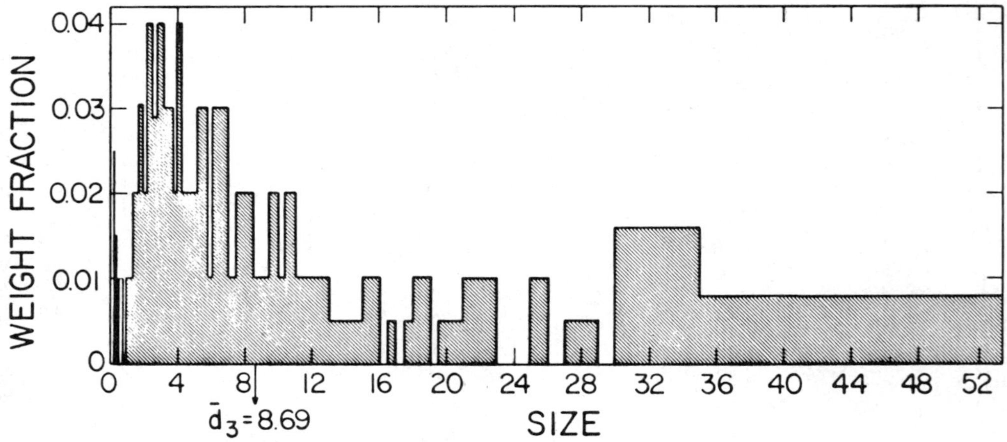

Figure 1.9. PRR 2557-2 data (80 intervals).

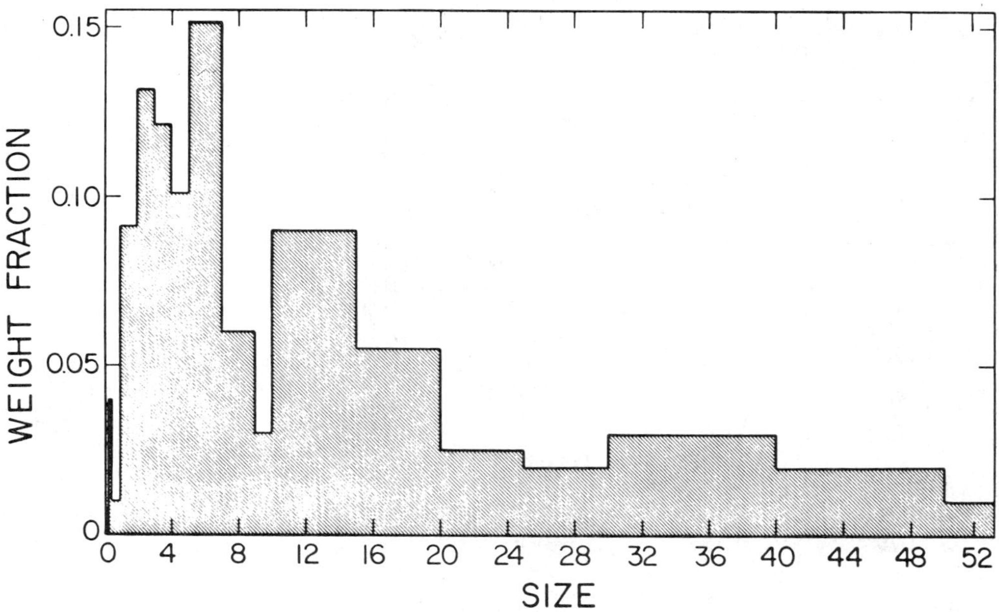

Figure 1.10. PRR 2557-2 data (18 unequally spaced intervals).

Subjectively, it looked like a good approximation; however, the continuity of the data prohibited the use of tests for log normality such as Wilks-Shapiro or chi square, and worse yet the lack of a "true" sample size invalidates the analysis of variance procedure for comparing the six different distributions. The next section discusses some qualitative solutions to the latter problem.

Methods for Data Comparisons

The inadequacy of subjective comparison of distribution curves is illustrated by the following example using the method of Chernoff faces.

Chernoff faces is a method that represents a multidimensional array as a cartoon with each variable depicted as a feature on a face. For example, curvature of the mouth, face width, and ear level would correspond to three components such as height, weight, and blood pressure. The purposes of examining a set of faces determined from many observations would include finding groups with similar characteristics, detecting data irregularities, or determining trends. Six different powders (see Figure 1.8) were broken into 17 intervals and weight fractions mapped by the Chernoff method. The results appear in Figure 1.11. There are 17 characteristics (weight fractions in intervals) incorporated into the drawing of each face including features such as face width, brow length, mouth curvature, pupil position, nose width, ear level, etc. Though the faces show quantitative characteristics of the curves of Figure 1.8, subjective comparison of these faces is no better than subjective comparison of the original curves in revealing "significant" differences. Therefore a more quantitative method for comparison must be developed. The authors' modification of the Kolmogorov method[28] provides a quantitative approach.

The Kolmogorov statistic D_n (Refs. 31, 32) is defined as the maximum absolute difference between two distributions. The n subscript is attached to indicate that its probability distribution is a function of sample size. The following example illustrates the modified method.

Example II.11. Using D_n to determine differences among distributions given in Figure 1.8, let $n = 2m$, where m is the "size" of each sample.

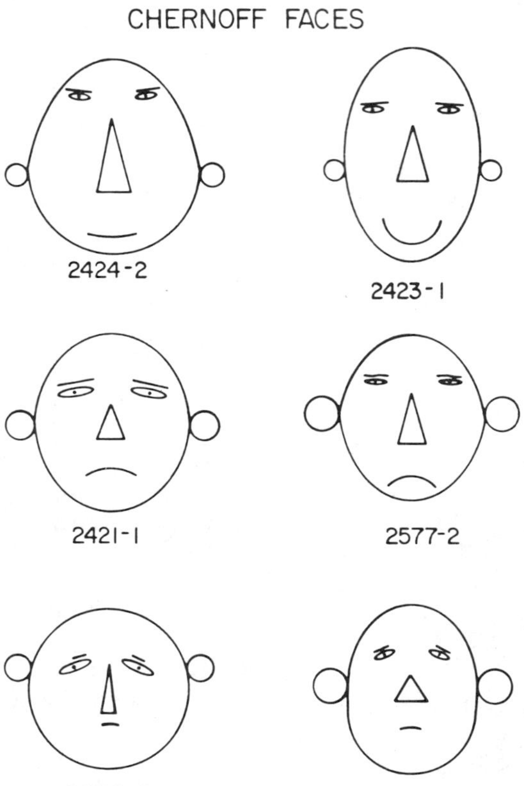

Figure 1.11. Chernoff faces.

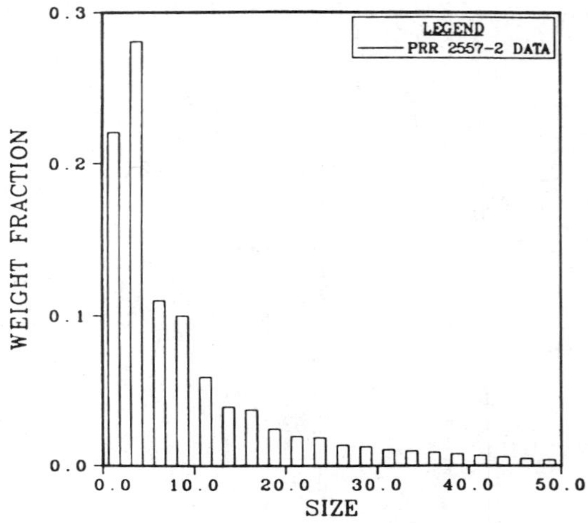

Figure 1.12. PR 2557-2 data partitioned into 20 equal intervals.

Step 1: Determine a critical value. Some critical values of α vs D_n are given.

α	.54	.39	.27	.20	.18	.15	.10	.08
D_n	$.8/\sqrt{n}$	$.9/\sqrt{n}$	$1/\sqrt{n}$	$1.07/\sqrt{n}$	$1.1/\sqrt{n}$	$1.14/\sqrt{n}$	$1.22/\sqrt{n}$	$1.25/\sqrt{n}$
α	.05	.04	.02	.01	.004	.001	.0007	
D_n	$1.36/\sqrt{n}$	$1.4/\sqrt{n}$	$1.5/\sqrt{n}$	$1.63/\sqrt{n}$	$1.75/\sqrt{n}$	$1.9/\sqrt{n}$	$2/\sqrt{n}$	

The value of $\alpha = .20$ is suggested instead of the typical $\alpha = .05$ simply because the chances of determining differences are enhanced.

The rationale for selecting n is based on the number of intervals required to determine the shape of the distribution and the number of intervals required to obtain effective sample statistics. Twenty equal sized intervals (Fig. 1.12) give an excellent pictorial description of a typical set of data. In addition, a total of 25 intervals seemed to give reasonable results for computing sample statistics.

A value of m between 20 and 25 seems appropriate. A value based on $\alpha = .20$, $m = 23$, $2m = n = 46$ gives a critical tabulated value of $1.07/\sqrt{46} = .158$. Any difference larger than .158 would indicate a difference in distributions; otherwise the hypothesis of equal distributions is accepted.

Step 2: Obtain values of D_{2m} by taking all possible absolute differences from the six curves in Figure 1.8 and look for values greater than .158.

	a	*c*	*d*	*e*	*f*	*g*
a	0	.0863	.1811*	.0839*	.0684	.1825*
c	.0863	0	.0994	.1702*	.0492	.0965
d	.1811*	.0994	0	.2526*	.1210	.0291
e	.0839	.1702*	.2526*	0	.1362	.2572*
f	.0684	.0492	.1210	.1362	0	.1269
g	.1825*	.0965	.0291	.2572*	.1269	0

The conclusion would be that *a* differs from *d* and *g*; *c* differs from *e*; *d* differs from *a* and *e*; *e* differs from *a*, *c*, *d*, and *g*; *f* differs from none of the others; and *g* differs from *a* and *e*.

Concluding Statement. The proof of the value of this or any other method for distribution comparison is in the application of the results of the method to predict process behavior of the particulate material.

REFERENCES

1. T. Allen, *Particle Size Measurement*, Chapman and Hall Ltd., London, (1968).
2. J. M. Dalla Valle, *Micrometrics*, 2d ed., Pitman Publishing Corp, New York (1948).
3. R. R. Irani and C. F. Callis, *Particle Size Measurement, Interpretation and Application*, Wiley, New York (1963).
4. G. Herdan, *Small Particle Statistics*, 2d ed., Academic Press, Inc., New York (1960).
5. H. D. Lewis and A. Goldman, "Theoretical Small Particle Statistics: A Summary of Techniques for Data Analysis with Recent Developments in Data Comparison, Notation, and Mixture Theory," Los Alamos Scientific Laboratory Report LA-3656 (1968).
6. H. D. Lewis and A. Goldman, "Proper Analysis of Coulter Counter Data," *Rev. Sci. Instrum.*, pp. 868–69 (June, 1965).
7. T. Allen, *Particle Size Measurement*, 2d ed., Chapman and Hall, London, (1977).
8. R. D. Cadle, *Particle Size, Theory, and Industrial Application*, Reinhold, New York (1965).
9. C. Orr and J. M. Dalla Valle, *Fine Particle Measurement*, Macmillan, New York (1959).
10. E. M. Wewerka and J. M. Dickinson, "Use of Gel Permeation Chronotography in the Development of High Quality Graphite," *J. of Chromatography* v*55*:25–31 (1971).
11. H. D. Lewis, "Small Particle Statistics: The Analysis of Particle 'Size Data'," in K. Beddow and T. P. Malloy, *Testing and Characterization of Powders and Fine Particles*, Hayden & Sons, Ltd. Spectrum House, Hill View Gardens, London NW4, pp. 136–137 (1980).
12. H. D. Lewis, "Small Particle Statistics–The Analysis of Particle 'Size Data'," Los Alamos Scientific Laboratory Report LA-5254-MS (1973).
13. S. S. Shapiro and M. B. Wilk, "An Analysis of Variance Test for Normality (Complete Samples)," *Biometrika* v. *52*:591–611 (1965).
14. Biomedical Computer Programs, Health Science Computing Facility, Dept. of Biomathematics, School of Medicine UCLA, Univ. of California Press (1975).
15. S. D. Wicksell, "The Corpuscle Problem," *Biometrika* v. *17*:84–89 (1925).
16. G. S. Watson, "Estimating Functions of Particle Size Distributions," *Biometrika* v. *58*:483–490 (1971).
17. S. A. Saltikov, "The Determination of the Size Distribution of Particles in an Opaque Material from a Measurement of the Size Distribution of Their Sections," in H. Elias (ed.), *Stereology*, Springer-Verlag, New York, pp. 163–173 (1967).
18. A. Goldman, H. D. Lewis, and Wm. M. Visscher, "Constrained Minimum Variance Estimation: A New Procedure for Estimating the Distribution of Spherical Particles from Plane Section," *Proceedings of the Technical Program, International Powder and Bulk Solids, Handling, and Processing*, pp. 151–160 (1979).
19. Wm. Visscher and A. Goldman, "Estimating the Distribution of Spherical Particles from Plane Sections An Optimal Algorithm for Solution of the Abel Integral Equation." To appear in SIAM Journal of Applied Computing (1983).
20. R. S. Anderson and A. J. Jakeman, "Abel-Type Integral Equations in Stereology," *Journal of Microscopy 105*:121–153 (1975).
21. J. W. Mellor, *Trans. Ceram. Soc. 9*:94 (1910).
22. M. E. Wise, "Converting a Number of Distribution of Particle Size into One for Volume or Surface Area," *Philps Res. Rep. 9*:231–237 (1954).
23. T. Allen and K. Marshall, "The Electrical Sensing Zone Method of Particle Size Measurement (The Coulter Principle)," University of Bradford, Bradford, England (Contact authors for article).
24. H. Lewis, K. Walters and K. Johnson, "Particle Size Distribution by Area Analysis: Modifications and Extensions of the Saltykov Method" *Metallography* 6:93–101 (1973).
25. J. Aitchison and A. C. Brown, "Lognormal Distributions," Cambridge University Press, New York (1957).
26. A. Goldman, H. D. Lewis, and Wm. M. Visscher, "Random Packing of Particles: Simulation With a One Dimensional Parking Model" *Technometrics 16*(2) (May, 1974).
27. H. D. Lewis and A. Goldman, "Theorems for Calculation of Weight Ratios that Produce Maximum Packing Density of Powder Mixtures," *J. Am. Cer. Soc. 49*:323–327 (1966).
28. A. Goldman and H. D. Lewis, "A Modified Kolmogorov Method for Comparison of Particle Size Distributions," Los Alamos Scientific Laboratory Report LA-UR-38-1361 (May, 1978).
29. M. Kendall and A. Stuart, *Advanced Theory of Statistics*, vol. 2, 3d ed., Hafner Press, New York (1976).
30. H. Chernoff, "Using Faces to Represent Points in k–Dimensional Space Graphically," *Journal of the American Statistical Association*, pp. 361–368 (June, 1973).
31. M. Kendall and A. Stuart, *Advanced Theory of Statistics*, vol. 3, 3d ed., Hafner Press, New York (1976).
32. W. J. Dixon and F. J. Massey, *Introduction to Statistical Analysis*, 3d ed., McGraw-Hill, New York (1969).

APPENDIX A1: TABLE OF COEFFICIENTS (a_{n-i+1}) USED IN W TEST FOR NORMALITY FOR n = 3(1)50

i \ n	3	4	5	6	7	8	9	10	11	12	13	14	15	16	17	18
1	0.7071	0.6872	0.6646	0.6431	0.6233	0.6052	0.5888	0.5739	0.5601	0.5475	0.5359	0.5251	0.5150	0.5056	0.4968	0.4886
2		.1667	0.2413	0.2806	0.3031	0.3164	0.3244	0.3291	0.3315	0.3325	0.3325	0.3318	0.3306	0.3290	0.3273	0.3253
3				0.0875	0.1401	0.1743	0.1976	0.2141	0.2260	0.2347	0.2412	0.2460	0.2495	0.2521	0.2540	0.2553
4						0.0561	0.0947	0.1224	0.1429	0.1586	0.1707	0.1802	0.1878	0.1939	0.1988	0.2027
5								0.0399	0.0695	0.0922	0.1099	0.1240	0.1353	0.1447	0.1524	0.1587
6										0.0303	0.0539	0.0727	0.0880	0.1005	0.1109	0.1197
7												0.0240	0.0433	0.0593	0.0725	0.0837
8														0.0196	0.0359	0.0496
9																0.0163

i \ n	19	20	21	22	23	24	25	26	27	28	29	30	31	32	33	34
1	0.4808	0.4734	0.4643	0.4590	0.4542	0.4493	0.4450	0.4407	0.4366	0.4328	0.4291	0.4254	0.4220	0.4188	0.4156	0.4127
2	0.3232	0.3211	0.3185	0.3156	0.3126	0.3098	0.3069	0.3043	0.3018	0.2992	0.2968	0.2944	0.2921	0.2898	0.2876	0.2854
3	0.2561	0.2565	0.2578	0.2571	0.2563	0.2554	0.2543	0.2533	0.2522	0.2510	0.2499	0.2487	0.2475	0.2463	0.2451	0.2439
4	0.2059	0.2085	0.2119	0.2131	0.2139	0.2145	0.2148	0.2151	0.2152	0.2151	0.2150	0.2148	0.2145	0.2141	0.2137	0.2132
5	0.1641	0.1686	0.1736	0.1764	0.1787	0.1807	0.1822	0.1836	0.1848	0.1857	0.1864	0.1870	0.1874	0.1878	0.1880	0.1882
6	0.1271	0.1334	0.1399	0.1443	0.1480	0.1512	0.1539	0.1563	0.1584	0.1601	0.1616	0.1630	0.1641	0.1651	0.1660	0.1667
7	0.0932	0.1013	0.1092	0.1150	0.1201	0.1245	0.1283	0.1316	0.1346	0.1372	0.1395	0.1415	0.1433	0.1449	0.1463	0.1475
8	0.0612	0.0711	0.0804	0.0878	0.0941	0.0997	0.1046	0.1089	0.1128	0.1162	0.1192	0.1219	0.1243	0.1265	0.1284	0.1301
9	0.0303	0.0422	0.0530	0.0618	0.0696	0.0764	0.0823	0.0876	0.0923	0.0965	0.1002	0.1036	0.1066	0.1093	0.1118	0.1140
10		0.0140	0.0263	0.0368	0.0459	0.0539	0.0610	0.0672	0.0728	0.0778	0.0822	0.0862	0.0899	0.0931	0.0961	0.0988
11				0.0122	0.0228	0.0321	0.0403	0.0476	0.0540	0.0598	0.0650	0.0697	0.0739	0.0777	0.0812	0.0844
12						0.0107	0.0200	0.0284	0.0358	0.0424	0.0483	0.0537	0.0585	0.0629	0.0669	0.0706
13								0.0094	0.0178	0.0253	0.0320	0.0381	0.0435	0.0485	0.0530	0.0572
14										0.0084	0.0159	0.0227	0.0289	0.0344	0.0395	0.0441
15												0.0076	0.0144	0.0206	0.0262	0.0314
16														0.0068	0.0131	0.0187
17																0.0062

i \ n	35	36	37	38	39	40	41	42	43	44	45	46	47	48	49	50
1	0.4096	0.4068	0.4040	0.4015	0.3989	0.3964	0.3940	0.3917	0.3894	0.3872	0.3850	0.3830	0.3808	0.3789	0.3770	0.3751
2	0.2834	0.2813	0.2794	0.2774	0.2755	0.2737	0.2719	0.2701	0.2684	0.2667	0.2651	0.2635	0.2620	0.2604	0.2589	0.2574
3	0.2427	0.2415	0.2403	0.2391	0.2380	0.2368	0.2357	0.2345	0.2334	0.2323	0.2313	0.2302	0.2291	0.2281	0.2271	0.2260
4	0.2127	0.2121	0.2116	0.2110	0.2104	0.2098	0.2091	0.2085	0.2078	0.2072	0.2065	0.2058	0.2052	0.2045	0.2038	0.2032
5	0.1883	0.1883	0.1883	0.1881	0.1880	0.1878	0.1876	0.1874	0.1871	0.1868	0.1865	0.1862	0.1859	0.1855	0.1851	0.1847
6	0.1673	0.1678	0.1683	0.1686	0.1689	0.1691	0.1693	0.1694	0.1695	0.1695	0.1695	0.1695	0.1695	0.1693	0.1692	0.1691
7	0.1487	0.1496	0.1505	0.1513	0.1520	0.1526	0.1531	0.1535	0.1539	0.1542	0.1545	0.1548	0.1550	0.1551	0.1553	0.1554
8	0.1317	0.1331	0.1344	0.1356	0.1366	0.1376	0.1384	0.1392	0.1398	0.1405	0.1410	0.1415	0.1420	0.1423	0.1427	0.1430
9	0.1160	0.1179	0.1196	0.1211	0.1225	0.1237	0.1249	0.1259	0.1269	0.1278	0.1286	0.1293	0.1300	0.1306	0.1312	0.1317
10	0.1013	0.1036	0.1056	0.1075	0.1092	0.1108	0.1123	0.1136	0.1149	0.1160	0.1170	0.1180	0.1189	0.1197	0.1205	0.1212
11	0.0873	0.0900	0.0924	0.0947	0.0967	0.0986	0.1004	0.1020	0.1035	0.1049	0.1062	0.1073	0.1085	0.1095	0.1105	0.1113
12	0.0739	0.0770	0.0798	0.0824	0.0848	0.0870	0.0891	0.0909	0.0927	0.0943	0.0959	0.0972	0.0986	0.0998	0.1010	0.1020
13	0.0610	0.0645	0.0677	0.0706	0.0733	0.0759	0.0782	0.0804	0.0824	0.0842	0.0860	0.0876	0.0892	0.0906	0.0919	0.0932
14	0.0484	0.0523	0.0559	0.0592	0.0622	0.0651	0.0677	0.0701	0.0724	0.0745	0.0765	0.0783	0.0801	0.0817	0.0832	0.0846
15	0.0361	0.0404	0.0444	0.0481	0.0515	0.0546	0.0575	0.0602	0.0628	0.0651	0.0673	0.0694	0.0713	0.0731	0.0748	0.0764
16	0.0239	0.0287	0.0331	0.0372	0.0409	0.0444	0.0476	0.0506	0.0534	0.0560	0.0584	0.0607	0.0628	0.0648	0.0667	0.0685
17	0.0119	0.0172	0.0220	0.0264	0.0305	0.0343	0.0379	0.0411	0.0442	0.0471	0.0497	0.0522	0.0546	0.0568	0.0588	0.0608
18		0.0057	0.0110	0.0158	0.0203	0.0244	0.0283	0.0318	0.0352	0.0383	0.0412	0.0439	0.0465	0.0489	0.0511	0.0532
19				0.0053	0.0101	0.0146	0.0188	0.0227	0.0263	0.0296	0.0328	0.0357	0.0385	0.0411	0.0436	0.0459
20						0.0049	0.0094	0.0136	0.0175	0.0211	0.0245	0.0277	0.0307	0.0335	0.0361	0.0386
21								0.0045	0.0087	0.0126	0.0163	0.0197	0.0229	0.0259	0.0288	0.0314
22										0.0042	0.0081	0.0118	0.0153	0.0185	0.0215	0.0244
23												0.0039	0.0076	0.0111	0.0143	0.0174
24														0.0037	0.0071	0.0104
25																0.0035

APPENDIX A2: PERCENTAGE POINTS OF W TEST FOR NORMALITY FOR $n = 3(1)50$

n	1	2	5	10	50
3	0.753	0.756	0.767	0.789	0.959
4	0.687	0.707	0.748	0.792	0.935
5	0.686	0.715	0.762	0.806	0.927
6	0.713	0.743	0.788	0.826	0.927
7	0.730	0.760	0.803	0.838	0.928
8	0.749	0.778	0.818	0.851	0.932
9	0.764	0.791	0.829	0.859	0.935
10	0.781	0.806	0.842	0.869	0.938
11	0.792	0.817	0.850	0.876	0.940
12	0.805	0.828	0.859	0.883	0.943
13	0.814	0.837	0.866	0.889	0.945
14	0.825	0.846	0.874	0.895	0.947
15	0.835	0.855	0.881	0.901	0.950
16	0.844	0.863	0.887	0.906	0.952
17	0.851	0.869	0.892	0.910	0.954
18	0.858	0.874	0.897	0.914	0.956
19	0.863	0.879	0.901	0.917	0.957
20	0.868	0.884	0.905	0.920	0.959
21	0.873	0.888	0.908	0.923	0.960
22	0.878	0.892	0.911	0.926	0.961
23	0.881	0.895	0.914	0.928	0.962
24	0.884	0.898	0.916	0.930	0.963
25	0.888	0.901	0.918	0.931	0.964
26	0.891	0.904	0.920	0.933	0.965
27	0.894	0.906	0.923	0.935	0.965
28	0.896	0.908	0.924	0.936	0.966
29	0.898	0.910	0.926	0.937	0.966
30	0.900	0.912	0.927	0.939	0.967
31	0.902	0.914	0.929	0.940	0.967
32	0.904	0.915	0.930	0.941	0.968
33	0.906	0.917	0.931	0.942	0.968
34	0.908	0.919	0.933	0.943	0.969
35	0.910	0.920	0.934	0.944	0.969
36	0.912	0.922	0.935	0.945	0.970
37	0.914	0.924	0.936	0.946	0.970
38	0.916	0.925	0.938	0.947	0.971
39	0.917	0.927	0.939	0.948	0.971
40	0.919	0.928	0.940	0.949	0.972
41	0.920	0.929	0.941	0.950	0.972
42	0.922	0.930	0.942	0.951	0.972
43	0.923	0.932	0.943	0.951	0.973
44	0.924	0.933	0.944	0.952	0.973
45	0.926	0.934	0.945	0.953	0.973
46	0.927	0.935	0.945	0.953	0.974
47	0.928	0.936	0.946	0.954	0.974
48	0.929	0.937	0.947	0.954	0.974
49	0.929	0.937	0.947	0.955	0.974
50	0.930	0.938	0.947	0.955	0.974

Appendixes A1 and A2 are reproduced from Ref. 13 with the kind permission of the publishers.

2

Particle Size Measurement: Experimental Techniques

Reg. Davies
E.I. DuPont deNemours
Wilmington, Delaware

2.1. INTRODUCTION

There are very few industries that do not at some stage in their manufacturing processes, produce, handle, characterize, and control matter in the particulate form. Our current lifestyle demands products in the form of powders, pastes, ointments, creams, tablets, suspensions, and aerosols, each exhibiting a specific chemical, biological, or physical property. These products have to be produced profitably, which means that they must have a low mill cost and high consumer acceptance. In these times of high inflation, rising energy costs and dwindling natural resources, profitability is geared to one's ability to constantly modify a product or process economically while maintaining its quality and end-use to the consumer. This demands an extensive knowledge of the characteristics of the raw materials and intermediates used in the process, and more important, a knowledge of how they interact in that unique and complex manner peculiar to particulate systems. In addition to the difficulties of production, industry now has to face additional problems relating to government regulations. If one prepares a particle balance analogous to making a heat balance in thermodynamics, one finds that in addition to the expected particulate products of the reaction, additional particles are continually being formed as a result of energy processes or mineral beneficiation, and many of these are continually lost to the environment via air and water discharges. The Environmental Protection Agency maintains a strict surveillance and control of these discharges to meet the environmental standards currently in force, and in order for industry to comply, it must design and install separation and collection devices at each discharge. These design criteria demand a knowledge of the particle characteristics and particle interaction behavior in the discharge fluids. More recently, concern has focused on the health and safety of workers, who use, manufacture, or otherwise have contact with particulate materials. It is known that both short- and long-term medical effects are dependent on the characteristics and interaction of particulate species that are inhaled or ingested. So in considering these three areas of industrial concern—production, pollution, health and safety—one finds that the need for particle characterization is the common thread that runs through them all. The countless, complex, and often unique particle products that are an integral part of our modern lifestyle can be best produced and used within the constraints of conservation and safety, when the chemical and physical properties of the particles have been measured, and their interactions comprehended, predicted, and accounted for.

However, as in most aspects of life, such general statements are easier to make than put into practice. Particle technology is a highly complex subject, and is not yet a science in the strict sense of the word. Table 2.1 lists many of the parameters important to the properties of particulate systems. All of them can be synthesized from four apparently basic particle parameters; size, shape, density and surface; and these four parameters act synergistically to various degrees. The difficulty arises when one attempts to define any one of them in isolation from the others. In analytical and physical chemistry, our building blocks are the atoms themselves, and a parameter such as the atomic number of an element is clearly defined and universally accepted. In particle technology such definitions are rare, and any statement that is made with respect to any one parameter has to be qualified, and, even then, measurement and interpretetation are always difficult. Such a listing as that given in Table 2.1 is naturally beyond discussion in this chapter, and, in consequence, only one basic parameter, particle size and size distribution is discussed here.

The purpose of preparing this chapter then is twofold. First, it is to help a prospective analyst to understand the practical difficulties of size measurement and to avoid the pitfalls. Second, it is to serve as a guide through the maze of instrumentation that is currently marketed by eager vendors, and to help in selecting the technique and instrument to best suit one's needs. No chapter of this type can be fully comprehensive, and so where further details are required, basic texts on the subject are recommended.

Table 2.1. Single Particle and Bulk Powder Parameters of Importance to Product Properties.

Single Particle Parameters
Size
Shape (macro and micro)
Surface (internal and external)
Density
Refractive index
Hardness
Melting point
Hygroscopicity
Light scattering and adsorption
Bulk Particle Parameters
Size distribution
Surface area
Bulk density
Tap density
Cohesion
Adhesion
Surface energy
Surface charge
Porosity
Pore size distribution
Moisture content
Conductivity
Dielectric strength
Tensile strength
Shear strength
Opacity

2.2. PARTICLE SIZE MEASUREMENT

An Overview of the General Problem

The particle size has been usefully defined as the state of subdivision of the material. For a smooth dense sphere, it can be accurately defined by a measurement of its diameter. In most applications, however, accurate particle size measurement is difficult, because most particles are irregular in shape; and therefore, the diameter is a function of the direction of measurement. It is also a function of the physical principle by which it is detected. Hence, a diameter calculated from data extracted from (a) an envelope-volume measurement of a powder in an electrolyte, (b) a light scattering measurement of a powder in aqueous media, or (c) a permeability measurement of the dry powder in air, will generally not be the same. The resulting differences, however, provide a valuable fingerprint of shape variations and can be used advantageously for this purpose. Typical methods of expressing size

Table 2.2. Several Methods of Expressing the Size of Irregular Particles.

Average thickness	The average diameter between the upper and lower surfaces of a particle in its most stable position of rest
Average length	The average diameter of the longest chords measured along the upper surface of a particle in the position of rest
Average breadth	The average diameter at right angles to the diameter of average length along the upper surface of a particle in its position of rest
Feret's diameter	The diameter between the tangents at right angles to the direction of scan, which touch the two extremities of the particle profile in its position of rest
Martin's diameter	The diameter which divides the particles profile into two equal areas measured in the direction of scan when the particle is in a position of rest
Projected area diameter	The diameter of the sphere having the same projected area as the particle profile in the position of rest
Particle sieve diameter	The width of the minimum square aperture through which the particle will pass
Specific surface diameter	The diameter of the sphere having the same ratio of external surface area to volume as the particle
Surface diameter	The diameter of the sphere having the same surface area as the particle
Stokes' diameter	The diameter of the sphere having the same terminal velocity as the particle
Volume diameter	The diameter of the sphere having the same volume as the particle

are shown in Table 2.2. The first six are obtained from microscopy and image analysis, the particle sieve diameter from sieving methods; the specific surface diameter from permeability measurements, the surface diameter from surface methods, the Stokes diameter from sedimentation methods, and the volume diameter from sensing zone methods. In selecting a method, therefore, one should choose one that correlates most closely with the process or problem at hand. No sample should ever be submitted to a physical testing laboratory simply requesting particle size analysis; neither should one use an instrument because it is conveniently located near one's workplace. The instruments available for the measurement of these diameters number several hundred, and their applicability to any particulate material will depend on the size range of particles present in the sample. Table 2.3 gives an approximate range of many particulate systems that find widespread use in today's applications.

Table 2.3. Range of Particle Size Common to Industrial Particulate Systems.

	SIZE RANGE						
PARTICULATE SYSTEM	1000	100	10	1	0.1	0.01	0.0001
Coarse sand	—						
Find sand		—					
silt			—				
clay				—	—	—	
Explosive fillers			—	—	—		
Paper fillers			—	—	—		
Paint fillers			—	—	—		
Paint pigments coarse		—	—	—			
Paint pigments, fine				—	—		
Paint pigments, ultrafine					—	—	—
Detergents	—	—					
Cosmetic powders		—	—	—			
Pharmaceutical powders		—	—	—			
Pesticides		—	—				
Metallurgical powders	—	—	—	—	—		

These span six orders of magnitude, and in view of this, before any decision is made with reference to the selection of a sizing method, an estimate of the largest and smallest particle present should always be made by a combination of optical and electron microscopy.

When one has decided on both the size parameter closest to the problem at hand and the approximate size range of the sample, a method for size analysis and an instrument can be selected. Before analysis can begin, however, the adequacy of the sampling statistics of the particulate should be examined. It is worthless to demand high accuracy and reproducibility of sizing technique if the sample has not been taken with at least the same degree of accuracy.

Finally, as many sizing techniques demand the suspension of a powder in a liquid, the effect of that liquid on particle interaction should be determined. For meaningful size analysis, each particle should be individually dispersed in the liquid and be stable with regards to coagulation. In addition, for submicron powders, solubility factors, however small on a microscopic basis, can account for the disappearance of small sized fractions during size analysis. The effect of pH, the nature and concentration of any additive and the energy required to disperse the powder in the liquid without affecting the particle structure should also be experimentally determined.

From the above general information, it can be seen that there are several basic steps that should be followed in sequence from the initial decision that size data is required to the examination of the final results. These steps are:

1. Definition of the particle processing problem
2. Definition of the type and extent of the particle size information required—if at all!
3. Selection of the most relevant measuring principle
4. Selection of an instrument that uses this principle
5. Acquisition of a reproducible sample
6. Preparation of the sample for analysis
7. Size measurement within the constraints of the method
8. Selection of the most relevant data-handling method.

Each of these steps is now discussed in sequence.

Definition of the Particle Processing Problem

This is clearly beyond the scope of this chapter, and obviously has to be decided individually, industry by industry, process by process, and problem by problem. It demands a basic knowledge of particle technology coupled with experience. In this chapter, we are dealing with particle size measurement and, therefore, assume that the problems considered demand particle size measurement.

Definition of the Type and Extent of Particle Size Information

Particle size measurement is most often required as:

a. A simple quality control technique
b. A method of investigating changes in R&D processes
c. A requirement in patent or product specification applications

In the above three categories, size information is generated for three different types of applications, and the corresponding requirements are slightly different. For quality control, a simple, reproducible, and robust plant method is usually needed. For R&D purposes an accurate, flexible laboratory technique has to be employed; and for patent purposes, an accurate and generally accepted technique has to be used and specified. Because each of these techniques will provide certain size information and limit the choice of instrument, the user should be sure that this information is truly what he needs. For example:

a. For film coating applications, the percentage of the largest size fraction is usually required.
b. For dustiness, the smallest size fraction is needed.

c. For inhalation, the fraction between one-half and five μm may be required.
d. For chemical reactivity, the surface diameter is usually required.
e. For permeability through porous media, the specific surface diameter is required.

In other words, it is important to measure the parameter of interest and significance to the problem at hand, and not to measure an unspecified size parameter.

In addition, most counting methods provide the option of reporting size data on a population, area, or weight basis; and it must be recognized that the mean sizes and other percentile values are widely different dependent on the basis selected. The one most sensitive to the problem at hand should be employed. Such factors have been discussed under theory and statistics in Chapter 1.

Selection of the Most Relevant Measuring Principle

As a general rule, it is wise to measure a particle by a method that closely related to its end-use properties and to measure it in the medium in which it is used. If a powder is to be coated onto a surface in air, sedimentation data from measurements in liquid will most likely bear no resemblance to the dust size distribution. Also, in estimating the tendency for a slurry to settle, sedimentation methods and not microscope counting should be considered. With some materials, an ASTM recommended method may already exist, so the user is advised to check the ASTM Standards before making any decision. If none exist, Table 2.4 can be used.

Table 2.4 shows some of the physical principles used to measure particle size. The one most relevant to the particle system should be selected. In order to do this, each method category, and each instrument contained in each category needs to be discussed in some detail.

Particle Size Measurement Methods

Microscopy. Table 2.5 shows a summary of particle size measurement equipment that utilizes microscopy. These methods can be

Table 2.4. Common Methods of Size Analysis.

METHOD	CONDITION	TECHNIQUE
1. Microscopy	Dry or wet	Optical
	Dry	Electron and scanning electron
	Dry	Automatic image analyzers
2. Sieving	Dry or wet	Wire woven and micromesh screens
	Wet	Automatic screening devices
3. Sedimentation	Dry/gravity	Micromerograph
	Wet/gravity	Pipette extraction; photosedimentation density difference; β back-scattering; balances sediment accumulation; x-ray sedimentation
	Wet/centrifugal	Pipette extraction; photosedimentation; x-ray sedimentation; sediment accumulation
4. Sensing zones	Wet	Electrical resistance change
	Wet or dry	Optical scattering; diffraction; obscuration
5. X-ray method	Dry	Absorption; low-angle scattering and line broadening
	Wet	β ray absorption
6. Surface methods	Dry	External surface–permeametry
	Dry	Total surface–gas adsorption volume or pressure change; weight changes; thermal conductivity change
	Wet	Adsorption from solution–fatty acids; isotopes; surfactants–dyes Heat of solution–heat of wetting
7. Miscellaneous	Dry and wet	Holography; ultrasonic attenuation momentum transfer; evaporative cooling of a hot wire

Table 2.5. Particle Size Measurement Equipment Microscopy.

	ANALYSIS TIME (MIN)	SIZE RANGE (μm)	APPROXIMATE COST ($)	AMOUNT OF SAMPLE NEEDED (g)
1. Imaging process				
Optical microscope	30	1–500	100–2000	0.1
Transmission electron microscope	60	0.01–10	30,000–250,000	0.1
Scanning electron microscope	60	0.01–100	30,000–250,000	0.1
2. Counting process				
2.1. Semi-automatic				
Zeiss–Endter Timbrell	90	1–500	5000–15,000	0.1
2.2. Automatic T.V. interfaced systems				
Cambridge–Imanco–Quantimet 720 Leitz–Classimat Zeiss–Microvideomat Millipore–πmc Bausch and Lomb–Ominicon Ealing Beck–Histotrak Hamamatsu–Multi–Analyzer –Poly–processor Leitz–Texture Analyzer	15–30	Optical interface 1–500 Photomicrograph interface 0.01–100	25,000–150,000	0.1
2.3. Automatic non-T.V. interfaced systems				
Carl Zeiss Jena–Epiquant Kontron/Wild–Digiscan AMEDA FIDAC and FIDACYS Photoscan Joyce Loebl Lemont Scientific–CESEMI	15–30	Optical interface 1–500 Photomicrograph interface 0.01–100	30,000–125,000	0.1

seen to consist of two parts, one concerned with imaging and the other with counting.

The physics of imaging techniques are not discussed here. For a good optical microscope, calibrated with a stage micrometer, particles can be counted from 0.5-50 μm, whereas particles of size 0.001-5 μm have to be counted from electron microscope photomicrographs. Sample preparation techniques are a critical factor in this method of analysis, but details are provided later. It is also important that particles measured optically be representative of the bulk powder or suspension; sampling details are also provided later. In this section, emphasis is given to measurement technique itself and the need for a statistically accurate count.

Particles can be measured one at a time using a linear scale. This method is slow and orientation dependent. Alternatively, an image-shearing eyepiece such as is used in the Timbrell microscope can be used. In such devices the light beam is split and separated to form two images by the rotation of two mirrors or two prisms. One sheared image is colored red and the other green. Under no shear, the two images coincide and the particle appears black. With a small amount of shear the images slightly overlap. In this instance one extremity appears red, the other green; and only the overlap appears black (Fig. 2.1). With greater shear, the red and green images separate leaving a bright gap between them. At the amount of shear equal to the particle diameter, the red and green images just touch, and exhibit no black area. By setting an amount of shear on the eyepiece, and then observing several fields, the number of particles just touching can be quickly estimated. By changing the amount of shear, and repeating the procedure, a particle count can be quickly established. This is, however, orientation dependent, and rotation of the field through 90° may significantly change the data. To minimize this, circular reticules were introduced, with which a particle can be compared to the circle of equal project area. To minimize optical illusions prevalent when open circles are compared to opaque images, some reticules contain both open and opaque

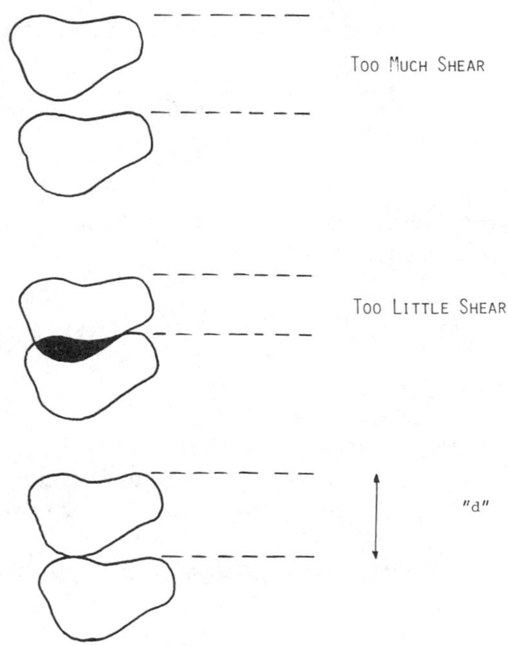

Figure 2.1. Image shearing.

circles for comparison. These are described by Allen.[1]

A semi-automatic size analyzer, the Zeiss-Endter, was developed to directly compare circles with particle projected areas. An adjustable spot of light from an iris diaphragm is focused underneath a particle image or a photomicrograph. The iris is adjusted until the areas of both the particle and the light spot are equivalent. The diameter of the equivalent circle is automatically recorded in a 48-channel register by the depression of a foot switch.

All of these methods are dependent on the statistical considerations. These change when frequency or weight distributions are required.

For frequency determinations the number of particles that have to be counted in any size range is given by Allen[1] to be:

$$\sigma(F_1) = \left\{ \frac{F_1\ (100 - F_1)}{n_t} \right\}^{1/2}$$

where

$\sigma(F_1)$ = Standard deviation expressed as a percentage of the total by frequency

F_1 = Frequency of particles in a given size range
n_t = Number of particles of all size ranges counted

For most purposes n_t is greater than 600.

For weight distributions, the number can be calculated from:

$$\sigma(Mq) = Mq/\sqrt{nd}$$

where

$\sigma(Mq)$ = Standard deviation expressed as a percentage of the total by weight
Mq = Percent by weight in the given size range
n_d = Number of particles counted in the size range

For weight distributions, the number of particles counted in the top range should generally lie between 15 and 30. Such counting procedures are slow and tedious. To eliminate this problem automatic image analyzers were developed. These are subdivided into TV camera interfaced devices, and non-TV camera interfaced systems.

Table 2.5 includes many of the TV interfaced systems commercially available. These instruments convert optical information taken picture point by picture point into an electrical signal via a TV camera. The features of an image are then detected by signal modulation, i.e., by a measurement of the grey level of the image using a signal discriminator. Projected lengths and Ferets diameter and other statistical diameters at angles to the horizontal are sometimes used. With many of these TV interfaced instruments, it is fair to comment that without software programming they are limited in scope and flexibility. However, the instruments are modular, fast, and reproducible; and many have software inputs for both data handling and processing and statistical calculations.

Many of the instruments in this class also offer the option of a light pen module. In this manual mode, the instruments can be classed as semi-automatic, but even in such a mode, the speed of data processing is significantly faster than the Timbrell or Zeiss-Endter.

Automatic non-TV interfaced image analyzers were the forerunners of the automatic microscope analyzers like the Mullard, Rank-Cintel Flying Spot Scanner, and Casella that are now no longer manufactured. These used optical scanning techniques coupled with photomultiplier signal converters. Newer variations using this technique include the Digiscan manufactured in cooperation of Wild Heerbrug, Ltd. It uses an optical beam with a photodiode converter to give a high grey value discrimination. Another variation, the Epiquant, manufactured by Carl Zeiss Jena, employs a combination of optical scanner and photomultiplier converter. Such a combination is claimed to give a higher grey value discrimination of low contrast images than TV interfaced systems. A dot-counting instrument, the AMEDA, developed by Femco, Ltd., was developed for use on metallographical and petrographic analysis. It directs a light spot onto a polished section and measures the optical reflectivity along scan lines. Points of similar reflectivity are stored in registers.

Ledley and Ruddle developed one such device for the examination of the size and shape of chromosomes. The film input to a digital automatic computer or FIDAC examines a film with a beam of light, and digitizes the intensity into seven grey levels. The data are stored in the memory of the FIDAC computer system. The system then establishes the grey value, which represents the points within the particle feature, and creates the silhouette of the image. This is, then, examined by comparing it to a criterion previously established for the accepted shape of a chromosome. A decision is made at this point, as to whether the image is indeed a chromosome or not. If so, the image is stored; if not, it is rejected. At the completion of the film scan, all established chromosomes are counted and their diameters, areas, perimeters, curvatures, etc., are calculated and recorded.

Simpler systems have been developed for particle analysis. The Photoscan has been

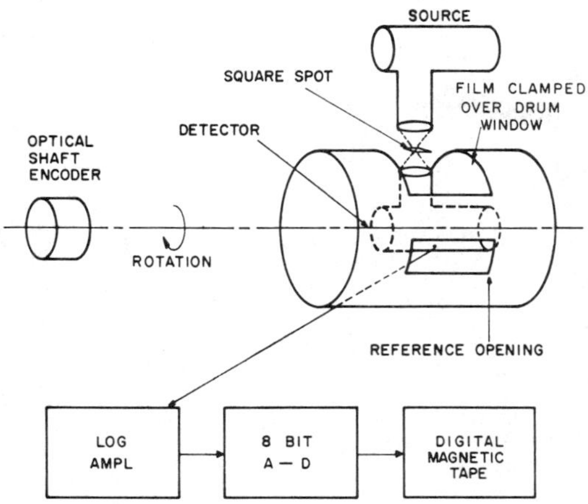

Figure 2.2. Photoscan.

widely used by Du Pont for particle size, shape, and texture analysis (see Fig. 2.2).

A scanning microdensitometer manufactured by Optronics, Inc., measures the optical reflection density of a 12 cm × 12 cm photomicrograph mounted on a revolving drum. The output is then recorded onto seven-track magnetic tape, and data is extracted using a size, form, and texture algorithm. Size analyses by frequency, area and mass, and area to perimeter^2 distributions are calculated and plotted. The number of asperities around the perimeter of the image, are then measured and incorporated into a texture factor. Joyce Loebl and colleagues have recently offered a similar capability using their microdensitometer systems.

With the advent of scanning electron microscopy (SEM), it became advantageous to bypass optical scanning of photomicrographs by taking the electrical signal already available from the scanning system and develop logic systems to process SEM images directly. E. White et al., at Penn State University developed a system for the computer evaluation of SEM images (CESEMI). The instrument is now commercially available through Lemont Scientific. The output from the SEM is recorded on digital magnetic tape and processed by the computer. Each feature, e.g., size, area, perimeter, etc., can be recorded as form factors by ratios. The minimum and maximum axes of fitted ellipses are used to yield shape data, which are tabulated and condensed into meaningful computer-plotted distributions. One additional feature of CESEMI and, in fact, other recent SEM interfaced analyzers, is the measurement of composition by x-ray spectroscopy on the SEM. Hence, shape can be related to composition.

Sieving. Table 2.6 shows some of the equipment used for laboratory sieving of particulates. Sieving is perhaps the most widely used method of size analysis. It can be performed wet or dry; but for most practical purposes, wet sieving is performed below 200 mesh, and dry sieving above this value. This is because the surface energy and charge associated with particles below 200 mesh induces agglomeration, and good separation becomes difficult. Deagglomeration can be achieved by brushing the particles through the mesh, but this is not recommended as it results in sieve distortion and damage. This is particularly true when sieves are cleaned. Particles of mesh size lodging in the mesh apertures are best removed in an ultrasonic bath. However, high frequency and low power input are essential if sieve damage is to be avoided. It must be realized that analyses of the same material by different sieves of the same nominal apertures made by the same manufacturer are subject to appreciable discrepancies. The coefficient of variance varies between 10% at small aperture widths to 5% at large aperture widths; and from this, it is appreciated that small aperture sieves cannot be reproduced as accurately as large ones. One study has shown that using tolerances specified in sieve standards, approximately two-thirds of sieves manufactured did not meet specification. This demands careful sieve calibration and for this purpose, the method originally suggested by Andreasen, later used by Anderson, and recently recommended by Leschonski is perhaps the best.[2] The basis for this method arises from studies on the rate of sieving. Whitby[3] showed that there are three regions during a

Table 2.6. Sieving.

	ANALYSIS TIME (MIN)	RECOMMENDED SIZE RANGE (μm)	APPROXIMATE COST ($)	AMOUNT OF SAMPLE NEEDED (g)
1. Dry (wirewoven screens) Rotap; Allen Bradley; Cenco; Alpine; Geoscience; Gilson; Endecotts; LaPine; Pascall;	15–30	+70	$100–1000	100
2. Wet				
2.1. Wirewoven: Geoscience; Tyler;	15–30	+44	$100–1000	100
2.2. Electroformed mesh	15–30	10–44		5

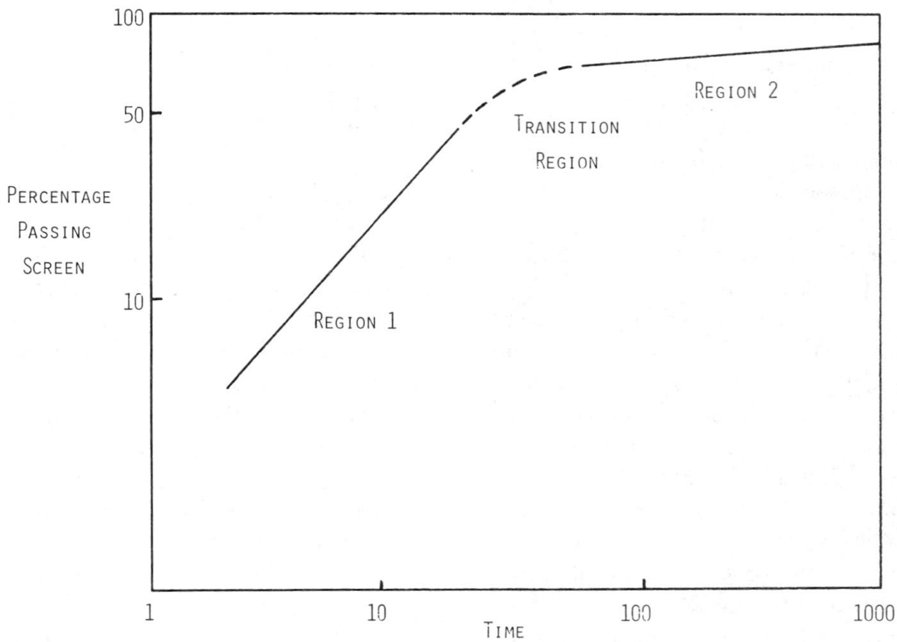

Figure 2.3. Particle sieving rate (after Whitby).

sieving process (see Fig. 2.3). Region 1 exists at the beginning of sieving when there are many particles much less than the mesh size still on the sieve. Region 3 exists when the particles on the sieve are near mesh size or larger. Region 2 is the transition zone between them. During Region 3, the percent passing becomes very small, and for most practical purposes sieving is stopped when the percentage passing is less than 0.1% of the sample during a one-minute sieving period. At this time, if the sieve is hand-tapped and the particles passing are weighed and counted, their average particle size can be taken to be the cut size of the sieve. Generally, this is calculated in terms of a volume diameter. Leschonski shows good agreement between such calibrated sieves and other methods.

Other factors affect the accuracy of sieving. For example, the probability of a particle passing through an aperture will depend on:

1. The characteristics of the particle
2. The particle load on the sieve
3. The method of shaking the sieve
4. The geometry of the sieve surface and its wear
5. The angle of presentation of the particle to the aperture

Of the instruments listed in Table 2.6, the first five illustrate the differences in sieving mechanisms. The Rotap is horizontally rotated while vertically tapped, the Allen-Bradley employs a pulsating air column, the Cenco is horizontally vibrated, the Alpine has a continually rotating air jet beneath the mesh, and the Geoscience is electromagnetically vibrated. In an independent study on test powders, using the same sieves on all the above machines, the results from the Rotap were found to most closely match the size distribution measured on ten powders of varying size distribution. For laboratory use, however, the Rotap is a noisy machine and has no variable controls for "tuning" the sieving action to any particular powder. This lack of variation is not always a disadvantage, as it makes the machine easy to operate by unskilled labor.

In the laboratory, the Alpine and Allen Bradley machines appeared more attractive; but the problem of fines loss due to electrostatics was experienced.

The Alpine utilizes the labor-saving qualities of a machine on a single sieving surface, the speed at which sieving is accomplished is rapid; and it has the advantage that fines are removed by a vacuum system, which makes it a clean instrument for laboratory use.

The Allen Bradley is perhaps the most rapid nest sieving machine commercially available. It is quiet in operation, but it does have a slight disadvantage in that standard eight-inch screens cannot be used. The smaller sieves manufactured by Allen Bradley are still U.S. standard woven or micromesh screens, but their smaller size limit the permissible load, which demands a more stringent control of sampling and analytical procedures. However, this sieve size does facilitate weighing, and the sieves can be weighed directly on precise analytical balances.

The dry sieving of fine powders on 10 μm and 20 μm micromesh screens was not recommended on eight of the two instruments, even though some success may have been achieved on certain powders by some operators.

For the vibratory machines, these are less noisy than the Rotap and have variable controls that have to be selected carefully for satisfactory fractionation.

However, these machines require some skill and a knowledge of the behavior of powders, if they are to be operated to full advantage, as a continual increase in amplitude of vibration does not guarantee a more effective separation. In addition, ideal conditions for one sieve size are not necessarily ideal for the next sieve in the nest. In the above study, this may explain why the analyses of vibrating machines were coarser than the analyses from other types of machines; There could be a resonant effect in which a laterally vibrating wire mesh oscillates about a mean aperture size which is effectively smaller than under nonvibratory conditions.

All the machines in Table 2.6, then are useful for size analysis provided that the limitations of each machine and sieving surface are recognized and accounted for.

Electron-formed or micromesh screens find some application, but sieving machines are not available that use them. Wet sieving in an ultrasonic bath is generally practiced and is the best method currently available when sieve aperture sizes of 10–25 μm are employed.

Automatic sieving machines have been developed. Hinde and Lloyd described the development of an automatic wet sieving machine for in-plant mill control. Single-point measurements on the size distribution were obtained by sieving a slurry through one screen using impeller agitation and ultrasonic energy, and weighing the residue by applying Archimedes' principle. In this way, drying of the screen and residue was not necessary. Schönert et al. described a unit equipped with seven screens which operated on a cyclic system. Screening, drying, weighing, and cleaning were performed in sequence; and weights were printed, punched, and processed using an electronic dataprocessing facility.

An automatic sieving device, the Autosieve, has been developed by the Micromeritics Instrument Company. This is a wet-sieving device that separates a particle sample into five size fractions. Particles are induced to pass each sieve in turn by upward-directed jets of water that fluidize the particles on the sieve during oscillation. Those particles that pass the sieve are directed to a microfine filter coated with a filter aid. This filter is totally immersed in water and is attached to a weighing transducer. When eight consecutive weighings deviate by no more than 1%, the data are stored and the next sieve is automatically brought into use. Weight percentage data from the total analysis are the final result. The unit is shown in Figure 2.4. The smallest sieve mesh used on the device is 325 mesh.

Sedimentation. There are perhaps more instruments available in this general area of size analysis than any other. In general, sedimentation methods involve an instrument that detects the settling rate of particles in a fluid by monitoring (a) a change in concentration of particles at a known depth in the fluid or (b) the rate at which particles settle out onto the bottom of the container in which they are confined. The parameter measured is the Stokes diameter, and calculation of this diameter assumes that Stokes Law is obeyed.

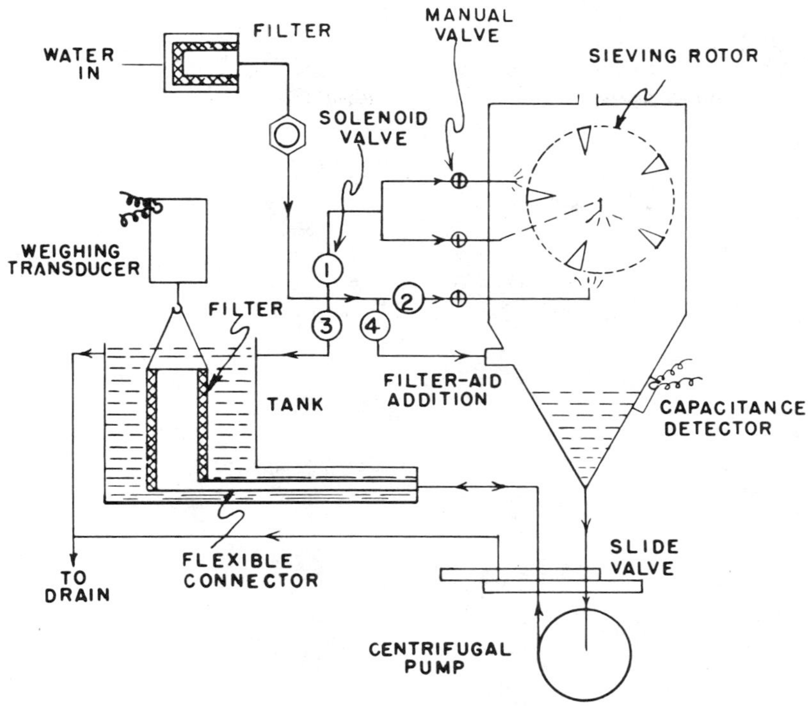

Figure 2.4. Micromeritics Autosieve.

Sedimentation under gravity is generally used for particle sizes greater than 2 μm, and sedimentation under centrifugal force is used for sizes less than this value. Table 2.7 shows the methods available.

Gravity systems are discussed first. One of the most labor-intensive but lowest capital investment techniques uses the Andreasen pipette. Here a homogeneous suspension of 1% by weight is allowed to settle, and at a predetermined time t a sample is extracted from a fixed depth h, in the suspension; and the particle concentration is then estimated by evaporating the sample to dryness and weighing. The ratio of this concentration to the initial concentration, expressed as a percentage, indicates the percentage finer than the Stokes diameter calculated using the fluid constants, the settling height h, and the time of settling t. A correction for surface active agents or dispersing aids present in the fluid has to be made. Analytical times are long; and in the conventional equipment, two errors arise, i.e., the dead space error and the density convection error (see Fig. 2.5). A special pipette was developed by Leschonski and manufactured by Sartorius to overcome these effects.

A less labor-intensive approach, but sometimes equally slow, is the use of photosedimentometers that measure the concentration change at a fixed depth in a very dilute suspension by monitoring the change in optical density of the suspension with time. Most of these devices display the optical density relationship on a strip chart recorder. To minimize long analytical times for powders containing particles less than 2 μm, some devices employ a scanning sedimentation cell system, which permits the monitoring of concentration at progressively shorter sedimentation heights, hence, speeding up the analysis. Typical devices are the Wagner, Fritsch, Ladal, and Micromeritics 5500. With the exception of the Ladal, Micromeritics 5500, and the Photomicron Sizer in Table 2.7, most of the photosedimentometers are relatively old designs. These newer instruments are worthy of some description.

The Micromentics Sedigraph 5500 determines

Table 2.7. Sedimentation.

	ANALYSIS TIME (MIN)	RECOMMENDED SIZE RANGE (μm)	APPROXIMATE COST ($)	AMOUNT OF SAMPLE NEEDED (g)
1. Gravity methods				
1.1. Pipette method	90	2–75	60	50
1.2. Optical photosedimentometers				
Microscal, Wagner, EEL, Micron Photo	60	2–75	1000–5000	1
Sizer, Boundbrook, Micromeritics				
1.3. X-ray sedimentometers	45	0.1–75	10,000–20,000	10
Micromeritics, Microscal				
1.4. Sedimentation balances	60	2–75	3000–8000	50
In liquid: Gallenkamp, Sartotius, Shimadzu				
In air: Micromerograph		5–75	15,000	50
1.5. Sedimentation tubes				
US Sedimentometer, Palo-Travis, Granulometer Size	60	2–75	100–1000	5
1.6. Density methods				
Fisher-Dotts, Hydrometers, Divers	60	2–75	100–500	50
1.7. β-backscattering sedimentometer	60	2–75	5000–10,000	50
1.8. Diffraction				
Talbot Disa		2–70	5000–10,000	5 liters
2. Centrifugal				
2.1. Weight method				
Simcar, M.S.A.	60	0.5–5	8000–15,000	10
2.2. Optical sensing				
Joyce Loebl, Kaye Disc	30	0.05–5	15,000–25,000	1
2.3. X-ray sensing				
Microscal; British Titan	30	0.05–5	10,000–20,000	50

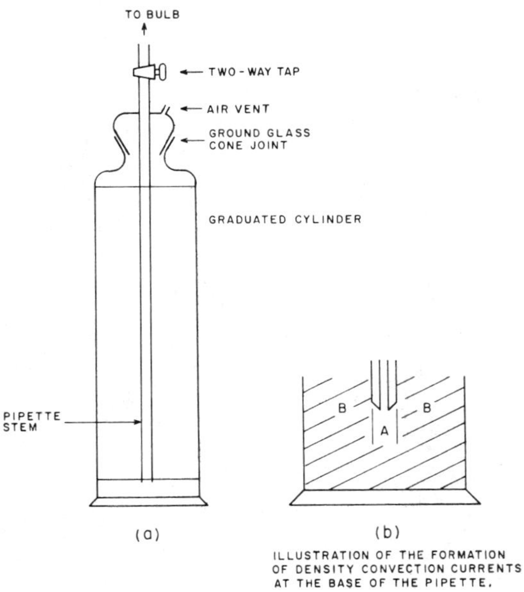

Figure 2.5. Basic system of fixed pipette equipment.

the concentration of particles settling under gravity in a cell at various sedimentation depth as a function of time. The logarithm of the transmitted light intensity is generated, scaled, and presented linearly as the cumulative projected area diameter distribution on the y axis of the recorder. The position of the cell continuously changes so that the effective sedimentation depth is inversely proportional to the elapsed time. The cell movement is synchronized by a mini-computer on the x axis of the recorder to indicate directly the equivalent area diameter corresponding to the sedimentation time and depth. The size information is presented logarithmically on a three decade graph. A conversion to mass diameters is then made using Mie theory. Concentration of 0.025% by volume are employed.

The Photomicron Sizer consists of two components, a gravity photosedimentometer and an analytical centrifuge. A suspension of particles in liquid is allowed to settle in a cuvette in the gravity sedimentometer while the optical density is monitored continually by measuring the attenuation of a parallel beam of light that passes through the suspension at a fixed height. To permit the rapid measurement of small particles, the cuvette is removed from the sedimentometer and placed in the centrifuge. It is then centrifuged for a predetermined time, taken out, replaced in the sedimentometer and the resulting drop in optical density measured at the same fixed point. From a combination of gravity and centrifugal measurement, a projected area diameter size distribution can be produced. Again, concentrations as low as .025% by volume are used.

In the Ladal instrument, a parallel white light beam passes through a sedimentation tank to a photocell; and a second similar beam passes to a reference photocell. These two cells are connected in opposition to form part of a Wheatstone network so that it is a simple matter to continually preserve a constant zero. To measure particle size, 1 g of powder is dispersed and pipetted into the sedimentation tank until an optical density of 0.5 is achieved. The resulting dilute suspension is stirred and when well mixed, stirring is stopped and the optical density is recorded with time. The sedimentation tank is then automatically lowered at a set rate to speed up the analysis. The optical density curve is plotted on a pen recorder; and when normalized, this cumulative percent surface undersize curve is integrated to yield the cumulative weight undersize distribution.

The above instruments can be used for all types of particles dispersed in fluids, but Allen points out a serious discrepancy in their data. As the laws of geometric optics do not hold, as size becomes small, photosedimentometers are biased towards the coarse end of a distribution. Allen, therefore, suggests their use be limited to comparative analyses only. For inorganic particles, when mass distributions are required directly, x-ray sedimentometers can be used. Typical examples are the Tioxide, Ladal, and Sedigraph 5000 gravity sedimentometers. These are used when the atomic number of the particle exceeds 14. The latter unit is the most widely used device of its type and operates identically to the 5500 above, except that polymeric cell windows are used with a collimated 0.002-inch x-ray beam. Particle con-

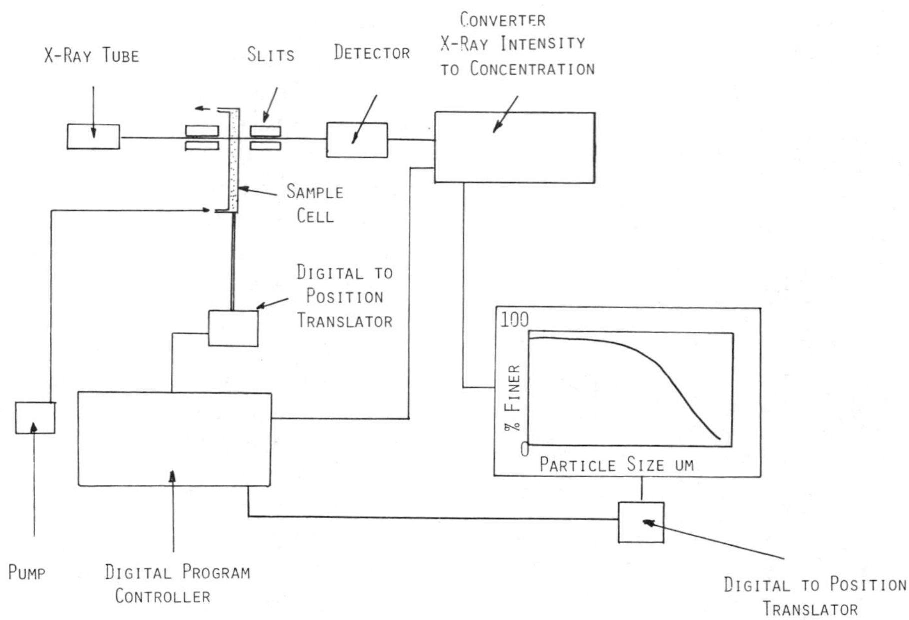

Figure 2.6. Essential features of the Sedigraph 5000.

centration of 0.2–4% depending on atomic number are commonly required, and the cumulative mass percent undersize is displayed on the y axis of its recorder (see Fig. 2.6).

In addition to the above techniques, concentration changes in suspensions have been measured as a function of time using small hydrometers sometimes termed divers, which monitor density changes. These suffer from long analytical times and are slowly being phased out of usage in industry with the advent of the newer techniques.

In contrast to the above techniques that can be termed incremental methods, several cumulative methods are available in which particles are permitted to settle completely. Sedimentation tubes are typical examples. A thin layer of suspension containing particles is laid on the surface of a long column of clean fluid and allowed to settle as a function of time. At its base, the column narrows so that particles collect and form a thin column of sediment with time. The settling velocities are calculated from the height of sediment buildup with time. A similar procedure is used with the β-Backscatter sedimentometer, except that the cumulative mass buildup is measured by β-ray absorption through a thin window in the sedimentometer base. Another example of cumulative methods is one that uses sedimentation balances. Here particles settle through a column of fluid to collect on a balance pan, where they are directly weighed. The simplest and most robust version is the Shimadzu balance. The balance pan is suspended from the arm of a beam balance that is depressed as particles settle onto the pan. When a certain displacement is reached, contacts mounted on the end of the beam complete control circuits, which when activated release a steel ball onto the counter balance pan. This is recorded. Unfortunately, 2% by volume suspensions have to be used, which is outside the range of Stokes free settling. Another version is the Bostock-Gallenkamp balance (Fig. 2.7) in which premixed homogenous suspension is quickly discharged into a sedimentation column and the settling weight recorded on a torsion balance. A sedimentation accessory is also available for use with the Cahn Electrobalance to convert this into a sedimentation balance. All the above units suffer from errors associated with

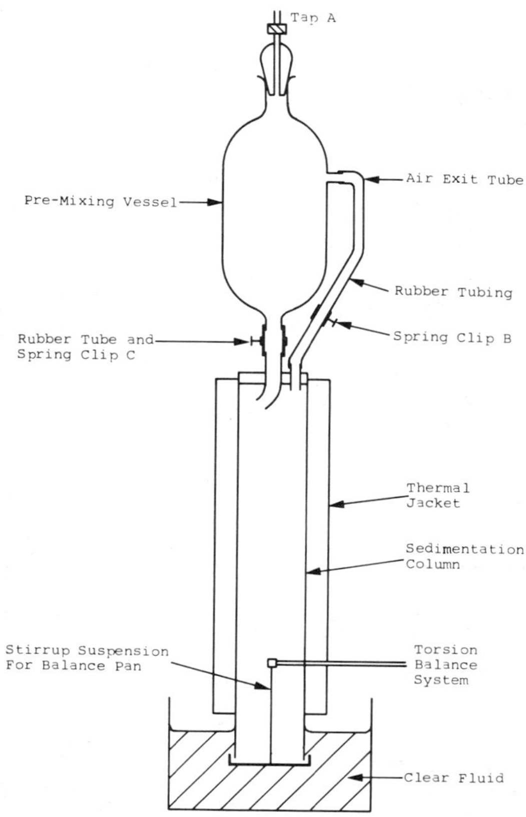

Figure 2.7. Bostock-Gallenkamp sedimentation balance.

particle leakage around the balance pan and air bubble attachment to the pan and frame. The Sartorius Balance was designed to eliminate these problems.

This is a commercially available sedimentation balance in which the weight of solids arriving at the base of the sedimentation column are continually recorded by a servo-mechanism-controlled, accurate, analytical balance. The balance pan onto which the particles sediment is surrounded by a reservoir of clear fluid as shown in Figure 2.8. The system shown in this figure is actually a modification of the original commercial system described by Leschonski.[4]

The balance pan is provided with a high rim to minimize suspension movement around the edge of the pan. The fluid below the pan is outgassed to prevent bubbles attaching themselves to the balance pan and causing error in the recorded weight of sedimented particles. The external rod system holds the balance pan against the rim of the sedimentation column as the prepared suspension of powder is poured into the column. The rod mechanism is then released and the pan allowed to swing freely. The size distribution is deduced from the curve of sediment weight against time as for other sedimentation balances.

A sedimentation balance for use in air is the Micromerograph. This uses a six-foot column and is useful for particles greater than 5 μm. Powder loss to the column walls due to electrostatic charge is one problem with this device. These devices are again slow and are limited to particles greater than 2 μm.

When the particle size is less than 2 μm, centrifugal sedimentation devised are employed. Cumulative methods for mass distributions include the Simcar, M.S.A., and Ladal pipette centrifuges. The M.S.A. is based on the sedimentation tube principle, and the rate of particle buildup with time is measured in the capillary of a centrifuge tube.

The Ladal/Fritsch pipette centrifuge was designed to extend the Andreasen gravitational pipette method into the submicron range. One hundred-sixty mls of homogeneous suspension are mixed and 10 mls extracted, dried, and weighed for an assessment of the initial concentration. The remaining 150 ml are then added to the centrifuge revolving in a horizontal plane at a speed of either 750 or 1500 rpm. Ten ml samples are then withdrawn at precalculated times, dried, and weighed. As the samples are extracted, the radial sedimentation height continually decreases. This has the effect of further speeding up the size analysis. A range of 0.3 μm to 2.5 μm is claimed, and volume concentrations of 1-2% by volume are employed (see Fig. 2.9).

Similar to gravity systems, incremental methods can be used in centrifugal fields. One system using optical density changes is the Joyce Loebl disc centrifuge. This instrument employs a hollow centrifuge disc which rotates in the vertical plane at a selected speed that can

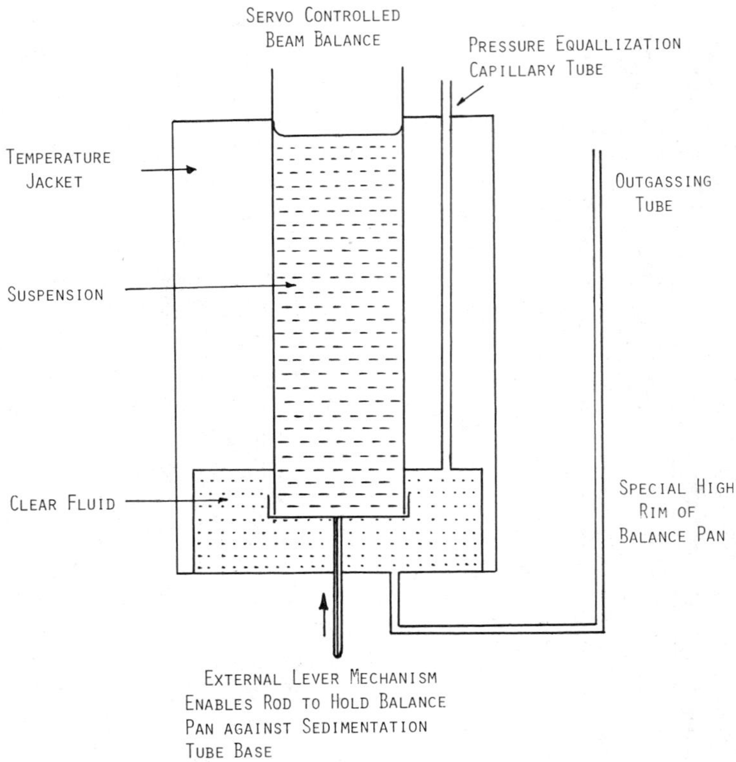

Figure 2.8. Modified Sartorius sedimentation balance as described by Leschonski.

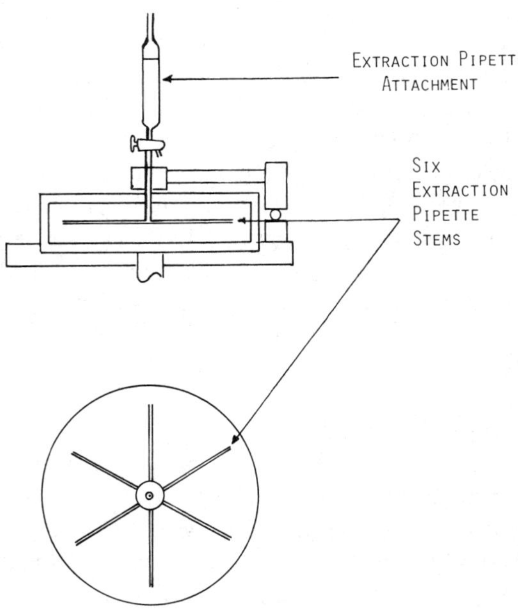

Figure 2.9. Ladal/Fritsch pipette centrifuge.

range from 1000-8000 rpm. A known volume of sedimentation fluid is injected into this rotating disc, and the particles to be analyzed are injected onto the fluid in a dilute suspension using a buffered line start method. Stokes Law modified for centrifugal fields is used as the equation for particle motion. This relationship can be used providing the system is hydrodynamically stable, i.e., no streaming occurs. This is minimized by the buffered line start system. In this method a line of buffer fluid is introduced onto the surface of the sedimentation fluid. The inner surface of this buffer annulus, i.e., the one which contacts the sedimentation fluid, is then blended into it by a brief disc acceleration followed by resynchronization to the selected speed. This produces a short density gradient through which injected particles have to pass in order to enter the sedimentation fluid. In the Mark I version, absolute cumulative weight vs Stokes diameter distribution curves are obtained by collecting all particles less than a predetermined Stokes diameter from the disc in a volumetric flask via a special sampling probe. The weight of each collected fraction is determined by suitable chemical or gravimetric analysis. Because of this analytical requirement, concentrations as high as 5% by volume have been used. In the Mark III, optical density vs Stokes diameter curves are automatically drawn, by measuring the attenuation of a beam of light passed through the disc cavity at the same radius as the probe collection point. Mass calibrations can then be computed using extinction coefficients. Using this alternative procedure, particle concentrations as low as 0.03% by volume are usually employed.

Cumulative mass distributions can be directly determined using one of three disc centrifuges fitted with x-ray beams. In the Tioxide, a perspex centrifuge disc is driven in the vertical plane by a constant speed 1000 rpm motor. The centrifuge assembly can be raised or lowered relative to a 0.5 mm collimated x-ray beam, to permit variations in the radial settling distance to be made. However, in each analysis, the disc does not scan; but measurements are recorded at one preselected settling distance. In contrast to the Joyce Loebl line start method, homogeneous suspensions of concentration 0.6% by volume for TiO_2 are employed in this unit. As settling height can be varied, an accurate knowledge of the x-ray beam below the surface has to be obtained prior to each analysis. The position of the liquid surface, e.g., meniscus, relative to the beam is first determined and is defined as that position where the x-ray count rate is midway between that through air and that through the sedimentation fluid. When this has been found, the disc is raised, a mechanical gauge graduated in 0.01 mm divisions recording the movement, and the beam is positioned at the required distance below the surface—usually 2 mm. Because of this small settling distance, analytical times are usually very short. Analysis is performed by quickly adding the homogeneous suspension and recording the x-ray count, until it reaches that previously recorded for clear sedimentation fluid. Mass distributions are then calculated using a desk computer (see Fig. 2.10).

In contrast, the Ladal disc centrifuge revolves in the horizontal plane. It uses a sealed radioactive source of low activity (100 m Ci) to generate a highly stable x-ray beam which, after passing through the sedimenting suspension, is monitored by a scintillation counter. Homogeneous suspensions are again used and concentrations of 1% by volume give adequate sensitivity for elements of atomic number of 12, and 0.2% by volume are adequate for many higher atomic number elements. Analysis begins when

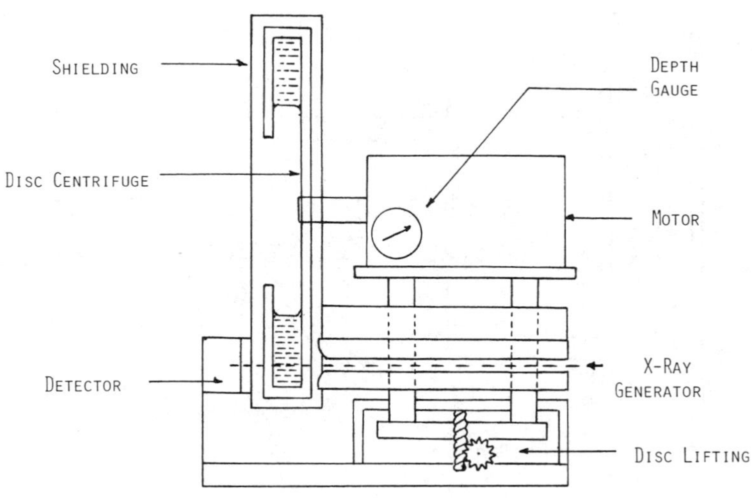

Figure 2.10. Tioxide disc centrifuge.

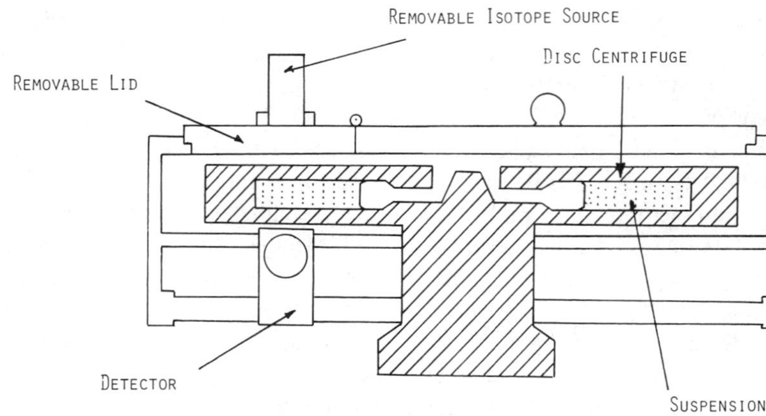

Figure 2.11. Modified Ladal x-ray centrifuge.

120 ml of homogeneous suspension are added to the disc, and ceases when the measured concentration falls to less than 5% of the initial value. The recorded curve is evaluated by extracting data points in a time progression of 2:1, and converting the deflections to mass % undersize using a set of linear equations. Two speeds of rotation, 750 and 1500 rpm, are available and an operating range of 0.01-5 μm is claimed (see Fig. 2.11).

The Fritsch Analysette 23 is a centrifugal sedimentometer in which particle concentration is monitored by x-rays from a Promethium 147 radioactive source (see Fig. 2.12). Here, the x-ray beam is made to scan radially inward to shorten the analytical time. Computer evaluation of the data is normally pursued due to the complexity of mass distribution data processing under conditions of gradually decreasing settling heights. The recorder, power supply, and amplifier can be used for the Fritsch Gravity instrument operating in the range of 0.01-100 μm. Both instruments use homogeneous suspensions.

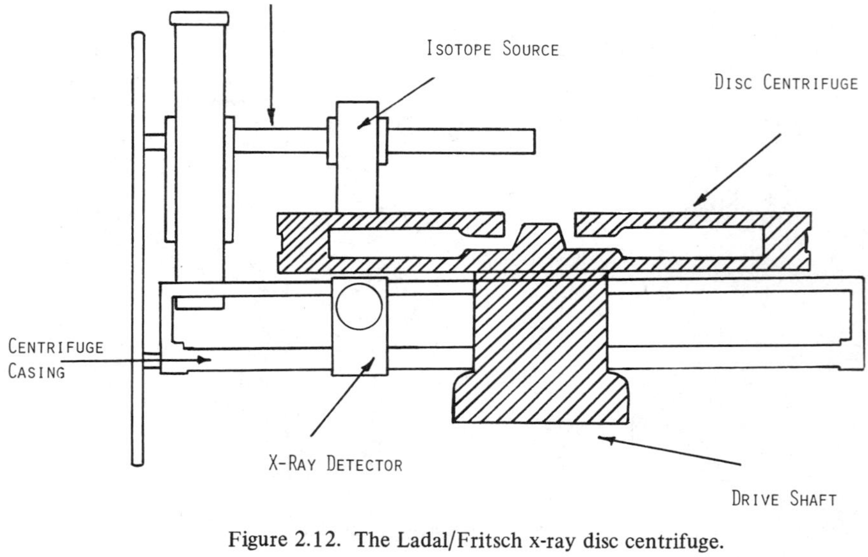

Figure 2.12. The Ladal/Fritsch x-ray disc centrifuge.

Table 2.8. Sensing Zone Methods.

	ANALYSIS TIME (MIN)	RECOMMENDED SIZE RANGE (μm)	APPROXIMATE COST ($)	AMOUNT OF SAMPLE NEEDED (g)
1. Electrical resistance change Coulter, Celloscope, Telefunken, Royco, Clay-Adams	5	0.5–1000	10,000–20,000	1
2. Optical (liquid-borne) HIAC; Climet; Prototron; Royco	5	2–75	10,000–20,000	1

Sensing Zone Methods. These are subdivided into two categories, electrical resistance change and optical methods (see Table 2.8).

Electrical resistance change methods have found application in the rapid sizing of solid and liquid particles contained in electrically conducting liquids. In the method, a dilute suspension of particles is made to flow through a small aperture having an immersed electrode on each side. The concentration of particles is such that the particles traverse the aperture one at a time. Each particle passage replaces electrolyte within the aperture, changing the resistance between the electrodes and producing a voltage pulse. The magnitude of this pulse is claimed to be proportional to the particle volume. The series of pulses resulting from many particles passing individually through the aperture is electronically scaled and counted.

In 1956, a device was marketed known as the Coulter counter and since that time many other instruments have appeared which use the same principle. Many models of Coulter counter have appeared over the past 25 years; one of the most recent and useful in particle size analysis is the TA II. This has 16 channels of readout with preset controls for a rapid push-button operation. A particle system can be analyzed and the results presented in digital and graphical form in 90 seconds. Five thousand particles per second are counted to within 1-2% reproducibility. Both a population and volume basis can be used in the analysis. Features include a digital readout selector that enables cumulative or differential volume, total count or analysis time to be obtained. It has a concentration indicator for the rapid evaluation of concentration, a curve normalizer, a data adjust control, a histogram display selector, and an X-Y plotter. The particle size range is from 0.7-500 μm.

A similar device manufactured by Particle Data is the computerized ELZONE™ system. A main advantage of the ELZONE system is the great variety of data processing/recording operations that are or can be made instantly available. When these are pre-organized in the minicomputer memory for auto-preparation of size analysis reports, maximum speed is achieved in providing the exact output wanted for a given measurement task.

Particle Data has been making the computerized system since 1970. The ELZONE™ basic particle counter provides several unique features which are as follows.

- Flexibility and accessibility of glassware and solid state electronics, housed in rugged cabinetry;
- An internal "standard particle" signal for validating overall performance and calibration;
- Effective monitoring of all ELZONE orifice conditions during data acquisition;
- An internal flushing system for providing easy care of glassware and removal of particles lodging in the aperture;
- A continuous flow-through sampling adaptor;
- A logarithmic conversion of particle-pulse amplitude spectrum, with log-scale span selectable from 16 : 1 to 16,000 : 1 in particle volume;
- Analog-to-digital conversion for digitizing of particle-pulse amplitudes for operating choice of 64, 128, or 256 channels.

Broadly, the Particle Data ELZONE program for size analysis operates in three major stages, and requires 16K or more of memory:

- Multi-channel pulse height analysis, producing a particle population distribution (histogram) as the primary size analysis, with "live" visual X-Y display;
- Size analysis data conditioning:
  a. application of the size scale calibration,
  b. normalization of the X and/or Y axis scale factors,
  c. blending of adjacent size spans of data on a given broad-distribution sample, and
  d. analysis editing, e.g., extrapolation, multimodal separation, extraneous data subtraction and least-square smoothing · · ·
- Computation and recording of pertinent parameters as selected from: percentiles,

statistical indices, graphing of curves both differential (histogram) and integral (cumulative), deviations from reference data, channel content tabulations · · ·

The ELZONE program also contains features such as storage and retrieval of data sets, optional details of the data display and recording, storage of calibration and identification information, overlay (comparison) display of second data sets.

Of great importance is the storage of preorganized routines for automatic, easy, uniform preparation of analysis reports. A given routine consists of the desired composite of text, data processing steps, results and formatting for the samples under consideration.

Though other devices are currently available, they find almost all their use in medical applications and not in particle technology. The Coulter and ELZONE™ systems, in contrast, find application in industrial quality control and the medical field and, in consequence, are the only two devices discussed in detail here.

The second sensing zone category is the one concerned with optical particle size measurements. When fine particles in gaseous or liquid suspensions interact with a light beam, some of the light is scattered in a forward direction, some at 90°, and some is backscattered. Further, if the particle passes through the light beam, a geometric shadow can be momentarily cast due to light obscuration. In this interaction, some light is also absorbed and some diffracted. Instruments have been developed to measure the magnitude of forward, 90°, diffracted, and obscured light, and through calibration relate it to the particle size. These have found use in airborne and liquid-borne single particle counters.

Royco Instruments manufactures the widest range of airborne particle counters, and manufacturers them as interchangeable modular units. Model 220 is a multirange counter that uses right-angle scatter optics. Particles of size 0.5 μm and larger are counted at a concentration of 10^6 particles/ft^3 and a flow rate of 0.1 ft^3/min. The Model 225 uses near-forward scatter for counting 10^8 particles/ft^3 at a flow rate of 0.01–0.1 ft^3/min. For high flow rate sampling, e.g., 1 ft^3/min the Model 245 is used. This again has near forward optics, but counts at 10^5 particles/ft^3. Special counters for automatic maintenance monitoring and for portable sampling are also available. The Model 345 liquid-borne counter is made to detect and count particles of size 5 μm and larger in liquids containing 100 particles/ml at flow rates of 100 ml/min. This uses light adsorption/total scatter optics. A new submicron particle counter has also been developed that uses a low power He–Ne laser to count and size particles in liquids. The laser beam is passed through a beam compressor to reduce the beam size. A portion of the direct beam is passed through the return mirror to a calibration fiber optic system, which passes chopped light pulses to the scattered-light detector. These pulses are used to define the light level from the laser through the folding optics and cell, to the scattered-light detector. As particles pass through the cell, the scattered-light pulse is produced and collected by the scattered-light detector. By virtue of the scatter mirror design, both forward- and back-scattered light is collected and sent through the folding prism to the scattered-light detector. An absorption signal is also produced by taking the return beam to the beam separator, collecting the separated beam in the lens at the periphery of the beam separator prism, and directing that beam to the extinction collection optics. Using a pulse-height analyzer, preliminary data taken from the scattered-light portion of the system have shown it to be relatively noise-free and capable of sensing particles 0.5 μm in diameter (see Fig. 2.13).

Climet Instruments produce single particle counters that employ a wide angle optical system that is claimed to collect 90% of the scattered light. The C1-208 airborne particle counter counts and sizes particles at concentrations of 10^7 particles/ft^3. The C1-250 model is a portable, digital multipurpose instrument, and for more versatile applications the C1-252 portable analog particle counter is used. These

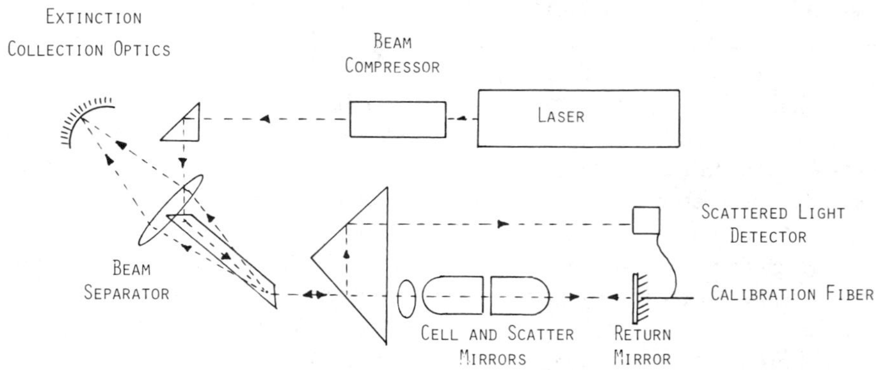

Figure 2.13. Royco liquid-borne submicron particle analyzer.

devices measure particles of diameter in excess of 0.5 μm. The Model C1-220 is the liquid-borne counter which uses the same optics, but can detect particles of size 2–200 μm. Concentrations of 10^0–10^8 at a flow rate of 500 ml/min can be measured. An on-line monitor for high-purity liquids is also available. This is the Model C1-221.

A counter was developed by Prototron for the rapid count of particulate in liquids, and in particular, the count of particles in sealed bottles or vials. A one milliwat He-Ne laser is spatially filtered and focused to provide a well-defined small illuminated volume inside the bottle. The beam is then rotated so that it travels in a circle at 900 rpm, and the digital readout displays the average number of particles with sizes above a threshold value. By taking sequential measurements at different thresholds, qualitative size distributions can be developed.

Light obscuration counters are typified by the HIAC PA-520 automatic size analyser and plotter. By projecting a constant beam of light through a sensing zone onto a photodiode, the instrument measures the amount of light blocked by each particle as it passes through the beam. Twenty-four size thresholds can be set and a microprocessor is used for data processing. Various sensors are available in a 1–45 range, i.e., 1–45 μm; 2–90 μm, etc.; up to a top size of 1 mm. The instrument finds wide application in the monitoring of hydraulic fluids for wear particles (see Fig. 2.14).

The Leeds and Northrup Company marketed the Microtrac particle size analyzer in 1975 for liquid-borne particles counting. In this instrument, the particles flow through the sample cell and are illuminated by a continuous laser beam. Suspended particles scatter light, which is collected by the primary lens, passed through a rotating optical filter, and focused by a secondary lens onto a solid-state detector. The resulting outputs are proportional to selected mathematical functions of the particle diameter, and are processed by software to provide the required size data. The size range of application is 2–200 μm.

Although used originally as a rapid-response batch apparatus, the instrument found application in the on-line monitoring of taconite. For such monitoring, the common diameters measured by the unit, e.g., average diameter of the area distribution MA and the average diameter of the volume MV, were found to correlate well with Blaine specific-surface and percent-passing standard sieves, respectively.

The instrument can be used to measure particles in air or liquid. In 1980 Leeds and Northrup developed a submicron particle size analyzer for the measurement of particles of diameter 0.17–20 μm in liquids. This instrument uses a dual optical system incorporating both a laser and a tungsten lamp. The optical system consists of two light paths, one for forward scatter and one for right-angle scatter. During part of the measurement cycle, the sample cell containing particles is illuminated by a helium–neon laser to produce 11 channels

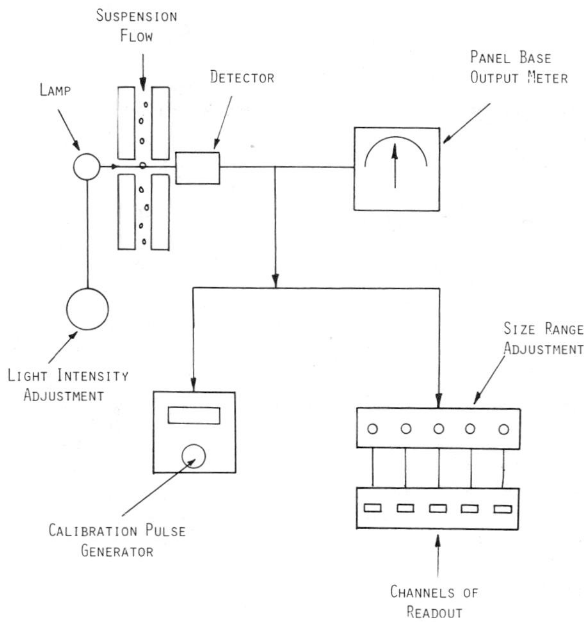

Figure 2.14. HIAC SS automatic particle counter.

of forward-scattered light. The other part of the cycle places several band pass filters and polarizers in the path of a tungsten-halogen lamp to produce three channels of 90° scattered light. The two groups of channels are combined and normalized by a microprocessor to display the size range above (see Fig. 2.15).

Another instrument designed on similar principles to the Microtrac developed in 1975, is the C1-Las. This is manufactured in France; and although differing in the design of the rotating mask, the instrument combines scattering and diffraction to measure particle size. It was originally designed for the analysis of cements, and is marketed under the name of Granulometer 226.

The combination of scattering and diffraction-pattèrn analysis is implemented in the size analyzer marketed by Malvern Instruments, Ltd., in England. In the Malvern ST 1800S and ST 1800P, light energy information is extracted from the particle diffraction pattern by a specially developed large scale integrated detector. This takes the form of concentric rings at increasing radii. The size range measured by one lens is approximately 100–1. The

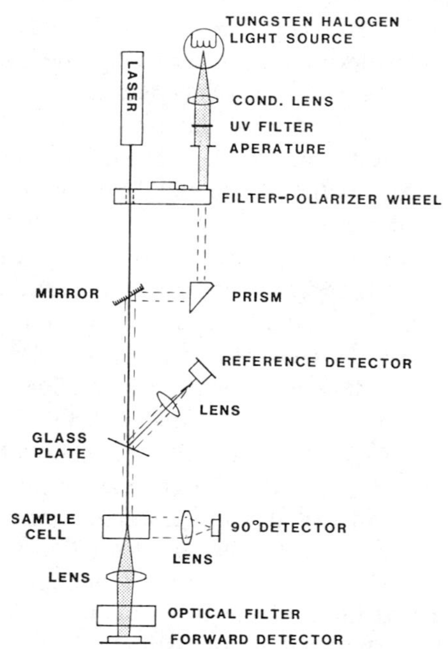

Figure 2.15. Leeds and Northrup small-particle analyzer.

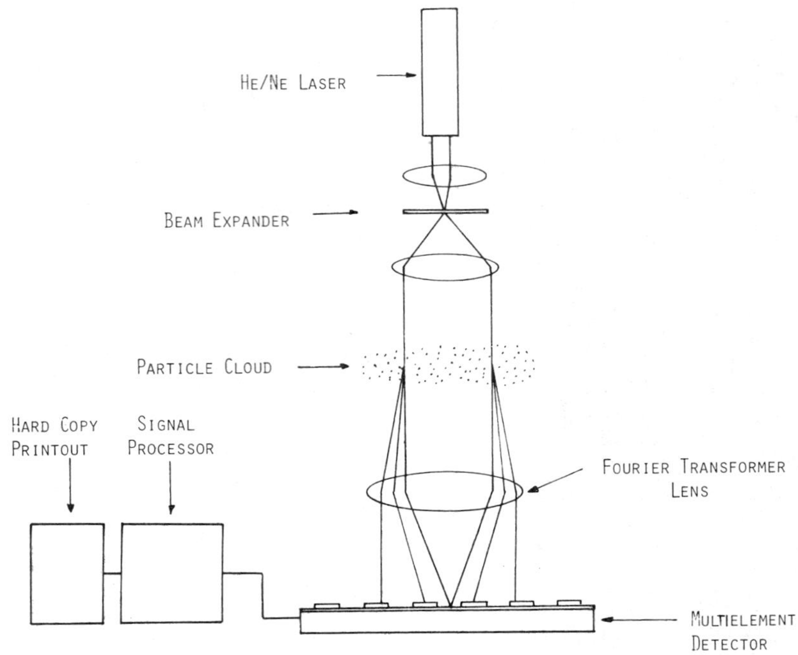

Figure 2.16. The Malvern ST 1800 particle and droplet analyzer.

output from the detector rings is scanned electronically and fed into a signal processor. Data are computed into 15 size bands and printed in hard copy. Both solid (*P*) and liquid (*S*) particles can be analyzed by the instrument. The advantage of the instrument is that light energy from a laser interacts with particles in an unrestricted working volume, and in consequence, particles need not be sampled and passed through tubes or cells (see Fig. 2.16).

The Malvern 4400 Photon Correlation Particle Analyzer is designed for 30 Å–3 μm particles in liquids. The particles in suspension are illuminated with a laser and scatter light. Due to Brownian motion, the scattered light fluctuates as the phases of the scattered light change relative to each other. Observation of these fluctuations with the Photon detector indicate the changes that take place with a time characteristic related to the size of particles under study and yield a correlation function. This can be used to compute the diffusion coefficient of the particle and, hence, its size.

The Coulter Nanosizer™ is based on a similar principle. Average particle size is computed in 1–8 minutes depending on the diameter.

Permeametry. Unlike most previous methods, permeametry provides a measure of the specific surface diameter—an average size. When considering the flow of fluid through a packed bed of particles, a linear relationship can be obtained between the flow rate and pressure drop. Apparatus for measuring specific diameters are therefore simple and are divided into constant flow and constant volume types. Allen points out that constant flow rate methods have some advantage for research purposes if provision is made for varying the flow rate, since the effect of flow-rate variation on bed permeability can be studied. Constant volume methods are simpler and more robust but less versatile. Table 2.9 shows the methods available.

Many of the devices developed for specific surface diameters measurement using constant pressure systems were designed years ago. The commercial version of this permeameter is the Fisher Sub-Sieve Sizer. In Figure 2.17, a steady head of pressure is maintained using a standpipe regulator. The use of two calibrated orifices enables the sensitivity of the measurements to be changed, i.e., make the pressure drop across the orifice and a packed powder bed to be the

Table 2.9. Permeametry.

	ANALYSIS TIME (MIN)	RECOMMENDED SIZE RANGE (μm)	APPROXIMATE COST ($)	AMOUNT OF SAMPLE NEEDED (g)
1. Atmospheric pressure				
1.1. Steady state				
Fisher; Rigden; Permaran	5	2–75	500–2000	5
1.2. Transient flow				
Blaine	5	2–75	500–1000	50
2. Low pressure				
Micromeritics-Knudsen Flow	45	0.1–75	5000–10,000	10

same. A potential calculator is mounted behind the pressure manometer, and is designed so that under steady conditions, the upper meniscus of the manometer fluid indicates the average particle size of the powder in the bed. A more recent variation of this device is the Permaran manufactured in Finland.

Constant-volume permeameters are more rare, and the only one of any note is the one devised by Blaine. This consists of a U-tube containing fluid. The powder bed is fitted to the upper end of one arm of the U-tube after the level of fluid has been raised so that the level of the meniscus is higher than in the arm open to the atmosphere. A tap joining the manometer to the bed is then opened so that air is drawn through the bed permitting the meniscus level to fall. The time required for the meniscus to fall from H_1 to H_2 is measured and from the use of basic equations, the specific surface area of the powder can be calculated, and the specific surface diameter estimated (see Fig. 2.18).

For fine particles, i.e., <5 μm, errors due to slip flow are experienced. To circumvent this problem, low pressure permeameters termed Knudsen flow permeameters have been developed. Knudsen flow prevails when the mean free path of gas molecules is approximately ten times the diameter of the passageways between particles. The system is flushed with helium to remove air and moisture, and a vacuum pump started and allowed to pump till the absolute pressure gauge indicates 0.1–1 mm of mercury. Flow is then allowed to stabilize at a maximum pressure of 15–22 mm of mercury. The flow rate is then determined by measuring the time required for 1 ml of gas to pass. Further values are taken at successively lower pressure and a linear plot developed. Extrapolation of the specific surface versus pressure data to zero pressure gives the specific surface

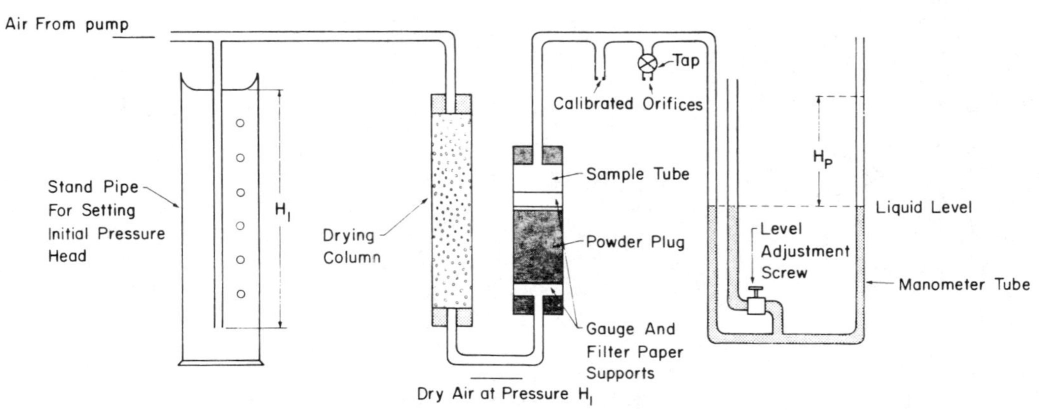

Figure 2.17. Fisher Sub-Sieve Sizer.

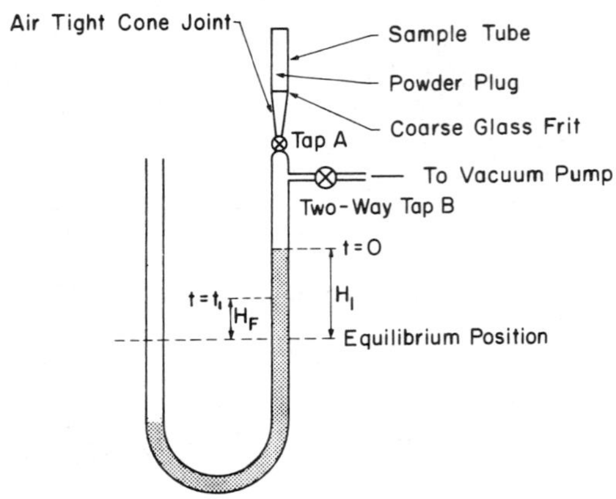

Figure 2.18. The Blaine permeameter.

area of the powder. This is termed the external surface area of the powder. A commercial version of this instrument is manufactured by the Micromeritics Instrument Company (see Fig. 2.19).

Surface Methods (see Table 2.10). In the previous paragraphs, particle size in terms of specific surface diameter, was discussed. It was stated that the surface measured is the external surface of the particles S. Knowing the volume of particles in the bed V_s, the surface-volume mean diameter d_{sv} can then be calculated from

$$\frac{S}{V_s} = \frac{6}{d_{sv}}$$

External surface implies that the surface detected is the envelope surface of the particles and no additional surface due to pore surface area is included in the measurement. When other specific surface methods are employed, i.e., gas adsorption of fatty acids, polymers, dyestuffs, electrolytes from solution, and heats of adsorption, varying amounts of internal surface due to macro, micro, and meso pores are measured in addition to the external surface. This significantly reduces the value of the surface-volume mean diameter, and introduces a sizing error. For this reason, total surface area measurements are not of much value to provide basic particle size analysis data but sometimes they are calculated and reported. In consequence, these methods are not discussed here, but are described in a later chapter under "Porosity, Permeability, Surface Area, and Pore Size Measurements."

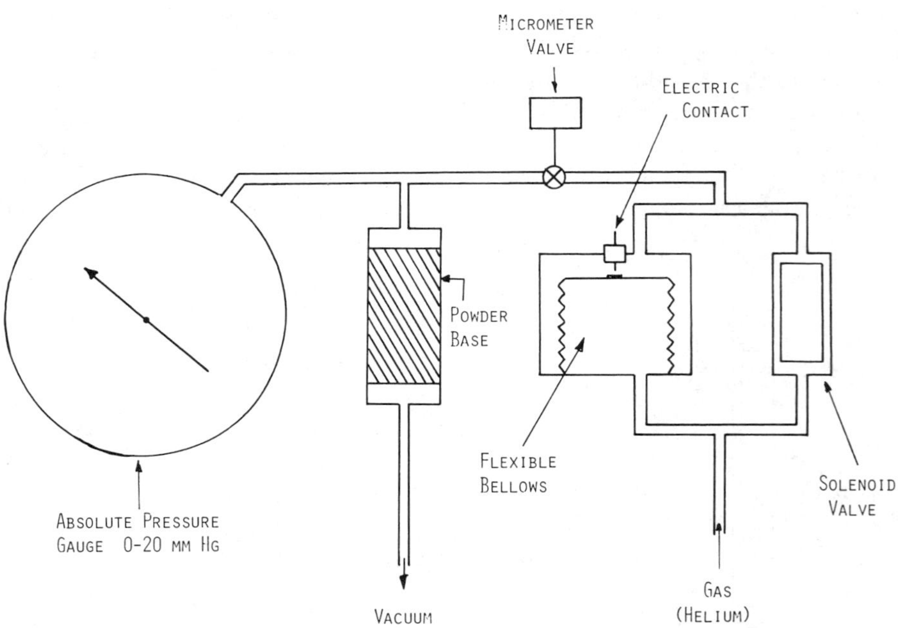

Figure 2.19. The Micromeritics Knudsen Flow permeameter.

Table 2.10. Surface Area.

	ANALYSIS TIME (MIN)	RECOMMENDED SIZE RANGE (μm)	APPROXIMATE COST ($)	AMOUNT OF SAMPLE NEEDED (g)
1. Multipoint BET				
1.1. Static pressure				
Adsorptomat, Micromeritics 2500	60	–	10,000–50,000	5
Micromeritics 2100				
1.2. Continuous flow				
Perkin Elmer, Quantasorb, Isorpta, Sorbet	60	–	10,000–25,000	20
2. Single point BET				
Strohlein Area Meter; Monosorb; Micromeritics 2200	20	–	5000–15,000	5

Table 2.11. Miscellaneous.

	ANALYSIS TIME (MIN)	RECOMMENDED SIZE RANGE (μm)	APPROXIMATE COST ($)	AMOUNT OF SAMPLE NEEDED (g)
1. β-ray attenuation Milltronies	20	10–500	30,000	100 liters
2. Ultrasonic attenuation Autometrics	10	10–500	30,000	100 liters
3. Differential air classification	10	10–500	–	2000
4. Pulsating gas flow	10	2–500	–	100
5. Low-angle x-ray scattering	30	0.01–1	–	1
6. X-ray line broadening	30	0.001–0.5	–	1

Miscellaneous Methods (see Table 2.11). Autometrics in Boulder, Colorado, have developed an onstream size analyzer for slurries. The lower limit of size detection is around 200 mesh and a solids concentration of 0–25% can be usefully handled. The system consists of an air eliminator and a sensor. The air eliminator utilizes a combination of centrifugal force and reduced pressure to remove air that would affect the accuracy of size and concentration measurements. The sample flow stream enters through a stationary pipe into a rotating impeller of elliptical cross-section, where pressure of air is reduced to 4 psi. The slurry then flows into the sensor section that contains two pairs of sensors, each of which consists of a transmitting and a receiving component. Pulses of high-frequency acoustic vibrational energy are transmitted through the sample stream at a rate of about 1000 pulses/second. The amount of transmitted acoustic energy that is received across the stream is dependent upon the number and size of particles in the slurry. Each pair of sensors operates at a different frequency and performs a different measurement. One determines size and the other concentration. The electronics section provides the signals to drive the transmitting sensors, and electronically conditions the electrical outputs of the receiving sensors so as to separate the measurements of size from percent solids. Changes in either parameter are claimed to have no effect on the accuracy of measuring the other (see Fig. 2.20).

Two instruments were developed in Japan for measuring the average size and size distribution of powders by elutriation and air classification. A stream of solid particles is fed into a classifier and separated continuously into two streams at a specific size termed the cut-off size. This value is a function of the air velocity. Thus, if the cumulative oversize fractional R = (Flow rate out)/(Flow rate in) is controlled so as to keep a preset value, e.g., 50%, then the cut-off size is the average size of the powder.

This is done by automatic adjustment of the air velocity in response to the flow rate signal of F_{in} and F_{out}. Using this method, the average size is obtained in 10–15 seconds.

If the air velocity is now changed continuously by a programmed system, e.g., a continuous divider and analog computer, the size distribution curve can be drawn. A typical analytical time for the determination of size distribution is one minute.

Flow rates of the oversize and feed powders are measured by particle flow meters that consist of a beam balance and a differential transformer. In this way the momentum measurements are converted to an electrical signal. Only particles greater than 100 μm are counted because cohesion at the low sizes limits the application.

In addition, the use of an air classifier or elutriator for on-line size analysis was discussed, developed, and reported by Metzger and Leschonski.[5] This was based on the principle of cross-flow classification. In this device the

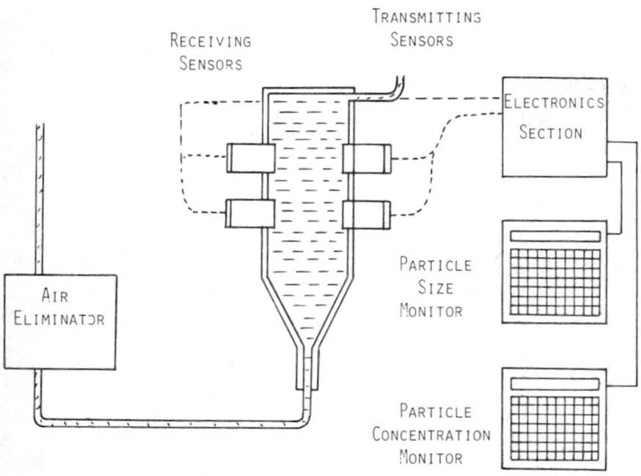

Figure 2.20. Autometrics ultrasonic attenuation size analyzer.

feed stream entering the classifying zone was divided according to size into eight fractions by means of seven blades. Extremely sharp classification cuts were obtained over the size ranges and compositions tested. One-line capability for size measurement was made possible by concentration sensors at the throats of each classified fraction container.

Continuing along the lines of particle analysis in gas streams, instruments have been developed to measure impact and momentum, and from these measurements, estimate size.

When a moving particle strikes a surface, its momentum is transferred to the surface. If the surface is suspended in such a way that it can vibrate on impact, the amplitude of vibration is proportional to the transferred momentum. Thermo-Systems in Minneapolis, Minnesota, manufacture a piezoelectric transducer to convert the mechanical vibration into an electrical signal. If the velocity and density of the particle is known, the size can be calculated.

Evaluation of the device as an on-line particle size analyzer for particles suspended in air has shown that accurate sizing of particles of size less than 30 μm is impossible; and, hence, the method has limited application.

However, a piezoelectric balance which has been developed can sense the impact of single particles. The actual weight of each particle on the sensor is recorded using a pulse height analyzer. The ultimate size resolution claimed is approximately one micrometer. It is obvious that for this system to accurately measure the mass of a particle, it must adhere to the sensor or contact; and then if the density is known, the size can be calculated.

KLD Associates market a droplet size analyzer based on evaporative cooling of a hot wire. This is for use in pesticide disbursement, fog and cloud analysis, and aerosol monitoring. Sizes from 1-600 μm were sized at a count rate of 500 counts/second.

A rapid response batch-type instrument using electrostatic ion capture has been developed by Thermo-Systems in Minneapolis, Minnesota, for the sizing of submicron particles in air. The aerosol is drawn into the instruments by a vacuum pump and receives a negative electrical charge from a sonic jet diffusion charger. The charged aerosol stream is then passed to the analyzer which consists of a cylindrical tube whose inner walls are protected by a sheath of air. A metal rod passes axially through the tube, and variable positive voltages can be applied to it. Particles smaller than a certain cut-off size which is a function of the positive voltage are drawn to the rod. Larger particles pass through the analyzer and are collected on a current collecting filter. The electrical charge on these particles drains off through an electrometer and a measure of the current is obtained (see Fig. 2.21).

By increasing the positive rod voltage, larger cuts of size can be selected, with a consequent decrease in electrometer current. This change in current is proportional to the number of particles present of size between the two cut-off values. By varying the positive voltage and measuring the electrometer current, a size distribution can be constructed.

So far, the methods that have been discussed have measured particle sizes down to approximately 0.1 μm, or 1000 Å. Many powdered materials are prepared with sizes of 100 Å; and at present, there are very few techniques capable of providing size data. X-ray scattering and line broadening offer two methods of measuring the average size of such small particulates. A similar procedure is followed as for light scattering but in order to measure the

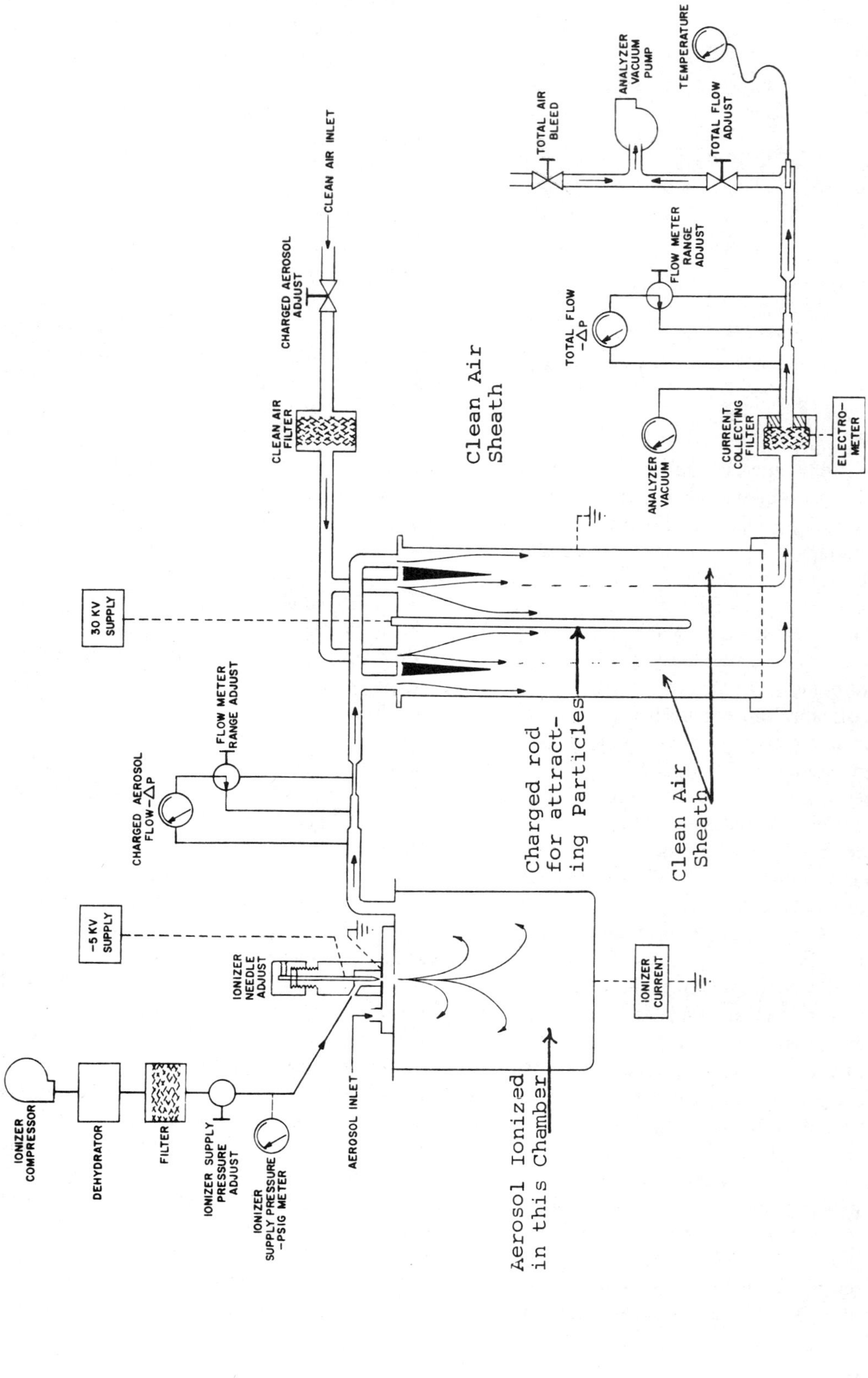

Figure 2.21. The Whitby aerosol analyzer.

diffraction pattern of x-rays scattered by particles, extremely low angles have to be observed. Unlike optical methods, the definition of diffraction rings requires optimum instrument design and alignment. In the most frequently used method, values of log (intensity) are plotted against the square of the diffraction angle, the slope of the straight line obtained is measured, and the root mean square radius calculated.

Selection of a Potential Measuring Principle

With a wide range of methods and instruments now in hand, let us now consider how we choose a potential measuring principle. Consider that we need to measure the size distribution of a 0.1-1.0 μm organic dyestuff, whose end-use application is in suspension or slurry form. This means that measurement should be made in liquid, and immediate sieving, elutriation, classification, and sensing zone methods are of no value because they are unable to detect and measure particles in the size range in question. X-ray methods cannot be used because organic particles do not absorb x-rays; and, hence, their masses would not be detected. Surface methods are of no value because they only yield average size data. Finally, the miscellaneous methods listed either cannot detect small particles or only yield mean particle data. Hence, we are left with microscopy, sedimentation, and light scattering. Microscopy is not desirable if alternative methods are available as it is slow and subject to error when the size range is wide. At this point, one should consider the end-use problem more specifically to determine if the problem is one of shelf-life, i.e., unstability in dispersion and settling; if so, sedimentation is the best choice. If it is more related to optical properties, then optical methods should be selected.

Selection of an Instrument that Uses this Principle

In sedimentation, instruments such as the Joyce Loebl Mark I, the Ladal pipette centrifuge, the Micromeritics 5500, or the Photomicron Sizer are possibilities. The former two instruments measure mass distributions directly, whereas the latter two do not.

For optical needs, the Leeds and Northrup Small-Particle Analyzer is a possiblity, but care should be taken in interpreting Mie scattering calculations for dyestuffs having complex refractive indices.

The actual instrument selected will now depend on such factors as availability, cost, sample size, analytical time, etc. With the Joyce Loebl Mark I, an instrument cost of \$25,000 can be expected; sample size is typically several milligrams, and analytical time several hours. With the Ladal pipette centrifuge, a cost of \$10,000 can be expected; a sample size of several grams is typical, and an analytical time of two hours is common. With the Leeds and Northrup SPA, the cost is approximately \$35,000; sample size several milligrams and analytical time one minute.

In the above dyestuff example, the choice of instrument was quite limited, alternative instruments few, and a selection not too difficult. However, if instead of a submicron organic particle, an inorganic particle of size 1-50 μm had been specified, then approximately 30 instruments could have done the job. Here a decision on which instrument to use would be more difficult, but chances are that one could be found more easily in the quality control laboratory that would be satisfactory. This decision on the instrument for the size analysis job now fixes the sample size required for analysis, and ways of taking a representative sample of that quantity are now described.

Acquisition of a Reproducible Sample

Sampling is usually divided into four necessary parts: bulk, gross, laboratory, and measurement sample acquisition. Bulk samples are usually taken from product flows of many tons/hour, and normally such samples constitute several hundred pounds. For example, a bulk sample could be represented by a tank car of slurry from a train or twenty 10-lb bags taken during a bagging operation, etc. These bulk samples are generally taken at the production site and need to be subdivided so that a gross sample can be taken representative of the production run. This gross sample would usually weigh a few kilograms. Preparation of gross samples from bulk samples is best achieved by the practice of two general rules. First, particulate samples whether wet or dry should always be sampled when in motion, and the whole stream should always be sampled using short-time increments. This procedure is sometimes called primary and secondary sampling. In consequence, a sample can be extracted from a tank car or truck only when it is being filled or emptied. Likewise, samples can be taken from the powder falling from a conveyor at a transfer point or from a chute or pipe in which particulate is moving. If heaps, tank cars, trucks, bags, etc., are sampled when the particulate is stationary, then a non-reproducible sample will almost always result. This is because of segregation of particles of different size, density, and shape in heaps. When particles are poured into a heap, a conical surface is quickly established down which large particles can roll and small particles cannot. These tend to percolate into sublayers of the structure resulting in a triangular cross-section having a higher concentration of fines in the central core and a higher concentration of coarse at the surface and at the bottom edges. Such conical or pyramidal shapes are formed in heaps, on conveyors, in containers, etc., so that representative sampling is impossible by any means except the one above. For this procedure automatic samplers are generally used. Figure 2.22 shows a primary sample cutter that moves through a sample stream at several inches/second. Commercial samples can travel 10 feet and longer to traverse conveyor belts or wide flow streams. A variety of cutters are available for wet and dry streams. The width of the cutter opening is a function of the largest particle present, and should always be larger than three times the maximum particle diameter. The primary cutter is designed to keep the sample within the cutter in motion so that it can be further subdivided by a secondary device. This usually takes the form of the Vezin sampler in Figure

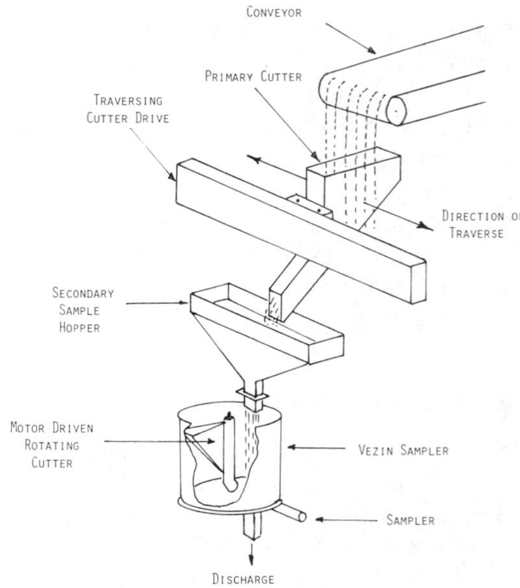

Figure 2.22. Primary and secondary sample dividers.

2.22. The stream from the primary cutter enters a cylindrical drum fitted with a secondary radial cutter. This revolves at a speed designed to extract particles from the stream without interference. Samplers of diameters from 6 to 60 inches are available and usually extract 5% of the sample stream. At this point, the sample exiting the Vezin is usually of the order of one or two kilograms. This is then packaged and sent to the laboratory. A laboratory sample is extracted from this package, by passing it through a smaller laboratory Vezin sampler or other rotary sample divider (see Fig. 2.23). The disadvantage of a laboratory Vezin sampler is that one subsample results from the sampling step and so a sampler such as the Ladal, Quantachrome, Leeds and Northrup, and Geos units that can separate the sample into sixteen subsamples is more often used for the final analytical sample. In this case, several representative samples can be analyzed for duplicate tests.

For example, let us assume that the primary cutter extracts 51,200 g of powder from a flowing stream and the secondary resin sampler extracts 5% of this to yield a 2560 g sample for the laboratory. A small laboratory Vezin sampler again operating at 5% removal rate

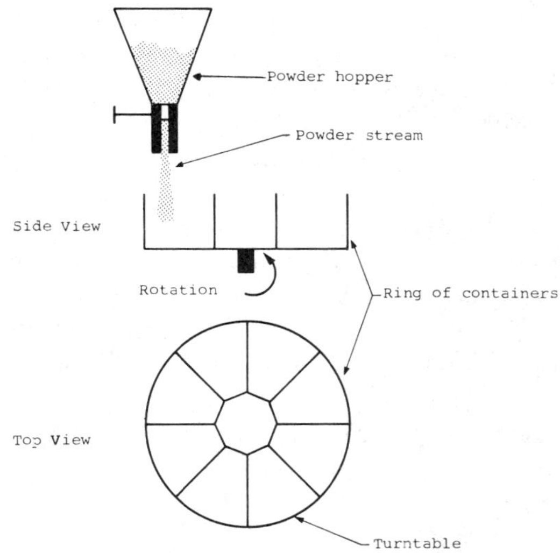

Figure 2.23. The rotary sample divider or spinning riffler has a ring of containers that rotates under the sampling outlet; each time a container passes under the stream of powder, it collects a small portion, which is used in the laboratory.

would cut 128 g from the sample for the quality control room. Using a rotary sample divider with a low feed rate and moderate rotational speed, 8 g would be subdivided into each of the 16 cups. For the Joyce Loebl, 2 cups would be removed from opposite sides of the sampling table, and the remaining 14 emptied. The first cup would then be run again through the rotary sample divider so that 0.5 g was obtained in each cup. Two cups would again be removed from opposite sides of the table and the particles deposited in small sample vials as Samples 1 and 2. The remaining cups would be emptied, and the second sample cup run. Two cups from opposite sides would again be taken, and deposited in vials as Samples 3 and 4. For analytical techniques requiring 2 g of sample, four of the 8 g subsamples would be passed through the rotary sample divider, depositing 2 g in each cup. Four samples would be taken one from each quadrant of the sampler table and labeled samples 1-4. Table 2.12 shows a comparison of the variances between sampling techniques according to Kaye.[6] Clearly, rotary sample dividing is the best method to use.

In the case of fine pigment powders or concentrated viscous slurries of dyestuffs or pigments, sampling can be achieved without fear of segregation or sedimentation. After good mixing, samples can usually be extracted with confidence by pipette or spatula. However, wherever good rotary sampling can be performed, the rule is to use it.

In those instances where sampling has to be performed in a gas stream, such approaches are of no value. Here, a sample has to be drawn from the gas via a sampling probe. This technique has been well documented by the Environmental Protection Agency, and need not be discussed in detail here. Suffice it to say that the sample probe should have sharp leading edges facing into the gas flow so as to minimize any disturbances. It should be aligned directly into the flow, and be pulling the sample at the same velocity as the gas stream. If the probe and sampling line have to be bent, care should be taken to estimate losses due to impaction within the sample line. Furthermore, in the event of temperature differentials, turbulence, dry gas, etc., sample line losses to effects of Brownian and Eddy Diffusion, thermophoresis, and electrostatic charge should be accounted for. Probe and sample line design can do much to minimize losses. Use of transverse flow of gas through porous walled sampling lines has been one approach.

Collection of the airborne sample can then be achieved using an impactor, an impinger, a thermal or electrostatic precipitator, and a cyclone.

Preparation of the Sample for Analysis

Regardless of the size analysis procedure used, the state of dispersion of the sample must

Table 2.12. Comparison of Sampling Method Variances.

SAMPLING METHOD	VARIANCE
Scoop sampling	2.33
Chute riffler	0.336
Spinning riffler	
Slow speed	0.058
High speed	0.023

always be considered. If the sample is not dispersed well enough initially or if it is allowed to fluctuate during the analysis, erroneous results will be obtained.

Imaging and Counting Requirements. For optical imaging and counting methods where the particles have to be mounted on solid surfaces with a minimum potential for movement, mixing of the sample with a few drops of dispersing agent on a clock glass with gentle shearing with a camel hair brush will usually disperse the sample. One drop can then be removed and placed on a microscope slide, a cover glass laid over it and sheared gently to exclude air bubbles. Alternatively, the sample can be mixed in a low-viscosity oil, and again sheared, and coated on a slide. Several slides should be prepared from drops taken from different places on the clock glass, to ensure representative sampling. For electron microscopy, films of collodion, carbon, or plastic are formed and mounted on electron microscope grids. Particles are usually dispersed in liquid, and a drop laid onto the grid by use of a hypodermic syringe. Dilution and spreading of the drop is then performed followed by vacuum drying. For scanning electron microscopy (SEM), a sample drop is placed on an SEM sample stub, vacuum dried and carbon- or gold-coated in a vacuum evaporator.

Observation of the dispersion is then made to determine the quality of the dispersion. Hence, good dispersion is interpreted as good particle separation. For some analyses, counting can then be performed directly. For others, such as image analysis, good quality photomicrographs must be taken. The requirements for these are identical, whether the photomicrograph is from an optical, electron, or scanning electron microscope. Good contrast between particles and the background, and uniform shading of the photomicrograph is essential. Consultation with the image analysis operator is therefore imperative, before photomicrographs are taken. Such precautions will prevent unnecessary and unsatisfactory photography.

Sedimentation and Sensing Zone Method Requirements. Dispersion in this instance demands dynamic stabilization. The sample has to be dispersed into bulk liquid, and must remain stable for time periods from 15 minutes to several hours. Dispersion is divided into two stages, both of which are usually performed simultaneously. In the first, the agglomerated and aggregated sample has to be separated into discrete particles and aggregates by the application of energy.

Energy can be and is applied by the use of mixers, blenders, and ultrasonic devices. Here, care must be exercised to ensure that particle

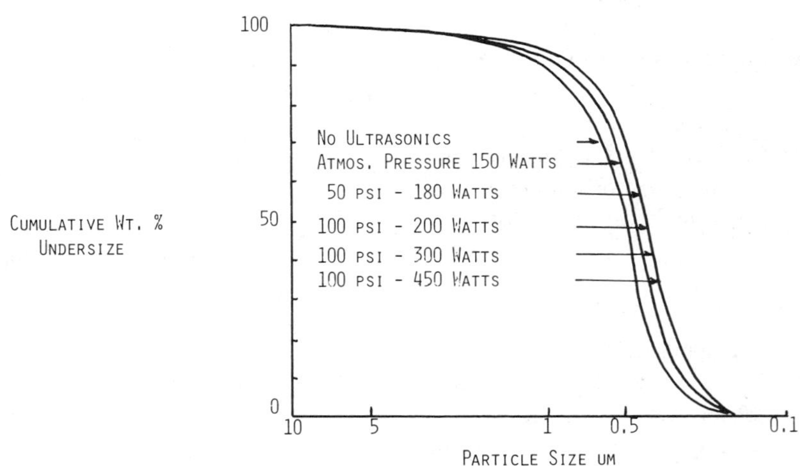

Figure 2.24. Size distributions of as-received powder under varying degrees of ultrasonic energy input.

breakage does not occur. Dispersion should always be accompanied by microscope examination of the sample after the application of energy to investigate this. Figure 2.24 shows the effect of ultrasonic power and applied pressure on a metal oxide dispersion. Figure 2.24 shows that the as-received powder yields a constant particle size distribution with increasing energy after a certain level, and SEM examination showed no tendency for breakage. Figure 2.25 shows a similar oxide slightly sintered, and here the application of increasing energy was found to continually break sinter bonds between particles, and a constant size distribution was never attained. Such behavior should be investigated before size analysis is reported. Obviously, in this case end-use should be correlated with energy input.

Once separated, the second stage of stabilization has to be implemented. This can be done by adjusting pH to give the optimum surface charge or by the use of additives. Phosphates and surfactants are the most widely used additives. Phosphates are particularly useful for dispersion of metal oxides and clays, and are a universal dispersing solution. Sedisperse has found wide application in sedimentation studies. The degree and state of dispersion is usually measured by optical microscopy during a sequence of (a) additive selection, (b) energy input, and (c) size analysis. Good appearance under the microscope and constancy of size data with time are usually taken as the indicators of a good dispersion.

If dilution of a dispersed sample has to be performed, it is important that the diluent contain the identical concentrations of dispersant as the original sample or flocculation can take place. Dispersion is discussed by Parfitt.[7]

Surface methods. For gas adsorption data, the powder must be dry and have no adsorbed gases or vapors on its surface. Degassing by heat treatment, followed by storage under high vacuum must precede all gas adsorption measurements.

Measurement of Particle Size and Size Distribution

Before any data are taken, it is recommended that the operator be asked the date of the last instrument calibration. If one week has elapsed since the last calibration, a measurement on a standard powder should be made. Such powders are available from the National Bureau of Standards or Duke Standards. When completed, the size measurement can be performed on the sample. Size analysis of duplicate or triplicate samples is made. For most methods reproducibility should be within 2% at the mean. One factor, mentioned previously, and

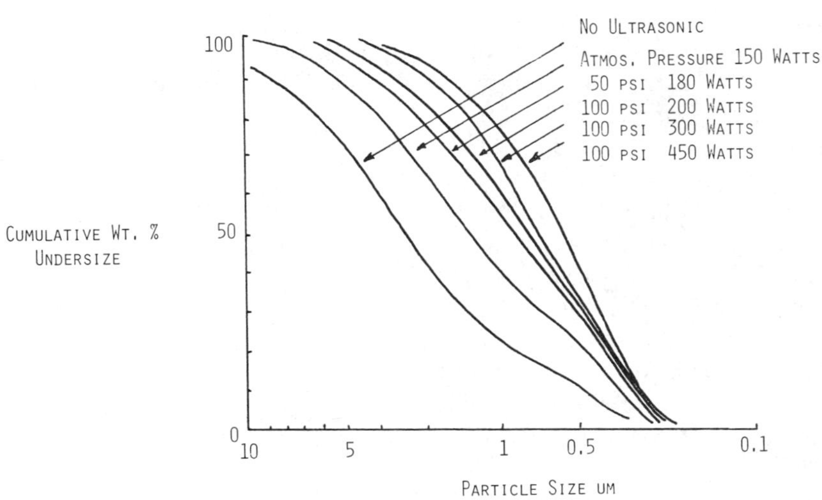

Figure 2.25. Size distributions of sintered powder under varying degrees of ultrasonic energy input.

emphasized here, is that of particle concentration. In all sedimentation and sensing zone method analyses, different degrees of dilution should be tried to satisfy the user that concentration effects are absent or minimal. In sedimentation, these effects are manifest by the clustering phenomena, and in sensing zone methods by coincidence counting. Both result in a coarser size analysis than reality. Concentration effects are observed by noting the shift of the mean size with increasing dilution. If concentration effects are present, the mean size becomes smaller with each increase in dilution until a constant value is reached. This may never be attainable with some instruments, because analytical detection and size measurement may become inaccurate on account of low sensitivity. An example is the Andreasen pipette and the gravity photosedimentometer. If concentration effects are present at 1%, increase in dilution will result in lower weights per sample extracted, and low weighing accuracy. Use of photosedimentometer at 0.005% by volume, will be acceptable.

Selection of Data Display Methods

There are many ways of representing size data. These were discussed in Chapter 1. The advantages of population, area, and mass distribution calculations have already been stated. Plotting any of these three distributions on linear graph paper produces the familiar bell-shaped histogram or "S" shaped cumulative distribution. Histogram plotting is extremely sensitive to detect bimodal or trimodal distributions, whereas cumulative plots are not. Both should be plotted if data has been taken on an unknown. Alternatively, the use of arithmetic probability and logarithmic probability paper to plot size distributions permit the size data to be represented as straight lines. From logarithmic probability plots, the geometric means appears at the 50% point of the distribution, and the ratio of the size at the 84.1%/50% point yield the geometric standard deviation.

One property of the log-probability plot is that population, area, and weight distributions appear as parallel straight lines. In consequence, a knowledge of the geometric mean and standard deviation of the population distribution permits one to use the Hatch-Choate equations to calculate and plot the corresponding area and weight distributions or vice versa.

In many instances, size distribution data are used to indicate skewness and kurtosis or other selected distribution parameters. Such data are usually reported in summary tables, and not as the graphical distribution.

2.3. SUMMARY

Size analysis is a complicated branch of science, and those requiring it should follow the steps outlined in this chapter. Time or space does not permit a thorough guide to be prepared for the user, but use of this approach will permit most users to avoid the common pitfalls experienced by those newly introduced to the field.

REFERENCES

For most needs in size analysis, sampling, and data representation the following reference is adequate:

1. T. Allen, *Particle Size Measurement*, 2 ed., Chapman and Hall, London, (1975).

Specific references reported were:

2. K. Leschonski, *Powder Technology*, **24**:115-124 (1979).
3. K. T. Whitby, Symposium on Particle Size Measurement, *ASTM Special Publication* **234**:3 (1958).
4. K. Leschonski, *Staub* **22**:475 (1962).
5. K. L. Metzger, and K. Leschonski, *Proceedings, 1st European Symposium on Particle Size Analysis*, Nurnberg, (September, 1975).
6. B. H. Kaye, London University, Ph.D. thesis, (1962).

For most needs in dispersion and sample preparation, the following reference is recommended:

7. G. D. Parfitt, *Dispersion of Powders in Liquids*, 2d ed., Wiley and Sons, New York, (1973).

3
Particulate Characterization: Future Approaches

Thomas P. Meloy
West Virginia University
Morgantown, West Virginia

3.1. INTRODUCTION

Particles are ubiquitous. They are in the air, the ocean, lakes, drinking water, soil, the ocean bottom and shores, space, the lunar surface, the Martian surface, the rings of Saturn, comets, vast clouds of the galaxy; in fact, particles pervade all of space, even deep space. Particles come from a number of sources: rocks through weathering, flacking, where flacks and sand are washed down the streams into the rivers, and finally from sandy beaches and their muddy bottoms. Particles are formed by condensates from the atmospheric pollutants; they are formed from evaporating droplets blown from waves in the ocean. Particles come from the breakup of rock by frost or entrainment and fracture after being entrained by winds. In the ocean, particles are formed by living matter forming a skeleton, dying, and leaving behind the skeleton; surprisingly, the earth creates a significant tonnage of particles daily from the solar system. One often forgets how extensive particles are, but one only has to look at beach sands, rocks, sediment, clays, and soil to see them. In fact almost the entire exposed surface of the earth is comprised of or covered with particles.

The greatest interest in particles, however, is industrial or man-made particles or man-handled particles. Yearly, over one cubic mile of particulates are created and manipulated by the iron, coal, copper, sand and gravel, cement, fertilizer, and chemical industries. Coal waste products alone are expected to reach between 300 and 500 million tons a year with disposal cost running $100 per ton because of their toxic nature. While we know a great deal about the chemical composition of many particles, we know next to nothing about their shape because until recently, we were unable to define particle shape.

3.2. CLASSICAL APPROACH

Characterizing particles shape is an ancient art. According to Bond,[1] there is a biblical reference to sizing ground wheat through woven cloth. Separating wheat from chaff by allowing the more compact, denser wheat grain to settle and the chaff to blow away is a typical form of sizing by sedimentation. Widespread use of screening and sedimentation is made by industry today to size particles. Laboratory sizing and sedimentation are also extensively used.

Particle size is one measure of particle shape, and there are a number of ways to size particles. Sizing by sieves or screens is a complex process by which particles are allowed to pass through a series of screens, each with a decreasing screen opening. Basically, the screens are made of woven wire cloth with reasonably rigid rectangular openings as shown in Figure 3.1. The particles to be sized readily pass through the larger screen openings until they reach a screen close to their "size." The screen openings look something like the grid in the lower left of Figure 3.1. The particle is jiggled by an external energy source until it up-ends and tries to go through the sieve opening. If it just sneaks through the opening as shown in Figure 3.1 in the bottom right, then it is said to be that size because it is retained on the screen below which has smaller openings. The length of the particle passing through the screen has little to do whether it will pass through the screen or not. The likelihood of the particle passing through the screen is a function primarily of its width and secondarily of its height. In Figure 3.1, lower right, one can readily see that the size of the particle is a complex function of its width and height. If the particle just passes through the screen, then it is said to be the size of an equivalent sphere of radius r, shown in the next grid opening, lower right of Figure 3.1. One parameter, the diameter of the equivalent sphere does little to define the particle shape even in a gross sense. For example, if the particle were very platy, the width w would dominate the probability of the particle passing through the sieve of a given size because h is very small compared to w. If w and h are roughly comparable dimensions, one can gain no information on the particle length. In fact, one of the major problems in characterizing the shape of a particle by size is that the approach only gives one parameter to describe particle shape.

To get a better handle on particle shape, a very tedious, microscopic method was introduced whereby the longest dimension of the particle l was measured and then the width w, which was the largest dimension at right angles to l. See Figure 3.2. By combining w and l, one could get a measure of the particle size and

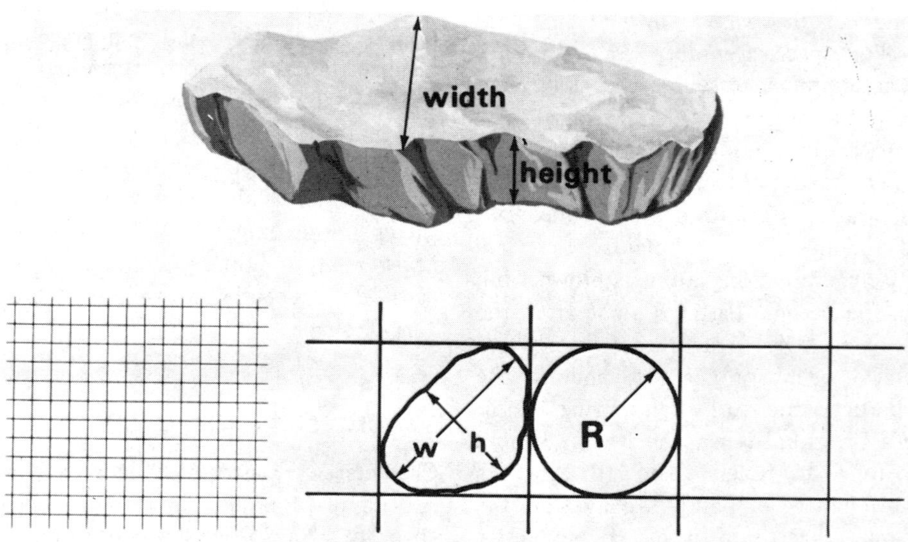

Figure 3.1. *Sizing by square screens.* Screens or sieves are widely used for sizing particles from 5 cm down to 40 μm. Most screens are made of woven wire and have square openings. The size of the particle that will just be retained on the screen is defined as that sphere which will just fit the square opening. Actually the size of the particle retained is a complex function of particle shape governed primarily by the particle width w and height h.

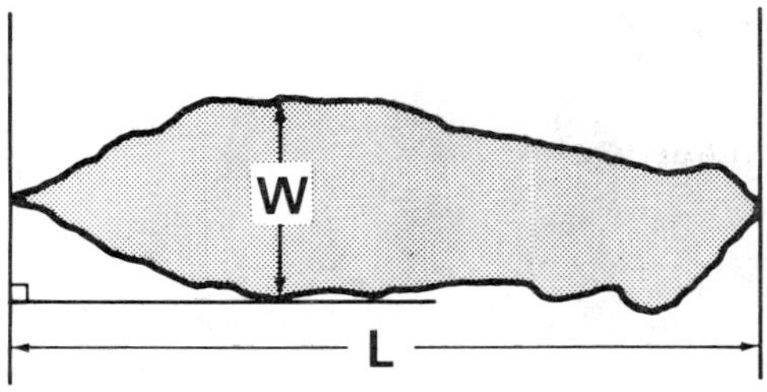

Figure 3.2. *Sizing by length and width.* Microscopic sizing is done by mesuring the longest dimension L and the maximum width W perpendicular to the length. These two measurements do not predict the screen size. Note the particle is characterized by two parameters.

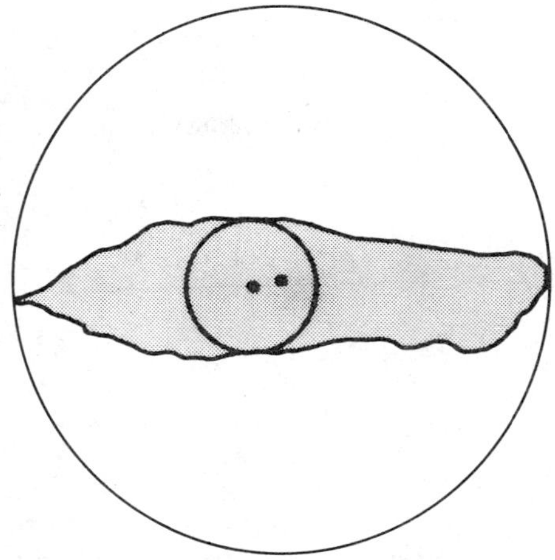

Figure 3.3. *Sizing by inscribed and circumscribed circle.* A faster method of characterizing a particle is to find the particle inscribed and smallest circumscribed circle for a particle. Again two parameters are used to describe the particle. These two parameters are similar to those in Figure 3.2 but will not predict the screen size.

by dividing a by w, one could get a measure of the particle aspect ratio. Because this method was both useful and tedious in characterizing particles, a faster system was developed.

By using a series of circles, the smaller of which was the largest circle that could be inscribed in the particle, and then by finding the smallest circle which could completely encompass the particle, as shown in Figure 3.3, a measure of the particle size and aspect ratio would be obtained. Between the two parameters obtained by the method shown in Figure 3.2 and those obtained by the method shown in Figure 3.3, there is a high correlation. For example, the length in Figure 3.2 is quite close to the diameter of the circumscribed circle in Figure 3.3. It is not the same because if one circumscribed a square particle, then the diameter of the circumscribed circle will be larger than the length. Thus the substitution of the faster, more economical inscribed/circumscribed circles method yields two parameters that describe reasonably well the gross appearance of the particle.

To improve the information one obtains from the relatively simple method of screening using square openings, particles can be run through slotted screens, shown in Figure 3.4, where the height of the particle is measured. Combining

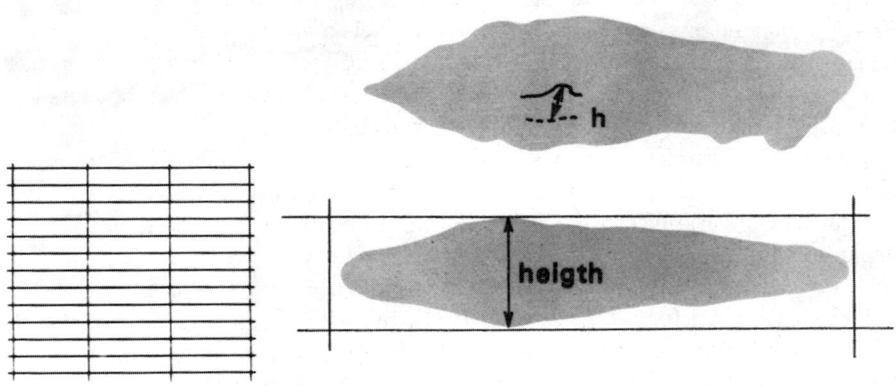

Figure 3.4. *Sizing by slot screens.* A different type of sieve is the slot screen which primarily measures the height of a particle. This is different from the screen in Figure 3.1 and from the parameters measured in Figures 3.2 and 3.3.

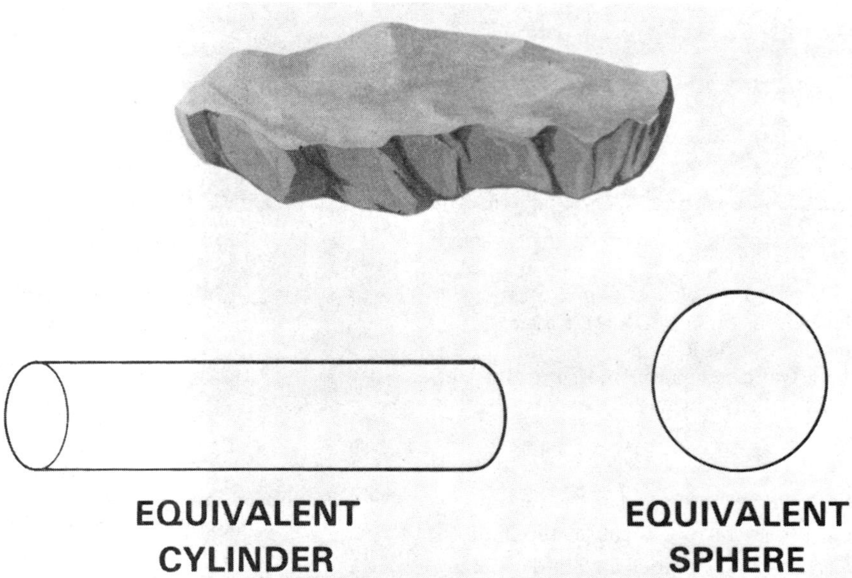

Figure 3.5. *Sizing by settling in a fluid.* Letting a particle fall in a fluid and measuring its terminal velocity yields the aerodynamic particle size. One says that the particle size is equivalent to the sphere or cylinder that will settle at the same rate. The aerodynamic diameters differ from all previous measurements.

this technique with rectangular screens (Fig. 3.1) yields information on particle shape and tends to define the thickness-to-width ratio. Once again, only two parameters are available for describing particle shape.

As described earlier, sizing by settling in a fluid has had long historical applications. In Figure 3.5, by taking the particle, letting it settle in the liquid, and then measuring the velocity of the settling rate, one can get an estimate of the particle size. If the particle is chunky, then one can take the settling rate and say it is equivalent in size to a sphere of the same density, which would settle at that rate. On the other hand, if the particle is long and has a high aspect ratio, then one can compare its size to that of a equivalent sphere with the same density that would settle at the same velocity. There are variations in the sedimentation technique in that one can settle particles by laminar flow or turbulant flow. Depending on the shape of the particle, two particles that settle in the laminar flow region with identical velocities may have different settling velocities in the turbulant region.

Rumpf[2] finds that there are 19 different methods of mechanically measuring particle size that are used to measure 19 different behavior characteristics of the particle. None of these 19 methods characterizes the particle shape by more than two parameters.

Recently, the flying spot microscope has been developed and is now widely used to characterize particles. By looking at the particles' silhouette and scanning across the particle, one obtains a measure of the silhouette's area by the total length of scan line that lies within the particle silhouette (see Figure 3.6). By a somewhat more complex measure, an estimate of the particle perimeter can be made. By dividing the area by the perimeter squared, a measure of the particle aspect ratio can be obtained. These methods of optical scanning are very fast and can handle a number of particles, but by the same token, can obtain only two independent measures of the particle shape. Unfortunately, it is widely implied that these optical scanning techniques give six measures of the particle shape rather than the actual two. The other four are linear combinations of the two independent measurements.

Two measurements are not sufficient to describe a particle's shape. Several attempts have been made to set up a series of standard particle

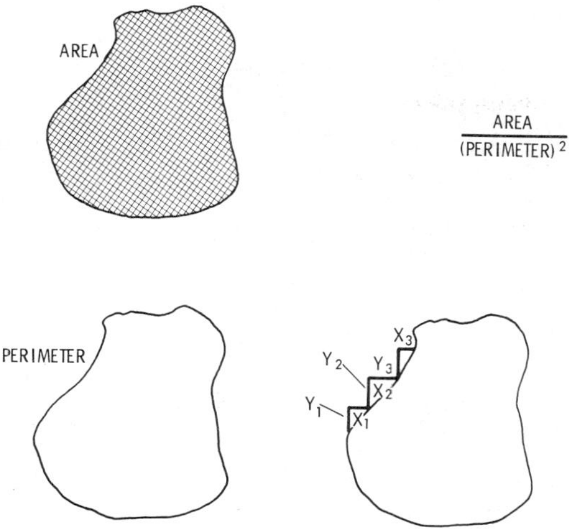

Figure 3.6. *Particle characterization.* Automated scanning optical microscope measure two new parameters: the projected area and the projected particle perimeter.

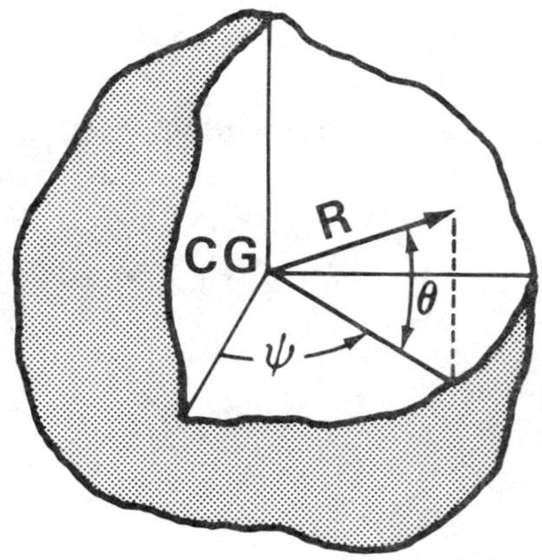

Figure 3.7. *Spherical coordinates.* The particle surface is traced out by a radius vector as a function of angles Θ and Ψ. Origin of spherical coordinates at particle center of gravity–*CG.*

shapes and then describe particles as belonging to one of these standard shapes. The ultimate in this approach is McCrone's[3] particle atlas, where several hundred typical particles are shown. For airborne particulates, this atlas has been successful. However, looking up a particle shape by comparing one's particle to a photomicrograph is not effective quantization.

3.3. MODERN CHARACTERIZATION TECHNIQUE

Three investigators[4,5,6] working independently developed a similar method for characterizing the shape of particles. A particle can be described by polar coordinates. Figure 3.7 shows a particle where the center of gravity is exposed by cutting away the surface. A radius vector is run from the center of gravity to the particle surface. By using the radius vector and the two polar coordinate angles, all points on the particle surface can be described. This technique then can describe to any desired degree of accuracy the shape of the particle surface.

A whole series of orthonormal functions can be used to interpret the information. However, the most familiar, and as it turns out, the most applicable, are the Fourier functions.

Figure 3.7 shows how a particle surface can be described by polar coordinates. Before describing how one composes a particle descriptor in the polar shape by use of the Fourier functions, one first must realize that all one normally sees of a particle is its silhouette or profile. Figure 3.8 is a transposition of the coordinate system in Figure 3.7 such that the polar coordinates have been converted directly to linear coordinates, with the radius vector being the height and the two polar coordinate angles being plotted along the x and y axis. The result is a surface floating above the x–y plane. This surface represents the surface of the particle. Any cut or sample of this surface will give information as to the roughness and physical parameters needed to describe the surface. Conversely, no two cuts will be the same. The edge of the silhouette will be one of the cuts along that particle surface. The information in that cut, the edge of the silhouette, describes the characteristic of that surface. Unless the cut misses a unique dominant feature on the particle surface, then the cut will be representative of the particle. By sampling that cut discretely, one can decompose it into a series of coefficients which can be used to reproduce that cut.

Figure 3.9 is a demonstration of how the silhouette is unrolled. The particle's silhouette

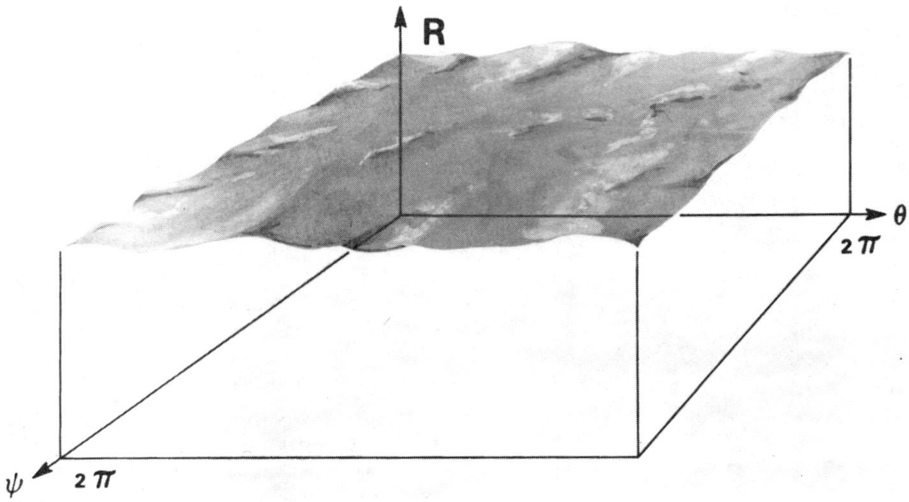

Figure 3.8. *Particle surface transformed to rectangular coordinates.* Particle surface in Figure 3.7 is plotted as a single valued surface in rectangular coordinates. Sampling the surface to obtain data for spectral analysis can be done at random points or along random cuts, such as the equator or silhouette.

in the upper left is sampled at various intervals along the profile. Radius vectors are run from the CG to the sample points and then the silhouette is literally unrolled and plotted as shown in the bottom right of Figure 3.9 as x vs. y in the rectangular graph.

As is well known, a signal may be analyzed or expanded into a Fourier series. In this case we will use a discrete Fourier series. The result will be a series of fourier coefficients and phase angles that can reproduce the particle shape with great accuracy. In the techniques used, which are described later, the sample points will be equally spaced such that the angular distance between sample points is the same. This is known as nyquist set.

Let us examine some of the limitations of this technique. Figure 3.10 shows a particle silhouette with the radius vector r going in and out of the particle to a region that can be called the reentry, an invagination, or an embayment. A particle with this type of characteristic is called nonanalytical because of the ambiguous value of the radius vector for a given angle. There are techniques developed by Zahn[8] that enable one to avoid this dilemma and are described later. Another type of problem encountered in using this technique, particularly when the optical scanning is done automatically, is how the date points are connected. Figure 3.11 shows three different ways to connect the seven intersecting points that the optical scan produces. In other words, as the scan goes along line 1, it first encounters point one, then point two. The scan retracts, steps down to line 2, encounters a third point, then steps down to line 3 and encounters point four on the left and point five on the right. These points are distinguished by rapid changes in grey level. How these points are connected is not determined by the scan, only the location of the points. Algorithms connecting the interface points can connect them in the wrong order, but the problems involved using the newer algorithms are small.

Another type of error encountered is known as the glitch. On the left of the drawing in Figure 3.12, there is a pyramid or tower that is a glitch. A glitch results when a speck of dirt is picked up and interpreted as a data point. More modern algorithms tend to minimize this type of error.

From the evenly spaced nyquist data, the discrete sampling of the particle profile has been taken. By using a discrete Fourier transform, the coefficient's characteristic of the Fourier

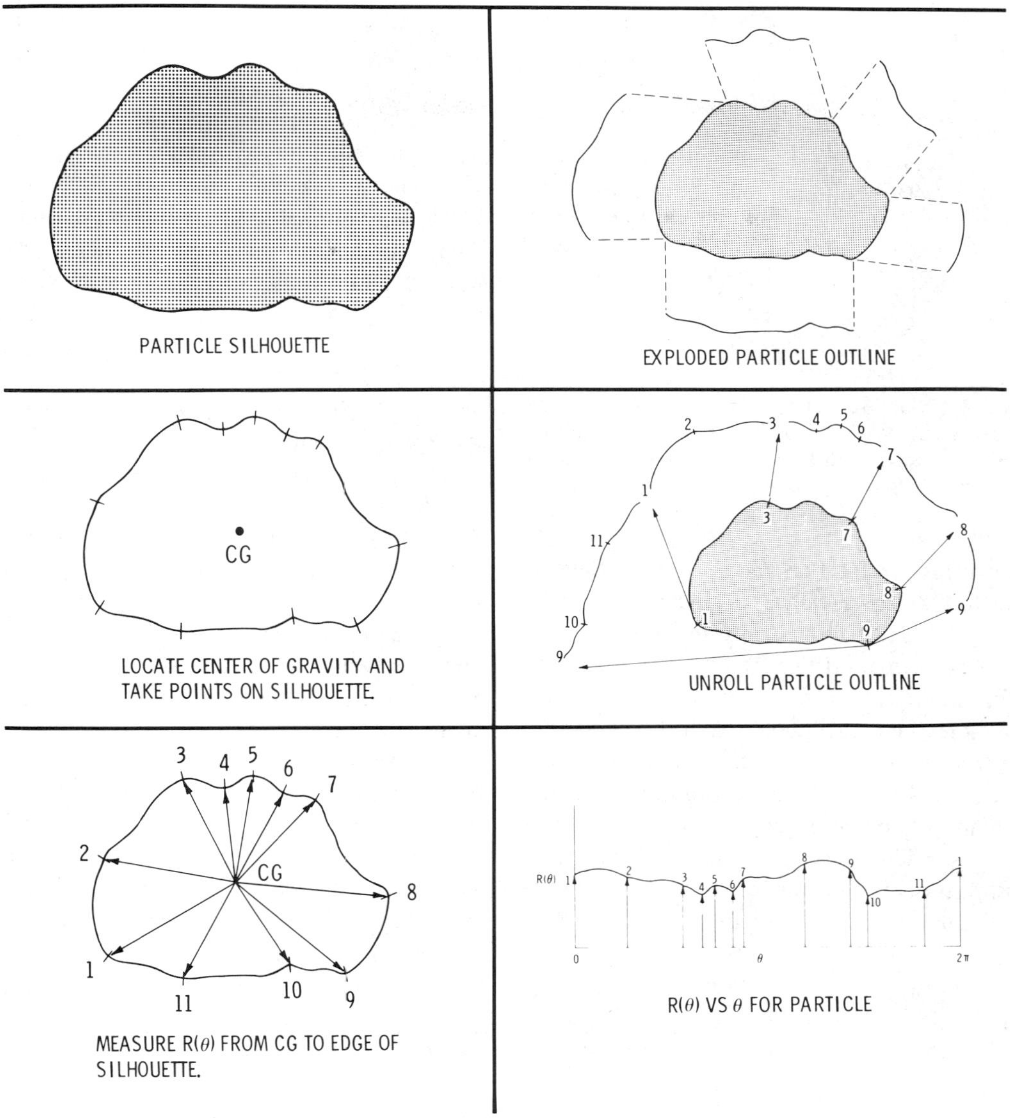

Figure 3.9. *Unrolling the particle silhouette.* To transform the particle outline into a single valued plot of particle radius $R(\Theta)$ vs polar angle Θ using the *CG* as the reference point, one takes the particle silhouette, finds the center of gravity, selects points on the edge of the particle silhouette, runs radius vector from *CG* to particle edge, explodes edge segments, and, finally, strings the segments together in a rectalinear plot of $R(\Theta)$ vs Θ. As the number of edge segments becomes more numerous the plot becomes more accurate. The R vs Θ plot can be approximated by a Fourier series.

series is obtained. One can either use a full Fourier or a fast Fourier to do the inversion. The full Fourier gives the amplitude of the chosen functions as well as the phase angles associated with each coefficient. Since the development of the Cooley-Tukey[9] algorithm, which uses significantly less computer time to compute the only amplitude coefficient of the discrete Fourier transform, it has been widely used. To reproduce the particle shape, one needs not only the amplitude coefficients, but also their phase angles. Thus the fast Fourier

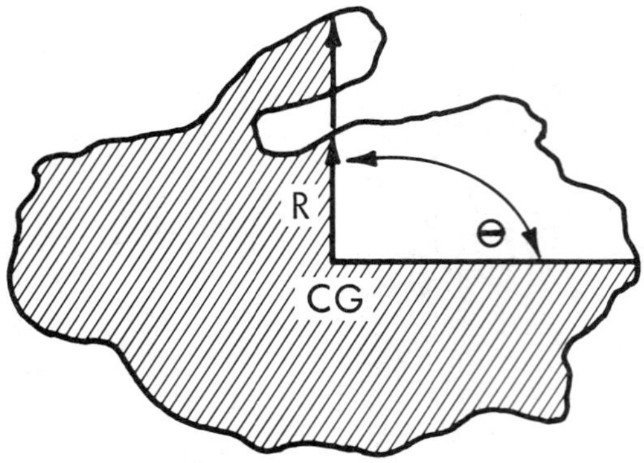

Figure 3.10. Holomorphic or analytic particles are single valued–$R(\Theta)$–for all Θ. The nonholomorphic particle is shown with $R(\Theta)$ having three values as the radius transverses the peninsula of the invagination.

transforms, or for that matter any of the fast transforms, lose the phase angle data, and while they give a great deal of information about the physical characteristics for the particle, the information set of the particle outline is not complete. By the same token, however, the physical significance of the amplitude coefficients is direct and unambiguous, whereas the meaning of the phase angle relationship is ambiguous. Thus one is left in the paradoxical position of being able to use the amplitude coefficient data readily, while needing, but by the same token, not being able to use the phase angle information.

Although most investigators agree that the Fourier is the best spanning set to use to describe the particle silhouette, there are other orthonormal spanning sets that can be used. These orthonormal sets are the Walsh and the Haar functions and the intergrals thereof. The Walsh functions are made up of a linear combination or Radamacher functions. Radamacher functions have the values of + 1 and - 1. Figure 3.13 shows the first five Radamacher functions. They are comprised of square waves and are very similar to cosin functions except that they are square. By linear combinations of the Radamacher functions, Walsh functions shown in Figure 3.14 are obtained. Unfortunately, the Radamacher functions are not complete, meaning that one can not describe a curve below any given error with these functions. However, the Walsh functions are complete and can be used to describe curves the same way that the Fourier functions can be used. Discrete Walsh transforms are available. In fact one of the available transforms is position invariant as is the Fourier transforms. Position invariant means that, irrespective of where one sets zero on the particle silhouette, the values of the Fourier coefficients are the

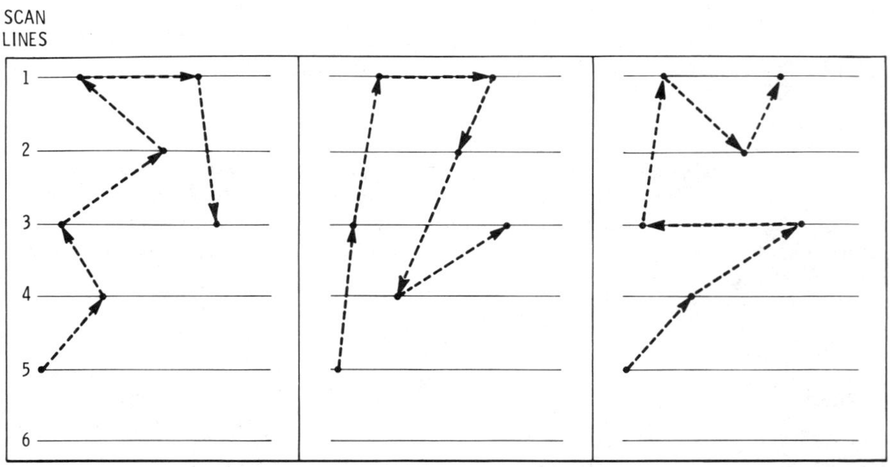

Figure 3.11. Determining the order in which the x,y digitalized coordinate points are connected is sometimes difficult because of ambiguities. Roughness causes the scan lines to pick up coordinate points in an order that differs from the order in which they appear in the particle silhouettes. This difficulty can be resolved if they are taken in the order of increasing Θ.

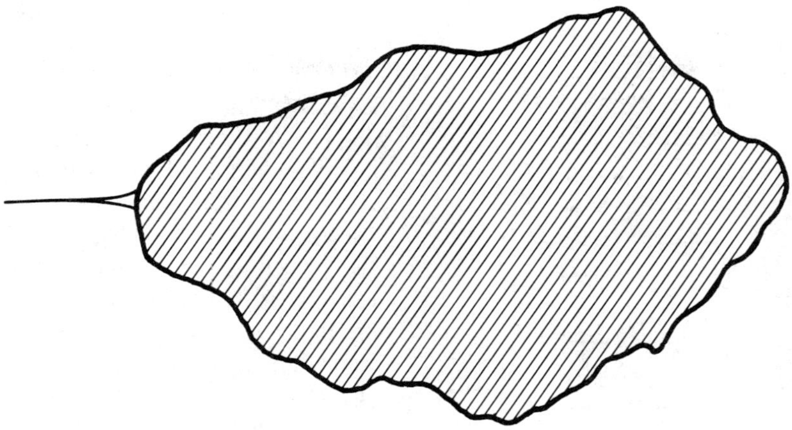

Figure 3.12. Glitch error occurs when a false reading is interpreted as being on the particle outline. This results in the apparent "spike" on the particle surface. To recreate this spike via the Fourier series the higher-frequency coefficients must be increased, and the associated angles become very important. Modern equipment reading well-prepared samples is free of this type of error.

same. As will be described later, conceptually, the Walsh functions are very useful for describing certain physiochemical properties of the particle surface. Another set of complete orthonormal functions are the Haar functions and their intergrals. Figure 3.15 shows the first 16 Haar functions, which is really the first 5 sets of functions. These functions have found their primary use in analyzing for time delays in electrical network. Meloy[10] suggested the use of these functions, and Beddow[11] has been exploring their applicability. Unfortunately, the number of Haar functions expands rapidly with the order of the Haar functions and thus the number of coefficients needed to give a

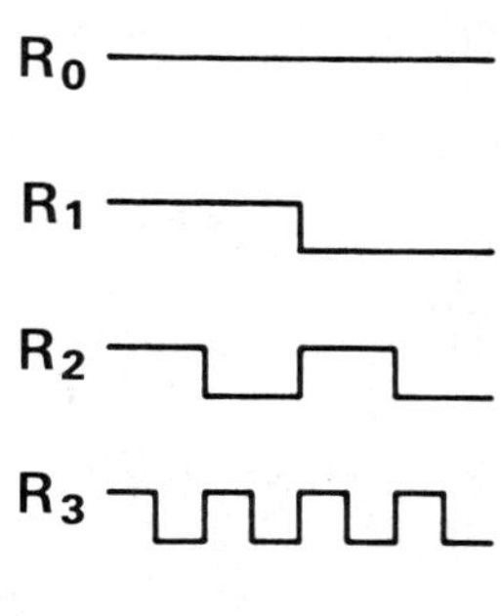

Figure 3.13. *Radamacher functions.* These have a value of +1 and are orthogonal over the interval 0 to 1. R_n has 2^{n-1} square waves associated with it.

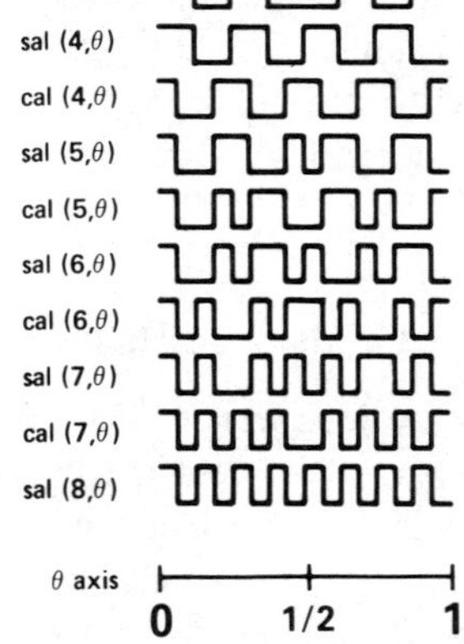

Figure 3.14. *Walsh functions.* Plot of the first 16 Walsh functions which is a set of complete orthonormal functions. Note the zero crossings and both the symmetry or inverted symmetry about the mid point. Walsh functions are about half as efficient as Fourier functions in approximating a given curve and have values of 1 or −1.

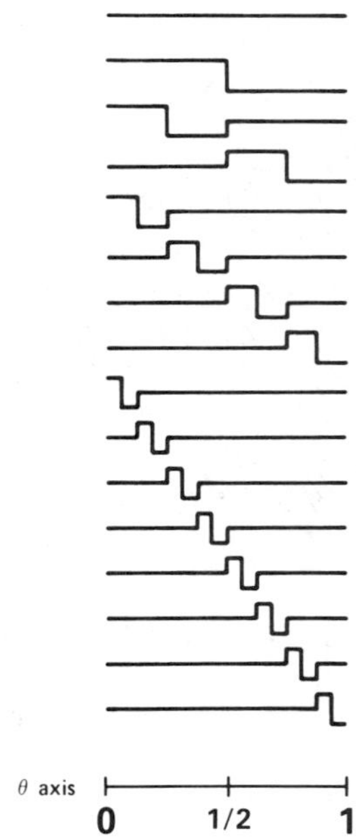

Figure 3.15. *Haar functions.* Plot of first four levels of Haar functions, another orthonormal set of functions having values of 1, 0, or −1. Note that the 4th level has 8 functions. The number of Haar functions increases very rapidly with the level of the function.

certain degree of error rises very rapidly compared to Walsh and Fourier. Meloy[12] speculates that for the same level of error for a 100 Fourier coefficients will give, requires 200 Walsh coefficients and perhaps over 5,000 Haar coefficients. On the other hand, for particles with distinct *asperities*, Haar functions, and particularly their integrals, may be very efficient.

3.4. PARTICLE SHAPE SIGNATURE

In previous sections wer developed a technique for finding a series of coefficients that describe the particle shape. We can take this information and recreate the particle shape below any given error. These coefficeints of the Fourier expansion are actually the signature of the particle, particularly if they are coupled with their phase angle.

If one plots the log of the amplitude of the nth coefficient versus log$n$, then one obtains a straight line as one can see from Figure 3.16. Meloy[13] has named this the law of morphological coefficients. Figures 3.17 and 3.18 are based on data obtained respectively by Schwartz[7] and Beddow[14] but plotted by Meloy. All three of these curves show decreasing coefficient amplitude with increasing n but with data scattered about what appears to be a straight line. There is a large amount of scatter. The percentage of scatter does not appear to decrease with increasing coefficient number, but there is a clear decrease in amplitude with decreasing coefficient number.

Figure 3.19 shows four different shaped particles and Figure 3.20 and the corresponding positions show the log-log plot of A-n versus log$n$ for each of the six particles shown in Figure 3.19. Once again, there is a straight line down to the 130th coefficient and a large amount of scatter above the straight line. The question is, What does the plot mean?

For white noise, one would expect the line to be horizontal. That would mean that the nth coefficient was equally likely to be the same magnitude as the second coefficient. On the other hand, in any Fourier approximation, one would expect the coefficient to decrease with increasing n. The question arises as to whether the straight line is an artifact or not.

3.5. RANDOM PARTICLES

In order to test the concept of signature as seen in Figures 3.16, 3.17, and 3.18, Meloy[15] defined the concept of random particles. A random particle is a particle that takes a straight line as its signature and makes certain assumptions about randomness. Figure 3.21 defines Type I and Type II random particles. Basically, what one is doing is generating particles from a signature such as a straight line as shown in Figures 3.15, 3.16, and 3.17. The radius from the particles' center of gravity R is the polar radius and is a function of the angle. As will be shown later, A is the particle diameter, and the sum of the An cosin functions is the typical Fourier function describing the particle radius. In Type I random particles, part of the polar angle is randomly oriented, and in Type II,

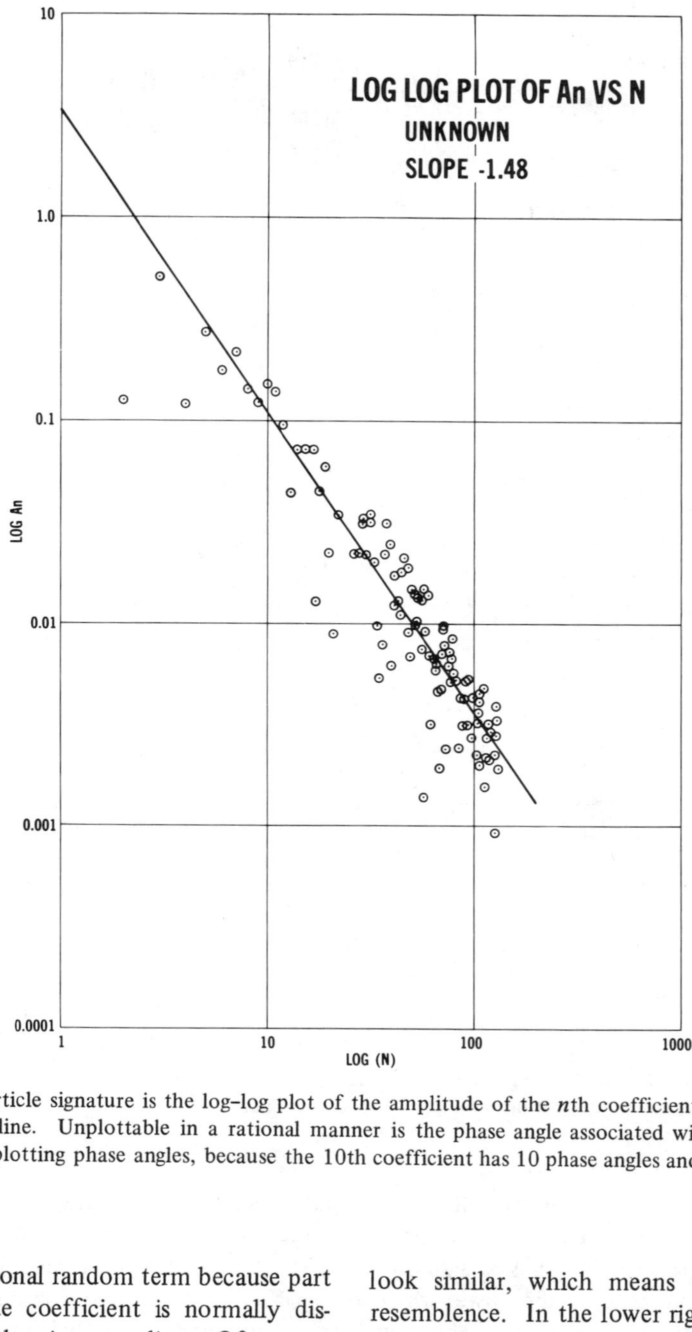

Figure 3.16. A particle signature is the log-log plot of the amplitude of the nth coefficient A_n *vs* N. Note the scatter about the line. Unplottable in a rational manner is the phase angle associated with each point. The problem arises in plotting phase angles, because the 10th coefficient has 10 phase angles and the nth coefficient has n coefficients.

there is an additional random term because part of the amplitude coefficient is normally distributed about the signature line. Of course, An is a function of $(1/n)^k$ and varies with each signature.

Figure 3.22 shows Type I random particles where the slope of the straight line on the log-log plot is -2 and the intercept of A_1 is equal to 0.6. The five particles shown in Figure 3.22 look similar, which means there is a family resemblence. In the lower right hand corner of Figure 3.6 is a blow-up of the surface between 0 and 12° showing the deviation of the radius vector from its average value. This plot would show up the roughness of the particle if there were any. Unfortunately, this microstructure is very smooth in this particle.

In Figure 3.23, five families of particles are

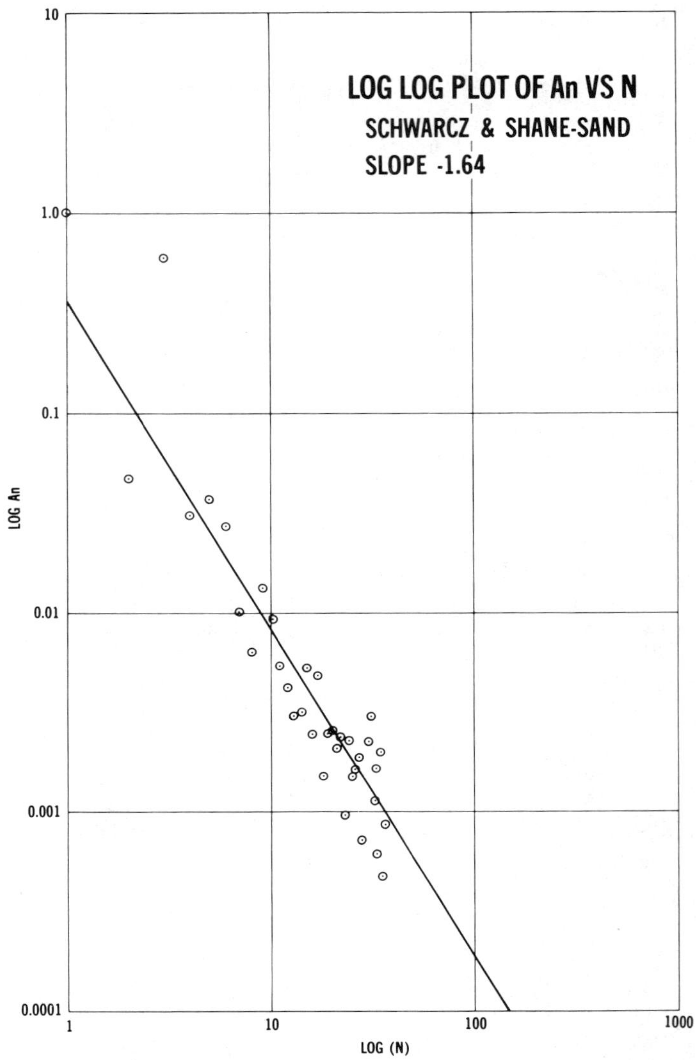

Figure 3.17. Data from Schwarcz[6] and Shane[7] is plotted in the same manner as the data in Figure 3.16, and it shows the same characteristic straight line and scatter about the line.

shown, each family differs from the others. In Figure 3.22, a family of particles that resembles one another was shown. In Figure 3.23, the sixth particle, appearing in the lower right, belongs to a family of one of the other five particles. Since it is readily identifiable as belonging to the family of the particle directly above it, it is easy to see that these reconstituted particles have an uniqueness about them.

Three types of random particles have been defined. They are Type I, Type II, and Type III. Figures 3.22 and 3.23 show Type I random partilces but Type II random particles look very similar to Type I and are difficult to distinguish. Type III random particles pin the phase angles of some of the coefficients, and they look quite different from Type I and Type II.

Another type of particle is the odd or even particle. In even particles, the even coefficients are significantly larger than the odd coefficients. A totally even particle, that is, one where all the odd coefficients are zero, is rotationally symmetrical in that one could cut the particle in half through the center of gravity, rotate it

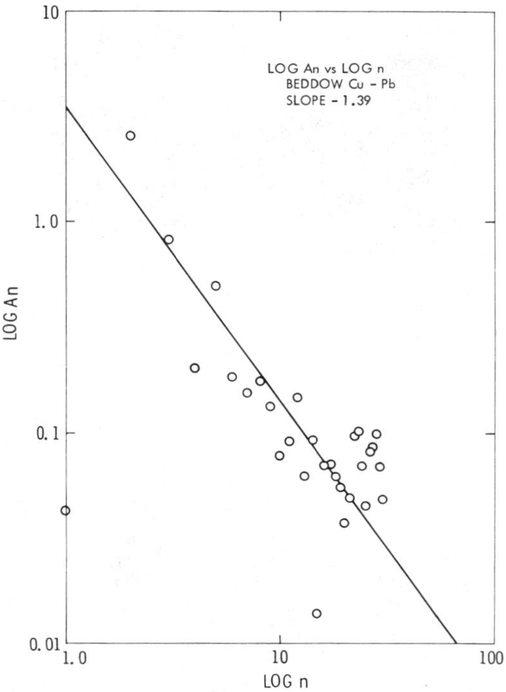

Figure 3.18. Data from Beddow plotted in the same manner as Figure 3.16 showing the characteristic straight line and scatter. Equations in Figure 3.21 are based on the data present in this figure and in Figures 3.16 and 3.17.

180°, and the two halves would superimpose. One can visualize this if one looks at Figure 3.29: cut it in half, rotate one half around, and it will superimpose on the other half. On the other hand, odd particles, that is, where the even coefficients are low valued or zero, are not symmetrical and thus can not be rotated 180° and superimposed. Figure 3.30 is a good example of an odd particle.

Although the five particles in Figure 3.22 look like real world particles, they almost certainly represent a very restricted segment of real world particles.

In making these plots, an assumption was made that the phase angle from the Fourier analysis did not contain any information, and therefore could be randomly distributed. Types I and II random particles make this assumption and while a number of realistic particle families appear, all the families appear to be stereotypes of a limited range. This indicates that the phase angles do play an important role in characterizing the particle.

Unfortunately, the phase angle is a very difficult parameter from which to extract information. The reason is that for the nth coefficient A_n, there are n different phase angles. What this means is that if one rotates the particle by $2\kappa/n$ radians, there is no change in particle shape. One does not know which of these n different identical positions is the important phase angle. Thus, if you are taking the 20th coefficient, there are 18° between the twenty different positions. Consequently, if the lowest phase angle is 10°, then the next lowest phase angle would be 10° + 18°, or 28°, the next one would be 10° + 2 X 18°, or 46°, and then 10° + 3 X 18°, and so on up to 10° + 19 X 18°. Meloy is currently working on techniques to find the significance of the phase angles.

It is important to realize that if one can show that the phase angles are not relevant to the analysis, then the computations can be done using the fast Fourier transforms, which are orders of magnitudes faster then the normal discrete transforms.

To get an idea of the difference between a random and a nonrandom particle profile, consider the nonrandom particles in Figure 3.24. These particles were formed by drying gel on a flat plate. The sharp angle and relatively straight edges are very unlikely. Since the signature or log-log plot of A_n versus n is not appreciably different from the random particles seen in Figures 3.22 and 3.23, then the phase angle must be closely related to the points. As mentioned before, it is how this phase angle relationship relates to these sharp points that is not understood. Figures 3.25 and 3.26 show the signature of nonrandom particles that will be discussed later.

Each of the coefficients in the Fourier or other orthonormal sets of functions has a physical interpretation. The zero coefficient, when translated back to the polar coordinate representation of Figure 3.9, yields a perfect circle about the center of gravity. Figure 3.27 shows this to be true both for Fourier and for Walsh theories. The first coefficient A_1 is a

Figure 3.19. Note the wide range of particle types that have been analyzed. Figures 3.19 and 3.20 correspond, so that the plot entitled *Block Long* in Figure 3.20 derives from data taken from the corresponding particle in Figure 3.19, and so on.

measure of the error in locating the center of gravity of the particle profile. It is the only coefficient that is not symmetrical about the center of gravity. The actual plot you see in Figure 3.28 is a combination of A_0 and A_1. Because of numerical methods used to compute these coefficients it is important that the first coefficient A_1 be small. The second coefficient A_2, which is even, is a measure of the aspect ratio of the particle. In Figure 3.29, a representation of $A_2 + A_0$ is shown. Note that all even functions are symmetrical about the origin. The third coefficient A_3, a measure of the triangularity of the particle, is not sym-

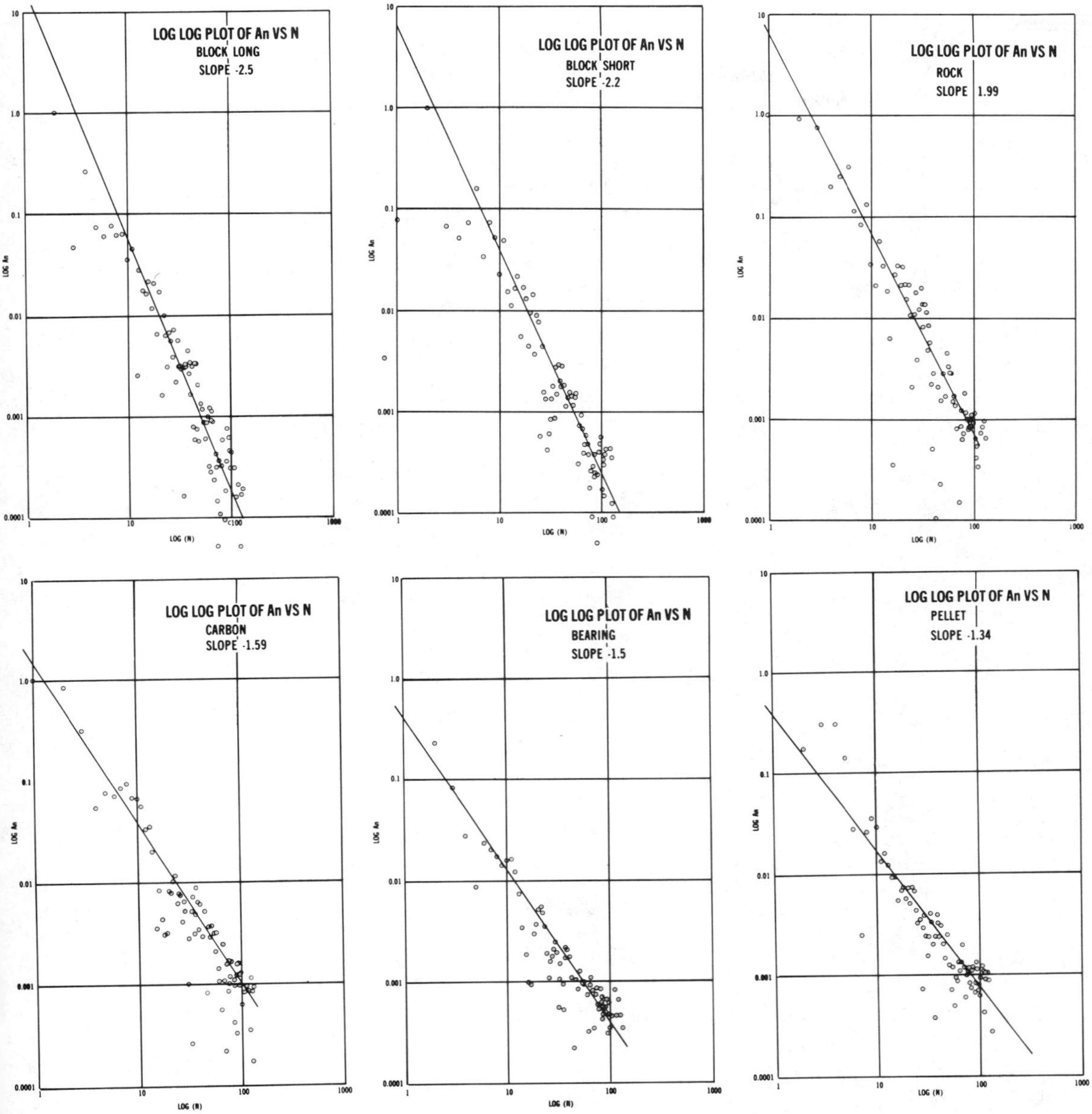

Figure 3.20. These six plots illustrate the law of morphological coefficients. Note in each plot the data can be fitted by a straight line. The line has a slope, an intercept, and scatter about the line. Therefore, 3 not 2 parameters characterize the data that fit the straight line.

metrical about the origin in the sense that it gives a mirror image. For example, one can not cut the particle in half through the center of gravity, rotate it 180°, and have the two halves superimpose themselves on each other. On the other hand, like all other coefficients except A_1, the center of gravity of the particle is also the center of gravity of Figure 3.30.

Figures 3.28 through 3.30 were all reconstructions that used the zero coefficient plus one other coefficient. To show how one can build up particle shape using several coefficients

TYPE I

$$R(\Theta) = A_0 + \sum_{n=2}^{\infty} A_n \cos(n\Theta + \Theta_{Rn})$$

Θ_{Rn} RANDOM COMPONENT (RECTANGULARLY DISTRIBUTED)

$A_n = f(1/n)$

TYPE II

$$R(\Theta) = A_0 + \sum_{n=2}^{\infty} (\bar{A}_n + A_{Rn}) \cos(n\Theta + \Theta_{Rn})$$

A_{Rn} RANDOM COMPONENT (NORMALLY DISTRIBUTED)

Figure 3.21. ***Random Particles.*** Shown here are equations for Type I and Type II random particles where the coefficients A_n are calculated from the particle signatures in the previous figure and the random component of the phase angle Θ_{Rn} is rectangularly distributed between 0 and 2π. In Type II random particles, an additional random component is added, A_{Rn}, which simulates the scatter about the straight line in Figure 3.20. Type III random particles pin some of the phase angles. All these random particles may be odd or even if the even or odd coefficients are set equal to zero.

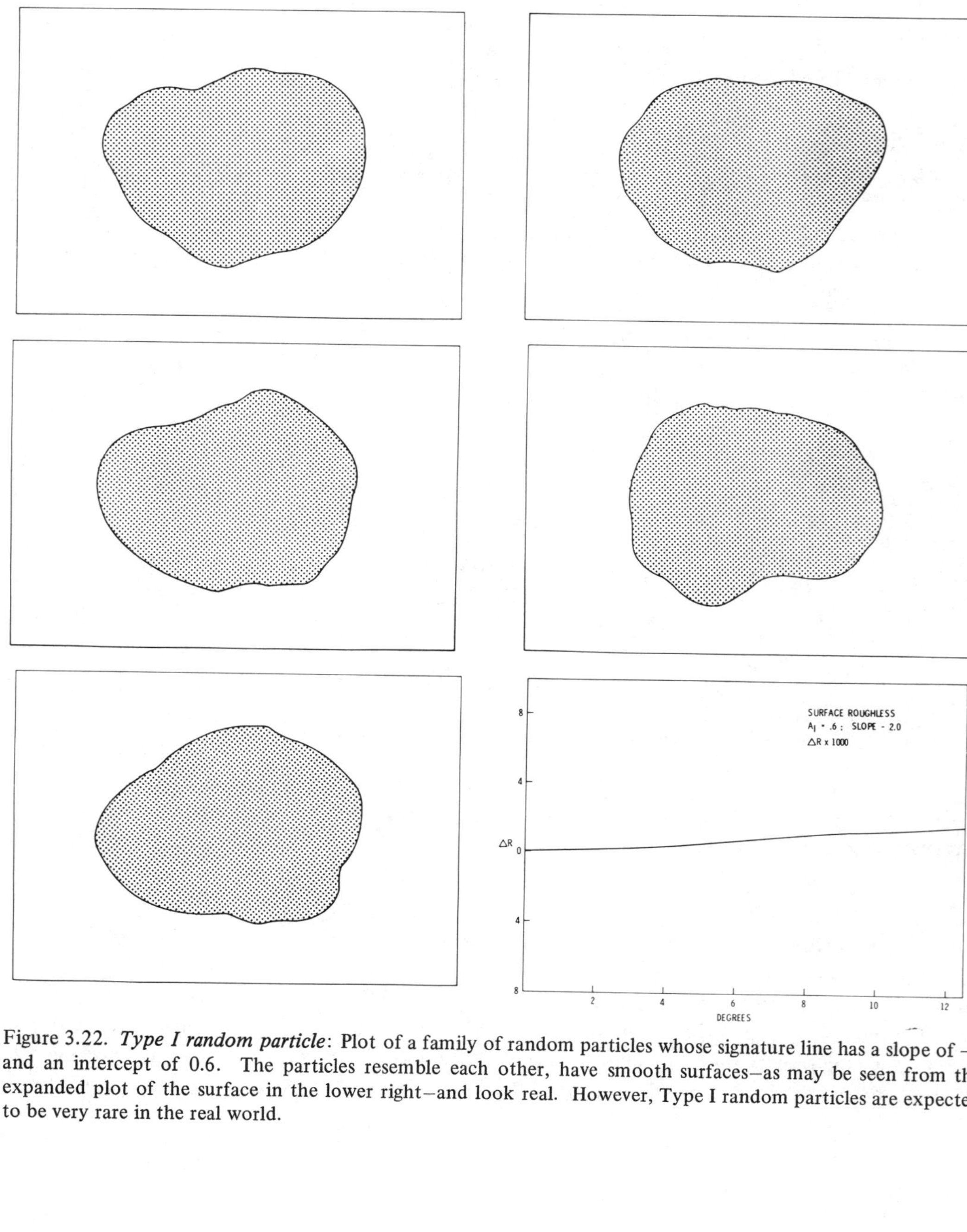

Figure 3.22. *Type I random particle*: Plot of a family of random particles whose signature line has a slope of −2 and an intercept of 0.6. The particles resemble each other, have smooth surfaces—as may be seen from the expanded plot of the surface in the lower right—and look real. However, Type I random particles are expected to be very rare in the real world.

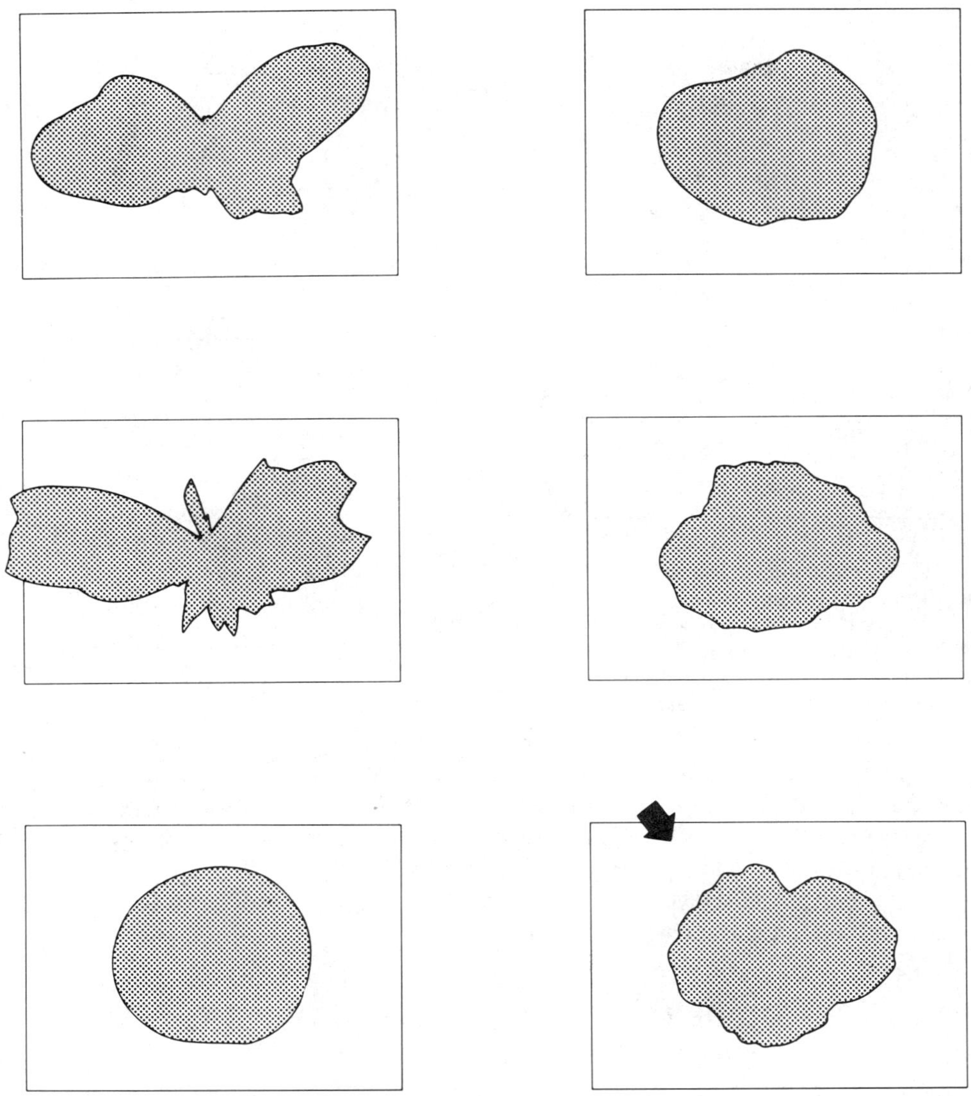

Figure 3.23. *Type I random particles*: Five families of random particles are shown. At the lower right is a particle that belongs to one of the five families of random particles. Can you tell which family it belongs to? It belongs to the family directly above it.

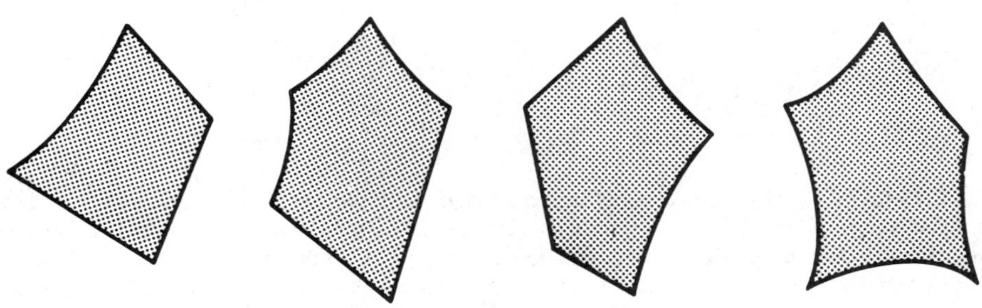

Figure 3.24. Nonrandom particles formed by drying a thin gel film. Note the sharp points and smooth sides.

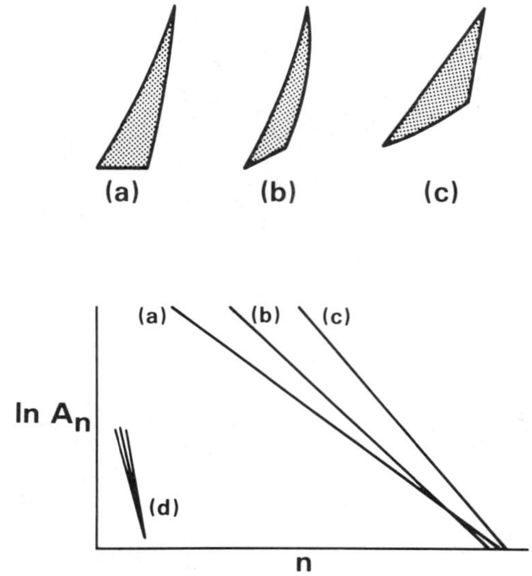

Figure 3.25. *Reconstructed particles using melamorphs.* Three non-random particles are formed by breaking window glass. In the lower right the signature of the three particles is drawn. In the lower left, labeled *d*, is one of the three signatures when the triangular melamorph is substracted out.

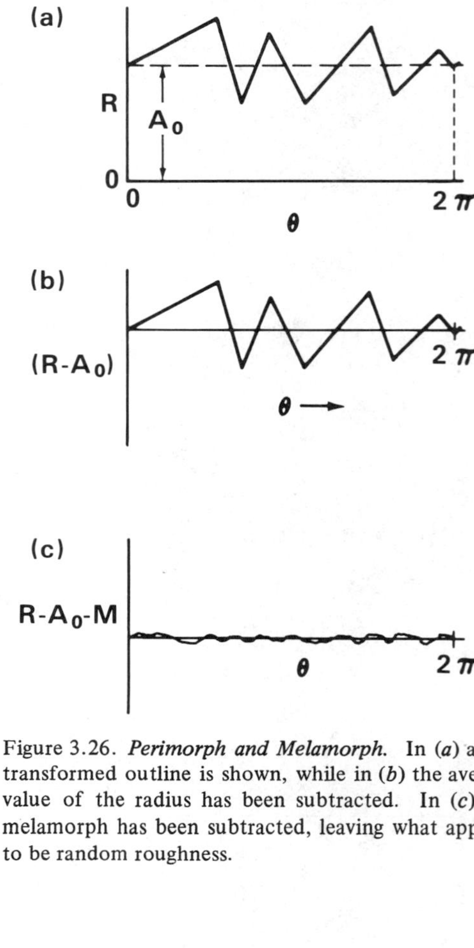

Figure 3.26. *Perimorph and Melamorph.* In (*a*) a raw transformed outline is shown, while in (*b*) the average value of the radius has been subtracted. In (*c*) the melamorph has been subtracted, leaving what appears to be random roughness.

and how these particle shapes can vary by changing the phase angles, two examples are given. In Figure 3.31, using Walsh functions because they are easier to plot, a combination of A_0 and A_2 and A_3 are shown. The amplitude of A_2 and A_3 is $\frac{1}{5}$ that of A_1. Note the

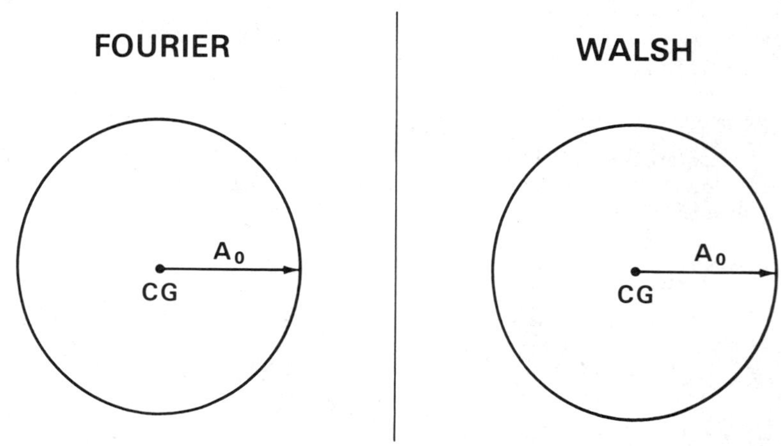

Figure 3.27. *Zero coefficient* A_0: *particle size.* Zero coefficient when plotted in polar coordinates yields a circle for both Fourier and Walsh functions. The subtle implications is that all particles are basically spheres. Higher-order coefficients merely show deviation from sphericity. It is possible to find another idealized shape such as a cube and measure deviation from that shape but it cannot be done with existing orthonormal functions without preparation.

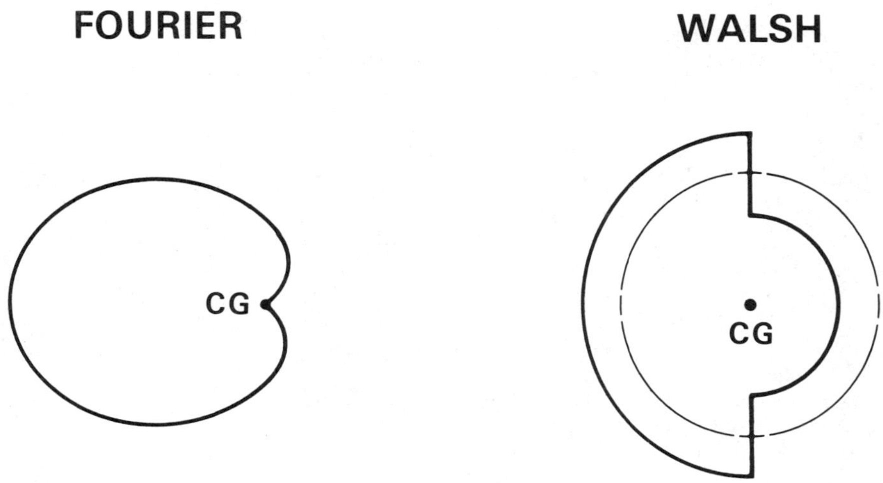

Figure 3.28. *First coefficient* A_1: *error.* The first coefficient A_1 is not symmetrical about the *CG*, as a plot of $A_0 + 0.2A_1$ shows. Thus the value of A_1 is the error in locating the *CG* correctly.

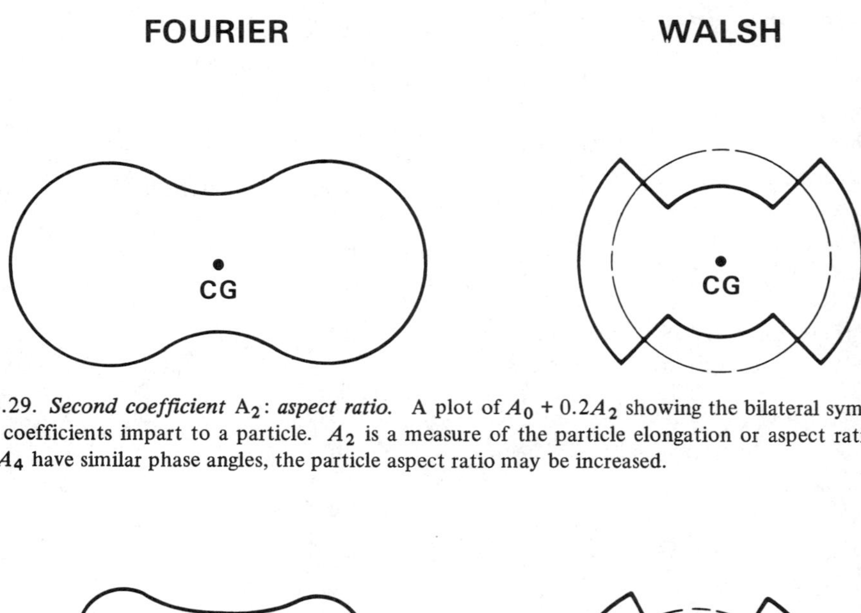

Figure 3.29. *Second coefficient* A_2: *aspect ratio.* A plot of $A_0 + 0.2A_2$ showing the bilateral symmetry which all even coefficients impart to a particle. A_2 is a measure of the particle elongation or aspect ratio, and when A_2 and A_4 have similar phase angles, the particle aspect ratio may be increased.

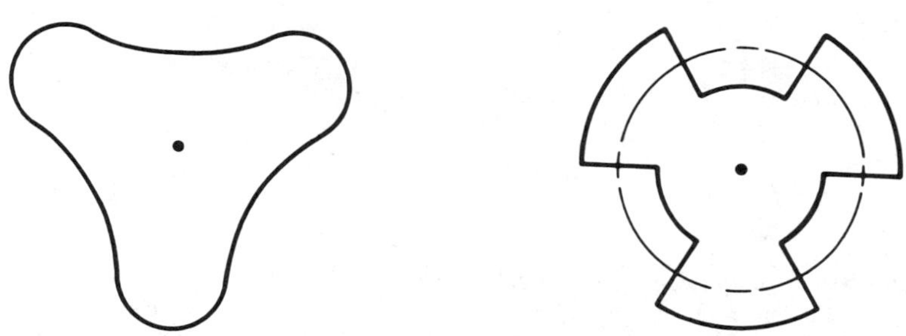

Figure 3.30. *Third coefficient* A_3: *triangularity.* Combined plot of $A_0 + A_3$ showing the triangularity of the particle. Note the lack of bilateral symmetry imparted by all odd coefficients.

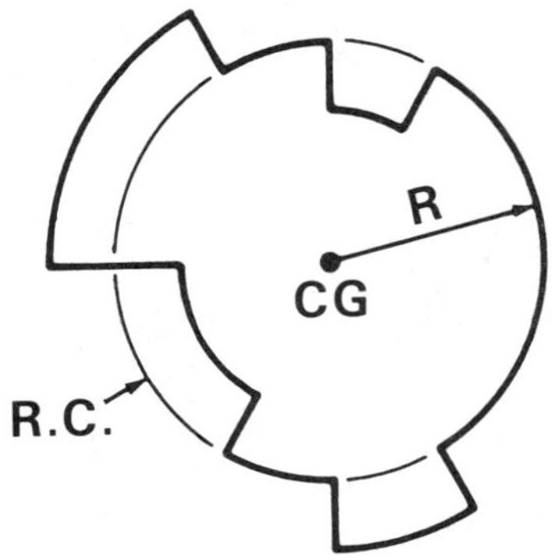

Figure 3.31. *Constructed morphology.* Using Walsh coefficients $R = A_0 + A_2 + A_3$ where $A_2 = A_3 = 1/6A_0$ showing a complex particle shape with just 3 coefficients. See Figure 3.32.

particle shape is rather unique and because Walsh functions are step functions, it looks artificial. Now examine Figure 3.32, which uses exactly the same coefficients, but rotates A_3 15°. Note now the very distinct difference in the shape of the two particles. If one now imagines the smoothing or rounding of the corners, one can begin visualizing naturally occuring particles. If one were to use 128 coefficients, one can see how very sophisticated particle shapes could be built up.

If one looks at Figure 3.32, one can ask the questions, What is the total perimeter of the particle, What is the length of the perimeter that moves in toward and out from the radius,

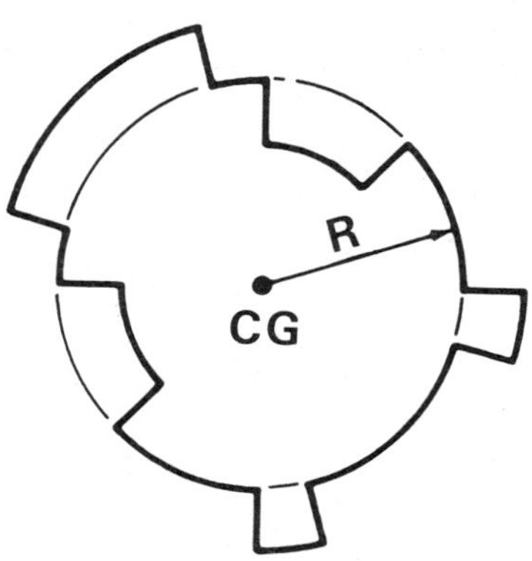

Figure 3.32. *Variation of morphology.* The values of A_0, A_2, and A_3 are the same as in Figure 3.31 but the position of the A_3 peaks have been shifted 15° clockwise. Note the difference in shape between Figure 3.31 and 3.32.

FOURIER SERIES

$$R(\Theta) = A_0 + \sum_{2}^{\infty} A_n \cos n\Theta$$

LINE INTEGRAL USING FOURIER SERIES

$$\int_0^{2\pi} dS = \int_0^{2\pi} \sqrt{R^2 + \frac{dR^2}{d\Theta^2}}\, d\Theta$$

LINE INTEGRAL USING WALSH FUNCTION

$$L = 2\pi A_0 + 4 \sum_{n=2}^{\infty} nA_n$$

Figure 3.33. *Line integrals using Fourier and Walsh functions.* To evaluate the line integral of a Fourier series in closed form is impossible and by numerical methods very difficult. For Walsh functions a simple summation is possible.

and What is the length of the perimeter that is curved or perpendicular to the radius? By knowing the radius of each point, one can very quickly make these computations. However, if one wants to make the same computations using the Fourier series, one has great difficulty. For example, in Figure 3.33, the line integral in discrete form is shown and is basically non-integrable. However, the line

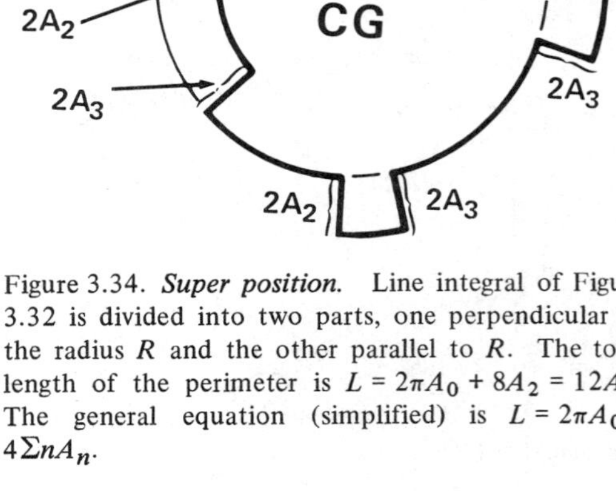

Figure 3.34. *Super position.* Line integral of Figure 3.32 is divided into two parts, one perpendicular to the radius R and the other parallel to R. The total length of the perimeter is $L = 2\pi A_0 + 8A_2 = 12A_3$. The general equation (simplified) is $L = 2\pi A_0 = 4\Sigma nA_n$.

integral using Walsh functions is a simple sum that can be done very quickly in close form.

The significance of each of the components is shown in Figure 3.34, where by superposition, one can add and substract the various components. Note now the importance of being able to separate the rounded part of the perimeter from the stair-step part. For the stair-step part, it is simply the sum of all the steps, and for the rounded part, since for each coefficient the number of up-steps is equal to the number of down-steps, the total rounded surface length becomes $2\pi A_0$. That is the meaning of the last equation in Figure 3.33. The number of other computations can be made in a like manner. There is one problem, however, and that is these wave trains, whether they be Fourier, Walsh, or other orthonormal sets are best adapted to flat surfaces, not solid ones.

Figure 3.35 shows the problems of cubing the sphere. Orderly arrays of waves can be put on flat surfaces, as are shown in Figure 3.35, but most particles are basically spherical (Figure 3.27). Somehow one must transform the cube into a sphere. Cartographers have done this for years by a series of projections. However, to have an orderly array on the surface of the sphere is not possible. Two golf balls in Figure 3.36 show how golf ball makers have tried to put orderly dimpling or orderly holes in their balls. Obviously, they have not succeeded. The easiest way to tell this from the photographs is to pick any hole or depression, and then count the number of nearest neighbors. On the golf ball at the right, one can see the six-fold coordinations outlined, but one can also find five-fold coordinations, and seven-fold coordination elsewhere on the golf ball. In essence, this means that you can not have an orderly array of waves on the surface of a sphere. That may be true both with golf balls and in reality; there is an old saying, however, that "when the physics doesn't fit the mathematics, then we will make the mathematics fit the physics." In essence, we assume that the surface of the sphere is flat, we set up our rectilinear wave trains, count the number of wave peaks that cross per unit surface area, and multiply this by the total surface area of the sphere. Once one gets eight or more waves on the surface of the sphere, the error becomes rather small. In Figure 3.37, a pair of two-dimensional waves are shown, the one on the left labeled A is a single wave form in two dimensions showing the peaks and the troughs. In B is shown two waves, one of a frequency n and the other of the frequency $2n$. To keep the computations straight a half wave length is χ/n, as shown. The height is shown as $2A_n$ where the amplitude of the wave is A_n. The peak to trough height is $2A_n$ because it goes from $-A_n$ to $+A_n$. What is shown is how a series of Walsh waves would look like on a particle surface. The more complex pattern B represents the surface up and down for a two-coefficient wave. There is a spherical component, which is the white top of the blocks and the valleys of the blocks. Then there is the block sides shown by the hatched lines. This is similar to the rounded edge and vertical edge in Figure 3.34. By like computational procedure, one can calculate the top and

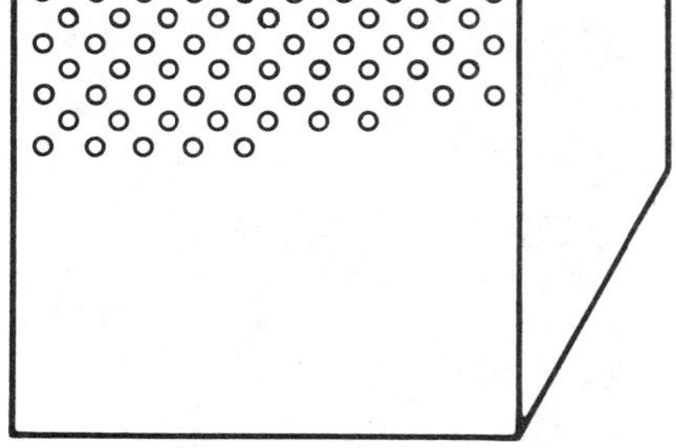

Figure 3.35. The cube shows how an orderly rectalinear array can be represented on the cube's surface but not on a sphere. See Figure 3.36. Also note that the perimeter of the cube is $4s_s$ a side, and thus on a planar surface such as Figure 3.8, the total surface area would be $(4s)^2$ or $16s^2$ but the surface of the cube is only $6s^2$.

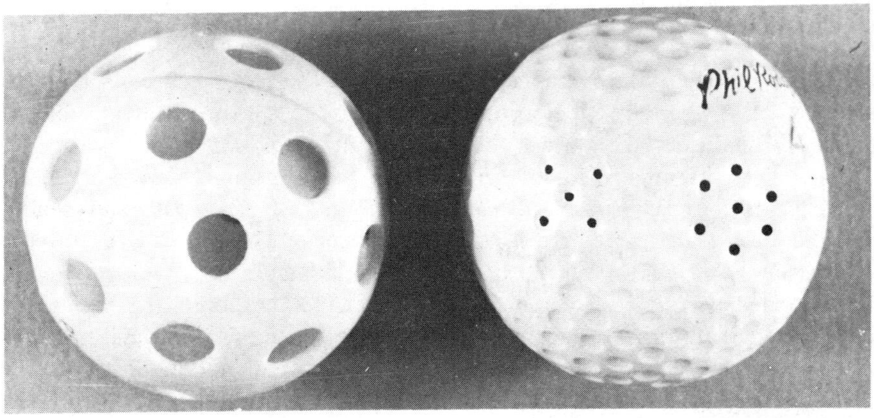

Figure 3.36. *Cubing the sphere.* Orderly arrays on a planar surface (Figure 3.35) are readily achieved but on a sphere they are impossible except in very special cases. Note that 4-, 5- and 6-fold coordination numbers appear on the golf ball.

bottom rounded surface area, the edge length, the side area, and the number of corners, and the like. This has been done by Meloy[10] and the results are shown in Figure 3.38, where there are some 25 different functions, some of which represent the excess side area, excess surface area, edge length, and the like.

Some of the functions in Figure 3.38 have physical significance. For example, the sum of n^2 is the number of corners. This, however, has no physical significance because the number of corners of every particle is independent on the shape of the particle, and thus this is a meaningless function. On the other hand, one can convert the corner problem into a more meaningful one by assuming the blocks are actually pyramids and asking how sharp is the pyramid. This is represented by Figure 3.39. After all the mathematical cranks have been turned, the corner function turns out to be that in row 3, column 5. The side area computation appears in row 2, column 2, and the edge length appears in row 2, column 3. Schwartz and Shane[7] found that the rollability of a particle was propor-

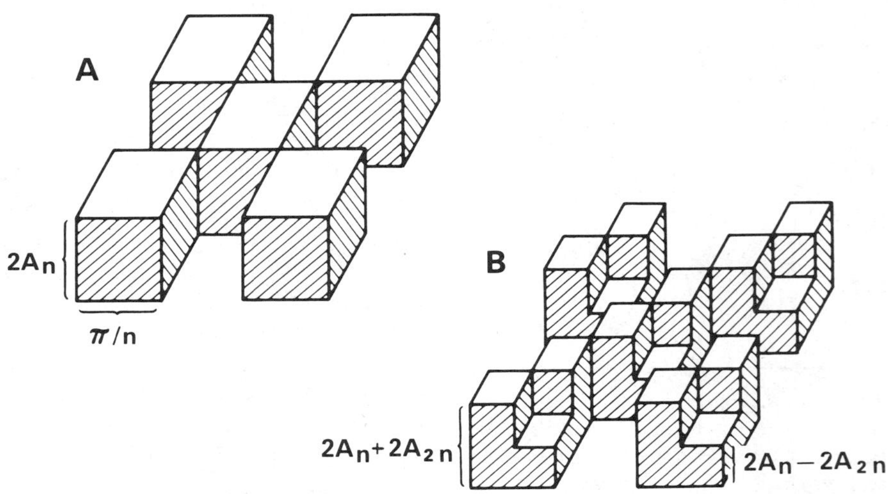

Figure 3.37. *Simulation of particle surface.* In part A, the two dimensional array of square-based blocks of height $2A_0$ is shown. In part B is shown the result of a second wave of twice the frequency superimposed on a block shown in part A.

A_0	$\sum N$	$\sum N^2$	$\sum N^3$	$\sum N^4$
$\sum A_N$	$\sum NA_N$	$\sum N^2A_N$	$\sum N^3A_N$	$\sum N^4A_N$
$\sum A^2_N$	$\sum NA^2_N$	$\sum N^2A^2_N$	$\sum N^3A^2_N$	$\sum N^4A^2_N$
$\sum A^3_N$	$\sum NA^3_N$	$\sum N^2A^3_N$	$\sum N^3A^3_N$	$\sum N^4A^3_N$
$\sum A^4_N$	$\sum NA^4_N$	$\sum N^2A^4_N$	$\sum N^3A^4_N$	$\sum N^4A^4_N$

Figure 3.38. Twenty-five functions that relate to the physical and physiochemical behavior of particles and their powders. The first row as it stands does not have physical significance. Further, some of the functions will not converge unless $A_n = f(1/n^{\alpha})$ such that $\alpha = 2$, 3, or 4.

tional to the sum of the coefficients squared, which is a function that appears in row 3, column 1.

Already, it has been shown that the number of corners, the sum of n^2, does not converge. The questions arises as to how many of the functions in Figure 3.38 do converge. Figure 3.40 is a second table that shows the convergence criteria for the 25 functions. The ones in white do not converge, whereas those hatched from upper right to lower left appear to converge for all particles experienced so far—those signature plots where the slope is -1.5 or greater. This brings up the very interesting observation that the pyramidic conversion of corners, which has the vertical hatching, converges only for very steep slope coefficients' signatures. This in turn brings about the clear possibility that corners, at least as analyzed so far, play no significant role. On the other hand, Schwartz and Shane's[7] data as well as data by Fong[16] indicate that the convergent areas often have physical significance. It is believed that these functions will turn out to represent the amount of work needed to compact the powders, the initial compaction, density, the angle of repose, the rate of fall in fluids and other physical properties which we will discuss. Currently, this field is being investigated by Beddow,[17] Fong, and others.

To understand the importance of these functional relationships to physical and physiochemical properties, one has to look at the quantum mechanical interpretation of solid matter. Let us consider six-fold coordinated

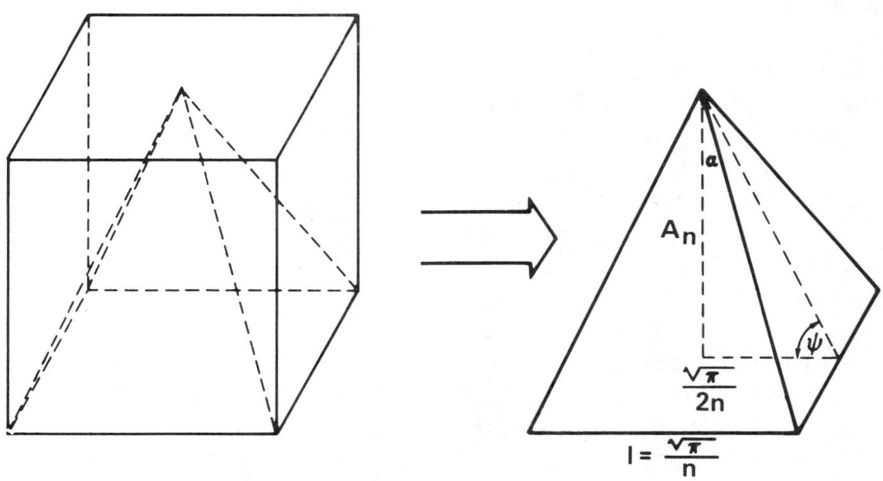

Figure 3.39. *Corner analysis.* A block is converted into a pyramid to obtain an estimate of the energy in the corners. The solid angle of the apex is considered to have zero energy when the height of the pyramid is zero ($A_n = 0$) and to be proportional to 2π – Solid angle. (2π Solid apex angle for zero height pyramids)

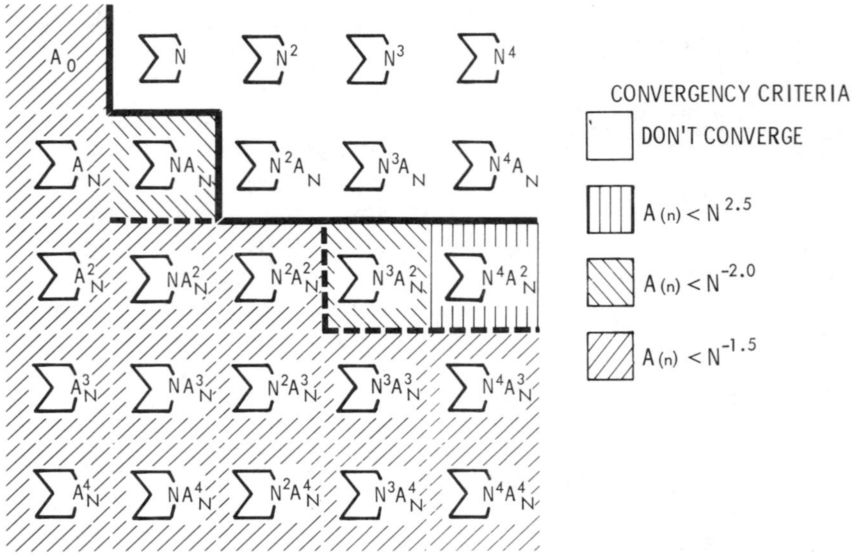

Figure 3.40. This is a tabular compilation of 25 functions, some of which in the hatched area are convergent and are believed to have physical significance. For example, Schwarcz and Shane[4] have shown the ΣAn^2 is a measure of the rollability of sand.

crystal matter because it is the easiest to look at from a numerical point of view. The crystal structure, however, is irrelevant to the final concept. Figure 3.41 represents a picture of quantum mechanical concepts of atoms that lie within the matrix, those that lie on the surface shown in A, those that lie on an edge, shown in B, and those that lie in a corner, shown in C. With six-fold coordination and the atom lying within the matrix, all six coordinate bonds are complete. On the other hand, if one takes an atom on the surface, as shown in A, where one of the bonds is broken (the dotted line) then the energy level of that atom is higher, and it turns out that generally the amount of energy in the surface is $\frac{1}{6}$ the total energy required to melt the particle. Looking now at the particle on the edge, one sees that there are two bonds broken, meaning that there is twice as much energy left unsatisfied by the particle on the edge as the particle in the surface. In a like manner, the particle in the corner has three bonds broken and therefore three times the amount of surface energy as the single particle on the surface. Thus, particles that are on an edge or corner are extremely reactive and are those most likely to go into solution, most likely to migrate while sintering, most likely to vaporize, and certainly most likely to react chemically. If one considers a particle on a freshly fractured coal surface in a furnace, the particle on the corner is likely to heat up very fast because it has a much wider solid angle

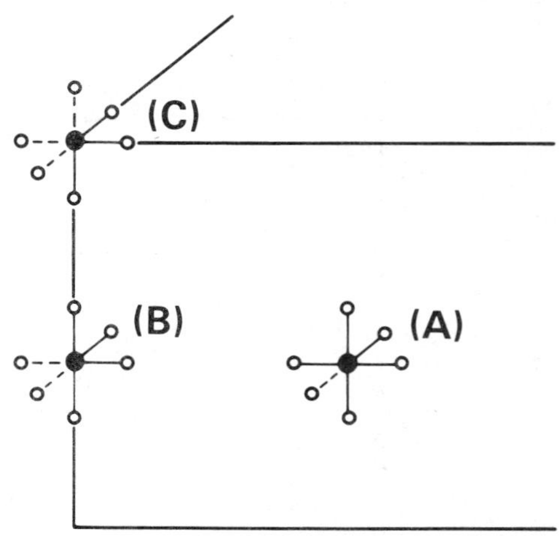

Figure 3.41. *Applications of quantum mechanics.* Atoms in surfaces, edges, and corners have higher energies because the number of bonds broken respectively are 1, 2, and 3, assuming six-fold coordination. These active sites influence sintering, catalysis, combustion and other physiochemical properties. With the proposed techniques the number of atoms in each type of site can be computed.

exposed to the radiant flame field and in addition is very reactive. One would expect these corner particles to initiate combustion. Paradoxically, they may be so reactive that before other particles have a chance to heat up, it may have already reacted, lowered the corners free energy, formed an ash layer, and actually hindered combustion initiation. In any event, an understanding of the detailed characteristics of the particle surface should help explain the behavior of particulates in physical and physiochemical reactions.

Until now, all the analytical work done on particles using this new technique has assumed that the particle was essentially a sphere and has measured the deviations from sphericity using orthonormal functions such as the Fourier set. In short, all particles have been considered as deviant spheres. But are particles deviant spheres? Meloy[10] has suggested another approach, that is, to find a melamorph for a given family of particles. A melamorph is a macroenvelope that instead of being a sphere, fits the natural shape of a particle. For example, galena particles tend to cleave into cubes. Therefore, the melamorph of galena particles instead of being a sphere would be a cube. All deviations from cubicity would be measured in a manner already described. It is believed that this approach would simplify the analysis in that the higher-order coefficients would play a much lower role, meaning their amplitudes would be much lower than if one assumes the particle's natural shape is a sphere.

Figure 3.25 shows three *shards* of a broken windowpane. The three signatures of shards *a*, *b*, and *c* are shown in the plot of the lower half of Figure 3.25. These signatures are based on the assumption that the melamorph is a sphere. Obviously these particles are much closer to triangles then spheres. If one assumes the melamorph is a triangle and then reanalyzes the data to obtain the coefficients in the classical manner and replots them in the lower half of Figure 3.25, one sees that they are moved to the left and downward, meaning that the higher-order coefficients have a much lower amplitude.

Another way to look at the same analysis is to look at the plot of the polar coordinates plotted rectilinearly as shown in Figure 3.26. In *a*, the curve is plotted with all the coefficients in it. In *b*, the curve is replotted without using the first coefficient. As can be seen, all that has happened is that the curve has been lowered by a factor A_0 from A. However, there must be considerable amplitude to the coefficients in order to reach the peaks and valleys shown in *b*. The melamorph used in the analysis so far has been that of a sphere. If now one subtracts out not the coefficient A_0, which describes a sphere, but another melamorph, which is more like the naturally occuring particle, then one sees a much flatter curve as shown in *c*. In other words, by using a melamorph and subtracting that information from the particle shape, one finds a much neater and more compact error deviation signal, shown in *c*. Unfortunately, the author is not aware of any simple mathematical tool to use this technique.

There are a number of other methods for particle analysis. Figure 3.10 depicts a particle with a center of gravity located such that the radius vector yields three values at some points. The classical technique proposed is unable to analyze such particles with modification. Zahn[8] has proposed a method whereby the vector describing the particle outline moves along the particle silhouette edge. In other words, a point is chosen on the particle surface and that is the base of the vector. Then the second point is chosen on the surface a short distance away. The vector is run from the first point to the second, and the angle and amplitude of the vector are analyzed in the same way as before. If one looks back at Figure 3.9 in the lower left hand corner and thinks of running one vector from points 1 to 2, and another from 2 to 3, another to 3 to 4, another from 4 to 5, one can see how the system can be set up. Looking now at Figure 3.10, one can see that if the length of the vector is kept short enough, one can walk around the reentry points, avoiding the problem of multivalues. Fong[16] and Beddow[17] have used this technique to study the dissolutionment of metal particles in acid.

$$[w\ (3)] = \begin{bmatrix} 1 & 1 & 1 & 1 & 1 & 1 & 1 & 1 \\ 1 & 1 & 1 & 1 & -1 & -1 & -1 & -1 \\ 1 & 1 & -1 & -1 & -1 & -1 & 1 & 1 \\ 1 & 1 & -1 & -1 & 1 & 1 & -1 & -1 \\ 1 & -1 & -1 & 1 & 1 & -1 & -1 & 1 \\ 1 & -1 & -1 & 1 & -1 & 1 & 1 & -1 \\ 1 & -1 & 1 & -1 & -1 & 1 & -1 & 1 \\ 1 & -1 & 1 & -1 & 1 & -1 & 1 & -1 \end{bmatrix} \begin{matrix} 0 \\ 1 \\ 2 \\ 3 \\ 4 \\ 5 \\ 6 \\ 7 \end{matrix}$$

$$[H\ (3)] = \begin{bmatrix} 1 & 1 & 1 & 1 & 1 & 1 & 1 & 1 \\ 1 & -1 & 1 & -1 & 1 & -1 & 1 & -1 \\ 1 & 1 & -1 & -1 & 1 & 1 & -1 & -1 \\ 1 & -1 & -1 & 1 & 1 & -1 & -1 & 1 \\ 1 & 1 & 1 & 1 & -1 & -1 & -1 & -1 \\ 1 & -1 & 1 & -1 & -1 & 1 & -1 & 1 \\ 1 & 1 & -1 & -1 & -1 & -1 & 1 & 1 \\ 1 & -1 & -1 & 1 & -1 & 1 & 1 & -1 \end{bmatrix} \begin{matrix} 0 \\ 7 \\ 3 \\ 4 \\ 1 \\ 6 \\ 2 \\ 5 \end{matrix}$$

Figure 3.42. *Walsh to Hadamard.* The Walsh matrix at the top is transformed into a Hadamard by rearranging of the rows in the Walsh functions. Using the symmetry and orthogonality thus obtained the Hadamard can be factored into a series of sparse matrices. Efficient algorithms for very fast computation have been developed.

The coefficients of the center of gravity, Fourier series, and the Zahn Fourier series are different but linearly related. As a matter of fact, the coefficients of the Walsh, Zahn, Fourier, and Haar functions are all linearly related. The linear relationship between various transformation techniques is pervasive in the discrete transforms. The matrix used to transform the raw data of Walsh functions is shown in Figure 3.42. As mentioned before, this matrix is comprised exclusively of +1 and −1. To convert to a Hadamard transform, there is an interchange of rows. Note that at the extreme right of the matrix, the rows on the Walsh are numbered 1, 0-7, whereas in the Hadamard, the rows have been rearranged from the Walsh matrix above. It is interesting to note how redundant the Hadamard transform is. If one takes any natural 2 X 2 matrix within the 8 X 8 shown of the Hadamard, one will see that it is identical to every other 2 X 2 matrix of the opposite sign. This property allows for very rapid computation.

As proposed, an analysis of a particle uses one of the orthonormal functions to obtain a large number of coefficients. A technique called the *Karhune Loue*[18] enables one to combine the coefficients to come up with optimum descriptors for a given type of particle. While not currently used, it is believed as particle systems are better understood that this type of analysis particularly will find widespread application for specifications.

Kaye[19] has come up with two other ways of analyzing particles. The first of these is to spin a particle silhouette about its center of gravity in front of a square matrix of light-sensing elements. Interruption of the light enables a binary signal to be obtained and when analyzed gives spectral information about the particle. Kaye's[20] second approach is to blow the particle up in a photograph and do a fractual analysis of the surface. This approach yields another parameter about the particle.

In the technique analyzing the surfaces by orthonormal functions, the question often arises as to how unique is the system, that is, how many different particles could it reasonably accommodate. With existing equipment, it is possible to get 4,000 coefficients describing a particle each having 100 different levels. This means that it is reasonable to expect the system to describe 10^{8000} different particles. Considering that there are only 10^{100} particles in the universe—electrons, protons, neutrons, and the like—the system seems more than adequate for current use.

One method being used to analyze particles depends on the analysis of the particle silhouette and not the total particle. Several questions arise as to whether this technique is satisfactory. First, since one is analyzing a two-dimensional particle, or the edge of a two-dimensional particle, is that enough information to analyze the three-dimensional particle? Second, do particles assume the same position every time they are thrown on a surface? Ehrlich[21] has studied this problem extensively by taking a particle, suspending it in a transparent block, and then studying the transform of a variety of projections of that particle. For each arbitrary position of that particle, the Fourier pattern changes, but not by that much. Meloy[22] has studied the number of different stable positions a particle can assume. He finds that most particles have two or three different positions, with seven being a very high number. Furthermore, Meloy finds that each position has an energy level associated with it in that the distance the center of gravity is raised above the plane is a measure of the energy level of that position. As might be expected, only one or two energy levels are low, and thus the likelihood of a particle falling into any of the higher energy levels becomes quite small. Ehrlich finds that particles assume a position such that their maximum area is projected.

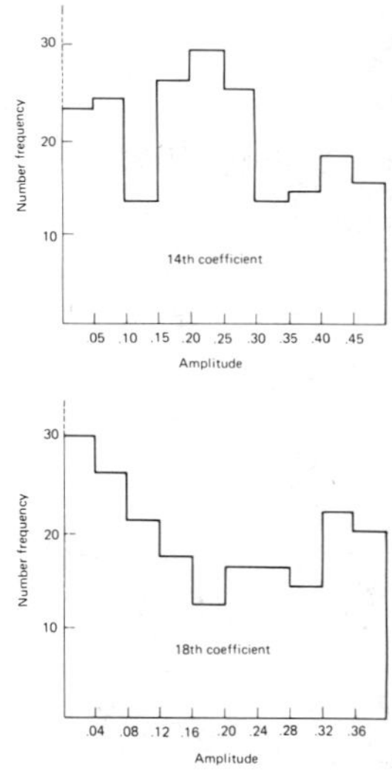

Figure 3.43. Histograms of the 14th and 18th coefficients of the first sample taken from powder A.

3.6. POWDER SIGNATURE

Hitherto, we have discussed the signature of individual particles. It has been shown that each particle shape can be represented by a series of terms in the Fourier series. By combining the terms, that is, the amplitude and phase angles, the individual particle shapes can be reproduced below any error designated. The coefficients constitute a spectrum for the individual particles. We are now going to use this spectrum of information from the individual particles to analyze powders. Powders even from homogeneous sources are represented by a variety of particle shapes. Thus a homogeneous powder may have a variety of repeating particle shapes within the powder. The spectrocomposite of these different shapes can not be used to reproduce the typical particle shape in the powder.

To use this information to tell one powder from another, a very simple statistical approach developed by Ehrlich[23] is used. A histogram is made for each coefficient. The histogram is a plot for a number of particles of the *n*th coefficient for each particle. For example, let us take the 14th coefficient and in the first block we will count the number of particles having an amplitude between 0.0 and 0.05. In the second block we will plot the number of particles having a 14th coefficient value of 0.05 to 0.1. Likewise, in the third block will be the number of particles having a coefficient between 0.1 and 0.15. This will be done not only for the 14th coefficient, but all the coefficients for powder A. In a like manner, a series of histograms will be made for all the coefficients in powder B. This type of histogram can be seen in Figure 3.43 where a histogram was made for the 14th and 18th coefficients of powder A and in Figure 3.44, where histograms were made for the 14th and 18th coefficients of powder B. Comparing now the histograms of the 14th coefficients from A and B, shown respectively

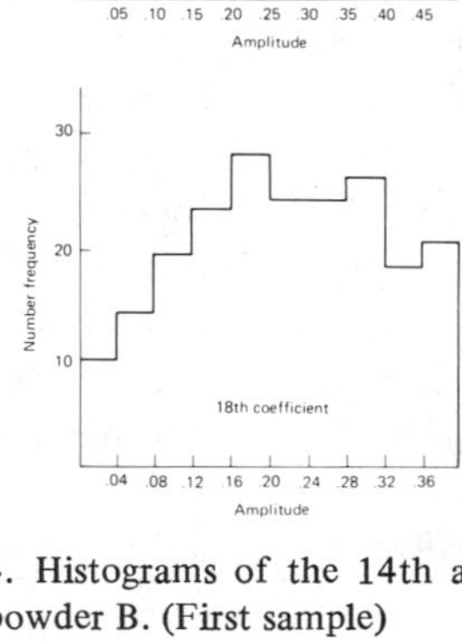

Figure 3.44. Histograms of the 14th and 18th coefficients of powder B. (First sample)

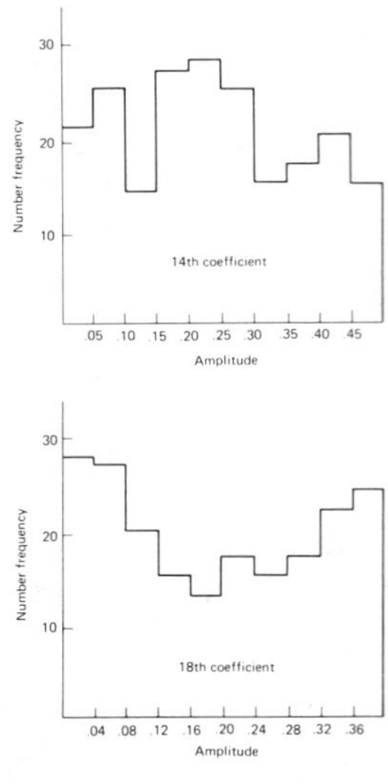

Figure 3.45. Histograms of the 14th and 18th coefficients of the second sample taken from powder A.

in Figures 3.43 and 3.44, yields a very clear insight that the two sets of coefficient histograms are not similar. In a like manner, if one compares the histograms of the 18th coefficients of powders A and B, respectively in Figures 3.43 and 3.44, one also comes to the conclusion that these histograms are not similar. Thus, powders A and B are not the same. If on the other hand, one takes a second sample of particles from powder A, runs them through the Fourier analysis technique, and then plots the histograms of the 14th and 18th coefficients as has been done in Figure 3.45, one can compare the histograms for the first sample taken from A with the second sample taken from A–Figures 3.43 and 3.45. By comparing histograms for powder A in Figure 3.43 and 3.45 one can see that the histograms are very similar and thus can draw the conclusion that the powder signature is repeatable.

To put the approach on a sound, analytical basis Ehrlich,[23] who developed the method of comparing histograms, decided to make a chi-square comparison for each bar in the histogram. Using the chi-square differentiater of the histogram, he[24] is able to distinguish very subtle differences in two powders. For example, in river systems, he can tell which sand source is contributing to a deposit. For another example, he has been able to tell whether the sand flowing along the beach is from the dunes behind the beach or from a nearby river. This is the first time powders have been identified by shape.

To do the analysis of a powder, one takes a sample of 200-300 particles from source A and another 200-300 particles from source B and then runs them through the Fourier analysis technique. The two histograms are then constructed and compared. Initially, each piece of data was taken in manually. As a result, about one particle could be done every four minutes. With the development of the Pagan Automatic Particle Analyzer System, a sample of 200-300 particles can be done in 10-15 seconds once the samples are properly prepared. The analysis and data manipulation are all done automatically. As a result, a great deal of the tedium that one would normally associate with this type of work is removed. Furthermore, it doesn't require skilled investigators.

It is believed that the method of comparing two different powders has a significant future in research but more likely, in industrial processing. Particle shape is very sensitive to conditions under which the particles are made. Slight variations, which may be unmeasurable in an industrial operation, will significantly alter the particle shape. Consequently, particle shape can become a powerful tool, but only for understanding industrial operations but for controlling them. Another, and in the long run, equally important application will be the use of these powder signatures in specifications.

Homogeneous powders have a number of interesting characteristics that are both unexpected and unexplained. For instance, homogeneous powders have not one or two predominent shapes, but usually five or more. Ehrlich[25] has recently developed analytical techniques that evaluate the Fourier coefficient signature for each particle from a homogeneous powder and then uses this signature to determine how many different generalized shapes there are. The surprising result is that there are so many different shapes. For example, in a common material from a homogeneous source such as quartz, one usually finds there are five or more different shapes. The question arises as to the relation of these predominent shapes to one another. There are, of course, transitional shapes, that is, shapes that are part way between two or more of the dominant shapes. Research is continuing in this area because not only is it interesting from a particle-formation point of view, but it raises some deep theoretical, mathematical questions.

3.7. SUMMARY

Recent rapid developments in the ability to analyze individual particles has led to new insight into the processes that form and modify particles. The ability to quantize powders promises to have widespread application in industry and research. It is believed that this new tool is at the same stage of development as the electron microscope was in the 1930s. As far as it is known, it is the only method of quantifying particle shape. It may well be that from an automatic scan of a powder and subsequent shape analysis one may be able to compute the initial and final packing density of a powder (see Figure 3.46).

REFERENCES

1. F. C. Bond, "Wet vs Dry Grinding," Europaeisches Symposium, "Zerkleinern," Frankfurt/M-1 97, 1962, *Aufbereitungs-Technik 3*(3):89–128 (1962).
2. H. Rumpf, "Problems of Scientific Development in Particle Technology Looked at from a Practical Point of View," *Powder Technol. 18*:3–17 (1977).
3. W. McCrone, "Experimental Techniques for the Shape Analysis of Particles," International Powder and Bulk Solids Conference, Rosemont, Ill. (1977).

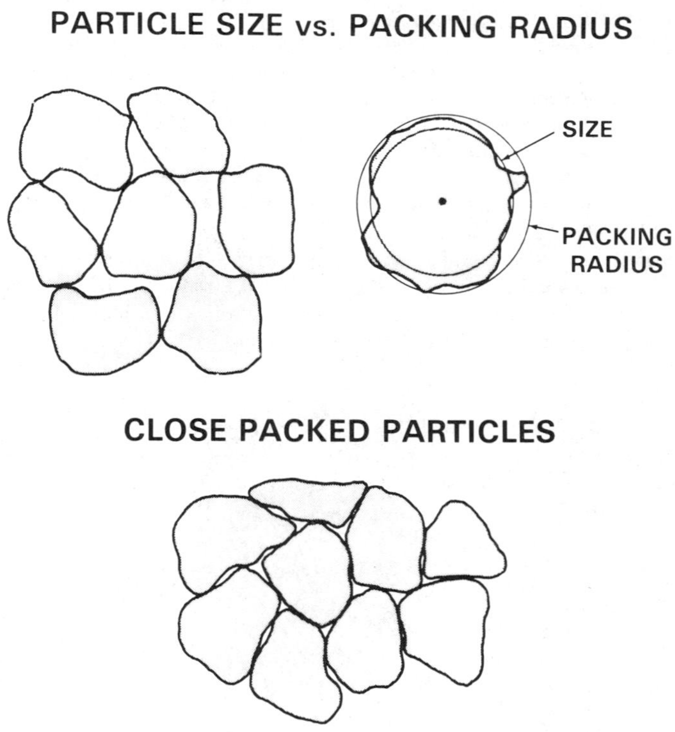

Figure 3.46. Particle's shape effects its behavior in powders. A particle's equivalent radius changes with the physical situation. Using the proposed techniques it appears possible to compute an equivalent particle radius for loose packing, dense packing, and the amount of work required to go from one to the other. Furthermore, it should be possible to compute a powder's resistance to penetration and its angle of repose.

4. T. P. Meloy, "A Hypothesis for Morphological Characterization of Particle Shape and Physiochemical Properties," *Powder Technol. 17*:233–253 (1977).
5. R. Ehrlich, and B. Weinberg, "An Exact Method for Characterization of Grain Shape," *J. of Sedimentary Petrology 40*(1):205–212 (1970).
6. J. K. Beddow, and G. Philip, "On the Use of a Fourier Analysis Technique for Describing the Shape of Individual Particles," *Planseeber 23*:1 (1975).
7. H. P. Schwarcz, and K. C. Shane, "Measurement of Particle Shape by Fourier Analysis," *Sedimentology 13*:213–231 (1969).
8. C. J. Zahn, and R. Z. Roskies, "Fourier Descriptors for Plane Closed Curves," *IEEE Trans Comput*, C-21:269–281 (March, 1972).
9. J. W. Cooley, and J. W. Tukey, "An Algorithm for the Machine Calculation of Complex Fourier Series," *Math. Comp. 19*:297 (1956).
10. G. Lew, J. K. Beddow, and A. F. Vetter, "Morphological Coefficients of Particle Profiles Derived from Walsh Coefficients," *CRC 47*:84 (1980).
11. J. K. Beddow, private communication (1978).
12. T. P. Meloy, "Characteristics of Particle Signatures," in J. K. Beddow and T. P. Meloy (eds.), *Testing and Characterization of Powders and Fine Particles*, Heyden and Son, Ltd., London, pp. 44–54 (1980).
13. T. P. Meloy, "Fast Fourier Transforms Applied to Shape Analysis of Particle Silhouettes to Obtain Morphological Data," *Powder Technol. 16*:27–35 (1977).
14. J. K. Beddow, private communication (1979).
15. T. P. Meloy, "Type I Random Particles," Proceedings of the 1978 Powder and Bulk Solids Conference, Industrial and Scientific Management Inc., Chicago, pp. 137–143 (1978).
16. S. T. Fong, J. K. Beddow, and A. F. Vetter, "A Technique to Show Effects of Chemical Reaction upon Particle Morphology" Proceedings of the 1978 Powder and Bulk Solids Conference, Industrial and Scientific Management Inc., Chicago, Illinois, pp. 214–218 (1978).
17. J. K. Beddow, private communication (1977).
18. Karhunen-Lo`eve Transform—sometimes called factor analysis.
19. A. G. Naylor, G. A. Robin, and B. H. Kaye, "Shape Analysis of Ceramic Particle Profile from Electron and Photo Micrographs, Fine Particles," in W. E. Kuhn (ed.), *Second International Conference*, Electrochemical Society, pp. 28–42 (1974).
20. B. H. Kaye, "Specification of the Ruggedness and/or Feature of a Particle Profile by its Fractal Dimension," *Powder Technol. 21*:1 (1978).
21. S. E. Tilman, "The Effect of Grain Orientation on Fourier Shape Analysis," *J. Sedimentary Petrology 43*:867–869 (1973).
22. T. P. Meloy, "Effect of Particle Projected Shape as a Function of Energy Level," Manuscript in preparation.
23. J. V. Mrakovich, R. Ehrlich, and B. Weinberg, "New Techniques for Stratigraphic Analysis and Correlation—Fourier Grain Shape Analysis—Louisiana Offshore Pliocene," *J. Sedimentary Petrology, 46*:226–233 (1976).
24. D. S. Van Nieuwenhuise, J. M. Yarus, R. S. Przygocki, and R. Ehrlich, "Source of Shoaling in Charleston Harbor—Fourier Grain Shape Analysis," *J. of Sedimentary Petrology 48*(2): 373–383 (1978).
25. R. Ehrlich, J. J. Orzeck, and B. Weinberg, "Detrital Quartz as a Natural Tracer—Fourier Grain Shape Analysis," *J. of Sedimentary Petrology 44*(1):145–150 (1974).
26. K. Gotoh, and J. K. Beddow, "Recent Progress in Particle Shape Analysis," *J. Society Powder Technology* (Japan) *15*(4):227–230 (1978).

4
Structural Properties of Packings of Particles

F. A. L. Dullien*
Institute de Mécanique des Fluides
Toulouse, France

INTRODUCTION

There is a wealth of literature on packed beds of particles, a study of which leads to the conclusion that it is difficult to meaningfully divide the great variety of known packs into different categories. The range of complexity of the structure of beds of particles is extremely wide: at one end of the spectrum are some natural soils, containing particles of various highly different shapes and sizes that are arranged in an infinite variety of irregular and unknown configurations, whereas at the other extremity of the scale of orderliness one finds the regular arrangements of monosized spheres. The latter are distinguished by a perfect order that facilitates a detailed description of their geometry. Owing to their great potential for analytical work, the structural properties of regular packings of equal spheres have been studied extensively. The results of these studies have been very interesting and enlightening, but sometimes they led to erroneous assumptions on the capillary and transport properties of irregular packs, containing particles of a wide range of different sizes and shapes. More recently, so-called random packs of monosized spheres have been also studied extensively, using statistical methods.

Although packs of particles of practical importance usually do not consist of arrangements of equal spheres, the amount and the significance of the work done on such systems make it a logical choice to place them in a separate category and to delegate all the rest of the packings to the other, second category of "general systems."

Of all the packs, probably soils and other unindurated sediments are of greatest significance in nature, and they occur also in the largest quantities. Other packed systems of great importance in technology include the so-called powders, usually products of crushing and grinding operations, the so-called bulk goods, such as various grains, coal, gravel, ores, etc., and the artificial packs of Raschig rings, Berl saddles, etc., used in contacting equipment of the chemical industry. In the present review

*This chapter was written while the author was on sabbatical leave from the University of Waterloo, Ontario, Canada.

no attempt is made to discuss systematically and separately the various different systems of irregularly packed particles.

In this chapter the structures of packings, as well as certain properties that are determined by the packing structure and the void geometry, are reviewed. The so-called macroscopic structure parameters include the porosity, the permeability, the specific surface, the reduced breakthrough capillary pressure and the resistivity factor.

4.1. MACROSCOPIC STRUCTURE PARAMETERS

All properties of packings of particles are influenced to a greater or lesser degree by the geometry of the void space which, in turn, depends on: (1) the type of packing, (2) the particle shape, and (3) the particle size distribution.[1] It is of great practical significance to identify those properties of the packings that are uniquely defined by geometry and, hence, do not depend on the nature of fluids contained in the void space unless, to be sure, the geometry itself is changed as a result of the action of the fluid (or fluids) present in the voids.

Mean Voidage or Porosity

The most obvious and most readily measurable macroscopic structure parameter is the *void fraction* ϵ usually called "mean voidage" in packed beds and "porosity" in general. Its definition is:

$$\epsilon = \frac{\text{volume of voids in packing}}{\text{bulk volume of packing}} \tag{1}$$

The various experimental methods used to determine ϵ have been adequately reviewed in the technical literature.[2,3]

The mean voidage is such a convenient basic parameter that it is common practice to use it as an *independent* parameter, separated from type of packing, particle shape, and size distribution,[1] even though it *depends* on these more fundamental, but also more baffling factors.

Specific Surface

Another simple geometrical parameter is the *specific surface* S_0 based on solids volume, i.e.,

$$S_0 = \frac{\text{internal surface of packing}}{\text{volume of solids in packing}} \tag{2}$$

which is altogether independent of the type of packing and the mean voidage. It is completely determined, in principle, by the particle shape, and size distribution. It can be readily combined with the mean voidage to give the specific surface S_v based on the bulk volume of the packing:

$$S_v = S_0(1 - \epsilon) \tag{3}$$

The major methods of determining specific surfaces have been reviewed in the literature.[2,3]

Permeability and Inertia Parameter

The *permeability* k often called specific permeability, is defined by Darcy's law:

$$Q = (kA/\mu)(\Delta \mathcal{P}/L) \tag{4}$$

Q is the volumetric flow rate, or "discharge," in sufficiently slow, unidirectional, flow of a Newtonian liquid of viscosity μ, through a sample of normal cross-sectional area A, and length L, in the macroscopic flow direction, under the influence of the pressure drop $\Delta \mathcal{P}$, where:

$$\mathcal{P} = P + \rho g z$$

where P is hydrostatic pressure, ρ the liquid density, g the acceleration due to gravity, and z is the distance measured vertically upward from an arbitrarily chosen datum level.

$\mathcal{P}$ is measured, in principle, by a pipe called the piezometer (see Fig. 4.1) and is indicated as the "piezometric head" ϕ (dimension of length):

$$\phi = \mathcal{P}/\rho g = (P/\rho g) + z \tag{6}$$

which is the sum of the elevation head z and the pressure head $P/\rho g$.

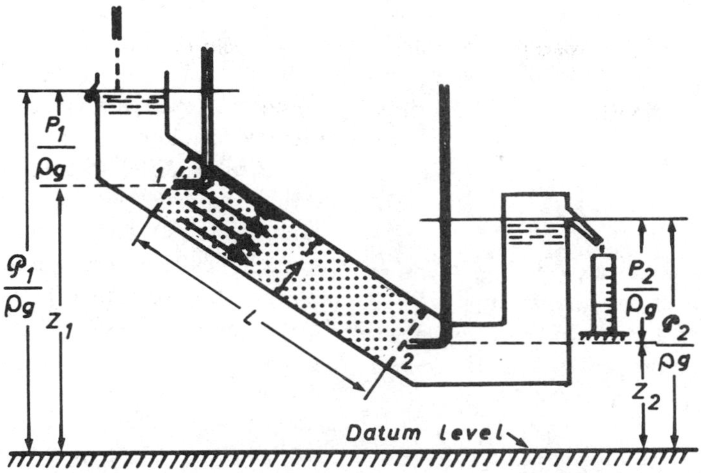

Figure 4.1. Illustration of the "piezometric head," "elevation head," and "pressure head."

Darcy's law is mostly used in differential form:

$$\mathbf{V} = -(k/\mu)\nabla \mathcal{P} = -(k/\mu)(\nabla P - \mathbf{g}) \qquad (7)$$

where $\mathbf{V} = (\delta Q/\delta A)\mathbf{n}$, the filter velocity or specific discharge, and $\mathbf{g}$ the gravitational acceleration vector. $\mathbf{n}$ is an outward unit normal vector.

When the only fluid of interest is water the *hydraulic conductivity* k_H defined as:

$$k_H = k\rho g/\mu \qquad (8)$$

is used. Darcy's law can then be written as:

$$\mathbf{V} = -k_H \nabla\phi \qquad (9)$$

In the case of gas flow, in most practical cases the elevation head may be neglected and $\Delta\mathcal{P} \approx \Delta P$. Owing to the compressibility of gases, however, both the volume flow and the filter velocity vary with pressure from one face of the sample to the other. In this case the correct integral form of Darcy's law is obtained by integrating the differential form, using the condition that at constant temperature and steady state the (pressure $\times$ velocity) product is constant throughout the sample. Thus, for gases:

$$\begin{aligned} V_2 &= (k/\mu)(P_2^2 - P_1^2)/(2P_2L) \\ &= (k/\mu)(P_m/P_2)(\Delta P/L) \end{aligned} \qquad (10)$$

where P_m is the arithmetic mean pressure.

It has been found that gas permeabilities sometimes vary with P_m due to a so-called slip effect. The equation taking the effect of "slip" into account is named after Klinkenberg[4]:

$$V_2P_2L\mu/(\Delta P P_m) = k[1 + (b/P_m)] \qquad (11)$$

where b is the constant characteristic of both the gas and the porous medium. The left side of Eq. (11) is plotted versus $1/P_m$, and a straight line is fitted to the data, whenever possible. The slope is bk and the intercept is k. Numerous methods have been proposed to measure permeabilities.[2,5,6,7]

As the filter velocity V is increased, increasing deviations from Darcy's law are observed, due to inertial effects. The appropriate modified form of Darcy's law is the so-called Forchheimer Equation, written in vectorial form as follows:

$$-\nabla\mathcal{P} = \mathbf{V}(\alpha\mu + \beta\rho V) \qquad (12)$$

where $\alpha = (1/k)$, i.e., the reciprocal permeability and β the so-called inertia parameter.

The Forchheimer equation has been used mostly in its one-dimensional form, which has been rearranged[8,9,10] into the following dimensionless form:

$$\frac{\Delta \mathcal{P}}{L\beta\rho V^2} = \frac{\alpha\mu}{\beta\rho V} + 1 \qquad (13)$$

All available data indicate (see Fig. 4.2) that Eq. (13) is a universal relationship.[18] The inertia parameter β appears to be also independent of the fluid properties and determined uniquely by the geometry of the pore space.

The Carman-Kozeny Equation and the Ergun Equation

S_V and ϵ are often used to calculate permeabilities based on a channel diameter characteristic of the packing by assuming[11,12,13,14] that the void space in packed beds is equivalent to a conduit, the cross section of which has an extremely complicated shape but, on the average, a constant area. This assumption is, strictly speaking, incorrect because it ignores the facts that (1) the void space in a packing is not a single conduit but the network consisting of a multitude of conduits, (2) each conduit has a variable area along its axis because it consists of an alternating sequence of "voids" and "windows" separating adjacent voids, and (3) different conduits consist of "voids" and "windows" of different sizes. Notwithstanding its shortcomings, this assumption, called the hydraulic radius theory or the Carman-Kozeny approach, has been very useful in random packings consisting of rotund particles of a narrow size distribution. Presumably the reasons for its success are that (1) in a packing of this type most of the conduits are of not very different sizes (i.e., most of the "voids" and the "windows" are contained in a narrow size range), and (2) the shapes of the conduits in different packings are not very different from each other. The effects of the axial variations in the conduit cross section are taken care of by a (fairly constant) empirical coefficient, the so-called Kozeny constant k'. Details of this theory follow below.

In analogy with established practice in hydraulics (although valid only in fully developed turbulent flow), the channel diameter D_H governing the flow rate through the conduit of uniform cross section is assumed to be four

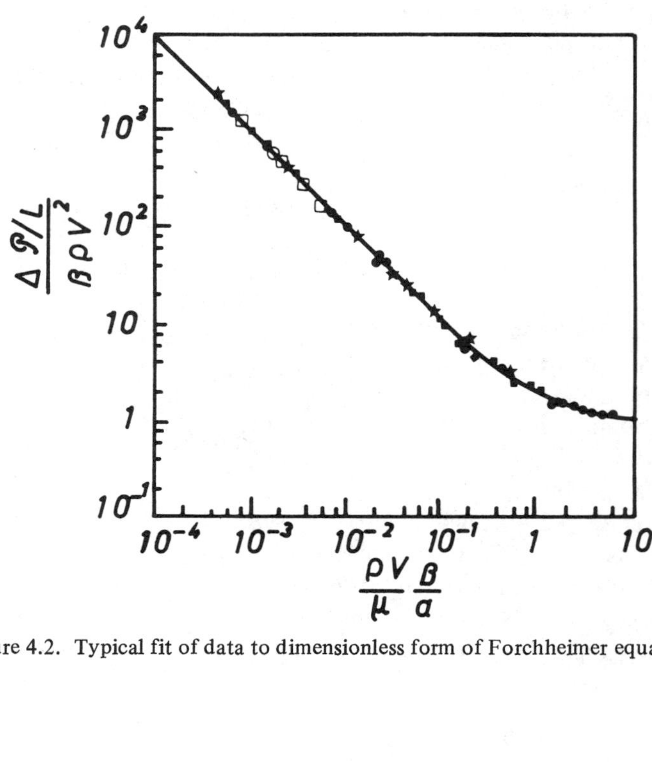

Figure 4.2. Typical fit of data to dimensionless form of Forchheimer equation.

times the hydraulic radius, which is defined as the flow cross-sectional area divided by the wetted perimeter, i.e.,

$$D_H = \frac{4 \times \text{void volume of medium}}{\text{surface area of channels in medium}} = \frac{4\epsilon}{S_0(1-\epsilon)} \tag{14}$$

It is noted that sometimes the pore diameter calculated from the breakthrough capillary pressure of a nonwetting phase (such as air through a system saturated initially with water), often called the bubbling pressure (for explanations see next section), is also called hydraulic diameter,[15] although this quantity is related to the volume-to-surface ratio of a porous medium only if all the conduits are equal and of uniform cross section (cf. Ref. 3, p. 59).

The average pore, interstitial, or seepage velocity $\overline{V}_p$ in the hypothetical single conduit is assumed to be given by a Hagen-Poiseuille type equation for a noncircular conduit:

$$\overline{V}_p = (\Delta \mathcal{P}/L_e)(D_H^2/16k_0\mu) \tag{15}$$

where L_e is the length of the conduit and k_0 is a "shape factor." It is of interest to note that in a conduit of uniform cross section Eq. (15) stays valid up to a Reynolds number of about 2100, whereas in a packing Darcy's law and, hence Eq. (15), start to break down at a Reynolds number of 0.2. This striking difference in behavior is due, in part, to variations in cross section of the conduits present in packings.[16,17]

The average "pore velocity" $\overline{V}_p$ and the filter velocity V [Eq. (7)] are assumed to be related as follows:

$$\overline{V}_p = (V/\epsilon)(L_e/L) = V_{DF}(L_e/L) \tag{16}$$

The division of V by ϵ is often referred to as the Dupuit-Forchheimer assumption, which defines the interstitial velocity V_{DF}, corresponding to flow through conduits of a net cross section equal to ϵA (A is the normal cross section of the system) in which all the microscopic streamlines are parallel to the macroscopic flow direction. The recognition of the fact that the microscopic streamlines follow a tortuous path is due to Carman, and it is reflected also in the fact that in Eq. (15) L_e was used, instead of L.

Combination of Eq. (7), (14), (15), and (16) gives k_{CK}, the Carman-Kozeny permeability:

$$k_{CK} = \frac{\epsilon^3}{k_0\left(\frac{L_e}{L}\right)^2 (1-\epsilon)^2 S_0^2} \tag{17}$$

The coefficient $(L_e/L)^2 \equiv T$ is usually called tortuosity factor or tortuosity. According to a great deal of data, the best empirical value of k', the Kozeny constant, i.e.,

$$k' = k_0(L_e/L)^2 \tag{18}$$

is in the neighborhood of 5.

Defining a mean particle diameter $\overline{D}_{p2}$ as the diameter of the hypothetical sphere with the same S_0 as the packing, i.e.,

$$\overline{D}_{p2} = 6/S_0 \tag{19}$$

the following popular form of the Carman-Kozeny equation is obtained from Eqs. (17)-(19):

$$k_{CK} = \frac{\overline{D}_{p2}^2 \epsilon^3}{180(1-\epsilon)^2} \tag{20}$$

At higher filter velocities, where Darcy's law breaks down and the Forchheimer equation [Eq. (12)] applies, a more general form of the Carman-Kozeny equation is the so-called Ergun equation:

$$-\nabla \mathcal{P} = \mathbf{V}\left[\frac{A\mu(1-\epsilon)^2}{\overline{D}_{p2}^2\epsilon^3} + \frac{B\rho V(1-\epsilon)}{\overline{D}_{p2}\epsilon^3}\right] \tag{21}$$

In the original form of this equation[9] $A = 150$ and $B = 1.75$. Macdonald et al.,[18] using all available data found that the best values are $A = 180$ and $B = 1.80$ for very smooth particles, but $B = 4.00$ for the roughest particles. Almost all experimental data lie within the ±50% envelope. For some of the data the following correlation holds:

$$A = 3.27 + 118.2\overline{D}_{p2}(10^{-3}\ \text{ft}) \qquad (22)$$

which improves the accuracy of Eq. (21) considerably.

It is customary to write the one-dimensional form of the Ergun equation in the form of a so-called friction factor versus Reynolds number correlation:

$$f_p \frac{\epsilon^3}{1-\epsilon} = \frac{180(1-\epsilon)}{Re_p} + 1.8 \qquad (23)$$

where

$$f_p = \frac{\overline{D}_{p2}\ \Delta \mathcal{P}}{\rho V^2 L} \qquad (24)$$

is a friction factor, and

$$Re_p = \frac{\overline{D}_{p2} V \rho}{\mu} \qquad (25)$$

is the "superficial" or "particle" Reynolds number.

At the end of this section some comments are in order regarding the method of obtaining the mean particle diameter $\overline{D}_{p2}$ of the packing, which is not usually obtained by a direct application of Eq. (19), because S_0 of a packing is seldom measured. Instead, one calculates $\overline{D}_{p2}$ from a particle size distribution $N(D_p)$ (or, n_i particles of diameter D_{pi}) of a sample. Subject to the condition:

$$D_{pi} = 6\frac{v_i}{s_i} \qquad (26)$$

there exist the following identities:

$$\overline{D}_{p2} = \frac{\sum_i D_{pi}^3 n_i}{\sum_i D_{pi}^2 n_i} = 6\frac{\sum_i v_i}{\sum_i s_i} = \frac{1}{\sum_i Y_i/D_{pi}} \qquad (27)$$

where v_i and s_i are the volume and the surface respectively, of a particle of diameter D_{pi}, and Y_i is the volume (or mass) fraction of the sample represented by particles of size D_{pi}.

In the case of wide particle size distributions, there is evidence[19] that $\overline{D}_{p2}$ is an inadequate mean particle diameter or, in other words, the Carman-Kozeny and the Ergun equations are inapplicable in the case of wide particle size distributions. The reason for this behavior is that even a relatively small fraction by volume of very small particles causes $\overline{D}_{p2}$ to be very small, whereas the resistance to flow of the sample is determined mainly by larger channels between larger particles.

For nonspherical particles, however, the customary methods of particle sizing and, in particular, sieve analysis do not determine particle diameters in conformity with Eq. (26), but yield values which need correcting to be equal to $6\ v_i/s_i$ for the particular shape under consideration.

Reduced Breakthrough Capillary Pressure

The capillary pressure P_c is defined[3,7] as the pressure difference in mechanical equilibrium across an interface separating two immiscible fluids nw and w:

$$P_c = P'' - P' = P_{nw} - P_w > 0 \qquad (28)$$

Here P'' and P' are the pressures on the concave and the convex sides of the interface, respectively, the nw and w refer to the nonwetting and the wetting fluid, respectively. *Wetting fluid* is defined as the one through which the "effective contact angle" $\theta + \phi$ is less than 90° (see Fig. 4.3). θ is the contact

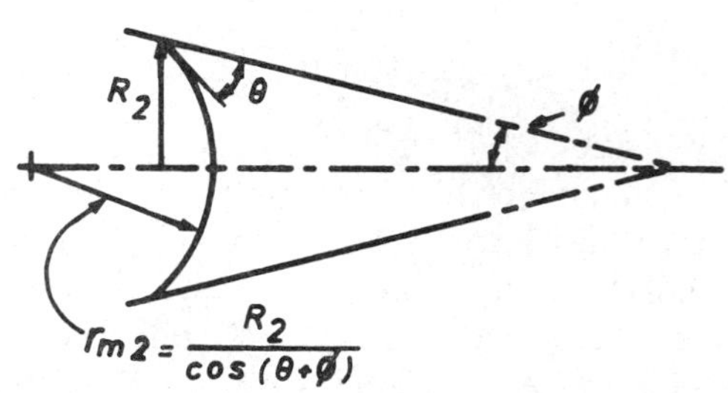

Figure 4.3. Meniscus in conical capillary.

angle and ϕ is half of the cone angle of the tapered capillary. P_c is related to the capillary radius R, the interfacial tension σ, and the angle $\theta + \phi$, by the following form of Laplace's equation:

$$P_c = \frac{2\sigma \cos(\theta + \phi)}{R} \tag{29}$$

A great deal of experimental research has been done on the so-called capillary pressure curves of packings.[20-32] Details of capillary pressure curve determination may be found in these references (see also Ref. 7). Here only its most essential features will be outlined with reference to the diagram in Figure 4.4. Refer to curve R_0, the so-called primary drainage curve. The experiment is started usually with the packing saturated 100% with a wetting phase, such as water. In an appropriate apparatus (there are many different types; see references) one face of the pack is contacted with a nonwetting phase, such as air, whereas at the other face a capillary barrier is used that permits the passage of water but prevents the penetration of air by virtue of its very fine

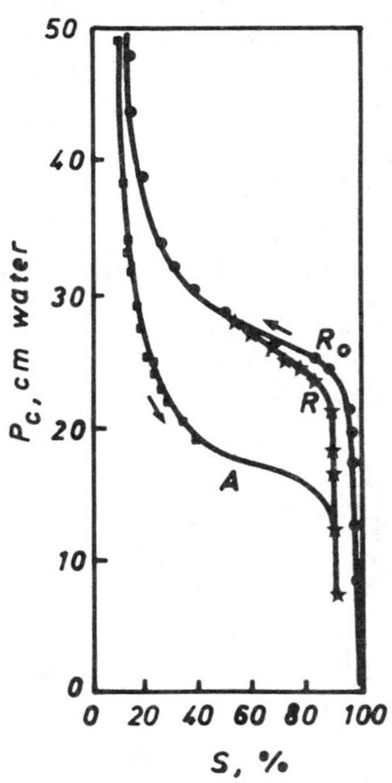

Figure 4.4. Capillary pressure curve with hysteresis loop.

pores. Starting at $P_c = 0$, as the value of P_c is increased in small steps, first there is penetration only into the surface irregularities of the sample. Later a value of P_c is reached (so-called entry pressure) which is high enough to force the air through the largest "windows" or "necks" at the surface of the sample and into the voids lying behind them. In regular packings of equal spheres (see section below) the same "window" size is repeated throughout the entire sample and, hence, once the entry pressure has been reached the air penetrates all across the sample and it "breaks through" at the other face. Hence, in this case, the entry pressure is also the breakthrough pressure. In random packs of equal spheres, and in general systems (see section below), however, somewhere along the path of the invading air, there always are windows that are smaller than the entry windows and, as a result, the penetration at the entry pressure is limited to a thin layer of the packing at the penetration face. The value of P_c must be raised to a higher value, permitting penetration through small windows (but *not the smallest windows* in the packing), before breakthrough is achieved. Hence, both in the "random" and the "general" packings the value of the breakthrough pressure P_{cb} exceeds the entry pressure. P_{cb} is at the first elbow of the curve R_0. The value of the breakthrough pressure P_{cb} is readily determined if no capillary barrier is used, because at this pressure the air will start bubbling across the sample. This accounts for the name *bubbling pressure.* It is an important parameter, characteristic of the packing and the fluid pair (in this case; water/air) used. Using Eq. (29) applied to a cylindrical capillary ($\phi = 0$) the *reduced breakthrough capillary pressure* P'_{cb} is defined:

$$P'_{cb} = \frac{P_{cb}}{2\sigma \cos\theta} = \frac{1}{R_b} \tag{30}$$

where R_b is the effective radius of the windows which it is necessary and sufficient to traverse in order to pass form one end of the system to the other. The two parameters P'_{cb} and ϵ correlate well with the permeability k and the resistivity factor F (see Ref. 15) (see below).

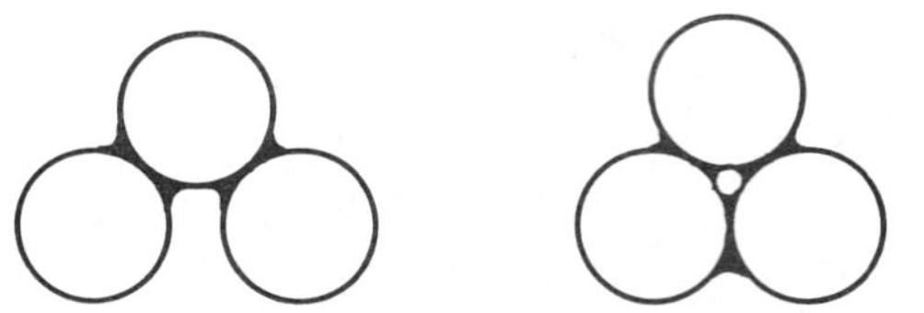

Figure 4.5. Pendular rings of water trapped between touching spheres.

As the pressure is increased past P_{cb}, (after replacing the capillary barrier) even smaller windows (and the voids behind them) are penetrated. The voids, however, are never completely filled because (1) as shown in Figure 4.5, there is water trapped between two touching particles in the form of so-called pendular or toroidal rings, and (2) there are also entire voids, and even systems of voids, containing water that are bypassed by the invading air, because they are "protected" by small windows all around. One way that this may happen is shown in Figure 4.6. As a result, even very high capillary pressures fail to reduce the water saturation below what is called the "irreducible" water saturation S_{wi}.

Reducing the capillary pressure will result, under appropriate experimental conditions, in reimbibition of water along the imbibition curve (refer to curve A in Fig. 4.4). The phenomenon that the curves R_0 and A do not coincide is called capillary hysteresis, which is due to (1) the nonuniform conduit cross section and (2) the hysteresis of the contact angle θ. Whenever one of the two phases is perfectly wetting, i.e., $\theta = 0$, such as water on clean glass and against air as the other fluid, only (1) is held responsible for capillary hysteresis.

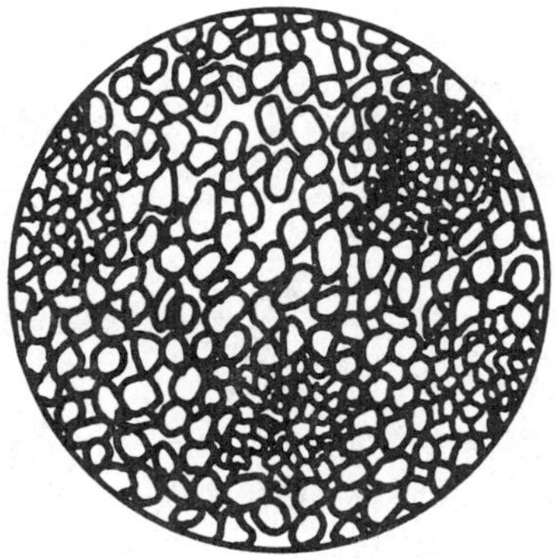

Figure 4.6. Microscopically heterogeneous distribution of grains illustrated as a monolayer. All the water between the small grains is trapped.

Even if the capillary pressure is reduced back to zero there is always some air trapped in the packing in voids which have been bypassed by the invading water. The corresponding saturation, S_{nwr}, is the "residual" nonwetting phase saturation.

If P_c is increased once more, the so-called secondary drainage curve (refer to curve R in Figure 4.4) will join up with curve R_0. The loop RA is called a "permanent" hysteresis loop because it can be traced reproducibly.

Resistivity Factor

The *resistivity factor* or formation resistivity factor F is defined as the ratio of the resistivity of the (electrically nonconductive) packing saturated with an ionic solution to the bulk resistivity of the same ionic solution. Thus:

$$F = R_0 / R_w \tag{31}$$

where R_0 is the resistivity of the saturated sample and R_w the resistivity of the electrolyte solution. It is evident that F, by definition, is always greater than one. The experimental technique of measuring F has been described in the literature.[33,34]

In the absence of surface conductivity, the value of F is uniquely determined by the geometry of void space. It is important to realize, however, an important difference between k and F as far as their dependence on the pore geometry is concerned. The reason for this difference is that the electric conductivity of a conductor of uniform cross section A_i is proportional to A_i, whereas the hydraulic conductivity in laminar flow is proportional to A_i^2. Hence, in two samples of identical mean voidage ϵ and geometrically similar void structure, but different void (pore) size, the values of F will be the same, whereas the value of k will be smaller in the sample with the smaller pore sizes. If all conduits in a packing had uniform cross sections along their axes, F

would not depend at all on the diameter of the conduits but only on their lengths and, of course, the mean voidage ϵ. Things are complicated, however, by the fact that the cross sections of the conduits vary periodically between windows and voids. Evidently if the sizes of all the windows were decreased to zero, the value of F would increase to infinity. This effect has been taken into account by several authors.[32,38] Owing to this, as well as to the tortuosity of the conduits, F is not simply in inverse proportion to the mean voidage ϵ, as one might first expect. The first empirical relationship for F was suggested by Archie.[117]

$$F = \epsilon^{-m} \tag{32}$$

where m is the cementation exponent. For random suspensions of spheres, cylinders, and sand in aqueous solutions of zinc bromide of approximately the same densities as the particles, De La Rue and Tobias[39] found Eq. (32) to hold with $m = 1.5$ with a high degree of precision over a range of ϵ extending from 0.25 to 0.55. Wyllie[40] used the following relationship:

$$F = X/\epsilon \tag{33}$$

which defines the electric tortuosity X. One has the following relationship:

$$X = TS' \tag{34}$$

where S' is a "constrictedness factor" that takes into account the axial variation of the conduit cross sections, referred to above.

4.2. PACKING STRUCTURES OF EQUAL SPHERES

Regular Packings

Important literature references on this subject include Haughey and Beveridge,[41] Fejes Toth,[42] Hrubisak,[43] and Graton and Fraser.[44]

It is convenient to think in terms of rows of touching spheres as the basic unit. The rows may be assembled parallel to and touching each other to form an infinite variety of different regular two-dimensional layers. There are two limiting forms, the square pattern characterized by an angle of 90° and the equilateral triangular (rhombic, hexagonal) pattern characterized by an angle of 60°. The layers are always thought of as being arranged in the horizontal plane. Considering only the action of the force of gravity, there are three stable ways in which two square or two triangular layers may be stacked one on top of the other.

1. Each sphere in the second layer is placed with its center exactly above that of a sphere in the first layer (sequence A). The number of layers in this sequence that is necessary to obtain spatial periodicity, the so-called order of the packing, is equal to 1. If this sequence is repeated (AA . . .) the simple cubic and the orthorhombic (cubic-tetrahedral) packings are obtained in the case of the square and the triangular layers, respectively.

2. Each sphere in the second layer is placed with its center exactly above the half-distance point between centers of touching spheres (i.e., the point of contact of touching spheres) in the first layer.

In the case of the square layer, depending on the direction of the displacement of the second layer relative to the first (i.e., whether this is in the x or in the y direction), there are two sequences possible (sequences B and C). If either sequence B or sequence C is repeated (BB . . . *or* CC . . .) equivalent orthorhombic packings of order 2 are obtained in which the sphere centers in layers i and $i + 2$ overlap, and there are clear vertical passages throughout the pack. When sequences B and C alternate (BCBC . . .), however, a different orthorhombic packing of order 3 is obtained where the sphere centers in layer $i + 2$ lie exactly above the centers of the holes between touching spheres in layer i, and there are no clear vertical passages throughout the pack (blocked passages).

In the case of the triangular layer each sphere touches six other spheres in the same layer. When placing the spheres in the second layer exactly above the points of contact of touching spheres in the first layer, the centers of two touching spheres in the second layer will lie above two of the six points of contact of a sphere in the first layer. Hence, there are three different ways of placing the second layer over the first one (sequences B, C, and D). When placing the third layer over the second one, however, there may result only two different packings. Sequences BB . . . , CC . . . , and DD . . . lead to equivalent tetragonal sphenoidal packings of order 2, where the centers of spheres in layers i and $i + 2$ overlap and that contain vertical passages throughout the pack. Sequences BCBC . . . , BDBD . . . , and CDCD . . . , on the other hand, result in equivalent tetragonal sphenoidal packings of order 3, where the sphere centers in layer $i + 2$ lie exactly above the points of contact between touching spheres in layer i, as well as those between touching spheres in layer $i + 1$. In this packing there are no clear vertical passages (blocked passages).

3. Each sphere in the second layer is placed with its center exactly above the center of the hole formed by touching spheres in the first layer.

For the square layers only one sequence (sequence E) is possible which if repeated (sequence EE . . .) results in the rhombohedral (dense cubic) packing of order 2, where the centers of spheres in layers i and $i + 2$ overlap and there are no clear vertical passages (blocked passages).

In the case of the triangular layers, however, only every second hole is covered, and thus there are two possible sequences (sequences E and F). When placing the third layer over the second one, two different packings may be obtained, depending on how one proceeds. Sequences EE . . . and FF . . . result in equivalent rhombohedral (close packed hexagonal) structures of order 2, where sphere centers in layers i and $i + 2$ overlap and there are clear vertical passages. Sequence EFEF . . . , however, results in a different rhombohedral, i.e., face-centered cubic packing of order 3 such that the sphere centers in layer $i + 2$ lie exactly above centers of holes in layer i, and there are no clear vertical passages (blocked passages).

Some important *structural properties* of regular packings of equal spheres are listed in Table 4.1.[41] *Unit cell* of the pack is defined as the smallest geometrical unit that will reproduce the pack if translated periodically, parallel to itself in the three spatial directions. The layer spacing β ($\beta = d/D_p$, where d is the distance between two consecutive parallel layers of the packing), in general, varies with the direction in the pack. The values contained in the table are the layer spacings in the vertical direction (cf. the description, given above, of the method of construction of regular packs). The coordination number n of a packing, is the number of spheres in contact with every sphere in the pack. It is seen to increase from 6 to 12 as the packing structure becomes more compact.

The coordination number of the pack n should be carefully distinguished from the *coordination number z of the network of void space* in the pack. Groupings of neighbouring spheres surround so-called voids that are separated from adjacent voids by so-called windows, access openings, or necks. The mean number of windows per void is the coordination number z of the pore or void network. From the point of view of transport phenomena taking place in the void space it is z, and not n that is of importance. The value of z is 6 in the simple cubic packing, as is evident by inspecting the unit cell shown in Figure 4.7. In the rhombohedral (close packed) structures, however, the "square" windows split into two "triangular" ones, as shown in Figure 4.8, and the rhombohedral unit cell contains two tetrahedral cells and a rhomboidal one, each of which contains a void inside. Each tetrahedral cell has 4 triangular windows, whereas the rhomboidal cell has 8, the average number of windows per void, i.e., $z = 5.33$. Another method of calculating the coordination number of a network which sometimes yields a different value will be discussed in the section void structures.

The mean free area fraction ϵ_A represents the fractional void area in a plane section of the unit cell. In regular packs the value of ϵ_A varies both with the orientation and the position of the sectioning plane. The values of ϵ_A shown in the table are taken in the horizontal plane (cf. the above discussion of the method of con-

Table 4.1. Structural Properties of Regular Packings of Equal-Sized Spheres.

PACKING GROUP	COMMON NAME	PACKING ORDER	PATTERN SQUARE LAYER	PATTERN TRIANGULAR LAYER	UNIT CELL VOLUME PER D_p^3	UNIT CELL VOID VOLUME PER D_p^3	MEAN VOIDAGE ϵ	LAYER SPACING PER D_p' β	COORDINATION NUMBERS (POINTS OF CONTACT PER SPHERE) n AND (WINDOWS PER VOID), z: n	z	FRACTIONAL FREE AREA, ϵ_A: MAX.	MIN.
Cubic	Cubic, cubic N° 1	1	AA . . .	–	1.000	0.476	0.4764	1.000	6	6	1.000	0.21
Orthorhombic	Orthorhombic N° 4 Cubic-tetrahedral	1		AA . . .	0.867	0.343	0.3954	1.000	8	5	1.000	0.09
	Orthorhombic N° 2 (clear passage)	2	BB . . . CC . . .					0.866			0.635	0.21
	Orthorhombic N° 2 (blocked passage)	3	BCBC . . .	–								
Tetragonal-sphenoidal	Tetragonal-sphenoidal N° 5 (clear passage)	2	–	BB . . . CC . . . DD . . .	0.750	0.226	0.3019	0.866	10	6	0.580	0.09
	Tetragonal-sphenoidal N° 5 (blocked passage)	3		BCBC . . . BDBD . . . CDCD . . .								
Rhombohedral	Rhombohedral N° 3 Pyramidal, dense cubic, Cubic close packed	2	EE . . .		0.708	0.184	0.2595	0.707	12	5.33	0.349	0.214
	Rhombohedral N° 6 (clear passage), close packed hexagonal, Tetrahedral	2		EE . . . FF . . .				0.816			0.455	0.093
	Rhombehedral N° 6 (blocked passage) Face centered cubic	3		EFEF . . .								

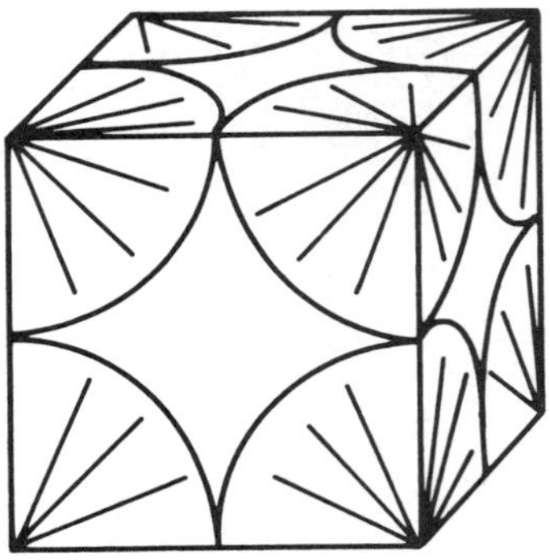

Figure 4.7. Unit cell of simple cubic packing.

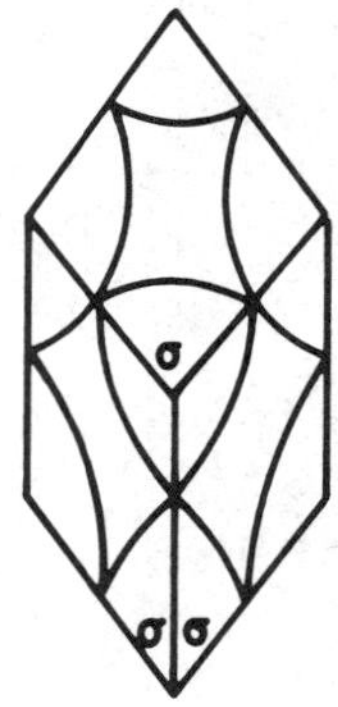

Figure 4.9. Unit cell after Mayer and Stowe.[46]

struction of regular packs) and they show the maximum and the minimum values of ϵ_A as the plane is moved across the unit cell. As a corollary, one may note that it is not permissible to try to infer the bulk porosity of regular packs from measurements performed in sections of the bed.

The surface areas per unit bulk volume of the packing multiplied by D_P ($S_v \times D_p$) can easily be shown to be a function of ϵ only, i.e.:

$$D_p S_v = 6(1 - \epsilon) \tag{35}$$

The number of spheres per unit bulk volume of the packing N_s multiplied by D_p^3 can readily be shown to be equal to:

$$D_p^3 N_s = 6(1 - \epsilon)/\pi \tag{36}$$

The *mathematical models* used to represent certain ranges of regular packs by Frevel and Kressley,[45] and by Mayer and Stowe[46,47] are useful. The first of these authors used triangular (close-packed) layers, which they stacked, one on top of the other. The relative position of the two layers was varied continuously between the two limits, corresponding to cubic-tetrahedral and rhombohedral (complete close-packed) structures. These arrangements include the tetragonal sphenoidal packing, but not the simple cubic packing. Mayer and Stowe, however, used a single packing angle σ, as shown in Figure 4.9, which they varied from 90° to 60° in order to obtain a continuous spectrum of regular packings, ranging from the simple cubic to the rhombohedral one, but bypassing the orthorhombic and the tetragonal sphenoidal structures.

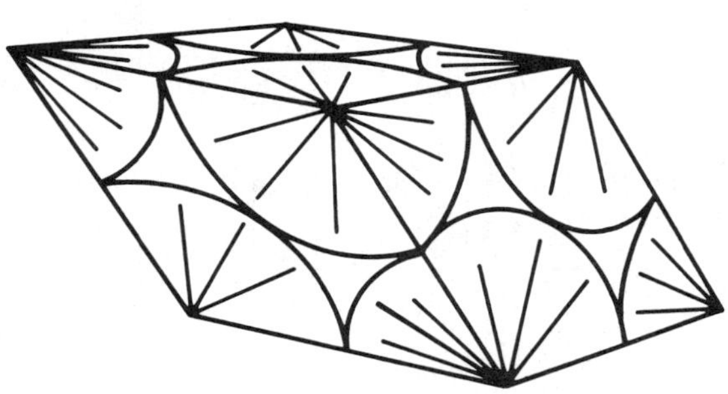

Figure 4.8. Unit cell of rhombohedral packing.

Random or Irregular Packings

It is easy to define a regular pack, but there does not seem to exist a completely satisfactory mathematical definition of a "random" pack.

A conceptual problem, inherent in "random" packs, is related to efforts to link the "random" distribution of particles to the mathematical abstraction of random distribution of points in a volume. Whereas points do not interfere with each other, particles do and, therefore, it may not be a fruitful idea to try to define the "random" distribution of particles in terms of the random distribution of points.

The definition of a random packing, used by Debbas and Rumpf[48] is as follows: "All particles of the same size and shape have the same probability to occupy each unit volume of the mixture." The following generally accepted theorems[49] follow from this definition.

1. The mean fractional free area ϵ_A is equal to the mean voidage ϵ of the packing,

2. The probability density function of diameters of circles appearing in any sufficiently large section plane through a random bed of spheres is identical to the theoretical distribution obtained by parallel sectioning a single sphere at infinitesimal, constant intervals, perpendicular to an arbitrarily selected radial direction.

Most details of the characteristics of "random" packings have been obtained experimentally. Some of these may be summarized by the so-called radial distribution function, shown for close random packings in Figure 4.10. The concept of radial distribution function shows the difference between regular packs, on the one hand, and "random" packs, on the other, most clearly.

Choosing the center of an arbitrary sphere as the origin of a coordinate system, in a regular pack the positions of *all the other sphere centers can be given exactly*. In a random pack, however, the position of *none of the other sphere centers can be given exactly*. In this case one can talk only about the *probability of finding sphere centers* at a given distance from the center of the base sphere. For the case of so-called uniformly random distribution one must, by definition, have the same probability density, independently of the distance. This is indeed the case in random packings of equal spheres at great distances from the center of the base sphere. Closer to the base sphere however it has been found experimentally[50,51] that the probability density varies with distance, as shown in Figure 4.10. As in the case of regular packs the radial distribution consists of a discrete set of Dirac delta functions (i.e., vertical lines) separated by gaps (i.e., regions of zero probability density) it is logical to interpret the peaks in the diagram at $R = 1$, etc.. ($R = r/D_p$, where r is the radial distance measured from the center of the base sphere) as an indication of a relative order over a short range around the base sphere. As expected, if the random packing is looser (greater mean voidage), the short range order becomes less pronounced.[52-54]

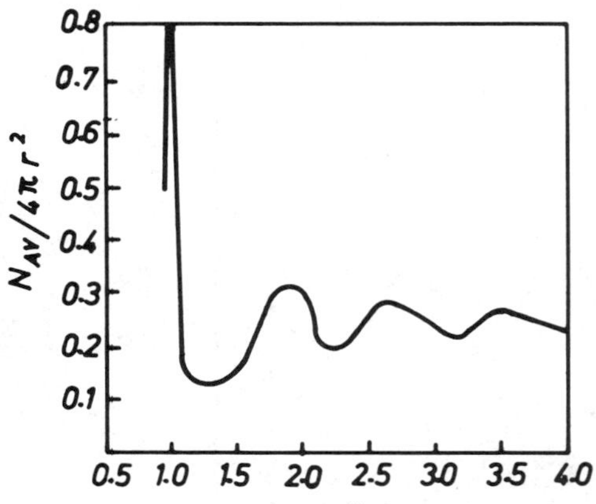

Figure 4.10. The radial distribution of the average number of sphere centers per unit area in a spherical shell.

For random packings, average values of the layer spacing β have been defined and expressed in dependence on the mean voidage, as follows:[55,56]

$$\beta = \sqrt{(2/3)}\{\pi/[3\sqrt{2}(1 - \epsilon)]\}^{1/3} \tag{37}$$

The mean coordination number n has been determined experimentally in random packs of equal spheres by several researchers.[57-61] Ridgway and Tarbuck[62] found the following correlation between n and ϵ in random packings of equal spheres:

$$\epsilon = 1.072 - 0.1193n + 0.00431n^2 \tag{38}$$

This correlation was based on data covering an extremely wide range of values of the parameters, i.e., $3 \leq n \leq 12$ and $0.27 \leq \epsilon \leq 0.78$, whereas in all common random packs of equal spheres $0.36 \leq \epsilon \leq 0.44$, approximately.

Considering the fact that there is an infinite number of different regular packings, it is not surprising to find also different kinds of random packs. These may be divided into the following four categories:[41]

Close Random Packing. These are obtained when the bed is vibrated or vigorously shaken down, and results in mean voidages of 0.359 to 0.375,[50,51,63–71] which is considerably in excess of the mean voidage of 0.26, corresponding to hexagonal close packing.

Poured Random Packing. Pouring spheres into a container, corresponding to a common industrial practice of discharging powders and bulk goods, results in mean voidages of 0.375 to 0.391.[11,42,73–75]

Loose Random Packing. Dropping a loose mass of spheres into a container, or packing spheres individually and randomly by hand, or permitting them to roll individually into place over similarly packed spheres, results in mean voidage values of 0.40 to 0.41.[52,57,60,64–70,76–78]

Very Loose Random Packing. The fluidized bed at the minimum fluidization has a mean voidage ϵ of 0.46 to 0.47.[79]

Slowly reducing the fluid velocity to zero in a fluidized bed,[80] or by sedimentation of spheres,[31] or by inversion of a bed container[82,83] a mean voidage of 0.44 is obtained.

In the case of random packs a so-called *wall effect* exists, because the proximity of a solid surface will introduce some local order into a random packing. Thus, the particles next to the solid surface tend to form a layer of the same shape as the surface. This so-called base layer is a mixture of clusters of square and triangular units. Randomness increases with increasing distance from the base layer, with a resultant disappearance of distinct layers.

Another important aspect of the *wall effect* is the existence of a region of relatively high voidage next to the wall due to the discrepancy between the radii of curvature of the wall and the particles.[44,84] A cyclic variation of the local voidage ϵ' with distance from a cylindrical wall has been measured,[62,85–90] extending some 3 to 4 particle diameters into the packing (see Figure 4.11). The local voidage ϵ' is the fractional free volume in a small or thin element of the bed dv', whereas the local mean voidage $\overline{\epsilon}$ is the mean value of ϵ' taken over a region of the bed, i.e.,

$$\overline{\epsilon} = \frac{1}{v'} \int_0^{v'} \epsilon' \, dv' \tag{39}$$

The *wall effect* has been studied as a function of the particle diameter ratio D_p/D_T. Various empirical formulas exist to correct for this effect.[41]

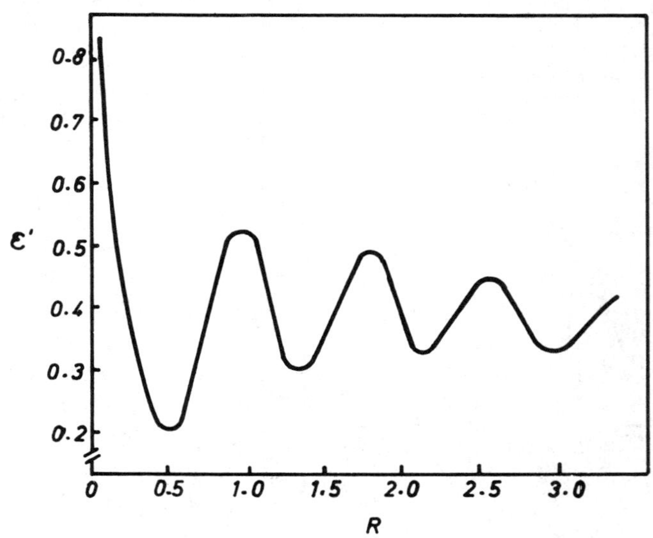

Figure 4.11. Variation of local volume fraction with distance from a cylindrical surface.

4.3. PACKING STRUCTURES OF GENERAL SYSTEMS

Almost all real packings of particles fall in this category. Unfortunately, it is not possible to say anything specific about the structure of the general systems, and one must be content to state a few principles regarding this subject.

There are two major factors to consider: (1) *particle shape* and (2) *particle size distribution*. Reviews (e.g., Ref. 41) offer very little useful information on the subject of the effect of *particle shape* on packing structure. One may note only that very special packing structures arise in the case of highly anisometric particles, such as platelets and needles, which can be packed in a great variety of very different ways, resulting in systems either of very high void fractions, when the orientations of the particles are random, or of very low void fractions, when the particles are stacked with their axes aligned. Systems of this latter type are also highly anisotropic, i.e., their physical properties are very different in the direction of the particle axes than in the perpendicular direction.

Particulate systems generally involve a range of particle size between 2- and 10^5-fold. When smaller particles are mixed into a bed of larger particles, then the former, on one hand, tend to increase the voidage, by forcing the larger particles apart but, on the other hand, also tend to decrease the voidage by filling the voids between the larger particles. The latter effect predominates for a size ratio greater than about 3:1. The net effect will depend not only on the particle size ratio (and particle shape), but also on the relative amounts of each size fraction present. Furnas,[91] and Sohn and Moreland,[92] showed that in binary mixtures of particles the bulk voidage always decreases below the values existing at both pure component ends of the composition range, and for any given ratio of particle diameters it has a minimum value at some composition.

Horsefield[93] showed that the filling of voids of a rhombohedral packing by five successive specified sizes would give a minimum voidage of 0.1481, which can be further reduced by the addition of even finer particles. Similar studies have been made also in other kinds of packings.[41]

A very important phenomenon that is directly related to particle size is the fact that *packings of smaller particles tend to result in relatively high voidages*, because small particles have a tendency to pack less closely than do larger particles. In beds of small particles there is frequent "arching" or "bridging" which yield larger voids.[65,94-96] This phenomenon is related to the smaller volume-to-surface ratio of small particles, resulting in weights which are insufficient to overcome a variety of surface resistances opposing attainment by the particles of positions of a minimum local potential energy by rolling into nearby wells or holes.

The *packing structures* of general systems vary widely and are seldom known in any detail. The only practicable technique, known to this author, which is available to explore the packing structures of "general systems" consists of filling the voids with some coloured or fluorescent plastic, or a low melting alloy, which make it possible to distinguish between the voids and the solid matrix, and study polished sections of the system.[7] Evidently both the packing structure and the void structure may be analyzed simultaneously by this technique, but only analysis of the latter has been attempted and that only in a few instances.

4.4. STRUCTURE OF INTERPARTICLE SPACE (VOID STRUCTURE)

General Characteristics

The geometry of the void structure of regular packings of equal spheres has certain characteristics that are shared, qualitatively anyway, by the "random" packs of equal spheres and also by most "general" systems:

Proceeding from one face of a unit cell in the direction of the opposite face, the channel cross section first increases and then decreases to the original value. It is convenient to refer to windows or access openings at the faces of the unit cell, and to voids or hollows on the inside (see Figures 4.7-9). The rhombohedral unit cell contains two kinds of voids, two tetrahedral ones and a rhomboidal one. In the general case, the coordination number $z = 6$, but it may be as low as 5. The size of a window or of a void can be conveniently measured by the diameter of the largest sphere that can be fitted inside. Qualitatively speaking, there is a *periodic* or, in the case of random or general packings, a *quasiperiodic variation in the cross section* of the channels traversing a packing. Quasiperiodic variation of the cross section simply means that the channel cross section first increases to a maximum value than it decreases to a minimum value, subsequently it increases to another maximum value, and so forth.

The shape of the channel cross section is always concave in packings of spheres and other rotund particles. In regular packings of equal spheres the general shape is a concave rhomboid, which becomes a concave square and a concave triangle in the limits of the simple cubic packing and the close-packed hexagonal packing, respectively. Qualitatively speaking, the *channel cross sections* in packings *are angular and irregular*, and *never circular*.

The channels are, in general tortuous. By definition, a channel is said to be *tortuous* when the axis, or centerline, of the channel connecting two arbitrary points in the packing along the shortest route possible, is not a straight line. In some regular packings of equal spheres there are certain directions along which the channels are not tortuous, such as the x and the y directions in a simple cubic packing, and the y direction in all those packings with clear vertical passages, listed in Table 4.1.

It is important to note that, owing to the periodic variations in the channel cross section, streamlines which do not lie on the channel axis are tortuous even in those channels whose axes are straight lines. Therefore, the tortuousity of the flow pattern across a packed bed is, in general, the sum of two effects, i.e., the tortuosity of the channel axis, and the tortuosity of the microscopic streamlines relative to the channel axis.[5]

The void space in all packings forms a three-dimensional interconnected network. In the case of regular packings of equal spheres the networks are built from repeating unit cells whose voids represent the nodes of a regular network of short capillaries. In a regular network all the windows lying in planes of identical orientation have the same size. The windows of the unit cells represent the portions with the smallest cross section of the capillaries. The average number of windows per void is the coordination number z of the network.

In random and general packings there are no repeating unit cells, nevertheless there is always present a network that can be divided into capillaries and nodes, and is characterized by an average coordination number z. In general packs both the windows and the voids are distributed over a range of sizes, a fact that has far reaching consequences for the capillary and transport phenomena taking place in the void space of the packing.

Methods of Investigation

There are a number of methods that have been used for the investigation of the void structure of packings, of which here only the two most important ones will be reviewed briefly. These are: (1) Capillary pressure curves, and (2) optical methods, using sections.

Capillary Pressure Curves. The discussion under "Reduced Breakthrough Capillary Pressure" has been limited to the important special case where the system originally is saturated with an incompressible wetting fluid such as water. Part of the water gets trapped when the system is invaded by the nonwetting phase. For the purpose of void structure studies, a widely used variant of the capillary pressure curve method is the so-called *mercury intrusion porosimetry*, first introduced by Ritter and Drake[97] (for an extensive literature review see, e.g., Modry et al.[98]).

Mercury Porosimetry. In a mercury porosimetry experiment, the primary drainage capillary pressure curve of an evacuated sample is determined with the help of mercury as the nonwetting fluid. The mathematical analysis of the results for the purpose of void structure investigation is facilitated by the absence of any trapped wetting phase in such an experiment.

Penetration of mercury under increasing pressures into regular packings of equal spheres was analyzed theoretically by Frevel and Kressly[45] and, in greater detail, by Mayer and Stowe,[46,47] who analyzed also the filling of the toroidal spaces, or pendular rings, after the breakthrough of mercury. Depending on the packing structure, there is different value of the quantity (reduced breakthrough capillary pressure x sphere radius), which is independent of the sphere size. In regular packs of equal spheres all the voids are penetrated simultaneously by the mercury, but they are not filled completely, owing to their angular shapes. So-called toroidal spaces or pendular rings are left empty, and these are filled gradually as the pressure is increased further. For a regular pack of known geometry and sphere size the theoretical results of Mayer and Stowe[46,47] offer a good quantitative interpretation of the relationship between the capillary pressure and the volume of mercury penetrating the pack.

For the case of random and general systems the interpretation of mercury intrusion porosimetry curves is done mostly in terms of a so-called pore size distribution $\alpha(D_e)$ of the sample, defined as follows:

$$dv = \alpha(D_e)\, dD_e \tag{40}$$

where dv is the volume of void space having entry diameters between D_e and $D_e + dD_e$. *Entry diameter* D_e, is defined by Eq. (29) where it is usually assumed, without justification, that $\phi = 0$. Hence, differentiation of $P_c D_e$ = cont., gives:

$$P_c\, dD_e + D_e\, dP_c = 0 \tag{41}$$

One should be careful to distinguish two completely different categories of the entry diameter D_e. First, entry diameter means the equivalent diameter of a window or neck to penetration by mercury. In addition to being the entry diameter of a window, the same is also the entry diameter of the divergent capillary on the other side of the window constituting a part of the void, and of the rest of the void, characterized by greater equivalent diameters than that of the window. The entry diameter bears absolutely no relationship to the size of the void penetrated through the window, however. Second, entry diameters are the equivalent diameters of convergent capillaries, such as the one shown in Figure 4.3. These correspond to equilibrium values of the capillary pressure from which the capillary dimensions may be calculated via Eq. (29). The latter entry diameters correspond to the dimensions of capillaries in the process preceding penetration of mercury through a window, and in the process of filling up the voids (the toroidal spaces) already penetrated by mercury when the capillary pressure is being increased further.

From Eq. (40) and (41) one obtains:

$$\alpha(D_e) = -\frac{P_c^2}{4\sigma\cos\theta}\,\frac{dv}{dP_c} \tag{42}$$

The pore size distribution calculated from Eq. (42) does not reflect the sizes of those spaces, such as voids, which are penetrated by the mercury through smaller openings, such as windows. Hence, it is not representative of the actual range of dimensions at the interparticulate spaces in the sample. Second, it assigns the volume of those spaces which are penetrated by the mercury through smaller openings to the dimensions characteristic of these openings and, therefore, grossly exaggerates the number of the openings or else makes them appear to be long capillary tubes. Third, it creates the illusion of the existence of very small pores, which in reality correspond to the toroidal spaces near the points of contact of touching particles.

Mason[99] used a tetrahedral pore model, i.e., a coordination number $z = 4$ to represent a random packing of equal spheres. Depending on the various different assumptions made, he obtained four different drainage capillary pressure curves and one imbibition curve.

Percolation theory. A promising approach to the interpretation of mercury porosimetry curves in terms of structure of interparticle space of general packings is offered by the percolation theory.[100-107]

A brief summary of this approach is given here for the benefit of research-oriented readers. The model used in this approach consists of a *capillary network.* A capillary network is conveniently divided into capillaries and nodes, the latter being the topological entities formed by the intersecting capillaries, which correspond to the voids in packings. There follows that a capillary connecting two nodes plays the same topological role as a window between two adjoining voids. The nodes can be made to correspond to the more general concept used in percolation theory, i.e. site, and the capillaries can be made to correspond to what are called bonds in percolation theory.

A network is characterized by the following parameters:

1. the dimensionality of the network,
2. the coordination number z, and
3. the microscopic topology of the network.

There are networks of one, two, and three-dimensions. The coordination number z, is the *average number of bonds meeting at a site.* The "average," however, can be defined in more than one way, which generally result in different values of z for one and the same lattice.

According to one definition, one writes:

$$z = \sum_r z_r f_r' \tag{43}$$

where z_r is the number of bonds meeting at a site of type r, and f_r' is the relative frequency of such sites in the lattice.

According to another definition, however, one writes:

$$Z = \sum_r Z_r f_r \tag{44}$$

where

$$Z_r = \frac{1}{2}\left(\sum_m\right)_r + 1 \tag{45}$$

with $(\Sigma_m)_r$ being the number of bonds connected to a bond of type r at both ends, and f_r the relative frequency of such bonds in the lattice. Hence, Z_r is the average number of bonds per site meeting at both ends of a bond of type r.

Usually, regular networks are used as models, because in these it is much easier to keep track of the calculations, but a regular network is equivalent to a statistically uniform irregular network characterized by the same values of the above parameters as the regular network.

One can distinguish between two basic types of percolation, i.e., percolation of *bonds* and percolation of *sites.* In the former all the sites are always "conductive" or "open" or, depending on one's viewpoint, they do not exist at all, and at the beginning of an "experiment" all the bonds are nonconductive or closed. An experiment consists of opening the bonds in the network, one-by-one, at locations chosen at random. It is convenient to assume that one face of the network is exposed to the invading fluid (mercury), the opposite face communicates with a reservoir that is empty (evacuated) at the beginning of the experiment, but it will be filled with mercury after breakthrough, and all the sides of the network are sealed with impervious walls. This set-up is shown for a two-dimensional square lattice in Figure 4.12.

At the beginning, all the network is empty (evacuated). As the bonds are getting opened, one-by-one, those open bonds which are at the exposed face of the network are penetrated by the invading fluid (mercury).

Using the terminology of mercury porosimetry and pore structure research, the first bond on the face of the network which is opened and is penetrated by mercury may be considered as the largest window or neck on the exposed face of the sample. Since the bonds are opened one-by-one, according to this viewpoint there are no two equal windows or necks in the sample. It is equally justified, however,

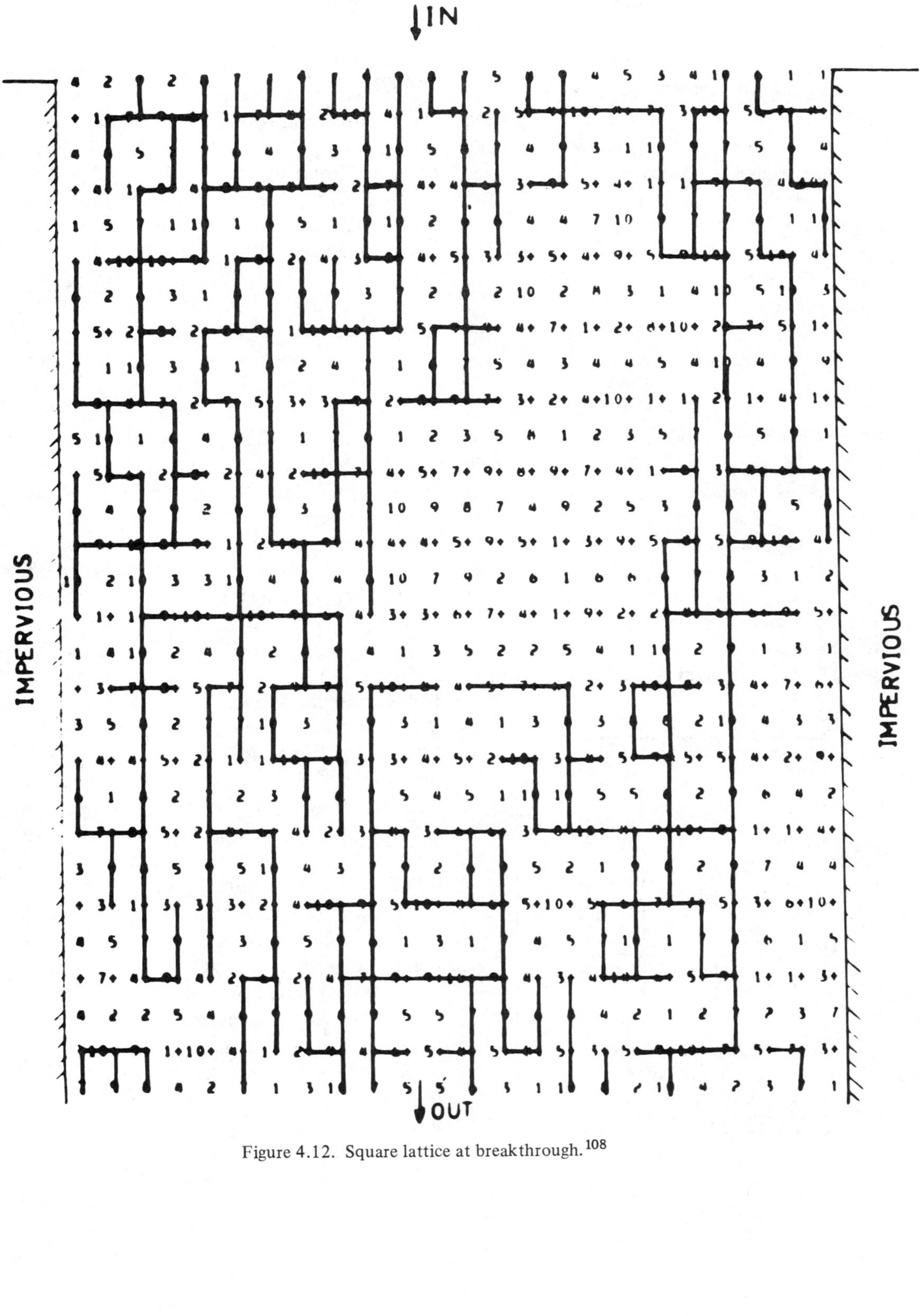

Figure 4.12. Square lattice at breakthrough.[108]

to consider the *first two open bonds to represent two windows of the same size* or, in general, *the first m_b open bonds to represent m_b windows of the same size*; similarly, *for the second m_b (or m_b') open bonds to represent another set of m_b (or m_b') smaller windows of the same size*, and so forth.

In order to obtain a sufficiently detailed information on the penetration history, however, one must choose values of m_b such that $m_b/N_b \ll 1$, where N_b is the total number of bonds present in the network.

The open bonds which are connected to the exposed face of the network via other open bonds are also penetrated by mercury. As a result, there is an increasing portion of the network which is penetrated by mercury, as the fraction of the number of open bonds p_b in the network increases. The depth of penetration (see Fig. 4.13) measured from the exposed face remains small, on the order of the first power of ten, in units of bond lengths, in the direction of penetration, until a certain, theoretically exact fraction p_{cb} of the number of bonds present in the network have been opened. At that point, suddenly the depth of penetration becomes equal to the depth of the network, i.e., *breakthrough is achieved, which persists even if the dimensions of the network are increased indefinitely, provided that the breadth and width of the network are at least as great as its depth*. The value of p_{cb} is characteristic of the network (see Table 4.2), and it corresponds to a generalized form of the breakthrough capillary pressure of a macroscopic sample discussed earlier.

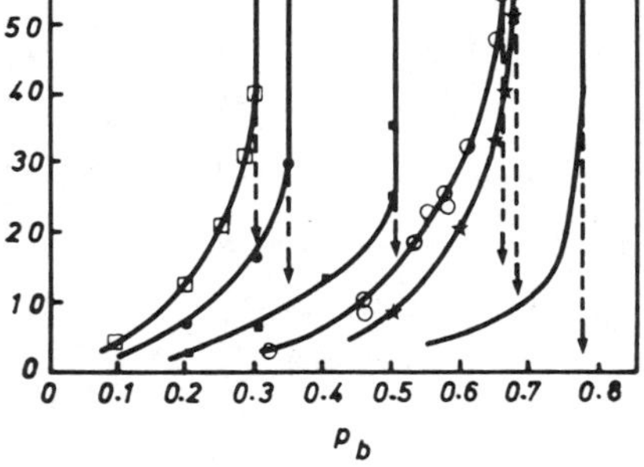

Figure 4.13. Depth of penetration in two-dimensional networks (curves form left-to-right: tetratriangular, hexatriangular, square, hexagonal, ditetragonal and tritetragonal).[108]

In order to simulate and interpret the mercury porosimetry curve of a general packing with the help of percolation of a network, one must count the number of bonds that have been penetrated by the fluid (mercury) at any given value of p_b. Looking at the process of penetration of network from the point of view of percolation of bonds, the open bonds represented by the fraction p_b of the N_b bonds present in the network are distributed in a random manner over the whole network. As explained above, initially only some of the open bonds form continuous paths connected to the exposed face of the network and are, therefore, penetrated by the fluid (mercury). Let us denote by Y_b the fraction of the N_b bonds present in the network that are connected to the exposed face of the sample at any value of p_b. It is evident from the above discussion that as the dimensions of the network are increased beyond any limit, for all values of p_b such that $p_b < p_{cb}$, i.e., "before breakthrough," $Y_b \to 0$.

The slope of the curve Y_b vs p_b at p_{cb}, i.e., $(dY_b/dp_b)_c$ increases with N_b because each bond rendered open immediately after the point of breakthrough results in the penetration of an increased number of additional bonds in a network of greater dimensions. Hence, $(dY_b/dp_b)_c \to \infty$ as $N_b \to \infty$. As a result, in an infinite network there appears to exist a range of values of Y_{cb} at $p_b = p_{cb}$, which is physically meaningless, because for a given value of p_b there can be only one value of Y_b.

On the basis of physical considerations, it is the upper end of this range which corresponds to reality. *The virtually vertical portion of the curve lacks a physical meaning because it is*

Table 4.2. Breakthrough Conditions for Various 2-D and 3-D Lattices.[108]

			I. BOND SIZES RANDOMLY DISTRIBUTED (BOND PERCOLATION)						II. BOND CORRELATED NET-WORKS WITH NODE SIZES RANDOMLY DISTRIBUTED (SITE PERCOLATION)					
			p_{cb}			$Z\,p_{cb}$			p_{cs}			$z\,p_{cs}$		
NETWORK	Z	z	*a*	*b*	*c*	*a*	*b*	*c*	*a*	*b*	*c*	*a*	*b*	*c*
Tri-tetragonal	2.66	2.4	0.78	-	-	2.07	-	-	-	-		-		
Di-tetragonal	3	2.66	0.68	-	-	2.04	-	-	0.72	-		1.92	-	-
Hexagonal	3	3	0.66	0.653	-	1.98	1.958	-	-	0.70	0.697	-	2.10	2.091
Square	4	4	0.50	0.500	0.499	2.0	2.0	1.996	0.59	0.590	0.591	2.36	2.36	2.364
Triangular	6	6	0.345	0.347	0.349	2.07	2.08	2.094	0.50	0.50	0.500	3.0	3.0	3.0
Tetra-triangular	6.66	6	0.30	-	-	2.0	-	-	-	-	-	-	-	-
Tetrahedral	4	4	-	0.388	0.390	-	1.55	1.56	-	0.425[d]	-	-	1.7	-
Simple cubic	6	6	-	0.257	0.254	-	1.54	1.52	-	0.307	0.32	-	1.842	1.92
Face-centred Cubic	12	12	-	0.119	0.125	-	1.43	1.50	-	0.195	0.208	-	2.34	2.496

[a]Chatzis and Dullien (1978)
[b]Sykes (1963)
[c]Dean (1963), Dean and Bird (1967)
[d]Domb and Dalton (1964)

practically impossible to distinguish between the sizes of windows which are penetrated over an infinitesimal range dp_b, whereas the approach in which the bonds are rendered open one-by-one presupposes that the sizes of all windows are distinct and different. Although it is theoretically possible to assign different sizes to all windows, physically speaking, in a macroscopic system a number of them is indistinguishable and will be assigned the same size. Hence, in a network of infinite size, instead of one bond, an infinite number of indistinguishable bonds will be opened simultaneously at breakthrough, resulting in a discontinuous increase of Y from zero to some finite value Y_{cb}. (It is helpful to think of dividing a network of infinite size into an infinite number of finite but large networks, each characterized by the same distribution of window sizes as the infinite network). The actual value of Y_{cb} found experimentally depends on the size of the step Δp_b, but it is always finite for it is impossible to proceed in infinitesimal steps dp_b.

In the absence of a theoretical method of calculating the entire Y_b vs p_b relationship for a network of infinite dimensions, one must be content with empiricism consisting of performing the same experiment with networks of the same type but of increasing sizes, until there is no more significant difference between the results obtained in two consecutive network sizes (see Fig. 14). With the exception of the virtually vertical portion of the curve, the value of Y_b obtained for a given p_b in a network of infinite size is independent of the path by which p_b is reached, i.e., whether each bond is opened individually, or all $p_b \times N_b$ bonds are opened at once, or whether they are opened in groups of m_i bonds per stage, such that $\Sigma_i\, m_i = p_b \times N_b$.

As is evident from Figure 4.14, after break-

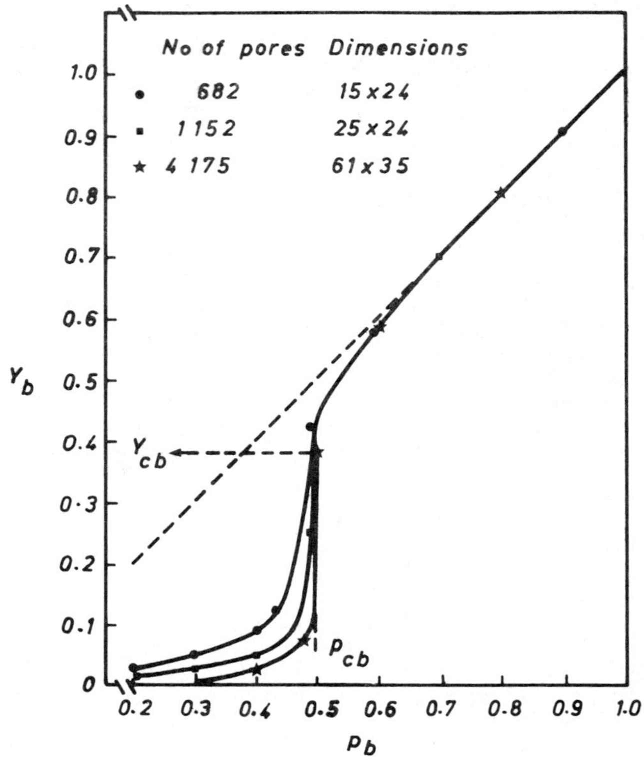

Figure 4.14. Y_b vs p_b curves for square networks of various sizes.[108]

through the Y_b vs p_b curve gradually becomes a straight line passing through the origin, i.e., $Y_b = p_b$, or all the open bonds are always penetrated by the fluid (mercury).

The percolation mechanism whereby first bonds at random locations in the network are rendered open, followed by identification of the open bonds with windows of various sizes, makes it easy to understand that the ratio $y_b = Y_b/p_b$, i.e., the fraction of the number of open bonds that are penetrated by the fluid (mercury), is the same value for windows of different sizes, and it depends only on the value of p_b.

As is evident from an inspection of Figure 4.12, certain parts of the penetrated network form dead-end channels, or communicate only through one bond with the rest of the penetrated network. The fluid contained in these parts of the network does not contribute to transport if a small potential difference is applied between the two faces of the network. The fraction of the number of penetrated bonds, q_b, belonging to this category is called "pseudodead-end bond fraction." The pseudodead-end fractions of windows of different sizes are identical, as q_b depends only on p_b. Before breakthrough $q_b = 1$, and its value decreases exponentially to zero for $p_b = 1$ (see Fig. 4.15).

It has been shown[7,108,109] that the Y_b vs. p_b curve of a network may be transformed into the mercury porosimetry curve of a sample if it may be assumed that the pore structure of the sample consists of a network of capillaries where the nodes are merely parts of the capillaries, the frequency distribution of capillary diameters is known, and reasonable assumptions may be made on the length-to-diameter ratios of the capillaries. It is evident that packings of particles, and most other porous media as well, very seldom conform to the above assumption. The void structure of packings of particles can be expected to be simulated much better by the bond-correlated site

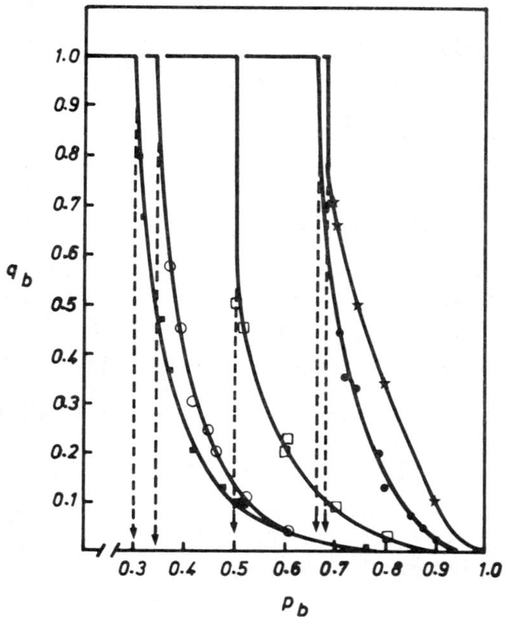

Figure 4.15. q_b versus p_b curves for two-dimensional networks (curves from left-to-right: tetra-triangular, hexa-triangular, square, hexagonal, di-tetragonal).[108]

percolation, the essence of which can be readily understood after having read the description of bond percolation presented above.

In *site percolation* the bonds do not exist or, depending on one's viewpoint, are all open, and sites of the network chosen at random are rendered open one-by-one. As the sites correspond to the nodes in a real network of capillaries, i.e., to the voids in a packing, site percolation cannot simulate the mercury intrusion porosimetry experiment the mechanism of which is controlled by the windows in the network.

In the case of *bond-correlated site percolation* sites are rendered open one-by-one in a random manner, and adjacent sites are assumed to be linked to each other by a bond which is, in effect, automatically opened when the smaller one of the two sites at its ends is open. In an experiment the numbers are counted of both the sites and the bonds such that are connected to the exposed face of the network via a pathway of open sites and bonds.

Bond-correlated site percolation may be used to simulate the mercury intrusion porosimetry experiment as follows.[109] A site which is rendered open first, is considered to represent a larger void than one that is rendered open subsequently. *The size of the window connecting two adjacent voids is considered to be an arbitrary fraction C of that of the smaller one of the two voids.*

Mercury penetration into a physical model built according to these specifications would proceed as follows. First, penetration takes place into the largest, then into the second largest void situated at the exposed face of the model, and so forth. It is possible that one of these voids has a neighbor void, not situated at the exposed face, which is even bigger. The size of the window between these two voids being a given fraction of the size of the void at the exposed face, this window will be penetrated first, after increasing the capillary pressure sufficiently. The void on the other side of this window, being bigger than the window, is also penetrated at the same pressure. If none of the penetrated voids at the exposed face has a bigger neighbor void, then it is the biggest of all voids adjacent to a penetrated void that must be considered. The window separating this void from the neighboring penetrated void has a size that is the same fraction of the size of the (smaller) empty void as in the case of all voids of the same size in the network. At this point of the experiment, this is the largest window which is accessible to mercury and it will be penetrated next, after increasing the capillary pressure to the necessary value, given by Eq. (29). The void on the other side of the window is also penetrated at the same pressure. Further penetration into the network consists of a repetition of the steps outlined above, and it is in all other respects identical to the case, discussed above in detail, of the penetration by bond percolation. *It is evident from the description of this experiment that penetration is controlled by the window sizes and, therefore, it may be used to simulate mercury intrusion porosimetry.*

In the case of bond-correlated site percolation, the fraction p_s of the total number N_s of sites is open and one enumerates the corresponding fractions Y_s of N_s and Y_b of N_b, such that are connected to the exposed face by a continuous path of open sites and bonds. Evidently, $p_b = p_s^2$.

Let $P_v(D_v)$ and $P_w(D_w)$, the cumulative distribution functions of void diameters D_v and window diameters D_w respectively, in a packing be assumed known. Then one can find D_v by letting $P_v(D_v) = p_s$, and D_w, by letting $P_w(D_w) = p_b = p_s^2$. Using D_w in Laplace's equation (Eq. 29), the corresponding value of the capillary pressure, P_c, can be calculated.

The calculation of the percentage of void space of the packing occupied by mercury, i.e., the saturation S at the capillary pressure P_c necessitates assumptions to be made about the shapes of both a window and a void. For example, one may assume that a window is a short tube, e.g., with a length-to-diameter ratio of unity, and of a triangular cross section. Thus, D_w is proportional to the size of the triangle. At the first penetration of a window only the core of the triangle D_w is filled, whereas the rest of the triangular tube is filled gradually afterwards as the capillary pressure is increased further. Similarly, it may be assumed that a void has, for instance, the shape of a regular tetrahedron and D_v is proportional to the size of tetrahedron. The fractions Y_s and Y_b may then be transformed into the customary saturation S of the packing by the following formula:

$$S = xy_sS_v^0 + (1 - x)y_bS_w^0 \tag{46}$$

where x is the percentage of interparticle space occupied by the (tetrahedral) voids, $(1 - x)$ is the percentage of this space occupied by the (triangular) windows, $y_s = Y_s/p_s$, and $y_b = Y_b/p_b$.

$$S_v^0 = \frac{\int_{D_v}^{D_{v,\max}} v(D_v, P_c)n(D_v)\,dD_v}{\int_{D_{v,\min}}^{D_{v,\max}} v(D_v)n(D_v)\,dD_v} \tag{47}$$

where $v(D_v)$ = volume of tetrahedron of "diameter" D_v, $v(D_v,P_c)$ = volume of mercury contained in a tetrahedron of diameter D_v at the capillary pressure P_c, and $n(D_v)$ = frequency of tetrahedra of diameter D_v, i.e.,

$$P_v(D_v) = \int_{D_v}^{D_{v,\max}} n(D_v)\,dD_v \tag{48}$$

Similarly,

$$S_w^0 = \frac{\int_{D_w}^{D_{w,\max}} v(D_w,P_c)n(D_w)\,dD_w}{\int_{D_{w,\min}}^{D_{w,\max}} v(D_w)n(D_w)\,dD_w} \tag{49}$$

where $v(D_w)$ = volume of triangular window of diameter D_w, $v(D_w, P_c)$ = volume of mercury contained in a triangular window of diameter D_w at capillary pressure P_c, and $n(D_w)$ = frequency of triangular windows of diameter D_w, i.e.,

$$P_w(D_w) = \int_{D_w}^{D_{w,\max}} n(D_w)\,dD_w \tag{50}$$

The calculations are performed analogously also for other, different assumed geometries of the voids and the windows.

Summing up, the mercury porosimetry curve, i.e., the relationship S versus P_c, of packing of particles may be calculated from the relationships Y_s vs p_s and Y_b vs p_b (see Fig. 4.16), if the size distributions, $n(D_v)$ and $n(D_w)$ are known or may be assumed. The void size distribution, $n(D_v)$, may be determined by optical methods, using sections through samples of the packing, as described in the next section.

Optical Methods Using Sections (Photomicrography). There is no obvious way of placing in every void of a packing a largest fitting sphere, whose size distribution $N(D)$ might serve as a measure of the void size distribution. It is worth noting however, that were there a way to do this, easy and rapid measurements performed on sections through the sample

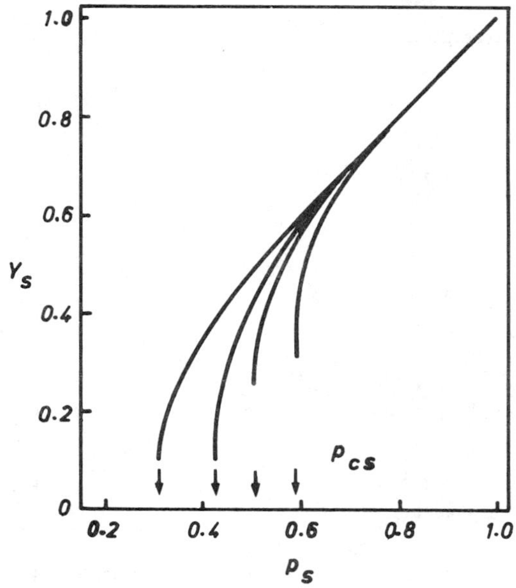

Figure 4.16. Y_s vs. p_s curves of bond-correlated networks after breakthrough (curves from left-to-right: cubic, tetrahedral, triangular, square).[108]

would provide the data from which the sphere size distribution $N(D)$ could be calculated *exactly*, using the methods of *quantitative stereology*[49,111] from the following integral equation:

$$n(\delta)\, d\delta = \delta d\delta \int_{D=\delta}^{D_{\max}} \frac{N(D)\, dD}{(D^2 - \delta^2)^{1/2}} \quad (51)$$

Here, $n(\delta)d\delta$ is the number of circles per unit of the plane of section with diameters between δ and $\delta + d\delta$. There are several approximate solutions of Eq. (51) described in the literature.

The ideally desirable method not being possible, the next best thing to do, following a suggestion by Rumpf[48] is to regard the interparticle space as a packing of irregular void particles that are in contact at the windows and thus form a continuum. Indeed, a "pore cast" can be made by injecting Wood's metal or plastic into the interparticle space,[112-114] and subsequently dissolving the solid matrix of the porous medium. There is no obvious method, however, to separate such a pore cast into the individual void particles that could be sized by any one of a number of available methods of particle size analysis. In the case of nonspherical particles all these methods are only approximate, of necessity. They include also approximate methods of *quantitative stereology* which are based on measurements made on sections through or projected images of the particles.[115-116] These methods have been applied using information obtained from measurements made in polished sections through samples saturated with Wood's metal. In this case, the generalized form of Eq. (51), i.e.,

$$n(\delta)\, d\delta = d\delta \int_{D=\delta}^{D_{\max}} F(D,\delta) N(D)\, dD \quad (52)$$

must be solved. Eq. (52) presupposes that a definite geometrical model ("object") of voids is assumed and that all voids have the same shape. The longest chord of a section in the chosen direction of measurement is δ and D is the magnitude of the chord between two points farthest away from each other in the "object" in a specified direction (see Fig. 4.17). $F(D,\delta)$ is a (shape) function that is uniquely determined by the shape of the object. It is defined as (see Fig. 4.17):

$$F(D,\delta) = \overline{\left(\frac{\partial z}{\partial \delta}\right)^{-}} - \overline{\left(\frac{\partial z}{\partial \delta}\right)^{+}} \quad (53)$$

where the average is taken over all possible orientations of the object. It is easier to obtain

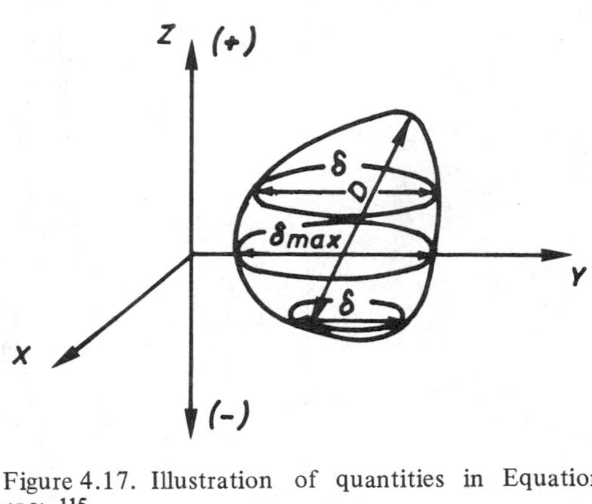

Figure 4.17. Illustration of quantities in Equation (52).[115]

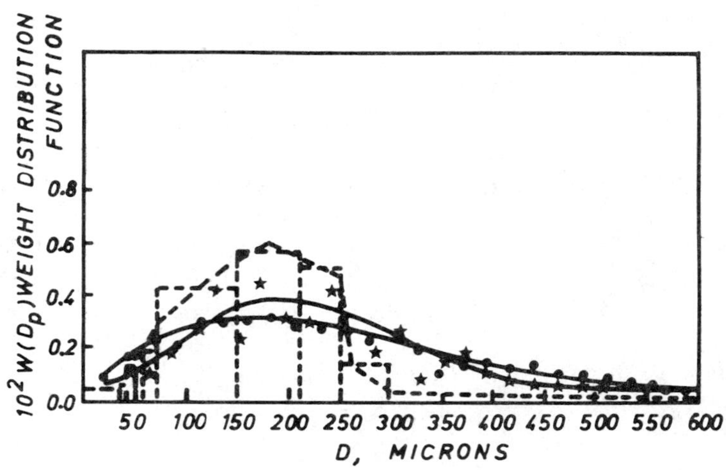

Figure 4.18. Particle size distributions obtained by sieve analysis (—), photo-micrography of sections of a packed bed (.) and of projection of a monolayer of particles suspended in oil (*).[112]

$F(D,\delta)$ by interchanging the averaging operation with the partial differentiation operation, i.e.,

$$\overline{\left(\frac{\partial z}{\partial \delta}\right)^-} = \left(\frac{\partial \overline{z}}{\partial \delta}\right)^-, \quad \overline{\left(\frac{\partial z}{\partial \delta}\right)^+} = \left(\frac{\partial \overline{z}}{\partial \delta}\right)^+ \tag{54}$$

which is permissible if the variation of z with δ is either monotone decreasing or monotone increasing. It is noted moreover that $\overline{z}(\delta,D_1)$ is similar to $\overline{z}(\delta,D_2)$, etc., because the objects of different sizes have the same shape.

Because in packings of particles the void particles touch each other, there is a certain, relatively low probability of regarding two (or

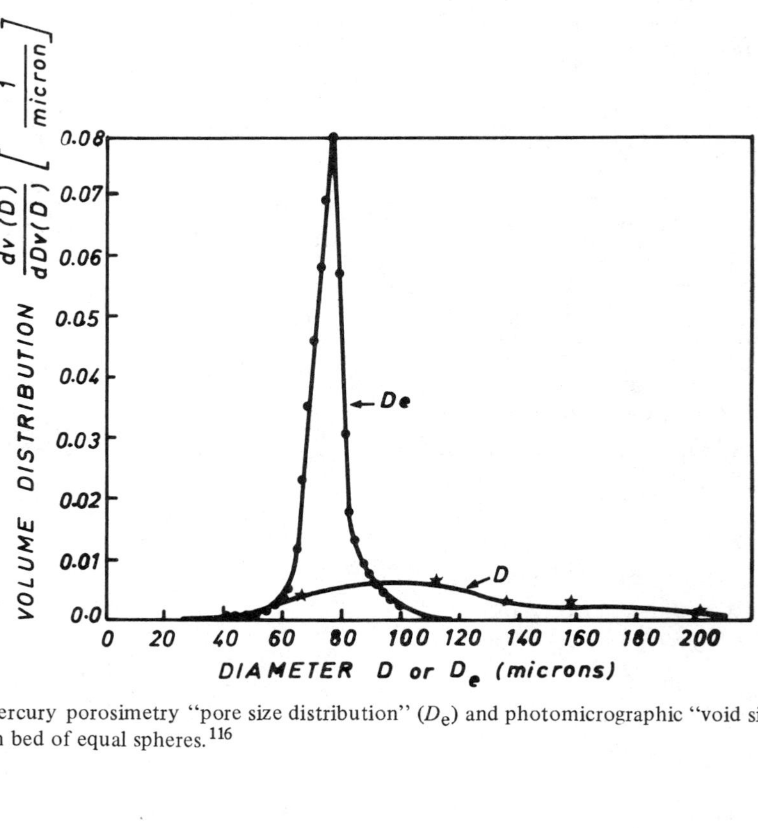

Figure 4.19. Mercury porosimetry "pore size distribution" (D_e) and photomicrographic "void size distribution" (D) of a random bed of equal spheres.[116]

more) adjacent void particles as a single void particle. Comparison of particle size distributions of the same common salt sample obtained (1) by sieve analysis, (2) by using sections through a packed bed of salt, and (3) by using projected images of the particles suspended in oil, are shown in Figure 4.18.

In Figure 4.19, the mercury porosimetry pore size distribution of a random pack of equal beads ($D_p = 250\ \mu m$) is compared with the void size distribution obtained from polished sections of the bed saturated with Wood's metal. In this case the deviation from spherical void shape was taken into account in the processing of the raw data, according to the method of Dullien and Dhawan.[113]

REFERENCES

1. H. Rumpf, and A. R. Gupte, "Einflüsse der Porosität and Korngrössenverteilung im Widerstandgesetz der Porenströmung," *Chem. Ing. Tech. 43*:367–375 (1971).
2. R. E. Collins, *Flow of Fluids through Porous Materials*, Van Nostrand Reinhold, New York (1961).
3. A. E. Scheidegger, *The Physics of Flow through Porous Media*, University of Toronto Press, Toronto, Canada (1974).
4. L. J. Klinkenberg, "The Permeability of Porous Media to Liquids and Gases," *A.P.I. Drill. Prod. Pract.*, pp. 200–213 (1941).
5. J. Bear, *Dynamics of Fluids in Porous Media*, American Elsevier, New York, (1972).
6. R. M. Barrer, *In The Solid-Gas Interface*, E. A. Flood (ed.), Vol. II, Dekker, New York, pp. 557–609 (1967).
7. F. A. L. Dullien, *Porous Media-Fluid Transport and Pore Structure*, Academic Press, New York (1979).
8. L. Green, and P. Duwez, "Fluid Flow through Porous Media," *J. Appl. Mech. 18*:39–45 (1951).
9. S. Ergun, "Fluid Flow through Packed Columns," *Chem. Eng. Progr. 48*:89–94 (1952).
10. N. Ahmed, and D. K. Sunada, "Nonlinear Flow in Porous Media," *J. Hyd. Div. Proc.* ASCE *95*(HY6):1847–1857 (1967).
11. P. C. Carman, "Fluid Flow through a Granular Bed," *Trans. Int. Chem. Eng. London 15*:150–156 (1937).
12. P. C. Carman, "Determination of the Specific Surface of Powders," *J. Soc. Chem. Ind. 57*:225–234 (1938).
13. P. C. Carman, *Flow of Gases through Porous Media*, Butterworths, London (1956).
14. J. Kozeny, "Concerning Capillary Conduction of Water in the Soil (Rise, Seepage and Use in Irrigation)," Royal Academy of Science, Vienna, *Proc. Class I 136*:271–306 (1927).
15. R. B. MacMullin, and G. A. Muccini, "Characteristics of Porous Beds and Structures," *AIChE J. 2*:393–403 (1956).
16. M. I. S. Azzam, and F. A. L. Dullien, "Application of Numerical Solution of the Complete Navier-Stokes Equation to Modelling the Permeability of Porous Media," *IEC Fundamentals 15*:281–285 (1976).
17. M. I. S. Azzam, and F. A. L. Dullien, "Flow in Tubes with Periodic Step Changes in Diameter: A Numerical Solution of the Navier-Stokes Equations," *Chem. Eng. Sci. 32*:1445–1455 (1977).
18. I. F. Macdonald, M. S. El-Sayed, K. Mow, and F. A. L. Dullien, "Flow through Porous Media—the Ergun Equation Revisited," *Ind. Eng. Chem. Fundamentals 18*:199–208 (1979).
19. A. R. Gupte, Dr.-Ing. Ph.D. dissertation, University of Karlsruhe, Germany (1970).
20. N. R. Morrow, "Physics and Thermodynamics of Capillary Action in Porous Media," *Ind. Eng. Chem. 62*:32–56 (1970). Reprinted as a chapter in *Flow through Porous Media*, American Chemical Society, Washington, D.C. (1970).
21. N. R. Morrow, and C. C. Harris, "Capillary Equilibrium in Porous Materials," *Soc. Petrol. Eng. J. 5*:15–24 (1965).
22. D. H. Everett, *The Solid-Gas Interface*," E. A. Flood (ed.), Vol. II, Dekker, New York, pp. 1055–1070, (1967).
23. F. A. L. Dullien, and G. K. Dhawan, "Bivariate Pore Size Distribution of Some Sandstones," *J. Interface Colloid Sci. 52*:129–135 (1975).
24. G. C. Topp, and E. E. Miller, "Hysteretic Moisture Characteristics and Hydraulic Conductivities for Glass-bead Media," *Soil Sci. Soc. Am. Proc. 30*:156–162 (1966).
25. A. Poulovassilis, "Hysteresis of Pore Water, an Application of the Concept of Independent Domains," *Soil Sci. 93*:405–412 (1962).
26. A. Poulovassilis, "Hysteresis of Pore Water in Granular Porous Bodies," *Soil Sci. 109*:5–12 (1970).
27. N. R. Morrow, and N. Mungan, "Mouillabilité et Capillarité en Milieux Poreux," *Revue I.F.P. 26*:629–650 (1971).
28. F. F. Craig, Jr., "The Reservoir Engineering Aspects of Water-Flooding," *Society of Petroleum Engineers of AIME*, Monograph, Vol. 3, Dallas, Texas (1971).
29. N. R. Morrow, "Capillary Pressure Correlations for Uniformly Wetted Porous Media," *J. Can. Pet. Technol. 15*:49–69 (1976).

30. J. M. Dumoré, and R. S. Schols, "Drainage Capillary Pressure Functions and the Influence of Connate Water," *Soc. Pet Eng. J. 14*:437–444 (1974).
31. M. Muskat, *Physical Principles of Oil Production*, McGraw-Hill, New York (1949).
32. W. O. Smith, "Minimum Capillary Rise in an Ideal Uniform Soil," *Physics 4*:184–193 (1933).
33. J. W. Amyx, D. M. Bass Jr., and R. L. Whiting, *Petroleum Reservoir Engineering*, McGraw-Hill, New York (1960).
34. M. S. El-Sayed, Ph.D. dissertation, University of Waterloo, Canada (1978).
35. E. E. Petersen, "Diffusion in a Pore of Varying Cross Section," *AIChE J. 4*:343–345 (1958).
36. A. S. Michaels, "Diffusion in a Pore of Irregular Cross Section: A Simplified Treatment," *AIChE J. 5*:270–271 (1959).
37. J. A. Currie, "Gaseous Diffusion in Porous Media, Part 2–Dry Granular Materials, *Brit. J. Appl. Phys. 11*:318–324 (1960).
38. F. A. L. Dullien, "Prediction of 'Tortusity Factors' from Pore Structure Data," *AIChE J. 21*:299, 820–822 (1975).
39. R. E. De La Rue, and C. W. Tobias, "On the Conductivity of Dispersions," *J. Electrochem. Soc. 106*:827–833 (1959).
40. M. R. J. Wyllie, *The Fundamentals of Electrical Log Interpretation*, Academic Press, New York (1957).
41. D. P. Haughey, and G. S. C. Beveridge, "Structural Properties of Packed Beds–A Review," *Can. J. Chem. Eng. 47*:130–140 (1969).
42. L. Fejes Toth, *Lagerungen in der Ebene, auf der Kugel und in Raum*, Springer, Berlin (1953).
43. J. Hrubisek, "Filtration–Geometry of Systems of Spheres," *Kolloid-Beihefte 53*:385–452 (1941).
44. L. C. Graton, and H. J. Frazer, "Systematic Packing of Spheres–with Particular Relation to Porosity and Permeability," *J. Geol. 43*:785–909 (1935).
45. L. K. Frevel, and L. J. Kressley, "Modifications in Mercury Porosimetry," *J. Anal. Chem. 35*:1492–1502 (1963).
46. R. P. Mayer, and R. A. Stowe, "Mercury Porosimetry–Breakthrough Pressure for Penetration Between Packed Spheres," *J. Coll. Sci. 20*:893–911 (1965).
47. R. P. Mayer, and R. A. Stowe, "Mercury Porosimetry: Filling of Toroidal Void Volume Following Breakthrough Between Packed Spheres," *J. Phys. Chem. 70*:3867–3873 (1966).
48. S. Debbas, and H. Rumpf, On the Randomness of Beds Packed with Spheres or Irregular Shaped Particles, *Chem. Eng. Sci. 21*:583–607 (1966).
49. E. E. Underwood, *Quantitative Stereology*, Addison-Wesley, Reading, Mass. (1970).
50. J. D. Bernal, J. Mason, and K. R. Knight, Letter, *Nature 194*:957–958 (1962).
51. G. D. Scott, "Radial Distribution of the Random Close Packing of Equal Spheres," *Nature 194*:956–957 (1962).
52. G. D. Scott, A. M. Charlesworth, and M. K. Mak, "On the Random Packing of Spheres, *J. Chem. Phys. 40*:611–612 (1964).
53. G. Mason, and W. Clark. "Distribution of Near Neighbours in a Random Packing of Spheres," *Nature 207*:512 (1965).
54. G. Mason, and W. Clark, "Fine Structure in the Radial Distribution Function From a Random Packing of Spheres," *Nature 211*:957 (1966).
55. W. O. Smith, P. D. Foote, and P. F. Busang, "Capillary Rise in Sands of Uniform Spherical Grains," *Physics 1*:18–26 (1931).
56. F. B. Hill, and R. H. Wilhelm, "Radiative and Conductive Heat Transfer in a Quiescent Gas–Solid Bed of Particles: Theory and Experiment," *AIChE J. 5*:486–496 (1959).
57. W. O. Smith, P. D. Foote, and P. F. Busang, "Capillary Rise in Sands of Uniform Spherical Grains," *Phys. Ref. 34*:1271 (1921).
58. E. Manegold, R. Hoffmann, and K. Solf, "Capillary Systems–XII Mathematical Treatment Ideal Sphere Packing and the Free Space of Actual Frame-like Structures," *Kolloid-Z 56*:142–159 (1931).
59. E. Manegold, and W. VonEngelhardt, "Capillary Systems XII (2). The Calculation of Content of Substance of Homogeneous Frame-like Structures. I. Sphere Planes and Sphere Layers as Structure Elements of Sphere Lattices," *Kolloid-Z 62*:285–294 (1933).
60. J. D. Bernal, and J. Mason, "Co-ordination of Randomly Packed Spheres," *Nature 188*:910–911 (1960).
61. W. H. Wade, "The Co-ordination Number of Small Spheres," *J. Phys. Chem. 69*:322–326 (1965).
62. K. Ridgway, and K. J. Tarbuck, "The Random Packing of Spheres," *Brit. Chem. Eng. 12*:384–388 (1967).
63. W. E. Ranz, "Friction and Transfer Coefficients for Single Particles and Packed Beds," *Chem. Eng. Progr. 48*:247–253 (1952).
64. H. Susskind, and W. Becker, "Random Packing of Spheres in Non-rigid Containers," *Nature 212*:1564–1565 (1966).
65. J. C. Macrae, and W. A. Gray, "Significance of the Properties of Materials in the Packing of Real Spherical Particles," *Brit. J. Appl. Phys. 12*:164–172 (1961).
66. J. D. Bernal, and J. L. Finney, "Random Packing of Spheres in Non-rigid Containers," *Nature 214*:265–266 (1967).
67. R. Rutgers, "Packing of Spheres," *Nature 193*:465–466 (1962).
68. G. D. Scott, "Packing of Equal Spheres," *Nature 188*:908–909 (1960).

69. G. Sonntag, "Einfluβ des Lückenvolumens auf den Druckverlust in Gasdurchströmten Füllkörpersäulen," *Chem. Ing. Tech. 32*:317–329 (1960).
70. R. Jeschar, "Pressure Loss in Multiple-Size Charges Consisting of Globular Particles," *Archiv. Eisenhüttenwesen*, p. 91 (1964).
71. W. H. Denton, "The Packing and Flow of Spheres," AERE-E/R-1095, Gt. Brit., Atomic Energy Research Establishment, Harwell, England, p. 22, (1953).
72. R. K. McGreary, "Mechanical Packing of Spherical Particles," *J. Am. Ceramic Soc. 44*:513–522 (1961).
73. J. Wadsworth, Nat. Res. Council of Canada, Mech. Eng. Report MT-41-(NRC N° 5895), (February, 1960).
74. J. Wadsworth, "An Experimental Investigation of the Local Packing and Heat Transfer Processes in Packed Beds of Homogeneous Spheres," *International Developments in Heat Transfer Proceedings*: American Society of Mechanical Engineers, pp. 760–769 (1963).
75. A. E. R. Westman, and H. R. Hugill, "The Packing of Particles," *J. Am. Ceram. Soc. 13*:767–779 (1930).
76. R. J. Parsick, S. C. Jones, and L. P. Hatch, "Stability of Randomly Packed Beds of Fuel Spheres," *Nucl. Appl. 2*:221–225 (1966).
77. O. K. Rice, "On the Statistical Mechanics of Liquids, and the Gas of Hard Elastic Spheres," *J. Chem. Phys. 12*:1–18 (1944).
78. N. Epstein, and M. J. Young, "Random Loose Packing of Binary Mixtures of Spheres," *Nature 196*, pp. 885–886 (1962).
79. S. Ergun, and A. A. Orning, "Fluid Flow Through Randomly Packed Columns and Fluidized Beds," *Ind. Eng. Chem. 41*:1179–1184 (1949).
80. D. E. Lamb, and R. H. Wilhelm, "Effects of Packed Bed Properties on Local Concentrations and Temperature Patterns," *I and EC Fund 2*:173–182 (1963).
81. H. H. Steinour, "Rate of Sedimentation, Nonflocculated Suspensions of Uniform Spheres," *Ind. Eng. Chem. 36*:618–624 (1944).
H. H. Steinour, "Rate of Sedimentation, Suspensions of Uniform-size Angular Particles," *Ind. Eng. Chem. 36*:840–847 (1944).
H. H. Steinour, "Rate of Sedimentation, Concentrated Flocculated Suspensions of Powders," *Ind. Eng. Chem. 36*:901–907 (1944).
82. J. Happel, "Pressure Drop Due to Vapor Flow Through Moving Beds," *Ind. Eng. Chem. 41*:1161–1174 (1949).
83. A. O. Oman, and K. M. Watson, "Pressure Drops in Granular Beds," *Natl. Petrol. News 36*:R795–802 (1944).
84. M. Rendell, J. Ramsey *Soc. Chem. Eng. 10*:31 (1963).
85. R. F. Benenati, and C. B. Brosilow, "Void Fraction Distribution in Beds of Spheres," *AIChE 8*:359–361 (1962).
86. L. H. S. Roblee, R. M. Baird, and J. W. Tierney, "Radial Porosity Variations in Packed Beds," *AIChE 4*:460–464 (1958).
87. G. Speck, Über die Randgängigkeit in Gasdurchströmten Füllkörperschüttungen, dissertation TH Dresden (1955).
88. J. T. Chiam, "Voidage and Fluid distribution in Packed Beds," Ph.D. thesis, Manchester College of Technology, 1962.
89. M. C. Thadani, and F. N. Pebbles, "Variation of Local Void Fraction in Randomly Packed Beds of Equal Spheres," *I and EC Process Design and Development 5*:265–268 (1966).
90. M. Kimura, K. Nono, and T Kaneda, "Distribution of Void in Packed Tubes, *Chem Eng.* (Japan) *19*:397–400 (1955).
91. C. C. Furnas, "Grading Aggregates I–Mathematical Relations for Beds of Broken Solids of Maximum Density," *Ind. Eng. Chem. 23*:1052–1058 (1931).
92. H. Y. Sohn, and C. Moreland, "The Effect of Particle Size Distribution on Packing Density," *Can. J. Chem. Eng. 46*:162–167 (1968).
93. H. T. Horsfield, "Strength of Asphalt Mixtures," *J. Soc. Chem. Ind. 53*:107–115T (1934).
94. B. S. Newman, *Flow Properties of Disperse Systems*, J. J. Hermans (ed.), *Interscience* Pub., Chapter 10 (1953).
95. J. C. Macrae, P. C. Finlayson, and W. A. Gray, "Vibration Packing of Dry Granular Solids," *Nature 179*:1365–1366 (1957).
96. J. Kolkuszewski, "Notes on the Deposition of Sand," *Research, Lond., 3*:478–483 (1950).
97. H. L. Ritter, and L. C. Drake, "Pore-size Distribution in Porous Materials. Pressure Porosimeter and Determination of Complete Macropore-size Distribution," *Ind. Eng. Chem. Anal. Ed. 17*:782–786 (1945).
98. S. Modry, M. Svata, and J. Jindra, "Bibliography on Mercury Porosimetry," in *House of Technology*, Prague, (1972).
99. G. Mason, "A model of the Pore Space in a Random Packing of Equal Spheres," *J. Colloids Interface Sci. 35*:279–287 (1971).
100. V. K. S. Shante, and S. Kirkpatrick, "Introduction to Percolation Theory," *Adv. Phys. 20*:325–357 (1971).
101. S. Kirkpatrick, "Percolation and Conduction," *Rev. Mod. Phys. 45*:574–588 (1973).
102. V. A. Vyssotoky, S. B. Gordon, H. L. Frisch, and J. H. Hammersley, "Critical Percolation Probabilities (Bond Problem)," *Phys. Rev. 123*:1566–1568 (1961).
103. M. F. Sykes, and J. W. Essam, "Some Exact Critical Percolation Probabilities for Bond and Site Problems in Two Dimensions," *Phys. Rev. Lett. 10*:3–4 (1963).

104. P. Dean, "A New Monte Carlo Method for Percolation Problems on a Lattice," *Proc. Camb. Phil. Soc. 59*:397–410 (1963).
105. P. Dean, and N. F. Bird, "Monte Carlo Estimates of Critical Percolation Probabilities," *Proc. Camb. Phil. Soc. 63*:477–479 (1967).
106. H. L. Frisch, J. M. Hammersley, and D. J. A. Wells, "Monte Carlo Estimates of Percolation Probabilities for Various Lattices," *Phys. Rev. 126*:946-951 (1962).
107. M. F. Sykes, D. S. Gaunt, and M. Glen, "Percolation Processes in Two Dimensions II, Critical Concentrations and the Mean Size Index," *J. Phys. A: Math Gen. 9*:97–103 (1976).
108. I. Chatzis, and F. A. L. Dullien, "Modelling Pore Structure by 2-D and 3-D Networks, with Applications to Sandstones," *J. Can. Pet. Technol. 16*:97–108 (1977).
109. I. Chatzis, and F. A. L. Dullien, "A Network Approach to Analyze and Model Capillary and Transport Phenomena in Porous Media," Symposium de l'A.I.R.H. sur les effets d'échelle en milieu poreux, Thessalonilea, Aug. 29–Sept. 1, 1978, 1.1–1.22.
110. P. G. De Gennes, and E. Guyon, "Lois Générales pour l'Injection d'un Fluide dans un Milieu Poreux Aléatoire," *J. de Mecanique 17*:403 (1978).
111. F. A. L. Dullien, E. Rhodes, and S. Schroeter, "Comparative Testing of Some Statistical Methods for Obtaining Particle Size Distributions," *Powder Technol. 3*:24–35 (1969–70).
112. F. A. L. Dullien, and P. N. Mehta, "Particle Size and Pore Size Distributions Determination by Photomicrographic Methods," *Powder Technol. 5*:179–193 (1971–72).
113. F. A. L. Dullien, and G. K. Dhawan, "Characterization of Pore Structure by a Combination of Quantitative Photomicrography and Mercury Porosimetry," *J. Interface and Colloid Sci. 47*:337–349 (1974).
114. N. C. Wardlaw, and R. P. Taylor, "Mercury Capillary Pressure Curves and the Interpretation of Pore Structure and Capillary Behaviour in Reservoir Rocks," *Bull. Can. Pet. Geol. 24*:225–262 (1976).
115. K. S. Chang, and F. A. L. Dullien, "Section Diameter Method for Size Distribution Determination of Non-Spherical Objects," *J. Microsc. 108*:801–808 (1976).
116. F. A. L. Dullien, and G. K. Dhawan, "Photomicrographic Size Distribution Determination of Non-spherical Objects," *Powder Technol. 7*:305–313 (1973).
117. G. E. Archie, "The electrical resistivity log as an aid in determining some reservoir characteristics," *Trans. AIME 146*:54–62 (1942).

5
Fundamental Properties of Powders

PART 1. RHEOLOGICAL PROPERTY OF PARTICULATE SOLIDS

Kunio Shinohara
Department of Chemical Process Engineering, Hokkaido University, Sapporo

There are several kinds of bulk properties of powders such as mechanical, thermal, electrical, magnetic, optical, acoustic, and surface physicochemical ones. Among them, the rheological property of particles is widely investigated in the applied fields. It is closely related not only to the material properties of a single particle but also to the unit operations in powder technology. Included here are the most fundamental characteristics of deformation and flow of particulate solids, that is to say, packing and strength, mainly in the dry system. Thus, the bulk properties are essential for describing various rheological behaviors in powder-handling processes, and they lie within the scope of current topics in the field of particle science.

5.1.1. PACKING CHARACTERISTICS OF PARTICLES

Each single particle has its own physical properties such as size, density, shape, restitution, etc., as mentioned in earlier chapters. Though powder consists of a number of individual particles, the bulk property of powder is not usually the simple summation of the physical properties of a single particle.

General characterization of the particle is not established yet, and it is difficult to prescribe the location of each particle under the influence of external and/or self-exerting forces in the open as well as closed systems. Information is, however, available on the packing structure of particle assemblage as a basis of the rheological properties of particles.[1]

Representative Parameters of Packing[2]

There are some fundamental and useful representations of the overall state of packing: Void fraction or fractional voidage ϵ is defined as the interstitial void volume in the unit bulk volume of particle assemblage.

Packing density or fractional solids content ϕ_p is then written by the void fraction as

$$\phi_p = 1 - \epsilon \qquad (5.1.1)$$

Bulk density ρ_b is the apparent density of powder mass, given by the weight per bulk volume of powder. Thus the following relation holds among the void fraction, the particle density ρ_p, and the bulk density.

$$\rho_b = \rho_p(1 - \epsilon) \tag{5.1.2}$$

Apparent specific volume V_s is the bulk volume of powder of unit weight, which is the inverse of bulk density as

$$V_s = \frac{1}{\rho_b} = \frac{1}{\rho_p(1 - \epsilon)} \tag{5.1.3}$$

Bulkiness ϕ_b indicates the bulk volume of solids in comparison with the unit volume of particles alone, that is, the inverse of packing density as

$$\phi_b = \frac{1}{\phi_p} = \frac{1}{1 - \epsilon} \tag{5.1.4}$$

Void ratio ϕ_v is written as the ratio of the voids volume to the net volume of particles.

$$\phi_v = \frac{\epsilon}{1 - \epsilon} \tag{5.1.5}$$

The coordination number N_c, which is defined as the number of contact points per each particle, is often referred to as relating to the rheological behaviors of particle assemblage.

Specific surface is the surface area of particles per unit weight S_w, per net volume of solids S_v, or per apparent volume of powder mass S_{av}. The relationship among them are

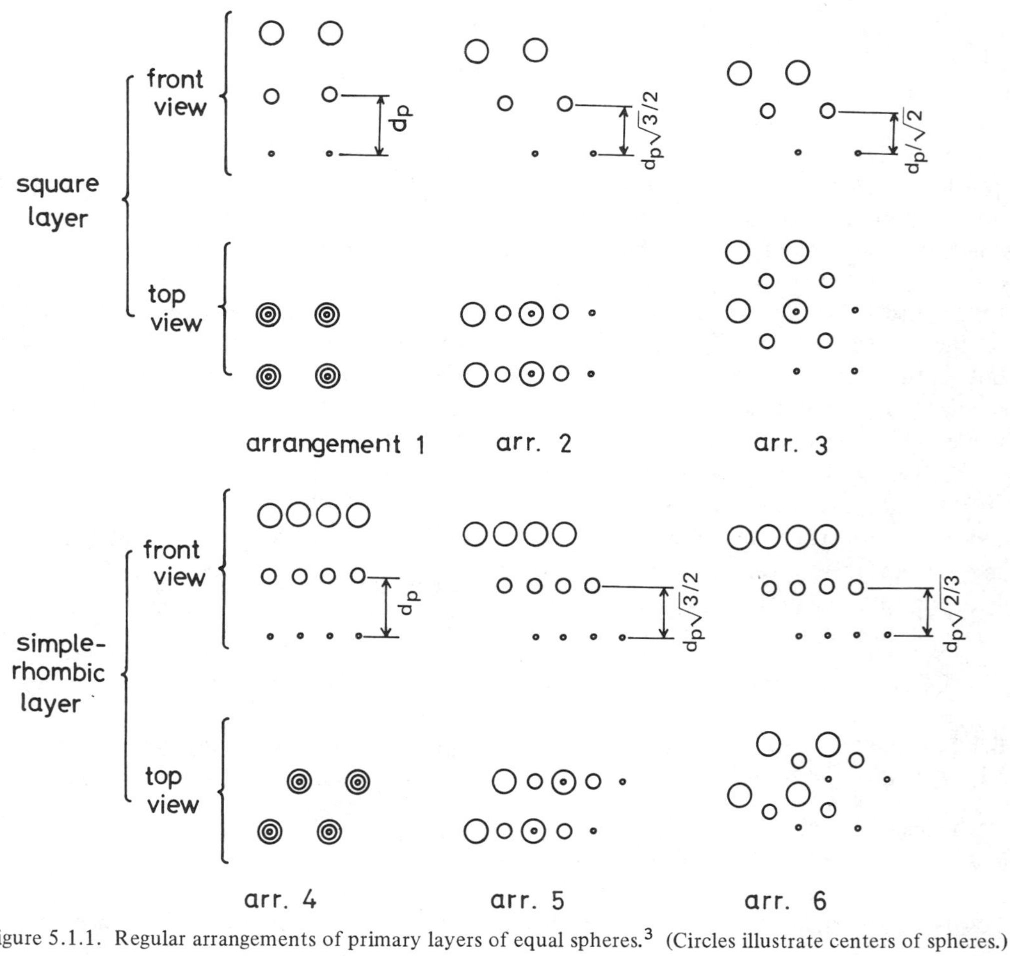

Figure 5.1.1. Regular arrangements of primary layers of equal spheres.[3] (Circles illustrate centers of spheres.)

$$S_w = \frac{S_v}{\rho_p} = \frac{S_{av}}{\rho_p(1-\epsilon)} = \frac{\phi_s}{\rho_p \cdot d_{p,av}} \tag{5.1.6}$$

where ϕ_s is the shape factor, and $d_{p,av}$ is the average particle diameter based on the specific surface.

Tortuosity is defined as the ratio of the length of the hypothetical curved capillary consisting of voids to the thickness of powder layer.

Other ways of representation are also available in connection with the distribution of voids, as mentioned later in the section on the random packing of equal spheres.

Regular Packing of Spheres

Though only limited cases are known of regular packing of spheres, the geometrical arrangements will be the basis of understanding the general state of packing of particles.

Packing of Equal-Sized Spheres. There are two kinds of primary layers of equal spheres, that is, square and simple rhombic layers. Four central points of spheres form the square arrangement on the same plane, and three centers of spheres in contact make a triangle, respectively. Thus, six variations are considered as stable geometrical arrangements in placing such a primary layer on the bottom one, as shown in Figure 5.1.1.[3] It is possible to introduce the concept of unit cell for these arrangements, as shown in Figure 5.1.2.[3] The corresponding packing characteristics are listed in Table 5.1.1. In fact, except for direction of the arrangement of voids, the second and the third type of arrangements in the square layer are the same as the fourth and the sixth ones in the rhombic layer, respectively.

It could be inferred in general that the void fraction decreases and the coordination number increases with increasing degree of deformation along four different kinds of regular arrangements of equal spheres.

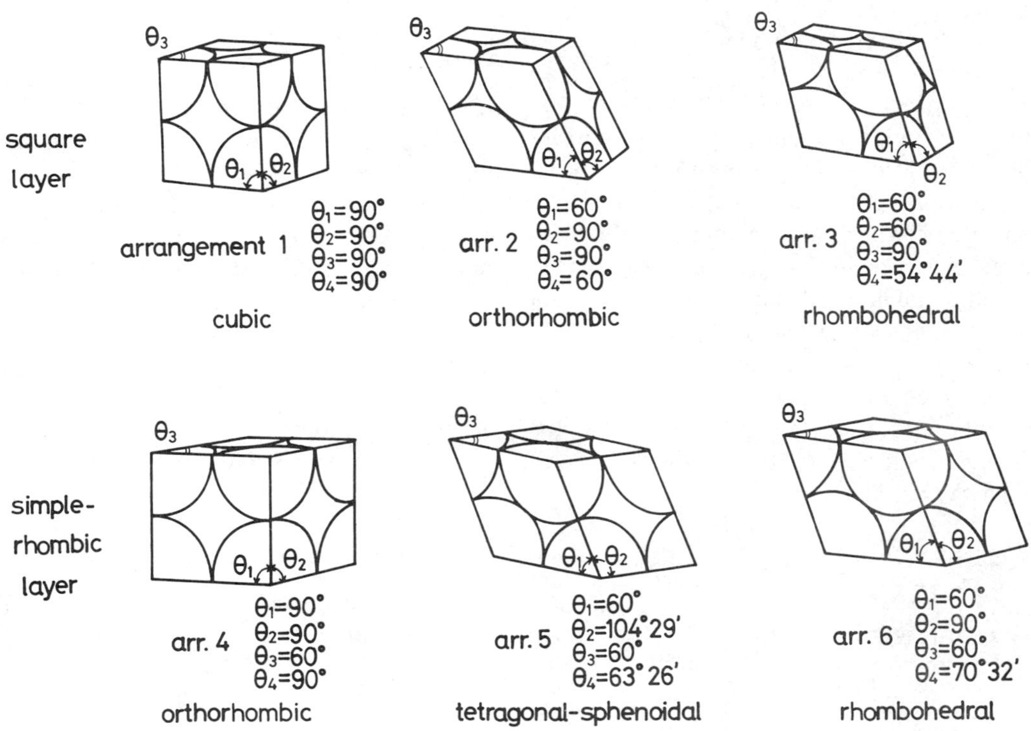

Figure 5.1.2. Unit cells of regular packing of equal spheres.[3] (θ_4 indicates inclined angle of right-side plane to horizontal one.)

Table 5.1.1. Packing Properties of Unit Cell.[3]

ARRANGEMENT NO.	BULK VOLUME	VOID VOLUME	VOID FRACTION	COORDINATION NO.
1	1	0.4764	0.4764	6
2	$\sqrt{3}/2$	0.3424	0.3954	8
3	$1/\sqrt{2}$	0.1834	0.2594	12
4	$\sqrt{3}/2$	0.3424	0.3954	8
5	3/4	0.2264	0.3019	10
6	$1/\sqrt{2}$	0.1834	0.2595	12

It is interesting to know that ellipsoidal particles yield the same void fractions for the stable and the unstable packing arrangements as those of the closest and loosest of the spheres, respectively.[4]

Packing of Different-Sized Spheres. Interspaces among equal spheres in the regular packing can ideally be filled with smaller spheres to attain higher density of the assemblage.

When only one small sphere is placed in each void, the diameter of which is the maximum to fit the void space, the packing characteristics are given in Table 5.1.2.[5]

After the square hole among six equal spheres in the rhombohedral arrangement is filled with one secondary sphere of maximum size, the triangular hole surrounded by four primary spheres is occupied by the third largest sphere. Furthermore, the fourth and the fifth largest spheres are packed into the interspaces between the secondary and primary spheres and between the third and primary spheres, respectively. All the remaining voids are finally filled up with considerably smaller spheres of equal size in the closest rhombohedral packing. Table 5.1.3 presents the characteristics indicating the minimum voidage of 0.039. This is so-called Horsfield packing.[6,7]

When more than one equal sphere are filled into interspaces of the closest rhombic arrangement, the void fraction varies with the size ratio of the smaller sphere to the primary one, as tabulated in Table 5.1.4. It appears that the void fraction decreases with each increase in the number of smaller spheres in the square hole, but this is not always true because of the intermittent number of the spheres in the triangular hole. A void fraction of 0.1130 is the minimum at the size ratio of 0.1716 on the basis of the triangular hole. This arrangement is called Hudson packing.[8]

Random Packing of Equal Spheres

Even with spheres of equal size the geometrical structure of random packing deviates far from that of regular packing. In other words, the characteristics of random packing are rather closer to the actual ones. Thus, the computer simulation of random packing is getting popular, as mentioned below.

Overall Packing Characteristics. In reality the randomness is always associated with the effects of particle properties, ways of filling, the dimension of the container, and its wall surface property.

Table 5.1.2. Packing Properties of Mixture with One Largest Sphere in Each Void.[5]

ARRANGEMENT	VOIDAGE	DIAMETER OF SMALL SPHERE	VOIDAGE OF MIXTURE	VOLUME RATIO OF SMALL SPHERE
cubic	0.4764	$0.723d_p$	0.271	0.391
orthorhombic	0.3954	$0.528d_p$	0.307	0.147
rhombohedral	0.2595	$0.225d_p$	0.190	0.019
		$0.414d_p$		0.070

Table 5.1.3. Properties of Horsfield Packing.[6,7]

SPHERES	SIZE RATIO	NUMBER OF SPHERES	VOIDAGE OF MIXTURE
Primary	1.0	–	0.260
Secondary	0.414	1	0.207
Ternary	0.225	2	0.190
Quaternary	0.175	8	0.158
Quinary	0.117	8	0.149
Filler	fines	many	0.039

By the usual packing experiments under gravity alone, the overall void fraction is approximately 0.39 and the coordination number is around 8 for relatively large spheres such as steel balls, round sand,[9] and glass beads.[5]

When spherical particles of about 3 mm diameter are poured without free falling, the datum value of voidage for the loose packing ranged from 0.393 to 0.409 for different particle density and surface friction.[10]

Without wall effects the void fractions in the loose and close packings were 0.399 and 0.363, respectively, which were extrapolated by filling dimpled copper cylinders of various heights and diameters. For the dense packing, steel balls of 3.18 mm were gently shaken down for about two minutes. Nonrigid balloons were also filled with the balls to confirm $\epsilon = 0.37$ for dense random packing with small peripheral error.[11]

On tapping vertically the same steel balls in the glass cylinder, the void fraction becomes 0.387 after 400 taps, which is close to that of the orthorhombic packing, whereas three-dimensional vigorous and prolonged shaking yielded the hexagonal close packing, whose void fraction is 0.26.[12] The structure was examined by removal of layers and arrays of balls and individual balls frozen in water as the thawing progressed.

According to the experiments with spherical lead shots of diameter 7.56 mm poured into a beaker,[13] the relationship between the average coordination and the voidage could be derived by assuming that the state of packing is represented by the mixture of the cubic and the rhombohedral packing in between the two.[14] The void fraction is written by using the fraction of the rhombohedral packing, R_r, as

$$\epsilon = 0.2595R_r + 0.4764(1 - R_r) \quad (5.1.7)$$

The average coordination number N_c is then given by

$$N_c = \frac{12\sqrt{2}\,R_r + 6(1 - R_r)}{\sqrt{2}\,R_r + (1 - R_r)} \quad (5.1.8)$$

because the volumes of unit cells for cubic and rhombohedral packings are 1 and $1/\sqrt{2}$, the numbers of spheres per unit volume are 1 and $\sqrt{2}$, and the coordination numbers are 6 and 12, respectively. Thus, eliminating R_r in Eqs. (5.1.7) and (5.1.8) leads to ϵ vs N_c as

$$\epsilon = \frac{0.414N_c - 6.527}{0.414N_c - 10.968} \quad (5.1.9)$$

The packing density of spherical particles, $1 - \epsilon$, increases with the diameter ratio of the vessel to the sphere up to about 10, over which the packing density gets nearly constant, 0.62.[15,16]

Purely by data correlation, the parabolic curve was found to fit well in the wide range of void fraction.[17]

$$\epsilon = 1.072 - 0.1193N_c + 0.00431N_c^2 \quad (5.1.10)$$

In the range of void fraction between 0.259 and 0.5 a model gives[18]

$$N_c = 22.47 - 39.39\epsilon \quad (5.1.11)$$

Table 5.1.4. Properties of Hudson Packing.[8]

SYMMETRICAL PACKS GOVERNED BY DIMENSIONS OF THE SQUARE INTERSTICE

SQUARE INTERSTICE					TRIANGULAR INTERSTICE		
$n_{p,s}$	STOICHIOMETRY	PACKING	d_{ps}/d_p	DENSITY INCREMENT	$n_{p,s}$	DENSITY INCREMENT	TOTAL DENSITY INCREMENT
1	Simple cube	tight	0.4142	0.07106	0	–	0.07106
2	Along cube diagonal	tight	0.2753	0.04170	0	–	0.04170
4	Crossed parallel face diagonals	tight	0.2583	0.06896	0	–	0.06896
6	Centers of cube faces	tight	0.1716	0.03028	4	0.04038	0.07066
8	Simple cube	tight	0.2288	0.09590	0	–	0.09590
9	Body-centered cube	tight	0.2166	0.09150	1	0.02034	0.11184
14	Face-centered cube	slack	0.1716	0.07074	4	0.04042	0.11116
16	Concentric simple cubes	slack	0.1693	0.07768	4	0.03882	0.11647
17	Concentric cubes, body-centered	slack	0.1652	0.07660	4	0.03605	0.11265
21	Hopper-faced cube, body-centered	slack	0.1782	0.11892	1	0.01132	0.13025
26	Hopper-faced cube, face-centered	slack	0.1547	0.09626	4	0.02962	0.12588
27	Simple cube	slack	0.13807	0.07108	5	0.02632	0.09740

SYMMETRICAL PACKS GOVERNED BY DIMENSIONS OF THE TRIANGULAR INTERSTICE

SQUARE INTERSTICE				TRIANGULAR INTERSTICE					
$n_{p,s}$	STOICHIOMETRY	PACKING	DENSITY INCREMENT	$n_{p,s}$	STOICHIOMETRY	PACKING	$d_{p,s}/d_p$	DENSITY INCREMENT	TOTAL DENSITY INCREMENT
8	Simple cubic	slack	0.09083	1	single	tight	0.22475	0.02271	0.11354
21	Hopper-faced cube, body centered	slack	0.10611	4	tetrahedral	tight	0.1716	0.04042	0.14653
26	Hopper-faced cube, face centered	slack	0.07457	5	body-centered tetrahedral	tight	0.14208	0.02868	0.10325

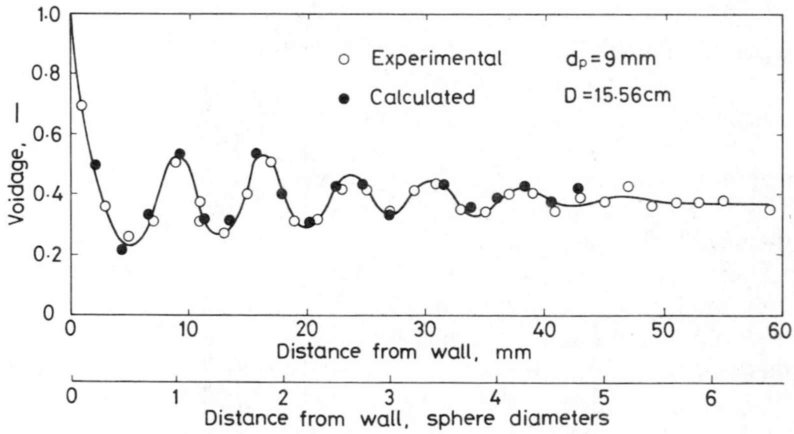

Figure 5.1.3. Voidage variation for randomly packed spheres in a cylinder.[23]

Local Packing Characteristics. The point of contact between rigid spheres is classified into close and near contacts, which are distinguished by a black paint ring with a clear center and a black spot, respectively. In the random close packing with the particle fraction of 0.62 the model coordination number lies between 8 and 9, the largest number having between 6 and 7 close contacts and between 1 and 2 near contacts. The means are 8.5 total and 6.4 close contacts. In the random loose packing of ϕ_p = 0.6, the modes were between 7 and 8 for the total coordination and between 5 and 6 for the close contacts, the means of which were 7.1 total and 5.5 close contacts.[19]

The local voidage of randomly packed spheres fluctuates in the vessel, as illustrated in Figure 5.1.3, according to the results of various ways of measurements.[20-23] The cyclic damping curve coincides irrespective of the sphere size, taking the distance by the sphere diameters, and the wall effect disappears at the distance of

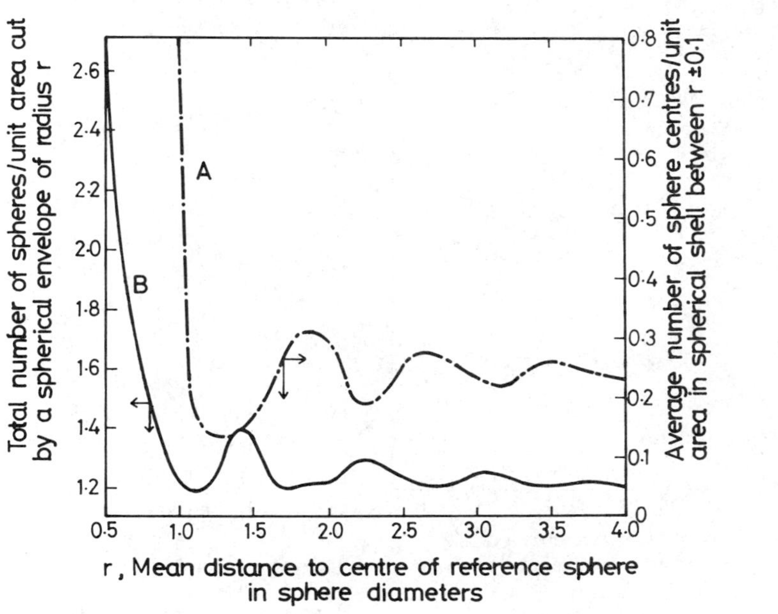

Figure 5.1.4. Variation of number of spheres in radial direction, A[18] and B.[24]

about 4 or 5. Similar fluctuation curves are also observed in the radial direction with respect to the average number of sphere centers per unit area in the spherical shell, curve A,[18] and the total number of spheres cut per unit area by a spherical envelope, curve B[24] (Fig. 5.1.4).

Based on the distribution function of the distance from a reference particle, the number of spheres within the spherical shell of differential thickness is theoretically calculated and compared with the packing data of an average number of neighboring particles, as shown in Figure 5.1.5.[25] The average number of spheres in contact is 6.0, irrespective of the packing density.[26,27]

Angular distribution of contacting spheres around a sphere could represent the packing structure of a randomly packed bed, as shown in Figure 5.1.6, which compares the experimental curve A[29] with the theoretical curve B.[30] Any given contact point is taken as a pole.

Analogously to the liquid structure,[31] the relationship between the coordination number and the packing density is approximated by

$$\phi_p = 0.1947N_c - 0.1301N_c^2 + 0.05872N_c^3 - 0.0128N_c^4 + 1.438 \times 10^{-3}N_c^5 - 8.058 \times 10^{-5}N_c^6 + 1.785 \times 10^{-6}N_c^7 \tag{5.1.12}$$

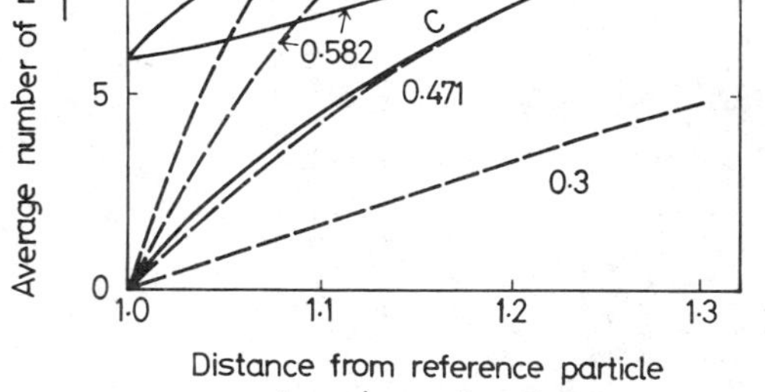

Figure 5.1.5. Average number of neighboring particles within a spherical shell.[25] (A,[26] B,[27] and C[28] are experimental)

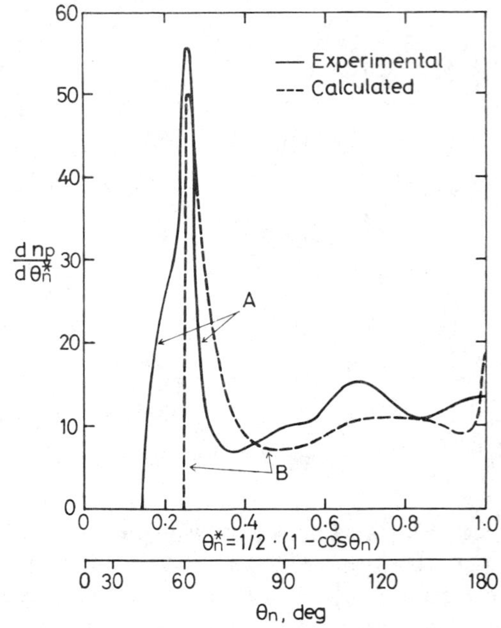

Figure 5.1.6. Angular distribution of contacting spheres around a sphere.[1] (A[29] is experimental, and B[30] is theoretical.)

with sufficient accuracy for $\phi_p \gtrsim 0.15$. The local mean packing-density distribution function f_p is derived on the basis of allocation of spheres to space cells according to a binomial probability mechanism[32]

$$f_p \, d\phi_{pl} = \frac{1}{\sqrt{2\pi}\zeta_p} \cdot \exp\left[-\frac{1}{2}\left(\frac{\phi_{pl} - \phi_p}{\zeta_p}\right)^2\right] d\phi_{pl} \tag{5.1.13}$$

where ϕ_{pl} and ϕ_p are the local mean and the bulk mean packing density, respectively. The standard deviation ζ_p becomes

$$\zeta_p = 4.75(0.7405 - \phi_p)^2 \cdot \phi_p / \sqrt{\pi/6} \tag{5.1.14}$$

Putting Eqs. (5.1.12) and (5.1.14) into Eq. (5.1.13) gives the distributions of the coordination number, as shown in Figure 5.1.7, which agree with the data.[19]

Packing of General Particles

Particles are not always spherical, packed regularly or perfectly at random. Thus, the

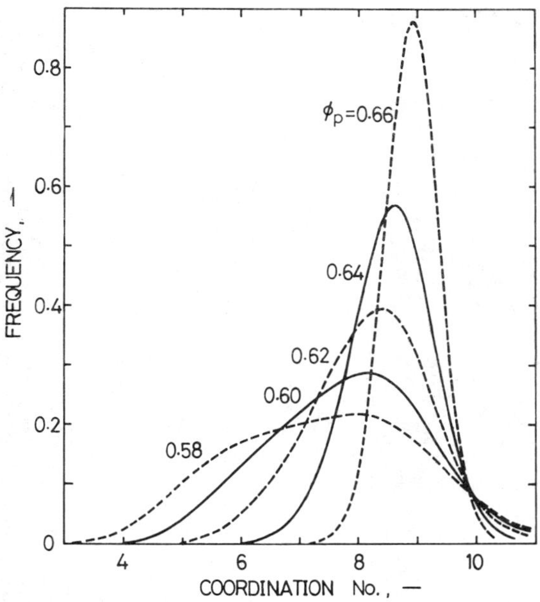

Figure 5.1.7. Distribution of coordination number.[31]

following packing characteristics are known to be useful in practice.

Overall Degree of Packing. The bulk density of general particles within the vessel decreases under gravity alone with decreasing diameter of the container and increasing height of the particles bed. Higher filling rate gives rise to smaller bulk density for coarse particles, but for fine cohesive powder like flour, dilute feeding yields the loose packing.[33]

The void fraction increases with decreasing sphericity in general, as shown in Figure 5.1.8.[34] Here the sphericity is defined as the ratio of the sphere surface to the irregular particle surface of the same volume. But it is also reported that angular particles have a larger void fraction in the loose packing, and it is inverse in the close packing.[35] Particles of higher surface roughness exhibit higher voidage, as shown in Figure 5.1.9.[35]

Smaller particles have higher voidage because of cohesion between particles, contrasting with the ideal independence of particle size of void fraction.[36,37] Thus, the apparent volume of moist powder grows larger with increasing water content.[38]

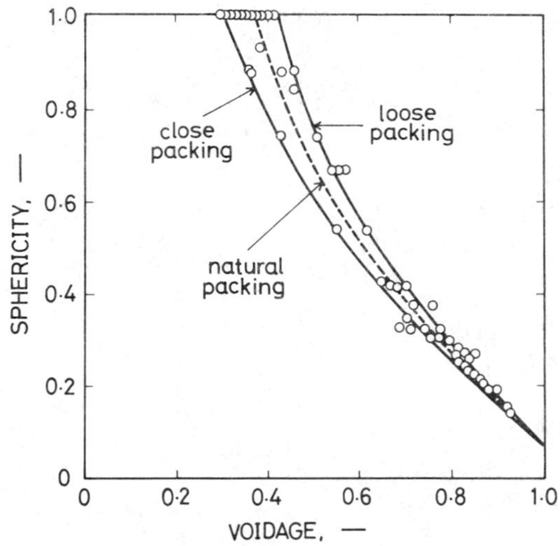

Figure 5.1.8. Relationship between void fraction and sphericity.[34]

Particles of distributed size tend to yield closer packing as spheres. But, in general, it is difficult to evaluate the void fraction corresponding to the mixing ratio and the particle-size ratio theoretically. Experimental relationship is available, as shown in Figure 5.1.10.[39]

Packing Model of Powder Mass.[40,41] In order to express uneven packing structure of general particles over the wide range of void fraction, a

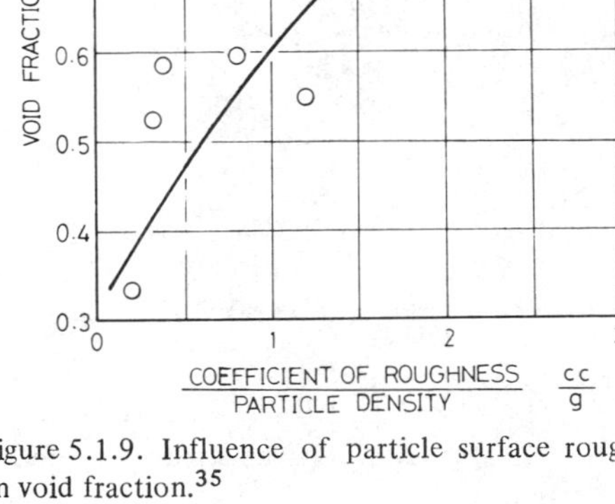

Figure 5.1.9. Influence of particle surface roughness on void fraction.[35]

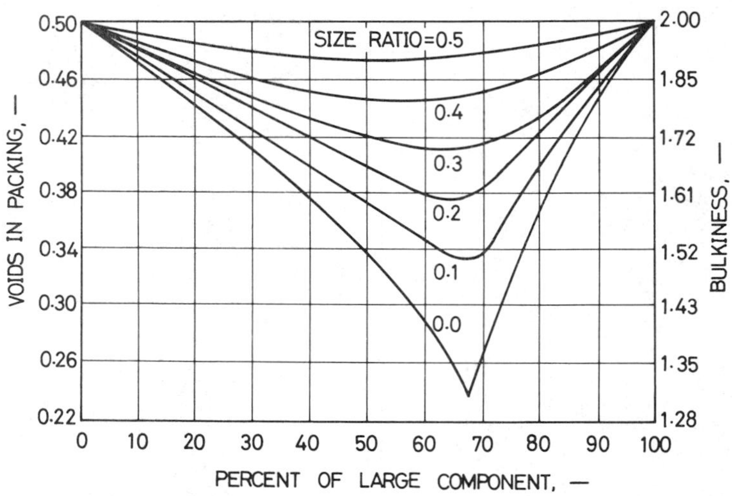

Figure 5.1.10. Relation between voids and size composition in two-component system of broken solids when the voids of single components are 0.5.[39]

mathematical model is proposed, (illustrated in Fig. 5.1.11). A unit cross-sectional area of powder mass is supposed to be composed of two kinds of regular packing portions of equivalent spheres, the cubic packing R_c and the rhombohedral packing R_r, and of the effective void ϵ_e that is independent of the particle packing portions. These three kinds of subdivisions are mixed at random over a certain cross-sectional area. The void fraction ϵ is then written as the sum of the voids by

$$\epsilon = 0.476R_c + 0.260R_r + \epsilon_e \quad (5.1.15)$$

Since

$$\epsilon_e = 1 - (R_c + R_r) \quad (5.1.16)$$

putting Eq. (5.1.16) into Eq. (5.1.15) gives

$$\epsilon = 1 - (0.524R_c + 0.740R_r) \quad (5.1.17)$$

The area ratio R of the loosest to the closest packing portion is assumed to increase with the void fraction along consolidation of powder mass by

$$\frac{dR}{d\epsilon} = \alpha R + \beta \quad (5.1.18)$$

where

$$R = R_c/R_r \quad (5.1.19)$$

and α and β are positive constants. Integrating Eq. (5.1.18) for the initial condition of $R = 0$ in $\epsilon = 0.260$, when the cubic packing portion disappears ($R_c = 0$) and all the portions are packed in the rhombohedral packing ($R_r = 1$), we find that

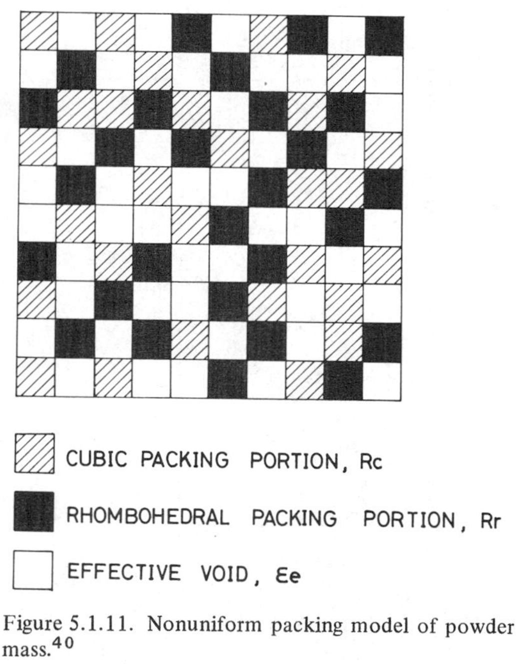

Figure 5.1.11. Nonuniform packing model of powder mass.[40]

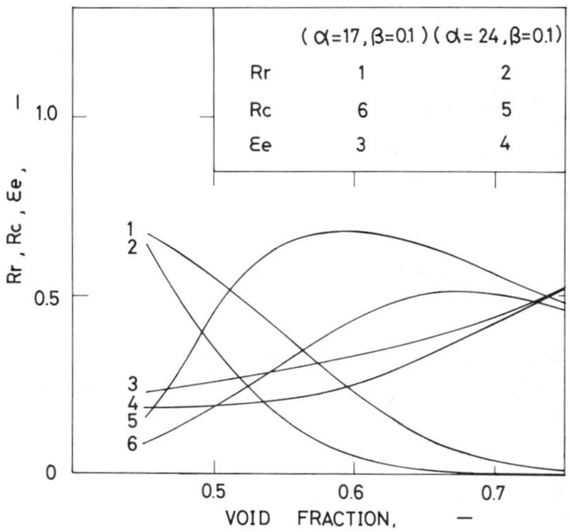

Figure 5.1.12. Variation of R_c, R_r, and ϵ_e with void fraction.[41]

$$R = \frac{\beta}{\alpha}\{e^{\alpha(\epsilon - 0.260)} - 1\} \qquad (5.1.20)$$

Hence, from Eqs. (5.1.16), (5.1.17), (5.1.19), and (5.1.20), each portion is obtained as a function of the void fraction, as depicted in Figure 5.1.12.

Figure 5.1.13 shows comparison of Eq. (5.1.24) with other empirical equations and data, including those for spheres mentioned before.

$$N_c\epsilon = \pi^{42} \qquad (5.1.25)$$

$$N_c = 19.3 - 28\epsilon^{43} \qquad (5.1.26)$$

$$N_c = 20.0(1 - \epsilon)^{1.7\ 41} \qquad (5.1.27)$$

Closest Packing of Different Sized Particles.[44] In the binary system of particles, interspaces among large particles are filled up with small ones to give the closest packing. The weights of large and small particles within the mixture of unit volume are written, respectively, as

$$W_l = 1 \cdot (1 - \epsilon_l) \cdot \rho_{pl} \qquad (5.1.28)$$

$$W_s = 1 \cdot \epsilon_l(1 - \epsilon_s) \cdot \rho_{ps} \qquad (5.1.29)$$

Thus, the weight fraction f_l of large particles is

$$f_l = \frac{W_l}{W_l + W_s} = \frac{(1 - \epsilon_l)\rho_{pl}}{(1 - \epsilon_l)\rho_{pl} + \epsilon_l(1 - \epsilon_s)\rho_{ps}} \qquad (5.1.30)$$

$$R_c = \frac{(1 - \epsilon)\{e^{\alpha(\epsilon - 0.260)} - 1\}}{0.524\{e^{\alpha(\epsilon - 0.260)} - 1\} + 0.740\alpha/\beta} \qquad (5.1.21)$$

$$R_r = \frac{(1 - \epsilon)\alpha/\beta}{0.524\{e^{\alpha(\epsilon - 0.260)} - 1\} + 0.740\alpha/\beta} \qquad (5.1.22)$$

$$\epsilon_e = \frac{(\epsilon - 0.476)\{e^{\alpha(\epsilon - 0.260)} - 1\} + (\epsilon - 0.260)\alpha/\beta}{0.524\{e^{\alpha(\epsilon - 0.260)} - 1\} + 0.740\alpha/\beta} \qquad (5.1.23)$$

These results are applied to analyses of shear and tensile strength, as mentioned later (see p. 155, 163).

Corresponding to the ideal maximum tensile strength, the relationship between the coordination and the voidage is obtained from the present packing model as

$$N_c = \frac{3}{2}\,\frac{e^{\alpha(\epsilon - 0.260)} - 1 + 3.99\alpha/\beta}{e^{\alpha(\epsilon - 0.260)} - 1 + 1.41\alpha/\beta} \qquad (5.1.24)$$

For the same solid material as the single component of equal voidage, that is, $\rho_{pl} = \rho_{ps}$ and $\epsilon_l = \epsilon_s = \epsilon$, the volume fraction of large particles becomes equal to f_l as

$$f_l = 1/(1 + \epsilon) \qquad (5.1.31)$$

where small particles should be completely involved in the matrix of large ones. Then the size ratio is lower than about 0.2.[8]

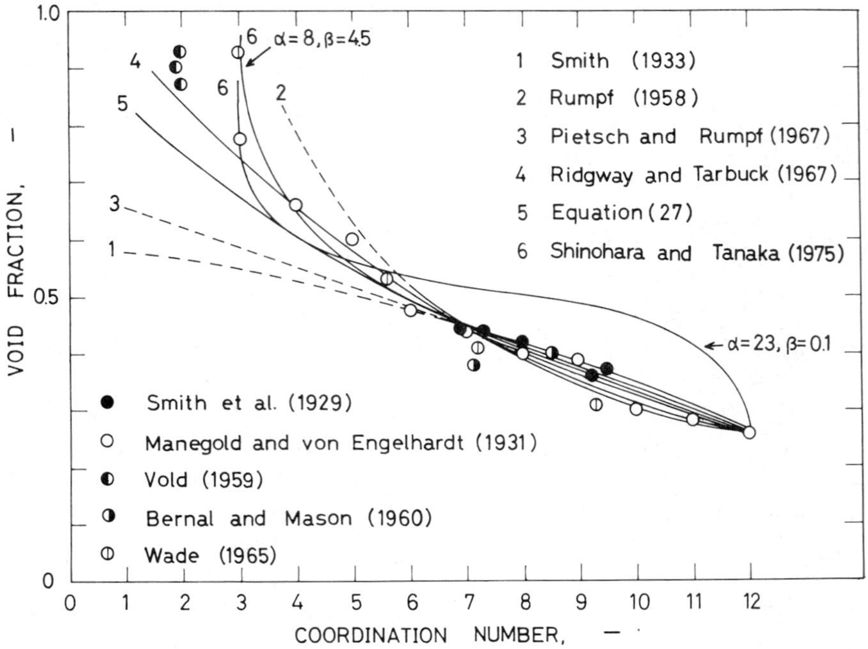

Figure 5.1.13. Relationship between coordination number and void fraction.[41]

In the multicomponent system of particles consisting of the same solid material, the interspaces among primary large particles are filled up with secondary small particles, the interspaces of which are packed by the tertiary small ones. Following the same way of packing by further smaller particles, the volume of each component V in the bulk volume of the mixture per unit particle volume, $V_m = 1/(1 - \epsilon^2)$, is given as

$$
\begin{aligned}
V_1 &= V_m(1 - \epsilon) = f_l \\
V_2 &= \epsilon V_m \cdot (1 - \epsilon) = 1 - f_l \\
V_3 &= \epsilon \cdot \epsilon V_m \cdot (1 - \epsilon) = (1 - f_l)\epsilon \\
V_4 &= \epsilon \cdot \epsilon^2 V_m \cdot (1 - \epsilon) = (1 - f_l)\epsilon^2 \\
&\vdots \\
V_n &= (1 - f_l)\epsilon^{n-2}
\end{aligned} \tag{5.1.32}
$$

where the sum of the particle volumes of the primary and the secondary particles are taken as unity just for computational convenience sake. Substituting Eq. (5.1.31) and summing up the volumes of all the components in Eq. (5.1.32) yields the following:

$$
\begin{aligned}
V_{ts} &= \frac{1}{1+\epsilon} + \left(1 - \frac{1}{1+\epsilon}\right) + \left(1 - \frac{1}{1+\epsilon}\right)\epsilon \\
&\quad + \left(1 - \frac{1}{1+\epsilon}\right)\epsilon^2 + \cdots \\
&\quad + \left(1 - \frac{1}{1+\epsilon}\right)\epsilon^{n-2} \\
&= \frac{1 - \epsilon^n}{1 - \epsilon^2}
\end{aligned} \tag{5.1.33}
$$

Hence, the volume fraction of each component is obtained by dividing $V_1, V_2, V_3, \ldots, V_n$ by V_{ts} Equation (5.1.33) is for the hypothetical case where each size acts as if it were infinitely small.

In the actual case of several different components uniformly mixed, the total volume of the mixed system V_{tm} is somewhat reduced as compared to the sum of the volume of separate layers of the components,

$$V_{tm} = \{V_{ts} - f_y(V_{ts} - f_l)\} \cdot (\rho_p/\rho_b) \tag{5.1.34}$$

where $f_y(V_{ts} - f_l)\rho_p/\rho_b$ indicates the volume decrease upon mixing and f_y is the factor ranging in value between 0 and 1.0 corresponding to separate layers of equal-sized particles and an ideal mixing with infinitely small particles. Thus, the bulk density of the mixed system is

$$\rho_{b,m} = \frac{\rho_b}{1 - \dfrac{f_y(V_{ts} - f_l)}{V_{ts}}} \tag{5.1.35}$$

For the closest packing or the maximum bulk density, the quantity $f_y(V_{ts} - f_l)/V_{ts}$ in the denominator should be maximum and thus differentiated with respect to the number of component sizes added to the original largest single size of the system, $n' = n - 1$.

$$\frac{f_l}{V_{ts}(V_{ts} - f_l)} \cdot \frac{dV_{ts}}{dn'} = -\frac{df_y}{f_y \cdot dn'} \tag{5.1.36}$$

Since the diameter ratio between particles of successive sizes for the maximum density is independent of the voidage and must be constant for the entire system,

$$\frac{d_{p,2}}{d_{p,1}} = \frac{d_{p,3}}{d_{p,2}} = \frac{d_{p,n'+1}}{d_{p,n'}} = K_s^{1/n'} \tag{5.1.37}$$

where K_s is defined as the size ratio of the smallest $d_{n'+1}$ to the largest particles, d_{pl}. Utilizing the experimental correlation between the total volume decrease f_y and the size ratios on binary systems,

$$f_y = 1.0 - 2.62K_s^{1/n'} + 1.62K_s^{2/n'} \tag{5.1.38}$$

Hence, differentiating Eqs. (5.1.33) and (5.1.38) with respect to n' and putting them into Eq. (5.1.36) gives the relationship among ϵ, K_s, and n', as

$$\frac{\epsilon^{n'} \cdot \ln \epsilon(1 - \epsilon)}{(1 - \epsilon^{n'+1})(1 - \epsilon^{n'})} = \frac{(2.62K_s^{1/n'} - 3.24K_s^{2/n'}) \ln K_s}{(1.0 - 2.62K_s^{1/n'} + 1.62K_s^{2/n'})n'^2} \tag{5.1.39}$$

According to the above equations, the minimum voidage is calculated from Eq. (5.1.35), for example, in the packings of two- to four-component systems, as shown in Figure 5.1.14 and listed in Table 5.1.5.

For varying voidages and particle densities, it is also possible to extend similar treatments.

$$W_t = f_1 + (1 - f_1) + (1 - f_1) \cdot \left(\frac{1 - f_2}{f_2}\right) + (1 - f_1)\left(\frac{1 - f_2}{f_2}\right)\left(\frac{1 - f_3}{f_3}\right) + \cdots \tag{5.1.40}$$

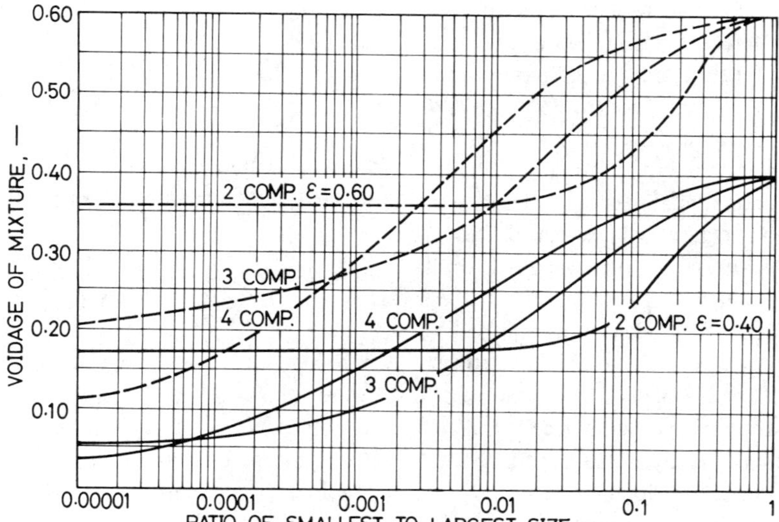

Figure 5.1.14. Minimum voidage for two to four component sizes for initial voidage of 0.40 and 0.60.[44]

Table 5.1.5. Composition of Packings for Minimum Voids.[44]

INITIAL VOIDAGE IN SINGLE COMPONENT	NUMBER OF COMPONENTS	VOLUME PERCENTAGE OF EACH COMPONENT			
		$d_{p,1}$	$d_{p,2}$	$d_{p,3}$	$d_{p,4}$
0.30	2	77.0	23.0	–	–
	3	72.0	21.5	6.5	–
	4	70.7	21.1	6.3	1.9
0.40	2	71.5	28.5	–	–
	3	64.2	25.6	10.2	–
	4	61.7	24.6	9.8	3.9
0.50	2	66.7	33.3	–	–
	3	57.2	28.5	14.3	–
	4	53.3	26.7	13.3	6.7
0.60	2	62.5	37.5	–	–
	3	51.0	30.6	18.4	–
	4	46.0	27.6	16.5	9.9

$$f_n = \frac{(1 - \epsilon_n)\rho_{p,n}}{(1 - \epsilon_n)\rho_{p,n} + \epsilon_n(1 - \epsilon_{n+1})\rho_{p,n+1}} \tag{5.1.41}$$

Compaction of Powders

As the essential characteristic that connects the packing structure with stress propagation within the powder mass, the compaction of powders has been extensively investigated for many years.[45]

From the operational viewpoint there are several kinds of compaction. Piston press and hydrostatic pressing are rather static ways of compaction, whereas tapping, vibration, hammering, and explosion belong to impact compaction. Other types of compaction are also available, such as roller press, vacuum pressing, multiaxis compression, and so forth.

The compaction will proceed along with the free-flowing region where aggregates of particles move mutually to reduce the bulk volume, the compaction region where the aggregates are broken to yield dense phase, the region where particles undergo plastic deformation, and the pure deformation region associated with the strain of crystal lattice.[46,47]

But no systematic analysis has generally been carried out to derive the detailed packing structure of particle assemblages that relate to the compaction pressure.

Variation of Density of Powder with Consolidating Pressure. Quite a few empirical equations are proposed regarding the piston press in one direction, as listed in Table 5.1.6.[48] These equations are rewritten and roughly classified into three types according to the differential variation of void fraction with respect to the compressive pressure.[49]

$$-\frac{d\epsilon}{dP} = c_n \cdot \epsilon^x \tag{5.1.42}$$

where $y = 1$ and $z \neq 1$ for Nutting's, and $y = 2$ equation.

$$\frac{d\epsilon}{dP} = c_n \cdot \frac{(1 - \epsilon)^y}{P^z} \tag{5.1.43}$$

where $y = 1$ and $Z \neq 1$ for Nutting's, and $y = 2$ and $z = 1$ for Terzaghi's equation.

$$-\frac{d\epsilon}{dP} = c_n \cdot \frac{\epsilon^x(1 - \epsilon)^y}{P^z} \tag{5.1.44}$$

where $x = 1, y = 1$ and $z = 0$ for Ballhausen's.

Some analytical equations lead to the relationship between the consolidating pressure and the volume of bulk solids: Based on the change in Mohr's circles before and after compaction, the difference in the angle of internal friction $\Delta\phi_i$ is derived as

Table 5.1.6. Various Equations on Compaction of Powders.[48]

Balshin	$\ln P = -c_1(V/V_P) + c_2$
Smith	$\frac{1}{V} - \frac{1}{V_0} = c_3 P^{1/3}$
Murray	$\ln\left(\frac{V}{V - V_P}\right) = c_4\left(\frac{V_P}{V - V_P}\right)^{1/3} + c_5 P$
Ballhausen	$\ln\left(\frac{V_P}{V - V_P}\right) = c_6 P + \ln c_7$
Konopicky	$\ln\left(\frac{V}{V_0 - V_P}\right) = c_8 P + \ln\left(\frac{V_0}{V_0 - V_P}\right)$
Jones	$\ln P = -c_9\left(\frac{V}{V_P}\right)^2 + c_{10}$
Athy	$\frac{V - V_P}{V} = \frac{V_0 - V_P}{V_0} e^{-c_{11}P}$
Nutting	$\ln\left(\frac{V_0}{V}\right) = c_{12} P^{c_{13}}$
Tanimoto	$\frac{V_0 - V}{V_0} = \frac{c_{14}P}{V_0} + \frac{c_{16}P}{P + c_{15}}$
Terzaghi	$\frac{V - V_P}{V_P} = -c_{17} \ln (P + c_{18}) - c_{19}(P + c_{19}) - c_{20}P + c_{21}$
Cooper	$\frac{V_0 - V}{V_0 - V_P} = c_{22} \cdot e^{-c_{23}/P} + c_{24} \cdot e^{-c_{25}/P}$
Gurnham	$P = c_{26} \cdot e^{c_{27}/V}$
Nishihara	$\ln\left(\frac{V_0}{V}\right) = -\left(\frac{P}{c_{28}}\right)^{1/c_{29}}$
Tsuwa	$\frac{V_0 - V}{V_0} = \frac{V_0 - V_P}{V_0} \cdot \frac{(1/c_{30})P}{1 + (1/c_{30})P}$
Kawakita	$\frac{V_0 - V}{V_0} = \frac{c_{31}c_{32}P}{1 + c_{32}P}$

$$\Delta\phi_i = \frac{\sqrt{K_r}}{1 + K_r} \cdot \frac{\Delta P}{P} \qquad (5.1.45)$$

where K_r is Rankine coefficient given by

$$K_r = \frac{1 - \sin \phi_i}{1 + \sin \phi_i} \qquad (5.1.46)$$

In case $\Delta\phi_i$ is proportional to the variables representing the state of packing such as voidage, strain, bulkiness, etc., various types of empirical equations are derivable for comparison.[49] Corresponding to the linear relationship between stress P and strain γ for general solids,

$$P = C_n \cdot \gamma^{1/(1 + \lambda)} \qquad (5.1.47)$$

was theoretically obtained for particulate matters,[51] which is comparable to Nutting's equation. By thermodynamic consideration,[52]

$$Q_d = \frac{1}{S_p} \cdot \log \frac{P - c_n}{P_0 - c_n} \qquad (5.1.48)$$

where Q_d and S_p are the displacement and the cross-sectional area of the piston, respectively; P_0 is the initial pressure and c_n is the constant throughout the compaction equations above. It is similar to Terzaghi's equation.

Compaction Due to Tapping and Vibration. Tapping and vibration are often adopted as an easy way of producing the compaction of power, though the bulk density achieved is not so high as that by piston press.

Experimental equations are available for the packing due to tapping.

$$\rho_{b,f_n} - \rho_{b,n} = (\rho_{b,f_n} - \rho_{b,0}) \cdot \exp(-k_1 n_{tp}) \qquad (5.1.49)$$

where ρ_{b,f_n} is the final bulk density attained to be constant after many times of tapping. $\rho_{b,0}$ is the initial bulk density before tapping, n_{tp} is the number of tapping, and k_1 is a constant.[53] ρ_{b,f_n} is not the material property but depends on the height of fall or the falling velocity of the tapped vessel. The straight line on the semilog plots sometimes indicates different slopes after a certain number of tappings in the series, especially with cohesive fine powders, as shown in Figure 5.1.15.[53,54]

$$\frac{V_0 - n_{tp}}{V_0 - V_{nt}} = \frac{1}{k_2 k_3} + \frac{n_{tp}}{k_2} \qquad (5.1.50)$$

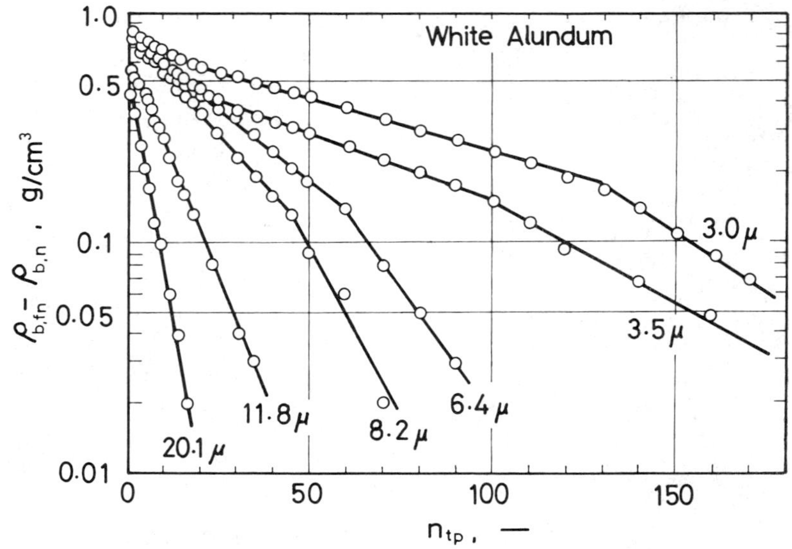

Figure 5.1.15. Relationship between bulk density and number of tapping.[54]

where V_0 is the initial bulk volume of powder, V_{nt} is that after n_{tp} times of tapping, and k_2, k_3 are constants.[55] For white alundum k_2 increases with decreasing diameter of particle and k_3 becomes minimum around 3 micrometers in particle diameter.[56] k_3 in Eq. (5.1.50) is found to be in direct proportion to k_1 in Eq. (5.1.49).

The degree of packing under sinusoidal vibration is correlated with a measure of compaction, $\Delta v/v_0$, by[57,58]

$$(1 - \epsilon) = (1 - \epsilon_0) + 0.073(\Delta v/v_0) \tag{5.1.51}$$

where ϵ and ϵ_0 are the void fractions of a particle bed in a cylinder during and before vibration, respectively. Δv is the impact velocity or the relative velocity of a single particle and the vibrating plate, and v_0 is the initial velocity of the particle as it leaves the vibrating plate. According to the intensity of vibration, $G = a_m(2\pi f)^2/g$,

$$G \leqq 1; \; \Delta v/v_0 \longrightarrow 0$$

$$1 < G \leqq 3.3; \; \Delta v/v_0 = \frac{\sqrt{G^2 + 1 - 2G \sin 2\pi n_1} + G \cos 2\pi n_1}{\sqrt{G^2 - 1}}$$

$$3.3 < G \leqq 3.92; \; \Delta v/v_0 = \frac{\sqrt{2G(\sin 2\pi n_1 - \sin 2\pi n_2) + G^2 \cos^2 2\pi n_1} + G \cos 2\pi n_2}{\sqrt{G^2 - 1}} \tag{5.1.52}$$

where a_m and f are the amplitude and the frequency of vibration, and n_n is the dimensionless time, $f \cdot t \cdot n_0$ corresponds to the point where a particle jumps from the plate, and n_1 and n_2 are the time when the particle falls on the plate, given by

$$n_0 = \frac{1}{2} \sin^{-1} \frac{1}{G}$$

$$n_1 = n_0 + \frac{1}{2\pi}\left(\sqrt{G^2 + 1 - 2G \sin 2\pi n_1} + \sqrt{G^2 - 1}\right)$$

$$n_2 = n_1 + 1 \cdots (n_1 < n_0 + 1)$$

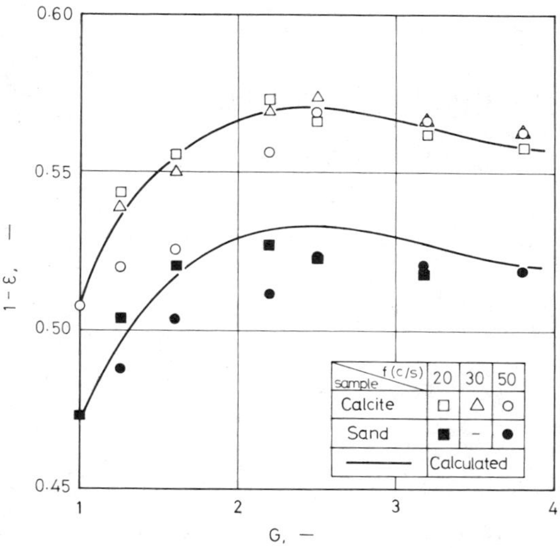

Figure 5.1.16. Variation of packing density with intensity of vibration.[58]

$$= n_1 + \frac{1}{2\pi}\{\sqrt{2G(\sin 2\pi n_1 - \sin 2\pi n_2) + G^2\cos^2 2\pi n_1} + G\cos 2\pi n_1\}\cdots(n_1 \geqq n_0 + 1) \qquad (5.1.53)$$

Figure 5.1.16 shows the data fitted by Eq. (5.1.51), which indicates the maximum packing density at $G = 2.5$.

Distribution of Bulk Density. The bulk density of powder varies within a large scale of a storage vessel, or even a small container under compression.

Assuming that the compressibility of powder is expressed by Eq. (5.1.54), as shown in Figure 5.1.17,[59]

$$\rho_b = aP^b \qquad (5.1.54)$$

the distribution of solids pressure, p_{s1}, within a cylindrical vessel is described after Janssen's[60] and Shaxby's[61] derivation as

$$P_{s1} = \left\{\left(P_0^{1-b} - \frac{aD}{4k_1\mu_w}\right)e^{-(1-b)4k_1\mu_w/D\cdot h} + \frac{aD}{4k_1\mu_w}\right\}^{1/1-b} \qquad (5.1.55)$$

where P_0 is the compressive pressure acting on the upper surface of powder bed, h is the depth from the upper surface, D is the diameter of the cylinder, μ_w is the frictional coefficient of wall surface, and k_1 is the ratio of lateral to vertical pressure assumed to be nearly constant throughout the vessel. Finally, a and b are coefficients

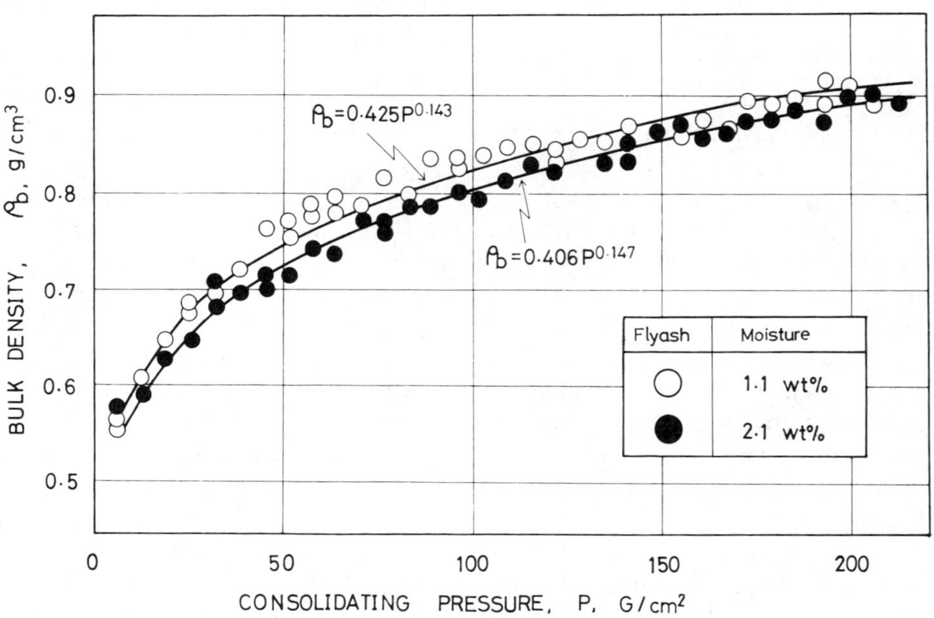

Figure 5.1.17. Relationship between bulk density and consolidating pressure.[59]

defined by Eq. (5.1.54). Hence, by putting P_{sl} from Eq. (5.1.55) into P in Eq. (5.1.54), the distribution of bulk density in the cylinder is drawn.

For a conical vessel, the density distribution is obtained in the same way by substituting Eq. (5.1.56) in Eq. (5.1.54).[59]

$$c(1-b) \neq 1; P_h = \left[P_0^{1-b}\left(\frac{Y}{h}\right)^{c(1-b)} + \frac{a(1-b)}{1-c(1-b)} \cdot Y\left\{\left(\frac{Y}{h}\right)^{c(1-b)-1} - 1\right\}\right]^{1/1-b}$$

$$c(1-b) = 1; P_h = \left\{P_0^{1-b}\left(\frac{Y}{h}\right) + a(1-b)Y \log_e\left(\frac{h}{Y}\right)\right\}^{1/1-b} \qquad (5.1.56)$$

where Y is the distance above the apex, θ is a half of the cone angle of the conical hopper and

$$c = 2\mu_w \cot\theta\,(k_1 \cos^2\theta + \sin^2\theta)$$

In the case of a bin consisting of a cylindrical silo over the conical hopper, P_0 in Eq. (5.1.56) is replaced by the bottom pressure in the cylinder P_{sl} given by Eq. (5.1.55).

For a cohesive powder the solids pressure distribution within a container is also derived under gravity alone,[62] tapping,[63] and aeration[64] in connection with the blockage criterion and a discharge rate of particles.

5.1.2. STRENGTH OF PARTICLE ASSEMBLAGE

The strength of powder is defined at the critical condition that the particle assemblage initiates flow from the stationary state. Two kinds of basic factors, friction and cohesion, act in the separation of solid bodies. They correspond to two types of strength, shear and tensile, according to the breakage mechanism of particulate materials. These strengths are directly based on the packing structure of particle assemblage through the degree of mechanical interlocking among particles and the coordination number, as mentioned below in close relation to the former section.

It is not too much an exaggeration to say that all the unit operations handling bulk solids are associated with the frictional and cohesive properties, and thus they are fundamental to the interpretation of particle behaviors, especially in storage, supply, transport, mixing, agglomeration, and so on. Some test devices and methods are proposed so far to evaluate them in a more comprehensive and reproducible manner, and an analysis of data obtained is attempted on a quantitative basis.

Interparticle Forces at a Contact Point

In principle, the strength of powder originates from the resistant forces at a contact point between two particles. Brief reference in general is, therefore, useful to frictional and cohesive forces between continuous solid bodies.

Friction Force Between Solid Surfaces.[65] By definition, the friction force is equivalent to the resistance exerted by one solid body against the motion of another in contact with it. The force is tangent to the contact surfaces. The coefficient of static friction is the ratio of the maximum friction of impending motion to the corresponding normal pressure. The coefficient of kinetic friction corresponds to the ratio for two surfaces moving relative to each other.

Provided Coulomb's empirical law of friction holds for comparatively dry and clean surfaces of solids: (1) the friction force is independent of apparent area of contact, and is proportional only to the normal load on the surface; and (2) the coefficient of kinetic friction is independent of the relative sliding velocity and is less than the coefficient of static friction. Experimentally the numerical value of the kinetic friction coefficient is found to increase gradually up to that of static friction as the velocity is decreased.

The coefficient of friction becomes large with a higher degree of vacuum, higher temperature of material, and thinner oxidation layer caused by a smaller quantity of molecules being adsorbed on or reacting with the solid surface. But the dry friction characteristics of the materials still act effectively through the boundary layer of lubricants. Coulomb's law is approximately applicable in such a wide variety of surface conditions.

Cohesive Force Between Solids. There are various kinds of attractive forces between solid materials. Among them, the following are the most basic and often encountered in cohesion phenomena of powders;

(1) *Van der Waals force* F_{vw} acts between molecules of solid surfaces within the shortest distance l of about 10^{-5} cm.[66] It is said that l is equal to 4×10^{-8} cm in close contact.[67] Between parallel planes of facing area s_f,

$$F_{vw} = \frac{A \cdot s_f}{6\pi l^3} \qquad (5.1.57)$$

Between sphere and plane,

$$F_{vw} = \frac{A \cdot d_p}{6l^2} \qquad (5.1.58)$$

Between different-sized spheres,[68]

$$F_{vw} = \frac{A}{12l^2} \cdot \left(\frac{d_{p1} \cdot d_{p2}}{d_{p1} + d_{p2}}\right) \qquad (5.1.59)$$

where A is a constant inherent to the material and is usually in the order of 10^{-12} erg.[69]

(2) *Electrostatic attractive force* F_e is given for two spheres apart whose electric charges are positive q_1 and negative q_2 in coulomb unit.[42]

$$F_e = \frac{q_1 \cdot q_2}{d_p^2} \cdot \left(1 - 2\frac{l}{d_p}\right) \qquad (5.1.60)$$

(3) *Solid bridge* due to chemical reaction, sintering, melting, and recrystalization gives rise to a strong bond between solids under the influence of temperature, pressure, humidity, water content, and so forth. The following is an example of the one analytical approach.[70]

Based on the rates of solid dissolution and of vaporization of bonding liquid between spherical particles in contact, the radius of the narrowest portion of the solid bridge r_n is approximately related to the initial liquid volume at the contact point V_{lq} as

$$\frac{r_n}{d_p} = 1.64 \cdot \frac{c_s V_{lq}}{\rho_p \cdot d_p^3/8} \cdot X^{\frac{1}{1-X}} \qquad (5.1.61)$$

where c_s is the saturated concentration of liquid in g/cm^3 and X is the dimensionless ratio of the rate of drying to the rate constant of dissolution in cm/s, which is the function of temperature. The bonding force F_b is then given by

$$F_b = \pi r_n^2 \cdot C_{rc} \qquad (5.1.62)$$

where C_{rc} is the strength of the bridge material formed by recrystalization of solid constituents.

(4) *Liquid bridge* between solids produces the bonding force F_{lq} as the sum of the forces due to the capillary suction pressure and the surface tension T of the liquid.

Assuming constant curvature of the liquid profile and perfect wetting ($\delta = 0$), the bonding force between different-sized spheres apart, as shown in Figure 5.1.18, is calculated at the narrowest portion of the liquid pendular ring as[71]

$$F_{lq} = \pi r_n^2 T\left(\frac{1}{r_{lp}} - \frac{1}{r_n}\right) + 2\pi r_n T$$
$$= \pi d_p T \cdot \frac{E}{4(m+n+1)} \cdot \left\{\frac{E}{4(m+n+1)} \cdot \frac{d_p}{r_{lp}} - 1\right\} \qquad (5.1.63)$$

where

$$E = \sqrt{\{4m + n(2m + n + 2)\} \cdot \{4(m + 1) \cdot \frac{2r_{lp}}{d_p} - n(n + 2m)\} + 2(n + 2)(n + 2m)\{8\left(\frac{r_{lp}}{d_p}\right)^2 - n\}}$$

Figure 5.1.19 depicts variations of the dimensionless cohesive force with constant curvature; for equal spheres, $m = 1$, and for a plane, $m =$ infinity, at different separation distances.

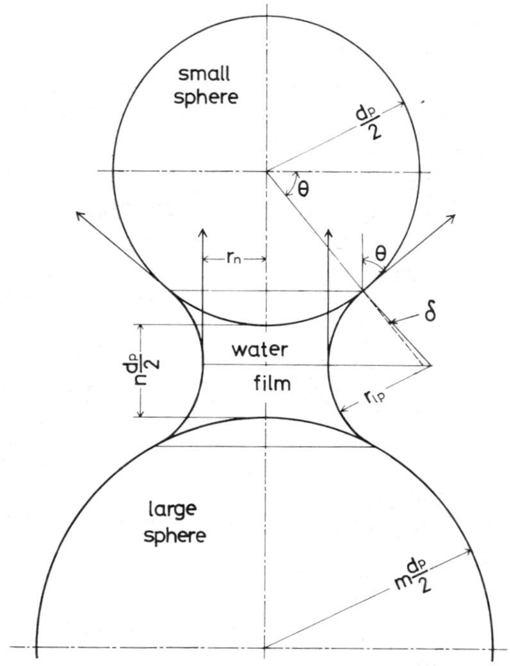

Figure 5.1.18. Model of cohesion due to liquid bridge between separate spheres of different size.[71]

The bonding force at the contact portion between the liquid and the equal sphere is given for the liquid volume V_{lq}, with different contact angles δ, by[43]

$$F_{lq} = \pi T d_p \cos\theta \left\{\cos(\theta - \delta) + \frac{d_p}{4}\left(\frac{1}{r_{lp}} - \frac{1}{r_n}\right)\cos\theta\right\} \quad (5.1.64)$$

$$V_{lq} = 2\pi\left[\{r_{lp}^2 + (r_{lp} + r_n)^2\} r_n \sin(\theta - \delta) - \frac{r_n^3 \sin^3(\theta - \delta)}{3} - (r_{lp} + r_n)\{r_{lp}^2 \sin(\theta - \delta) \cdot \cos(\theta - \delta) + r_{lp}^2(\theta - \delta)\} - \frac{d_p^3}{24}(2 + \sin\theta) \cdot (1 - \sin\theta)^2\right] \quad (5.1.65)$$

F_{lq} at the contact portion is always greater than that at the narrowest portion, and it becomes maximum in the close contact of solids. With the exception of the close contact, F_{lq} increases with V_{lq} at a certain separation distance and passes through the maximum point, as shown in Figure 5.1.19.

In the case of a cone and a sphere, as shown in Figure 5.1.20, the bonding force is obtained

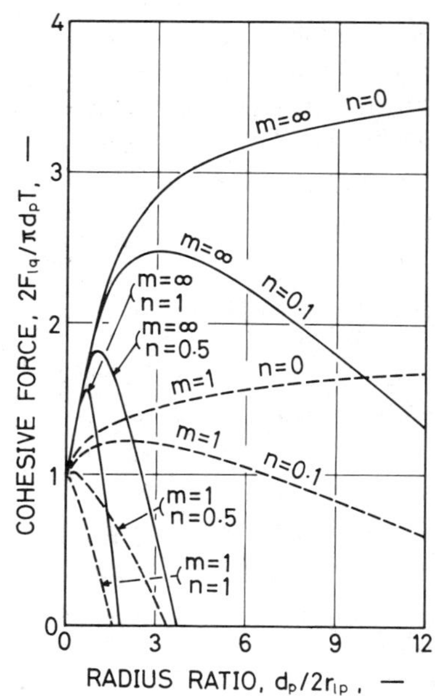

Figure 5.1.19. Cohesive force due to water bridge between equal spheres and between a sphere and a plane.[71]

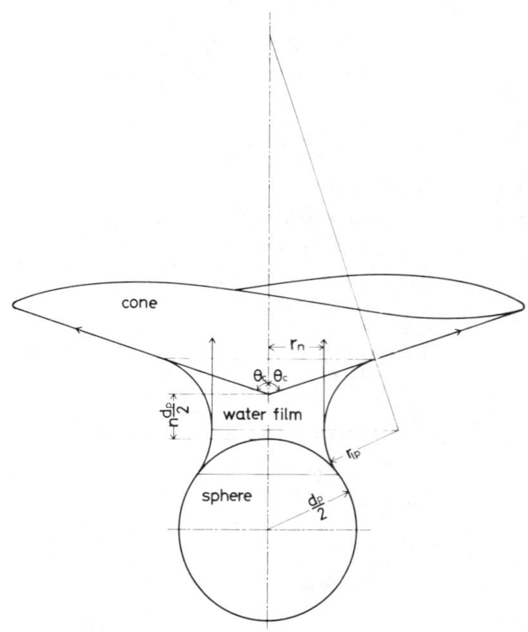

Figure 5.1.20. Model of cohesion due to liquid bridge between separate cone and sphere.[71]

in a similar way to the force with spheres at the narrowest portion.[71]

$$F_{1q} = \frac{\pi T d_p}{2} \cdot \left(\frac{B}{2} \cdot \frac{d_p}{r_{1p}} \cdot \cos\theta_c - 1 \right) B \cdot \cos\theta_c \qquad (5.1.66)$$

where

$$B = \frac{2r_{1p}}{d_p} + \left\{ \sqrt{1 - n(n+2)\tan^2\theta_c + \frac{4}{d_p} r_{1p} \left(\frac{2}{1 - \sin\theta_c} + n\frac{\tan\theta_c}{\cos\theta_c} \right)} - (n+1) \right\} \sin\theta_c$$

Figure 5.1.21 illustrates that the cohesive force increases with a larger cone angle, shorter distance, and with greater curvature of the liquid profile. d_p = infinity leads to F_{1q} for a cone and a plane as[71, 72]

$$F_{1q} = \pi T r_n \left(\frac{r_n}{r_{1p}} + 1 \right) \qquad (5.1.67)$$

$$V_{1q} = \pi r_{1p}^3 \left[(1 + \sin\theta_c) \left\{ \left(\frac{r_n}{r_{1p}} + 1 \right)^2 + (1 + \sin\theta_c) - \frac{1}{3}(1 + \sin\theta_c)^2 \right\} - \left(\frac{r_n}{r_{1p}} + 1 \right) \left(\sin\theta_c \cos\theta_c + \theta_c + \frac{\pi}{2} \right) - \frac{1}{3} \left(1 + \sin\theta_c - \frac{n d_p}{2 r_{1p}} \right)^3 \tan^3\theta_c \right] \qquad (5.1.68)$$

where

$$\frac{r_n}{r_{1p}} + 1 = \frac{1 + \left(1 - \frac{n d_p}{2 r_{1p}} \right) \sin\theta_c}{\cos\theta_c}$$

The above conical cases suppose angular particles, the surfaces of which have the conical projection at the contact point between particles.

Measurements of Cohesive Force Between Two Solids. The spring balance method is the direct way to evaluate the cohesive force between a particle and a flat plane through the displacement of the spring[73] or the elastic beam[74], as sketched in Figure 5.1.22. The automatic electrobalance method is a modification where the variation of interparticle force with the distance between two sample particles is measured under various atmospheres with high sensitivity of 10^{-8} grams.[75, 76] Particle diameter is usually in the order of several hundreds micrometers, and the corresponding cohesive force is in the order of several dynes.

The pendulum method is a way of separation using a gravitational component to move the wall around the center where the particle is suspended with a piece of fiber string, as shown in Figure 5.1.23a, or to expand two suspending points of nylon strings, as shown in Figure 5.1.23b.[77] The angles or the amplitudes of the particle pendulum are measured in separation.

The centrifugal method adopts a rotating cell,[78-80] for example, inside of which equal spheres are bonded on spheres fixed in a line on the razor edge, as shown in Figure 5.1.24a,[81] or particles of distributed size are spread in mono-layer over the plane, as shown in Figure 5.1.24b.[82] Changing the rotation speed ω, the cohesive force between two spheres is detected in separation, and the residual percentage of particles on the wall surface ψ is calculated from the centrifugal force, F_c,

$$F_c = \frac{\pi}{6} d_p^3 \cdot \rho_p \cdot r_c \omega^2 \qquad (5.1.69)$$

The latter results[82] indicate that the cohesive force at a contact point is distributed and that ψ vs F_c follows the log-normal function as

$$\psi = \int_0^\infty \frac{1}{\sqrt{2\pi}\log\zeta} \cdot \exp\left[-\frac{(\log F_c - \log F_{c,50})^2}{2\log^2\zeta} \right] d(\log F_c) \qquad (5.1.70)$$

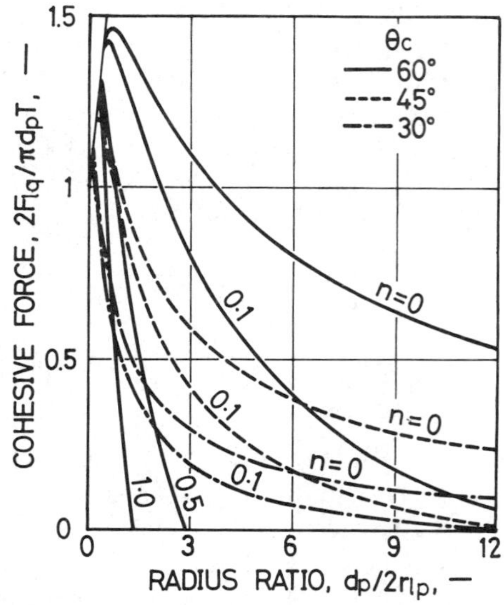

Figure 5.1.21. Cohesive force due to water bridge between separate cone and sphere.[71]

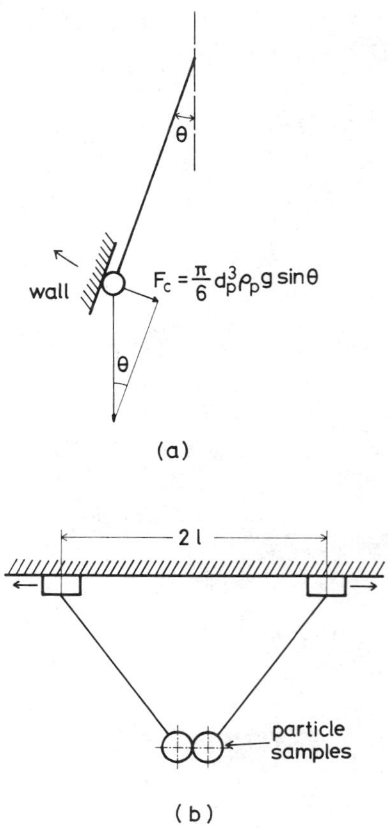

Figure 5.1.23. Pendulum method.[77]

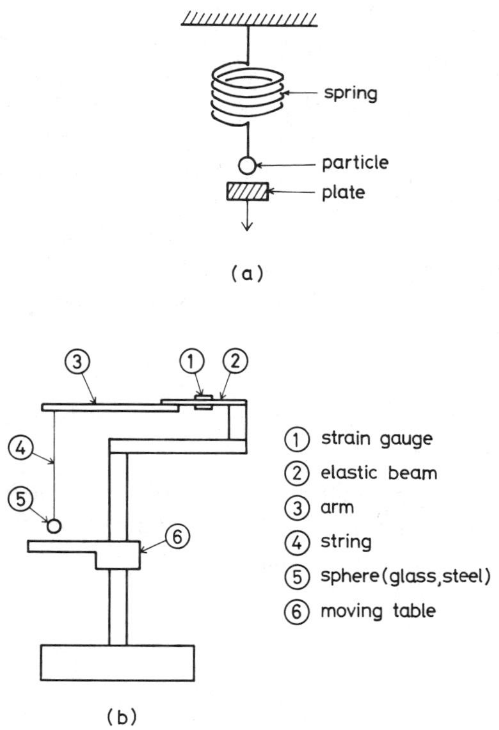

Figure 5.1.22. Spring balance method, (a)[73] and (b).[74]

This method is applicable to a large number of finer particles around ten micrometers and yields smaller values of cohesive force by 10 to 100 dynes, as compared with those by the balance methods.

Tensile Strength of Powder Mass

Based on the cohesive force at a contact point between particles and the number of contact points over the yield plane of powder mass, the ultimate tensile strength is evaluated without frictional effects among particles by means of some tensile test devices.

Devices for Measuring Tensile Strength. The methods of tensile testing are classified into two kinds of direct tension and three kinds of indirect tension by their ways of compaction and breakage.

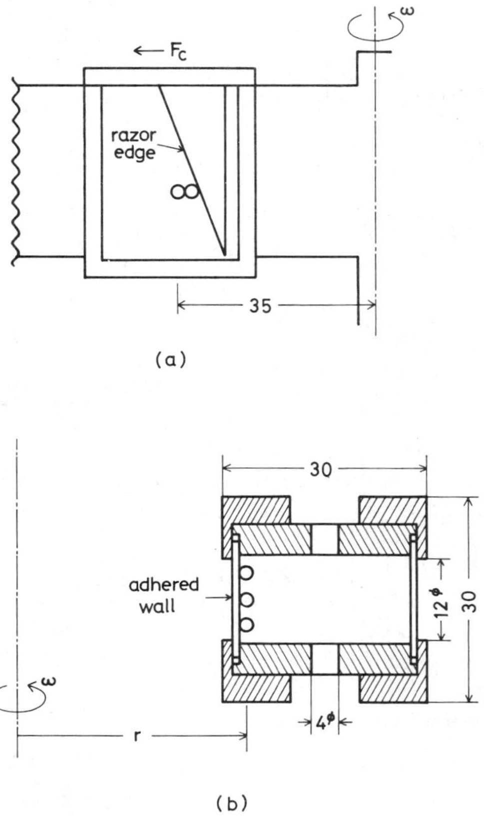

Figure 5.1.24. Centrifugal method.[81,82]

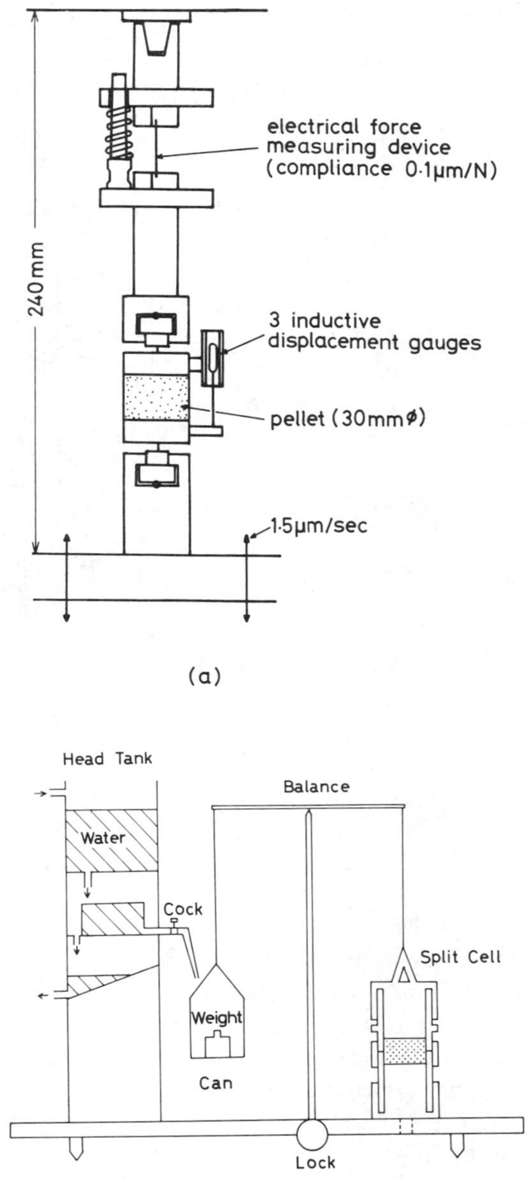

Figure 5.1.25. Devices for vertical tensile test, (a) adhesive method[83] and (b) wall clamping method.[84]

1. *Vertical tensile test*. Compacted powder is subjected to a vertical tensile load, and thus the rupture plane reveals horizontally or at right angles to the direction of compaction. Two types of testing apparatus have been devised, for example, as shown in Figure 5.1.25.[83,84] Modifications of these are available, especially in the manner of application of tensile load and the detector.[84–86] The main difference between the present apparatuses lies in the way of clamping the particle specimen.

(a) One way is the *adhesive method*,[85] in which the cylindrical pellet of particles prepared under high pressure (tons per cm^2) is glued to a pair of adaptors with a strong adhesive and set in the standard material-testing machine for vertical loading, as shown in Figure 5.1.25a.[83] The device gives the tensile stress-strain relationship at the same time. The pellet must be strong at both end planes without damage in order not to be separated from the adaptors in tension. Thus, it is not adequate for a loosely packed powder mass in the usual condition of handling.

(b) The second way of clamping the particle specimen is the *wall clamping method*, in which compacted powder is clamped due to friction

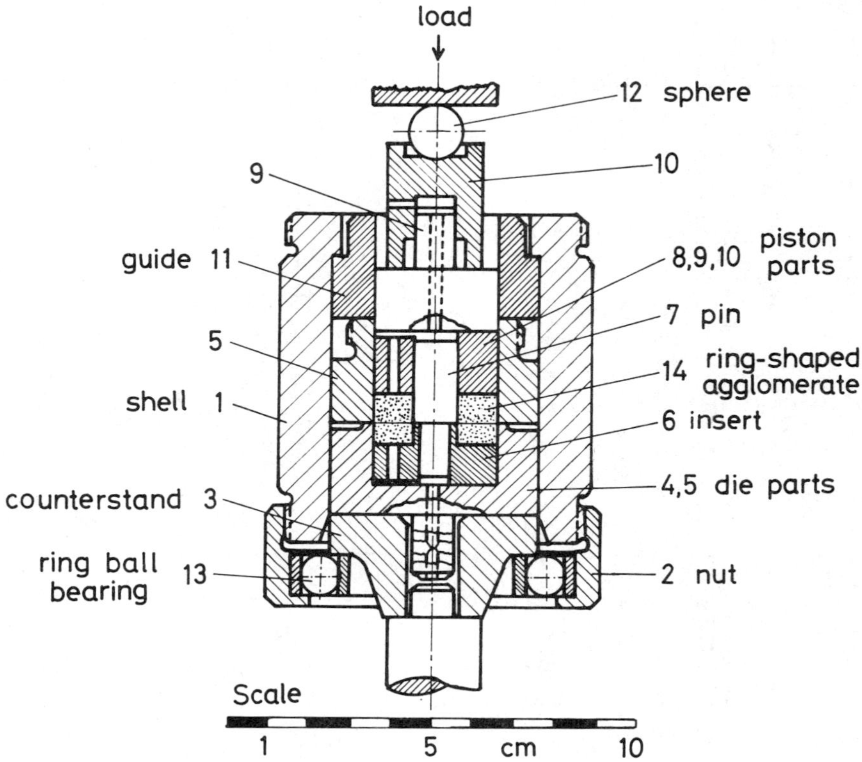

(a)

Figure 5.1.26. Split cells for wall clamping, (a) annular cell[87] and (b) cylindrical cell.[84]

and cohesion between particles and the walls of pistons and cylindrical cells. This method much improves the adhesive method in the range of compaction pressure or voidage by employing such adaptors as shown in Figure 5.1.26.[84,87] Part *a* of the figure shows the central pin inserted to increase the contact area or the resistant forces of the ring-shaped agglomerate (prepared at about 70 kg per cm^2) and to eliminate the inhomogeneous core of the cylindrical pellet.[87] Part *b* illustrates the joined cells and pistons, the internal wall surfaces of which are roughened by screw cuts for the cylindrical compact (up to about $\epsilon = 0.75$) not to slide down easily during tensile testing.[84] In contrast to the one-directional piston press in Figure 5.1.26a, powder is compressed in the joined cells by screwing up both pistons simultaneously, which produces shear loading with turning and piston pressing from both sides of the cylinder at the same time. As a result, the stress state becomes uniform and the void fraction of the sample is the largest at the joint section of the cells, that is, the yield plane is always prepared at the joint under tensile load. The void fraction over the failure plane is estimated from the weight of the powder slice at the joint section. Figure 5.1.27 presents the typical results of tensile strength by means of two kinds of wall clamping method.

2. *Horizontal tensile test.* Powder is compacted in the shallow cylinder under vertical loads and is diametrically split into two semicircular blocks by the horizontal tensile load. The vertical fracture plane is yielded at the joint of two half cells, one of which is fixed and the other of which is mounted on the traction table, as shown in Figure 5.1.28.[88]

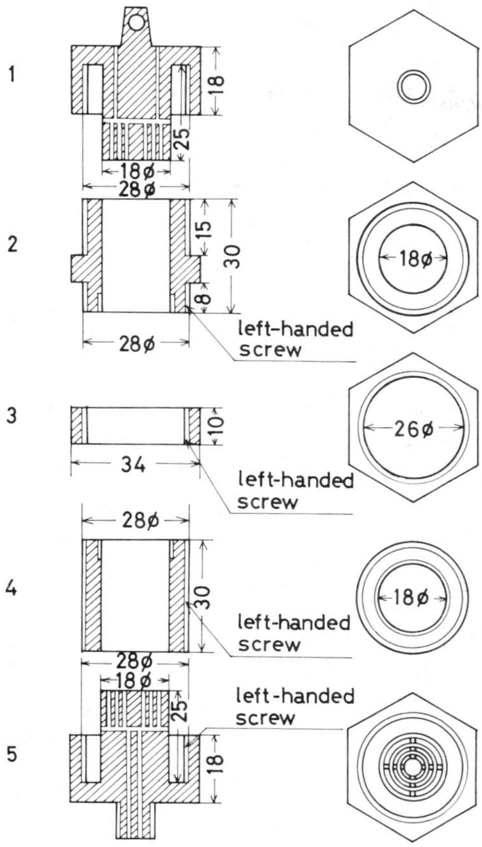

(b)

Figure 5.1.26. *Continued*

The similar split-plate apparatus is used, in which two movable cells are attached on both sides of the central fixed plate.[89] This method is intended to measure precisely the low tensile strength of loosely packed powder. It is, however, difficult in practice not only because of unstable guides for the movable half cell on ball bearings but also because of unavoidable inhomogeneity of stress and voidage along the powder depth. Thus, extrapolation of measured strength to zero bed height, if possible, could suggest the most appropriate value of tensile strength.[89] Figure 5.1.29 shows some examples of the data obtained by the horizontal tensile test.[86,89,90]

3. *Diametral compression test.* A discoidal or cylindrical agglomerate of particles is compressed across the diameter between two platens, as shown in Figure 5.1.30. In the case of ideal line loading the uniform distribution of tensile stress develops along the vertical diameter. The direction of the stress is at right angles to the vertical load. Assuming that the particle agglomerate is homogeneous and behaves like an elastic and brittle material up to the yield stress, this method gives a direct evaluation of tensile strength of powder by the maximum tensile stress in the agglomerate as

$$\sigma_{t,\max} = \frac{2\,L_t}{\pi D\,T_k} \tag{5.1.71}$$

where L_t is the total load and T_k is the thickness of the disk.

However, in the actual case with particle agglomerates, the load distributes over a finite area at the point of load application. Thus, the maximum tensile stress deviates from Eq. (5.1.71), and fracture occurs at a certain degree of deformation that relates to the compressibility of powder and the variation of tensile strength with consolidation. The general approach is given later in the analysis of tensile strength. Figure 5.1.31 shows one of the examples of the data.[91]

4. *Break-off test.* A powder sample consolidated in a tube with a plunger is extruded in the horizontal direction, to be broken off because of its own weight, as shown in Figure 5.1.32.[42] The weight of the powder column fallen down is divided by the cross-sectional area of the cylinder to give the cohesive strength. Then the powder slice of discrete thickness is scraped off at the failure portion to know the void fraction. Though indirect, it is a quite simple way of measuring cohesiveness of powders. Figure 5.1.33 shows the results.

5. The *penetration method* is sometimes proposed to measure tensile strength in the liquid phase.[93] It is not advantageous and attempts have been made to improve it by the hollow cylinder method.[94]

Analysis of Ultimate Tensile Strength. In order to investigate tensile strength, knowledge of the

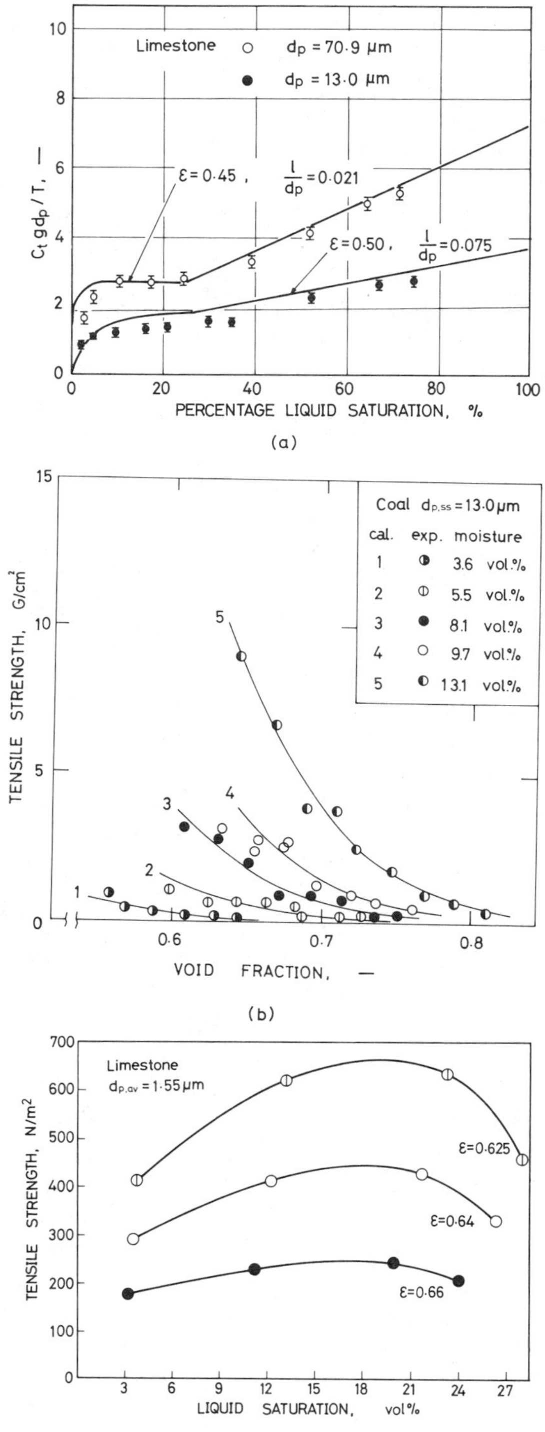

Figure 5.1.27. Tensile strength by vertical test, (a) with annular cell,[87] and (b)[41] and (c)[86] with cylindrical cell.

stress-strain behavior of particulate materials would be necessary, as is the case with continuous solid bodies. But only a few qualitative data are available as yet.[83,84,95]

On the other hand, several analytical equations have been proposed so far on the ultimate tensile strength of particle agglomerate in three kinds of liquid state.

1. *In the pendular state*, where only liquid bridges exist between the individual particles. Basically, the tensile strength is given as a product of the total contact points over the unit area of yield plane and the cohesive force at a contact point between two particles. The number of contact points is obtained from the number of reference particles multiplied by the coordination number of a single particle, which relates to the particle size and the packing structure of particle assemblage. Thus, the tensile strength of cohesive powder, C_t, is written in general as

$$C_t = k_1 \cdot \frac{1-\epsilon}{\phi_{sp} d_p^2} \cdot N_c \cdot F \qquad (5.1.72)$$

where k_1 is a proportional constant, ϕ_{sp} is the shape factor to represent an effective projected area of particle, and F is the cohesive force at the contact point.

(a) *Uniform distributions* of voidage and interparticle force give rise to an ideal tensile strength that is usually the maximum and does not change so much with the void fraction. The coordination number N_c is represented as a function of the void fraction ϵ, for instance, by Eqs. (5.1.10), (5.1.11), (5.1.23), (5.1.24), and (5.1.25). Among them, the simple relationship of Eq. (5.1.10) is often adopted to give the tensile strength as

$$C_t = k_2 \cdot \frac{1-\epsilon}{\epsilon} \cdot \frac{F}{d_p^{\,2}} \qquad (5.1.73)$$

where $k_2 = \frac{9}{8}$[85] is now revised to be unity.[96]

(b) An *increase in interparticle force with increasing contact area* due to consolidating pressure will interpret the rapid change in the tensile strength data more than that described

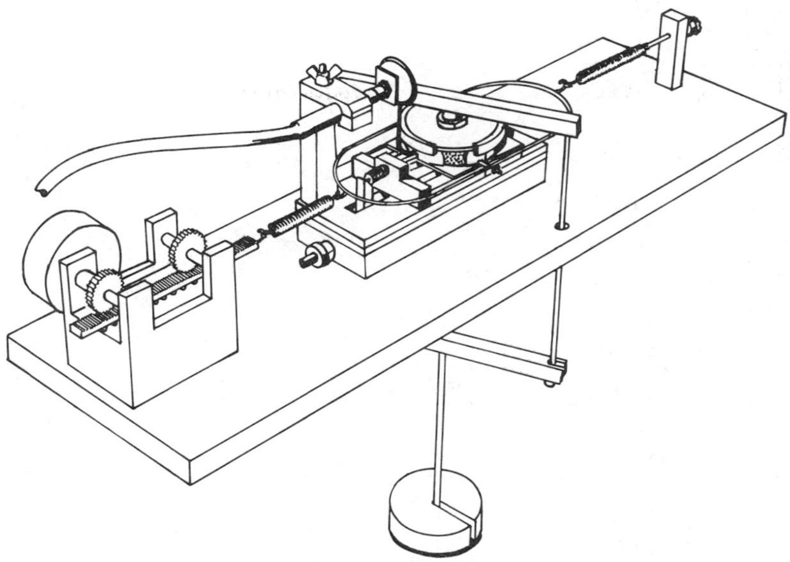

Figure 5.1.28. Apparatus for horizontal tensile test.[88]

by the voidage function $(1 - \epsilon)/\epsilon$ in Eq. (5.1.73), as shown in Figure 5.1.34.[95,97]

The cohesive force over the contact area between the particle and the plane F_d is derived as the sum of the force without deformation F and that due to the pressing force on the particle, F_p.[98]

$$F_d = F + F_p \cdot \frac{F_{vw}}{H_p \left(1 + \frac{2}{3} \cdot \frac{s_{el}}{s_{pl}}\right)} \qquad (5.1.74)$$

$$= F + F_p \cdot \frac{F_{vw}}{H_p} \quad \text{for } s_{el} << s_{pl} \qquad (5.1.74)$$

where F_{vw} is the Van der Waals force, H_p is the Hertz hardness of the particle, s_{el} and s_{pl} are contact areas attributing to the elastic and the plastic deformations, respectively.

The deformation of spheres in the bed by the external force F_{ex} is analyzed on the contact area as[99]

$$s = k_3 \cdot F_{ex}^q \qquad (5.1.75)$$

where q is $\frac{2}{3}$ for elastic deformation, 1 for plastic, and lies in between $\frac{1}{2}$ and $\frac{2}{5}$ in other cases.

In principle, it is almost possible to derive the relationship between F_d and ϵ by combining these two equations.

(c) *A nonuniform packing structure* of the particle assemblage, mentioned above, results in a pronounced increase in the ultimate tensile strength with a slight decrease in the void fraction.[41] Provided that only particles in the closest packing portion govern predominantly the ultimate strength,

$$C_{t,r} = n_{p,r} \cdot 4 \cdot F \qquad (5.1.76)$$

where 4 is the number of contact points of a single particle concerned with the breakage and $n_{p,r}$ is the number of particles in the rhombohedral packing portion given by

$$n_{p,r} = \frac{R_r(1 - 0.260) \cdot (d_p/\sqrt{2})}{\pi d_p^3/6} \qquad (5.1.77)$$

Thus, putting Eqs. (5.1.22) and (5.1.77) into Eq. (5.1.76),

$$C_{t,r} = \frac{F(1-\epsilon)}{\pi d_p^2/6} \cdot \frac{3.99(\alpha/\beta)}{e^{\alpha(\epsilon - 0.260)} - 1 + 1.41(\alpha/\beta)} \qquad (5.1.78)$$

Then, the tensile strength in the cubic packing

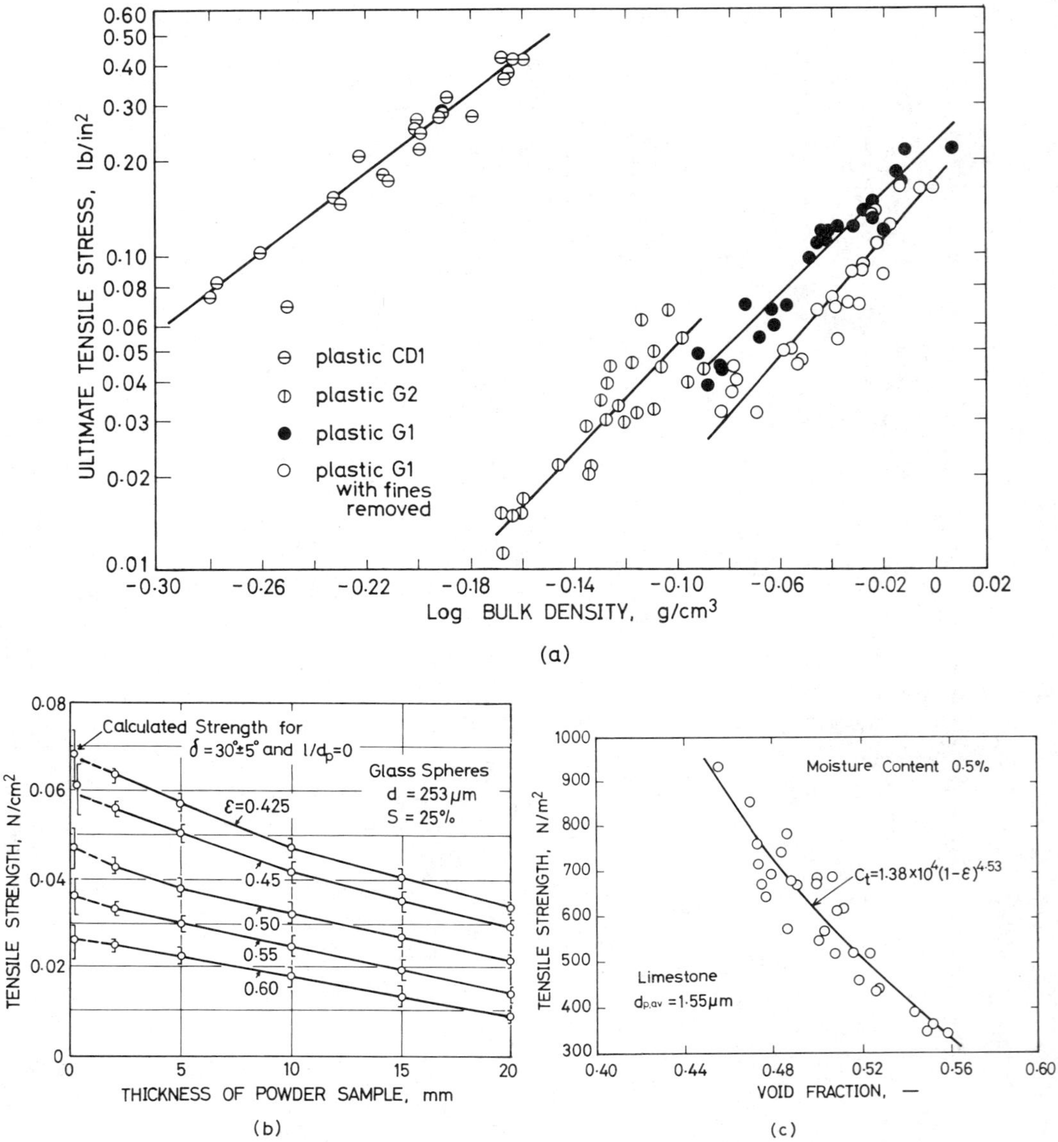

Figure 5.1.29. Tensile strength by horizontal test, (a),[90] (b),[89] and (c).[86]

portion $C_{t,c}$ and the maximum tensile strength in both particle packing portions $C_{t,max}$ are also derived and compared with Eq. (5.1.78) in Figure 5.1.35.

$$C_{t,c} = n_{p,c} \cdot 1 \cdot F = \frac{F(1-\epsilon)}{\pi d_p^2/6} \cdot \frac{e^{\alpha(\epsilon - 0.260)} - 1}{e^{\alpha(\epsilon - 0.260)} - 1 + 1.41\,(\alpha/\beta)} \tag{5.1.79}$$

$$C_{t,max} = n_{p,c} \cdot 1 \cdot F + n_{p,r} \cdot 4 \cdot F = \frac{F(1-\epsilon)}{\pi d_p^2/6} \cdot \frac{e^{\alpha(\epsilon - 0.260)} - 1 + 3.99\,(\alpha/\beta)}{e^{\alpha(\epsilon - 0.260)} - 1 + 1.41\,(\alpha/\beta)} \tag{5.1.80}$$

where

$$n_{p,c} = \frac{R_c(1 - 0.476)d_p}{\pi d_p^3/6} \tag{5.1.81}$$

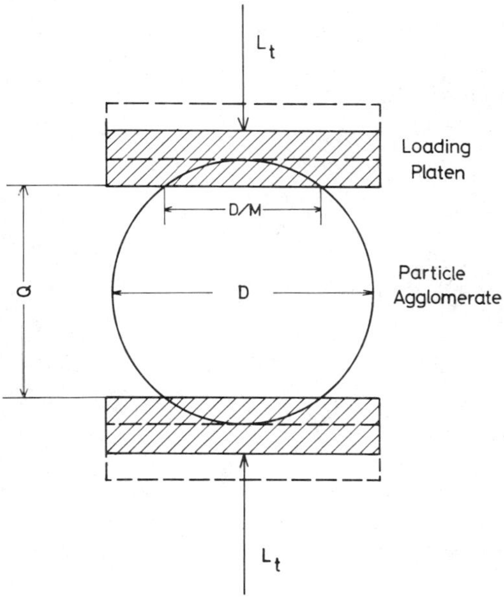

Figure 5.1.30. Diametral compression test.

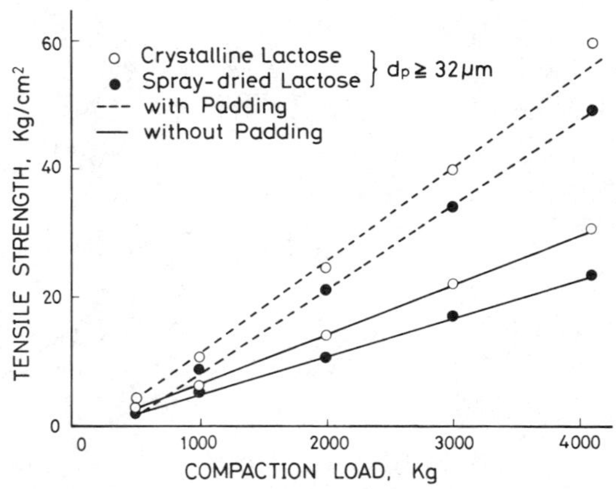

Figure 5.1.31. Tensile strength of agglomerate by diametral compression test.[91]

All the lines in Figures 5.1.27b and 5.1.36 are calculated from Eq. (5.1.78).

There are other types of representation of tensile strength;

(d) *For the powder compact of a single component of distributed size,*[100]

$$C_t = \frac{r_a \cdot r_b \cdot N_c}{2} \cdot S_v \cdot (1 - \epsilon) \cdot F(l) \quad (5.1.82)$$

where r_a is the ratio of the number of particle pairs per unit area of failure to that per unit volume; r_b is the ratio of overall area of contact per particle pair to the surface area of the smaller particle of the pair, and $F(l)$ is the interparticle force per unit overall area of contact as a function of the surface separation distance l of the particle pair, which is given by

$$l = l_{vn} - \frac{d_{p,av}}{3} \cdot \frac{\epsilon_{vn} - \epsilon}{1 - \epsilon_{vn}}$$

where l_{vn} is the effective range of the attractive interparticle force, and ϵ_{vn} is the corresponding voidage at which the tensile strength vanishes. For the binary mixture of particles, Eq. (5.1.82) is modified into

$$C_{t,mix} = \frac{r_a \cdot r_b}{2} N_c (1 - \epsilon) \cdot M_i \{S_{v,1} \cdot f_1^2 + 2S_{v,12} \cdot f_1 (1 - f_1) + S_{v,2} (1 - f_1)^2\} \cdot F(l) \quad (5.1.83)$$

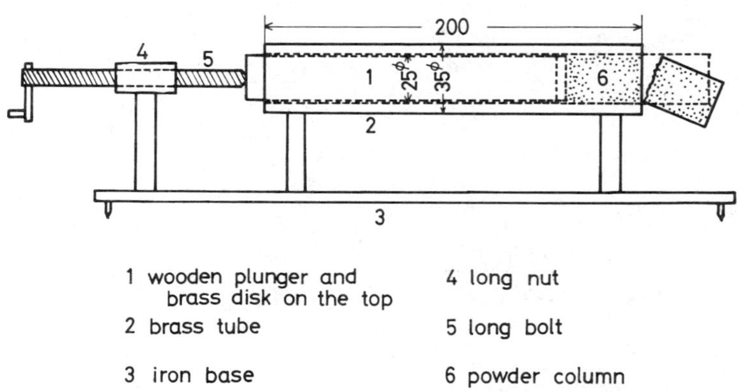

Figure 5.1.32. Device for break-off test.[92]

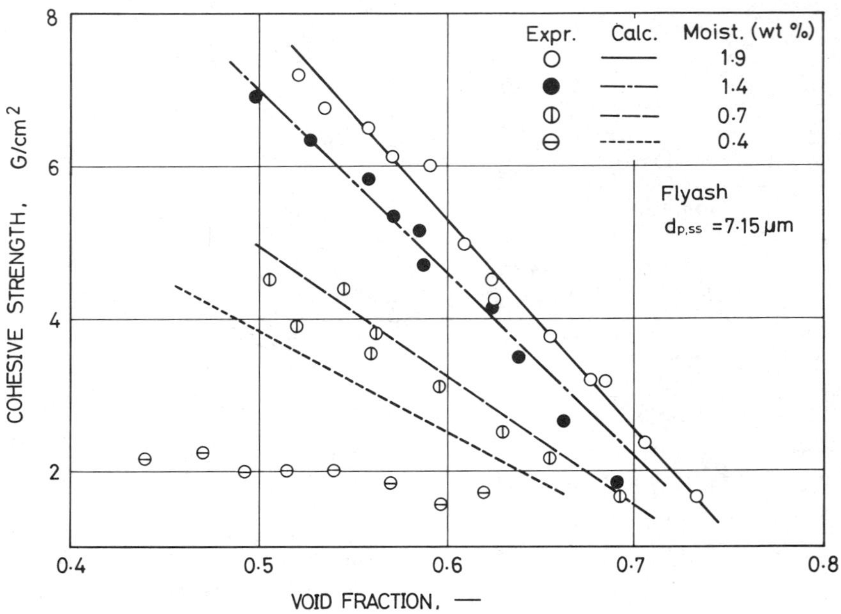

Figure 5.1.33. Cohesive strength by break-off test.[92]

where f_1 and f_2 are the weight fraction of components 1 and 2, respectively, $S_{v,12}$ is the specific surface for the particle pair of the mixture, and M_i is the index of mixing.[101]

(e) As an *experimental correlation*,

$$C_t = k_a (1 - \epsilon)^{m_a} \tag{5.1.84}$$

where k_a and m_a are fitting parameters.[102]

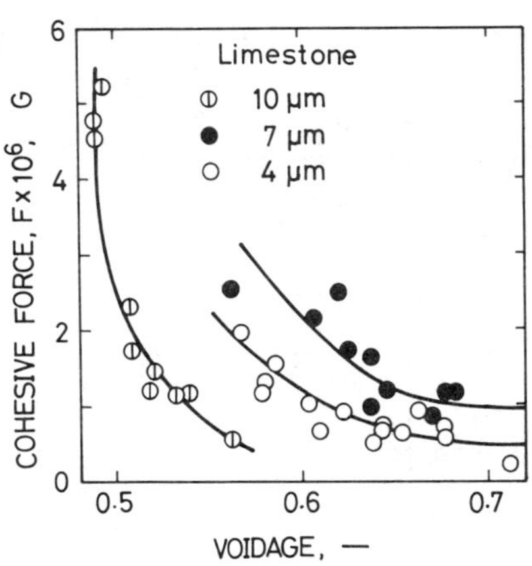

Figure 5.1.34. Variation of cohesive force at a contact point with void fraction.[97]

2. In the *capillary state*,[103] where all the capillaries composed of voids in the powder are completely filled with the binding liquid, the tensile strength is given by the capillary pressure, p_c.

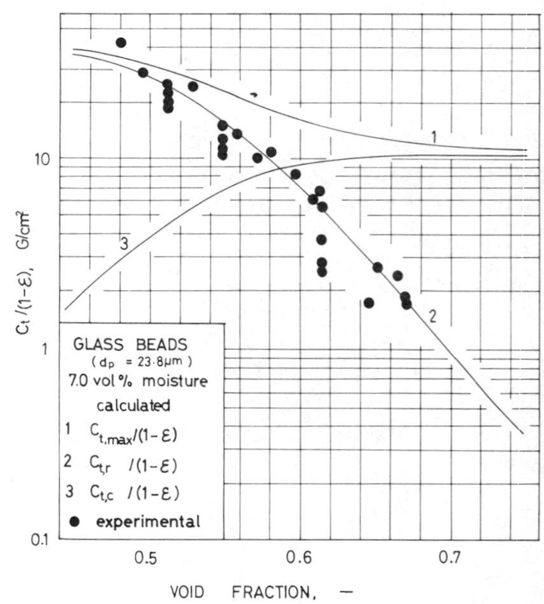

Figure 5.1.35. Comparison of $C_{t,\max}$ and $C_{t,c}$ with $C_{t,r}$.[41]

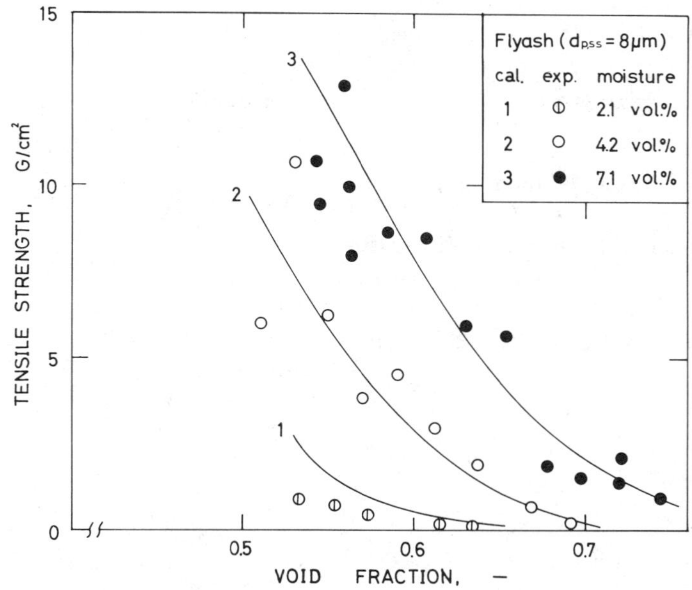

Figure 5.1.36. Variation of tensile strength with void fraction by wall clamping method.[41]

$$C_t = p_c = k_4 \cdot \frac{T}{d_{ps}} \cdot \frac{1-\epsilon}{\epsilon} \cdot f(\delta) \tag{5.1.85}$$

where d_{ps} is the surface equivalent diameter, and $f(\delta)$ is the contact angle function. $f(\delta = 0°) = 1$.

3. In the *funicular state*,[43] $S^* < S < 100$, where both liquid bridges and capillaries filled with liquid are present,

$$C_t = C_t^* + \frac{S - S^*}{1 - S^*} \cdot (p_c - C_t^*) \tag{5.1.86}$$

where S is the percentage liquid saturation, and C_t^* is the tensile strength evaluated by Eq. (5.1.72) at the critical saturation S^*, where liquid bridges begin to touch each other,

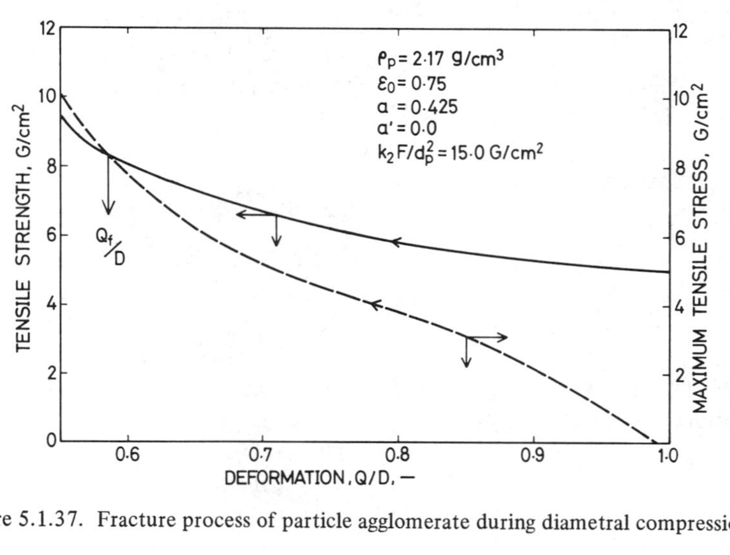

Figure 5.1.37. Fracture process of particle agglomerate during diametral compression.[105]

given by

$$S^* = 3 \cdot \frac{1-\epsilon}{\epsilon} \cdot \frac{N_c}{\pi} \cdot \frac{V_{t,lq}}{d_p^3}$$

where $V_{t,lq}$ is the volume of liquid bridges.

Analysis on Diametral Compression Test. In the general case of distributed loading over the discoidal agglomerate of particles, the maximum tensile stress always occurs at the center of the disk as[104]

$$\sigma_{mj,\,max} = -\frac{2L_t}{\pi D T_k} \cdot J(M, b', n) \quad (5.1.87)$$

where

$$J(M, b', n) = \sum_{i=1}^{n+1} \left[\frac{1 - 2^{b'}|I_n|^{b'}}{(n+1) - 2^{b'} \sum_{i=1}^{n+1} I_n^{b'}} \cdot \left(\frac{16}{M^2} \cdot I_n^2 - 1 \right) \sqrt{1 - \frac{4}{M^2} I_n^2} \right]$$

and

$$I_n = \frac{i-1}{n} - 0.5$$

Here, $n + 1$ is the number of concentrated loads, and the loads distribution expressed by Eq. (5.1.88) is adopted.

$$L_i = -a'|x_i|^{b'} + c' \quad (5.1.88)$$

where a', b' and c' are constants and x_i is the x coordinate of the point of load, L_i. Thus,

$$x_i = \frac{D}{M} \cdot I_n \quad (5.1.89)$$

where D/M is the contact width between the disk and the compressing plate, as denoted in Figure 5.1.30.

By simple geometry,

$$\frac{1}{M} = \sqrt{1 - \left(\frac{Q}{D}\right)^2} \quad (5.1.90)$$

$$\rho_b = \frac{4W}{D^2 T_k} \cdot \frac{1}{\pi - 2\cos^{-1}\frac{Q}{D} + 2\frac{Q}{D}\sqrt{1 - \left(\frac{Q}{D}\right)^2}} \quad (5.1.91)$$

where Q is the distance between two platens and W is the weight of the agglomerate sample.

Incorporating the compaction characteristics of powders mentioned above, for example, Eq. (5.1.54),

$$\rho_b = aP^b \quad (5.1.92)$$

equating C_t by Eq. (5.1.72), (5.1.84), and (5.1.85), and $\sigma_{mj,\,max}$ by Eq. (5.1.87) leads to the failure condition by the bulk properties of powder as[105]

$$\left(\frac{2L_t}{\pi D T_k}\right)_f = \frac{2P_f}{\pi} = -\frac{C_t(\epsilon_f, F)}{J_f(M_f, b', n)} \quad (5.1.93)$$

where subscript f denotes failure condition, and M_f in Eq. (5.1.90), ϵ_f in Eq. (5.1.91) and P_f in Eq. (5.1.92) are all represented by Q_f/D. Hence, the deformation at failure Q_f/D is obtained from Eq. (5.1.93) at the intersection of two curves, $\sigma_{mj,\,max}$ and C_t, as illustrated in Figure 5.1.37. In other words, since the bulk properties of powder give the value of Q_f/D, the ultimate tensile strength is predicted by Eq. (5.1.87)

Shear Strength of Particles

In contrast to the tensile strength in the normal direction to the failure plane, shear strength is yielded along the parallel plane to the breaking force. It arises from friction and interlocking in addition to cohesion between particles, as analyzed later.

The shear stress τ is written as a function of the normal stress σ by the following equations;

For Coulomb powder with a linear relationship,

$$\tau = \mu_i \sigma + C_s \tag{5.1.94}$$

where μ_i is the coefficient of internal friction, and C_s is called the stickiness corresponding to τ at $\sigma = 0$.

For general powder

$$\tau = \frac{\partial \tau}{\partial \sigma} \cdot \sigma + C_s(\sigma) \tag{5.1.95}$$

where $\mu_i = \partial\tau/\partial\sigma$.[106]

$$\left(\frac{\tau}{C_s}\right)^{q_{id}} = \frac{\sigma - \sigma_0}{\sigma_0} \tag{5.1.96}$$

where q_{id} is one sort of flowability indices, and σ_0 is the normal stress σ extrapolated to $\tau = 0$[102] or the apparent tensile strength.

$$\tau = k_6\ (\sigma - \sigma_0)^q \tag{5.1.97}$$

where k_6 and q are fitting constants, and Eq. (6.1.97) coincides with Eq. (5.1.96) by taking $k_6 = C_s/-\sigma_0^{1/q\,id}$ and $q = 1/q_{id}$.[107]

In the critical stress state, these equations describe the yield loci that are tangential to Mohr's circle, as illustrated in Figure 5.1.38. Usually the particle bed is not isotropic before flowing, because, in the case of Coulomb powder, the following relation holds between the major σ_{mj} and the minor σ_{mn} principal stresses as

$$\frac{\sigma_{mn} - \sigma_0}{\sigma_{mj} - \sigma_0} \geqq \frac{1 - \sin\phi_i}{1 + \sin\phi_i} \tag{5.1.98}$$

where it is in the elastic state up to the plastic equilibrium indicated by the equal sign.

After the failure, isotropic flow of continuous slipping is represented by the straight line through the origin with the effective angle of friction δ_e that is tangential to Mohr's failure circles passing through the terminal points of the yield loci, as shown in Figure 5.1.38. Thus, Eq. (5.1.99) holds simultaneously with Eq. (5.1.98), and equating them gives Eq. (5.1.100).

$$\frac{\sigma_{mj}}{\sigma_{mn}} = \frac{1 + \sin\delta_e}{1 - \sin\delta_e} \tag{5.1.99}$$

$$C_s = \frac{\sigma_{mj}(\sin\delta_e - \sin\phi_i)}{(1 + \sin\delta_e)\cos\phi_i} \tag{5.1.100}$$

In order to get such fundamental flow factors as ϕ_i, C_s and δ_e, shear and compression tests are conducted to draw the yield loci.

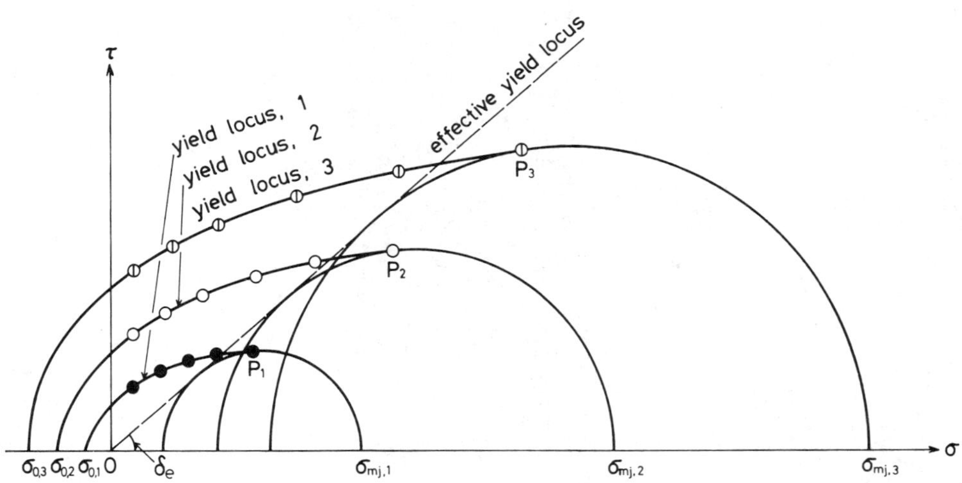

Figure 5.1.38. Relationship between yield loci and Mohr's circles.

Methods of Shear and Compression Tests.

1. *Direct shear test.* Powder mass consolidated in the circular or square cells under vertical loads is subjected to horizontal shear in three ways, as shown in Figure 5.1.39.

One of the cells is horizontally moved against the fixed cells to measure the shearing force in the initial but steady state of slipping. Repeated measurements with samples of constant voidage under various loads give a line of σ–τ relation. As contrasted with one- or two-plane shear by the first two methods in Figure 5.1.39, the last method of ring shear employs annular cells.[108,109] Its advantage lies in the constant area of shear plane during test and the possible measurement of the coefficient of dynamic friction of powder.

2. *Compression test.* Cylindrical powder agglomerate of height about 2 to 3 times the diameter is compressed in two ways to measure the yield strength, as schematically shown in Figure 5.1.40, from which the σ–τ relation is obtained.

The uniaxial compression test without lateral pressure gives the major principal stress at fracture in the axial direction $\sigma_{mj,f}$, and thus the Mohr's circle passes through the origin. Measuring the inclination angle of the slip plane from the vertical direction of the major principal stress θ_{uc}, the angle of internal friction ϕ_i is calculated from

$$\theta_{uc} = \frac{\pi}{4} - \frac{\phi_i}{2} \qquad (5.1.101)$$

As the test presents only one circle for cohesive powder ($C_s \neq 0$), the yield locus is drawn at the point on the circle whose tangent is tan ϕ_i. In

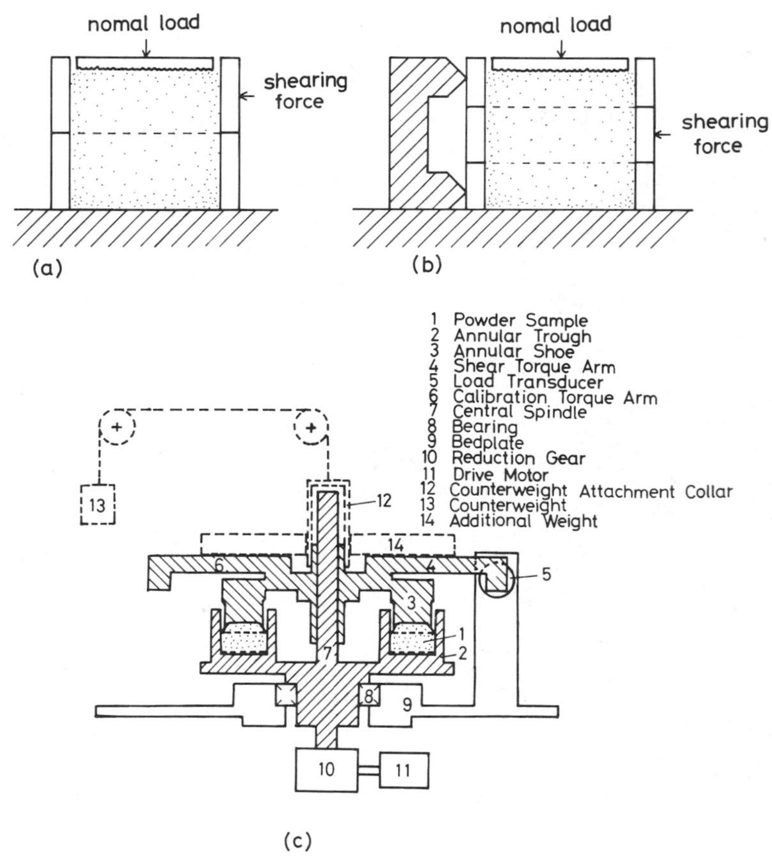

Figure 5.1.39. Direct shear tests, (a) one-plane shear, (b) two-plane shear, and (c) ring shear.[108]

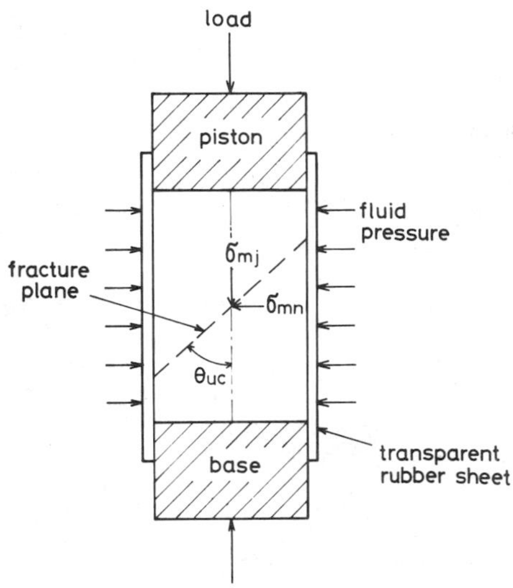

Figure 5.1.40. Uni- or tri-axial compression test.

the case of general powder, measuring the tensile strength or σ_0, the *power* q_{id} in Eq. (5.1.96) is given as

$$q_{id} = \frac{\cos\phi_i \cdot \cot\phi_i}{(1 - \sin\phi_i) - 2\dfrac{\sigma_0}{\sigma_{mj \cdot f}}} \tag{5.1.102}$$

Figure 5.1.41 shows an example of the data by this test.[110]

The triaxial compression test employs cylindrical powder that is enveloped by a thin rubber sleeve and pressed laterally with fluid. When one compresses the cylinder with a piston up to the point of failure, the major principal stress $\sigma_{mj,f}$ is then obtained in the vertical direction together with the minor one $\sigma_{mn,f}$ in the horizontal. Pairs of these stresses form the Mohr's circles, which are to be tangential to the yield locus or the shear strength under normal stress for constant voidage.

Analysis of Shear Strength. There are several kinds of angle properties of powders, such as the angle of internal friction, the angle of slide or wall friction, and the angle of repose. These are not primary material properties but vary with the packing conditions. Based on the nonuniform packing model mentioned above, they are analyzed as a function of void fraction.[40]

1. *Angle of internal friction* ϕ_i. The total shearing force required over the unit area of the rupture plane τ is given as the sum of the breaking forces in both particle packing portions by

$$\tau = (\mu_p \sigma_c + n_{p,c} \cdot 1 \cdot F) + \left(\frac{1 + \sqrt{2}\,\mu_p}{\sqrt{2} - \mu_p} \cdot \sigma_r + \frac{4\sqrt{3}}{\sqrt{2} - \mu_p} \cdot n_{p \cdot r} \cdot F\right) \tag{5.1.103}$$

where the first parentheses indicate the sum of the frictional force and the cohesive one in the cubic packing portion, and the term in the second parentheses is derived from the force balance in the horizontal direction of minimum shear force for a square arrangement of the rhombohedral packing. σ_c and σ_r are normal loads in each portion allotted with respect to the number of particles supporting the total load σ, as

$$\sigma_c = \frac{n_{p,c}}{n_{p,c} + n_{p,r}} \cdot \sigma, \quad \sigma_r = \frac{n_{p,r}}{n_{p,c} + n_{p,r}} \cdot \sigma \tag{5.1.104}$$

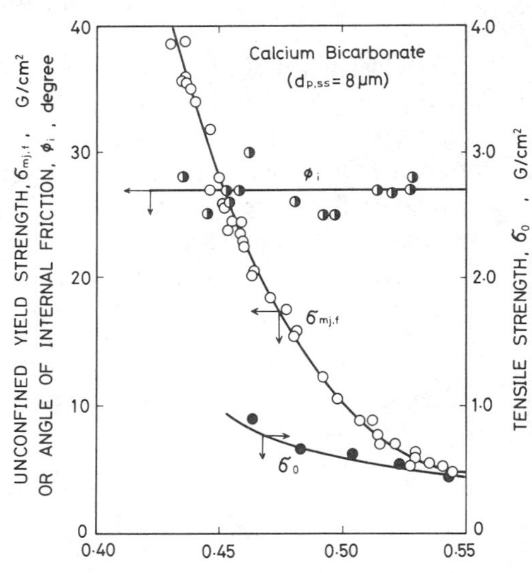

Figure 5.1.41. Yield strength and angle of internal friction by uni-axial compression test.[110]

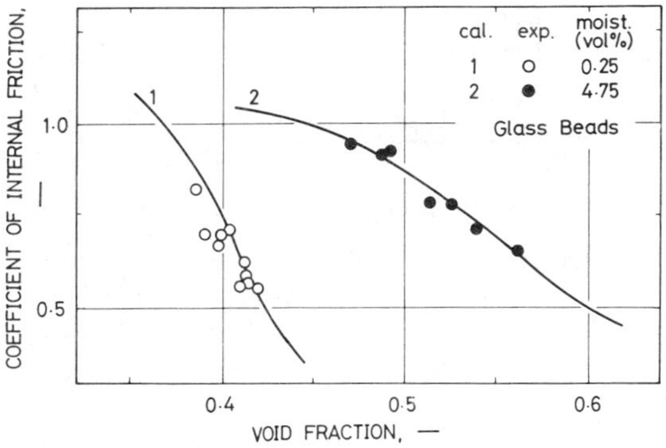

Figure 5.1.42. Variation of coefficient of internal friction with void fraction.[40]

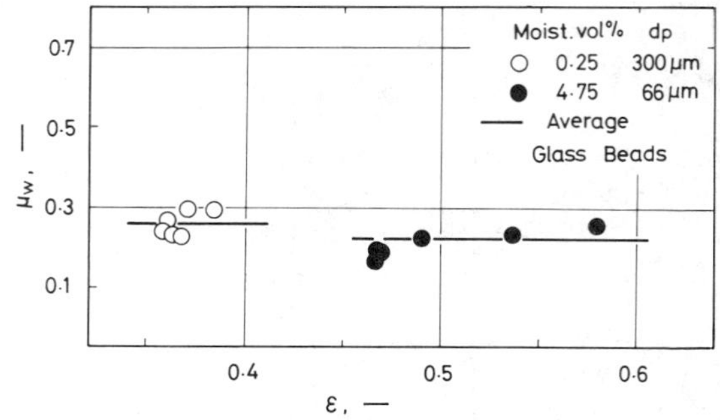

Figure 5.1.44. Independency of coefficient of wall friction of void fraction of powder.[40]

Substituting Eq. (5.1.104) and Eqs. (5.1.21), (5.1.22), (5.1.77), and (5.1.81) for n_p into Eq. (5.1.103) leads to the shear strength of powder in the general form as

$$\tau = \left\{\mu_p + \frac{1 + \mu_p^2}{\sqrt{2} - \mu_p} \cdot \frac{(\alpha/\beta)}{e^{\alpha(\epsilon - 0.260)} - 1 + (\alpha/\beta)}\right\} \cdot \sigma + \frac{F(1 - \epsilon)}{\pi d_p^2/6} \cdot \frac{e^{\alpha(\epsilon - 0.26)} - 1 + \{6.93/(\sqrt{2} - \mu_p)\}(\alpha/\beta)}{e^{\alpha(\epsilon - 0.260)} - 1 + 1.41(\alpha/\beta)} \quad (5.1.105)$$

In a comparison of Eq. (5.1.105) with Eq. (5.1.94) of Coulomb's, the second term on the right hand side indicates the physical content of the stickiness on the basis of the cohesive force between particles and the number of contact points, which increases with decreasing void fraction. The first term except σ corresponds to the coefficient of internal friction as the resistant effects of solid surface friction and geometrical interlocking among particles, which decreases with increasing voidage, as shown in Figure 5.1.42. The similar experimental trend is also reported.[111]

2. *Angles of wall and slide friction.* When the lower cell is replaced by a plane wall in the direct shear test, the shear force required per unit area of powder mass τ_w is obtained by the same considerations as is the case with τ:

$$\tau_w = \mu_w \sigma + \frac{F_w(1 - \epsilon)}{\pi d_p^2/6} \cdot \frac{e^{\alpha(\epsilon - 0.26)} - 1 + 1.0(\alpha/\beta)}{e^{\alpha(\epsilon - 0.26)} - 1 + 1.41(\alpha/\beta)} \quad (5.1.106)$$

where μ_w is the coefficient of wall friction between particles and the wall, and thus $\tan^{-1} \mu_w$ is the angle of wall friction.

By the force balance of powder on the inclined plane, as shown in Figure 5.1.43,[40]

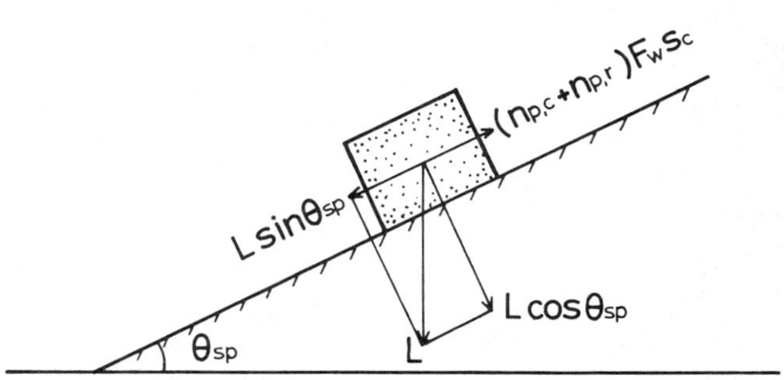

Figure 5.1.43. Force balance of powder block on an inclined plane.[40]

$$L \sin \theta_{sp} = L \cos \theta_{sp} \cdot \mu_s + (n_{p,c} + n_{p,r}) \cdot 1 \cdot F_w \cdot s_c$$

or

$$\theta_s = \theta_{sp} - \sin^{-1} \left\{ \frac{s_c}{L} \cdot \frac{F_w(1-\epsilon)}{(\pi d_p^2/6)\sqrt{1+\mu_s^2}} \times \frac{e^{\alpha(\epsilon - 0.26)} - 1 + 1.0(\alpha/\beta)}{e^{\alpha(\epsilon - 0.26)} - 1 + 1.41(\alpha/\beta)} \right\} \quad (5.1.107)$$

where L and s_c are the weight and the cross-sectional area of powder block, θ_{sp} is the angle of inclined plane, and μ_s is the coefficient of slide friction that defines the angle of slide friction as $\theta_s = \tan^{-1} \mu_s$. For noncohesive particles, $\theta_s = \theta_{sp}$.

These two coefficients of friction, μ_w and μ_s, are independent of void fraction, as shown in Figure 5.1.44, and depend only on the surface properties of particle and plane wall.

REFERENCES

1. D. P. Haughey, and G. S. G. Beveridge, *Can. J. Chem. Eng. 47*:130 (1969).
2. K. Kubo, E. Suito, Y. Nakagawa, and S. Hayakawa, *Powder–Theory and Application*, Maruzen Co., Tokyo, p. 208 (1962).
3. L. C. Graton and H. J. Fraser, *J. Geol. 43*:785 (1935).
4. T. Uematsu, K. Tsuchiya, and S. Okamura, *J. Mech. Eng. Japan 17*:72 (1951).
5. E. Manegold, R. Hofmann, and K. Solf, *Kolloid-Z. 56*:142 (1931).
6. H. T. Horsfield, *J. Soc. Ind. 53*:108 (1934).
7. H. E. White and S. F. Walton, *J. Am. Cer. Soc. 20*:155 (1937).
8. D. R. Hudson, *J. Appl. Phys. 20*:154 (1949).
9. A. E. R. Westman and H. R. Hugill, *J. Am. Cer. Soc. 13*:767 (1930).
10. J. C. Macrae and W. A. Gray, *Brit. J. Appl. Phys. 12*:164 (1961).
11. G. D. Scott, *Nature 188*:908 (1960).
12. T. G. Owe Berg, R. L. McDonald, and R. J. Trainor, Jr., *Powder Technol.*, *3*:183 (1969;70).
13. W. O. Smith, P. D. Foote, and P. F. Busang, *Phys. Rev. 34*:1271 (1929).
14. W. O. Smith, *Physics 4*:425 (1933).
15. R. K. McGeary, *J. Am. Cer. Soc. 44*:513 (1961).
16. M. Leva and M. Grummer, *Ind. Eng. Chem. 40*:415 (1948).
17. K. Ridgway, and K. J. Tarbuck, *Brit. Chem. Eng. 12*:384 (1967).
18. D. P. Haughey, and G. G. Beveridge, *Chem. Eng. Sci. 21*:905 (1966).
19. J. D. Bernal, and J. Mason, *Nature 188*:910 (1960).
20. L. H. S. Roblee and J. W. Tierney, *A. I. Ch. E. Journal 4*:460 (1958).
21. R. F. Benenati and C. B. Brosilow, *A. I. Ch. E. Journal 8*:359 (1962).
22. M. C. Thadani and F. N. Peebles, *Ind. Eng. Chem. Proc. Des. Develop. 5*:265 (1966).
23. K. Ridgway and K. J. Tarbuck, *Chem. Eng. Sci. 23*:1147 (1968).
24. G. D. Scott, *Nature 194*:956 (1962).
25. K. Gotoh, *J. Soc. Powder Tech. Japan 15*:223 (1978).
26. J. L. Finney, *Proc. R. Soc. London A319*:479 (1970).
27. E. M. Tory, B. H. Church, M. K. Tam, and M. Ratner, *Can. J. Chem. Eng. 51*:484 (1973).
28. J. A. Barker and D. Henderson, *Molecular Phys. 21*:187 (1971).
29. G. D. Scott and D. L. Mader, *Nature 201*:382 (1964).
30. M. M. Levine and J. Chernick, *Nature 208*:68 (1965).
31. K. Gotoh, *Nature Phy. Sci 231*:108 (1971).
32. K. Gotoh, *Ind. Eng. Chem. Fundam. 10*:161 (1971).
33. R. Aoki, *Handbook for Transport Engineering*, edited by T. Uematsu, K. Ikemori, Y. Mori, and S. Itoh, Asakura Shoten, Tokyo, p. 34 (1966).
34. G. G. Brown, *Unit Operations*, John Wiley and Sons Inc., New York, p. 214 (1950).
35. E. J. Crosby, *Kagaku Kogaku 25*:124 (1961).
36. P. S. Roller, *Ind. Eng. Chem. 22*:1206 (1930).
37. I. Shapiro and I. M. Kolthoff, *J. Phys. Colloid Chem. 52*:1020 (1948).
38. R. Sakata, *Oyo-butsuri 21*:24 (1952).
39. C. C. Furnas, *Bur. Mines Bull. 307*:74 (1929).
40. K. Shinohara and T. Tanaka, *Kagaku Kogaku 32*:88 (1968).
41. K. Shinohara and T. Tanaka, *J. Chem. Eng. Japan 8*:50 (1975).
42. H. Rumpf, *Chem. Ing. Techn. 30*:144 (1958).
43. W. Pietsch and H. Rumpf, *Chem. Ing. Techn. 39*:885 (1967).
44. C. C. Furnas, *Ind. Eng. Chem. 23*:1052 (1931).
45. D. Train and C. J. Lewis, *Trans. Instn. Chem. Eng. 40*:235 (1962).
46. E. Shotton and D. Grauderton, *J. Pharm. Pharmacol. 12*:87T (1960).
47. B. E. Kurtz, *Chem. Eng. Progress 56*:67 (1960).
48. K. Kawakita and K. H. Ludde, *Powder Technol. 4*:61 (1970/71).

49. K. Kawakita and S. Taneya, *Powder Technology*, Plant Kogaku Sha, Tokyo, p. 71 (1967).
50. T. Wakabayashi, *Powder Metall. Japan 10*:83 (1963).
51. A. Watanabe and K. Umeya, *Zairyo 13*:237 (1964).
52. T. Mogami, *Trans Japan Soc. Civil Engrs. 129*:39 (1966).
53. H. Kuno, *Proc. Fac. Eng. Keioh Univ. 11*:(41)1 (1958).
54. M. Arakawa, T. Okada, and E. Suito, *Zairyo 15*:151 (1966).
55. K. Kawakita, *Zairyo 19*:574 (1970); *19*:579 (1970).
56. M. Arakawa, T. Okada, and E. Suito, *Zairyo 14*:764 (1965).
57. T. Kohata, K. Gotoh, and T. Tanaka, *Kagaku Kogaku 31*:55 (1967).
58. A. Suzuki, H. Takahashi, and T. Tanaka, *Powder Technol. 2*:72 (1969).
59. K. Shinohara, K. Tamura, K. Gotoh, and T. Tanaka, *Kagaku Kogaku 31*:287 (1967).
60. H. A. Janssen, *VDI. Z. 39*:1045 (1895).
61. J. H. Shaxby and J. C. Evans, *Proc. R. Instrum. 19*:742 (1910).
62. K. Shinohara and T. Tanaka, *Chem. Eng. Sci. 29*:1977 (1974).
63. K. Shinohara and T. Tanaka, *Ind. Eng. Chem. Proc. Des. Develop. 14*:1 (1975).
64. K. Shinohara and T. Tanaka, *Chem. Eng. Sci. 30*:369 (1975).
65. N. Soda, *Friction and Lubrication*, Iwanami Shoten, Tokyo, p. 40 (1953).
66. J. Th. Overbeek, *Colloid Science*, Elsevier Pub. Co., Amsterdam, 1:245 (1952).
67. H. Rumpf, *J. Res. Assoc. Powder Techn. Japan 9*:3 (Special autumn issue, 1972).
68. R. S. Bradley, *Trans. Faraday Soc. 32*:1088 (1936).
69. J. Overbeek and M. J. Sparnay, *Disc. Faraday Soc. 18*:12 (1954).
70. N. Kudoh, M. Kuramae, and T. Tanaka, *Kagaku Kogaku Ronbunshu 2*:625 (1976).
71. K. Iinoya and H. Muramoto, *Zairyo 16*:70 (1967).
72. H. Tsunakawa and R. Aoki, *Kagaku Kogaku 36*:281 (1972).
73. M. Corn, *J. of Air Poll. Contr. Assoc. 11*:528 (1961); *11*:566 (1961).
74. K. Hotta, K. Takeda, and K. Iinoya, *Powder Technol. 10*:231 (1974).
75. M. Arakawa and S. Yasuda, Preprint of 37th Symposium by Tokai Branch of Soc. of Chem. Engrs. Japan, p. 7 (1977).
76. M. Chikazawa, W. Nakajima, and T. Kanazawa, *J. Res. Assoc. Powder Tech. Japan 14*:18 (1977).
77. H. Schubert, *Chem. Ing. Techn. 40*:745 (1968).
78. M. S. Kordecki and C. Orr, *Arch. of Environ. Health 1*:1 (1960).
79. G. Boehme, H. Krupp, H. Rabenhort, and G. Sandstede, *Trans. Instn. Chem. Engrs. 40*:252 (1962).
80. B. V. Deryagin and A. D. Zimon, *Kolloidn. Zh. 23*:544 (1961).
81. H. Emi, S. Endoh, C. Kanaoka, and S. Kawai, *Kagaku Kogaku Ronbun-shu 3*:580 (1977).
82. S. Asakawa and G. Jimbo, *Zairyo 16*:358 (1967).
83. H. Schubert, W. Herrman, and H. Rumpf, *Powder Technol. 11*:121 (1975).
84. K. Shinohara and T. Tanaka, Preprint of Third Congress CHISA, Marianske Lazne, 1969; *J. Chem. Eng. Japan 8*:46 (1975).
85. H. Rumpf, *Agglomeration* ed. by W. A. Knepper, Interscience, New York, p. 379 (1962).
86. G. A. Turner, M. Balasubramanian, and L. Otten, *Powder Technol. 15*:97 (1976).
87. W. Pietsch, E. Hoffman, and H. Rumpf, *Ind. Eng. Chem. Prod. Res. Develop. 8*:58 (1969).
88. M. D. Ashton, R. Farley, and H. H. Valentin, *J. Sci. Instrum. 41*:763 (1964).
89. H. Schubert and I. W. Wibowo, *Chem. Ing. Techn. 42*:541 (1970).
90. R. Farley and F. H. H. Valentin, *Trans. Instn. Chem. Engrs. 43*:T193 (1965).
91. J. T. Fell and J. M. Newton, *J. of Pharm. Sci. 59*:688 (1970).
92. K. Shinohara, H. Kobayashi, K. Gotoh, and T. Tanaka, *J. Res. Assoc. Powder Tech. Japan 2*: 352 (1965).
93. R. M. Griffith, *Chem. Eng. Sci. 20*:1015 (1965).
94. H. Schubert, *Powder Technol. 11*:107 (1975).
95. R. Aoki and H. Tsunakawa, *Zairyo 18*:497 (1969).
96. H. Rumpf, *Chem. Ing. Techn. 42*:538 (1970).
97. G. Jimbo, S. Asakawa, and N. Soga, *Zairyo 17*: 540 (1968).
98. H. Schubert, K. Sommer, and H. Rumpf, *Chem. Ing. Techn. 48*:716 (1976).
99. T. Nagao, Bulletin of *J. S. M. E. 10*:775 (1967); *Kikai Gakkai Ronbun-shu* (2)33:229 (1967).
100. D. C. H. Cheng, *Chem. Eng. Sci. 23*:1405 (1968).
101. S. Kocova and P. Pilpel, *Powder Technol. 7*:51 (1973).
102. M. D. Ashton, D. C. H. Cheng, R. Farley, and F. H. H. Valentin, *Rheol. Acta 4*:206 (1965).
103. W. Pietsch, *Nature 217*:736 (1968).
104. K. Shinohara and C. E. Capes, *Powder Technol. 24*:179 (1979).
105. K. Shinohara, C. E. Capes, and A. Fouda, Powder Technol. 32:163 (1982).
106. R. Aoki and K. Yamahuji, *J. Res. Assoc. Powder Tech. Japan*, Special Issue, p. 33 (February, 1966).

107. K. Umeya, N. Kitamori, M. Araki, and H. Miwa, *Zairyo 15*:166 (1966).
108. J. F. Carr and D. M. Walker, *Powder Technol. 1*:369 (1967/68).
109. S. Kocova and N. Pilpel, *Powder Technol. 5*:3?9 (1971/72).
110. R. Aoki, *Methods of Measuring Powder Properties*, ed. by S. Hayakawa, Asakura Shoten, Tokyo, p. 85 (1973).
111. T. Ohtsubo, *J. Res. Assoc. Powder Tech. Japan 1*:97 (1964).

LIST OF SYMBOLS

A	material constant in Equations (5.1.57)-(5.1.59), dyne. cm.
a_m	amplitude of vibration, cm.
a, b	coefficients defined by Equation (5.1.54).
a', b', c'	coefficients defined by Equation (5.1.88).
C_{rc}	strength of solid bridge material formed by recrystalization of constituents, dyne/cm^2.
C_s	stickiness in Coulomb's equation, (5.1.94), G/cm^2.
C_t	tensile strength of powder mass, G/cm^2.
c	factor used in Equation (5.1.56).
c_n	coefficients in compaction equations.
c_s	saturated concentration of liquid, g/cm^3.
D	diameter of cylinder or disk, cm.
d_p	particle diameter, cm.
F	cohesive force at a contact point, dyne.
$F(l)$	interparticle force per unit overall area of contact for surface separation, l, G.
F_c	centrifugal separating force, dyne.
F_{ex}	external force acting on spheres in the bed, dyne.
F_p	pressing force on a particle, dyne.
f	frequency of vibration, s^{-1}.
f_l	weight fraction of large particles.
$\mathrm{f_p}$	distribution function of local mean packing density.
f_y	factor in between 0 and 1 used in Equation (5.1.34) or total volume decrease.
G	intensity of vibration.
g	acceleration due to gravity, cm^2/s.
H_p	Hertz hardness of particle, dyne.
h	depth from the upper surface of solids bed, cm.
i	integer.
K_r	Rankine coefficient given by Equation (5.1.46).
K_s	size ratio of smallest to largest particle in the mixture.
k_a	fitting parameter in Equation (5.1.84), G/cm^2.
k_n	constants.
k_l	ratio of lateral to vertical solids pressure.
L	load, G.
l	shortest distance between two solid surfaces, cm.
l_{vn}	effective range of interparticle force, cm.
M	number of contact width equivalent to disk diameter.
M_i	mixing index.
m_a	fitting parameter in Equation (5.1.84).
m, n	integer or numerical value.
N_c	number of contact points per each particle (also called *the coordination number*).
n'	number of component sizes.
n_n	dimensionless time of vibration (=ft).
n_p	number of particles.
$n_{t,p}$	number of tappings.
P	compaction or solids pressure, G/cm^2.
P_0	solids pressure acting on the upper surface of powder bed, G/cm^2.
P_c	capillary pressure, G/cm^2.
Q	distance between two platens, cm.
Q_d	displacement of piston, cm.
q	exponent in Equations (5.1.75) and (5.1.97).
q_{id}	exponent in Equation (5.1.96).
q_1, q_2	electric charges with two spheres apart, Coulomb.
R	area ratio of cubic-packing portion to rhombohedral one.
R_c	volume fraction of cubic-packing portion of powder mass.
R_r	volume fraction of rhombohedral-packing portion of powder mass.

r radius, cm.

r_a ratio of number of particle pairs per unit area of failure to that per unit volume.

r_b ratio of overall area of contact per particle pair to surface area of smaller particle of the pair.

r_c radius of centrifuge, cm.

r_{lp} radius of liquid profile, cm.

r_n radius of narrowest portion of solid or liquid bridge, cm.

S saturation of liquid among voids, vol%.

S_{av} specific surface based on apparent volume of particles, cm^2/cm^3.

S_v specific surface based on net volume of particles, cm^2/cm^3.

S_w surface area of particles per unit weight, cm^2/g.

s contact area of particle, cm^2.

s_c cross-sectional area of powder block, cm^2.

s_f facing area of parallel planes, cm^2.

T surface tension of liquid, dyne/cm.

T_k thickness of discoidal agglomerate, cm.

V volume or bulk volume of particles, cm^3.

V_{lq} liquid volume at a contact point, cm^3.

V_p net volume of particles, cm^3.

V_s apparent specific volume of powder.

$V_{t,s}$ total volume of separate layers of particles, cm^3.

Δv impact velocity of a particle against vibrating plate, cm/s.

v_0 initial velocity of a particle leaving vibrating plate, cm/s.

W weight of particles, g.

X ratio of rate of drying to rate constant of dissolution.

Y distance above apex of conical hopper, cm.

x, y, z exponents in Equations (5.1.42)-(5.1.44).

x_i x-coordinate of loading point, L_i, cm.

α, β coefficients defined by Equation (5.1.18).

γ strain of particulate matters.

δ contact angle of liquid, degree.

δ_e effective angle of friction, degree.

ϵ void fraction of powder mass.

ϵ_e effective void fraction defined by Equation (5.1.16).

ϵ_{vn} void fraction at which tensile strength vanishes.

ζ_p standard deviation of packing density given by Equation (5.1.14).

ζ_s standard deviation of separation force due to centrifuge, dyne.

θ angle, degree.

θ_c half of cone angle, degree.

θ_s angle of slide friction, degree.

θ_{sp} angle of inclined plane, degree.

θ_{uc} angle of fracture plane to vertical direction by uniaxial compression test, radian.

θ_n angle around a sphere, degree.

λ power constant in Equation (5.1.47).

μ_i coefficient of internal friction.

μ_p frictional coefficient of particle surface.

μ_s coefficient of slide friction.

ρ_b bulk density of powder, g/cm^3.

ρ_p particle density, g/cm^3.

σ internal stress, G/cm^2.

σ_0 normal stress extrapolated to $\tau = 0$, G/cm^2.

τ shear strength of powder mass, G/cm^2.

ϕ_b bulkiness of particles.

ϕ_i angle of internal friction of particles, rad.

ϕ_p packing density of particles.

ϕ_s shape factor of particles based on specific surface.

ϕ_{sp} shape factor for effective projected area of a particle.

ϕ_v void ratio of particles.

ψ residual percentage of particles.

ω angular velocity, rad/s.

Subscripts

av average

b bonding due to solid bridge

c cubic packing

d	deformed particle	mn	minor
e	electrostatic attractive	0	initial
el	elastic deformation	pl	plastic deformation
f	at failure	r	rhombohedral packing
f_n	final	s	small particles
i	integer, index or internal	sl	cylindrical silo
l	large particles	ss	specific surface
l_q	liquid bridge	t	total or tensile
m	mixture	tp	n_{tp} times of tapping
max	maximum	vw	Van der Waals
mj	major	w	between wall and particle

PART 2. BEHAVIOR OF POWDERS IN A CONDENSED STATE

John Bridgwater
The University of Birmingham
Birmingham

INTRODUCTION

The purpose of this chapter is to provide insight into the physical mechanisms, in so far as these are understood, that control the behavior of powders and granular materials when these are in a condensed state. By the condensed state is meant that each particle has contact with other particles and that it is possible to proceed from one particle in an array to another anywhere in the array by passing through the point contacts between particles, there being a very large number of possible routes between the two particles. Thus, systems to which the concepts discussed are relevant include solids storage vessels, moving beds, spouted beds, the continuous phase of fluidized beds, filter cakes, materials processing equipment, some ore roasters and kilns, but systems such as pneumatic conveyers are excluded. In describing the mechanisms, an introduction will be provided toward mathematical ways of describing behavior.

5.2.1. TOTAL STRESS AND EFFECTIVE STRESS

The first point to make is that although each individual particle is in contact with others, it does not necessarily follow that powder behavior is controlled by the properties of the solid matrix alone. The behavior of the fluid in the pores, the interstitial fluid, is significant or controlling in many applications. In others, if the dominant fluid is a gas, for example, accumulation of liquid about the points of solid contact can arise. The first of these two instances is of particular significance and leads to the concepts of total stress, effective stress, and pore fluid pressure. These ideas will now be discussed.

Consider initially a powder that is an equilibrium, there being no motion of the solid particles nor of the interstitial fluid. Any interstitial fluid may be present; there is no restriction to the two most common ones found in processing industries, namely air and water. In Figure 5.2.1 is shown a unit cross-section of a powder that will be considered to be acted upon by a normal stress σ acting in the whole of the powder section. Now the effective stress σ' and pore fluid pressure p are related to σ by

$$\sigma = \sigma' + p \tag{5.2.1}$$

The total pore fluid pressure acts on the pore spaces and is, by hydrostatics, also uniformly superimposed on the powder matrix. The effective stress is the force borne by the powder skeleton per unit area of the powder itself plus the interstitial fluid. Equation (5.2.1) is thus a force balance, the overall stress σ being opposed by a stress in the solid and a stress in the pore fluid, each being taken to act over the whole section of the powder. The local solid stresses will vary from point to point in the section depending on the local packing and the position of a point with respect to contacts between particles. We shall take σ and σ' to be positive for compressive stresses.

The significance of Eq. (5.2.1) lies in the fact that the voids ratio (i.e. the ratio of the volume of pores or interstices to the volume of the solid) is affected by the effective stress σ', not by the total stress σ. A change in σ' leads to a displacement of the soil skeleton and to a change in the voids ratio. By way of example, consider the behavior of a body of powder held in a container with porous walls surrounded by a fluid at constant pressure. If the normal force is changed, σ is immediately changed. However, the powder matrix is not able to move until the interstitial fluid has been squeezed out or sucked in under the action of the pressure gradient in the pore fluid. Immediately the change in σ is balanced by a change in the pore water pressure p, and σ' will be unchanged. As the boundaries of the system permit the ingress or egress of fluid, the pressure gradient in the pore fluid induces flow of the pore fluid. This flow permits the voids ratio to change and σ' to alter until the pore pressure returns to its original value. In many applications in processing industries, concern will often be with the steady-state flow of fluid and solids, but the importance of the relationship between σ' and the voids ratio remains. The conclusion from this illustration is that both the flow of the pore fluid and the behavior of the powder matrix itself may be important. Each of these areas will now be considered.

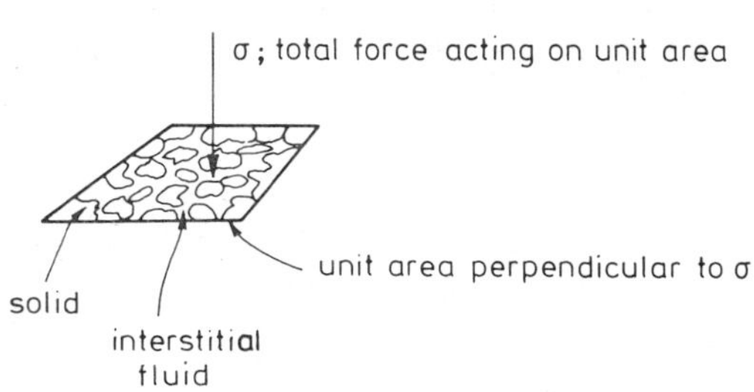

Figure 5.2.1. Stresses on a plane cut through a powder.

5.2.2. PORE FLUID FLOW

If the fluid flow through a matrix of stationary powder particles is sufficiently low, the pressure gradient will be controlled by viscous effects alone, provided the molecular mean free path is much less than the pore size. Flow due to Knudsen diffusion in the pores will not be considered, because even if it does occur, it will be for systems in which the total and effective stresses are virtually equal. Consider, for the present, one-dimensional fluid flow in a horizontal direction x. If u denotes the superficial velocity, i.e., the volumetric flow per unit cross-sectional area of solid plus pores, then it is found that

$$u = -k\frac{dp}{dx} \tag{5.2.2}$$

the negative sign arising from mathematical convention since the fluid flows from a point of high pressure to one of low pressure. Thus, if the pressure rises as x increases, fluid will flow in the direction of negative x. If flow is in the vertical direction y, there will be no flow if the pressure distribution is hydrostatic. Note that the fluid phase is completely connected and that there is no correction due to the fluid only

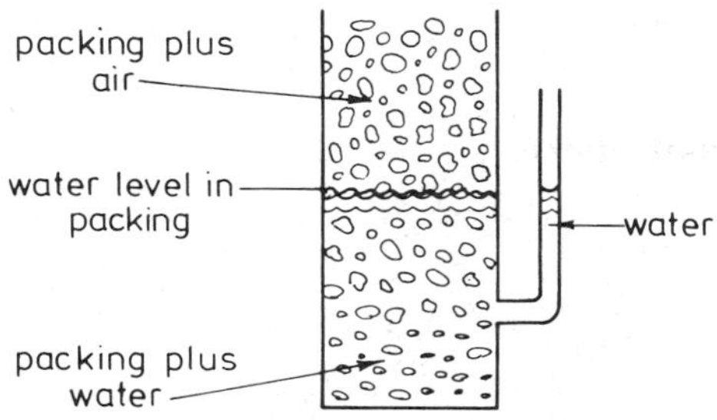

Figure 5.2.2. Hydrostatic pressure generation in a granular material.

occupying a portion of the cross section. This is illustrated in Figure 5.2.2.

If fluid is flowing in a vertical direction (Fig. 5.2.3) and if P now denotes the pressure at position y, the superficial velocity of the fluid at position y, the superficial velocity of the fluid is given by

$$v = -k\frac{dp}{dy} \tag{5.2.3}$$

where $p = P - \rho g y$, ρ being the fluid density and g the acceleration due to gravity. This equation shows that the extent to which the pressure gradient exceeds the hydrostatic pressure gradient determines the flow of pore fluid.

If the pore fluid is incompressible, then conservation of mass during flow leads, in Cartesian coordinates, to

$$\frac{\partial u}{\partial x} + \frac{\partial v}{\partial y} + \frac{\partial w}{\partial z} = 0 \tag{5.2.4}$$

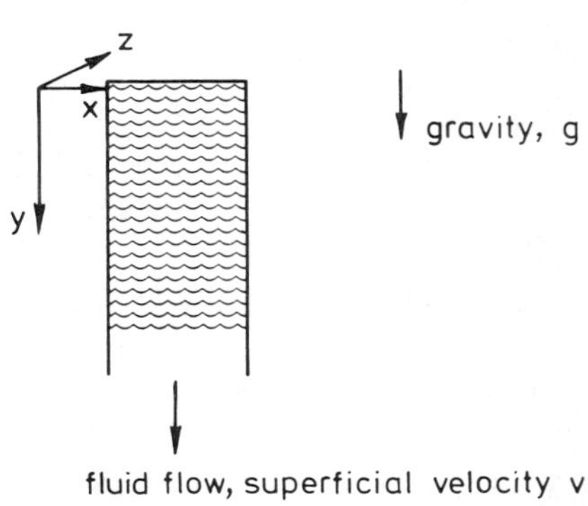

Figure 5.2.3. Cartesian coordinate frame with fluid downflow.

w being the superficial velocity in the direction z (Fig. 5.2.3). This may be written in the more general form

$$\text{div } \mathbf{v} = 0 \tag{5.2.5}$$

Combining Eq. (5.2.4) with equations such as (5.2.2) and (5.2.3) gives for an isotropic flow i.e. k independent of direction,

$$\frac{\partial^2 p}{\partial x^2} + \frac{\partial^2 p}{\partial y^2} + \frac{\partial^2 p}{\partial z^2} = 0 \tag{5.2.6}$$

which is Laplace's equation, an equation arising in many branches of science and engineering. For example, in heat conduction problems, it describes temperature distributions, temperature replacing p in Eq. (5.2.6).

Writing Eq. (5.2.2), which is often known as Darcy's law, in a more general form, we get the velocity vector $\mathbf{v}$ related to the pressure gradient by

$$\mathbf{v} = -k \text{ grad } p \tag{5.2.7}$$

The general result corresponding to Eq. (5.2.6)

$$\nabla^2 p = 0 \tag{5.2.8}$$

then arises from vector identities.

Now it is possible that the permeability k may be dependent upon direction. If the coefficients are designated k_x, k_y, k_z for the x, y, and z directions, then

$$k_x\frac{\partial^2 p}{\partial x^2} + k_y\frac{\partial^2 p}{\partial y^2} + k_z\frac{\partial^2 p}{\partial z^2} = 0 \tag{5.2.9}$$

Here we assume that the orientation of axes x, y, and z is such that a pressure gradient in one direction does not produce flow in another direction. Note that Equation (5.2.9) can be reduced to the same form as Equation (5.2.6) by making the substitutions $X = x/(k_x)^{1/2}$, $Y = y/(k_y)^{1/2}$, $Z = z/(k_z)^{1/2}$ yielding

$$\frac{\partial^2 p}{\partial X^2} + \frac{\partial^2 p}{\partial Y^2} + \frac{\partial^2 p}{\partial Z^2} = 0 \tag{5.2.10}$$

The present state of knowledge does not often warrant the use of three-dimensional analyses. Many engineering problems are one-dimensional or two-dimensional. For example, the form of Laplace's equation with constant coefficients k in cylindrical form is

$$\frac{\partial^2 p}{\partial r^2}+\frac{1}{r}\frac{\partial p}{\partial r}+\frac{1}{r^2}\frac{\partial^2 p}{\partial \theta^2}+\frac{\partial^2 p}{\partial z^2}=0 \qquad (5.2.11)$$

using the (r, θ, z) coordinate system shown in Figure 5.2.4. This equation reduces to

$$\frac{\partial^2 p}{\partial r^2}+\frac{1}{r}\frac{\partial p}{\partial r}+\frac{\partial^2 p}{\partial z^2}=0 \qquad (5.2.12)$$

if angular effects may be ignored, and to

$$\frac{d^2 p}{dr^2}+\frac{1}{r}\frac{dp}{dr}=0 \text{ or } \frac{d}{dr}\left(r\frac{dp}{dr}\right)=0 \qquad (5.2.13)$$

if axial effects may be ignored.

In spherical coordinates (Fig. 5.2.5), the general form of the equation is

$$\frac{1}{r^2}\frac{\partial}{\partial r}\left(r^2\frac{\partial p}{\partial r}\right)+\frac{1}{r^2\sin\theta}\frac{\partial}{\partial\theta}\left(\sin\theta\frac{\partial p}{\partial\theta}\right)+\frac{1}{r^2\sin^2\theta}\frac{\partial^2 p}{\partial\psi^2}=0 \qquad (5.2.14)$$

which reduces to

$$\frac{d}{dr}\left(r^2\frac{dp}{dr}\right)=0 \qquad (5.2.15)$$

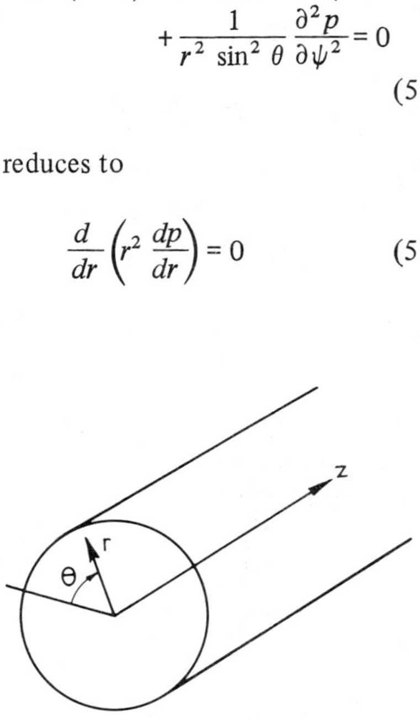

Figure 5.2.4. Polar coordinate system.

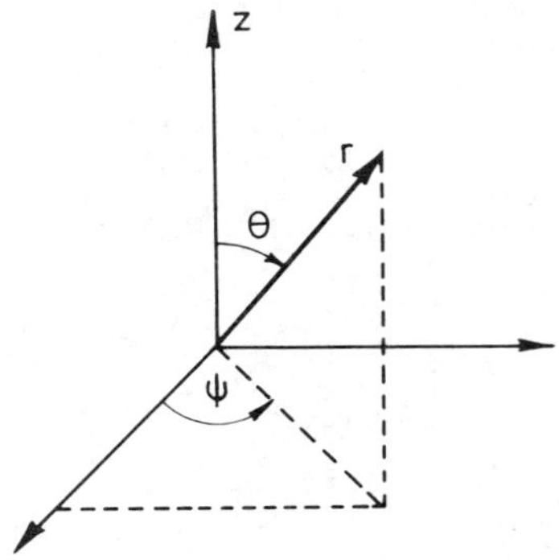

Figure 5.2.5. Spherical coordinate system.

if the behavior is spherically symmetrical and is independent of the angular coordinates.

Once the pressure distribution is known, the fluid flow rate can be found by employing equations such as Eq. (5.2.2); this now demands a knowledge of k. Once one is given a powder or granular material it is not difficult to determine its value experimentally in a permeability cell. However, the designer would like to know in advance the influence of particle size on fluid flow rate. Kozeny (1927) and Carman (1937) suggest that the pores may be represented by a number of equivalent passages with a cross-section independent of axial position, these being inclined at some angle λ to the direction of the applied pressure gradient. Davidson and Harrison (1963) provide a full discussion of the approach. Consideration of viscous flow in the equivalent passages enables the pressure drop Δp over length L of the bed to be given by

$$\frac{\Delta p}{L}=\left[\frac{72}{\cos^2\lambda}\right]\frac{\mu U(1-\epsilon)^2}{d^2\epsilon^3} \qquad (5.2.16)$$

where d is the particle diameter (the material having been taken to be made up of monosized spheres), ϵ is the bed voidage, the volumetric fraction of bed that is occupied by fluid, μ the fluid viscosity, and U the superficial velocity of the fluid. Experimentally, the factor $[72/\cos^2\lambda]$ is found to be 162 to 184, implying that λ lies between 48° and 51°, a physically reasonable range. Equation (5.2.16) may be rearranged to give

$$f = \frac{H}{Re} \tag{5.2.17}$$

where $184 > H > 162$, f is a friction factor given by $f = \Delta p d \epsilon^3 / L \rho U^2 (1 - \epsilon)$ and Re is a Reynolds number given by $Re = \rho U d / \mu (1 - \epsilon)$.

At higher Reynolds number, inertial effects become significant and compete with viscous effects. The empirical correlation according to Ergun (1952) is then the means of determining pressure gradient. His correlation is

$$f = \frac{150}{Re} + 1.75 \tag{5.2.18}$$

At low Reynolds numbers, Eq. (5.2.18) reduces to Eq. (5.2.17) except that the numerical constant is slightly different. At high Reynolds numbers, f becomes constant and inertial effects control behavior.

Note, however, that Eqs. (5.2.6) and (5.2.8) are only true if the flow is such that the first term on the right hand side of Eq. (5.2.18) is dominant. If the Reynolds number is about 10 or greater, the error in f, due to neglecting the second term on the right hand side of Eq. (5.2.18), exceeds 10%, and Laplace's equation will no longer provide a sufficiently accurate guide to the pressure distribution. Previously at low Reynolds numbers where Darcy's law applied [Eq. (5.2.3)]

$$v = -k \frac{dp}{dy}$$

Now

$$-\nabla p = \mathbf{v} (f_1 + f_2 |v|) \tag{5.2.19}$$

where f_1 and f_2 are estimated by the Ergun equation or have been determined empirically. The analysis is then more complicated and proceeds via introduction of the stream function. Stanek and Szekely (1974) provide a discussion of these matters.

If the particles constituting a bed are not spherical, then the friction factor f and Reynolds number Re appropriate to Eq. (5.2.18) are given by

$$f = \Delta p (\phi_s d) \epsilon^3 / L \rho U^2 (1 - \epsilon)$$

$$Re = \rho U (\phi_s d) / \mu (1 - \epsilon)$$

where ϕ_s is a shape factor defined by

$$\phi_s = \frac{\text{surface area of sphere of equal volume to the particle}}{\text{surface area of particle}}$$

The appropriate particle diameter d for inclusion is the diameter of the sphere having the same volume as the particle.

When the particles are distributed in size, Kunii and Levenspiel (1969) recommend for use in the expressions for f and Re the surface area mean diameter, which is given by

$$\left[\sum_{i=1}^{n} \frac{X_i}{d_i} \right]^{-1}$$

where X_i is the mass fraction of solids of diameter d_i, n denotes the number of size fractions (say, sieve cuts).

The discussion has considered the powder matrix to be stationary; often this will be the case or else a sufficiently good approximation of the physical situation. If the solid is in motion, the flux of fluid will be that which would occur if there were no pressure gradient and the pore fluid moved along with the solid, plus that due to pressure gradient.

The literature on the solution of Laplace's equation is immense. Many analytical solutions are to be found in Carslaw and Jaeger (1959) and in Crank (1975). Numerical methods are often used, and reference may be made to many textbooks for procedures. Jenson and Jeffreys (1977) and Bayley et al. (1972) provide useful introductions for the beginner.

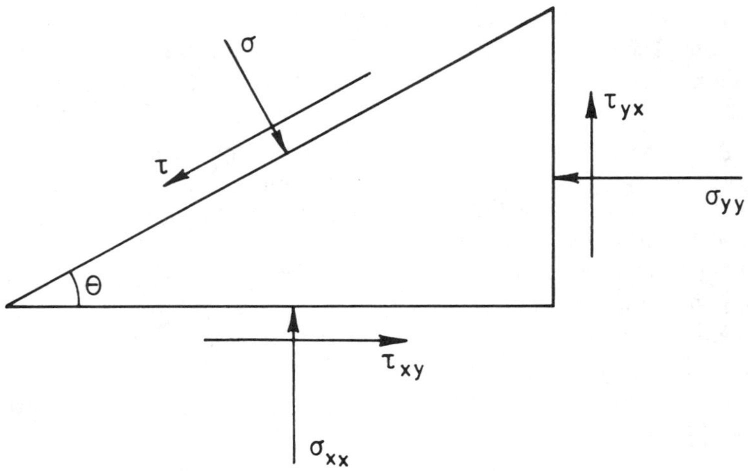

Figure 5.2.6. Equilibrium of a wedge of powder under the action of normal and shear stresses.

5.2.3. EQUILIBRIUM OF A MASS OF POWDER

Consider a mass of particles, and let interstitial fluid be absent. At static equilibrium, the states of stress in two- and three-dimensional problems may be shown as illustrated in Figures 5.2.6 and 5.2.7. For a given set of applied normal stresses and shear stresses, there exists a unique set of mutually perpendicular planes upon which no shear stress exists. These planes are termed principal planes, and it is normally desirable to conduct analysis using the principal stresses that are acting on the principal planes.

In two dimensions, the relationship between the stresses and angle may be found using the well-known Mohr's circle construction (Fig. 5.2.8), here using the convention that com-

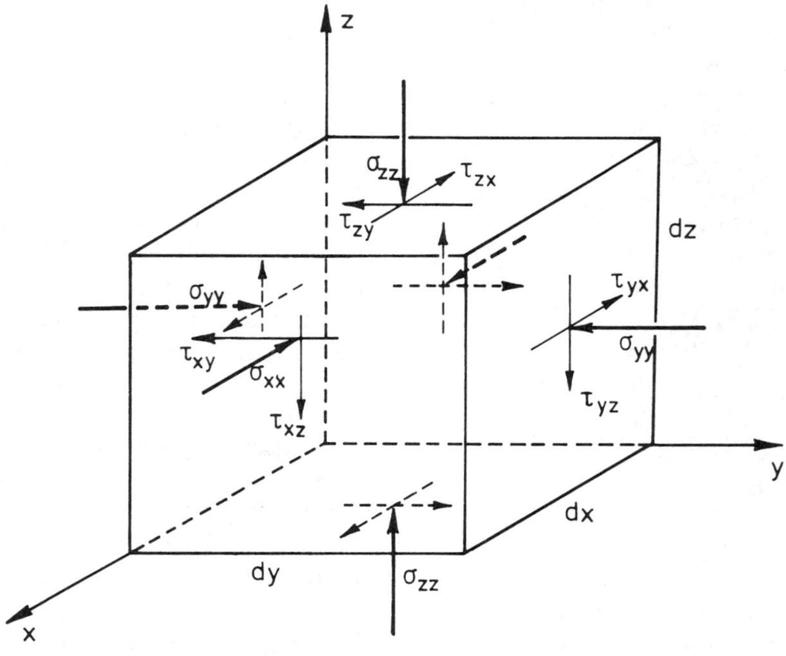

Figure 5.2.7. Equilibrium of a block of powder.

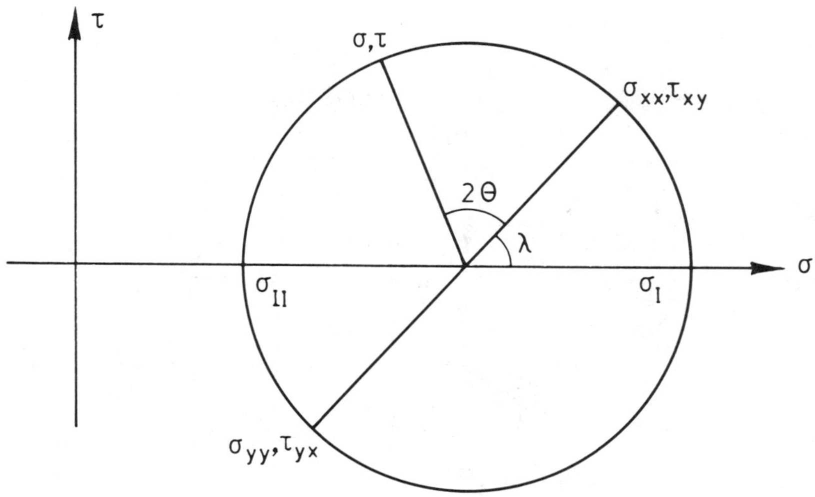

Figure 5.2.8. Mohr's circle construction.

pressive normal stresses and counterclockwise shear stresses are positive. Formal force balances as an element of the powder yield

$$\sigma = \frac{\sigma_{xx} + \sigma_{yy}}{2} + R\cos(2\theta + \lambda) \qquad (5.2.20)$$

Note that the total stress is equal to the effective stress for the purposes of the present discussion, and also

$$\tau = R\sin(2\theta + \lambda) \qquad (5.2.21)$$

where R is the radius of Mohr's circle given by

$$R = \left\{ \tau_{xy}^{\ 2} + \frac{(\sigma_{xx} - \sigma_{yy})^2}{4} \right\}^{1/2} \qquad (5.2.22)$$

and λ is twice the angle between the major principal plane and the plane containing (σ_{xx}, τ_{xy}).

In three dimensions, three principal stresses may be found, for instance, σ_1, σ_2, and σ_3, and related procedures can be used to allow for rotation of axes (Harr, 1966). There are three invariants I_1, I_2, I_3 that control such behavior, namely

$$\begin{aligned} I_1 &= \sigma_1 + \sigma_2 + \sigma_3 \\ I_2 &= \sigma_1\sigma_2 + \sigma_2\sigma_3 + \sigma_3\sigma_1 \qquad (5.2.23) \\ I_3 &= \sigma_1\sigma_2\sigma_3 \end{aligned}$$

In general, the stresses in a powder will vary with position. In the absence of pore fluid flow, differential force balances may be carried out in an element of the powder such as that shown in Figure 5.2.9, which shows only those forces acting in the z direction. These balances yield

$$\frac{\partial \sigma_{xx}}{\partial x} + \frac{\partial \tau_{yx}}{\partial y} + \frac{\partial \tau_{zx}}{\partial z} = 0 \qquad (5.2.24a)$$

$$\frac{\partial \tau_{xy}}{\partial x} + \frac{\partial \sigma_{yy}}{\partial y} + \frac{\partial \tau_{zy}}{\partial z} = 0 \qquad (5.2.24b)$$

$$\frac{\partial \tau_{xz}}{\partial x} + \frac{\partial \tau_{yz}}{\partial y} + \frac{\partial \sigma_{zz}}{\partial z} + \gamma = 0 \qquad (5.2.24c)$$

where ρ_s is the bulk density of the solids and $\rho_s g = \gamma$

5.2.4. FAILURE OF POWDERS

The manner in which a powder deforms depends on the initial voidage of the powder system and upon the applied stresses. If a powder is subjected to a constant normal stress, two possbile trajectories of the relationship between shear stress and strain may be found. If the material is loosely packed initially, the shear stress increases with the strain and ulti-

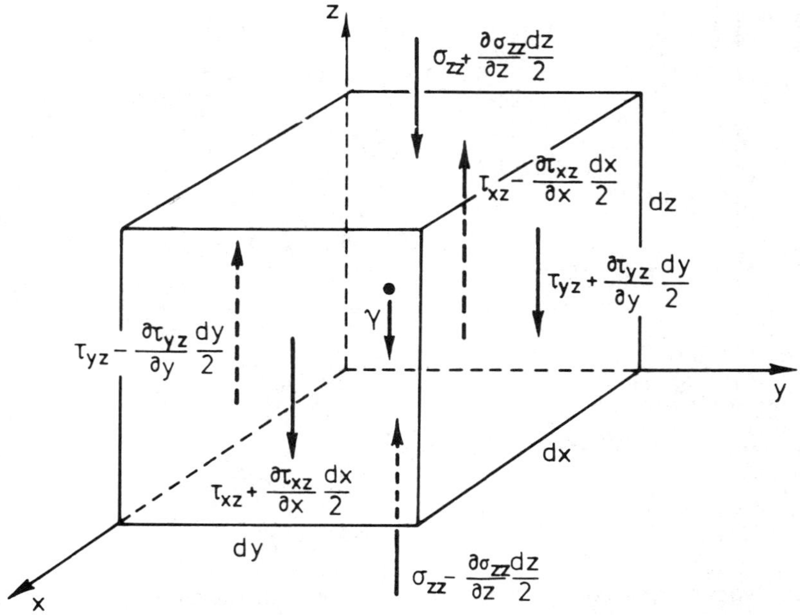

Figure 5.2.9. Equilibrium of a block of powder, stresses varying with position.

mately reaches an asymptotic value. If the material is densely packed initially, then the shear stress passes through a maximum value but then settles out at the same asymptote (Fig. 5.2.10). The relationship between applied stress and the specific volume of the powder v is formalised in the three-dimensional presentation termed a Hvorslev surface, illustrated in Figure 5.2.11. This figure is shown, for conceptual simplicity, as a relationship between normal stress, shear stress and v, although a more fundamental set of coordinates is called for (Atkinson and Bransby, 1978). In this figure, the line shown on the plane $\tau = 0$ gives the relationship between consolidating normal stress and specific volume. The curved plots drawn on planes of constant specific volume show the relationship between normal

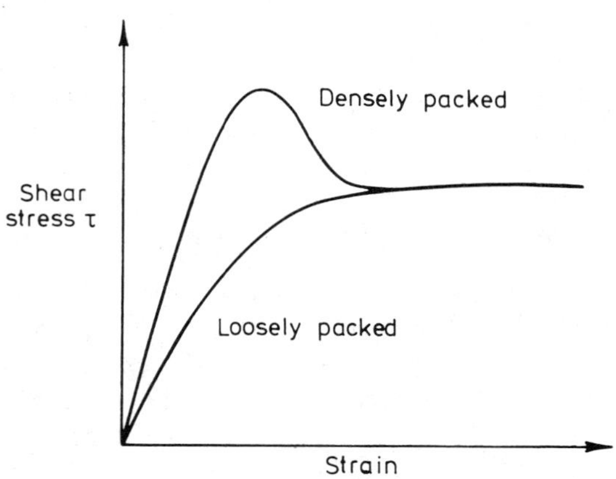

Figure 5.2.10. The relationship between stress and strain for a powder.

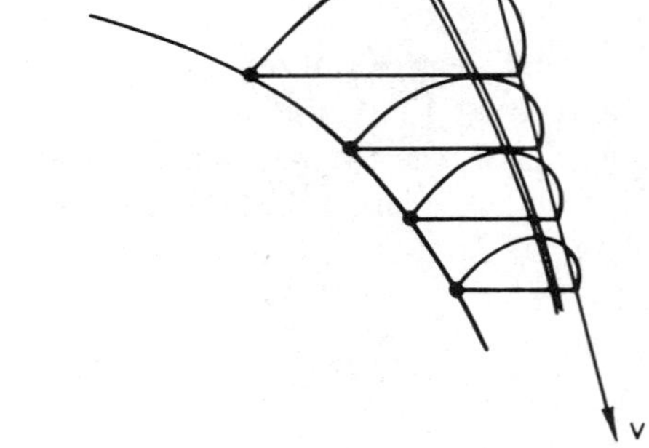

Figure 5.2.11. Illustration of the Hvorslev surface and of the critical state.

stress and shear stress for tests conducted at constant volume. The double line shows the critical state condition, i.e., that at which the powder suffers indefinite strain at the specified conditions of shear stress, normal stress, and specific volume. The curves shown in Figure 5.2.11 are seen on sections drawn on planes of constant specific volume v. Whether a powder is cohesive or cohesionless has no effect on the general behavior.

From powder testing, it is common to determine a material failure locus. At low stresses this is commonly found to be a straight line, although the fundamental reason why remains obscure. If the vertical normal stress on a plane is σ_z (Fig. 5.2.12), then it is possible to draw two Mohr's circles that are tangential to the material failure locus, the powder having a cohesion c and angle of internal friction ϕ, the failure stress τ_f being given by

$$\tau_f = c + \sigma \tan \phi \qquad (5.2.25)$$

taking the failure locus to be linear. A Mohr's circle may not penetrate above the failure locus since, on some planes, the powder would be called upon to sustain a shear stress greater than it is able to do. For the left-hand circle, the soil fails by horizontal expansion $\sigma_z > \sigma_x$, and the failure is said to be active. For the right-hand circle failure is by horizontal compression $\sigma_x > \sigma_z$ and the failure is said to be passive. From the geometry of the circle, in the active case, slip planes form at an angle of $\pi/4 + \phi/2$ from the σ_z plane and in the passive case $\pi/4 - \phi/2$ from the same plane.

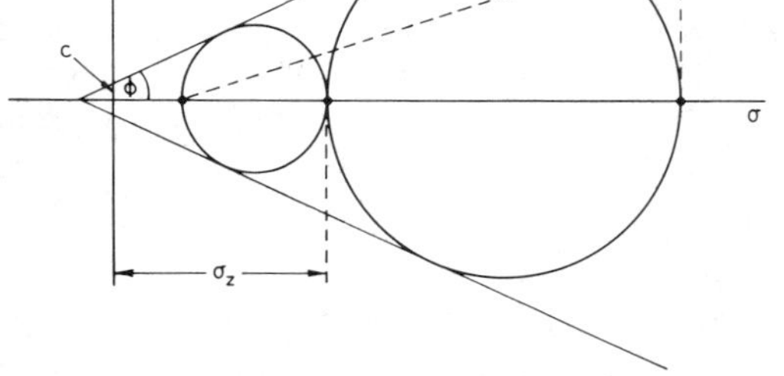

Figure 5.2.12. Material failure; active and passive states.

For a cohesionless powder, the minimum horizontal stress, termed the fully active stress, is given by

$$\left(\frac{1 - \sin \phi}{1 + \sin \phi}\right) \sigma_z$$

and the maximum horizontal stress by

$$\left(\frac{1 + \sin \phi}{1 - \sin \phi}\right) \sigma_z$$

Many problems are solved by first deciding upon the stress state that exists and then solving the force equations governing the system.

The partial differential equations (Eqs. (5.2.24b,c)] may be solved with a failure criterion such as that illustrated in Figure 5.2.12, by the method of characteristics developed by Sokolovskii (1965); a useful recent reference is that of Horne and Nedderman (1976) but such methods of solution have not been developed greatly. Use is probably restricted since the material usually does not fail as a whole, failure being confined to the narrow regions.

5.2.5. VARIATION OF POWDER PROPERTIES

In practice, during processing, powder is not homogeneous. The voidage will be different both adjacent to vessel walls and also vary throughout the system in an approximately periodic manner that is associated with the occurrence of failure zones. This is shown schematically in Figure 5.2.13. The significant shear strain occurs in these failure zones, the voidage in the failure zones being governed by the condition of unlimited strain on the Hvorslev surface. The exact position on this curve will be determined by the stresses existing in the zone.

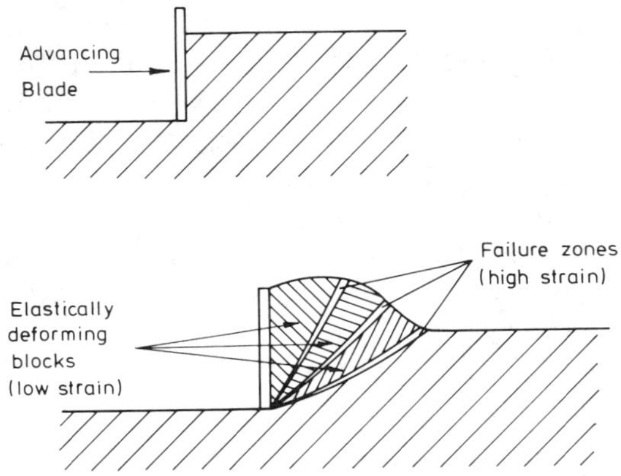

Figure 5.2.13. Development of failure zones as a blade advances into a powder.

For powders that are of diameter greater than 100 μm, size and density segregation of material may occur. Such separation processes have been shown to take place in failure zones because of inter-particle percolation caused by free-surface separation in tumbling regions near the surface and segregation and by fluid forces during any free-flight in the gas space above the powder.

The factors responsible for separation have been reviewed at length by Bridgwater (1976). The work of Szekely (1976), concerned with the influence of particle maldistribution on gas distribution as applied primarily in the metallurgical industries, shows the consequence of powder inhomogeneity.

5.2.6. ELASTIC EFFECTS

The behavior of a mass of particles prior to failure has not been explored very greatly in chemical processing problems, analyses of those situations in which unlimited strain due to plastic failure being those of much practical significance and upon which most attention has been lavished. However, elastic behavior is of significance, say, in the compression of filter cakes or in the penetration of jets of fluid into a bed of particles, as in a spouted bed. Such areas have not yet been explored fully for problems in the chemical industry. In practice, of course, a powder is not strictly elastic in that whereas stress and strain are related prior to failure, the behavior is not strictly linear and the unloading behavior is different from the loading behavior. Nonetheless, it is possible to introduce a Young's modulus for a given part of a process and follow the formalism of elastic analysis. This is reviewed well by Harr (1966). A recent example of work in the civil engineering field concerned with the development of cavities in soil caused by flow of interstitial fluid has been given by Carter (1979). These ideas can be integrated into the critical state concept (Atkinson and Bransby, 1978).

5.2.7. INERTIAL EFFECTS IN POWDER FLOW

In previous sections, the elastic and plastic properties have been noted but inertial (momentum) effects have not been included. Very full equations for the behavior of powders have been given by Goodman and Cowin (1972); a recent review of continuum mechanical and statistical approaches to the mechanics of granular materials has been given by Cowin and Satake (1979). In general, as for elastic problems, few problems of significance for chemical processing have yet been solved in which inertial terms are significant.

Rather than examining general equations that have yet to lead to general methods of solution, we shall follow the specific problem studied by Davidson and Nedderman (1973). They analyzed the flow of material in a conical convergent hopper, as shown in Figure 5.2.14. Material is assumed to flow radially towards the point 0. It is thus being extended in the r direction and compressed in the θ direction, which is shown and the ψ direction, the other spherical coordinate, which is not shown. The material motion is taken to be radial only, and the granular solid leaves the hopper at $r = r_0$. The solid, with a constant angle of internal friction ϕ and with walls that are smooth, exerts no shear stress on the moving particles.

Now the shear stress $\tau_{r\theta}$ is zero on the center line of the hopper by symmetry and is zero at the hopper wall, which has been taken to be

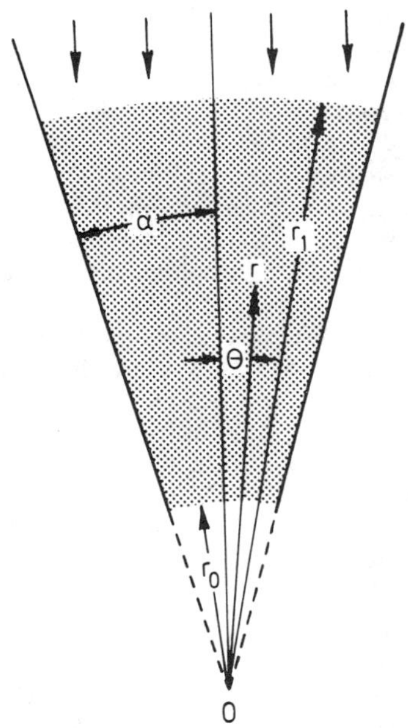

Figure 5.2.14. Flow of powder out of a conical hopper.

smooth. $\tau_{r\theta}$ is taken to be zero for all values of θ, an assumption strengthened by the experimental observation that the radial velocity depends little on θ. Furthermore, since the motion is symmetrical about the vertical axis, $\tau_{r\psi} = \tau_{\theta\psi} = 0$. These statements about shear stresses enable us to deduce that σ_r, σ_θ, and σ_ψ are principal stresses. Furthermore, in view of the deformation behavior already described, σ_r must be less than both σ_θ and σ_ψ. Hence

$$\sigma_\theta = \sigma_\psi = K\sigma_r \tag{5.2.26}$$

where

$$K = \frac{1 + \sin\phi}{1 - \sin\phi} \tag{5.2.27}$$

Now the radial velocity of the particles is related to the mass flow-rate of particles W by

$$v = \frac{W}{2\pi\rho_s r^2(1 - \cos\alpha)} \tag{5.2.28}$$

where ρ_s denotes the bulk density. A radial force balance, including the effects of momentum, gives

$$\frac{dv}{dr} = -\frac{1}{r^2}\frac{d}{dr}(r^2\sigma_r) + \frac{(\sigma_\theta + \sigma_\psi)}{r} - \rho_s g\cos\theta \tag{5.2.29}$$

which when θ is small can be rearranged to yield

$$2\rho\frac{V^2}{r^5} = \frac{d\sigma_r}{dr} - 2(K - 1)\frac{\sigma_r}{r} + \rho_s g \tag{5.2.30}$$

where

$$V = \frac{W}{2\pi\rho_s(1 - \cos\alpha)} \tag{5.2.31}$$

Upon integration, letting the radial stress fall to zero at the hopper outlet, $\sigma_r = 0$ at $r = r_0$,

$$V^2 = \left(\frac{1 + K}{2K - 3}\right)\left(\frac{1 - y^{3-2K}}{1 - y^{-2-2K}}\right) g r_0^5 \tag{5.2.32}$$

Hence the mass flow rate of material is proportional to (outlet diameter)$^{5/2}$, an experimental result found by many workers. It should be noted that the predicted rate of flow is about twice than observed, and this is ascribed by Davidson and Nedderman (1973) to the wall shear generating a velocity profile in the material.

It may be seen that inertial effects may be important in solids flow and a method for introducing such effects in one particular case has been shown.

5.2.8. CONCLUSION

Some of the important aspects of powder behavior have been outlined. The references given in the test permit access to the more detailed structure of the subject. The principles have been applied to a number of important problems but much remains to be done. Much also needs to be done to introduce the ideas into undergraduate engineering curricula.

REFERENCES

1. J. H. Atkinson, and P. L. Bransby, 1978. *The mechanics of soils: An introduction to critical state soil mechanics*, New York: McGraw-Hill.
2. F. J. Bayley, J. M. Owen, and A. B. Turner, 1972. *Heat transfer*. London: Nelson.
3. J. Bridgwater, 1976. Fundamental powder mixing mechanisms. *Powder Technol. 15*:215–236.
4. P. C. Carman, 1937. Fluid flow through granular beds. *Trans. Instn. Chem. Engrs. Lond. 15*:150.
5. H. S. Carslaw, and J. C. Jaeger, 1959. *Conduction of heat in solids*. 2nd edition. Oxford University Press.
6. J. P. Carter, M. F. Randolph, and C. P. Wroth, 1979. Stress and pore pressure changes in clay during and after the expansion of a cylindrical cavity. *Int. J. Num. Anal. Meth. Geomechanics 3*:305–322.
7. S. C. Cowin, and M. Satake, 1979. Continuum mechanical and statistical approaches in the mechanics of granular materials. *J. Rheology 23*:243–256.
8. J. Crank, 1975. *The mathematics of diffusion*, Oxford: Clarendon Press.
9. J. F. Davidson, and D. Harrison, 1963. *Fluidised particles*. Cambridge, England: Cambridge University Press, pp. 9–12.
10. J. F. Davidson, and R. M. Nedderman, 1973. The hour-glass theory of hopper flow. *Trans. Instn. Chem. Engrs. 51*:29.
11. S. Ergun, 1952. Fluid flow through packed columns. *Chem. Engng. Progr. 48*:89.
12. M. A. Goodman, and S. C. Cowin, 1972. A continuum theory for granular materials. *Arch. Rat. Mech. Anal. 44*:249.
13. M. E. Harr, 1966. *Foundations of theoretical soil mechanics*. New York: McGraw-Hill.
14. R. M. Horne, and R. M. Nedderman, 1976. Analysis of the stress distribution in two-dimensional bins by the method of characteristics *Powder Technol. 14*:93–102.
15. V. G. Jenson, and G. V. Jeffreys, 1977. *Mathematical methods in chemical engineering*. London: Academic Press.
16. J. Kozeny, 1927. Über Kapillare Leitung des Wassers im Boden. *S. B. Akad. Wiss. Wien*, Abt. IIa, *136*:271.
17. V. V. Sokolovskii, 1965. *Statics of granular media*. Oxford: Pergamon Press.
18. V. Stanek, and J. Szekely, 1974. Three-dimensional flow of fluids through non-uniform packed beds. *A.I.Ch.E.J. 20*:974.
19. J. Szekely, J. W. Evans, and H. Y. Sohn, 1976. *Gas–solid reactions*. New York: Academic Press.

6
Vibration of Fine Powders and its Application

A. W. Roberts
The University of Newcastle
N.S.W., Australia

INTRODUCTION

The effect of mechanical vibrations on the physical characteristics and dynamic behavior of cohesive and noncohesive powders and granular materials is a subject of broad engineering interest. For instance, it is well known that under certain conditions the application of mechanical vibrations will cause bulk powdered materials to consolidate or to compact, this process being accompanied by an increase in strength. On the other hand, given an alternative set of conditions, the application of vibrations will cause bulk materials to dilate and undergo a reduction in strength. It is important, therefore, that these two apparently diverging characteristics be understood. As is usually the case, these characteristics are used to advantage in order to achieve certain desired objectives in powder handling and processing. However, situations may occur in practice where the presence of vibrations will have a detrimental effect on the handling and processing of bulk materials. It is important, in such cases, that the full significance be understood in order that adequate steps may be taken for preventative action at the design stage or for corrective action during the process or plant operation.

The industrial applications involving vibrations of powders and bulk solids (The term "bulk solid" is used to describe the general case of cohesive powder or bulk materials) are extremely wide and varied. For instance, many industries rely, to a considerable extent, on the need to handle materials in bulk form and it is important, therefore, that handling systems be designed to operate as efficiently and effectively as possible. One bulk handling function of particular significance concerns the design and operation of storage bins and their associated discharge equipment. Where it is usual to rely on gravity flow to discharge the contents of a bin, often there is a need to promote and control the flow, particularly where the storage bin has not been designed to operate under mass flow conditions. Where flow promotion is necessary, devices that impart mechanical vibrations to the bulk solid are used extensively.

The principle embodied here is one in which the vibrations decrease the strength of the bulk material, thereby increasing its ability to flow.

Perhaps the more obvious and widely used application of vibration of powders is concerned with increasing compaction and density. This field of usage is of importance to many industries, such as those concerned with powder metallurgy, casting and foundry practice, ceramics, cements and silicates, and plastics. The pharmaceutical industry, with its heavy reliance on packaging and pelletizing of powders, has a dominant interest in this field of application; so too does the food industry.

The diversity in range of application is further shown in the use of vibration to fluidize granular materials and the associated application of vibrated beds in convective drying of granular materials. On other occasions, the damping characteristics of such materials as sand have been used to reduce transmitted vibration and noise.

The nature and characteristics of wave propagation through powdered or granular media are of importance to industries involved, in some way, with the general field of geophysics. This is true, for example, for oil exploration; it is equally true for civil engineering, where great reliance is given to the correct design of foundations and footings for building, dams and other large structures. With respect to the latter, earthquakes or tremors, blasts, traffic, and machines are all sources of vibration that may propagate through soils. It is not surprising, therefore, that the general field of soil mechanics has given much attention to this subject.

The objective of this chapter is to identify and discuss some important basic concepts of the vibration of powders and bulk solids. Following a general overview of some relevant research in this field, the body of the material presented is devoted to the study of the influence of mechanical vibrations on shear strength of consolidated bulk solids and the determination of stiffness, damping, and resonance characteristics. A vibrating shear cell apparatus, which permits these properties and characteristics to be determined, is described. A failure criterion is presented that relates the shear strength during vibration under given normal pressure with the voidage and vibration velocity on the plane of failure. The effect of vibrations in reducing friction of powders and bulk solids in contact with boundary surfaces such as the metal walls of a hopper is discussed. In view of its importance to bulk materials handling operations, the application of vibrations to promote the gravity flow of bulk solids from storage bins is briefly reviewed. Other topics discussed include the applications of random vibrations of bulk solids and some basic principles concerning the role of vibrations in compacting or consolidating powders.

6.1. LITERATURE REVIEW

The overall field of mechanical vibrations is quite extensive and has been the subject of a great deal of research, in the area, for instance, of vibrations of rigid and elastic bodies, machines, and structures. The general theory is now well established and thoroughly documented. The particular area of vibration of powders and bulk solids has, by comparison, received less attention, and this may be due largely to the extreme difficulty of accurately modeling and theoretically analyzing such materials under dynamic conditions. The complexities due to such parameters as particle shape and size distribution, moisture content, temperature, consolidation, and loading conditions have meant that much of the research to date has relied heavily on experimental investigation and, consequently, the published results are somewhat empirical.

So that some appreciation may be gained of the research associated with the vibration of powders, the following overview of the published literature is presented. This overview is by no means exhaustive, but it is based on a selection of published works deemed relevant to the present topic.

Developments in Bulk Solids Research

It is important to recognize the foundation work in the study of particulate solids. The origins of this area of study can be traced back to such people as Coulomb[1] and Rankine[2] who studied the frictional behavior of sand and Reynolds[3,4] who observed the dilatancy effect of sand while undergoing deformation. In Reynold's work, two areas of development are significant: that associated with the storage, flow, and handling of bulk solids and that associated with soil mechanics.

The early work associated with bulk solids was carried out in the latter part of the nineteenth century and the beginning of the present century. Of particular note are the contributions of Janssen,[5] Airy,[6] Jamieson,[7] and Ketchum,[8] who were concerned with, essentially, the behavior of granular material under static conditions.

However, it was not until the 1950s that any significant progress in this field took place. The modern developments are almost entirely due to the pioneering work of Jenike. Although Jenike and his colleague Johanson have published many papers in this field, the three University of Utah bulletins[9,10,11] lay the foundations for the modern theory of bulk solids storage and flow. The work of Jenike has precipitated a great deal of research in the field of bulk solids handling.

The Contribution of Soil Mechanics

Prior to the work of Jenike and study of particulate solids was mainly associated with soil mechanics. Since soil mechanics is mainly concerned with retaining walls and foundation design, the internal stresses are much higher than those encountered in bulk solids handling. Furthermore the main concern of soil mechanics is with the conditions existing within soils before failure, whereas the main interest in bulk solids handling is with the conditions under which failure and flow can occur. Nevertheless, the general similarities between the two fields of study permit some qualitative comparisons to be made.

Although the general area of soil mechanics, like the overall field of mechanical vibrations, is quite vast, only the work of certain people is particularly relevant to the present discussion and needs to be highlighted. The distinctive work of Hvorslev,[12] who established the fundamental principle of failure, is of particular importance to understanding the mechanism of failure in bulk solids as induced by mechanical vibrations. Hvorslev, who studied the stress condition in cohesive soils, showed that the peak shear stress at failure is a function of the effective normal stress on and the voids ratio (or density) in the plane of failure; this condition is independent of the stress history of the sample. The work of Hvorslev was further extended by Roscoe et al.[13] who established the concept of a failure surface in the three dimensional space of shear stress τ, normal stress σ, and voids ratio e. They also showed the existence of a critical void ratio boundary at which unlimited deformation could take place without change in the three variables τ, σ and e.

The fundamental concepts developed by Hvorslev and later developed by Roscoe et al. stimulated further research in the same general vein. For instance Ashton et al.[14] proposed a similar three dimensional surface that extended into the tensile region. Their results showed a functional relationship between bulk density and strength for a given powder at constant humidity. Williams and Birk[15] proposed that the Jenike consolidation procedure was aimed at the attainment of the critical density for a given normal pressure, since for a correctly consolidated sample in the direct shear cell test, there is virtually no density change during shear.

Other researchers such as Palma,[16] Wroth and Basset,[17] and Rowe et al.[18] have made valuable contributions by deriving stress-strain relationships for sand and clay, utilizing energy concepts. In essence, the external deformation energy was equated to the internal energy changes caused by frictional loss and stored elastic or potential energy.

In the area of soil mechanics considerable attention has been given to the study of dynamic loadings on soils and foundations. For instance Mitchell[19] used the theory of rate processes to relate the shearing resistance of soils in triaxial compression to frictional and cohesive properties, effective stress, soil structure, rate of strain, and temperature. Lysmer and Richart[20] studied the dynamic response of footings to vertical loading. They showed that all vertically loaded footing-soil systems are strongly damped because of wave propagation in the subsoil. Mitchell et al.[21,22] followed the early work of Mitchell in using the fundamental theory of rate processes to study the time-dependent deformation of soils. Funston and Hall,[23] in carrying out research into footing vibrations with nonlinear subgrade support, used a simple, single degree-of-freedom model to represent the dynamic motion of a foundation. While recognizing the limitations of the model, they showed that it had considerable merit in facilitating investigation of soil properties and foundation parameters as well as permitting estimates to be made of foundation motions.

D'Appolonia[24] presented criteria for the placement and improvement of soil required to sustain dynamic loadings. Timmerman and Wu[25] studied the behavior of sands under cyclic loading. Their results lead to some qualitative conclusions concerning the stress-strain relationships under dynamic loading. For the range of stress and acceleration studied, the soil deformation was primarily due to shear. In the earlier research of Greenfield and Misiaszek[26] concerning the vibration compaction of Ottawa sand, the existence of a resonant frequency at approximately 30 Hz was shown to occur. This frequency affected the maximum change in void ratio and minimum final void ratio.

Instrumentation and Experimental Techniques

The area of soil mechanics has also given rise to the development of instrumentation and experimental techniques for the determination of the dynamic properties of soils. For example, Yong Suk Chae[27] employed four different methods to determine certain dynamic material constants of soil. The four methods were the resonance column method, amplitude ratio method, elastic half-space method, and wave propagation method. The dynamic shear modulus and wave velocity can be determined with respect to changes in frequency, amplitude, confining pressure, and water content. Similar testing techniques were also used by Drnevich et al.[28] and de Graft-Johnson[29] to measure the damping capacity, shear modulus, and elastic modulus of soil.

In the research of Youd[30] a direct shear apparatus was developed in order to investigate the effects of vibration on the shear strength and void ratio of dry granular materials. The results showed that both the critical void ratio and coefficient of internal friction are reduced considerably by the application of vibration.

Fundamental Studies in Powder Mechanics

A number of other researchers working in the general area of powder mechanics have contributed significantly to the study of vibrations of powders. For example, Scarlett and Eastman[31] studied the propagation of shock waves through a bed of granular materials. Their apparatus consisted of a vertical cylinder filled with granular materials, the cylinder having a centrally located loading transducer in the base. By applying single-impulse shock loading via this transducer the velocity and attenuation of the shock wave reaching the surface was measured.

It was found that the intensity of the disturbance reaching the surface was strongest vertically above the energy source and decreased along inclined lines until no disturbance was detected outside a 45° cone emanating from the energy source. The velocity through the bed was independent of the height of the bed and the size fraction of the sand used, but was a maximum in the vertical direction, while decreasing along lines of increasing angle to the vertical. The total impulse at a plane in the bed decays exponentially with the height of the bed. The velocity of propagation (around 140 m/s for dry sand) is substantially slower than that in a solid and decreases as it deviates from the original input direction.

Gray and Rhodes[32] examined the mechanism of energy transfer during the vibratory compaction of powders. They modeled the bed as plastic bodies on the one hand and as viscoelastic bodies on the other. When a powder is compacted by vibration, the final density is a function of the energy transferred to the bed from the vibrator. The authors found that at frequencies less than 150 Hz the powder acts as a coherent mass and is projected from the container base subsequently colliding with it. The motion is then nonsinusoidal. Both models predicted a decrease in the energy transferred to the powder as the acceleration is increased at constant frequency, but the models fail when the powder ceases to behave as a coherent mass. This occurs at frequencies above 150 Hz, accelerations greater than 10 g or when the bed is fluidized by vibration. Within the limits of the models the visco-elastic model was shown to be superior.

The application of vertical vibrations to fluidize beds of granular materials has been the subject of some research, such as that reported by Nicklin and Hopkins[33] and by Chlenov and Mikhailov.[34] Vibration frequencies between 2 and 100 Hz have been used to achieve fluidization in this way with corresponding vibration amplitudes being such that the accelerations were in the range 0–10 g. The effect of the compressibility of the air in the gap between the container and the bed of granular material or the percolation of air through the material is shown to be of particular interest and importance.

This problem has been analyzed in some detail by Gutman[35] who improved the earlier model of Kroll[36] with the inclusion of the air compressibility effect. Gutman performed a very comprehensive analytical and experimental study, and in addition to providing a more fundamental insight into the behavior of vibrated beds with air percolation, he also demonstrated the basic mechanism of heat transfer in, and energy dissipation of a vibrated bed. He showed that when a vertical plate heater is immersed in a vibrated bed, the heat transfer from the plate is enhanced by the scouring action of the gas sublayer as a result of the motion of the particles in the vertically vibrated container.

Harwood,[37] in studying the vibration segregation of lactose, used radioactive pills to trace the movements in a vertically vibrated tube. The results showed that at a critical vibration condition (100 Hz and 10 g) the powder was induced to take on a semifluidized state.

A fundamental understanding of the nature and mechanism of vibration of powders and bulk solids must ultimately be greatly assisted by a more thorough appreciation of the behavior of powders as conglomerations of discrete particles. The linking together of theories developed by such analyses with the already widely established and utilized continuum theory would be very beneficial. Some pioneering work in the mechanics of packings of discrete spherical particles was performed by Deresiewicz.[38] More recently Molerus[39] used the continuum and particulate approaches to formulate a theory on the yielding of cohesive powders. A simple model was proposed for the transmission of external stresses through a lattice of solid particles whereby contacting points on the surface of the particles were assumed to transmit force. This concept resulted in an expression relating the macroscopic externally applied stress to the microscopic interparticle force. The cohesive forces were shown to be a result of plastic deformation of particle contacts under external load. The concept embodied in this behavior is important in determining the resonance frequencies of bulk granular materials.

A number of other papers of fundamental importance to the understanding of the deformation and flow of powders and bulk solids have been published. By way of example mention is made of the work of Becker and Lippman,[40] Nova and Wood,[41] Blinowski,[42] Cowin,[43] Goodman and Cowin,[44] Passman,[45,46] Passman and Thomas,[47] Nemat-Nassar,[48] and Nunziato et al.[49] In the latter paper, the authors studied the behavior of one-dimensional accleration waves in an inhomogenous granular solid. They found that the average wave speed varied significantly for short propagation distances, but approached an equilibrium value for large distances. In general the amplitude of the waves attenuated with propagation distance. They concluded that the acceleration wave behavior was dominated by the elasticity of the granules and the dispersion caused by the initial nonuniformity of the material.

Vibrations of Bulk Solids

An important application of mechanical vibrations is in the promotion of flow of bulk solids from storage bins. The papers by Wahl[50] Myers[51] and Carroll and Colijn[52] emphasized the practical applications of vibrations to the solution of flow problems. The types of vibrations and live bottom bins were discussed with some rule-of-thumb indications of appropriate frequencies and amplitudes to be used.

Shinohara, Suzuki, Tanaka, and Takahashi[53-55] studied the gravity flow of noncohesive and cohesive bulk solids from vibrating hoppers. In these processes, vibrations are applied to the whole hopper in the vertical direction and the behavior analyzed by a block flow model. Discharge occurs when impact forces caused by the vibrations, together with the weight of the materials within the block, exceed the shear force within the material and that at the hopper wall. A critical value of vibration intensity can then be determined. The experimental studies show that for the cohesive materials low frequency or high intensity of vibration gave the best flow ratio with good theoretical agreement. These results are in contrast with the results for the noncohesive materials, which show that higher frequency give the best discharge rates.

Some preliminary work reported by Arnold et al.,[56] Croft,[57] and Roberts et al.[58] referred to experiments performed on a model plane-flow hopper with a horizontally vibrating insert. Both discrete frequency sinusoidal vibrations and broad band random vibrations were examined. The results indicated the significant influence of moisture content on the ability of the bulk material, in this case sand, to flow. The experiments also indicated the sensitivity to flow of the input frequency. Consistent performance of the bin was obtained by using broad band random vibrations. Narrow band vibrations had little influence in promoting flow. The need to examine the frequency and amplitude dependence of bulk solids through physical testing procedures was realized, and some preliminary work in this field was performed by Kaaden[59] and Arnold et al.[56] A vibrating shear cell apparatus was used to examine the influence of vibrations on the consolidation strength of bulk solids. Furthermore, the significant influence of mechanical vibrations on reducing hopper wall friction was shown, a result of particular importance to bin and hopper design.

The research of Roberts and Scott,[60] Li,[61] and Roberts, Scott, and Li[62] extended this earlier work in order to show the influence of mechanical vibrations on reducing the strength of bulk solids, thus increasing their ability to flow. An important aspect of this work was the development of a special vibrating, direct shear testing apparatus. The contribution by Li[61] in establishing a theory for the failure of vibrated bulk solids when subjected to shear loading is of particular significance. The work conducted by these authors is discussed in greater detail in subsequent sections of this chapter.

The design of bulk handling equipment requires a knowledge of boundary friction characteristics between a granular material or bulk solid and an adjacent surface such as a wall, or conveyer flight or casing. Although it is common to employ the Jenike type direct shear tester for this purpose, other test equipment and associated procedures have been developed. Rademacher[63] proposed a method that permitted accurate determination of the kinetic coefficient of friction between a surface and a granular mass. Sharma et al.[64] investigated the high frequency vibrational effects on soil–metal friction. Although this particular study was directed at the design and operation of tillage tools, the results are of interest to the field of bulk solids handling.

The application of laboratory scale experiments to full scale operations often requires that the necessary laws for dynamic similarity be established. These laws must be kept in mind when developing laboratory models. The final application of mechanical vibrations as an aid to flow will require careful consideration of scaling laws to ensure that correct dynamic conditions are obtained in the prototype. Some progress on modeling has already been made. For example Molerus and Schöneborn[65] have demonstrated the application of a bunker centrifuge to examine, on a model basis, the performance of actual bunkers. Johanson[66] and Sharma et al.[64] have used dimensional analysis in an attempt to generalize the results of experiment. Nonetheless, in the application of mechanical vibrations to flow promotion of bulk solids, no satisfactory modeling laws have yet been developed. However, the work already published may be considered as a guide to future work in this area.

Compaction of Powders

The application of mechanical vibrations to the compaction of powders has been studied indirectly by several authors already mentioned. For instance Gray and Rhodes[32] and Kaaden[59] include some consideration of the influence of mechanical vibrations on the packing of powdered materials.

A comprehensive overview of the principles and methods of vibration compaction is given by Shatolova et al.,[67] McGeary,[68] Hauth,[69] and Evans and Millman.[70]

6.2. MEASUREMENT OF DYNAMIC SHEAR

Introductory Remarks

A knowledge of the dynamic shear properties of powders, soils, and bulk solids is of importance to particular areas of application. For example, in soil mechanics these properties will have a significant influence on the design of foundations and retaining walls. In developing testing equipment to measure these properties, it is important to give careful consideration to the manner in which consolidation and loading stresses act as well as to the way in which vibrations may be propagated.

One area where a knowledge of dynamic shear properties is of considerable importance is in the design and operation of flow promotion devices for bulk solids handling. Mechanical vibration, as an aid to flow promotion in storage bins, may be applied in several ways. In some cases the bin may be vibrated in the vertical direction, and in this respect the distribution of acceleration within the mass is important. In other cases the vibrations may be applied in the horizontal direction or more generally in an oblique plane. Much depends on the design of the flow promotion device. For instance, some flow promotion devices apply vibrations directly to the hopper walls; another type consists of a vibrating or gyrating insert as part of a live bottom bin. An important factor is the manner in which the vibration energy is transmitted through the bulk mass in a way to ensure a reduction in shear strength in the critical regions of the flow obstruction.

The theory of bin design as developed by Jenike[9,10] is now well established and widely used. So that the application of mechanical vibrations as an aid to flow may become more effective in the design sense, there is a need for research on bulk solids vibrations, to be directed in a way that is compatible with the concepts of flowability and bin design as developed by Jenike. The research of Roberts, Li, and Scott[58,61,62] had this objective. Primarily the research was aimed at providing an insight into the mechanisms of shear failure and flow of bulk solids excited by vibration.

A significant aspect of the research concerned the development of a dynamic shear testing apparatus and associated measurement technique that was entirely consistent with the bulk solids testing procedures established by Jenike.[10,71] The apparatus, which is described in the following subsection, was designed to determine the effect of vibration frequency, amplitude, and energy transfer on the shear strength of bulk solids, as well as the wall frictional characteristics.

The dynamic shear cell apparatus developed by Roberts et al. followed the earlier work of Arnold et al.[56,72] in which a Jenike direct shear tester, fitted with vibration excitor, was used to determine the effects of vibration on the consolidation strength of bulk solids as well as wall friction. In this case the effects of vibration in reducing shear strength were not examined.

A vibrating direct shear test was used by Youd[30] in his research into the effects of vibration on the shear strength and void ratio of dry granular materials. In this case the shear cell was mounted on a shaker table, the cell being fitted with a pneumatic loading device which permitted normal pressures to be applied to the test samples in a way which could not be influenced by inertia effects due to the vertical vibrations. As will be explained in the next subsection the vertical mode of vibration employed by Youd differed from the horizontal mode used by Roberts et al.

A comment concerning the suitability of the Jenike direct shear tester for bulk solids analysis needs to be made. As is widely known, several other methods exist for determining the strength of bulk solids. These include ring shear tests, torsion shear tests, and triaxial tests. A complete review of all such methods and associated test equipment is given by Schwedes.[73] On balance, the Jenike direct shear tester is the most widely used and offers several advantages, in particular the ability to perform time consolidation tests. It is the most readily adaptable to the determination of vibration shear tests. Care needs to be taken in using the direct shear tester, since errors may occur in the measurements, as has been discussed by Rademacher and Haaker.[74]

Dynamic Shear Apparatus

The dynamic shear cell apparatus used by Roberts et al. is essentially a modified version of the Jenike, direct shear cell apparatus with provision for the application of mechanical vibrations in the horizontal plane parallel to the plane of shear. The horizontal plane was chosen in order to isolate from the measurements any influence of vertical acceleration in the gravitational field. The apparatus is arranged so that either the whole cell or the top half of the cell may be vibrated. These two arrangements are shown in Figures 6.1 and 6.2. As can be observed, the vibrations are applied transversely to the direction of shear in order to minimize any interaction effects between the vibration excitation and the recorded shear force.

The test arrangement for vibrating the whole cell during shear is shown schematically in Figure 6.1.

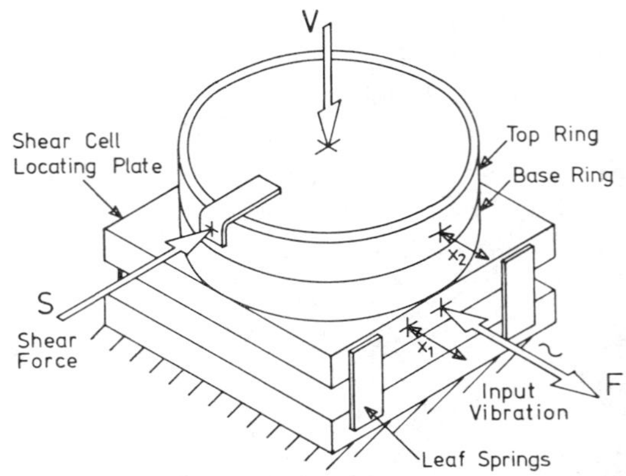

Figure 6.1. Vibrating shear cell apparatus–whole cell vibrated.

The cell is located on a vibrating platform mounted on vertical leaf springs, as indicated. Provision is made for variations in shear cell size, but basically the testing is performed on a cell of 95 mm internal diameter. Vibration of the whole cell is used primarily to examine the resonance characteristics of consolidated bulk solids.

When only the top half of the cell is vibrated, the test arrangement shown in Figure 6.2 is used. Tests performed in this way permit a greater insight into the mechanism of failure, particularly with respect to vibration frequency and amplitude. The same arrangement as shown in Figure 6.2 is used in the determination of the influence of vibrations on the shear of bulk solids in contact with bin or hopper walls. In this case a sample of the hopper wall material is located in place of the lower half of the shear cell. In all cases the amplitude of

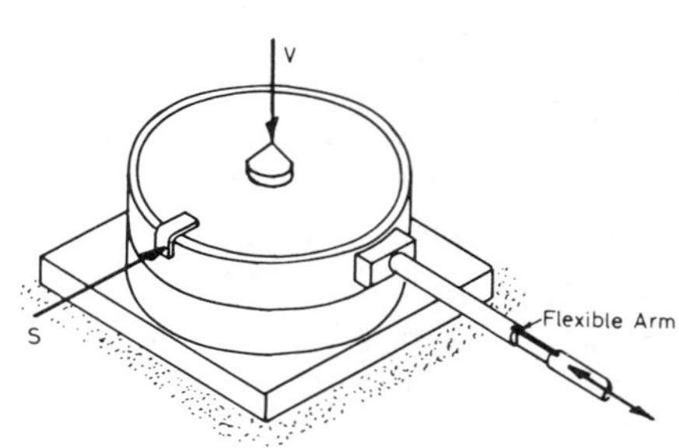

Figure 6.2. Shear cell arrangement for vibration of top half of cell.

vibration needs to be kept very small in order to prevent premature shearing of the test sample by vibration alone.

The test instrumentation is shown in block diagram form in Figure 6.3. Provision is made for either discrete frequency sinusoidal or random vibration excitation to be applied. In the case of the latter the experimental work of Roberts et al. incorporated the use of either pseudo random binary or Gaussian signals. Vibrations are applied via the force transducer, with the associated instrumentation to provide consistent amplitude control under load.

For the tests when the whole cell is vibrated, it is necessary to tune the apparatus, so that the natural frequency and hence cut-off frequency of the vibrating cell does not influence the dynamic measurement being made. It has been found advisable to use flexible springs for discrete frequency excitation such that the natural frequency of the cell and its mounting is well below the resonance frequencies of the bulk solid contained in the cell. On the other hand, for broad band random vibration excitation, the spring stiffness should be sufficiently high to provide the necessary band width. That is, since the vibrating platform behaves as a mechanical low-pass filter, its cut-off frequency must be higher than the highest frequency of interest in the random vibration analysis.

Included in the instrumentation are transducers to measure the input displacement as well as the displacement of the top ring of the shear cell in both the lateral and vertical directions. As indicated in Figure 6.3, signal processing equipment such as a correlator and spectrum analyser are desirable for analyzing the measured signals in the time domain and frequency domain, respectively. This equipment has the added advantage of permitting separate identification analysis to be performed in association with the determination of the dynamic characteristics of the particular bulk solid under test.

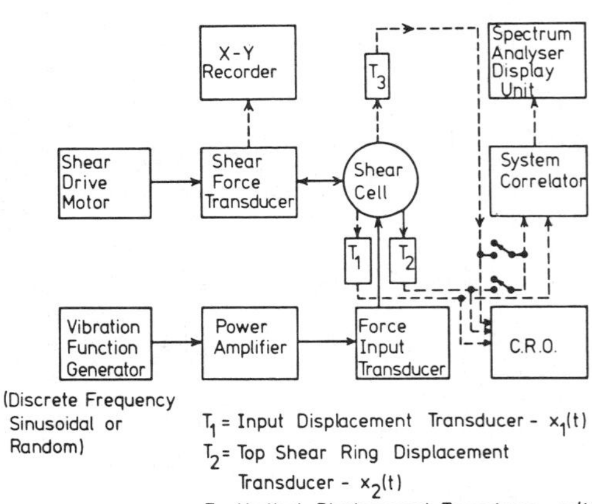

Figure 6.3. Block diagram showing test instrumentation.

Since the overall objective is to determine the effect of applied vibrations on the shear strength and flow functions of bulk solids, the standard procedure as recommended by Jenike[71] for the preparation and consolidation of the samples is adopted. The samples are first preconsolidated by a twisting procedure under the applied normal load, followed by consolidation under shear to obtain the steady state condition. Normally no vibrations are applied during the consolidation phase. Once the samples are consolidated, the vibrations are then applied during the shear test under the predetermined normal load. The apparatus does, however, permit the effects of vibration on the consolidation strength of the sample to be determined. This may be achieved by vibrating the cell under the consolidation load prior to performing the shear test. Such information is of particular value in assessing, for example, the effects on the bin performance of the vibrator being accidentally switched on before the discharge gate is open. The information is also of relevance in determining the effects a vibratory feeder may have on the performance of a bin.

6.3. DYNAMIC SHEAR CHARACTERISTICS—SINUSOIDAL VIBRATION EXCITATION

Bulk Solid Strength and Flow Parameters

The purpose of the test program using the apparatus described in the previous section is to determine the relevant strength and flow properties under applied vibrations for bin design and evaluation. For a given bulk solid the dynamic shear strength τ_f under applied sinusoidal vibrations is dependent on a number of parameters. In functional form

$$\tau_f = f_1(\sigma_1, \bar{\sigma}, f, x, \rho, H, d, T) \qquad (6.1)$$

where

σ_1 = major consolidating stress
$\bar{\sigma}$ = normal stress corresponding to shear failure
f = vibration frequency
x = amplitude of vibration
ρ = bulk density of the material
H = moisture content
d = average particle size
T = temperature of material

The major consolidation pressure is defined for the yield locus by the applied normal consolidating pressure σ and corresponding shear stress τ for which consolidation in the shear cell is perfomed. That is, for a given set of material parameters

$$\sigma_1 = f_2\ (\sigma, \tau) \qquad (6.2)$$

In bin design the Flow Function, as defined by Jenike,[71] is an important parameter. For a given set of material properties the functional form of the Flow Function is given by

$$\text{FF} = f_3\ (\sigma_1, \sigma_{cf}) \qquad (6.3)$$

where σ_{cf} = unconfined compressive stress under vibration, and is obtained from the yield function given by (6.1).

A series of tests have been performed to examine the interrelation of vibration frequency and amplitude on the shear strength for a range of material parameters. Some typical results are presented here by way of example. The results apply to pyrophyllite, a material used in the manufacture of refractories.

Shear Force versus Shear Deformation

It is interesting to observe the shear force versus shear deformation records obtained during an actual test. A typical set of results for the material pyrophyllite are given in Figure 6.4. Curve A_1 shows the consolidation graph during shear deformation under the applied normal stress τ = 8.33 kPa, while curve A_2 is the shear versus shear deformation record for the unvibrated condition under the applied normal stress σ = 2.84 kPa.

Curve B_1 is a repeat of the consolidation condition as in A_1 for a fresh sample, while curve B_2 is the corresponding shear record for the applied normal stress σ = 2.84 kPa, this time the shear force being applied simultaneously with an impressed sinusoidal vibration of frequency f = 200 Hz and amplitude X_r = .01 mm measured at the shear plane. As can be seen, the sample sheared at a value of S = 25 N compared with a value of S = 29.3 N in the case of curve A_2 for the unvibrated sample.

The reduction in shear strength as a result of the impressed vibration is immediately apparent. The behavior of the second sample beyond the point b of the shear force versus shear deformation second is interesting to note. At point b the vibration was switched off and, as indicated, the sample immediately reconsolidated and sheared at a value similar to that of the unvibrated sample depicted in curve A_2.

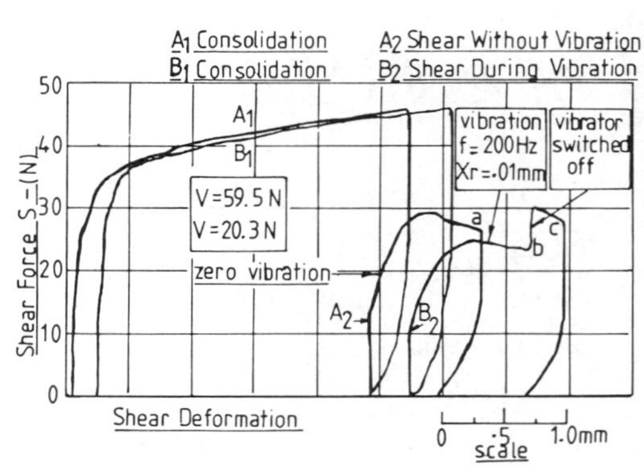

Figure 6.4. Shear force versus shear deformation. −1 mm pyrophyllite m.c. = 5% (d.b.).

Curves A_2 and B_2, which are typical of the shear force versus deformation curves obtained for all test conditions, permit some general conclusions to be drawn concerning the dynamic characteristics of a bulk solid. During loading under the action of an applied deformation rate $\dot{x}_1$ there is an instantaneous "step" change or increased in the shear force S. This increase, denoted by F_f, is due to internal Coulomb frictional characteristics of the bulk solids. Following this initial step change the shear force then increases linearly indicating elastic deformation. As loading further proceeds the shear force versus deformation characteristics become nonlinear as plastic flow occurs. Ultimately, the point of failure occurs, and this is accompanied by a reduction in shear force as the top half of the shear cell slides relative to the bottom half.

The behavior during unloading is interesting to observe. If at any time such as at points a and c of Figure 6.4 the load deformation input is stopped and the sample is held under constant load, there is shown to be practically no relaxation of the shear force S; apart from observed reductions over expanded periods of time of around 2 to 4%, the shear force remains substantially constant. When the shear force actuator of the testing machine described in Figure 6.1 is retracted, the shear force versus deformation characteristics due, in essence, the mirror image of those obtained during loading. That is, there is an immediate "step" reduction in force due to the Coulomb frictional resistance followed by a "ramp-like" elastic deformation as the elastic energy is recovered. As indicated in Figure 6.4, the slopes of the elastic deformation graphs during loading and unloading are slightly different with the stiffness of the material being lower during unloading. This is no doubt due to the change in internal characteristics of the material resulting from the shear failure.

The characteristic behavior of the bulk material during shear is shown by Figure 6.4 to be similar for the case of shear both during and in the absence of applied vibrations. The effect of the applied vibrations is to cause a slight reduction in the internal stiffness, otherwise the general characteristics are of the same form.

Dynamic Shear—Whole Cell Vibrated

A set of typical results depicting the influence of consolidation stress, applied normal stress, frequency, particle size, and moisture content (m.c.) are presented in Fig. 6.5. In order to show the magnitude of the reduction in shear stress with frequency, the shear stress ratio τ_f/τ_{f0} is plotted against frequency in all cases, τ_{f0} being the shear stress under zero frequency.

Effect of Consolidation Stress—Applied Normal Stress. Figure 6.5a shows τ_f/τ_{f0} versus frequency f for the major consolidating stress σ_1 = 17.4 kPa (corresponding to σ = 7.9 kPa and τ = 7.6 kPa) for -1 mm pyrophyllite at 5% moisture content (d.b.). The whole cell was vibrated with constant impressed amplitude X_i = ± 0.008 mm. The shear stress τ_f was determined for the three normal stress $\bar{\sigma}$ = 5.47 kPa, $\bar{\sigma}$ = 4.23 kPa, and $\bar{\sigma}$ = 2.98 kPa.

The significant reduction in shear stress with increase in frequency is clearly shown. As indicated for each applied normal stress $\bar{\sigma}$, the shear stress reaches a minimum value at a particular frequency, this characteristic being indicative of a resonance effect. Similar results were obtained for other major consolidating pressures. The resonance frequency ω_n is shown to increase with increase in major consolidating stress σ_1 (and hence σ); for each major consolidating stress the resonance frequency also increases with increase in applied normal stress $\bar{\sigma}$. These results support the results obtained in the study of X_r/X_i in the inertia model investigation. (See below, Amplitude Ratio—Whole Cell Vibrated.)

Effect of Particle Size. Figure 6.5b shows the shear stress ratio τ_f/τ_{f0} for four particle size ranges of pyrophyllite at 5% moisture content (d.b.). As indicated there is a well-defined resonance frequency in all cases, with the greatest reduction in shear stress occurring in the case of the largest particle size.

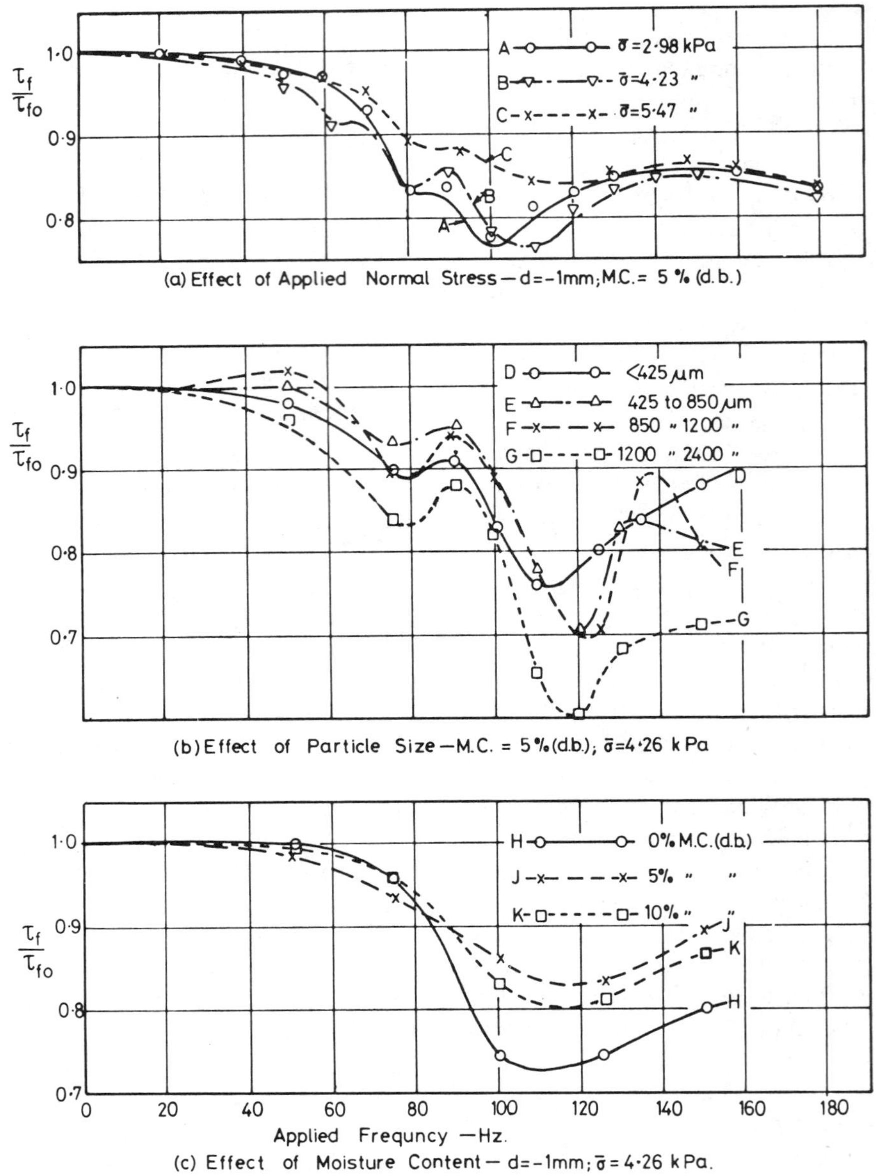

Figure 6.5. Effect of consolidation stress, applied normal stress, particle size, and moisture content on τ_f/τ_{f0} during sinusoidal vibration of whole shear cell. Pyrophyllite; σ = 7.9 kPa, x_i = ± 0.008 mm.

Effect of Moisture Content. Figure 6.5c shows the effect of moisture content on the τ_f/τ_{f0} versus frequency characteristics for −1 mm pyrophyllite. Although there is little difference between the 5% and 10% moisture content, the stress reduction at lower moisture levels is quite significant. It is clear that moisture within a bulk solid affects its dynamic characteristics, particulary with respect to stiffness and damping.

Amplitude Ratio—Whole Cell Vibrated

Measurements have been made of the absolute amplitude X_2 of the top half of the shear cell while the whole cell was being vibrated with input amplitude X_1 and frequency $\omega = 2\pi f$.

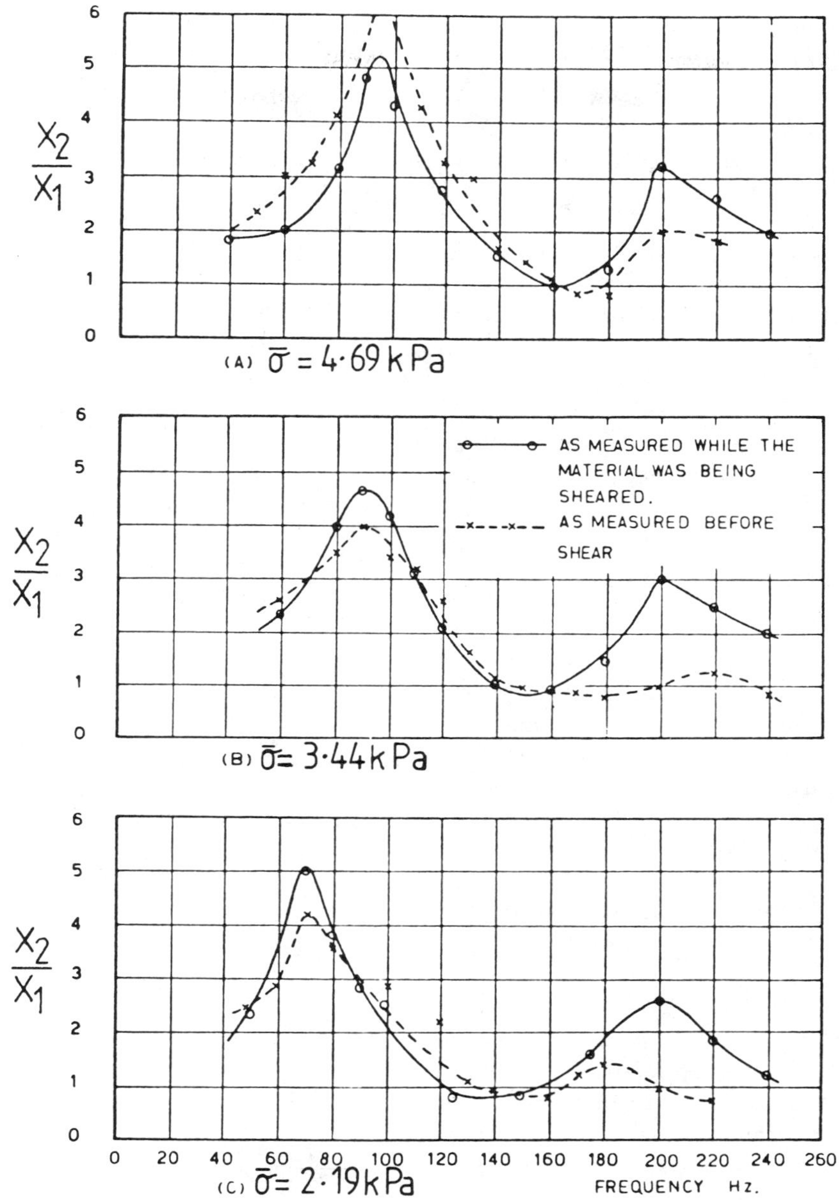

Figure 6.6. Effect of vibration on amplitude ratio X_2/X_1. -1 mm pyrophyllite, 5% m.c. (d.b.), 0 = 7.81 kPa, X_1 = .006 mm.

These measurements were made both before the samples were sheared and during shear. Figure 6.6 shows the amplitude ratios X_2/X_1 for pyrophyllite for the consolidation condition σ = 7.81 kPa and for the three applied normal stresses $\bar{\sigma}$ = 4.69 kPa, $\bar{\sigma}$ = 3.44 kPa, and $\bar{\sigma}$ = 2.19 kPa.

The effect of a resonance condition for maximum X_2/X_1 is clearly pronounced, the fundamental resonance frequency occuring at values similar to that for the corresponding minimum τ_f/τ_{f0}. Also, as indicated, there appears to be a second resonance point occurring at a frequency of approximately 200 Hz, which can be observed from the trend of the curves in Figure 6.5. It is believed that this frequency is due to the measuring equipment and not directly attributable to the bulk material.

The amplitude X_2/X_1 measured before and during shear shows an appreciable difference particularly in the region of resonance. A major influence is the effect of the loading stem (see Fig. 6.1) on the movement of the top ring. Furthermore, during actual shear, the shear deformation causes the bonds between adjacent particles in contact to fail and reform in a cyclic manner. This action may also affect the dynamic properties of the material particularly with respect to increasing the internal damping.

Dynamic Shear—Top Half of Cell Vibrated

When the top half of the cell is vibrated, the impressed amplitude X_1 and amplitude X_r on the shear plane are the same. Consequently there is no resonance effect, and the shear stress reduction with increase in frequency has a smooth decaying characteristic shape approaching, asymptotically, a limiting value, as illustrated in Figure 6.7.

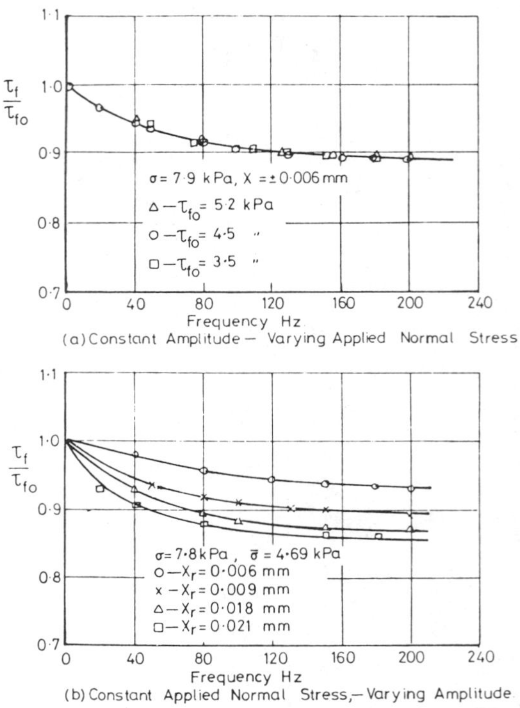

Figure 6.7. Effect of consolidation stress, applied normal stress, and amplitude on τ_f/τ_{f0} during sinusoidal vibration of top half of shear cell. −1 mm pyrophyllite, 5% m.c. (d.b.).

Effect of Consolidating Stress and Applied Normal Stress. Figure 6.7a shows τ_f/τ_{f0} versus frequency f for −1 mm pyrophyllite at 5% moisture content (d.b.). The normal consolidation stress is σ = 7.9 kPa, and the amplitude of vibration is X_r = 0.006 mm. Three applied normal pressures are examined; however, as the results indicate, the τ_f/τ_{f0} variation with frequency is virtually the same for all three normal stresses.

Effect of Amplitude. Figure 6.7b shows the influence of an increase in amplitude on the reduction in shear strength. Analysis shows that for a given frequency there is a limiting amplitude beyond which the shear stress at failure does not decrease any further.

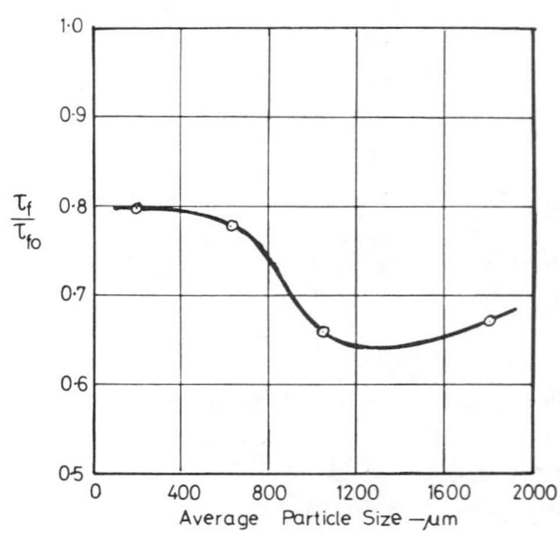

Figure 6.8. Effect of particle size on shear strength, top half of shear cell vibrated. f = 200 Hz, x_i = ± 0.01 mm, pyrophyllite, 5% m.c. (d.b.).

Effect of Particle Size. The effect of particle size on the shear ratio is illustrated in Figure 6.8. The results correspond to the impressed

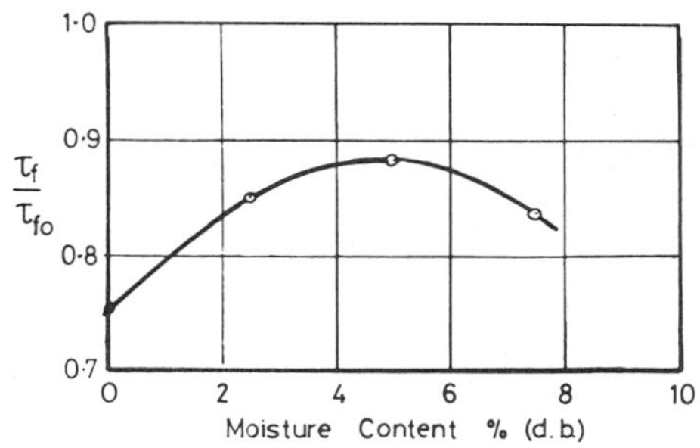

Figure 6.9. Effect of moisture content on shear strength, top half of shear cell vibrated. f = 200 Hz, x_i = ± 0.01 mm, –1 mm pyrophyllite.

frequency f = 200 Hz and amplitude X_1 = ± 0.01 mm. The reduction in strength with increase in particle size confirms the results presented in Figure 6.5b for vibration of the whole cell.

Effect of Moisture Content. Figure 6.9 shows some characteristic results for the reduction in shear strength obtained by vibrating the top half of the cell at a frequency f = 200 Hz and amplitude X_i = ± 0.01 mm. The trend of the results is similar to that indicated by Figure 6.5c for the case when the whole cell is vibrated.

Yield Loci and Flow Functions

From the viewpoint of assessing the effects of vibration on the strength reduction of bulk solids, the relevant information is best obtained from the yield loci and Flow Function. By way of illustration the yield loci for –1 mm pyrophyllite at 5% moisture content for the frequencies 0, 150, and 200 Hz and amplitude X_1 = 0.01 mm are shown plotted in Figure 6.10. These results are based on the vibrations of the whole cell, the consolidation condition being defined by the major consolidating stress σ_1 = 16.0 kPa. The reduction in shear strength is clearly evident.

The Flow Function of a bulk solid is a plot of the unconfined compressive strength σ_c against the major consolidating stress σ_1. For the single consolidation condition depicted by Figure 6.10 the reduction in unconfined compressive strength as a result of applied mechanical vibrations is readily observable. To enable a complete set of Flow Function graphs to be drawn, a minimum of three sets of yield loci corresponding to three consolidation conditions is needed. On this basis the Flow Function graphs shown in Figure 6.11 have been obtained. The static zero vibration (0 Hz) or instantaneous Flow Function provides a

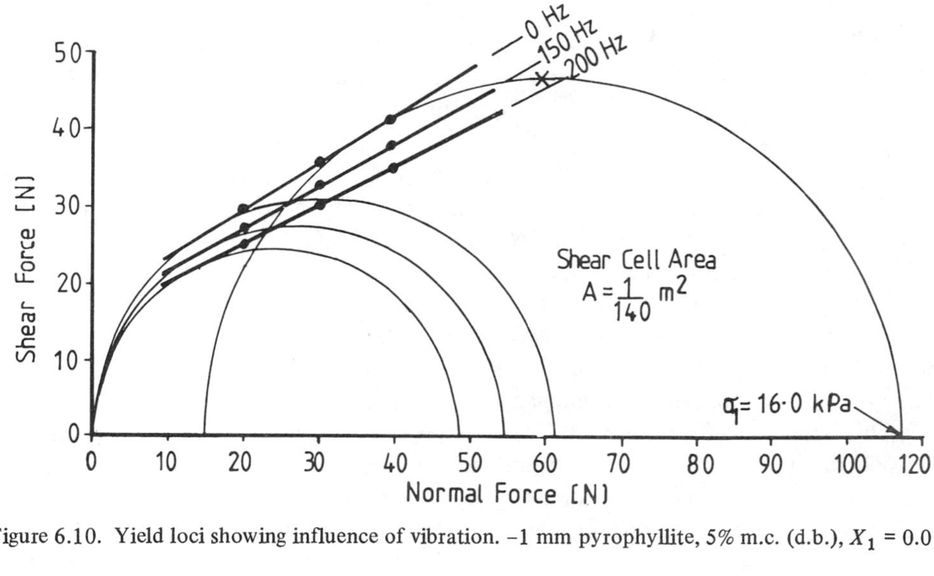

Figure 6.10. Yield loci showing influence of vibration. –1 mm pyrophyllite, 5% m.c. (d.b.), X_1 = 0.01 mm.

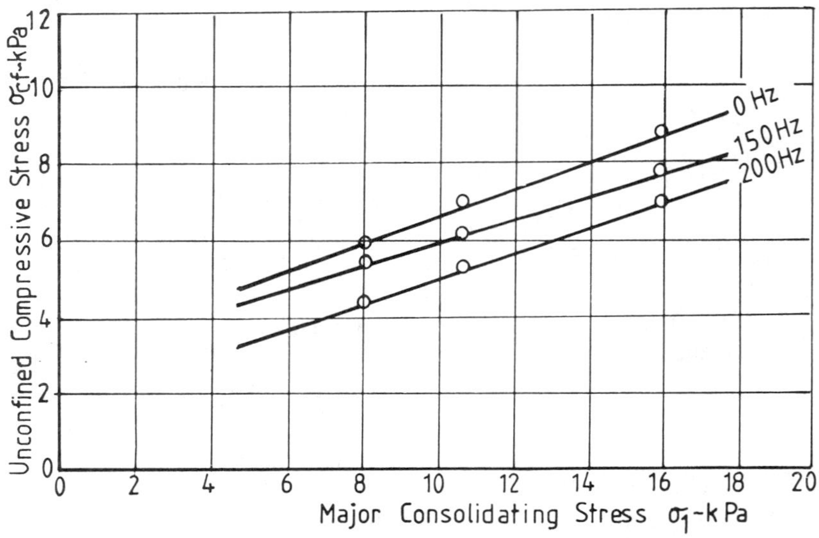

Figure 6.11. Flow Functions showing influence of vibration. −1 mm pyrophyllite, 5% m.c. (d.b.), x_1 = 0.01 mm.

reference for comparisons to be made with the vibrated Flow Functions. The reduction in strength under the applied vibrations is quite significant and clearly indicates the relevance of vibrations as an aid to flow.

Similar results may be obtained by vibrating the top ring of the shear cell only. The significance of this together with the implications for storage bin design is discussed in Section 6.9.

6.4. AN INTERTIA MODEL FOR VIBRATION OF WHOLE SHEAR CELL

Introductory Remarks

In view of the complex structure of cohesive and noncohesive granular materials, difficulties are encountered when attempts are made to analyze the deformation and flow characteristics of these materials. Broadly speaking two approaches to modeling such materials may be attempted, namely the continuum approach and the particulate or discrete particle approach. Much of the work to date has been based on the continuum approach, and already a great deal of progress and development have taken place.

The continuum model assumes that the properties of the bulk materials may be represented by continuous functions of positions in both time and space. This procedure is equivalent to assuming that the material may be subdivided indefinitely into smaller components or constituents without any change in its properties. A lumped model can therefore be used to determine quantitatively the behavior of the whole aggregate or mass. So far, this approach is most commonly used principally because it lends itself to easier experimental work and subsequent analysis. It suffers from the disadvantage, in many cases, of leading to results that are often empirical in nature. However, provided the bounds on the solutions, particularly with respect to the limits of experimental and theoretical analysis, are clearly enunciated and understood, then the continuum approach is invaluable in providing data, theories, and procedure for ready application to design and development.

In the particulate approach, a conglomeration or array of finite size particles is considered. Often it is necessary to make some simplifying assumptions, for instance, that the particles behave as rigid spheres. A physical model may be formulated by investigating the behavior of the particles due to their interactionary effects.

On this basis attempts can then made to deduce the behavior of the entire aggregate. This approach may provide a qualitative insight into the behavior of the material but may not be adapted to provide quantitative results. Over recent years it has become evident that the fundamental behavior of granular materials in the constitutive sense is receiving considerable attention. A selection of references[39-47,75] is included in the list at the end of this chapter. For the purpose of the present discussion only simplified lumped parameter models are considered.

Stiffness of Bulk Material

In the shear force versus shear deformation graphs of Figure 6.4; as indicated the characteristic "step" plus "ramp" response records are similar for the case of shear both during, and in the absence of, applied vibrations. The effect of the vibrations is to cause a slight reduction in the internal stiffness; otherwise the general behavior is the same.

The shear versus shear deformation behavior within the linear region may be represented by the simple model depicted in Figure 6.12. $\dot{x}_r$ is the applied rate of deformation and S the measured shear force. As indicated the shear force is given by

$$S = F_f + kx_r \tag{6.4}$$

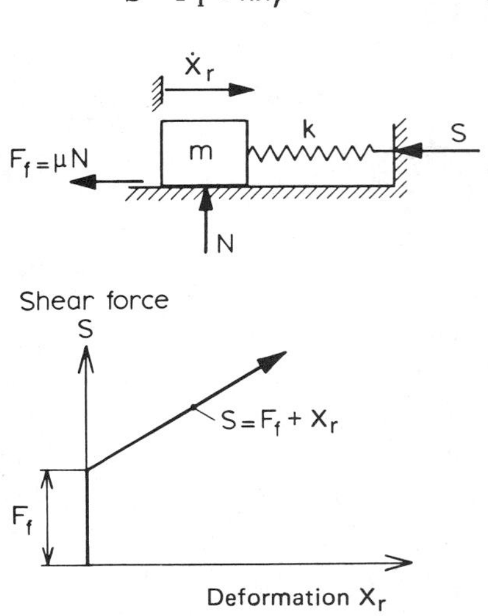

Figure 6.12. Simple Coulomb-elastic model.

where

F_f = Coulomb force
x_r = Relative deformation on shear plane
k = stiffness

The variations in stiffness may be readily observed from the shear force versus shear deformation results obtained in the actual shear tests on the bulk solid (see Fig. 6.4). The stiffness is shown to vary with both consolidation pressure and applied normal stress as illustrated in Fig. 6.13. The stiffnesses are average values based on significant deformations in the elastic range.

Resonant Frequencies

As indicated in Figures 6.5 and 6.6, shear tests performed on the direct shear apparatus show pronounced resonant frequencies when the whole cell is vibrated. The experimental studies indicate that the resonant frequencies vary with both the major consolidation stress σ_1, or applied normal consolidating stress σ, which acts during the consolidation phase, and the applied normal stress $\bar{\sigma}$ which acts during shear. Li,[61] in following the analysis of Molerus,[39] considered the contact forces and elastic defor-

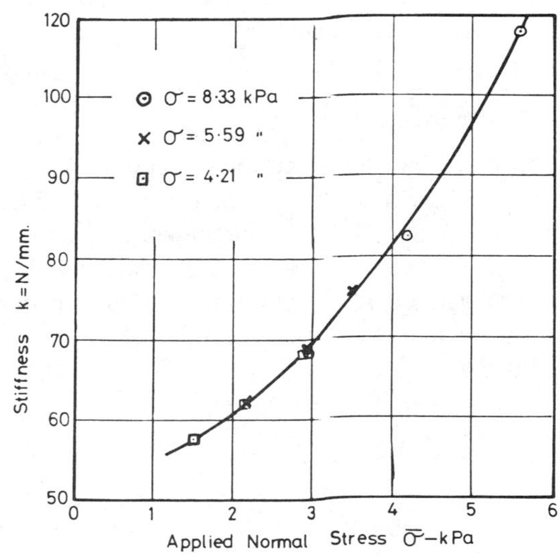

Figure 6.13. Average stiffness as a function of consolidation and applied normal stress. –1 mm pyrophyllite, 5% m.c. (d.b.).

mation between individual particles and related the load versus deformation behavior of the particles to that of the whole mass of material in the shear cell. Li deduced that the natural frequency may be expressed by

$$\omega_n = a_0 (N + C) \tag{6.5}$$

where

a_0 = constant
N = Inter-particle normal force due to the external applied load
C = Cohesive force generated by the external load.[57]

Molerus derived a relationship between the external compressive load and the resulting normal load. The equation is

$$\left(\frac{1-e}{e}\right)\frac{N}{d^2} = \frac{V}{A} \tag{6.6}$$

where

e = void ratio
$\overline{V}$ = compressive force acting on the plane of interest (equivalent to the normal force acting on the shear cell)
A = area of the plane of interest (equivalent to area of the shear cell)
d = particle diameter

The void ratio e depends on the degree of consolidation. However, the percentage change in e as a result of the changing consolidation is usually small. Thus, for the present discussion, it will be assumed that the bulk density, and hence voidage e, is constant. It will also be assumed that the particle diameter is constant. Hence Eq. (6.6) becomes

$$N = a_1 \overline{V} \tag{6.7}$$

where $a_1 = \text{constant} = \dfrac{ed^2}{(1-e)A}$

During consolidation the particle contact also receives plastic deformation and according to Molerus these deformations are responsible for the cohesive forces in bulk materials. The following relationship was derived.

$$C = a_2 N_c + C_0 \tag{6.8}$$

where

C = total cohesive force
C_0 = initial cohesive force before consolidation
N_c = interparticle normal force due to consolidation.

Using a similar argument to that embodied in Eq. (6.7) it follows that

$$N_c = a_3 V \tag{6.9}$$

From Equations (6.5), (6.7), (6.8), and (6.9) the expression for ω_n, as derived by Li, becomes

$$\omega_n = a_0(a_1 \overline{V} + a_2 a_3 V + C_0) \tag{6.10}$$

or in terms of stress units

$$\omega_n = c_2 \sigma + c_1 \sigma + c_0 \tag{6.11}$$

where

$\sigma = V/A$
$\sigma = \overline{V}/A$
$c_0 = \dfrac{C_0 a_0}{A}$, $c_1 = \dfrac{a_0 a_1}{A}$, $c_2 = \dfrac{a_0 a_2 a_3}{A}$
A = shear cell area

The validity of the relationships given by Eq. (6.11) has been verified experimentally for pyrophyllite at 5% moisture content (dry basis). The results which confirm Eq. (6.11) are plotted in Figure 6.14.

For the pyrophyllite, under the conditions examined, the constants are

$c_0 = 263 \quad (s^{-1})$

$c_1 = 48.2 \times 10^{-3} \quad (m^2/Ns)$

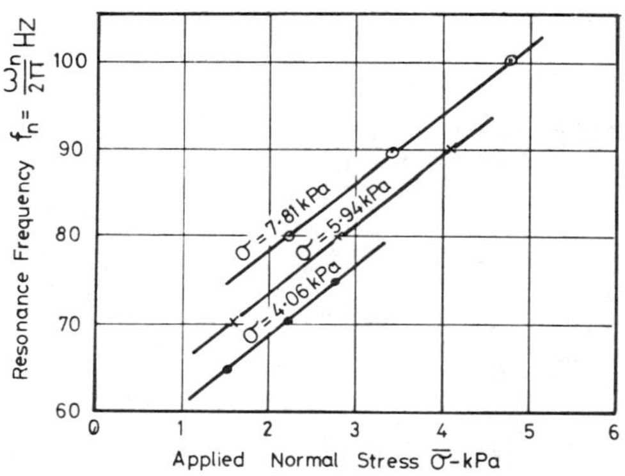

Figure 6.14. Resonance frequency. -1 mm pyrophyllite, 5% m.c. (d.b.).

$$c_2 = 16.5 \times 10^{-3} \qquad (m^2/Ns)$$

The lower bound of the resonant frequency is $\omega_n = c_0$ which occurs when both σ and $\bar{\sigma}$ equal zero. As both the consolidation stress σ and applied stress $\bar{\sigma}$ increase, the material approaches a solid condition and the resonant frequency increases.

Embodied in Eq. (6.11) are the variations in stiffness and mass that result from changes in σ and $\bar{\sigma}$. The variations in stiffness may be readily observed from the shear force versus deformation results obtained in the actual shear tests on the bulk solid. (see Fig. 6.4). The stiffness is shown to vary with both consolidation pressure and applied normal stress as illustrated in Figure 6.13. The stiffness and resonant frequencies will also be influenced by moisture content and particle size distribution.

Shear Cell Vibration Model

The graphs of Figures 6.5 and 6.6 show the existence of dominant resonance frequencies. Since these frequencies are of fundamental importance, the possibility of higher and less significant resonance frequencies will be ignored. This follows the reasoning of Li[61] who shows that a one degree-of-freedom lumped vibration model provides a satisfactory analogy for the purposes of examining the dynamic characteristics of the shear cell test. Li assumes a linear model with viscous damping which, for the small vibration amplitudes involved, gives good correlations between the analytical predictions and experimental results.

Nevertheless, for the general case, the presence of the Coulomb resistive force cannot be ignored and for this reason the model presented by Roberts et al.[62] is proposed. This model is shown in Figure 6.15a while the corresponding forces acting on the top half of the cell are shown in Figure 6.15b. The displacement coordinates are

x_1 = absolute displacement of the lower half of shear cell or input displacement

x_2 = absolute displacement of top ring

$x_r = x_2 - x_1$

= relative displacement between top and bottom halves of shear cell

The force S transmitted by the shear force actuator acts perpendicular to the direction of vibration.

The formulation of the model is based on the following assumptions

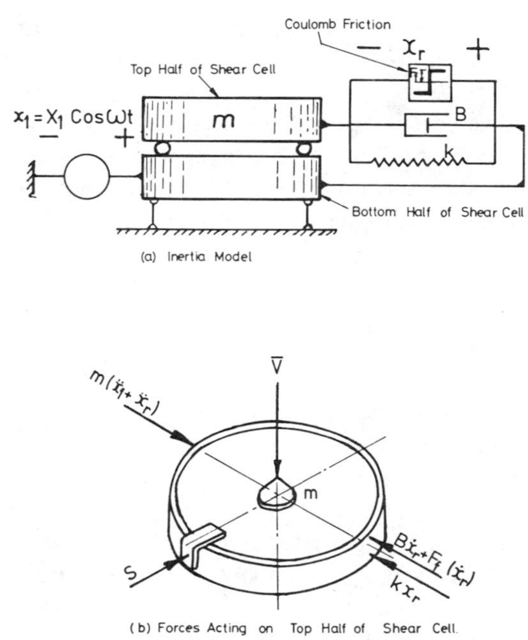

Figure 6.15. Inertia model for vibration of whole shear cell.

1. The material is isotropic across the shear plane.
2. The material behaves as a cohesive mass with one dominant natural frequency.
3. During shear the properties of the material in the direction of vibration remain unchanged.
4. The applied vibration is not of sufficient intensity to shear or fluidize the material. Hence there is little loss due to friction and plastic deformation.
5. The amplitude of vibration on the shear plane is small relative to the particle size.
6. Under the influence of the applied vibration the particles in the shear cell move substantially in the direction of the applied force. Interactionary effects are negligible.

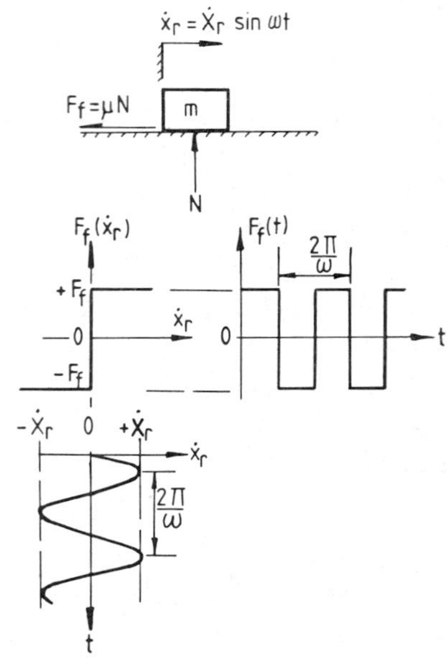

Figure 6.16. Coloumb frictional drag force during sinusoidal excitation.

While the above assumptions are satisfactory for the present analysis it is recognized that they are not strictly correct. For instance, Molerus[39] has shown that anisotropy is inherent in the Jenike direct shear test in view of the fact that the orientation of the principal axes during consolidation may not necessarily coincide with the principal axes during incipient yield. Further, the material in the shear cell may not be completely homogeneous as a result of uneven loading and variations in particle size distribution.

The development of the dynamic model may proceed as follows: During an applied sinusoidal vibration the lower half of the cell has an impressed vibration

$$x_1 = X_1 \cos \omega t \tag{6.12}$$

where

X_1 = impressed amplitude (mm)
$\omega = 2\pi f$ = frequency (rad/s)
f = frequency (Hz)

The Coulomb frictional resistive force has a magnitude $F_f = \mu N$ and always acts in a direction opposing the motion. Thus the Coulomb resistive force is linked directly to the velocity but always acts in the opposite sense to the velocity. Since the top half of the shear cell vibrates relative to the bottom half with frequency ω, the Coulomb resistive force, as a function of time, will be a rectangular wave of amplitude $\pm F_f$ and frequency ω. This is illustrated in Figure 6.16. The Coulomb frictional force may be represented by the Fourier series.

$$f(t) = \frac{4F_f}{\pi}\left(\sin \omega t + \frac{1}{3}\sin 3\,\omega t + \cdots\right) \tag{6.13}$$

This is an odd function with half wave symmetry. Hence only the sine terms and odd harmonics appear.

As in describing function analysis, a technique used in control systems analysis, a single nonlinearity such as that in the present example is approximated by the first term or fundamental frequency component of the Fourier series. Neglecting the higher harmonic terms does not introduce serious error owing to the diminishing amplitude effect associated with these terms. Thus, in this case, the nonlinear frictional resistive force may be approximated by an equivalent linear viscous drag force F_E

$$F_E = B_E\,\dot{x}_r \tag{6.14}$$

where

$$B_E = \frac{4F_f}{\pi \dot{X}_r} \tag{6.15}$$

or

$$B_E = \frac{4F_f}{\pi X_r \omega} \tag{6.16}$$

$\dot{x}_r$ = relative sliding velocity

$\dot{X}_r$ = amplitude of sliding velocity

X_r = amplitude of relative displacement

Here B_E is the describing function for the frictional resistive force and may be regarded as the equivalent damping coefficient. Of particular interest in the study of the shear cell model is the relationship for the relative motion between the top and bottom sections of the shear cell. It may readily be shown that the differential equation is

$$\frac{d^2x_r}{dt^2} + 2\zeta\omega_n \frac{dx_r}{dt} + \omega_n{}^2 x_r = X_i \omega^2 \cos \omega t \tag{6.17}$$

where

$\omega_n = \sqrt{k/m}$ = Natural frequency (rad/s)
ζ = Damping ratio
x_r = Relative displacement
m = Total mass in motion with top half of cell
k = Dynamic stiffness of bulk solid
$\omega = 2\pi f$ = impressed frequency

The absolute motion of the top ring is also of interest since it is often easier to measure than the relative motion, particularly in the case of small impressed amplitudes. The differential equation in this case is

$$\frac{d^2x_2}{dt^2} + 2\zeta\omega \frac{dx_2}{dt} + \omega_n^2 x_2 = \frac{X_1\omega^2}{\sqrt{1 + (2\zeta r)^2}} \cos(\omega t + \theta) \tag{6.18}$$

where

r = frequency ratio ω/ω_n

$\theta = \tan^{-1}(2\zeta r)$

The total system damping is the sum of the actual viscous damping denoted by B and the equivalent damping B_E due to Coulomb friction.

Thus the damping ratio is

$$\zeta = \frac{B + B_E}{B_{cr}} \tag{6.19}$$

where

$B_{cr} = 2m\omega_n$ = critical damping factor.

From Eq. (6.16) it follows that when Coulomb damping is present, ζ will not be constant but will vary with the impressed frequency ω and amplitude X_r. The amplitude X_r is of particular importance in that it relates directly to the deformation on the shear plane.

Before discussing the general case of damping given by Eq. (6.19) it is important to first consider the two particular cases of viscous damping and Coulomb damping when each is present separately.

Viscous Damping

Here, ζ is constant and the steady state solutions of Eqs. (6.17) and (6.18) may be expressed in terms of the amplitude and phase relationships. For relative motion:

$$\frac{X_r}{X_1} = \frac{r^2}{\sqrt{(1 - r^2)^2 + (2\zeta r)^2}} \tag{6.20}$$

The phase angle ψ_r between the relative motion and impressed motion is given by

$$\psi_1 = \tan^{-1} \frac{2\zeta r}{1 - r^2} \tag{6.21}$$

where

r = frequency ratio, ω/ω_n

A resonance condition corresponding to maximum X_r/X_1 is obtained at the frequency

$$\omega_{re} = \frac{\omega_n}{\sqrt{1 - 2\zeta^2}} \tag{6.22}$$

For absolute motion

$$\frac{X_2}{X_1} = \sqrt{\frac{1 + (2\zeta r)^2}{(1 - r^2)^2 + (2\zeta r)^2}} \tag{6.23}$$

The phase angle ψ_a between the absolute motion and impressed motion is given by

$$\psi_a = \tan^{-1}\left[\frac{2\zeta r^3}{1 - r^2 + (2\zeta r)^2}\right] \tag{6.24}$$

Coulomb Damping

Using the approximated equivalent viscous damping coefficient B_E given by Eq. (6.16), it follows from Eq. (6.19) that when $B = 0$, the damping ratio is given by

$$\zeta = \frac{2F_f}{\pi m X_r \omega \omega_n}$$

or

$$\zeta = \frac{2F_f}{\pi k X_r r} \tag{6.25}$$

Substituting for ζ in Eq. (6.20), the amplitude ratio for the relative motion becomes

$$\frac{X_r}{X_1} = \pm\sqrt{\left[\frac{r^2}{1 - r^2}\right]^2 - \left[\frac{4F_f}{\pi F_s(1 - r^2)}\right]^2} \tag{6.26}$$

and the phase angle ψ_1 is

$$\psi_r = \tan^{-1}\left[\frac{4F_f}{\pi \frac{X_r}{X_1} F_s(1 - r^2)}\right] \tag{6.27}$$

where

$F_s = kX_1$
= equivalent static restoring force

Referring to Eq. (6.26), it can be seen that X_r has a real value only when

$$\frac{F_f}{F_s} < \frac{\pi r^2}{4} \tag{6.28}$$

When small frictional forces are involved, as is usually the case, the condition is easily satisfied. On the other hand, the condition will not be satisfied at low frequencies where $r << 1.0$. For all cases when Eq. (6.28) is satisfied, Eq. (6.26) shows that the amplitude X_r becomes infinite at resonance, that is when $\omega = \omega_n$. Timoshenko et al.[76] explain this condition in terms of the energy dissipated per cycle being less than the energy input.

A more rigorous solution of the Coulomb damping vibration problem with a harmonic force acting on the system has been presented by Den Hartog.[77] The procedure he followed may be readily adapted to the case where the system is disturbed by a harmonic displacement, as occurs in the shear cell model of Figure 6.15. The differential equation is

$$\frac{d^2 x_r}{dt^2} + \omega_n^2 x_r = X_1 \omega^2 \cos \omega t \pm \frac{F_f}{k} \omega_n^2 \tag{6.29}$$

where the + sign refers to the friction force under the condition that the mass moves in the positive direction and vice versa.

It can be shown that the amplitude ratio obtained from a solution of Eq. (6.29) is given by

$$\frac{X_r}{X_1} = \sqrt{\left[\frac{r^2}{1 - r^2}\right]^2 - \left[\frac{F_f}{F_s}\right]^2 Q^2} \tag{6.30}$$

where $r = \omega/\omega_n$

$$Q = \frac{\sin\left(\frac{\pi}{r}\right)}{r\left[1 + \cos\left(\frac{\pi}{r}\right)\right]} \tag{6.31}$$

It may be observed that Eq. (6.30) is similar in form to the approximate solution given by Eq. (6.26). In the exact solution given by Eq.

(6.30) the Coulomb damping function Q, given by Eq. (6.31), varies with the frequency ratio as indicated. The phase angle from the rigorous solution is

$$\psi_r = \tan^{-1}\left[-\frac{F_f Q}{F_s \frac{X_r}{X_1}}\right] \quad (6.32)$$

Diagrams of the amplitude ratio X_r/X_1 and phase angle ψ_r given by Eqs. (6.30) and (6.32), respectively, are represented in Figure 6.17.

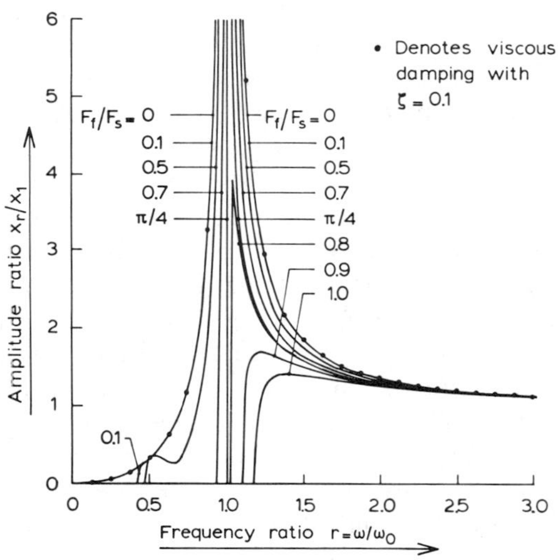

Figure 6.17a. Amplitude ratio.

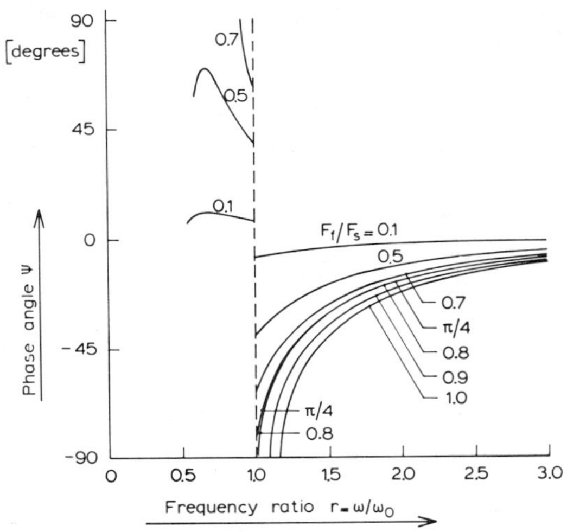

Figure 6.17b. Phase angle.

The various curves illustrate the influence of the force ratio F_f/F_s. Referring to Figure 6.17a, for comparison purposes, the plotted points indicate the curve for viscous damping with damping ratio $\zeta = 0.1$. As can be seen, this curve follows closely the results for very small Coulomb frictional forces when $F_f/F_s = 0.05$ or lower. It is readily observed that for the zero damping case, Eq. (6.30) for Coulomb damping and Eq. (6.20 for viscous damping become identical. From a practical point of view the equivalent constant viscous damping factor of $\zeta = 0.1$ provides a convenient approximation of the lower bound for the Coulomb damping case.

Combined Viscous and Coulomb Damping

Using the approximated equivalent viscous damping coefficient B_E for the Coulomb Damping then the combined damping factor as given by Eq. (6.19) may be written in the form

$$\zeta = \zeta_v + \frac{2F_f}{\pi X_r k r} \quad (6.33)$$

where

$\zeta_v = \dfrac{B}{2m\omega_n}$ = damping ratio for the viscous component of the damping.

Following a similar procedure to the approximate Coulomb damping case it can be shown that the amplitude ratio X_r/X_1 is given by the quadratic

$$\left(\frac{X_r}{X_1}\right)^2\left[(1-r^2)^2 + (2r\zeta_v)^2\right] + \left(\frac{X_r}{X_1}\right)\left[\frac{16\zeta_v r F_f}{\pi F_s}\right] + \left[\left(\frac{4F_f}{\pi F_s}\right)^2 - r^4\right] = 0 \quad (6.34)$$

and phase angle ψ_r is

$$\psi_r = \tan^{-1}\left[\frac{2\zeta_v r}{1-r^2} + \frac{4F_f}{\frac{\pi X_r}{X_1} F_s(1-r^2)}\right] \quad (6.35)$$

Solutions for the real values of X_r/X_1 can be obtained from (6.34) for given values of ζ_v and F_f/F_s.

A rigorous analysis of the influence of combined viscous and Coulomb damping has been presented by Den Hartog.

Verification of Model

The actual damping characteristics of cohesive bulk solids may be a combination of a number of factors such as inter-particle friction, plastic deformation at contact points, and interfacial fluid damping. The characteristics are certainly highly nonlinear and extremely difficult to analyze in a rigorous way. However, it is reasonable to assume that the combination of viscous and Coulomb damping, as previously described, provides a satisfactory approximation for modeling purposes. As to which of the Coulomb or viscous components of the damping is dominant will depend, to some extent, on the amplitude of vibration. Certainly if the amplitude is large in relation to average particle size, then Coulomb friction will have a major influence. On the other hand, for very small amplitudes Li[61] argues that Coulomb friction is minimal and that the particles simply undergo small oscillatory motions about their pinning points or points of contact.

For the experimental work using the vibration shear cell apparatus, Li concludes that because of the small amplitudes involved, the damping is dominantly viscous. Following extensive tests he established that for pyrophyllite and iron ore the damping was not influenced significantly by the consolidation stresses and applied normal stresses during vibration. The following viscous damping factors were shown to fit the data quite well.

for -1 mm pyrophyllite at 5% moisture content (d.b.), $\zeta = 0.1$

For iron ore at 5% moisture content (d.b.), $\zeta = 0.125$ (see Tables 6.2 and 6.3 for more detailed information).

As an indication of the degree of fit given by the assumption of viscous damping with $\zeta = 0.1$, the ratio of the absolute to impressed amplitude X_2/X_1 computed using Eq. (6.23) is compared with the corresponding experimentally obtained results for pyrophyllite. The two curves are shown in Figure 6.18, and the agreement is considered satisfactory. It is to be noted that the absolute amplitude rather than the relative amplitude is used, since the former was easier to obtain experimentally. Further, as previously stated, the model was developed to predict the fundamental natural frequency; no attempt has been made to analyze the presence of the second and higher natural frequency shown in the experimental results, since this frequency is of lower significance in affecting the behavior of the material during shear.

Whether the reasoning given by Li in favor of viscous damping being dominant in this case is correct, is difficult to say. Certainly the very small amplitudes used in the experimental work lend weight to his argument. On the other hand, the general form of the shear load versus deformation characteristic of Figure 6.4 favors Coulomb damping as being dominant, particularly when the amplitude is of a reasonable order. For the present results, reference to Figure 6.17 indicates that a viscous damping ratio of $\zeta = 0.1$ is equivalent to a low value of F_f/F_s in the case of Coulomb Damping, that is $F_f/F_s < 0.1$. While this is feasible, it is difficult to quantify; it implies a low value of F_f and a high value of the initial stiffness k. It should be noted that the stiffness values plotted in Figure 6.13 are average values for large deformations obtained under very low (almost static) deformation rate conditions. Nonetheless, the shear force versus shear deformation graphs of Figure 6.4 indicate initial values of k substantially

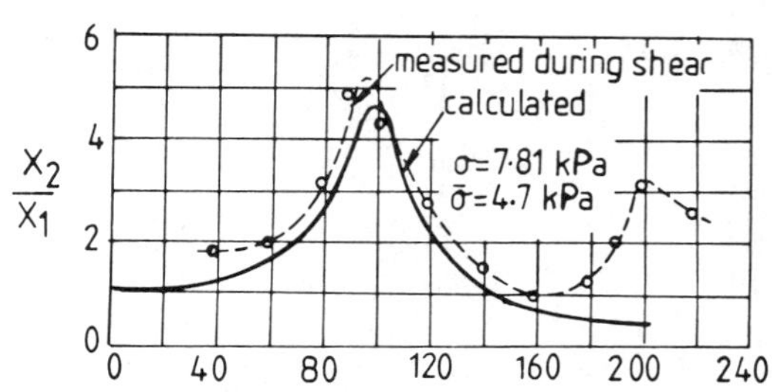

Figure 6.18. Comparison of measured and calculated amplitude ratios. -1 mm pyrophyllite, 5% m.c. (d.b.).

higher than the average values plotted in Figure 6.13.

The overriding result of this study is that the shear cell model adequately depicts the behavior of the sample during vibration. It is clear that the damping is of very low order and for Coulomb damping it is equivalent to $F_f/F_s \simeq 0.05$ for the pyrophyllite. The assumption of viscous damping with $\zeta = 0.1$ fits the data sufficiently well for practical purposes and provides a simple model for analysis.

Concept of Resonance

The concept of resonance in the case of a bulk solid is somewhat complex, particularly in view of the difficulty of defining, with any degree of certainty, the mass and stiffness contributions in the actual vibrating system. Even in the shear cell apparatus there are problems in establishing the actual vibrating mass, but the problems are compounded when the same exercise is attempted for a bulk solid in an actual hopper. However, as discussed later, the significant parameters affecting the dynamic shear failure are vibration frequency and relative amplitude on the shear plane. Thus the dynamic properties of bulk solids are important in view of their influence on the transmission of vibrations, particularly with respect to the possible amplification of vibration amplitude.

6.5. A FAILURE CRITERION

Correlation of Dynamic Shear Stress with Sliding Velocity

In order that the shear cell test data be applied to the design or bin activators and flow promotion devices, consideration of the appropriate parameters for dynamic similarity is required. Such parameters need to take into account the relevant information on vibration frequency, amplitude, and energy transfer. As an initial step in this investigation studies have been undertaken to ascertain the degree of compliance of the shear stress during vibration with the relevant vibration parameters, notably the velocity and acceleration. Which one of these two parameters is the more appropriate will depend, to some extent, on such factors as the intensity of the vibration, the manner in which the vibration is applied, and the design features of the particular vibration system concerned. For instance, in many cases the material is vibrated in a gravitational field either in the vertical direction or at some angle to the vertical. In such cases the acceleration ratio $\omega^2 X_i/g$ is a significant parameter.

However, if the material is vibrated horizontally, as in the shear cell apparatus of Figure 6.1, then the effects of gravity and acceleration become less important.[60-62] The experimental results show that the shear stress correlates well with both the vibration velocity and acceleration. However, as shown by Li[61] the weight of evidence favors the vibration velocity (and hence energy) as being the parameter with the most significant influence on the dynamic shear behavior. The correlation with sliding velocity is shown, by way of example, with the graphs of Figure 6.19. These graphs apply to -1 mm pyrophyllite at 5% moisture content. Figure 6.19a corresponds to the vibration of the whole cell, and it is clear that the plotting of τ_f against the maximum relative sliding velocity $2\pi f X_r$ on the shear plane has resulted in a smoothing out of the curve into a simple decaying shape with the resonance effect, as indicated in Figure 6.5, being no longer visible. This characteristic results from the fact that amplitude X_r in the maximum sliding velocity is the relative amplitude on the shear plane and is related to the impressed amplitude X_i as shown by Figure 6.6.

Figure 6.19b corresponds to the vibration of the top of the cell only. Here $X_r = X_i$ and the τ_f plot shows a decaying shape similar to that of Figure 6.19a. The similarity in shape between the two curves of the figure is clearly evident. In fact, analysis shows they are virtually identical in shape. It is clear, therefore, that resonance is only meaningful in terms of the amplitude transmitted to the plane of shear.

Concept of Failure

Li based his studies on the concept of failure originally developed by Hvorslev[12] and later extended by Roscoe et al.[13] Hvorslev, whose work was concerned with cohesive soils, showed that the shear stress at failure is a function of the effective normal stress on, and voids ratio (or density) in the plane of failure at the instance that failure occurs; failure, defined in this way is independent of the stress history of the material. Rosco et al. showed the relationship between shear stress τ, normal stress σ and void ratio e, which, for a given bulk material defines a failure surface as illustrated in Figure 6.20. The state of a sample defined by any point below the surface indicates a condition of no failure. States of stress defining a point on the surface indicate a condition of yield, and hence failure, or incipient yield. States of stress defining points above the surface are not possible. Roscoe et al. also showed the existence of the critical void ratio line (CVR) which defines a boundary of the Hvorslev failure surface. The critical void ratio indicates the limiting state of a sample at which a further increment of shear distortion will not result in any change in the void ratio. The CVR is a unique line to which all loading paths in the e-σ-τ space converge.

It is proposed that the Hvorslev failure surface applies equally well to the dynamic case as to the static. Referring to Figure 6.20, the specimen in the shear cell is originally consolidated with normal stress σ_0 and shear stress τ_0, the corresponding void ratio being e_0. The condition of the sample is given by D'. In the static case, reduced normal stresses $\bar{\sigma}_1$ and $\bar{\sigma}_2$ will lead to points C' and B' on the yield surface. The projection of $B'C'D'$ onto the σ–τ

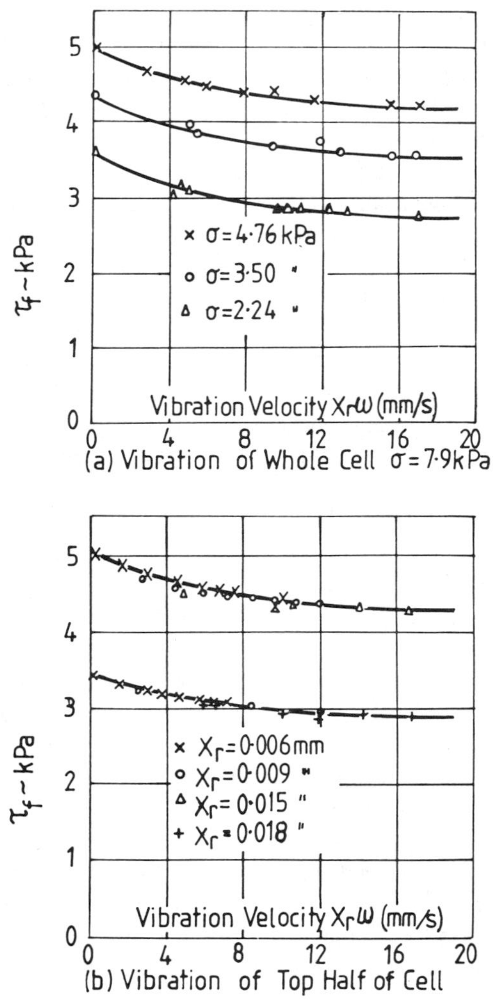

Figure 6.19. Correlation of dynamic shear force with maximum vibration velocity. –1 mm pyrophyllite, 5% M.C. (d.b.).

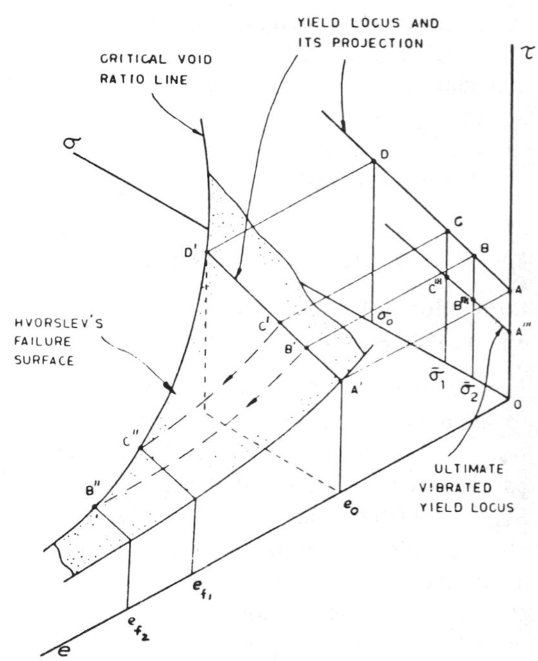

Figure 6.20. Behavior of material under vibration on the Hvorslev's failure surface.

plane gives the yield locus for the material for the given consolidation condition. If vibration is applied during the application of the normal stress $\bar{\sigma}_1$, the material dilates on the shear plane, causing a localized increase in the void ratio at the shear plane where relative motion exists. Elsewhere in the cell the material is in fact consolidating as the voids are reduced. The increase in voidage on the shear plane causes a change of state along the line $C'C''$. A similar argument holds for the applied stress $\bar{\sigma}_2$ where a change of state occurs along $B'B''$.

For each applied normal stress the application of mechanical vibrations may cause the void ratio to increase until a limiting value is reached. The limiting void ratios for the two stresses $\bar{\sigma}_1$ and $\bar{\sigma}_2$ are respectively e_{f_1} defined by point C'' and e_{f_2} defined by point B''; the points B'' and C'' lie on the critical void ratio line. The projection of points B'' and C'' onto the σ–τ plane specifys two points on the ultimate vibrated yield locus which define the lowest possible values of the shear stress. A further increase in the intensity of vibration will not cause any greater reduction in shear stress. As will be readily apparent these arguments apply only to the case of horizontally applied vibration; in the case of vertical vibrations, failure of the material may take place under conditions of high vibration acceleration.

During the application of mechanical vibrations, yielding may occur anywhere on the Hvorslev surface between the boundaries defined by the static condition and the ultimate or critical voidage condition. The condition for failure defined by points on the failure surface directly depends on the impressed velocity, frequency, and applied normal stress. For this reason the actual vibrated yield loci may not be straight or parallel lines. They may even cross each other as indicated in Figure 6.21. For a given normal stress $\bar{\sigma}$ the reduction in shear stress has been shown by Roscoe et al.[13] to follow a negative exponential curve expressed by

$$\tau = \mu\bar{\sigma} + \beta \exp(-Be_f) \qquad (6.36)$$

where μ, β, and B are material constants and e_f is the void ratio in the plane of failure.

Referring to Figure 6.19 it is apparent that the shear strength τ_f during vibration follows a

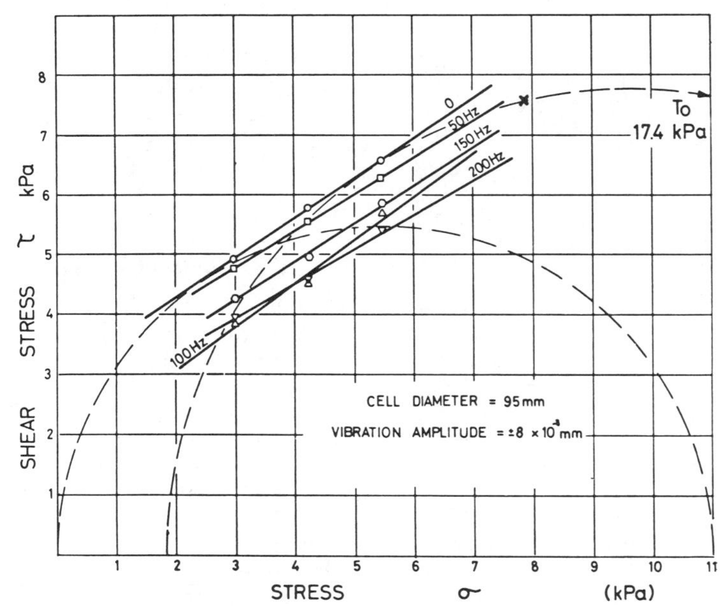

Figure 6.21. Yield loci for pyrophyllite illustrating the effect of sinusoidal vibration.

similar negative exponential curve that suggests that the vibration velocity on the shear plane is directly correlated with the void ratio e. Consequently Li proposed a decaying exponential function of the form.

$$\tau_f = \tau_\infty + \beta \exp\left(\frac{-2\pi X_r f}{\gamma}\right) \qquad (6.37)$$

where $f = \omega/2\pi$ (Hz) and τ_∞, β, and γ are constants which for a given material, depend on the consolidation and normal stress during shear. The form of equation is indicated in Figure 6.22. τ_∞ represents the limiting shear stress, which is an indication of the effectiveness of the vibrations on the shear reduction. The static value of the shear stress τ_{f0}, which occurs when the input frequency $f = 0$, is

$$\tau_{f0} = \tau_\infty + \beta \qquad (6.38)$$

As an alternative, Eq. (6.37) may be formulated in terms of a relationship for the shear stress ratio τ_f/τ_{f0}. It may readily be shown that

$$\frac{\tau_f}{\tau_{f0}} = 1 - \frac{\beta}{\tau_{f0}}\left\{1 - \exp\left(-\frac{2\pi X_r f}{\gamma}\right)\right\} \qquad (6.39)$$

The constant γ in Eqs. (6.37) and (6.39) is called the vibration velocity constant and indicates the rate of decay of the shear stress.

Table 6.1. Shear Stress for Multiple Valves of γ.

VIBRATION VELOCITY $U = 2\pi X_r f$	SHEAR STRESS τ_f
0	$\tau_\infty + \beta$
γ	$\tau_\infty + 0.368\beta$
2γ	$\tau_\infty + 0.135\beta$
3γ	$\tau_\infty + 0.05\beta$
4γ	$\tau_\infty + 0.015\beta$
5γ	$\tau_\infty + 0.007\beta$
10γ	$\tau_\infty + 0.00005\beta$
∞	τ_∞

Table 6.2. Typical Dynamic Properties. (Material: Pyrophyllite with 5% moisture content (d.b) input amplitude $X_i = 0.006$ mm.)

σ kPa	$\bar{\sigma}$ kPa	f_n Hz	ζ	τ_{f0} kPa	τ_∞ kPa	β kPa	γ
7.81	4.69	100	0.11	4.86	3.92	0.94	7
7.81	3.47	90	0.10	4.21	3.28	0.94	7
7.81	2.20	80	0.095	3.36	2.42	0.94	7
5.94	4.06	90	0.105	4.07	3.28	0.80	7
5.94	2.81	80	0.10	3.42	2.62	0.80	7
5.94	1.57	70	0.10	2.59	1.79	0.80	7
4.06	2.81	75	0.10	2.94	2.24	0.70	7
4.06	2.19	70	0.10	2.59	1.89	0.70	7
4.06	1.57	65	0.10	2.22	1.52	0.70	7

By way of example the decay in the shear stress for multiple values of γ is indicated in Table 6.1:

Thus for $U = 3\gamma$, for example, the shear stress decreases by 95% of its total possible reduction. The velocity $U = 5\gamma$ causes the shear strength to decrease some 99.3% of its total possible reduction.

Prediction of Vibrated Shear Strength

For the purpose of this discussion attention is drawn to the various experimentally obtained parameters for pyrophyllite and iron ore listed in Table 6.2 and 6.3, respectively. With respect

Table 6.3. Typical Dynamic Properties. (Material: Iron ore with 5% moisture content (d.b) input amplitude $X_i = 0.0075$ mm.)

σ kPa	$\bar{\sigma}$ kPa	f_n Hz	ζ	τ_{f0} kPa	τ_∞ kPa	β kPa	γ
8.54	4.41	110	0.15	5.35	4.44	0.91	10
8.54	3.04	100	0.125	4.44	3.53	0.91	10
6.44	3.72	100	0.125	4.35	3.54	0.81	10
6.44	2.35	95	0.125	3.47	2.66	0.81	10

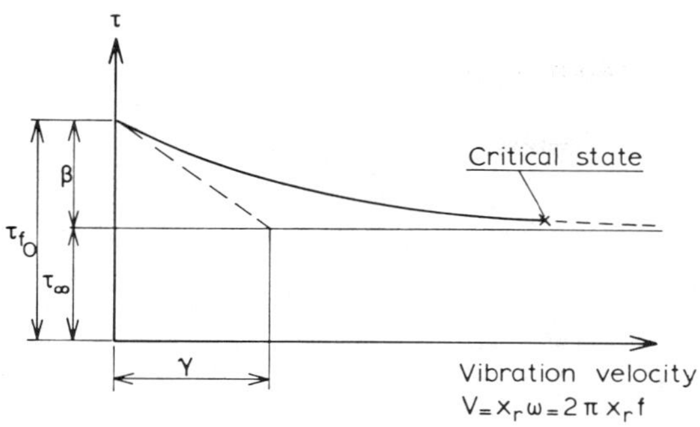

Figure 6.22. Shear stress failure criterion.

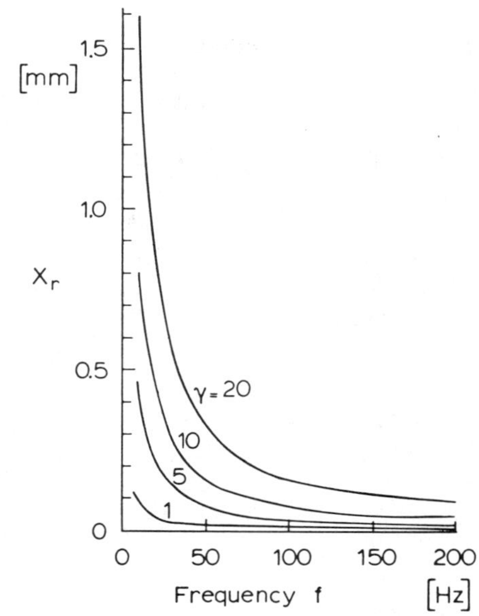

Figure 6.23. Relationship between X_r and f to obtain maximum reduction in shear strength.

to Eq. (6.37) it can be seen that the parameter β, which represents the maximum possible reduction in shear stress, depends only on the consolidation condition given by σ; for each σ, β does not change with the applied normal stress $\bar{\sigma}$. The vibration velocity parameter γ is shown to be a constant for the particular material and independent of the consolidation stress and applied normal stress.

Thus, by vibrating the top ring only, an estimate of the shear strength during vibration may be obtained as follows:

1. For each consolidation condition, determine τ_{f_0} for the chosen applied normal stress.

2. For each consolidation condition and a nominal applied normal stress $\bar{\sigma}_1$, estimate the maximum possible strength reduction β. This value may be checked for other applied normal stress values. For this test it will be necessary to use a high vibration velocity excitation. The properties of the exponentially decaying curve of Figure 6.22 will enable the values of both β and γ to be determined. As a guide, the relative amplitude X_r versus frequency f curves of Fig. 6.23 will allow various alternative valves of X_r and f corresponding to $U = 5\gamma$ to be determined.

The curves are plotted for a range of possible values of γ. The values of $U = 5\gamma$ correspond to the shear stress reduction of 99.3% of β.

Where resonance effects are known to be present, such as when the whole shear cell is vibrated, the resonant frequencies may be estimated using Eq. (6.11) with the coefficients c_0, c_1, and c_2 determined for the particular bulk material. The damping characteristics need to be estimated, but for powders similar to iron ore or pyrophyllite for which the properties are known, the assumption of an equivalent viscous damping factor ζ in the range 0.1–0.125 would seem to be satisfactory. Using linear theory, the relative amplitude X_r as a function of impressed amplitude X_1 may be estimated. This enables the shear stress as a function of frequency for various consolidation and loading conditions to be determined.

Li[61] used this general procedure to check the validity of the parameters determined by experiment. A typical set of values for pyrophyllite are shown in Fig. 6.24. The agreement between the calculated and experimental values is considered to be quite satisfactory.

6.6. BOUNDARY SHEAR AND WALL FRICTION

Introductory Remarks

There are many applications in industry involving powdered or bulk materials where it is necessary to obtain low frictional drag forces

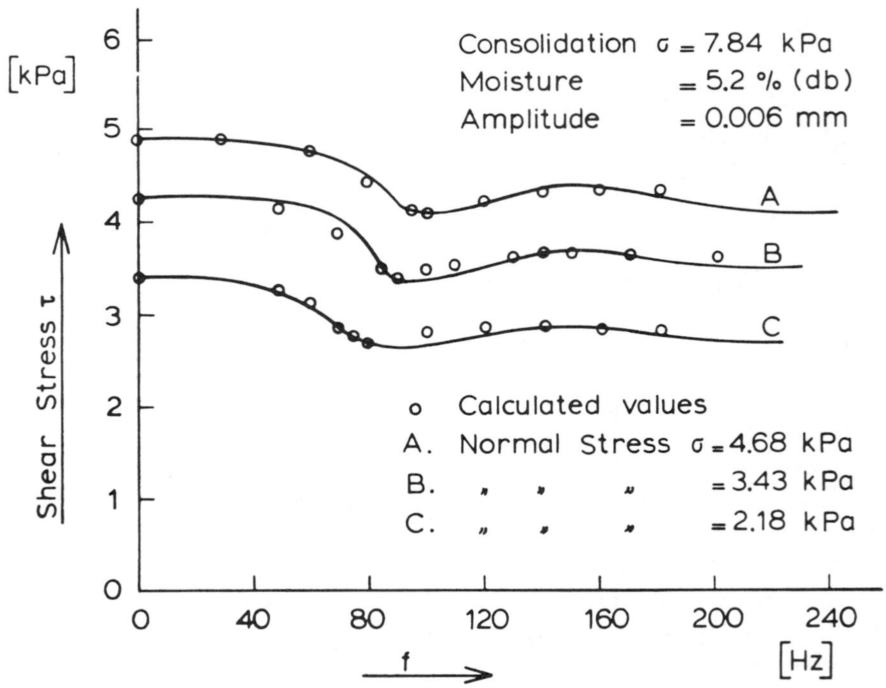

Figure 6.24. Comparison between experimental and calculated shear stress for vibration of whole shear cell. −1 mm pyrophyllite, 5% m.c. (d.b.).

between the bulk material and an adjacent metal or plastic surface. For instance, in the design and operation of gravity mass-flow storage bins, a low friction coefficient at the hopper walls is essential. To achieve this condition, it is often necessary to employ polished stainless steel or an appropriate plastic coating as a lining for the hopper. The significance of this statement may be gauged from the example given in Figure 6.25, which shows the wall yield loci for a typical coal in contact with several wall surface materials. However, it has also been established by several workers,[56,58,61,64,72,79] that mechanical vibrations can significantly reduce the frictional drag between a bulk material and a solid surface.

Much work has been done to examine the effect of vibration on the friction between solid surfaces.[79-81] Although the sliding bodies in such cases may be somewhat different from that of a bulk material in contact with a solid surface, as far as the frictional drag is concerned, there are many similarities. From a macroscopic point of view, the force to overcome friction in solid contacts is equivalent to the force to shear a Coulomb powder. Microscopically both processes involve the shear deformation of asperities between adjacent contact surfaces and particles.

In studying the effect of vibration on the reduction of friction between metal surfaces, Lenkiewizc[79] employed vibration amplitudes which ranged from 0.005 to 0.2 mm. He showed that the sliding velocity due to the

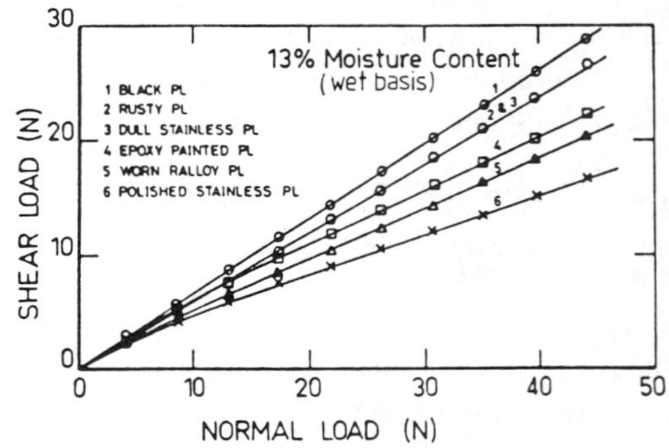

Figure 6.25. Wall yield loci for a typical coal (from Arnold et al.[78]).

imposed vibration was a critical factor in the process although the correlations did not generally apply at large amplitudes.

Similar experiments were performed by Tolstoi,[81] who attributed the reduction in friction between metal surfaces to the increase in sliding velocity. The reasons for this behavior are summarised below:

1. The higher the velocity, the shorter is the time during which adjacent asperities compress each other, and hence the separation between contact surfaces can be maintained at a maximum.
2. The increase in sliding velocity increases the upward component of the impulsive forces exerted on the asperities as they collide with those of the adjacent surfaces. In this way the amplitude of the natural vibration is increased. The effect is the same as in (1) above.
3. The higher the sliding velocity due to vibration, the shorter is the time available to squeeze the boundary lubricant between the asperities. This aids the reduction in the frictional drag force.

It is interesting to observe the similarity between the results and observations of Lenkiewizc[79] and Tolstoi[81] with the criterion for shear failure of a vibrated bulk solid presented in Section 6.5. In the case of sliding between two adjacent solid surfaces, the increase in separation between the surfaces resulting from an increase in vibrational sliding velocity is equivalent to the effect of dilation during shear of a consolidated bulk solid.

Sharma et al.[64] studied the influence of high-frequency vibration on soil-metal friction. A dimensional analysis approach was used, and several dimensionless parameters were derived. Of these, the following three are of interest to the present discussion:

$$\pi_1 = \frac{\omega Xi}{U_s}$$

$$\pi_2 = \frac{N_c \omega^2 X_i}{g}$$

$$\pi_d = \frac{F_f}{N} \tag{6.40}$$

where

π_d = dependent term
F_f = average friction force
U_s = velocity of sliding of contact surfaces
N_c = inertial coefficient
X_i = amplitude.

The apparent coefficient of soil-metal friction given by π_d decreases as π_1 increases as illustrated in Figure 6.26. However, the variation of π_d with π_2 was within 3.07% of the mean at the 95% probability. Hence the inertia number π_2 was considered insignificant. The significance of the velocity number π_1 adds further weight to argument that the vibrational velocity on the plane of failure being the important parameter in causing dilation or separation leading to a lowering of the frictional drag force.

Experiments Using Vibrating Shear Cell Apparatus

In the work of Kaaden[59] and Arnold et al.[56, 72] the effect of vibration on the wall yield loci for bulk solids was determined using a modified Jenike direct shear test apparatus. Vibrations were applied in the same direction as the direc-

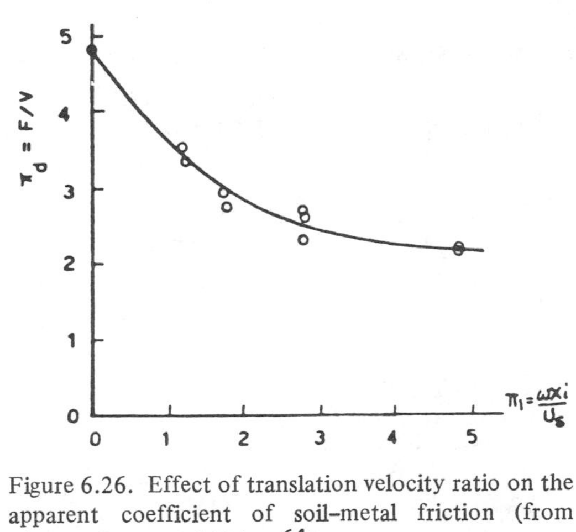

Figure 6.26. Effect of translation velocity ratio on the apparent coefficient of soil-metal friction (from Sharma, Drew, and Nelson[64]).

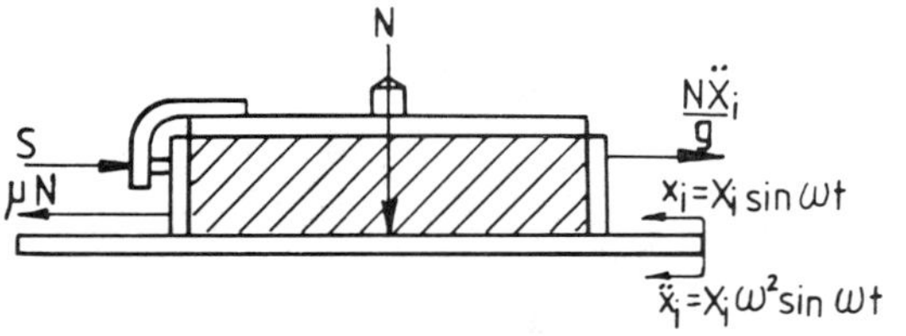

Figure 6.27. Wall yield locus analysis (from Kaaden.[59])

tion of sliding as indicated in Figure 6.27. In order to faciliate experimental measurements, the metal plate was vibrated, while the bulk solid sample contained within the shear cell ring was pushed relative to the plate.

Referring to the forces shown on the shear cell in Figure 6.27 and following the simplified analysis given by Kaaden, it can easily be seen that the shear force S can be reduced by an amount not exceeding

$$\Delta S = \frac{NX_i\omega^2}{g} \tag{6.41}$$

where

N = normal force on shear plane
X_i = amplitude of vibration
ω = frequency
ΔS = maximum reduction in shear force.

Reference to Figure 6.28 will permit the reasoning behind this analysis to be seen.

Figure 6.29 shows the results obtained from tests on Mt. Newman iron ore fines at 9.3% moisture content (w.b.). For this test a constant frequency of 25 Hz and amplitude X_i = 0.23 mm was used and, as is readily observable,

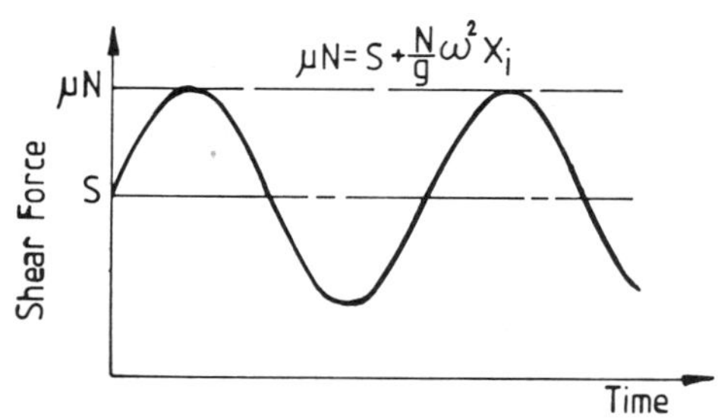

Figure 6.28. Oscillation of total shear force at limiting condition.

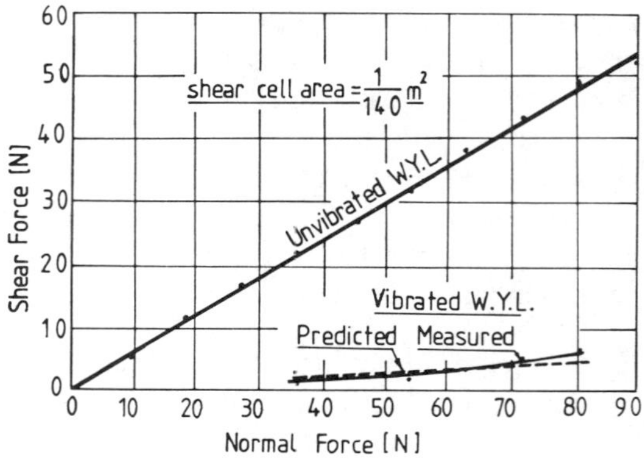

Figure 6.29. Vibrated versus normal wall yield locus, Mt. Newman fines. 9.3% m.c. (w.b.) on black steel (Kadden[7-29]–Arnold et al.[56]).

shear force values of from 5 to 10% of the static values have been obtained. Also, as indicated by Figure 6.29 the correlation between the measured and predicted vibrated wall yield loci is very reasonable.

In order to isolate the effect of the inertia force in lending assistance to the reduction in shear force as in the previous case, Roberts et al.[58,62] and Li[61] used the vibrating shear cell apparatus describing in Section 6.2 of this chapter. Vibrations were applied in the horizontal plane in a direction perpendicular to the direction of shear. In this way the results obtained are consistent with those presented in Section 6.3.

By way of illustration, a set of wall yield loci curves for -1 mm pyrophyllite (5% m.c.) on mild steel is presented in Figure 6.30. A constant amplitude of X_i = 0.006 mm was used.

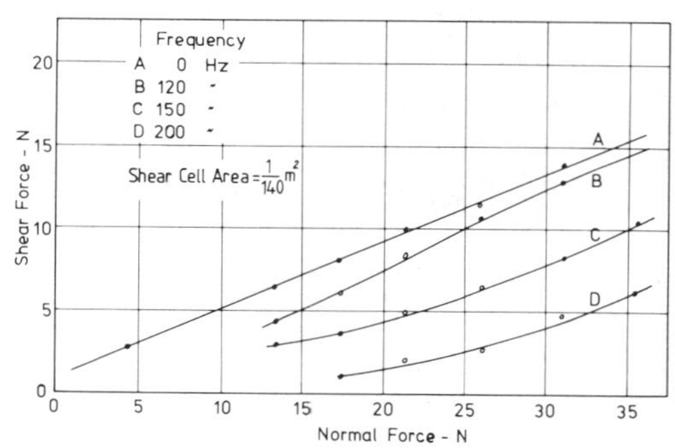

Figure 6.30. Vibrated wall yield loci for pyrophyllite on mild steel plate. X_r = ±0.006 mm; m.c. = 5% (d.b.).

The reduction in wall friction with increase in frequency is quite evident. The reduction in friction is considerably less than would be obtained when vibrations are applied in the direction of shear.

6.7. RANDOM VIBRATION EXCITATION

Application of Random Vibrations

The application of broad-band random vibrations to promote the flow of bulk solids has been discussed by Arnold et al.[56] and Roberts et al.[58,60] In view of the frequency dependence of many bulk solids with respect to strength reduction, broad-band random vibrations have certain advantages. Since the vibration energy is distributed uniformly across all frequencies within a given bandwidth, then it is certain that a resonant frequency of the particular material and flow promotion device would be excited, provided the resonant frequency lies within the bandwidth of the random vibration input. This will not be the case when discrete frequency sinusoidal vibration excitation is used when the input frequency differs significantly from the resonant frequency of the system.

The influence of random vibrations on the shear strength of bulk solids has been examined using the shear cell apparatus of Figure 6.1, some typical results being shown in Figure 6.31 (See Roberts and Scott[60]). Here a comparison is made between two cell diameters, namely 57 and 95 mm and two band widths 40 and 150 Hz; the bulk material is pyrophyllite. The samples were consolidated under the applied normal stresses $\sigma = 7.9$ kPa in the 95 mm cell and $\sigma = 9.5$ kPa in the 57 mm cell; in each case the samples were sheared under the normal stress of $\overline{\sigma} = 4.23$ kPa during the application of random vibrations.

Figure 6.31 shows the variation of τ_f/τ_{f_0} for pyrophyllite as a function of RMS amplitude. Curves A and B apply to the 95 mm cell, whereas curve C applies to the smaller 57 mm cell. For curves A and C the bandwidth is 90 Hz; for curve B the bandwidth is 150 Hz.

The comparison between curves A and B clearly shows the advantages, in terms of reduction in shear strength, in using the wider bandwidth. These findings support the results shown in Figure 6.5, which indicate minimum shear stresses occurring around 100 Hz under sinusoidal excitation. The comparison between curves A and C of Figure 6.31 shows that the

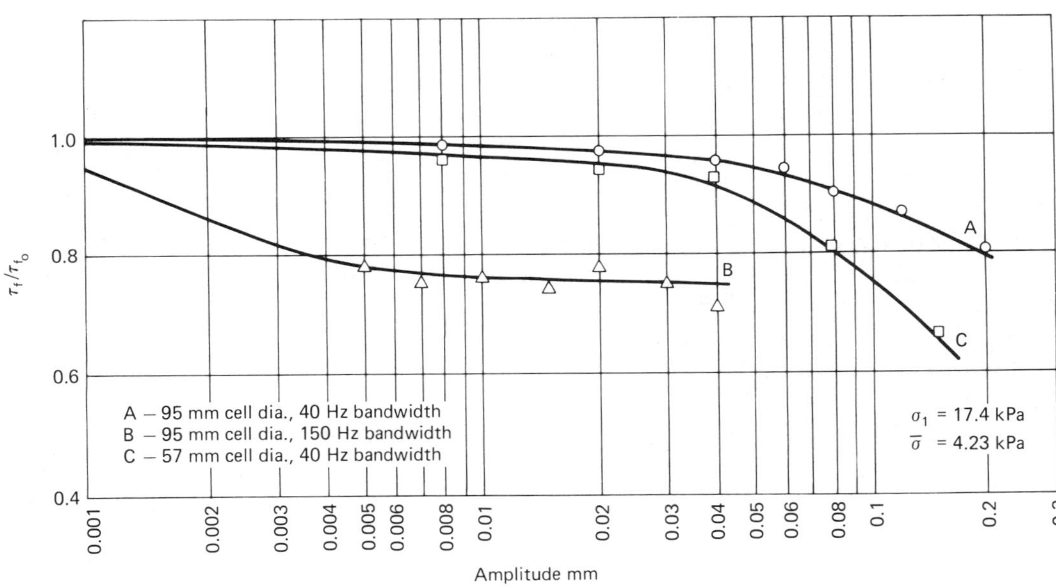

Figure 6.31. Effect of random vibration excitation on shear stress ratio τ_f/τ_{f_0} for pyrophyllite.

shear stresses for the same bandwidth are lower in the smaller cell for corresponding amplitudes. This may be explained by the fact that the vibration energy input per unit mass is higher in the case of the smaller cell.

The results of this study clearly indicate that, in order to be most effective, the bandwidth in the case of random excitation should span the resonant frequencies. This does not happen, in this case, for the 40 Hz bandwidth.

The results may be viewed another way. For most bulk solid storage systems, the major reduction in strength occurs at, or in the vicinity of, the resonant frequency in view of the amplitude, and hence velocity amplification, that this provides. Since the resonant frequency will vary from one material to another, then the general application of wide-band random excitation in flow promotion has some merit. The continuous range of frequencies within a given bandwidth would render this type of excitation applicable to a wide range of bulk materials.

Dynamic System Identification

The application of system identification by random excitation and cross-correlation and spectral analysis to bulk materials handling systems has been fully described by Roberts et al.[82-84] The method has also been used by Roberts and Scott[60] to determine the dynamic characteristics of bulk solids in the shear cell test.

The test procedure consists, essentially, of applying a pseudo-random binary coded displacement signal $x_1(t)$ to excite the whole shear cell and obtaining the cross-correlation function and cross-spectral density of the two signals $x_1(t)$ and $x_2(t)$ where $x_2(t)$ is the dynamic displacement of the top half of the shear cell. The cross-correlation function $R_{x_1x_2}(\lambda)$ and cross-spectral density $S_{x_1x_2}(f)$ are, respectively, defined as

$$R_{x_1x_2}(\lambda) = \lim_{T \to \infty} \frac{1}{T} \int_0^T x_1(\lambda)x_2(t+\lambda)\,dt \tag{6.42}$$

and

$$S_{x_1x_2}(f) = \int_{-\infty}^{\infty} R_{x_1x_2}(\lambda)e^{-j2\pi f\lambda}\,d\lambda \tag{6.43}$$

That is $S_{x_1x_2}(f)$ is the Fourier Transform of $R_{x_1x_2}(\lambda)$.

Under the condition that the input signal bandwidth is significantly greater than that of the system, in this case the shear cell and bulk solid, it can be shown that the impulse response or weighting function $h(\lambda)$ is given by

$$h(\lambda) \simeq KR_{x_1x_2}(\lambda) \tag{6.44}$$

where K = constant under the same conditions.

$$H(f) \simeq KS_{x_1x_2}(f) \tag{6.45}$$

where $H(f)$ is the system transfer function.

In performing the system identification analyses on the bulk solids, the samples were first prepared and consolidated in the normal way. The random excitation was then applied with the selected normal pressure σ applied to the shear cell, and measurements of $R_{x_1x_2}(\lambda)$ and $S_{x_1x_2}(f)$ were obtained.

To illustrate the application of this method, results are given for pyrophyllite and iron ore. Figure 6.32 shows the cross-correlation functions for these two materials while Figure 6.33 shows the corresponding cross-spectral densities. The results were obtained using the 95 mm diameter shear cell, the samples being consolidated under the normal stress $\sigma = 7.9$ kPa. The shear cell was subjected to pseudo-random excitation under the applied normal stress $\bar{\sigma} = 4.23$ kPa.

It is interesting to compare the two $S_{x_1x_2}(f) \simeq H(f)$ curves of Figure 6.33 with the corresponding τ_f/τ_{f_0} curves obtained from sinusoidal excitation. These latter curves for pyrophyllite and iron ore are shown in Figure 6.34. Referring first to Figure 6.33 it may be seen that the $S_{x_1x_2}(f)$ characteristic increases to a maximum around 90 Hz, which indicates that the peak vibration energy is concentrated at

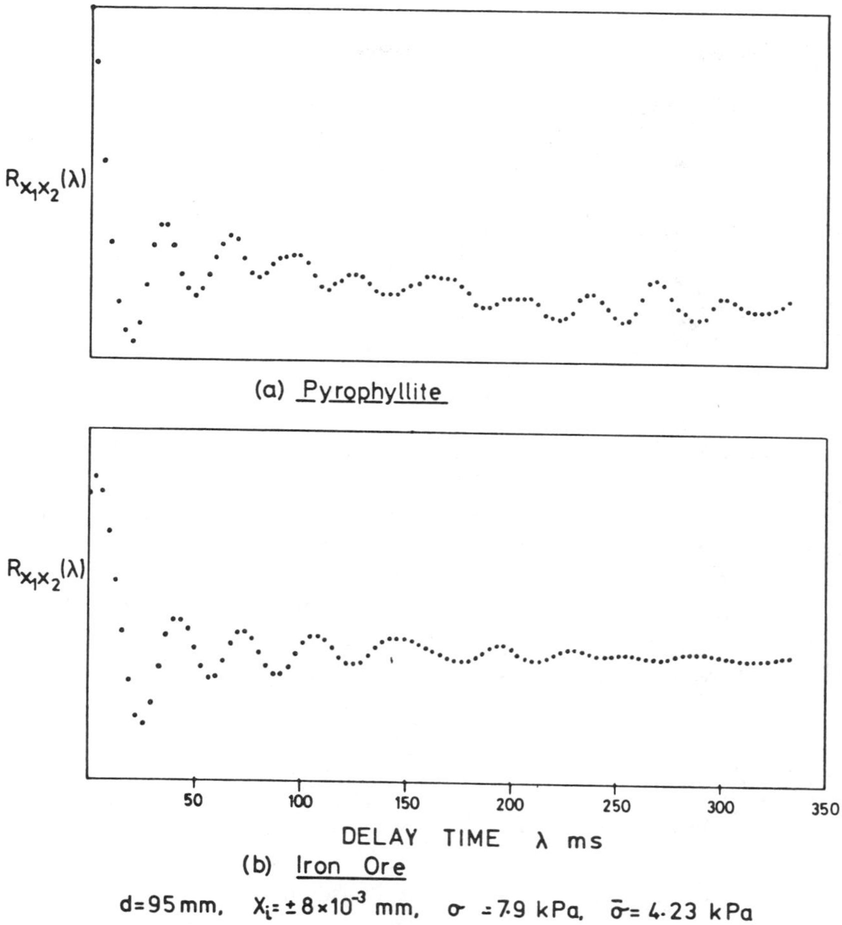

Figure 6.32. Cross-correlation functions for pyrophyllite and iron ore.

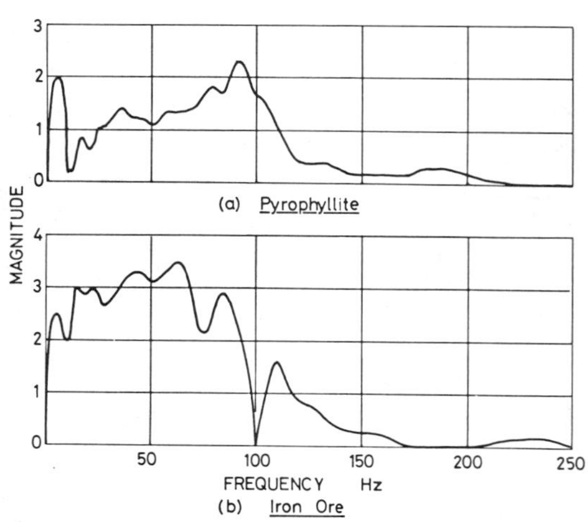

Figure 6.33. Cross-spectral densities $S_{x_1x_2}(f) \simeq H(f)$ for pyrophyllite and iron ore.

this particular frequency. On the other hand, the "flatter" $S_{x_1x_2}(f)$ characteristic for iron ore indicates that the vibration energy is more uniformly distributed over the frequency range 0–100 Hz. This explains why the τ_f/τ_{f_0} graph of Fig. 6.34 for iron ore shows a greater reduction than the corresponding graph for pyrophyllite over the 0–100 Hz frequency range.

6.8. COMPACTION OF POWDERS AND BULK SOLIDS

Mechanical vibrations are employed extensively to compress and compact powders, the range of applications in this respect being wide and varied. Whereas vibrations are used to advantage in such cases, there are often situations where

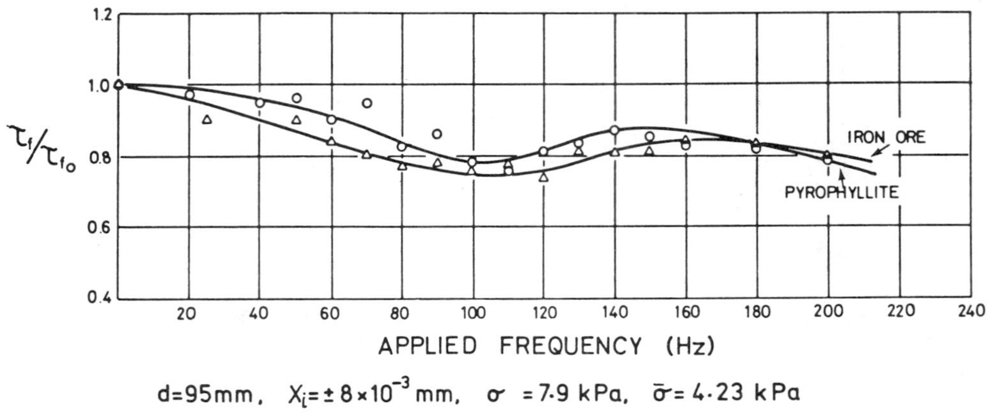

Figure 6.34. Comparison of effect of sinusoidal vibration on shear ratio τ_f/τ_{f0} for iron ore and pyrophyllite.

compaction of powders or bulk solids may have detrimental effects. This is particularly the case in bulk handling operations where increased strength due to vibratory compaction may cause flow interruptions. This section consists of a brief review of some salient aspects of the effects of vibrations on strength and bulk density.

Effect of Vibration on Shear Strength and Bulk Density

Kaaden[59] used a Jenike direct shear tester to examine the effect of sinusoidal vibrations on the consolidation and shear strength of bulk solids. In the standard shear test the consolidation of the sample in the shear cell comprises two phases, a preconsolidation phase involving a twisting procedure under the applied normal load followed by the consolidation under shear phase to obtain a steady state condition. To assess the effect of vibration on the shear strength, Kaaden used the same consolidation procedure as in the standard test, but prior to shearing, the samples were further consolidated by vibration for a predetermined period of time.

Figure 6.35 shows a typical set of yield loci results for Mt. Newman iron ore at 9% moisture content (w.b.); in this case the samples were consolidated by horizontal vibrations being applied to the top ring. For the range of frequencies examined, the increase in shear strength due to vibration of the confined samples was most affected by the increase in time of vibration. Presumably there is a limiting time of vibration consolidation when the bulk material approaches a limiting density and any further vibration would have little effect. For comparison purposes the time yield locus for a 48 hour storage time is also shown and it is readily observable that the increased strength due to vibration is far more significant than that due to time storage. The results presented in Figure 6.35 are in sharp contrast to the results given in Section 6.3; in that case the shear strength was considerably reduced when vibrations were applied during the actual shearing operation.

It is well known that a correlation exists between the shear strength and density of a bulk solid; the more a sample is consolidated, the higher will be its density and the higher its

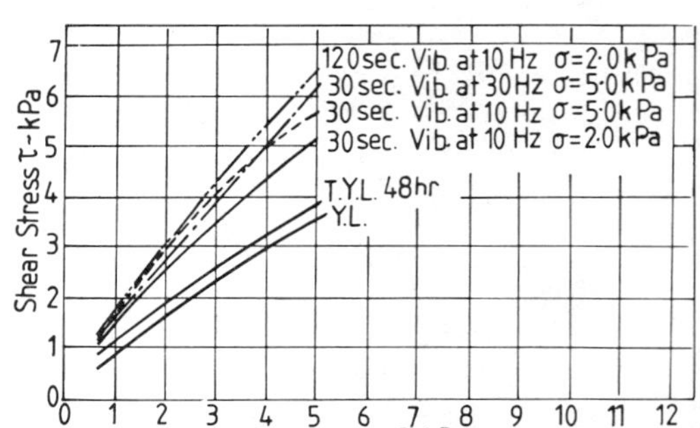

Figure 6.35. Effect of vibration consolidation on yield loci, Mt. Newman iron ore. 9% m.c. (w.b.).

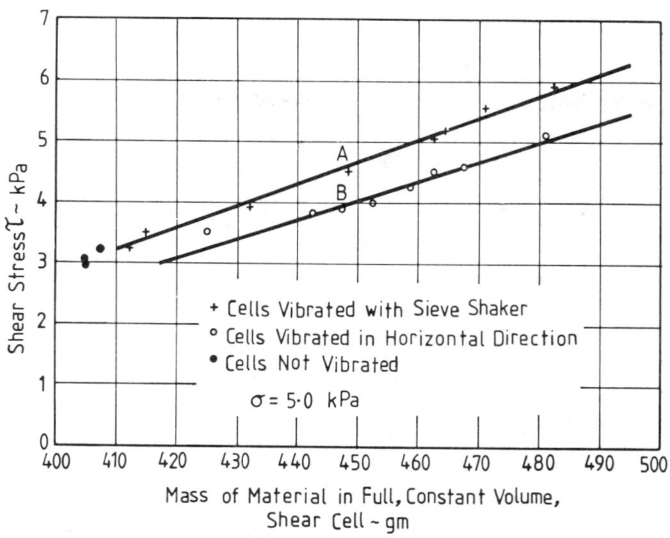

Figure 6.36. Effect of density on shear strength, Mt. Newman iron ore. 9% m.c. (w.b.).

shear strength. By way of example Figure 6.36 shows the correlation between shear stress τ and bulk density for Mt. Newman iron ore fines. For curve *A* the complete cell was vibrated using a sieve shaker, whereas for curve *B* the top ring of the shear cell was vibrated in the horizontal direction using a electromagnet shaker.

The sieve shaker produced samples that were more uniformly packed and hence of greater strength than the samples produced by the electromagnetic shaker. These results indicate a 20% density increase being accompanied by an 80–90% increase in shear strength.

Vertical Compressive Deformation during Lateral Vibration

To gain a better insight into the effects of vibration on the compressibility of bulk solids, Roberts and Scott[60] examined the vertical compressive deformation of samples in the shear cell when subjected to laterally applied vibrations. The shear cell apparatus shown in Figure 6.1 was used for this test.

Samples were prepared and consolidated under shear in the usual way, the consolidating stress being $\tau = 7.9$ kPa. They were then subjected to lateral vibrations under reduced normal stress, during which records of the compressive deformation were made. For this series of tests −2 mm pyrophyllite was used instead of the −1 mm pyrophyllite, as used in the work presented in Section 6.3.

Figure 6.37 shows a set of compressive deformation versus time records obtained during sinusoidal vibration under the applied normal stress $\tau = 5.3$ kPa. The amplitude of vibration was 0.012 mm. As indicated, the major transient deformations occur within the first minute; after this the response curves approach asymptotically steady-state deformations.

The steady-state deformations as a function of frequency are shown in Figure 6.38. For the range of frequencies examined, the minimum deformation occurs around $f = 140$ Hz, with a maximum value being approached at $f = 100$ Hz. Although the evidence is somewhat inconclusive, the maximum consolidation occurring around the value $f = 100$ Hz lends weight to the notion of this frequency being a resonance frequency similar to that depicted in the results presented in Section 6.3. A plausible argument that relates the results given here with the reduction in shear strength shown in Figure 6.5 might proceed as follows: It is apparent that the maximum reduction in shear strength occurs when the dilation, and hence voidage,

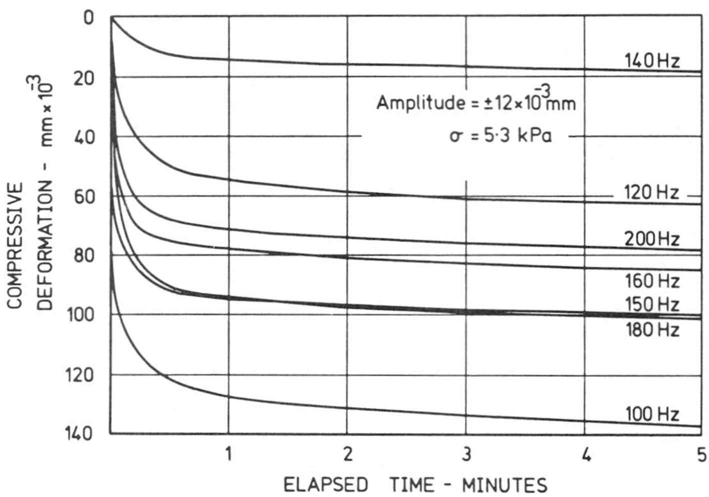

Figure 6.37. Vertical compressive deformation of pyrophyllite during lateral vibration. 95 mm shear cell.

on the shear plane is a maximum. This occurs at reasonance. Hence, when the sample is vibrated while being sheared, the resonance condition implies that the maximum dilation on the shear plane is accompanied by a maximum compressive deformation of the material elsewhere in the shear cell.

Some General Remarks

The application of vibrations to the compaction of powders has received wide attention. A comprehensive review of the underlying principles and associated research in this topic is given by Shatolova et al.,[67] McGeary,[68] and Evans and Millman.[70]

It is clear that the application of vertical vibrations to powders and bulk materials held within containers provides an efficient mode of compaction, and consequently this field of application has been studied in some detail. For instance Gray and Rhodes[32] point out that the final density of vibrated powder is a function of the energy transferred to the bed from the vibrator. They modeled the bed as (1) plastic bodies and (2) visco-elastic bodies; both models predict a decrease in energy transferred to the powder as the acceleration is increased at constant frequency. For frequencies below 150 Hz and accelerations less than 10 *g*, the powder behaves as a coherent mass, compaction taking place as a result of the block type motion of the mass being projected from the base of the container and subsequently collecting with it. Above the frequency and acceleration limits indicated, the bed becomes fluidized.

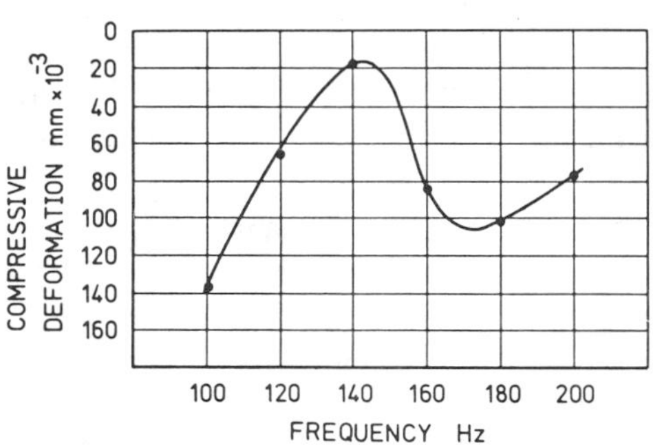

Figure 6.38. Steady-state vertical compressive deformation during lateral vibration. 95 mm shear cell.

Other work of relevance to vibratory compaction is that of Shinohara et al.[53] and Suzuki and Tanaka.[55] Particular attention is drawn to the work of Gutman,[35] who studied the effect of air compressed between a vertically vibrating bed and the bottom of the container.

For more details on the effect of vibrations on the compaction of powders, the reader is referred to the literature cited.

6.9. APPLICATION OF VIBRATIONS IN FLOW PROMOTION

Reference has been made several times throughout this chapter to the application of vibrations in promoting the flow of bulk solids from storage bins. When used correctly, vibrations can significantly reduce both the strength of bulk solids and the wall friction, and as a result, greatly increase the ability of the material to flow.

There are a variety of ways in which mechanical vibrations are used in practice in association with flow promotion in storage bins, and these have been reviewed in the papers by Carroll and Colijn,[52] Myers,[51] and Wahl.[50] Although a great deal of practical knowledge has been gained, the information available in the past for design purposes has been largely empirical. There has been a general lack of information concerning the relevant frequency, amplitude, and inputs that should be used for each particular bulk solid.

The manner in which the information presented in earlier sections of this chapter may be integrated with the general philosophy of bin design is discussed in this present section.

Bin Design Philosophy

The theory of storage bin design, as developed by Jenike,[9,10] is well documented and widely used. However, in order that the effect of vibrations on the flow of bulk solids may be better understood, the salient aspects of the bin flow characteristics and design philosophy are briefly reviewed.

Flow Patterns. As indicated in Figure 6.39, there are two basic modes of flow, namely mass-flow and funnel-flow.

In mass flow the bulk material is in motion at substantially every point in the bin whenever material is drawn from the outlet. The material flows along the walls with the bin and hopper (that is, the tapered section of the bin) forming the flow channel. Mass-flow is the ideal flow pattern and occurs when the hopper walls are sufficiently steep and smooth and there are no abrupt transitions or inflowing valleys.

Funnel-flow, on the other hand, occurs when the material sloughs off the surface and discharges through a vertical channel that forms within the material in the bin. This mode of

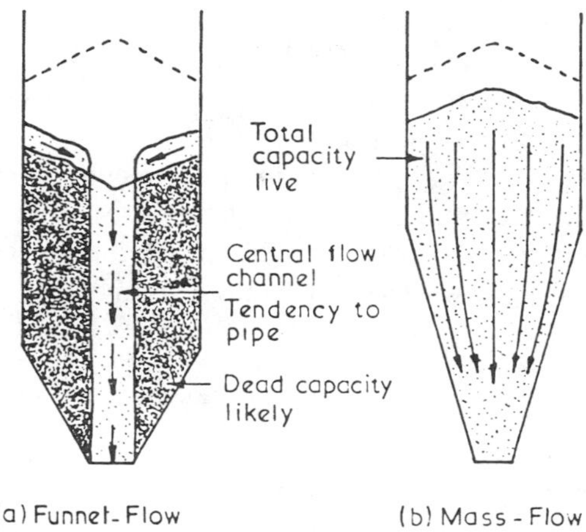

Figure 6.39. Bin flow characteristics.

flow occurs when the hopper walls are rough and the slope angle α is too large. The flow is erratic, with a strong tendency to form stable pipes which obstruct the bin discharge. When flow does occur, segregation takes place, there being no remixing during flow. It is an undesirable flow pattern.

The limits for mass-flow depend on the hopper half angle α, the wall friction angle ϕ, and the effective angle of internal friction δ. In the case of conical hoppers the limits for mass-flow are clearly defined as illustrated in Figure 6.40. On the other hand plane-flow or wedge-shaped hoppers have similar limits for mass-flow, but these are less critical.

Hopper Geometry for Mass Flow. Basically the aim in mass-flow design is to determine the

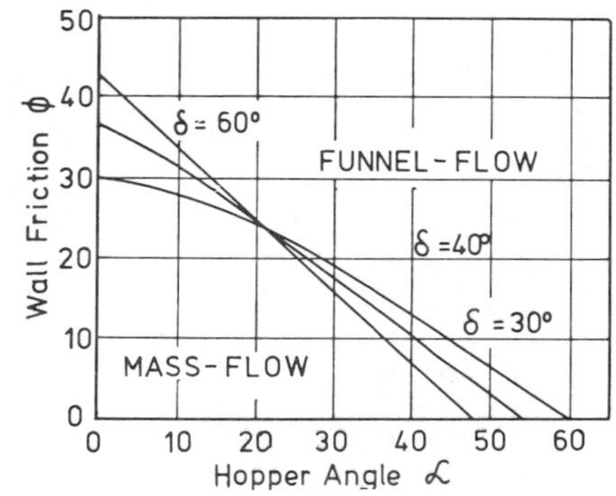

Figure 6.40. Funnel flow versus mass flow limits for conical hoppers.

hopper geometry, in particular the hopper half angle α and opening size B, so that a stable cohesive arch cannot form over the bin outlet. Two parameters are important: first, the *Flow Function FF* representing the strength of the material, and second, the *flow factor ff* representing the stress condition in the hopper during flow. The flow factor is given by

$$ff = \frac{\sigma_1}{\bar{\sigma}_1} \tag{6.46}$$

where

$\bar{\sigma}_1$ = stress that can develop in an arch
σ_1 = major consolidating stress

The flow factor is a linear function and is shown together with a typical flow function in Figure 6.41. The flow factor depends on the wall friction angle ϕ, the hopper half angle α, and the effective angle of internal friction δ. That is,

$$ff = f(\phi, \delta, \alpha) \tag{6.47}$$

The determination of flow factors is described by Jenike[10] and Arnold et al.,[78] who also give the associated flow factor design charts.

With reference to Figure 6.41, the critical condition for flow is defined by $\bar{\sigma}_1 = \sigma_c$ where σ_c is the unconfined compressive strength. The

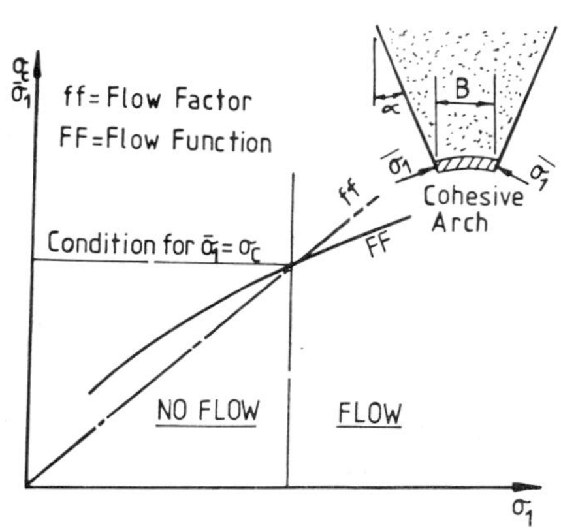

Figure 6.41. Flow/no-flow critierion for mass flow hopper design.

minimum opening size B is defined by

$$B = \frac{\bar{\sigma}_1 H(\alpha)}{\rho g} \tag{6.48}$$

where

ρ = bulk density
g = gravitational acceleration.

The function $H(\alpha)$ depends on the outlet shape and hopper half angle and is plotted by Jenike[10] and Arnold et al.[78]

In practice the opening size is made larger than the minimum value given by Eq. (6.50) in order to ensure a satisfactory flow rate. In the case of fine powders of low permeability, the determination of opening size to achieve a prescribed flow rate is far more complex, and air permeation may be necessary.[78, 85]

Influence of Mechanical Vibrations

As previously discussed, two significant parameters in bin and hopper design and performance evaluation are the Flow Function and wall yield locus. In Figure 6.11 the Flow Functions from -1 mm pyrophyllite at 5% moisture content were presented for the three frequencies 0 Hz (instantaneous, static condition), 150 Hz, and 200 Hz. However, on the basis of the failure criterion presented in Section 6.5, it is really only necessary to determine, for the vibration case, the limiting Flow Function corresponding to the maximum vibration velocity for maximum shear strength reduction. The Flow Function determined in this way for the pyrophyllite is shown in Figure 6.42. This graph is based on shear tests in which the top half of the shear cell was vibrated at a frequency of 200 Hz and amplitude ±0.01 mm; the corresponding velocity on the shear plane is 12.6 mm/s, which is the value approaching that for maximum shear strength reduction as indicated in Figure 6.19. The unvibrated Flow Function is shown, for comparison purposes, in Figure 6.42c. The reduction in strength due to vibration is quite significant and has a major influence on bin design.

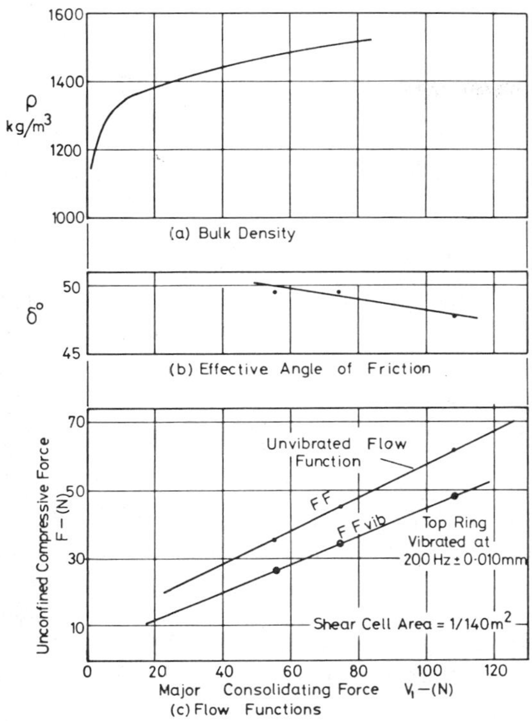

Figure 6.42. Flow properties of vibrated and unvibrated samples. −1 mm prophyllite, 5% m.c. (d.b.).

Figures 6.42a and 6.42b show, respectively, the bulk density ρ and effective angle of internal friction. Both these parameters are required for bin design.

The vibrated wall yield loci for the pyrophyllite are presented in Figure 6.30. For the present discussion reference will be made to the instantaneous and 200 Hz curves. The reduction in wall friction due to vibration at 200 Hz will permit a significant increase in hopper half angle α for the case of a conical hopper, as may be observed from an inspection of Figure 6.40.

The influence of the vibrations in improving the flow characteristics may be seen by reference to the mass-flow hopper proportions given in Table 6.4. Here a comparison is made between the hopper half angle and opening sizes for conical hoppers determined on the basis of the instantaneous and the vibrated properties of pyrophyllite.

The increase in hopper half angle α and the reduction in opening size B show the significance of using mechanical vibrations as an aid to flow. The increase in α can be even greater if the vibrations at the wall are applied in the direction of flow. The results of this study show that a funnel-flow bin can operate under mass flow giving uniform discharge, provided the vibrations are applied correctly. It is important that the bin and hopper be in a potential flow mode with the flow control gate open when the vibrations are applied; otherwise the vibrations will have the adverse effect of increasing the consolidation and strength of the bulk solid and reducing its ability to flow. This condition is indicated in Section 6.8. A flow mode will exist when an arched stress field occurs in the hopper.[78]

In order that the shear cell data and corresponding design information of the type given in Table 6.4 be applied to the design of bin activators, consideration needs to be given to the appropriate dynamic scaling parameters. Such parameters would need to translate the information on frequency and amplitude relating to the shear cell tests to equivalent data for a full-scale bin. The theory of failure presented in Section 6.5 shows that the vibration velocity on the plane of failure is the significant parameter causing local dilation and a corresponding reduction in shear strength.

Table 6.4. Comparison of Mass Flow Conical Hoppers for Unvibrated and Vibrated Conditions. (Material: −1 mm pyrophyllite, 5% moisture content (d.b.))

HOPPER GEOMETRY	INSTANTANEOUS PROPERTIES (UNVIBRATED)	VIBRATED PROPERTIES $f = 200$ Hz $x_r = 0.006 - 0.01$ mm
Half angle α	19°	33°
Opening size B	0.9 m	0.6 m

Vibration velocity $X_r\omega$ is directly related to the vibration energy E_v per unit volume in the following way

$$E_v = \rho X_r^2 \omega^2/2 \qquad (6.49)$$

It follows that the energy transfer characteristics of a bulk solid have a direct influence on the vibration velocity, and hence shear strength reduction, at the critical region of a flow obstruction. Based on this reasoning it seems, therefore, that the required vibration energy per unit mass at a flow obstruction needs to be the same as that determined for the shear cell.

In practice, the vibration flow promotion device may, if necessary, be located some distance from the actual flow obstruction. For this reason the vibration propagation characteristics of the bulk solid, as discussed by Scarlett and Eastman,[31] need to be taken into account. Fortunately, as already indicated, a reduction in wall friction can significantly improve the flow characteristics and, for this reason, a vibrator suitably placed on the hopper wall may be very effective. Such a device has been described by Carroll and Colijn.[52]

Another effective flow promotion device is the bin activator described by Wahl.[50] In this case the bin activator consists of a domed baffle plate located just above the outlet which, together with the rounded hopper bottom, is vibrated in the horizontal plane.

The domed baffle plate is said to eliminate overhead bridging by transmitting vibrations vertically up the hopper to cause flow to occur at least in the area vertically above the activator. An effect not mentioned is the reduction of friction on the surface of the baffle that enhances material flow off the edge of the baffle. In addition, the transmission of vibrations to the hopper walls also reduces friction and further enhances the flow.

Other types of vibratory-induced flow systems involve the vertical vibration of the whole bin or hopper. The comprehensive research of Shinohara, Suzuki, Takahashi, and Tanaka[53-55] has presented data indicating the conditions under which flow of both noncohesive and cohesive bulk solids can be best obtained by such systems. A difficulty of employing vertically vibrating hoppers in practice, particularly in large installations, is the design complexity and energy requirements to vibrate the whole hopper-bin system.

Performance of Bin with Vibrating Insert

Reference was made above to the vibrating baffle plate flow promotion device described by Wahl.[50] Studies performed by Croft[57] and Roberts et al.[58] have focused attention on some performance characteristics of this type of flow promotion. Salient aspects of their research are briefly reviewed.

The test rig used for this work is drawn schematically in Figure 6.43. The model bin is of rectangular cross section and operates under plane-flow, this shape and flow mode being chosen for ease of manufacture and testing. The bin is clear plastic, and the vibrating insert located inside the bin discharge section can be fitted with flat or convex baffle plates of various geometrical proportions.

Transverse vibrations are applied to the insert by the force transducer driven by either a discrete-frequency or random-signal generator through a power amplifier. The vibrator provides the driving force and may be set at a constant force amplitude, while the input displacement of the insert and output material flow rate are recorded continuously. Signal processing instrumentation permit information on correlation functions and power spectral densities to be obtained.

By way of example, one series of tests was conducted with moist sand using a flat plate baffle. The applications of sinusoidal forced excitation of varying frequencies and pseudo-random binary forced excitation of varying bandwidths were examined. In all cases the effective force amplitudes were kept approximately the same.

The results for the wet sand indicate that the low frequencies (in the order of 10 Hz) do not readily initiate flow. Quite often considerable time may elapse before flow commences, if at

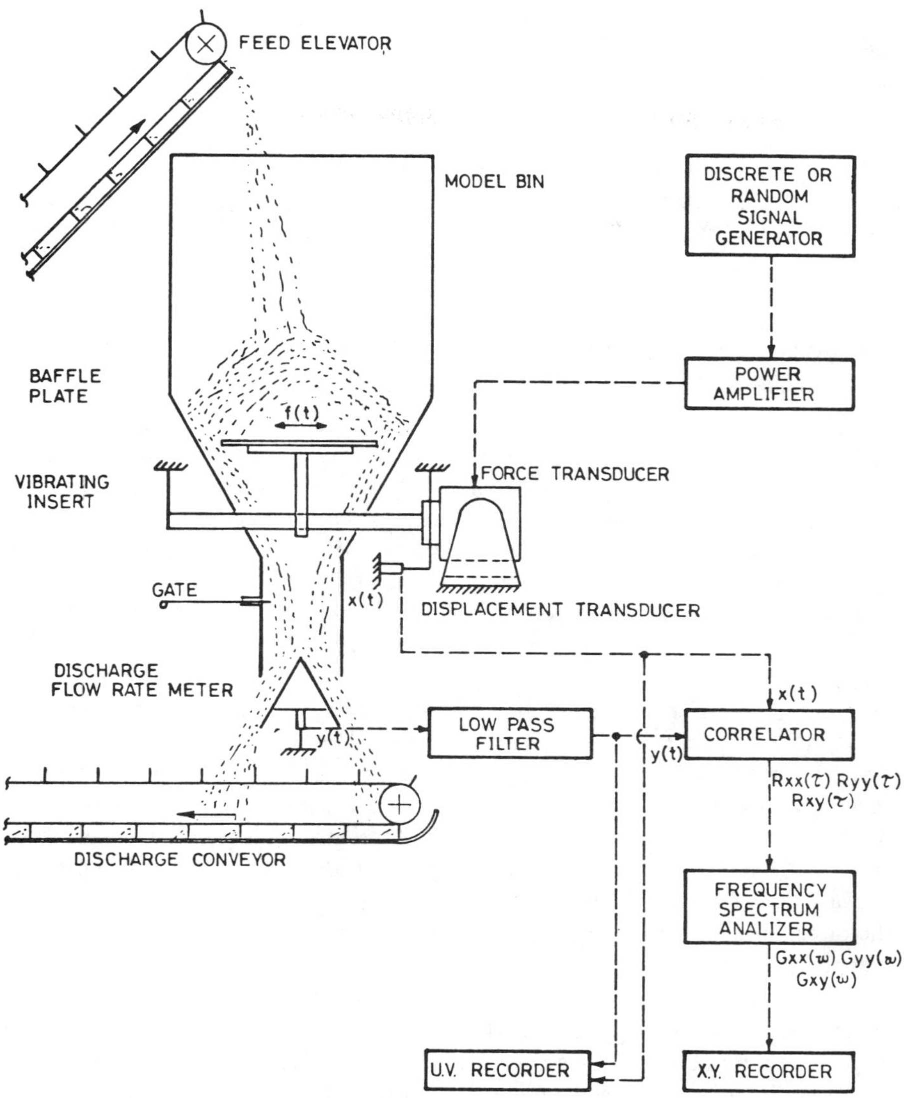

Figure 6.43. Model bin experimental test rig.

all. On the other hand with higher frequencies in the order of 100-300 Hz, flow commences almost immediately and continues at a steady rate.

The results also indicate that the excitation frequency is dependent on material properties, which confirms the results of the shear cell tests. For this reason, broad-band random excitation, in which there are an infinite number of frequencies represented in the bandwidth, will have distinct advantages over discrete-frequency sinusoidal type excitation. Table 6.5 gives a typical set of results for the wet sand, using pseudorandom binary force excitation. Although the results relate specifically to the model bin, nevertheless they provide an indication of comparative performance.

6.10. CLOSURE

The application of mechanical vibrations with respect to powders and bulk solids is wide and

Table 6.5. Typical Performance Results for Wet Sand using Pseudorandom Binary Force Excitation.

BANDWIDTH	ELAPSED TIME (s) BEFORE FLOW COMMENCED	ELAPSED TIME (s) TO EMPTY BIN ONCE FLOW COMMENCED	AVERAGE FLOW RATE m^3/s	COMMENTS
10 Hz	79[a]	35	.0023	Flow not uniform
100 Hz	18	23	.0031	Uniform flow
300 Hz	13	17	.0042	Uniform flow

[a]This is a typical figure—much longer times have been recorded at this bandwidth.

varied, and a basic understanding of the effect of vibrations on strength and flow properties is of fundamental importance to all areas of application. On the basis of the research reviewed in this chapter, some general conclusions may be drawn:

1. The shear strength of a powder on bulk solid undergoing shear deformation may be reduced significantly by the application of mechanical vibrations. For a given vibration energy input, the reduction in strength is a function of the degree of consolidation, the applied normal stress during shear, and the frequency. Experimental evidence suggests that the strength reduction increases with increase in particle size and reduction in moisture content.

2. In the direct shear test using the vibrating shear cell apparatus, vibration of the whole cell during shear deformation shows the presence of critical or natural frequencies at which the shear strength is a minimum. The presence of these natural frequencies has also been observed in measurements of the amplitudes of vibration on the shear plane. The natural frequency for a particular bulk solid is shown to increase with increases in both consolidation stress and applied normal stress. While the exact nature of the damping characteristics of powders and bulk solids is difficult to determine, there is evidence to suggest that damping due to Coulomb friction has a major influence, particularly when the amplitude of vibration relative to the average particle size is not insignificant. For very small amplitudes the influence of viscous damping is believed to be more pronounced. Since the damping is of low order, it has been shown that for simplicity an equivalent viscous damping factor may be assumed. Tests on pyrophyllite and iron ore, both at 5% moisture content (d.b.), show values of $\zeta = 0.1$ and $\zeta = 0.125$, respectively, that fit the experimental data quite well.

3. In the direct shear test when the top half of the cell is vibrated, no resonance effect is observed. The shear stress reduces exponentially with frequency, approaching asymptotically a limiting value that depends on the consolidation and applied normal stress.

4. The shear failure of a vibrated powder or bulk solid may be directly related to the influence of the relative vibration velocity on the shear plane. The vibration velocity may be correlated with the voidage on the shear plane as indicated by the Hvorslev failure criterion. For a given consolidation and applied normal stress, the shear strength is a decaying exponential function of the maximum vibration velocity, approaching asymptotically a limiting value.

5. The frictional drag forces between powders or bulk solids and adjacent boundary surfaces such as the wall of a hopper may be considerably reduced by the application of mechanical vibrations. The extent of this reduction is a function of the intensity or energy level of the vibration applied. The direction of the vibration in relation to the direction of shear deformation is shown to be of importance; when these two directions coincide the greatest benefit in the lowering of the drag forces may be achieved because of the assistance rendered by the inertial forces.

6. Reductions in shear strength of powders or bulk solids may also be accomplished by the use of broad-band random vibrations. To be effective, the frequency band width should span the critical frequency or frequencies of the material.

7. System identification studies employing cross-correlation and spectral analysis may be used to obtain the dynamic characteristics of a bulk solid in both the time and frequency domains. The latter provides useful information on the energy versus frequency distribution.

8. In the application of vibrations to the compression or compaction of powders, the degree of compaction achieved is directly related to the vibration energy. Vibratory compaction greatly increases the shear strength.

9. Mechanical vibrations may be used to advantage in promoting the flow of bulk solids from storage bins. When correctly applied, the use of vibrations will significantly lower the strength of the bulk solid and its wall friction coefficient. In this way storage bins that may operate as funnel flow bins in normal circumstances can be made to operate as mass-flow bins under applied vibration, using appropriate flow promotion devices. It is most important that a flow stress field exist within the material in the bin when the vibrations are applied.

ACKNOWLEDGMENTS

Much of the material presented in this chapter is based on research into vibrations of powder conducted at The University of Newcastle N.S.W., Australia. The contribution of staff and students to the development of this work is gratefully acknowledged. In particular special mention is made of the research of Mr. Li Kin Wah, a former research student of the university, for his valuable contribution to the theory of vibrations of powder and bulk solids and to Mr. O. J. Scott for his assistance both with the development of the experimental equipment and the conducting of experimental work.

The final writing of the material for this chapter was performed while the author was on a period of study leave in the Department of Mechanical Engineering at the University of Twente, Enschede, The Netherlands. The invaluable assistance rendered by the staff of this University in typing the manuscript and preparing diagrams is acknowledged with appreciation. In particular the assistance rendered by Mr. H. J. Rietman is gratefully acknowledged. Special mention is made of the support and encouragement given by the author's colleague Prof.dr.Ir. F. J. C. Rademacher.

LIST OF SYMBOLS

A	Area of shear plane (m^2)
a_0, a_1, a_2	Constants
B	Damping coefficient (N.s/mm)
B_E	Equivalent damping coefficient (N.s/mm)
B_{cr}	Critical damping coefficient (N.s./mm)
C	Cohesion
C_0	Initial cohesive force before consolidation (N)
c_0, c_1, c_2	Constants
d	Average particle size (mm)
E	Constant
E_v	Energy per unit mass
e	Void ratio
f	Frequency (Hz)
f_n	Natural frequency (Hz)
FF	Flow Function
F_f	Friction force (N)
F_s	Equivalent static restoring force (N)
ff	Flow factor
g	Gravitational Acceleration (m/s^2)
H	Moisture content (also abbreviated as m.c.)
$H(f)$	System transfer function
$h(X)$	System impulse response or weighting function
k	Stiffness (N/mm)
K	Constant
m	Mass (kg)
N	Interparticle normal force (N)
N_c	Inertial coefficient
N_e	Interparticle normal force due to consolidation (N)

$Rx_1x_2(\lambda)$	Cross-correlation function
r	Frequency ratio
Q	Coulomb damping function
S	Shear force (N)
$S_{x_1x_2}(f)$	Cross-spectral density
T	Temperature (°C)
U	Vibration velocity (mm/s)
U_s	Velocity of sliding (mm/s)
V	Normal force (N)
$\overline{V}$	Compressive force on plane of interest (N)
X	Amplitude of vibration (mm)
x_1, x_i	Input displacement (mm)
X_1, X_i	Input displacement amplitude (mm)
x_r	Relative displacement (mm)
X_r	Relative displacement amplitude (mm)
x_2	Absolute displacement of top half of shear cell (mm)
X_2	Absolute displacement amplitude (mm)
α	Hopper half angle
δ	Effective angle of internal friction
ϕ	Kenematic angle friction at hopper wall
β	Reduction in shear stress (kPa)
γ	Vibration velocity constant
ρ	Bulk density
σ_1	Major consolidating stress (kPa)
σ	Normal stress during consolidation (kPa)
$\overline{\sigma}$	Applied normal stress (kPa)
σ_c	Unconfined compressive strength (kPa)
σ_{cf}	Unconfined compressive stress during vibration (kPa)
Ψ	Phase angle
τ_∞	Steady state shear stress (kPa)
τ	Shear stress during consolidation (kPa)
τ_f	Shear stress plotted against frequency
$\overline{\tau}$	Measures shear stress
ω	Frequency (rad/sec)
ω_n	Natural frequency (rad/sec)
ζ	Damping ratio
ζ_v	Damping ratio, viscous component
μ	Friction coefficient

REFERENCES

1. C. A. Coulomb, "Essai Sur une Application des Règles des Maximis et Minimis à Quelques Problèmes de Statique Relatifs de l'Architecture." *Mem. Math. Phys.*, pp. 343–381 (1773).
2. W. Rankine, "On the Stability of Loose Earth." *Phil. Trans.* (1857).
3. O. Reynolds, "On the Dilatancy of Media Composed of Rigid Particles in Contact." *Phil. Mag. 20*:469–481 (1885).
4. O. Reynolds, "Experiments Showing Dilatancy, a Property of Granular Material, Possibly Connected with Gravitation." *Proc. Roy. Inst. Gr. Brit. 11*:354–363 (1887).
5. H. A. Janssen, "Tests on Grain Pressure Silos." *Z. Ver. Dtsch. Ing. Beih. 35*:1045–1049 (1895).
6. W. Airy, "The Pressure of Grain." *Proc. Inst. Civ. Eng.* Paper 3049 (1897).
7. J. A. Jamieson, "Grain Pressures in Deep Bins." *Trans. Can. Soc. Civ. Eng. 17*:554–654 (1903).
8. M. S. Ketchum, *The Design of Walls, Bins and Grain Elevators*, 3d ed., McGraw-Hill, New York (1919).
9. A. W. Jenike, "Gravity Flow of Bulk Solids." *Utah Agric. Exp. St. Bull.* 108 (1961).
10. A. W. Jenike, "Storage and Flow of Solids." *Utah Agric. Exp. St. Bull.* 123 (1964).
11. J. R. Johanson, "Stress and Velocity Fields in Gravity Flow of Bulk Solids." *Utah. Agric. Exp. St. Bull.* 116 (1962).
12. M. J. Hvorslev, "On the Physical Properties of Distributed Cohesive Soils." *Ingeniorvidensk. Skr. 45* (1937).
13. K. H. Roscoe, A. N. Schofield, and C. P. Wroth, "On the Yielding of Soils." *Geotechnique 8*:22–53 (1958).
14. M. D. Ashton, D. C. H. Cheng, R. Farley, and F. H. H. Valentin, "Some Investigations into the Strength and Flow Properties of Powders." *Rheol. Acta. 4*(3):206–218 (1965).
15. J. C. Williams and A. H. Birks, "The Preparation of Powder Specimens for Shear Cell Testing," *Rheol. Act. 4*(3):170–180 (1965).
16. A. C. Palma, "Stress-Strain Relation for Clay, an Energy Theory." *Geotechnique 17*:348–358 (1967).
17. C. P. Wroth and R. H. Bassett, "A Stress-strain Relationship for the Shearing Behaviour of Sand." *Geotechnique 15*:32–56 (1965).
18. P. W. Rowe, L. Barden, and I. K. Lee, "Energy Components during the Triaxial Cell and Direct Shear Tests." *Geotechnique 14*:247–261 (1964).
19. J. K. Mitchell, "Shearing Resistance of Soils as a Rate Process." *Proc. Am. Soc. Civ. Eng., J. Soil Mech. and Foundations Div.*, pp. 29–61 (1964).
20. J. Lysmer and F. E. Richart, "Dynamic Response of Footings to Vertical Loading." Jnl. of Soil

Mechanics and Foundations Division, *Proc. Am. Soc. Civ. Eng., Soil Mech. and Foundations Div.*, pp. 65–91 (1966).

21. J. K. Mitchell, "Soil Creep as a Rate Process." *Proc. Am. Soc. Civ. Eng., J. Soil Mech. and Foundations Div.*, pp. 231–254 (1968).
22. J. K. Mitchell, A. Singh, and R. G. Campanella, "Bonding, Effective Stresses and Strength of Soils." *Proc. Am. Soc. Civ. Eng., J. Soil Mech. and Foundations Div.*, pp. 1219–1246 (1969).
23. N. E. Funston, "Footing Vibration with Non Linear Subgrade Support." *Proc. Am. Soc. Civ. Eng., J. Soil Mech. and Foundations Div.*, pp. 191–211 (1967).
24. E. D'Appolonia, "Dynamic Loadings." *Proc. Am. Soc. Civ. Eng., J. Soil Mech. and Foundations Div.* (1970).
25. D. H. Timmerman and T. H. Wu, "Behaviour of Dry Sands Under Cyclic Loading." *Proc. Am. Soc. Civ. Eng., J. Soil Mech. and Foundations Div.*, pp. 1097–1112 (1969).
26. B. J. Greenfield and E. T. Misiaszek, "Vibration-Settlement Characteristics of Four Gradations of Ottawa Sand." *Proc. Int. Symp. on Wave Propagation and Dynamic Properties of Earth Materials*, University New Mexico, pp. 787–795 (1967).
27. Suk Chae Yong, "The Material Constants of Soil as Determined from Dynamic Testing." *Proc. Int. Symp. on Wave Propagation and Dynamic Properties of Earth Materials*, University of Mexico, pp. 759–770 (1967).
28. V. P., Drnevich, J. R. Hall, and F. E. Richart, "Effect of Amplitude of Vibration on the Shear Modulus of Sand." *Proc. Int. Symp. on Wave Propagation and Dynamic Properties of Earth Materials*, University New Mexico, pp. 189–199 (1967).
29. J. W. S. de Graft-Johnson, "The Damping Capacity of Compacted Kaolinite under Low Stresses," *Proc. Int. Symp. on Wave Propagation and Dynamic Properties of Earth Materials*, University New Mexico, pp. 771–780 (1967).
30. L. T. Youd, "Reduction of Critical Void Ratio during Steady-State Vibration." *Proc. Int. Symp. on Wave Propagation and Dynamic Propagation and Dynamic Properties of Earth Materials*, University of New Mexico, pp. 737–744 (1967).
31. B. Scarlett and I. E. Eastman, "Stress in Granular Materials due to Applied Vibration." *I. Chem. Eng. Symp.*, ser. 29., Inst. Chem. Eng., London, pp. 45–68 (1968).
32. W. A. Gray and G. T. Rhodes, "Energy Transfer during Vibratory Compaction of Powders." *Powder Technol. 6*:271–281 (1972).
33. D. Nicklin and W. Hopkins, "Fluidization by Vibration." Fluidization Symposium, Melbourne, Australia (1967).
34. V. Chlenov and N. Mikhailov, "Vibrofluidised Beds." *Izdatelstvo Nauka*, Moscow (1972).
35. R. G. Gutman, "Vibrated Beds of Powders." *Trans. Inst. Chem. Eng. 54*:174–183 (1976).
36. W. Kroll, "Forschung aut der Gebiete des Ingenieur wesen," 20, Ed A (1), 2. (1954).
37. C. F. Harwood, "Powder Segregation due to Vibration." *Technol. Powder Technol. 16*:51–57 (1977).
38. H. Deresiewicz, *Mechanics of Granular Matter Advances in Applied Mechanics*, vol. 5, Academic Press, New York (1958).
39. O. Molerus, "Theory of Yield of Cohesive." *Powder Technol. 12*:259–275 (1975).
40. M. Becker and H. Lippman, "Plane Plastic Flow of Granular Model Material: Experimental Set-up and Results." *Arch. Mech. Stosow. 29*(6):829–846 (1977).
41. R. Nova and D. M. Wood, "A Constitutive Model for Sand in Triaxial Compression." *Int. J. Numerical and Analytical Methods in Geomechanics 3*:255–278 (1979).
42. A. Blinowski, "On the Dynamic Flow of Granular Media." *Arch. Mech. Stosow. 30*(1):27–34 (1978).
43. S. C. Cowin, "A Theory for the Flow of Granular Materials." *Powder Technol. 9*:61–69 (1974).
44. M. A. Goodman and S. C. Cowin, "A Continuum Theory for Granular Materials." *Arch. Ration. Mech. Anal. 44*:249–266 (1972).
45. S. L. Passman, "Balance Equations for Mixtures of Granular Materials." MRC Technical Summary Report No. 1390, January 1974 University of Wisconsin.
46. S. L. Passman, "Mixtures of Granular Materials." *Int. J. Eng. Sci. 15*:117–29 (1977).
47. S. L. Passman and J. P. Thomas, "On the Linear Theory of Flow of Granular Media." *Dev. Theor. Appl. Mech. 9* (1978).
48. S. Nemat-Nasser, "On Behaviour of Granular Materials in Simple Shear." Earthquake Res. and Eng. Lab., Tech. Rep. No. 79-6-19. Dept. of Civ. Eng., Northwestern University, Evanston, Ill. (June, 1979).
49. J. W. Nunziato, J. E. Kennedy, and E. K. Walsh, "The Behaviour of One-Dimensional Acceleration Waves in an Inhomogeneous Granular Solid." *Int. J. Eng. Sci. 16*:637–648 (1978).
50. E. A. Wahl, "Bin Activators–Key to Practical Storage and Flow of Solids." Symposium on Storage and Flow of Solids, Chicago, Ill., ASME Paper 72-MH 29 (1972).
51. J. I. Myers, "Vibrating Hoppers." *Mech. Eng.*, pp. 27–31 (1970).
52. P. J. Carroll and H. Colijn, "Vibration in Solids Flow." *Chem. Eng. Prog. 71*(2):53–65 (1975).
53. K. Shinohara, A. Suzuki, and T. Tanaka, "Gravity

and Vibration Effects on Flow of Cohesive Materials from Hopper." Paper presented at ASME Symp. on Flow of Solids, Boston, Mass. (October, 1968).

54. A. Suzuki, H. Takahashi, and T. Tanaka, "Behaviour of a Particle Bed in the Field Vibration, Part II: Flow of Particles through Slits in the Bottom of a Vibrating Vessel." *Powder Technol.* *2*:72–77 (1969).
55. A. Suzuki and T. Tanaka, "Behaviour of a Particle Bed in the Field of Vibration, Part IV: Flow of Cohesive Solids from Vibrating Hoppers." *Powder Technol.* *6*:301–308 (1972).
56. P. C. Arnold, A. S. Kaaden, and A. W. Roberts, "Effects of Vibration on the Flow of Bulk Solids." Paper presented at Symp. Am. Inst. Chem. Eng. (1976).
57. A. J. Croft, "Investigation of the Trends of Behaviour of the Flow Parameters of Cohesive Bulk Materials when Subjected to Vibration." M. Eng. Sc. thesis, University of Newcastle (1977). Australia.
58. A. W. Roberts, O. J. Scott, and Kin Wah Li, "Effects of Mechanical Vibration on the Flow of Bulk Solids." Paper presented to Conf. on Agric. Eng., Inst. Aust. Eng., Toowoomba, Australia August (1978).
59. A. S. Kaaden, "Gravity Flow of Some Steelmaking Raw Materials with Particular Reference to the Effects of Vibration," M. Eng. Sci. thesis, University of Wollongong (1975). Australia.
60. A. W. Roberts and O. J. Scott, "An Investigation into the Effect of Sinusoidal and Random Vibrations on the Strength and Flow Properties of Bulk Solids." *Powder Technol.* *21*:45–53 (1978).
61. Kin Wah Li, "Effect of Vibration on the Shear Strength and Flow Properties of Bulk Solids." Unpub. M. E. thesis, Dept. of Mech. Eng., The University of Newcastle, Australia (1978).
62. A. W. Roberts, O. J. Scott, and Kin Wah Li, "The Influence of Mechanical Vibrations on the Strength and Flow Properties of Bulk Solids." *Proc. Int. Conf. on Powder and Bulk Solids*, Philadelphia, Pa. (May, 1979).
63. F. J. C. Rademacher, "Accurate Measurement of the Kinetic Coefficient of Friction Between a Surface and a Granular Mass." *Powder Technol.* *19*:65–77 (1978).
64. V. K. Sharma, L. O. Drew, and G. L. Nelson, "High Frequency Vibrational Effect on Soil Metal Friction." *Trans. ASAE 20*(1):46–51 (1977).
65. O. Molerus and P. R. Schöneborn, "Bunker Design Based on Experiments in a Bunker-Centrifuge." *Powder Technol.* *16*:265–72 (1977).
66. J. R. Johanson, "Modelling Flow of Bulk Solids." *Powder Technol.* *5*:93–99 (1971).
67. I. G. Shatalova, N. S. Gorbunov, and V. I. Likhtman, "Physichemical Principles of Vibratory Compacting." *Perspectives in Powder Metallurgy*, vol. 2, *Vibratory Compacting*, Plenum Press, New York (1967).
68. R. K. McGeary, "Mechanical Packing of Spherical Particles." *Perspectives in Powder Metallurgy*, vol. 2, *Vibratory Compacting*, Plenum Press, New York (1967).
69. J. J. Hauth, "Vibrational Compaction of Nuclear Fuels." *Perspectives in Powder Metallurgy*, vol. 2, *Vibratory Compaction.* Plenum Press, New York (1967).
70. P. E. Evans and R. S. Millman, "The Vibratory Packing of Powders." *Perspectives in Powder Metallurgy*, vol. 2, *Vibratory Compacting.* Plenum Press, New York (1967).
71. A. W. Jenike, "Determination of Flow Properties of Powders by Shear Cell Measurements." *Particulate Matter*, *4*(2):11–14 (1973).
72. P. C. Arnold and A. S. Kaaden, "Reducing Hopper Wall Friction by Mechanical Vibration." *Powder Technol.* *16*:63–66 (1977).
73. J. Schwedes, "Vergleichende Betrachtungen zum Einsatz von Schergeraten zur Messung von Schüttguteigenschaften" [Shear Testers for Measuring the Flow Properties of Bulk Solids], Institute fur Mechanische Verfahrenstechnik, Technische Universitat, Braunschweig (1979).
74. F.J.C. Rademacher and G. Haaker, "Analysis of the Possible Errors Caused by the Loading Mechanism of Both the Original Jenike Shear Cell and Modified Version." *Proc. Int. Conf. on Powders and Bulk Solids*, Philadelphia (May, 1979).
75. S. C. Cowin and M. A. Goodman, "A Variational Principle for Granural Materials." *Zeitschrift Fur angewandte Mathematic und Mechanik* 56:281–286 (1976).
76. S. Timoshenko, D. H. Young, and W. Weaver. *Vibration Problems in Engineering.* 8th. ed., Wiley, New York (1974).
77. J. P. Den Hartog, "Forced Vibrations with Combined Coulomb and Viscous Friction." *Trans. A.S.M.E. 53*:107–115 (1931).
78. P. C. Arnold, A. G. McLean, and A. W. Roberts, "Bulk Solids: Storage Flow and Handling." *TUNRA*, The University of Newcastle, (1979). 2nd Edition (1981).
79. W. Lenkiewicz, "The Sliding Friction Process–Effect of External Vibration." *Wear 13*:99–108 (1969).
80. D. Godfrey, "Vibration Reduces Metal-to-Metal Contact and causes an Apparent Reduction in Friction." *Trans. ASLE. 10*:183–192 (1967).
81. D. M. Tolstoi, "Significance of the Normal Degree of Freedom and Natural Normal Vibration in Contact Friction." *Wear 10*:199–213 (1967).
82. A. W. Roberts and W. H. Charlton, "Application

of Pseudo-Random Test Signals and Cross Correlation to the Identification of Bulk Handling Plant Dynamic Characteristics." *Trans. ASME*, *J. Eng. Ind.* *95*:31–36 (1973).

83. A. W. Roberts and G. S. Montagner, "Identification of Transient Flow Characteristics of Granular Solids in a Hopper Discharge Chute System." Paper presented at the Symp. on Solids and Slurry Flow and Handling in the Chemical Process Industries, AICHE 77th National Meeting, Pittsburgh, Pa. (1974).
84. A. W. Roberts and W. H. Charlton, "Determination of Natural Response of Mechanical Systems Using Correlation Techniques." *Expt. Mech.*, *15*(1):17–22 (1975).
85. A. G. McLean and P. C. Arnold, "Two Phase Flow in Converging Channels." Proc. European Symp. Particle Technology Amsterdam (3–5 June 1980).

7
Size Enlargement Methods and Equipment

PART 1. INTRODUCTION

C. E. Capes
National Research Council
Ottawa, Ontario

Size enlargement can be defined as any process in which small particles are gathered into larger masses. Such processes are used to improve the usefulness of fine materials either in a downstream processing step or as a final product. The substitution of granular material for fine powders yields a number of possible benefits, as indicated in Table 7.1.1.

Methods used to agglomerate fine particles may be broadly classified into two main categories. On the one hand are the forming-type processes in which the properties of the *individual agglomerates* (such as shape, size, composition, density, etc.) are carefully controlled. On the other hand are the size enlargement methods in which a coarser granular material is created from fine powders. The properties of the *bulk material* are controlled in this case, and the characteristics of the individual agglomerates are important only insofar as they affect the properties of the whole or bulk product. As a result of these differences, the forming-type methods are usually of low capacity, often measured in pieces per hour, whereas the methods to beneficiate bulk material are of much larger capacity, usually measured in tons per hour.

In this chapter, some of the forming-type agglomeration processes as practiced in the pharmaceutical and other industries are dealt with in Sections 3 and 4. The balance of the chapter concerns size enlargement to improve the bulk properties of particulates. These latter techniques are classified here according to the principal methods used to bring particles together into agglomerates, including thermal methods (sintering), tumbling (balling) and other agitation methods, dispersion methods (prilling and other spray processes), and agglomeration from liquids.

Table 7.1.1. Benefits of Size Enlargement and Some Representative Applications

BENEFIT	EXAMPLES OF APPLICATION
1. Production of useful structural forms and shapes	Pressing of intricate shapes in powder metallurgy; manufacture of spheres by planetary rolling
2. Preparation of definite quantity units	Metering, dispensing, and administering of drugs in pharmaceutical tablets
3. Reduced dusting losses	Briquetting of waste fines
4. Creation of uniform, nonsegregating blends of fine materials	Sintering of fines in the steel industry
5. Better product appearance	Manufacture of fuel briquets
6. Prevention of caking and lump formation	Granulation of fertilizers
7. Improvement of flow properties	Granulation of ceramic clay for pressing operations
8. Greater bulk density to improve storage and shipping of particulates	Pelleting of carbon black
9. Reduction of handling hazards with irritating and obnoxious materials	Flaking of caustic
10. Control of solubility	Production of instant food products
11. Control of porosity and surface-to-volume ratio	Pelleting of catalyst supports
12. Increased heat transfer rates	Agglomeration of ores and glass batch for furnace feed
13. Removal of particles from liquids	Pellet flocculation of clays in water using polymeric bridging agents
14. Fractionation of particle mixtures in liquids	Selective oil agglomeration of coal particles from dirt in water

PART 2. AGGLOMERATE BONDING AND STRENGTH

Wolfgang B. Pietsch
Maschinenfabrik Köppern GmbH & Co.KG.
Hattingen/Ruhr

7.2.1. BINDING MECHANISMS

To obtain agglomerates from particulate matter, binding forces must act between the individual particles. According to Rumpf,[1] who first published a classification, the possible mechanisms can be divided into five major groups.

Table 7.2.1. Binding Mechanisms of Agglomeration[1]

1. Solid bridges
2. Interfacial forces and capillary pressure at freely movable liquid surfaces
3. Adhesion and cohesion forces at not freely movable binder bridges
4. Attraction forces between solid particles
5. Form-closed bonds (interlocking)

Solid Bridges. At elevated temperatures, solid bridges may develop by diffusion of molecules from one particle to another at the points of contact. Heat can be introduced from an external, secondary source or created during agglomeration by friction and/or energy conversion. Solid bridges can also be built up by chemical reaction, crystallization of dissolved substances, hardening binders, and solidification of melted components.

Interfacial Forces and Capillary Pressure at Freely Movable Surfaces. Capillary pressure and interfacial forces in liquid bridges can create strong bonds that disappear if the liquid evaporates and no other binding mechanisms take over.

Adhesion and Cohesion Forces in not Freely Movable Binders. Highly viscous bonding media such as tar and other highly molecular organic liquids can form bonds very similar to those of solid bridges. Thin adsorption layers are immobile and can contribute to the bonding of fine particles under certain circumstances.

Attraction Forces Between Solid Particles. The typical short-range forces of the van der Waals, electrostatic, or magnetic type can cause solid particles to stick together if they approach each other closely enough. Decreasing particle size clearly favors this mechanism.

Form-Closed Bonds. Fibers, little platelets, or bulky particles can interlock or fold about each other resulting in "form-closed" bonds.

Another classification into only two groups[2] distinguishes between the presence of material bridges between the primary particles in the agglomerate and attraction forces (Fig. 7.2.1).

7.2.2. THEORY OF AGGLOMERATE BONDING AND STRENGTH

The most important characteristic of all forms of agglomerates is their strength. For the determination of agglomerate strength, real stresses are often simulated experimentally. In addition to the usually applied crushing, drop, and abrasion tests, methods for the determination of impact, bending, cutting, or shear strength are employed. All values obtained by these methods are strictly empirical and cannot be predicted by theory, since it is not known which stress component causes the agglomerate to fail. For the same reason, the experimental results from different methods cannot be compared with each other.

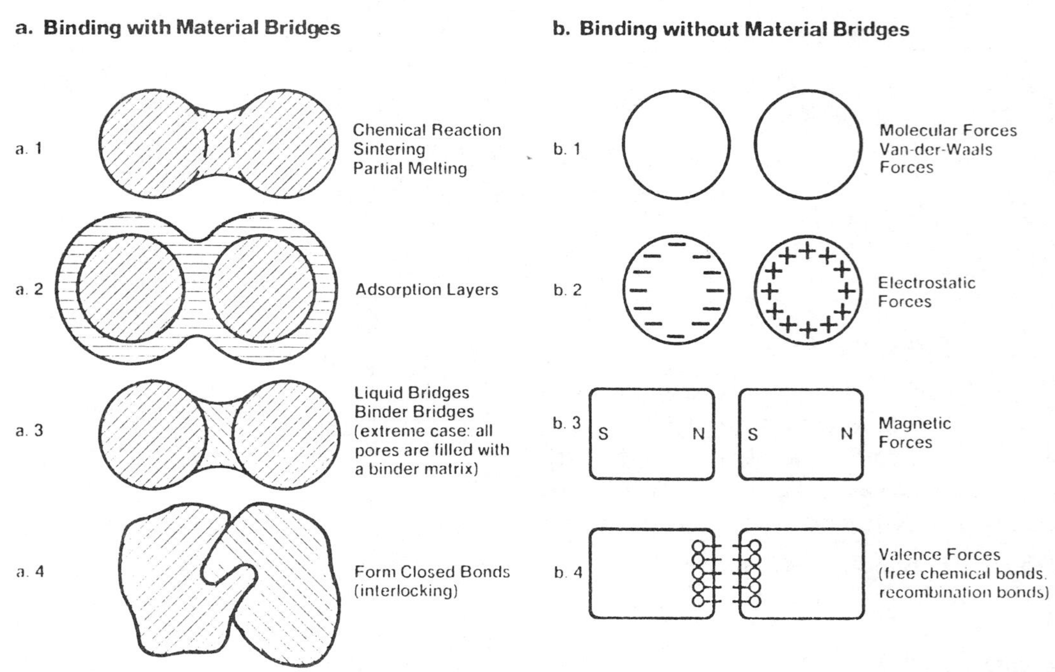

Figure 7.2.1. Alternative classification of the binding mechanisms.[2]

Therefore, Rumpf[1] proposed to determine the tensile strength of agglomerates. It is defined as the tensile force at failure divided by the cross section of the agglomerate. Because with high probability failure occurs as the result of the highest tensile stress in all stressing situations, this proposal is justified. Moreover, the tensile strength can be approximated by theoretical calculations.

All binding mechanisms listed above can be described by one of three models:

1. The entire pore volume of the agglomerate is filled with a substance that can transmit forces and thereby causes strength.
2. The pore volume of the agglomerate is entirely filled with liquid.
3. Binding forces are transmitted at the contact and coordination points of the primary particles forming the agglomerate.

Maximal Tensile Strength if the Pore Volume of the Agglomerate is Filled with a Stress-Transmitting Substance

If the pore volume of the agglomerate is completely filled with a stress-transmitting substance, for example a hardening binder, three strength components must be distinguished:

1. $\sigma_{t\epsilon}$ (Pore volume strength) = Strength of binder substance
2. σ_{ta} (Grain boundary strength) = Strength caused by adhesion between binder and solids
3. $\sigma_{t(1-\epsilon)}$ = Strength of the solids forming the agglomerate

The relative lowest strength component determines the agglomerate strength. If the pore volume strength or, respectively, the strength of the solids forming the agglomerate are the determining factors and if they are everywhere the same, then the cross section of the respective material defines the agglomerate strength. A theoretical approximation is possible using the same assumptions described on p. 234 for solid bridges between particles. If the agglomerate strength is caused by the grain boundary strength, it can be approximated by calculating the adhesion forces (see below, Theoretical Approximation of Adhesion Forces).

Maximal Tensile Strength if the Pore Volume of the Agglomerate is Filled with a Liquid

If the entire pore volume of the agglomerate is filled with a liquid such that concave minisci are formed at the agglomerate surface a negative capillary pressure p_c develops in the interior. Because the membrane forces at the surface are negligibly small in relation to the capillary pressure, the tensile strength σ_{tc} of the agglomerates filled with a liquid can be approximated by the capillary pressure:

$$\sigma_{tc} \approx p_c \qquad (7.2.1)$$

Assuming that the pore diameter is characterized by the mean half hydraulic radius of the pore system, further assuming complete wetting and spherical monosized particles, the following equation is obtained:

$$\sigma_{tc} \approx p_c = a' \cdot \frac{1-\epsilon}{\epsilon} \cdot \alpha \cdot \frac{1}{x} \qquad (7.2.2)$$

Therefore, the maximal tensile strength of agglomerates filled with a liquid is proportional to a porosity function $(1-\epsilon)/\epsilon$ and the surface tension of the liquid α; it is inversely proportional to the grain size x of the particles forming the agglomerate. The factor a' has a value between 6 and 8.

To correctly describe the capillary pressure, and thereby the tensile strength, a function of the wetting angle $f(\delta)$ would have to be included in the above formula. Experience teaches, however, that this function equals one if the liquid completely wets the solid.

Normally the particles forming agglomerates are not monosized. Comparisons between experimental results and the theory showed that a mean grain size x_0 calculated as the diameter equivalent to the specific surface of the particles describes the relations well.[3]

Maximal Tensile Strength if Binding Forces are Transmitted at the Contact and Coordination Points

The model used for agglomerates, the strength of which is caused by solid bridges, assumes that the entire solid binder material is uniformly distributed at all contact and coordination points and, there, forms bridges with constant strength σ_B. Then, the relative cross section of that material defines the agglomerate strength. In a random packing the cross-sectional area of one component (area porosity ϵ_a) is approximately equal to the relative volume of that same component (i.e., volume porosity $\epsilon_v = \epsilon_a = \epsilon$). Thus, the tensile strength σ_{tB} of agglomerates with solid bridges can be approximated by:[4]

$$\sigma_{tB} = \frac{M_B}{M_p} \cdot \frac{\rho_p}{\rho_B} \cdot (1 - \epsilon) \cdot \sigma_B = \psi_B \cdot \epsilon \cdot \sigma_B \tag{7.2.3}$$

where M_B is the mass of the bridge building solid, M_p the mass of the particles forming the agglomerate, ρ_B and ρ_p the density of the respective solids, $(1 - \epsilon)$ the relative unit volume of the solid, σ_B the tensile strength of the bridge building solid, ϵ the specific void volume (porosity) of the agglomerate, and ψ_B the degree of filling, i.e., the fraction of the void volume filled with the bridge building material.

Equation (7.2.3) is only valid if, in addition to the restrictive assumptions mentioned above, the failure occurs only through solid bridges. In reality all these conditions are never fulfilled. Particularly the uniform distribution and the constant strength of the binder material are seldom realized.

Often, the strength is caused by adhesion forces A acting at the coordination points of the particles forming the agglomerate. Based on statistical considerations Rumpf[5] developed a general formula for the tensile strength of such agglomerates. Assuming that the particles forming the agglomerate are monosized and spherical the tensile strength σ_t can be approximated by:

$$\sigma_t = \frac{1 - \epsilon}{\pi} \cdot k \cdot \frac{A}{x^2} \tag{7.2.4}$$

where ϵ is the specific void volume (porosity) of the agglomerate, $\pi = 3.14$, k the average coordination number, and x the size of the particles forming the agglomerate.

After a small correction[5] Eq. (7.2.4) can also be applied for nonspherical particles. Then the estimated elementary tensile strength σ_{te} becomes:

$$\sigma_{te} = (1 - \epsilon) \cdot k \cdot A/O_p \tag{7.2.5}$$

with O_p the particle surface. Equation (7.2.5) is valid for agglomerates formed by approximately isodisperse, convex, and monosized particles. With the third moment M_{30} of the number density distribution $n(x)$ and a shape factor f_0, a formula can be derived which is valid for distributions of similar, approximately isometric, and convex particles:

$$\sigma_{te} = \frac{1 - \epsilon}{f_0 M_{30}} \int_0^\infty k(x) \cdot A\,[x, n(x)] \cdot x \cdot n(x)\, dx \tag{7.2.6}$$

This equation can only be integrated if the relationships are known between coordination number and particle size $k(x)$ as well as between adhesion force and particle size and distribution $A\,[x, n(x)]$. In most instances this is not the case. To measure σ_{te}, an agglomerate free of cracks must be uniformly stressed by tensile forces. This requires very sophisticated methods and experimental care (see section 7.2.3).

Theoretical Approximation of Adhesion Forces

Adhesion Force of a Liquid Bridge.[6,7] The maximal tensile force that can be transmitted by a liquid bridge between two monosized spheres consists of two components:

(a) An adhesion force component A_c caused by the negative capillary pressure in the bridge:

$$A_c = p_c \cdot \frac{\pi}{4} \cdot x^2 \cdot \sin^2\beta$$

(b) An adhesion force component A_b caused by the boundary force at the contact line solid-liquid-gaseous, which is determined by the surface tension of the liquid, α:

$$A_b = \alpha \cdot x \cdot \pi \cdot \sin\beta \cdot \sin(\beta + \delta)$$

By adding the two parallel adhesion force components A_c and A_b and after introducing the dimensionless function F_A a formula for the effective adhesion force A_L of a liquid bridge is obtained:

$$A_L = \alpha \cdot x \cdot F_A = \alpha \cdot x \cdot f_1\left(\beta, \delta, \frac{a}{x}\right) \quad (7.2.7)$$

Where α is the surface tension of the liquid, x the diameter of the spherical, monosized particles, β the angle according to Fig. 7.2.2, and a the distance of the particle surfaces at the coordination point.

Therefore, the adhesion force of a liquid bridge is proportional to the surface tension α, the particle diameter x, and a function of the angle β, the angle of contact δ and the dimensionless quotient a/x. β defines the size of the liquid bridge and can be substituted by ϕ, the liquid volume divided by the volume of the solid particles:

$$\phi = \frac{V_b}{2 \cdot \pi \cdot x^3/6} = \frac{3}{\pi} \cdot f_2\left(\beta, \delta, \frac{a}{x}\right)$$

with V_b the volume of the liquid bridge.

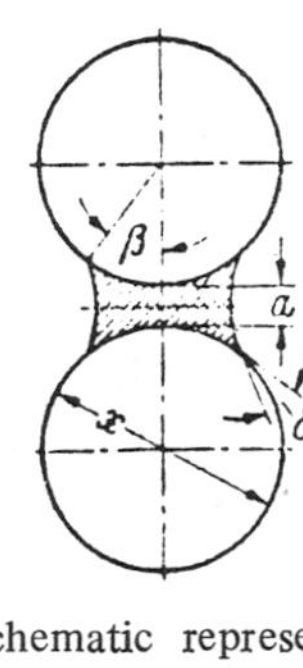

Figure 7.2.2. Schematic representation of a liquid bridge between two spherical, monosized particles.

By inserting the adhesion force A_L [Eq. (7.2.7)] into the basic formula, Eq. (7.2.4), and assuming that $k \cdot \epsilon \approx \pi$,[8] the maximal tensile strength σ_{tb} of agglomerates with liquid bridges becomes:

$$\sigma_{tb} = \frac{1-\epsilon}{\epsilon} \cdot \frac{\alpha}{x} \cdot F_A \quad (7.2.8)$$

Adhesion due to van der Waals Forces. Depending on the geometrical model (Fig. 7.2.3) being used and on the theoretical approach taken, different relationships exist for the approximation of adhesion by van der Waals forces. The best-known equations are those developed by Hamaker[9] based on the microscopic theory of London-Heitler. For the model sphere/plane, Figure 7.2.3a, a distance $a < 1000$ Å, and a particle diameter x, the adhesion force A_{v_1} is:

$$A_{v_1} = \frac{H}{12 \cdot a^2} \cdot x \quad (7.2.9)$$

For the model sphere/sphere, Fig. 7.2.3b, and the same limitations as mentioned above, Hamaker calculates an adhesion force A_{v_2}:

$$A_{v_2} = \frac{H}{24 \cdot a^2} \cdot x \quad (7.2.10)$$

H, the "Hamaker Constant", which depends on the material characteristics, has values in the order of 10^{-13} to 10^{-12} erg.

More recently, Krupp[10] has developed a formula for the model sphere/plane (Fig. 7.2.3a) which is based on the macroscopic calculations of Lifshitz-Landau:

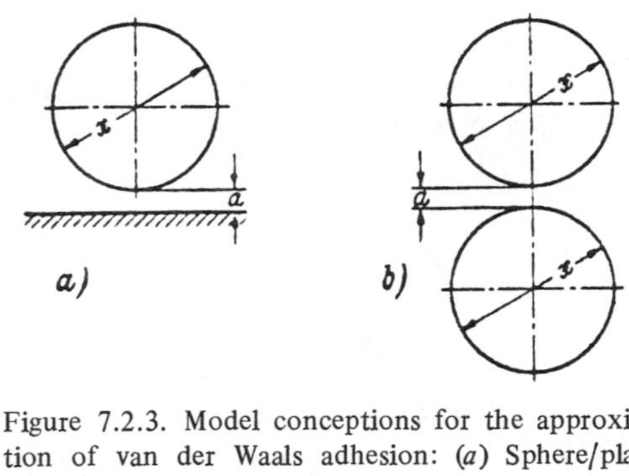

Figure 7.2.3. Model conceptions for the approximation of van der Waals adhesion: (*a*) Sphere/plane: (*b*) Sphere/sphere.

$$A'_{v_1} = \frac{\hbar\overline{\omega}}{16 \cdot \pi \cdot a^2} \cdot x \qquad (7.2.11)$$

$\hbar\,\overline{\omega}$ is the "Lifshitz–van der Waals Constant," which, depending on the material characteristics, has values between 0.2 and 9 eV (1 eV = $1.6 \cdot 10^{-12}$ erg). All equations for the approximation of van der Waals forces only differ in the constants. The adhesion force A_v is always proportional to the particle diameter x and inversely proportional to the squared distance a:

$$A_v = c \cdot \frac{x}{a^2} \qquad (7.2.12)$$

By inserting Eq. (7.2.12) into the basic formula, Eq. (7.2.4), and assuming that $k \cdot \epsilon \approx \pi$,[8] the maximal tensile strength σ_{tv} of agglomerates bound by van der Waals forces becomes:

$$\sigma_{tv} = \frac{1-\epsilon}{\epsilon} \cdot \frac{c}{a^2} \cdot \frac{1}{x} \qquad (7.2.13)$$

Adhesion due to Electrostatic Forces. In the case of electrostatic forces, one must distinguish between an excess charge and the electrical double layer (equilibrium).

The strength due to excess charges can be estimated if it is assumed that positively and negatively charged particles are arranged in a uniform pattern.[4] The basis for the derivation is Coulomb's formula for the attraction force between two spherical, nonconducting particles of equal size, the distance between which is much smaller than their diameter. If the charges $Q = \gamma \cdot \pi \cdot x^2$ are uniformly distributed on the surfaces the adhesion force A_c can be approximated by:

$$A_c = \frac{10\gamma^2}{(1 + a/x)^2} \cdot x^2 \qquad (7.2.14)$$

For quartz the maximal charge density per unit area, γ, was estimated to be $\gamma^2_{max} \approx 2.5$ dyn/cm^2.[1] If it is assumed that the charged particles forming an agglomerate are arranged like an ion lattice, then the attraction force between two adjacent, oppositely charged particles is approximately a factor 0.3 smaller because of the repulsion of neighboring particles with the same charge.

By inserting Eq. (7.2.14) into the basic formula, Eq. (7.2.4), and assuming that $k \cdot \epsilon \approx \pi$,[8] the maximal tensile strength σ_{tc} of agglomerates due to excess charges is:

$$\sigma_{tc} = \frac{1-\epsilon}{\epsilon} \cdot \frac{3\gamma^2}{(1 + a/x)^2} \qquad (7.2.15)$$

Because of the field character of this binding mechanism, the tensile strength is independent of the particle size. Also, the strength due to excess charges is very small, and the charges tend to equalize with time. Therefore, this mechanism is most often only important for the initial formation of agglomerates.

Much more important, however, are adhesion forces due to electrical double layers. This phenomenon can develop if the particles touch each other and is permanent. According to Krupp[10] the "attraction pressure" due to electrical double layers between two semi-infinite bodies is in the order of $P_{el} = 10^5$ to 10^8 dyn/cm^2 ($\approx 10^{-1}$ to 10^2 kp/cm^2). In comparison, the van der Waals attraction pressure between two semi-infinite bodies is $P_{vdW} = 2 \times 10^8$ to 3×10^9 dyn/cm^2 ($\approx 2 \times 10^2$ to 3×10^3 kp/cm^2).

It may seem as if the two mechanisms exclude each other. However, since P_{vdW} decreases with $1/a^3$ and P_{el} stays almost constant even over macroscopic distances, the electrical double layer will contribute to the adhesion of particles, particularly if the contact surfaces are rough.

A theoretical approximation for specific systems is still not yet possible, since little is known about the distribution of charges in different materials. The effect of magnetic particles in agglomerates corresponds to that of excess charges and is subject to the same limitations.

7.2.3. EXPERIMENTAL DETERMINATION OF AGGLOMERATE BONDING AND STRENGTH

The most important techniques for the experimental determination of agglomerate strength known today measure crushing, shear, and tensile strengths. Sketches *A*-*F* in Figure 7.2.4 show schematically the methods for measuring the strength of agglomerates and particle conglomerates.

Figure 7.2.4*A* shows the determination of the crushing strength.[11] This method is a very simple one. Individual agglomerates are placed between two parallel plates and loaded with a uniformly increasing force *P* until failure occurs. Usually the "agglomerate strength" is defined as the mean statistical force at failure of a larger number of agglomerates tested by this method. Sometimes a crushing strength is calculated by dividing the force at failure by the projection area of the agglomerate; however, from a physical point of view this is not acceptable.

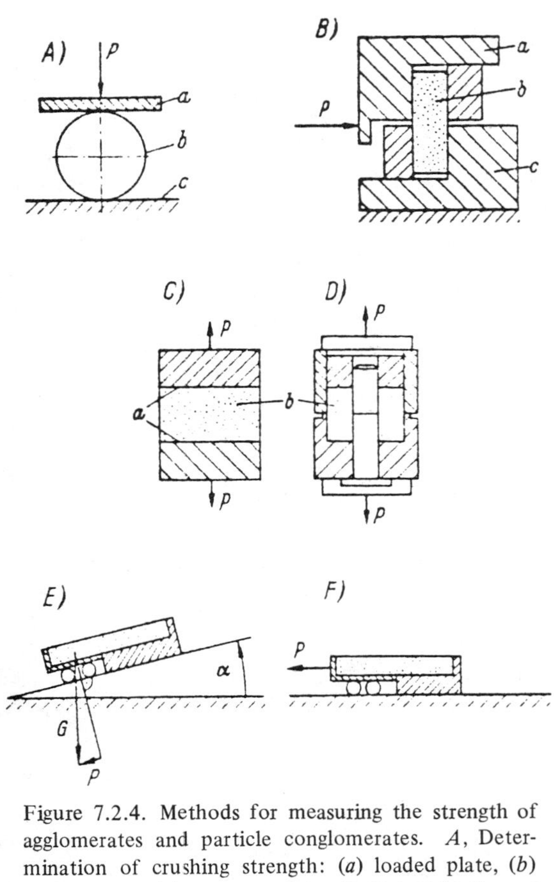

Figure 7.2.4. Methods for measuring the strength of agglomerates and particle conglomerates. *A*, Determination of crushing strength: (*a*) loaded plate, (*b*) agglomerate, (*c*) support plate, *B*, Determination of shear strength: (*a*) upper receptable, (*b*) compact or briquet, (*c*) lower receptacle; *C* and *D*, Determination of tensile strength of strong agglomerates: (*a*) adhesive, (*b*) agglomerate (eventually machined); *E* and *F*, Determination of tensile strength of weak agglomerates and of particle conglomerates.

The results of this test method are very rarely comparable. For spherical pellets the stressing is only uniform from test to test if all agglomerates are absolutely globular. In the case of agglomerates with flat ends the transverse expansion is blocked by friction between pellet and plate; thereby uncontrolled stress concentrations build up which can be the true cause for failure.

In Figure 7.2.4*B*, an apparatus is sketched for the investigation of shear strength. Originally this method was used in soil mechanics for the determination of shear curves of cohesive bulk solids. The "strength" of the conglomerate caused by internal friction can be determined graphically from the shear curves. The agglomerate must have two parallel surfaces, which may have to be produced by machining. The test specimen is fastened in the apparatus and stressed by the force *P*. The shear strength is defined by the shear force at failure divided by the shear plane.

Figure 7.2.4*C* shows in principle the "adhesive" method for the determination of tensile strength. Cylindrical agglomerates with two parallel and flat ends are centrically cemented between two so-called adaptors. To eliminate bending stresses it is necessary to machine spherical or nonsymmetrical agglomerates into cylindrical specimens using a special method (Fig. 7.2.5).[12] This sample is fastened to two thin wires (Fig. 7.2.6) and subjected to tensile forces in a conventional testing machine (Fig. 7.2.7*a* and *b*). The tensile strength of the agglomerate is defined as the tensile force *P* at failure divided by the cross section of the cylindrical specimen.

Figure 7.2.4*D* sketches the determination of

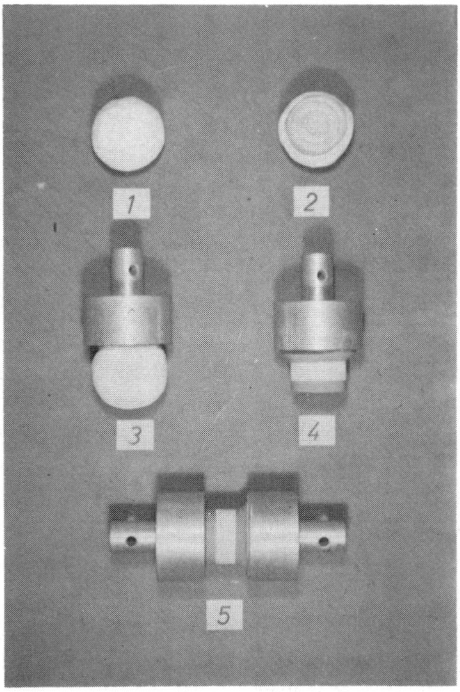

Figure 7.2.5. Preparation of "spherical" agglomerates for the determination of tensile strength by the "adhesive" method.[12]

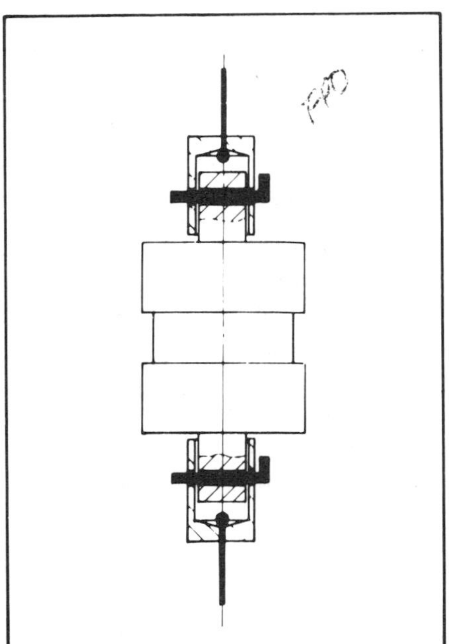

Figure 7.2.6. Schematic representation of the fastening method with thin wires.

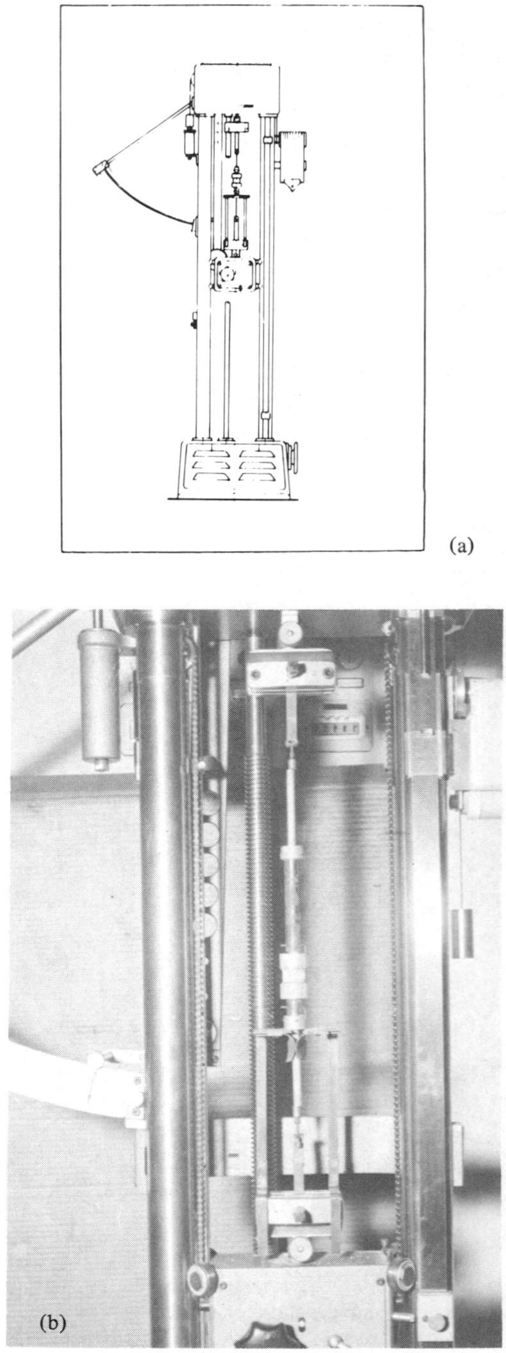

Figure 7.2.7. Testing machine adopted for the determination of the tensile strength of agglomerates: (*a*) schematic overall view, (*b*) Close-up during an actual tensile test.

the tensile strength of model agglomerates by means of the wall friction method.[7] For this test a cylindrical pellet –potentially with a central pin– is produced in a press. After removing the specimen from the press, it is stressed directly in the die shell. The tensile force is transmitted by adhesion between the end surfaces and the "pistons" as well as on the circumference and the die walls. Again, the tensile strength is defined by the quotient rupture force P divided by the cross section of the cylindrical or ring-shaped sample.

Figure 7.2.4*E* shows a method that is particularly suitable for the determination of low conglomerate strengths. The rupture stress is measured in a model arrangement. The powder to be investigated is filled into a flat, often rectangular split mold and densified by vibration or compaction. The upper half of the mold is fastened to a tiltable plate while the lower half is supported frictionless by rolls or spheres on the same plate. For the determination of the rupture force, the plate is slowly lifted at the end carrying the fixed mold until the particle conglomerate separates. The strength is defined as the force at failure divided by the cross section of the agglomerate.

Another method based on the same principle uses a slowly increasing horizontal force P to pull apart the specimen (Fig. 7.2.4*F*). In this case the arrangement remains horizontal during the entire test. The strength is defined as described above. Figure 7.2.8 shows a recent design of this experimental apparatus.[7] It is constructed such that two tests can be carried out on the same sample, and, if moist agglomerates are being investigated, the capillary pressure p_c can be measured simultaneously by means of a U-tube manometer.

7.2.4. RESULTS OF THE DETERMINATION OF AGGLOMERATE STRENGTH

Theoretical Approximation of Agglomerate Strength

With the theories of Section 7.2.2 the orders of magnitude of agglomerate strength can be defined for the different binding mechanisms.

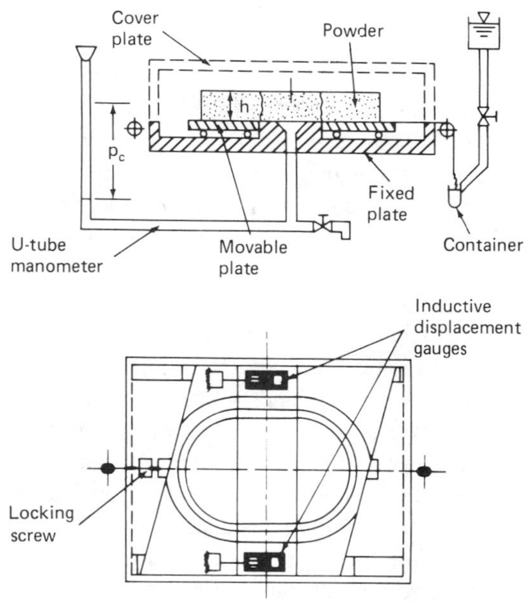

Figure 7.2.8. Diagrammatic representation of the split plate apparatus, according to Schubert.[13]

They often depend on the size of the particles forming the agglomerate and predict the maximum strength of conglomerates caused by a particular binding mechanism. Figure 7.2.9 shows the tensile strength σ_t versus particle size x in a double logarithmic plot.

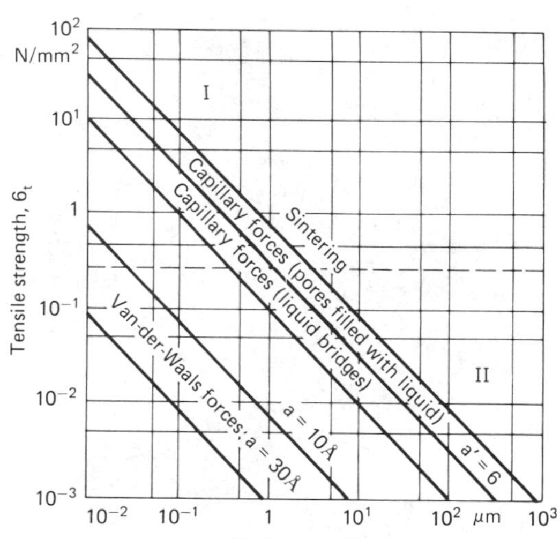

Figure 7.2.9. Approximation of the maximal theoretical tensile strength of agglomerates. Porosity: ϵ = 0.35. Region I: Briquetting, hardening binders. Region II: Crystallizing soluble substances, e.g., salts.

The horizontal dotted line divides the entire field into two regions, I and II. These represent binding mechanisms that are independent of particle size. Region I describes briquetting. This technology uses high compaction pressures causing considerable deformation of particles and interparticle contact. Agglomerates with highly viscous and hardening binders are also included in this region. The lower region II describes the much weaker bonds caused by recrystallization of dissolved substances.

The diagonal lines define maximal tensile strengths, which depend on particle size. The van der Waals lines were calculated using the model sphere/sphere and particle distances of a = 30 and 10 Å. Assuming a distance of a = 4 Å (equilibrium distance) and using the model sphere/plane, the line would be pushed higher, close to the one representing capillary forces. If, in addition, plastic deformation of particles is considered, still higher agglomerate strengths can be obtained.

A narrow region characterizes the effect of liquid bridges. Somewhat higher is the line for the strength of agglomerates that are completely filled with a liquid. For this diagram it was calculated assuming water and the constant $a' = 6$. The strength of agglomerates with sinter and adhesion bridges can be expected above this line.

In each case, the predictions of Figure 7.2.9 are only valid for certain assumptions. In the following a few examples shall demonstrate the variability of the correlations if individual parameters are changed.

Figure 7.2.10*a* and *b* show salt bridges that were obtained at different drying temperatures[14] during a model experiment. The visual examination indicates that the drying temperature must play an important role in the development of agglomerate strength even if all other parameters are kept constant.

As already explained (see p. 233 and pp. 242–243, 248–250), capillary pressure and tensile strength of moist agglomerates are associated with each other. To a great degree they are influenced by the amount of liquid that is present in the pore volume of the agglomerate.

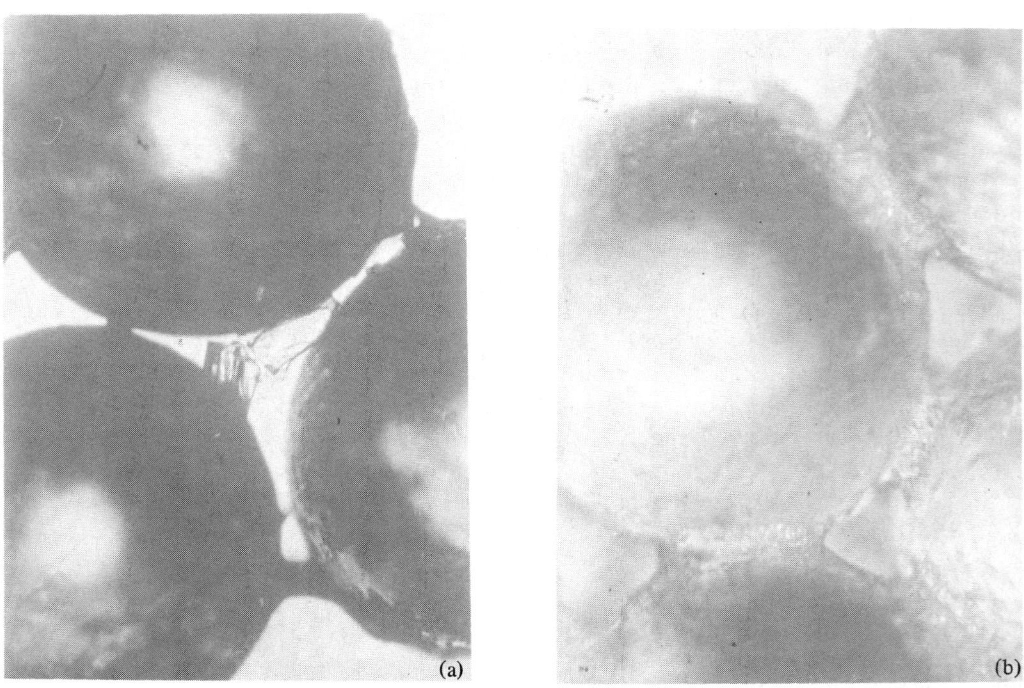

Figure 7.2.10. Common salt (NaCl) crystallized between glass spheres (model experiment): (*a*) drying at room temperature, (*b*) drying at 110° C.

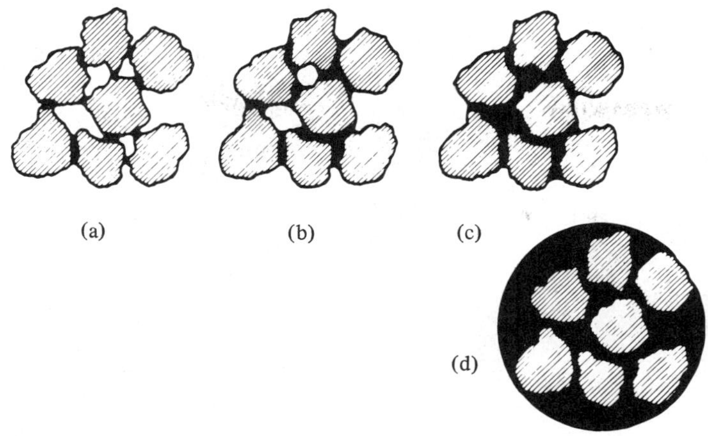

Figure 7.2.11. Different models of liquid distribution in moist agglomerates: (*a*) pendular state, (*b*) funicular state, (*c*) capillary state, (*d*) liquid droplet with particles inside or at its surface.

Assuming that the liquid wets the solid particles ($\delta = 0$), a classification as shown in Figure 7.2.11 can be defined. It is valid for three-phase systems consisting of a disperse solid material and two immiscible fluid phases. The dark colored area represents the wetting fluid phase.

A small quantity of liquid causes liquid bridges between the particles forming the agglomerate (Fig. 7.2.11*A*). This region is called the pendular state. By increasing the amount of liquid, the funicular state is obtained (Fig. 7.2.11*B*) where both liquid bridges and pores filled with liquid are present. The capillary state (Fig. 7.2.11*C*) is reached when all pores are completely filled with the liquid, and concave menisci develop at the surface of the agglomerate. The last state (Fig. 7.2.11*D*), a liquid droplet with particles inside or at its surface, is an important mechanism for wet scrubbing but has no relevance for agglomerate strength. Corresponding to the two patterns, Figures 7.2.11*A* and *C*, different models exist for the theoretical determination of agglomerate strength with a transition range in between (Fig. 7.2.11*B*).

Formerly, mathematical approximations were used for estimating the adhesion forces that can be transmitted through a liquid bridge. More recently, Schubert[15,16] developed exact equations for all rotationally symmetric liquid bridges. In Figure 7.2.12 the nondimensional force $F_A = A_L/\alpha \cdot x$ [Eq. (7)] is plotted versus V_b/V_s, where V_b is the bridge volume, V_s the volume of the solid (sphere), x the diameter of the spheres, α the surface tension of the liquid, the wetting angle $\delta = 0$, and a the distance at the coordination point. As the value of V_b/V_s increases, the attraction forces increase for planes and cones and decrease for spheres.

Under normal atmospheric conditions and with wetting solids it must be expected that liquid bridges are developing by capillary con-

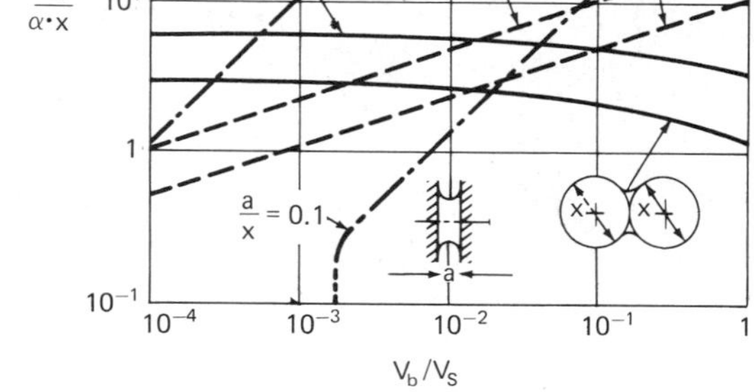

Figure 7.2.12. Computed adhesion forces resulting from liquid bridges for various geometric situations.[15]

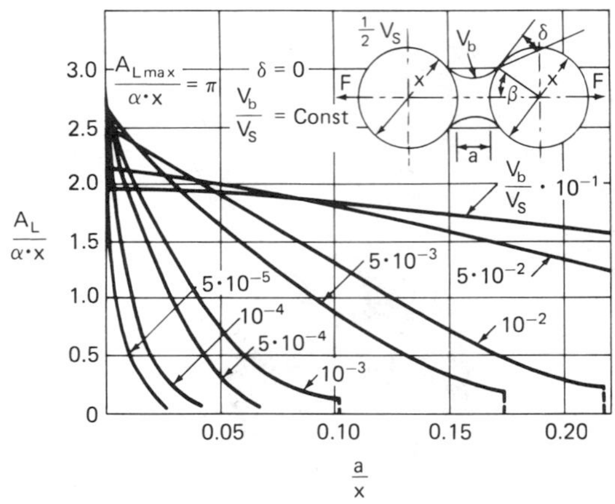

Figure 7.2.13. Nondimensional adhesion force $F_A = A_L/\alpha \cdot x$ as a function of the dimensionless distance at the coordination point a/x.[17]

densation at contact points ($a = 0$). Depending on the contact geometry involved, the attractive forces resulting from this mechanism may exceed the van der Waals forces. For example, Figure 7.2.13 shows the nondimensional force $F_A = A_L/\alpha \cdot x$ [Eq. (7)] as a function of the dimensionless distance at the coordination point a/x for the model sphere/sphere.[17] At $a/x = 0$ and $\beta \rightarrow 0$, the maximum value $A_L = \pi \cdot \alpha \cdot x$ is obtained. If the bridge is stretched the attraction force decreases the more, the smaller the liquid volume.

Assuming complete wetting ($\delta = 0$), the correlation between the nondimensional force $F_A = F_L/\alpha \cdot x$ [Eq. (7)] and the filling angle

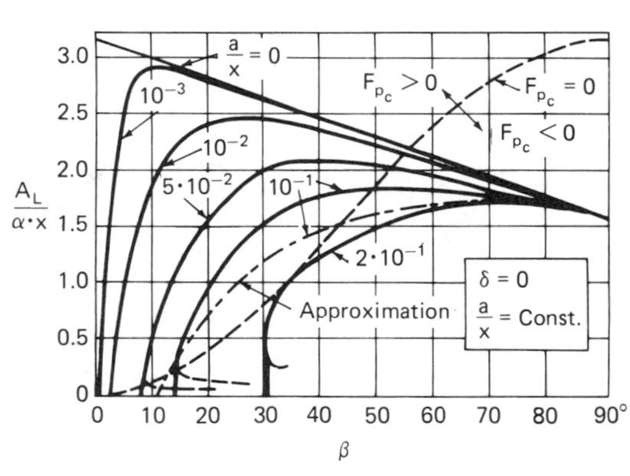

Figure 7.2.14. Correlation between the dimensionless adhesion force $F_A = A_L/\alpha \cdot x$ between two equal spheres and the filling angle β. Assumptions: $\delta = 0$ and a/x = constant.

β (Fig. 7.2.2) at different values of a/x is presented in Figure 7.2.14. For spheres in contact ($a/x = 0$) the value of $F_A = A_L/\alpha \cdot x$ decreases from π (at $\beta = 0$) as the liquid content increases. For finite values of a/x, however, the curves pass through a maximum or, respectively, increase with higher liquid saturation.

The dotted curve in Figure 7.2.14 divides the graph into a field of capillary excess pressure ($F_{p_c} < 0$) and a field of capillary suction ($F_{p_c} > 0$). For comparison, two curves were calculated for $a/x = 0.1$, one representing the exact theory of Schubert[15,16] and the other an earlier approximation by Pietsch and Rumpf.[6]

For the determination of the maximally transferrable uniaxial tensile stress Eq. (8) can be rewritten:

$$\frac{\sigma_{tb} \cdot x}{\alpha} = \frac{1 - \epsilon}{\epsilon} \cdot F'_A(\beta, \delta, a/x) \qquad (7.2.16)$$

For the transition range (Fig. 7.2.11*B*), the easily measurable liquid saturation S is introduced:

$$S = \pi \cdot \frac{1 - \epsilon}{\epsilon} \cdot \phi \qquad (7.2.17)$$

It is defined as the ratio of liquid volume to pore volume of the agglomerate. Thus, Eq. (7.2.16) yields:

$$\frac{\sigma_{tb} \cdot x}{\alpha} = \frac{1 - \epsilon}{\epsilon} \cdot F''_A(\epsilon, S, \delta, a/x) \qquad (7.2.18)$$

Equation (7.2.18) can be used to compute the correlation between the value $\sigma_{tb} \cdot x/\alpha$ and the liquid saturation S. Results for two different porosities ϵ and different distance ratios a/x (assuming complete wetting [$\delta = 0$]) are presented in Figure 7.2.15. Again, for comparison, the curves for $\epsilon = 0.35$ and $a/x = 0.1$ representing the exact theory and the approximation, respectively, have been included in Figure 7.2.15.

In the capillary state the strength of the agglomerate is determined by the capillary suction. Since only those pores filled with the

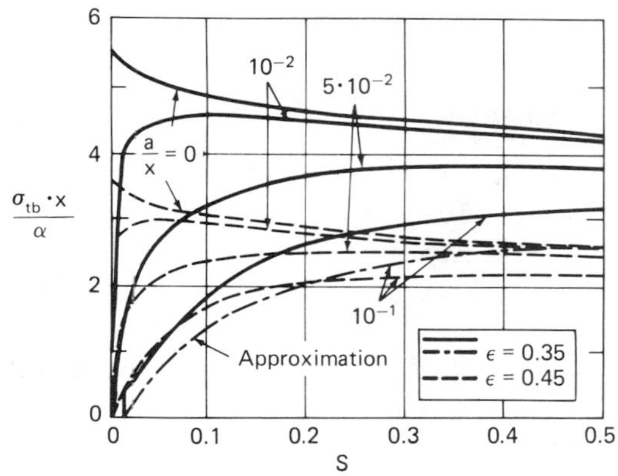

Figure 7.2.15. Maximally transferable uniaxial tensile stress $\sigma_{tb} \cdot x/\alpha$ as a function of the liquid saturation S of moist agglomerates. Assumption: complete wetting ($\delta = 0$).

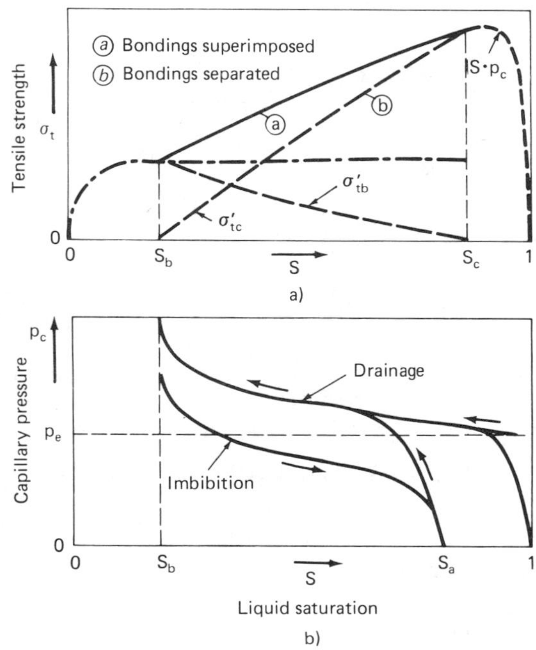

Figure 7.2.16. (*a*) Maximally transferable tensile stress σ_t and (*b*) capillary pressure p_c as a function of the liquid saturation S.

liquid contribute to the strength, Eq. (7.2.1) must be rewritten:

$$\sigma_{tc} = S \cdot p_c \tag{7.2.19}$$

Equation (7.2.19) assumes that the liquid is uniformly distributed in the agglomerate. The product $S \cdot p_c$ can be calculated from the capillary pressure/saturation curve (Part *b* of Figure 7.2.16). By definition, the starting point of capillary pressure curves is $p_c = 0$ at $S = 1$. During drainage of the agglomerate the capillary pressure follows the curve marked in Figure 7.2.16*b* until it reaches a point at which only isolated capillaries exist. If, starting at that point, liquid is reintroduced into the agglomerate, $p_c = 0$ is reached at $S < 1$, as not all pores can be filled with liquid by imbibition. The remaining air pockets are blocked off by adjacent pores that are already filled. Repeated drainage/imbibition tests lead to the typical hysteresis loop shown in Figure 7.2.16*b*. It is explained by the existence of pore bulges and pore necks as well as by the contact angle hysteresis.[18]

Part *a* of Figure 7.2.16 shows schematically the maximally transferable tensile stress σ_t as a function of the liquid saturation S. The capillary pressure curve (Figure 7.2.16*b*) is used to calculate $\sigma_{tc} = S \cdot p_c$ for $S_c \leq S \leq 1$. The capillary state ends when liquid bridges between the particles start forming ($S \leq S_c$). The funicular state exists in the region $S_b \leq S \leq S_c$. For the pendular state ($S \leq S_b$) the already discussed Eq. (7.2.18) is valid.

In the transition range, the funicular state ($S_b \leq S \leq S_c$), in which liquid bridges coexist with liquid-filled pores, two cases can be constructed which follow a model published first by Rumpf.[1]

1. Both bonding mechanisms can be superimposed.
2. Each of the bonding mechanisms acts alone.

Assuming that the ratio of the liquid in the bridges to the total liquid diminishes linearly from 1 at S_b to 0 at S_c, one obtains the following for the individual bonding mechanisms (see Fig. 7.2.16*a*):

$$\sigma_{tb}' = \sigma_{tb}\,(S_c - S)/(S_c - S_b) \tag{7.2.20}$$

$$\sigma_{tc}' = p_c \cdot S_c\,(S - S_b)/(S_c - S_b) \tag{7.2.21}$$

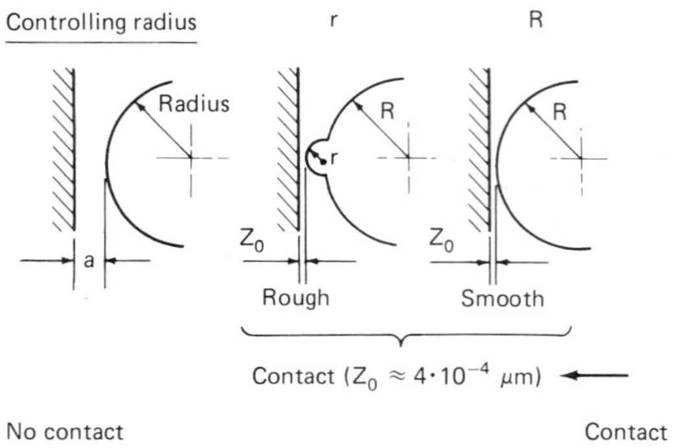

Figure 7.2.17. van der Waals model sphere/plate with and without surface roughness.[17]

If both mechanisms act alone curve *b* in Figure 7.2.16*a* represents the expected results. If the bonding mechanisms can be superimposed curve *a* results from the sum σ_{tb}' plus σ_{tc}'.

In all those cases where adhesion is caused by van der Waals or electrostatic attraction or by liquid bridges, surface roughness reduces the maximally transferable adhesion force. For the model sphere/plane and van der Waals attraction, Figure 7.2.17 shows the controlling radii for the calculation of the adhesion force according to Eq. (7.2.11). The shape of particles with surface roughness can be approximated by superimposing two spheres. The large radius R is considered the equivalent radius of a sphere of same volume as the particle, whereas the small radius r represents the surface roughness.[17]

Considering the model of Figure 7.2.17 and Eq. (7.2.11), the highest attraction forces A_{max} must be obtained if the adhesion partners are in contact and have smooth surfaces. Contrary to the indication of Eq. (7.2.11), the attraction force on contact is in reality finite. Therefore, an adjustment parameter Z_0 must be introduced:

$$A''_{v_1} = \frac{\hbar\overline{\omega}}{8\pi(a + Z_0)^2} \cdot R \qquad (7.2.22)$$

Krupp[10] has defined $Z_0 = 4 \cdot 10^{-8}$ cm (4 Å) as a measure for the atomic distance. For Figure 7.2.18*b*, A_{max} was calculated for different adhesion mechanisms using this value of Z_0, although it does not represent a true atomic distance. Rather, it is an approximate or adaptive parameter, the accurate value of which needs still to be determined.

Figure 7.2.18*b* shows for the model sphere/plate the correlation between the maximum adhesion force A_{max} on contact and the diameter $x = 2R$ of smooth particles for different adhesion mechanisms. The highest attraction forces are caused by liquid bridges assuming complete wetting ($\delta = 0$) and water as the liquid. Van der Waals forces are smaller by almost an order of magnitude, although a relatively high Lifshitz–van der Waals constant ($\hbar\overline{\omega} = 5$ eV) was chosen. If two different materials contact, an electrostatic attraction force develops that is caused by the contact potential. The latter depends on the characteristics of the two contacting materials and their surface conditions. Again, the potential chosen ($U = 0.5$ V) represents a relatively high value. For conductors the electrostatic attraction force is higher than for nonconductors with the same contact potential because the charge is concentrated at the surface. Electrostatic attraction forces can also result from excess charges originating from friction, crushing, or electron and, respectively, ion adsorption. The highest possible excess charges are around 10^2 elementary charges $e/\mu m^2$.

Figure 7.2.18*b* indicates that for smooth spheres with sizes below 100 μm the electrostatic adhesion is negligible compared with van der Waals forces and even more so in relation to forces caused by liquid bridges.

Figure 7.2.18*a* describes the influence of surface roughness, represented by r (abszissa), on the attraction force A for different adhesion mechanisms. The curves were calculated for spheres with constant diameter $x = 2R = 10$ μm. The corresponding values of A_{max} can be determined in Figure 7.2.18*b*. Only for van der Waals forces two further curves for $R = 0.5$ μm and $R = 50$ μm were plotted since—because of their short-range character—the influence of roughness on van der Waals forces is very pronounced.

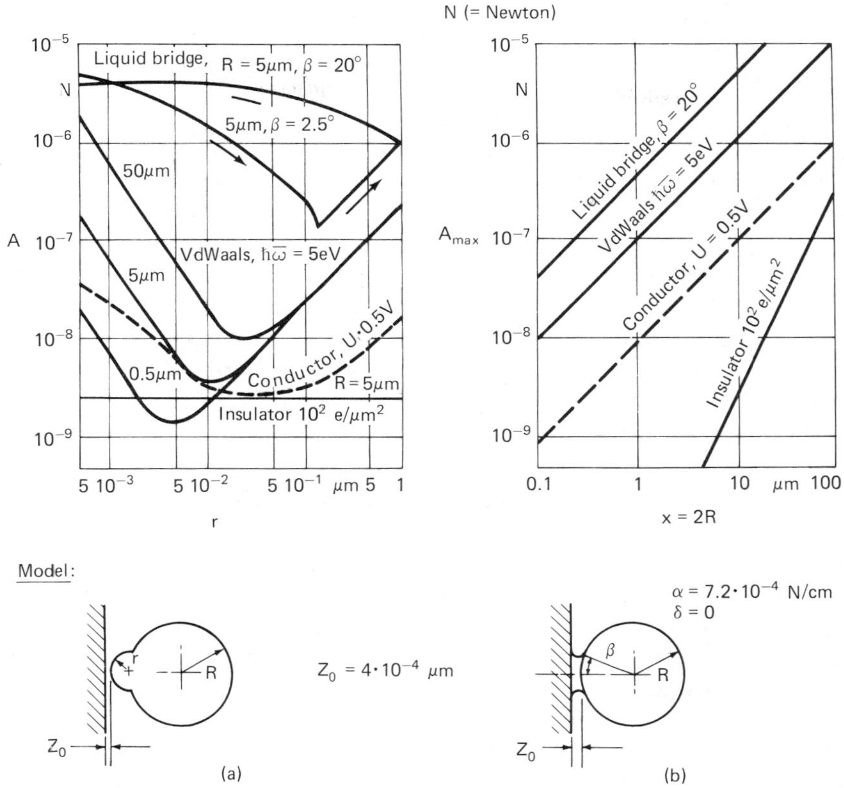

Figure 7.2.18. Attraction forces caused by different adhesion mechanisms for the model sphere/plate. Contact: $a = 0$, $Z_0 = 4$ Å.[17] (*a*) Influence of the roughness radius r on the attraction force A. (*b*) Influence of the diameter R of smooth spheres on the attraction force A_{max}.

Investigating the curve for $R = 5\ \mu m$ and van der Waals attraction, the following observations can be made. At $r = 0$ (not shown in Fig. 7.2.18*a*) the maximum adhesion force $A = A_{max} = 10^{-6}$ N is obtained (Fig. 7.2.18b). With increasing r, the distance of the larger sphere from the plane grows, and the adhesion force decreases proportional to r^{-2}. Later the influence of the large sphere diminishes, and a minimum is reached at which both attraction forces act simultaneously. A_{min} is only approximately $1/250\ A_{max}$. If r grows further, the influence of the large sphere disappears. Then, only the attraction force of the small sphere remains, which increases proportional to r. With growing R, r_{min} and the corresponding A_{min} increase but not at the same rate as A_{max}. At $R = 50\ \mu m$, A_{min}/A_{max} equals 1/1000.

Liquid bridges are much less sensitive to surface roughness. If the angle β is not too small, e.g., $\beta = 20°$, the roughness is immersed in the liquid and merely increases the distance a. The smaller β, the more pronounced is this influence. At $\beta = 2.5°$ the liquid bridge breaks off from the large sphere at $\sim r = 10^{-1}\ \mu m$ and remains on the small sphere. Then, the attraction force increases proportional to r. $\beta = 2.5°$ can already correspond to the capillary condensation. If β is still smaller, the transition to the line proportional to r occurs at smaller values of r, e.g., for $\beta = 1°$ at $r \approx 5 \cdot 10^{-2}\ \mu m$. In any case, the attraction force due to capillary condensation at the roughness peaks is always larger than the van der Waals force.

The electrostatic attraction forces of electrical conductors and of insulators with excess charges are smaller than the van der Waals forces. However, the influence of roughness is less pronounced and disappears completely for nonconducting particles facing a plane with an opposite charge of the same density.

For a long time the opinion existed that compared with van der Waals forces electrostatic attraction is always negligible. Based on the knowledge of the independence of roughness on electrostatic adhesion, this must now be corrected. Since the electrostatic attraction force increases with R^2, the 50 μm sphere with a charge density of 10^2 $e/\mu m^2$ would generate a higher value than could be obtained with the van der Waals model in the entire range of roughnesses shown in Figure 7.2.18*a*. Therefore, in dry agglomerates formed by relatively large particles the electrostatic attraction forces due to excess charges–which do not depend on surface roughness–may have an important share in agglomerate strength.

The following results are generally valid:

1. For adhesion, forces caused by liquid bridges are most important and, normally, represent the highest share. Even in dry systems –due to capillary condensation– liquid bridges may be the controlling mechanism.
2. Van der Waals forces are extremely sensitive to surface roughness but should always be larger than forces caused by the contact potential.
3. Excess charges can also be a controlling factor for adhesion, particularly if relatively large particles form a dry conglomerate.
4. At distances in excess of 1 μm or, respectively, $a/x > 0.2$, only electrostatic forces are effective. They cause particle attraction before adhesion takes over and forms agglomerates.

Results of Experimental Determinations of Agglomerates Strength.

In the following, some results of experimental investigations will show that if suitable model materials, agglomeration techniques, and experimental stressing methods are used, the respective theoretical expectations according to Sections 7.2.2 and 7.2.4 (see Theoretical Approximation of Agglomerate Strength) can be well approximated.

Crystallization of Dissolved Substances During Drying.[14] If drying of agglomerates that are filled with a salt solution starts at high liquid saturation S, evaporation begins at the surface of the agglomerate. The liquid flows to the surface by means of capillary suction. There, the dissolved substance crystallizes, forms a crust, and decisively controls further drying of the porous body. The crystal structure can therefore be influenced by either the drying temperature or the presence of a crust. In the latter case, the tensile strength of the agglomerate changes due to the varying drying conditions even if the strength is measured on the core after removing the crust.

A typical example is shown in Figure 7.2.19. The tensile strength, σ_t, is plotted versus the

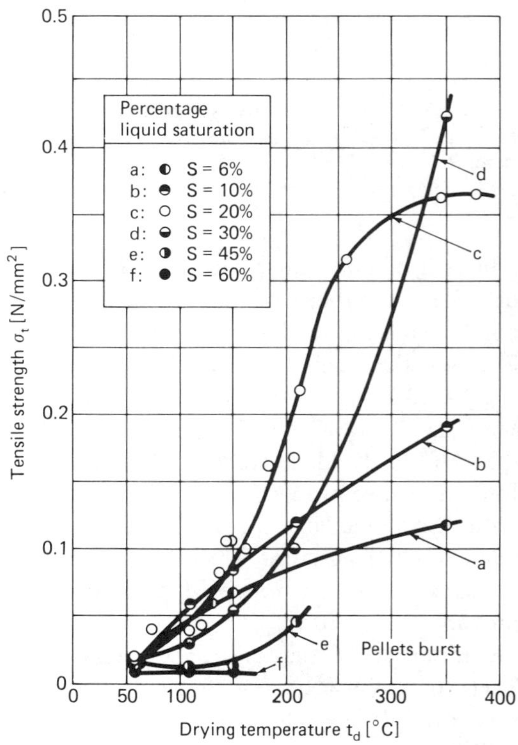

Figure 7.2.19. Tensile strength σ_t of the core of agglomerates (crust removed) with salt bridges as a function of the drying temperature t_d at different liquid saturations S before drying.[14]

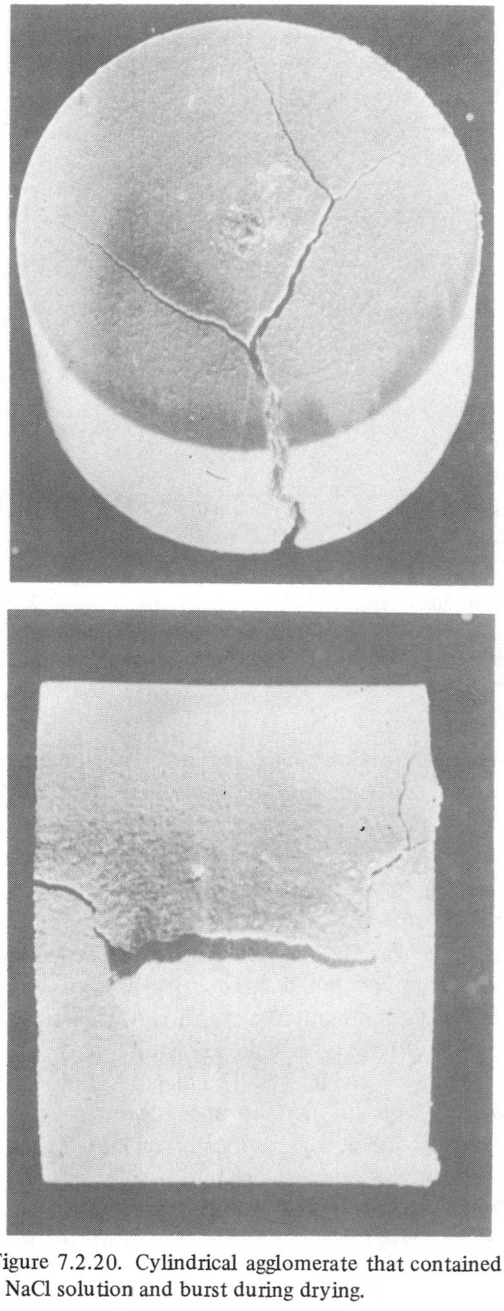

Figure 7.2.20. Cylindrical agglomerate that contained a NaCl solution and burst during drying.

drying temperature, t_d. The parameter is the liquid saturation S. This diagram was obtained using a nearly saturated salt (sodium chloride) solution and a narrow limestone powder fraction. At very small liquid saturations ($S \leq 20\%$, curves *a* and *b*) no crust is built up, and the tensile strength increases with the drying temperature. This rise is caused first by the increasing crystallization velocity and second by the amount of salt forming bridges that changes almost proportionally to the liquid saturation. At a liquid saturation of 20% the formation of a thin crust starts to influence the tensile strength slightly. At the highest liquid saturations examined ($S \geq 45\%$, curves *e* and *f*), the dense crust is at all temperatures the deciding factor for drying and for the crystallization velocity. Their core tensile strength is low and remains almost constant. Such agglomerates burst at high drying temperatures much like a pressure vessel (Fig. 7.2.20). The high tensile strength of agglomerates with $S = 30\%$ dried at a temperature of 350°C (curve *d*) is caused by small hair fractures in the crust that did not cause the agglomerate to disintegrate but raised the drying rate and thus increased the tensile strength of the dry agglomerate core.[19]

Figure 7.2.21 shows the same set of results as

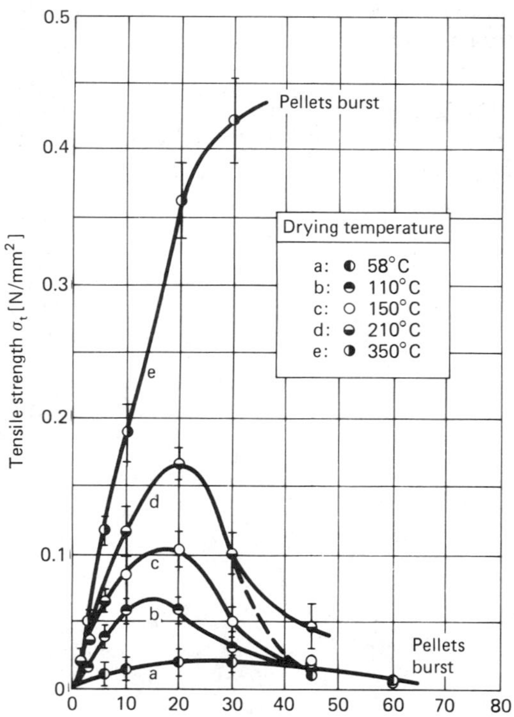

Figure 7.2.21. Tensile strength σ_t of the core of agglomerates (crust removed) with salt bridges as a function of the liquid saturation S before drying at different drying temperatures t_d.[14]

shown in Figure 7.2.19 but plotted in a different way. This time the tensile strength σ_t is presented as a function of the liquid saturation S before drying. The parameter is the drying temperature t_d. This graph confirms that normally the highest strength is obtained at $S = 20\%$ if it is measured after removing the crust. However, an optimum drying temperature exists whereby an agglomerate dries quickly but does not build-up enough inside pressure to cause cracking or disintegration. Then, the tensile strength can also be determined with crust.

Tests to investigate this optimum drying temperature were carried out in an instrumented drying channel.[20] Figure 7.2.22 shows a result. Below $S = 0.2$ the strength values for agglomerates with and without crust are identical. At saturations above 0.2 the strength of agglomerates with crust increases proportionally.

For practical applications the following conclusion can be drawn: To obtain high strength, drying should be carried out at the highest possible (without cracking) temperature using a saturated solution. Charé[20] found further that the air velocity does not substantially change the drying rate and, therefore, the agglomerate strength.

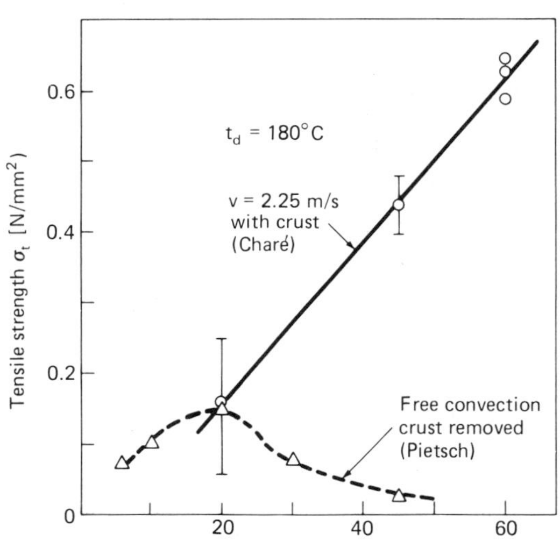

Figure 7.2.22. Tensile strength σ_t of agglomerates with salt bridges as a function of the liquid saturation S before drying with and without the crust.[21]

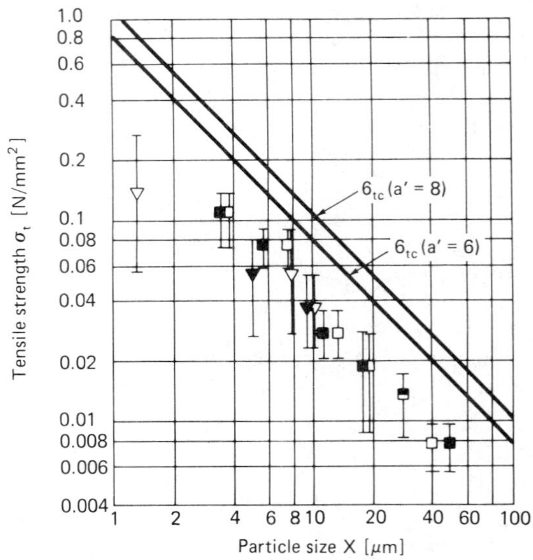

Figure 7.2.23. Tensile strength of moist agglomerates with high liquid saturation as a function of particle size. $\epsilon = 0.35$. Quartz powder: ▽x_0, ▼x_l. Limestone powder: □x_0, ■x_l.

Strength of Moist Agglomerates. Agglomerates which are being built-up by balling, that is the snowball-like forming of pellets in drums or discs, are nearly saturated with liquid. Figure 7.2.23 shows results of the determination of tensile strength plotted versus particle size. x_0 is the surface equivalent diameter and x_l is the maximum of the diameter distribution. The diagonals represent the maximally transferable tensile strength calculated with Eq. (7.2.2) using $a' = 6$ and $a' = 8$. The diagram shows that $\sigma_t \sim 1/x$ is fulfilled. Values lower than theoretically predicted are mostly due to the fact that the agglomerates were not fully saturated when the tensile strength was determined.

The relationship $\sigma_t \sim \alpha$ was confirmed by Conway-Jones[11] with compression tests on spherical agglomerates (Fig. 7.2.24) and $\sigma_t x/\alpha \sim (1 - \epsilon)/\epsilon$ was checked by Schubert[15] who confirmed this correlation, too (Fig. 7.2.25).

It can be assumed that up to saturations of $\approx$ 20–40% the liquid in moist agglomerates is present in the form of discrete liquid bridges at the contact and coordination points between the particles forming the agglomerate. The tensile strength of such an agglomerate is pre-

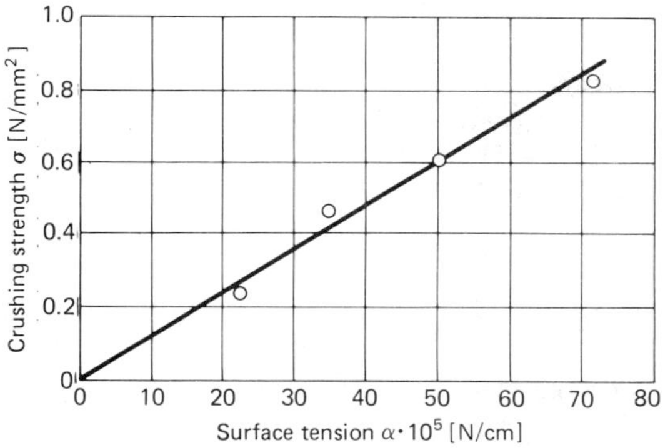

Figure 7.2.24. Relationship between crushing strength σ of moist agglomerates and surface tension α of the liquid.[11]

ALCOHOL/WATER MIXTURES

VOL. %		α	σ
ALCOHOL	WATER	N/cm	N/mm^2
0	100	72.2×10^{-5}	8.31×10^{-1}
10	90	50.2×10^{-5}	6.02×10^{-1}
30	70	35.0×10^{-5}	4.53×10^{-1}
100	0	22.2×10^{-5}	2.42×10^{-1}

dicted by Eq. (7.2.8). Experimentally it was investigated with the wall friction method (Figure 7.2.4*D*) using pellets made of narrowly distributed limestone powder and distilled water. In Figure 7.2.26 the experimental results are shown in comparison to the theory. The curves were approximated by varying the distance a (respectively, a/x). They seem to fit the experimental results well, although the

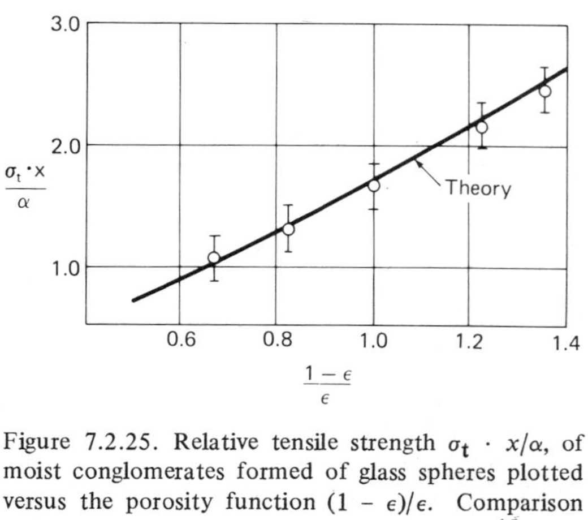

Figure 7.2.25. Relative tensile strength $\sigma_t \cdot x/\alpha$, of moist conglomerates formed of glass spheres plotted versus the porosity function $(1 - \epsilon)/\epsilon$. Comparison between theory [Eq. (7.2.8)] and experiment.[15]

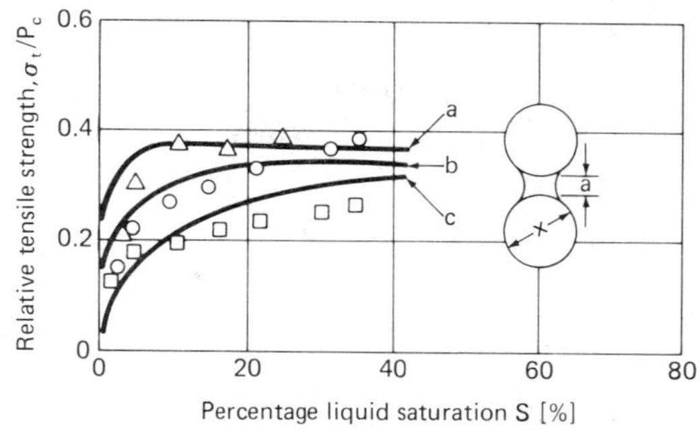

Figure 7.2.26. Relative tensile strength σ_t/p_c of agglomerates with liquid bridges as a function of the liquid saturation S. Limestone powder; Δ: $x_0 = 71$ μm; ○: $x_0 = 35$ μm; □: $x_0 = 13$ μm.
a $\epsilon = 0.45; a/x = 0.02$ ($a = 1.4$ μm)
b $\epsilon = 0.45; a/x = 0.04$ ($a = 1.4$ μm)
c $\epsilon = 0.50; a/x = 0.1$ ($a = 1.3$ μm)

parameters a/x are purely empirical. However, considering the surface roughness of the particles the a/x values may be in the correct order of magnitude.

At higher liquid saturations more and more pores fill up and the models *liquid bridges* and *saturated pores* coexist. The theories described in Section 7.2.4 (Fig. 7.2.11, 7.2.15, and 7.2.16) were checked by Schubert.[15] Figure 7.2.27 shows the results. The tensile strength is plotted versus the liquid saturation and compared with the theory. At $S = 1$ the capillary

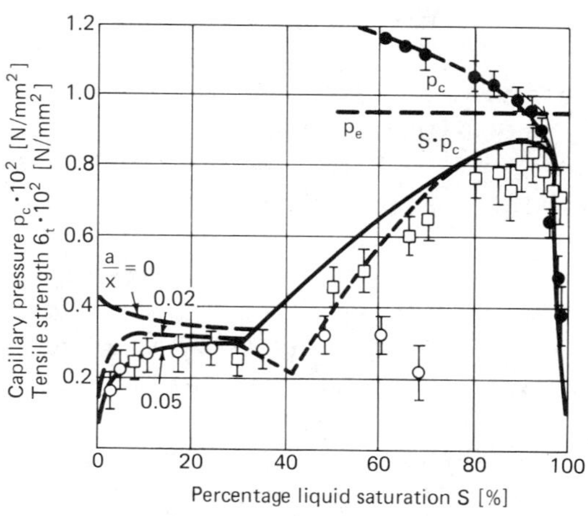

Figure 7.2.27. Tensile strength σ_t and capillary pressure p_c as a function of liquid saturation S. Limestone: $x_0 = 71$ μm, $\epsilon = 0.415$. ●: p_c; □: σ_t drainage; ○: σ_t imbibition.

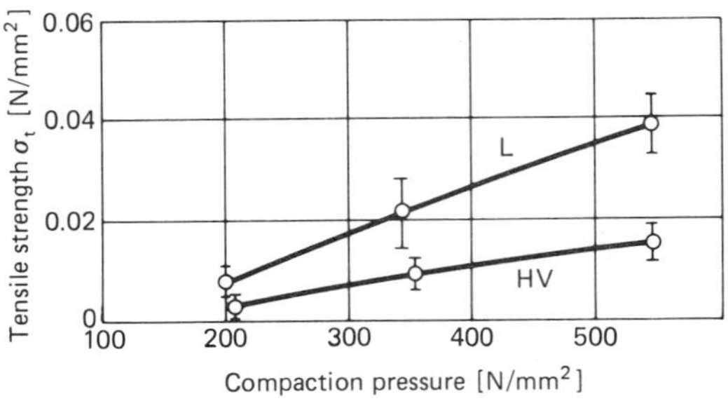

Figure 7.2.28. Tensile strength σ_t of barium sulfate pellets with (L) and without (HV) adsorption layers as a function of compaction pressure. Particle size of starting material: 50–100 μm.[22]

pressure is zero. If, starting at this point, the agglomerate is drained, the capillary pressure raises steeply and then turns into a flatter curve. The point where the two tangents meet is defined as the so called entry suction pressure p_e. It is generally located near $S = 0.9$. Approximately at this point the maximun tensile strength of moist agglomerates exists. At lower and higher saturations the strength decreases. The results show that between $0.3 < S < 0.9$ both mechanisms contribute to the strength of agglomerates.

The capillary pressure and, therefore, the tensile strength are much larger if the liquid is drained than after imbibition. This knowledge can be very important for agglomeration and certain other technologies, e.g., filtration (strength of filter cakes).

Strength due to Van Der Waals Forces. Because of the short range of van der Waals forces, particles forming an agglomerate must be brought closely together to cause significant attraction. To investigate the influence of pressure and van der Waals attraction on agglomerate strength, cylindrical pellets were produced in a hydraulic press using barium sulfate as the model substance. This material excludes other binding mechanisms.

The influence of adsorption layers on agglomerate strength was demonstrated using, respectively, air-dry material and powder, which was dehydrated at a temperature of 600°C prior to pressing it into pellets at 10^{-5} mbar and room temperature. Figure 7.2.28 shows the results.[22] With increasing compaction pressure, i.e., smaller particle distance, the tensile strength increases. Pellets which were produced at atmospheric conditions from air-dry material (*L*) and, therefore, contain adsorbed water exhibit higher strength than those pressed at high vacuum from desorbed barite (HV). This is in general agreement with the expected influence of water adsorption on adhesion discussed above.

Herrmann[23] investigated in more detail the influence of water adsorption on the tensile and shear strengths of barium sulfate briquets. Figure 7.2.29 shows some results. The tensile and shear strengths were determined on briquets produced and stressed in a high vacuum and at varying levels of relative humidity of the surrounding atmosphere. The normal relative humidity lies between 60 and 80%. Therefore, in the common sense, the powder must be considered dry. The strength is plotted in both parts of Figure 7.2.29 versus the relative water

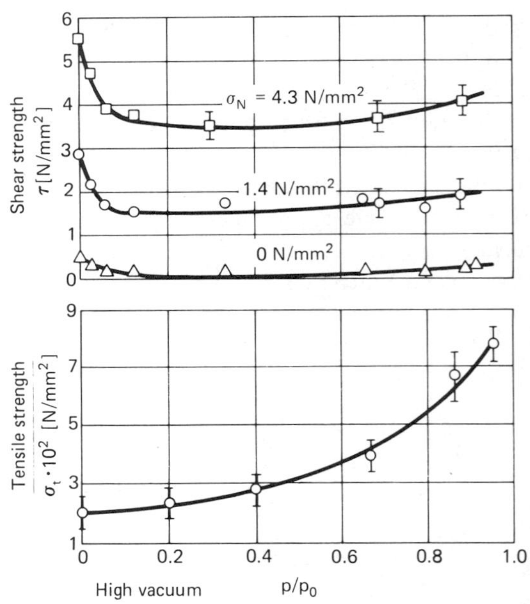

Figure 7.2.29. Shear strength τ and tensile strength σ_t of barite briquets as functions of the relative water vapor pressure p/p_0.[23] Particle size of starting material: 50–100 μm. Compaction pressures: 2.2×10^2 N/mm² (shear strength graph); 4.8×10^2 N/mm² (tensile strength graph).

vapor pressure p/p_0 (with p_0 = water vapor pressure at saturation). The following conclusions can be drawn:

1. The tensile strength σ_t increases with growing relative water vapor pressure p/p_0. Responsible for this rise is the capillary condensation. Van der Waals forces participate only to a small extent.
2. Due to interparticle friction the shear strength τ is high in a high vacuum. This explains, for example, the extremely well-developed footprints that were visible during the first moon landing of man in the loose dust (Apollo II). Interparticle friction is highest at space conditions. With increasing relative humidity the shear strength τ first decreases rapidly. This is due to the fact that liquid films "lubricate" the particles and the friction decreases. Later, liquid bridges develop by capillary condensation, and the strength increases again slowly.
3. At a normal load of $\sigma_N = 0$ the tensile strength σ_t of barite briquets is smaller than the shear strength τ. Therefore, the tensile strength is the critical strength, and failure occurs under tensile load.
4. The shear strength τ depends on the normal load σ_N which acts upon the specimen during shearing.

Other Investigations

A large number of other, specific investigations was carried out by various researchers, confirming still more theories of agglomerate bonding and strength. However, since this chapter is only meant to introduce some basic theoretical and experimental information, further results should be obtained from the respective scientific and technical literature.

REFERENCES

1. H. Rumpf, "The strength of Granules and Agglomerates," *Agglomeration*, edited by W. A. Knepper, John Wiley, New York, pp. 379–418, (1962).
2. W. Pietsch, "Granulieren durch Kornvergrösserung," *CZ-Chemie-Technik 1*(3):116–119 (1972).
3. H. Rumpf, und E. Turba, "Über die Zugfestigkeit von Agglomeraten bei verschiedenen Bindemechanismen," *Ber. dtsch. keram. Ges. 41*(2):78–84 (1964).
4. W. Pietsch, "Die Festigkeit von Agglomeraten," *Chem. Techn. 19*(5):259–266 (1967).
5. H. Rumpf, "Zur Theorie der Zugfestigkeit von Agglomeraten bei Kraftübertragung an Kontaktpunkten," *Chemie-Ing. Techn. 42*(8):538–540 (1970).
6. W. Pietsch, and H. Rumpf, "Haftkraft, Kapillardruck, Flüssigkeitsvolumen und Grenzwinkel einer Flüssigkeitsbrücke zwischen zwei Kugeln," *Chemie-Ing. Techn. 39*(15):885–893 (1967).
7. W. Pietsch, E. Hoffman, and H. Rumpf, "Tensile Strength of Moist Agglomerates," *I + EC Product, Research & Development 8*(3):58–62 (1969).
8. W. O. Smith, P. D. Foote, and P. F. Busang, *Phys. Rev. 34*:1271–1274 (1929).
9. H. C. Hamaker, "The London–van der Waals Attraction between Spherical Particles," *Physica 4*:1058–1072 (1937).
10. H. Krupp, "Particle Adhesion, Theory and Experiment," *Advances Colloid Interface Sci. 1*(2) (1967).
11. J. M. Conway-Jones, "An Investigation into the Mechanism of the Unit Operation of Granulation," Ph.D. thesis, University of London, 1957.
12. H. Rumpf, "Das Granulieren von Stäuben und die Festigkeit der Granulate," *Staub 5*(5):150–160 (1959).
13. H. Schubert, "Tensile Strength of Agglomerates," *Powder Technology 11*:107–119 (1975).
14. W. Pietsch, "The Strength of Agglomerates Bound by Salt Bridges," *Can. J. Chemical Engng. 47*:403–409 (1969).
15. H. Schubert, "Untersuchungen zur Ermittlung von Kapillardruck und Zugfestigkeit von feuchten Haufwerken aus körnigen Stoffen, Ph.D. thesis, University of Karlsruhe, 1972.
16. H. Schubert, "Kapillardruck und Zugfestigkeit von feuchten Haufwerken aus körnigen Stoffen," *Chemie-Ing. Technik, 45*(6):396–401 (1973); and *VDI-Bericht Nr. 190*, pp. 190–194, (1973).
17. H. Rumpf, "Die Wissenschaft des Agglomerierens," *Chemie-Ing. Technik 46*(1):1–11 (1974).
18. N. R. Morrow, "Physics and Thermodynamics of Capillary Action in Porous Media," *Ind. Eng. Chem. 62*(6):32–65 (1970).
19. W. Pietsch, "Festigkeit und Trockungsverhalten von Agglomeraten, deren Festigkeit durch bei der Trockung auskristallisierende Salze bewirkt wird," Ph.D. thesis, Universität (TH) Karlsruhe, 1965.
20. I. Charé, "Trockung von Agglomeraten bei Anwesenheit auskristallisierender Stoffe. Festigkeit und Struktur der durch die auskristallisierten Stoffe verfestigten Granulate," Ph.D thesis, Universitat (TH) Karlsruhe, 1976/77.

21. H. Rumpf, "Particle Adhesion," in "Agglomeration 77," edited by K. V. S. Sastry, *Proc. 2nd Int. Symp. Agglomeration*, Atlanta, March 6–10, 1977, *1*:97–129 (1977).
22. E. Turba, "Die Festigkeit von Briketts aufgrund von van der Waals Kräften und der Einfluß von Adsorptionsschichten," Ph.D. thesis, Universität (TH) Karlsruhe, 1963.
23. W. Herrmann, "Die Adsorption von Wasserdampf in Schwerspat-Preßlingen und ihr Einfluß auf deren Festigkeit," Ph.D. thesis, Universität (TH) Karlsruhe, 1971/72.

PART 3. TABLETING AND PELLETIZATION IN THE PHARMACEUTICAL INDUSTRY

J. T. Carstensen
*University of Wisconsin,
Madison, Wisconsin*

Compressed tablets are the most common pharmaceutical dosage form. The reasons for this are: (1) they are convenient, compact, easy to carry and ship, and (2) they are usually more chemically stable than other dosage forms, since most drugs decompose by hydrolysis.

The first tablet machines were introduced in the nineteenth century, and have by now developed into sophisticated, high-precision tools. The writing to follow will describe machines, manufacturing procedures, formulation parameters, and basic principles on which formulation principles are based. An extensive list of machine specifications is also included, since such a compilation has been deemed useful for both the reader and the practitioner.

7.3.1. TABLET MACHINES

The process of tableting is one in which (Fig. 7.3.1) (1) a powder (or granulation) is transported into the cylindrical cavity in a die, (2) retained there by a lower punch, (3) compressed by the lowering of an upper punch, and (4) ejected by the raising of both punches.

Tablet machines used for this may be either single-punch (excentric) machines (Fig. 7.3.1) or rotary presses (Fig. 7.3.2). In the excentric machine, powder flows from the shoe into the die in position 1. The shoe then swings away, and the upper punch is lowered to compress the powder (position 2). Both punches then are raised (position 3) lifting the tablet out of the die, and the hopper then comes back into its original position (and knocks the ejected tablet onto the discharge chute). The powder level in the shoe is maintained by gravity feed from the hopper.

The fill weight can be adjusted by the (low) position of the lower punch. The lower it is, the higher the fill weight. The fill weight, as shall be seen, is also a function of the apparent density and the flow rate of the powder. The compression pressure (and hence, as described later, the tablet hardness and porosity) can be adjusted by the (low) position of the upper punch.

In a *rotary press* there is a series of dies positioned circularly on a die table (Fig. 7.3.2a). The upper and lower punches glide on cams (Figs. 7.3.2b, c, d). An evoluted picture is shown in Figure 7.3.2b. The filling takes place between points *A* and *B*, i.e., under the feed frame. This in turn is fed by the hopper. The powder is leveled (scraped) at point *B*, so that the fill is a function of the level of the lower

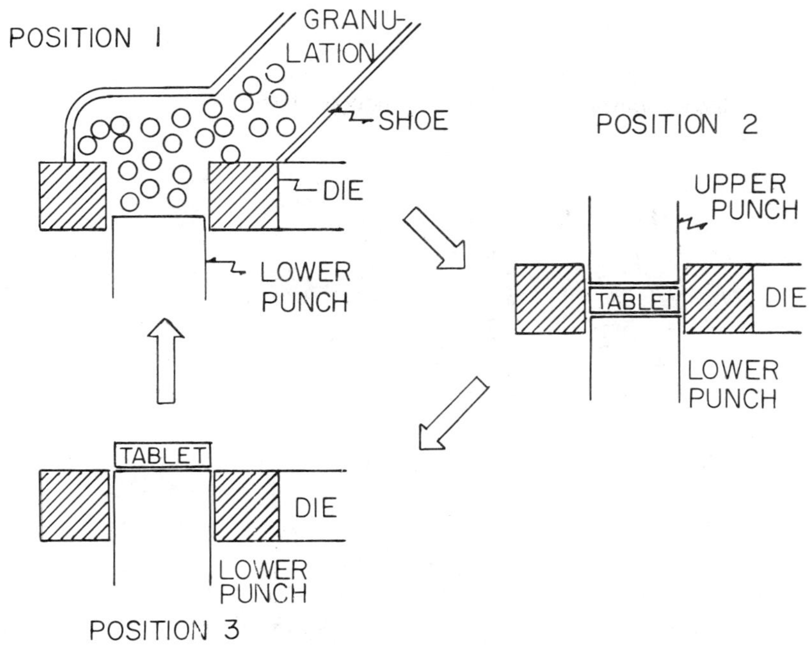

Figure 7.3.1. Steps in the formation of a tablet on a single punch machine.

punch at this point. As the table rotates (goes from right to left in Figure 7.3.2b), the die passes the feed frame, and the lower punch drops a small amount. (No powder is introduced into the die after it has passed the feed frame.) At the point of the pressure wheels, the upper punch is brought down and the lower punch raised to form the tablet. Both are then raised (by the cam contour), and the tablet is ejected. Point A' corresponds to A (the back end of the feed frame), and this serves as an ejection bar for the tablet, which in this fashion is guided into the ejection chute. It is obvious from the drawing that tablet weight can be adjusted by screw E, ejection by screw F (where the ejected tablet must be flush with the table) and compression pressure by the relative position of the pressure wheels.

Photographs of a single-punch machine and a rotary press are shown in Figure 7.3.3, and typical characteristics of several types and production capacities of various machines are listed in Tables 7.3.1-7.3.19. The range is as follows: the single-punch machine produces about 100 tablets per minute, whereas high-speed presses produce up to 10,000 tablets per hopper per minute; that is (since there are, usually, two hoppers and two cycles of compression per rotation), a total of about one million tablets per hour is the upper limit of capacity. The principal machines in use in North America are Stokes machines (Sharples-Stokes division of Pennwalt Corp., 955 Mearns Rd., Warminster, Pa., 18974.) and Manesty machines (Thomas Engineering, Hoffman Estates, Ill. 60195, and Manesty Machines Ltd., Speke, Liverpool L24 9LQ, England).

In the simplest case of a rotary machine there is one hopper and a certain number of "stations" (as few as four) on the die table. In other words, one rotation produces the number of tablets given by the number of dies (and punch sets) on the machine. Examples of this are the Stokes model 512 rotary press, the Stokes model B-2 (Table 7.3.7), and the Manesty B3B press (Table 7.3.8). The next most complicated model has two hoppers, as is the case with, for instance, the Stokes model 513-2 rotary press (Table 7.3.9) and the Manesty Rotapresses (Tables 7.3.10 and 7.3.11).

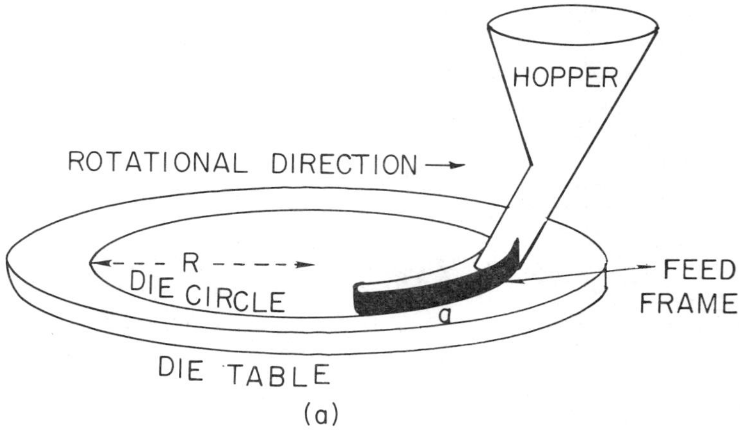

Figure 7.3.2a. Schematic of a rotary machine.

It shall be seen that expulsion of entrapped air from a granulation or (particularly) a powder mix is important since it reduces lamination and capping of the produced tablets. Machines such as the Stokes model 580 Precompression Press (Table 7.3.12) and the Manesty Express (Table 7.3.13) and Betapress (Table 7.3.14) as well as the high-speed machines of the type of Stokes model 610 and Manesty Mark III are equipped with a precompression feature. Solids for tableting are of three types:

a. noncompressible powders
b. compressible powders possessing poor flow
c. compressible powders possessing good flow

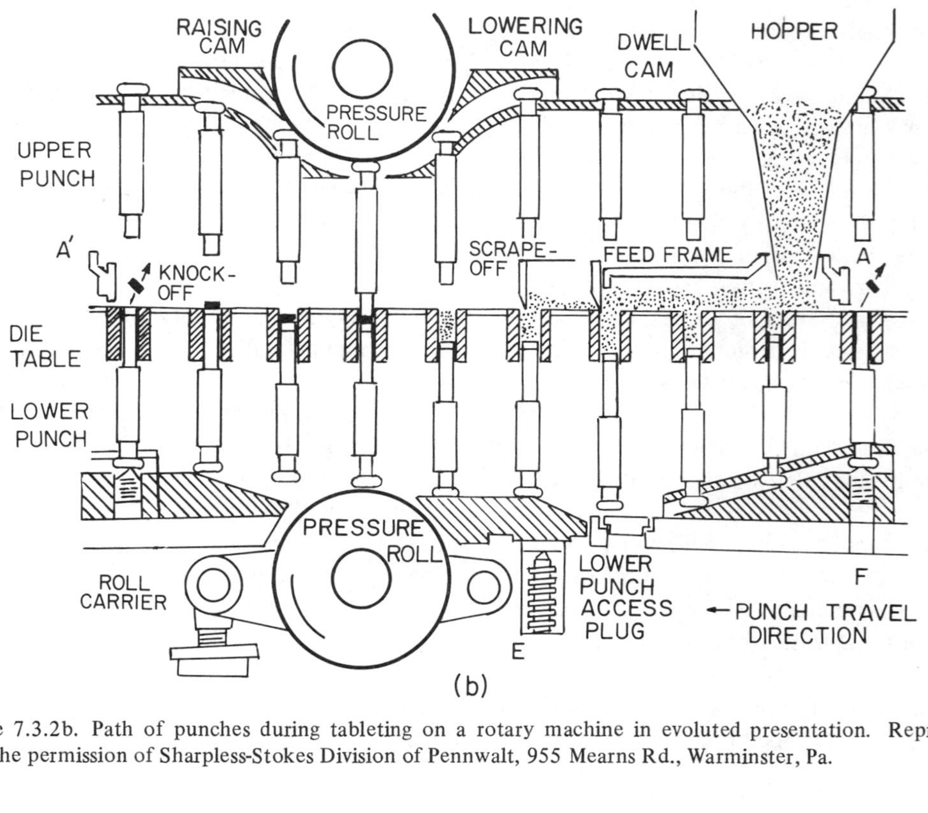

Figure 7.3.2b. Path of punches during tableting on a rotary machine in evoluted presentation. Reproduced with the permission of Sharpless-Stokes Division of Pennwalt, 955 Mearns Rd., Warminster, Pa.

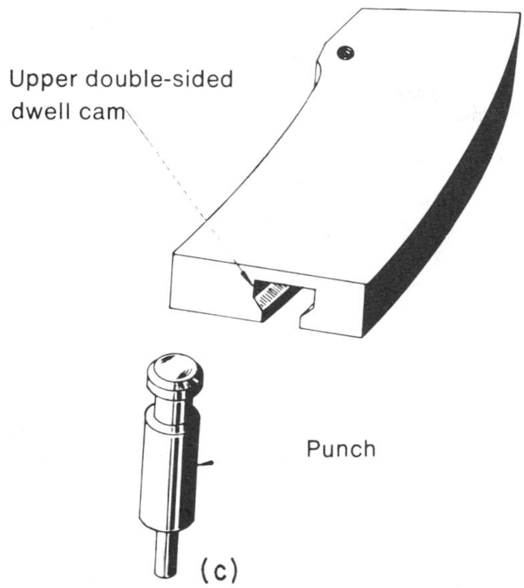

Figure 7.3.2c. Double-sided upper cam, used in Stokes GTP tableting press. Reproduced with the permission of Sharpless-Stokes Division of Pennwalt, 955 Mearns Rd., Warminster, Pa.

(d)

Figure 7.3.2d. Photograph of punch being installed on a Stokes GTP tableting press. Reproduced with the permission of Sharpless-Stokes Division of Pennwalt, 955 Mearns Rd., Warminster, Pa.

Noncompressible powders are either wet granulated (which adds a binder, which makes them compressible) or (if they are of sufficiently low dosage level) mixed with a powder excipient of type *c*, so that the mixture is compressible and free flowing.[1]

When powders are granulated, particle enlargement ensues, a feature which augments flow rates. These flow rates are usually superior to those of naturally free-flowing powders; hence direct compression powders (i.e., mixtures of type *c*) are usually aided in the filling step by artificial means, the so-called forced feeders. The Manesty Betapress for instance (Table 7.3.14) is equipped with a two paddle Rotaflow feeder, which is fitted with a controlled flow inlet. The Stokes series 580 presses contain a pin-type transit-level tunnel feeder, which directs the powder directly in over the die cavity (Table 7.3.12). Flow in the hopper, of course, can be of concern also (and if not uniform will cause inconsistent tablet weights). The Stokes 580 presses are equipped with a metering hopper, which allows materials that tend to pack to be fed in a uniform manner. Automatic punch adjustment during operation to adjust tablet weights to within the limits of a control chart is made possible by the use of the Thomas Engineering Sentinel system.

Powders of type *b* (especially if moisture sensitive) will form tablets, but because of inconsistent flow they can not be compressed directly to produce tablets with uniform tablet weights. The flow in these cases is often improved by particle-size enlargement effected by first making large tablets (slugs, boluses) on a heavy-duty machine (e.g., Stokes model 515-1 or 516-1 [Table 7.3.16] or a Manesty RS3 [Table 7.3.17]). These slugs do not (necessarily) have good weight control. They are broken up by milling through a suitable screen, to form fragments of a larger particle size than the parent powders. Hence flow is better, and tablets can be produced that have satisfactory weight variation.

It should be noted that a fair amount of development of tablet formulas is done at a stage where only small amounts of drug are available (the so-called stage I and II in the

Table 7.3.1. Specification Parameters for Stokes Models Laboratory Single-Punch Presses. (Reproduced with permission of Stokes-Pennwalt, 955 Mearns Rd., Warminster, Pa., Bulletin 365A)

MODEL	519	525	511
Production Rate (compacts/min)	60–95	20–60	40–75
Maximum compact diameter (in.)	$\frac{3}{4}$	3	$\frac{1}{2}$
Maximum depth of fill (in.)	$\frac{11}{16}$	2	$\frac{7}{16}$
Maximum operating pressure (tons)	4	20	1.5
Ejection pressure (tons)	1	5	
Power required (hp)	1	5	$\frac{1}{4}$
Tool length (in.) (upper)	2.187	5.500	1.625
Tool length (in.) (lower)	2.250	4.375	1.625
Die diameter (in.)	1.500	4.500	1.062
Die depth (in.)	0.750	2.250	0.625

Table 7.3.2. Specification Parameters for Stokes DD-2 Rotary Tableting Press. (Reproduced with permission of Stokes-Pennwalt, 955 Mearns Rd., Warminster, Pa., Bulletin 358)

Output (tablet/min)	720
Number of tool statins	23
Fill depth (in.)	$\frac{1}{8}$–$1\frac{3}{8}$
Max. Tablet diameter (in.)	$1\frac{3}{16}$
Max. operating pressure (tons)	
Standard	15
Optional	20

clinical progression of drug development), and that scale-up difficulties into high-speed equipment can be anticipated. Because of government requirements that eventual production formulas be identical to those tested in the clinic, severe pilot problems always exist in the pharmaceutical industry.

The general dimensions and physical characteristics are as follows: Tablets, in general, have diameters of $\frac{1}{4}$ to 1 inch and are usually cylindrical with a circular cross section (Fig. 7.3.4). The crowns can be either flat with a beveled edge, standard concave, or deep concave. Large tablets of this geometry are difficult to swallow, and tablets weighing more than 800 mg are usually tableted into an oblong shape (Fig. 7.3.4c and d), allowing easier swallowing "the long way." A host of other shapes exist (Fig. 7.3.5).

The deep concave, oval, and capsule-shaped tablets are usually sugar coated. In the case of the deep concave (or extra deep concave) tablets, this shape is specifically selected for coating reasons, since it is nearly spherical, and hence lends itself better to rolling and interaction with added liquid in a coating pan. The shapes of the latter two cases, the oval and

Table 7.3.3. Specification Parameters for the Stokes *Pacer* Rotary Tableting Presses. (Reproduced with permission of Stokes-Pennwalt, 955 Mearns Rd., Warminster, Pa., Bulletin 555)

MODEL	PACER 1	PACER 2
Output (tablets/min)	4,200	3,200
Number of tool sets	45	35
Max. tablet diameter (in.)	$\frac{7}{16}$	$\frac{5}{8}$
Fill depth (in.)	0–$\frac{11}{16}$	0–$\frac{11}{16}$
Operating pressure (tons)		
Maximum	4	4
with standard release spring	1.5–4	1.5–4
with light release spring	0–1.5	0–1.5

Table 7.3.4. Specification Parameters for Stokes Model 328 Rotary Tableting Presses. (Reproduced with permission of Stokes-Pennwalt, 955 Mearns Rd., Warminster, Pa., Bulletin 374)

Number of tooling stations	27	33	45
Output (tablets/min)	1,000–2,700	1,200–3,300	1,600–4,500
Max. tablet diameter (in.)	$1\frac{3}{16}$	$1\frac{1}{16}$	$\frac{3}{4}$
Fill depth (in.):			
Standard	$0–\frac{11}{16}$	$0–\frac{11}{16}$	$0–\frac{11}{16}$
Optional	$\frac{11}{16}–1\frac{3}{8}$	$\frac{11}{16}–1\frac{3}{8}$	$\frac{11}{16}–1\frac{3}{8}$
Max. operating pressure (tons)	10	10	10
Pressure release adjustment (tons)	0–10	0–10	0–10
Upper punch entrance (in.)	$\frac{3}{16}–\frac{7}{16}$	$\frac{3}{16}–\frac{7}{16}$	$\frac{3}{16}–\frac{7}{16}$

Table 7.3.5. Specification Parameters for Stokes Model 541 and 551 Rotary Tableting Presses. (Reproduced with permission of Stokes-Pennwalt, 955 Mearns Rd., Warminster, Pa., Bulletin 372 A)

Model	541	551
Output (tablets/min)	1,500–4,100	1,800–5,100
Number of tool stations	41	51
Max. tablet diameter (in.)	$\frac{5}{8}$	$\frac{7}{16}$
Depth of fill (in.)	$\frac{1}{8}–\frac{11}{16}$	$\frac{1}{8}–\frac{11}{16}$
Max. operating pressure (tons)	4	4

Table 7.3.6. Specification Parameters for Stokes Model 560 *Versapress*™ Tableting Press. (Reproduced with permission of Stokes-Pennwalt, 955 Mearns Rd., Warminster, Pa., Bulletin 560)

Model	560-1	560-2
Output, tablets/min		
Two-layer or precompression	2,100	1,630
Single-component operation	4,200	3,260
Number of tool stations	45	35
Max. tablet diameter (in.)	$\frac{7}{16}$	$\frac{5}{8}$
Fill depth (in.)		
Two-layer, first fill	$0–\frac{7}{16}$	$0–\frac{7}{16}$
Two-layer, second fill	$\frac{1}{8}–\frac{3}{8}$	$\frac{1}{8}–\frac{3}{8}$
Single component or precompression	$0–\frac{11}{16}$	$0–\frac{11}{16}$
Operating pressure, precompression (tons)		
Standard release	1.5–4	1.5–4
light release (optional)	0–1.5	0–1.5
Upper punch entrance (in.)	$\frac{1}{8}–\frac{3}{8}$	$\frac{1}{8}–\frac{3}{8}$
Upper punch stroke (in.)	$1\frac{1}{8}–1\frac{3}{8}$	$1\frac{1}{8}–1\frac{3}{8}$
Lower punch stroke (in.)	$\frac{3}{4}$	$\frac{3}{4}$

Table 7.3.7. Specification Parameters for Stokes B-2 Rotary Tableting Press. (Reproduced with permission of Stokes-Pennwalt, 955 Mearns Rd., Warminster, Pa., Bulletin 368)

Output (tablets/min)	1,050
Number of tool stations	16
Fill depth (in.)	0–$\frac{11}{16}$
Max. tablet diameter (in.)	$\frac{5}{8}$
Maximum operating pressure (tons)	4
Upper punch entrance (in.)	$\frac{1}{8}$–$\frac{3}{8}$
Upper punch stroke (in.)	$1\frac{1}{4}$
Lower punch stroke (in.)	$\frac{3}{4}$

capsule-shaped tablets, are chosen to facilitate administration, and the coat is employed to make the surface more smooth. Sugar coating is expensive, and in recent years has been replaced in many cases with film coating.[1]

To assure proper alignment the oval and capsule-shaped punches (as well as punches with embossing) are keyed or restrained (Fig. 7.3.6). This prevents rotation of the punch in its path along the cam.

In many cases there exist incompatibilities[2] among drugs, and such solids are kept apart from one another by special means, notably by double- or triple-layer tablets or by compression-coated tablets (tablet within a tablet). In the triple-layer tablet, compression takes place in several stages, requiring a special press. There are three hoppers, 120° apart on the die table. Filling takes place in three steps. In the first stage, the low position of the lower punch dictates the fill weight (of the first layer), in the next stage the "bottom" of the die is the top layer of the first filled powder, and in the last stage, the surface of the second layer is the "bottom" of the die. Intermediate "tamping" is possible, and this, for instance, improves the

Table 7.3.8. Specification Parameters for B3B Rotary Machines. (Reproduced with permission of Manesty Machines Ltd., Speke, Liverpool, England and Thomas Engineering Inc., Hoffman Estates, Ill., Bulletin 190 177)

Machine	BB3B	BB3B	BB3B	BB3B	B3B	B3B	D3B
Series	27	33	35	45	16	23	16
Number of stations	27	33	35	45	16	23	16
Max. pressure (tons)	6.5	6.5	6.5	6.5	6.5	6.5	10
Max. tablet diameter (in.)	$\frac{5}{8}$	$\frac{7}{16}$	$\frac{6}{8}$	$\frac{7}{16}$	$\frac{5}{8}$	$\frac{7}{16}$	1
Maximum depth of fill (mm)	17.4	17.4	17.4	17.4	17.4	17.4	20
Output (tablets/min)	760–1,520	924–1,848	1,490–2,980	1,913–3,826	350–700	5,000–1,000	260–520

Table 7.3.9. Specification Parameters for Stokes Series 513 Chord-Loc Rotary Tableting Machine (Reproduced with Permission of Stokes-Pennwalt, 955 Mearns Rd., Warminster, Pa., Bulletin 513)

Model	513-2	513-3
Output (tablet/min)	4,200	3,260
Number of tool stations	45	35
Max. tablet diameter (in.)	$\frac{7}{16}$	$\frac{5}{8}$
Fill depth (in.)	0–$\frac{11}{16}$	0–$\frac{11}{16}$
Operating pressure (tons)		
Maximum	4	4
With standard release spring	1.5–4	1.5–4
With light release spring	0–1.5	0–1.5
Upper punch entrance (in.)	$\frac{1}{4}$	$\frac{1}{4}$
Upper punch stroke (in.)	$1\frac{1}{4}$	$1\frac{1}{4}$
Lower punch stroke (in.)	$\frac{3}{4}$	$\frac{3}{4}$

Table 7.3.10. Specification Parameters for a High-Performance, High-Speed Rotary Press, the Manesty *Rotapress* Mk 11. (Reproduced with permission of Manesty Machines Ltd., Speke, Liverpool, England, and Thomas Engineering Inc., Hoffman Estates, Ill., Bulletin 250275)

Series	37	45	55	61
Number of stations	37	45	55	61
Max. operating pressure (tons)	10	6.5	6.5	6.5
Max. depth of fill (in.)	$\frac{13}{16}$	$\frac{11}{16}$	$\frac{11}{16}$	$\frac{11}{16}$
Max. tablet diameter (in.)	1	$\frac{5}{8}$	$\frac{7}{16}$	$\frac{7}{16}$
Output (tablets/min)	888–3,552	2,050–8,200	2,500–10,000	2,775–11,100

Table 7.3.11. Specification Parameters for a High-Output Tablet Production Machine, the Manesty *Rotapress* MK IIa. (Reproduced with Permission of Manesty Machines Ltd., Speke, Liverpool, England and Thomas Engineering Inc. Hoffman Estates Ill., Bulletin 281077).

Series (number of stations)	37	45	55	61
Max. operating pressure (tons)	10	6.5	6.5	6.5
Max. depth of fill (in.)	$\frac{13}{16}$	$\frac{11}{16}$	$\frac{11}{16}$	$\frac{11}{16}$
Max. tablet diameter (in.)	1	$\frac{5}{8}$	$\frac{7}{18}$	$\frac{7}{16}$
Output (tablets/min)	710–3,550	1,640–8,200	2,000–10,000	2,220–11,100

Table 7.3.12. Parameters for Stokes Series 580 Tri-Pact™ Rotary Tableting Presses. (Reproduced with Permission of Stokes-Pennwalt, 955 Mearns Rd., Warminster, Pa., Bulletin 362).

Output, (tablets/min)	2,100	1,600
Number of punches	45	35
Max. tablet diameter (in.)	$\frac{7}{16}$	$\frac{5}{8}$
Fill depth (in.)	$0-\frac{11}{16}$	$0-\frac{11}{16}$
Number of compression stations	3	3
Operating pressures (tons)		
First precompression station	0–1	0–1
Second precompression station	0–2	0–2
Third (final) compression	0–4	0–4

precision of the fill of each layer. Examples of multilayer tablet presses are the Stokes 580 (Table 7.3.12) and the Manesty Bicota (Table 7.3.18).

In the case of a compression coating (e.g., Manesty Drycota, Table 7.3.19), a tablet is first manufactured on one press (which constitutes one half of the total press assembly), and then transferred into the half-filled larger die, with the "outer" granulation on the other half of the machine. This is then filled to the top and compressed. In both compression coating and multiple-layer tablets, the intergranulation and

Table 7.3.13. Specification Parameters for a Single-Hopper Medium Production Rotary Press, the Manesty *Express*. (Reproduced with Permission of Manesty Machines Ltd., Speke, Liverpool, England and Thomas Engineering Inc., Hoffman Estates Ill., Bulletin 100877)

Series	X20	X25	X30
Number of stations	20	25	30
Max. pressure (tons)	10	6.5	6.5
Max. precompression pressure (tons)	1	1	1
Max. tablet diameter (in.)	1	$\frac{5}{8}$	$\frac{7}{16}$
Max. depth of fill (in.)	$\frac{13}{16}$	$\frac{11}{16}$	$\frac{11}{16}$
Output (tablets/min)	800–2,000	1,000–2,500	1,200–3,000

Table 7.3.14. Specification Parameters for a Single-Hopper High-Speed Machine with Forced Feed (Rotaflow Feeder), the Manesty *Betapress*. (Reproduced with the Permission of Manesty Machines Ltd., Speke, Liverpool, England and Thomas Engineering Inc., Hoffman Estates, Ill., Bulletin 231,171)

Number of stations	16	23
Maximum operating pressure (tons)	6.5	6.5
Max. depth of fill (in.)	$\frac{11}{16}$	$\frac{11}{16}$
Max. tablet diameter (in.)	$\frac{5}{8}$	$\frac{7}{16}$
Diameter of die (in.)	1.187	0.945
Output of tablets per min, 1500 rpm.	600–1,500	860–2,160
Output of tablets per min, 1800 rpm.	750–1,500	1,050–2,160

Table 7.3.15. Specification Parameters for Stokes GTP Presses (Reproduced with permission of Stokes-Pennwalt, Sharpless-Stokes Division, 955 Mearns Rd., Warminster, Pa., Bulletin 610-378-3M-1P)

Model	610-1	610-2	610-3
Output (tablets/min)	3,500–10,000	2,800–8,000	2,150–6,150
Number of tooling stations	65	53	41
Number of pressure stations			
(Precompression)	2	2	2
(Final compaction)	2	2	2
Operating pressure, max. (tons)			
Final compression	10	10	10
Pressure release adjustment (tons)	0–10	0–10	0–10
Standard punch stroke (in.)			
Upper	$1\frac{11}{14}$	$1\frac{11}{14}$	$1\frac{11}{14}$
Lower	$\frac{11}{16}$	$\frac{11}{16}$	$\frac{11}{16}$

layer bonding and the amount of moisture are exceedingly important parameters.

In a triple-layer tablet, the precision of fill is less than in a conventional tablet, as far as the individual layers are concerned. Defects are primarily (1) insufficient interlayer bonding, giving rise to separation of layers, (2) unevenness of the layers (which can be seen directly if multicolor schemes are employed). In the compression-coated tablet the defects are (1) missing core, (2) poorly centered core, which can be seen from the "outside" of the tablet, and (3) splitting caused by inadequate bonding in the outer layer. The formulation of these types of products is difficult.

Table 7.3.16. Specification Parameters for Stokes 515 Rotary Presses. (Reproduced with Permission of Stokes-Pennwalt, 955 Mearns, Rd., Warminster, Pa., Bulletin 372 A)

Model	515	515-16
Output (tablets/min)	335	500
Number of tool stations	15	16
Fill depth (in.)	$\frac{1}{8}$–$1\frac{1}{16}$	$\frac{1}{8}$–$\frac{13}{16}$
Max. tablet diameter (in.)	$1\frac{3}{16}$	$\frac{15}{16}$
Max. operating pressure (tons)	10 (standard)	

7.3.2. TABLET FORMULATIONS

To formulate a tablet one must first know the desired size as well as shape and approximate thickness. In this manner one may estimate the approximate weight. The sum of all ingredients is, of course, the tablet weight, and estimates are then made of the required amounts of the necessary ingredients. The amount is then

Table 7.3.17. Specification Parameters for a Slugger, the Manesty RS3, High-Performance Heavy-Duty Rotary Press. (Reproduced with the Permission of Manesty Machines Ltd., Speke, Liverpool, England and Thomas Engineering, Inc., Hoffman Estates Ill., Bulletin No 101072)

	15 TON MODEL			20 TON MODEL		
Number of stations	14	16	21	14	16	21
Max. diameter of tablet (in.)	2.75	2.5	1.94	2.75	2.5	1.94
Depth of fill (in.)	0.25–2.5	0.25–2.5	0.25–2.5	0.25–2.5	0.25–2.5	0.25–2.5
Output (tablets/min)	84–225	95–256	126–336	84–224	96–256	126–336
Max. pressure (tons)	15	15	15	20	20	20
Max. precompression pressure (tons)	5	5	5	5	5	5

Table 7.3.18. Specification parameters for the Manesty *Bicota* Compression Coating Machine. (Reproduced with the Permission of Manesty Machines Ltd., Speke, Liverpool, England and Thomas Engineering Inc., Hoffman Estates, Ill., Bulletin 010973). The Machines Produce over 150,000 Ordinary Tablets Per Hour, over 50,000 Single-Coated, Double-Coated, or Two- or Three-Layer Tablets Per Hour.

Model	900	600
Output per min, bicoated	320–900	224–624
Output per min, core + 1st coat	320–900	224–624
Output per min, uncoated	960–2,700	224–624
Max. diameter, bicoated (in.)	$\frac{7}{8}$	$\frac{13}{16}$
Min. shell thickness (in.)	$\frac{1}{32}$	$\frac{1}{32}$
Max. core diameter (in.)	$\frac{1}{2}$	$\frac{1}{2}$
Max. diameter, single coated (in.)	$\frac{5}{8}$	$\frac{13}{16}$
Max. diameter, single coat (in.) core	$\frac{9}{16}$	$\frac{9}{16}$
Max. diameter, uncoated (in.)	$\frac{5}{8}$	1

Table 7.3.19. Specification Parameters for a Compression Coating Machine, the Manesty *Drycota*. The Machine Produces 100,000 Tablets Per Hour, 50,000 Compression (Dry-) Coated Tablets Per Hour, and 50,000 Layer Tablets Per Hour. (Reproduced with the Permission of Thomas Engineering Inc., Hoffman Estates, Ill., Bulletin 010973)

Model	500	600	900
Output (coated tablets/min)	176–496	224–624	323–897
Output (uncoated tablets/min)	352–992	448–1,248	646–1,794
Max. diameter of coated tablet (in.)	$\frac{13}{16}$	$\frac{13}{16}$	$\frac{5}{8}$
Max. diameter of core (in.)	$\frac{9}{16}$	$\frac{9}{16}$	$\frac{9}{16}$
Max. thickness of core (in.)	$\frac{3}{16}$	$\frac{3}{16}$	$\frac{3}{16}$

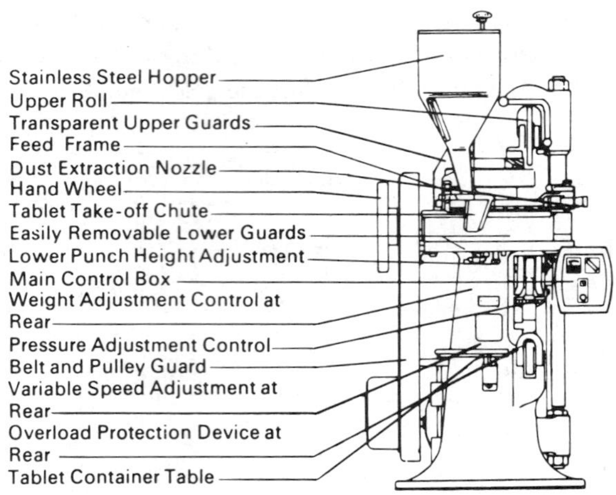

(a)

Figure 7.3.3A. Schematic view of a B3B Manesty Rotary Tablet Machine. Reproduced with permission of Manesty Machines Ltd., Speke Liverpool, England and Thomas Engineering Inc., Hoffman Estates, Ill.

brought to the desired weight (q.s.) with filler. A list of ingredients and approximate concentration ranges is shown below:

DRUG	EXAMPLE	RANGE
Disintegrant	Cornstarch	0–8%
Lubricant	Magnesium stearate	0–2%
Glidant	Talc	0–1%
Binder	Cornstarch	0–5%
Filler	Lactose	q.s.

Except when placebo tablets are made, the drug is present, and in an amount dictated by its nature. When a tablet is administered to a patient, it must *disintegrate* in the gastric (and intestinal) fluids. On contact with biological fluids, swelling substances such as starch, certain resins (Amberlite XE88), alginic acid, and modified polyvinylpyrrolidone will expand sufficiently to "blow" apart the tablet.

When a tablet is compressed in the die, a residual force exists against the die wall (Fig. 7.3.7). This force P is perpendicular to the ejection force E, exerted by the lower punch during the ejection phase of the tableting. The two forces are related by the frictional equation:

$$E = FP \tag{7.3.1}$$

where F is the frictional coefficient. The function of the lubricant is to reduce the value of F. Improperly lubricated formulations will, in milder cases, give rise to tablets that are prone to cap (or that actually do cap), that is, the crown separates from the rest of the tablet. Hairline cracks in the walls of the tablet are usually indicative of this condition. In more severe cases the formulation will "bind up" in the die, and the tablet machine will stop operating. There are formulation reasons for capping as well; for instance, a too large quantity of fines will give rise to capping. The actual capping often occurs as the tablet is being ejected (i.e., actually outside the die), because at this point the tablet expands. Lubrication, machine speed, and reduction of fines are usually the remedies employed in the case of "capped tablets."

To obtain a good tablet the powder or granulation must flow well. *Glidants* are sometimes added to improve flow, but most frequently, as shall be discussed shortly, flow is controlled by particle size and surface. The property affected by poor flow is the consistency of the tablet weight. The United States Pharmacopeia XVI states the following requirements for weight: of 20 individually weighed tablets only two may differ from the mean by

Figure 7.3.3B. Photograph of a B3B Manesty Rotary Tablet Machine. One model has 16 stations and is capable of producing tablets up to 15.8 mm diameter. Capacity is 700 tablets per minute. Another model has 23 stations. Filling depth of both machines is 17.4 mm and highest pressure is 6.5 tons. Reproduced with permission of Manesty Machines Ltd., Speke Liverpool, England and Thomas Engineering, Inc., Hoffman Estates, Ill.

Figure 7.3.3C. Photograph of a single punch machine, Stokes Model 519 (F) Press. Reproduced with the permission of Sharpless-Stokes Division of Pennwalt, 955 Mearns Rd., Warminster, Pa.

more than the stated percentage, and no tablet may differ by more than twice the stated percentage:

Tablets weighing 13 mg or less	15%
Tablets between 13 and 130 mg	10%
Tablets between 130 and 324 mg	7.5%
Tablets more than 324 mg	5%

The *binders* are added to tablet formulations to produce granules or powders that will bind together to make a good compact in the tablet die. To describe binders, it is necessary to briefly classify manufacturing methods. These are (1) wet granulation, in which the binder is added as a paste (i.e., water is added to the granulation in the process), and (2) dry methods, in which powders are blended and compressed (direct compression); or compressed, reground, and recompressed (slugging). The pastes used in wet granulation[3] are mostly: Cornstarch paste (0–10%), sucrose (usually added dry, water being the granulating liquid), povidone (polyvinylpyrrolidone) (10% alcoholic solution), acacia (10% aqueous solution), and gelatin (5–13% aqueous solution).

The *fillers* are usually sugars, sugar alcohols, or inorganic substances. Lactose, dicalcium phosphate, sucrose, and mannitol are common tablet fillers. All nondrug substances in a tablet are denoted excipients.

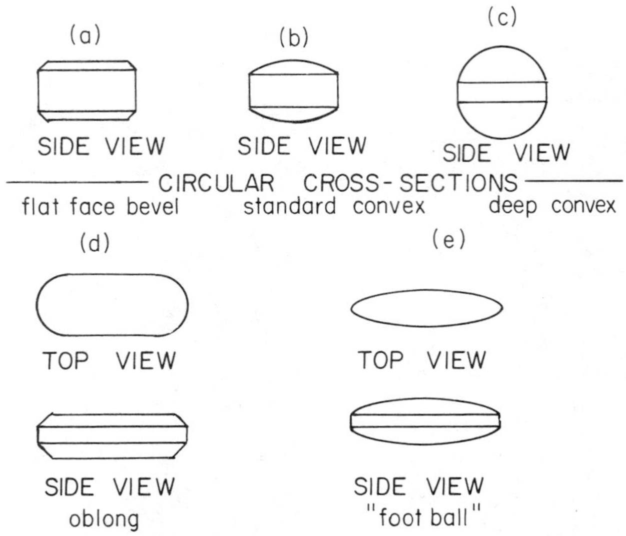

Figure 7.3.4. Tablets shapes. *a*, flat faced beveled (ffb); *b*, standard concave (sc); *c*, deep concave (dc); *d*, capsule shaped; and *e*, oval.

7.3.3. FACTORS AFFECTING FLOW

Flow rates of powders affect tableting in two ways: the flow from the hopper to the feed frame must be adequate, and the flow from the feed frame to the die must be adequate.

Powder flow is a function of

1. Particle size
2. Particle shape
3. Roughness of surface
4. The chemical nature of the compound (cohesion)
5. Moisture

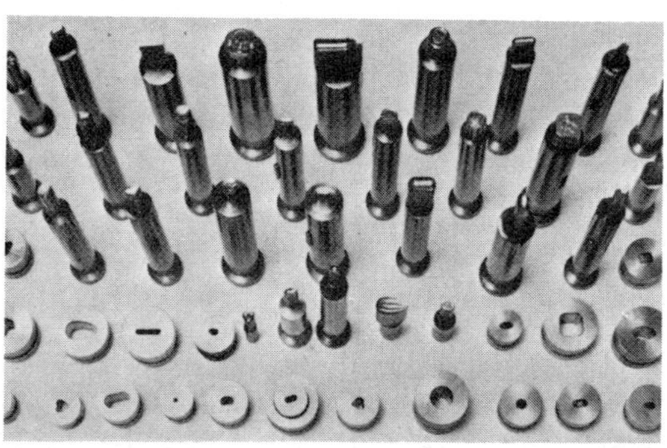

Figure 7.3.5. Punches of various shapes. Reproduced with the permission of Sharpless-Stokes Division of Pennwalt, 955 Mearns Rd., Warminster, Pa.

In general, flow versus particle diameter is a parabolic function, such as shown in Figure 7.3.8.[3] The maximum (d_m, W_m), where d and W are diameter and flow rate, respectively, occurs at faily large diameters (400–1000 μm), so that flow problems associated with fineness

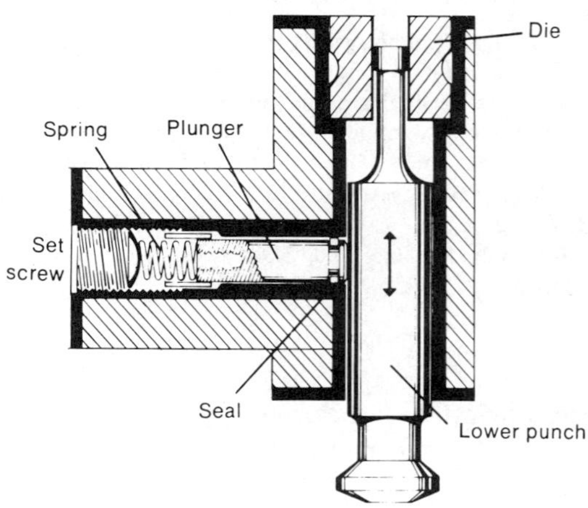

Figure 7.3.6. Punch restraining system used in Stokes GTP tableting press. This assures lower punch positioning by use of a spring-loaded plumber. Rotation of punches is hence prevented, which, for instance, avoids double impressions (with embossed punches) and also assures level positioning so that fill weight is better controlled. Reproduced with the permission of Sharpless-Stokes Division of Pennwalt, 955 Mearns Rd., Warminster, Pa.

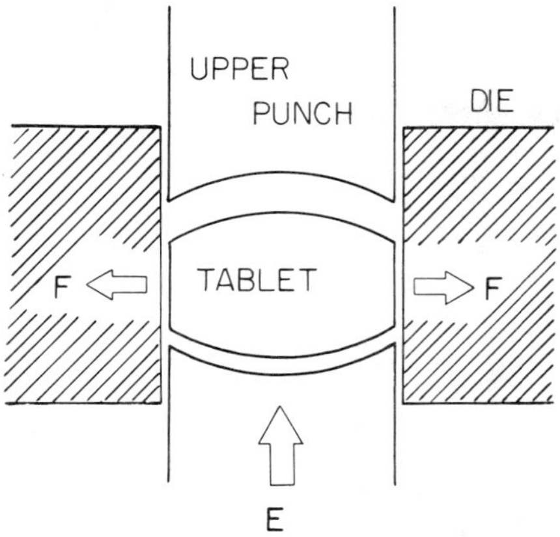

Figure 7.3.7. Residual die wall force F and ejection force E.

and cohesiveness of powders is usually solvable by particle enlargement. The general method employed is either wet granulation or slugging, both of which will be discussed shortly.

The *effect of the particle shape* has been described by Ridgway and Rupp.[4] They define a quantity for describing particle shape (shape factor) in the following fashion: If d denotes the projected mean diameter of the particle, it is possible to express the surface A and the volume V of the particle as $A = q_1 d^2$ and $V = q_2 d^3$, and the shape factor is then $G = q_1/q_2$. In general the effect of the shape factor on flow amounts to a 20% drop in flow rate with a doubling (from, e.g., 7.5 to 15) of shape factor. The effect is *most* prominent with larger orifices.

The effect of orifice diameter on flow is described by the Brown and Richards Equation[5]

$$(4W/\pi\rho g)^{0.4} = mD + C \qquad (7.3.2)$$

The effect of the *addition of fines* to a monodisperse powder has been described for instance by Danish and Parrott.[6] The general effect of this step is shown in Figure 7.3.9.

Equation (7.3.2) shows that flow rates increase with a 2.5 power of the orifice opening diameter. Exponents of this general order of magnitude have been reported in the literature.[7]

7.3.4. EFFECTS OF FLOW RATE ON TABLET PROPERTIES

The amount of material that can be filled into a tablet die is the apparent density ρ' (g/cm^3) multiplied by the volume V(cc) of the die cavity. If the contact time between the die and the feed frame of length a (cm) is t (seconds), then on a die table of radius R (cm) and rotational speed Ω (rotations per second),

$$t = \frac{a}{\Omega 2\pi R} \qquad (7.3.3)$$

In general, as long as the flow rate has a value over a critical value W' given by:

$$W' = V\rho'/t = D\Omega 2\pi R/a \qquad (7.3.4)$$

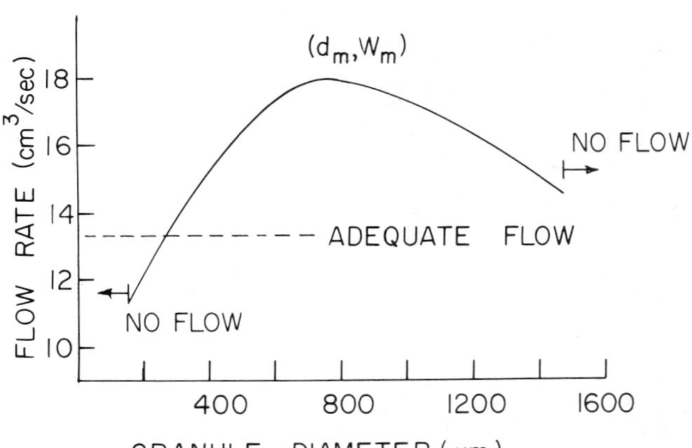

Figure 7.3.8. Flow rates as a function of particle diameter.[3]

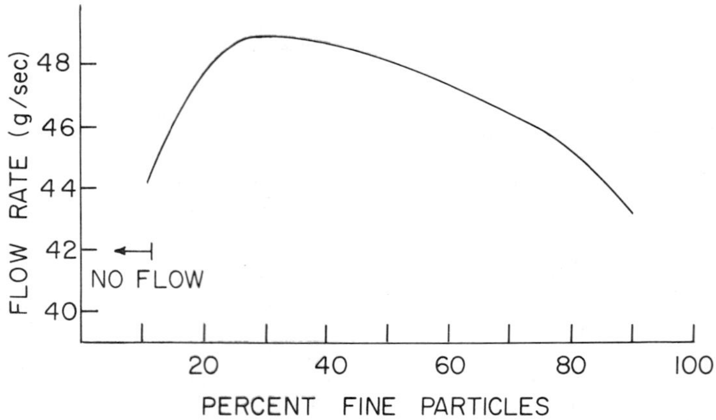

Figure 7.3.9. Flow rate as a function of percent fines in a granulation.[6]

the fill weight will be D(g). However, for values of $W < W'$, this is not the case, and here the tablet weight becomes a function of flow rate:

$$D = Wa/(\Omega 2\pi R) \tag{7.3.5}$$

These relations are shown in Figure 7.3.10. There is no sharp break between the two linear portions predicted by Eqs. (7.3.4) and (7.3.5), and on high-speed machines, the situation is frequently in the transitional region (the curve in Fig. 7.3.10.)

7.3.5. FACTORS AFFECTING THE DIMENSIONS OF TABLETS

The thickness h (cm) and the hardness H (kg) of a tablet are functions of the pressure P (Pascals) applied in the formation of the tablet. This, of course, is a function of the relative distance between the two punches at their closest point of approach. The thickness h follows the Fell-Newton law[8]

$$\ln \frac{h - h_\infty}{h_0 - h_\infty} = -k\,(P - P_i) \tag{7.3.6}$$

and this relation is shown (in linear fashion) in Figure 7.3.11. h_∞ is a function of the true density of the tablet ρ (g/cm^3), in that the (nonporous) mass of the compact is given by:

$$D = h_\infty \pi (D/4)^2 \rho \tag{7.3.7}$$

h_0 is given by the apparent density (ρ') in a similar expression:

$$D = h_0 \pi (D/4)^2 \rho' \tag{7.3.8}$$

Inspection of Figure 7.3.11a shows that Eq. (7.3.6) applies beyond $P = P_i$ (i.e., to the portion AB). The reason for this is that the sequences in the tablet formation are as shown in Figure 7.3.11b: deposition of a powder bed of cascaded apparent density; b, consolidation to a closest packing; c, deformation; d, fragmentation; and finally e, fusion. It is assumed that Eq. (7.3.6) applies only to the last steps;

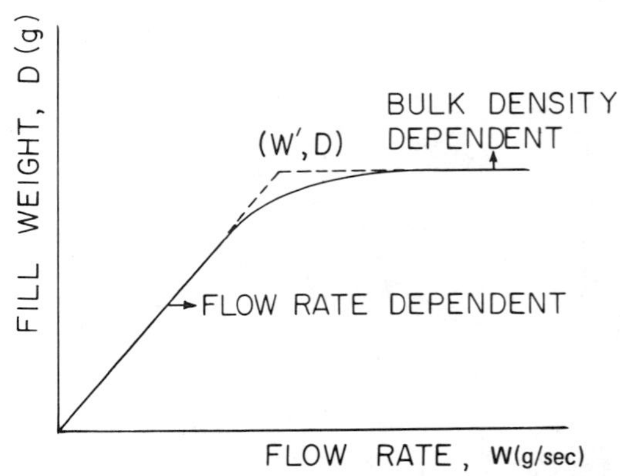

Figure 7.3.10. Fill weight as function of flow rate of a granulation or powder. D is dose and W' is the critical flow rate.

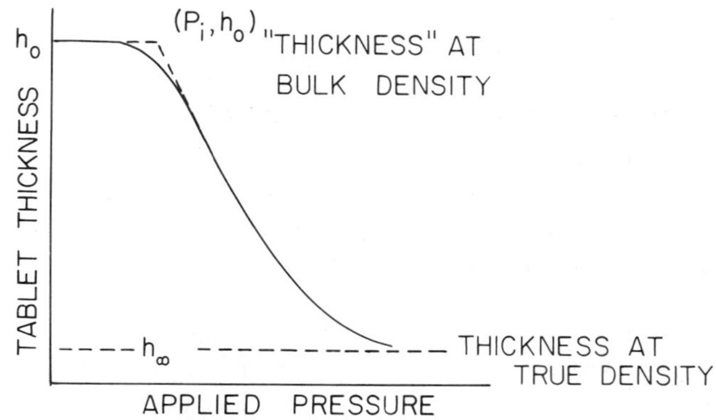

Figure 7.3.11a. Tablet thickness h as a function of applied pressure P.

hence P_i somehow relates to the elastic limit beyond which deformation no longer gives rise to the same (1) shape or (2) size of the particle when the pressure is released.

It is apparent from Figure 7.3.11b, that the rate at which a powder or a granulation can consolidate may be critical when high-speed machines are used, and that, therefore, consolidation rates play a part in compression physics. Takieddin et al.[9] have checked the fundamentals of consolidation rates by cascading a monodisperse granulation into a graduate and recording the volume V_0 (cm^3), then tapping it through a certain height n times and recording the volume V (cm). The volume will approach that of a close packing V_∞ (cm) after a large number of taps. The type curve obtained adheres to a best fit of:

$$\frac{V - V_\infty}{V_0 - V_\infty} = (1 + \mathrm{k} \cdot n)^{-1/4} \qquad (7.3.9)$$

where k is a constant dependent on material, particle shape, and diameter.

7.3.6. PRESSURE RELATIONS IN THE COMPRESSION CYCLE

Leigh and Carless[10] have treated the pressure relations in the compression cycle by comparing the tablet with a solid (a Mohr's body), and the cycle in Figure 7.3.12 is suggested. Here the radial stress σ (Pascals) is plotted as a function of the axial stress τ (Pascals). The point B

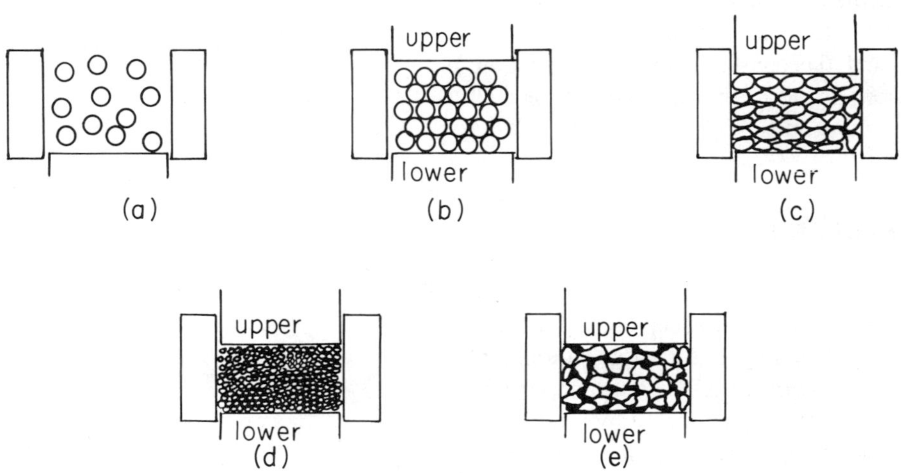

Figure 7.3.11b. Steps in the formation of a tablet. (a) powder in cascaded apparent density, (b) consolidation to tapped density, (c) deformation, (d) fragmentation and (e) fusion.

is interpreted as the point where elastic recovery has its limit, and plastic flow prevails beyond this value. In a manner of speaking this corresponds to the point P_i in Figure 7.3.11a, but it should be stressed that the analogy is but a similarity, because the tablet mass is not nonporous.

The residual stress (AE) is the pressure exerted by the tablet on the die wall after removal of the upper punch. It follows that an equation similar to Eq. (7.3.1) holds

$$\tau = \mu \sigma \tag{7.3.10}$$

where stresses replace forces and where μ is the frictional coefficient. One of the functions of a lubricant in a tablet is to reduce the value of μ.

The lubricant also manifests itself in how well the compression pressure is propagated through the solid mass. Tablet machines are frequently instrumented[11,12] by strain gauges or piezoelectric cells, so that the pressure exerted on the upper punch P_u (Pascals) and the lower punch P_l can be monotored. The closer to unity the ratio P_l/P_u is the better the tablet is lubricated. This is obviously (partly a kinetic problem, since its severity is increased with increasing speed of the tablet punches. The consolidation rate plays a part in the process, and if the time for complete consolidation does not exist, then fragmentation will take place in a structure that is not completely closely packed, and the consolidation, fragmentation, and fusion portions of the compression will occur simultaneously.

7.3.7. TABLET DURABILITY

The tablet produced must have the desired physical durability to withstand the vicissitudes of packaging, handling, and transportation. In these aspects, *hardness* is the most important quality. This is usually measured by means of a diametral hardness test: The tablet is placed (diametrally) between two anvils, and the force necessary to cause mechanical failure (breaking) is measured. This can be measured in Newtons or in arbitrary units (SCU). The latter is derived from the original instrument used in the pharmaceutical industry (the Strong Cobb Hardness tester, e.g., model no. B-48, Strong-Cobb-Arner, Inc., Cleveland 4, Ohio.). Sophisticated instruments of this type (the Heberlein hardness tester, for instance.) are in wide use today.

The systematic investigation of the diametral-compression test for pharmaceutical tablets is in great part due to the studies of Newton and co-workers.[13-16] With line loading under ideal circumstances, the values of compressive, tensile, and shear stresses can be calculated by elastic theory (assuming the tablet to be a nonporous solid). The derived maximal tensile stress σ' is:[14]

$$\sigma' = \frac{2F}{\pi d h} \tag{7.3.11}$$

where F is the load (Newton), d is the tablet diameter (cm), and h is the tablet thickness (cm). Some authors note that the tablet is not a nonporous solid and include a porosity term in Equation (7.3.11).

The test gives different types of failure,[15] and there is a sizeable scatter in results. Fell and Newton[14] have shown that the fracture strength of tablets made under identical conditions can give rise to either tensile failure (in which case

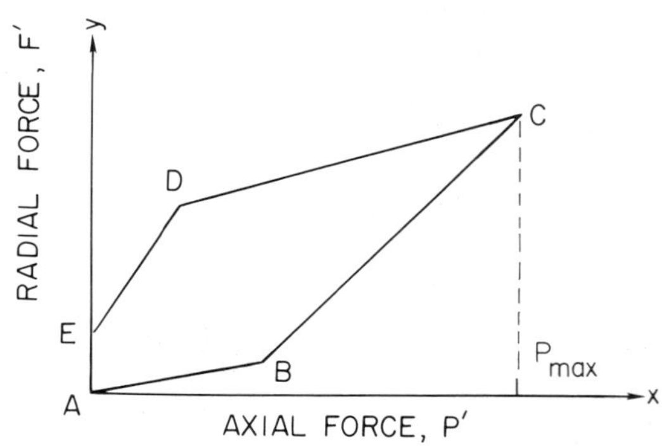

Figure 7.3.12. Radial force (or stress) as a function of axial force (or stress).

the tablet splits cleanly in two parts) or to shear or compressive failure (in which case the tablet falls into many smaller parts). If limited to tensile failure, Newton and Stanley[15] have shown that the scatter, statistically, adheres to a Weibull function.

If Eq. (7.3.11) were correct, then a plot of σ' versus P would be linear through the origin. The data of Fell and Newton[15] when plotted this way are quite linear, but require a small adjustment due to the nonzero intercept.

REFERENCES

1. J. T. Carstensen, *Pharmaceutics of Solids and Solid Dosage Forms*, Wiley, New York, p. 161 (1977).
2. J. T. Carstensen, J. B. Johnson, W. Valentine, and J. J. Vance, *J. Pharm. Sci. 53*:1050 (1964).
3. J. T. Carstensen and P. Chan, *J. Pharm. Sci. 66*:1235 (1977).
4. K. Ridgway and R. Rupp, *J. Pharm. Pharmacol. 21*:30S (1969).
5. R. L. Brown and J. C. Richards, *Trans. Inst. Chem. Eng. 38*:243 (1960).
6. F. Q. Danish and E. L. Parrott, *J. Pharm. Sci. 60*:550 (1971).
7. J. T. Carstensen and S. Laughlin, *Powder Technology 23*:79 (1979).
8. J. T. Fell and J. M. Newton, *J. Pharm. Sci. 60*: 1428, 1868 (1971).
9. M. Takieddin, F. Puisieux, P. Touré, D. Duchêne, and J. R. Didry, *Proc. First. Int. Conf. on Pharm. Technology*, APGI, Paris, *1*:248 (31 May–2 June, 1977).
10. S. Leigh, J. E. Carless, and B. W. Burt, *J. Pharm. Sci. 56*:888 (1967).
11. T. Higuchi, E. Nelson, and L. W. Busse, *J. Am. Pharm. Assoc.*, 2d ed., *43*:345 (1954).
12. E. Shotton, J. J. Deer, and D. Ganderton, *J. Pharm. Pharmacol. 15*:106T (1963).
13. J. M. Newton, P. Stanley, and C. S. Tan, *J. Pharm. Pharmacol. 29*:40P (1977).
14. J. T. Fell and J. M. Newton, *J. Pharm. Sci. 59*:688 (1970).
15. J. M. Newton and P. Stanley, *J. Pharm. Pharmacol. 26*:60P (1974).
16. J. M. Newton and D. J. W. Grant *Powder Technol. 9*:295–297 (1974).

PART 4. ROLL PRESSING, ISOSTATIC PRESSING, AND EXTRUSION

Wolfgang B. Pietsch
Maschinenfabrik Köppern GmbH & Co.KG.
Hattingen/Ruhr

7.4.1. INTRODUCTION

Pressure agglomeration, like all agglomeration techniques, participates heavily in the growth and diversification of applications. Literally all particulate matter can be or are being agglomerated by pressure agglomeration equipment. As a result, interdisciplinary contacts take place to utilize know-how and the results of research and development for design improvements. An example for the latter is the application of rolling mill and calender technology for roll presses.

Because of the wide diversification of applications, recent improvements and future developments have centered and will center around improved designs, increased capacity, better materials of construction (particularly of wear parts), improved feeding, instrumentation and automation, safer operation, and optimization of auxiliary equipment.

There is a great variety of agglomeration methods, each one corresponding to one or more binding mechanisms[1, 2] (see also part 2, section 7.2.1). According to the mechanisms involved, two categories can be distinguished.

They are: binderless agglomeration and agglomeration with the addition of binders.

Binderless agglomeration comprises all size enlargement processes that use one or more of the following binding mechanisms: Solid bridges caused by sintering; chemical reaction and/or partial melting; adsorption layers; molecular, electrostatic, magnetic, and valence forces; recombination bonds; and interlocking.

Binderless high strength agglomerates are obtained by applying high compaction forces, the main binding mechanisms being: van der Waals forces; sorption layers; solid bridges due to partial melting and solidification; recombination bonds; and interlocking. Some materials contain binders naturally, for example, the bituminous components in many coals. Since this "binder" was not added purposefully, agglomeration methods treating such materials are considered binderless.

Much particulate matter does not exhibit inherent binding tendencies; therefore, a binder must be added to secure adhesion of the solid particles. The main binding mechanisms are: Bridges of highly viscous bonding media; capillary pressure at the surface of aggregates filled with liquid; and freely movable liquid bridges.

Normally, agglomerates with low or medium strength are produced. To obtain strong bonds, subsequent drying, firing, or curing is necessary. Then the following binding mechanisms can apply: Solid bridges caused by sintering; chemical reaction; partial melting and solidification; crystallizing of dissolved substances; and hardening binders.

The simplest agglomeration method, which usually requires the addition of binders, is the buildup of aggregates during an appropriate state of motion in the particulate matter system (Fig. 7.4.1). Relatively loosely bound spherical agglomerates with, in most cases, a broad range of sizes and often uneven surfaces are formed from particulates during tumbling in rotating drums, discs, or cones and mixers, movement in vibrating or pulsating machine troughs, and turbulent motion in fluid beds or suspensions. Strength and quality of the green (freshly produced) agglomerates can be improved to some extent by specific handling procedures (rolling) and curing or firing (sintering).

A very versatile method is pressure agglomeration. Some techniques use high forces (Fig. 7.4.2) to compact particulate matter into tablets, compacts, and briquets of defined size and shape or into dense, flat, or corrugated sheets. Under the acting pressure the solid particles approach each other closely resulting in strong bonds; even fusion may take place at the grain boundaries. Normally, high pressure compaction produces agglomerates that exhibit high strength instantaneously. Binders are not needed in many cases.

Another pressure agglomeration method is a relatively low-force application which is often called "pelleting" (Fig. 7.4.3). During this process, materials with sufficient binding tendency or those containing binders are forced through open dies or holes in dies of different shapes. Normally, the pellets or briquets formed this way are cylindrical with a defined diameter and variable length. Often such products require curing to gain final strength.

7.4.2. MECHANISM OF COMPACTION

The production of a powder tablet, compact, or briquet can be carried out by a number of techniques, the purpose of which is usually to form the powder into a more or less well-defined shape. Within each method many routes are possible, each resulting in the manufacture of different types of products with respect to size, shape, and physical properties. However, all have in common a basic compaction mechanism.

When a particulate solid is placed into a die and pressure is applied, a reduction in volume will occur due to the following mechanisms (Fig. 7.4.4):

1. At low pressures, rearrangement of the particles takes place, leading to a closer packing. At this stage, energy is dissipated mainly in overcoming particle friction, and the magnitude of the effect depends on the

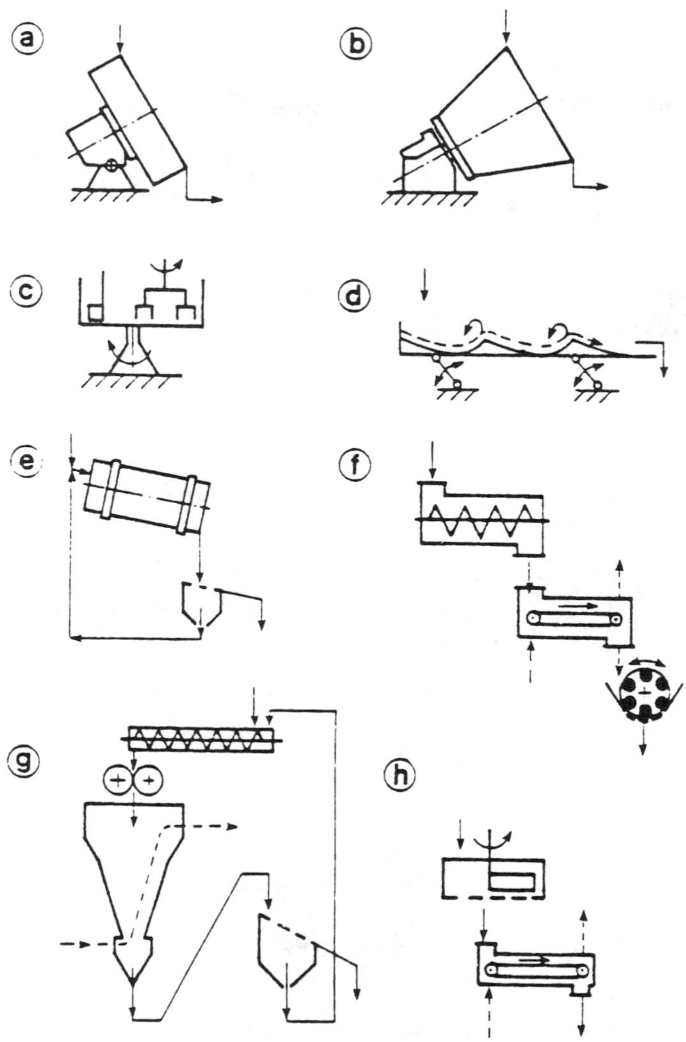

Figure 7.4.1. Schematic representation of different balling techniques: *a*, balling disc; *b*, balling cone; *c*, batch mixer; *d*, vibrating trough conveyor; *e*, balling drum (closed circuit); *f*, continuous mixer, dryer, and granulator (mill); *g*, prilling tower; *h*, sieve granulator and dryer (curing).

coefficient of interparticle friction. In the case of fine powders, cohesive arches may collapse at this stage.

2. At higher pressures, elastic and plastic deformation of the particles may occur, causing particles to flow into void spaces and increasing the area of interparticle contact (A). Interlocking of particles may also occur. For materials of low thermal conductivity and low melting point, the heat generated at points of contact may be sufficient to raise the local temperatures to a point where increased plasticity and even melting facilitate particle deformation.

With brittle materials, the stress applied at interparticle contacts may cause particle fracture followed by rearrangement of the fragment to give a reduced volume (B).

3. Stages A and B continue until the compact density approaches the true density of the material. Elastic compression of the particles

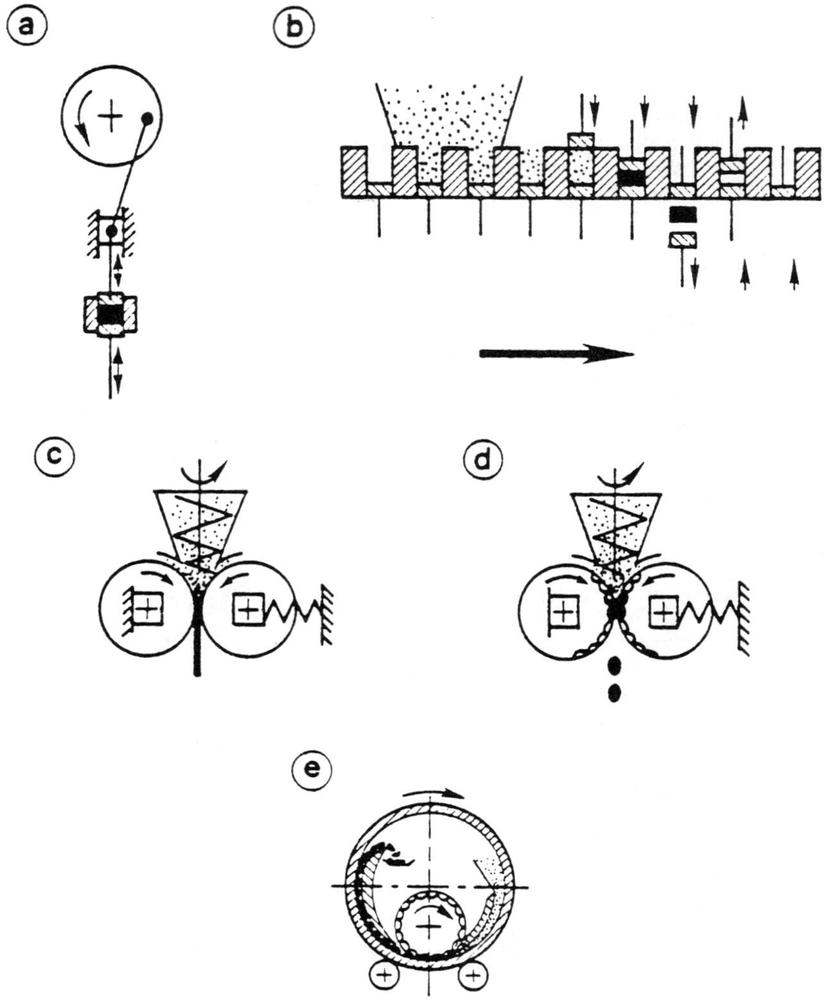

Figure 7.4.2. Schematic representation of different pressure agglomeration methods: *a*, reciprocating piston press; *b*, rotary tableting machine; *c*, roll press (compacting); *d*, roll press (briquetting); *e*, ring roll press.

and entrapped air will be present at all stages of the compaction process.

The mechanisms discussed may occur simultaneously. The relative importance of the various mechanisms and the order in which they occur depend on the properties of the particles and on the speed of pressing.

The aim of compaction is to bring small particles into sufficiently close contact so that the forces acting between them are large enough to produce a product that has sufficient strength to withstand subsequent handling. Therefore, it is often necessary to carry the compaction into the bulk compression stage, in which the stressing is hydrostatic in character. Broken or deformed particles are no longer able to change position because of the few remaining cavities, and a certain amount of interparticle conformity has been achieved. With increasing pressure the apparent density will gradually approach that of the theoretical density. The rate of this approach depends on the yield point of the material. Brittle materials are more difficult to densify to a high degree by pressure only because fragmentation decreases due to the hydrostatic pressure conditions and higher strength of smaller particles. When

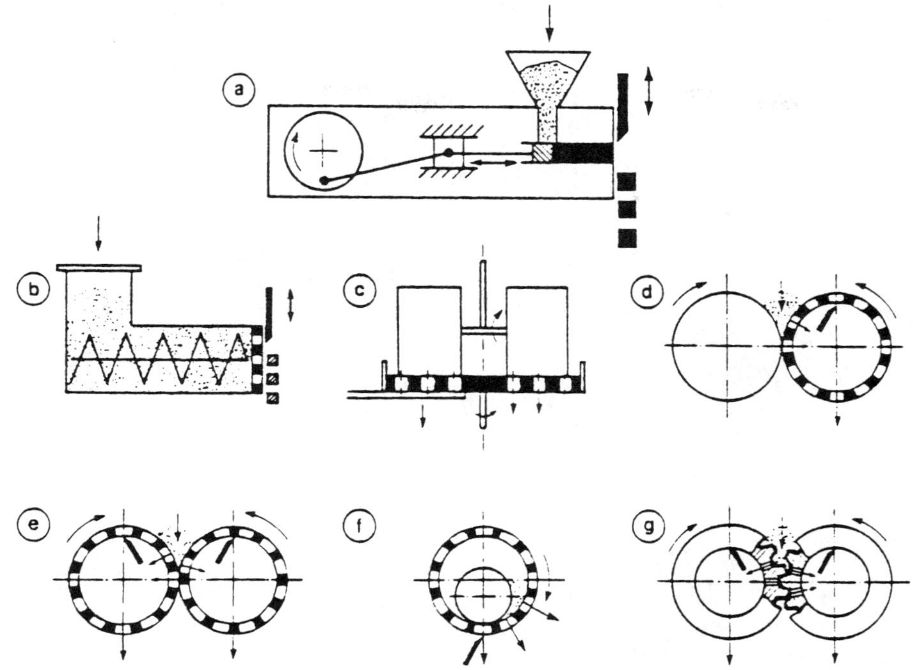

Figure 7.4.3. Schematic representation of different pelleting machines: *a*, extrusion press; *b*, screw extruder; *c*, muller with flat perforated die; *d*, roll pelleting machine (one hollow roll); *e*, roll pelleting machine (two hollow rolls); *f*, roll pelleting machine (internal press roll); *g*, gear type pelleting machine.

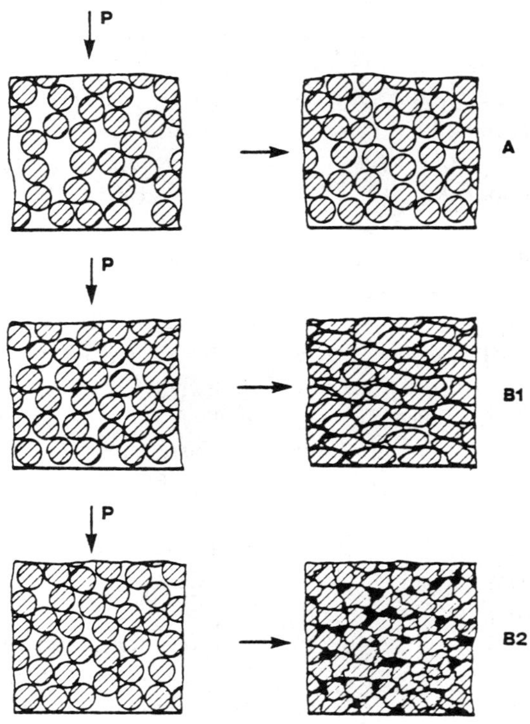

Figure 7.4.4. Mechanisms of compaction.

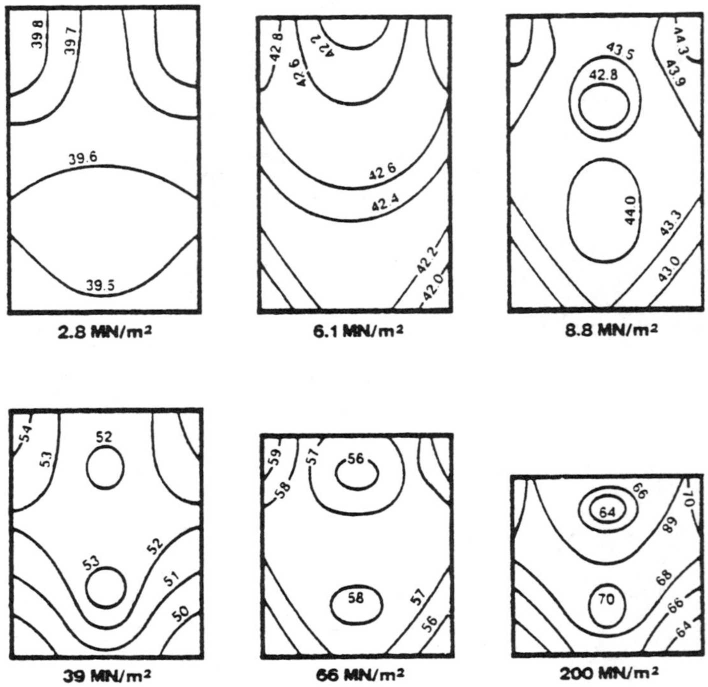

Figure 7.4.5. Density distributions in cylindrical compacts.[3]

porosity becomes fully disconnected, the isolated pores may set up considerable internal gas pressures which, together with stored elastic energy, can contribute to the disintegration of compacts if the pressure is released too quickly.

If a particulate solid were compacted in a cylindrical die with frictionless walls, it is expected that the pressure exerted by the piston would be transmitted throughout the material giving uniform pressure and, therefore, uniform density throughout the compact. In practice, the presence of frictional shear forces at the wall leads to a nonuniform pressure distribution causing variations in the density of the compact (Fig. 7.4.5). These variations are present in products from all pressure agglomeration techniques and lead to weakening of the compact. If a sintering step follows, distortion is possible due to differences in the amount of contraction occuring at the positions of different density.

Figure 7.4.5 shows the highest density at the top edge of the compact and the lowest density at the bottom edge. A region of high density occurs near the axis a short distance above the bottom of the compact. In some cases the density in this position is higher than that observed near the axis at the top of the compact.

The general conclusion from investigations into the effects of operating conditions of pressure agglomeration equipment are that density variation:

- increases with the applied pressure and with the height of the specimen for constant diameter,
- decreases with increasing diameter even at constant height-to-diameter ratio,
- is slightly reduced by the addition of a lubricant to the powder, and
- is considerably reduced by lubricating the die walls or tools.

Segregation during feeding and filling also leads to density variations due to local changes in size distribution and, in the case of mixtures, to differences in the plasticity and friability

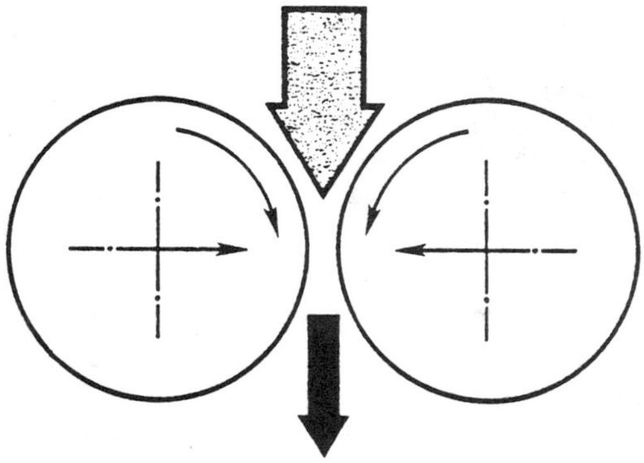

Figure 7.4.6. The principle of roll pressing.

of the component materials. Since there is evidence that radial flow of powder during compaction is negligible, it is expected that variations in density before compaction have an appreciable effect on the uniformity and quality of the compact.

A knowledge of the relationship between compacting pressure and density is important because pressure or force, more than any other factor, controls the attainment of high density, high strength, and low porosity in green compacts and markedly influences the same properties in the final product. A number of empirical formulas has been proposed to describe the pressure-porosity relationship; however, none of these formulas is universally applicable, giving acceptable results over a limited range of pressures only.

7.4.3. ROLL PRESSING[4]

General

Roll pressing[4] of particulate matter is a pressure agglomeration technique that is of growing interest for all industries in which large quantities of finely divided solids, both valuable and worthless, must be handled. Originally developed as an economical method to agglomerate coal screenings, today this technology is being applied for the treatment of a large number of materials in the chemical, pharmaceutical, food processing, mining, mineral, and metallurgical industries. This versatile technology lends itself to such different uses as compaction and granulation of highly heat- and pressure-sensitive pharmaceutical materials such as pancreatin or penicillin, briquetting of extremely corrosive and poisonous materials such as sodium cyanide, compaction and granulation of large tonnage materials such as fertilizers, and briquetting of crude hot materials such as metal chips and turnings, ores, or "sponge iron" at temperatures of up to 1,000°C.

Historical Background

The roll type briquet machine or "Belgian" press was developed in Europe during the last half of the past century. The object of its invention was to find an economic method for agglomerating coal screenings.

The first successfully operating machine was constructed by the Belgian Loiseau and was installed in a briquetting plant at Port Richmond, USA, in the late 1870s. Similar machines were built shortly thereafter by Fouquemberg, Zimmermann & Hanrez, Schüchtermann & Kremer, and others. The first American roll press was constructed by G. J. Mashek.

Although some of the first machines and several others built shortly thereafter are still in operation, the construction and application of roll-type machines are again of growing interest as a means for compacting and/or briquetting a great variety of chemical and metallurgical products. For simple applications, today's machines use the original design principles whereas others employ sophisticated features to cope with extremely difficult process conditions.

The particular merit of the roll-type machine is its ability to compact large quantities of material at low cost. Other considerations may be the ease with which it can be made dusttight for briquetting toxic chemicals or the great variety of materials from which the rolls can be constructed for those cases where corrosive or abrasive resistance is of importance. Of more recent importance is its usefulness for compacting hot materials with temperatures up to 1,000°C.

Roller presses do not produce compacts having the same fine detail and uniformity as those made with tableting machines or other machines that compress materials between punches in a closed die. The flashing or web that is usually around the edges of the briquets may be objectionable in some cases, too. Because of that, normally, roll presses find their natural field of application where low capital and operating costs are of more importance than the uniformity of the product.

Roll Pressing Equipment

The most widely used machines are double roll presses that achieve compaction by squeezing material between two rolls (Fig. 7.4.6), much in the manner of a rolling mill. Pockets or indentations cut into the surfaces of the rolls form briquets or compacts. Between smooth rolls, material is compacted into dense sheets. In the early machines, and in many application today, the material to be briquetted is gravity fed into the nip of the rolls and distributed over the roll width by simple mixing—that is, by distributing devices mounted on top of the press.

The process occuring during compaction of particulate matter in roll presses is described and interpreted by different authors in a rather similar way. The feed mechanism is characterized by the pressure caused by gravity or a force feed system and the friction between material and roll surface. Compaction between two rolls may be explained by dividing the roll nip area into two zones: the feed zone and the compaction zone.

As depicted in Figure 7.4.7, showing a smooth roll press, the feed zone is defined by the two angles α_E' and α_E. In the feed zone, the material is drawn into the nip by friction on the roll surface. Densification is solely due to the rearrangement of particles (Fig. 7.4.4A). The density of the feed is characterized by the bulk density ρ_0 and reaches the tap density ρ_t at the point α_E. The peripheral speed w of the rolls is higher in this zone than the velocity u of the material to be compacted. α_0 is the so-called angle of delivery which is defined by the width h_0 of the feed opening above the rolls.

The compaction zone follows after the heavy solid line (Fig. 7.4.7). α_E is the angle of rolling, the gripping angle, or angle of compaction. In

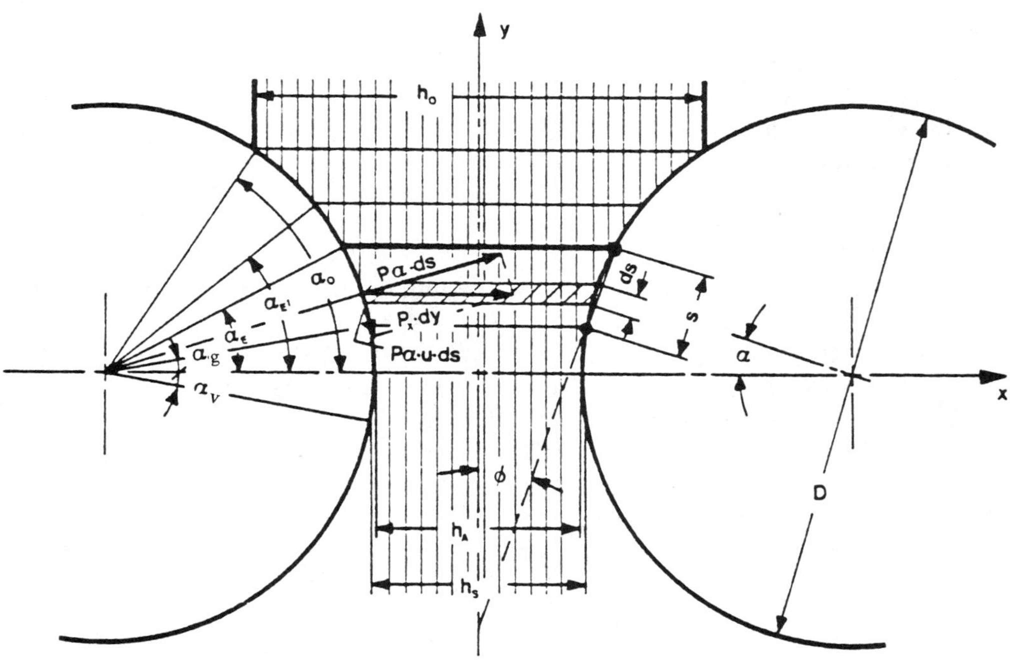

Figure 7.4.7. Compaction of particulate matter in the nip of a smooth roll press.

the compaction zone the pressing force becomes effective and the powder particles deform plastically (Fig. 7.4.4B1) and/or break (Fig. 7.4.4B2). α_g is the neutral angle where the sign (direction) of the friction force changes. At this point, the pressure in the material and the density have their highest values.

α_V is the angle of elastic compression of the rolls that determines the thickness h_s of the compacted sheet. α_V becomes zero and the sheet thickness h_A if the elastic deformation of the rolls can be ignored. However, in most cases the strip is even thicker than h_s due to elastic recovery of the compacted material. The angle corresponding to this actual outlet plane is called angle of release α_R.

Much of this work is phenomenological in character. A comprehensive theory of roll pressing of particulate matter is not yet available.

In the ring roll press, an important alternative to the double roll press has been developed for high-pressure work. The particulate matter, normally powdered coal, is pressed between a roll and the inner surface of a ring (Fig. 7.4.8). Thus, a very narrow angle of entry is achieved, and with it, of course, considerable drag, which obviates forcible feeding of the powdered coal. Such a system has many advantages, but also some disadvantages that have not yet been completely overcome.

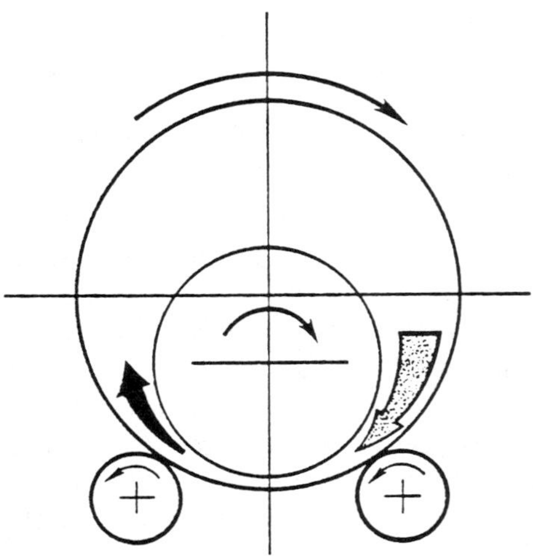

Figure 7.4.8. Operating principle of the ring roll press.

Present Position and Future Trends

Although the technique of roll pressing is over a hundred years old, it is still more an art than a science. Many details of the process are not yet fully understood. A basic understanding of the roll pressing process in terms of the fundamental mechanisms involved is indispensable in guiding steps to improve the performance of existing plants, for optimizing the design of future plants processing conventional materials, and for extending the process to materials currently considered not capable of being handled by roll presses. Considerable effort has already been expended in obtaining this understanding; however, much of this work has been directed to specific materials and situations and may not have general application. Consequently, there is still a need for extensive fundamental work.

Such work falls into two areas. First, attaining a knowledge of the general deformation characteristics of materials under various conditions of stress and, second, to explain the behavior of materials with different deformation characteristics in geometrically different presses and under various operating conditions.

More practical problems that need investigation are improved methods to calculate roll force, roll torque, and roll power, the minimum roll diameter, and the maximum roll speed. This is particularly true for pocketed rolls and for machines applying force feeders. Especially the latter case needs research because, so far, the optimum force feeding pressure and the feeder design for each application are determined by specific experimental laboratory tests, using a sample of the actual feed material and different feeder arrangements, which are limited by their availability in the respective laboratory.

The economics of many roll press applications depend on the operating cost and here mainly on the wear cost, particularly if pocketed rolls are being used. So far, the only reliable way to check the wear situation is to run a commercial unit under continuous production conditions for a prolonged time. This method is expensive and time consuming. Also, during such a test there is only a limited possibility of investigating a variety of roll materials and different roll surface configurations simultaneously. Therefore, a simulation method should be developed that makes it possible to test the wear characteristics of different roll materials and roll configurations under processing conditions, but with a small amount of material and in a short period of time.

7.4.4. ISOSTATIC PRESSING[5]

General

Isostatic or hydrostatic pressing is the compaction of a powdered material into predetermined shapes by the application of pressure via a fluid through a flexible mold. The arrangement may be such that the flexible tool contracts or dilates by the application of the pressure. Isostatic pressing covers liquids and gases as the pressure transmitting medium, whereas hydrostatic pressing is best reserved for liquids. However, the two terms are used freely to cover both aspects. Depending on whether the flexible tool is an integral part of the press or removed from the pressure vessel after each compaction cycle one distinguishes between the "dry" and "wet bag" process.

Isostatic pressing using gases as pressure-transmitting medium is still in development and practiced by only a few. It is particularly attractive at high temperatures where compaction and sintering are combined into one operation, i.e., isostatic hot pressing. The preform produced by cold isostatic pressing is in most cases further consolidated by sintering, forging, extrusion, rolling, etc. When the economics of isostatic pressing are considered, it must be in relation to the final product and not for the shaping operation alone. The advantages often lie in a better product and reduction in final machining requirement.

Historic Background

In 1913, Madden first described an isostatic pressing technique in a U.S. patent assigned to the Westinghouse Lamp Co., the method clearly being developed to overcome the limitations of die-compacted billets. Madden claimed that isostatically pressed billets were uniformly compacted, devoid of strata, and possessed sufficient green strength to permit handling. Further patents were taken out on the isostatic pressing of refractory metal powders by Coolridge in 1917 (for tubes of tungsten and molybdenum), and by Pfanstiehl in 1919, while Fehse described the wet bag isostatic pressing of tungsten tubes in 1928. Almost no subsequent development of the isostatic pressing of refractory metal powders appears to have occurred and little further interest was shown in isostatic pressing until the 1930s and early 1940s when a series of isostatic techniques was described by Jeffery (1932–1942) and Daubenmeyer (1934) in patents assigned to the Champion Spark Plug Company.

During the same period, Fessler and Russell patented a technique for pressing spark plug insulators by direct compression isostatic pressing. These workers cited the low number of rejects, rapidity, and the need for only a limited amount of equipment as economic advantages of isostatic pressing.

By 1942, most of the advantages of isostatic pressing had been recognized, and the basic principles in common use today had been established, i.e.,

- The wet bag pressing of large or complex shapes in which the flexible tool is filled externally and subsequently immersed in the fluid,
- The dry bag pressing of smaller, regular shapes in which the tool forms an integral part of the pressure vessel.
- The use of rigid formers to produce accurate internal or external surfaces, and
- Pressurization by pumped systems or by direct compression with punches in a die.

Materials that had been pressed included ceramics, metals, and cermets.

In recent years, fully automatic dry bag presses for producing small ceramic components have been developed, while semiautomatic wet bag presses have been used to manufacture large and sometimes complex components with reasonable dimensional accuracy and requiring only minor trimming to produce the final form. The size of pressure vessels has increased greatly. Additional materials now being isostatically pressed include plastics (particularly PTFE), explosives, and chemicals. It has been reported that isostatic pressing is being developed for the food and pharmaceutical industries.

Hot isostatic pressing, including so-called gas pressure bonding, has only been seriously developed during the last twenty years. This technique has been developed for two main research applications: the solid-state diffusion bonding of components of various metals and cermets, and the hot compaction of metal, ceramic, and cermet powders. However, hot isostatic pressing has remained confined to special applications for which the high operating costs and low rates of production are of less importance.

Isostatic Pressing Equipment

Isostatic powder compaction equipment consists of a pressure vessel, pumps to generate the necessary hydraulic pressure, and related equipment to enable effective and safe machine operation.

Many pressure vessels used for isostatic pressing fall into the high-pressure range and would be classified as thick walled pressure vessels. The literature on this subject is quite considerable but, unfortunately, in many countries there are no rules covering minimum requirements for the design, fabrication, operation, inspection, and certification of such vessels. Design and fabrication of high-pressure vessels are very much in the domain of specialists who are relying to a great extent on their past experience.

The time to reach the required pressure depends on a number of factors, i.e., volume of the cavity, volume and compaction ratio of powder and tool, compressibility of the fluid, and delivery rate of the pumping system. In order to speed up pumping, it is possible to use a number of pumps in parallel. Alternatively, a pump system using different types of pumps to reach different pressure levels may be designed.

Air-driven and hydraulically driven pumps can be built easily in a variety of modules for various demands. It is simple, therefore, to change the pumping requirements by changing the intensifier (Fig. 7.4.9) or increasing the number of intensifiers.

Most isostatic presses operate satisfactorily up to 400 MN/m^2 on an oil/water emulsion or hydraulic oil; for higher pressures special fluids may have to be used, but the tools used must be compatible with these liquids. Problems can arise when it is necessary to dispose of contaminated fluid after each pressing operation. Such contamination may originate from powder adhering to the external tool walls or from a tool bag failure.

To be effective an isostatic press must be joined with equipment that fulfils some or all of the following functions: filling and consolidating the powder in the tool; loading and unloading the tool set into the vessel; handling, i.e., insertion and removal of the vessel's closure; controlling the pressure in the vessel; and stripping the compact from the tool.

The difference between wet bag and dry bag pressing is illustrated in Figure 7.4.10. In the dry bag process, the flexible tool is fixed in the pressure vessel, and the powder can be loaded without the tool having to be removed from the vessel. The tool thus forms a membrane between fluid and powder; optionally, the tool being used can be placed inside a primary diaphragm so that it never comes into contact with the fluid.

Dry bag tooling is used for the production of small components at a fast rate. It is common to make provisions for loading the powder

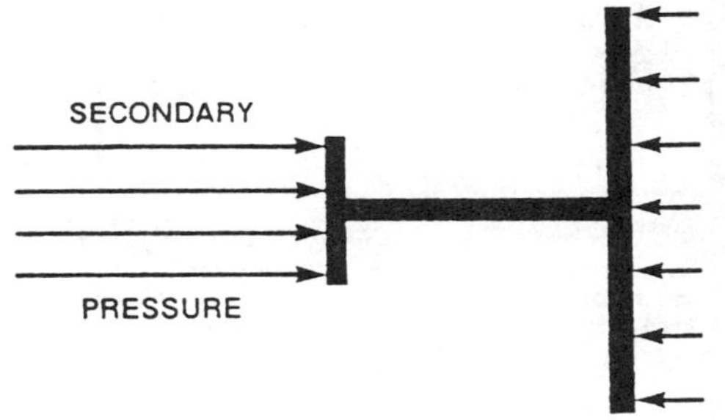

Figure 7.4.9. The principle of intensifiers.

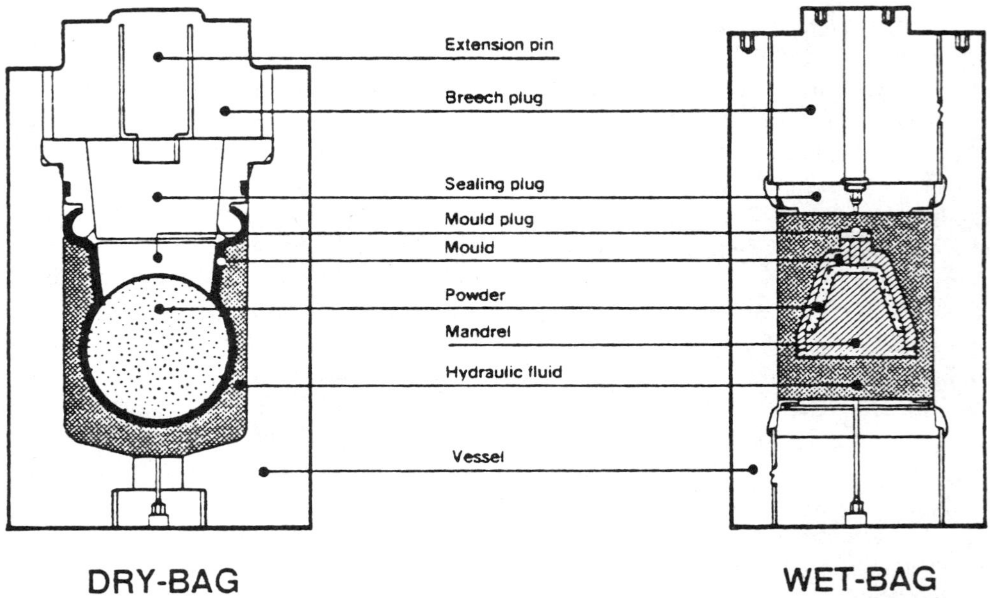

Figure 7.4.10. Schematic representation of the difference between dry bag and wet bag pressing.

automatically into the tool, dispensing an accurately premeasured quantity. The automatic filling, the permanent location of the tool, and the smaller fluid volume required make for faster operation. Dry bag tooling has also the advantage that the fluid cannot be contaminated with powder. However, because the tool has to stand up to many pressing cycles and since tool changing is time consuming, it has to be made of a very durable material.

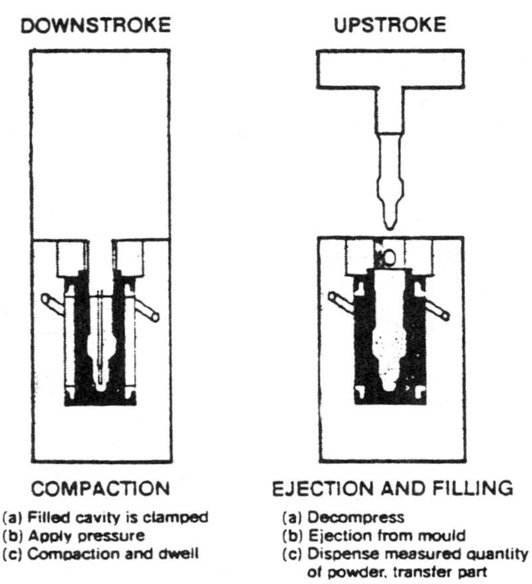

Figure 7.4.11. Operational sequence of a "densomatic" press (Olin Energy Systems Ltd.).

Where mass production of simple small powder compacts (e.g., spark plug insulator blanks, grinding media, carbide tools, electrical insulators) is required, the equipment usually takes the form of a battery of small presses generally similar to and operationally having a great deal in common with a conventional hydraulic press.

It has proved relatively easy to make such a machine for automatic operation with production rates up to 90 components per minute being claimed. The compaction, ejection, and filling is demonstrated on a spark plug insulator in Figure 7.4.11. The production rates depend on maximum pressure, size of component, powder properties, and number of tool cavities.

The larger automatic units that have been developed include a rotating pressure chamber system where loading of powder, pressurizing, depressurizing, and unloading of the compact are carried out automatically at various work stations during the complete cycle. It has also proved possible to automatically operate certain functions on large isostatic presses. For instance, loading and unloading of the tooling, insertion and closure of the breech, pressurization depressurization of the vessel, all have been carried out automatically by ingenious arrangements of mechanisms and controls. Figure 7.4.12 shows the operational sequence of a three-station rotary automatic press capable of pressing parts at rates up to 300 parts per hour.

Present Position and Future Trends

In general, the development of isostatic pressing has been comparatively slow, particularly for metal powders, and even today the technique is still regarded as an alternative to be used only when the technical limitations of conventional methods are too restrictive. Until recently, only the ceramic manufacturers have commercially exploited isostatic pressing, but only to a limited extent and mainly in the United States. In addition, isostatic pressing is suitable for producing high-purity ceramics and long ceramic tubes, for which there is an increasing demand. In contrast, the common metals can be readily formed by long-established methods such as casting, rolling, forging, or extrusion and only recently have metal fabricators begun to look more closely at the feasibility of isostatic pressing.

Isostatic pressing also shows great promise of becoming an established production technique for the fabrication of components from PTFE and high-molecular-weight polyethylene. PTFE, for instance, although a thermoplastic, has a very high melt viscosity, which precludes satisfactory processing by established injection moulding and extrusion techniques. This has led to the adoption of techniques used in powder metallurgy, which involve initial cold compaction and subsequent sintering, of which isostatic moulding is the latest.

7.4.5. EXTRUSION

General

The term *extrusion* has a wide range of meanings. In the present context, it extends from compaction such as in coal briquetting to pelleting of particulate matter with inherent binding characteristics, from forming pastes to shaping hard ceramics such as alumina using binders. Extrusion uses open-ended dies in which the wall friction replaces the resistance of a closed die. In some cases, the column of

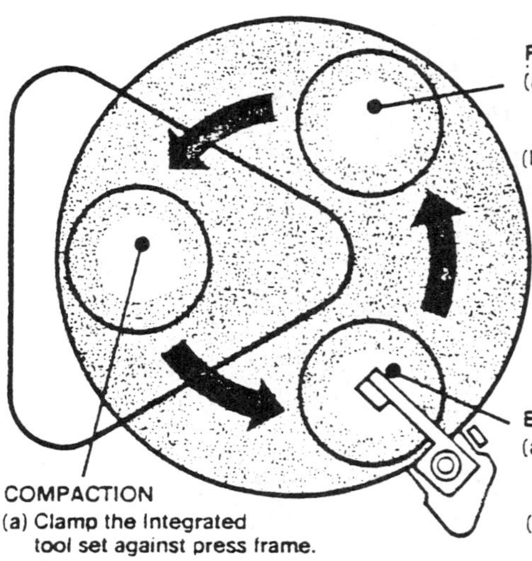

Figure 7.4.12. Operational sequence of an automatic, rotating isostatic press (Olin Energy Systems, Ltd.).

already finished extrudates or briquets provides an additional back pressure. The differences between these forms of extrusion and their relation to powder compaction in closed dies are illustrated diagrammatically in Figure 7.4.13.

Powder consolidation by extrusion takes two main forms: In one the powder is shaped dry, whereas in the other the process is assisted by the presence of a liquid "lubricant" or binder. In the latter case, the materials are limited to those in which the liquid content is never more than sufficient to fill the pores between the solid particles.

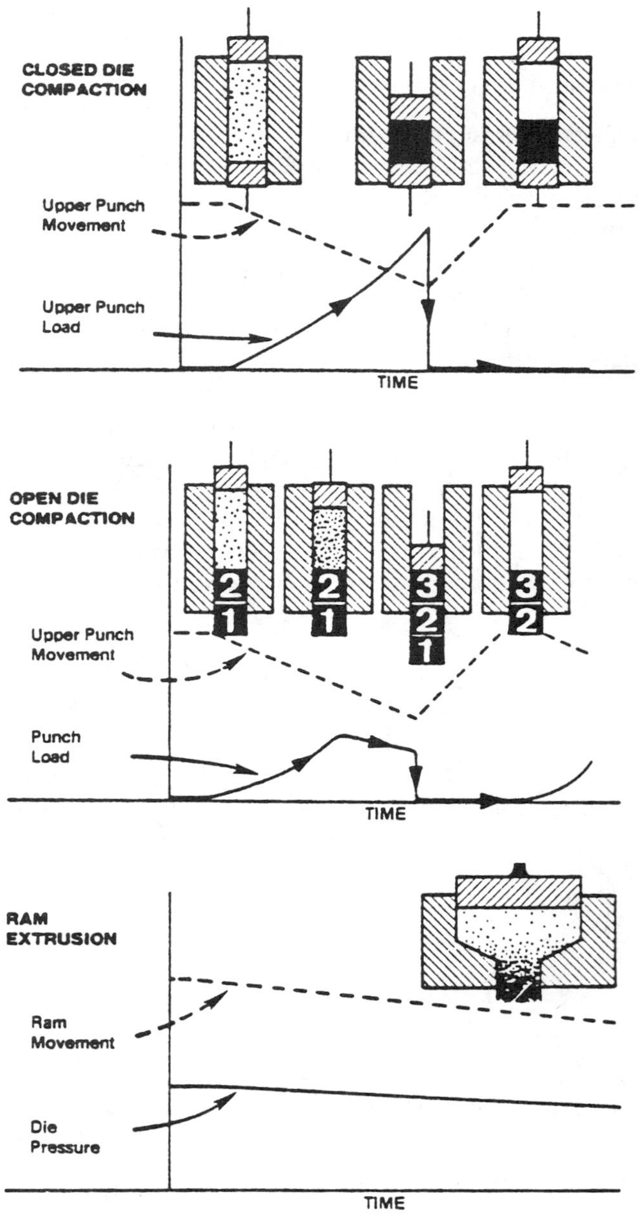

Figure 7.4.13. Differences between two forms of extrusion and compaction in closed dies.

Extrusion has the advantage over die compaction of being able to continuously shape uniform, mostly cylindrical or polygonal but sometimes also more complex cross sections from rigid particles with little loss of porosity. It does not generally impart much strength but relies on subsequent steps such as cementing, drying, or sintering to do this.

Historic Background[6]

The modern history of pressure agglomeration by extrusion begins in the United States as early as 1848. At that time, Easby obtained a patent for a method of converting coal into briquets by pressure alone. In Germany, the industry was established in the second half of the nineteenth century. The press used had been developed by Exter originally for the manufacture of peat briquets in Ireland. It was introduced into Germany for the briquetting of brown coal in 1858, and although modifications have been made, essentially the same press is being used today to produce approximately 90 percent of the world's output of brown coal briquets. In spite of this, unlike the situation in the bituminous coal briquetting industry, where roll presses are being applied, a great deal of research has taken place that has added much to the technique of briquetting by extrusion and to an understanding of the mechanisms.

The different types of so-called pelleting machines were developed during the first part of this century to shape often precooked and/or enriched mixtures of animal feed into optimal feed quality and sizes.

Extrusion Presses

To illustrate the extrusion compaction of soft materials, the "extrusion briquetting" process employed in the coal industry may be taken as typical. The (Exter) press consists essentially of a channel through which coal is pushed by means of a reciprocating ram (Fig. 7.4.14). Each backward stroke of the ram permits powdered coal to fall into the channel, and each forward stroke compacts the powder and pushes the compact or briquet along the channel. The frictional resistance of the die and the column of extruding or extruded material provides the necessary back pressure.

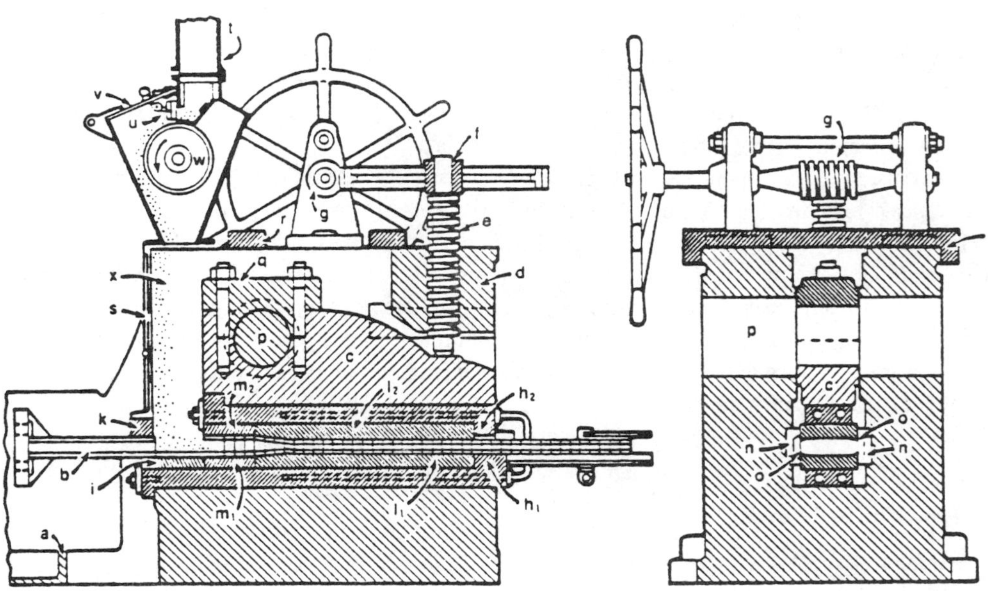

Figure 7.4.14. Cross sections through the "Exter" press.

In a closed mold, the development of a predetermined pressure presents no difficulty, but in an extrusion press, the situation is complicated. The peak pressure developed at each stroke depends not only on the power of the ram, but also on the restriction imposed by the channel to the forward movement of the briquetted coal. This movement depends on many things: the shape and length of the channel or die, its smoothness, the nature of the coal, its temperature and the type and length of the open cooling channel. The rate of development of the pressure is also important and depends on the frequency of the stroke and its length and on the rather complicated relationship between the movement of the ram and the magnitude of the resisting frictional force between the briquet and the channel. This frictional force changes with both the state of compaction and the rate of movement.

Soft and hard materials that are not easily compressed are extruded as pastes. Those sufficiently plastic are generally auger extruded, but less easily deforming substances often require the higher pressures obtained with hydraulic rams or other effective mechanical means as used in the different types of pelleting machines (Fig. 7.4.3). The working principle of this equipment is similar to that of the extrusion briquetting process described above. In most cases the materials have sufficient plasticity and binding characteristics that pellets must be cut to length by fixed or moving cutting blades. In contrast to closed die presses, extrusion presses operate at substantially constant peak load irrespective of the die fill. Therefore, accurate powder metering equipment is not required provided the variation in compact length can be tolerated. By using long stroke and/or long channel presses, very low density material can be handled. In extreme cases, auger precompactors are used to feed the press die.

Present Position and Future Trends

The particular advantage of extrusion presses is that, if desired, materials can be shaped from soft and even pasty mixtures with very little change in density. For this reason, pelleting machines, originally developed for the production of animal feeds, find increasing applications in manufacturing catalyst carriers, in shaping food stuffs, and in granulating pastes. The extrusion briquetting process finds new applications in the reclamation of wood wastes (saw dust) and in processes seeking new forms of energy based on coal.

7.4.6. GENERAL STATE OF THE ART AND DEVELOPMENT TRENDS

The basic principle of pressure agglomeration is very old. Although not specifically documented, the first application was most probably in medicine for the production of pills. The "machines" were simple, hand-operated die presses. Because the compaction forces were low, the mixtures to be tableted were moistened and bonding resulted from liquid bridges in the green (fresh) state and from recrystallized soluble matrix materials after curing (hardening).

A diversification of pressure agglomeration techniques occurred in the nineteenth century from piston and die presses to other forms of compaction of particulate matter. Except for more recently included modern design features, the basic principles of all pressure agglomeration equipment are at least a century old. In spite of that, most are still more an art than a science.

The best researched field is die pressing where the most important application is still in the pharmaceutical industry. The nature of the product, pharmaceutical formulations, calls for extremely accurate weight control and high compact quality. This together with the demand for higher outputs resulted in considerable research and development efforts, which led to very sophisticated controls and the design of high-accuracy, high-speed tableting machines.

Most other pressure agglomeration methods do not produce compacts having the same fine detail and uniformity as those made with tableting machines or certain isostatic presses.

Often very old machines are still in use with only the drives modernized and many new ones resemble in their basic design machines that were built 50 to 100 years ago.

With modern solids-processing technologies, finely divided particulate matter is obtained as the intermediate or final product in an ever increasing proportion. For example, in many areas the rich mineral resources are depleted and the poorer ores require beneficiation. In most cases, fine grinding is a step in the upgrading of such ore. Before the fine concentrates can be used, they must be agglomerated to improve handling and processing and to reduce losses. Another example is the increasing requirement for ultrapure materials that often precipitate or crystallize from solutions. After drying, such powders must be agglomerated before they can be further processed. With those technologies producing or using finely divided particulate matter, environmental control systems must be employed to prohibit or reduce contamination of air and water. New antipollution regulations often require agglomeration of dust from dust collection systems or sludge from water treatment plants for safe handling or disposal even if they are worthless wastes.

All these general trends in solids processing have greatly stimulated the importance of the unit operation agglomeration. In addition, many other new developments require the step agglomeration in the processing cycle. One example is the trend to higher capacities and automation of industrial plants. Often, the accurate and reliable automatic feeding of solids at high rates is only possible if powder materials have been agglomerated into a suitable, free flowing, granular, or particulate material.

One result of this fast growing, very diversified need for agglomeration, is the interdisciplinary exchange of results of research and development and the application of certain agglomeration techniques in, so far, rather foreign fields. Two examples illustrate these trends.

The development of ultrahigh-speed tableting presses was hampered by the requirement of accurately feeding the dies with the required amount of material. Because the doses of active medicine in the mixture are often very small, the particle size of the matrix material must be small, too, to avoid segregation. However, such powders do not flow easily. Originally, this problem was overcome by compacting the powder on slow piston presses into "slugs" (large tablets) which then were granulated (crushed) and screened into a free-flowing granular material suitable for feeding ultrahigh-speed tableting machines. Because of the high capacity of the tableting machine and the relatively small yield of granular material after crushing, several "slugging presses" were often necessary to feed one single tableting press. A modern solution is to use modified roll presses, which were originally developed and used for very crude applications in the coal and minerals industry, to continuously compact the pharmaceutical mixture. The compacted sheet is granulated into the free-flowing tableting machine feed and the undersized material is recirculated to the compaction. Because of better feeding characteristics, the continuous operation of roll presses, and easy cleaning, one such system can serve a large number of high-speed tableting machines.[7]

On the other hand, tableting machines are now being used in the fertilizer industry to produce accurately dosed, special fertilizer rations for flower and vegetable gardening and farming.

REFERENCES

1. H. Rumpf, "The Strength of Granules and Agglomerates," *Agglomeration*, edited by W. A. Knepper, Wiley, New York, pp. 379–418 (1962).
2. W. Pietsch, "Granulieren durch Kornvergrößerung," *CZ-Chemie-Technik 1*(3):116–119 (1972).
3. D. Train, "Transmission of Forces through a Powder Mass during the Process of Pelleting," *Trans. Inst. Chem. Engrs. 35*(4):258–266 (1957).
4. W. Pietsch, *Roll Pressing*, Heyden, London (1976).
5. P. Popper, *Isostatic Pressing*, Heyden, London (1976).
6. D. C. Rhys Jones, "Briquetting," *Chem. of Coal Util.*, Supplementary Volume, edited by H. H. Lowry, Wiley, New York, pp. 675–753 (1963).
7. W. Pietsch, "Granulation Techniques for Pharmaceutical Applications," *Drugs Made in Germany 13*(2):58–66 (1970).

PART 5. AGITATION METHODS

C. E. Capes and A. E. Fouda
National Research Council
Ottawa, Ontario

7.5.1. INTRODUCTION

Virtually all solids mixers are capable of forming agglomerates when processing fine powders mixed with a wetting liquid. The agitation methods of size enlargement most often used for large tonnage applications make use of the tumbling, rolling, cascading action produced in disc, drum, and cone devices. This equipment has been discussed briefly in Appendix 7.5.7. In this part, alternative methods of mixer agglomeration are considered.

Mixer agglomeration can be broadly classified into two major groupings according to the size, density, and state of wetting of the agglomerates produced. In the first group, the agglomerates are similar in physical characteristics to those produced by tumbling. Dense, capillary-state agglomerates are formed by using agitator/internals within the mixing vessel to provide a positive rubbing and shearing action. Hignett[1] claims that the agglomerates made in this way are harder and stronger than those produced by tumbling. Among other advantages over tumbling methods is the ability to process plastic, sticky materials and greater tolerence in accommodating variations in operating conditions. Less wetting phase can be used in a mixer than in a tumbling device.[2] Disadvantages include generally higher maintenance and power requirements and an irregular product that may require further shaping, as in a tumbling dryer.

In the second group of mixer agglomerators, powders are moistened to a lesser degree than the wet capillary state. The product is in the form of weak clusters and the technique is suitable, for example, to produce "instantized" food products.

Specialized equipment has been developed for each of these two major groupings. Some mixers, however, are suitable for both methods. Some of the most common equipment in mixer agglomeration are discussed in the following examples.

7.5.2. PAN MIXERS

The horizontal pan mixer shown in Figure 7.5.1 was used primarily as a batch mixer-granulator in the early development of fertilizer granulation. A typical pan might be 2.3 m in diameter and 0.5 m deep and contain a 0.5 to 0.9 Mg batch of material. Mixing blades rotate in a direction opposite to the rotation of the pan maintaining the charge in a constant state of agitation. If sufficient water is added, a certain degree of plasticity is created, and agglomerates are formed in 2 to 3 minutes of agitation. In modern fertilizer technology, horizontal pan mixers have been replaced by rotary cylinders that are better suited to continuous processing.

7.5.3. PADDLE MIXERS

These devices, also known as pugmills, pug mixers, and blungers, consist of a horizontal trough containing a mixing shaft with attached mixing blades of various designs. Single or double trough designs are used, although the latter type is most popular (see Fig. 7.5.2). Twin shafts rotate in opposite directions, throwing the materials forward and to the center of the machine as the pitched blades on the shaft pass through the charge. Incoming material may be added at various locations along the length of the mixer to ensure that the entire mixing length is used and to add versatility to the processing. Table 7.5.1 gives

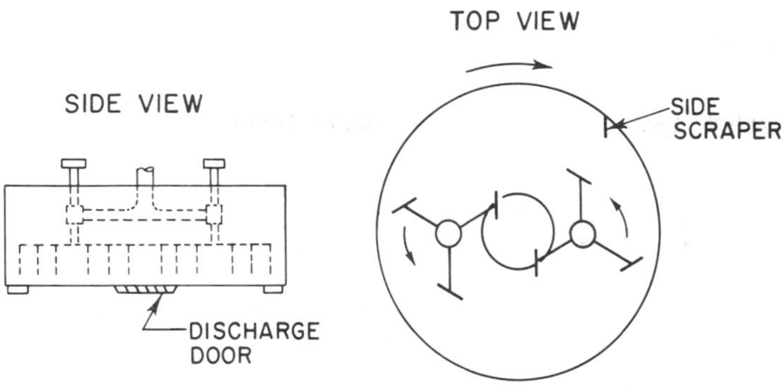

Figure 7.5.1. Schematic of a horizontal pan mixer.[12]

the general characteristics of the range of pug mixers offered by one manufacturer for fertilizer granulation.

7.5.4. HIGH-SPEED MIXERS

Shaft mixers operating at high rotational speeds provide a more intensive mixing-granulating action than that obtained with conventional paddle mixers. These machines are generally single-shaft devices that may be operated either vertically or horizontally. They find application in granulating extreme fines that may be highly aerated when dry and plastic or sticky when wet. The intensive mixing action may achieve agglomeration with short residence times, leading to very compact continuous flow-through designs.

Typical examples of high-speed mixer-agglomerators are the peg granulator[3] used to treat ceramic clays in the china clay industry (see Fig. 7.5.3) and the pin mixer[4] used to densify carbon black into pellets (see Fig. 7.5.4). These machines are similar in design, consisting of a metal cylinder housing a rotating shaft carrying a number of pins or pegs arranged in a helix. Wet feed or dry feed, which is immediately moistened, enters the machine at one end and emerges as pellets at the opposite end.

Figure 7.5.2. Double trough pugmill for fertilizer granulation. (Courtesy Edw. Renneburg & Sons Co.)

Table 7.5.1. Characteristics of Pug Mixers for Fertilizer Granulation. (Courtesy Feeco International, Inc.)

MODEL	MATERIAL BULK DENSITY LB/FT^3	APPROXIMATE CAPACITY (TONS/HR)	SIZE (WIDTH × LENGTH) (FT)	PLATE THICKNESS (IN.)	SHAFT DIAMETER (IN.)	SPEED (RPM)	DRIVE (HP)
A	25	8	2 × 8	1/4	3	56	15
	50	15	2 × 8	1/4	3	56	20
	75	22	2 × 8	1/4	3	56	25
	100	30	2 × 8	1/4	3	56	30
B	25	30	4 × 8	3/8	4	56	30
	50	60	4 × 8	3/8	4	56	50
	75	90	4 × 8	3/8	4	56	75
	100	120	4 × 8	3/8	5	56	100
C	25	30	4 × 12	3/8	5	56	50
	50	60	4 × 12	3/8	5	56	100
	75	90	4 × 12	3/8	6	56	150
	100	120	4 × 12	3/8	6	56	200
	125	180	4 × 12	3/8	7	56	300

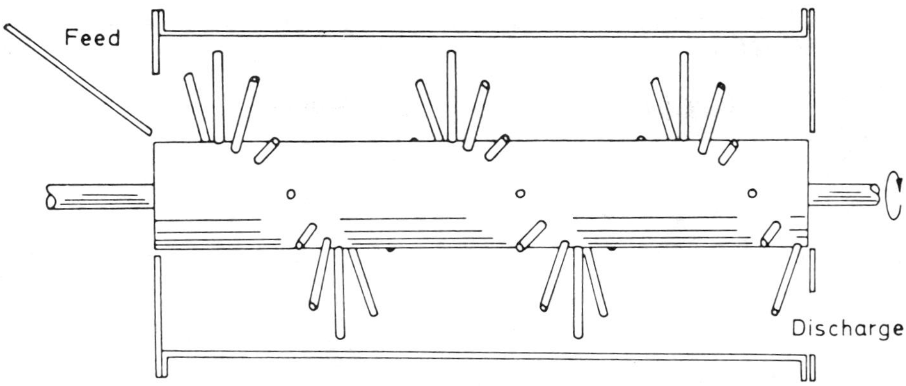

Figure 7.5.3. Horizontal peg granulator for ceramic clay preparation.[3]

As illustrated in Figure 7.5.4, the pelletizing of carbon black in a pin mixer is considered[4] to occur in three stages:

1. The mixing zone is roughly 15-20% of the total length. In this stage small droplets of binder are brought into intimate contact with the powder by the interaction of mechanical and aerodynamic forces produced by the agitator.
2. Agglomeration begins in the pelletizing zone, which is roughly 35% of the effective length. Moist solid particles introduced into the pelletizing zone are eventually combined into a number of nuclei granules and grow into spheroidal pellets fairly uniform in size and density.
3. The densifying zone comprises the final 50% of the effective machine length. The granules formed in the previous zones require very little additional mass but are hardened, densified, and polished through the action of the pins and interaction with each other.

Table 7.5.2 shows pelletization test results using the pinmixer with a furnace oil carbon black.

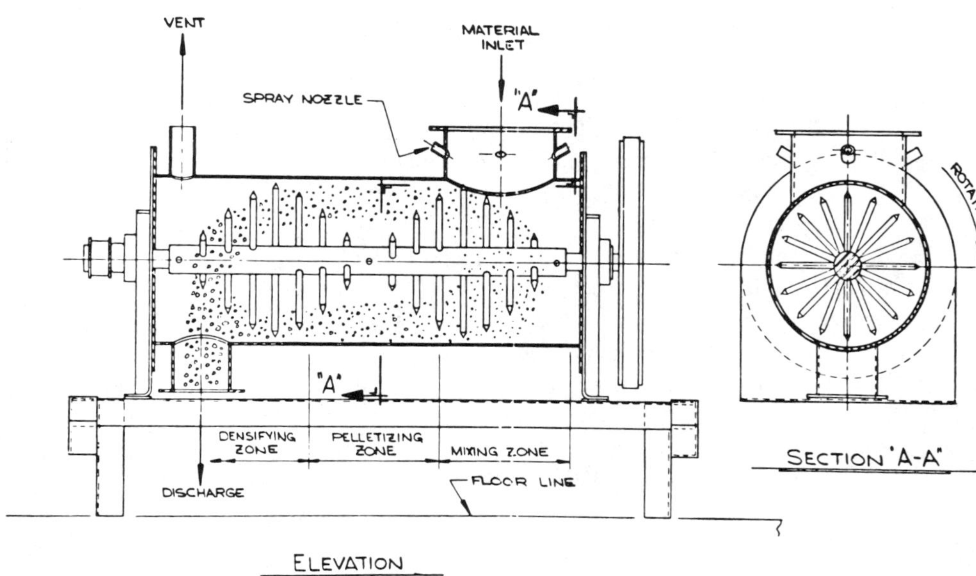

Figure 7.5.4. Pinmixer used in pelleting carbon black.[4]

Table 7.5.2. Pinmixer Used for Pelleting Carbon Black. Test Results with a Furnace Oil Carbon Black Using a 0.67 × 2.54 m Stainless Steel Unit. (Reproduced from Reference 4)

Carbon black feed	
Rate, Mg/day	26.3[a]
Bulk density, kg/m^3	51.3
Pellets produced	
Wet basis	
Production rate, kg/h	2108.3
Bulk density, kg/m^3	562.3
Densification ratio	11.0
Dry basis	
Production rate, kg/h	1096.3
Bulk density, kg/m^3	394.1
Binder	
Specific gravity	1.05
Injection rate, kg/h	1011.5
Use ratio, weight of binder to weight of wet pellets	0.92
Power Consumption[b]	
Rate, kW	18.5
Per Mg of wet pellets, kWh	15.0[c]
Production quality	
Rotap test (5 min), %	1.4 (avg. of 45 samples)
Crushing strength, g	25 (avg. of 73 samples)

[a]Average from 5-day test, plus subsequent production.
[b]Ammeter readings.
[c]Cold shell.

7.5.5. POWDER BLENDERS AND MIXERS

In applications such as the preparation of tableting feeds and the manufacture of detergent powders, the aim is to produce small agglomerates (usually 2 mm diameter and less) with improved flow, wetting, dispersing, or dissolution properties. Agglomeration takes place by wetting the feed powders in a relatively dry state in standard or specialized powder mixers.

In the standard wet-granulating method used to produce tablet feed in the pharmaceutical industry, sigma blade or heavy duty planetary mixers are often employed.[5] These machines may handle 100 or 200 kg batches and employ 5 to 7.5 kW drives to knead and mass the moistened charge. Mixing times from 15 minutes to an hour may be necessary, depending on the formulation. The mass is then wet screened or milled, dried and rescreened to the required size that is dictated by the size of the tablets to be produced.

The time-consuming wet-milling step can be omitted and the agglomerates sent directly to drying, provided an appropriate granular texture can be formed in the mixer. This can be achieved by the use of specialized intensive powder mixers such as the Littleford-Lodige unit shown in Fig. 7.5.5. Powder is fed through the filler-opening at the top of the mixer while the product is discharged through a contour door at the bottom. The working level is normally 50% of the total volume, and cleaning is easily accomplished through the two wide access doors at the front. The material is subjected to a dry mix cycle to eliminate any lumps that might have formed during storage. The granulating solution is introduced to the mixer through liquid injectors mounted over high-speed blending choppers. Spray nozzles are not needed since the high-speed blending choppers quickly disperse the granulating liquid. Plows intermingle the powder and drive material into the high-speed choppers, which

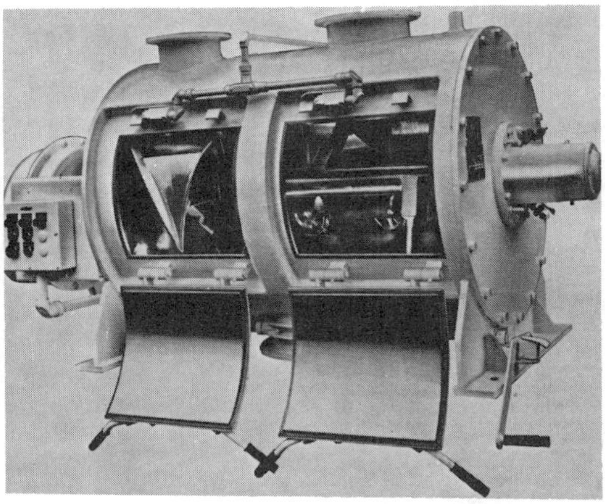

Figure 7.5.5. Littleford-Lodige mixer-granulator for tablet feed preparation. (Courtesy Littleford Bros., Inc.)

are independently powered. The choppers also control the upper size of lumps and agglomerates formed. Standard mixers with working capacities up to 4.8 m^3 are available in this design. Relatively short batch granulation cycles of less than 10 minutes are claimed for this equipment.

7.5.6. OTHER CLUSTER-TYPE AGGLOMERATION PROCESSES

Two other applications requiring small, cluster-type agglomerates with improved flow, wetting, dispersing and or dissolution properties are the manufacture of home dishwashing detergents and "instantized" powdered food products.

Powdered detergent ingredients can be gathered together into a homogeneous granular product by the application of a liquid silicate spray as the bonding agent. A unique design of agglomerator has been developed[6,7] for this application in which the liquid spray is applied to a falling curtain of powder ingredients of constant thickness. The curtain is generated in a rotary drum containing an internal cage of bars separated from the drum walls by a spiral ribbon (Fig. 7.5.6). The cage, together with inertial and centrifugal forces, holds the powder bed against the shell until it falls through the cage to form a constant density curtain. The spiral serves to recirculate fine material toward the feed end. The curtain of powder absorbs the liquid spray before it can impinge on internal agglomerator surfaces while the free-floating action of the internals keeps all surfaces free of build-up, both of which prevent lump formation and encourage uniform granulation. An agglomerator 1.5 m in diameter by 4.9 m long typically produces 4.5 Mg of dishwasher detergent per hour.

In the food industry, continuous-flow mixing systems are used to bring together powder and moistening liquid to form clustered products with "instant" properties. Several types of moistening-agglomerating devices are possible,[8-10] including rotating cones, powder

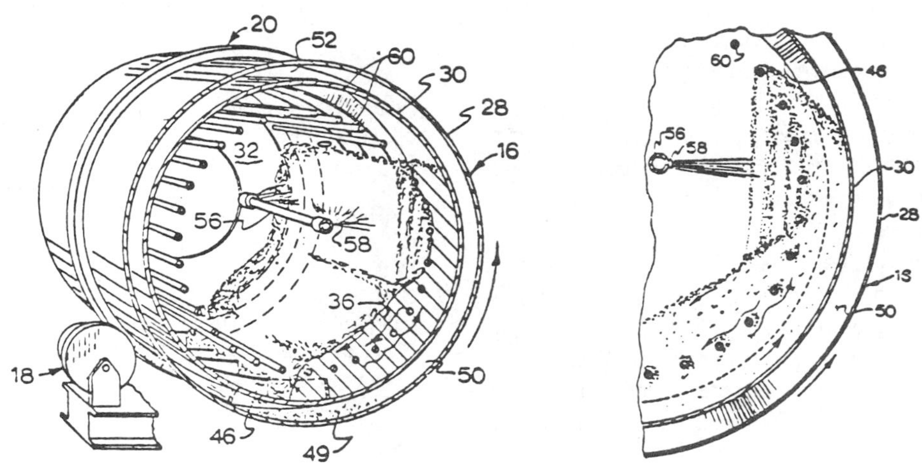

Figure 7.5.6. Constant density falling-curtain agglomerator.[7]

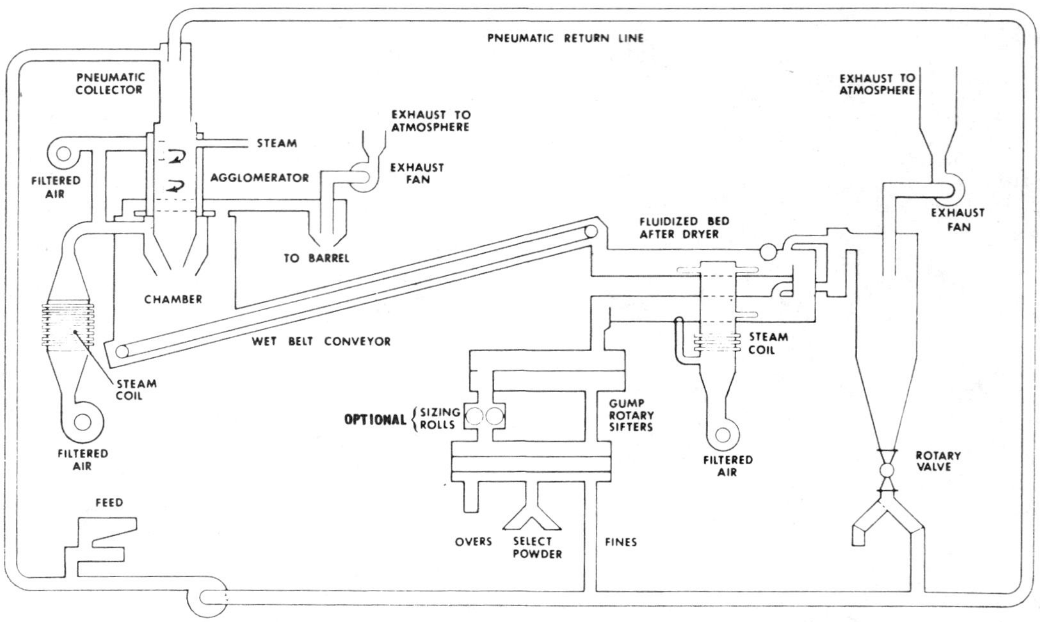

Figure 7.5.7. Flow diagram of the Blaw-Knox Instantizer. (Trademark of Blaw-Knox Food & Chemical Equipment, Inc.)

funnels, and vortex tube mixers. An illustrative example of this type of system is the Blaw-Knox Instantizer Agglomerator[11] depicted in Figure 7.5.7. Feed powder, at a rate controlled by a rotary valve, is introduced to the wetting section via a pneumatic conveying line. The powder falls as a narrow stream between two jet tubes that inject the wetting liquid in a highly dispersed state. Steam is often used but water, other solvents, or a combination of these may be used. Air at ambient temperature is introduced through radial wall slots in the moistening chamber to produce a vortex motion. The resulting lower particle temperature condenses fluid onto the particles while the vortex motion enhances particle-particle collisions. The clustered material then drops through an air-heated chamber onto a conditioning-conveyor where it is allowed sufficient time to reach a uniform moisture distribution. The material then passes to an after-dryer, cooler, and sifter followed by bagging of the selected product.

REFERENCES

1. T. P. Hignett, "Manufacture of Granular Mixed Fertilizers." In V. Sanchelli (ed.) *Chemistry and Technology of Fertilizers*, Reinhold, New York (1960).
2. P. J. Sherrington, "The Granulation of Sand as an Aid to Understanding Fertilizer Granulation." *Chem. Eng. (London) No. 220*, CE 201-CE 215 (1968).
3. R. E. Brociner, "The Peg Granulator," *Chem. Eng. (London) No. 220*, CE 227-CE 231 (1968).
4. J. A. Frye, W. C. Newton, and W. C. Engelleitner, "The Pinmixer–a Novel Agglomeration Device." *Proc. Inst. Briquet. Agglom. Bien. Conf. 14*, pp. 207–217 (1975).
5. L. Lachman, H. A. Lieberman, and J. L. Kanig (eds). *The Theory and Practice of Industrial Pharmacy*, Lea and Febiger, Philadelphia (1970).
6. C. A. Sumner, "Agglomeration of Dishwasher Detergents." *Soap Chem. Spec.* (July, 1975).
7. C. A. Sumner, and E. O'Brien, "Constant Density Falling Curtain Agglomeration of Detergents and Other Materials." In K. V. S. Sastry (ed.) *Agglomeration 77*, AIME, New York (1977).
8. J. D. Jensen, "Some Recent Advances in Agglomerating, Instantizing and Spray Drying." *Food Technol.*, Chicago, pp. 60–71 (June, 1975).

9. R. Wood, "Getting to Grips with Granulation." *Mfg. Chem. Aerosol News*, pp. 23–27 (June, 1975).
10. K. Masters, and A. Stoltze, "Agglomeration Advances." *Food Eng.*, pp. 64–67 (February, 1973).
11. J. G. Moore, W. E. Hesler, M. W. Vincent, and E. C. Dubbels, "Agglomeration of Dried Materials." *Chem. Eng. Prog. 60*(5):63–66 (1964).
12. J. O. Hardesty, "Granulation." In *Superphosphate: Its History, Chemistry and Manufacture*, U.S. Dept. of Agriculture, Washington, 1964.

APPENDIX 7.5.7—AGGLOMERATION BY TUMBLING; BALLING

Tumbling agglomeration is known by several names. Fertilizers are granulated with solution phase and hardened by drying. Ores are balled or wet-pelletized to form a green agglomerate that may be sintered or heat hardened to form pellets. Full details of these processes are available elsewhere[1-4]. Only a brief description of the two most common pieces of equipment, the rotary drum and the inclined disc, is given here.

INCLINED DISKS

These machines consist basically of a tilted rotating plate equipped with a rim to contain the agglomerating charge. Solids are fed continuously onto the central part of the disc and discharge as agglomerates over the rim. Moisture or other binding agents can be added at various locations on the plate surface by means of sprays provided. To promote lifting and rolling of the charge and to protect the disc surfaces, a lining of expanded metal or an abrasive coating is provided. Adjustable scrapers and plows which may oscillate mechanically maintain a uniform protective layer of product over the surface and also control the flow pattern of material on the disc. Pan angle is adjustable from 40° to 70° to the horizontal to obtain the best results. Both constant-speed and variable-speed motors are available as disc drives. Dust covers are fitted when required.

In a survey[4] of machines offered in North America, capacity data were found to range between 0.5 D^2 Mg/h and 1.2 D^2 Mg/h where D is disc diameter in metres. Installed power ranged from 0.6 D^2 kW to 2.3 D^2 kW. These quite broad variations in capacity and power requirements are due to variations in rotational speed, pan depth and inclination, feed density and different specifications for agglomerate properties such as size and porosity.

DRUM AGGLOMERATORS

These machines consist of an inclined rotary cylinder powered by a fixed- or variable-speed drive. The pitch of the drum, which may be up to 10° from the horizontal, assists material transport down the length of the drum. A retaining ring is often fitted to the feed end to prevent spill-back of feed. A dam ring may also be used at the exit to increase the depth of material and residence time in the drum.

Feed material, consisting of fresh and recycled particles, may be introduced by a belt conveyor, a chute, pipe or similar devices. Liquid phase may be introduced either before or immediately after the solids enter the cylinder. In iron ore balling, premoistened wet filter cake usually constitutes the feed but water sprays may also be located inside the drum for moisture addition to aid balling control. Although water is the usual wetting agent, various solutions and slurries may be used in mixed fertilizer granulation. These are introduced as sprays or through other distributors on the bed surface or through spargers under the rolling bed of solids.

As with inclined discs, the character of the internal surface is important in producing the correct tumbling action in the drum. In iron ore balling the steel drum is commonly lined with a layer of concrete (approx. 2.5 cm thick) or by expanded metal. Various types of internal scrapers are used to limit build-up of material on the inside surface and to provide a uniform layer to promote the correct rolling, tumbling action in the drum. Rubber flaps and liners as well as external knockers are used to limit build-up in fertilizer processing.

Characteristics of some typical drum agglomerators are given in the following table, based on information from Ref. 4.

REFERENCES

1. D. F. Ball, J. Dartnell, J. Davison, A. Grieve and R. Wild, *Agglomeration of Iron Ores*, Heinemann, London (1973).
2. P. J. Sherrington and R. Oliver, *Granulation*, Heyden, London (1981).
3. K. Meyer, *Pelletizing of Iron Ores*, Springer-Verlag, Berlin (1980).
4. C. E. Capes, *Particle Size Enlargement*, Elsevier, Amsterdam (1980).

APPLICATION	DIAMETER (FT)	LENGTH (FT)	INSTALLED POWER HP	RPM	APPROXIMATE* CAPACITY TONS/H
Fertilizer granulation	5	10	15	10–17	7.5
	8	16	75	8–14	40
Iron ore balling	9	31	60	12–14	54
	12	33	75	10	98

*Capacity excludes recycle; actual drum throughput may be much higher.

PART 6. PRILLING AND OTHER SPRAY METHODS

C. E. Capes and A. E. Fouda

National Research Council

Ottawa, Ontario

7.6.1. INTRODUCTION

In these methods of size enlargement, feed in a liquid or semiliquid form is sprayed into a gas to produce granular solids through heat and/or mass transfer. A variety of process equipment may be used, including spray dryers, prilling towers, spouted or fluidized beds, and pneumatic conveying dryers. Agglomerates are formed by the direct conversion of feed droplets into solid particles, the layering of solids deposited from the feed onto existing nuclei, and/or the sticking together of small particles into aggregates by deposition of binding solids from the spray.

Features common to all these spray and dispersion techniques include the following:

1. The feed liquid must be pumpable and dispersible.
2. The processes are usually amenable to continuous, automated large-scale operation.
3. Attraction and fines carryover are often a problem, and the systems are designed to recover and/or recycle them.

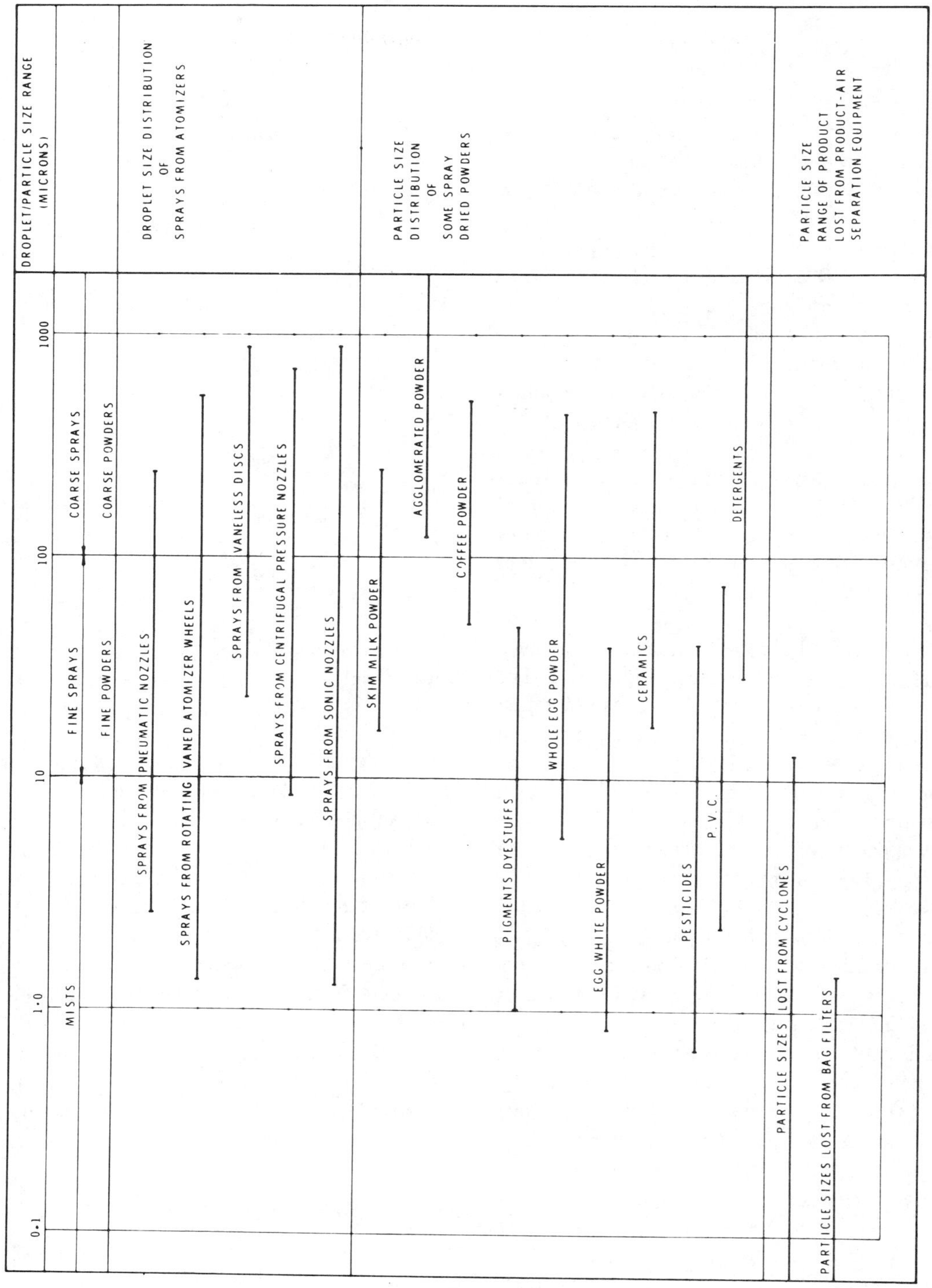

Figure 7.6.1. Particle size range of some spray-dried products.[1]

4. Product size is limited to about 5 mm diameter particles and is often much smaller.

7.6.2. SPRAY DRYING[1]

In this process, feed material is dispersed in droplet form into a drying chamber where it contacts a large volume of hot gas. The liquid carrier is evaporated, and the dry product is recovered. Control of the operating variables can lead to rounded product particles varying from quite fine powders to relatively coarse granular materials (see Fig. 7.6.1).

Spray drying represents an attractive alternative to traditional granulation and feed preparation methods used, for example, in the ceramics and pharmaceutical industries. This procedure is illustrated in Figure 7.6.2 where the unit operations associated with the conventional wet preparation of ceramic tilebodies is compared with the spray-drying alternative. The latter method eliminates a number of processing steps between slip preparation and finished press-body.

The four fundamental unit processes involved in spray drying are shown in Fig. 7.6.3. The liquid feed is dispersed into droplets in the first stage, mixed with the gas stream, and then introduced to the drying chamber. The moisture is evaporated from the droplets, which form solid granules. The dried particles are separated from the gas stream in the fourth stage. Control of the properties of spray dried products requires close attention to the design of each of these four unit processes.

Atomization of the liquid feed and contacting the spray with air are the critical features of spray dryers. Dispersion of the feed into droplets is accomplished with either rotary devices or with nozzles. In rotary atomization (Fig. 7.6.4a), feed is introduced centrally to a wheel (with vanes or bushings) or a disc (vaneless plates, cups, inverted bowls) and is flung off at the periphery where it disintegrates into droplets. Nozzles used can be either single-fluid pressure (Fig. 7.6.4b) or two-fluid pneumatic (Fig. 7.6.4c). Thus atomization of the feed can use centrifugal, pressure, or kinetic energy.

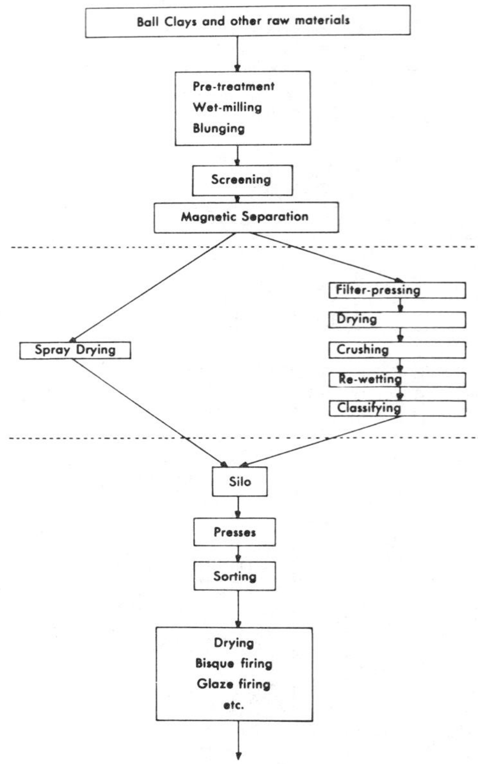

Figure 7.6.2. Operations used in wet preparation as compared with spray drying of ceramic press feeds. (Courtesy Niro Atomizer, Inc.)

Other design features include the solid-gas flow system and the chamber shape. Solid-gas flow can be cocurrent, countercurrent, or a combination of these. Chamber shape is chosen to accommodate the type of atomization. Because of the narrow cone pattern produced, a tall tower is required when nozzles are used, whereas chambers of smaller height-to-diameter ratios are suitable when droplets spun horizontally from a centrifugal atomizer are used.

Spray dryers are available with water evaporative capacities up to 40,000 lb/h (18,100 kg/h) or more. Details of the theoretical and empirical design of spray dryers is given elsewhere.[1]

Coarse Products by Spray Drying

Many different properties of spray-dried products (for example, density, friability, reactivity, etc.) are of interest depending upon the

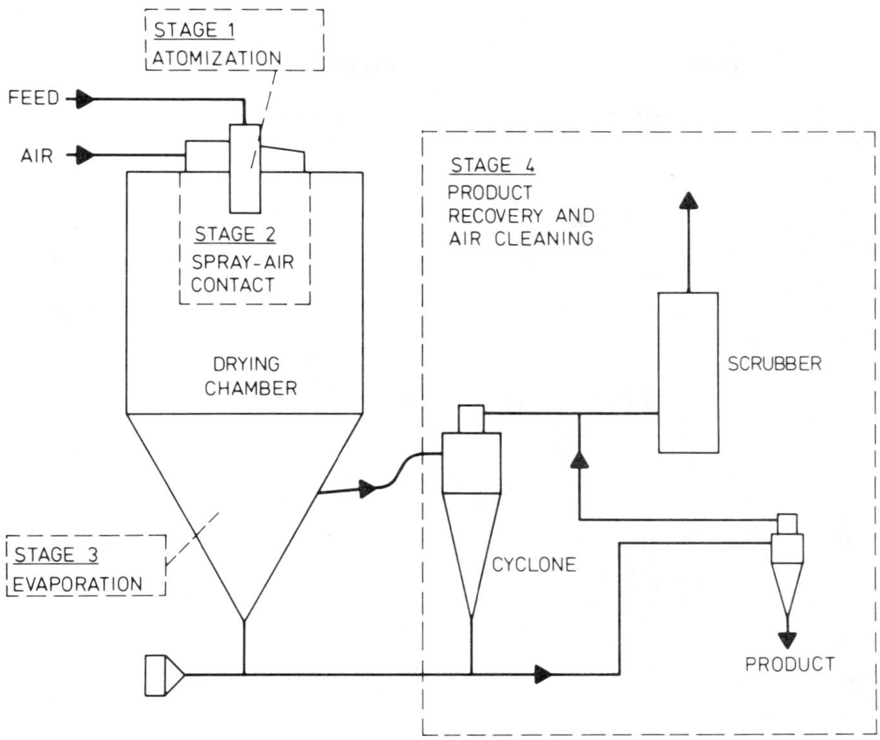

Figure 7.6.3. The four fundamental unit processes associated with spray drying.[1]

application under discussion. When size enlargement to a coarse granular structure is a primary objective, however, particle size and its distribution are of greatest concern.

Although the variables of spray dryer design and operation all interact to influence product characteristics, a number of these have important effects on product size and size distribution. Decreased intensity of atomization and of spray-air contact and lower exit temperatures from the dryer all tend to increase the particle size obtained. Higher liquid feed viscosity and feed rate as well as the presence of natural or added binders that lend tackiness also favor larger product size.

The flowsheet in Figure 7.6.5 gives one example of a system designed to yield agglomerated products. Coarse spray-dried food powders with "instant" properties are produced directly from liquid in this system. Two stages of agglomeration are involved. The initial stage occurs in the atomization zone where relatively cool air is passed in order to retard the evaporation rate and enhance the agglomeration of

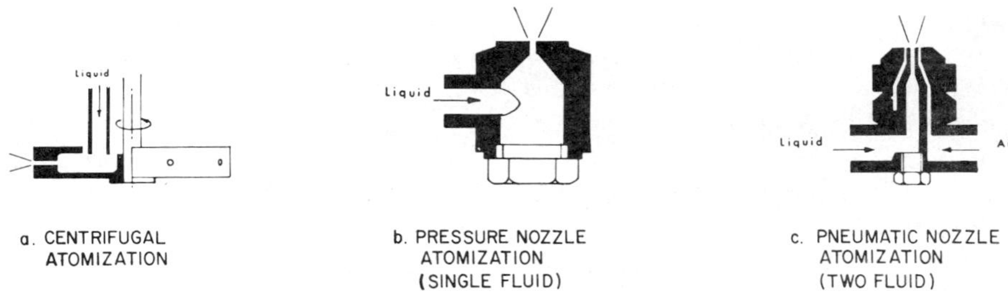

Figure 7.6.4. Feed atomization methods used in spray drying. (Courtesy Anhydro, Inc.)

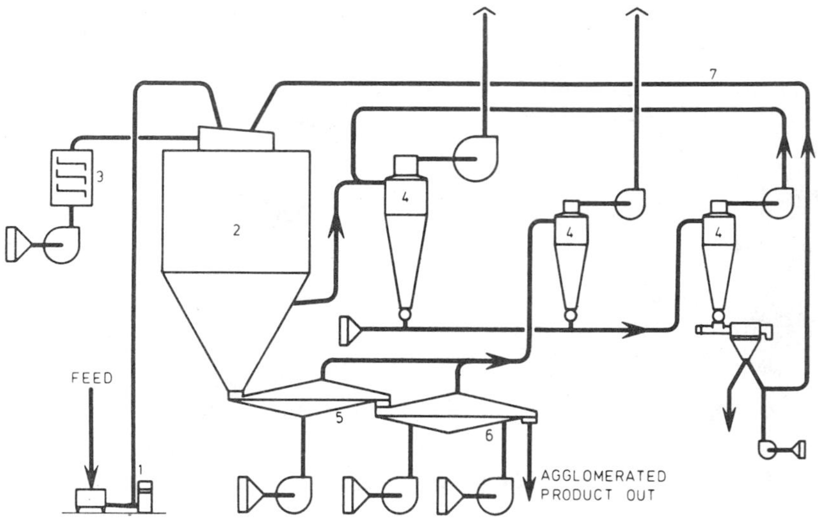

Figure 7.6.5. Flow sheet for production of coarse food powders with "instant" properties by spray drying:[16] (1) liquid feed system; (2) spray dryer chamber; (3) drying air heater; (4) cyclones for fines recovery; (5) vibrofluidizer as after-dryer; (6) vibrofluidizer as after-cooler; and (7) fines return to drying chamber.

fines. Further agglomeration is achieved by operating the spray dryer so that the powder is still moist on leaving the drying chamber. The agglomerated powder passes out of the bottom of the drying chamber to a vibrating fluid bed where drying is completed, then into a second fluid bed for cooling.

7.6.3. PRILLING

This process is also known as spray cooling, spray solidification, spray congealing, and shot forming. It is similar to spray drying in that liquid feed, dispersed into droplets at the top of a chamber, solidifies into a granular product as it falls down the chamber. It differs from spray drying in that the liquid droplets are formed from a melt that solidifies primarily by cooling in the chamber with little drying. Product size is also generally larger (up to about 3 mm diameter). As a result of this relatively large prill size, the process is generally carried out in narrow but very tall towers to ensure that the prills are sufficiently solid when they reach the bottom. Prilling is limited to materials of low melting point that do not decompose on fusion. Urea and ammonium nitrate fertilizers are traditionally treated by prilling.[2,3] Other prillable materials are listed in Table 7.6.1.

In most cases air is used as the cooling medium but other gases or liquids may also be used. In the manufacture of lead shot, molten lead is prilled into water. Prilling into a liquid medium requires shorter towers because of the improved cooling by liquids, but an extra step to remove the liquid cooling medium is necessary.

Some Design Features of Prilling Towers

In principle, a prilling tower is extremely simple (see Fig. 7.6.6.). Melt is fed to a dispersing or prilling device at the top of the tower. The droplets produced travel countercurrent to cooling air as they fall, and the solidified prills are conveyed away from the bottom of the tower to appropriate downstream handling.

In the fertilizer industry, the two most commonly used droplet-forming devices are nozzles or spray pipes and rotating perforated buckets.[4] Examples are shown in Figure 7.6.7. For viscous, high-melting and sticky materials such as resins, a hot chamber method, the Thermaprill System,[3] has been developed. Prilling nozzles are mounted atop the prilling

Table 7.6.1. Some Typical Prillable Materials. (Courtesy HPD Incorporated, Glen Ellyn, Ill.).

Adhesives
Adipic Acid
Alpha naphthol
Ammonium nitrate and additives
Asphalt
Bisphenol-A
Bitumen
Carbon pitch
Caustic soda
Cetyl alcohol
Coal-derived waxes
Coal tar pitch
Dichloro-benzidine
Fatty acids
Fatty alcohols
Epoxy resins
Hydrocarbon resins
High-melting inorganic salts
Ink formulations
Lauric acid
Myristic acid
Myristryl alcohol
Paraffins
Pentachlorophenol
Petroleum wax
Phenolic resins–Novalak resin
Pine rosin
Polyethylene resins
Polystyrene resins
Polypropylene–maleic anhydride
Potassium nitrate
Resins
Sodium glycols
Sodium nitrate
Sodium nitrite
Sodium sulphate
Stearic acid
Stearyl alcohol
Substituted aliphatics
Substituted amides
Sulphur
Urea and additives
Urea–sulphur mix
Wax–resin blends

tower in a radiant-heat thermal conditioning chamber through which no cooling air is allowed to flow. The arrangement prevents the nozzles from plugging or freezing and provides a zone in which streams from the orifices can break into droplets before solidification begins. Fewer fines are claimed as is

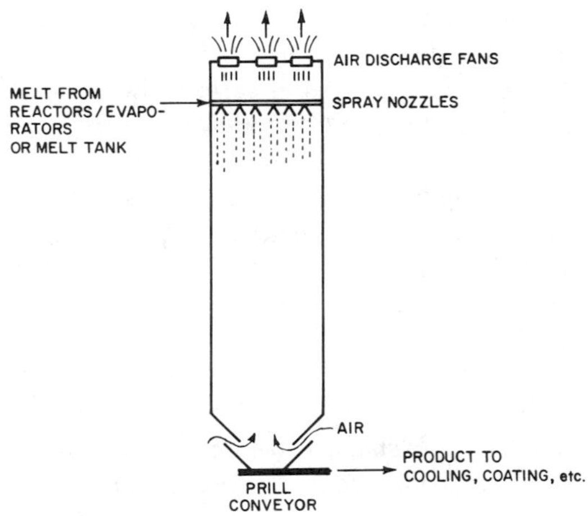

Figure 7.6.6. The basic features of a prilling tower.

the ability to prill many materials not treated this way in the past.

The two principal tower dimensions to be determined are the cross-sectional area and the free-fall height. Restraints that must be considered in sizing the *cross-sectional area* include

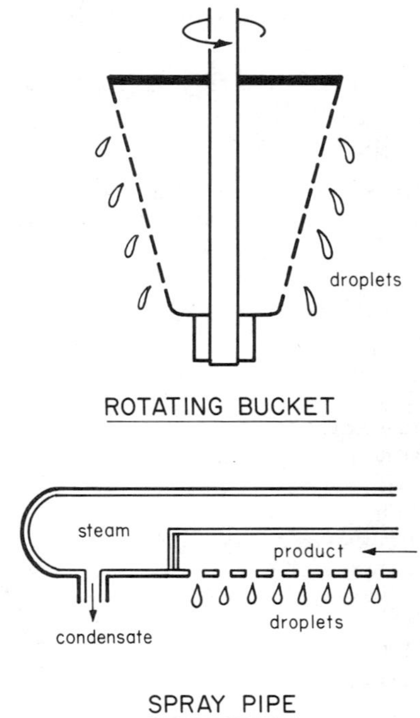

Figure 7.6.7. Two droplet-forming devices used in prilling towers.[4]

the production rate and the number of prilling nozzles required to achieve this production and the minimum area required to allow sufficient air through the tower to satisfy the cooling load at a velocity that is sufficiently low to prevent excessive prill carry over. The *free-fall height* is determined by the retention time necessary to cool the desired prill size to the required temperature at the existing air velocity. The effect of the countercurrent air flow, which may be of the order of 0.9–2.1 m/s, is considered to be negligible by Vernède.[4] Theoretical calculations are possible to determine tower height with reasonable accuracy. Details are available in References 5–8. Some characteristics of typical prilling systems for two fertilizer materials are given in Table 7.6.2.

7.6.4. FLUID BED GRANULATION

In this process, simultaneous drying and particle forming are carried out by spraying liquid feed onto a fluidized layer of essentially dry particles. Particle growth occurs either by particle coalescence or by layering of solids from the feed liquid onto the surface of bed particles.

Because of their ability to deposit multiple layers of solids on a given particle or cluster of particles, fluidized bed (and spouted bed) systems can produce larger granules than spray dryers. The product is thus less dusty, and the longer residence times possible mean that larger dryer loads with more dilute feed liquors can be handled. Since the drying particles are less dispersed in fluid beds, smaller equipment is needed.[9]

A typical fluid bed spray granulation unit is shown in Figure 7.6.8. The fluidizing gas is heated externally and introduced to the base of the unit through a suitable distributor plate. In addition to product support, the distributor ensures a uniform distribution of the fluidizing medium over the cross section of the granulator. Any poorly fluidized region which is subjected to the feed spray might cause the formation of large lumps. The liquid is sprayed by an atomization nozzle centered in the expansion section. The solids to be granulated are fed into the unit below the expansion section.

Air leaving the fluidized bed passes through a cyclone collector, which removes the entrained solids. The solids are returned to the fluidization section and the air is passed through a scrubber for further cleaning. Granulated product is removed near the bottom of the bed through an outlet pipe located slightly above the distributor plate. Pressure drop measurements indicate the weight of solids in the bed and can be used to control the rate of product removal.

A number of important design factors should be emphasized. Often the fluidization chamber

Table 7.6.2. Some Characteristics of a Typical Prilling Operation. (Courtesty HPD Incorporated, Glen Ellyn, Ill.).

Tower size			
Prill tube height, ft		130	
Rectangular cross section, ft		11 by 21.4	
Cooling air			
rate, lb/h		360,000	
inlet temperature		ambient	
temperature rise, °F		15	
Melt			
Type	Urea		Ammonium Nitrate
rate, lb/h	35,200 (190 lb H_2O)		43,720 (90 lb H_2O)
inlet temperature, °F	275		365
Prills			
outlet temperature, °F	120		225
size, mm	approximately 1 to 3		

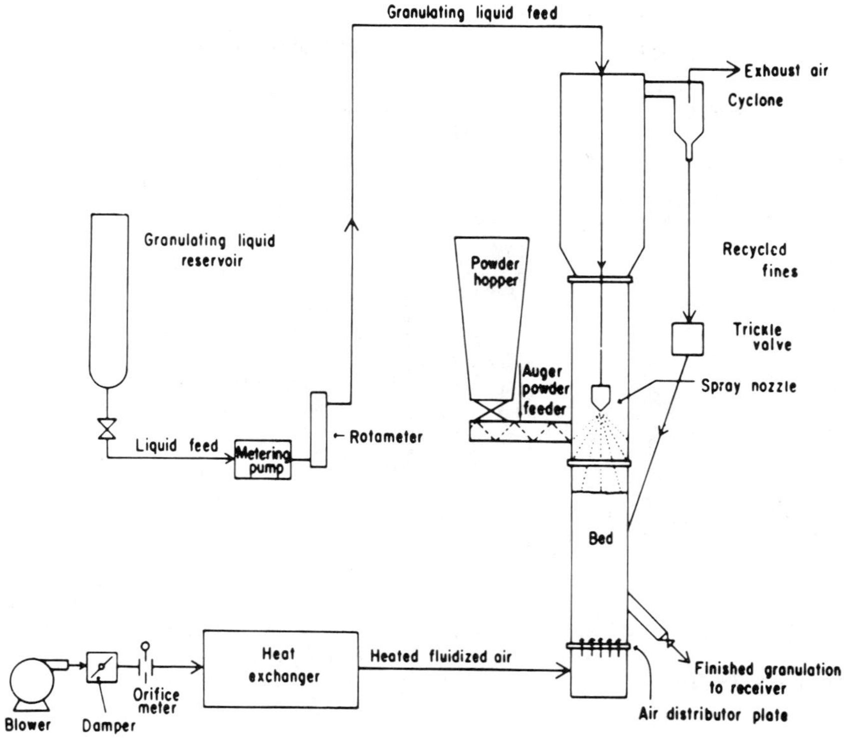

Figure 7.6.8. A typical fluid bed spray granulation unit. (Reproduced from Scott et al.[17] with permission from the copyright owner)

is conical in shape, so that the gas velocity is highest near the distributor. In this way the larger granules which tend to segregate to the bottom of the bed are kept in motion, and overheating is prevented. Fluidizing velocity in general should be selected so that the bed surface, where the feed spray is deposited, is maintained in vigorous movement. Under these conditions, carryover of the smaller particles can occur and a de-entrainment section in the upper part of the bed is necessary. Product discharge takes place through an opening below the bed surface, often relatively close to the distributor. In this way buildup of larger granules and lumps at the distributor plate is avoided.

The fluidizing chamber can consist of more than one compartment (see Fig. 7.6.9). This provides different process conditions (e.g., temperature, moisture level, gas velocity, etc.) as material flows through the bed and encourages conditions closer to plug flow for the granular solids, leading to a more uniform product size distribution.[10]

Control Parameters[11,12]

As in all size enlargement processes, the control of granule nucleation is essential to stable operation. In continuous operations such as those in Figures 7.6.8 and 7.6.9, the rate of production of stable new seeds must equal the rate of production of product size granules. New seeds are generated by a number of mechanisms, including drying of liquid feed to solid before contacting the bed, by attrition and fracture of bed particles, by recycling of crushed oversize product and by introducing new solid particles as part of the feed.

Some general guidelines on the effect of various operating parameters can be given, but these require experimental verification for each spray granulation application.

Increase in the rate of liquid feed addition

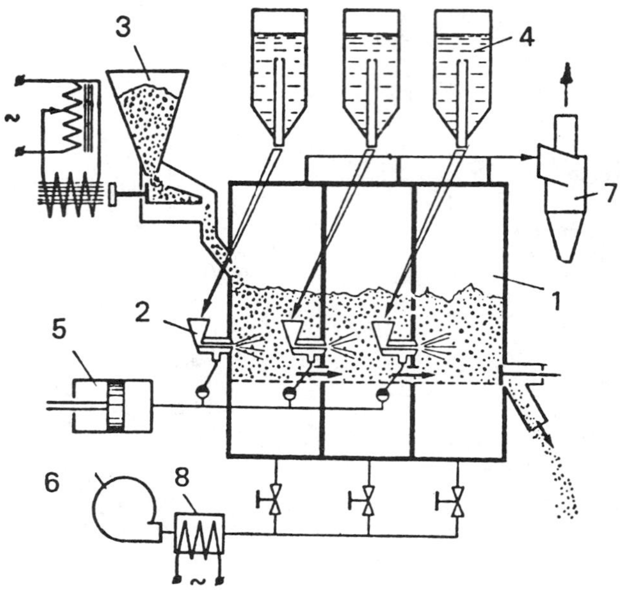

Figure 7.6.9. A nulticompartment fluid bed granulator:[10] (1) fluid beds; (2) compressed air operated injectors for introducing solution into the fluid bed; (3) vibratory feeder for introducing the solid phase; (4) solution tanks; (5) compressor; (6) blower; (7) cyclone; and (8) heater.

and in its solids or binder content generally produces larger, stronger and more dense granules. In some cases, as the solids content of the feed increases, the feed may tend to spray dry in the space above the bed, forming new seed particles and smaller particle size in the bed.

Large agglomerates can be obtained by decreasing the intensity of feed atomization. This effect is lessened as the granule/droplet size ratio increases. Increase in the fluidizing gas rate and bed temperature decreases the ability of the spray to penetrate and wet the bed material and hence smaller particle size is obtained. The geometry of the spray plays an important part in the product size. For example, a narrower, more concentrated spray angle wets a smaller fraction of the bed material and would be expected to yield larger granules.

It is often found that the rate of agglomera-

Table 7.6.3. Characteristics of Batch Fluid Bed Spray Granulators to Produce Tablet Granulations in the Pharmaceutical Industry. (Flowsheet Given in Fig. 7.6.10.)

	APPROXIMATE RANGE		
Batch load, dry basis, lb	20	to	400[a]
Volume of container for static bed, ft^3	2	to	15
Fluidizing air fan, hp	5	to	25
Air (Steam) heating capacity, Btu/h	70,000	to	600,000
Drying air temperature, °C	40	to	80
Granulating liquid spray[b]	Two fluid nozzle		
Air volume	$\frac{1}{2}$	to	2 SCFM
Liquid volume	500	to	1500 cm^3/min
Batch processing time, min	30	to	50
Average granule size	24	to	8 mesh

[a]Batch capacity exceeds 1500 lb in the largest modern units.
[b]Typical granulating liquids are gelatin or sodium carboxymethyl cellulose solutions.

Table 7.6.4. Granular Products from Fluidized Bed Incineration.[15]

TYPE OF SLUDGE	INCINERATOR SIZE	BED TEMPERATURE	CAPACITY	GRANULAR PRODUCT COMPOSITION
Oil refinery waste sludge (85–95% water)	40 ft high; 20 ft ID at base increasing to 28 ft at top	1330°F	31×10^3 lb/hr of sludge	Start-up material was silica sand; replaced by nodules of various ash components such as $CaSO_4$, Na, Ca, Mg silicates, Al_2O_3 after operation of incinerator.
Paper mill waste liquor[a] (40% solids)	20 ft ID at top	1350°F	31×10^3 lb/hr	Sulfur added to produce 90–95% Na_2SO_4 and some Na_2CO_3

[a] Flowsheet, Figure 7.6.11.

tion increases as the gas velocity decreases. This is due to a less rapid exchange of particles within the wetted zone of the fluidized bed. The extent to which the gas velocity can be decreased is limited by the formation of lumps and eventually by termination of the fluidization process.

As noted above, recycled particles are an important source of new seeds for larger granules. The extent to which recycled particles are milled has a profound effect on granule size. For a constant rate of spray addition, increased grinding of recycled material increases the seeding effect and reduces the size of the bed material.

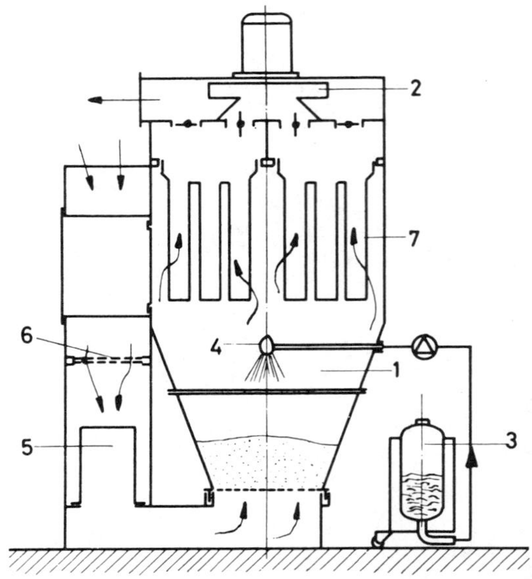

Figure 7.6.10. Batch fluid bed spray granulator used to produce tablet granulations in the pharmaceutical industry. (Air-flow necessary for fluidization is generated by a suction-fan (2) mounted in the top portion of the unit, directly driven by an electric motor. The air being used is heated to the desired temperature by an air heater (5). Prefilters remove all impurities at the air inlet (6). The material to be processed has been loaded into the material container (1). The container bottom consists of a perforated plate above which a fine mesh stainless steel retaining screen is fitted. Exhaust filters (7) mounted above the product container retain fines and dust. The granulating liquid (3) is sprayed as a fine mist through a mechanical or pneumatically actuated nozzle onto the finely dispersed, fluidized material to form the desired agglomerates.) (Courtesy Aeromatic AG.)

Performance Data

Performance data for two industrial versions of the fluid bed spray granulation technique are given in Tables 7.6.3 and 7.6.4. Corresponding equipment diagrams are found in Figure 7.6.10 and 7.6.11, respectively. In Table 7.6.3, data are listed for a range of batch spray granulators available for the production of tablet granulations in the pharmaceutical industry. In this application, the fluidized bed granulator combines into one step several of the individual operations (e.g. size control, drying, blending) normally used in other granulation techniques. Table 7.6.4, contains data on the fluid bed incineration process. Although the main objective of this process is disposal of waste sludges, the granular ash product may often be a salable chemical byproduct. In this secondary aspect, fluid bed incineration can be considered as a size enlargement process.

Spouted Bed Granulation

This technique differs from the fluidized bed process in the method used to agitate the growing bed particles. As shown in Figure 7.6.12 hot spouting gas is injected as a single jet into the conical base of the granulation chamber, causing the bed material to circulate much like a water fountain. Particles are carried up the central spout as a dilute phase until they lose their momentum and fall back onto the top of the bed around the outer periphery. They recirculate back down the column as a dense moving bed and are directed back into the gas stream by the conical base of the apparatus. Liquid feed, injected as a spray into the base together with the hot spouting gas, deposits a thin layer of liquid onto the recirculating seeds. Feed solids deposit by drying from the liquid onto the particles as they cycle up the spout and down the annulus.

The gas-solids contacting efficiency of fluidized systems becomes impaired at particle sizes larger than, say, 1 mm diameter when more and

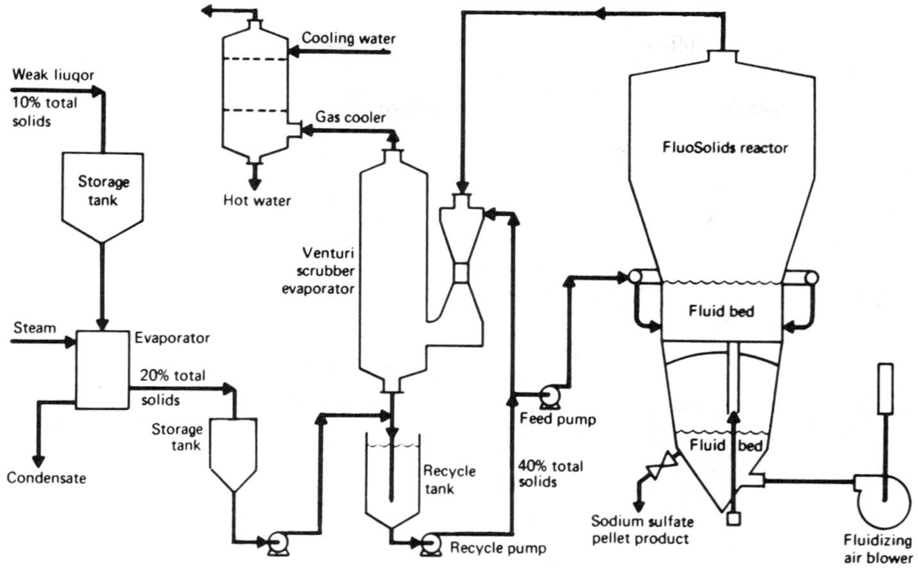

Figure 7.6.11. Flowsheet of fluid bed incinerator used to treat paper mill waste liquor.[15]

more gas bypasses the solids in the form of large bubbles. Spouted beds avoid this problem and are thus suited to the formation of larger granules than those produced in fluid beds.

In their recent book, Mathur and Epstein[13] have noted other advantages of spouted beds when compared with fluid beds:

1. Higher permissible inlet gas temperatures since the spray liquid rapidly cools the gas when injected into the high-velocity region at the base of the spout.
2. Layer-by-layer growth mechanism favors well-rounded and uniform granules.
3. A classification effect at the top of the bed allows the largest particles to be removed through the outlet pipe, yielding a narrow product size distribution.
4. There is no gas distribution plate to become scaled and plugged.

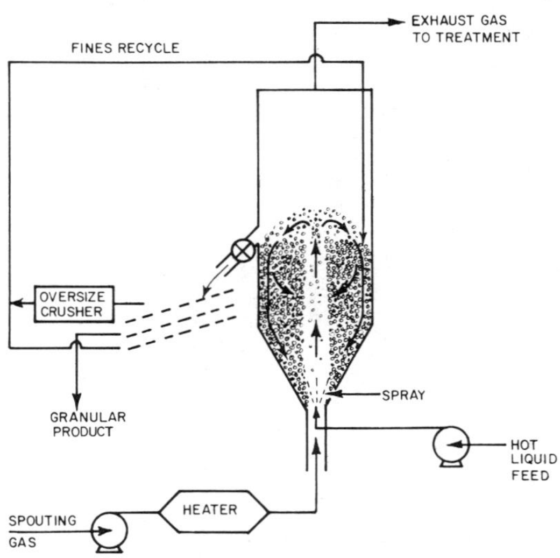

Figure 7.6.12. A typical spouted bed granulator.

Table 7.6.5. Spouted Bed Granulation Data for Some Agricultural Chemicals.[13] (Performance Data Reported by Berquin (Ref. 14).)

MATERIAL	FEED SOLUTION		PRODUCT		AIR TEMPERATURE (°C)		AIR FLOW RATE (M^3/S)	CAPACITY (MG/H OF PRODUCT)	WEIGHT OF BED OF SEED GRANULES (MG)
	MOISTURE (%)	TEMPERATURE (°C)	SIZE (MM)	MOISTURE (%)	INLET	OUTLET			
Complex fertilizer (nitro phosphorus)	27	Cold	3–3.5 (90%)	2.4	170	70	13.9	4	–
Potassium chloride	68	Cold	4–5 (oversize <5%)	–	200	60–75	13.9	1	1
Ammonium nitrate	4	175	2.5–4	0.2	Cold	55	13.9	9.5	1.5
Sulfur	0	135	2–5	0	Cold	–	0.011[a]	0.04	0.008

[a]Injecting 1 liter/h water as spray into the spouting air reduced the air requirement to 0.007 m^3/s for the same product output.

Performance data for the spouted bed granulation of some agricultural products are given in Table 7.6.5.

REFERENCES

1. K. Masters, *Spray Drying Handbook*, 3d ed., George Godwin London; Halsted Press, New York (1979).
2. "Explosives and Fertilizers." *Can. Chem. Process Ind. 28*:299–303 (1944).
3. Horton Thermaprill Process. Bulletin No. 8305P, HPD Incorporated.
4. J. Vernède, "Prilled or Granulated Fertilizers?" *Nitrogen No. 60*, pp. 29–33 (July–August, 1969).
5. R. W. R. Carter and A. C. Roberts, "The Production of Ammonium Nitrate Including Handling and Safety." *Proceedings of the Fertilizer Society, No. 110*, London (1969).
6. H. A. Becker, "The Effects of Shape and Reynolds Number on Drag in the Motion of a Freely Orientated Body in an Infinite Fluid." *Can. J. Chem. Eng. 37*:85–91 (1959).
7. H. Kramers, "Heat Transfer from Spheres to Flowing Media." *Physica 12*:61–80 (1946).
8. J. F. Richardson and P. Ayers, "Heat Transfer between Particles and a Gas in a Fluidized Bed." *Trans. Inst. Chem. Eng. 37*:314–322 (1959).
9. J. W. D. Pictor, "Solids from Solutions in One Step." *Process Eng.*, pp. 66–67 (June, 1974).
10. N. A. Shakhova, B. G. Yevdokimov, and N. M. Ragozina, "An Investigation of a Multi-Compartment Fluid-Bed Granulator." *Process Technol. Int. 17*:946–947 (1972).
11. W. L. Davies, and W. L. Goor, "Batch Production of Pharmaceutical Granulations in a Fluidized Bed." *J. Pharm. Sci. 60*(12):1869–1874 (1971); ibid. *61*:618–622 (1972).
12. S. Mortensen and S. Hovmand, "Particle Formation and Agglomeration in a Spray Granulator." In D. L. Keairns (ed.) *Fluidization Technology*, Hemisphere Pub. Corp., Washington (1976).
13. K. B. Mathur and N. Epstein, *Spouted Beds*, Academic Press, New York (1974).
14. Y. F. Berquin, *Method and Apparatus for Granulating Melted Solid and Hardenable Fluid Products*. U.S. Patent 3, 231, 413 (January 25, 1966).
15. C. J. Wall, J. T. Graves, and E. J. Roberts, "How to Burn Salty Sludges." *Chem. Eng. 82*(8):77–82 (1975).
16. K. Masters, and A. Stoltze, "Agglomeration Advances." *Food Eng.*, pp. 64–67 (February, 1973).
17. M. W. Scott, H. A. Lieberman, A. S. Rankell, and J. V. Battista, "Continuous Production of Tablet Granulations in a Fluidized Bed." *J. Pharm. Sci. 53*(3):314–320 (1964).

PART 7. SINTERING—IRON ORE

R. A. Limons,
Bethlehem Steel Corporation
Bethlehem, Pennsylvania

7.7.1. INTRODUCTION

By 1910, the steel industry in the United States was generating in excess of three million tons per year of flue dust from blast furnace operations. In addition, millions of tons of hematite ore fines from the Mesabi range were being accumulated, because their size consist was far too small for direct charging to the blast furnaces. Magnetite concentrates, roasted ore residues, and beneficiated limonitic or "brown" ore fines, all finely divided oxides of iron, were also being stocked in huge quantities. The same situation existed in those countries throughout the world that were producing basic iron. Necessity being the mother of invention, the sintering process was born. The infant process grew rapidly through the years in conjunction with the worldwide expansion of the steel industry. Growth of the steel industry, with its new methods of refining and finishing steel, generated additional tonnages of revert materials for the sintering process to reclaim. Depletion of screened ore fines and the production of tremendous tonnages of concentrates further complicated the raw material feed to the process. The development of fluxed and superfluxed sinter—of great value to blast furnace operations—was still another complication of the process. Today these raw material complexities are being capably handled by the still-maturing sintering process, a process of great versatility that produces high-quality blast furnace burden material.

7.7.2. HISTORY AND DEVELOPMENT OF SINTERING

Technical publications from the early part of this century record the basic thinking of the investigators and the criteria they proposed for solutions of problems relating to the potential use of stockpiles of fine ores and flue dust.

One of these investigators, Felix Vogel, at an AIME meeting in 1912, said the following:[1]

> Flue-dust, to most blast-furnace operators, means a troublesome by-product, the formation of which should be curtailed, if not prevented entirely. However, with the increasing use of fine ores, larger furnaces, and high-pressure blast, the production of flue-dust is constantly increasing and amounts annually in the United States to from 3,000,000 to 3,500,000 tons, an exceedingly high tonnage, of which a large part has been discarded as valueless.
>
> As a result of greater economy in the iron industry, the attention of our furnace-men has been directed toward the utilization of this enormous amount of waste material, a problem which has also been given due consideration by metallurgists abroad.
>
> To recover, in the blast-furnace, all the values represented by the material contained in the flue-dust, the following conditions should be complied with:
>
> 1. The dust should be agglomerated into lumps about the size of furnace-coke, so that it will help to carry the burden and facilitate the flow of the gases.
> 2. The agglomerated material should be strong enough to carry the burden without disintegrating; it should be heavy, so as to decrease its volume, and it should be sufficiently porous to permit the furnace-gases to penetrate fully. Under no condition should the surface be glazed.
> 3. It should contain all of the valuable constituents of the dust, such as coke, iron-ore, lime, etc.
> 4. It must stand handling, without undue breakage, and should not produce more than 5 per cent of dust.
> 5. It must stand the weather.
> 6. It must not disintegrate in the blast-furnace before being greatly or totally reduced.
> 7. It must submit to easy reduction without requiring additional fuel.
> 8. It must not contain substances detrimental to blast-furnace operations.
> 9. Its cost of production must be low.
>
> Blast-furnace operations, by the use of such agglomerated material will result in
>
> 1. Regular steady operation
> 2. Increased burden, with increase in the metal produced
> 3. Decrease in the consumption of coke
> 4. Decrease in the production of flue-dust
> 5. Decrease in the cost of producing pig-iron

Thus, not only the need for an agglomerate was recognized, but also the value of such a material to the blast furnace operation was postulated.

In the decade preceding 1910, investigators had developed a number of agglomeration processes such as briquetting, nodulizing, compacting, and sintering. Static bed sintering as a method of agglomeration had been developed both in the United States and abroad. Arthur S. Dwight and Richard L. Lloyd had been working together in Cananea, Mexico, between 1903 and 1906 on the development of a downdraft continuous sintering system for the production of blast furnace charge material from copper sulfide ore fines. Following their achievement in Mexico, the Dwight-Lloyd team returned to the United States in 1906, and by 1908 the first commercial continuous sintering plant was built in Colorado for a copper smelter. Dwight and Lloyd then immediately turned their attention to the ferrous industry—again successfully. In 1911 James Gayley described the "Type E" Dwight and Lloyd sintering machine shown in Figure 7.7.1.[2]

In 1912 B. G. Klugh reported on the Dwight and Lloyd operation of the Brooke Iron Company in Birdsboro, Pennsylvania.[3] This first continuous sintering machine in the United States for ferrous materials went into operation in October of 1911 and by February of 1912 had made several thousand tons of sinter for blast furnace use. Klugh reported that the sinter had proved to be a superior and beneficial material for the blast furnace charge. In his words, "Materials otherwise regarded as unfit for the blast furnace may be readily and cheaply converted into a product most desirable for that purpose. For instance, as considerable flue-dust carries an amount of carbon in excess of that required for sintering its own iron-bearing constituents, this may be used by mixing the flue dust with blue billy (hematite sludge) or magnetite, thus giving a product of uniform, cellular composition without cost for fuel." Further, Klugh proposed that sinter mixes should contain fine ores, flux for the entire blast furnace burden (crushed to pass 10 mesh), and the proper carbon. He suggested that the lime in the sinter be calcined and combined with the other gangue materials to form useful slag constituents. Thus, the ideas for making fluxed and superfluxed sinter and for utilizing revert iron oxides for sintering had been publicly porposed at least as early as 1912.

Figure 7.7.2 illustrates the growth of the

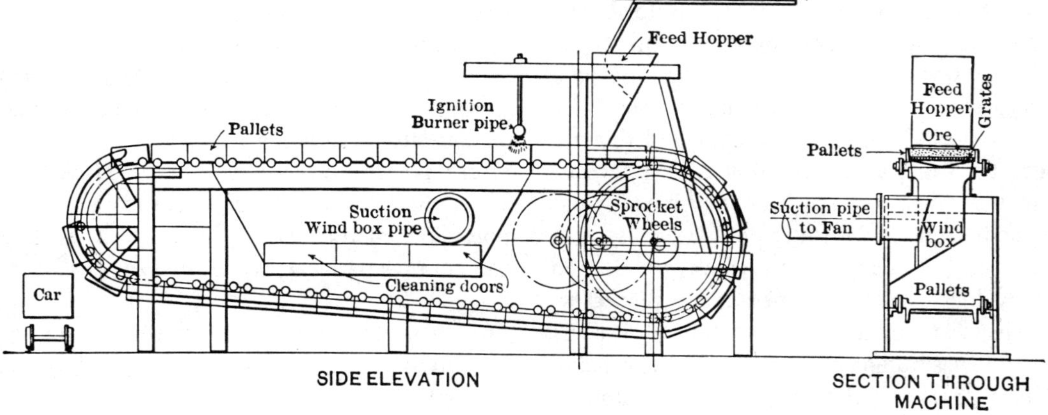

Figure 7.7.1. Dwight and Lloyd sintering machine, conveyor type.

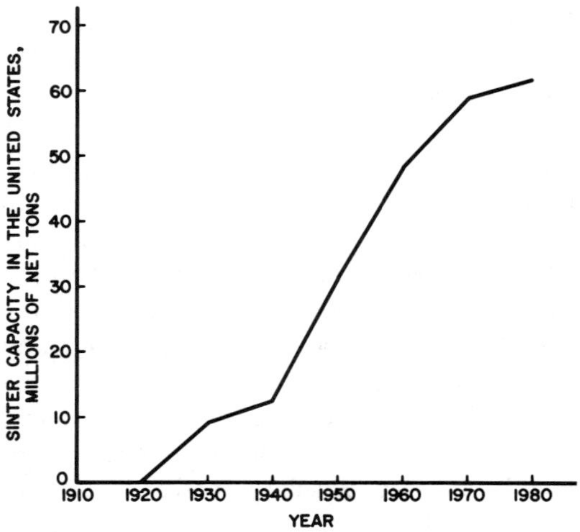

Figure 7.7.2. Sinter capacity in the United States.

sintering industry in the United States. Prior to World War II sintering plants were characterized by small machines—most often 1.82 meter wide by 30.04 meter long (6 ft by 100 ft), installed in multiples of two. Feed materials were not usually bedded and blended and were proportioned by gate openings in bins and by adjusting the speed of table feeders. Preparation of the mix was accomplished in small pug mills. Moisture control relied on the feel and appearance of the mix. Mix was delivered to the machine by a swinging spout without the benefit of a hearth layer. Ignition time was short, and limited fan capacity did not usually draw more than 508 mm (20 in.) of vacuum under the bed. Incandescent sinter often fell into an open pit, and this was followed by water spraying or quenching. Screening of the product was often inefficient, resulting in a carryover of fines to the blast furnace bins. Relatively low product strength added more fines for furnace consumption. Blast furnace flue dust rates were normally very high. Fluxing and sizing of the sinter product was seldom practiced. Windbox exhaust gases were cleaned by cyclones and/or electrostatic precipitators that had varying degrees of efficiency. Process control was rudimentary at best. The many changes that took place in the process following World War II will be outlined when we look at current sintering practice.

7.7.3. RAW MATERIALS AND THEIR PREPARATION FOR SINTERING

Before dealing with the sintering process itself, we must first outline the raw materials and their preparation for sintering. These include:

Iron Ores

The oxides of iron, i.e., hematite, magnetite, limonite, geothite, and, to a lesser extent, the carbonate of iron, siderite, comprise most of the virgin iron ores used for sintering. Ilmenite, the titania-bearing oxide, and pyrite, the sulfide of iron, may be used in sintering but only after beneficiation, particularly in the case of the pyrite. Characteristics of iron ore that have a direct effect on sintering rates and sinter properites are: (1) size consist, (2) chemisty, and (3) mineralogy.

The size consist of iron ores for sintering is established by two factors. The first is, screening practice within the basic iron operation of the steel plant where run-of-mine ore is screened to various sizes for blast furnace, steel plant, and sinter plant use. Screening practice should limit the top size of the sinter fraction to minus 9.5 mm ($\frac{3}{8}$ in.). However, such screening practice is often disrupted by the vagaries of weather such as extreme cold and excess moisture. The second factor is purchase contracts of large tonnages of sintering fines. Here, specifications may be established by operations and purchasing departments for the most effective size consist and chemistry at each operation. Sized and washed sintering fines that limit oversize and undersized fines deleterious to sintering are being more sought after on a worldwide basis. On the other hand, some ultrafines may be desirable from a standpoint of micropelletizing characteristics that can improve permeability of the bed.

The chemistry of iron ores is very important to sintering. For example, magnetite ore, because of the exothermic reaction of Fe_3O_4 upon ignition, is desirable since it reduces the solid-fuel requirement of the sinter mix. However, adequate air and on-strand residence time must be provided to oxidize most of the ferrous iron since sinter product is less reducible as the ferrous iron content increases.[4] On the other hand, hematite ore fines generally sinter faster than magnetite ore fines. The elements sulfur, phosphorous, potassium, sodium, and zinc are particularly undesirable above certain limits. With the exception of sulfur, these elements are largely retained by the sinter product after processing. Alkali and zinc in sinter can cause serious problems in the operation of blast furnaces, and the surlfur and phosphorus content must be reckoned with in steel refining.[5]

The mineralogy of iron ores also plays an important part in sintering. The crystal size and related surface area, the crystal structure (bridged, bonded, or loose matrix), and the gangue constituents and their dispersion all have an effect on the sintering rate and the physical-chemical characteristics of the resulting product. The British Steel Corporation has recently published a model for predicting sinter productivity from ore properties.[6]

Revert Materials

Recycled materials, also called reverts, have long been a major portion of the raw materials used for sintering. These materials, as we have already noted, were one of the bases for developing the sintering process. In recent years changes in methods of refining steel, changes in finishing steel, and, possibly of greater importance, legislated regulations for environmental control have greatly increased the percentages of revert materials. Reverts originating outside of the sinter plant, which often form a major portion of the sinter-mix raw materials, are blast furnace flue dust and filter cake, basic oxygen dust and slag, scarfing scale, hot and cold mill scales, pit and ladle scrap, and so on. For purposes of classification we will also include here the hot and cold return fines generated during the sintering process and during sinter product preparation as well as waste gas and water treatment. All of these reverts have the same type of effects on sintering as do the virgin ore fines described in the previous section. They contain, to some degree, desirable elements such as ferrous, ferric, and metallic iron, carbon, lime, magnesia, silica, and alumina. The undesirable elements, as in the case of iron ores, are sulfur, phosphorus, sodium, potassium, zinc, and of course those oils associated with mill scales of various types. There are in operation today plants, called reclamation plants, that treat only recyclable materials in the production of blast furnace sinter.

Fluxes

Dolomite and calcite are the most generally used minerals for the fluxing and the superfluxing of sinter. As with the iron-bearing raw materials, sintering characteristics are affected by the size consist, chemistry, and mineralogy of the fluxstone used. Some of these effects, which have been reported by Bethlehem Steel Corporation, show that the crystal size as well as the size consist of the dolomite are major factors influencing the sintering rate.[7] Other flux materials successfully used in sintering are finely divided burned lime, hydrated lime, and partially calcined particles and dusts from bulk lime-burning operations. The most dramatic effect on sintering rate is the use of burned lime as 2 or 3% of the raw mix.[8] Increases in sinter rate of up to 20% along with high product strength have been reported.

Fuels

Solid fuels that can be used in sintering are beehive coke fines, byproduct coke breeze, anthracite fines, bituminous fines and petroleum coke breeze. Of these, coke breeze fines are by far the most efficient and are the most generally used for sintering. Research work by United States Steel and others established the optimum size consist of coke breeze at minus 3.2 mm plus 100 mesh for sintering purposes.[9] Because of logistics and economics, large tonnages of anthracite fines No. 4 and No.

Table 7.7.1. Comparison of Sinter Mix Compositions, Bethlehem Steel Corporation Plants (October 1976).

ITEM	LOCATION BETHLEHEM	BURNS HARBOR	JOHNSTOWN	LACKAWANNA	SPARROWS POINT
Virgin ore fines (%)	37.6	12.7	43.3	23.2	39.7
Hot and cold return fines (%)	38.6	27.6	26.9	29.7	28.0
Revert materials, steel plant generated (%)	4.2	42.9	10.1	19.2	11.5
Fluxes (%)	14.4	14.1	15.1	24.9	17.3
Fuel, solid (%)	5.2 (a)	2.7 (c)	4.6 (c)	3.0 (c)	3.5 (c)
Total (%)	100.0	100.0	100.0	100.0	100.0

(a) Anthracite fines
(c) Coke breeze

5 are used for sintering is the eastern United States.

Sinter Mix Composition

Table 7.7.1 shows the diversity of the materials making up the sinter mix at each of five Bethlehem Steel Corporation plants. The variations in the proportions of ore fines, fluxes, and revert materials used in the mixes reflect differences in ore scheduling, steel mill practice, and blast furnace operating practice among the five plants.

7.7.4. SINTER PLANT BEDDING AND BLENDING SYSTEMS

Sinter plant production rate and coke breeze consumption as well as sinter product chemical and physical properties are adversely affected by uncontrolled swings in the chemistry and size consist of materials making up the raw mix. In its turn, the efficient operation of the blast furnace depends to a large degree on the chemical and physical consistency of the burden materials. That is, loss in hot metal production and increased coke consumption both result from swings in the chemical and physical properties of sinter charged to the furnace.

The increasing use of revert materials such as basic oxygen furnace slag and dust creates an increased chance of chemical variations in the raw mix because such materials fluctuate widely in their chemistry. Iron ore fines for sintering also show excursions in chemistry, notably in the silica and alumina contents. These factors underline the need for chemical-physical control of the raw sinter mix before conversion to product.

Bedding and blending plants for sinter mix have been around for a long time. These systems have been used to a much greater extent in many European countries and Japan than in North America. Very recently, however, the trend in the United States has been to install sophisticated computer-controlled bedding and blending systems. Bethlehem Steel Corporation has recently installed bedding and blending systems at Burns Harbor, Indiana, and Sparrows Point, Maryland, in support of the newly constructed sinter plants at these two locations. These recently installed systems are designed to prepare a homogeneous, consistent mix from a variety of ores, fluxes, and reverts, many of which have variations in chemical and physical properties.

Figure 7.7.3 illustrates a traveling stacker that takes materials from a storage bin system and deposits them in predetermined amounts as thin layers down the length of the bed. This process continues—often 500 layers of material are stacked—until completion of the bed. One bedded pile may be of sufficient size for six to eight days of sinter plant operation. Bed-building practice varies, but, generally speaking, all the ores, revert materials, and the major proportion of flux stone are bedded. After completion of the bed, a bucket wheel reclaimer and harrow, shown in Figure 7.7.4, are positioned to blend and retrieve the mix. The harrow moves into the face of the pile and rakes material to the base of the pile. Here, a bucket wheel picks up the partially blended material and deposits it onto a traveling conveyor. Blending is complete by the time the mix is transferred by conveyor into the sinter plant raw-mix bins.

Bedding and blending systems are efficient when rapid, accurate chemical analysis of the individual feed material is available for compu-

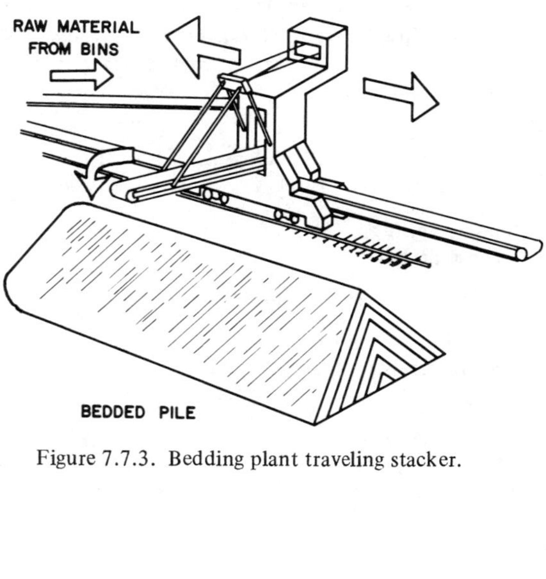

Figure 7.7.3. Bedding plant traveling stacker.

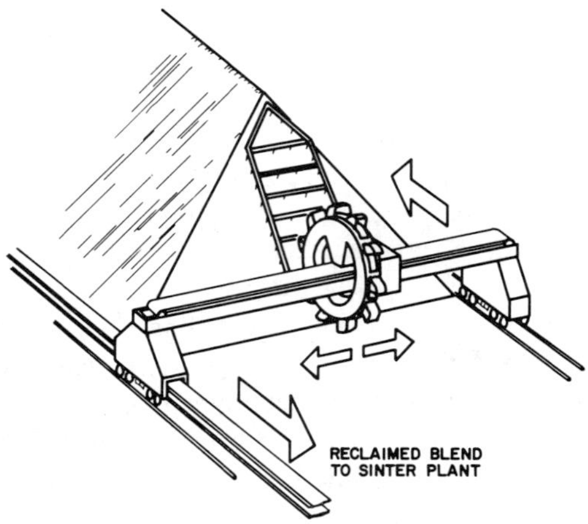

Figure 7.7.4. Bedding plant bucket-wheel reclaimer.

tation of pile chemistry. Extensive raw material sampling programs and statistical analyses are usually employed to determine the frequency and extent of analysis required for computing bed chemistry. Automatic sampling and rapid analysis of the reclaimed mix prior to sinter plant storage provides a further check on predicted pile chemistry. A computer system is effective in correcting feed rates of raw materials to the bedding system.

A striking example of the effective use of a bedding and blending system has been published by Hoogovens in Ijmuiden, Holland.[10] Table 7.7.2 summarizes the improvements made through installation of a bedding and blending system for their sinter plant. Hoogovens concluded that the savings in blast furnace operating costs justified the investment.

7.7.5. LABORATORY SINTERING IN A GLASS CONTAINER

Some time ago an apparatus was developed in Bethlehem Steel Corporation's laboratories that permits observation and visual study of the downdraft sintering process. The sinter mix container is a Pyrex glass tube 40.6 cm (16 in.) long, 10.2 cm (4 in.) in diameter, and with a 3.2 mm ($\frac{1}{8}$ in.) wall thickness. The glass container is mounted in a steel adapter that replaces the cast-iron pot of our laboratory sintering unit. The grate is a 6.4 mm ($\frac{1}{4}$ in.) thick stainless steel plate perforated on a tripod to provide a clear view of the area beneath the grate. Ignition of the mix is initiated by a 15.2 mm (6 in.) diameter gas burner.

Visual phenomena associated with the downdraft sintering process are:

- Segregation of coarse and fine particles occurring during charging of the raw sinter mix into the container
- Compaction of the raw sinter mix by as much as 38.1 mm (1 $\frac{1}{2}$ in.) in the 30.5 cm (12 in.) deep bed as suction is applied before ignition

Table 7.7.2. Improvements Realized From Installation of a Sinter Plant Bedding and Blending System, Hoogovens, Ijmuiden, Holland.

	WITHOUT BEDDING-BLENDING	WITH BEDDING-BLENDING
Sinter production	–	7% increase
Sinter plant fuel consumption	–	15% decrease
SiO_2 content of sinter	6% α = 0.27%	6% α = 0.051
Blast furnace slag basicity	0.95 α = 0.043	0.93% α = 0.023
Blast furnace hot metal sulfur	0.07% α = 0.0149%	0.028% α = 0.0085%

- Calcination of the zone immediately below the flame front, as indicated by the lighter color of the calcination zone
- Condensation of moisture on individual ore particles of the mix during ignition and throughout sintering
- Advancement of the flame front in an irregular fashion, forming a jagged zone across the section
- Escape of incandescent fines through the grate as sintering approaches completion
- Continual shrinkage of the sinter product during residence on the grate

The use of this glass pot has been valuable in studying the sintering process and for making quick, inexpensive tests for go or no-go situations involving additives to the mix and changes in types of ore fines and ore size consist, mix moisture levels, etc.

Figure 7.7.5 illustrates key stages of a typical test run in the glass sintering pot.

7.7.6. CURRENT SINTER PRACTICE

We have reviewed raw material properties and the bedding and blending of these raw materials prior to sintering. Now we are ready to describe current sinter plant practice. Current sintering practice means the pyrochemical conversion of finely divided, iron-bearing materials to produce a strong, chemically consistent, closely sized burden material for the blast furnace.

Figure 7.7.6 illustrates this sinter practice in flow-diagram form. Reclaimed mix from the bedding plant is transferred to the sinter-mix bins. An automatic sampler has cut this stream for analysis as a check on the computed chemistry of the bed. Flux and coke breeze as well as hot and cold return fines are also transferred to the bin building. To ensure a continuous flow of materials from the bins to the belt weigh-feeders, each of the bins is equipped with a vibrating bin-bottom. The weigh-belt proportional feeders deliver sinter mix to the conveyor at a preset rate. Trim fluxstone is added as a constant ratio of trim fluxstone to mix.

The moisture of the coke breeze is determined by an analyzer mounted on the bin, and coke is added on a dry basis as a ratio of the new raw feed. Return fines are then metered to the mix as a fixed ratio to the weight of the new raw feed.

The total sinter mix is conveyed to the mixing-reroll drum for conditioning. The major proportion of water is added to the first half of the drum, where lifters may be installed for thorough blending of the raw materials and water. Final moisture addition is made in the second half of the drum, where the mix is rolled to promote micropelletizing of the fines to develop optimum mix permeability. A downstream moisture gage provides control of the final mix moisture.

Figure 7.7.7 illustrates the feed-end arrangement of the sinter strand. Approximately 25.4 mm of hearth layer, screened from product sinter and sized to about 12.7 X 25.4 mm, is added to the grate of the machine. The hearth layer protects the grate bars from excessive temperature, reduces the amount of fines drawn through the grate bars, and helps to prevent "stickers," i.e., molten sinter adhering to the grate. The hearth-layer bin is mounted on load cells for control of the material requirement to the grate.

Sinter mix is delivered to a hopper by an oscillating conveyor belt and deposited on top of the hearth layer by a roll feeder. The roll-feed hopper is mounted on load cells to control the volume of mix in the hopper. The roll-feeder speed is ratioed to machine speed by means of probes extending into the mix behind a cutoff plate that maintains a flat level of mix on the machine. Mix level behind the cutoff plate is critical, since bed compaction can result from too much mix being piled behind the plate or from an uneven bed for ignition if not enough mix is present. An adjustable gate at the discharge of the roll feeder maintains the proper level of mix on the bed.

The sinter mix on the traveling grate enters the ignition zone. Temperature in the ignition hood is graded from initial fusion of the surface to final annealing of the surface. Temperatures in the hood are controlled to prevent slagging,

Figure 7.7.5 Sintering in a glass pot.

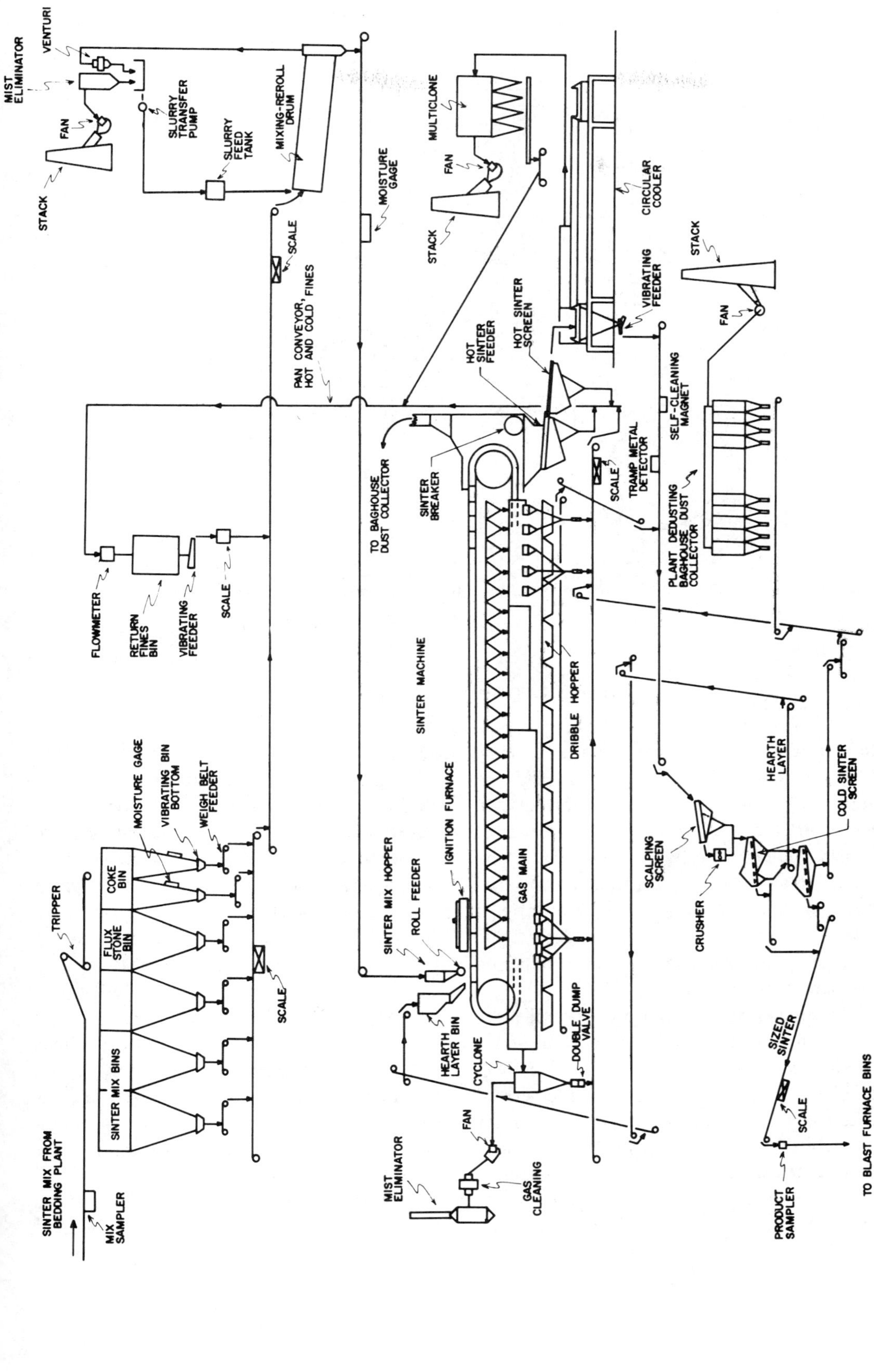

Figure 7.7.6. Current sinter practice flow diagram.

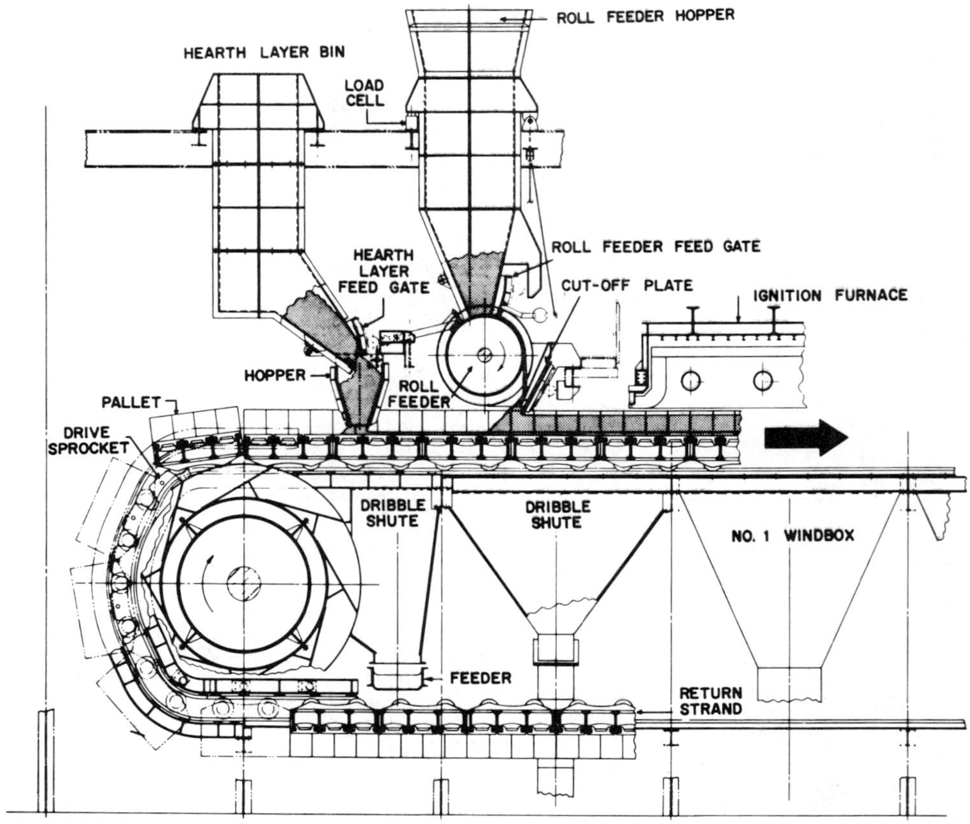

Figure 7.7.7. Sinter strand feed end arrangement.

which restricts air flow, and to prevent cooling, which can shatter the surface. Sintering continues as the strand moves from the ignition zone over the windboxes. Completion of sintering, determined by peak windbox temperature, is automatically controlled to the first section of the last windbox. Burn-through control prevents unsintered material from being discharged and maximizes sinter production per square meter of grate area.

Sinter discharged from the grate is crushed by a revolving toothed roller against fixed grizzly bars to a nominal 15.24 cm or 20.32 cm top size. The minus 6.4 mm fines thus generated are discharged to a pan conveyor for transfer to the return-fines bin. Oversize hot sinter goes to a hot sinter screen for further scalping. Oversize sinter is fed to the sinter cooler. The cooler, mounted on load cells, is speed-automated to control the feed-hopper level and maintain a full bed of sinter. Cooled sinter is discharged into a hopper for transfer to the cold screening and crushing station. This hopper is also mounted on load cells to balance product-to-screening capacity. A metal detector and magnet are installed to remove tramp metal, thus protecting the crusher and conveyor system. A scalping screen separates the oversize product for crushing. The rejoined streams are then screened to provide either one or two sizes of sinter product and to extract hearth layer as required for the process. Minus 12.7 mm fines are sent to the return-fines bin by conveyor. The final product-sized sinter is conveyed to the blast furnace bins. An automatic sampler cuts representative samples for physical testing and chemical analysis.

The environmental quality control systems shown on the flow diagram will be discussed later.

7.7.7. SINTERING PROCESS CONTROL

The most efficient sintering operation is achieved by precise control of the process. The sintering process may be considered as consisting of a sequence of segments covering feed, moisture, ignition, burnthrough, and cooling. Control loops of these segments, when interconnected, will provide an essentially automated process, and this objective is being approached by some of the installations in our industry. Each segment of the process may require a series of two or more control loops in order to provide coarse and then fine control. That is, upstream loops control to a coarse output range, and refining is then achieved to a precise set point by the final loop.

Feed Control

Figure 7.7.8 illustrates the feed-control system. Several interconnected proportioning control loops are utilized. Sinter-feed components are delivered to the collecting belt as preset ratios as the total feed varies. The feed rate of recycle fines is balanced with the amount of fines returning from the screening station. Coke breeze feed rate is set as a ratio of the recycle-fines rate, thereby maintaining sinter quality with an optimum amount of recycled fines. The level of raw mix in the surge hopper above the roll feeder is maintained at an optimum point as a trim control to balance material required for sintering with the amount of mix fed. The amount of sinter mix delivered to the strand is further controlled as a function of machine speed and bed depth. These controls automatically balance the mix feed rate with the sintering rate for maximum production and optimum quality.

Moisture Control

Figure 7.7.9 illustrates the moisture-control system, which consists of two control loops. One meters a major portion of the required water into the mixing area of the mix-reroll drum as a function of the weight of feed material. The second trims the quantity of water sprayed into the reroll section as a function of the amount of feed plus a feedback signal from a nuclear sensor that measures the moisture content of the mix after discharge from the drum.

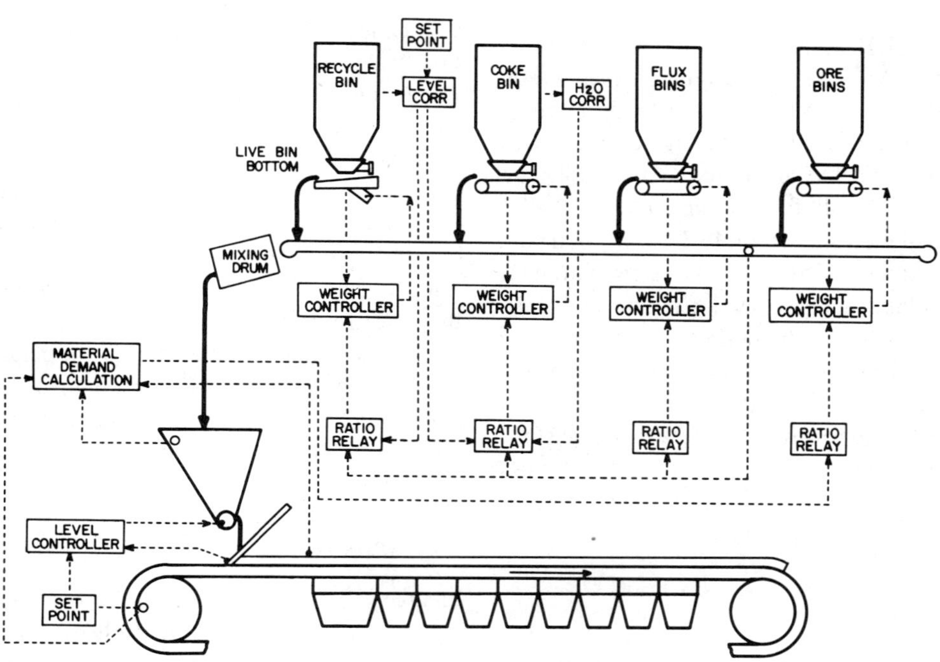

Figure 7.7.8. Control of sinter feed materials. (Courtesy Dravo Corp., Pittsburgh, Pa.)

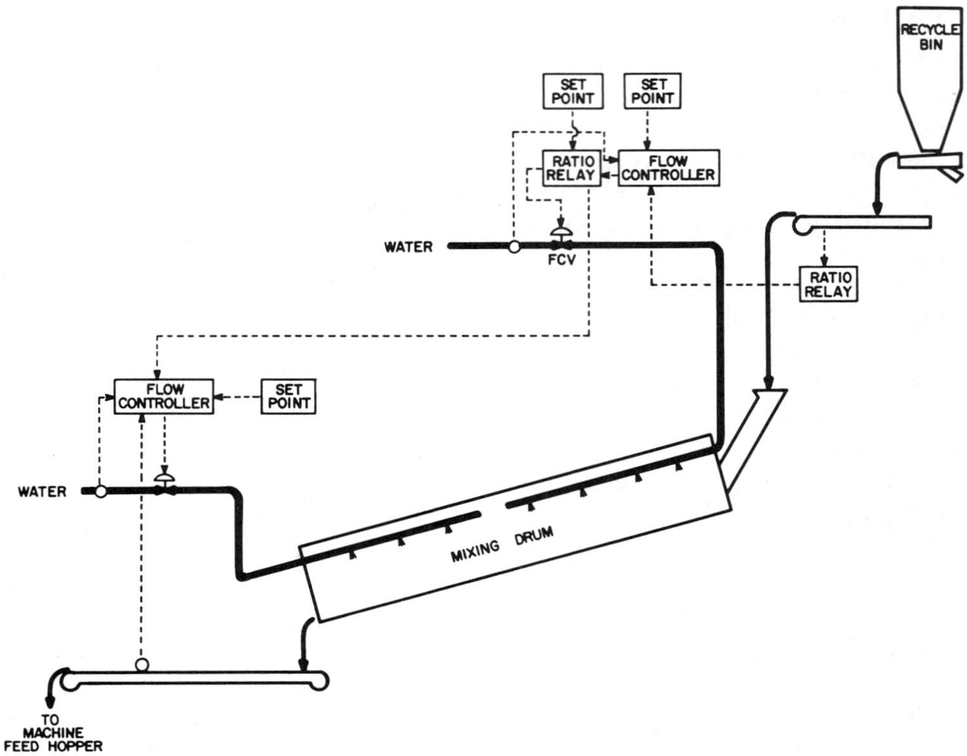

Figure 7.7.9. Control of sinter mix moisture. (Courtesy Dravo Corp., Pittsburgh, Pa.)

Ignition Control

Figure 7.7.10 illustrates the ignition-control system. An array of thermocouples in the ignition furnace programs a declining temperature gradient profile in the ignition zone by adjusting the fuel-air ratio at the individual burners. Overall fuel-air ratio in the ignition zone is preset and controlled by ignition-zone temperatures. The control system automatically adjusts the fuel and air flow in a fixed ratio to maintain the desired temperature profile. A zero-speed signal from the machine drive automatically shifts the temperature control system to low-fire when the machine stops. Dampers in the windboxes under the furnace automatically maintain a constant pressure within the ignition hood.

Burnthrough Control

Figure 7.7.11 illustrates the burnthrough-control system. This system continuously adjusts machine speed to maintain peak windbox temperature at a predetermined point on the machine. Two control loops are utilized. The second loop maintains the peak windbox temperature in the next to last windbox. The anticipatory loop, which operates from a selected windbox two-thirds of the distance down the machine, anticipates changes to the burnthrough point as the result of mix, moisture, etc. and adjusts the machine speed to keep the burnthrough point at a desired location. A selector switch is included in the anticipatory control loop, because the optimum windbox from which to control may vary as the composition of the sinter mix changes.

Cooler Control

Figure 7.7.12 illustrates the cooler-control system. Cooler speed is controlled by a combination of the speed and bed level of the sinter machine plus a bias signal set by the operator.

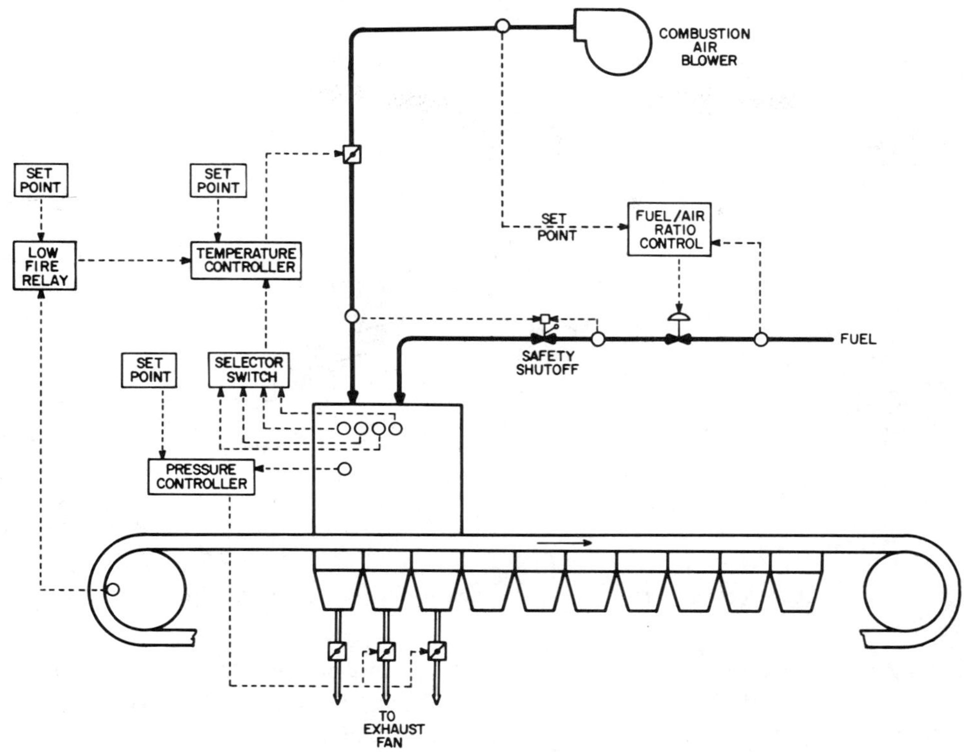

Figure 7.7.10. Control of sinter mix ignition. (Courtesy Dravo Corp., Pittsburgh, Pa.)

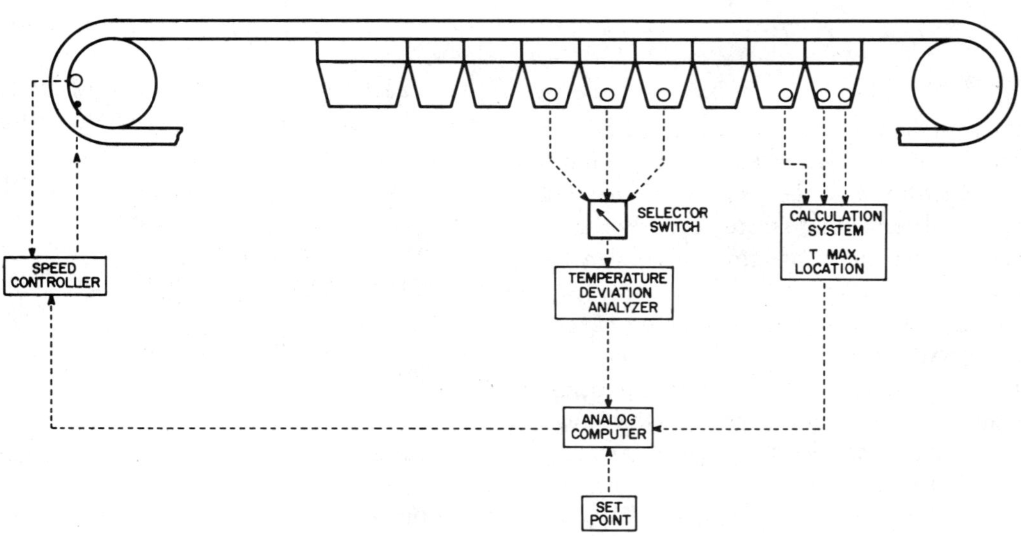

Figure 7.7.11. Control of sinter burnthrough. (Courtesy Dravo Corp., Pittsburgh, Pa.)

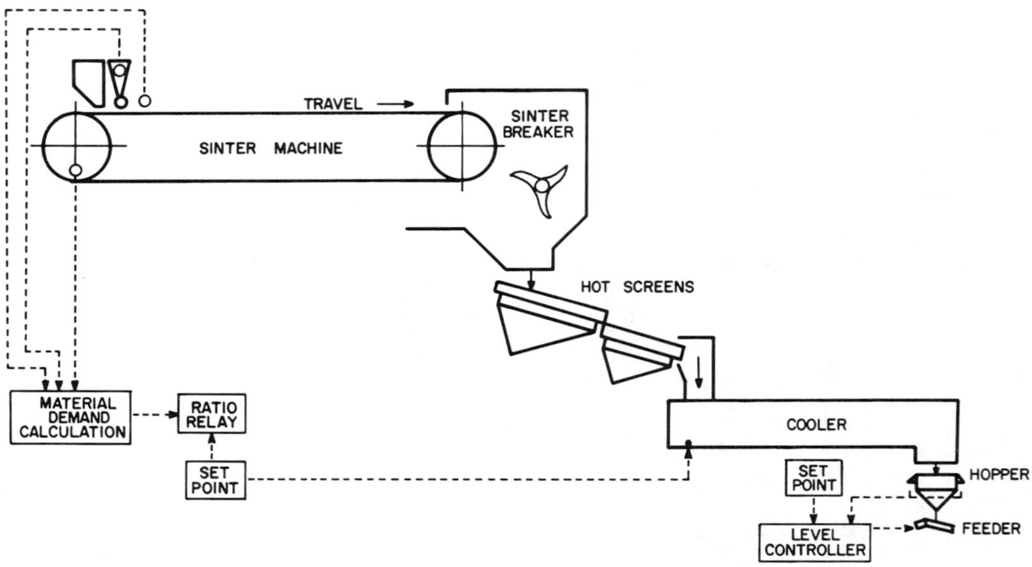

Figure 7.7.12. Control of sinter cooler operation. (Courtesy Dravo Corp., Pittsburgh, Pa.)

This arrangement allows for changes in the yield and for changes in the return fines generated by different mixes. Close control between sintering rate and cooling maintains a full bed of sinter on the cooler. A constant feed of cooled sinter to the screening station is achieved by using a variable-rate feeder discharging from a surge hopper mounted on load cells.

7.7.8. TYPES OF SINTER

Categories

Sinter product may be divided into these three general mineralogical categories: acid, fluxed, and superfluxed. The basicity of the sinter product determines its category. The definitions of basicity and stone equivalent are illustrated in Figure 7.7.13.

Of the two concepts, stone equivalent is more useful for the blast furnace operator. This term defines the raw-stone replacement capability of fluxed sinter going to the furnace. Conversely, a negative stone equivalent defines the raw-stone addition required by acid sinter going to the furnace. Specifically, stone equivalent is the kilograms of raw flux that can be replaced by the addition of 100 kg of sinter to the ore burden. It is a measure of the lime and magnesia in excess of the requirement for fluxing the sinter itself.

Mineralogy of Sinter

The mineral assemblage of sinter is important, because it is a major factor in establishing sinter product strength and sinter reducibility. Figure 7.7.14 illustrates the mineral assemblage and structure of acid, fluxed, and superfluxed sinters.

The following is a brief mineralogical description of sinter made in the laboratory using normal ores and revert materials from Bethlehem Steel Corporation's Lackawanna, New York, plant. At the 1.0 base-to-acid ratio described, the sinter product contained 6.7% CaO + MgO and an equal percentage of SiO_2 plus Al_2O_3, or a total of about 13.5% gangue. Total carbon in the mix at the 1.0 basicity was 5.8%.

Acid Sinters. By definition, sinter is acid at base-to-acid ratios of less than 1.0. The mineral constituents are predominantly magnetite and ferrous silicate glass. The major strength mechanism is magnetite crystal intergrowth and

BASICITY

Basicity = Percent total basic oxides divided by percent total acid oxides

$$B/A = \frac{(\%CaO + \%MgO)}{(\%SiO_2 + \%Al_2O_3)}$$

STONE EQUIVALENT

Stone Equivalent = Kg of equivalent stone per 100 Kg of material

$$S.E. = \frac{100[(\%CaO + \%MgO)_m - (B/S)[(\%SiO_2)_m - 2.14(\%Fe)_m(\%Si)_{hm}/(\%Fe)_{hm}]]}{[(\%CaO + MgO)_s - (B/S)[(\%SiO_2)_s - 2.4(\%Fe)_s(\%Si)_{hm}/(\%Fe)_{hm}]]}$$

subscripts: m = material
s = stone being charged to blast furnace
hm = blast furnace hot metal

CaO = Calcium Oxide
MgO = Magnesium Oxide
SiO_2 = Silicon Dioxide
Fe = Iron
2.14 = Conversion Factor, Si to SiO_2
B/S = basicity of blast furnace slag: (%CaO + MgO)/(%SiO_2)

Figure 7.7.13. Sinter basicity and stone equivalent

glass matrix. Sinter product strength is high with these strength factors.

Self-Fluxing Sinters. At base-to-acid ratios of 1.0 to 1.5, sinter is in the self-fluxing range. The mineral assemblage shows a pronounced change in character. Some of the magnetite oxidizes to hematite. The glass matrix is reconstituted to a lime-iron silicate glass and crystalline dicalcium silicate. Sinter strength falls off appreciably in this range, because magnetite intergrowth is less developed, the bridging of

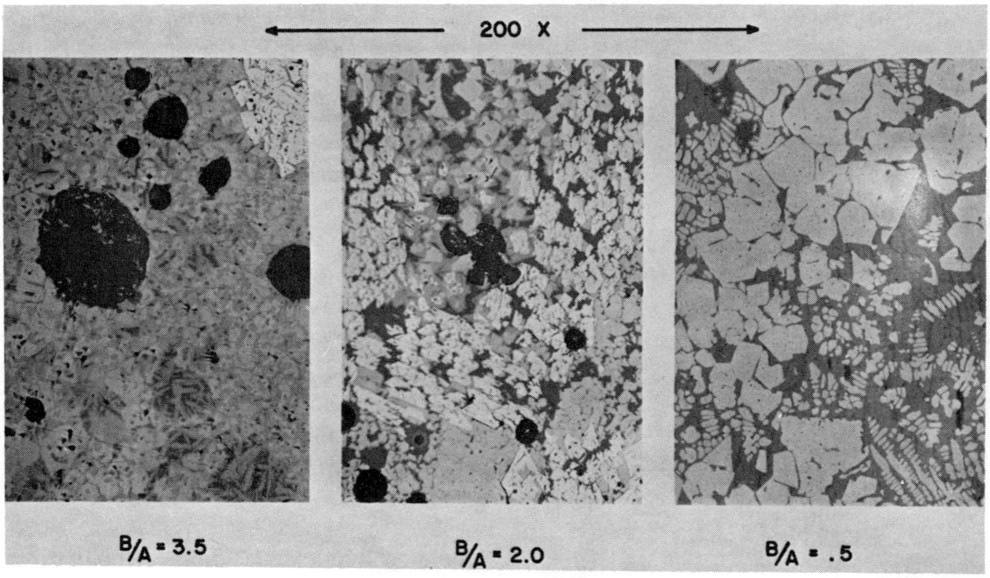

Figure 7.7.14. Mineral structure of acid, fluxed, and superfluxed sinter.

hematite is not yet well-developed, and dicalcium silicate crystals disrupt the glass matrix.

Superfluxed Sinters. At base-to-acid ratios of 2.0 to 4.0, sinter is well within the superfluxed range. Dicalcium ferrite is now a major constituent, whereas the glass and dicalcium silicate are minor. Bridging of the iron oxides is enhanced and in fact approaches a monolithic bond where the iron oxide grain boundaries are no longer evident as in the original bridging. Because of the monolithic nature of the dicalcium ferrite matrix, sinter strength becomes very high at 4.0 basicity.

Figure 7.7.15 illustrates the characteristic curve for sinter strength as the product goes from acid to self-fluxed to superfluxed. Sinter strength was determined by the ASTM standard tumbler test for sinter, this being the test most generally used in the U.S. As for reducibility, because ferric iron reduces at a faster rate than ferrous iron, the reducibility of sinter increases as the base-to-acid ratio increases. Thus, superfluxed sinter, which contains hematite and monocalcium and dicalcium ferrite, is more easily reduced than acid sinter, which contains magnetite and lime-iron-silicate glass.

The Range Between Self-Fluxed and Superfluxed. At base-to-acid ratios of 1.5 to 2.0, sinter is approaching the superfluxed range.

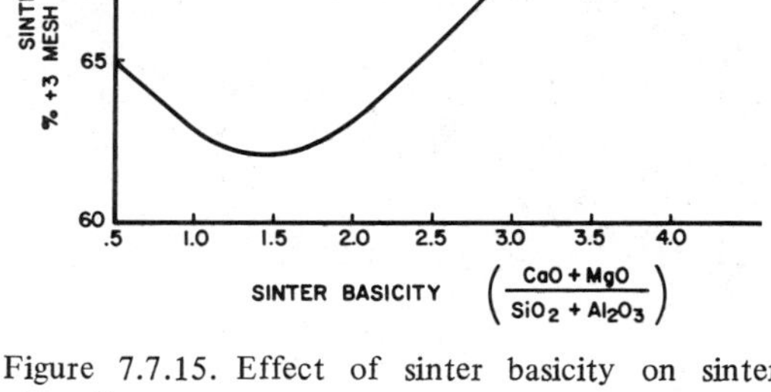

Figure 7.7.15. Effect of sinter basicity on sinter strength.

More of the magnetite has oxidized to hematite. Bridging of the hematite grains is more pronounced. Lime-iron silicate glass is less apparent, but dicalcium silicate is more pronounced. Dicalcium ferrite is now present and, along with iron oxide bridging, plays a major role in strength of the product, which is now on an upward curve.

7.7.9. ENVIRONMENTAL QUALITY CONTROL OF SINTERING OPERATIONS

Environmental quality control systems command an increasingly larger proportion of the capital and operating costs of our sintering operations. All operating sinter plants emit particulate and gaseous pollutants and, to some degree, liquid pollutants. These pollutants are generated during: (1) the handling and preparation of raw materials prior to sintering, (2) the sintering process itself, (3) cooling of the sinter, (4) sizing and handling of the sinter product, and finally (5) processing of the large volume of gas used for sintering.

Particulate emissions are: (1) dusts entrained from ores, fluxes, solid fuel, and the hot and cold return fines as they are prepared for sintering, (2) dust entrained in the windbox system that consists of sintered and unsintered materials, and (3) dusts entrained during cooling, sizing, and handling of the sinter product.

Gaseous emissions are carbon monoxide, carbon dioxide, sulfur oxides, nitrogen oxides, hydrocarbons, water vapor, and possibly fluorides. These pollutants are generated during the sintering process as products of combustion and high-temperature chemical reactions. All of these gaseous pollutants are contained within the windbox exhaust system, with the possible exception of minor amounts generated in the cooler if sintering has not been completed on the strand.

Liquid pollutants are those soluble sulfates, chlorides, fluorides, condensible hydrocarbons, and other particulate solids that are contained in scrubber liquor after cleaning of the windbox exhaust gas.

Up until the mid-1960s nearly all of the operating sinter plants, both here and abroad, used some combination of cyclone mechanical collectors and dry electrostatic precipitators, or baghouses, to collect particulate emissions. Although a great deal of research was being conducted on control of fine particulates and gaseous pollutants, very few full-scale sintering operations included facilities capable of capturing these constituents. Two factors greatly accelerated the work on the effective cleaning of gaseous emissions as well as improved particulate control. The first factor was the legislation, at all governmental levels, of strict regulations for sinter plant environmental quality control. The second factor was a pronounced change in the sintering process itself when superfluxed sinter with basicities of 2.0 and often as high as 4.0 was introduced. The use of superfluxed sinter resulted, in many instances, in reduced coke consumption and increased hot metal production in blast furnace practice. However, superfluxed sinter dust has a much higher resistivity than acid or fluxed sinter dust.[11] High-resistivity dust greatly decreases the efficiency of dry electrostatic precipitators, and, as a result, most plants so equipped found they could no longer meet emission regulations.

The steel industry, sinter plant design and construction firms, and pollution control equipment firms embarked on intensive programs with the objective of meeting environmental quality control regulations. A number of systems have been piloted or tried at full scale, such as:

- Afterburning the windbox exhaust gas
- Recycling the windbox exhaust gas through the hottest portion of the sinter strand
- Gravel-bed filters for the windbox exhaust gas
- Chemical additives sprayed into the exhaust gas
- Baghouse filters
- High-energy wet scrubbers
- Wet electrostatic precipitators
- Removal from the sinter mix of materials that generate hydrocarbons such as mill scale

Some degree of success or failure was found for each of these systems. There remains, however, the challenge of designing a system that will effectively clean windbox exhaust gas streams containing high-resistivity dust, ultrafine particulate matter, condensible hydrocarbons, and gaseous pollutants for the lowest possible capital and operating costs. This challenge cannot be avoided, because in many cases we must recycle the oily mill scales and other pollutant-bearing ores, fluxes and revert materials, and also because superfluxed sinter is a valuable blast furnace burden material.

7.7.10. USE OF SINTER IN THE BLAST FURNACE

Over a period of 65 years the use of sinter in the blast furnace has evolved from charging everything that came off the end of the sinter strand, including all the fines, to charging a very strong closely sized, chemically tailored product.

Replacing Direct Ore

Recently reported trials carried out on No. 4 blast furnace at the Cornigliano plant of Italsider demonstrated the effect of replacing sized ore with fluxed sinter.[12] Figure 7.7.16 shows the gains in hot metal production and the decreased coke consumption as fluxed sinter replaced sized ore. Sinter basicity, defined by Italsider as $CaO \div SiO_2$, was about 1.58 for the various levels of sinter-sized ores shown on the graph. The blast furnace, with a working volume of 644 m^3 was able to use nearly 10% more wind at the maximum replacement of sinter for sized ore.

Sized Sinter

There are many well-documented reports of the improvements in blast furnace operation from the use of sized sinter, but three of these from

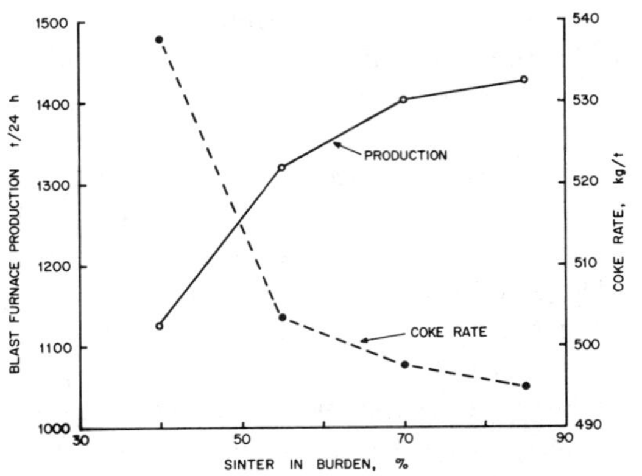

Figure 7.7.16. Improvement in blast furnace operation with sinter replacing sized ore. (Cornigliano plant, Italsider.)

three different countries will suffice to establish the economic benefits of this practice. An American example is taken from experience at the Bethlehem, Pennsylvania, plant of Bethlehem Steel Corporation. Two sizes of sinter, "nut" 38.1 mm × 12.7 mm ($1\frac{1}{2}$ in. × $\frac{1}{2}$ in.) and "pea" 12.7 mm × 6.4 mm ($\frac{1}{2}$ in. × $\frac{1}{4}$ in.), were produced by careful screening practice and tested in "B" furnace.[13] The use of nut-sized sinter resulted in a hot-metal production gain of 16% and a 3% decrease in coke rate as compared with the base period. The use of the pea-sized sinter resulted in a hot-metal production gain of 10% and a 6% decrease in coke rate compared to the base.

At the Mizue works of Nippon Kokan a crushing and screening plant was installed to size all burden materials, primarily sinter, to 41.28 mm × 4.8 mm.[14] As a result of the use of 66% sized sinter in the blast furnace compared to 73% unsized sinter in the base period, production rose from 2971 to 3377 NTHM per day, and the coke rate decreased from 434 to 423 kg per NTHM per day. The $CaO \div SiO_2$ ratio of the sized and unsized sinter burdens used was 1.10.

In 1971 at the Peine-Salzgitter Ag plant in Germany a new sinter crushing and screening plant was put into operation to size sinter at 45 mm × 5 mm.[15] A 30-day test comparing sized to unsized sinter showed a production increase from 1640 to 1960 NTHM per day and a decrease in coke consumption from 461 to 434 kg per ton.

Use of Fluxed and Superfluxed Sinter

In a review of blast furnace burden materials used in the United Kingdom the author states, "the benefits of self- and super-fluxed sinters are realized both at the sinter plant, where higher productivities, higher strength, and higher reducibility are obtained, and at the blast furnace, where the removal of raw flux from the burden and the improved sinter result in marked improvements in furnace productivity and coke rate."[16]

A unique approach to sinter fluxing was developed by Republic Steel in 1968. Sinter was produced with the objective of forming as much of the mineral dicalcium ferrite as possible.[17] This objective was based on laboratory tests that showed that dicalcium ferrite sinter is characterized by high strength, high productivity, and excellent reducibility. Subsequently, large tonnages of dicalcium ferrite sinter, which had a basicity of 4.5, were tested in the blast furnace. Hot-metal production increased from 1,366 to 1,567 NT per day, and the coke rate decreased from 703 to 605 kg per NTHM per day. These improvements were accomplished with 22% of the dicalcium ferrite sinter in the furnace burden replacing sized ore.

At the Lackawanna, New York plant of Bethlehem Steel Corporation, the sinter basicity in February 1966 was increased from 1.0 to 2.3.[18] The percentage of this superfluxed sinter on "J" furnace was increased from 12 to 53% while the acid pellet burden was reduced from 85 to 47%. During this period the hot-metal production increased from 2,828 to 3,035 NT per day, the coke consumption decreased from 427 to 415 kg per NT of hot metal. Raw stone charged to the furnace was reduced to 0 kg per ton of hot metal from a previous 241.

Sinter Versus Pellets

One of the world's largest blast furnaces, Fukuyama No, 5, has averaged over 11,000 NTHM per day with a burden of about 75%

fluxed sinter and 25% screened ore. Generally speaking, the use of high-sinter burdens is practiced to a greater extent in Asia and Europe than in North America.

Logistics and economics have resulted in high-pellet availability in North America, and a number of operators have first-hand experience with high-pellet burdens and the benefits that can be obtained from this practice.

Nevertheless, rather than compete with each other, sinter and pellets should complement one another. For example, Inland Steel reported an average daily production of 1909 NTHM per day, or 5.65 NTHM per ft^2 of hearth area for October 1965.[19] This level of production was achieved with a burden of 73% acid pellets and 23% superfluxed sinter having a basicity of 3.35. The coke rate was 479 kg per NTHM, the raw stone to the furnace was reduced to 37 kg per NTHM. The campaign of the previous year, without use of superfluxed sinter, averaged 1,774 NTHM per day and 498 kg of coke per NTHM. Thus, a gain of 135 NTHM per day was achieved and a reduction of 19 kg per NTHM in coke rate was realized by employing a burden in which both superfluxed sinter and acid pellets played their parts.

The majority of North American blast furnace operators deal with some combination of sinter and pellets for blast furnace use. By practicing the sinter technology that is available, i.e., the production of strong closely sized, chemically consistent sinter with high flux capability, our blast furnaces will respond favorably.

7.7.11. INDUSTRY TRENDS IN SINTERING

In discussing emerging trends in industrial sintering practice, both here and abroad we will briefly look at:

- Machine size
- On-strand cooling
- Retrofitting of sinter plants
- Automation and computer control
- Research and development

Machine Size

Sinter machine size is going the same route as blast furnace size. Bigger and bigger single units are replacing existing smaller units or multiples of smaller units. Sinter machines 4 m wide × 100 m long are no longer uncommon, especially in Europe, Russia, and Japan.[20] For example, presently in Japan two machines with pallets 5 m wide × 120 m long are under construction. Each of these big machines is designed to produce 27,000 tons per day or about 9.45 million tons per year of sinter for blast furnace consumption. To meet this design capacity, these plants must be available 96% of the time. Carefully programmed mechanical-electrical maintenance programs are scheduled to meet this availability. Replacement of high-wear units such as the toothed sinter breakers, grizzly screen, and hot sinter screen must be made on a standby basis with spare-units availability. Computer control of the operation is required.

On-Strand Cooling

Complete cooling of sinter on the strand was practiced by Dwight and Lloyd in their original work. Many of the original machines were built to accomplish cooling on the strand, but the ever-increasing push for more hot metal soon accelerated sinter machine speeds to produce more tons. In the last several years on-strand cooling has gained considerable interest, and a number of such units have been and are being built in Europe.[21]

Advocates of strand cooling claim that the system has the advantages of:

Less pollution
Less mechanical wear due to no handling of hot sinter
Less return fines because cooled sinter crushes easier than hot sinter
More uniform size of sinter product due to crushing characteristics
Savings in sinter-mix solid-fuel requirements

Those opposed say that the process is more difficult to control because of:

Changes in burnthrough point with changes in mix composition

Increase in maintenance of the strand as a result of its greater length

Increased difficulty of pallet tracking because of the increased number of pallets

Limited flexibility of ore fines receipts to the plant

Both parties claim capital cost advantages, but this is certainly a point of contention. On a positive note, if a long-term stable ore supply is indicated and the sintering-to-cooling ratio of the strand is about 1 to 1, then there is a good chance that on-strand cooling will be economical. There is also the prospect that if blast furnace burden requirements are increased, the means are at hand to nearly double sinter production by adding a cooler to the end of the strand. Figure 7.7.17 illustrates one presently used method of on-strand cooling.

Retrofitting of Sinter Plants

In terms of utilizing existing installations, the fixing up of old sinter plants, i.e., retrofitting, is an economically important route. This route applies to a great many of the steel companies presently operating and can be implemented quite effectively.[22] Our objectives have not changed in that we still seek to produce strong chemically consistent, closely sized sinter for the blast furnace. Each operation in the plant must be examined and recommendations made for improvements. For instance, table-feeder rates can be more closely controlled by installing ratchet gates whose positioning is controlled on the basis of frequent weight samples collected with a quarter-circle pan. Pug mills may be replaced with roll-feed drums, perhaps not as large as desired, given the space restrictions, but an improvement nevertheless. Moisture content of the mix can be controlled by installing electrical-resistance or nuclear devices. Swinging spouts can be replaced by roll feeders. Ignition furnaces can be improved and tempering zones added. Induced-draft fans can be increased in size to provide increases in production. Improvement in cooling, crushing, and sizing of the final product can all be engineered. Windbox exhaust gas-cleaning and environmental control can be engineered for compliance with regulations. These are only the highlights, and the list of units for improvement could go on.

Automation and Computer Control

Computer control of the bedding and blending operation and computer control of the sintering process have been described earlier. Integrating these control segments into one programmable computer provides the basis for a fully automated sinter plant. The technology of sensing devices, control systems, and coordinating systems is available, and new green-field instal-

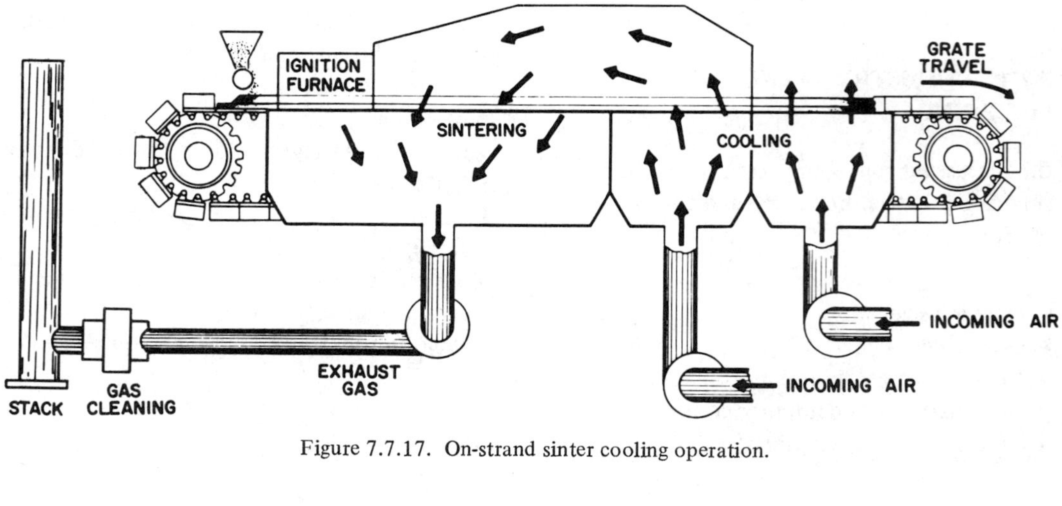

Figure 7.7.17. On-strand sinter cooling operation.

lations are already utilizing or are planning to utilize automated and computer-controlled sinter production.[23]

Research and Development

Sinter plant engineering, design, and construction companies and most steel corporations have ongoing sinter research programs. Laboratory sintering facilities employed by both of these groups are capable of a wide range of both fundamental and applied research. There are many areas of sintering research and development that are currently under way. A few of the more active areas are: (1) raw-mix chemistry control, (2) control of the ferrous content of sinter product for maximum reducibility, (3) mathematical modeling of the process, (4) computer control of the process, and (5) automation of the process. The publication of about 850 technical articles on iron ore sintering since 1960 is evidence of the worldwide interest in the subject. The rising costs of raw materials, utilities, and labor dictate that those involved in sintering seek out the most economical methods of producing superior quality sinter for the blast furnace.

7.7.12. THE ECONOMICS OF SINTER

The economics of sinter, as in the case of all blast furnace burden materials, is important because of its effect on the cost of producing hot metal.

Table 7.7.3 illustrates sinter plant capital and operating costs.

Capital costs

The economics of scale must be reckoned with. For example, there may be local conditions where construction of excess sinter capacity today will be more economic than waiting until additional tons of burden material are needed in the foreseeable future because of projected plant expansion.

The high cost of meeting environmental regulations is reflected in the percentage of the total capital costs designated for this purpose.

Operating Costs

Operating costs are presented in the manner shown in Table 7.7.3 so that local wages, utilities, and supplies may be computed to give

Table 7.7.3. Sinter Plant Capital and Operating Costs.

CAPITAL COSTS

1.0 $\overline{\text{M}}$ NT/year product plant ~ $46/annual NT product
2.0 $\overline{\text{M}}$ NT/year product plant ~ $32 annual NT product

Notes: (a) Estimates are for green field site, within battery limits of plant.
(b) No bedding-blending facilities included.
(c) Estimates include a high-energy wet scrubber system.
(d) Environmental quality control costs total 25 to 30% of the capital cost estimate.

OPERATING COSTS

Supervisory labor	.015 man-month/NT product
Operating labor	.060 man-month/NT product
Maintenance labor	.050 man-month/NT product
Supplies	90 ¢/NT product
Coke breeze	22.7 to 45.4 kg/NT product
Ignition fuel	250,000 BTU/NT product
Power	65 kWh/NT product
Water	1514 liter/NT product
Chemicals (for water treatment system)	22 ¢/NT product
Laboratory, freight, misc.	15 ¢/NT product

Notes: (a) Power costs include high-energy wet scrubber operating at 60 in. pressure drop.
(b) Costs do not include ores, fluxes, additives, plant depreciation.

a final operating cost, less the cost of ores. The delivered-ore costs and flux costs per ton of product added to the "cost above" will give the final plant cost of sinter product for the blast furnace.

Justification of Capital for Sinter Plants

Capital expenditures have to be justified on some designated rate-of-return basis. This principle applies to the refitting of older sinter plants as well as to new installations at green-field sites. Justification must be based on the increased value of the new or of the improved sinter product to the blast furnace. Some of these considerations are:

1. Higher flux capacity of sinter can reduce coke consumption in the blast furnace.
2. Additional tons of sinter can decrease the higher cost of pellet units.
3. Sized sinter with minimum fines can increase furnace production and reduce coke rates.
4. Increased efficiency of operation can reduce sinter plant maintenance cost.

A widely used method of computing savings from improving both the burden materials and the blast furnace practice is based on articles published by R. Flint.[24]

7.7.13. SUMMARY

Early in this century, continuous-downdraft sintering emerged as a viable process for agglomerating those iron-bearing materials that were too finely divided for direct charging to the blast furnace. The growth of sintering throughout the world closely followed the growth of the steel industry. Likewise, sintering technology has advanced, often by necessity, with the advancing technology of blast furnace practice. Major advances in the product were the development of sized, fluxed, and superfluxed sinter. The past 25 years have seen other advances in the bedding and blending of sinter mix, machine design and construction, ignition, sinter cooling, process control, including the use of computers, and, last but not least, environmental quality control. Increased hot-metal production and reduced coke rate are the benefits derived from using high-quality sinter in the blast furnace. Presently, major concerns for both new and old facilities are the capital and operating costs of the sintering operation, including those for protecting the environment. Today, sintering is the one agglomeration process that has the versatility to accept large varieties of physically and chemically variable materials and to convert them into a strong chemically consistent, closely sized product for use in the blast furnace.

ACKNOWLEDGMENT

This chapter is based on a lecture, "Iron Ore-Sintering," that is part of the Blast Furnace Ironmaking course sponsored by McMaster University, Hamilton, Ontario, Canada. The author thanks the organizing committee of the course, in particular D. W. R. George of The Steel Company of Canada, Chairman and Dr. W-K Lu, Department of Metallurgy, McMaster University for their kind permission to publish this material.

REFERENCES

1. F. Vogel, "Sintering and Briquetting of Flue-Dust," *Transactions of the American Institute of Mining Engineers*, vol. XLIII, pp. 381–386 (1913).
2. J. Gayley, "The Sintering of Fine Iron-Bearing Materials," *Transactions of the American Institute of Mining Engineers*, vol. XLII, pp. 180–190 (1911).
3. B. Klugh, "The Sintering of Fine Iron-Bearing Materials by the Dwight & Lloyd Process," *Transactions of the American Institute of Mining Engineers*, vol. XLIII, pp. 364–375 (1913).
4. Y. Doi and K. Kasai, "The Making of Self-Fluxing Sinter and the Blast Furnace Practice With Its 100 Per Cent Burden," Blast Furnace, Coke Ovens, and Raw Materials Committee, *AIME*, vol. 18, pp. 182–217 (1959).
5. N. Standish and W-K Lu, "Alakalis In Blast Furnaces," Proceedings of Sympsium On Alkalis In Blast Furnaces–State of the Art, (1973).
6. P. Seaton and F. Traice, "Development of a Model for the Prediction of Sinter Plant Productivity from Ore Properties," *Ironmaking Proceedings 35*:539–564 (1976).
7. R. Limons and H. Kraner, "An Evaluation of the Properties of Dolomite Affecting Self-Fluxing

Sinter Rates," *Agglomeration*, Interscience Publishers, New York, pp. 117 (1962).

8. R. Limons, et al., "The Use of Burned Lime as Partial Replacement for Dolomite Flux in Sinter Mixes," *Ironmaking Proceedings*, *23*:303–323 (1964).
9. T. Myron, et al., "Fuels for Sintering," Proceedings, Blast Furnace, Coke Oven, and Raw Materials Committee, *AIME 17*:282–293 (1958).
10. G. Bruynel, et al., "The Influence of Ore Bedding on the Sintering Operation," International Iron and Steel Congress, Düsseldorf, vol. 1 (1974).
11. J. Varga, "Control of Reclamation (Sinter) Plant Emissions Using Electrostatic Precipitators," Battelle Columbus Laboratories, Columbus, Ohio (1976).
12. G. Sironi, "Influence of Burdening and Charging Practices on Blast Furnace Performance and Control," International Iron and Steel Congress, Brussels, pp. 3.2.2. (1976).
13. J. Breitenstein, et al., "Operating Practice and Production Benefits from Sized Burdens on the Bethlehem *B* Furnace," *Ironmaking Proceedings 28*:471–477 (1969).
14. S. Horie, "Blast Furnace Operation to Make More Than 3300 T/D Hot Metal With Coke Rate Less Than 950 Lbs.," Annual Meeting of AIME, Chicago (Feb. 1965).
15. E. Teichert and V. Peome, "Change in the Operating Characteristics of a Blast Furnace Due to Charging Sized Sinter," *Stahl und Eisen 94*(4): 133–136 (1974).
16. J. Darnell, "Effect of Burden Quality on Furnace Productivity and Efficiency," *Journal of the Iron and Steel Institute* (March 1969).
17. M. Fedock, et al., "Dicalcium Ferrite (C_2F) Sinter–Its Development, Production and Use," *Ironmaking Proceedings 27*:144–149 (1968).
18. R. Limons and C. Felton, "Superfluxed Sinter Practice at Bethlehem Steel Corporation's Lackawanna Plant," *Ironmaking Proceedings 25*:78–81 (1966).
19. J. Kapitan, "Use of High Basicity Sinter and Pellets for High Rates of Hot Metal Production," *Ironmaking Proceedings 28*:217–224 (1969).
20. H. Beer, et al., "New Developments in Sintering Iron Ore," *Stahl und Eisen 96*(7):8.4 (1976).
21. F. Kinsey and D. Kestner, "Design Parameters for Strand Cooling of Sinter," *Ironmaking Proceedings*, vol. 34 (1975).
22. H. Nuessel and J. Otto, "Possibilities of Improvements to Existing Sinter Plants," *Ironmaking Proceedings 35*:565–581 (1976).
23. R. Greenawalt and S. Sephton, "Automation for Superfluxed Sinter Production," *Ironmaking Proceeding 29*:89–96 (1970).
24. R. Flint, "Effect of Burden Materials and Practices on Blast Furnace Coke Rate," American Iron and Steel Regional Technical Meeting, Chicago (1961).

SELECTED BIBLIOGRAPHY

Ball, D. F., Dartnell, J., Davison, J., Grieve, A., and Wild, R., *Agglomeration of Iron Ores*, American Elsevier, New York (1973).

Cappel, F., and Wendeborn, E. H., *Sintern von Eisenerzen*, Verlag Stahleisen, Dusseldorf (1973).

Developments in Ironmaking Practice, The Iron and Steel Institute, London (1973).

Knepper, W. A., *Agglomeration*, Interscience Publishers, New York (1962).

McGannon, H. E., *The Making, Shaping and Treating of Steel*, 8th ed., United States Steel Corporation, Pittsburgh (1964).

Ore Mining and Materials Handling, Special Report No. 82, The Iron and Steel Institute, London (1963).

Roe, L. A., *Iron Ore Beneficiation*, Minerals Publishing Co., Lake Bluff, Ill. (1957).

Second Symposium International Sur L'Agglomeration, IRSID, St.-Germain-en-Laze (1957).

Strassburger, J. H., *Blast Furnace–Theory and Practice*, Gordon and Breach Science Publishers, Inc., New York (1969).

Symposium on Sinter, Special Report No. 53, The Iron and Steel Institute, London (1955).

PART 8. AGGLOMERATION IN LIQUID SYSTEMS

C. E. Capes and A. E. Fouda
National Research Council
Ottawa, Ontario

7.8.1. INTRODUCTION

Although fine *dry* powders present difficulties such as dusting losses and other handling hazards, finely divided materials *in liquids* are also difficult to deal with. The size of the individual particles is often so small that methods to capture them (such as filtration) are difficult unless some form of size enlargement is applied.

Traditional procedures for agglomerating fine particles in liquids, such as flocculation, rely on relatively small interparticle bonding forces to form rather weak cluster-type agglomerates which occupy a large volume. Often the objective is simply to remove the fines from the liquid medium. In contrast, the present discussion deals with those techniques in which stronger bonding and specialized equipment are used to form generally larger and more permanent agglomerates in liquid suspensions. In addition to separation of particles from suspensions, these latter methods have other broad objectives as shown in Table 7.8.1, including the production of granular (often spherical) materials, maximum displacement of suspending liquid from the product, and the selective agglomeration of one or more components of a multiparticle mixture.

7.8.2. MECHANISMS OF LIQUID-PHASE AGGLOMERATION

In the growth of agglomeration technology in liquid systems, three broad types of processes have evolved. Two of these rely on different bridging mechanisms to pull suspended particles together into larger agglomerates. The third involves conversion of the solid feed into a liquid form, which is then dispersed as droplets in a second liquid, followed by solidification to the final particulate product.

Wetting by Immiscible Liquids[1]

Fine particles in liquid suspension can readily be formed into large dense agglomerates of considerable integrity by adding suitable amounts of a second or bridging liquid under appropriate agitation conditions. This second liquid must be immiscible with the suspending liquid and must wet preferentially the solid particles that are to be agglomerated. A simple example is the addition of oil to an aqueous suspension of fine coal. The oil readily adsorbs preferentially on the carbon particles and forms liquid bridges between these particles by coalescence during collisions under the agitation conditions. Inorganic impurity (ash) particles are not wetted by the oil and remain in unagglomerated form in the aqueous slurry.

The agglomeration phenomena that occur as progressively larger amounts of bridging liquid are added to a solids suspension are depicted in Figure 7.8.1. The general relationships shown are not specific to a given system. Figure 7.8.1

Table 7.8.1. Some Important Agglomeration Processes Carried Out in Liquid Systems.

PROCESS OBJECTIVE	MATERIAL TREATED AND PROCESS USED	REFERENCES
Sphere formation and production of coarse granular products	Nuclear fuel and metal powder production by sol-gel processes	9, 10
	Manufacturing of small spheres from refractory and high-melting-point solids (e.g., tungsten carbide) by immiscible liquid wetting	7, 8
Removal and recovery of fine solids from liquids	Removal of soot from refinery waters by wetting with oil	17
	Recovery of fine coal from preparation plant streams to allow recycling of water	13, 14
Displacement of suspending liquid	Dewatering of various sludges by flocculation followed by mechanical drainage on filter belts, in revolving drum, etc.	4, 15
	Displacement of moisture from fine coal by wetting with oil	13, 14
Selective separation of some components in a mixture of particles	Removal of ash-forming impurities from coal and from tar sands by selective agglomeration	14, 18

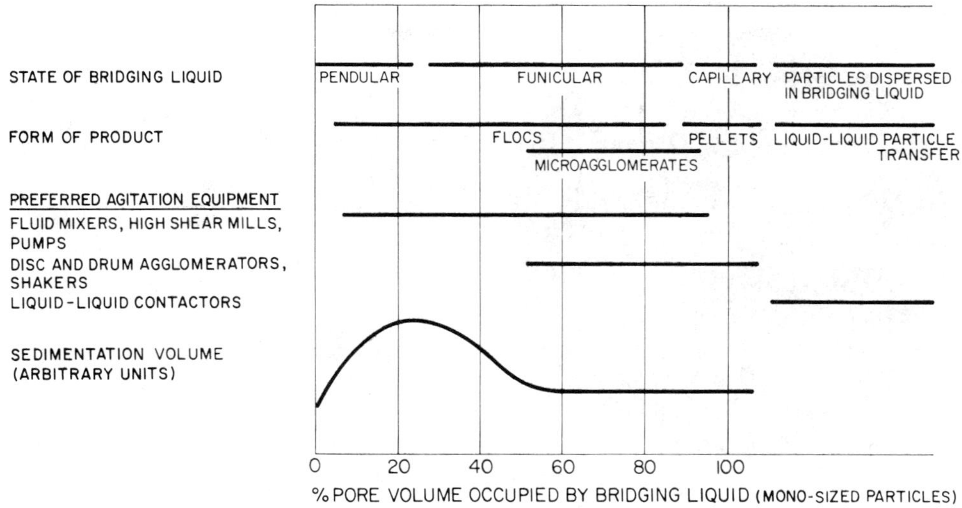

Figure 7.8.1. "Phase diagram" for agglomeration by immiscible liquid wetting.[1] The effect of increasing amounts of bridging liquid on the process.

relates equally well, for example, to siliceous particles suspended in oil and collected with water, or to coal particles suspended in water and agglomerated with oil.

At low levels of bridging liquid, only pendular bridges can form between the particles, with the result that an unconsolidated floc structure exists. As seen in the lower part of Figure 7.8.1, a loose settled mass of volume larger than that of the unflocculated particles results. As the funicular region of bridging-liquid levels is reached, the flocs consolidate somewhat, and lower settled volumes are recorded. Some compacted agglomerates appear and increase in number, until about midway in the funicular region the whole system has been formed into "Microagglomerates." As the amount of bridging liquid is increased, the agglomerates grow in size and reach a peak of strength and sphericity near the capillary region. Beyond this region the agglomerates exist as pasty lumps; the solids are then essentially dispersed in the bridging liquid. Figure 7.8.2 shows these different stages in the development of coal agglomerates.

Bridging With Polymeric Flocculants

A wide range of polymeric agents[2,3] (e.g., polyacrylamides) is available today to aid the aggregation and subsequent removal of fine particles from water. These polymeric flocculants, due to their large molecular size, cause aggregation of particles by a bridging mechanism in which several particles are united by adsorption onto one molecule of flocculant. The agglomeration of particles into a floc structure results in faster settling of the suspension and allows the supernatant liquid to be recovered more quickly. Flocculated particles, however, tend to stick to each other as they settle and form a loose, bulky layer. Although the pores in the settled layer are relatively large and its filtration and drainage rate is thus enhanced, the high porosity of the settled layer means that a larger proportion of suspending medium is often retained by the flocculated material than is the case with the unflocculated particles.

Techniques have been developed to form more compact sediments (agglomerates) of reduced liquid content in flocculated systems. These techniques, sometimes known as "pellet flocculation,"[4] combine relatively large amounts (a few pounds per ton of solids) of polymeric flocculants with gentle agitation, such as a rolling tumbling action, to reduce the moisture content of the separated solids. The agglomerates thus formed contain more interparticle

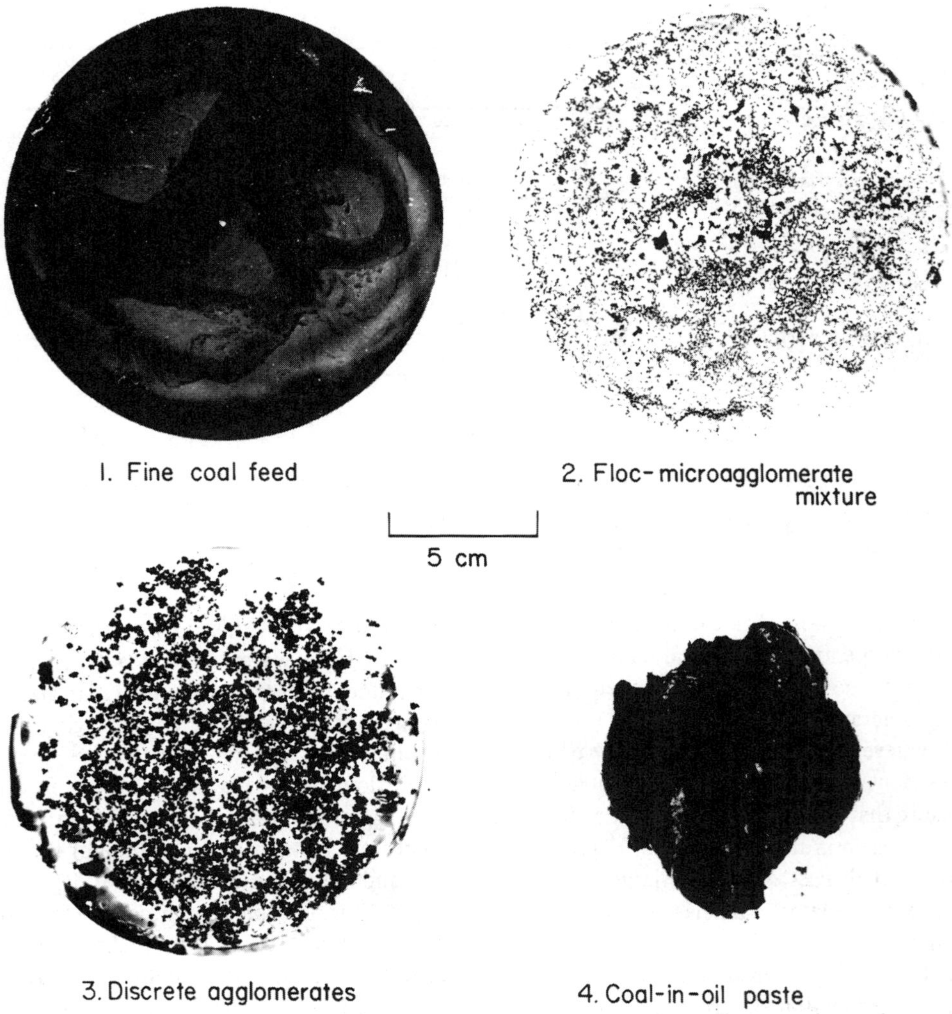

Figure 7.8.2. Coal agglomerates formed with increasing amounts of bridging oil.

bridges than with lower polymer levels, are able to grow to a larger size permitting easier separation from the liquid phase, and are strong and pliable enough to allow entrapped liquid to be squeezed out under mechanical working.

Dispersion in Liquids

A number of processes exist in which solid materials are converted into a liquid form, dispersed as droplets in a second liquid by some suitable means and solidified to a final particulate product. When the starting material is a massive solid, the process is then one of size reduction.[5] When a powder feed is used, size enlargement results.

Many variations are possible depending on the method used to disperse the liquid phase and on the procedure used to harden the droplets. For example, the feed liquid can be prilled into a quiescent column of the second liquid or dispersed by mechanical agitation. Hardening the droplets can be accomplished by chemical reaction, by cooling, by solvent extraction, by evaporation, or by combinations of these methods.

7.8.3. PROCESSES AND EQUIPMENT

A number of specific processes will now be described in which the various mechanisms of liquid-phase agglomeration discussed above are utilized.

Spherical Agglomeration Processes

Spherical shapes are required for a variety of applications, many of which are associated with the field of powder metallurgy. While it is relatively easy to produce spheres by conventional techniques such as shot or prilling towers, refractory solids in general and high melting point metals do not readily respond to these methods. If the solid is available in powder form, however, various techniques are available to agglomerate the powder into highly spherical shapes.

One attractive method uses immiscible liquid wetting to pull the powder together into agglomerates while suspended in a second fluid under highly energetic agitation. Such methods are part of the family of techniques generally known as "Spherical Agglomeration Processes."[6] In this operation, the interparticle bonds formed by the immiscible liquid bridges, especially in the capillary region, are very rugged and are readily replaced if they become dislocated. Consequently, intense mechanical energy may be used to produce dense, highly spherical agglomerates. Compaction and rounding is facilitated by multitudinous collisions between the agglomerates themselves and the container walls. Where unit batch size is small, high-energy shaking devices may be utilized to optimize sphericity, size distribution, and density of the spherical agglomerates. The advantage of such a process is that a finishing operation, such as lapping and grinding after a preliminary sintering step, is much reduced when compared to that necessary for spherical powder compacts made by other techniques, for example by press molding.

A typical example[7] of this technique is the manufacture of small tungsten carbide spheres for use as blanks in the manufacture of ball-point pens. The closely sized spheres, 1 mm diameter, are prepared by agitating tungsten carbide and cobalt powders, in a closed Teflon container with hemispherical ends, on a high-speed reciprocating shaker (see Fig. 7.8.3). Halogenated solvents are used as the carrier liquid and water as the bridging liquid. The addition of about 6% cobalt to the tungsten carbide powder is required to reduce sintering temperatures to more acceptable levels. Co-agglomeration of the mixed powders from liquid suspension tends to reduce segregation of the powders, ensuring a more homogeneous spherical product.

Energy-intensive batch agitators such as shown in Figure 7.8.3 are most suited to pro-

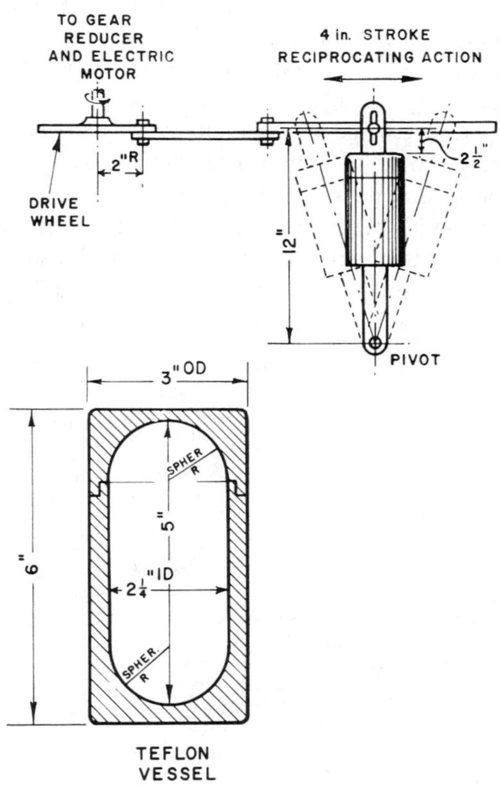

Figure 7.8.3. Teflon cylinder with hemispherical ends mounted in reciprocating shaker used to form small spheres by the Spherical Agglomeration Process[7,20] Typical conditions include 75 cm^3 carbon tetrachloride containing 200 g tungsten carbine powder, which is agglomerated with about 10 cm^3 water. Shaking speed 300 cpm for 5–7 minutes. (Reprinted with permission from Ref. 20. Copyright 1967 by the American Chemical Society)

ducing small spheres (1-2 mm diameter) since a large number can be made in a single small batch. Where comparatively high production rates are required, rotating drums and disc agglomerators are better suited to the process. In these tumbling agglomerators, the presence of a liquid slurry of feed is useful in reducing the dust nuisance that may be a problem, especially with toxic powders, when conventional dry granulation is used. The blanket of suspending liquid is helpful in at least two additional aspects. First, the phenomenon of "snowballing" is much reduced since the suspending liquid helps to disperse the bridging liquid uniformly throughout the agglomerating mass and the liquid turbulence opposes rapid agglomeration, allowing particles to layer into larger entities in a controlled manner. The liquid blanket also aids in developing a desirable tumbling, cascading motion in the equipment since the charge is more voluminous and better interparticle lubrication prevails than would be the case if no suspending liquid were present. The solids tend to be carried with the fluid. Indeed, this semifluid nature of the agglomerating mass has been used[8] to operate successfully a balling drum with an internal screen classifier.

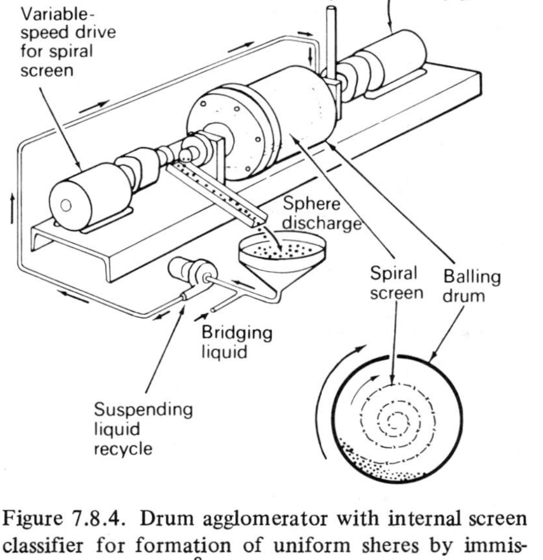

Figure 7.8.4. Drum agglomerator with internal screen classifier for formation of uniform sheres by immiscible liquid wetting.[8]

As shown in Figure 7.8.4, the horizontal, rotating drum contains a spiral screen rotating at a slower speed. This screen continually passes through the pellet charge. Undersize pellets return to the drum through the screen, while the onsize material progresses along the spiral until it reaches the axis of the drum. A hollow tube at the axis then directs the pellets to a discharge point outside the drum. Highly spherical products with very uniform size distributions have been produced in this equipment. For example, carbonyl iron spheres with a median diameter of 3.9 mm and a size spread of ±0.3 mm about this median were produced at a rate of about 40 g/min using a drum 280 mm diameter by 280 mm long. A light petroleum solvent was used to suspend the 10 μm powder and water at a rate of about 4 cm^3 min. was added as the agglomerating agent. Drum speed was in the range 60-100 rpm while the spiral screen revolved in the same direction at 4-6 rpm.

Sol-Gel Processes

In agglomeration by immiscible liquid wetting, small amounts of a bridging phase adsorbed on the particles coalesce to draw the particles into larger entities. In the sol-gel process, fine particles are initially suspended in an excess of a bridging phase; the suspension is formed into spherical droplets, and the excess bridging phase is removed to solidify the droplets into a particulate product.

The sol-gel process has been actively developed[9,10] for the preparation of spherical oxide fuel particles up to about 1000 μm diameter for nuclear reactors. The following operations are involved in converting the initial aqueous sol of colloidal particles into calcined microspheres:

1. Dispersion of sol into droplets
2. Suspension of droplets in an immiscible liquid that will extract water to cause gelation
3. Separation of gel microspheres
4. Recovery of immiscible liquid for reuse
5. Drying, calcining, and sintering of microspheres

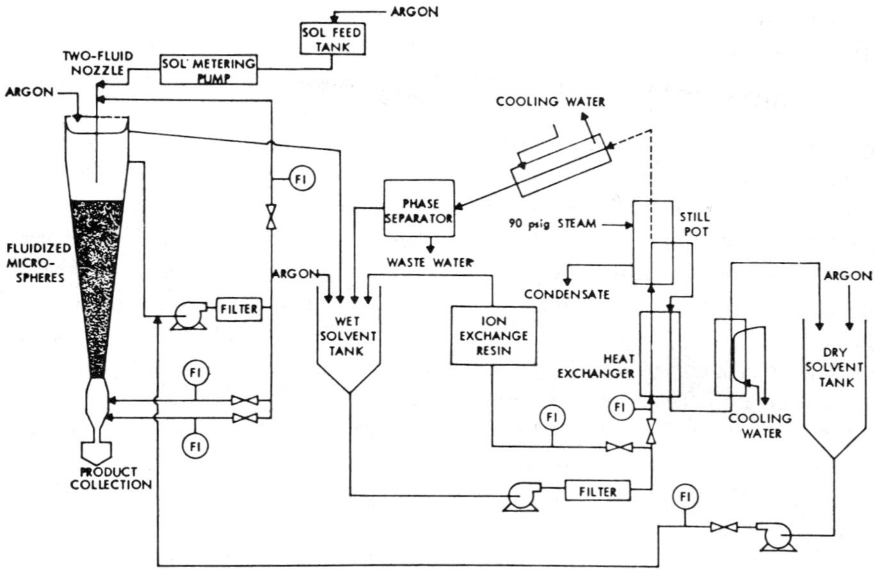

Figure 7.8.5. A flow diagram for microsphere formation by the sol-gel process. (Reprinted with permission from Ref. 9. Copyright 1966 by the American Chemical Society)

Equipment used[9] to accomplish steps 1 to 4 in a continuous operation is shown in Figure 7.8.5. The aqueous sol of colloidal particles (e.g., thoria, ThO_2) is dispersed into drops at the enlarged top of a tapered forming column. The droplets are fluidized by an upflowing stream of the immiscible water-extracting fluid, such as 2-ethyl-1-hexanol. Interfacial tension holds the drops in a spherical shape, but the maximum size is limited since large drops are more susceptible to distortion. A surfactant is dissolved in the immiscible liquid to prevent coalescence of the sol droplets with each other, their adhesion to the walls of the vessel, and/or sticking together of partially dehydrated drops. As the water is removed and the sol is converted to a gel, the particles become denser and their settling velocity increases. Column design and flow rates are controlled so that the densified particles drop out continuously to the product receiver, while new sol droplets are added at the top of the column. The extracting liquid is separated from the product and a portion of it is sent continuously to the distillation recovery system for purification to maintain a sufficiently low water concentration in the fluidizing liquid. The purified extracting liquid is then recycled to the column.

Typical capacity of a 76 mm ID (minimum) column is 9 kg/h of sintered oxide spheres using concentrated sols.

Agitation in baffled vessels can also be used to disperse and suspend the sol drops in the extracting liquid. Compared with the fluidized system described above, this more vigorous agitation produces smaller microspheres less than 100 μm in diameter.

Other Liquid-Phase Dispersion Processes

Many other granulation methods based on liquid-phase dispersion are possible depending on the way in which the feed material is converted into liquid form, the method used to disperse it into droplets, and the procedure used to harden the droplets. The prilling process described in Part 6 normally uses air as the cooling medium for the molten feed droplets. In the manufacture of the lead shot, however, molten lead is prilled into water for solidification. The prilling of calcium nitrate into mineral oil has been described.[11] Compared with air-cooled processes, prilling into a liquid medium requires shorter cooling times and smaller equipment because of the improved cooling by liquids, but an extra stage to remove the liquid phase is necessary.

Another variant involves agitation of powders in nonsolvent liquid above their melting temperatures to form droplets that are then cooled below their melting temperatures to produce solid enlarged particles. For example, small spheres of naphthalene required in the preparation of hollow metallic spheres[12] can be produced by agitating naphthalene in water at about 80°C to form emulsion droplets, which are then quenched at lower temperatures to yield the solid form.

A further example involves dissolution of appropriate particulate feeds in sufficient organic solvent to make it fluid, dispersion of the liquid into water by agitation or spray followed by steam-distillation of the solvent to yield solid enlarged particles.[5]

Fine Coal Cleaning Using Oil Agglomeration[13, 14]

Selective agglomeration is readily accomplished with processes based on wetting by immiscible liquids. One or more components of a complex solids mixture can be selectively agglomerated and removed from suspension (for example, by screening) while other components not wetted by the bridging phase remain in suspension. Where the natural wetting properties of a particulate component do not allow its separation from a suspending liquid and/or from other particles of a mixture, surface conditioning agents may be used to modify its surface properties and allow the desired separation.

The recovery of fine coals from aqueous waste suspensions is a problem of great current interest and will be used to illustrate the selective agglomeration process. Increasing quantities of fines in water suspension must be processed today during coal preparation. These fines result from natural degradation, increasingly mechanized mining methods, and the finer grinding necessary to liberate impurities from lower-quality coals. Coal particles are readily agglomerated and recovered from aqueous suspension upon agitation with many different oils as collecting liquids. Inorganic or ash-forming constituents remain in suspension and are thus rejected.

A simplified flow diagram for the recovery of fine coal by selective oil agglomeration is given in Figure 7.8.6. Standard equipment well known in the chemical and mineral industries can be applied in the process. Agitation serves initially to disperse the bridging oil and secondly to contact the oil droplets and coal particles so that bonds are formed between oil-coated particles. The coal agglomerates thus formed are readily recovered on a screen of suitable mesh size while the impurity particles pass through the screen to waste. The agglomerates recovered in this way are typically in the diameter range 0.5–1 mm and may be suitable in some cases for direct shipment with the coarser coal products from a preparation plant. Alternatively, further processing may be required such as centrifugal dewatering or balling with binder in a disc or drum pelletizer.

Oil-coal contact is the most critical step in the oil agglomeration process. The required intensity and duration of mixing are determined by the oil and coal characteristics and by the solids

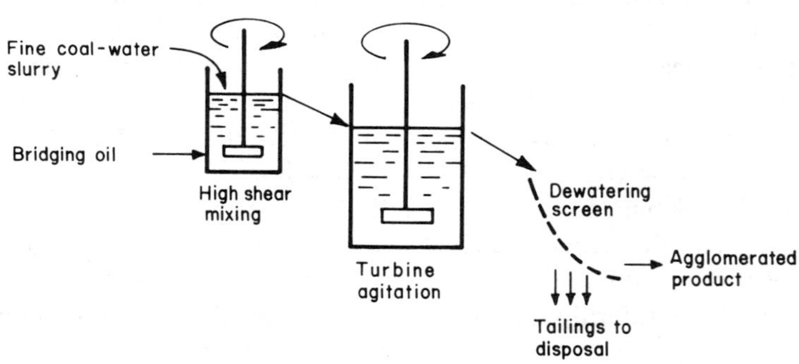

Figure 7.8.6. Flow diagram for the selective oil agglomeration of coal fines.

concentration and oil usage. Predispersion of the oil as an emulsion appears to be helpful. The range of operating conditions that have been used[14] in the oil agglomeration of fine coals includes:

Coal-water feed slurry	
Wt % solids	3-50
Particle size	typically, minus 200 mesh
Ash content, wt % dry basis	10-50
Oil usage (light fuel oil), % of solids wt	2-30
Turbine agitator	
Tip speed, m/s	~10-30
Mixing time	30 s to several min
Power consumption, kW/m^3	~10-40
Product agglomerates	
Wt % recovery of solid combustible matter	>90
Ash content, wt % dry basis	5-10

A most important characteristic of the oil agglomeration technique is its ability to recover extremely fine coal particles (for example, a few micrometers in diameter) even in the presence of clay slimes. A second important benefit of the oil agglomeration method is its ability to dewater fine coal without thermal drying. During agglomeration, the collecting oil is adsorbed on the surface of the particles and displaces the moisture to the surface of the agglomerates. The amount of moisture held by the coal then depends primarily on the surface area of the agglomerates formed. This is illustrated in Figure 7.8.7, in which the surface area is represented by the reciprocal of the agglomerate diameter. The moisture contents shown relate to the simple gravity drainage of the agglomerates on a 100 mesh screen.

It is evident that all the data in Figure 7.8.7 lie approximately on one line and as the diameter of the agglomerates or the coal particles

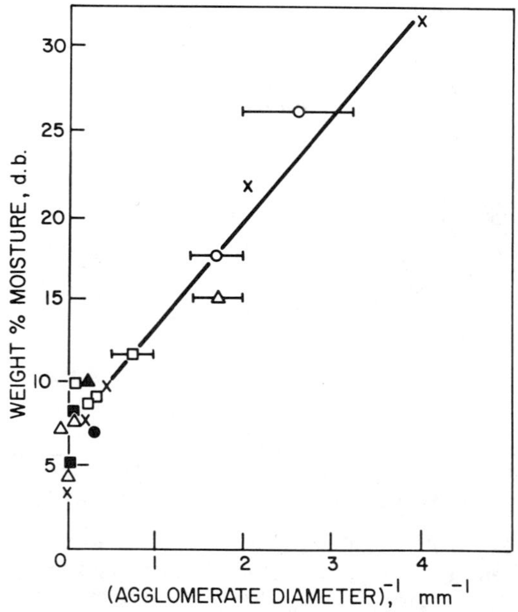

Figure 7.8.7. Moisture content for coal agglomerates and for unagglomerated coal as a function of the reciprocal of diameter.[19] (Data refers to gravity drainage on screens. X refers to unagglomerated coal of various sizes; △, □, ○, ▲, ■, ● refer to coal fines wetted by oil to form agglomerates of various sizes.)

increases, the moisture content decreases. These data indicate that the moisture content may be reduced to less than 10% by agglomeration and drainage on a screen without the need for thermal drying, provided that the size of the agglomerates is larger than about 2 mm.

The amount of oil necessary to form these large agglomerates that will drain to low moisture levels may be prohibitive in cases where extreme fines are being treated. When small agglomerates (less than 1 mm) are produced with lower oil levels, mechanically assisted dewatering in a centrifuge can be employed to attain low moisture levels without thermal drying. Some results are given in Figure 7.8.8 for a nominal minus 28 mesh metallurgical coal. With this particular feed material, the small agglomerates formed with only a small percentage of oil dewater to low levels in a centrifuge.

Recovery of fine waste coals from preparation plants obviously reduces the load on the tailings handling system. For example, if a

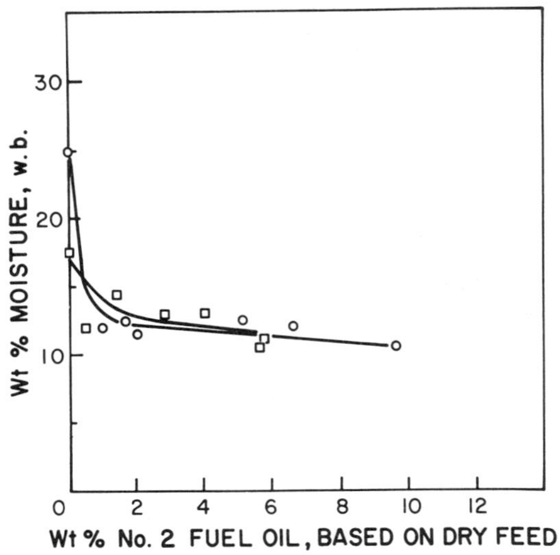

Figure 7.8.8. Effect of oil content on the moisture content of centrifuged (160G) agglomerates formed from two different samples of minus 28 mesh coal fines.

tailings slurry containing 70% coal is agglomerated prior to thickening and 90% of the coal is recovered, then the solids feed to the thickener and the tailings pond decreases by 63%. Thus, not only is the settling rate improved because of reduced solids concentration, but tailings pond life almost triples. Agglomeration of coal in existing tailings ponds recovers lost coal values, extends pond life, and in some cases may eliminate the need for new ponds.

Pellet Flocculation and Other Sludge Dewatering Systems

These techniques combine relatively large amounts of polymeric flocculants with gentle rolling mixing to consolidate settled flocs into compact agglomerate-like sludges of low liquid content.

Choice of flocculant is of prime importance in these processes. Criteria for choosing a flocculant include the degree of floc formation and effect on water clarification, the amount of water contained in the settled and dewatered flocs, and the dosage required in terms of cost per unit weight of dry solids. Organic polyelectrolytes provide the best results with many materials, and it can be anticipated that a cationic flocculant will be most useful with organic sludges, whereas an anionic or nonionic flocculant will be best for inorganic and mineral sludges.[15]

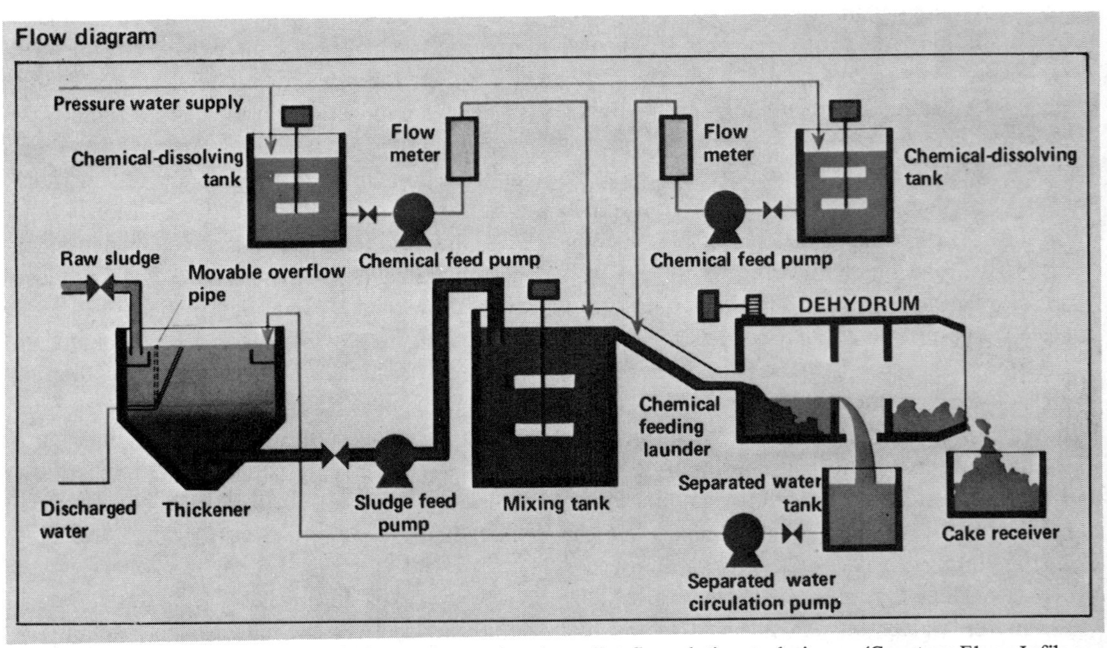

Figure 7.8.9. Flow diagram for sludge treatment by the pellet flocculation technique. (Courtesy Ebara-Infilco Co., Ltd., Tokyo)

Table 7.8.2. Performance Data for Treatment of Various Sludges Using "Dehydrum" of Figure 7.8.9. (Courtesy Ebara - Infilco Co. Ltd., Tokyo.)

	TYPE OF SLUDGE	WATER WORKS SLUDGE	BENTONITE SLURRY SLUDGE IN SHIELD TUNNELING PROCESS	SLUDGE IN MIXED WASTE EFFLUENTS FROM AUTOMOBILE SHOPS (CONTAINING OILS AND ACTIVATED SLUDGE)	SLUDGE IN GRAVEL-WASHING WASTE WATER	SEWAGE SLUDGE (MIXED RAW SLUDGE)	SLUDGE FROM DREDGING WASTE WATER
Raw sludge	Solid concentration (g/l)	30–40	150–300	50–70	100–200	40–75	200–250
	Ignition loss in solid matter (%)	15–30	–	40–45	–	45–50	12–21
	Oil content in solid matter (%)	–	–	7–10	–	–	–
Chemicals	Amount of polymeric flocculant/amount of solid (%)	0.1–0.2	0.07–0.1	0.25–0.3	0.04–0.05	0.25–0.35	0.05–0.1
	$Ca(OH)_2$ (%)	0	3	0	0	0	2–4
Moisture content in cake (%)		65–80	46–48	82–86	38–47	79–82	46–64
Turbidity of separated water (ppm)		50	50	100	100	100	100

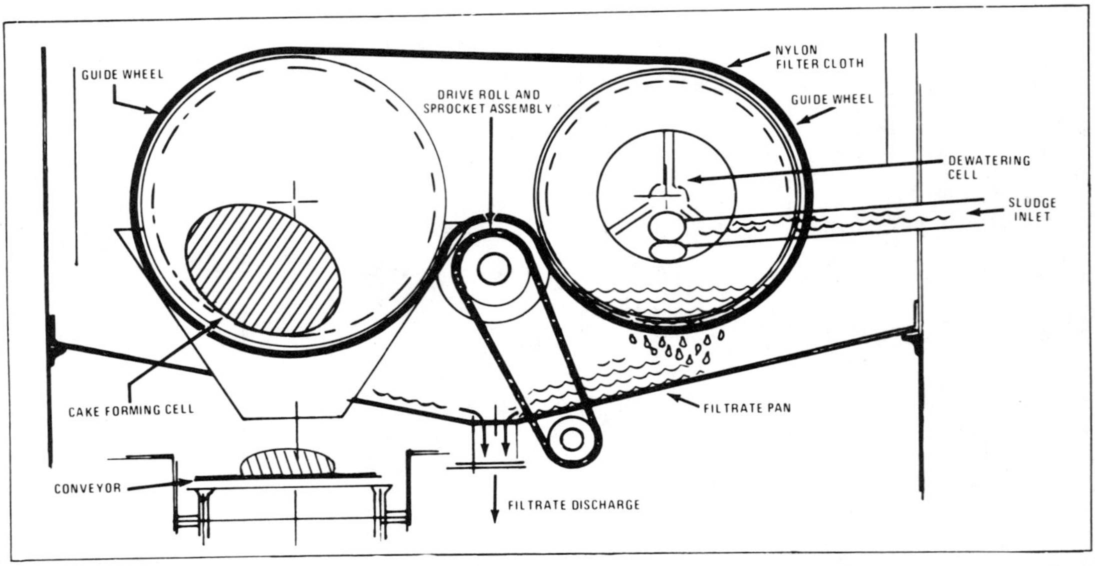

Figure 7.8.10. Dual-cell gravity sludge dewatering unit. (Courtesy The Permutit Company, Paramus, N.J.)

One effective piece of equipment to accomplish "pellet flocculation" has been developed in Japan and is depicted in the flow diagram of Figure 7.8.9. This process makes use of a slowly revolving (1 m/min peripheral speed) horizontal drum (called a "Dehydrum") to dewater sludge.[4] The drum interior is made up of three sections for successively pelletizing, decanting, and consolidating the solids. Polymeric flocculant is added to the suspension upstream of the drum, together with auxillary agglomerating agents such as calcium hydroxide or sodium silicate. Voluminous flocs formed ahead of the drum are rolled into denser sediment in the pelletizing section. These are then pushed into the decanting section by a guide baffle where water is removed through intermittent slits in the drum wall. In the final consolidating section, the agglomerates are gently tumbled and rolled into a

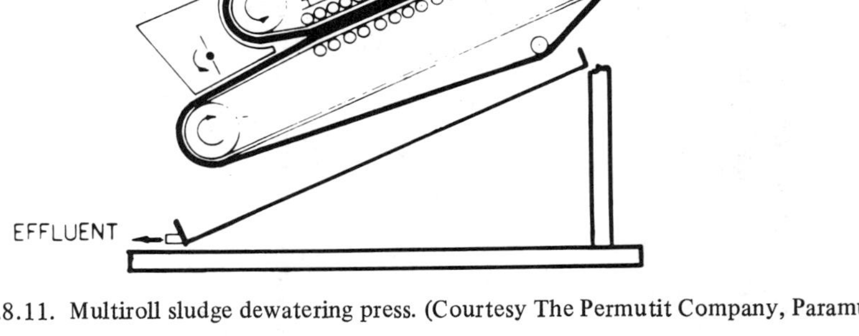

Figure 7.8.11. Multiroll sludge dewatering press. (Courtesy The Permutit Company, Paramus, N.J.)

Table 7.8.3. Operating Results for the Sludge Dewatering Units Shown in Figures 7.8.10 and 7.8.11. (Courtesy The Permutit Company, Paramus, N.J.)

TYPE OF SLUDGE	WT % FEED	DRY SOLIDS DCG CAKE[a]	DRY SOLIDS MRP CAKE[b]	APPROX. CAPACITY FOR DCG 200[a] LB/H DRY BASIS
Raw primary sanitary	3	12–17	20–23	300
Iron clarification	6	19	34	600
Aluminum hydroxide	1.2	8–13	15–19	52
Iron hydroxide	3.5	13–15	25–30	150
Paper mill waste	2.4	5–10	14–25	280
Latex	4	12–14	–	400

[a]Dual cell unit shown in Figure 7.8.10. DCG 200 consists of 2 of the modules in Figure 7.8.10.
[b]Multiroll press in Figure 7.8.11.

denser form and water again escapes through wall slits. Product solids then discharge as a low-water-content cake. These cylindrical vessels are available[16] in standard sizes up to 3.4 m diameter by 9.2 m long with a 5.5 kW drive. Typical sludge treating capacities for a 2.4 m diameter unit are 6 to 9 Mg/h for gravel waste sludge, 1.4 to 2.2 Mg/h for a dredged mud sludge and 0.4 Mg/h for a mixed waste sludge from an automobile factory. Table 7.8.2 provides performance data for the treatment of a number of suspensions by this technique.

Another approach to the dewatering of flocculated sludges is through the use of filter belt units. One type of unit is shown in Figure 7.8.10. The unit contains two cylindrical cells, each about 400 mm in diameter, formed by a fine-mesh nylon filter cloth that travels continuously over two guide wheels. Initial dewatering occurs in the first cell, and the solids are then carried over the drive roll, which separates the cells, into a second cake-formation cell. Here the solids are compacted by rolling action to form a cake of low moisture content before discharge from the unit. A multi-roll press, using dual endless filter belts and shown in Figure 7.8.11, may be used as a secondary dewatering device in conjunction with the unit in Figure 7.8.10. In this way, the volume and moisture content of treated sludges is further reduced. Typical operating and capacity data for these machines are given in Table 7.8.3.

REFERENCES

1. C. E. Capes, A. E. McIlhinney, and A. F. Sirianni, "Agglomeration from Liquid Suspension–Research and Applications." In K. V. S. Sastry (ed.). *Agglomeration 77*, A. I. M. E. New York (1977).
2. R. Akers, *Flocculation*, Inst. Chem. Engrs., London (1975).
3. J. A. Kitchener, "Principles of Action of Polymeric Flocculants." *Br. Polym. J. 4*:217–229 (1972).
4. M. Yusa, H. Suzuki, and S. Tanaka, "Separating Liquids from Solids by Pellet Flocculation." *J. Am. Water Works Assoc. 67*:397–402 (1975).
5. R. H. Perry and C. H. Chilton, (eds.), *Chemical Engineers' Handbook*, 5th Ed., section 8, McGraw-Hill, New York (1973).
6. J. R. Farnand, H. M. Smith, and I. E. Puddington, "Spherical Agglomeration of Solids in Liquid Suspension." *Can. J. Chem. Eng. 39*:94–97 (1961).
7. A. F. Sirianni and I. E. Puddington, "Forming Balls from Powder." U.S. Patent 3,368,004 (Feb. 6, 1968).
8. C. E. Capes, R. D. Coleman, and W. L. Thayer. "The Production of Uniformly Sized Spherical Agglomerates in Balling Drums and Discs." *Int. Conf. Compact. and Consolid. of Part. Matter, Proc.*, 1st, London (1972).
9. P. A. Haas and S. D. Clinton, "Preparation of Thoria and Mixed Oxide Microspheres." *Ind. Eng. Chem. Product Res. Dev. 5*(3):236–246 (1966).
10. M. E. A. Hermans, "Sol-gel Processes–A Curiosity or a Technique? *Powder Met. Int. 5*(3):137–140 (1973).
11. P. J. van den Berg and G. Hallie, "New Developments in Granulation Techniques." *Proceedings of the Fertilizer Society, No. 59*, London (1960).
12. J. R. Farnand and A. F. Sirianni, "Hollow Article

Production." U.S. Patent 3,528,809 (Sept. 15, 1970).

13. C. E. Capes, A. E. McIlhinney, R. E. McKeever, and L. Messer, "Application of Spherical Agglomeration to Coal Preparation." *Int. Coal Prep. Conf., Proc.*, 7th, Sydney, Australia (1976).
14. C. E. Capes and R. L. Germain, "Selective Oil Agglomeration in Fine Coal Beneficiation." In Y. A. Liu (ed.). *Physical Cleaning of Coal*, Markel Dekker, New York (1982).
15. *Flocpress*, Bull. DB845, Infilco Degremont Inc. (Sept. 1976).
16. *Dehydrum Continuous Pelletizing Dehydrator*, Ebara-Infilco Co., Ltd., Tokyo, Japan.
17. F. J. Zuiderweg and N. van Lookeren Campagne, "Pelletizing of Soot in Waste Water of Oil Gasification Plants–The Shell Pelletizing Separator." *Chem. Eng. (London) 220*:CE 223–227 (1968).
18. W. Campbell, "Oil from Alberta's Tar Sands." *Science Dimension 8*(1):10–15 (1976). Published by the National Research Council of Canada, Ottawa.
19. C. E. Capes, A. E. McIlhinney, A. F. Sirianni, and I. E. Puddington, "Agglomeration in Coal Preparation." *Proc. Inst. Briquet. Agglom. 12*:53–65 (1971)
20. C. E. Capes and J. P. Sutherland, "Formation of Spheres from Finely Divided Solids in Liquid Suspension." *Ind. Eng. Chem. Process Design Develop. 6*:146–154 (1967).

8
Mixing of Powders

Sherman S. Weidenbaum
United States Coast Guard Academy
New London, Connecticut

8.1. INTRODUCTION

The section on solid-solid mixing which follows this introduction was originally prepared for the Chemical Engineers Handbook (5th ed.).[15] At the time of preparation, a major revision in approach to summarizing the essentials of the field of solids mixing was made, so that the material would be of an enduring nature. Consequently it is reproduced in its entirety. Any equipment cost estimates should be reupdated in accordance with the change in the economic index since the time the figures were given.

References: A comprehensive bibliography is available in References 9 and 17. Reference 17 is a fairly recent article that carries, in addition to the materials indicated from the title, an extensive list of references arranged in categories as follows; general mixing; theories of mixing and mixing quality; statistical analysis of mixing–quality investigation; problems of phase separation; investigation of mixing quality using radioactive substances; uniformity and mixing quality of concrete and building materials; investigations of mixers and mixing systems; mixing in the glass industry; mixing in the ceramic industry; mixing of lime sandstone masses; mixing in the pharmaceutical industry; and mixing of colors. A summary emphasizing methods for scaling-up tumbling mixers is given in Reference 16.

Equipment photographs and details are available in References 2, 4, 5, 7, 8, and 10. References 3 and 6 give excellent theoretical work. Reference 5 gives a tabulation and summary of many mixer types and applications. References 8 and 9 are book chapters dealing with mixing of solids and cover both the theoretical and equipment aspects. Interpretive summaries of the literature in various areas (state of mixedness, theoretical frequency distributions, rate equations, and equipment) are included in Reference 9. Reference 1, which was updated and published in 1978, gives a procedure for testing solids-mixing equipment.

Equipment: Additional equipment for dry solids blending have been developed and manufactured during the last few years. These are:

- Static Mixer units as shown in Fig. 8.1 *a* and *b*. The units have been used to blend dry segregating particles as well as nonsegregating particles. If the particles are free flowing, they can be mixed either by gravity feed or pneumatically fed through the mixer. Figures

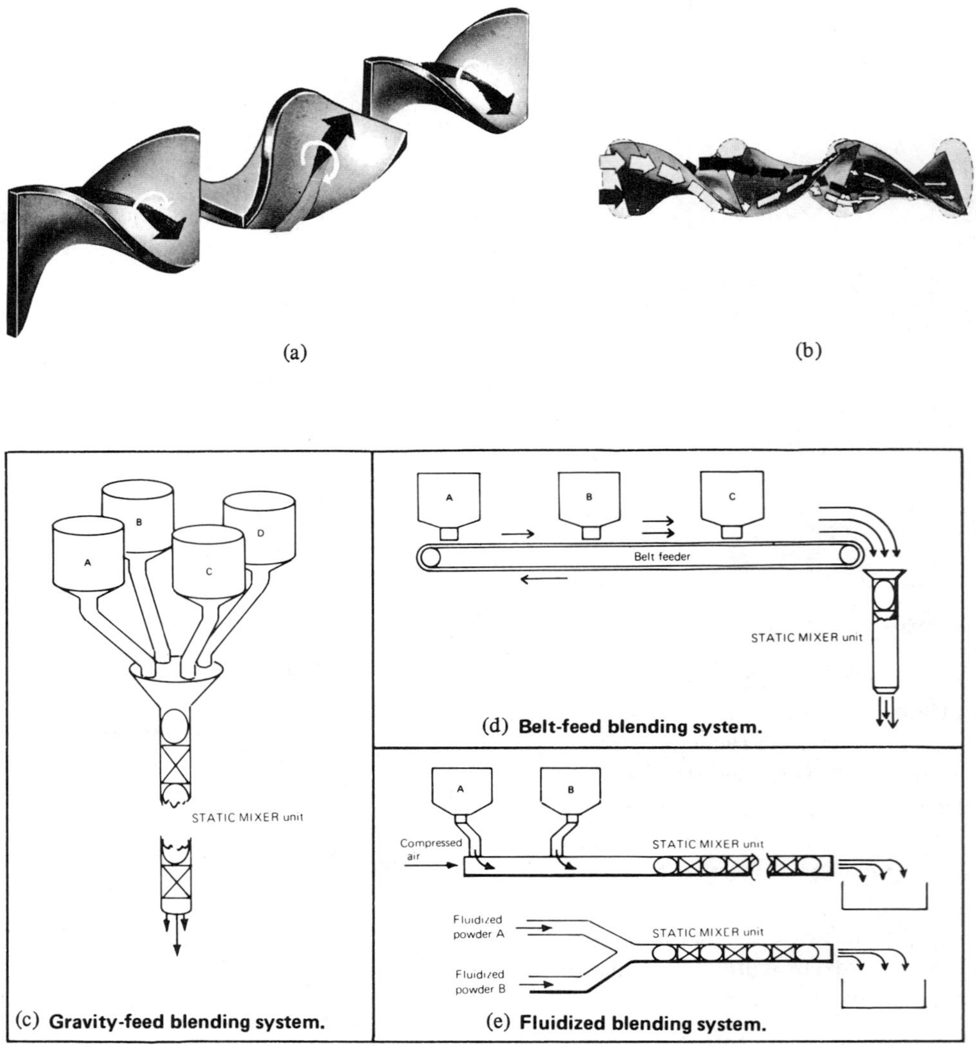

Figure 8.1. Static mixer systems

8.1, *c*, *d*, and *e* are the schematics of three solids blending systems that incorporate Static Mixer units.

- A mixer employing a twin rotary shaft, which is called the "Fluidized Zone" mixer (Fig. 8.2).
- A recent modification of the Twin-Shell blender in which the two legs are of unequal length (Principle of Unequal Displacement (Fig. 8.3*a*) used in the Cross Flow blender (Fig. 8.3*b*)).

8.2. FUNDAMENTALS

Objectives of Solid-Solid Contacting

Equipment in which solid materials are mixed may be used for a number of operations. Blending of ingredients may be the main objective as, for example, in the preparation of feeds, insecticides, fertilizer, glass batches, packaged foods, and cosmetics. Other objectives may include cooling or heating such as in the cooling of limestone or sugar, or the preheating of

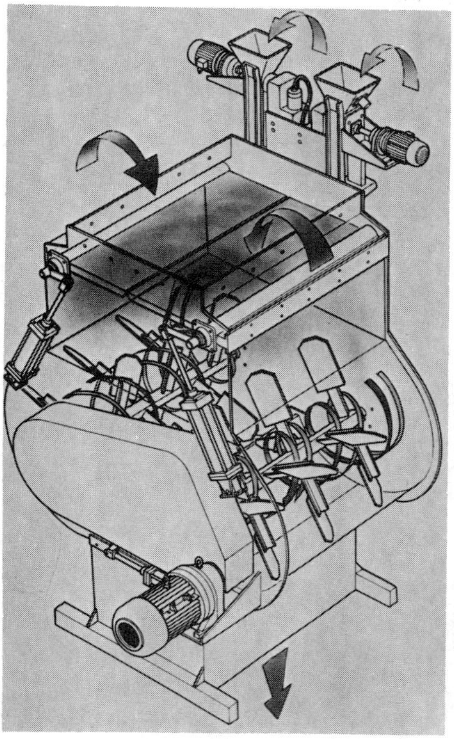

Figure 8.2. Fluidized Zone mixer.

plastic prior to calendering. Drying or roasting of the solids is sometimes desired. In some applications, such as polymerization of plastics, catalyst manufacture, or the preparation of cereal products, the solids mixture may be reacted. Coating is desired in some cases as in the manufacture of pigments, dyes, minerals, candy, and other food products, and the preparation of feeds. In certain of these cases, small amounts of liquid may be added, but the end product is a solids mixture. Sometimes agglomerates are desired as in the preparation of food products, pharmaceuticals, detergents, and fertilizer. Often size reduction is desired while solids are being mixed. In all cases, the mixing of solids occurs. However, in some of these operations, the details of the equipment to accomplish operations other than pure blending may become a major problem. This chapter will deal with equipment whose major function is to give a thorough mixture of solids. Specialized equipment to perform the other functions is discussed elsewhere in this book and will not be dealt with here. Thus, for example, chapters 7 and 13 are devoted to size enlargement and reduction, although equipment mentioned there may also accomplish mixing.

Batch Ingredient Properties Affecting Solids Mixing

Wide differences among properties such as particle-size distribution, density, shape, and surface characteristics (such as electrostatic charge) may make blending very difficult. In fact, the properties of the ingredients dominate the mixing operation. The most commonly observed characteristics of solids are: (1) *Particle-size distribution.* This tells the percentages of the material in different size ranges. (2) *Bulk density*, which is weight per unit of volume of a quantity of solid particles; this is usually expressed in $lb/ft.^3$ It is not a constant and can be decreased by aeration and increased by vibration or mechanical packing. (3) *True density* of the solid material; usually expressed in $lb/ft.^3$ This, divided by the density of water, equals specific gravity. (4) *Particle shape.* Some types are pellets, egg shapes, blocks, spheres, flakes, chips, rods, filaments, crystals, or irregular shapes. (5) *Surface characteristics* such as surface area and tendency to hold a static charge. (6) *Flow characteristics. Angle of repose* and *flowability* are measurable characteristics for which standard tests are available (e.g., A.S.T.M. Test B213-48 on Flow Rate of Metal Powders, etc.). A steeper angle of repose would indicate less flowability. The term "lubricity" has sometimes been used for solid particles to correspond roughly to viscosity of a fluid. (7) *Friability.* This is the tendency of the material to break into smaller sizes in the course of handling. There are quantitative tests specially devised for certain materials such as coal which can be used to estimate this property. Abrasiveness of one ingredient upon another should also be considered. (8) *State of agglomeration.* This refers to whether the

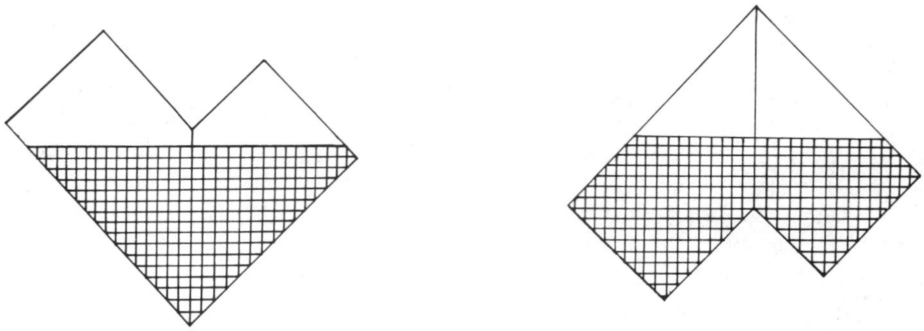

In the upright position, each blender leg holds an equal amount of material. As the blender rotates, the uneven legs force the material to shift continuously from one leg to the other. This axial-flow pattern and the radial mixing action — the intermeshing of the material within the two inclined cylinders — produce a rapid, uniform blend.

Figure 8.3a. Principle of unequal displacement.

Figure 8.3b. Cross-Flow blender.

particles exist independently or adhere to one another in clusters. The kind and degree of energy employed during mixing and the friability of the agglomerates will affect the extent of agglomerate breakdown and particle dispersion. (9) *Moisture or liquid content of solids.* Often a small amount of liquid is added for dust reduction or special requirements (such as oils for cosmetics). The resultant material may still have the appearance of a dry solid rather than a paste. (10) *Density, viscosity, and surface tension* at operating temperature of any liquid added. (11) *Temperature limitations of ingredients.* Any unusual effects due to temperature changes which might occur (such as heat of reaction) should be noted.

A look at the above properties for the ingredients to be mixed is a first step toward selecting mixing equipment.

Batch Homogeneity

Measuring Uniformity. Except for cases where a coating of one ingredient with another takes place, the theoretical end result of mixing will not be an arrangement in which one type of particle is directly next to a different type. Rather, the theoretical end result where *random tumbling* takes place will be a random mixture along the lines shown in Figure 8.4. With easily distinguishable particles which can be counted, the variation between spot samples of a known size (i.e., number of particles) can be theoretically predicted for a random mixture and used as a guide to determine how closely random blending of the ingredients is approached. Where individual particles cannot be easily distinguished and particle counts are not practical, various types of analyses can be made on spot samples to determine batch uniformity. Recent advances in instrumental analysis have made it much easier to give rapid and numerous analyses which are of great benefit for statistical analyses. Some of these methods are X-ray fluorescence, flame spectrometry, polarography, and emission spectroscopy. Also, radioactive-tracer methods have been used. Regardless of the analytic methods chosen, whether gravimetric, volumetric, electrometric, particle counts, optical or other, it is very important that the data be objectively analyzed via statistical methods should there be any question as to the adequacy of the mixture. The analytical error should be very small compared

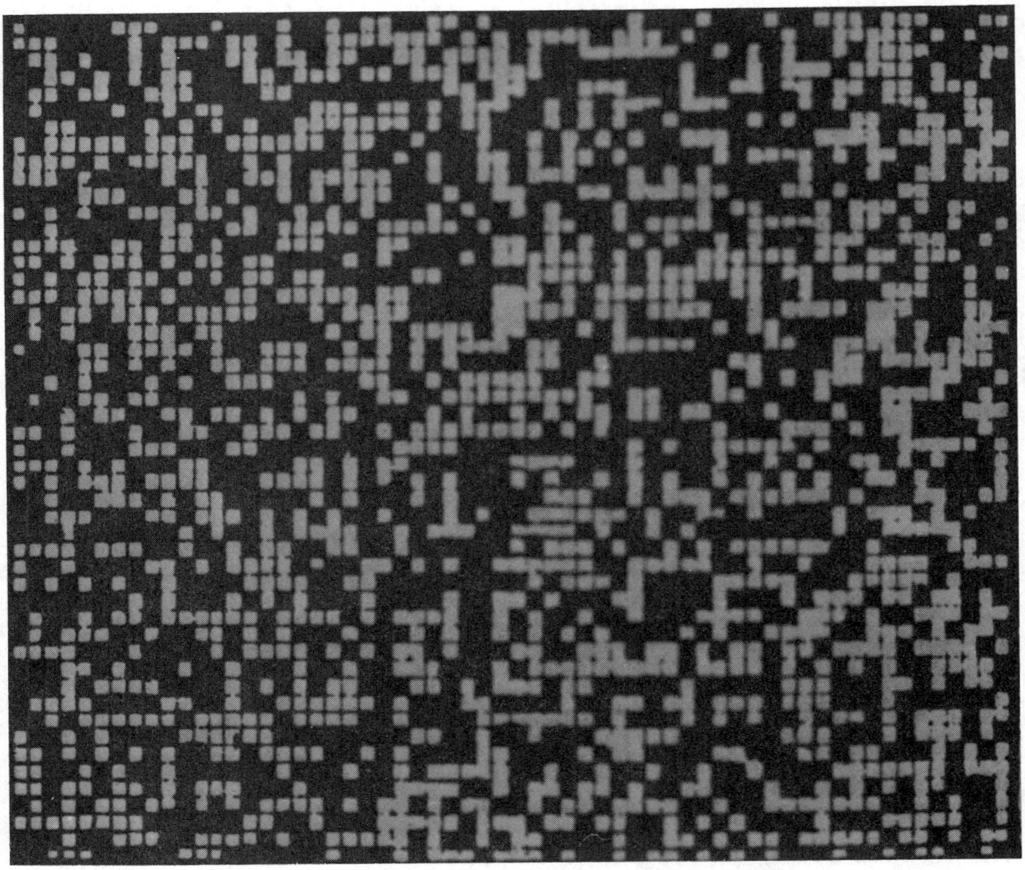

Figure 8.4. A random arrangement of black and white particles (From Ref. 19).

with the variation in the composition (or other property) between spot samples.

Reference 9 describes many different ways of measuring uniformity.

Evaluation. Whether the desired end product is satisfactory can be used as a practical criterion of the adequacy of the solids mixture. A further consideration is the effect of the solids mixture on the over-all economics of the manufacturing process. Studies of the type mentioned in the previous paragraph may be part of such an evaluation. Where the solids mixture is made directly into a product, as in the case of feed pellets or pharmaceutical tablets, uniformity tests on these items will speak for themselves. In cases where the solids mixture must be further processed, such as in the manufacture of glass or plastics, the efficiency and costs of the subsequent operations can often be related to the starting solids mixture. In such cases, knowledge of the homogeneity of the solids mixture is needed to determine its effect on the manufacturing process.

Regardless of the method of evaluating the solids mixture, the sampling procedure is vital. Often a sampling thief, or other special device, is used to remove samples from the mixture without excessive disturbance of the batch. If an easier method of sampling is obvious and will bring less contamination to the batch, it should be used.

Method of sampling, location, size and number of samples, method of sample analysis, and fraction of the batch removed for sampling all contribute to how well the sampling study reflects the actual conditions.

A standard testing procedure for solids-mixing equipment is available (Ref. 1). This contains details and references pertaining to sampling from solids mixtures for both batch and continuous mixing.

Segregation Problems. Previously it was pointed out that wide differences among properties may make blending very difficult. For example, natural segregating tendencies will be observed with extreme differences in specific gravity, size, or shape. The heavier, smaller, or smoother and rounder particles tend to sink through the lighter, larger, or jagged ones, respectively. In some cases, preparation of the materials to avoid extreme differences in such ingredient properties can avoid segregation problems.

There are also other factors which can cause segregation.

Electrostatic charges may cause particles to repel each other. In cases where continued blending may cause such charges to build up, it is important to determine the precise blending time required and not to overblend.

Loss of material as dust must be considered as a possible means of segregation and should not be aggravated by too strong suction in the dust-collection apparatus.

If there are smeary particles which have an almost pastelike behavior and barely flow (high angle of repose), frictional anchorage of these onto the other particles in the mixture may be necessary in order to achieve good mixing.

Where a batch ingredient is in agglomerate form, some device to break up the agglomerates should be used to prevent them from segregating from the rest of the mixture and to ensure the intimate dispersion of this ingredient throughout the mixture.

The use of a liquid such as water (possibly with a surface-active agent) can have remarkable effects in overcoming segregation which may appear inevitable otherwise.

Although the above applies to the actual solids-mixing operation, thought must be given to the subsequent processing steps. Thus, the solids-mixing operation must be checked from the point of view of delivering a well-mixed batch to a certain point. The system must be scrutinized for possible segregating points such as transfer points, long drops, flow-through silos, and vibratory equipment. Where a liquid is used, the amount that can be added without getting into caking problems which may upset the later processing of the solids mixture should be determined.

8.3. EQUIPMENT

Mixing Mechanisms

There are several basic mechanisms by which solid particles are mixed. These include small-scale random motion (diffusion), large-scale random motion (convection), and shear.

Motions which increase the mobility of the individual particles will promote diffusive mixing. If there are no opposing segregating effects, this diffusive mixing will in time lead to a high degree of homogeneity. Diffusive mixing occurs when particles are distributed over a freshly developed surface and when individual particles are given increased internal mobility. A plain tumbler gives the former while an impact mill gives the latter.

For most rapid mixing, in addition to diffusive (fine-scale) mixing, there should be a means by which large groups of particles are intermixed. This can be accomplished by either the convective or shear mechanism. A ribbon mixer illustrates the former whereas a plain tumbler gives the latter.

Types of Solids-Mixing Machines

There are several types of solids-mixing machines. In some machines the container moves. In others a device rotates within a stationary container. In some cases, a combination of rotating container and rotating internal device is used. Sometimes baffles or blades are present in the mixer. Table 8.1 classifies solids-mixing machines via the characteristics given in the column headings. Illustrations of several of the machines listed there are shown in Figure 8.5. The various types listed in Table 8.1 will be briefly discussed below, with paragraph numbers referring to the columns.

1. Tumbler. Suitable for gentle blending; capable of handling large volumes; easily cleaned; suitable for dense powders and abrasive materials. Not for breaking up agglomerates.

In Figure 8.5, parts *a* and *b* (without broken-line portions) show some unbaffled tumblers.

Parts *c* and *d* show some baffled tumblers.

2. Tumbler with Agglomerate Breaker. The ball mill, rod mill, and vibratory pebble mill will accomplish mixing along with size reduction.

Several tumblers are available with separately driven internal rotating devices for breaking up agglomerates. The tumbler itself can be used for gentle blending if agglomerate breakdown is not required.

The broken-line portions of Figure 8.5*a* and *b* show some types of agglomerate-breaking devices for tumblers.

Table 8.2 includes impact velocities for some internal rotating devices in tumblers as well as other mixers. Contamination and wear problems of internal rotating devices are discussed under Performance Characteristics.

3. Stationary Shell or Trough. There are a number of different types of mixers in which the container is stationary and material displacement is accomplished by single or multiple rotating inner mixing devices.

Ribbon Mixer (Fig. 8.5*e*). Within this subgroup there are several types. Ribbon cross section and pitch, clearances between outer ribbon and shell, and number of spirals on the ribbon are some features which can be varied to accommodate materials ranging from low-density finely divided materials that aerate rapidly to fibrous or sticky materials that require positive discharge aid. Other construction variations are either center or end discharge, and the mounting of paddles or cutting blades on the center shaft. A broad ribbon can be used for lifting as well as conveying while a narrow one will cut through the material while conveying. The ribbon is adaptable to batch or continuous mixing.

Vertical Screw Mixer. This subgroup also has several variations. One type is shown in Figure 8.5*f*. In this type, the screw rotates about its own axis while also orbiting around the center axis of the conical tank. In another variation, the screw does not orbit but remains in the center of the conical tank and is tapered so that the swept area steadily increases with increasing height. In another type, the central screw is

Table 8.1. Types of Solids-Mixing Machines*

TUMBLER (1)	TUMBLER WITH INTERNAL AGGLOMERATE BREAKER (2)	STATIONARY SHELL OR TROUGH (3)	BOTH SHELL AND INTERNAL DEVICE ROTATE (4)	IMPACT MIXING (5)	PROCESS STEPS WHICH CAN AFFECT SOLIDS MIXING‡ (6)
Without baffles:					
Drum, either horizontal or inclined	Ball mill Pebble mill	Ribbon	Countercurrent, muller turret and pan rotate in opposite directions	Hammer mill	Filling of hoppers
Double cone	Rod mill	Stationary pan, rotating muller turret†		Impact mill	Fluidization
Twin shell Cube	Vibratory pebble mill	Vertical screw	Planetary types	Cage mill	Screw feeders
Mushroom type	Double cone Twin shell Cube	Single rotor Twin rotor		Jet mill Attrition mill	Conveyor-belt loading Elevator loading
		Turbine Paddle mixer			Pneumatic conveying Vibrating
With baffles:					
Horizontal drum Double cone revolving around long axis		Sifter (turbosifter)			

*Diagrammatic sketches of many of these machines are shown in Fig. 8.5.
†There is also a muller in which the turret is stationary but the pan rotates.
‡Although these steps, when carefully selected, can aid mixing, caution must be exercised with pneumatic conveying and vibrating as they may tend to separate materials.

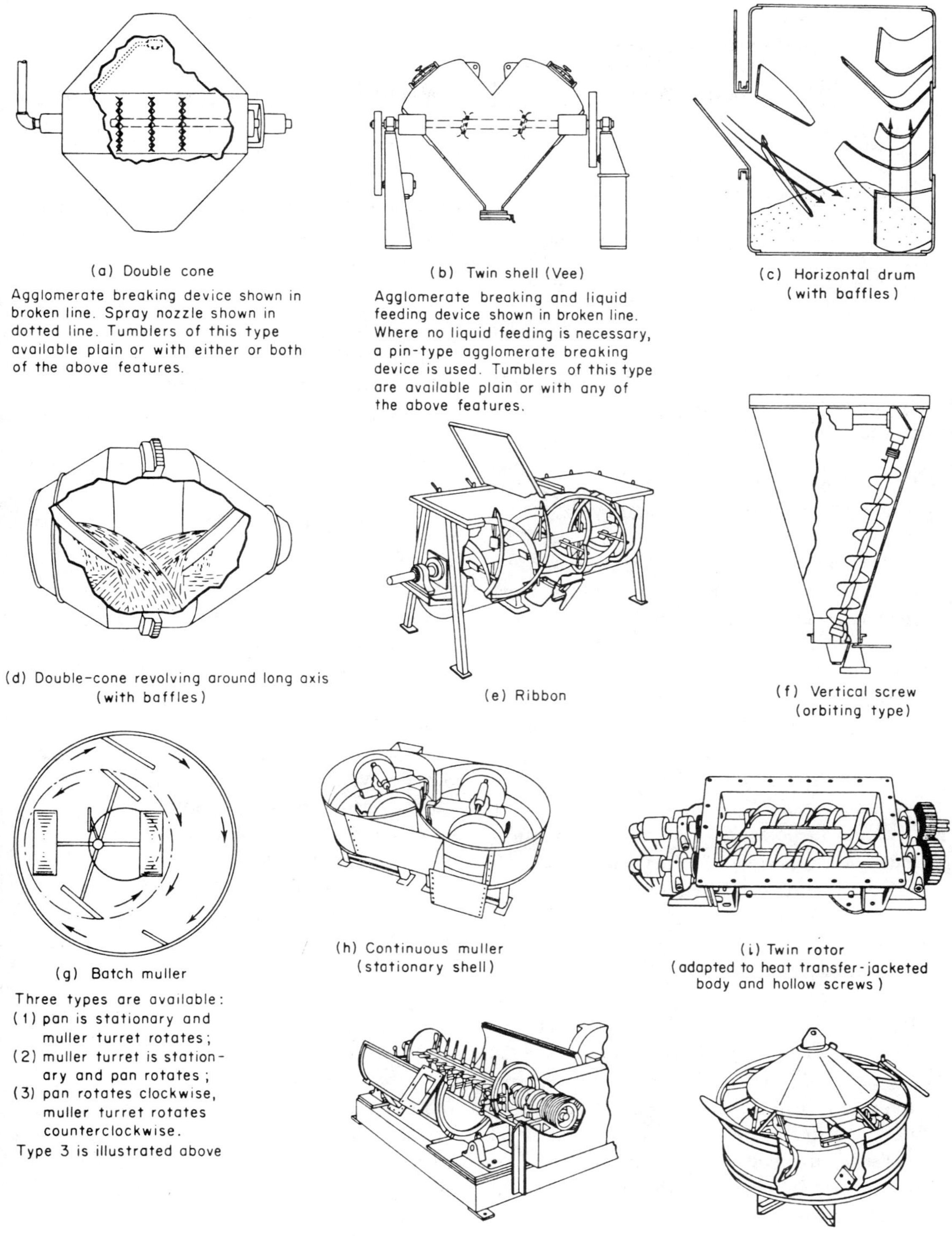

Figure 8.5. Several types of solids mixing machines (see Table 8.1).

Table 8.2. Approximate Impact Velocities of Some Rotating Internal Devices in Mixers

TYPE OF MIXER (SEE TABLE 8.1)	TIP SPEED, FT/MIN
Ribbon	280
Turbine	600
Twin-shell tumbler with:	
Pin-type intensifier	1700
Liquid-feed bar	3300
Twin rotor	Up to 1300
Single rotor	6000–9000
Mills of various types	2500–20,000

contained in an inner cylindrical casing. This type of mixer is primarily suitable for free-flowing dry solids.

Muller Mixer. The stationary-pan muller with rotating turret is one of several types. Other muller types are the countercurrent, in which the pan and muller turret rotate in opposite directions, and the rotating-pan type, in which the muller turret is stationary.

The heavy, wide roller rides over the material. There is some skidding action where the rollers engage the mass of materials. This gives local shearing plus coarse scale mixing which is aided by the plows and scrapers.

The muller is useful for mixing problems requiring certain types of aggregate breakdown, frictional anchorage of particles to one another, and densification of the final mix. Materials which are excessively fluid or sticky should be avoided. The muller mixer is generally used for batch operations (Fig. 8.5*g*) although Figure 8.5*h* shows a continuous muller.

Twin Rotor (Fig. 8.5*i*). This consists of two shafts which either paddles or screws encased in a cylindrical shell. There are various types available with shaft speeds ranging from moderately low to relatively high (see Table 8.2). It is useful for continuously mixing non-free-flowing solids; liquids can be added; there is minor product attrition; materials can be added beyond the inlet. Easily adaptable to heating or cooling. Some machines are specifically designed for heat transfer during mixing. The pug mill is one type of twin rotor.

Single Rotor (Fig. 8.5*j*). This consists of a single shaft with paddles, encased in a cylindrical shell. This type is available with relatively high speeds (see Table 8.2) although in certain cases lower speeds are used. A high-speed single rotor gives the maximum impact short of a grinding mill. It is used for intensive dispersion and disintegration; available with split casing; suitable for heating or cooling and for small amounts of liquid addition.

Turbine Mixer (Fig. 8.5*k*). This is a circular trough with a housing in the center around which revolves a spider or a series of legs with plow shares or mold boards on each leg. The mold boards spin around through the circular trough; suitable for free-flowing dry materials or semiwet materials which do not flow well; also adaptable to liquid-solid mixing and coating problems.

4. Shell and Internal Device Rotate. The countercurrent muller (Fig. 8.5*g*), which is in this category, is mentioned under Muller Mixer. This machine has a clockwise rotating mixing pan with a counterclockwise rotating mixing tool head mounted off center of the pan, thus providing a planetary mixing pattern. For mixing of free-flowing solids not requiring the shearing and compressive action of the mullers, plows are sometimes used alone. When used with mullers, the plows deflect material into their path. Special mixing tools are also available.

5. Impact Mixing. (includes size reduction.)

6. Process Steps. The process steps listed in Table 8.1 can sometimes be used to promote mixing (see Ref. 10). However, they are primarily for functions other than solids mixing. (Note precautions for pneumatic conveying and vibrating in Table 8.1.)

Since paste mixing is not within the scope of this section, such widely used paste mixers as the sigma blade and banbury types will not be covered here.

Performance Characteristics

Before selecting solids-mixing equipment, a careful study should be made of various performance characteristics. These are given below.

Uniformity of Mixture. The proper type of mixer should be chosen to assure the desired degree of batch homogeneity. This cannot be compromised for other conveniences. Information is given under Types of Solids-mixing Machines about the special abilities of various kinds of machines to blend different types of materials.

Care should be taken to avoid mixing too long, as in some cases this will cause a poorer blend to result. A graph of degree of mixing vs time should be made to quantitatively select the proper mixing time.

Mixing Time. The actual time during which the batch is being mixed is usually less than 15 min if the proper types of machine and working capacity have been chosen. In some cases much more lengthy mixing times are tolerated so as to avoid the cost of purchasing more efficient equipment. However, there is usually a machine that can properly homogenize almost any type of mixture in less than 15 min provided one is willing to pay the price. In fact, proper mixer design in most instances will produce the desired blend in a few minutes.

Besides actual mixing time, however, the total cycle time should be optimized.

Charging and Discharging. The total handling system must be considered in order to obtain optimum charging and discharging conditions. This includes the efficient use of weigh hoppers and surge bins, minor ingredient premixing, location of discharge gates, etc.

Power. In general, power requirements are not a major consideration in choosing a solids mixer since other requirements usually predominate. However, sufficient power must be supplied to handle the maximum needs should there be changes during the mixing operation. Also, where a variety of mixes may be required, power must be sufficient for the heaviest bulk-density materials. If the loaded mixer is to be started from rest, there should be sufficient power for this. Where speed variation may be desirable, this should be taken into account in planning power requirements.

Horsepower requirements of several types of mixers are listed in Table 8.3.

Cleaning. The ease, frequency, and thoroughness of cleaning may be crucial considerations where incompatible batches are to be mixed at different times in the same machine. Plain tumbling vessels are easy to clean provided that adequate openings are available. Areas that may present cleaning problems are (1) seals or stuffing boxes, (2) crevices at baffle supports, (3) any corners, and (4) discharge arrangement. Where cleaning between different batches may be time-consuming, several small mixers should be considered. Special sanitary construction can usually be provided at extra expense.

Agglomerate Breakdown and Attrition. The two methods of producing agglomerate breakdown and attrition are: (1) *Impact.* The major factor is the peripheral speed of the rotating internal device. Table 8.3 gives impact-velocity data for various mixers. (2) *Shearing and compressive action*, which in mullers depends upon the clearance between muller and pan, and the muller weight or spring load, respectively.

Where an attrition device is necessary to break down aggregates but may also produce too much size reduction on other batch ingredients, the tolerable attrition should be determined by tests.

Dust Formation. Loss of dust can seriously affect batch composition, particularly where vital minor ingredients are lost. Methods of minimizing dust formation are: (1) Use of less dusty but equally satisfactory batch ingredients. Sometimes a pelletized form of an extremely dusty material is available. (2) Proper venting so as to enable filtering of displaced air rather than unregulated loss of dust-laden air. (3) Dust-tight arrangements for loading and unloading the mixer. (4) Addition of liquids where tolerable. Water is not only effective in minimizing dust upon discharging from the mixer, but if properly added, it will also render the batch less dusty in subsequent handling steps. The addition of a small quantity of surface-active agent will improve the penetration of the water throughout the batch and enable it to wet even such materials as coal dust. The method of adding water is important (see Method of Adding Liquids).

Care should be taken to avoid powerful suction or air flow on the mixer or the weight hopper from which the ingredients feed into the mixer. If the dust-collection suction on the mixer is too strong, vital ingredients may be sucked out. If the dust-collection suction on the weighing system is too strong, errors in weighing may result.

Electrostatic Charge. Certain batch materials such as plastics tend to accumulate a charge easily. Work input will affect the charge on the batch. Coating of the inside of the mixer shell or rotating elements may occasionally result because of electrostatic charge. This can present a cleaning problem. Possible aids in overcoming this are (1) addition of special solid materials with very high surface area to weight ratios, (2) addition of liquids (see Dust Formation and Method of Adding Liquids), (3) proper choice of material of construction of the mixer, (4) controlling humidity, (5) preparation of the batch ingredients so as to minimize accumulated charge.

Equipment Wear. Simple tumbling mixers give the least wear. Attrition devices in tumblers may present serious abrasion problems with certain materials such as sand and abrasive grinding-wheel grains. Abrasion-resistant coating such as rubber coating, special alloys, or platings should be considered for these cases. An internal agitator device may wear even though its speed is low. Particularly where highly abrasive materials are to be mixed, the benefits of an agglomerate-breaking device must be weighed against the potential contamination and replacement and maintenance costs.

Contamination of Product. This has been partially covered under Cleaning and Equipment Wear. Other sources of contamination are lubricants and repair materials. Types which are not compatible with the batches to be mixed should be avoided.

Heating or Cooling. Nearly all commercial mixers can be heated or cooled. Some can be provided with heated or cooled agitators. Where temperature rise during mixing is detrimental, cooling facilities should be provided. The various manufacturers can provide details on the means of heating their machines. Most common heating means are (1) water or steam in the jacket and in hollow-screw or paddle-type internal agitator, (2) hot oil, (3) Dowtherm liquid or vapor, (4) electric heaters, contact or radiant, (5) hot air in direct contact with product (suitable only for revolving-drum-type mixers), (6) exterior heating of drum by direct or indirect firing. For cooling, the most common means are (1) water or refrigerated fluid in the jacket and in hollow-screw or paddle-type internal agitator, (2) an evaporant such as liquid ammonia, (3) direct air contact (for rotating-shell mixers), (4) oil or Dowtherm (or equivalent) for cooling high-temperature materials.

Flexibility. Where batches of widely different size must be mixed, flexibility of operating capacity may enable use of fewer mixers. Certain features may require a nonflexible capacity requirement. For example, ordinarily an internal agitating device in a tumbling mixer does not function effectively unless the batch is loaded to a certain level. The need for such features must be weighed against the limitations imposed by a narrow operating-capacity range when choosing equipment for an operation where batch size will vary considerably.

In general, the effect of the percentage of mixer volume occupied by the batch on the adequacy of mixing should be borne in mind, particularly when any change from the recommended volume percentage is considered.

Vacuum or Pressure. Most tumbling mixers can have provision for vacuum or pressure.

Mixers which cannot be adapted to these conditions are mullers with rotating pans. Continuous mixers introduce problems of sealing the charge and discharge ends.

Method of Adding Liquids. Where the addition of liquids may be desirable (see Dust Formation and Electrostatic Charge), this should be considered when designing the mixing system rather than hastily improvised. The purpose of the liquid should be considered, whether for (1) dust suppression; (2) product; (3) heating and cooling. If a viscous liquid must be well distributed, this should be considered when choosing the mixer.

Liquid should be directed into the batch materials and not onto bare mixer surface since this could cause build-up. Nozzle spray pressure should be sufficient to penetrate the batch but not so high as to cause heavy splashing. The liquid should be added to the well-mixed batch. In particular, where premature addition of liquid could impair the adequacy of blending, both the time it is added during the mixing cycle and the time taken to add the liquid are important.

Automated equipment for the addition of liquids can be worked into the over-all mixing plant where necessary. For dust-reduction purposes, a volumetric method of metering is satisfactory. However, should a critical batch ingredient be added in liquid form, a more precise method of metering may be necessary.

Other Considerations. (1) Proper ventilation and discharge enclosures, (2) provision for relief of internal explosion, (3) vibration isolation (shock mounts), (4) remote operation of charge and discharge, (5) noise during operation.

Equipment Selection

General Comments. Types of mixers and performance characteristics are given in Table 8.3. Segregating tendencies among solid materials were described previously. A sound approach to solids-mixer selection starts with a careful examination of these areas. However, mixer selection should also involve consideration of the mixer's place in the over-all process. Possible consolidation of many solids-processing steps, or the opposite (splitting up one operation into several), deserves scrutiny at this time. If no one standard machine has all the necessary requirements, thought should be given as to which machine can best be modified to achieve the most desirable combination of features. One should look at the over-all process objectives as well as the equipment details when selecting a solids mixer.

Pilot Tests. In some cases, it is possible to perform pilot tests on a small-scale version of the equipment to be used in production. Much useful information can be found here but the following must be borne in mind:

1. In general, the larger the pilot unit, the more reliable the prediction of large-scale performance. The pilot unit should be a prototype with all dimensions properly scaled down.

2. Published solids-mixing scale-up data are rare. Equipment suppliers can provide scale-up information for their particular type of equipment based on past experience. With geometrically similar tumblers, if the speeds are adjusted to give comparable motion and the mixer volume fraction occupied by the charge is the same, scale-up of results will be straightforward. The presence of a rotating internal device presents problems in the scaling up of clearances, blade area to mixture volume, and sizes and speeds of the rotating devices. For agglomerate breakers, the key factor in scaling up is impact velocity. Scale-up in cylinders is discussed on pp. 290–292 of Reference 9. Solids-processing scale-up is discussed in Reference 18 and the scale-up of tumbling mixers is discussed in Reference 16.

3. The actual process materials should be used if possible. If substitute materials must be used, they should have the same mixing characteristics. Tests with differently colored but otherwise identical beads can be misleading and so can tracers. The reason is that the flow properties of the specific materials to be mixed in the plant may not be the same as these

Table 8.3. Horsepower Requirements and Speeds of Rotation for Some Commercial Solids Mixers (Approximately 50 ft^3 Working Capacity).

TYPE OF SOLIDS-MIXING MACHINE	APPROX. WORKING CAPACITY, CU FT	HORSEPOWER		ROTATIONAL SPEED, RPM		COMMENTS
		SHELL	INTERNAL DEVICE	SHELL	INTERNAL DEVICE SHAFT SPEED	
1. Tumbler						
Without baffles:						
Double cone	54	$7\frac{1}{2}$		18		Based on 100 lb/cu. ft material
Twin shell	50	5		13.7		Max. bulk density of material = 55 lb/cu. ft^3
With baffles:						
Horizontal drum						
Manufacturer E	50	20		11.1		Heavy duty (material 100 lb/cu. ft). For extremely heavy duty (150–200 lb/cu. ft material), the maximum working capacity with 20-hp motor is 35 cu. ft.
Manufacturer F	50	10		14		For material of 40 lb/cu. ft max. bulk density
Double cone revolving about horizontal axis	56	25		11.5		Mixer can be tilted. Rear end charger. Capacity based on mixed concrete
2. Tumbler with agglomerate breaker:						
Double cone	54	$7\frac{1}{2}$	See Comment	18	See Comment	Hp requirement for internal device depends on character of material, type and speed of agitator. These are to be determined by adequate testing
Twin shell	50	5	5 (pin-type intensifier bar) $7\frac{1}{2}$ (liquid-solids intensifier bar)	13.7	945 (1730 ft/min tip speed) 1055 (3320 ft/min tip speed)	Max. bulk density of material = 55 lb/cu. ft.
3. Stationary shell or trough						
Ribbon:						
Manufacturer C	50		12		28	Hp required based on material of 50–60 lb/cu. ft bulk density, medium free-flowing, using 10 hp/ton for average mix cycle of 3–10 min (depending on material, range can be 3–18 hp/ton)

Manufacturer A	46		10		37	Based on material of 30 lb/cu. ft bulk density
Manufacturer D	50		15		45	Based on material of 40–50 lb/cu. ft bulk density
3-shaft ribbon	50		Blender shaft 20 Feeder shaft $7\frac{1}{2}$ (total)		Variable-speed drives on all shafts	This blender is rated at 300 cu. ft/h on batch-mixing basis; 900 cu. ft/h on continuous-mixing basis. Materials rated at 70 lb/cu. ft bulk density
Vertical screw	52.9		5		Screw, 64.4 Orbit, 2.2	Hp. based on 37 lb/cu. ft bulk density. This may vary with different materials. Max. hp = 10, max. weight = 4410 lbs.
Muller:						
Batch; stationary pan, rotating turret	40		60		24 (turret speed)	Based on material of 60–75 lb/cu. ft bulk density
Continuous; stationary pan, rotating turret	Basically, the continuous mullers are merely two batch mullers joined together at the cribs, making a figure 8 design. Thus, the 40 cu. ft batch muller rated at 60 hp becomes an 80 cu. ft working capacity continuous muller requiring 125 hp. This would give 125 tons/hr with a $2\frac{1}{2}$-min residence time. Turret speeds are 24 rpm.					
Single rotor	See Comments					In this *continuous* unit, the output can range from 25–600 lb/min with hp from 5–100 and rpm of 500–4000 depending on the materials mixed
Double rotor	See Comments					In this *continuous* unit the output can range from 200–500 lb/min with hp from 5–40 and rpm from 200–300 depending on the materials mixed
Twin-rotor heat-exchanger mixer	49.2		5–15		20–100	Amount of conveying and mixing action affected by amount of pitch and type of ribbons mounted on exterior of hollow screws
Turbine	50		50		Peripheral speed of 600 ft/min	
4. Both shell and internal device rotate						
Countercurrent muller	45 60–90†	* 20	* 25	6.75–8.75 5.65	28–35 20	

*One 25-hp motor drives both the shell (mixing pan) and internal device (mixing star).
†Batch capacity range depends on nature of materials to be mixed.

Table 8.4. Costs of Several Types of Solids-Mixing Machines (See Table 8.1).

The figures given below are derived from log-log graphs of cost (\$) on the ordinate vs working capacity, cu. ft, on the abscissa. The cost in dollars = c_2 of a mixer of working capacity = Q_2 (cu. ft) can be estimated from the equation $c_2 = c_1(Q_2/Q_1)^n$. The column entitled "Range" gives the working-capacity range over which this equation holds. Where several ranges are listed for any type of machine, care must be taken to use the proper c_1 and n for the range shown. c_1 is the ordinate intercept at the lowest working capacity Q_1; n is the slope of the log-log plot and represents the exponent of the relationship: capacity^n = cost. This type of relationship is commonly used in order-of-magnitude estimation. Further information on it can be found in Aries and Newton, "Chemical Engineering Cost Estimation," Chemonomics, New York, 1950, and Chilton, "Cost Engineering in the Process Industries," McGraw-Hill, New York, 1960.

The following qualifications must be borne in mind when using this method of estimating solids-mixer costs:

1. The cost figures pertain to 1960. Therefore adjust estimate accordingly.
2. Special features will increase the cost. Similar machines may differ in cost depending upon construction and other differences (type of drive, maximum bulk density which can be handled, etc.). Therefore, careful attention should be paid to the comments.
3. In order to simplify the tabulation of costs, the best straight line was drawn for each log-log graph, even though there was scatter about these lines and sometimes a slight curvature.

Therefore, the cost figures should be used only for order-of-magnitude estimations and not as exact quotes. This point cannot be stressed too heavily, since virtually every manufacturer who was contacted expressed misgivings about the advisability of including any cost figures because of the ease with which these can be misused so as to give misleading impressions. Despite these misgivings, the manufacturers did cooperate to help the engineer to obtain rapid "guesstimates" of solids-mixer costs.

When a definite quote on a solids mixer is needed, this should be obtained from the manufacturer after the type of machine which will be needed has been discussed with him in detail.

Many machines are available in much larger capacities than those included in the ranges in this table. For example, 700 cu. ft working capacity ribbon mixers and 1000 cu. ft working capacity tumbling mixers are available; even larger-capacity machines can be built.

TYPE OF SOLIDS-MIXING MACHINE	RANGE, CU. FT	c_1	n	GENERAL COMMENTS; CHECK WITH EQUIPMENT MANUFACTURER IF PRECISE FIGURE IS NEEDED
1. Tumbler				
Without baffles:				
Double cone				
Manufacturer A:				
Mild steel	13–40	1,500	0.36	For working-capacity range of 13–40 cu. ft, construction is for bulk density of 100 lb/cu. ft
	54–325	2,700	0.65	
304 stainless steel (reg. commercial finish)	13–40	2,300	0.32	For working-capacity range of 54–165 cu. ft., construction is for bulk density of 75 lb/cu. ft
	54–325	3,800	0.77	
316 stainless steel (reg. commercial finish)	13–40	2,600	0.30	For working capacity range of 240–325 cu. ft, construction is for bulk density of 50 lb/cu. ft
	54–325	4,400	0.77	Prices do not include motor drives
Twin shell (vee)				
Manufacturer B (Note: These figures apply also to double cone for Manufacture B):				
Carbon steel	1–200	600	0.47	Above 30 cu. ft working capacity, bulk densities of 50–65 lb/cu. ft.
Stainless steel	1–200	800	0.49	Below 30 cu. ft working capacity, some bulk densities may go as high as 135–165 lb/cu. ft although most are in the 65–75 lb/cu. ft range. Check with manufacturer. Type of drive also has bearing on this.

Table 8.4. (*Continued*).

TYPE OF SOLIDS-MIXING MACHINE	RANGE, CU. FT	c_1	n	GENERAL COMMENTS; CHECK WITH EQUIPMENT MANUFACTURER IF PRECISE FIGURE IS NEEDED
With baffles:				
Horizontal drum carbon steel				Electric motors included
Without liner	15–70	3,000	0.52	Light duty, bulk densities to 70 lb/cu. ft
With liner	13–110	3,800	0.73	Heavy duty, bulk densities of 100 lb/cu. ft
	35–170	5,950	0.72	Extremely heavy duty, bulk densities of 150–200 lb/cu. ft
Double cone rotating around long axis	56–210	12,000	0.65	Includes motor without starter, liners and blades, and rear-end charger
2. Tumbler with agglomerate breaker				
Twin shell:				
Pin-type intensifier (carbon steel)	1–75	1,200	0.41	Above 30 cu. ft working capacity, bulk densities of 55–65 lb/cu. ft
Pin-type intensifier (stainless steel)	1–75	1,400	0.45	Below 30 cu. ft working capacity, bulk densities as high as 115 lb/cu. ft with smaller sizes but mostly in the 65–75 lb/cu. ft range
Liquid-solids feed-bar type of intensifier (carbon steel)	1–50	1,900	0.39	Bulk densities in 50–65 lb/cu. ft range
Liquid-solids feed-bar type of intensifier (stainless steel)	1–50	2,100	0.43	
3. Stationary shell or trough ribbon				
Manufacturer A:				
Mild steel	7.5–186	840	0.54	Does not include motor drive. Double spiral blade. 30 lb/cu. ft bulk density material
304 stainless steel	7.5–186	1,560	0.57	
Manufacturer C:				
Standard carbon steel	25–300	1,500	0.52	Double ribbon, inner and outer; motor and drive not included. Capacities based on 50–60 lb/cu. ft bulk density material although mixer can be used with other bulk densities with necessary power adjustment
Standard 304 stainless steel	25–300	3,100	0.59	
Manufacturer D:				
Carbon steel without drive and motor	11–94	1,930	0.56	Batch capacity and range based on bulk density of 40–50 lb/cu. ft. Other densities will affect capacities accordingly
With drive and motor	11–94	2,650	0.56	
3-shaft ribbon (includes motor and variable drives):				
Carbon steel, heavy duty	25–50	8,900	0.31	May be used for wetted materials with angle of repose up to 60 deg. and weighing up to 60 lb/cu. ft or for free-flowing materials weighing up to 70 lb/cu. ft.
Stainless steel, heavy duty	3–50	3,800	0.48	
Carbon steel, medium duty	These are available in 25 and 50 cu. ft. models. The difference in price as compared with heavy duty would be due to the use of smaller motors			May be used for wetted materials with angle of repose of 60 deg. and weighing up to 30 lb/cu. ft and for free-flowing materials weighing up to 35 lb/cu. ft.
Stainless steel, medium duty				

Table 8.4. *(Continued).*

TYPE OF SOLIDS-MIXING MACHINE	RANGE, CU. FT	c_1	n	GENERAL COMMENTS; CHECK WITH EQUIPMENT MANUFACTURER IF PRECISE FIGURE IS NEEDED
Vertical-screw mixer with orbiting screw:				
Mild steel	3.5–12.5	3,100	0.11	Motor and motor controls (not mounted) are included. Will handle materials up to 189 lb/cu. ft depending on volume to be mixed.
304 stainless steel	3.5–12.5	4,300	0.12	Does not have universal joint at bottom of screw
Mild steel	7–100	4,100	0.10	Does have universal joint at bottom of screw
304 stainless steel	7–100	6,000	0.14	
Muller:				
Rotating muller turret, stationary shell carbon steel	6–60	5,100	0.69	Stainless steel approximately 200–240% of carbon-steel prices
Continuous muller with rotating turret and stationary shell	6–120 (actual batch capacity) 11–170 tons/h continuous mixing through-put for $2\frac{1}{2}$ min residence time	8,700	0.66	Based on material bulk density of 60–75 lb/cu. ft and continuous mixing comparable with approximately $2\frac{1}{2}$ min batch mulling cycle (residence time). Residence time will affect tons/hr; it can vary from 1–5 min.
Twin rotor–heat exchanger, mixer				
Carbon steel	2.7–49 (actual batch capacity)	5,100	0.44	Used where heat transfer is needed along with mixing. Does not include motor and drive; hollow screws with mixing ribbons; heavy-duty
Stainless steel	2.7–49 (actual batch capacity) (cap. in tons/h = 0.6 to 10 assuming 10 min residence time)	9,300	0.34	construction; mixing action depends on pitch and type of ribbon mounted on exterior of hollow screw. Material density assumed = 70 lb/cu. ft Heat-transfer capacities 20,000 to 1,000,000 Btu./h assuming 300°F heat-transfer medium, over-all heat-transfer coeff. of 8–30 and solids over-all coeff. of 8–15 Btu/(sq. ft)(h)(°F)
Turbine	$2\frac{1}{2}$–100	1,770	0.70	Includes basic mixer with motor with one manually operated discharge gate. $12\frac{1}{2}$ cu. ft and above have replaceable liners as standard equipment
4. Both shell and internal device rotate				Batch capacity range established on basis of material bulk density of 100 lb/cu. ft
Countercurrent mixer				Other densities will affect capacities accordingly
Carbon steel, plows only	5.5–45	4,000	0.50	(Not including motor drive)
With mullers	5.5–45	4,300	0.55	(Not including motor drive)
Stainless steel:				
Plows only	5.5–45	6,400	0.54	(Not including motor drive)
With mullers	5.5–45	7,200	0.60	(Not including motor drive)

Table 8.4. *(Continued).*

TYPE OF SOLIDS-MIXING MACHINE	RANGE, CU. FT	c_1	n	GENERAL COMMENTS; CHECK WITH EQUIPMENT MANUFACTURER IF PRECISE FIGURE IS NEEDED
Larger capacity				
Carbon steel:				
Plows only	90	24,500		(Not including motor drive)
With mullers	60	26,500		(Not including motor drive)
Stainless steel:				
Plows only	90	44,000		(Not including motor drive)
With mullers	60	53,000		(Not including motor drive)

demonstration materials. Regardless of how the mixer contents appear to be moved around, the properties of the actual batch ingredients may cause segregation or other problems.

4. Differences in materials of construction between the pilot unit and the production unit should be considered. These may have a bearing on caking, abrasion, and electrostatic effects.

Continuous Mixing. Although batch mixing has been the predominant method of mixing of solids, consideration is being given to the use of continuous mixing in many industries. There are two types of continuous-mixing operations. The first type has a low holdup volume and will provide fine-scale blending of the particles via impact and shear elements such as are used in grinding machines. Some machines of this type are the hammer, impact, cage, and jet mills. It is essential that the feed to these machines be properly proportioned and premixed in order to achieve a uniform product.

The second type of continuous mixer involves high holdup machines which contain agitating and conveying mechanisms. These rearrange the individual particles and also displace large volumes of material and move the batch through the machine. Mixers of this type can produce both fine-scale and coarse-scale blending. The ribbon-type mixer is frequently used for continuous mixing although this is also used for batch mixing. A continuous muller mixer has been developed as shown in Figure 8.5*h*.

The average composition of the stream leaving a continuous mixer is the same as the average of the added entering streams. Variations in proportions of the entering streams will be damped out by the mixing action of a continuous mixer. These effluent-stream variations will become smaller as average solids residence time is increased and the frequency of the variations becomes larger.

Certain general criteria can be used to determine whether continuous flow will be beneficial. Continuous flow is worth consideration if (1) a single formulation can be run for an extended period, (2) the fluctuations of the outgoing product are within process requirements, (3) sufficiently accurate metering of ingredients can be achieved, (4) the rest of the process warrants continuous mixing. Continuous flow is of doubtful benefit if (1) frequent changes of formulations are anticipated, (2) fluctuations of product composition will be outside the permitted range, (3) the ingredients cannot be metered with the necessary level of accuracy, (4) complex temperature or pressure cycles are involved.

Sometimes a system of mixing and dispersing is composed of one or more batch units providing a feed to a continuous intensive dispersion unit. Another possibility would be a batch mixer and surge bin which provide a continuous feed to a final dispersion unit. Various combinations of this type with adequate sampling at the proper points may be used where continuous flow would be beneficial provided that certain features could be overcome.

Cost

Costs for several types of solids mixing machines are given in Table 8.4.

REFERENCES

1. A. I. Ch. E. Dry Solids, *Past and Dough Mixing Equipment Testing Procedures*, 2d ed., American Institute of Chemical Engineers, New York (1978).
2. Bullock, H. L. *Chem. Eng. 66*:177 (1959).
3. Danckwerts, Research (London) *6*:355–361 (1953).
4. Fischer J. *Chem. Eng. 67*:107 (1960).
5. Kirk and Othmer (eds.), *Encyclopedia of Chemical Technology*, vol. 9, Reinhold, New York (1952).
6. Lacey, *J. Appl. Chem. 4*:257–268 (1954).
7. Quillen, *Chem. Eng.* 61:178 (1954).
8. Scott, Chapter on Mixing of Solids in *Chemical Engineering Practice*, Cremer and Davies (eds.), vol. 3, p. 362, Butterworth, London, (1957).
9. S. S. Weidenbaum, Chapter on Mixing of Solids in *Advances in Chemical Engineering*, Drew and Hoopes (eds.), vol. II, Academic Press, New York, (1958).
10. Work, *Chem. Eng. Progr. 50*(5): September, 1954).
11. Vance, "Statistical Properties of Dry Blends," *Ind. Eng. Chem. 58*:37–44 (1966).
12. Gren, "Solids Mixing: a Review of Theory," *Brit. Chem. Eng. 12*:1733–1738 (1967).
13. Ashton and Valentin, "Mixing of Powders and Particles in Industrial Mixers," *Trans. Inst. Chem. Engrs.* (London) *44*:t165–t188 (1966).
14. Valentin, "Mixing of Powders and Pastes: Basic Concepts," *Chem. Engr.* (London) *45*:CE99–CE106 (1967).
15. S. S. Weidenbaum, "Solid–solid Mixing" in *Chemical Engineers Handbook*, 5th ed., J. H. Perry (ed.), McGraw-Hill Co., New York, pp. 21–30 (1973).
16. R. H. Wang and L. T. Fan, "Methods for Scaling-Up Tumbling Mixers," *Chem. Eng., 81*(11):88–94 (1978).
17. H. B. Ries, "Mixing Quality, Problems, Test Methods, and Results," *Int. Chem. Eng., 18*:426–442. (1978).
18. K. E. Sterrett, "Scale-up for Solids Processing," *Chem. Eng.*, pp. 155–158 (September 21, 1959).
19. Lacey, *Trans. Inst. Chem Engrs.* (London) *21*:52 (1943).

9
Storage of Particulate Solids

Fred M. Thomson
E. I. Du Pont De Nemours & Co., Inc.
Wilmington, Delaware

9.1. INTRODUCTION

For many years, the structural design of bins for storage bulk solids has been based on computation methods derived in the early 1900s. Little was known about the flow properties of bulk solids, or even how to determine them. As a result, functional design of bins was concerned mainly with providing a storage structure having a specified gross storage volume and having adequate protection for the stored material.

Within the past few years, a number of workers in the rapidly growing powder mechanics field have turned their attention to storage and flow of bulk solids. Much of the new information coming from their work has modified or changed the early structural design methods and is providing a sound basis for functional design of bins.

Because the flow of new information appearing in the literature is continuing to escalate, this chapter can only provide a summary of the state-of-the-art in storage and flow of bulk solids.

9.2. DEFINITIONS

The following definitions are commonly accepted and will be used in this chapter:

Bin: Any upright container for storing bulk solids.

Silo: A tall bin, where $H > 1.5D$ (H is the height of the vertical and D is the diameter of a round bin or the dimension of the short side of a rectangular bin). A tall bin is sometimes described in some structural engineering texts as a bin where the "plane of rupture" of the contained material, determined by Coulomb's theory, intersects the side walls. There is disagreement among engineers regarding the actual location of this plane of rupture in bins having hopper bottoms and this definition is becoming less used.

Bunker: A shallow bin, where $H < 1.5D$ or, as above, where the "plane of rupture" intersects the top surface of the stored solids.

Hopper: A converging sloping wall section attached to the bottom of a silo. If a converging section stands alone as an independent bin, it is called a bunker.

Solids Flow Patterns: As solids flow from a bin, the boundaries between flowing and nonflowing regions define the flow pattern. Three common patterns, funnel flow, mass flow, and expanded flow, are defined in Section 9.5.

Flow Obstructions: It is assumed that interruption of solids flow in a bin can be caused by either of two types of obstructions: an arch (sometimes called a bridge) formed across a flow channel or bin opening, or a rathole (sometimes called a pipe) formed when the flow channel empties, leaving the surrounding stagnant material in place. These obstructions are defined in more detail in Section 9.5.

9.3. TYPES OF BIN CONSTRUCTION

Bins and silos can be categorized as either agricultural or industrial-type construction. The general descriptions that follow apply to either type.

Metal Construction

Shop-Welded. These are welded as a complete assembly in the shop (with roof in place) and then shipped as a complete unit to the site. The maximum width or diameter is normally limited to 3.6 to 4.0 m to accommodate rail and road clearances encountered during transportation. The maximum volume accommodated by shop-welded bins is about 1,700 m^3.

Field Assembly by Welding. Preformed parts are shipped to the site, fitted together, and assembled by welding. Elevated silos and bins with hopper bottoms have been built up to 15 m diameter. Flat bottom silos and bins resting on concrete slabs have been built up to 48 m diameter. Shop- or field-welded bins can be fabricated of any desired material. Carbon steel, aluminum, and stainless steel are all commonly used. A wide variety of wall coatings are available to protect carbon steel surfaces although sophisticated coatings that require curing, heating, etc., are best applied on shop-fabricated parts.

Field Assembly by Bolting. Bolted cylindrical bins are formed with rolled steel staves, normally 2.5 m high with flanged ends and sides (Fig.

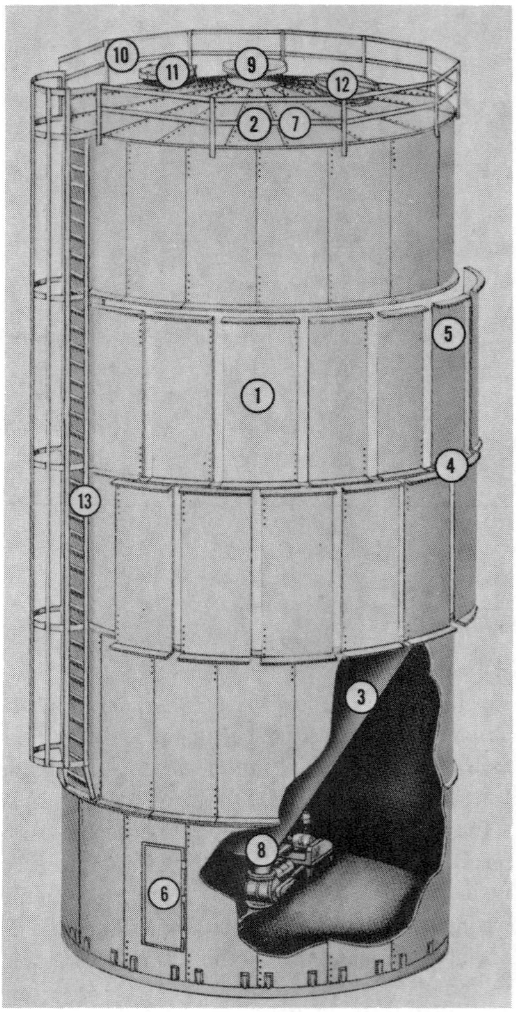

Figure 9.1. Assembly of bolted cylindrical steel bin. (1) prefabricated cylindrical shell, (2) circular deck, (3) conical hopper, (4) flanges joining shell rings, (5) rings formed by bolting sections (staves), (6) access door in silo skirt, (7) preshaped deck plates, (8) hopper formed by joining hopper sections that are equipped with compression bars that attach to the interior of the shell for support, (9) fill opening, (10) guard rail, (11) ventilator, (12) deck manhole (13) ladder. (Courtesty of Peabody Tec Tank, Inc., South Industrial Park, Parsons, Kans.)

9.1). These are gasketed and bolted together to form body rings (Figs. 9.2 and 9.3) and then stacked to form the bin or silo cylinder. The bin is anchored to a concrete slab with stirrups (Fig. 9.4).

Elevated cylindrical bolted bins with conical

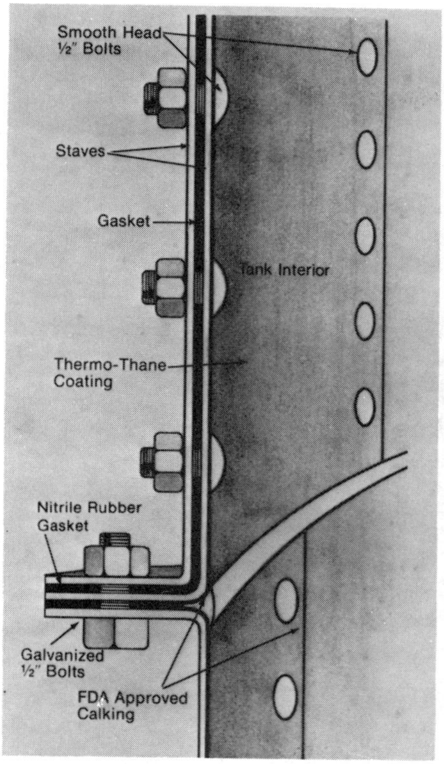

Figure 9.2. Detail of vertical and horizontal seams. (Courtesy of Peabody Tec Tank, Inc., South Industrial Park, Parsons, Kans.)

hopper bottoms, supported on a steel structure are available up to about 8 m diameter. Flat bottom bins supported on a concrete slab are available up to about 17 m diameter. These bins are usually fabricated of carbon steel with various paint, epoxy, or glass coatings.

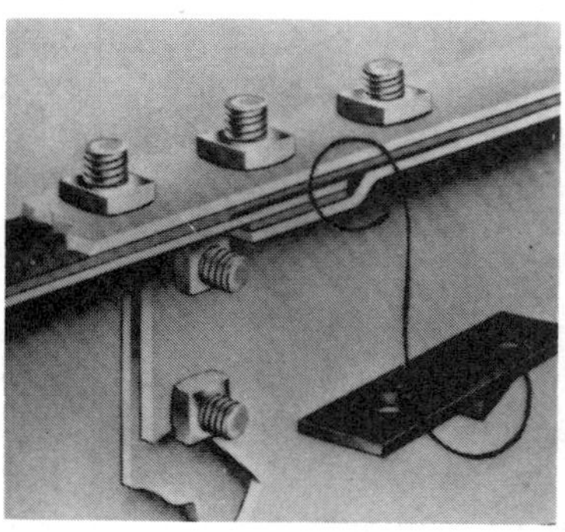

Figure 9.3. Detail of chime-lap gasket at sectional joints. (Courtesy of Peabody Tec Tank, Inc., South Industrial Park, Parsons, Kans.)

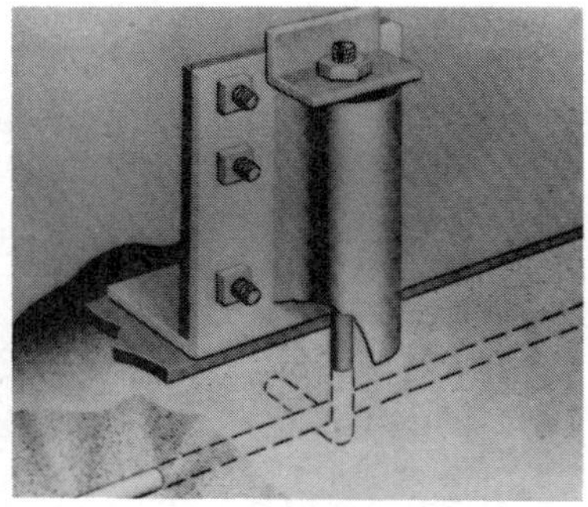

Figure 9.4. Detail of stirrup bolted to bottom ring and hook-end to foundation. (Courtesy of Peabody Tec Tank, Inc., South Industrial Park, Parsons, Kans.)

A common method of assembly of bolted circular bins is shown in Figure 9.5. In the case shown, a glass-coated steel bin is being assembled by forming the plates or staves into a ring on the foundation and jacking each section vertically as it is completed. Details of the bolted seam of a glass-coated steel bin are shown in Figure 9.6.

Bins are also formed from preformed flanged plates that can be assembled into rectangular or octagonal cross sections. These can be grouped to form large capacity storage systems (Fig. 9.7). Details of a typical rectangular storage structure are shown in Figure 9.8.

Very large silos with capacities of 1,000 to 60,000 m^3 have been built by Eurosilo.[1] These are constructed as a vertical structure steel frame supporting an outer and inner shell. The bottom is flat. Loading and discharge is accomplished by a special rotating device that levels the material while filling and draws the material to the center when unloading to underground conveyors. Other bolted agricultural silo designs are described by Reimbert.[2]

Spiral-Wound Coil Construction. With this unique fabrication method, aluminum or steel strip is unwound from a coil into a circular roll-forming machine that joins the edges of the strip in a continuous double-crimped spiraling seam. This generates a continuous rigid water tight cylinder (Fig. 9.9). Field-assembled bins are made by mounting coil and forming machine on the silo pad and continuously unwinding and seaming (Fig. 9.10), producing a

Figure 9.5. Assembling a glass-coated steel bin. (Permission Koppers Co., Inc., Sprout Waldron Div.)

continuous vertical cylinder. A detail of the double-crimped seal is shown in Figure 9.11. Prefabricated roof sections are put in place close to ground level after several revolutions have formed the initial cylinder; then unwinding and seaming is continued until the desired height is reached. These bins can also be shop-fabricated horizontally, then shipped to the site for erection.

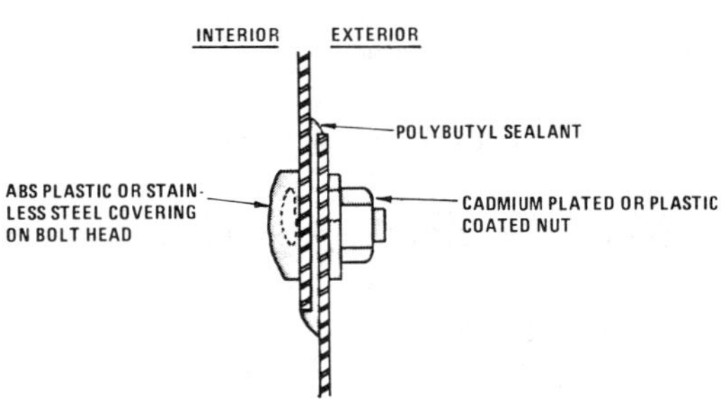

Figure 9.6. Details of bolted seam. (Permission Koppers Co., Inc., Sprout Waldron Div.)

Concrete Construction

Concrete bins become economically competitive with metal structures when diameters exceed 3.5 to 4.5 m. Concrete bins have been built up to 30 m diameter and larger units up to 46 m diameter are being designed. These bins can be constructed as a single circular or rectangular storage cell or in multiple cells. Common concrete bin constructions are shown in Figure 9.12. Various techniques used in concrete silo construction include the following:

Precast Construction

Concrete Staves. The bin is formed by assembling the staves in a circle and stacking them to form a cylinder. Circumferential steel hoops usually consisting of 3 or more rods connected by steel or malleable iron lugs, are spaced at intervals along the outside of the staves and posttensioned to place the staves in compres-

Figure 9.7. Exterior view of a bolted preformed rectangular steel bin. (Permission Leach Manufacturing Co., Inc [Lemanco])

sion (Fig. 9.13). Staves are usually about 250 to 300 mm wide by 500 to 750 mm long and 60 to 100 mm thick (Fig. 9.14). They are pressed in forms and then cured in high-pressure steam kilns. Staves can be made with hollow cores or can be made of lightweight aggregate to provide a measure of insulation to protect solids from the effects of sudden temperature changes and/or to minimize moisture condensation on the inner walls. Solid staves can be used for less critical applications and can be cast in heavier duty construction for storage of high-density solids. The staves are fitted together with a tongue-and-groove fit. Each stave has a tongue cast on top and one side with grooves cast on the bottom and opposite side.

The exterior of the structure is coated with either of several coatings as required by the application. These include sand and cement slurry coatings, waterproof agents combined with the slurry coatings, or special paint or epoxy coatings. The coatings provide weather protection, a water drip over the hoops, and improve the appearance of the silo. The interior of the structure can be finished to provide a smooth monolithic appearance using several coats of a brush and trowelled cement plaster or several coats of epoxy coatings applied with spray equipment or trowels.

A flat bottom silo mounted on a concrete slab at ground level will discharge from the side. When steel hoppers are used for discharge, they are fabricated with compression ring girders and supported on steel columns from grade. The

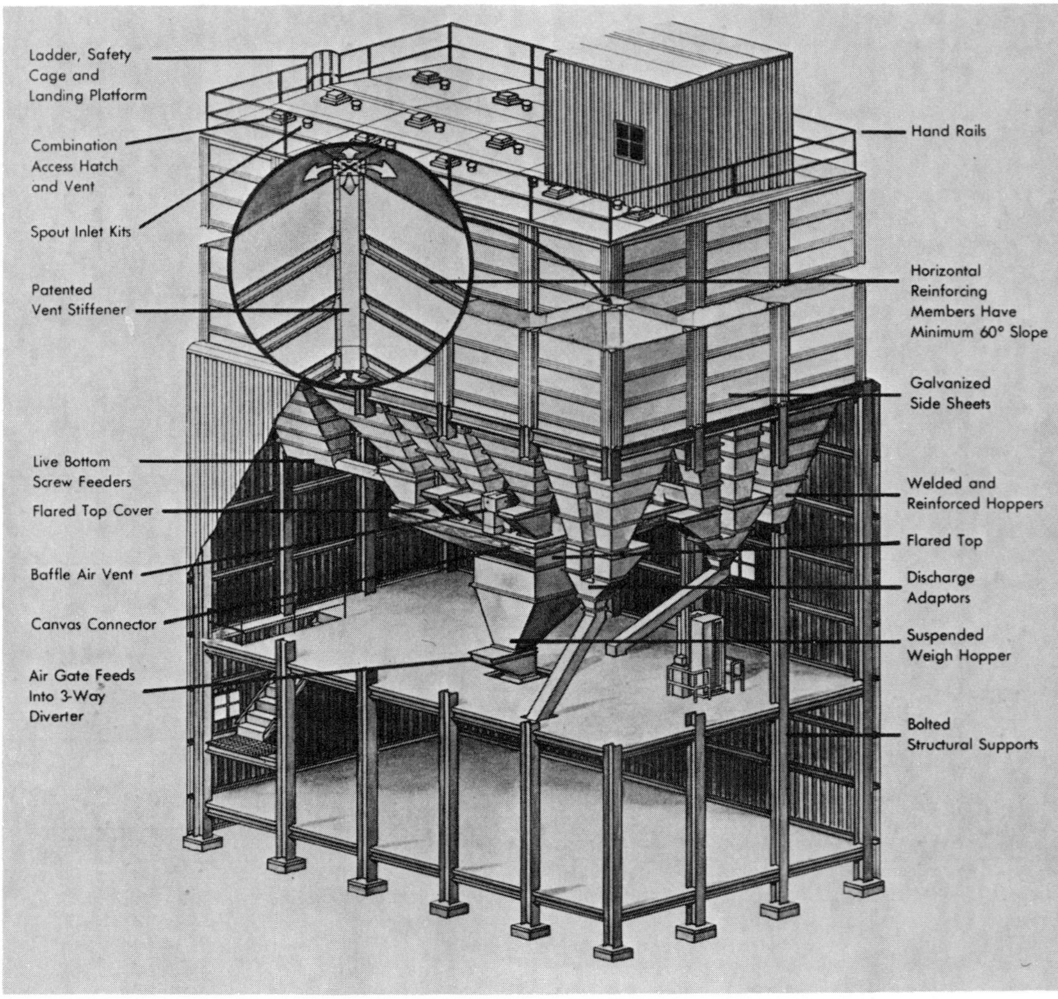

Figure 9.8. Typical construction details of a bolted preformed rectangular steel bin. (Permission Leach Manufacturing Co., Inc. [Lemanco])

silo roof normally consists of a reinforced concrete slab, sometimes mounted on a bar joist or structural steel beam support.

Where applicable, concrete slave silos are the lowest cost concrete construction. They have been built up to 12 m diameter and 30 m high. Specification and standards for concrete stave silos have been published by the American Concrete Institute.[16]

Posttensioned Rings. These structures are assembled by stacking circular precast concrete sections vertically on an elevated slab, capping with a precast roof slab, and then vertically posttensioning them together with wire rope. Details of the design of a group of 4.9 m diameter × 9 m deep bins are described in Reference 2.

Prefabricated Reimbert Silo. This construction is described in detail by the Reimberts in Reference 3. Shaped, precast reinforced concrete slabs about 4.5 m long by 0.5 m high are used as the basic structural element. These are stacked in a horizontal position with each end fastened to vertical concrete posts. These units can be used to form storage cells of any shape—rectangular, hexagonal, etc. Storage of up to

Figure 9.9. Spiral-wound bin. (Permission Conair, Inc.)

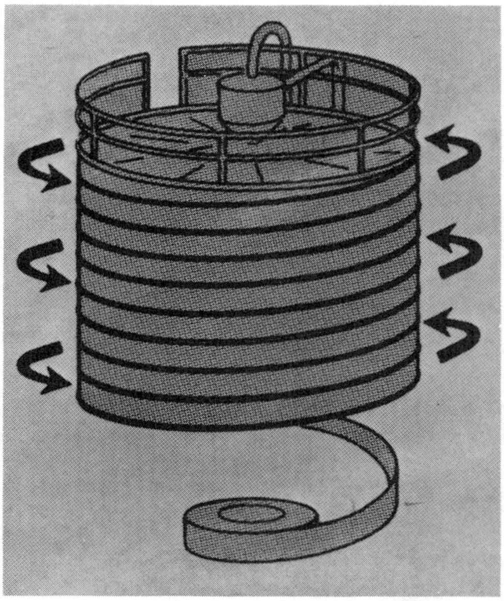

Figure 9.10. Forming a continuous spiraling seam to construct a spiral bin. (Permission Conair, Inc.)

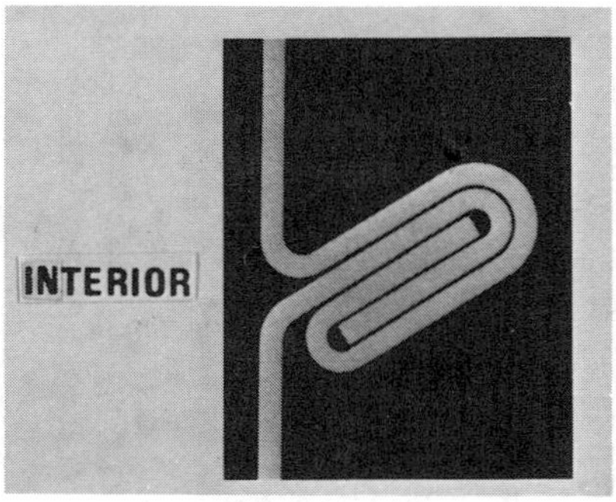

Figure 9.11. Detail of the double crimped seam. (Permission Conair, Inc.)

30,000 tons of grain and other agricultural products has been reported.

Cast In Place. Concrete silos or bunkers can be cast in stationary forms, slip forms, or jump forms in various configurations (Fig. 9.15). Detailed descriptions of these silos constructions are given by Safarian and Harris[5] and the Reimberts.[2] Slip forms and jump forms are well suited to silo fabrication and are the most commonly used method for large silos.

Slip Form. In this method, forms erected on the silo foundation are continuously raised by hydraulic screw jacks spaced about the periphery of the wall, and guided in the lift by vertical rods (Fig. 9.16).

Concrete is cast into the forms on a continuous basis. The speed of the upward movement of the form is determined by the setting time of the concrete. Continuous pouring assures that the concrete does not set before the following layer is cast, thus providing a monolithic structure. The slip form moves from 200 to 380 mm per hour (with 300 mm per hour a good average) and continues around the clock until the walls are complete. Steel reinforcing is placed in the forms as they reach predetermined positions in the pour. Conventional reinforcing bars or posttensioned steel strands or wire tendons may be used for reinforcing.

Slip form structures can provide greater load-bearing capacities than precast storage structures, provide smooth monolithic wall construction, and can be built in a variety of single cell

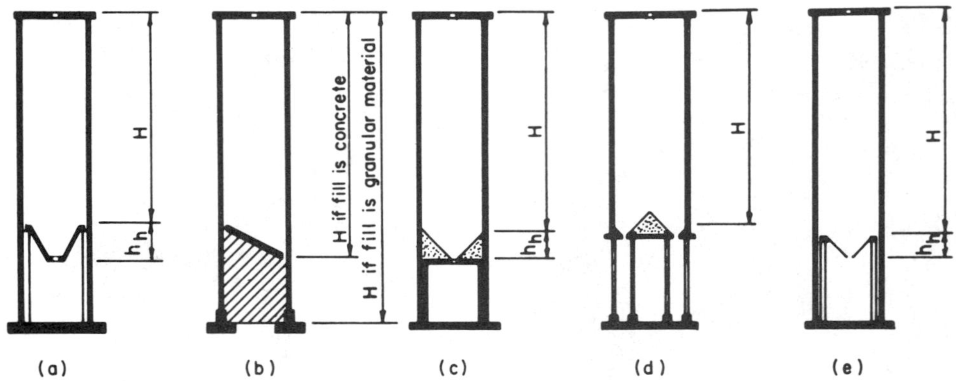

Figure 9.12. Typical Concrete Silo Constructions: *a*, silo on raft foundation, independent hopper resting on pilasters attached to wall; *b*, silo with wall footings and independent bottom slab supported on fill; *c*, silo with hopper-forming fill and bottom slab supported by thickened lower walls; *d*, silo with multiple discharge openings and hopper-forming fill resting on bottom slab, all supported by columns. Raft foundation has stiffening ribs on top surface; *e*, silo on raft foundation, with hopper independently supported by a ring-beam and column system. (Reprinted from Ref. 16 with permission of American Concrete Institute)

Figure 9.13. Concrete stave silos. (Permission First Colony Corp. and the Nicholson Co.)

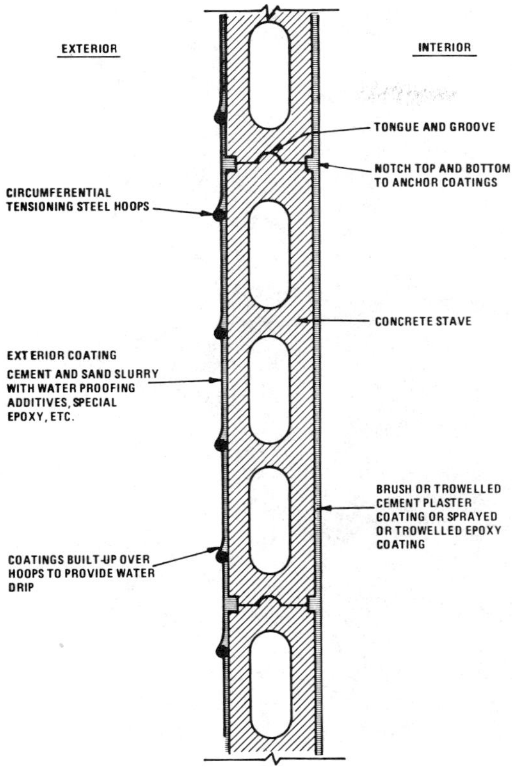

Figure 9.14. Detail of stave-wall. (Permission First Colony Corp. and the Nicholson Co.)

geometries, as well as grouped cells (Fig. 9.17). Slip form construction becomes most economical when silo diameters are about 12 m and larger. They have been built up to 37 m diameter and 52 m high. There are proposals for larger diameters on the drawing boards.

Jump Form. With this method, the forms are made up of sections about 1.8 m long by 1.3 m high, fastened together to form a continuous circular form. After two vertical layers have been cast and set, jump forming takes place. The forms from the previous pour are removed and hoisted into place above the top form and a new pour is made. Pouring need not be continuous and thus can be done on a day shift only, in contrast to slip forming, where casting must continue until the silo walls are completed. Jump form construction provides a storage system intermediate in size and cost compared to precast and slip form structures and is used mostly for cylindrical silos from about 9 to 18 m diameter and up to 46 m high.

9.4. NATIONAL CODES USED FOR STRUCTURAL DESIGN OF SILOS

Solids stored in a bin exert horizontal forces that result in tensile stresses in the walls. Also, part of the solids weight is transferred to the

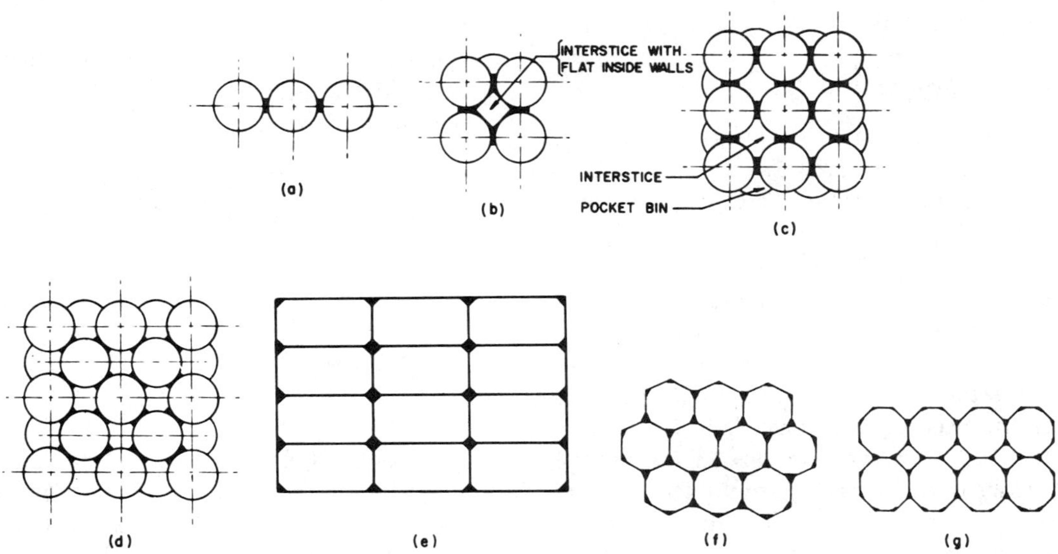

Figure 9.15. Typical silo or bunker grouping (From Ref. 5) (Permission Van Nostrand Reinhold Co.)

Figure 9.16. Slip-form concrete silo under construction. (Permission First Colony Corp. and the Nicholson Co.)

walls through friction, resulting in vertical compressive or buckling stresses. The Jannsen and Reimbert method for calculating wall pressures imposed by contained solids have been in general use for structural design of bins.

Jannsen's Method

For many years, Jannsen's pressure (The terms *pressure* and *stress* have the same units, force (N) per unit area $(m)^2$, and are often used interchangeably in the codes and by different writers.) theory[6] has been the basis for computing the stresses on bin walls caused by stored solids and has been incorporated in most national silo design codes. This theory is based on the assumption that vertical pressures are uniform over any horizontal bin cross section and these pressures vary only in the vertical direction. The average vertical static pressure q(kPa) at depth Y(m) below the bulk solids surface (Fig. 9.18) is given by:

$$q = \frac{\gamma R}{\mu k}(1 - e^{-\mu k(Y/R)}) \qquad (9.1)$$

The ratio of vertical to horizontal pressure k is assumed to be constant, and independent of the magnitude of the pressure, so that the magnitude of the pressure, so that the lateral static unit pressure p(kPa) given by:

$$P = kq = \frac{\gamma R}{\mu}(1 - e^{-\mu k(Y/R)}) \qquad (9.2)$$

where

$$k = \frac{1 - \sin\phi}{1 + \sin\phi} \qquad (9.3)$$

μ = coefficient of friction between material and silo wall ($\tan\phi'$ where ϕ' is the static angle of wall friction given in Section 9.7)

γ = bulk density (expressed as kN/m^3)

Figure 9.17. Aerial view–group of slip-form silos with silos in foreground under construction. (Permission First Colony Corp. and The Nicholson Co.)

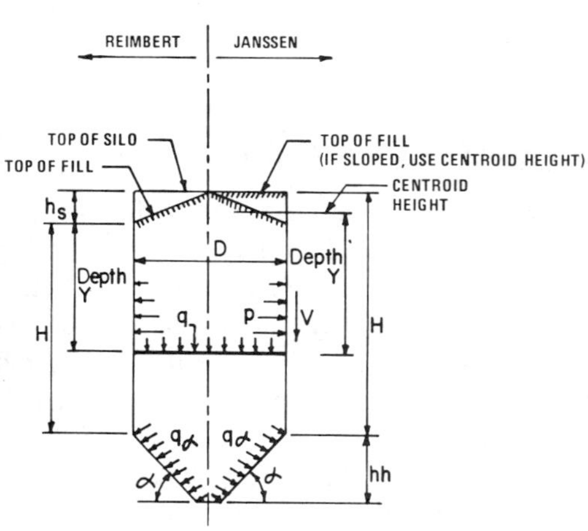

Figure 9.18. Storage bin dimensions used in Jannssen and Reimbert theories. (Reprinted from Ref. 16 with permission of American Concrete Institute)

ϕ = angle of internal friction of the solids. This can be obtained from the shear test described in Section 9.7. (In Jannsen's work, this value was assumed to be closely approximated by the angle of repose of the material and, therefore, angle of repose has often been used by many in this calculation. This assumption is not always correct, particularly at high values of ϕ.)

R = hydraulic radius, area of bin cross section/perimeter of bin (m)

$R = D/4$ for circular silo, where D = bin diameter

$R = D_e/4$ for regular polygonal bins where D_e is the diameter of a circle of area equivalent to the inside of the bin.

$R = a/4$ for square bins where a is the length of a side.

For computing pressures next to the short side of a rectangular silo with a = length of short side and b = length of long side, use $R = a/4$, and for pressures next to the long side use $R = a'/4$ where a', the length of side for an imaginary square silo, is:[3]

$$a' = 2ab/a + b \tag{9.4}$$

As depth Y increases, the lateral static pressure on the wall (p) asymptotically approaches:

$$p = \gamma R/\mu \tag{9.5}$$

In the vertical portion, the frictional stress v(kPa) is related to lateral pressure (p) by:

$$v = \mu p \tag{9.6}$$

The vertical frictional force V, per unit width of wall perimeter (kN/m) at depth Y, is given by:

$$V = (\gamma Y - q)R \tag{9.7}$$

(Note: The Jannsen method usually calculates higher vertical pressures than the Reimbert method (as given next), and therefore calculates lower frictional stresses. Equation (9.7) is therefore sometimes given as $V = (\gamma Y - 0.8q)R$, where q is computed by the Jannsen method and $V = (\gamma Y - q)R$, where q is computed by the Reimbert method so that the vertical stresses are approximately equal when computed by either method.[16])

The static unit pressure q_α(kPa) normal to the inclined hopper wall at depth Y (Fig. 9.18) is given by:

$$q_\alpha = p \sin^2 \alpha + q \cos^2 \alpha \qquad (9.8)$$

The hydraulic radius R is usually assumed constant and equal to that of the bin in this calculation but some designers reduce the pressures in accordance with the reduction in R within the hopper.

Reimbert's Method

A second method of wall stress computation has come into use based on the work by the Reimberts.[2,3] The static unit pressure at depth Y (Fig. 9.18) is computed as:

$$q = \gamma \left[Y \left(\frac{Y}{C} + 1 \right)^{-1} + \frac{h_s}{3} \right] \qquad (9.9)$$

where h_s is the height of the conical heap formed at the surface due to the angle of respose.

The lateral static unit pressure at depth Y is:

$$p = p_{max} \left[1 - \left(\frac{Y}{C} + 1 \right)^{-2} \right] \qquad (9.10)$$

For circular bins:

$$p_{max} = \frac{\gamma D}{4\mu} \qquad (9.11)$$

$$C = \frac{D}{4\mu K} - \frac{h_s}{3} \qquad (9.12)$$

k is given by Equation (9.3)

For regular polygonal silos with more than four sides:

$$p_{max} = \frac{\gamma R}{\mu} \qquad (9.13)$$

$$C = \frac{L}{\pi} \left(\frac{1}{4\mu k} \right) - \frac{h_s}{3} \qquad (9.14)$$

where L = perimeter of bin (m).

For rectangular silos, on the shorter wall a:

$$p_{max} = \frac{\gamma a}{4\mu} \qquad (9.15)$$

$$C = \frac{a}{\pi \mu k} - \frac{h_s}{3} \qquad (9.16)$$

For rectangular silos, on the longer wall b:

$$p_{max} = \frac{\gamma a'}{4\mu} \qquad (9.17)$$

$$C = \frac{a'}{\pi \mu k} - \frac{h_s}{3} \qquad (9.18)$$

a' is as defined in Equation (9.4).

In the above, R is as defined previously for the Jannsen method.

The vertical frictional force (V) per unit width of wall perimeter at depth Y is given by Equation (9.7). The static unit pressure q_α(kPa) in the hopper is calculated by Equation (9.7) using the Reimbert values for p and q.

Comments on Jannsen's Method

In recent years, as the knowledge of the strength of material properties of materials used in bin construction have become better known, design safety factors have been reduced. A number of bin failures occurred during this time. Theimer[7] and Sadler[8,9] described reinforced concrete silo failures. Many of these failures can be attributed to dynamic stresses that were not anticipated by the designers. Steel silo failures also occurred, including those caused by eccentric discharge.[10] These incidents have prompted comprehensive reviews of existing methods of calculating stresses.[10,11,13]

Some investigators have examined Jannsen's work and attempted to redefine his equations. Jenike and Johanson[95] pointed out that Jannsen's formula is a *lower* bound on the average pressures, and not necessarily the *actual* pressures on a silo wall. Sundaram and Cowin[14] reached the same conclusion and presented an "improved formula" that interpreted the coefficient k as the ratio of the horizontal pressure averaged over the perimeter to the vertical pressure averaged over the horizontal cross section of the silo. Bagster[15] points out that the common interpretation of Jannsen's k factor as being equal to $1 - \sin \phi / 1 + \sin \phi$ is erroneous and that the correct value lies between that and $1 - \sin^2 \phi / 1 + \sin^2 \phi$. Walker[91] also stated a similar conclusion.

It is known that wall pressures are very often not uniform over any horizontal cross section in a bin, that bulk density is not constant throughout the bin, and that the Jannsen ratio k is not constant.

The Reimberts[2,3] described extensive tests with a variety of noncohesive, agricultural grains in model and full-size bins. Their test data showed:

1. The angle of internal friction of a particular grain sample was not constant but varied with pressure. Particle shape, size, and surface characteristics of the grain significantly affected this value.
2. The angle of repose formed when pouring the grains onto a pile was only approximately equal to the angle of internal friction as measured under low stress. The angle of internal friction measured at high stress differed significantly from angle of repose.
3. Friction of solid on walls varies considerably with materials and wall roughness. In some cases, solid wall friction did not significantly increase when rough walls were substituted for smooth walls—contrary to what should be expected. In some cases, wall friction reduced with time; walls were polished from sliding of the grain, or oil from the grain coated the walls.
4. The ratio of vertical to lateral pressure was not constant. Fill pressures were dependent on speed and method of filling. During emptying, wall pressures were significantly higher.
5. Eccentric unloading caused significant localized low wall pressures where the flow channel was adjacent to the wall.

Use of "Overpressure Factors" and Changes in National Codes

The work by the Reimberts, Walker, Jenike, Johanson, and others involved in designing and constructing bins has clearly shown that the dynamic pressures occurring during filling and emptying of a silo can, under certain circumstances, substantially exceed the static pressures calculated by the Jannsen or Reimbert method. Flow theories that account for these pressures have been developed (Section 9.11). These theories are of recent origin and, consequently, there has not yet been time for them to receive widespread study and acceptance in the industry. However, the various silo design codes at this time recognize that these pressures can occur and they recommend the use of certain minimum safety factors to account for them.

American,[16] French,[17] and Russian Silo Codes.[18] These codes recommend that wall pressures be calculated by Jannsen's or Reimbert's method and that the pressures thus calculated be multiplied by an "overpressure factor" c_d to produce a design pressure, p_{des} for calculating wall stresses:

$$p_{des} = \text{Jannsen or Reimbert} \times c_d \tag{9.19}$$

It is known from work reviewed in Section 9.11 that localized stress peaks occur all along the length of the silo walls during flow, and the magnitude of the peak varies depending on silo geometry. The codes cited above, therefore, recommend that an envelope enclosing the

SEE ALSO COMMENTARY FOR SECTION 4.4.2.2. FOR MASS FLOW DISCUSSION

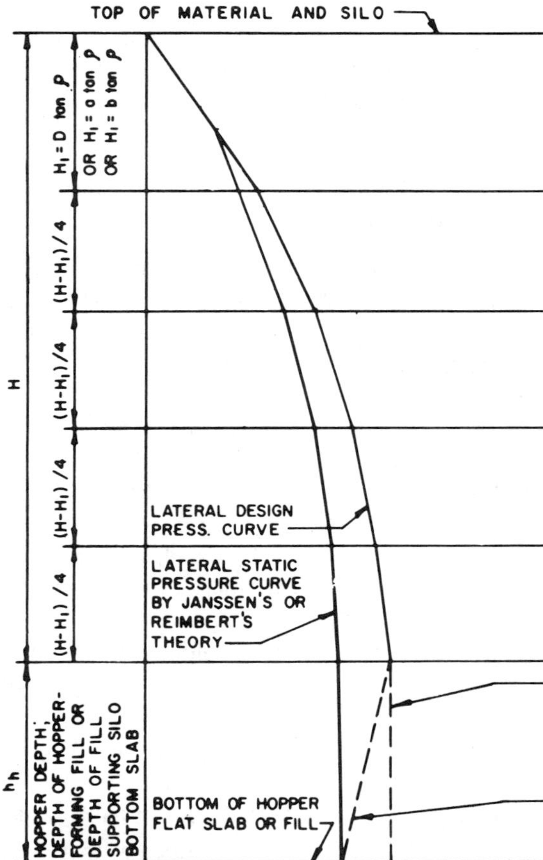

OVERPRESSURE FACTOR Cd	$\frac{H}{D} < 2$		$2 \leq \frac{H}{D} < 3$		$3 \leq \frac{H}{D} < 4$		$4 \leq \frac{H}{D} < 5$		$\frac{H}{D} \geq 5$	
	JANSSEN	REIMBERT	JANSSEN	REIMBERT	JANSSEN	REIMBERT	JANSSEN	REIMBERT	JANSSEN	REIMBERT
$H_1 = D \tan\rho$ OR $H_1 = a \tan\rho$ OR $H_1 = b \tan\rho$	1.35	1.10	1.45	1.20	1.50	1.25	1.60	1.30	1.65	1.35
$(H-H_1)/4$	1.45	1.20	1.55	1.30	1.60	1.35	1.70	1.40	1.75	1.50
$(H-H_1)/4$	1.55	1.45	1.65	1.55	1.75	1.60	1.80	1.70	1.90	1.75
$(H-H_1)/4$	1.65	1.65	1.75	1.75	1.85	1.85	1.90	1.90	2.00	2.00
$(H-H_1)/4$	1.65	1.65	1.75	1.75	1.85	1.85	1.90	1.90	2.00	2.00
h_h: HOPPER DEPTH; DEPTH OF HOPPER-FORMING FILL OR DEPTH OF FILL SUPPORTING SILO BOTTOM SLAB	USE SAME PRESSURE WITHIN HOPPER HEIGHT OR, IF DESIRED, REDUCE PRESSURES IN ACCORDANCE WITH HYDRAULIC RADIUS CHANGE. IF DESIRED, PRESSURES MAY BE REDUCED FROM TOP OF FILL TO TOP OF FLAT SLAB AS SHOWN									
OVERPRESSURE FACTOR Cd FOR USE IN CALCULATING DESIGN BOTTOM PRESSURES IN SILOS — CONCRETE BOTTOM	1.35	1.50	1.35	1.50	1.35	1.50	1.35	1.50	1.35	1.50
STEEL BOTTOM	1.50	1.75	1.50	1.75	1.50	1.75	1.50	1.75	1.50	1.75

Figure 9.19. Concrete overpressure factors as given in the American Concrete Institute *Recommended Practice for Design and Construction of Concrete Bins, Silos, and Bunkers for Storing Granular Materials*, ACI-313-77. Cd values given in this table are inadequate for the higher loads associated with mass flow. Notes: (1) Cd factor for lateral pressures is given for the bottom of each height zone shown, (2) In the region of a flow-correcting insert (e.g., Buhler-Nase) lateral pressures may be many times greater than static, and the Cd values above are not sufficient, (3) Bin bottom pressures need not be considered larger than the pressure caused by 100 percent of weight of silo contents, (4) The Cd values do not compensate for circumferential (horizontal) bending. Whenever such bending is present, it shall be computed. (From Ref. 16) (Courtesy of Committee 313 and the American Concrete Institute)

Ratio of volume dumped in one load to total silo capacity		1:2	1:3	1:4	1:5	1:6 and less
Impact factor C_i	Concrete bottom	1.4	1.3	1.2	1.1	1.0
	Steel bottom	1.75	1.60	1.5	1.35	1.25

Figure 9.20. ACI recommended minimum values of impact factor C_i (from Ref. 16).

probable high-stress peaks be determined by applying minimum values of c_d for different regions of the silo.

A typical example is shown in Figure 9.19 from the American Concrete Institute ACI 313-77 "Recommended Practice for Design and Construction of Concrete Bins, Silos, and Bunkers for Storing Granular Materials."[16] The static pressures calculated by Jannsen and Reimbert give different values, but the "over-pressure" factors c_d specified in the standard are intended to produce similar results whichever method is used. A multiplying factor, c_i for the effect of impact from loads imposed during high-volume filling is also included in this ACI standard (Fig. 9.20). Whichever is the higher value, c_d or c_i, is used for design. Overpressure factors for concrete and steel silos based on the Russian code are given in a similar diagram by Safarian and Harris.[5]

German Code DIN 1055, Part 6, *Loads in Silo Bins 1965* and the *Supplement* of 1977.[17] This code does not specify overpressure factors. Jannsen's method is recommended, but a value of solids-wall friction (μ_f) and Jannsen's (k_f) is used to computing filling pressures and another value of solids-wall friction (μ_e) and (k_e) is used for computing emptying pressures, where:

k_f (filling) = 0.5

k_e (emptying) = 1

μ_f (filling) = tan 0.75 ϕ for granular material (avg. dia. > 0.2 mm)

μ_f (filling) = tan ϕ for powder material (avg. dia. < 0.06 mm)

μ_e (emptying) = tan 0.60 ϕ for granular material

μ_e (emptying) = tan ϕ for powder

ϕ = Static angle of internal friction.

The angle of wall friction is thus assumed to be fixed as a fraction of the angle of internal friction. This does not take into account varying wall roughness and, therefore, is known to be an incorrect assumption.

Filling pressures are the Jannsen values:

$$q_f = \frac{R}{\mu_f k_f}(1 - e^{-\mu_f k_f Y/R}) \tag{9.20}$$

$$p_f = k_f q_f \tag{9.21}$$

Emptying pressures are Jannsen values multiplied by C:

$$q_e = C\left[\frac{R}{\mu_e k_e}\left(1 - e^{-\mu_e k_e Y/R}\right)\right] \tag{9.22}$$

$$p_e = C\,(k_e q_e) \tag{9.23}$$

where

$$C = 1 + 0.2\left(K + \frac{eL}{1.5A}\right) \tag{9.24}$$

where

e = eccentricity of outlet measured from the centroid
L = perimeter of the silo
A = cross-sectional area of the silo

K = 1 for organic materials
K = 2 for inorganic materials (for sugar the supplement specifies using C = 1)

The 1977 supplement excludes cohesive organic materials since the "pressures caused by these materials do not appear to be adequately described by this method." It also specified that calculated pressures occurring during flow should be multiplied by 30% for corn because of recent experience with imported corn containing fines and broken pieces caused by multiple handling.

Eccentric Discharge

It is known that eccentric discharge from a single or from multiple discharge openings can induce high localized pressures in silo walls and must be considered in design. When an eccentric flow channel intersects the wall on only one side of a bin while the opposite side remains submerged in stagnant material, the bins walls are exposed to horizontal bending stresses. In metal bins, this can lead to flattening of the wall surface adjacent to the flow channel[10] and in concrete silos to vertical cracks.[7] The DIN 1055 Standard attempts to account for this by the use of the eccentricity term in Eq. (9.22). Safarian and Harris[5] propose a correction factor to be added to the design pressure and assume that the value of this factor varies from a value of zero at the top of the silo to a maximum value at material depth Y. The American Concrete Institute Standard suggests a similar approach. Garfinkel[20] presents a method for accounting for eccentric discharge based on work by Pieper.

Filling and Emptying With Aeratable Materials

When a bin is rapidly filled with a fine powder that tends to aerate, pressures on the wall can assume hydrostatic properties. Recommendations for calculating wall pressures during rapid filling are given by Garfinkel[20] based on recommendations by Pieper. After the powder has deaerated, the pressures are determined as with any consolidated bulk material.

Air is often introduced along hopper bottoms to assist in the discharge of powders. If the imposed air pressure exceeds the emptying pressures, it acts on the bin walls for a certain height above the hopper and must be considered in the design. This height, determined by Pieper, is also given by Garfinkel.

Air is often used for homogenizing bin contents. For design purposes, wall pressures are considered to be almost hydrostatic. The ACI 313-77 Standard recommends that the pressure p_{air} during homogenization be calculated as:

$$p_{air} = 0.6\,\gamma H \tag{9.25}$$

where 0.6 is an empirical factor to account for the fact that particles are in partial suspension and not in contact, γ is the bulk density, and H is the distance below the surface of the powder.

Summary

With the increasing understanding of solids flow behavior has come a continuing review of the national codes. Wenzel[21] and Theimer[7] reviewed and compared the present German, American, Russian, and French versions. Deutsch[22], Safarian et al.,[23] and Appleton et al.[24] commented on the recently issued American ACI 313-77 Standard.

The German DIN 1055, Part 6 is being revised and was scheduled for reissue in 1983. This revision is expected to include, among other changes, recommended Jannsen k values for a number of materials, in place of the constant value given in the 1965 version; a method of multiplying calculated filling pressures to determine design emptying pressures; wall friction values to be used for various wall roughnesses and new factors to account for pressures induced by eccentric discharge, by filling with aeratable materials, and by mass flow at the transition between vertical and sloping hopper walls in a bin.

Codes of practice for design of bins and silos are being considered in a number of countries. A code of practice for steel bunkers is being formulated in the United Kingdom. The Japan Society of Civil Engineers has begun research work on the design requirements for steel silos to form a basis for a Japanese code.

It is recognized that some understanding of solids flow behavior is necessary for those structural engineers involved in bin design and for those who use the silo design codes. The distribution and magnitude of wall pressures in a funnel flow bin will be very different than those in a mass flow bin. The overpressure factors given in some of these codes may be conservative in the case of funnel flow but may be marginal or insufficient for mass flow. The designer must have an understanding of the difference. Typical tabular listings of "properties of materials" in the existing codes include bulk density, angle of repose, and coefficient of friction of material and silo wall. There is no reference given as to the conditions under which the measurements were made, moisture content, temperature, or pressure under which bulk density was determined. Wall friction with common silo materials other than concrete and clean steel are not given. Important bulk solids properties now being used in powder mechanics, including angle of internal friction and effective angle of internal shear, are not listed. These aspects are reviewed in the following sections.

9.5. FLOW PATTERNS IN BINS AND HOPPERS

A knowledge of flow patterns occurring in a bin are fundamental to any understanding of the forces acting on the material or on the bin walls. Wall pressures are determined not only by frictional forces caused by sliding of solids along the wall, but by the flow patterns that develop during filling and withdrawal.

Types of Flow Patterns

Three basic flow patterns have been identified:[55]

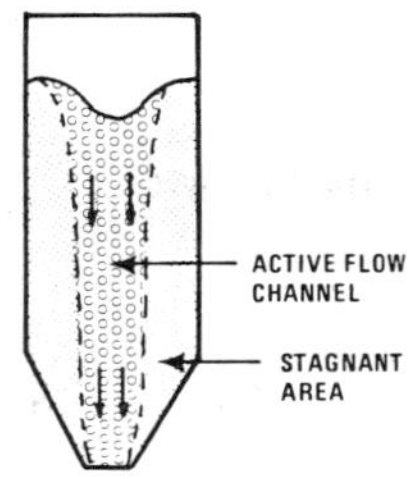

Figure 9.21. Funnel flow through an entire bin.

Funnel Flow. This is sometimes also called "core flow." It occurs in bins with a flat bottom or with a hopper having slopes too shallow or too rough to allow solids to slide along the walls. Solids flow to the outlet through a channel within a stagnant mass of material. This channel is usually conical in shape, with its lower diameter approximately equal to the largest dimension of the active area of the outlet. It usually increases in size as it extends from the outlet, up into the bin (Fig. 9.21). Serious flow problems can occur if the material compacts and exhibits poor flow properties when consolidated under solids head pressures. Material in the stagnant areas may gain strength with time and remain in place when the active flow channel empties out, forming a rathole or pipe (Fig. 9.22). In severe cases, the material can form a bridge or arch over the discharge opening (Fig. 9.23).

The flow channel may not be well defined. Material surrounding the channel may be unstable, and in this condition will cause stop-and-start flowing, pulsating or "jerky" flow. High pressures within the channel are often muffled by the stagnant material and may not

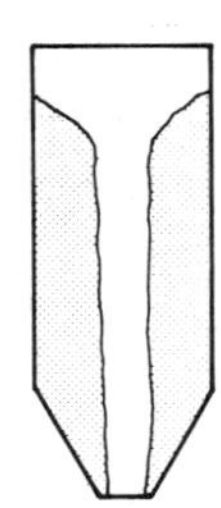

Figure 9.22. Rathole, formed when stagnant material gains sufficient strength to remain in place as flow channel empties.

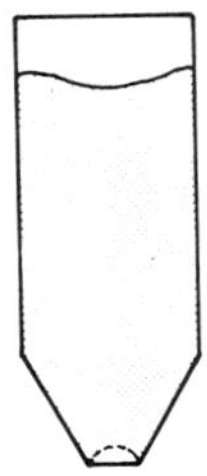

Figure 9.23. Arch or bridge, formed across a flow channel.

reach the walls. At high discharge rates, however, these pulsations could lead to structural damage. As the bin is emptied (assuming the material does not compact to form a stable rathole) solids continually slough off the top surface into the channel. If solids are simultaneously charged into the top and withdrawn from the bottom, the incoming solids will pass immediately through the channel to the outlet.

In tall bins or silos, the channel boundaries may expand to intersect the cylinder walls at a point defined as the effective transition (Fig. 9.24). Material above this intersection may move in plug flow and stresses developed within the flow channel reach the bin walls.

A storage bin having a funnel flow pattern is the most common in industry and many have been designed to provide a maximum volume for storage without considering that the actual discharge capacity may be much less due to accumulation of stagnant material.

The funnel flow bin is usually the least costly design. However, it has several disadvantages when handling certain materials:

- Flow rate from the discharge opening cán be erratic because arches tend to form and break and the flow channel becomes unstable. Powder density at discharge will vary widely because of the varying stresses in the flow channel. This can render ineffective any volumetric feeder installed at the silo discharge.
- Fine powders can become *aerated and flush uncontrollably when arches or ratholes collapse*. Positive sealing-type discharge devices or feeders are *mandatory* when these conditions exist.
- Solids can degrade or cake solid when left under consolidating stresses in the stagnant areas.
- A stable rathole or pipe can form if the stagnant material gains sufficient strength to remain in place after the flow channel drains out.
- Indicators mounted along the length of the bin wall to detect bin level will remain submerged in the stagnant areas and will not correctly signal solids level in the lower regions of the bin.

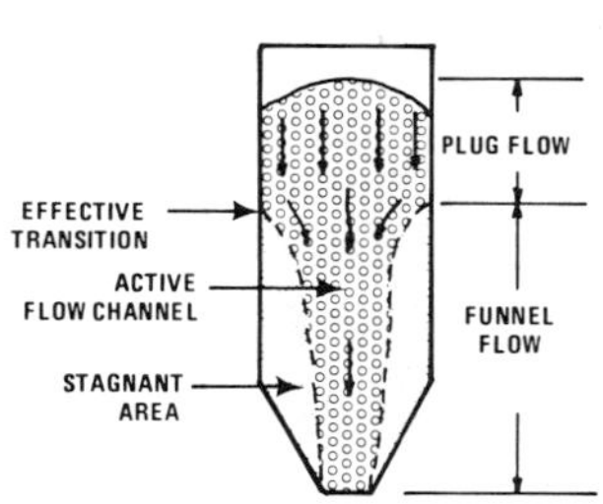

Figure 9.24. Funnel flow below an effective transition.

Funnel flow bins are entirely adequate for the many cases where noncaking or nondegradable materials are to be stored and the discharge openings are adequately sized to prevent bridging or ratholing. Many commercial, mechanical, and aerating devices, described in Section 9.15 are available for promoting flow. However, if these devices are considered for new installations, their investment and maintenance costs, as well as the probability of success in maintaining reliable flow should be calculated against the cost of a mass flow design.

Mass Flow. This occurs in bins with sufficiently steep and smooth hoppers, and where material discharges from the entire area of the outlet (the outlet *must* be fully active for mass flow to occur). In mass flow, the flow channel coincides with the bin and hopper walls; all material is in motion and sliding against the walls of the vertical section and the converging hopper (Fig. 9.25). Material in the vertical part moves down uniformly in plug flow as long as the level is above some critical distance above

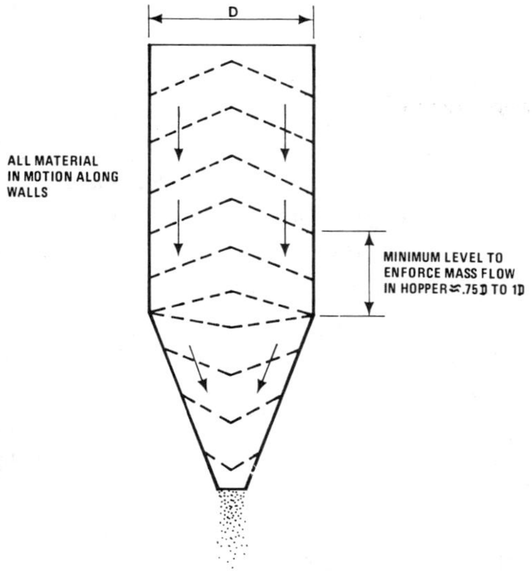

Figure 9.25. Mass flow.

the hopper-cylinder transition. If the level drops below that point, the material in the center of the channel will flow faster than the material at the walls. The height of this critical level has not been exactly defined but it is apparently a function of material angle of internal friction, material-wall friction, and hopper slope. The height shown in Figure 9.25 is approximate for many materials. In mass flow, stresses caused by the flow act on the entire wall surface of the hopper and vertical part.

Mass flow offers significant advantages over funnel flow:

Erratic flow, channeling, flooding of powders are avoided.
Stagnant regions within the silo are eliminated.
First-in first-out flow occurs, minimizing the problem of caking, degrading, or segregation during storage.
Particle segregation is considerably reduced or eliminated.
The material in the silo can act as a gas seal.

Solids density at discharge is constant, essentially unaffected by solids head. This makes it possible to get good solids controls with volumetric feed devices, and it improves the performance of gravimetric feeders.

Since flow is well controlled, pressures will be predictable and relatively uniform across any horizontal cross section: solids consolidation and permeability will be uniform, flow boundaries will be predictable and, therefore, the analysis based on steady state flow conditions (described in Section 9.10) can be used with a high degree of confidence.

Mass flow bins are usually more expensive than funnel flow bins of the same volumetric capacity because of the extra height required for the more steeply sloped hoppers. This is often not true if the cost of flow-promoting devices added to funnel flow bins are considered in the total cost.

Expanded Flow. Flow in a funnel flow bin can be improved by terminating the bin in a mass flow hopper section as shown in Figure 9.26. The mass flow section will expand the active flowing volume, minimize stagnant areas, and provide reliable discharge without bridging and may permit using a smaller discharge opening. This arrangement can be used to modify existing funnel flow bins to correct arching, ratholing, or flushing problems. It can be used in bins with multiple outlets (Fig. 9.27).

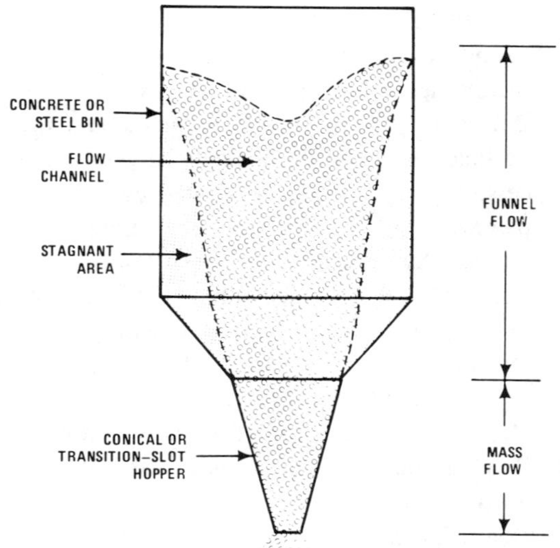

Figure 9.26. Expanded flow through a single outlet.

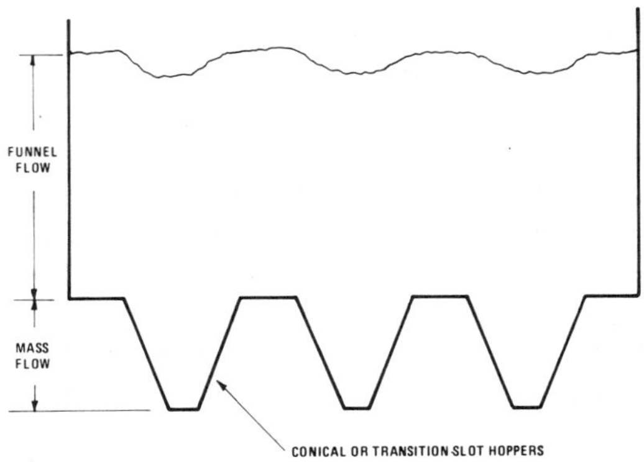

Figure 9.27. Expanded flow through multiple outlets.

Studies of Flow Patterns

A number of techniques have been used to study flow patterns in model bins.[27,28,29,30] A comprehensive summary of these techniques is given by Resnick.[29] Included are:

1. Observing the passage of tracer layers before, during, and after flow in transparent wall "thin slice" models.
2. Immobilizing the entire model bin contents with molten wax or polyester casting resin, then slicing the model longitudinally to study the flow patterns shown by tracer layers.
3. Sequentially photographing and tracking the passage of particles as viewed through a hopper with a transparent wall.
4. Photographing particles through a transparent wall using stereo-photogrammetric techniques. In this method developed by Butterfield et al.,[28] photographs of moving particles are taken by a single camera. Successive photographs are viewed as a stereo pair with each eye viewing one of the photographs. An optical, three-dimensional model can then be provided and from that iso-velocity contour maps can be constructed to reproduce the displacement field. An example is given in Reference 29.
5. Measuring gamma radiation absorption in flowing beds to determine variation in porosity.
6. Measuring x-ray densities during and after flow in a model hopper.
7. Tracing the flow with radioactive or colored markers deposited on various parts of the bed during flow.
8. Using tracer layers in a model that can be separated longitudinally. After the flow pattern is developed, the model is laid on its side, the top half removed, and excess material brushed off to reveal the tracer layer patterns on a plane across the center.

Sketches of flow patterns observed during model tests have been given in a number of papers. Deutsch and Clyde[12] and Deutsch and Schmidt[13] presented patterns observed during flow of granular material in semicylindrical 305 mm model bins having a height varying up to 1500 mm (Figs. 9.28 and 9.29). Sugita[33] showed similar patterns occurring with 177–250 μm beads in a 200 mm diameter $\times$ 800 mm high semicylindrical model (Figs. 9.30 and 9.31). He also observed flow patterns into an eccentric discharge opening (Fig. 9.32). Lenczner[35] and McCabe[34] studied the flow of sand in model bins. McCabe tested sand in a 2450 mm high model with diameters varying from 80 mm at 460 mm. Data from his test is shown in Figure 9.33. All these tests show that the flow patterns are very sensitive to degree of consolidation of the material. Nguyen et al.[36,37] studied the flow of dry granular sand, polystyrene, glass beads, and rice in model conical and wedge-shaped hoppers and showed that the height of the free surface of the material in the vertical part of the bin can have a significant effect on flow pattern.

These studies showed that the boundaries of funnel flow cannot yet be predicted with certainty. They are a function of not only the hopper configuration, but the solids level in the vertical portion of the bin, and the angle of internal friction of the material, which in turn is sensitive to changes in bulk density caused by changes in consolidating pressures during filling and discharge.

Blair-Fish and Bransby[27] observed variations in density in flowing material in mass and funnel flow using radiographic techniques with lead shot tracers in a model bunker. The velocity

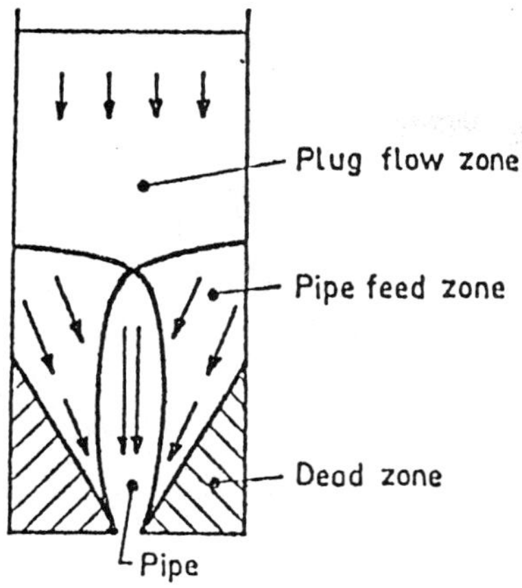

Figure 9.28. Flow zones during steady flow of sand from a model bin after loose filling, with no compaction. (From Ref. 12)

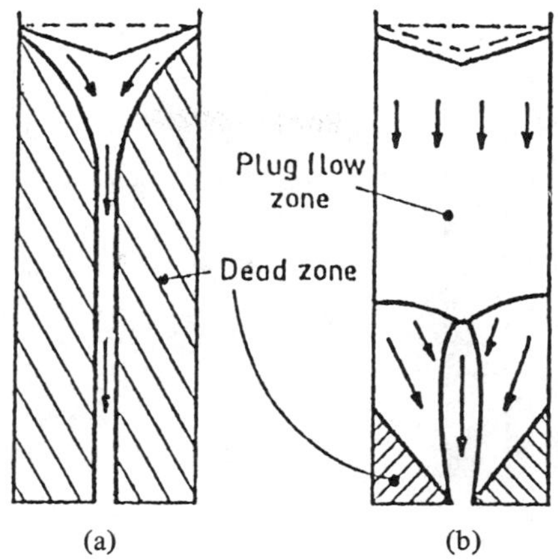

Figure 9.29. Flow zones during flow of sand from a model bin after compaction by vibration: (a) initial flow pattern: Central pipe extends to surface forming conical crater, (b) steady flow. (From Ref. 12)

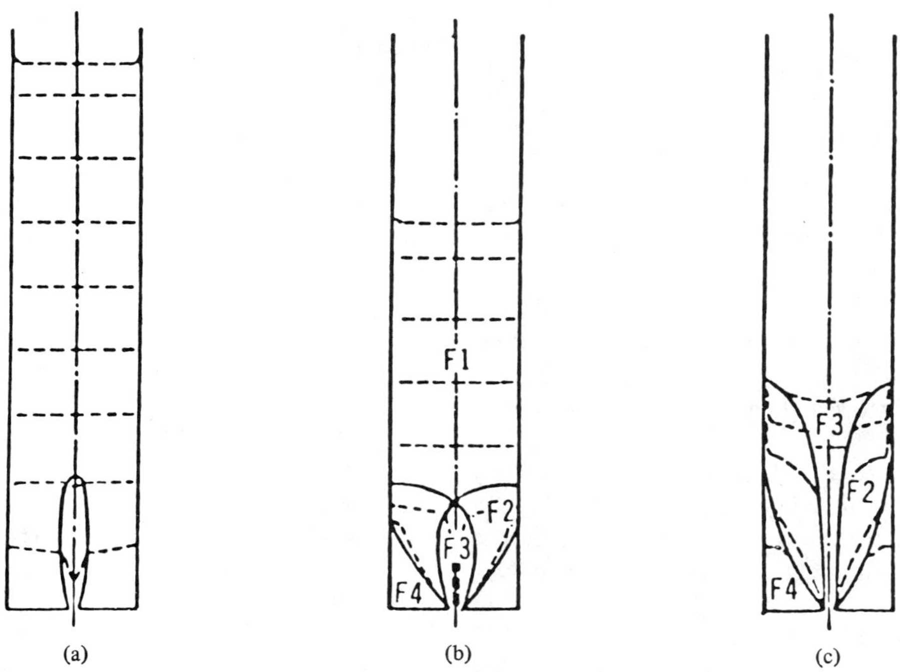

Figure 9.30. Flow zones during free discharge of glass beads from a model bin after loose filling with no compaction: (a) Immediately after discharge begins, the zone of flow extends vertically above the oriface. Height varies with oriface diameter and material properties. (b) During steady state flow, the following zones appear: *F1*, Material sinks uniformly with steady velocity. *F2*, Material entering this zone then flows into zone F3 with radial velocity. Material at boundary of F1 and F2 reaches state of failure and flows plastically. *F3*, Material falls vertically with a high radial velocity. *F4*, stagnant zone. (c) In the final state of discharge when the free surface moves down to a certain height, the boundary between *F1* and *F2* makes a slow ascent. Eventually this boundary rises up to the falling free surface, *F1* disappears, and a crater is formed. (From Ref. 33)

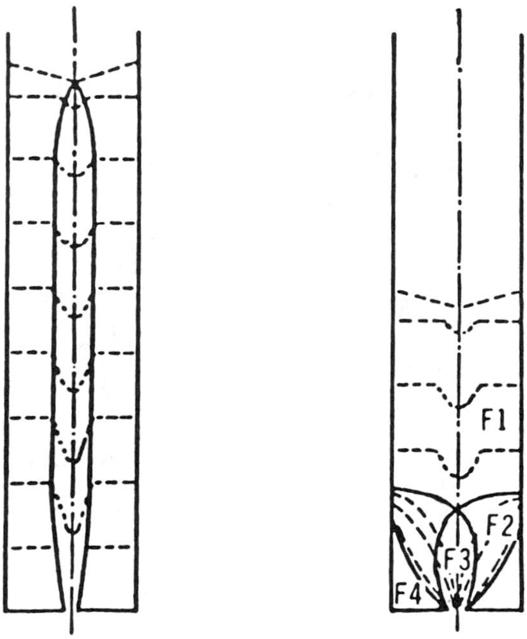

(a) (b)

Figure 9.31. Flow zones during free discharge of glass beads from a model bin after compaction by tapping after filling: (a) Immediately after discharge begins, the zone of flow extends from the oriface to the free surface of the material. A smaller crater is formed at the surface. (b) steady state flow (see Fig. 9.30 for identification of zones). (From Ref. 33)

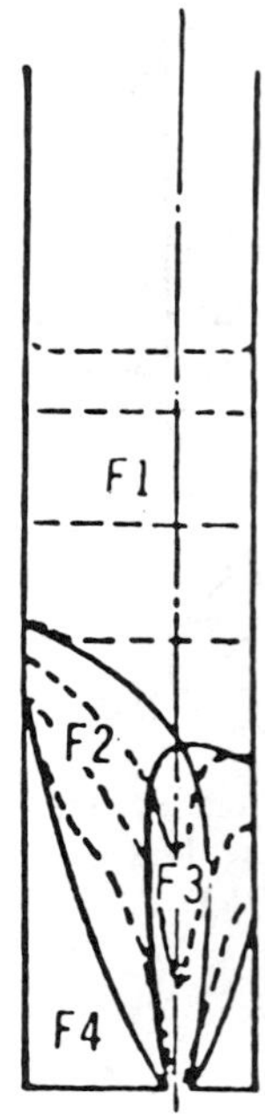

Figure 9.32. Flow zones during steady-state flow of glass beads from an eccentric opening (see Fig. 9.30 for identification of zones). (From Ref. 33)

fields they detected were similar to those reported by Deutsch, and their photographs clearly showed flow direction and rupture surfaces.

Chatlynne and Resnick[29] used gamma ray absorption in a flowing bed and found a dilation wave that moves up the bed as flow as initiated, similar to that shown in Figure 9.33. They reported that the porosity of their test bed was 0.41 at rest. It increased to 0.47 during flow, which was also the value they obtained at minimum fluidization conditions.

Giunta[38] studied the flow of fine powders (starch, pulverized coal, and iron concentrate) with spaced trace layers of marking material in an 457 mm dia. × 610 mm high split model, similar to Johanson's. He found the flow pattern to be a function of effective angle of internal friction, opening size, and head of material in the bin. He proposed the following equation for determining the boundaries as shown in Figure 9.34:

$$2\overline{Y} = D + 2\tan\overline{\theta}\,\frac{H - A\dfrac{D}{2}}{1 + A\tan\overline{\theta}} \qquad (9.26)$$

where

D = diameter of discharge opening (ft) (where D is large enough to prevent arching or ratholing)
$\overline{\theta}$ = angle of flow pattern boundary at edge of opening (degrees)
H = head of material in bin (ft)
$\overline{Y}$ = maximum radius of boundary between flowing and nonflowing material (ft)

Angle $\overline{\theta}$ and factor A are dependent on the angle of internal friction as given in Figure 9.35. This angle is defined in Section 9.7.

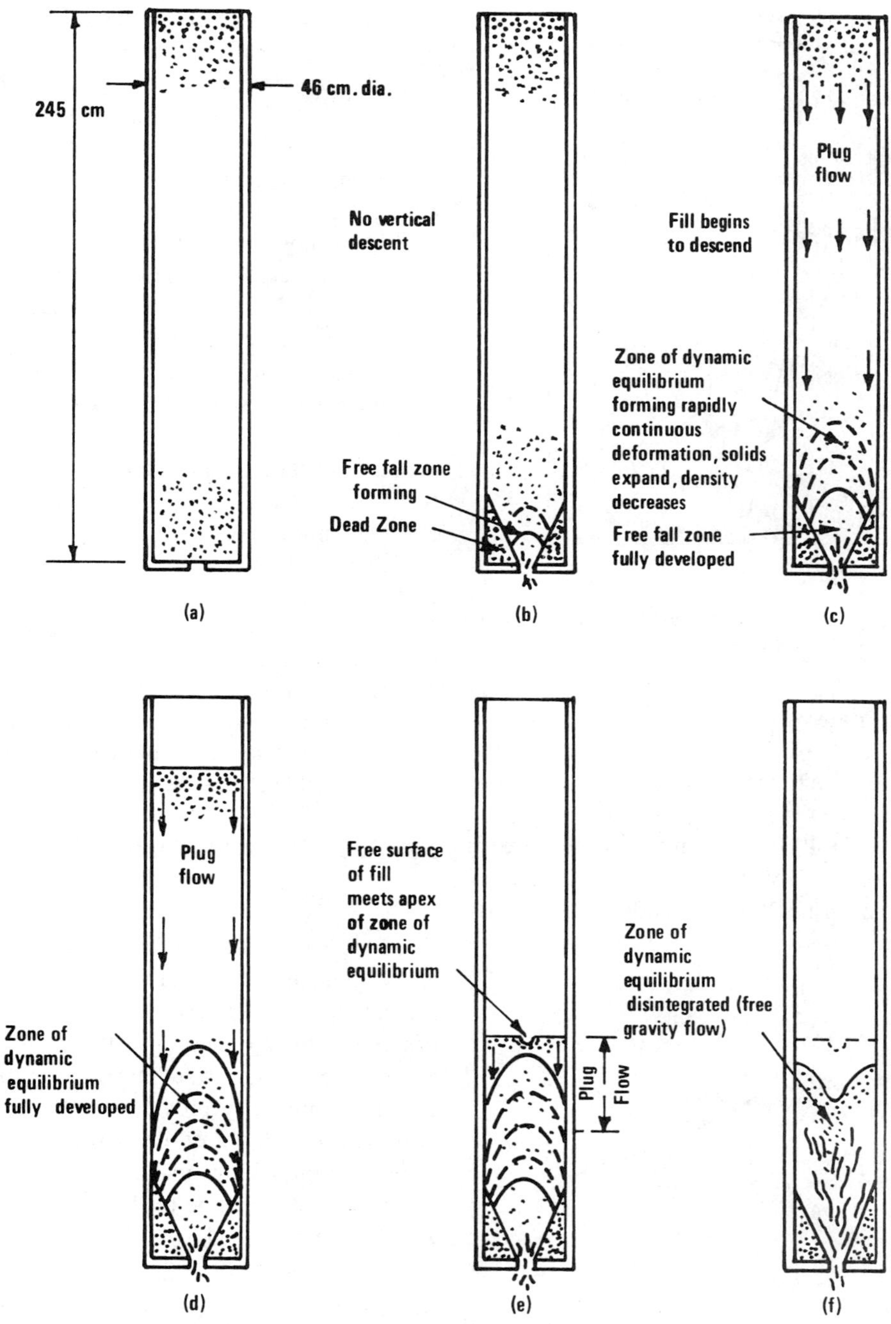

Figure 9.33. Flow zones during free discharge of sand from a model bin: (a) Bin is full, discharge port closed. (b) Discharge port opens, discharge begins. (c)–(f) Discharge continues. (From Ref. 34, permission of the Institute of Civil Engineers)

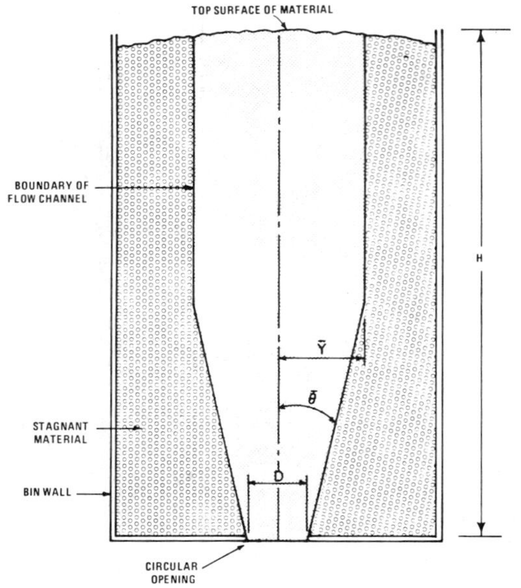

Figure 9.34. Giunta's predicted flow boundary in funnel flow. (From Ref. 38)

Guinta states that Equation (9.26) is valid only if $H > AD/2$. If $H \leqslant AD/2$, the diameter of the flow boundary will remain the same as the opening ($2\overline{Y} = D$).

A study of PVC and sand flow in a 1.5 m bunker reported by Van Zanten, Richards, and Mooij[87] confirmed the central flow region close

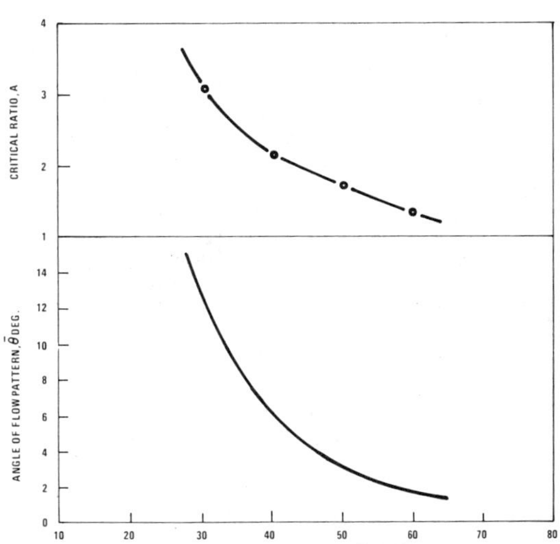

Figure 9.35. A and $\overline{\theta}$ versus δ by Giunta. (From Ref. 38)

to Giunta's prediction but also found a large cylindrical slow flow zone surrounding the central "fast flow" core. They identified four flow zones (Fig. 9.36a). The number of zones were found to be different depending on flow properties of the material (Fig. 9.36b) and the configuration of the bin. In some sloped hoppers, only a conical and cylindrical fast flow, with stagnant zone were present. The angle O_T was found to be close to that predicted by Jenike[52] and shown in Figure 9.36c.

Johanson[39] studied the zones in funnel flow. He identified the steady flow zones (Fig. 9.37a) as a function of angle of internal friction as in Jenike, but he further defined the surrounding region of unsteady flow that occurs with free flowing, noncohesive, frictional solids (Fig. 9.37b). He attributes formation of this secondary channel to pressure changes along the steady flow channel walls during flow that causes loosening of the adjacent material and causes the outer region to become unstable. These unstable zones in funnel flow cause pulsations or jerky flow that can lead to vibration or possible structural failure.

9.6. FLOW OF BULK SOLIDS

Introduction

Good engineering design of structures for storage of particulate solids can best be achieved if the design data includes the solids flow properties expressed in quantitative terms. A number of empirical tests to define these properties are in use, but they provide mostly qualitative information. The most useful will be described in this section. Flow analysis and testing techniques that do provide qualitative information will then be described.

Empirical Tests to Determine Bulk Solids Properties

In the past years, a number of empirical tests have been developed to described the properties of bulk solids. These include:

CEMA. The Conveyor Equipment Manufacturer's Association (CEMA) has published a

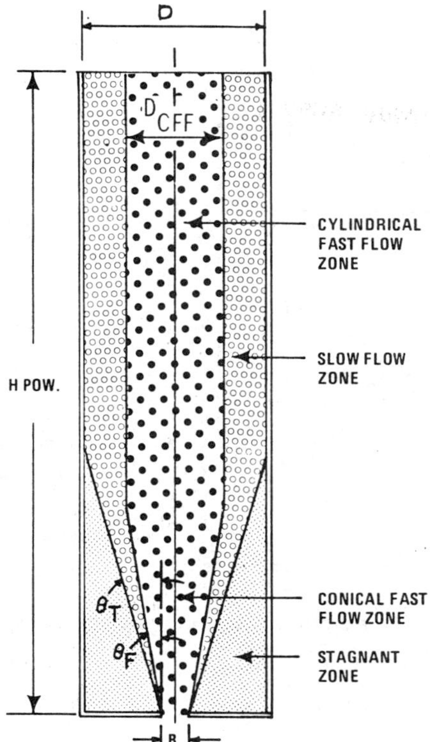

(a)

Properties of materials

Property	PVC grade X	PVC grade Y	Sand
ρ (kg/m³)	495	595	1525
δ (deg)	39.5	35.0	37.5
ϕ' steel/zinc compound (deg)	25.0	24.0	23.5
ϕ' aluminum (deg)	27.0	24.0	18.0

(b)

Summary of measured and predicted flow patterns in funnel flow

Material	Angle of cone to vertical, deg	D, m	B, m	H_{pow}/D	θ_T, deg	θ_F, deg	$D_{CFF,m}$	θ_F (Jenike (b)) deg	θ_F (Giunta (c)) deg	$D_{CFF,m}$ (c) [4])
PVC grade X	25	1.50	0.15	4.70	8.5	6	0.90	8	6	1.33
	25	1.50	0.15	3.63	9	5	0.70	8	6	1.05
	25	1.50	0.10	4.73	9	6	0.85	8	6	1.34
	20	2.00	0.15	2.57	<10	5	—(d)	8	6	1.00
PVC grade Y	15/25	1.50	0.10	5.00	11–12.5	7	0.80–0.90	11	8.5	1.76
	20	2.00	0.15	2.77	<9.5	6	—(d)	11	8.5	1.32
	90[a]	1.50	0.15	4.00	13.5	10	1.05	11	8.5	1.42
	90[a]	1.50	0.15	2.67	12	9.5	0.90	11	8.5	0.98
	90[a]	1.50	0.10	4.00	9–12.5	9	1.00	11	8.5	1.38
Sand	90[a]	1.50	0.10	4.00	12.5	—	0.75	9.5	8	1.26

(c)

Figure 9.36. Funnel flow patterns determined from test on model bins: (a) flow zones, (b) properties of the materials used in tests, (c) summary of measured and predicted flow patterns. (From Ref. 86)

guide to the classification and definition of bulk materials that is used by manufacturers of materials handling equipment.[40] This guide proposes establishment of terminology for describing various properties and characteristics of bulk materials that effect design of handling equipment. Two general classes are established. The first class is those physical characteristics that can be determined by test.

Test A-1, Abrasiveness. Four characteristics are considered; particle hardness (see Test A-16), particle shape (see Test B-17), bulk density (see Test A-8), and particle size (see

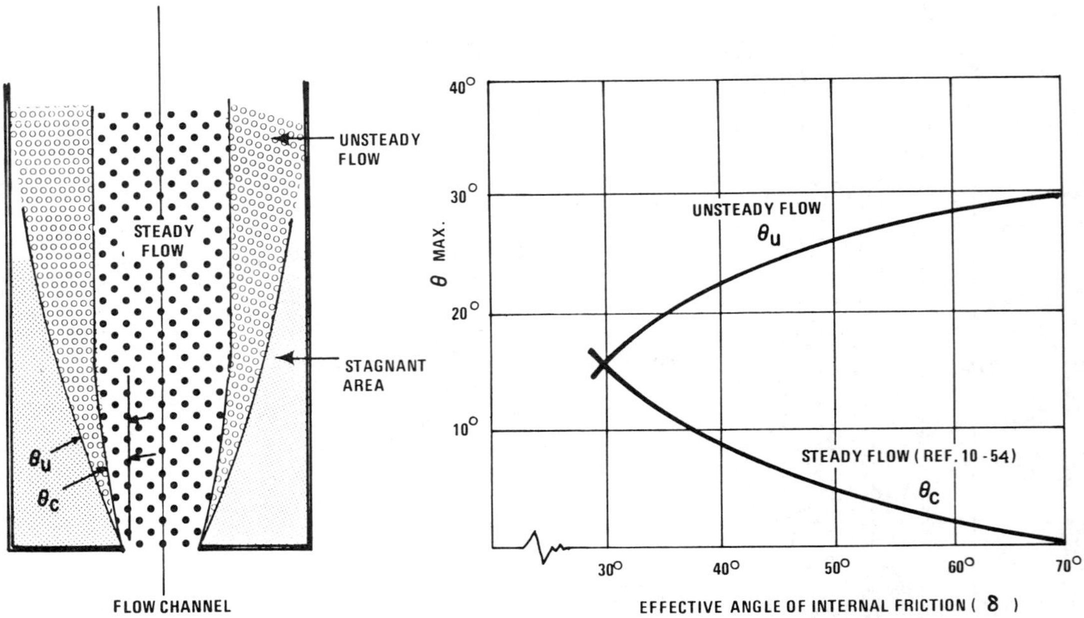

Figure 9.37. Predicted flow zones with free-flowing noncohesive frictional solids. (From Ref. 39) (Permission *Chemical Engineering*)

Tests A-17 and A-18). Each of these characteristics is assigned a factor as shown in Table 9.1.

To determine relative particle abrasiveness, multiply the appropriate CEMA factors to determine the abrasive index numbers and compare to the CEMA abrasive code shown in Table 9.2.

Test A-2, Angle of External Friction. This is the same as the kinematic angle of wall surface friction ϕ', as described in Section 9.7. A description of the type and surface finish of the material on which the test is performed must be included with the data.

Test A-3, Angle of Internal Friction. This is the angle ϕ and is determined as described in Section 9.7.

Test A-4, Angle of Maximum Inclination. This is the maximum angle measured from the horizontal plane, at which a belt conveyor can be inclined when conveying that particular material. It is determined through experience or by test.

Test A-5, Angle of Repose. In this test, the angle of repose is stated as the angle formed between a horizontal plane and the slope line extending along the face of a heap formed by pouring the material onto a horizontal surface. Angle of repose has limited usefulness. This property cannot be stated as a unique single value for a particular material, but must be defined in terms of the method by which the heap is formed. Several commonly used terms shown in Figure 9.38 include:

Loose or Poured Angle of Repose (Fig. 9.38a): Where material is poured onto a heap over a short distance without significant impact. The heap can also be formed by depositing through a pipe and withdrawing the pipe as the heap is formed to minimize impaction effects.

Impacted Angle of Repose (Figs. 9.38b, 9.38c): Where material drops from a height and consolidates due to impact. With some wet or cohesive materials, the heap can assume a shape with several angles of repose as shown on Figure 9.38b. Powders that readily fluidize, can

Table 9.1. CEMA Factors (Reprinted with permission of Conveyor Equipment Manufacturer's Association Ref. 40).

		MOHRS NO.	CEMA FACTOR
Hardness		1	1
		2	4
		3	9
		4	16
		5	25
		6	36
		7	49
		8	64
		9	81
		10	100
		LB/CU FT	
Density		0-60	1.0
		61-120	1.1
		121-180	1.2
		181-240	1.3
		241-300	1.4
		TYPE	
Shape	Rounded		1.0
	Subround or Subangular (approach rounded or angular shape but well-rounded edges)		1.5
	Angular		2.0
		CEMA SIZE CODE (FROM A-17 AND A-18)	
Size		A	1.0
		B	1.1
		C	1.2
		D	1.3
		E	1.3

Table 9.2. CEMA Abrasive Index (Reprinted with permission of Conveyor Manufacturer's Association, Ref. 40.

CHARACTERISTICS	CEMA CODE NUMBER	ABRASIVE INDEX RANGE
Mildly abrasive	5	1-17
Moderately abrasive	6	18-67
Extremely abrasive	7	68-416

assume the heaped shape shown in Figure 9.38a when poured on top of the heap but will assume the shape shown in Figure 9.38c when dropped from a height.

Drained Angle of Repose (Fig. 9.38d): Where material is withdrawn from an existing pile through an opening under the central portion of the pile. This angle can be steeper than the loose angle of repose due to consolidation of the material in the center of the pile, segregation of less free-flowing finer material in the center during pile formation and more resistance to flow inherent in an in-flowing converging conical surface.

Sliding Angle of Repose (Fig. 9.38e): Frydman[41] defines such an angle as that where material resting on a horizontal surface that is being slowly inclined, starts to move along the face of the pile.

Compacted Angle of Repose (Fig. 9.38f): This angle is sometimes quoted as the angle that defines the shape of a flow channel around a forced discharge feeder, for example a screw or drag chain feeder.

Aerated Angle of Repose (Fig. 9.38g): Angle assumed by a heap when aerated from a permeable membrane on which the material rests—sometimes used as a measure of "fluidity" or to determine if material can be transported or moved by or on aerated panels.

Rolling Angle of Repose (Fig. 9.38h): The angle assumed by material in a rotating drum, mixer, etc.

Discharge From a Feeder (Fig. 9.38i): The angle of repose formed at the discharge of a feeder can have two values: static angle when the feeder is stationary and dynamic angle when the feeder is operating.

A comprehensive review of various methods used for determining angle of repose is given by Brown and Richards.[42]

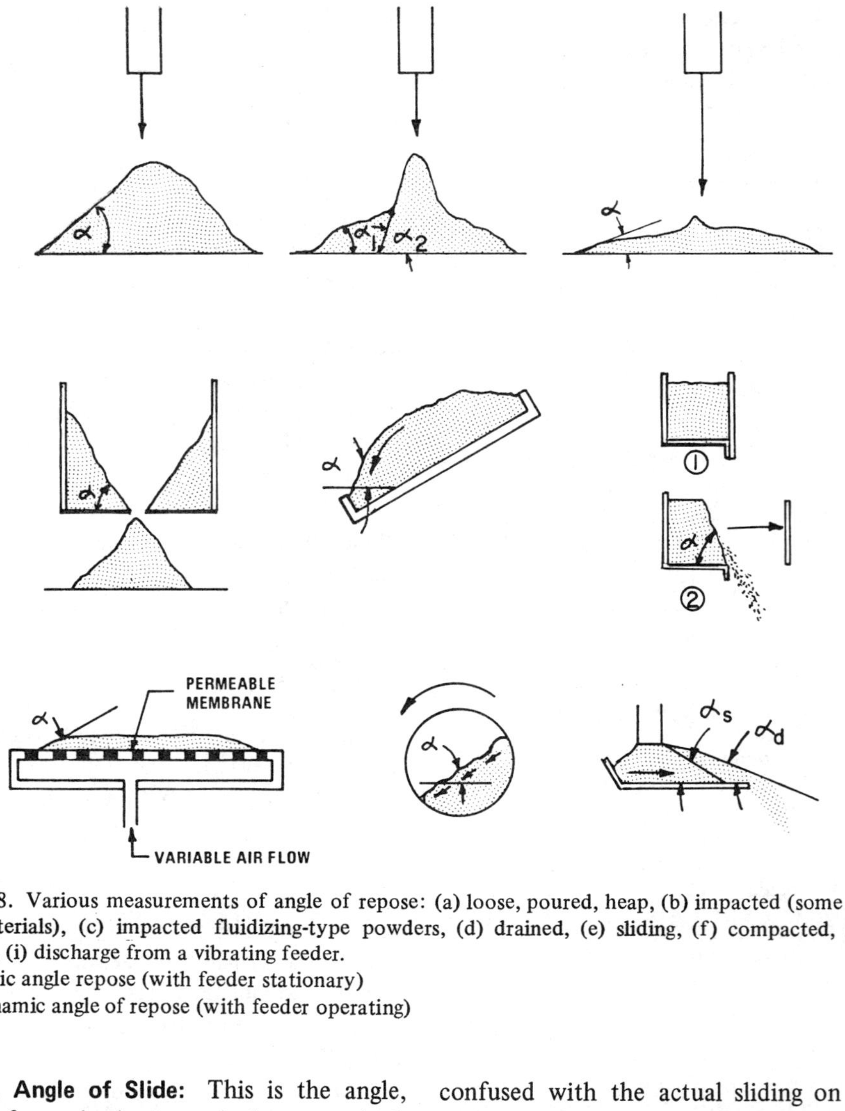

Figure 9.38. Various measurements of angle of repose: (a) loose, poured, heap, (b) impacted (some wet and cohesive materials), (c) impacted fluidizing-type powders, (d) drained, (e) sliding, (f) compacted, (g) aerated, (h) rolling, (i) discharge from a vibrating feeder.

α_s = static angle repose (with feeder stationary)

α_d = dynamic angle of repose (with feeder operating)

Test A-6, Angle of Slide: This is the angle, measured from the horizontal plane, to which an inclined surface must be tipped so that material will slide downward along the surface (Fig. 9.39). The test data should include the type and surface roughness of the plate. The test should be done with care so that the angle where material begins to slide upon itself is not confused with the actual sliding on the plate surface. At least five individual tests should be made and the results averaged.

Angle of slide is useful to determine slopes of chutes required for discharge of solids, to determine slopes of hoppers required for complete discharge of free flowing solids, and to determine if existing hoppers will require vibration

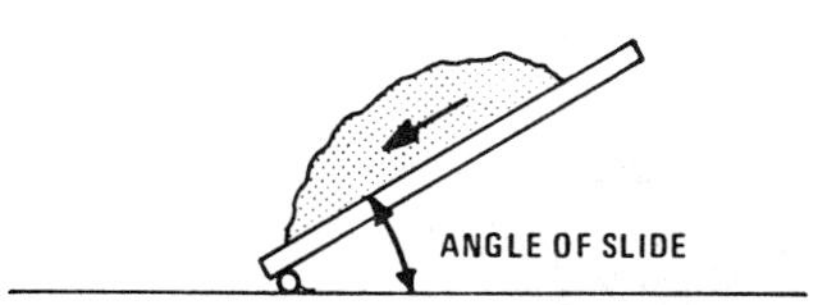

Figure 9.39. Angle of slide on a surface.

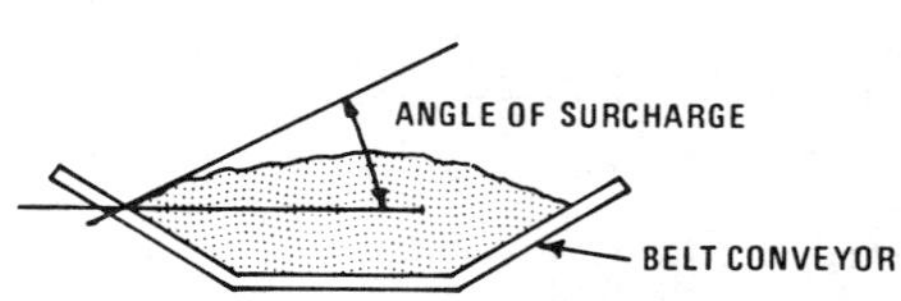

Figure 9.40. Angle of surcharge on a belt conveyor.

for final cleanout. This angle is *not* the same as the hopper slope angle required for mass flow.

Test A-7, Angle of Surcharge. This is the angle of surcharge of a bulk material on a moving conveyor (Fig. 9.40) and is used for calculating conveyor capacity. This is less than the loose angle of repose and is influenced by the type and speed of the conveyor. When handling very free-flowing materials, movement of the conveyor (a belt conveyor moving over troughing idlers for instance) will agitate the material and cause the angle of surcharge to become less than that at the loading point and could cause spillage on the belt after loading. This change in angle is similar to the angle of difference proposed by Carr and described later. An accurate determination of angle of surcharge is made by actual test on a belt conveyor.

Test A-8, Bulk Density (Loose). A measuring container mounted under an open filling mold

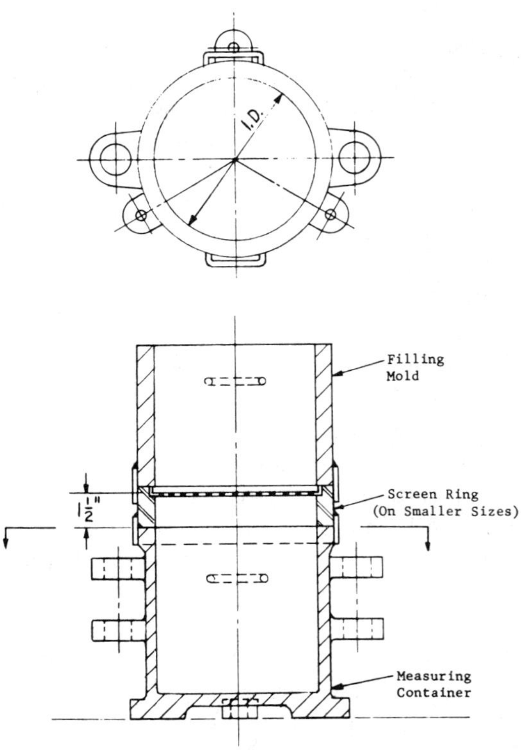

Figure 9.41. CEMA bulk density measuring container. (From Ref. 40) (Permission Conveyor Equipment Manufacturer's Association)

and separated by a screen is used for measuring loose bulk density (Fig. 9.41). The CEMA standard recommends a screen size at least four times the largest particle size. Material is gently deposited on the screen with a scoop allowing it to fall through the screen to the box below. No vibration or tapping is allowed. When sufficient material has been added to fill to the level of the screen, the top mold section and screen are gently removed. The container is then struck level with a straightedge and placed on a scale and weighed. The standard recommends that the minimum weight of sample be related to particle size as shown in Table 9.3.

Test A-9, Bulk Density (Vibrated). The same filling mold and container as described above is used here. The screen is omitted. The material is loaded to fill the mold about two thirds full and then the complete assembly is placed on a vibrating table and vibrated for one minute. A vibrating table having a frequency of 50–60 cps, a vertical amplitude variable between 0.05 and 0.60 mm under load is recommended. After vibration, the top mold section is removed, the container struck level with a straightedge, placed on a scale, and weighed. Percent compressibility is determined from this test as: $\gamma_{\text{vibrated}} - \gamma_{\text{loose}}/\gamma_{\text{vibrated}} \times 100$.

Test A-10, Cohesiveness. This property, characterizing the degree to which individual particles cling together is determined from data obtained with a direct shear tester as described later in this section.

Test A-11, Elevated Temperature. The standard merely states that materials to be handled at elevated temperatures should be classified as A-11 handling conditions, with the temperature stated.

Test A-12, Flowability—Flow Function. This is the Jenike flow function described later in this section.

Test A-13, Lumps—Size and Weight. This section of the standard specifies that large lumps over about 180 mm (7 inches) should be in-

Table 9.3. CEMA Procedure—Minimum Weight of Sample Versus Maximum Particle Size (Reprinted with permission of Conveyor Manufacturer's Association, Ref. 40.

MAXIMUM PARTICLE SIZE	MIN. WEIGHT OF SAMPLE		RECOMMENDED SIZE OF MEASURING BOX			
			VOLUME		DIA.	HGT.
	GRAMS	POUNDS	LITERS	CU FT	MM	MM
Class A finer than U.S. #40 sieve	3,000	6.6	1.0	0.035	102	123
Class B finer than U.S. #6 sieve	9,070	20	2.82	0.1	152	155
Class C $\frac{1}{2}''$ maximum lump	45,350	100	14.1	0.5	279	231
1″ maximum lump	90,700	200	28.2	1.0	337[a]	318
Class D 3″ maximum lump	272,100	600	84.6	3.0	489[b]	453

Alternate measuring box: [a]305 mm × 305 mm × 305 mm high.
[b]457 mm × 457 mm × 406 mm high.

dividually measured and weighed. Smaller lumps may be defined by the square mesh through which cubical lump theoretically could pass with the maximum dimension stated if the particle shape is very long in one direction.

The remaining tests describe methods that are well known and described in many engineering texts. These tests are:

- Test A-14, specific gravity
- Test A-15, moisture content
- Test A-16, particle hardness (MOHS hardness scale)
- Test A-17, screen analysis
- Test A-18, sized and unsized material

The second class of tests in the CEMA standard are those describing properties more difficult to test or to determine qualitatively and are described as hazards effecting conveyorability.

Test B-1, Aeration and Fluidity. To test a finely divided material for tendency to aerate or "flood" while handling, a glass container is filled about half full with the material and sealed with a tightly filled cover. The bottle is then shaken vigorously for one or two minutes. The cover is then quickly removed and the bottle slowly tilted to dump the contents into another container. If the material has become aerated, it will flow very nearly as a fluid at a small angle of tilt and will not form a conical pile in the second container.

The remaining tests describe additional properties that can best be determined by experience or by bench tests:

- Test B-2, becomes plastic or tends to soften
- Test B-3, builds up and hardens
- Test B-4, corrosive
- Test B-5, generates static electricity
- Test B-6, degradable—size breakdown
- Test B-7, deteriorates in storage—decomposition
- Test B-8, dusty
- Test B-10, explosiveness
- Test B-11, flammability
- Test B-12, harmful dust, toxic gas or fumes
- Test B-13, hygroscopic
- Test B-14, interlocks, mats, agglomerates
- Test B-15, oils or fats present
- Test B-16, packs under pressure
- Test B-17, particle shape
- Test B-18, stickiness—adhesion
- Test B-19, contaminable
- Test B-20, very light, fluffy—may be windswept

After all tests are completed, the material is assigned a code designation for each of the characteristics tested. These characteristics are listed as shown on the CEMA material classification chart (Table 9.4). In this classification, the bulk density used is usually the loose bulk density determined from Test A-8. Where a

Table 9.4. CEMA Material Classification Code Chart (Reprinted with permission of Conveyor Manufacturer's Association, Ref. 40.

MAJOR CLASS	MATERIAL CHARACTERISTICS INCLUDED		DEFINITION AND TEST REFERENCE	CODE DESIGNATION
Density	Bulk density, loose		A-8	Actual lbs/cf
Size	Very fine	No. 200 sieve (.0029″) and under	A-17	A_{200}
		No. 100 sieve (.0059″) and under		A_{100}
		No. 40 sieve (.016″) and under		A_{40}
	Fine	No. 6 sieve (.132″) and under		B_6
	Granular	$\frac{1}{2}$″ and under		$C_{1/2}$
		3″ and under		D_3
		7″ and under		D_7
	Lumpy	16″ and under		D_{16}
		Over 16″ to be specified X-actual maximum size		D_X
	Irregular	Stringy, fibrous, cylindrical, slabs, etc.		E
Flowability	Very free-flowing-flow function > 10		A-12	1
	Free-flowing-flow function > 4 but < 10			2
	Average flowability-flow function > 2 but < 4			3
	Sluggish-flow function < 2			4
Abrasiveness	Mildly abrasive–Index 1-17		A-1	5
	Moderately abrasive–Index 18–67			6
	Extremely abrasive–Index 68–416			7
Miscellaneous Properties of Hazards	Builds up and hardens		B-3	F
	Generates static electricity		B-5	G
	Decomposes–deteriorates in storage		B-7	H
	Flammability		B-11	J
	Becomes plastic or tends to soften		B-2	K
	Very dusty		B-8	L
	Aerates and becomes fluid		B-1	M
	Explosiveness		B-10	N
	Stickiness-adhesion		B-18	O
	Contaminable, affecting use		B-19	P
	Degradable, affecting use		B-6	Q
	Gives off harmful or toxic gas or fumes		B-12	R
	Highly corrosive		B-4	S
	Mildly corrosive		B-4	T
	Hygroscopic		B-13	U
	Interlocks, mats or agglomerates		B-14	V
	Oils present		B-15	W
	Packs under pressure		B-16	X
	Very light and fluffy–may be windswept		B-20	Y
	Elevated temperature		A-11	Z

range of values are shown, an average is used in the code designation.

As an example of this coding system, alumina with a bulk density of 50–65 lb per cu. ft, fine particle size less than no. 6 sieve, free flowing, and extremely abrasive, can become aerated, windswept, and dusty, is assigned a material code designation of $58B_6 27MY$.

Carr's Method of Classification

R. L. Carr developed a system to characterize bulk solids with respect to "flowability" and "floodability."[43,44] In the Carr method, the following properties are determined:

Bulk Density Loose. Determined when material is poured gently through a screen into a container and measured before settlement takes place, similar to CEMA Test A-8.

Bulk Density Packed. Measured after vibrating the container for a specified time to settle and deaerate the material, similar to CEMA Test A-9.

Compressibility. This is expressed as a percent difference between loose and packed density and computed as (packed density − loose density)/packed density × 100. The more compressible a material, the less flowable it will be. Carr states that the borderline between a free-flowing and a nonfree-flowing material is a compressibility of about 20%.

Angle of Repose. This is determined by free fall onto a conical pile from a funnel similar to CEMA Test A-5. Flowability materials are characterized by lower angles of repose as determined by this method.

Angle of Fall. This is determined by dropping a small weight on the platform on which a loose or poured angle of repose has been formed. This jarring causes a "slump" or decrease in angle of repose, forming a new angle identified as angle of fall. The more free-flowing or floodable a material, the lower the angle of fall.

Angle of Difference. This is the difference between the poured angle and angle of fall. The greater the angle of difference, the greater the potential for flooding or fluidizing.

Angle of Spatula. This is the angle (measured from the horizontal) that a material heap assumes on a flat spatula that has been inserted into a pile then withdrawn vertically. An average value is taken between the angle determined when the spatula is first withdrawn and after it has been tapped gently. The higher the angle of spatula, the less flowable the material.

Cohesion. This test is used with very fine powders. Material is passed through three vibrating sieves in series. The amount of material retained on each screen is weighed and a cohesion index determined from relative amounts retained.

Dispersibility. This is determined by weighing the amount of material that disperses and falls outside a given area when a sample of material is dropped from a distance of 600 mm (24 inches) to impact on a horizontal plate. The greater the amount of material that disperses, the more likely the material will flush or flood.

In Carr's method, results of each of the tests are assigned a numerical value, which is summed to produce a relative "Flowability Index" and "Floodability Index" for that particular bulk material. A commercial testing machine incorporating Carr's method is now being manufactured by Hosokawa Iron Works, Osaka, Japan (Fig. 9.42).

Other Tests

Methods of determining bulk density of specific materials are described in various ANSI/ASTM Standards. Most of these recommended methods are similar to loose and packed density as described in CEMA Test A-8. ANSI/ASTM Test B527(9) describes a method for density measurement of metal powders and compounds using a Tap-Pak Volumeter (Manufactured by J. Engelsmann, A. G. Ludwigshafen, West

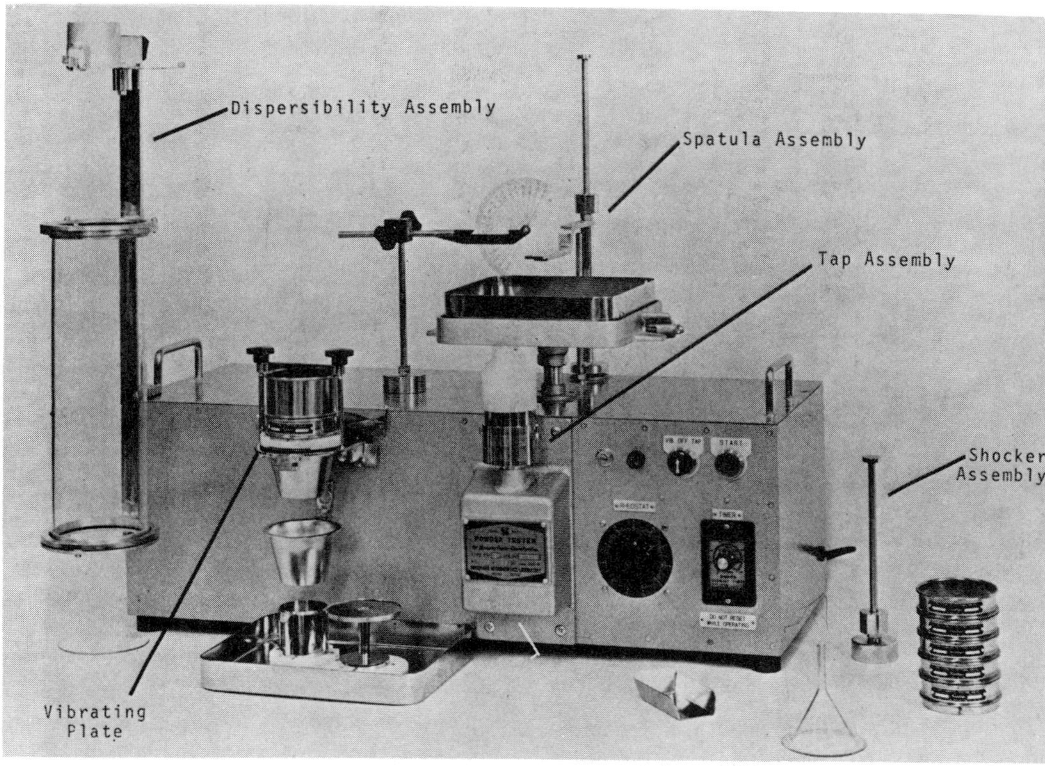

Figure 9.42. Hosokawa powder characteristics tester. (Permission VibraScrew Corp.)

Germany). In the test, a sample of material is subjected to a predetermined number of vertical drops, and a compacted density is obtained. Flowability testing of metal powders is described in ANSI/ASTM B213(9). The rate of flow of a powder sample through a calibrated funnel orifice (Hall Flowmeter) is determined and compared to other powders to obtain a relative measure of flowability. ANSI/ASTM B3292(36) describes flowability of TFE-Fluorocarbon molding materials by comparing flow rates through a stationary and vibrating funnel and orifice.

A simple cone flow test for sand is described in Reference 46, where the "roughness" of sand particles was related to flow rate through a calibrated funnel.

Material that will flood will exhibit a long dearation time. A test to identify powders that exhibit flushing or flooding tendencies is often made by filling a tube about half full of powder, sealing the tube, and shaking vigorously to aerate. After shaking is stopped, the solids level is measured over a period of time and then plotted. This test is also made by fluidizing a column of material (Fig. 9.43) and plotting column height with time after fluidizing air is stopped (Fig. 9.44). Sutton and Richmond[47] describe a more sophisticated test developed by Warren Springs Laboratory where the level of the top of the material is monitored by a video camera and can be analyzed in detail.

Bruff[48] described a shaking test with powdered anthracite. He also determined that flow and deaeration properties of powdered anthracite could be related to the profile of a heap formed on horizontal plate when an open-bottom tube filled with powder was slowly lifted free of the plate. Other empirical tests are described by Zenz[49] and Brown and Richards.[25]

All of the empirical tests described can be useful for comparing properties of a bulk material to a known material or to predict

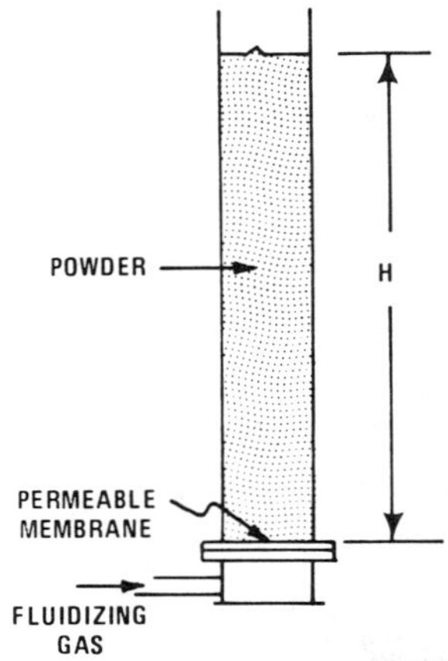

Figure 9.43. Dearation settling test apparatus.

possible flow problems. They do not provide the quantitative information necessary to design bins to avoid these problems.

Solids Flow Analysis and Testing

Attempts to develop an analytical model for predicing the behavior of bulk solids during flow have been based on either the *particulate* or the *continuum* approach. As defined by Goodman and Corwin,[50] in the particulate approach, the properties of discrete particles of finite size (idealized rigid or elastic spheres) are used to deduce the laws governing the behavior of the entire mass. In the continuum approach, the properties of the mass are assumed to be a continuous function, and the mass may be divided indefinitely without losing any of its defining properties. The discrete particles' properties are not considered.

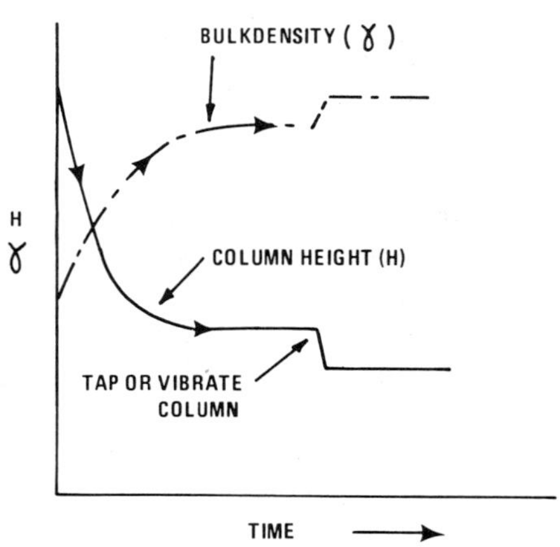

Figure 9.44. Plot of solids bulk density (γ) or solids head (H) vs time.

A considerable amount of work is underway to develop a theory of flow based on particle properties. But real life particulate granular and powder solids are nonhomogeneous, may have an infinite combination of particle sizes, shapes, and interstitial voids. The continuum approach, being less complex, has yielded the most useful information for purposes of engineering design and has been responsible for accelerating the development of the field of powder mechanics as it evolved from the soil mechanics work of Coulomb, Rankine, and others. There are, however, important differences between soil and powder mechanics.[55] Cohesion is usually not important in soil mechanics but is in powders. Stresses in powders stored in bins can be up to 1,000 times smaller than those normally encountered in soil and are not detectable in soils' mechanics tests; boundary conditions in powder mechanics are usually not the same as in soil mechanics, since powders are usually stored in bins; powders can be subjected to much larger deformations than is common in soil mechanics.

Continuum plasticity-type models for powders have been proposed by a number of workers,[51,52,53] including Jenike and Shields. Jenike was the first to use the concepts of plastic failure with the Mohr-Coulomb failure criteria in analyzing the flow of solids in bins and hoppers to develop the concept of a flow-no-flow criterion. This has produced an extremely useful quantitative method for designing storage bins for gravity flow of solids. Since this method has been proven in engineering practice, the information that follows is based on Jenike's original work.[54,55,60,61]

Jenike assumed that a bulk solid can be closely approximated by a rigid-plastic Coulomb solid. From soil mechanics, such a solid is characterized by a yield locus that defines the limiting shear strength under any normal stress (Fig. 9.45). A Coulomb solid has a linear yield locus. Plotting shear stress τ and normal stress σ, the yield locus for a Coulomb

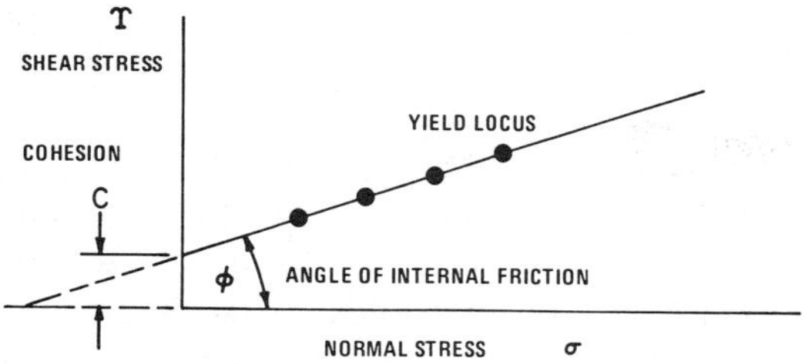

Figure 9.45. Yield locus of a Coulomb solid.

powder intersects the τ axis at a value of τ defined as cohesion C at an angle ϕ, defined as the angle of internal friction.

After many experimental measurements, Jenike found that with real bulk solids, at low pressures, the locus deviates significantly from a straight line (Fig. 9.46); the locus does not increase indefinitely with increasing values of σ but terminates at some point E; and the position of the locus is a function of the degree of consolidation of the material. During flow the stresses in the plastic regions of the solid are continuously defined by point E.

The yield locus for a cohesive solid is shown in Figure 9.46. The yield locus for a free-flowing material such as dry sand would have a locus as shown in Figure 9.47.

In his analysis, Jenike assumed that in the plastic region, solids properties at a point are the same in all directions (isotropic), and frictional, cohesive, and compressible. During incipient failure, the bulk solid expands and during steady-state flow it can either expand or contract, stress at any point does not change with time, and stresses are not significantly effected by velocity changes.

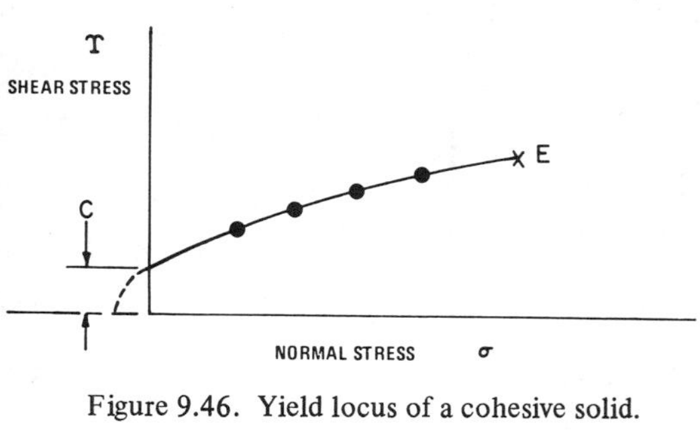

Figure 9.46. Yield locus of a cohesive solid.

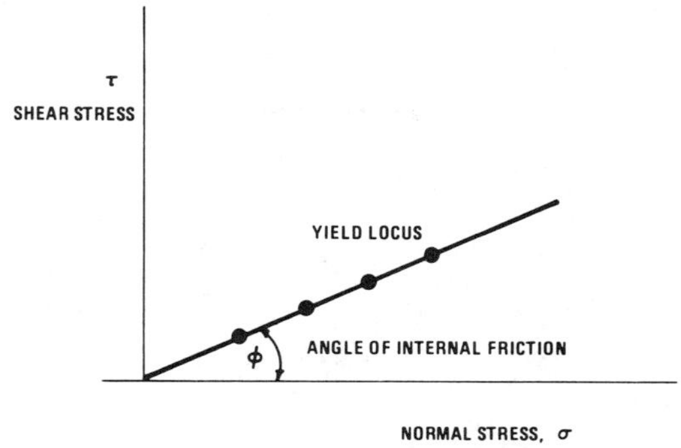

Figure 9.47. Yield locus of free-flowing sand.

Stress-Strength Relationships

As an element of material flows through a channel, the major consolidating stress σ_1 and minor consolidating stress σ_2 on the element change (Fig. 9.48) and continuous shear deformation occurs, causing slip planes as the elements slide on one another or on the bin wall. During flow the "strength" (resistance to shear failure) and density are a function of the last set of stresses and when flow stops, it is assumed these stresses remain. As the material remains stationary under these stresses, it may gain in strength, and resist flow when the bin outlet is reopened.

The shear stress–normal stress relationships for a powder are determined with a shear tester described in the following paragraphs. A set of

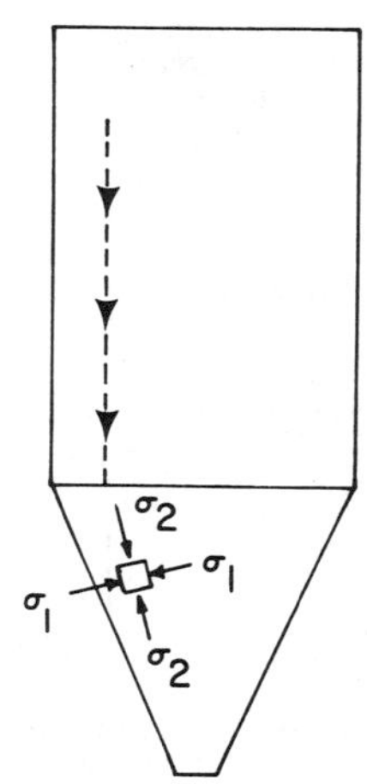

Figure 9.48. Stresses on an element flowing through a channel in a bin.

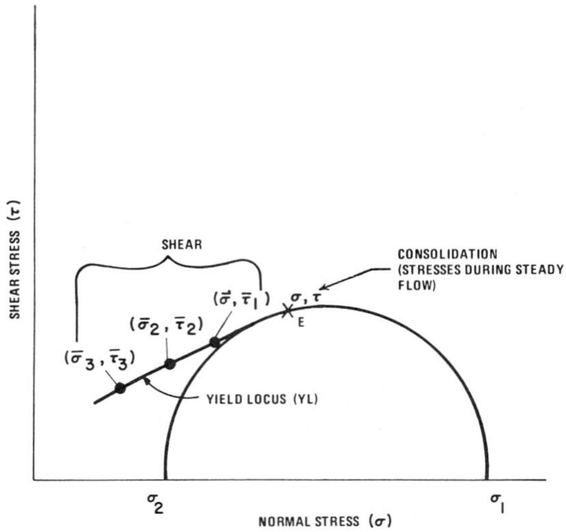

Figure 9.49. Construction of a yield locus.

samples are consolidated under the same applied normal stress σ and then each is sheared to failure while under a different applied normal stress. The yield locus for material in that particular state of consolidation can be drawn as shown in Figure 9.49. Point E represents shear stress τ and normal stress σ at initial consolidating conditions, and $(\bar{\sigma}_1, \bar{\tau}_1)$, $(\bar{\sigma}_2, \bar{\tau}_2)$, $(\bar{\sigma}_3, \bar{\tau}_3)$ are the values obtained during shear.

The state of stress on any plane within the bulk solid can be represented by a Mohr circle. For any stress condition represented by a Mohr semicircle tangent to the locus, the bulk solids will be at a yield, and the major and minor consolidating stresses at this condition will be defined by the intersection of the semicircle with the σ axis. Point E (located between the top of the semicircle and the yield locus) depicts conditions during consolidation (stresses during steady-state flow). The yield locus terminates at the point of tangency of the Mohr circle through E. This circle intersects the σ axis at the major stress σ_1, and minor stress σ_2 under which the specimen has been consolidated.

Testing Apparatus—Shear Cell

The pressure-strength relationships are measured in a shear cell. Several shear test cells have been developed to perform these tests. A design for a rotary cell is described in Reference 58. A rotary cell developed for industrial use by I. A. Peschl is described in Reference 59.

The direct shear cell, developed by Jenike at the University of Utah has become the most commonly accepted and widely used in industrial practice. For this reason, it is the device described here.

The cell consists of a shear ring, base, and cover (Fig. 9.50). The ring used most frequently has an inside diameter of 95.3 mm (3.75 in.).

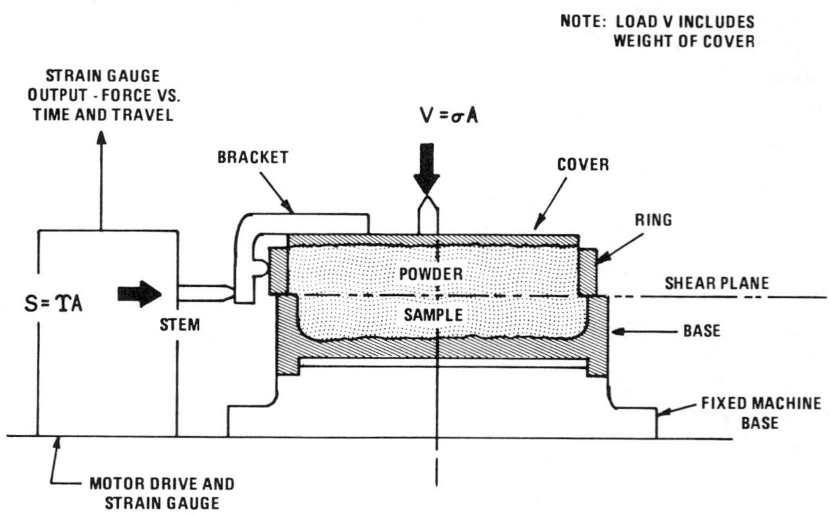

Figure 9.50. Jenike shear cell.

A 63.5 mm (2.5 in.) diameter ring is sometimes used to reduce handling of heavy weights when high consolidating loads are used. The bottom of the cover and inside of the base are roughened to increase solids adhesion. A bracket is attached to the top of the cover. The base and ring are filled with powder and the cover put in place. A vertical force is applied to the cover. A horizontal shearing force is applied to the bracket by a motor-driven stem. Part of the shearing force is transferred to the ring by a loading pin attached to the cover bracket. This helps to ensure more uniform distribution of shearing force across the cell during shear. With 60 Hz electrical supply, the shearing force is applied at a constant rate of 0.91 mm/min (0.036 in./min) in older machines, and 2.7 mm/min (0.108 in./min) in newer machines. With 50 Hz electrical supply, the rates are 0.76 mm/min (0.03 in./min) and 2.3 mm/min (0.09 in./min). The shearing force is transmitted through the pin to a load cell and displayed as shear force vs time and displacement.

The flowability and yield strength of a mixture of coarse and fine particles are most dependent on the properties of the fine fraction since shear occurs across the fine fraction during flow. Therefore, when testing such a mixture, particles greater than about 3 mm are screened and removed from the shear test sample.

Sample Preparation and Testing

To simulate stress conditions occurring during steady flow, the consolidation of the sample is done in two stages, preconsolidation and consolidation, followed by shearing of the sample.

Preconsolidation (Fig. 9.51a). This is done to ensure uniformity between samples. A packing mold is placed on top of the ring and powder packed to the top of the mold. A twisting top is placed over the material and a selected load V_t is applied to the top which is then given a number of oscillating twists of about 30° amplitude. The load V_t and top are then removed and the powder scraped level with the shear ring (Fig. 9.51b).

Consolidation (Fig. 9.51c). This reproduces flow with a given stress under steady-state conditions (point E). The cover is placed over the material, the consolidating load V applied and the shearing force S is applied through the pin. Shearing continues until the shearing force reaches a steady value indicating plastic flow across a layer of the specimen. At this point, consolidation is considered complete, the shearing force is interrupted, and the pin retracted. The stress in the solid at this point is plotted as τ versus σ at point E in the Mohr semicircle, as shown in Figure 9.49. For convenience, the data can be plotted so that stresses applied to the shear cell can be plotted in force coordinates as: $V = A\sigma$; $S = A\tau$, where A = area of the shear cell, as shown in Figure 9.52.

Shearing (Fig. 9.51d). The shear ring, sample, and cover remain in position. The vertical load is replaced by a smaller load noted as $\overline{V}$, and the shearing force reapplied until the stress–strain curve peaks and falls off, indicating development of a failure plane and a point $\overline{V}_1\overline{S}_1 = \overline{\sigma}_1, \overline{\tau}_1$, on the yield locus (Fig. 9.49). The specimen is examined after failure to ensure that the failure plane roughly coincides with the shear plane of the cell. If the planes do not, the test is repeated.

Several other points are required to construct the yield locus. Intermediate values of $\overline{V}$ are chosen by inspection of the data or by judgment, so that at least three well-spaced points along the locus are obtained. The lowest selected value of $\overline{V}$ should be not less than $\frac{1}{3}$ the value of the maximum consolidating load V. Each point on the locus is obtained by first consolidating a sample to point V, S representing steady-state flow conditions, then shearing to failure at lower values of $\overline{V}(\overline{V}_2, \overline{V}_3$, etc.). The range for acceptable values is shown in Figure 9.52. The yield locus through the $\overline{V}_1\overline{S}_1, \overline{V}_2\overline{S}_2 \ldots$ points is usually slightly convex shaped but for analyzing the areas of interest in industrial-size bins it is often approximated by a straight line. The Mohr semicircle drawn tangent to the yield locus and passing through the consolidation point $E, (V, S)$ intercepts the

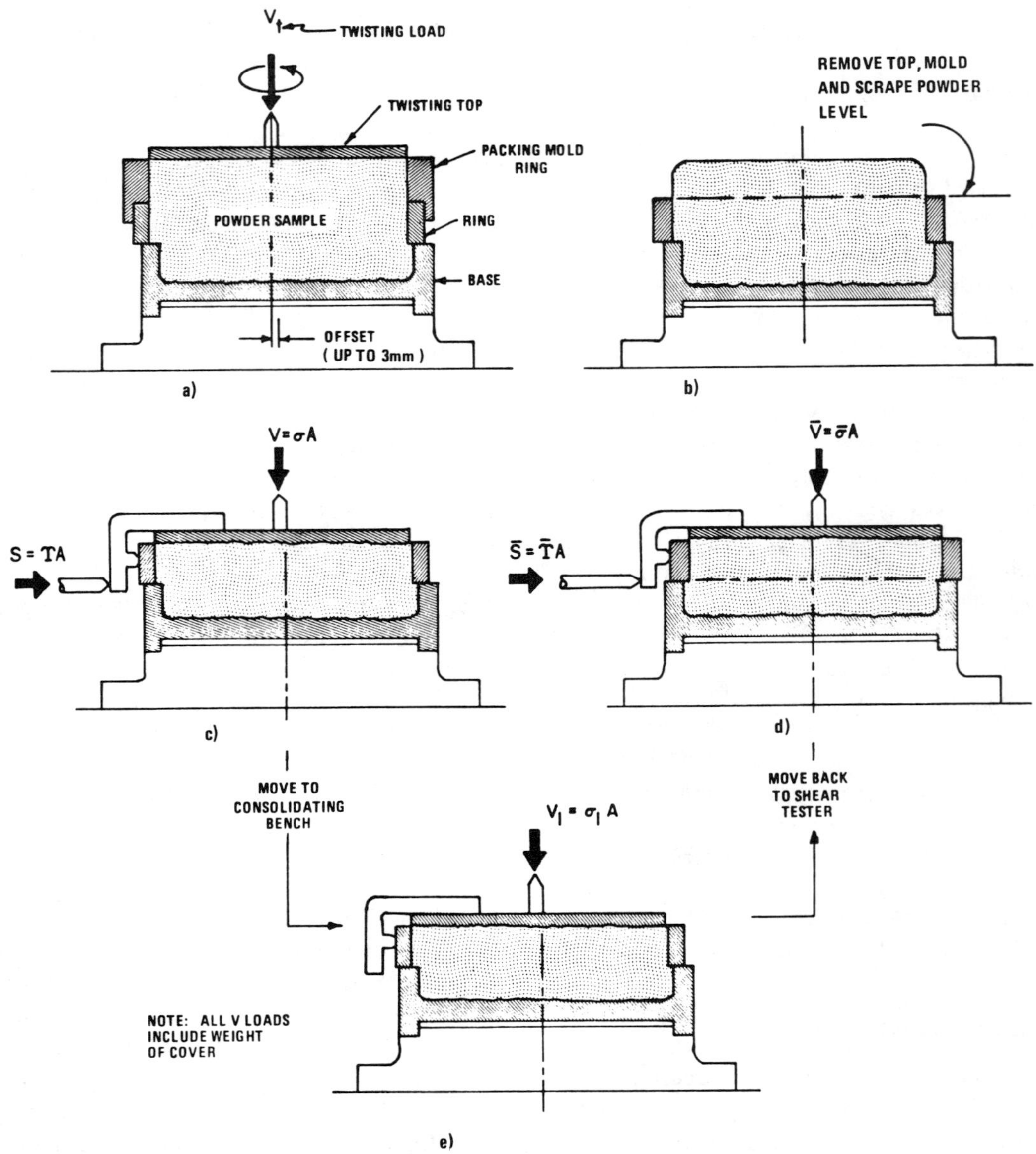

Figure 9.51. Jenike shear test sequence: (a) preconsolidation, (b) Removal of twising top and packing mold ring, (c) consolidation, (d) shear, (e) time consolidation.

V axis at the value of the major consolidating force $V_1 = A\sigma_1$.

A family of yield loci can be determined by consolidating the sample to different levels of normal stress before shearing. An example is shown in Figs. 9.53 and 9.54. The $(YL)a$ (Fig. 9.53) and $(YL)b$ (Fig. 9.54) represent the yield loci obtained after consolidation under loads $(V)a$ and $(V)b$, respectively, and are shown plotted in σ, τ coordinates.

Inspection of Shear-Strain Data

The direct shear cell has a limited horizontal movement. Specimens must be properly prepared during consolidation such that consolidation and shear are accomplished within this limit. If in a group of samples it is found that the value of S determined during steady-state consolidation varies more than 10% from the intermediate values of the group, that sample should be retested.

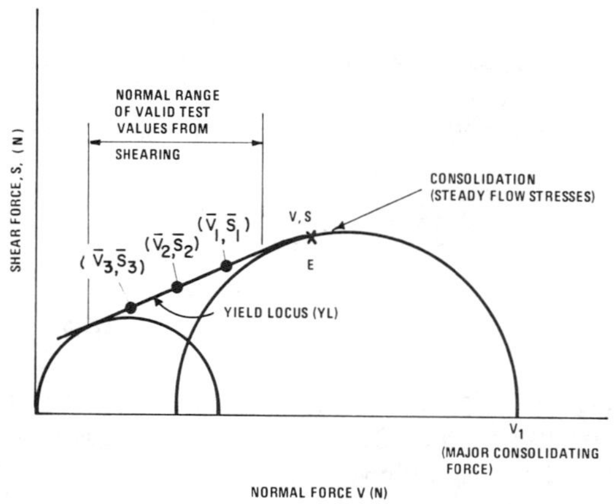

Figure 9.52. Yield locus plotted in force units.

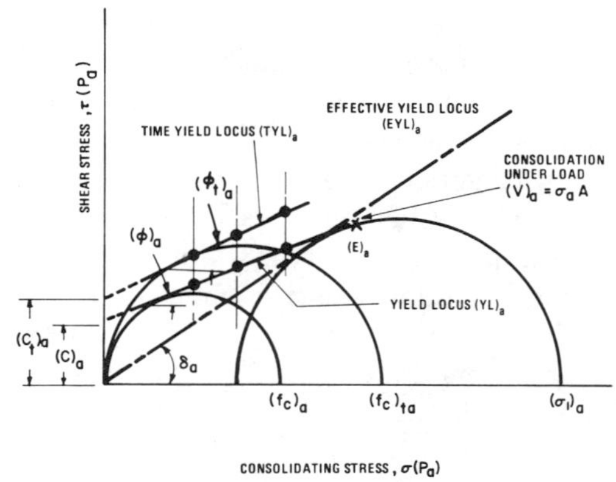

Figure 9.53. Yield locus under consolidation stress σ_a.

9.7. PROPERTIES USED IN FLOW ANALYSIS

The following properties of bulk solids, most of which are determined from the shear test, are used in the analysis of flowability and for the design of storage bins (refer to Figs. 9.53 and 9.54).

Effective Yield Locus (EYL)

Effective Angle of Internal Friction (δ)

These are a measure of conditions in a flowing material. The effective yield locus is defined by a straight line through the origin, tangent to the steady-flow Mohr stress circle as shown in the figure (9.53 and 9.54). The ratio of major and minor consolidating pressures during flow can be expressed by the effective yield function:

$$\frac{\sigma_1}{\sigma_2} = \frac{1 + \sin \delta}{1 - \sin \delta} \tag{9.27}$$

The EYL can be defined in the equation:

$$\sin \sigma = \frac{\sigma_1 - \sigma_2}{\sigma_1 + \sigma_2} \tag{9.28}$$

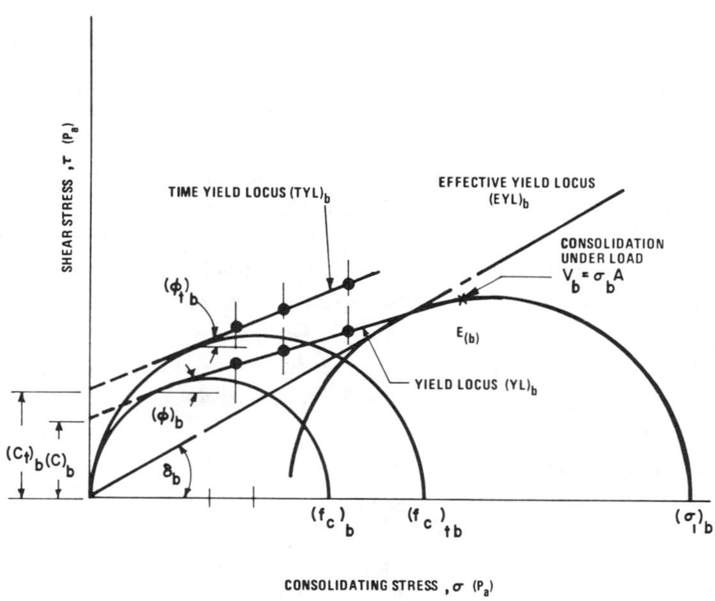

Figure 9.54. Yield locus under consolidation stress σ_b.

The angle δ is called the *effective angle of internal friction* and is related to the angle of internal friction of a material. It is a measure of resistance to flow while the material is in a flowing condition. As δ increases, the flowability of a solid decreases. With a given material, it usually increases with decreasing consolidating stress and may approach 90° at very low consolidating stresses. The value of δ ranges from 25 to 70° for most materials that have been tested.

Unconfined Yield Strength (f_c)

At a free surface formed when an arch is about to collapse, the minor consolidating stress σ_2 acting normal to the surface is equal to zero, and the major consolidating stress σ_1 is tangent to the surface. Therefore, a Mohr circle drawn through the origin, tangent to the yield locus defines the largest stress σ_c the material can withstand at a free surface (when $\sigma_2 = 0$). The value of σ_c at this point defines the unconfined yield strength, f_c(Pa) [expressed in force units, $F(N) = f_c A$, where A = area of shear cell (m^2)]. For each value of consolidation stress there is a corresponding value of unconfined yield strength f_c, and as consolidation stress increases, f_c increases.

Kinematic Angle of Internal Friction ϕ

This is the angle between the yield locus and σ axis. It can vary somewhat with increasing normal stress if the YL is convex shaped.

Time Yield Locus (TYL)

The yield locus for a material is determined under "instantaneous" flow conditions when the specimen is placed under stress and sheared immediately. In a storage bin, the stresses acting during flow remain after flow stops, exposing the solids to continuous consolidation. The yield locus determined after exposing a sample to this time consolidation will produce a yield locus representing the stresses required to *initiate* flow and reestablish the conditions represented by the instantaneous yield locus. The time yield locus is determined by consolidating the specimen to a steady-state condition (point E) as before and transferring the cell from the shear tester to a consolidating bench, where it is placed under major stress σ_1 (using load V_1) as determined from the Mohr circle for the instantaneous yield function. After the specified time, the V_1 load is removed, the cell is transferred to the shear tester, and the specimen sheared to failure as before (Fig. 9.51d) A time yield locus is constructed and a value for $(f_c)_t$ is determined for these conditions as shown in Figures 9.53 and 9.54.

Static Angle of Internal Friction ϕ_t

This is the angle between the time yield locus and the σ or V axis at the point of tangency of the locus and the Mohr circle through the origin. It is essentially constant but may vary at low consolidating stresses. This is the angle of internal friction used in the analysis of funnel flow.

Wall Yield Locus (WYL)

Kinematic Angle of Surface Friction (ϕ')

Solids flow along slip lines that form boundaries between flowing and static solids described by the YL, or they can flow along a rigid bin wall. Stresses along the wall during this type of flow lie along the wall yield locus (WYL).

This locus is determined by measuring the shear force required to slide the solids horizontally across a plate that reproduces the bin wall surface. In the test (Fig. 9.55) a normal load is applied to the specimen, and when a steady force required to move the cell is determined, the normal load is reduced and a new steady force determined. After a number of values are recorded, the wall yield (WYL) locus is constructed and the kinematic angle of friction ϕ' determined (Figs. 9.56 and 9.57).

In many cases the WYL is linear (Fig. 9.56), but it also can be convex or concave (Fig. 9.57). In those cases, ϕ' must be determined for stress conditions at the particular point under

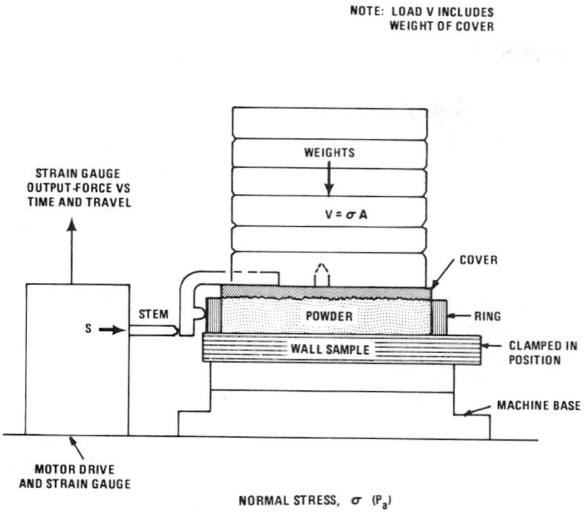

Figure 9.55. Jenike wall friction test.

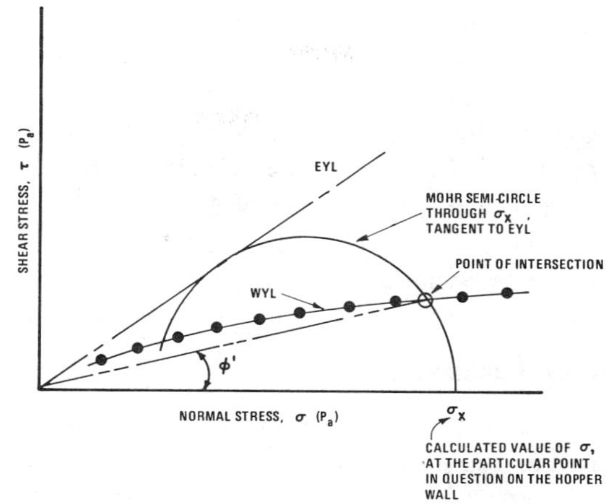

Figure 9.57. Convex wall yield locus.

investigation in a hopper. This is done by extending a straight line through the origin to the intersection of the WYL and the Mohr circle representing the stress conditions at the point in question and measuring ϕ' (Fig. 9.57).

At times, the shear force recorder pen will oscillate in a slip-stick type motion. The peak values are used. Sometimes the initial shear force required to start movement peaks then assumes a lower steady-state value. The peak value is used.

Adhesion or ϕ_t' Static Angle of Surface Friction

The kinematic angle of surface friction ϕ' occurs during flow. Some solids will stick or adhere to bin walls if allowed to remain at rest under consolidation and in contact with the wall. When this happens, a higher stress will be required to initiate flow and restore the WYL to steady-flow conditions.

This can be predicted by a wall adhesion test. Steady-state shear across a plate under a vertical load is established as described for the WYL test. The machine is then stopped, the specimen left under the load for a specified time, and then the shearing force is reapplied. The force required to initiate movement can be compared to the forces required in the kinematic WYL test. If higher values are found, these can be identified as ϕ_t' or static angle of surface friction. To facilitate testing a number of samples can be transferred to the consolidating bench to remain at rest under its particular consolidating load; then all can be transferred back in succession to the shear tester.

If adhesion is severe, smoother wall surfaces or alternate wall coatings or linings should be tested to find one that will eliminate or minimize the problem.

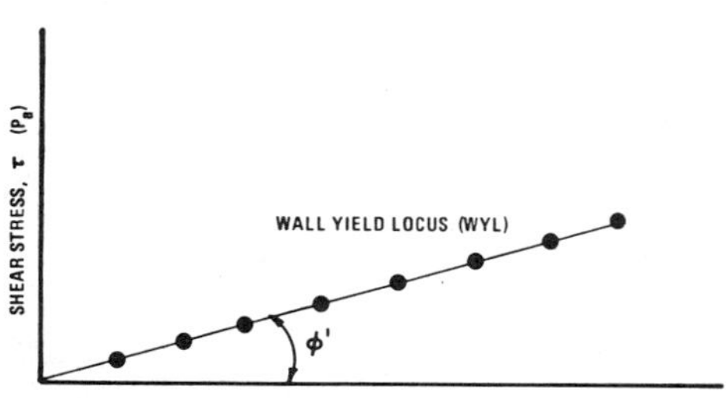

Figure 9.56. Linear wall yield locus.

Cohesion

Cohesion is the sticking together of particles of a bulk solid. A relative measure of cohesion can be determined by approximating the yield locus with a straight line across the low-stress region to intercept the shear stress axis (Figs. 9.53 and 9.54). The intercept is a measure of shear stress at zero normal stress.

Cohesion C increases and decreases with consolidating pressure during flow. Free-flowing materials such as sand have no cohesion, (C = 0), and the yield locus is through the origin ($\phi = \delta$). Cohesion values obtained with the shear test method can be of use in comparing various powders, but are not used in analysis of flow described in this chapter.

Flow Function (FF)

The flow function (FF) sometimes called the failure function, was conceived by Jenike to characterize the "flowability" of a bulk solid. The flow of a bulk solid is dependent on the strength it develops as a result of consolidation. The unconfined yield strength (f_c) is a measure of this strength and is a function of the major consolidating stress (σ_1). The flow function therefore is found by plotting values of f_c versus σ_1, (or F_c versus V_1) for each yield locus (Fig. 9.58). As an example, the values for the yield loci shown in Figs. 9.53 and 9.54 are plotted in Figure 9.58 to obtain the instantaneous time flow functions. *Instantaneous flow functions* are determined under conditions of zero consolidation time: *Time flow functions* are determined under conditions occurring after time consolidation. Many materials gain strength with time consolidation.

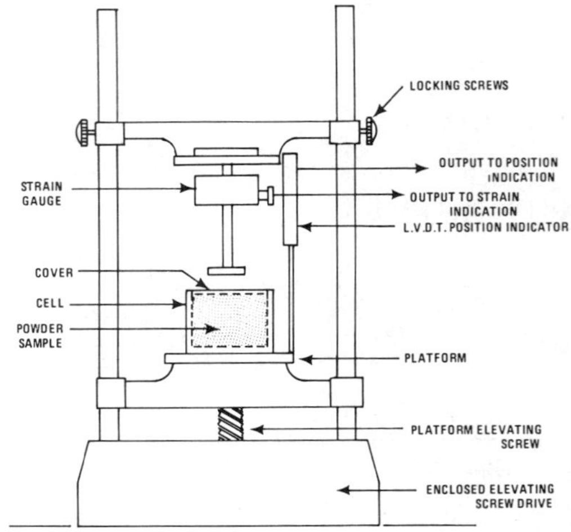

Figure 9.59. Bulk-density compression test apparatus.

In Figure 9.58. The highest curve represents the solid with the greatest strength and greatest ability to support an arch or rathole.

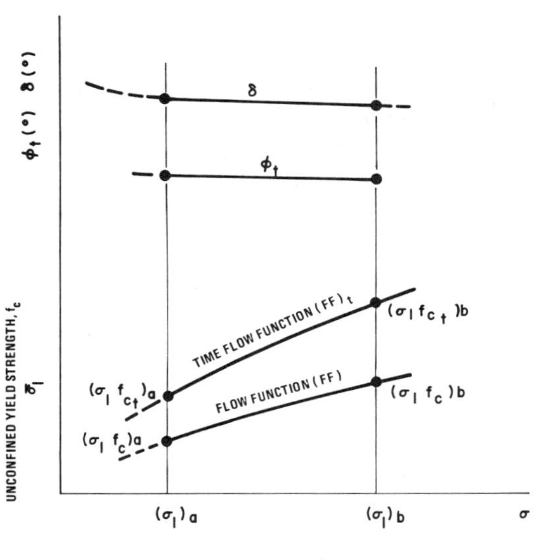

Figure 9.58. Flow functions. (From Figs. 9.53 and 9.54)

Bulk Density

Bulk density of a solid is a function of the consolidation stress and during flow, it changes as the stresses change. Bulk density as a function of these stresses can be determined by applying vertical loads to a bulk solids sample of known mass and recording compression of the sample, with a dial indicator, scale, or electronic position indicating equipment as shown in Figure 9.59. Since the mass, consolidating load, and volume are known, the relationship can be

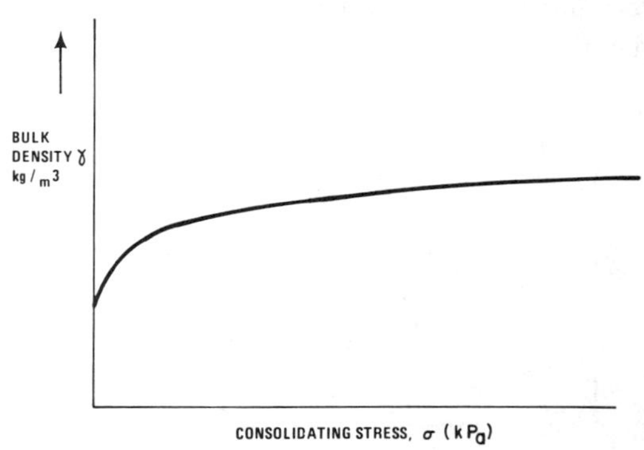

Figure 9.60. Typical plot of bulk density vs consolidating stress.

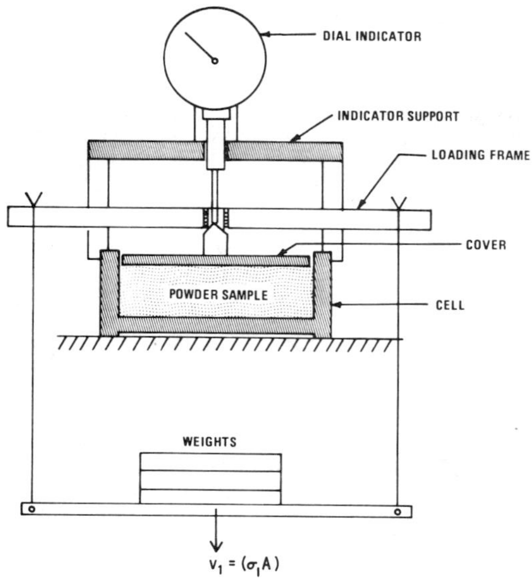

Figure 9.61. Jenike and Johanson Inc. compressibility tester. (From Ref. 61)

plotted as shown in Figure 9.60. To minimize wall effects, the cylinder used for the consolidating test should have a length to diameter ratio (L/D) not exceeding 1.

Compressibility

Jenike and Johanson have proposed a compressibility constant β as a measure of powder compressibility.[54] To obtain this, a powder sample is compressed in a shear cell ring (Fig. 9.61) and

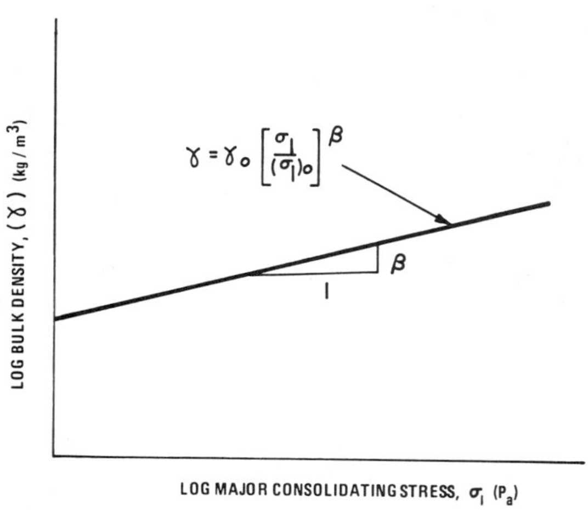

Figure 9.62. Determination of compressibility constant. (From Ref. 62)

the displacement recorded. The results are presented on a logarithmic plot (Fig. 9.62) and a straight line fitted through the points. The compressibility constant β is determined by assuming the relationship:

$$\gamma = \gamma_0 \left[\frac{\sigma_1}{(\sigma_1)_0}\right]^\beta \tag{9.29}$$

where

γ = bulk density at consolidating stress σ_1 or load V_1

γ_0 = bulk density at consolidating stress $(\sigma_1)_0$ or load $(V_1)_0$ (arbitrary chosen base point).

The slope of the line β is a measure of the compressibility of the bulk solids and will vary between 1.0 for very hard materials to close to 0 for very compressible materials. Most powders will have values ranging between 0.1 to 0.01. It can be assumed that when comparing powders, the sample having the higher value will be more likely to require longer times to settle and deaerate when loaded into a bin and will be more likely to flush or flood when discharged.

Permeability

A measure of powder permeability is useful for predicting potential problems in powder storage. Powders with low permeability can remain fluidized for long periods of time after charging to a bin. During this period, the powders can impose high lateral pressures on the silo walls and can be semifluid when withdrawn. On the other hand, these powders can deaerate with time and pressure and become flow-rate limited when withdrawn from mass flow hoppers, as described in Section 9.12.

A permeability factor has been proposed by Jenike and Johanson[62] to characterize this type of powder behavior. The factor is determined by testing a powder specimen in the apparatus shown schematically in Figure 9.63. Powder is placed in the cylinder, the bed height

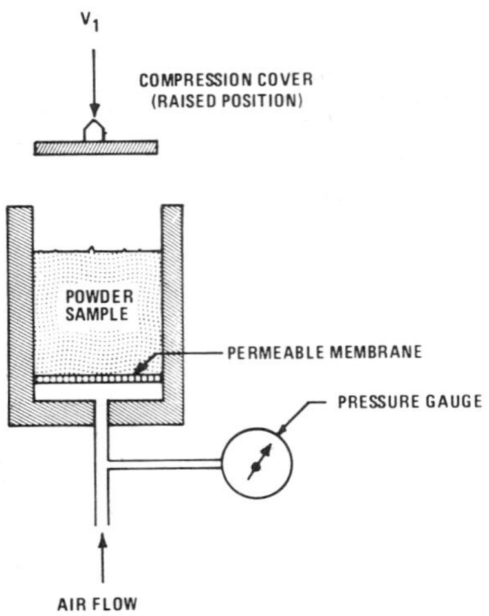

Figure 9.63. Jenike and Johanson Inc. permeability tester. (From Ref. 62)

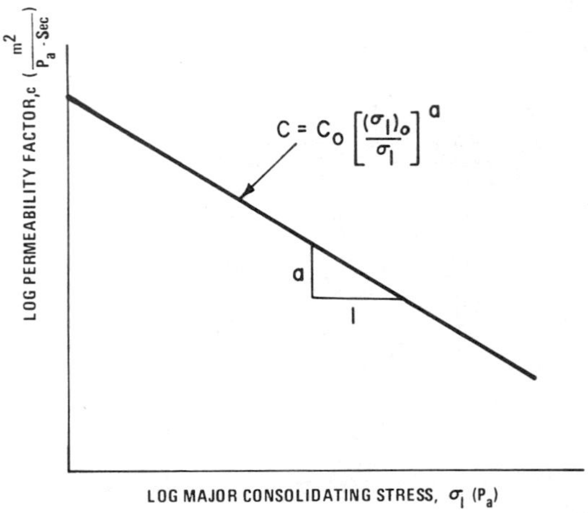

Figure 9.64. Determination of permeability factor. (From Ref. 62)

measured with a dial indicator, and then air is passed into the bottom through a distribution membrane and up through the sample. The flow is slowly increased until the pressure is slightly less than that required to lift the sample and the pressure and air flow are recorded. The powder sample is then compacted to increasingly higher densities and the test repeated several times.

Assuming laminar flow and constant gas viscosity, a permeability factor c is determined from a form of the D'Arcy equation:

$$V = c\,\Delta p/h \tag{9.30}$$

where

v = superficial gas velocity, m/s
c = permeability factor m^2/Pa-s
c = permeability factor
p = pressure drop across sample column, Pa
h = sample colume height, m

Therefore, the permeability factor:

$$c = \frac{V}{\Delta p/h} \tag{9.31}$$

A value of c is computed for each of the test points and plotted on a logarithmic scale as shown in Figure 9.64. The relationship between c and consolidation pressure σ as shown in this figure can be expressed as:

$$c = c_0\left[\frac{(\sigma_1)_0}{\sigma_1}\right]^a = c_0\left[\frac{(V_1)_0}{V_1}\right]^a \tag{9.32}$$

where

c = permeability factor at consolidating stress σ_1 or load V_1
c_0 = permeability factor at consolidating stress $(\sigma_1)_0$ or load $(V_1)_0$ (arbitrary chosen base point),
a = permeability constant for the powder specimen

The permeability constant (a) becomes a measure of the change in powder voidage with change in consolidating stress and is also a relative measure of powder compressibility.

9.8. OTHER FACTORS AFFECTING FLOW PROPERTIES

Impact During Loading

When a silo is initially filled, the stresses on the material at the point of impact may be higher than those that occur during flow. Assuming initial velocity of the solids to be zero, this impact stress σ_p (kPa) may be estimated as:

$$\sigma_p = \frac{Q}{gA}(2gh)^{0.5} \tag{9.33}$$

where

Q = Flow rate into bin, kN/s
g = gravitational constant, 9.8 m/s^2
A = estimated area of impact, m^2
h = height of fall, m

If a tall bin is to be designed, the yield strength f_c of the solid developed under the impact stress ($\sigma_p = \sigma_1$) should be checked to make sure it is less than f_c during time consolidation or flow. If not, the dimensions of the discharge opening should be calculated on the basis of the higher solid strength.

The impact stresses can be reduced by: installing a deflector plate at the bin inlet to distribute the material over a wider impact area, directing the incoming solids stream at the bin wall, or by maintaining a minimum solids head in the hopper so that impact occurs well above the solids outlet. Stresses at the outlet can also be reduced by withdrawing material (at a low rate if necessary) for a short time during initial filling. This will cause a passive (arched) stress field to develop immediately over the outlet as described in Section 9.11.

Temperature and Chemical Changes

Solids may agglomerate or soften at high temperature or may undergo phase changes when cooled. All these can affect flowability. Temperature changes can occur, for instance, when solids are dried, then loaded hot into storage silos and allowed to consolidate and cool at rest or when solids are loaded into silos, trucks, or rail cars that are subsequently exposed to cyclic temperature extremes occurring between day and night. In many cases, these conditions have been found to significantly increase solids strength (f_c).

Flow properties of these solids should be determined by duplicating *these conditions* (including cyclic changes) in the consolidating bench. Many startup problems can be avoided by such testing. Testing may indicate the need to cool and solids before storage or during storage or to insulate storage bins or rail cars to protect the material from ambient temperature changes.

Moisture

Moisture in a bulk solid may vary due to changes in operation of a process dryer or from exposure to weather when stored in outside piles. Changes also occur during storage if moisture enters the bin through open atmospheric vents, is blown in during pneumatic conveying of hot moist materials and condenses on cold bin walls, or migrates from the stored material. Moisture can effect yield strength (f_c) and wall friction ϕ' and can cause solid-wall adhesion (ϕ_t'). To properly access these effects, the expected storage conditions must be duplicated in the shear test. If the flow properties are time dependent they can only be determined by making time consolidation tests. Migration of moisture from large quantities of material in a storage bin, particularly if it is accompanied by chemical changes, may not be detected in shear tests since the sample size is small. Several useful studies of caking due to these mechanisms are described in References 66–70. The use of flow conditioning agents to improve flow are reviewed in References 71 and 72.

Particle Size

Flowability usually decreases, and wall friction ϕ' tends to increase as particles become finer. Permeability also decreases, resulting in an increasing potential for flooding or flow limitation from the hopper outlet.

If materials are to be handled in equipment that will cause particle attrition before storage (pneumatic conveyors for instance) the samples used for shear testing must be representative of material after handling.

Future possibilities of producing and storing finer particles should be anticipated in the initial design. Coarse particles may break and generate finer particles during storage and flow. Run crush tests to determine potential breakage with the anticipated consolidating stresses in the bin. If breakage cannot be tolerated, consider storage in smaller bins where the stresses will be lower. If breakage will occur and can be tolerated, shear test the minus 3 mm fraction since the flowability of the mixture will be most affected by these particles. Particles larger than this are considered coarse granular and are usually free flowing.

Vibration

Vibration can be induced into storage structures by nearby moving equipment, by transport in over-the-road vehicles, and by vibrating devices used to promote flow.

Many bulk solids, particularly those containing fine particles will tend to compact very rapidly when vibrated. Those solids that have high instantaneous flow functions are particularly susceptible to flow stoppages caused by compaction during vibration. Use of vibrators to assist flow should be restricted to *only* the time that material is flowing in the hopper. Vibration should be stopped immediately when the silo outlet is closed or the discharge feeder stopped.

The application of vibration to assist flow is strictly empirical at this time. Many manufacturers have developed rules of thumb regarding proper locations for their particular vibrator on a particular hopper geometry.

If the shear test data (or other experience) indicate that vibration will be required to start flow, (i.e., reduce the time flow function to the instantaneous flow function), the silo outlet should be designed with a safety factor to allow for possible solids over compaction due to vibration (see Section 9.10). Roberts describes the effects of vibration in more detail in Chapter 6.

9.9. FLOW-NO-FLOW CRITERION

Jenike's flow–no-flow criterion[55] provides an extremely useful means of predicting gravity flow or no-flow of solids in a bin. It has been found to provide a basis for design that is within normal engineering tolerances.

This criterion states that gravity flow will occur in a channel if the yield strength which a solid develops in that channel is not sufficient to support an obstruction to flow. (This obstruction is assumed to be an arch or a rathole.) It can be illustrated with reference to Figure 9.65 as follows:

Assume material flowing in a mass flow bin where solids continually flow into the top. As an element of material flows downward, it becomes compacted under the major consolidating stress σ_1 acting within the bin, and develops an unconfined yield strength, f_c.

The consolidating stress increases, then levels off in the vertical part of the bin, abruptly changes at the transition, then decreases toward zero at the vertex, while at the same time strength (f_c) changes approximately as shown.

The major stress $\overline{\sigma}_1$ that acts on the abutments of a stable arch has been shown to be directly proportional to the span B of the arch and, therefore, would vary as shown in Figure 9.65. Jenike[60] has shown that the stress, $\overline{\sigma}_1$(kPa), acting at the abutments of an arch, can be expressed by:

$$\frac{\overline{\sigma}_1}{\gamma B} = \frac{1}{H(\theta)} \tag{9.34}$$

where

γ = bulk density expressed as kN/m^2
B = diameter of circular opening, m
$H(\theta)$ = function given in Figure 9.73

The lines denoting the values of f_c and $\overline{\sigma}_1$ intersect at a critical value from which the arch dimension B is determined. according to the flow–no-flow criterion, below the point of intersection the material has developed sufficient strength to support an arch, and flow will stop. Above that point, the material has insufficient strength, no arch will form, and gravity flow will take place.

The unconfined yield strength f_c is determined by shear testing the material under appropriate consolidating stresses and from this a flow function (FF) for the material can be constructed as described earlier.

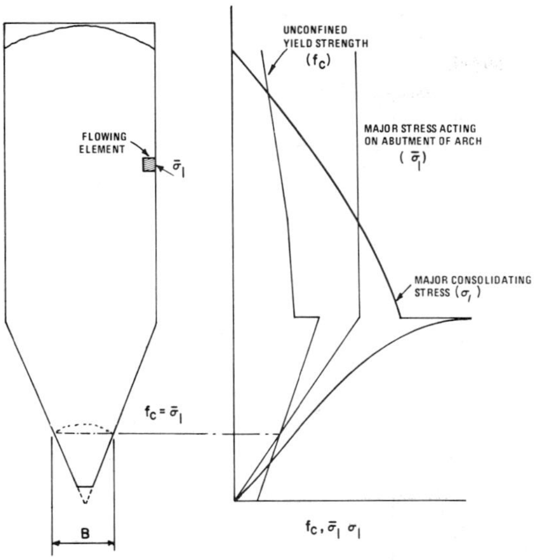

Figure 9.65. Stress on an element flowing through a mass flow bin.

The ratio $\sigma_1/\bar{\sigma}_1$ is defined as the *flow factor* (*ff*) describing the flowability of a *channel* or hopper. The higher the consolidating stress σ_1 acting in a channel and the lower the stresses $\bar{\sigma}_1$ acting on the arch, the lower will be the flowability of the channel or hopper. Jenike has obtained flow factors by solving the equations describing stress and velocity fields in flow channels. The solutions required for a wide range of solids and boundary conditions were found to be too complex and unwieldy for routine design. Jenike showed that the general solutions converged to radial stress field at the vertex of the hopper. Johanson in further studies shows that radial stresses were closely approximated in regions further from the outlet.[30,31] Perry and Jangda[32] reported pressure measurements within a bed of sand in a model bin using pressure-sensitive radio pills. Their data showed evidence of radial stress within the bed.

Assuming a radial stress field, the stress distribution during mass flow of the bulk solids were shown to be approximated by a straight line through the hopper vertex in accordance with:

$$\sigma = r\gamma S(\theta') \qquad (9.35)$$

where

σ = average of vertical and lateral stresses
γ = bulk density during flow
r = distance from the vertex
$S(\theta')$ = stress function

The stress function $S(\theta')$ has been computed and presented in chart form in Reference 54.

The assumption of radial stress has been found to be sufficiently accurate for engineering purposes. Stresses in the region of the outlet during steady mass flow therefore can be predicted on the basis of effective angle of internal friction δ, slope of hopper walls θ', and wall friction ϕ'. It has been determined that when boundary conditions are not compatible with radial stress, mass flow does not occur in a hopper, and a funnel flow pattern develops. Boundary conditions are shown for conical mass flow hoppers in Figure 9.66 and symmetric plane flow hoppers in Figure 9.67.

Combining Equation 9.32 with the equation for major consolidating stress σ_1:

$$\sigma_1 = \sigma(1 + \sin\delta) \qquad (9.36)$$

results in the equation for the *flow factor* (ff):

$$\text{ff} = \frac{S(\theta')(1 + \sin\delta)}{2\sin\theta'}(1 + m) \qquad (9.37)$$

where

m = 0 for plane flow channels (slot outlet)
m = 1 for conical hoppers

Flow factors for various values of effective angle of internal friction δ, angle of solids wall friction ϕ', and angles of hopper wall slope θ' have been calculated by Jenike and are presented in the form of charts in Bulletin 123 published by the University of Utah.[55] The charts are for solids having a δ of 30, 40, 50, 60, and 70°. The charts for symmetric plane flow and axi-summetric conical flow for solids for which $\delta = 40°$ are reproduced in Figures 9.68 and 9.69.

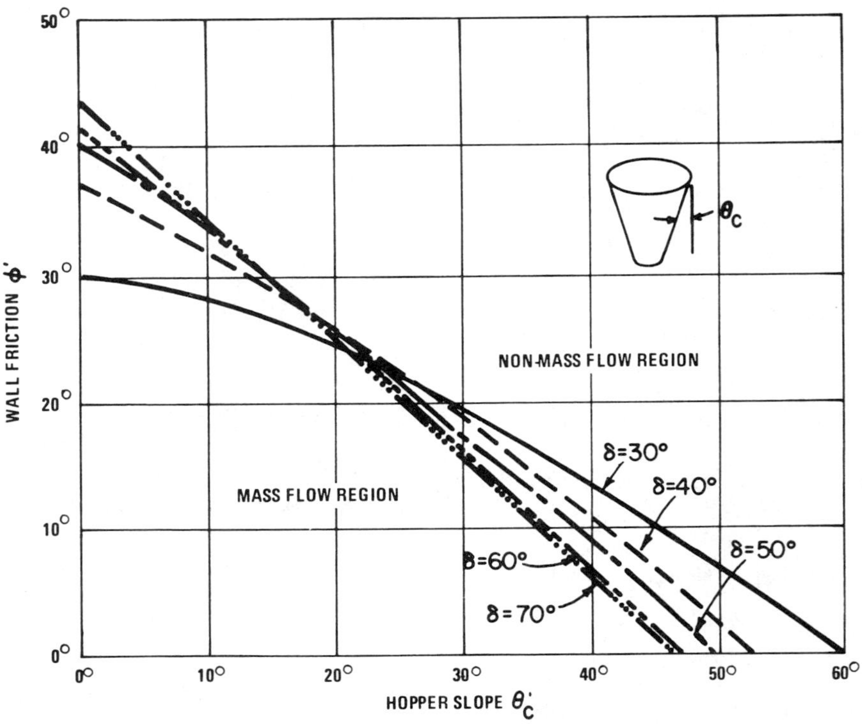

Figure 9.66. Critical wall slope for conical mass flow hoppers for δ values from 30 to 60°.

The flow function (FF) and flow factor (ff) can be superimposed as shown on Figure 9.70. In that part of the ff line that lies above the FF line (ff > FF) the flow criteria are satisfied; the stresses set up in the arch exceed the solid strength, and flow will occur. In that part of the ff line that lies below the FF line, the stresses are not sufficient to cause failure and arching will occur. The intersection of the two lines represents the critical value for calculating the minimum hopper opening size as described in the next section.

9.10. FUNCTIONAL DESIGN OF BINS FOR GRAVITY FLOW

Mass Flow Bins

Hoppers in mass flow bins must be sufficiently steep to cause flow along the walls and must have outlets large enough to prevent formation of an arch; in addition, any discharge device must withdraw solids uniformly *across the complete discharge opening.* If the feeder or connecting chutes cause the solids to flow preferentially from only one portion of the outlet, the mass flow pattern will be destroyed and funnel flow will result. Typical mass flow hopper geometries are shown in Figure 9.71. Design of the mass flow hopper begins with the shear test and from the data, plotting the instantaneous and time flow function, flow factor, δ, and ϕ_t (Fig. 9.72) and the wall yield locus. The required hopper slopes and discharge openings required for mass flow are then determined as follows:

Slope of Hopper Walls. Hopper slope θ' is a function of the effective angle of internal friction δ and the wall friction ϕ'. To determine the required hopper slope, δ and ϕ' are first determined by a shear test.

The required hopper slopes and appropriate flow factor are determined by entering the

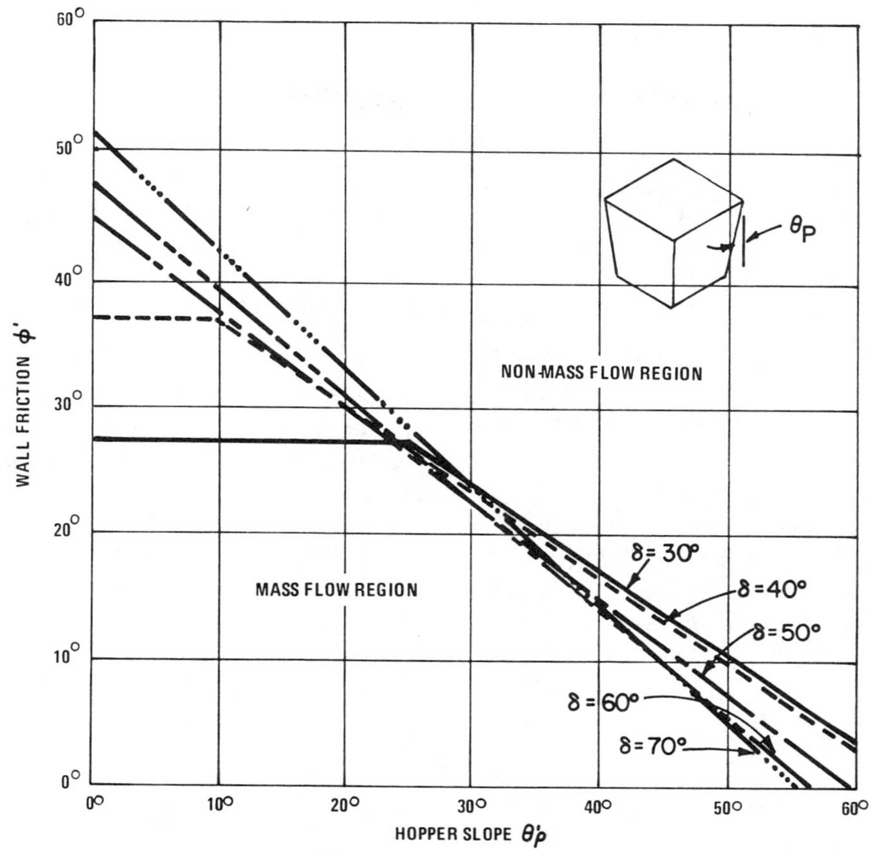

Figure 9.67. Critical wall slope for symmetric plane flow (slot opening) hoppers for δ values from 30 to 60°.

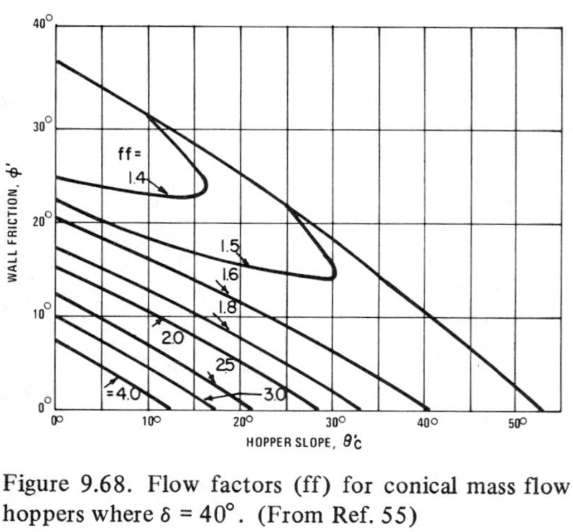

Figure 9.68. Flow factors (ff) for conical mass flow hoppers where δ = 40°. (From Ref. 55)

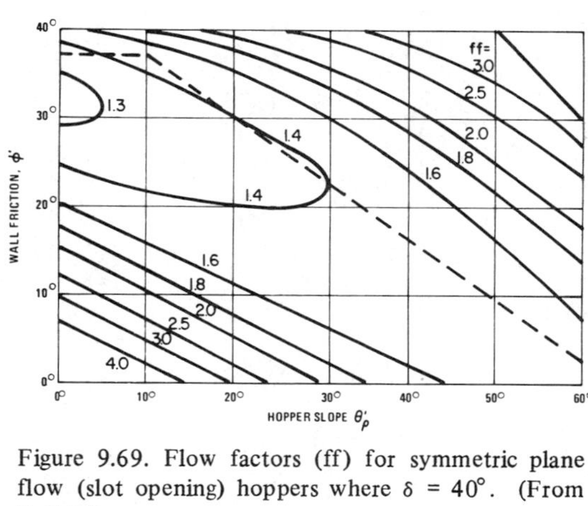

Figure 9.69. Flow factors (ff) for symmetric plane flow (slot opening) hoppers where δ = 40°. (From Ref. 55)

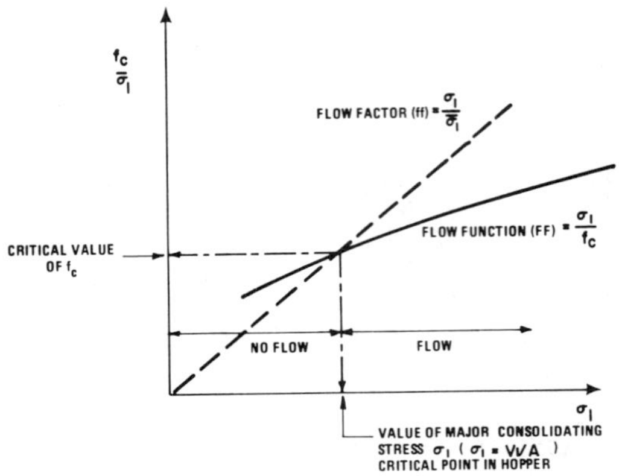

Figure 9.70. Flow factor–flow function relation.

measured value of ϕ' on the appropriate chart for the measured δ value. (Typical charts are shown in Figs. 9.68 and 9.69). Axisymmetric conical hopper slopes are noted as θ'_c, plane flow hopper slopes as θ'_p, and both are as measured from the vertical plane as shown in Figure 9.71.

It is recommended that the maximum hopper angle θ'_c used for design of conical hoppers be 4° less than the angle obtained from the boundary on the chart, as a safety factor to allow for differences in wall friction as measured in the shear tester and that which might occur under actual conditions in the plant. The limits on the region of mass flow in a plane flow hopper are much wider than in conical flow, so the safety factor is not used when selecting a value of θ'_p.

Minimum Discharge Opening. The hopper discharge opening must be large enough such that any arch that tends to form will fail. This opening dimension B(m), is found by:

$$B \geqslant \bar{\sigma}_1 H(\theta')/\gamma \tag{9.38}$$

where

A = Area of the shear cell ring (m^2)

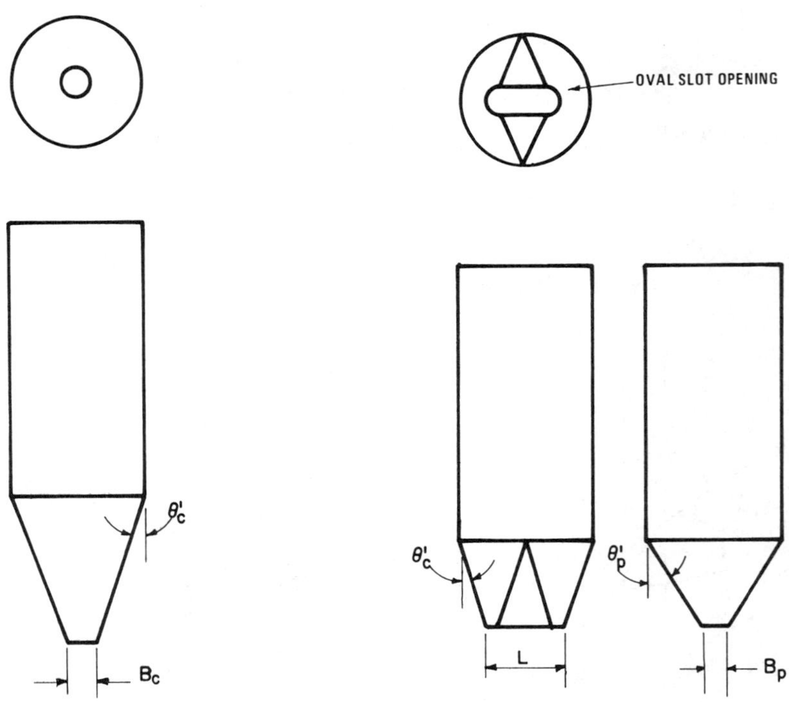

Figure 9.71. Various mass flow hopper geometries: (a) conical hopper, (b) transition hopper, (c) conical with transition hopper, (d) chisel hopper, (e) wedge hopper.

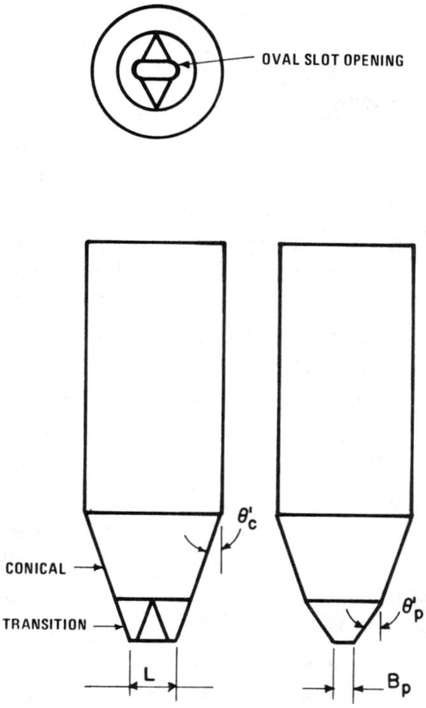

(c)

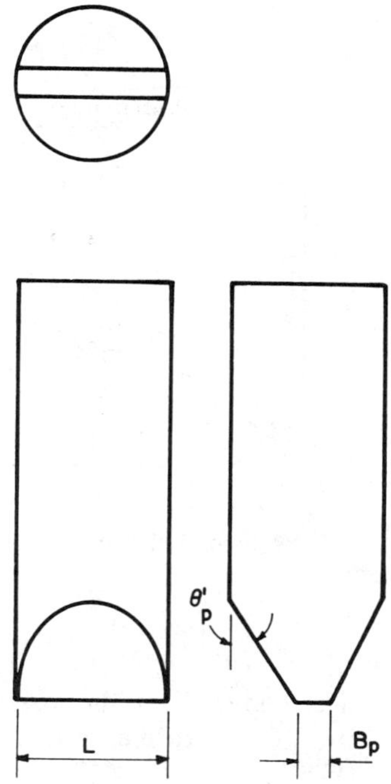

(d)

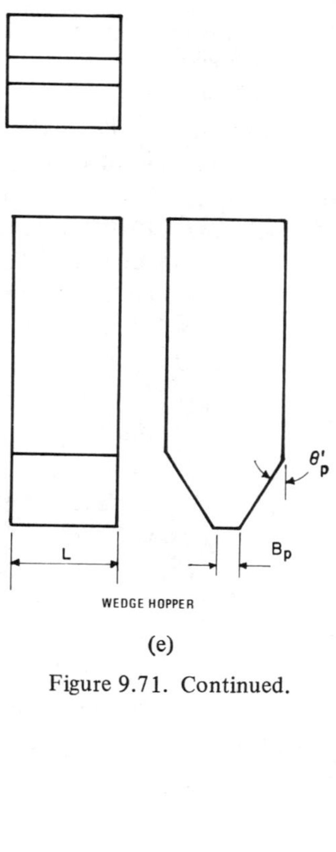

(e)

Figure 9.71. Continued.

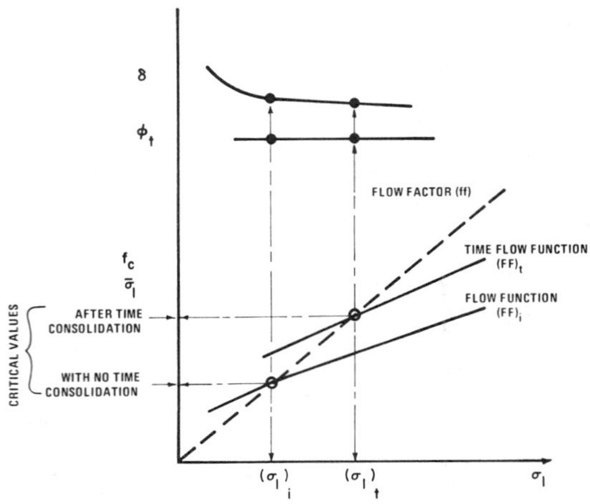

Figure 9.72. Flow factor with instantaneous and time-flow function.

$\overline{\sigma}_1$ = stress acting on the obstruction, determined from the intersection of the FF and ff lines as in Figure 9.72 (kPa).

γ = bulk density (kN/m^3)

$H(\theta')$ = function as given in Figure 9.73 for conical, rectangular and square openings.

B = diameter of a circular opening or width of a rectangular opening (m). Where a rectangular opening is used, the ratio of length to width should not be less than 3 to assure flow at the sides of the channel. For full slot wedge bins this ratio should be at least 6 to insure against funnel flow higher within the hopper.

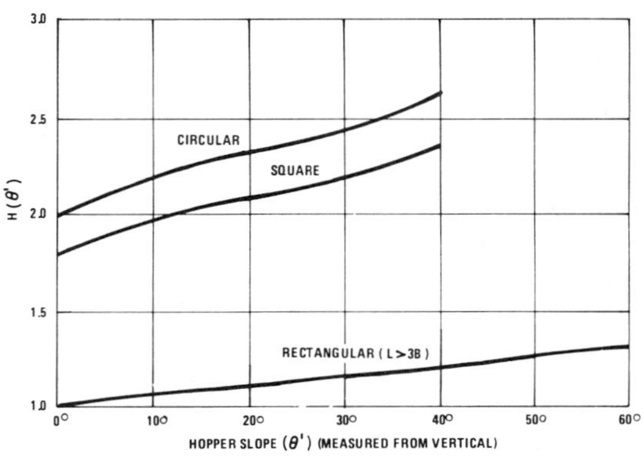

Figure 9.73. Function H(θ'). (From Ref. 55)

If FF lies completely below ff, the hopper opening can be very small and is not detectable with the flow analysis. If the instantaneous FF lies below, and the time FF lies above, then it is usually possible to use vibration or other means to start flow after time consolidation and thus return the material to the instantaneous flow condition. If vibration is to be used to initiate flow, a safety factor is applied to allow for possible compaction during vibration. Select a value of $\overline{\sigma}_1 = 1.5f_c$ to be used for the calculation of dimension B (where f_c is determined from the intersection of the ff and FF$_i$). If the instantaneous FF lies completely above ff, gravity flow is not feasible and mechanical aids will be required.

If the wall yield locus is pressure dependent and curved convex upward, then ϕ' will be higher in the lower regions of the hopper during flow. The value of ϕ' for the region near the hopper opening should be used to determine the flow function. To determine ϕ' at the hopper opening, first estimate a value for the discharge opening B and calculate the value of $\overline{\sigma}_1$, from formula 9.38

$$\overline{\sigma}_1 = \frac{B\gamma}{H(\theta')} \qquad (9.39)$$

From the intersection of $\overline{\sigma}_1$, and the estimated ff line, determine σ_1 and estimate δ (Fig. 9-72). Construct the EYL, note point σ_1, construct the Mohr semi-circle through this point tangent to the EYL (Fig. 9-57). The line through the origin and the point of intersection of the WYL and the Mohr semi-circle determines the value of ϕ'. A flow factor is determined using this value. If it varies by more than about 10% from the assumed value, a new estimate for B is made and the calculation repeated.

Surface Finish of Hopper Wall. A low solid-wall friction ϕ' is preferred for the *hopper* section of a mass flow bin. It permits larger hopper slope angles θ' and thereby reduces the overall height of the hopper.

A variety of hopper wall lining materials can be used to reduce wall friction and adhesion. These include Teflon®, glass, various epoxy paints, stainless steel, and ultra-high molecular weight (UHMW) polyethylene. The last four are the most commonly used.

It is not possible to make generalizations such as use of "smoother" surfaces will always result in lower solids-wall friction. Each solid must be tested. Hopper wall materials with the same apparent smoothness may exhibit widely differing kinematic angles of friction on dry solids.[2,74] Particle shape, size, and hardness, wall surface hardness, and surface profile all interact.

It should be recognized that abrasive materials sliding over the surface of mass flow hoppers may cause wear. Abrasive resistant plates, replaceable metal liners, and glass blocks have been used for this reason when handling abrasive ore, minerals, coal, and so on.

Care should be taken to ensure that a wall finish specified in the design is reproduced in the actual bin.

- Substitute "similar" paint or coated surfaces only if samples have been submitted and tested.
- When using unpainted carbon steel, use the WYL for rust-coated steel unless precautions are taken to present rusting. Mass flow may not start at all if the rusting is sufficient to prevent sliding of the material during initial use. If mass flow is established, the rust film will be those calculated on the basis of clean steel WYL.
- Specify surface finish of stainless steel sheets, and plates can be obtained in several different surface finishes (smoothness). The standard commercial finish usually described as a No. 1 finish has a surface profile ranging from about 150-500 μAA (micro inches, arithmetic average) and is standard on most sheet and plate. Sheets up to $\frac{3}{16}$ inch thickness can be furnished in a 2B finish, having a 5-15 μAA profile. Plates can sometimes be furnished with a 2D surface (40-60 μAA profile). Flat sheets and plates can also be polished to any desired surface finish before forming and fabrication.
- Construct the hopper so no ledges are presented to the flowing material. With lap welded construction, overlap plates in the direction of flow. Grind circumferential welds flush. Fasten any interior liners with countersunk or shallow-head fasteners.

Surface Finish of Vertical Section. Smooth walls on the vertical part of a silo may *not* be desirable. As the solids-wall friction in the vertical part of the silo is reduced, more of the consolidating stress from the stored material is transmitted directly to the material in the converging hopper below. This could cause arching across the silo at the transition between cylinder and hopper.

An inspection of Figures 9.68, 9.69, and 9.73 will show that as θ_c' decreases to zero (vertical wall) the function $H(\theta')$ decreases and the flow factor (ff) increases, indicating a decreasing flowability of the channel. The vertical wall flow factor can be superimposed on the flow function as before to determine if the material will gain sufficient strength to arch across the bin at the transition (the value of ϕ' appropriate to this area of the bin must be determined if the WYL is pressure sensitive).

If the time flow function FF_t continues to rise steeply at high pressures, even though it lies below the flow function (ff) (and indicates a small or zero arching diameter) arching may still be possible at the cylinder cone transition. This condition is described in Reference 9-109.

If it is suspected that smooth walls in the vertical portion of a silo will cause arching at the transition, the vertical wall specification should call for a rough surface, or the surface should be roughened by attaching internal bars or ledges for a distance of about one diameter above the transition.

Funnel Flow Bins

To insure gravity flow from a funnel flow bin, the discharge opening must be large enough to prevent a rathole or arch from forming. Outlets used in funnel flow bins are shown in Figure 9.74.

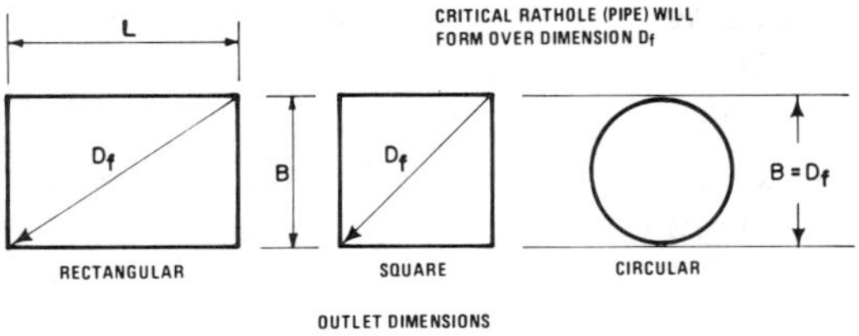

Figure 9.74. Outlet dimensions used in the analysis of funnel flow: (a) rectangular, (b) square, (c) circular.

Ratholing. The bin or hopper opening must be larger than the critical rathole (piping) dimension D_f. At this critical dimension, the stress imposed on the material will exceed the yield strength, and any rathole that tends to form will continually collapse.

The critical flow properties of materials determined from the shear test as described so far are based on steady-state flow conditions. As pointed out earlier, and in Section 9.11, initial pressures caused by filling a bin, without withdrawing solids, result in an active, peaked stress field and higher consolidating pressures on the material near the outlet. The material at the bottom of the bin, therefore, gains greater strength during initial filling than during steady flow.

In a mass flow bin, any obstructions to flow caused by initial filling will fail, and the critical flow properties are determined from steady-state conditions. This is not true in a nonmass flow bin. Therefore, the strength of the solids, the ability to support a rathole, and the minimum rathole diameter must be calculated on the basis of initial filling as well as steady-state flow.[56]

Under initial filling conditions (with no withdrawal) the critical rathole diameter D is determined first calculating the consolidation stresses on the material at the hopper opening. In bins where the height to diameter ratio exceeds 1, use Janssen's equation (Equation 9.1) to calculate q, using Janssen's k value = 0.4 and assume:

$$q = \sigma_1 \tag{9.40}$$

Transposing this value of σ_1 to the time flow function chart, determine a value of $\bar{\sigma}_1$ from the intersection of σ_1 and FF_t. Calculate the critical piping diameter D_f(m) as:

$$D_f = \bar{\sigma}_1 G(\phi_t)/\gamma \tag{9.41}$$

where

$G(\phi_t)$ = function of ϕ_t as given in Figure 9.75.
γ = bulk density (kN/m^3)

If the bin height to diameter ratio is 1 or less, use:

$$q = H\gamma \tag{9.42}$$

where H = height of material surface above the opening (m)

Since the major consolidating stress is lower in the upper part of the bin, the solids strength is correspondingly lower. Therefore, it is possible to have material flow in the upper part without piping but when the level drops to a certain height, a stable pipe can form as shown in Figure 9.76. The height at which this occurs can be estimated as follows. Restate formula (9.38):

$$\bar{\sigma}_1 = D_f\gamma/G(\phi_t) \tag{9.43}$$

Enter $\bar{\sigma}_1$ on the time flow function chart and determine the corresponding value of σ_1. Use this value of σ_1 to calculate H_1, the distance measured from the top of the material over

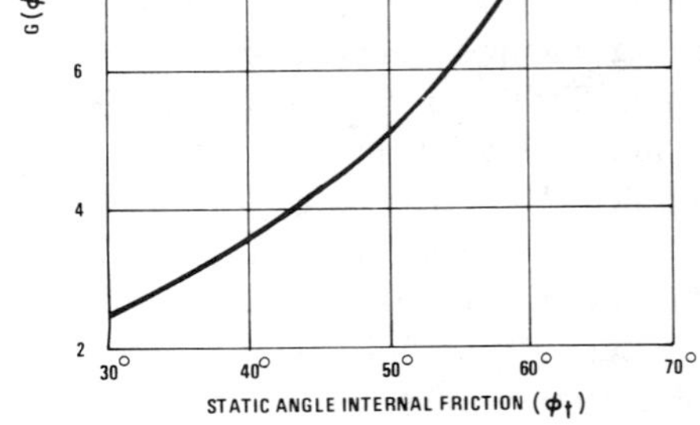

Figure 9.75. Function G(ϕ_t). (From Ref. 55)

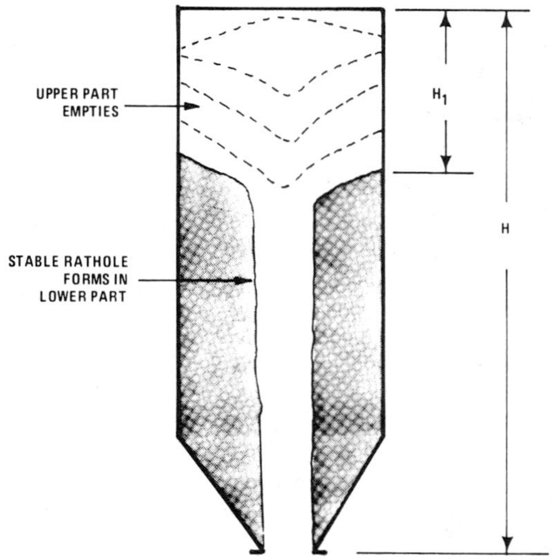

Figure 9.76. Formation of a stable rathole after upper part of bin empties to a critical level.

which flow will occur without ratholing:

$$H_1 = \frac{\sigma_1}{\gamma} \tag{9.44}$$

When the material level drops below this level, a rathole may form.

In the case where the solids are being continuously removed from the flow channel in a nonmass flow bin, while it is being simultaneously filled, the surrounding material is not subjected to high initial filling pressures. A flow factor for ratholing, ff_p, based on steady-state flow conditions, calculated by Jenike and presented in Reference 55, is reproduced in Figure 9.77. Using the ϕ_t value determined from the shear test, the appropriate ff_p is determined from the chart, and superimposed on the FF_t chart as described earlier for mass flow. The critical piping dimension D_f (m) is determined by:

$$D_f = \bar{\sigma}_1 G(\phi_t)/\gamma \tag{9.45}$$

where

$\bar{\sigma}_1$ = stress determined from the intersection of the FF and ff_p lines as in Figure 9.71.

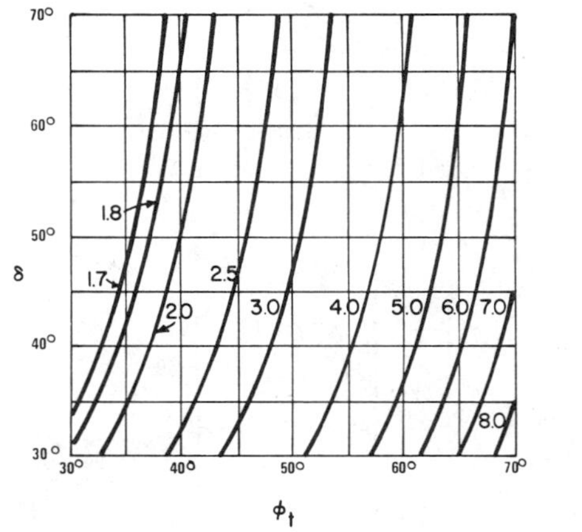

Figure 9.77. Flow factor for no-piping (ff_p). (From Ref. 55)

$G(\phi_t)$ = a function of ϕ_t as given in Figure 9.75.

γ = bulk density (kN/m³)

To ensure against ratholing, the outlet dimension, D_f is normally calculated on the basis of the largest critical rathole that can occur. This is usually that caused by *filling conditions* rather than steady-state flow.

Arching. Arching will not occur over a square or circular outlet with a D_f dimension sufficient to prevent piping (Fig. 9.74). Where D_f is the diagonal dimension of a square or rectangular opening, the minor dimension B must be sized so that arching does not occur. It is calculated as:

$$B \geq 1.15\, \bar{\sigma}_1/\gamma \tag{9.46}$$

where $\bar{\sigma}_1$ is determined from the intersection of the flow function and a flow factor of 1.7

Expanded Flow

The major diameter of the mass flow section must be at least larger than the critical piping diameter for the material; the outlet must be larger than the minimum arching dimension. Typical expanded flow bin geometries are shown in Figure 9.78.

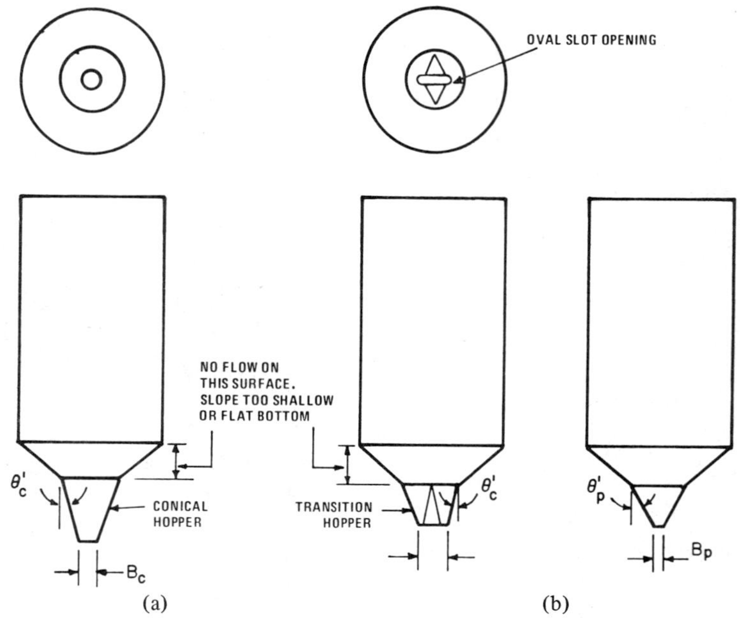

Figure 9.78. Expanded flow bin geometries: (a) conical hopper, (b) transition hopper.

Expanded flow can be achieved with an increased slope at the lower end of a hopper, but it is not the only way. A funnel flow hopper can be converted to expanded flow by applying a low-friction surface material, epoxy coating, plastic linings, glass coating, polished stainless steel, etc, to the lower wall surfaces. This technique can be particularly useful where ϕ' is pressure sensitive. The proper wall coating is selected by evaluating the WYL for the particular surface and the stresses expected in the low pressure regions near the outlet where the coating will be used.

When a funnel flow bin is converted to mass flow, the effect of higher wall stresses must be considered. The stable or unstable portions of the flow channel may expand to reach the walls in the vertical portion and impose higher wall stresses, as described in Section 9.11.

Summary

A number of workers have evaluated the Jenike method with model bins. Wright,[75,76] testing a number of different iron ores in wedge-shaped hoppers with various slopes, reported that this method provided a sound basis for functional design of mass flow bunkers handling ore under dynamic conditions and provided a reasonable safety factor for engineering design. The predicted critical outlet size agreed closely with experimental results, but the predicted hopper slopes were 5-10° steeper than those found by test to be required for mass flow. The same conclusion regarding plane flow hopper slopes was reported by Eckhoff and Leversen.[79] These results are consistent with theory, since it is known that while the regions of mass flow are quite restrictive in conical-flow channels, they are much wider in plane-flow channels. Accordingly, the Jenike flow factor charts for plane flow contain a safety factor to allow for variations in solid head in the vertical portion of the bin. Wright also confirmed that consolidating stresses caused by impact during initial fill must be considered in sizing the discharge opening and reported that arching at the vertical, sloping wall transition can occur.

Richards[86] tested wet and dry sand in symmetrical conical hoppers and reported that the critical hopper slope and outlet for mass flow and the critical outlet for funnel flow determined by test agreed very closely with values predicted by the Jenike method.

Enstad[78,79] pointed out that if the flow function is determined in a region of high stresses where it may be linear and this linear function is extrapolated to regions of low stress, it will intersect the flow factor at a higher, incorrect value. This overestimates the strength of the material in the region of the hopper outlet and will predict an opening considerably larger than that required. Eckhoff and Leversen[77] report similar results when the yield locus was extrapolated into a region of very low normal stress. Jenike, however, suggests that at low normal stresses, the powder in the cell is exposed to tensile stress components that produce measurements at shear stresses at failure that are actually too low. This is avoided by following a test procedure whereby normal stresses imposed on the cell during shear are greater than $\frac{1}{3}$ of the normal stress used for consolidation.

Flow of Large Granular Particles

Noncohesive large granular particles will bridge or arch by mechanical interlocking of particles. They develop very little, if any, unconfined yield strength and cannot by analyzed by existing powder mechanic's theory. Minimum hopper opening size to prevent bridging has been based mainly on rules-of-thumb. Orifice tests with noncohesive granular material reported by Reisner[80] indicate a theoretical *minimum* limit of 3 for the ratio of hopper opening diameter/maximum particle dimension (D/D_p).

The most commonly quoted minimum ratio for design is 5. Schwedes[81] recommends a ratio of 10 to provide a larger factor of safety to assure no mechanical block in a hopper outlet. Peschl[82] studied the flow of coarse granular materials in model bins. He described the flow as being characterized by constant formation and collapse of successive arches, which he termed "dynamical" arches. This is similar to observations reported with flow of granular material by early experimenters. Peschl concluded that the probability of arch formation with coarse granular materials cannot be predicted theoretically, but can be predicted by making a small number of repetitive tests with model hoppers, and statistically, analyzing the data.

9.11. STRESSES ON BIN WALLS

Static and Dynamic Conditions

Stresses on the bin walls are due to combinations of static and dynamic flow conditions that occur during loading and discharge. Extensive bibliographys on this subject are presented in References 11 and 101.

It is now accepted by most workers that wall stresses are influenced by at least three conditions: initial filling, flow, and switch from initial filling to flow. Each of these conditions can be described as follows:

Initial Filling. When a bin or hopper is initially filled, the material contracts vertically or near vertically under the pressure of the solids head with very little horizontal deformation. The major principle stresses therefore are assumed to be aligned along this near vertical direction and form what is defined as an active or "peaked" stress field, as shown in Figure 9.79a, for a mass flow bin and hopper. As the solids settle slightly, slip occurs along the walls and a friction stress develops [Equation (9.5)].

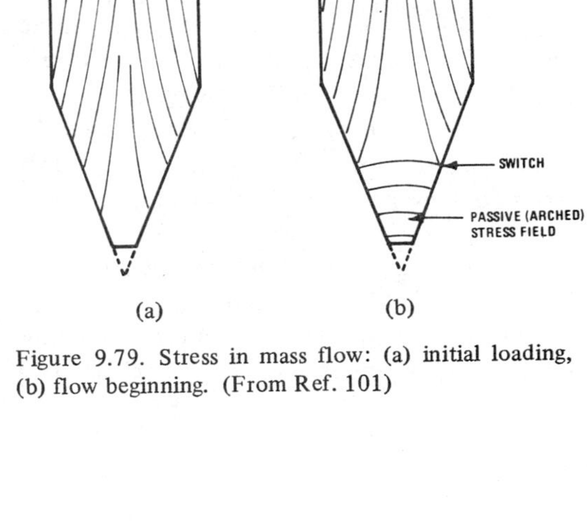

Figure 9.79. Stress in mass flow: (a) initial loading, (b) flow beginning. (From Ref. 101)

Flow Conditions. When the outlet gate is opened, the solids move in a flow channel that converges downward toward the outlet. In order to flow, the solids must expand vertically and contract laterally as the channel forms. This causes the stress field to realign, the major stress now acting in a close to horizontal direction, forming what is defined as a passive or "arched" stress field, as shown in Figure 9.79b.

Switch Conditions. Nanninger[97] observed that at the transition level between active and passive fields, equilibrium of the mass requires an overpressure to occur. Jenike and Johanson[101] and Walters[98,99] postulate that a large transient "switch" stress is developed when flow is initiated in a bin that has been filled without having any solids withdrawn. An active stress field is generated during filling. When discharge begins, the unsupported solids just above the outlet expand downward and a passive stress field develops in that region. As flow continues, the interface between the two fields moves very rapidly upward, to a point where the flow channel intersects the vertical section of the bin.

In a mass flow bin, this point would be at the transition between the hopper and vertical section. In a funnel flow bin, it may not intersect the walls, but if it does, it would be at a point in the vertical cylinder defined as the "effective transition" (Fig. 9.24).

The stress conditions during a switch in a mass flow bin at a point where the switch has reached a certain height are shown approximately in Figure 9.79b. Below the switch, the stresses are in a passive (dynamic) state and the hopper wall stresses are less than those in the (active) state above. The material at the transition between the stress fields is no longer supported by the flowing solids below and the equilibrium of forces results in an additional stress (or "overpressure") at the region of the switch.

The movement of the transitory switch stress, if present, would be very rapid and thus could be very difficult to detect on pressure measuring devices, on models, or full size silos. As a result, very little test data has been published to substantiate the presence of switch stresses. Walker[83] measured high overpressures in a bin which had been partially emptied, then refilled with the outlet closed. Van Zanten and Mooij[87] describe flow pressure variations on mass flow model bins which they attribute to switches in the stress field.

Wall stresses will be higher in a mass flow bin than in a funnel flow bin that has a flow channel that does not intersect the walls. The walls are isolated from these stresses by the stagnant material. The Reimbert antidynamic tube described in Section 9.14 has been used in bins to insure funnel flow and thereby protect the structure from mass flow stresses.

Stress Theories

Very little is known regarding stress conditions within the flowing mass. All useful stress measurements published to date have been made at the bin walls.

Test data from model or full-size bins that confirm the presence of varying wall stress conditions during filling and flow are given in References 83 through 95. It is generally agreed that stresses or wall pressures in *funnel* flow bins have a distribution about as shown in Figure 9.80 and *mass* flow bins have a stress distribution as shown in Figure 9.81.

A number of theories have been proposed to describe these stress distributions. These include those by Walker,[96] Walters,[98,99] Clague,[100] Jenike et al.,[101-104] Enstad,[78,79] and Takahashi et al.[105,106] A summary of the theories of the first five noted above are given by Arnold, McLean, and Roberts.[107]

There is yet no general agreement on a theory that best describes stress conditions in bins under all conditions of flow and as a result, they have not been incorporated in any of the national silo codes.

The theories of Walker and Walters and Jenike and Johanson have been the most widely quoted. Details can be found in the referenced papers. The information they have published is concerned with *concentric* discharge. It is recognized that much of the silo overstressing and failures have been caused by *eccentric* flow patterns induced by eccentric single or multiple discharge openings. These conditions impose severe, unbalanced lateral stresses along the horizontal cross sections of a bin, particularly where the flow channel intercepts the wall. The points of interception are often difficult to determine. This area of study remains the least explored and requires the most caution on the part of bin designers.

Walker and Walters. Walker proposed an approximate theory to describe stresses and arching in hoppers and bins, and presented a considerable amount of data derived from tests with wet and dry coal in a 1.8 m diameter and 1.8 m square bins. His data confirmed that the stress fields developed during filling and dis-

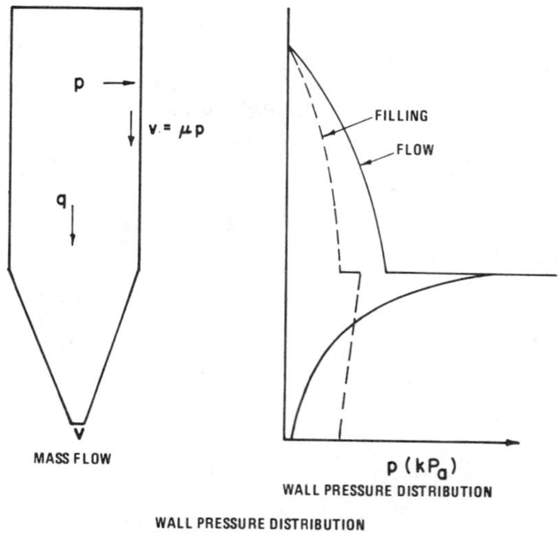

Figure 9.81. Wall pressure distribution–mass flow: (a) mass flow bin, (b) wall pressure distribution.

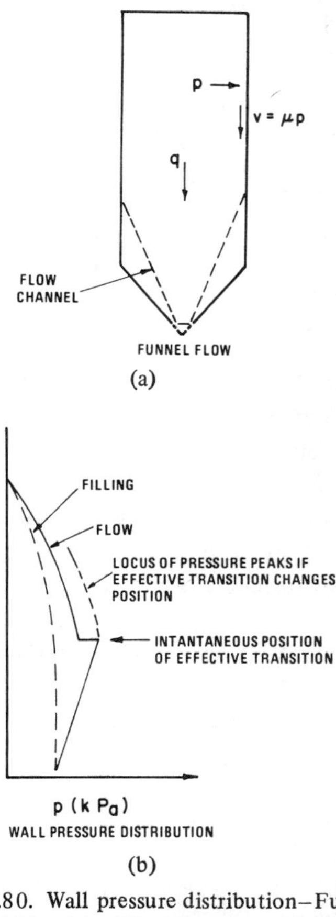

Figure 9.80. Wall pressure distribution–Funnel Flow: (a) funnel blow bin, (b) wall pressure distribution.

charge are very different; withdrawing a small amount of material while filling a bin significantly reduces the high initial pressures found near the hopper apex; during loading with no withdrawal, initial pressures increased with depth of fill; flow pressures in the lower region of a mass flow bin were linearly proportional to height above the hopper apex and independent of depth of fill (evidence of a radial stress field). Flow pressures were independent of flow rate and once established by withdrawal they remain, even when withdrawal is interrupted for a period of time.

Walters extended the Walker theory to distinguish between the stresses developed during filling and flow in conical hoppers and in conical hoppers with vertical sections above. He also proposed an approximate method of calculating the "switch stress." Clague extended it to plane flow bins. Arnold and Roberts[107,108] integrated these theories with Jannsen to propose a generalized theory for predicting wall stresses in mass flow bins.

Jenike, Johanson, and Carson. The authors measured wall stresses on 300 mm diameter model bins handling sand and coke. They found widely varying pressure fluctuations in the cylinder portion during flow, which they

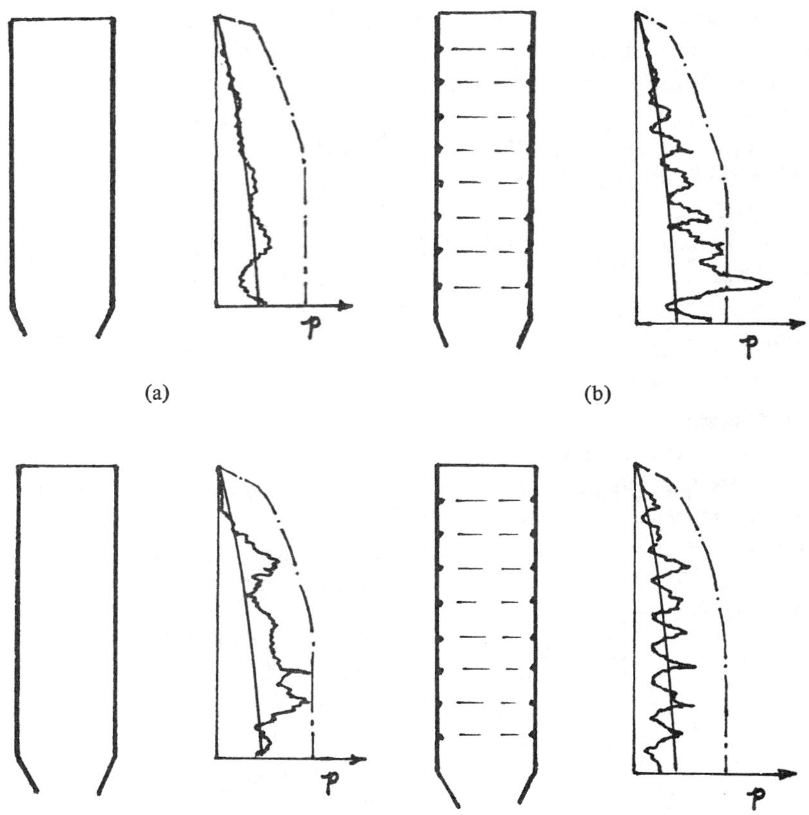

Figure 9.82. Wall pressures on model bin handling sand: (a) diverging 1/2° no ledges, (b) diverging 1/2° with ledges, (c) converging 1/2° no ledges, (d) converging 1/2° with ledges. *Pressures calculated by Jenike, Johanson and Carson strain energy theory. †Pressures calculated by Janssen theory. (From Ref. 102)

attributed to very slight deviations from perfect uniformity in the shape of the bin cross section. (These results and conclusions were confirmed by Van Zanten et al.).[87,88] Wall pressures were measured on models with nondiverging cylinders, with cylinders having surface imperfections caused by weld shrinkage on girth seams, with continuously converging cylinders, and with cylinders having internal ledges or constrictions. Typical results are shown in Figures 9.82 and 9.83. The authors concluded that deviations in cylinder geometry cause changes in pressure and strain energy in the bulk solids during flow, and they used strain energy as a basis for a theory for predicting wall pressures.[102-104]

In 1980, Jenike[109] proposed a design method for calculating wall loads in cylinders and circular hoppers. This method is based on Janssen's equation for the cylinder with upper and lower bounds placed on values of k and μ to take into account irregularities in cylinder construction and a modification of Jannssen's equation for calculating pressures at the transition.

The locus of pressures on bin walls during initial filling and during flow in cylindrical bins, according to the Jenike (1980) method, is shown in Figure 9.84.

Mass Flow. During filling without withdrawal, a Janssen pressure distribution (Formula 9.1) is used with a Janssen $k = 0.4$ (see Fig. 9.84b).

Since all cylinders converge and diverge along

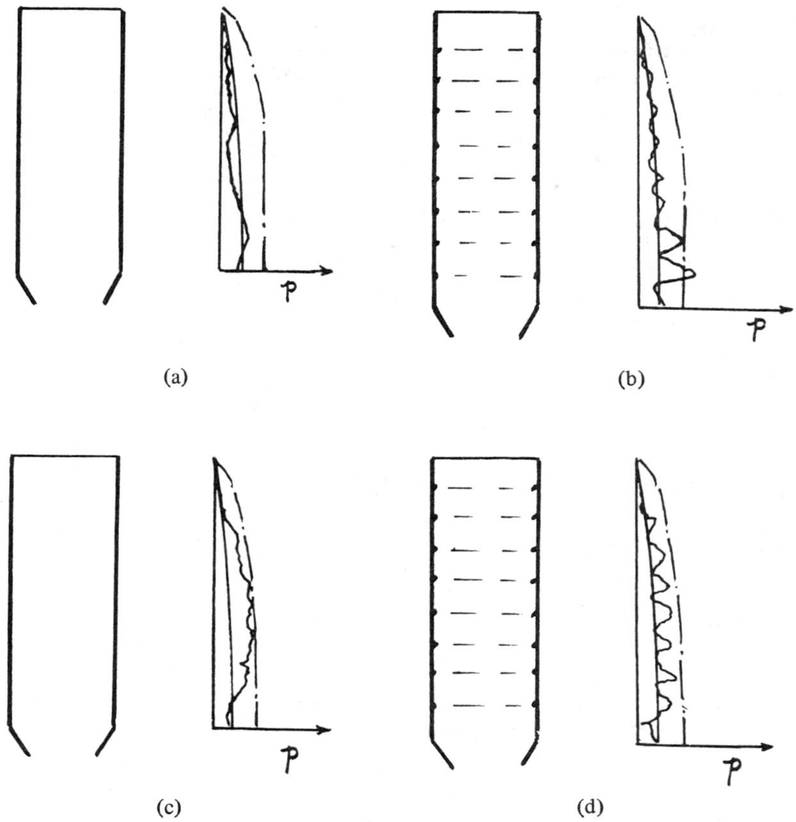

Figure 9.83. Wall pressures on model bins handling coke: (a) diverging 1/2° no ledges, (b) diverging 1/2° with ledges, (c) converging 1/2° no ledges, (d) converging 1/2° with ledges. (From Ref. 102)

their length due to manufacturing tolerance, Jenike describes the flow conditions as those where the pressure p increases to contract a solid laterally when a layer passes a convergence and decreases when the layer passes a divergence. This produces a varying value for Janssen's k. In addition, wall friction may vary between kinematic conditions (μ_k = tan ϕ') and static conditions (μ_s = tan ϕ_t').

A Janssen pressure distribution is therefore assumed for the cylinder, using the following bounds:

$$0.25 > k > 0.6$$

$$(\mu_k - 0.05) \leq \mu \leq (\mu_s + 0.05)$$

Therefore, for design purposes, the minimum product of $k\mu$ = 0.25 (μ_k - 0.05) gives the maximum value of q, and a value of $k = 0.6$ gives the maximum value of p.

As the flowing layer passes the transition, the pressure increases, taking the form shown in Figure 9.84c. Jenike theorizes that the overpressure at the transition reflects onto the opposite side of the cylinder to cause an overpressure in this area. (Fig. 9.49b) In the reference paper, Jenike gives suggestions for calculating this transition overpressure and its reflected overpressures.

Funnel Flow. In the stagnant areas, the cylinder deviations do not affect the wall pressures because there is no sliding along the walls. The wall pressures are calculated from Janssen, using $k = 0.4$.

If the flow channel intersects the cylinder wall, an "effective transition" is formed, and

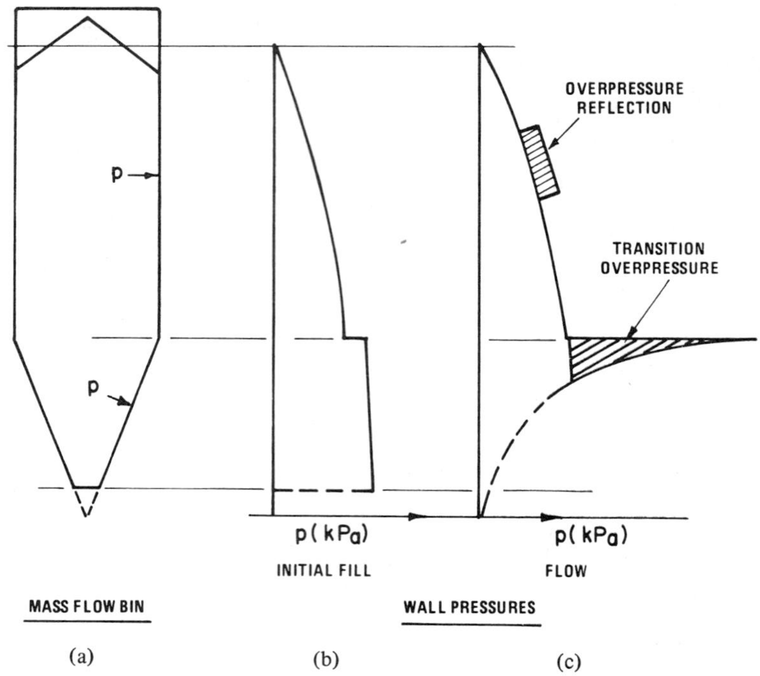

Figure 9.84. Wall pressures–Jenike pressure theory: (a) mass flow bin, (b) wall pressure, initial filling, (c) wall pressure during flow. (From Ref. 109)

an overpressure occurs due to the conveying flow channel. This reflects to the opposite side of the cylinder as an overpressure. Suggested methods of calculating these pressures and locating the reflected overpressure are given by Jenike in the referenced paper.

9.12. EFFECT OF THE GAS PHASE

The previous paragraphs on flow in silos were based on single phase flow under gravity forces, and the effect of the gas phase was not considered.

During loading and subsequent settling, the gas entrapped in a solids bed can have a significant effect on silo wall pressures and solids flow behavior, a fact recognized in the silo design codes. Entrapped gas in fine powders can be retained for an appreciable time and can cause a significant increase in lateral wall pressures in a silo. A mathematical analysis of this problem and proposals for solutions together with data from tests on model silos are presented in References 62 and 110.

Interstitial gas pressures within a solids bed change during flow and affect the discharge of fine particles from hoppers.[111,112] Wlodarski and Pfeffer[113] showed that air pressures within a flowing solids bed in a 90 mm tube exhibited a significant axial gradient. In their test, the solids head was maintained constant and air pressures along the axis of the tube were measured through hypodermic needles inserted through the tube walls. They found that a low pressure region developed just above the orifice (Fig. 9.85). When the lower needle was vented to atmosphere, the flow rate through the tube *increased* as air entered to restore this region to atmospheric pressure.

As mass flow silos have come into widespread use, it has been found that discharge of fine powders from these silos can become feed rate limited due to changes in pressure of the gas entrapped in the powder interstices. Figure 9.86 illustrates conditions occurring in a mass flow silo during flow of a fine powder. As a solid element flows through the silo, pressures on the element change, increasing with depth, reaching a maximum, then decreasing towards zero as the element approaches the hopper

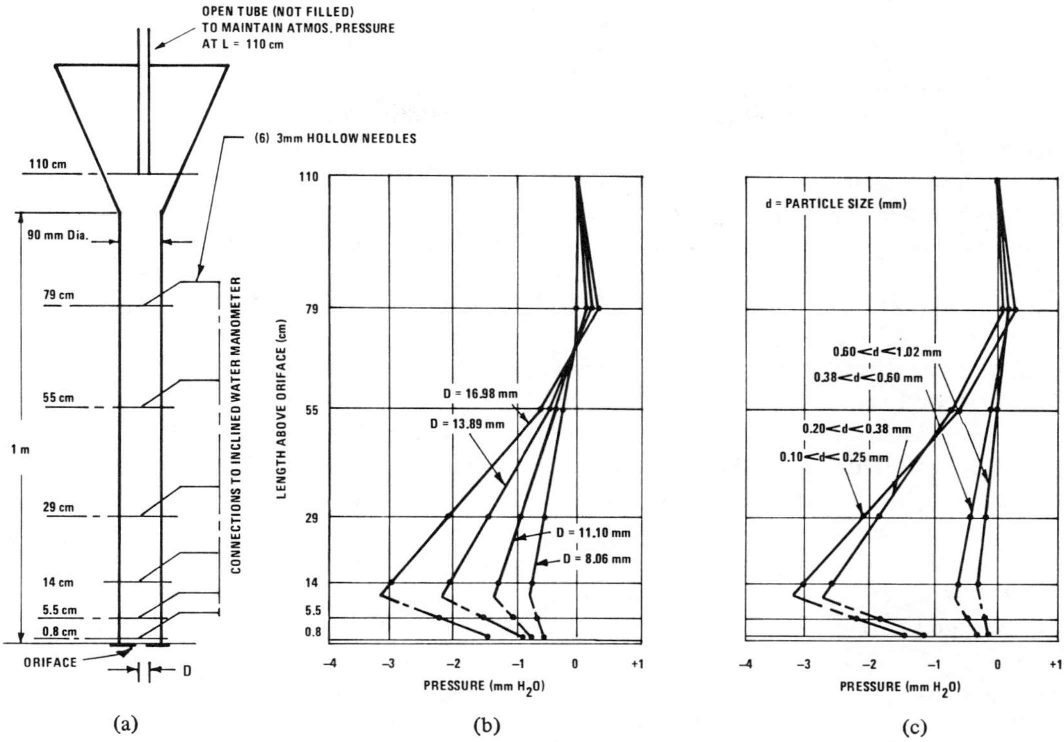

Figure 9.85. Air pressure measured during flow of sand through a model bin: (a) arrangement of test apparatus, (b) air pressure measured (in center) at varying heights, with sand particles having a size distribution d of 0.10 < 0.25 mm, and oriface diameters D varying from 8.06 mm to 16.98 mm, (c) air pressure measured (in center) at varying heights, with an oriface diameter D = 16.98 mm and various sand particle size distributions d. (From Ref. 113)

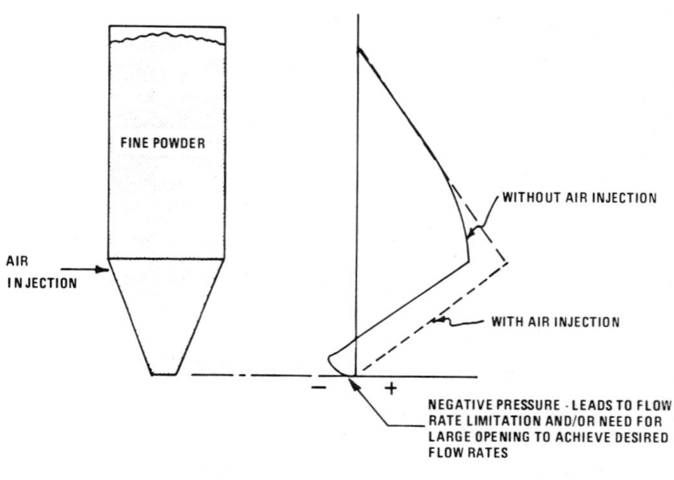

Figure 9.86. Air pressure gradient that may occur in a mass flow bin: (a) mass flow bin, (b) air pressure in flowing mass.

vertex. Bulk density changes with pressure. As the pressure increases, the element is compressed, reducing the voidage and squeezing air out through the top of the mass. As the solid element moves down through the converging hopper, the pressure decreases and the element expands as it approaches the outlet. The voidage volume increases and the interstitial air pressure decreases below atmospheric. As solids discharge from the outlet, air flows up through the outlet into the low-pressure region. This counter-flowing air significantly slows the free discharge of the solids.

This phenomena was pointed out by Bruff and Jenike in describing a design of a mass flow hopper for ground anthracite[114] in which air injection was used to overcome a powder flow limitation at high discharge rates. It has been found that injection of very small amounts of gas at appropriate points in a silo (Fig. 9.86) provides some support for the solids in the vertical portion of the silo, thereby reducing compaction and subsequent expansion in the lower regions of the hopper and supplies gas to the intertices preventing the pressure from dropping below atmospheric.

Several recent papers describe examples of air injection used to significantly increase the rate of solids flow under gravity force from mass flow silos and hoppers.[115, 116, 117, 118] It should be stressed that the gas injection is accomplished with very small gas flows (.03–0.3 m^3/min), which are substantially less than that required to fluidize the solids bed.

A recent paper by Johanson[119] discusses the use of continuum mechanics and test model data to predict the effect of the solid-gas interactions on the flow characteristics of solids.

9.13. PARTICLE SEGREGATION DURING STORAGE AND FLOW

Whenever particulate solids are moved, deposited on piles, or withdrawn from silos, there is a tendency toward segregation. In many cases, this is undesirable, and the storage and handling system must be designed to minimize the possibility of its occurring.

Segregation occurs most frequently in free-flowing granular materials having a wide size distribution and seldom in fine powders where particle size is about 70 μm or less.

Cohesive powders usually do not segregate during handling. Powders containing cohesive and noncohesive components can segregate. The more cohesive components tend to move together in relatively thick unsegregated layers or patches when sliding in a chute or on a pile and will form rivulets of nonsegregated cohesive material extending down the face of the chute or pile.

A review of segregation of particulate materials is given by Williams[120] and Johanson.[121] Particle properties that cause segregation are due to differences in particle:

- Size
- Density
- Shape
- Resilience
- Angle of repose
- Cohesiveness

Many workers[122-127] have confirmed that differences in particle size is by far the most important cause of segregation, with differences in particle density and shape (assuming not gross shape difference) being comparatively unimportant.

Mechanisms

The mechanisms leading to segregation of noncohesive particles include:

Percolation of Fine Particles. Fine particles can percolate through the voids of larger particles as they rearrange themselves during a disturbance. This can occur, for example, during shear induced by stirring, shaking, or pouring the particles in a heap, or during flow through a silo.

A well-known example of surface percolation occurs during filling of a silo (see Fig. 9.87a). The particles striking the heap form a thin layer of rapidly moving material. The finer particles in the moving layer percolate into the stationary

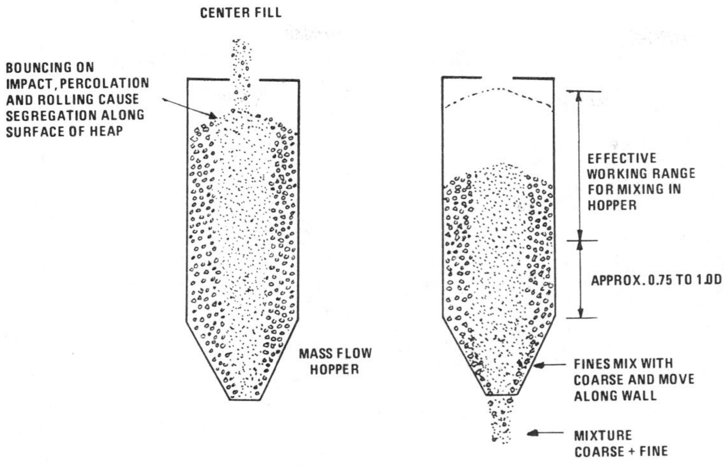

Figure 9.87. Typical segregation and mixing during mass flow: (a) bin filling with no discharge, (b) discharge.

layer below and become locked in position. The large particles do not penetrate and continue to roll or slide to the outside perimeter of the heap. This has been compared to a sieving or screening mechanism. The result is considerable radial inhomogenity.

When the silo discharges, particle rearrangement again takes place. In a *mass* flow silo, remixing occurs as the segregated material leaves the vertical section and enters the mass flow hopper (Fig. 9.87b) where the fine fraction mixes with the coarse fraction.[121]

In a *funnel* flow bin, particle segregation occurs during filling as the particles fall onto a heap (Fig. 9.88a). A central core of finer material is deposited during filling just as it does in the mass flow hopper. However, the mixture ratio exiting the hopper can vary, depending on the rate of refill as shown in Figures 9.88b through 9.88g. If the hopper is drained, the last material to exit the hopper will be mostly coarse. If the level is lowered and then refill is begun, a short-term increase in coarse fraction will be noted at the discharge, until the new incoming material has reestablished the central core flow.

If refill continues at the same rate as discharge and a narrow flow channel has formed, segregation at the outlet will be reduced. This condition will continue until a change in silo level takes place. In time, if the silo is not emptied, fines can percolate into the coarse fraction in the stagnant region and this could cause stable ratholes to form.

Percolation also occurs when a mixture of particles are vibrated or agitated during conveying. This effect can be noticed in vibrating conveyors and chutes and small hoppers that are vibrated to promote flow.

Vibration. Williams[120] describes a condition other than percolation that can cause even a single large particle in a vibrated bed to rise to the surface. Each vertical movement of the bed allows fines to run in under the large particle. As the fine material accumulates and compacts, it supports the large particle, causing it to rise to the surface.

Ahmad and Smalley[125] studied the movement of a single 12,700 μm diameter lead ball in a vibrated bed of 500–600 μm dry sand particles. They reported that at a constant frequency of vibration, segregation increased as acceleration increased, but at a constant acceleration, segregation was reduced as frequency increased. Acceleration was the most critical variable affecting segregation.

Harwood[122] studied the behavior of cohesive and noncohesive powders subjected to vertical vibration using tracer powders comparable in size to the powder bed to determine segregation. He reported that particle size was the major

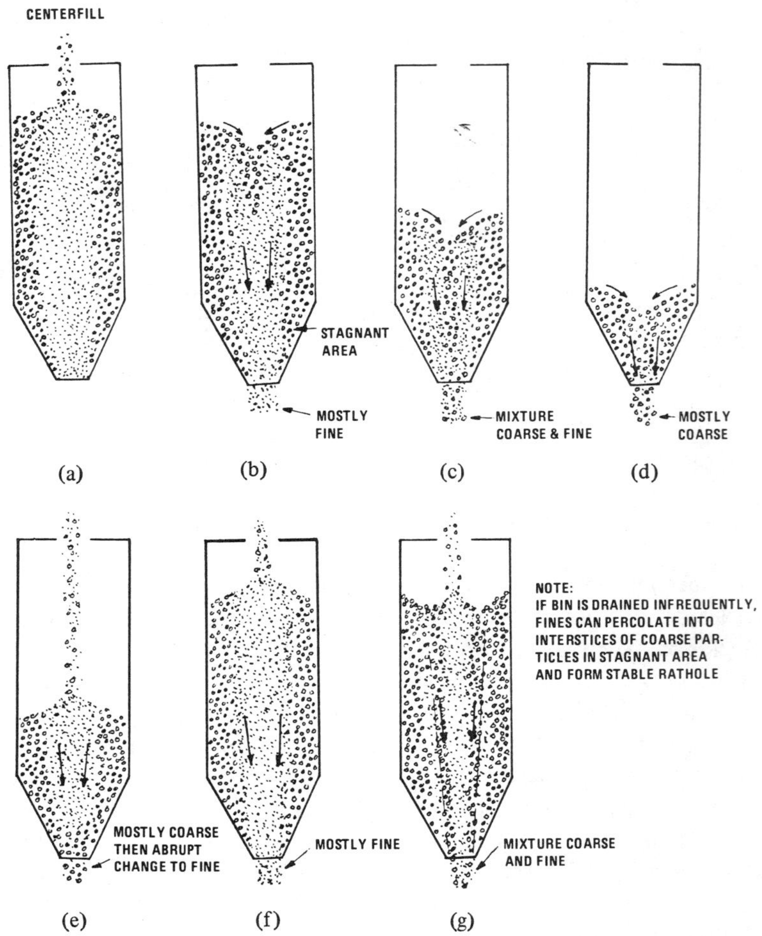

Figure 9.88. Typical segregation and discharge patterns during funnel flow: (a) center filling, no discharge, (b) discharge begins, (c) discharge continues, level in bin dropping, (d) level continues to drop, heel discharging, (e) start to refill before heel is completely discharged, (f) level rising (g) discharging at the same rate as filling, level remains unchanged. Note: Flow patterns shown are typical for a funnel flow bin when any free flowing (segregating or nonsegregating) material is stored.

controlling factor for segregation. In a binary system of free flowing and cohesive powders, segregation was very limited once the powder bed had become compacted, but if vibrational energy was sufficiently high to induce a semi-fluidized state in the bed, segregation was significantly increased.

Storage silos do not usually experience vibration with sufficient intensity to cause segregation, but small feed hoppers and chutes can.

Trajectory of Falling Particles. Material projected from a conveyor or chute onto a heap can segregate before impact due to differences in particle size, particle density and occasionally because of air drag effects.

Usually, if the material has segregating tendencies it has already occurred on the incoming conveyor or chute because of the mechanisms described above, and the trajectory of discharge serves to preserve this separation (Fig. 9.89).

Guidance on estimating the trajectory of material discharged from the head pulley of a belt conveyor can be found in Reference 128. Calculated trajectories of a single fine particle usually have no practical significance in this case since it is not possible to account for the effects of air turbulence and particle to particle contact in a dense falling stream of material.

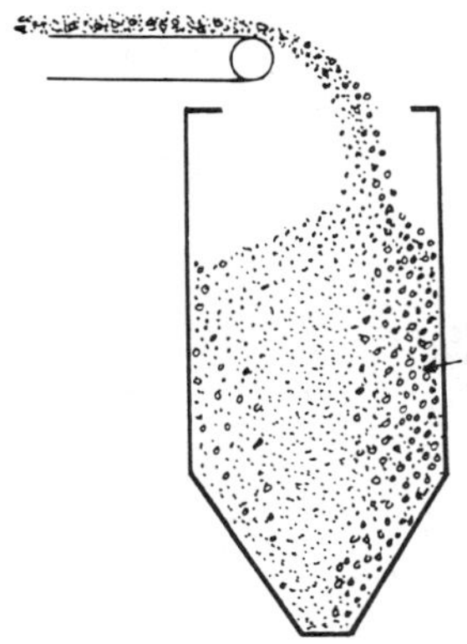

Figure 9.89. Trajectory segregation.

Impact on the Heap. After impacting on a pile, large coarse particles will tend to roll or slide over smaller coarse particles to concentrate on the outside. The more resilient larger particles will tend to bounce and also concentrate along the outside of the pile, while the smaller, less resilient particles will tend to concentrate in the center.

If the mixture contains sufficient moisture, the fine fraction will tend to stick on impact, and large particles will roll or bounce away.

If a mixture of fine powder and coarse particles are impacted directly onto a heap after discharge from an air slide,® or pneumatic conveyor, the fines fraction can become aerated during a free-fall and, on impact, will assume a very shallow or zero angle of repose. The heavier, coarse particles will concentrate in the impact area.

If the silo is loaded with a pneumatic conveyor impacting on the bed, the fine material can remain fluidized or can remain entrained in the moving air stream above the bed until loading is completed and the pneumatic conveyor shut down. If the silo is then completely emptied, the fine fraction that has settled at the top of the silo will discharge in a mass.

Angle of Repose. When a mixture of uniformly sized granular particles consisting of components with different angles of repose is poured on a heap, the particles having a steeper angle of repose tend to concentrate in the center of the heap.

Theoretical Analysis

Segregation is usually studied by sampling from a model bin or from a full-size bin and by reporting the results on a statistical basis. No theoretical basis for analyzing segregation mechanisms has yet been formulated, although some work in this area is beginning. Theoretical models to describe segregation by percolation have been proposed by Shinohara et al.[129, 130] and by particle size and density by Tanaka.[131] Matthee[132] has proposed an approach to modeling all aspects of segregation.

Minimizing Segregation

Particle rearrangement and segregation will occur each time a material is dumped onto a conveyor, a chute at the loading or transfer points. Knowing the likely segregation mechanisms that will be present, the probable distribution of coarse and fine particles across and along the length of the conveyor incoming to the silo can be predicted with reasonable certainty. This incoming stream then must be redistributed or mixed in the silo if a nonsegregated mixture is required at the silo discharge.

As long as the solids level in a mass flow bin remains above the transition, a distance equivalent to about three quarters the diameter, the material moves down the vertical section in plug flow with radial segregation unchanged. However, radial mixing will occur in the hopper section before discharge, as noted earlier in Figure 9.87.

Mixing will not occur in a funnel flow bin. To minimize segregation in these bins, some means of redistributing the incoming material and/or changing and internal flow pattern are required. A moving fill spout (Fig. 9.90a), a fixed deflector, flow spitter, or multiple loading spouts (Fig. 9.90b) have been used to distribute incoming material on the heap. A patented rotating device for this purpose is shown in Figure 9.91.

Devices to reduce segregation by changing the flow pattern are, in essence, designed to simulate mass flow as much as possible. An insert

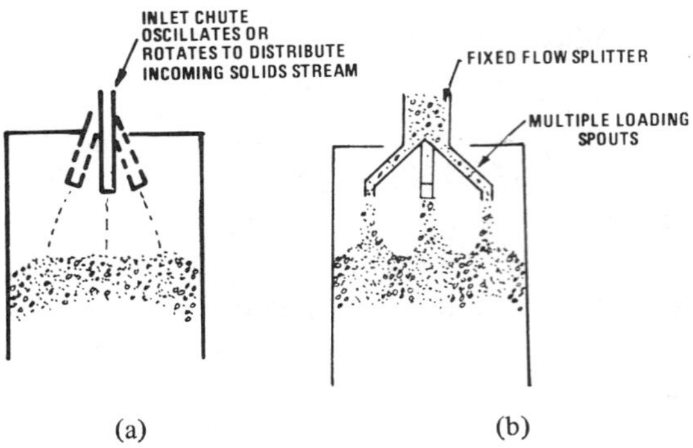

Figure 9.90. Devices to minimize segregation during filling of a bin: (a) moving fill spout, (b) flow splitter or spreader-deflector.

mounted high in the hopper section can widen the flow channel and assist in remixing (Fig. 9.92a). Multiple discharge pipes (Fig. 9.92b) have been used to extract material from different segregated areas of the bin and recombine them at the discharge point. A patented device for similar use is shown in Figure 9.93. A device patented by Fisher[133] is used for such a purpose. The Stock conical distribution chute is used to feed coal to stokers. The chute remains full at all times as the coal drains from a bin above to a spreader stoker below. Under these conditions, this device will produce very little segregation.

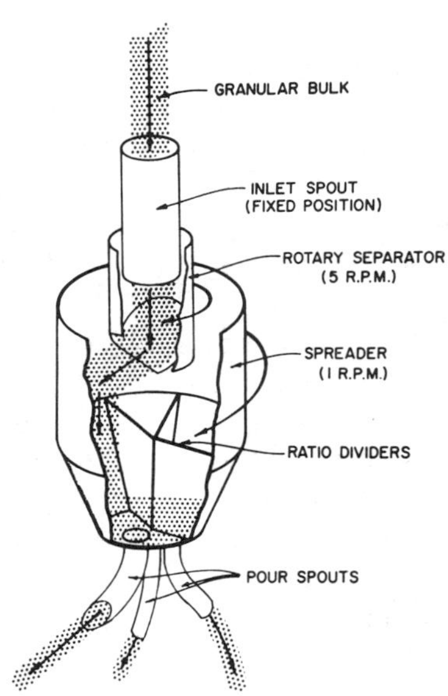

Figure 9.91. Rotary spreader (U.S. patent 3,285,438). (From Ref. 134)

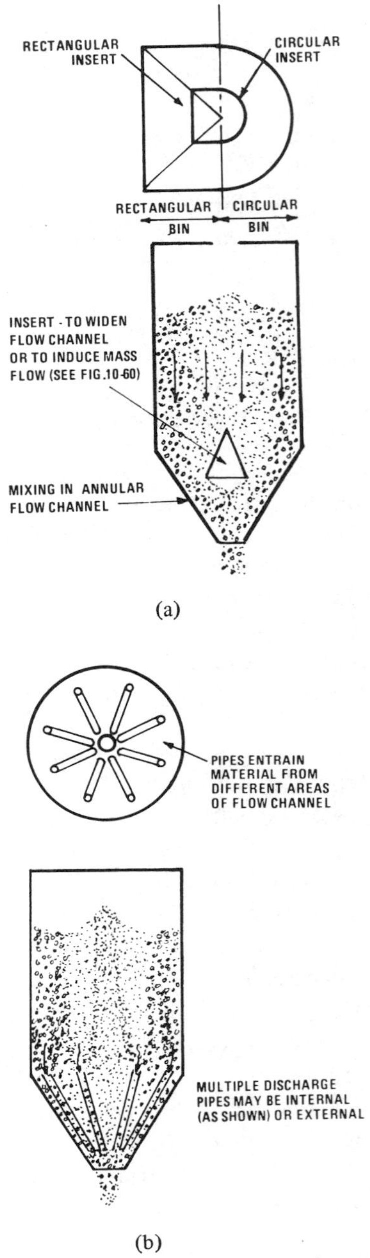

Figure 9.92. Devices to assist in mixing during discharge: (a) inserts, (b) multiple discharge ducts.

Some commercial devices described in Section 9.15 can also enforce mass flow.

Van Denberg and Bayer[123] reported on studies of segregation of granular particles in model bins as the bins flowed from full to empty. They obtained quantitative data by filling the

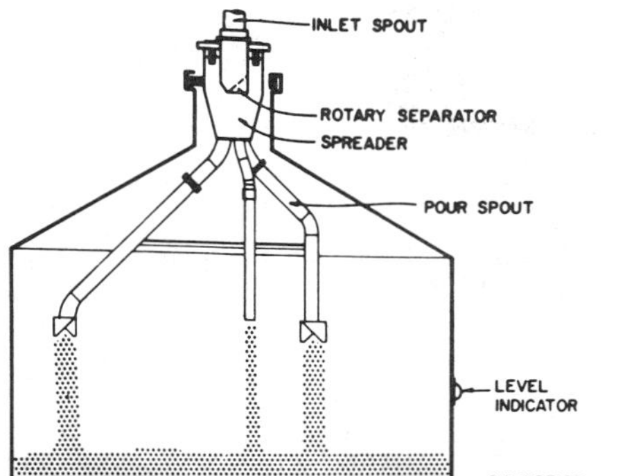

Figure 9.93. Metering rolls (U.S. patent 3,285,438). (From Ref. 134)

models with well-mixed material, then discharging and sampling continuously as the bin emptied. Each sample was analyzed with conventional sieving techniques and the results plotted as shown in Figure 9.94 with sample screen analysis as ordinate values vs sample order in terms of percent weight removed as abscissa values. Figure 9.94a shows segregation patterns typical of center filled funnel flow bins. Figures 9.94b and 9.94c show potential improvements with a well-placed insert or with multiple point fill. Although flow properties of the materials are not reported in Van Denberg and Bayer paper, it is apparent that the bin geometry and discharge arrangement shown in Figure 9.94d produced mass flow and segregation at discharge was markedly reduced. Figure 9.94e shows the serious segregation effects that can be induced by filling and discharging near a vertical baffle.

9.14. STATIC DEVICES TO PROMOTE GRAVITY FLOW FROM FUNNEL FLOW SILOS

BCR Easy-Flo Bin

Large discharge outlets are usually required to prevent bridging over the outlet in bins used to store fine coal, particularly when surface moisture is present. Bituminous Coal Research Inc. (BCR) has developed the Easy-Flo bin[135] for promoting flow of this material (Fig. 9.95). BCR reports that the interior double cone insert in this device controls pressures and flows such that the solids can be made to converge from the large silo opening to the small opening on the bottom. By reducing the coal outlet, the size and cost of associated valves and feeders can be reduced. The silo hopper opening to which such a device is to be attached must be larger than the critical opening for arching or ratholing.

Reimbert "Antidynamic Tube"

This is a vertical perforated tube developed by the Reimberts and is mounted above the discharge opening in a bin (Fig. 9.96). During solids discharge, only the top layer of material moves, sliding down the surface into the top perforation. Since the tube is full, the material below this point is prevented from entering through the lower perforations and thus remains stationary. As discharge continues, each top layer is successively discharged as another tube perforation is exposed. This device, of course, can only be used with noncohesive materials that will flow freely through the perforations.

The antidynamic tube was developed as a response to structural failures of bins caused by large overpressures for which they were not designed. These overpressures were thought to be caused by mass flow conditions that were not anticipated and/or not understood by the

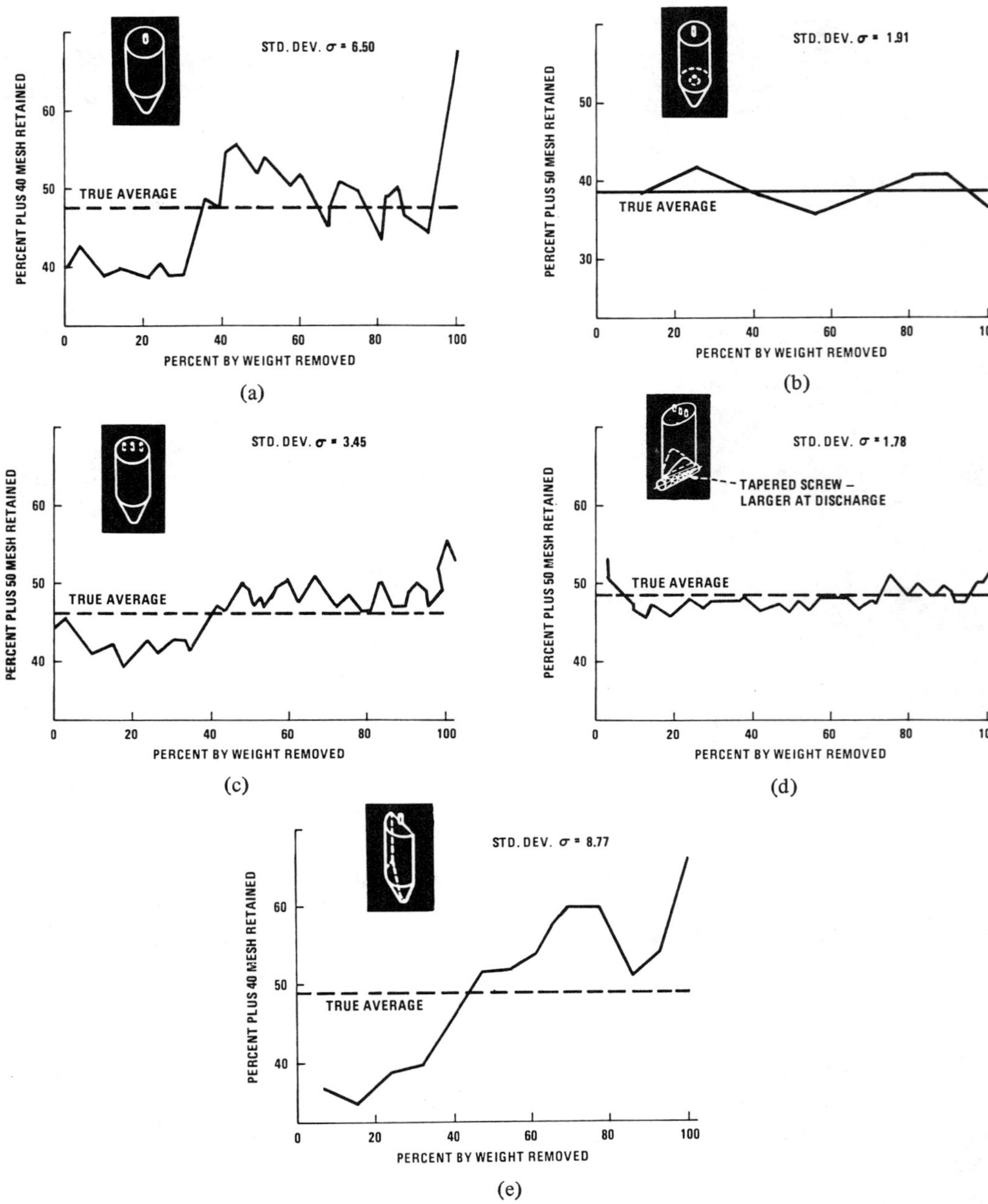

Figure 9.94. Segregation patterns in model bins handling granular material: (a) cylindrical unit with 60° cone bottom; axially filled, then axially discharged completely, (b) cylindrical unit with 60° cone bottom, well-sized and located insert, axially filled and then axially discharged completely, (c) cylindrical unit with 60° cone bottom, filled through three points, then axially discharged completely, (d) cylindrical unit with symmetrical wedge bottom, filled through three points, then discharged completely by uniform withdrawal across the slot discharge opening, (e) cylindrical unit with 60° bottom and vertical partition, filled and discharged completely through openings adjacent to the partition. (From Ref. 123) (Excerpted by special permission from *Chemical Engineering*, copyright © 1964 by McGraw-Hill, Inc., New York, N.Y.)

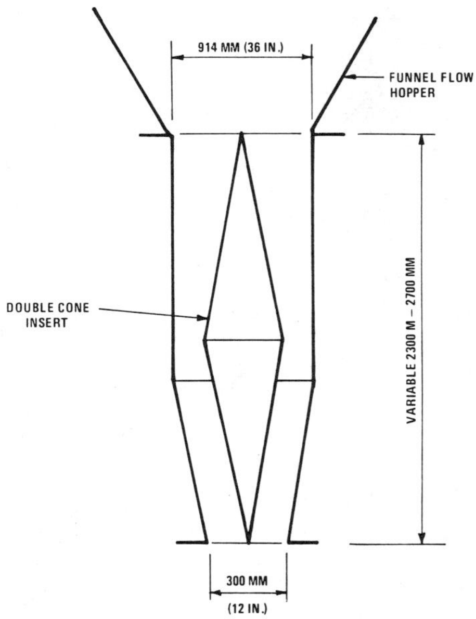

Figure 9.95. BCR Easy-Flow® cone. (Ref. 136) (Permission Bituminous Coal Research Association)

silo designer. These include "effective transitions" occurring during flow, and off-center discharge. The Reimbert tube enforces a funnel flow pattern so that most of the dynamic flow pressures will not extend to the silo walls. The tube has been installed as part of the repair of damaged bins and in new installations to ensure funnel flow.

Inserts

It has been known for a number of years that correctly placed inserts can solve flow-related problems in silos. Newton[136] in 1945 described the use of perforated trays and inclined pipes to provide even distribution (mass flow) of granular catalyst in a moving bed, Morse[137] described the sizing and placement of inserts installed on the vertical axis of a vessel, near the junction of the cone and vertical shell to cause mass flow in a moving bed when shallow sloped hoppers are used.

Sizing and placement of inserts generally have been based on rules-of-thumb or have been found by trial-and-error. Johanson has proposed a method for doing this based on consideration of the bulk solids flow properties and hopper geometry. This work, summarized in the following paragraphs, is the most specific guide to date on the design of inserts.[138, 139, 140]

Johanson reasoned that since an insert forms an annular opening that approaches a long slot opening (Fig. 9.97) a plain-strain wedge-shaped hopper is closely approximated in the area around the insert. Since flow will occur along the walls of wedge-shaped hoppers at relatively shallow slopes, an insert can change a funnel flow pattern to flow along the walls in the region influenced by the insert as shown in Figure 9.97. The critical ratio of the dimen-

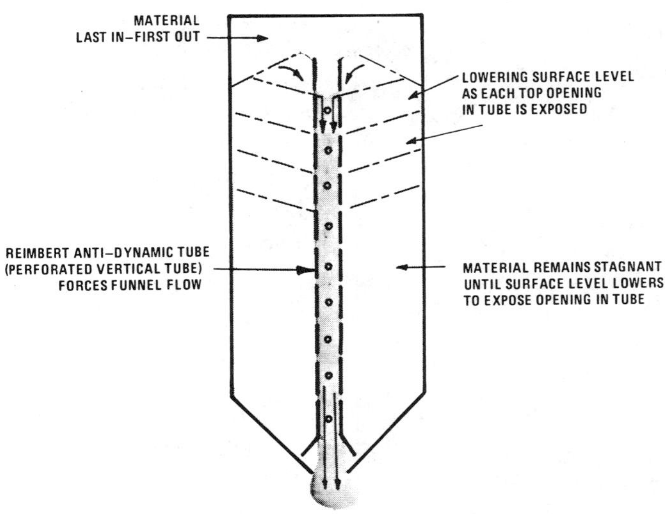

Figure 9.96. Reimbert antidynamic tube (From Ref. 2)

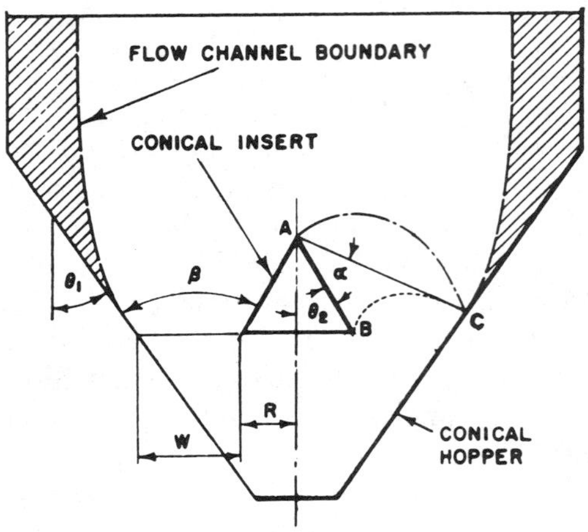

Figure 9.97. Insert geometry and placement (From Ref. 139)

sions W/R given by Johanson and shown graphically in Figure 9.98 were calculated for the case where $\theta_2 = \theta_1$. Johanson states that the values shown in the figure will give a good approximation for any insert slope angle. The angle between the insert wall AB and the line AC to the point at which flow occurs along the hopper wall is presented in graphical form (Fig. 9.99) as a function of the total included angle

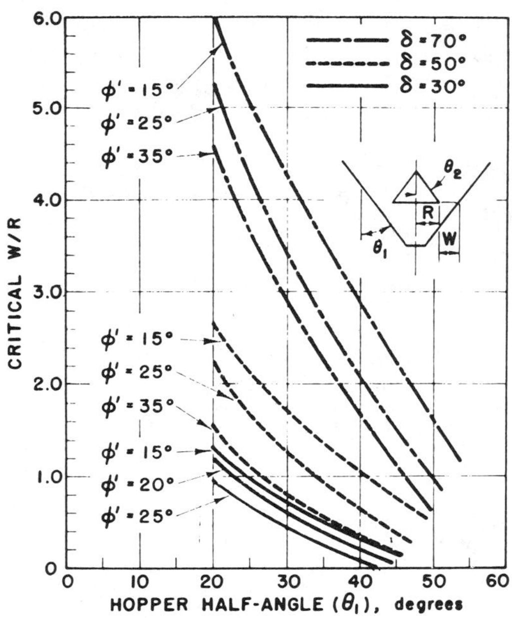

Figure 9.98. Approximate critical W/R for inserts and hoppers having the same slope.

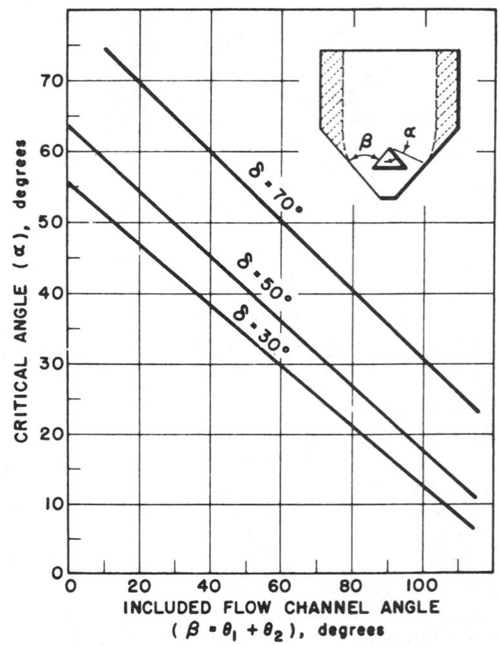

Figure 9.99. Approximate angle α to determine limit of flow along hopper walls.

$\beta = \theta_1 + \theta_2$, assuming a symmetric channel. (The value of α is approximately the same for nonsymmetrical channels where θ_1 and θ_2 are not equal.)

Inserts to Minimize Segregation. An insert designed to eliminate segregation during withdrawal from a funnel flow bin must be placed high enough in the bin near the junction of vertical wall and cone so as to cause the entire mass in the vertical section to move uniformly.

Such an insert is shown in Figure 9.100. With materials that will not arch or rathole, the insert is designed by Johanson's method as follows:

1. Select an insert slope angle θ_2. A horizontal flat plate can be used if cleanout is not a consideration.
2. Determine critical W/R and α values using solids flow properties and Figures 9.98 and 9.99. Johanson suggests adding a safety factor to the design by reducing the critical W/R by 10%.

3. On a sketch of the silo, draw the line AB having as its slope the angle $(\pi/2 - \alpha - \theta_2)$ from the horizontal. Draw line CD through the vertex at angle a, where $\tan a = \tan \theta_1 / (1 + WR)$. Points on this line represent critical values of W/R.
4. Draw line BE at slope angle θ_2 to determine point E, the bottom of the insert.

Insert to Widen the Flow Channel. The same procedure can be used to determine insert size and placement low in the hopper to widen the flow channel and reduce stagnant areas in a funnel flow silo (assuming the materials will not bridge or rathole). The diameter of the desired flow channel locates the approximate point C (Fig. 9.97).

Inserts With Cohesive Materials. An insert can be used to prevent ratholing by providing a vertical flow channel greater than the critical rathole diameter. To design this insert, first determine the minimum opening required to prevent an arch forming over a circular opening in a funnel flow bin, as described earlier. The critical dimensions W should be no less than three-fourths of this minimum opening. If this insert is placed high in the hopper, a rathole may form below if the angle of repose of the material allows a depth of material above the

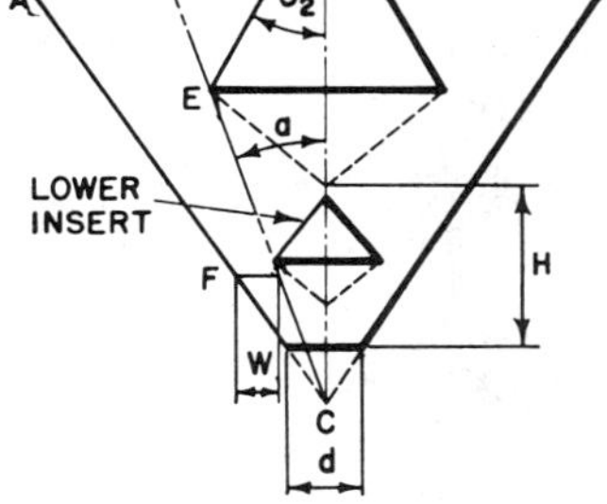

Figure 9.100. Placement of inserts.

hopper opening greater than the diameter of the opening. Johanson suggests a second, lower insert may be necessary to prevent this (Fig. 9.100).

Inserts placed in the hopper section usually do not cause higher overpressures (stresses) on the hopper wall. Inserts that project up into or are located in the cylinder portion can cause overpressure (stresses) on the cylinder walls because of the presence of the flow channel transition at the insert.

Special Hopper Geometries

Theimer[141] describes a variety of silo and hopper geometries, developed through trial and error, that have proven useful for promoting gravity flow of bulk solids. Most of the examples cited refer to storage of poor-flowing grain and food products in large concrete silos. In these large concrete silos, mass flow hoppers in many cases are prohibitively expensive. However, by taking into account the flow properties of the solids and by judicious shaping and proportioning of the silo bottoms, the structures described in this paper are successfully storing and discharging poor-flowing materials.

The design criteria for shaping bin and hopper geometry to improve flow include the following:

1. When handling poor-flowing powders in hoppers having a rectangular or square cross section, avoid sloping walls that intersect to form a valley angle. These materials will not flow in the region of the valley angles and will cause ratholes to form. Design so that inclined hopper surfaces intersect with vertical wall surfaces.
2. To promote flow of cohesive materials in large silos provide a design that reduces the consolidating pressures and allows expansion of the material as it flows through the lower hopper area. This can be accomplished by the inserts described previously, by pressure relief "noses," or by expansion of the hopper cross section at the junction of hopper and vertical section.

Other hopper geometries are reviewed by Reisner and Eisenhart.[142]

9.15. FLOW-PROMOTING DEVICES AND FEEDERS FOR CONTROL OF FLOW

Selection and design of feeders or other flow control devices to be installed at a bin outlet must be considered to be an integral part of the storage bin design. Feeders should be designed to withdraw material uniformly from the entire area of the discharge opening. This will ensure the largest possible flow channel in a funnel flow bin. It is a mandatory requirement for a mass flow bin. If the entire opening is not active in such a bin, mass flow will not occur.

The minimum opening size and shape required to insure flow from a bin must be determined before selecting a discharge device or feeder. It is not always correct to select the feeder and then match the hopper opening to it. Feeders are usually rated by manufacturers on the basis of volumetric capacity. If the feeder selected on this basis has an inlet smaller than the minimum required hopper opening size, it is unacceptable. Too small an opening could result in bridging, ratholing, and erratic flow. Selecting the feeder on the basis of opening size, therefore, may require a unit that is considerably "oversized," and operates at low speed.

Basic Feeder Types

There are many types of feeders described in the literature. The important criteria for any of these devices when used at the discharge of a bin is that it be configured and mounted so that material is withdrawn from the entire opening. Several feeders commonly used for control of powder and granular flow from bins are described in Reference 143. These include:

Rotary Feeders (also called rotary vane feeder or star valve). These can be used as a volumetric feeder and/or a gas pressure seal (air lock) to pass solids from one pressure environment to another (Fig. 9.101). It can be used under a circular, rectangular, or slot opening

Figure 9.101. Drop-through rotary feeder.

(Fig. 9.102). These valves are well suited for feeding materials that tend to flush or aerate in funnel flow bins, since they can be machined with close clearances between rotor and housing. The pockets of a rotary valve fill on the rising side of the rotor when the rotor is under a head of solids. Therefore, when the valve is mounted under a bin, withdrawal across the bin opening can be made more uniform by either of two arrangements. Separate the valve from the bin opening by a connecting spout having at least the same cross section of the opening. The length of the spout should be at least two times the bin opening as viewed along the axis of the valve rotor, to allow the solids flow channel into the rotor to diverge upward to meet the full opening of the bin. An alternative arrangement would be to direct the material to the rising side of the rotor, similar to that shown for slot openings in Figure 9.102.

Where the material properties dictate a large bin opening to prevent arching or ratholing, a

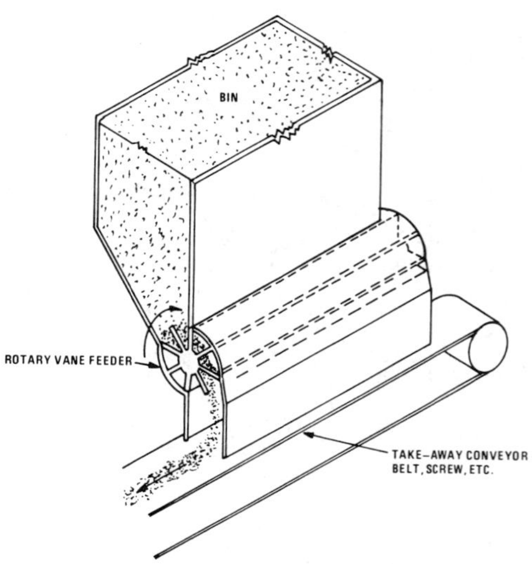

Figure 9.102. Slot-type rotary vane feeder.

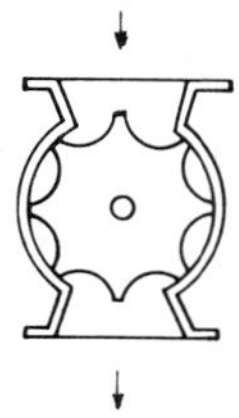

Figure 9.103. Drop-through rotary feeder with filled rotor.

shallow-pocket (filled-pocket) rotor is often used instead of the standard rotor (Fig. 9.103). When the valve opening is oversized to meet material property requirements, the volumetric capacity of the standard rotor becomes so large that very high drive speed reductions are required. In most cases, it is less expensive to reduce the capacity of the rotor with the shallow-pocket design and operate at speeds that require a more moderate, less costly drive.

Where a rotary valve is to be used to feed into a pneumatic conveyor there are additional considerations. Close clearances between rotor and body are required. Pellets or granular material can jam in these clearances. This can be prevented by using a side-entry (pellet) valve (Fig. 9.104) or a pellet shield with a flow control gate to meter the material and prevent filling of the rotor pockets. When feeding material into pressure pneumatic conveyors, the gas leakage past the rotor clearances will greatly exceed the pocket displacement and will pass up into the incoming material. In some cases, this gas "fluffs" the material in the bin and assists flow. In most cases, however, this gas impedes solids flow and must be *vented*. It can be vented through a connection to the inlet feed section or through a connection in the valve housing.

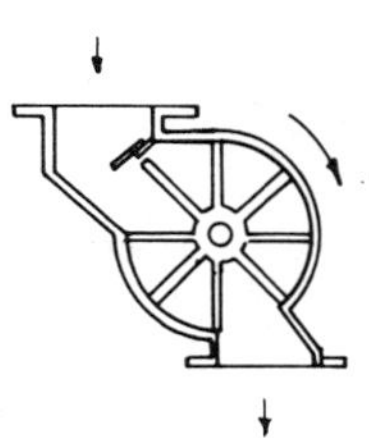

Figure 9.104. Side-entry rotary feeder.

Screw Feeders. Screw feeders handle a wide range of materials from lumps to powder, are relatively inexpensive, are easily enclosed to be dust tight, and easily accommodate slot openings. They will not seal against an uncontrolled flow of "flooding" fine powders and normally operate with a zero or low-pressure differential between outlet and inlet. Special designs have been made for feeding certain materials at pressure differentials up to 100 kPa.

If the required bin discharge opening determined from solids flow properties is very large, it may be necessary to use several parallel screws in a slot opening. No matter how many screws are used, they must be designed to promote uniform withdrawal from the hopper above. Nonuniform withdrawal can lead to solids arching, ratholing, or, if a rathole collapses, to *flushing through the screw* (Fig. 9.106).

A standard screw feeder has a pitch-to-diameter ratio of 1. This ratio is satisfactory only for withdrawing uniformly from openings where the maximum dimension does not exceed 1-1-$\frac{1}{2}$ pitches. If the hopper opening exceeds this, solids flow will occur at the back end of the screw, causing a channel or funnel flow to occur at that point (Fig. 9.105). There are several screw configurations that can be used to promote uniform withdrawal from slot openings.

Increasing Pitch (Fig. 9.106a). Each pitch progressively increases in the direction of flow, until the flight is out into the conveying section, away from the hopper opening. On long slot openings, exceeding about 5 to 6 screw diameters in length, successive increases in pitch may not be completely effective in achieving uniform withdrawal along the slot.

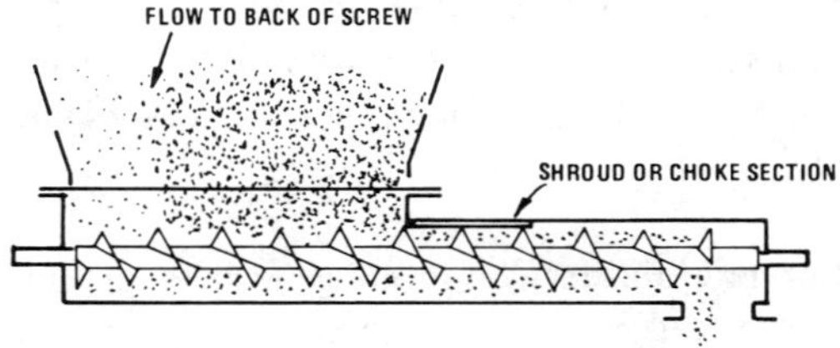

Figure 9.105. Screw feeder–uniform pitch–produces a poor flow pattern under a slot opening.

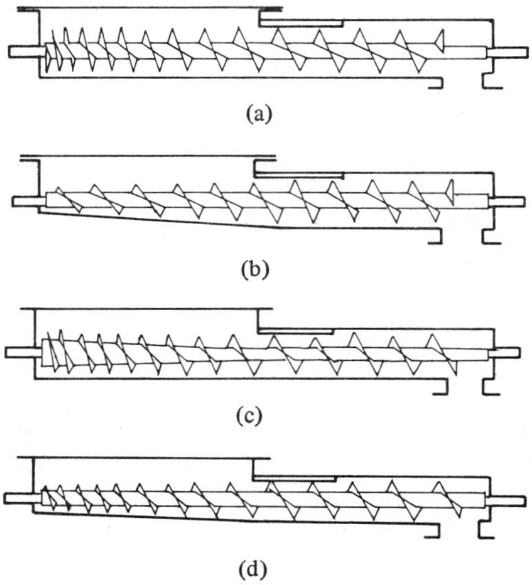

Figure 9.106. Various screw feeder geometries producing improved flow patterns under slot openings: (a) increasing flight pitch, (b) increasing flight diameter, (c) increasing pitch with decreasing shaft diameter, (d) increasing pitch with increasing flight diameter.

Increasing Flight Diameter (Fig. 9.106b). The diameter of the screw flight increases in the direction of flow. Some manufacturers offer this as a standard preengineered single or twin screw assembly to reduce cost. It can be very effective, but the powder properties must be such that they will not bridge over the smaller opening at the rear end of the screw.

Increasing Pitch With Decreasing Shaft Diameter (Fig. 9.106c). The flight pitch increases in the direction of flow, while the shaft diameter decreases in the direction of flow. This design is effective over long slot opening.

Increasing Pitch With Increasing Flight Diameter (Fig. 9.106d). The flight pitch increases in the direction of flow as does the flight diameter. This is effective with long slot openings but is usually more costly than Figure 9.106c.

Vibrating Feeders. Vibrating feeders provide precise feed control, handle material gently, are self-cleaning, and can handle hot materials. They normally operate at frequencies from 12

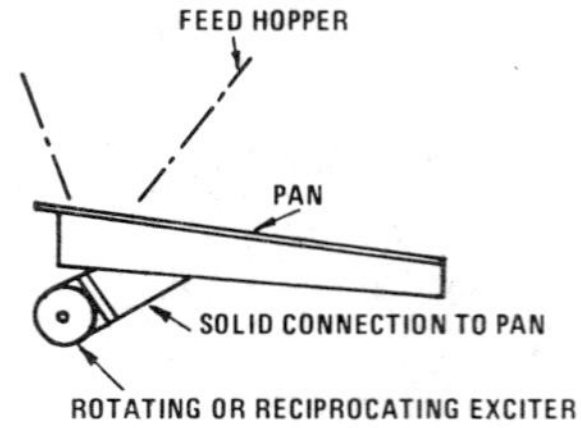

Figure 9.107. Direct force (single mass) vibrating feeder.

to 60 cps and strokes to about 10 mm. There are two general types of feeders: the direct force (single mass) machine (Fig. 9.107) and the indirect force (tuned two-mass) machine (Fig. 9.108).

A rotating counterweight or reciprocating piston causes the motion in a direct force feeder. Essentially a constant rate machine, it is low cost and can handle a wide range of particle sizes from lumps to damp fines, but does not provide precise flow control.

The vibrating forces from an exciter mass are amplified by a spring mass system to vibrate the trough of an indirect force feeder. This design is most commonly used since it provides the best control of solids flow, uses the least power, and normally requires less maintenance than the direct force machine. Two excitation sys-

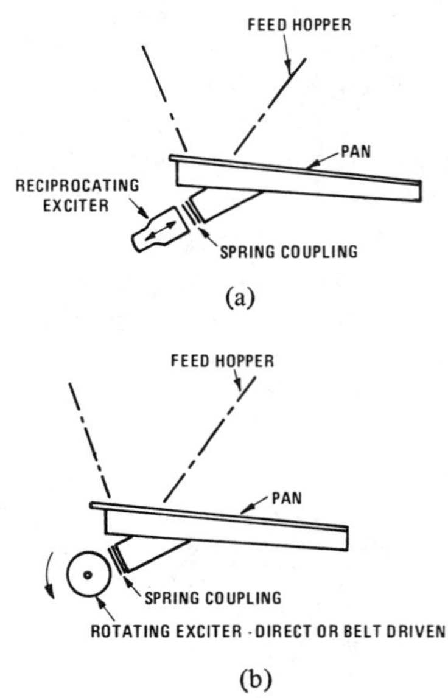

Figure 9.108. Indirect force (tuned two mass) vibrating feeders: (a) electromagnetic feeder, (b) electromechanical feeder.

tems are in common use: electromagnetic (Fig. 9.108a) and electromechanical (Fig. 9.108b).

In the electromagnetic feeders, an alternating or pulsating direct current drives a vibrator that is coupled to the pan through metal or fiberglass leaf springs. These machines have a short stroke (approx. 0.08 mm) and high frequency (50–60 Hz). Feed rate is adjusted by voltage control using a rectifier and rheostat, or variable voltage transformers. Very precise control from 0 to 100% and almost instantaneous shutoff of solids is possible with this machine.

In electromechanical feeders, an electric-motor-driven eccentric weight, coupled through mechanical, elastomer, or pneumatic springs, drives the pan. These machines can have strokes up to 0.10 mm and run at frequencies varying from 12 to 60 cps. Feed control can be accomplished by varying speed of the motor through a variable voltage control circuit, automatic changing of eccentric weight loading, or varying air pressure in pneumatic couplings. Some of these feeders can control solid flow from 20 to 100% of rated capacity; other models are capable of 5 to 100%.

Hoppers above vibrating feeders must be designed correctly to properly deliver the solids to the feeder trough. Improperly designed hoppers can put unnecessary loads on the feeder pan and can cause compaction of the material in the hopper opening, all of which can significantly reduce feeder capacity. Suggested methods of hopper outlet design for delivering to vibrating feeders are given in References 144 and 145, Figure 9.109 for circular openings, and Figure 9.110 for rectangular or slot openings.

The performance of a vibrating feeder is more sensitive to particle properties than any of the feeders discussed. Special consideration should be given to fine powders and powders that tend to aerate. These powders often move at very low rates on vibrating pans. They can deaerate on vibrating surfaces and only the top layer of material will move. They can also flush through an improperly designed inlet hopper. It would be prudent to test fine powders on a vibrating feeder before specifying their use.

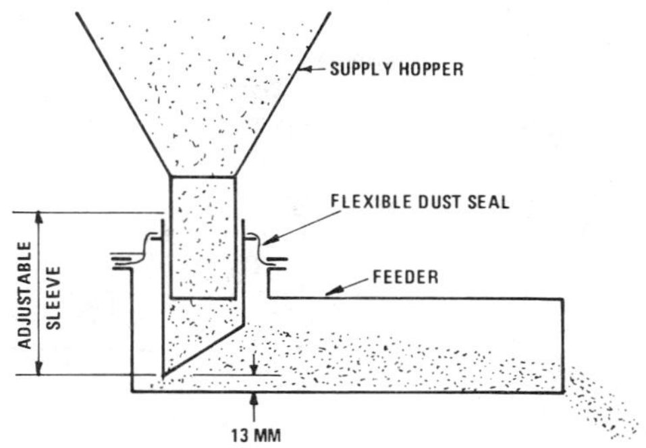

Figure 9.109. Circular feed opening to a vibrating feeder.

Belt Feeders. Belt feeders can withdraw material from very long slot openings in bins, can be designed to take very heavy impact and solids loads from the hopper above, can be combined with weigh decks or weigh idlers to gravimetrically meter flow and will handle practically any solid. The belt is usually a fabric or elastomeric covered, fabric reinforced band, riding on a slider bed or rollers.

Improperly designed feed hoppers over the belt can cause solids compaction, belt wear, and high horsepower demand (Fig. 9.111). Belt feeders having hoppers designed so that the opening diverges in the direction of flow have proven successful for handling a variety of granular and powdered material through long slots (Fig. 9.112).

When handling very abrasive materials or large lumps, an apron feeder may be used in place of a belt feeder. In this device, the carrying surface is made up of over-lapping metal pans, sup-

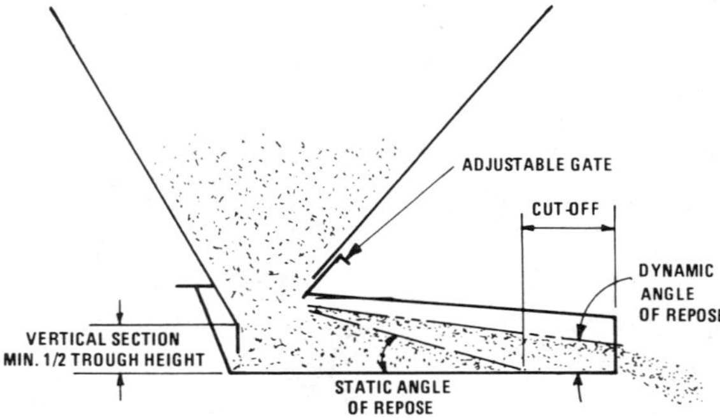

Figure 9.110. Rectangular or slot feed opening to a vibrating feeder.

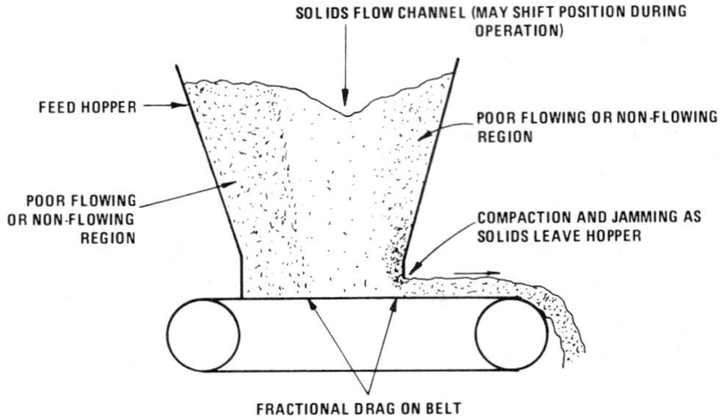

Figure 9.111. Poorly designed feed hopper over a belt feeder.

ported on each side by a driven roller chain, riding on steel tracks.

Other Feeders and Flow-Promoting Devices

A number of flow-promoting devices and special feeders have been developed for specific applications. Several commonly used devices classified by the principle method used to induce flow are described below.

Vibratory-Type Devices

Vibrating Bin Bottom or Bin Discharger (Fig. 9.113). This is a conical hopper mounted beneath the opening in a bin or hopper and suspended from the bin or hopper by elastomeric-bushed links. Elastomeric bands connect and seal the inlet to the bin above and to the feed device or chute below. Motor-driven eccentric weights, mounted on the vibrating hopper, cause it to gyrate in an elliptical path on a horizontal plane. The frequency is fixed by the rotational speed of the weights: amplitude is varied by positioning of the weights. Frequencies vary from 15 to 50 hZ, but 15 to 30 Hz are most commonly used. Weight positioning (amplitude) is determined by solids flow characteristics, density, and amount of material in the bin, and is based on experience with similar materials or from tests on small hoppers.

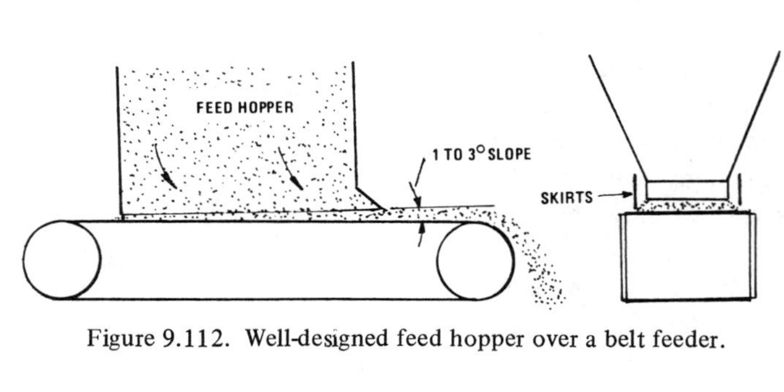

Figure 9.112. Well-designed feed hopper over a belt feeder.

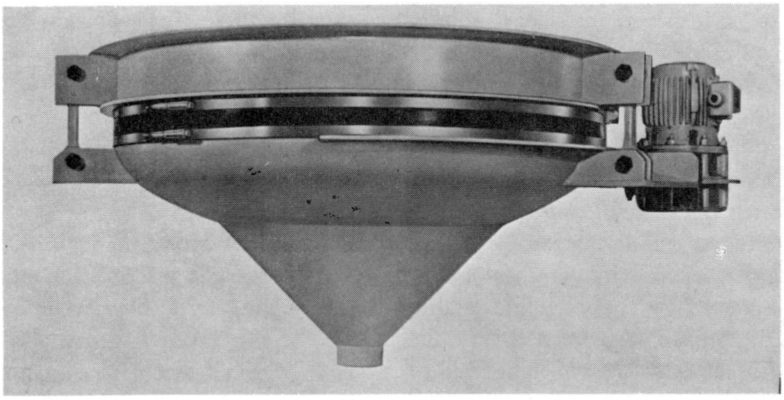

Figure 9.113. VibraScrew bin activator–vibrating hopper for attachment to hopper opening. (Permission VibraScrew Corp.)

A pressure cone or baffle mounted axially within the unit vibrates with the hopper and serves two purposes: It reflects the vibratory motion up into the material in the bin above and prevents solids compaction at the outlet by shielding the outlet from direct pressure from the solids.

In the Whirlpool vibrating hopper configuration (Fig. 9.114), two motor-driven vibrators with their axis of rotation inclined to the horizontal plane are mounted 180° apart on the hopper. The action of the vibrators impart a twisting and lifting motion that can be effective in inducing flow of very sticky or cohesive materials.

Vibrating hoppers can be very effective in promoting flow of a variety of powders, including those that agglomerate and form friable lumps, those that must be deaerated to prevent flooding, and cohesive powders that will not

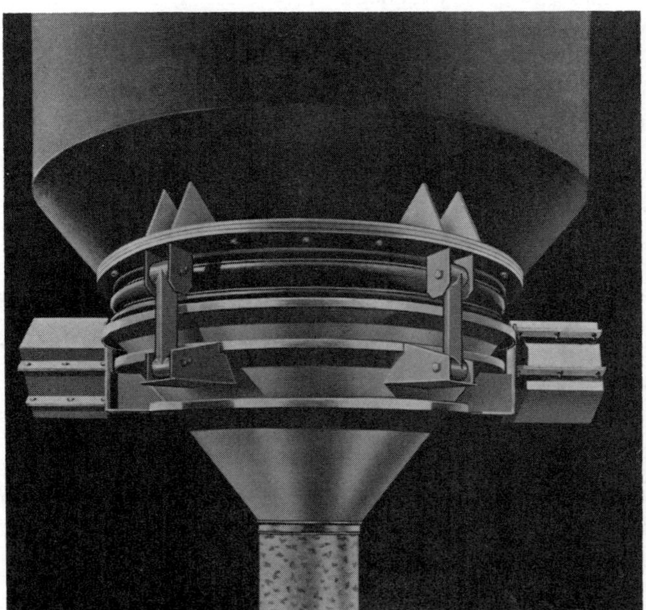

Figure 9.114. Whirlpool vibrating hopper. (Permission Carman Mfg. Co.)

flow by gravity. Because of their heavy rugged design, they will accept very high head loads from material in the bin. Selection of the hopper inlet diameter is dependent on the solids flow pattern desired in the bin above. If mass flow is required, the vibrating hopper can be sized to match the full cylinder diameter, or it can be mounted at the discharge of a conical mass flow hopper. This latter arrangement is useful where a mass flow hopper would require inordinately large openings for gravity flow. The vibrating hopper in that case can be used to converge this flow to a smaller outlet.

If funnel flow in the bin above is acceptable, the vibrating hopper can be sized large enough to expand the flow channel to the desired size, and converge to a small discharge opening.

The inlet dimension of a vibrating hopper must be sized to be larger than the minimum hopper outlet required to prevent bridging or ratholing in the hopper or bin under which it is mounted. This dimension can be determined by first making allowances for possible compaction due to vibration and using the techniques described in Section 9.10.

If vibrating hoppers are improperly applied, bridging can occur above the hopper or powders can be overcompacted and flow at very reduced rates from the outlet. Conversely, flooding can occur if ratholes form and collapse in the bin above, or if hopper flow rate, discharge nozzle size, and hopper amplitude are not properly matched. The Metalfab bin discharger (Figure 9.115) features a secondary adjustable baffle designed to prevent overcompaction at the outlet.

Since it is a vibrating device, care should be taken to prevent transmission of the vibration into building structures when designing supports for bins having large bin dischargers.

An additional feed device must be installed at the vibrating hopper outlet to achieve accurate flow control.

Vibra Screw Feeder (Fig. 9.116). In this feeder, a screw and trough assembly are mounted on an elastomeric isolation system. Motor-driven eccentric weights cause the entire assembly to vibrate or oscillate in a rocking motion. This keeps the material in the hopper section above the screw in motion, preventing bridging or channeling, and a more consistent solids density is achieved as the material flows into the vibrating screw. This action also permits feeding into very small screw feeders at low rates. The rate is controlled by screw size and speed. Frequency of vibration is fixed and

Figure 9.115. Metal Fab bin activator, vibrating hopper for attachment to hopper opening. (Permission Metal Fab Inc.)

Figure 9.116. VibraScrew feeder. (Permission VibraScrew Corp.)

amplitude, determined by test or experience, is set by eccentric weight positioning.

Acrison R-P (Recipro-Pivotal) Hopper (Fig. 9.117). The hopper is mounted on pivots at the upper end. An eccentric driving mechanism imparts a balanced oscillating force to the lower end. This action, combined with the effect of the hopper slope, induces flow toward the outlet.

Silleta Feeder (Fig. 9.118). In this device, a vibrating feed tray is formed by a series of parallel sloping plates spaced so as to form a

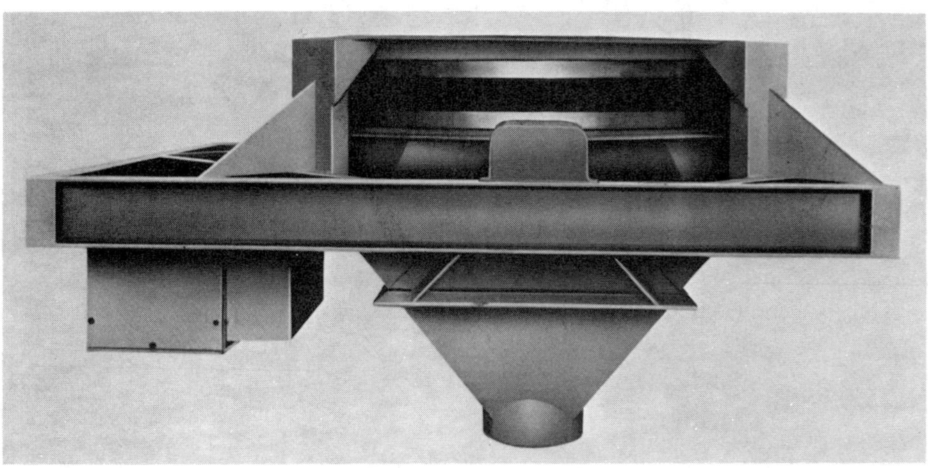

Figure 9.117. Acrison R-P (recipro-pivotal) hopper. (Permission Acrison, Inc.)

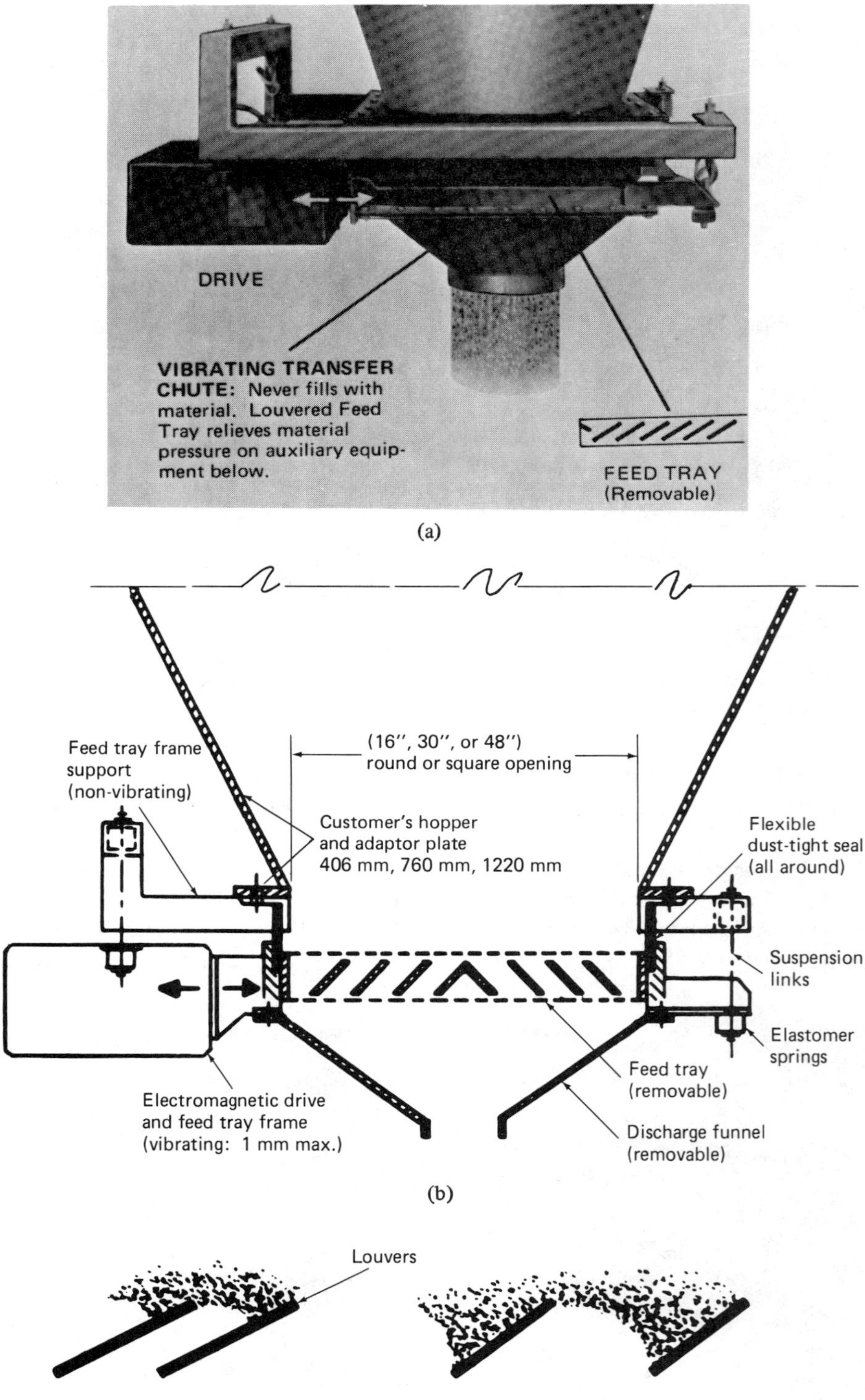

Figure 9.118. Silleta feeder: (a) assembly, (b) section through feeder, (c) typical feed tray for noncohesive powders, (d) typical feed tray for cohesive powders. (Permission Solids Flow Control Corp.)

louver. When the feeder is turned off, the solids rest on the tray and bridge over the slot openings and no flow occurs. When the unit is on, the tuned oscillator vibrates the tray assembly in a horizontal plane at amplitudes to 1.5 mm and a frequency of 50–60 Hz. This vibrating action destroys the bridges and induces flow. Flow rate is controlled by tray amplitude, which is varied by changing voltage to the oscillator. This machine is a combination bin discharger and feeder. It can be effective in controlling the flow of fine powders, and powders with friable lumps, and is capable of 0–100% flow regulation.

Thayer "Bridge Breaker" (bin discharger) (Fig. 9.119). Expanded metal or perforated metal screens in this device are positioned inside the hopper and parallel to the walls. They are attached by studs to externally mounted, low frequency, high amplitude air vibrators. The studs pass through and are supported by resilient elastomeric wall mounts. When activated, the vibrators agitate the screens in a reciprocating motion almost parallel to the plane of the hopper wall. Since this motion puts most of the energy directly into the material instead of the hopper walls, this device uses less energy and makes less noise, compared to standard bin vibrators.

Agitation-Type Devices

Acrison Bin Discharger (Fig. 9.120). In this unit, helical agitators turn at 1–2 rpm to prevent consolidation and maintain the solids in a flowable condition. Solids discharge from one or more openings on the bottom, with no control of rate. A variety of feeders can be mounted at the discharge to control feed rate. The size of the unit is governed by the maximum opening required to prevent bridging in the hopper opening under which it is mounted.

Acrison Bin Discharger Feeder (Fig. 9.121). This device combines the fixed-speed bin discharger to induce flow to a variable-speed screw. Feed rate is controlled by screw speed.

Acrison Feeder (Fig. 9.122). A slow-moving concentric ribbon (or agitator) "conditions" the solids as they enter the feed screw. This controlled agitation maintains density at a consistent level, reduces the tendency to arch or bridge over the feed opening, and permits feeding into very small screw feeders at low rates. Rate is controlled by screw speed and/or screw size.

Metal Fab Feeder (Fig. 9.123). Specially configured agitators mounted on the feed screw loosen the material in the feed hopper, prevent arching, and induce a consistent flow through the feed screw. This also permits feeding into small screws at low rates.

K-Tron Twin Screw Feeder (Fig. 9.124). The feeder uses a mechanical agitator to prevent bridging as solids flow into wiped-surface co-rotating screws. Specially designed flights on these screws intermesh in close proximity to provide a wiping action that aids in discharge of sticky or cohesive materials. The feeder offers good metering capability with a variety of cohesive and noncohesive powders.

Agitated Bin Unloader (Fig. 9.125). A single shaft-mounted ribbon or agitator can be used to prevent bridging and to promote flow from a bin outlet. The agitator can be installed in the hopper, or it can be mounted in a separate housing that is installed at the bin outlet. This device can handle a variety of powders, flake and fibrous materials; it can be mounted on bins that are susceptible to arching because process needs dictate the use of a small discharge opening. Control of flow requires adjustable gates or feed devices be installed at the outlet.

Koppers Live Center Bin (Fig. 9.126). In this bin, vertical screws agitate and loosen stored material to prevent compaction under a solids head; horizontal screws discharge the material. Material characteristics determine the direction of vertical screw rotation. They may be rotated in unison to lift material or, in certain parts of

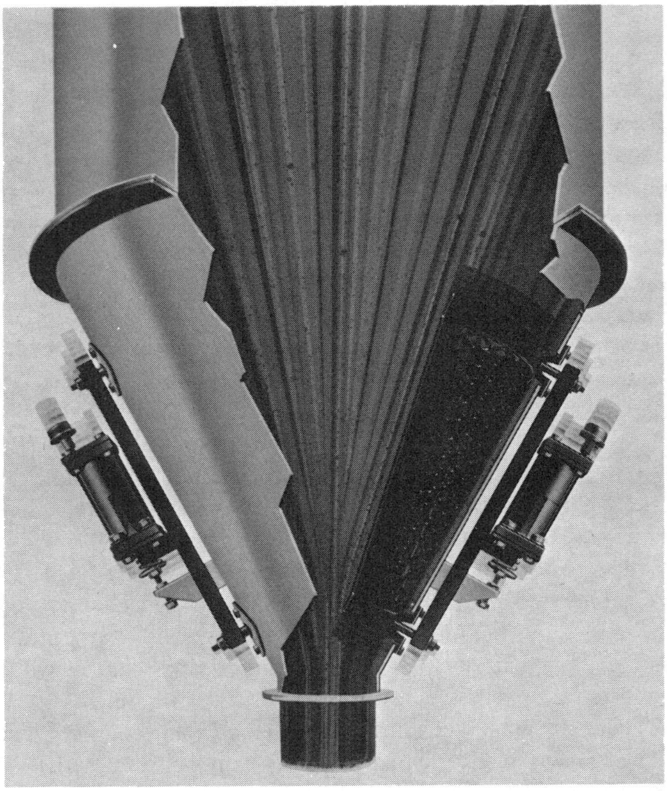

(a)

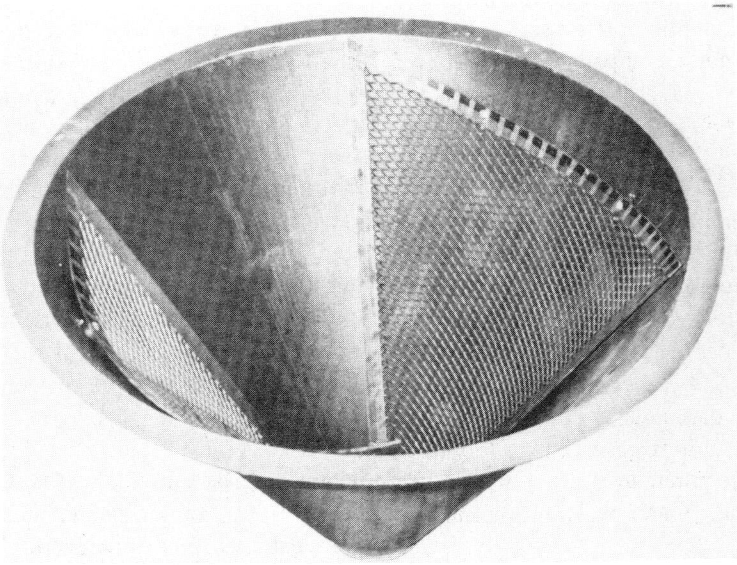

(b)

Figure 9.119. Thayer Bridge Breaker: (a) assembly of two units on a conical hopper, (b) internal view into conical hopper. (Permission Thayer Scale Co., Hyer Industries, Inc.)

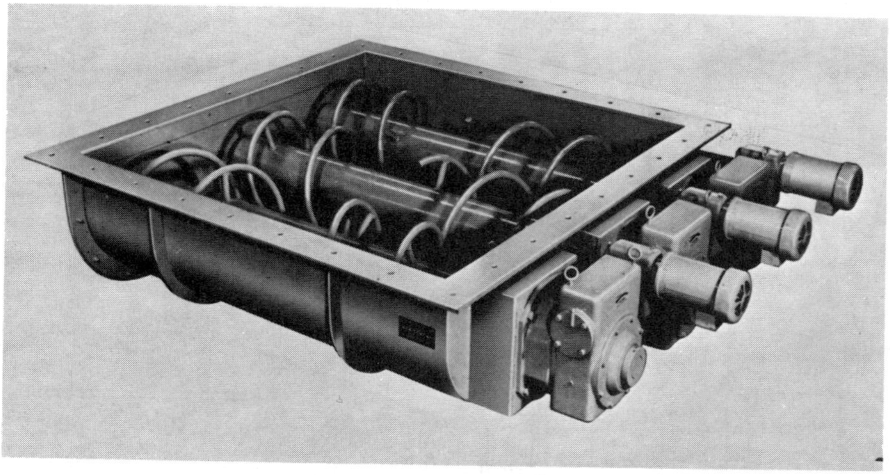

Figure 9.120. Acrison bin discharger. (Permission Acrison, Inc.)

the bin, may be rotated to drive material downward. This bin is designed primarily for storing and feeding low-density, poor-flowing, compressible materials such as chopped sheet, paper, and fibrous materials.

Force-Extraction Devices

Stephens-Adamson Circular Bin Discharger with Arch Breaker (Fig. 9.127). In the discharger section, rotating fingers extract a layer of material and move it toward a gravity discharge chute. A centrally mounted arch breaker extends up through the discharger into the conical portion of the bin.

The arch breaker, driven from below through a universal joint, rotates slowly on its axis to break up arches that may occur, and also induces flow to the discharger. This device works well on many powdery granular, flaky, and fibrous materials, as long as bridging does not occur in the cylindrical section of the bin above the arch breaker. This unit can operate with the discharge completely full since material not

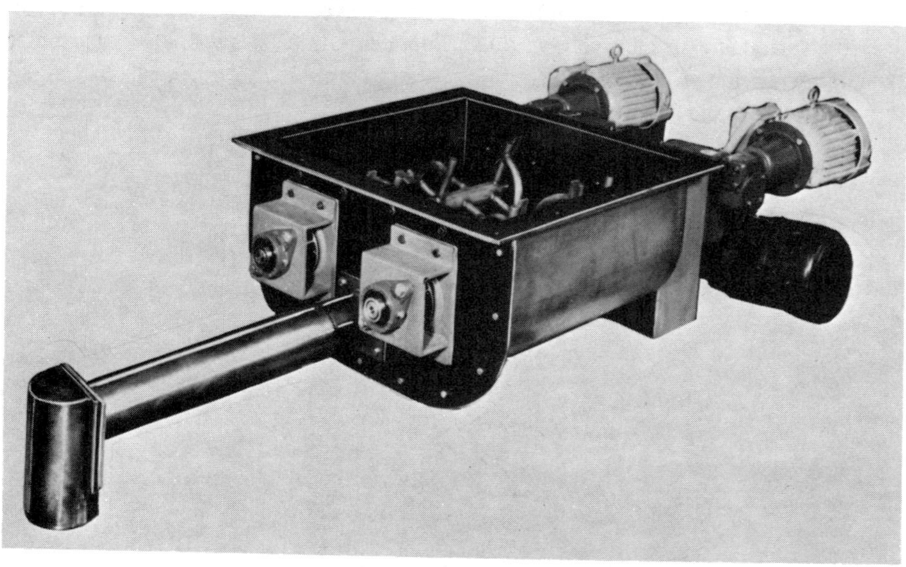

Figure 9.121. Acrison bin discharger–feeder. (Permission Acrison, Inc.)

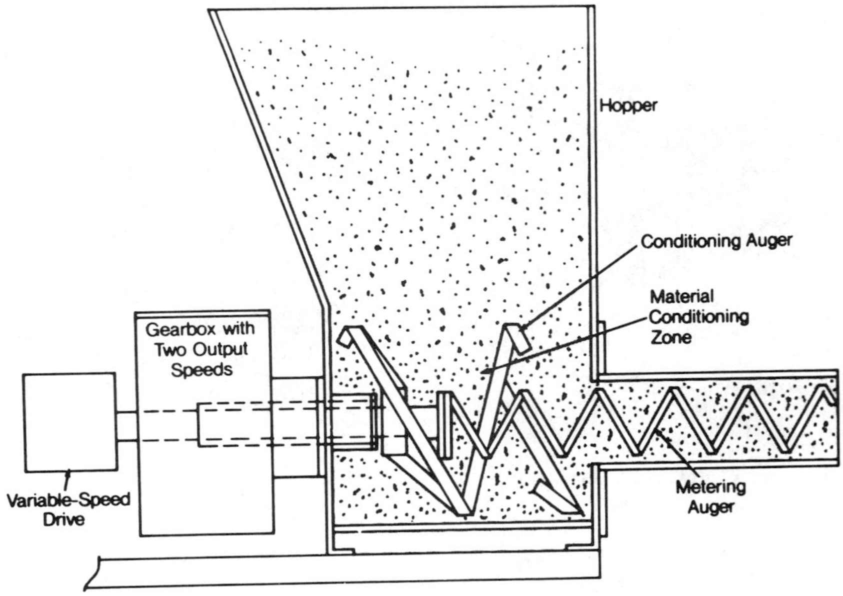

Figure 9.122. Acrison feeder. (Permission Acrison, Inc.)

removed will recycle through the discharger. Multiple discharges can be furnished. This is a combination bin unloader and rough feeder and can be equipped with a variable speed drive.

A. O. Smith Sweep-Arm Unloader (Fig. 9.128). The sweep arm, a chain equipped with teeth, rotates to sweep the flat bottom of a bin, dragging material to a center opening where it discharges to a second traveling chain that transports it to a discharge chute. This combination bin unloader and rough feeder has been used with powders and granular and flaky materials. It can be damaged by abrasive materials and flaky materials that tend to build up between chain and sprocket. It can be equipped with a variable speed drive, although it is not intended for precise feed control.

Some manufacturers offer a horizontal screw (in place of the chain) that rotates about the bin centerline to sweep the flat bottom and draw material into a central discharge opening

Figure 9.123. Metal Fab feeder. (Permission Metal Fab, Inc.)

Figure 9.124. K-Tron twin-screw feeder. (Permission K-Tron Inc.)

(Fig. 9.128). These sweep-type unloaders are used primarily with solids that tend to bridge, or to gain increased volumetric capacity at reduced cost by allowing a flat-bottom in place of a conical-bottom hopper.

Multi-Screw Unloader/Feeder (Fig. 9.129). A series of parallel screws, *proportioned for proper slot flow*, as described earlier, can be used to provide a large, fully active discharge. This device is particularly useful for feeding and discharging sticky, very cohesive or compacting-type solids that require very large hopper open-

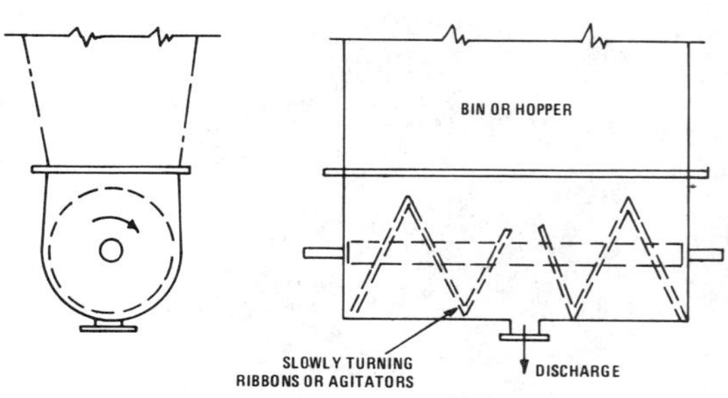

Figure 9.125. Agitated bin discharger.

Figure 9.126. Koppers Live Center bin. (Permission Koppers Co., Inc., Sprout Waldron Div.)

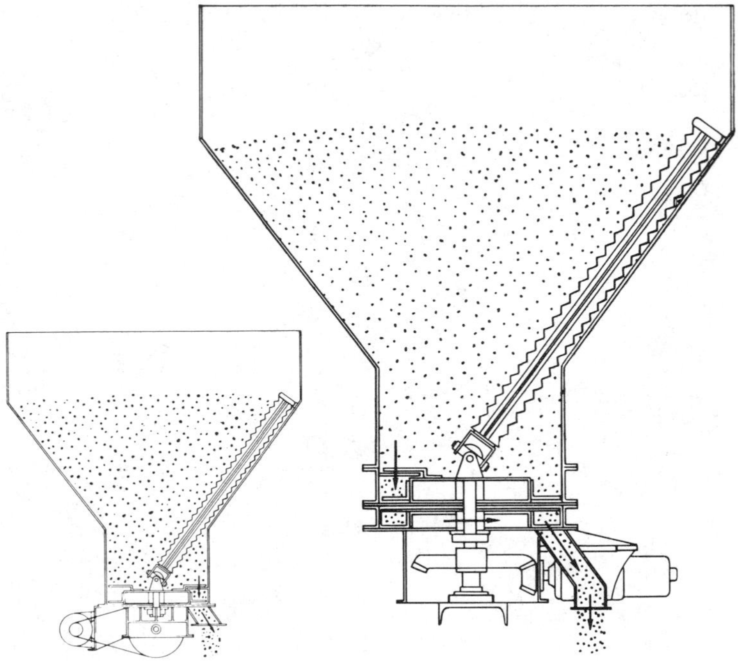

Figure 9.127. Stephens-Adamson circular bin discharger with arch breaker: (a) single stage feeder, (b) two-stage feeder. (Permission Stephens-Adamson Div., Allis Chalmers Co., Aurora, Ill.)

ings, or those that require vertical or negatively sloped hopper walls.

Rotary Table Feeder (Fig. 9.130). The table, a circular plate, rotates at about 2–10 rpm below a hopper opening. Material flowing onto the plate is discharged over the edge by a fixed plow. Flow can be regulated by changing the height of an adjustable feed collar or the speed of the plate.

For best flow control and to keep the hopper opening fully active, the feed collar should be

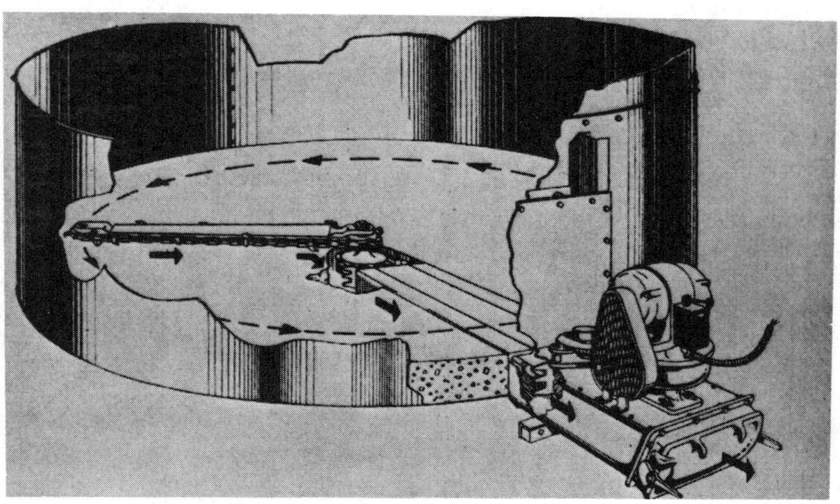

Figure 9.128. A. O. Smith sweep-arm unloader. (Permission Koppers Co., Inc., Sprout Waldron Div.)

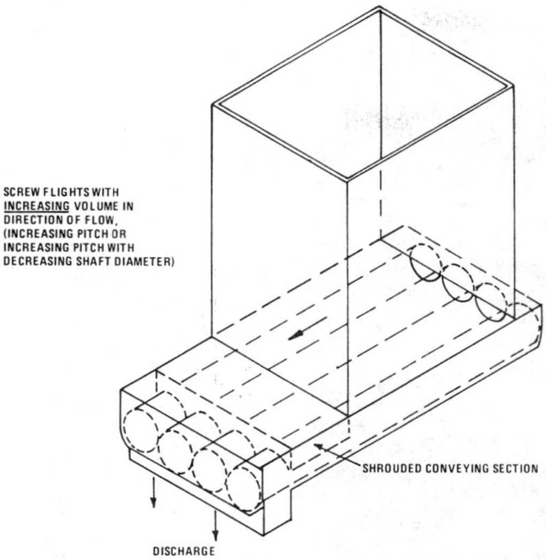

Figure 9.129. Multiscrew unloader.

high enough to allow flow from under the entire perimeter of the collar onto the rotating plate. The device is not intended for precise feed control. The table feeder is selected for materials that require large bin openings to eliminate arching, such as wood chips, sticky or wet granular materials, and for abrasive materials such as minerals and sand.

Com-Bin Feeder (Fig. 9.131). This feeder is designed for a variety of solids and is particularly effective for damp, oily, or sticky materials. It resembles a table feeder except that the shell and contained solids rotate with the plate. A stationary plow strips off solids from a gap between rotating shell and plate. Flow rate can be controlled by plot thickness, gap height, and rotational speed.

Figure 9.130. Rotary table feeder.

Flow Star Feeder (Fig. 9.132). In this device, specially configured wiping blades draw material from an annular space formed by a hopper wall and a stationary flow cone into a discharge opening. This action promotes mass flow in the vicinity of the hopper opening. This unit can be classified as a combination discharger and feeder. Feed control is achieved by varying drive-motor speed.

Disc Feeder. This is a small-scale version of the table feeder. The table is grooved to extract a fixed volumetric amount of material. It is used for very low feed rates, (about 1 to 2 cu ft/hr), with fine free-flowing or cohesive powders.

Rotary Plow Feeder (Fig. 9.133). In this device, a self-propelled carriage, supporting power-operated rotating arms, travels parallel to a slot opening in a bin and above a conveyor. The rotating arms plow material from a continuous shelf onto the conveyor below. This arrangement permits the design of a very long, fully active slot opening in a hopper or bin having a minimum overhead space requirement.

Flexible Wall Devices

Pneu-Bin Panels (Fig. 9.134). These elastomeric panels are fastened along the inner wall of hoppers. Periodic inflation with air expands the panel and forces solids into the flow channel. Pressure control of the air prevents overinflation. The panels are sized and spaced to suit the storage hopper geometry. Since they are elastomeric, they are temperature limited. The sequence and timing of inflation of single or multiple panels is determined by material characteristics and flow rate. Inflation is most effective in promoting flow if there is a void to accept the displaced material. Inflation will be

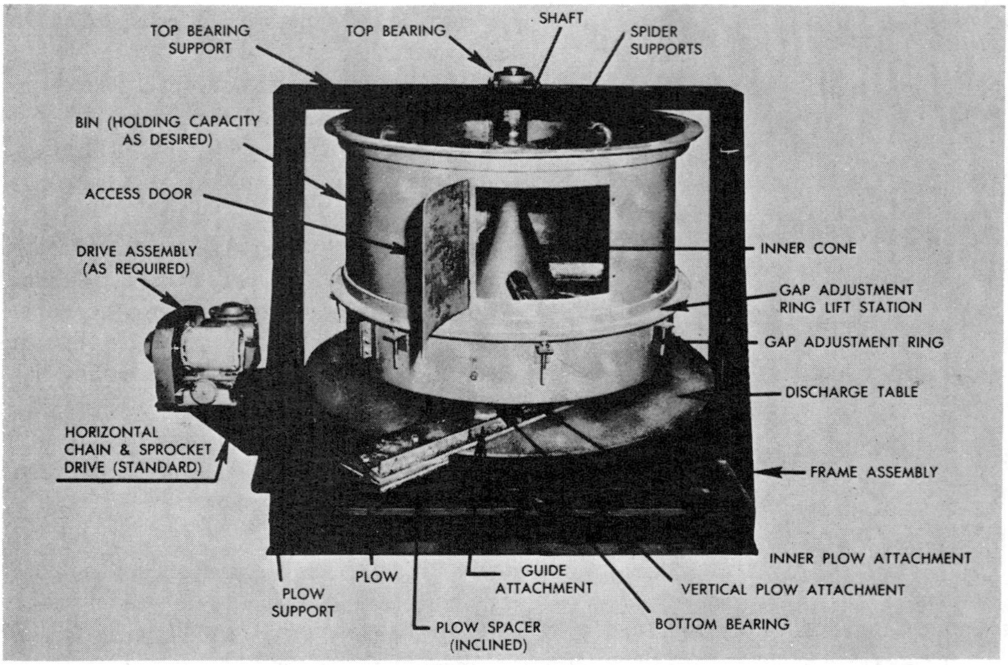

Figure 9.131. Com-Bin Feeder. (Permission Pulva Corp.)

ineffective if it packs cohesive material into a filled channel or if the cohesive material forms a void around the panel. The hopper walls must be sufficient to withstand the reaction forces generated during inflation. The panels are useful for powders as well as sticky materials and are for aiding flow, but provide no control of flow rate.

Accu-Rate Feeder (Fig. 9.135). This is a low-cost volumetric feeder combining a flexible wall hopper for loosening material in the hopper

Figure 9.132. Flow star feeder. (Permission Merrick Scale Mfg. Co., Passaic, N.J.)

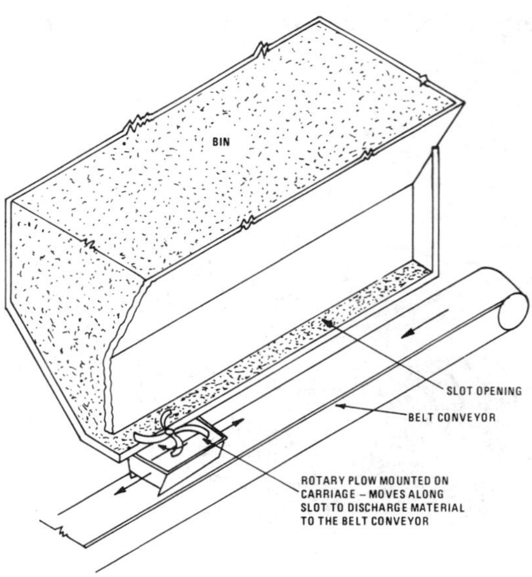

Figure 9.133. Rotary plow feeder.

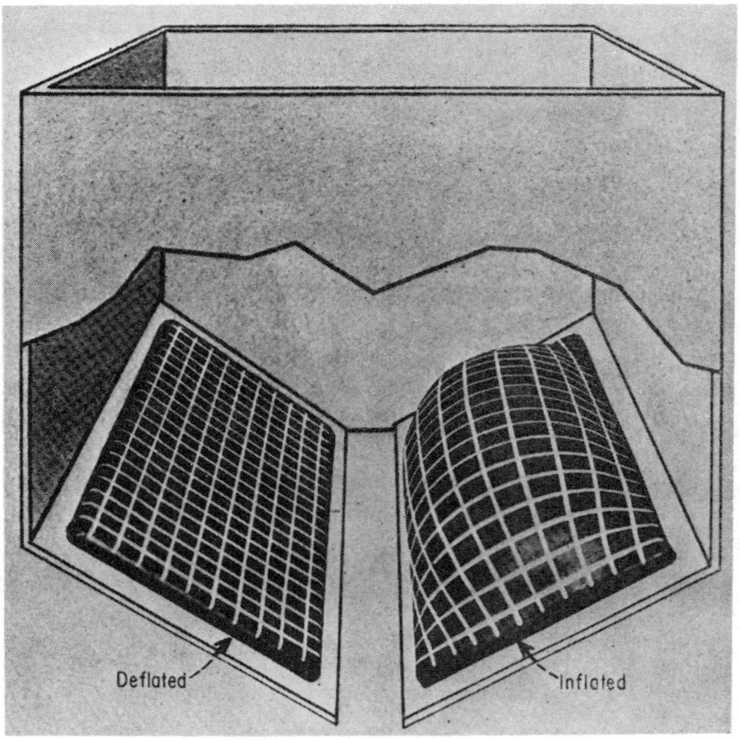

Figure 9.134. Pneu Bin panels. (Permission PneuBin Corp.)

with a variable speed screw feeder. Motor-driven mechanical agitators distort or agitate the walls of a one-piece molded flexible-vinyl hopper during operation to prevent bridging or ratholing and to provide a constant solids feed to the screw.

Figure 9.135. Accu-Rate feeder. (Permission Accu-Rate Div., Moksnes Mfg. Inc.)

Aeration-Type Devices

Certain solids can be aerated easily by controlled gas injection and are readily discharged from hoppers or fed to process by a variety of aeration devices. These devices operate at low noise levels, require little maintenance, are relatively low cost, and can handle large volumes of solids with low gas flows. If well distributed and controlled, most of the injected gas will exit with the powder. At low solids heads, however, more of the gas may exit up through the top of the bin.

For reliable operation, it is necessary to control dust, using dust collectors as necessary, and to provide dry clean air to prevent fouling of membranes and prevent entry of moisture into the material, which could reduce flowability.

The following are the major types of aeration-type flow-promoting equipment. There are a variety of aeration jets, impulse tank jets, and pads that are not discussed here.

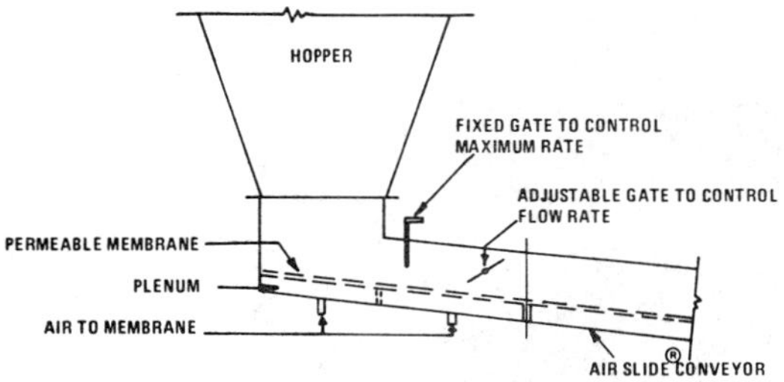

Figure 9.136. Air feeder. (Permission Air Slide Conveyor Reg. Trademark Fuller Co.)

Air Feeder (Fig. 9.136). Air introduced into the solids through an inclined permeable membrane causes solids flow. Flow rate is controlled by an adjustable gate.

Air Hopper (Fig. 9.137). This can take the form of an air feeder under a rectangular or transition slot hopper (Fig. 9.137a), a flanged, dished head surrounding a conical membrane, (Fig. 9.137b) or a conical or rectangular hopper fully lined with a permeable aerating membrane, or a hopper having individual spaced aerating panels or nozzles (Fig. 9.137c). Air (or other gas) is introduced through the membrane in sufficient quantities only as required to reduce the particle-particle and particle-wall friction in the immediate area of the membrane wall surface and exits with the solids. The upper dimension of the aerated portion of the unit is determined by the maximum opening required to prevent arching or ratholing in the hopper above. In determining the opening dimension it may be necessary to add a safety factor to account for the supporting forces caused by air passage up through the material

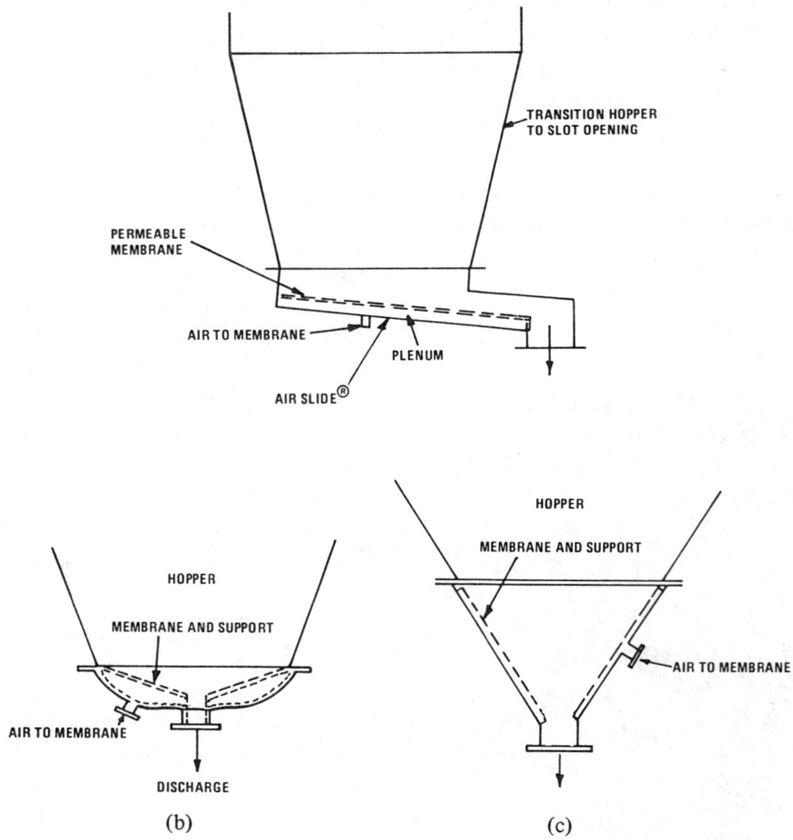

Figure 9.137. Air hoppers: (a) Airslide mounted under slot opening, (b) dished head with fully aerated interior surface, mounted under circular bin opening, (c) conical hopper with fully aerated interior surface, with radial aeration strips or individual aeration pads.

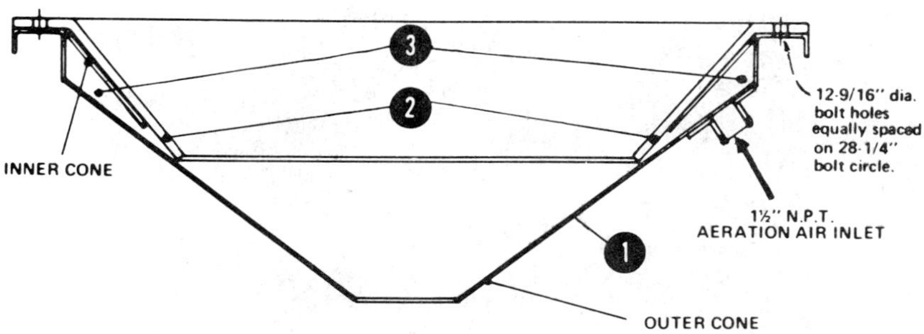

Figure 9.138. Aerated bin discharge cone: (1) steel discharge hopper, (2) neoprene inner seal hopper, (3) aeration plenum. (Permission Monitor Mfg. Co.)

that subtract from the consolidating forces tending to collapse an arch. These devices are essentially bin dischargers. Precise feed control requires a valve or feeder at the outlet.

Aerated Bin Discharge Cone (Fig. 9.138). In this device, air is directed to the solids, under an elastomeric conical insert in a steel hopper. Pulsed air flow can be used to agitate and loosen material. Currently manufactured only as a 762 mm (30 inch) diameter flanged cone it has been effective in promoting flow from bins into small (100-300 mm) diameter discharge openings. It has also been used for aeration during pressure differential unloading or bulk trucks.

Air Blasters (Fig. 9.139). In this device, a volume of compressed air is stored in a tank with its exhaust port sealed by a linear or spherical piston. When air pressure on one side of the piston is exhausted through a quick acting valve, the compressed air in the tank is released almost instantaneously into a mass of stored solids. Location of these devices on a silo, and the orientation of the air release nozzle is determined by the probable location of flow obstructions. The expanding air pocket can break down bridges or ratholes in the material. Single or multiple units can be activated sequentially at time intervals through electrical controls, to maintain the solids in a flowable condition.

Caution is advised if using blasters where serious bridging or ratholing may occur in large silos. The silo must be capable of withstanding the stresses caused by a sudden collapse of these flow obstructions, and the discharge device must be capable of sealing against a sudden rush of solids.

9.16. RECENT DEVELOPMENTS

Borg[147] presented an interesting paper summarizing a statistical evaluation of 500 shear tests on (unidentified) solids commonly used in the chemical industry. Hopper wall friction with various wall surfaces, hopper slopes, and critical outlet diameters required for mass flow were calculated for more than 200 bulk solids having varying degrees of cohesiveness. The critical outlet diameter exceeded 1.2 m for 35% of all products. The percentage of products for which mass flow could be achieved with the wall surface having the lowest friction (ϕ'), was plotted against the hopper angle θ_c for which mass flow will occur. Eighty percent of these products would mass flow at $\theta_c = 10°$; only 25% would mass flow at $\theta_c = 30°$. The curve between these points was almost linear. Confirming comments made earlier in this chapter, the author reported that polishing a wall surface does not always reduce solid wall friction.

Tsunakawa[148] described an experimental demonstration showing that placement of a circular cone insert (supported from below by vertical partition plates fastened to the walls and located along the axis of a conical hopper above the discharge opening) will significantly reduce vertical stresses on the particulate material below. This in turn reduces the yield strength (f_c) developed by the material in that

Figure 9.139. Air Blasters shown attached to conical hopper. (Permission Martin Engineering Co.)

region and will help prevent bridging or arching and/or allow the use of smaller discharge outlets on conical hoppers.

The author proposes a method for calculating static stresses within the hopper when this type insert is used with noncohesive, granular material. This method is shown to compare favorably with results of measurements made on a 300 mm diameter model hopper.

U.S. patent 4,286,883 has recently been issued to J. R. Johanson for a unique insert offered commercially by Jenike and Johanson Inc. as the Binsert.™ This insert consists of a hopper (having the minimum slope required to achieve mass flow) mounted concentrically within an outer hopper. It is claimed that the outer hopper can have a much less steeper slope, and, by proper proportioning of the annular space and discharge region, mass flow can be achieved in conical or wedge-shaped funnel flow hoppers.

REFERENCES

1. Eurosilo Holland, *Wormerveer*, The Netherlands.
2. M. Reimbert and A. Reimbert, *Silos–Theory and Practice*, Clausthal, Germany: Trans. Tech. Publications (1976).
3. M. Reimbert and A. Reimbert, *Silos–Traité Théoretique et Pratique*, Editions Eyrolls, Paris (1961).

4. J. M. Haeger and S. S. Safarian, "A New Concept of Storage Bin Construction," *9 Proceed V* 64: *J. Am. Conc. Inst. 9*:575–597 (Sept., 1967).
5. S. S. Safarian and E. C. Harris, "Silos and Bunkers," M. Fintel (ed)., *Handbook of Concrete Engineering*, Van Nostrand Reinhold Company, New York (1974).
6. H. A. Jannsen, *Versuche Über Getreidruck in Silozellen*, VD1 Zeitschrift, Düsseldorf, *39*: 1045–1049 (1895).
7. O. F. Theimer, "Failures of Reinforced Concrete Silos," *Trans. ASME, J. Eng. Ind.* (B)*91*(2):460–477 (May, 1969).
8. J. E. Sadler, "More Research Needed in Coal Silo Technology," *Coal Mining and Processing*, pp. 70–72 (May, 1976).
9. J. E. Sadler, "Silo Problems," *International Conference on Design of Silos for Strength and Flow*, Univ. of Lancaster, U.K. Powder Advisory Center, London (Sept., 1980).
10. A. W. Jenike, "Denting of Circular Bins with Eccentric Draw Points," *J. Struct. Div., Am. Soc. Civil Eng. 93*(ST1):27–35 (1967).
11. F. Turitzin, "Dynamic Pressure of Granular Material in Deep Bins," *J. Struct. Div., Am. Soc. Civil Eng. 89*(ST2):49–73 (1963).
12. G. P. Deutsch and D. H. Clyde, "Flow and Pressure of Granular Material in Silos," *J. Struct. Div., Am. Soc. Civil Eng.* (EMC):103–123 (1967).
13. G. P. Deutsch and L. C. Schmidt, "Pressures on Silo Walls," *Trans. ASME, J. Eng. Ind.* (B)*91*(2): 450–459 (1969).
14. V. Sundaram and S. C. Cowin, "A Reassessment of Static Bin Pressure Experiments," *Powder Technol. 22*:22–32 (1979).
15. D. F. Bagster, "A Note on the Pressure Ratio in the Janssen Equation," *Powder Technol. 4*:235–237 (1970/71).
16. "Recommended Practice for Design and Construction of Concrete Bins, Silos and Bunkers for Storing Granular Materials," (AC1-313-77) *J. Am. Conc. Inst. 10*:529–548 (1975).
17. "Règles de Conception et de Calcul des Silos en Beton," *Syndicat National du Beton Arme et des Techniques Industrialisées*, no. 189 (1975).
18. *Soviet Concrete and Reinforced Concrete Code*, Silo Code SN-302-65, Moscow (1965).
19. "Lastanahmen für Bauten-Lasten in Silozellen," DIN 1055 Blatt 6 (1964), *Erganzende Bestimmungen ZU.* DIN 1055 Teil 6 (1977).
20. G. Garfinkel, "Reinforced Concrete Bunkers and Silos: Steel Tanks," R. S. Wozniak, E. H. Gaylord, and C. N. Gaylord, (eds)., *Structural Engineering Handbook*, 2d ed., McGraw-Hill Company, New York (1979).
21. F. A. Wenzel, "A Critical Comparison of Different Standards and Methods of Computation of Silo Loads," *Am. Soc. Chem. Eng.*, 70th Annual Meeting, New York Paper No. 71b (1977).
22. G. P. Deutsch, "Bin Wall Design and Construction, Discussion," vol. 65, *J. Am. Conc. Inst. 3*:211–218 (1969).
23. "Commentary on Recommended Practice for Design and Construction of Concrete Bins, Silos and Bunkers for Storing Granular Materials," *J. Am. Conc. Inst. 10*:549–565 (1975).
24. "Discussion of Proposed AC1 Standard, Recommended Practice For Design and Construction of Concrete Bins, Silos and Bunkers for Storing Granular Materials." V. 73: *J. Am. Conc. Inst. 6*:345–361 (1976).
25. R. L. Brown, and J. C. Richards, *Principles of Powder Mechanics*, Pergamon Press, London (1970).
26. W. Resnick, "Flow Visualization Inside Storage Equipment," *Intern. Conference on Bulk Solids Storage, Flow and Handling*, Stratford-Upon-Avon, England, Powder Advisory Center, London (1976).
27. P. M. Blair-Fish and P. L. Bransby, "Flow Patterns and Wall Stresses in a Mass Flow Bunker," *Trans. ASME, J. Eng. Ind.* (B)*95*(1):17–26 (1973).
28. R. Butterfield, R. M. Harkness, and K. Andrews, "A Stereo-Photogrammetric Method for Measuring Displacement Fields," *Géotechnique 20*(3): 308–314 (1970).
29. C. J. Chatlynne and W. Resnick, "Determination of Flow Patterns For Unsteady-State Flow of Granular Solids," *Powder Technol. 8*:177–182 (1973).
30. J. R. Johanson, "Stress and Velocity Fields in Gravity Flow of Bulk Solids," Ph.D. thesis, University of Utah, Mech. Engrg. Dept. (1962).
31. J. R. Johanson, "Stress and Velocity Fields in the Gravity Flow of Bulk Solids," *Trans. ASME, J. Appl. Mech.*, (E)*31*(3):499–605 (1964).
32. M. G. Perry and H. A. S. Jangda, "Pressures in Flowing and Static Sand in Model Bunkers," *Powder Technol. 4*:89–96 (1970–1971).
33. M. Sugita, "Flow and Pressures in Non-cohesive Granular Materials in Funnel Flow Bins," ASME, *2d Symposium on Storage and Flow of Solids*, Chicago, Ill. Paper no. 72-MH-20 (1972).
34. R. P. McCabe, "Flow Patterns in Granular Materials in Circular Silos," *Géotechnique 24*(1):45–62 (1974).
35. D. Lenczner, "An Investigation into the Behavior of Sand in a Model Silo," *Struct. Eng. 41*(12): 389–398 (1963) and (7):243–246 (1963).
36. T. V. Nguyen, C. E. Brennen, and R. H. Sabersky, "Gravity Flow of Granular Material in Conical Hoppers," *Trans. ASME, J. App. Mech. 46*(3): 529–535 (Sept., 1974).
37. T. V. Nguyen, C. E. Brennen, and R. H. Sabersky,

"Funnel Flow in Hoppers," *Trans. ASME J. App. Mech. 47*:(4):729-735 (Dec., 1980).

38. J. S. Giunta, "Flow Patterns of Granular Materials in Flat Bottom Bins," *Trans. ASME, J. Eng. Ind.* (B)*91*(2):406-413 (1969).
39. J. R. Johanson, "Know Your Material–How to Predict and Use the Properties of Bulk Solids," *Chem. Eng. 85*(24):9-17 (Oct., 1978).
40. *Classification and Definitions of Bulk Materials*, Conveyor Equipment Manufacturers Association, Washington, Book no. 550 (1970).
41. S. Frydman, "The Angle of Repose of Potash Pellets," *Powder Technol. 10*:9-12 (1974).
42. R. L. Brown and J. C. Richards, *Principles of Powder Mechanics*, Pergamon Press, London (1970).
43. R. L. Carr, "Evaluating Flow Properties of Solids," *Chem. Engrg. 72*(2):163-168 (Jan., 1965).
44. R. L. Carr, "Particle Behavior Storage and Flow," *Brit. Chem. Engrg. 15*(12) (Dec., 1970).
45. R. L. Carr, "Classifying Flow Properties of Solids," *Chem. Engrg. 72*(3):68-72 (Feb., 1965).
46. R. E. Tobin, "Flow Cone Sand Tests," *J. Am. Conc. Inst. 1*:13-21 (1978).
47. H. M. Sutton and R. A. Richmond, "Improving the Storage Conditions of Fine Powders by Aeration," *Trans. Instn. Chem. Engrs. 51*:97-104 (1973).
48. W. Bruff, "Some Characteristic Qualities of Powder Materials," *Trans. ASME, J. Engrg. Ind.* (B)*91*(2):323-328 (1969).
49. F. A. Zenz and D. F. Othmer, *Fluidization and Fluid-Particle Systems*, Reinhold Publishing Corp., New York (1970).
50. M. A. Goodman and S. C. Cowin, "Two Problems in the Gravity Flow of Granular Materials," *Trans. ASME J. Fluid Mech. 45*(part 2):321-339 (1971).
51. K. Terzaghi, *Theoretical Soil Mechanics*, John Wiley, New York (1943).
52. D. C. Drucker, R. E. Gibson, and D. J. Henkel, "Soil Mechanics and Work Hardening Theories of Plasticity," *Trans. ASCE 122*:338-346 (1957).
53. A. W. Jenike and R. T. Shield, "On the Plastic Flow of Coulomb Solids Beyond Original Failure," *Trans. ASME, J. Appl. Mech.* (E)*26*(4): 599-602 (Dec., 1959).
54. A. W. Jenike, *Gravity Flow of Bulk Solids*, Bul. 108, University of Utah, Utah Engineering Station (1961).
55. A. W. Jenike, *Storage and Flow of Solids*, Bul. 123, University of Utah, Utah Engineering Station (1964) (revised 1976).
56. J. R. Johanson, "Effect of Initial Pressures on Flowability of Bins," *Trans. ASME, J. Engrg. Ind.* (B)*91*(2):395-399 (1969).
57. J. R. Johanson and H. Colijn, "New Design Criteria for Hoppers and Bins," *Iron and Steel Engineer.* (Oct., 1964).
58. J. F. Carr and D. M. Walker, "An Annular Shear Cell for Granular Materials," *Powder Technol. 1*:369-373 (1967/68).
59. M. U. Eiserhart-Rothe and I. A. S. Peschl, "Powder Testing Techniques for Solving Industrial Problems," *Chem. Engrg. 84*(7):97-102 (1977).
60. A. W. Jenike, P. J. Elsey, and R. H. Woolley, "Flow Properties of Bulk Solids," *Proc. ASTM 60*:1168-1181 (1960).
61. A. W. Jenike, *Flow Properties of Bulk Solids*, Bul. 95, University of Utah, Utah Engineering Station (1958).
62. A. W. Jenike and J. R. Johanson, "Settlement of Powders in Vertical Channels Caused by Gas Settlement," *Trans. ASME J. Applied Mech.* (E)*39*(4):863-868 (Dec., 1972).
63. J. R. Adams and W. Ross, "Relative Caking Tendency of Fertilizers," *Ind. and Eng. Chem. 33*(1):121-127 (1941).
64. J. Whetstone, "Solution to the Caking Problem of Ammonium Nitrate and Ammonium Nitrate Explosives," *Ind. Eng. Chem. 44*(11):2663-2667 (Nov., 1952).
65. A. L. Whynes and T. P. Dee, "The Caking of Granular Fertilizers: An Investigation on a Laboratory Scale," *J. Sci. Food Agric.*, *8*:577-590 (Oct., 1957).
66. W. B. Pietsch, "Adhesion and Agglomeration of Solids During Storage, Flow and Handling–A Survey." *Trans. ASME, J. Eng. Ind.* (B)*91*(2): 435-449 (May, 1969).
67. W. B. Pietsch, "The Strength of Agglomerates Bound by Salt Bridges," *Can. J. of Chem. Eng. 47*:403-409 (Aug., 1969).
68. J. Silverberg, J. R. Lehr, and G. Hoffmeister, "Microscopic Study of the Mechanisms of Caking and Its Prevention in Some Granular Fertilizers," *Agric. Food Chem. 6*(8):442-448 (1958).
69. R. R. Irani, H. L. Vandersall, and W. W. Mergenthaler, "Water Vapor Sorption in Flow Conditioning and Cake Inhibition," *Ind. and Eng. Chem. 53*(2):141-142 (1961).
70. R. R. Irani, C. F. Callis, and T. Liv, "How to Select Flow Conditioning and Anti-Caking Agents," *Ind. and Eng. Chem. 51*(10):1285-1288 (1959).
71. M. Peleg and C. H. Mannheim, "Effect of Conditioners on the Flow Properties of Powdered Sucrose," *Powder Technol. 7*:45-50 (1973).
72. M. Peleg, C. H. Mannheim, and Passy, "Flow Properties of Some Food Powders," *N. J. of Food Sci. 38*:959-964 (1973).
73. A. W. Jenike and T. Leser, "A Flow-No-Flow Criterion In the Gravity Flow of Powders in Converging Channels," *Fourth International*

Congress on Rheology, Brown University, Providence, R.I. (1963).

74. P. L. Bernache, "Flow of Dry Bulk Solids on Bin Walls," *Trans. ASME, J. Eng. Ind.* (B)*91*(2):489-496 (1969).
75. H. Wright, "An Evaluation of the Jenike Bunker Design Method," *Trans. ASME, J. Eng. Ind.* (B)*95*(1):48-54 (1973).
76. H. Wright, "Bunker Design for Iron Ores," Ph.D. thesis, Dept. of Chem. Engrg., University of Bradford (1964).
77. R. K. Eckhoff and P. G. Leversen, "A Further Contribution to the Evaluation of the Jenike Method for Design of Mass Flow Hoppers," *Powder Technol. 10*:51-58 (1974).
78. G. Enstad, "On the Theory of Arching in Mass Flow Hoppers," *Chem. Eng. Sci. 30*:1273-1283 (1975).
79. G. Enstad, A. "Note on Stresses and Dome Formation in Axially Symmetric Mass Flow Hoppers," *Chem. Eng. Sci. 32*:337-339 (1977).
80. W. Reisner, "Behavior of Granular Materials in Flow Out of Hoppers," *Powder Technol. 1*:257-264 (1967/68).
81. Schwedes, Jr., "Dimensionierung Von Bunkern" (Dimensioning of Bins), *Aufber Technol. 10*: 535-541 (1969).
82. L. A. S. Z. Peschl, "The Theory of Formation of Arches in Bins," *Trans. ASME, J. Eng. Ind.* (B)*91*(2):423-434 (May, 1969).
83. D. M. Walker and M. H. Blanchard, "Pressures in Experimental Coal Hoppers," *Chem. Eng. Sci. 22*:1713-1754 (1967).
84. R. Aoki and H. Tsunakaw, "The Pressures in a Granular Material at the Wall of Bins and Hoppers," *J. of Chem. Eng.* (Japan) *2*(1):126-129 (1969).
85. H. Tsunakaw and R. Aoki, "Lateral Pressures of Granular Materials in a Model Bin," *J. of Chem. Eng.* (Japan) 7(2):131-134 (1974).
86. P. C. Richards, "Bunker Design, Part 1: Bunker Outlet Design and Initial Measurement of Wall Pressures," *Trans. ASME, J. Eng. Ind.* (B)*99*(4): 809-813 (1977).
87. D. C. Van Zanten and A. Mooij, "Bunker Design, Part 2: Wall Pressures in Mass Flow," *Trans. ASME, J. Eng. Ind.* (B)*99*(4):814-818 (1977).
88. D. C. Van Zanten, P. C. Richards, and A. Mooij, "Bunker Design Part 3: Wall Pressures and Flow Patterns in Funnel Flow," *Trans. ASME, J. Eng. Ind.* (B)*99*(4):819-823 (1977).
89. R. Everts, D. C. Van Zanten, and P. C. Richards, "Bunker Design, Part 4: Recommendations," *Trans. ASME, J. Eng. Ind.* (B)*99*(4):824-827 (1977).
90. R. Moriyama and T. Jotaki, "An Investigation of Wall Pressures in Flowing and Static Bulk Materials in Model Bins," *Intern. Conf. on Design of Silos For Strength and Flow*, Univ. of Lancaster, U.K., Powder Advisory Center, London (Sept., 1980).
91. C. G. Tattersall and L. C. Schmidt, "Model Studies of a Plane Converging Hopper," *Intern. Conf. On Design of Silos For Strength and Flow*, Univ. of Lancaster, U.K. Powder Advisory Center, London (Sept., 1980).
92. P. G. Murfitt and P. L. Bransby, "Pressures in Hoppers Filled With Fine Powders," *Inter. Conf. on Design of Silos For Strength and Flow*, Univ. of Lancaster, U.K. Powder Advisory Center, London (Sept., 1980).
93. U. S. Mukhopadhyah and K. N. Srivastava, "Static and Dynamic Pressure Distribution in Steel Storage Bins Handling Fertilizer Raw Material," *Intern. Conf. on Design of Silos For Strength and Flow*, Univ. of Lancaster, U.K. Powder Advisory Center, London (Sept., 1980).
94. G. E. Blight and D. Midgley, "Pressure Measured in a 20 m Diameter Coal Load-out Bin," *Intern. Conf. on Design of Silos For Strength and Flow*, Univ. of Lancaster, U.K. Powder Advisory Center, London (Sept. 1980).
95. W. S. Patterson, "Measurement of Pressures in Hoppers and Silos," *Inter. Conf. on Design of Silos For Strength and Flow*, Univ. of Lancaster, U.K. Powder Advisory Center, London (Sept., 1980).
96. D. M. Walker, "An Approximate Theory for Pressures and Arching in Hoppers," *Chem. Eng. Sci 21*:975-997 (1960).
97. N. Nanninga, "Gibt die übliche Berechtungsart der Drücke auf die Wände und den Boden von Silobauten sichere Ergenbnisse?" *DeIngenieur 68* (Nov., 1956).
98. J. K. Walters, "A Theoretical Analysis of Stresses in Silos with Vertical Walls," *Chem. Eng. Sci. 28*:13-21 (1973).
99. J. K. Walters, "A Theoretical Analysis of Stresses in Axially Symmetric Hoppers and Bunkers," *Chem. Eng. Sci. 28*:779-789 (1973).
100. K. Clague, "The Effect of Stresses in Bunkers," Ph.D. thesis, Univ. of Nottingham (1973).
101. A. W. Jenike and J. R. Johanson, "Bin Loads," *J. Struct. Div. Am. Soc. Civil Eng. 94*(ST4): 1011-1041 (1968).
102. A. W. Jenike, J. R. Johanson, and J. W. Carson, "Bin Loads, Part 2: Concepts," *Trans. ASME, J. Eng. Ind. 95*(1):1-5 (1973).
103. A. W. Jenike, J. R. Johanson, and J. W. Carson, "Bin Loads, Part 3: Mass-Flow Bins," *Trans. ASM, J. Eng. Ind. 95*(1):6-12 (1973).
104. A. W. Jenike, J. R. Johanson, and J. W. Carson, "Bin Loads, Part 4: Funnel Flow Bins," *Trans. ASME, J. Eng. Ind. 95*(1):13-16 (1973).
105. H. Taksahashi, H. Yanai, and T. Tanaka, "An Approximate Theory for Dynamic Pressure of Solids in Mass Flow Bins," *J. of Chem. Eng. Japan 12*(5):369-375 (1979).

106. H. Takahashi, H. Yanai, and T. Tanaka, "An Approximate Theory for Dynamic Pressures of Solids in Funnel Flow Bins," *J. of Chem. Eng. of Japan 12*(5):376–382 (1979).
107. P. C. Arnold, A. G. McLean, and A. W. Roberts, *Bulk Solids: Storage Flow and Handling*, The Univ. of Newcastle, New South Wales, Tunra Limited (1979).
108. P. C. Arnold and A. W. Robert, "A Useful Procedure For Predicting Stresses of the Walls of Mass-Flow Bins," *AICHE 80th National Meeting*, Boston. Paper No. 49B (1975).
109. A. W. Jenike, "Effect of Solids Flow Properties and Hopper Configuration on Silo Loads," in *Unit and Bulk Materials Handling*, F. J. Loeffler and C. R. Proctor (ed.) Presented at the Materials Handling Conference, ASME Century 2, Emerging Technology Conferences, San Francisco, Calif. (August, 1980).
110. P. G. Murfitt and P. L. Bransby, "Deaeration of Powders in Hoppers," *Powder Technol. 27*:149–162 (1980).
111. J. E. P. Miles, C. Schofield, and F. H. H. Valentin, "The Rate of Discharge of Powders From Hoppers," *Inst. Chem. Eng. Symp. Ser. 29* (1968).
112. B. J. Crewdson, A. L. Ormond, and R. M. Nedderman, "Air Impeded Discharge of Fine Particles from A Hopper," *Powder Technol. 16*:197–207 (1977).
113. A. Wlodarski and A. Pfeffer, "Air Pressure in the Bulk Granular Solid Discharge from a Bin," *Trans. ASME, J. Eng. Ind.* (B)*91*(2):382–384 (1969).
114. W. Bruff and A. W. Jenike, "A Silo for Ground Anthracite," *Powder Technol. 1*:252–256 (1967/68).
115. A. W. Jenike and J. R. Johanson, U.S. Patent 3,797,707, *Bins For Storage and Flow of Bulk Solids* (March 19, 1974).
116. G. E. Reed and J. R. Johanson, "Feeding Calcine Dust with a Bulk Feeder at Falconbridge," *Trans. ASME, J. Eng. Ind.* (B)*95*(1):72–74 (Feb., 1973).
117. L. G. Laszlo, L. Williams, and J. W. Carson, "Brookfield Solves Fine Limestone Feed Problem," *Proc. Int. Bulk Solids Handling* and *Proc. Conf.*, Phila., Pa., International and Scientific Conference Management, Inc. (May, 1979).
118. M. Turco, C. Gaffney, and J. R. Johanson, "Feeding Dry Fly Ash without Flooding and Flushing," *Proceedings Inten. Proc. Int. Bulk Solids Handling* and *Proc. Conf.*, Phila., Pa., International and Scientific Conference Management, Inc. (May, 1979).
119. J. R. Johanson, "Two-Phase Flow Effects in Solids Processing and Handling," *Chem. Eng. 86*(1):77–86 (Jan., 1979).
120. J. C. Williams, "The Segregation of Particulate Materials: A Review," *Powder Technol. 15*:245–251 (1976).
121. J. R. Johanson, "Particle Segregation and What to do About It," *Chem. Eng. 85*(2):183–188 (May, 1978).
122. C. F. Harwood, "Powder Segregation Due to Vibration," *Powder Technol. 16*:51–57 (1977).
123. J. F. Van Denburg and W. C. Bauer, "Segregation of Particles in Storage of Materials," *Chemical Engrg. 71*(2):135–142 (Sept., 1964).
124. J. F. G. Harris and A. M. Hildon, "Reducing Segregation in Binary Powder Mixtures with Particular Reference to Oxygenated Washing Powders," *Ind. Eng. Chem., Proc. Des. Devel. 9*(3):363–367 (1970).
125. K. Ahmad and I. J. Smalley, "Observations of Particle Segregation in Vibrated Granular Systems," *Powder Technol. 8*:69–75 (1973).
126. "Matching Size Eliminates Fertilizer Segregation," *Chem. & Eng. News* (Sept. 24, 1962).
127. K. Clague and H. Wright, "Minimizing Segregation in Bunkers," *Trans. ASME, J. Eng. Ind.* (B)*95*(1):81–85 (1973).
128. *Belt Conveyors for Bulk Materials* (CEMA), Cahners Publishing Company, Boston (1979).
129. K. Shinohara, K. Shojik, and T. Tanaka, "Mechanism of Size Segregation of Particles in Filling a Hopper," *Ind. Eng. Chem., Proc. Des. Devel. 11*(3):369–376 (1972).
130. K. Shinohara, K. Shojik, and T. Tanka, "Mechanism of Segregation and Blending of Particles Flowing Out of Mass Flow Hoppers," *Ind. Eng. Chem., Process Des. Dev. 9*(2):174–180 (1970).
131. T. Tanaka, "Segregation Models of Solid Mixtures Composed of Different Densities and Particle Sizes," *Ind. Eng. Chem, Proc. Des, Dev. 3*:332–340 (1971).
132. H. Matthee, "Segregation Phenomena Relating to Bunkering of Bulk Materials: Theoretical Considerations and Experimental Investigations," *Powder Technol. 1*:265–271 (1967/68).
133. G. W. Fisher, U.S. Patent 3,575,321, *Solid Particulate Material Blender* (April 20, 1971).
134. C. A. Lawler, "New Method Controls Particle Segregation," *Materials Handling Eng.*, pp. 105–108 (Nov., 1968).
135. F. J. Loeffler and C. R. Proctor, (ed.), "Unit and Bulk Materials Handling," presented at the Materials Handling Conference, ASME Century 2, Emerging Technology Conferences, San Francisco, Calif. (August, 1980).
136. Bituminous Coal Research Institute, Pittsburg, Pennsylvania.
137. R. H. Newton, G. S. Dunham, and T. P. Simpson "The TCC Cayalytic Cracking Process for Motor Gasoline Production," *Trans. A.I.ChE. 41*:215–18 (1945).

138. H. H. Morse, U.S. Patent 2,555,052 *Method of Effecting Contract in a Pebble Heater* (May 29, 1951).
139. J. R. Johanson, "The Use of Flow Corrective Inserts in Bins," *Trans. ASME, J. Eng. Ind. 88*:224–230 (1966).
140. J. R. Johanson and W. K. Kleysteuber, "Flow Corrective Inserts in Bins," *Chem. Eng. Prog. 62*(11):79–83 (Nov., 1966).
141. J. R. Johanson, "The Placement of Inserts to Correct Flow Problems," *Powder Technol. 1*: 328–333 (1967/68).
142. O. F. Theimer, "Ablauf fördernde Trichterkonstruktion von Silozellen" (Discharge–Prompting Hopper Construction of Silos), *Powder Technol. 3*:253–248 (1969/70).
143. W. Reisner and M. Eisenhart Rothe, *Bins and Bunkers for Handling Bulk Materials*, Trans. Tech. Publications, Clausthal, Germany (1971).
144. F. M. Thomson, "Smoothing the Flow of Materials Through the Plant: Feeders," *Chem. Eng. 85*(24):77–87 (Oct. 1978).
145. P. J. Carroll and H. Colijn, "Vibrations in Solids Flow," *Chem. Eng. Progress 71*(2):53–65 (Feb., 1975).
146. P. J. Carroll, "Hopper Designs with Vibratory Feeders," *Chem. Eng. Progress 66*(6):44–49 (June, 1970).
147. Lambertus ter Borg, "Evaluation of Shear Test Results on Bulk Solids in Chemical Industry," *German Chem. Eng. 5*:59–63 (1982).
148. H. Tsunakawa, "The Use of Partition Plates and Circular Cones to Reduce Stresses on Particulate Solids in Hoppers," *Intern. Chem. Eng. 22*(2): 280–286 (April, 1982).

10
Fluidization Phenomena and Fluidized Bed Technology

Frederick A. Zenz
Manhattan College, Riverdale, New York

The term *fluidization* is used to designate the gas–solid contacting process in which a bed of finely divided solid particles is lifted and agitated by a rising stream of process gas. At the lower end of the velocity range, the amount of lifting is slight, the bed behaving like a boiling liquid (hence the term *boiling bed*). At the other extreme, the particles are fully suspended in the gas stream and are carried along with it; the terms *suspension*, *suspensoid*, and *entrainment* contact have all been used to designate this action.

10.1. HISTORICAL DEVELOPMENT

The fluidized technique as it is known today was born from the pioneering work of Standard Oil Development Co., The M. W. Kellogg Co., and Standard Oil of Indiana in their efforts to find a better catalytic cracking process than the fixed-bed process that was introduced commercially in 1937. The fixed-bed process was a major improvement over the earlier thermal cracking methods. It yielded more gasoline of higher octane rating and less low-value heavy fuel oil byproduct.

Initial experimentation in developing a still superior process began along the lines of the fixed-bed method. Oil vapor was passed through one of a pair of beds until the catalyst became fouled with carbon formed in the reaction; then the oil vapor was fed to an adjacent fresh bed while air passed through the fouled material to burn off the carbon and regenerate the catalyst. It was soon appreciated that some innovation would be desirable to avoid the complexity and cost of such intermittent operations.

Placing the regenerating and reacting beds in series and continuously moving the catalyst mechanically from one to the other appeared to be an obvious method of approach. Initial experiments indicated that such a system might suffer considerable loss from catalyst attrition unless pneumatic rather than mechanical conveying methods were adopted. Thus, experimentation turned to studies of pneumatic transport of catalysts.

It was soon discovered that, in order to avoid severe erosion as well as attrition, relatively lower gas velocities would be required. This led to the investigation of powder-form catalysts and eventually to the observation that dense beds of powder could be maintained with relatively low carryover losses even at superficial gas velocities that were orders of magnitude greater than the calculated settling velocity of the individual particles making up the bed. At these gas velocities the particles were observed to be considerably agitated, as gas bubbles passed upward through the bed in a manner analogous to the boiling of liquids. Simultaneously, it was observed that the pressure drop through such a boiling or fluidized bed was equal to the weight of the bed charge; the bed was in effect heterogeneously buoyed by the gas stream and thus took on effective flowing properties similar to liquids. These simple experiments gave birth to the present-day fluid-bed concepts.

Before such processing techniques could be applied commercially, considerable further work had to be done to develop satisfactory solids recovery systems, proper aeration techniques, instrumentation, methods of line sizing, minimizing of erosion problems, reactor conversion correlations, heat-transfer data, regeneration rates, and numerous other matters. The first commercial fluid-bed catalytic cracking plant was put in operation in 1942 followed by 31 additional plants during the war years alone.

During the past succeeding 35 years the application of the fluidized technique has spread rapidly to metallurgical ore roasting, limestone calcination, synthetic gasoline, petrochemicals, and even to the design of nuclear reactors. A realization of the scope of applications and number of organizations with a vested interest in fluidization is some indication of its importance and its rate of growth. Table 10.1 gives a representative sampling of fluid-bed applications that have been investigated over the past 40 years. This list is far from exhaustive; yet it records substantial evidence of a lively pace of interest and activity.

Each application listed in Table 10.1 may represent a host of operating units. Table 10.2, for example, lists a number of metallurgical processing installations and Table 10.3 indicates the level of activity in fluidized bed combustion. When it is realized that none of these tables are exhaustive, that equal if not greater activity exists in fluidized bed gasification than in fluidized bed combustion, and that nearly 200 fluidized bed catalytic cracking plants are in operation, it becomes apparent that fluidization as a unit operation has touched almost every process industry and every related corporate body.

Though the commercial development of the fluidized technique was a direct outcome of the work of the major petroleum process development companies, scattered references bearing on the fluidized technique can be found as far back as 1878.

In all processes using the fluidized-solid technique, it is common to handle the solid material in one or more stages or steps and to transfer it from step to step through pipe lines in much the same manner as with a liquid. To raise the material to a higher level, it is carried as a suspension in a gas stream; to take it to a lower level or to a region of higher pressure, the settled material is allowed to flow by gravity down a pipe line to the desired point, the weight of material in the pipe more than equaling the differential pressure.

10.2. ADVANTAGES AND DISADVANTAGES OF THE FLUIDIZED TECHNIQUE

Advantages

Temperature Control. The ability of the fluidized-solid bed to approach isothermal conditions is the outstanding advantage of this method over other methods of carrying out reactions. This factor is vital to nearly all applications, the other advantages generally being of lesser importance.

Close control of reaction variables is well known to be important in obtaining maximum yields of desired products. Of the several variables, temperature is one of the most important, for reaction rates change exponentially with temperature (often doubling for a 10°C change). In the common case in which several competing reactions may occur, a temperature change of a few degrees may shift the balance between the several rates from a favorable one to an unfavorable one.

The relatively close control of temperature that is possible in a fluidized-solid bed is due to

Table 10.1. Some Applications of Fluidization.

APPLICATION	ORGANIZATION	REFERENCE
Acetone recovery	British Celanese Ltd. Wrexham, Wales	*C. & E. N.*, 10/25/65, p. 56
Acetylation of polyformaldehyde	Sci. Res. Inst., Sofia, Bulgaria	*Int. Chem. Eng. 5*, No. 3, 415 (1965)
Acrylonitrile from C_3H_6 and NH_3	E. I. duPont de Nemours & Co., Beaumont, Texas; Standard Oil of Ohio; Montedison	U.S. Pat. 2,736,739 (Feb. 28, 1956); Chem. Eng. *68*, 122 (Jan. 23, 1961); Chem. Week *88*, 40 (Jan. 28, 1961); Hydr. Proc., 144–146 (11/72)
Activated charcoal manufacture	A. Godel	*Chem. Eng. 55* (7), 110 (July 1948)
Activated Carbon regeneration	Battelle Columbus Labs. (Ohio)	*Env. Sci. & Tech., 4*, No. 5, 432–437 (May 1970)
Adsorption with fluid char	L. D. Etherington, et. al.	*Chem. Eng. Progr. 52* (7), 274 (1956)
Adsorption in fluid beds	Food Machinery & Chemical Corp.	*Chem. Eng. 68* (11), 87 (May 29, 1961); Chem. Tech. *11*, 647 (1964)
Adsorption separation of gases in a fluid bed	D. L. Campbell, et al.	U.S. Pat. 2,446,076 (July 27, 1948)
Agglomerates from fines	The Pillsbury Co. Krimo-Ko Corp., Calif. and Hawaiian Sugar Refin. Corp.	*Chem. Wk.*, 6/27/64 p. 96
Alkylation and dehydrogenation of aromatic hydrocarbons	Mamedaliev Petrochemical Inst. ASSR	*Int. Chem. Eng. 5*, No. 3, 467 (1965)
Production of AlF_3	Montedison–Lurgi	*C. E. P., 67*, No. 2, 58–63 (1971)
AlF_3 from HF plus aluminum hydroxide	Kaiser Aluminum and Chem. Corp., Gramercy, La.	*Chem. Eng.*, 1/18/65, p. 92
Calcining aluminum hydroxide	Metals Research Institute, Budapest Hungary	*Brit. Chem. Eng. 10*, No. 10, 710 (1965)
Ammonium chloride and similar condensations of sublimable matl's.	Ivanov Chem. Engrg. Inst. (USSR)	*Int. Chem. Engrg., 8*, No. 4, 651–653, Oct. 1968; *2*, No. 1, 105–108, (Jan., 1962)
Ammoniation of superphosphate	Inorg. Chem. Res. Inst., Czechoslavakia	*Brit. Chem. Eng. 10*, No. 11, 756 (1965) *Chem. Eng. 68* (13), 74 (June 26, 1961)
Aniline via fluid-bed process	American Cyanamid Co.	*Chem. Week 85* (13), 68 (Sept. 26, 1959): U.S. 2,891,094
Aniline from nitrobenzene	M. S. Murthy et. al.	*Chem. Age of India 14*, No. 9, 653 (Dec., 1963)

Table 10.1. (*Continued*).

APPLICATION	ORGANIZATION	REFERENCE
Roasting of arsenopyrites	Piritas Espanolas Auxini S. A., Madrid	*I. & E. C. Proc. Des. & Dev. 2*, No. 3, 214 (1963)
$BaCl_2$ from Cl_2 + $BaSO_4$	Central Salt & Marine Chemicals Res. Inst. Bhavnagar, India	"Fluidization and Related Processes" CSIR, New Delhi, India
Barium oxide by reduction of $BaCO_3$	Barium Reduction Co., South Charleston, West Va.	*Chem. Eng. 67* (9), 107 (May 2, 1960)
Benzotrichloride from Toluene	Western N.Y. Nuclear Res. Ctr.	*Nuclear Technology, 18*, 29–45, (1973)
Bisulfite acid from SO_2 and limestone	Univ. of Naples Italy	*Pulp & Paper Mag. of Canada*, (Oct., 1965)
Blending solids in a fluid bed	Fuller Co.; Wilmot Castle Co.; General Electric N.C.	Fuller Co. Bull. B-1, Catasaugua, Pa.; Chem. Week *85* (19), 73 (Nov. 7, 1959)
Conversion of ethanol to butadiene	Indian Inst. of Tech. Kharagpur	*I. & E. C. Proc. Des. & Dev. 2*, No. 1, 45 (1963)
Calcination of phosphate rock	San Francisco Chemical Co. Montpelier, Idaho	*Chem. Eng. Progr. 55* (12), 77 (Dec. 1959)
Calcination of phosphates, Limestone, and magnesite	Krupp; J. R. Simplot; Door-Oliver	*Brit. Ch. E. & Proc. Tech.*, p. 381, May 1972 Chem. Processing, p. 78, 8/77
Coke calcination	Alum. Co. of Canada	*Chem. Week*, p. 69, 70, 10/19/63, Cdn. Jnl. of Chem. Engrg., 94–96 (April 1965); 146–149 (June 1965)
Calcium carbide nitration	Süddeutsche Kalkstickstoffwerke A. G., Trostberg, Bavaria	*Chem. Eng. 67* (3), 66 (Feb. 8, 1960)
Spent carbon recovery	Westvaco Corp.	*Chem. Engrg.*, p. 97 9/12/77 *Env. Sci. & Tech., 10* No. 5, 454–456 (1976)
Carbon disulfide production	R. P. Ferguson Kurashiki Rayon Co. Japan	U.S. Pat. 2,443, 854 (June 22, 1948); U.S. Pat. 3,402,021
CS_2 adsorption on carbon in 38 ft. diam. fluid bed	Courtlauds, Ltd., Holywell, Wales	*Chem. Eng.*, p. 92, 4/15/63; Brit. Chem. Eng., *8*, No. 3, 180 (1963)
Carbontetrachloride from Char	C. S. I. R. O., Clayton Victoria, Australia	*Brit. Chem. Engrg., 17*, No. 4, 319-322 (April, 1972)
Carbonization and drying by fluidization	V. Charvat	*Paliva 34*, 179 (1954)

Table 10.1. (*Continued*).

APPLICATION	ORGANIZATION	REFERENCE
Cement via fluid-bed process	R. Pyzel	*Chem. Week 80* (7), 108 (Feb. 16, 1957)
MFC fluidized cement calciner	Mitsubishi Hvy. Ind's. Ltd.	*Chem. Engrg.*, pp. 102–106, 6/24/74 C. E. P., pp. 36–44, (August, 1977)
CCl_3F and CCl_2F_2 production	Montedison, Italy	*Chem. Engrg.*, p. 40 2/23/70; pp. 75–77, (6/14/71)
Trichloroethane (methychloroform)	Ethyl Corp.	*Chem. Week*, pp. 30–31, (8/9/69)
Trichlor and perchlor-ethylene	PPG Industries	*Chem. Engrg.*, pp. 90–91, (12/1/69)
Chloromethane from HCl, natural gas, and oxygen	E. I. duPont de Nemours & Co., Inc. Orange, Texas	*Chem. Eng. 68* (15) 126 (July 24, 1961) and *68* (3), 33 (Feb. 6, 1961)
Reactivation of clays	Shri Ram Inst. for Ind. Res., Delhi, India	"Fluidization and Related Processes" CSIR, New Delhi, India
Coal Refinery	W. F. Coxon	*Gas World 144*, 148 (1956)
Coal Gasification	Philadelphia and Reading Corp., New York; Hydrocarbon Research, Inc. Inst. of Gas Tech., FMC Corp., M. W. Kellogg Co., Consolidation Coal Co., etc.	*C & E N*, 4/28/66, p. 68; C. E. P. *60*, No. 6, 35, 58, 69 (1964); Chem. Eng. 12/6/65, p. 78
Fluid Bed boiler	CEGB, England	*Elec. Rev. 176*, 39, 1/8/65
Coal combustion	Babcock and Wilcox (see also Table 2)	*Chem. Engrg.*, p. 114–127, 8/14/78
Coating drug tablets	Abbott Laboratories	*C. E. P. 62*, No. 6, 107 (1966)
Coating particulates	D. E. Marshall	U.S. Pat. 2,579,944 (Dec. 25, 1951)
Coating particles by the Wurster process	Abbott Laboratories; Merck Co.; Smith, Kline & French Laboratories	*Chem. Eng. 66* (15), 55 (July 27, 1959)
Coating in fluid beds	Knapsack-Griesheim A. G., Cologne; Whirl-clad Div., The Polymer Corp.; General Electric Co.; Rockwell Mfg. Co.	*Chem. Week 87* (10) 56 (Sept. 3, 1960); *Chem. Week 89* (2), 48 (July 8, 1961); *Chem. Eng. Progr. 56* (7), 75 (1960); *Chem. Eng. 66* (28), 100 (Dec. 28, 1959) *C. & E. N.*, p. 37, 1/25/71; *Chem. Engrg.* pp. 36–38, 7/12/71

Table 10.1. *(Continued).*

APPLICATION	ORGANIZATION	REFERENCE
Coating particles with ceramics	Battelle Memorial Institute; American Metal Products Co.; Mallinckrodt Nuclear Corp.; 3M Co.; High Temperature Materials (Union Carbide Corp.)	*Chem. Week 87* (19), 59 (Nov. 5, 1960); *Chem. Eng. News*, p. 41 (June 12, 1961); p. 25 (Nov. 21, 1960)
Coating Mb and Cb with silicon in 1900°F fluid bed	Boeing Co., The Pfaudler Co.	*Chem. Eng.* 6/24/63 p. 40; *A. I. Ch. E.*, paper 4E, Dallas Mtg. 2/6-9/66
Fluid coking of coal	Atlantic Refining Co.	*Chem. Eng.* 8/31/64, p. 22
Coking process	Esso Research & Engrg. Co.	*Chem. Eng. 67* (8), 79 (April 18, 1960); *67* (10), 112–115 (May 16, 1960)
coking of pelleted coal fines	U.S. Fuel Co. (U.S. Smelting, Refining & Mining Co.)	*Chem. Eng. 67* (2) 43 (Jan. 25, 1960)
Columbium chloride reduction	Battelle Memorial Institute	*Chem. Eng. 67* (13), 77 (June 25, 1960)
Condenser using fluidized solids	C. B. Beck et. al.	M. S. thesis, M. I. T. (June, 1952)
Cornstarch depolymerization to dextrins	R. Frederickson (A. E. Staley Mfg. Co.)	*Chem. Eng. 66* (18) 80 (Sept. 7, 1959)
Cracking with fluidized sand	Erdölchemie, GmbH, Dormagen; Lurgi, Frankfurt; Ruhrgas, Essen; Bayer; Leverkusen, Germany	*Chem. Eng. 66* (17) 66 (Aug. 24, 1959); *World Petrol. 30* (6) 62 (June 1959); *Petrol. Refiner 40* (10), 137 (1961)
Cristobalite from quartz in 2800°F fluid bed	A. D. Little, Inc.; Wedron Silica Co. Chicago	A. I. Ch. E. Symposium Series No. 62, Vol. 62, 56 (1966)
Cumene dealkylation	J. F. Mathis and C. C. Watson	*A. I. Ch. E. J., 2*, 518 (1956)
Oil decontamination of sand	Univ. of California, Santa Barbara, Calif.	*Chem. Engrg.*, p. 58 8/10/70
Desulfurization of coke-oven aromatics	U.S. Industrial Chemicals Co.	*Chem. Eng. 68* (31), 72 (June 26, 1961); *Chem. Week 88* (23), 64 (June 10, 1961)
Desulfurization of petroleum coke	Institute of Petroleum, Zagreb, Yugoslavia	*Chem. Eng. 68* (2) 126 (Jan. 23, 1961)
Dimethyl terephthalate from toluene	Bergwerksverband A. G.; Toyokoatsu Ind., Japan	*Hydrocarbon Processing 44*, No. 11 275 (1965)
Distillation with fluidized solids	R. W. Krebs and C. N. Kimberlin	U.S. Pat. 2,758,073 (Aug. 7, 1956)
Drying of fluidized solids	W. W. Niven	U.S. Pat. 2,715,282 (Aug. 16, 1955)
Drying with a plurality of fluid beds	W. N. Lindsay	U.S. Pat. 2,676,668 (April 27, 1954)

Table 10.1. *(Continued).*

APPLICATION	ORGANIZATION	REFERENCE
Paper and Textile drying	The Shirley Institute Dunlop Textiles Man-Made Fibers Res. Association	*Brit. Chem. Engrg.*, *17*, No. 5, p. 383 (1972); C. & E. N., p. 37, 12/6/65; *Chem. Enrg.*, p. 74, (4/30/62)
Synthetic fiber heat stretching and heat setting	Dunlop Textiles; Court aulds Ltd. England	*C. & E. N.*, 12/6/65 p. 37
Drying of heat- sensitive materials	Door-Oliver, Inc.	*Chem. Eng. 68* (15), 57 (July 24, 1961)
Electrolysis	Akzo Zout Chem. Netherlands	*Chem. Engrg.*, p. 72, 77, (8/14/78)
Plating and similar electrochemistry	Centre Nat. de La Rech. Sci., Nancy France; Central Electrochem. Res. Inst., India	*Ind. & Eng. Chem.*, *61*, No. 10, 8–17, Oct. 1969 *Ind. & Eng. Chem. Prod. Des. & Dev.*, *9*, No. 4, 563–567 (1970)
Electrostatic ore beneficiation	Univ. of Western Ontario	C. E. Sym. Ser., *66*, No. 105, 236–242 (1970); *Revue de L'industrie Minerale*, pp. 442–449 (June, 1967)
Electrothermal fluidized beds for Zr, P, etc.	Battelle Columbus Labs	*C. E. P.*, *61*, No. 2, 63–68 (1965) *Battelle Tech. Review*, *13*, No. 11, 3–9, (Nov., 1964)
Ethylene manufacture	R. P. Cahn	U.S. Pat. 2,752,407 (June 26, 1956)
Polyethylene, LD & HD	Union Carbide Corp.	*Hydrocar. Processing*, pp. 130–136, (Nov., 1972), *Chem. Proc.*, p. 18, (February 1978) *Chem. Engrg.*, p. 72–73, 11/26/73; pp. 25–27, 1/2/78
Fuel oil from polypropylene residue	Procedyne, New Brunswick, N.J.	*Chem. Engrg.*, p. 57, (12/4/78)
Ethylene oxide from ethylene	National Research Council, Ottawa, Ont., Canada	*Can. J. Chem. Eng. 38* (4), 108 (Aug. 1960)
Extraction countercurrently in fluidized beds	D. E. Weiss and E. A. Swinton	U.S. Pat. 2,765,913 (Oct. 9, 1956)
Feeder for solids	Hanna Furnace Corp. (National Steel Corp.)	*Chem. Eng. 68* (19) 64 (Sept. 18, 1961); *Chem. Week 89* (11), 126 (Sept. 9, 1961)
Fertilizers from oxidation and ammoniation of coal	Central Fuel Res. Inst. of India	*Brit. Chem. Eng.*, August, 1966), p. 799

Table 10.1. *(Continued).*

APPLICATION	ORGANIZATION	REFERENCE
Fire protection with fluidized solids	Anonymous	*Chem. Eng. 58* (2), 160 (Feb., 1951)
Fischer-Tropsch and related processes	H. C. Anderson et. al.	*U.S. Bur. Mines Bull.* No. 544 (1955)
Flowmeter using fluid bed	C. F. Gerald	*Ind. Eng. Chem. 44*, 233 (1952)
Freezing fruits and vegetables; grilling meats and potatoes	Swedish Food Processor	*Brit. Chem. Eng. 8*, No. 12, 800 (1963); ibid. *11*, No. 1, 2 (1966); *Food Processing*, (Nov. 1963)
Roasting chocolate beans	General Foods Corp.	*Chem. Wk.*, (8/31/63), p. 37
Formaldehyde from methane	Bergbau A. G.	*Chem. Week 85* (13), 76 (Sept. 26, 1959)
Gaseous diffusion	P. W. Garbo	U.S. Pat. 2,637,625 (May 5, 1953)
Gasoline from methanol	Mobil Oil Corp.	*C. & E. N.*, pp. 26, 28, (1/30/78) *Chem. Week*, pp. 35–37, (1/25/78)
Grain seed inoculation	Northrup King & Co.	*A. I. Ch. E.*, paper 16e, 57th Nat'l. Mtg., Minneapolis, (9/26–29/65)
Granulation and drying	Aeromatic, Inc.; Niro Atomizer	*Chemical Processing*, p. 53, Mid-Nov. (1978) *Chem. Engrg.*, p. 39, (11/6/78) *Brit. Chem. Engrg.*, p. 811, *12*, No. 6, (June, 1967)
Glycol production	D. F. Othmer and M.S. Thakar	ACS meeting, New York (Sept. 8–13, 1957)
Grinding	Southwestern Engrg. Co.	*Chem. Process* (Chicago) p. 132 (Sept., 1961)
Heat treating and stress relieving	Boeing Co., Seattle, Wash.; The Electric Furnace Co.	*Chem. Eng. News*, p. 80 (April 10, 1961); *Missiles Rockets*, p. 28 (Oct. 23, 1961)
High-temperature bath	C. E. Adams et. at.	*Ind. Eng. Chem. 46*, 2458 (1954)
Hot air for peak shaving	Stal-Laval (Gt. Brit.) Ltd.	*C. & E. N.*, pp. 33, 34, (9/3/73)
Humic acid by fluidized oxidation of Tandur coal	D. P. Agarwal and M. S. Iyengar	*J. Sci. Ind. Res.* (India) *12B*, 443 (1953)
Hydrogen production	Pittsburgh Consolidation Coal Co.	Brit. Pat. 673,332 (June 4, 1952)
Hydrogen from steam	Inst. of Gas Tech.	*Popular Science*, p. 91–94, (Jan., 1977)
HCl and Fe_2O_3 from waste pickling liquors	Nittetu Chem. Engrg. Ltd. American Lurgi Corp.	*Chem. Engrg.*, p. 107 (8/14/78); p. 102–103, (11/13/72)

Table 10.1. *(Continued)*.

APPLICATION	ORGANIZATION	REFERENCE
Regenerating HCl pickle liquor	Lurgi, GMBH; Hilgers AG at Rhein-Brohl, Germ.	*Chem. Eng.*, p. 32, (8/29/66)
HCN from C_3H_3 + NH_3; "Fluohmic" processes for titanium tetrachloride, CS_2, and desulfurizing coke	Shawinigan Products, Ltd., Montreal	*Chem. Eng. 68* (19), 72 (Sept. 18, 1961); *Chem. Week 89* (12) 104 (Sept. 16, 1961); *Ind. Chem. Eng. 53* (1), 19A (1961); *Chem. Eng. News*, p. 55 (Nov. 21, 1960)
HCl oxidation	A. J. Johnson and A. J. Cherniavsky	U.S. Pat. 2,644,846 (July 7, 1953)
Hydrogen sulfide removal	L. Jequier	Brit. Pat. 708,972 (May 12, 1954)
Hydrogenating residual petroleum oils	Hydrocarbon Research, Inc.; British Gas Council, Solihull, England	*Chem. Eng. 67* (19) 69 (Sept. 24, 1960); *Petrol. Refiner 39* (10), 151 (1960); *Chem. Eng. 67* (9), 115 (May 2, 1960)
H-Coal and H-Oil (Liquid fluidized beds)	Hydrocarbon Res., Inc. Cities Service Oil, Co.	*C. E. P., 67*, No. 8, 81–85, 8/71; *Brit. Chem. Engrg. 16*, No. 12, 1117–1119, (Dec., 1971); *Hydro. Proc., 45*, No. 5, 153–158, (May, 1966)
Ilmenite chlorination	L. K. Doraiswamy, et. al.	*Chem. Eng. Progr. 55* (10), 80 (1959)
Incineration of spent caustic & other refinery wastes	Amoco-Dorr-Oliver; Nichols Engrg.	*Chem. Processing*, pp. 8, 9, (Sept., 1973) *Chem. Engrg.* pp. 87–94, (1/2/78); p. 60, (8/14/78); pp. 71, 72, (10/4/71)
Econ-Abator catalytic incineration of organic emissions	Air Resources, Inc. Harshaw Chem. Co. Nipro, Inc.; Dutch State Mines	*C. E. P.*, pp. 31–35 (August, 1977) *Chem. Processing*, p. 13, (August, 1976)
Indole and benzofuran	SNAM Progetti, Italy	*Chem. Engrg.* p. 78, (5/17/71)
Ionic mass transfer	G. J. V. J. Raju	D. Sc. thesis, Andhra Univ., Waltair, India (1959)
Ion exchange	Himsley Engrg Ltd., Toronto; Liquitech, Inc., Houston; Farbenfabri ken Bayer (Mobay)	*C. & E. N.*, p. 23, 24, (Aug. 2, 1976) *Chem. Week*, p. 37, (Aug. 2, 1972) *Chem. Engrg.*, p. 60–62 (Jan. 8, 1973)
Chlorination of pyrites cinder to FeO	Montedison S. p. A.- A. G. McKee	*Chem. Engrg.*, p. 45, (4/26/76)

Table 10.1. (*Continued*).

APPLICATION	ORGANIZATION	REFERENCE
FeS to iron oxide in 14′ diam. bed at 1900°F	Lurgi, GMBH; Outokumpu Oy, Finland	*Chem. Eng.*, (2/14/66) pp. 122–124
Iron ore reducer (Wicke process)	Phoenix–Rheinrohr A. G.	*Chem. Week 85* (17), 42 (Oct. 24, 1959)
Iron ore reduction (multistage)	Arthur D. Little, Inc.	*Trans. Met. Soc. AIME 218* (1), 12 (Feb., 1960)
Iron ore reduction in a self-agglomerating fluid bed	Battelle Memorial Institute	Blast Furnace, Coke Oven and Raw Materials Conference., Apr. 4–6, 1960, Chicago, Ill.
Iron ore reduction by "Nu-Iron" process	U.S. Steel Corp.	*J. Metals 12*, 317 (Apr. 1960); *Chem. Eng. 67* (7), 64 (Apr. 4, 1960)
Iron ore reduction (FIOR)	Exxon Res. & Engrg.; A. G. McKee	*Chem. Wk.*, p. 32, (9/5/73) p. 66, (3/7/64); *Chem. Engrg.* p. 49, (5/28/73)
Iron ore reduction by the "H-Iron" process	Hydrocarbon Res., Inc.; Bethlehem Steel Corp.	*Chem. Eng. 67* (3), 96 (Feb. 8, 1960)
Magnetic Fe roasting	French Iron and Steel Res. Inst.	*A. I. Ch. E.* Symposium Series No. 62, Vol. 62, 15 (1966)
Isomerization of paraffin hydrocarbons	P. Tristmans	*Het Ingrablad Techn. Wetenschapp. Tijdschr. 22*, 269 (1953)
Isophthalonitrile	Badger (Boston); Mitsubishi (Japan)	*Chem. Week*, p. 39, (5/3/72)
Reforming and desulfurization in a magnetically stabilized bed	Exxon Res. & Engrg. Co.	*Chem. Engrg.*, p. 95, (10/23/78)
Maleic Anhydride	Shakhtakhtinskü, et. al.; Mitsubishi Chem. Industries	*Azerb. Khim. Zh.*, No. 2, 91–94 (1965) *Brit. Ch. E. & Proc. Tech.*, p. 13, Aug. 1974, *Chem. Engrg.*, pp. 107–109, (9/20/71)
Maleic anhydride from benzene and butylenes	Inst. for Gen'l. Chem. Warsaw, Poland, Petrochem. Processes Inst. ASSR Acad. of Science	*Brit. Chem. Eng. 10*, No. 10, 710 (1965); *Internat'l. Chem. Eng. 6*, No. 4, 674 (1966)
Maganese ore chloridization	M. L. Skow et. al.	U.S. Bureau Mines Rept. Invest. No. 5271 (1956)
Fluidized Mattress	Medical Univ. of So. Carolina	*Readers Digest* (1970) Dr's. Thos. Hargest and C. P. Artz
Melamine from Urea	BASF, Ludwigshafen,	*C. & E. N.*, p. 62–63,

Table 10.1. *(Continued).*

APPLICATION	ORGANIZATION	REFERENCE
	W. Germany U.S. Steel, Pitt. Coke Co., Chemico	9/22/69; *Chem. Engrg.*, 101–103, (10/19/70) Cdn. Patents 737,475; 737,476
Melamine from Urea and ammonia	Osterreichische Stickstoffwerke, AG; Power Gas Corp.; Uhde, GMBH	*Chem. Wk.*, 3/19/66, p. 87; Chem. Eng. (10/11/65), pp. 180–182
Mercury adsorption	S. F. Yavorovskaya, et. al.	*Khim. Prom.* (1955), 91
Metal oxide reductions	H. G. McGrath and L. C. Rubin	U.S. Pat. 2,671,765 (March 9, 1954)
Metal powders of high purity	J. E. Drapeau and R. J. Halsted	U.S. Pat. 2,758,021 (Aug. 7, 1956)
Reclamation of scrap metals	USI (Engrg.) Ltd.	*Brit. Chem. Engrg.*, *14*, No. 8, p. 1041, Aug. 1969
Metal carbide and nitrides	AERE (Gt. Brit.)	*Chem. Engrg.*, 134–136, 11/11/63
Methane reduction with cupric oxide	W. K. Lewis, et. al.	*Ind. Eng. Chem.*, *41*, 1227 (1949)
LPM process for Fluidized bed Methanation of H_2 + CO	Chem. Systems, Inc.	*C. & E. N.*, p. 30, 1/16/78
Catalytic cracking of methylcyclohexane	S. Tone et. al.	*J. Chem. Eng. Japan*, 7, No. 1, p. 44 (1974)
Molybdenum trioxide reduction	J. M. Dunoyer	Proc. Intern. Symp. Reactivity Solids, Gothenburg, (1952), p. 411
Recovery of metals from dilute solutions	Elec. Council Res. Ctr., Capenhurst, U. K.; Rockwell Intl.; Constructors John Brown	*Chem. Engrg.*, p. 44, (12/18/78); p. 78, 5/11/73, *Brit. Chem. Engrg.*, *15*, No. 9, 1191, (Sept., 1970)
Condensation of naphthalene	J. Ciborowski	*International Chem. Eng.* *2*, No. 1, 105 (1962)
Neutralization of liquids	H. W. Gehm and L. T. Purcell	U.S. Pat. 2,642,393 (June 16, 1953)
Nickel oxide reduction	A. Kivnick and N. Hixson	*Chem. Eng. Progr.* *48* (8), 394, (1952)
NiO and Nickel chloride	Falconbridge Nickel Mines, Quebec	*Chem. Week*, p. 52, (2/10/71); Can. Min. & *Met. Bul.* (August, 1961), p. 601
Niobium pentachloride reduction with hydrogen	Battelle Memorial Inst.; National Steel Corp.; Nova Beaucage Mines, Ltd., Lake Nipissing, Ont., Canada	*Intern. Congr. Chem. Eng.*, June 19–22, 1960, Mexico City, Mexico; *Chem. Eng.* *68* (2), 124 (Jan. 23, 1961); *Chem. Eng. News*, p. 51, (July 4, 1960)

Table 10.1. *(Continued).*

APPLICATION	ORGANIZATION	REFERENCE
Nitric acid by the "Wisconsin process"	Food Machinery & Chemical Corp.	*Chem. Eng. Progr. 52* (11), 483 (1956)
Nuclear liquid fluidized-bed reactor	Martin Co.	*Chem. Week 85* (18) 59 (Oct. 31, 1959)
Numerous processes	Society of Chem. Industry, England	*Fluidisation*, (1964)
Olefin polymerization	J. A. Carver	U.S. Pat. 2,686,110 (Oct. 10, 1954)
Olefins from crude oil on fluidized coke	BASF, Ludwigshafen, West Germany	*Chem. Eng. News 37* (45), 50 (Nov. 9, 1959); *Chem. Engrg.*, p. 17, (11/30/70)
Oxidation of aromatic hydrocarbons in fluid beds	H. K. Pargal	Ph.D. thesis, Univ. of Colorado, (1954)
Oxygen production	P. J. Gaylor	*Petrol. Process. 5*, 1211 (1950)
Generation of ozone	Iowa State Univ.	*I. & E. C. 59*, No. 3 64 (1967)
Packaging of solids	St. Regis Paper Co.	*Chem. Process. 21*, 182 (March 1958)
Paper mill black liquor recovery	Container Corp. of America; Copeland Process Corp.; Green Bay Packaging Inc.; Dorr-Oliver	*Tappi 47*, No. 6, 175A, 1964; *Chem. Wk.*, (7/31/65) p. 31
Continuous particle size separation	Baskakov et. al.; SWECO, Inc.	*Khim. Prom.*, No. 6, p. 59 (1974); *Chem. Processing*, p. 60, (mid-Nov., 1978)
Perchlorethylene	Diamond Alkali; Columbia-Southern Chemical Corp. (Pittsburgh, Plate Glass Co.)	*Hydrocar. Proc. 46*, No. 11, 210, 1967; U.S. Pats. 2,914,575 2,914,576, (Nov. 24, 1959); 2,951,103 (Aug. 30, 1960); 2,952,714 (Sept. 13, 1960); 2,957,924 (Oct. 25, 1960)
Phenanthraquinone	U.S. Steel Corp.	Brit. Pat. 771,085 (March 27, 1957)
Production of phosphorus	Goldberger, W. M. (Battelle)	*C. E. P., 61*, No. 2, p. 63, (February, 1965)
CaO and phosphorus from tricalcium phosphate in a plasma bed at 2000°F	Battelle Memorial Inst.	A. I. Ch. E. Symposium Series No. 62, Vol. *62*, 42 (1966)
Phthalic anhydride by air oxidation of phenathrene in a fluid bed	Central Fuel Research Institute of India	*Ind. Eng. Chem. 53* (7), 14A (1961)
Phthalic anhydride	American Cyanamid Co.	*Chem. Eng. 66* (25), 78 (Dec. 14, 1959); *Chem. Week 85* (3), 37 (July 18, 1959)
Phthalic anhydride from o-xylene	Badger (Cambridge, Mass.)	*Chem. Wk.*, (11/28/64) p. 63; *C. E. P., 66*, No. 9, 49–58, (1970)

Table 10.1. *(Continued).*

APPLICATION	ORGANIZATION	REFERENCE
Prilling fertilizers	Fisons Fertilizers, Ltd.	Brit. Pat. 1,187,372; *Brit. Chem. Engrg., 15,* No. 5, 585 (May, 1970)
Fluidized solids propellants	Bell Aerospace Div. of Textron Corp.	*Chem. Engrg.*, p. 54, (5/15/72)
Radioactive waste solidification	Newport News Ind. Corp.; Aerojet Energy Corp.; Atlantic-Richfield Hanford Co.	*C. & E. N.*, p. 21, (10/16/78); *Chem. Engrg.*, p. 39, (5/3/71)
Hydrocarbon Reforming		*Brit. Chem. Engrg.*, p. 27, (Dec., 1971)
Hydrogenation of waste rubber tires	Hydrocarbon Res., Trenton, N.J.	Paper presented at 165th ACS Nat'l. Mtg. Dallas, Texas, (4/9-13/73)
Scouring in sea water desalination	Brookhaven Nat'l. Lab.	*C. & E. N.*, (3/29/65), p. 42
Scrubbing of gases	Aerotec Industries; Aluminum Co. of Canada Ltd.	*Chem. Eng. 66* (5), 106 (Dec. 14, 1959)
Shale preheater	Oil Shale Corp. (TOSCO)	*C. & E. N.*, p. 24, (5/14/73)
Silicon-organic compounds	S. Nagata, et. al.	*Chem. Eng.* (Japan) *16*, 301 (1952)
Si and Zr tetrachlorides	Stauffer Chemical Co.	*Chem. Eng.*, p. 90, (9/3/62)
$HSiCl_3$ from Si + HCl	K. A. Andrianov, USSR Acad. of Sciences	*Brit. Chem. Eng., 11* No. 9, 927 (1966)
Styrenes from aromatic hydrocarbons	Mamedaliev Inst. ASSR	*Int. Chem.* Eng. *4*, No. 3, 382 (1964)
Sugar, process liquor treating	Spreckles Sugar Co.; American Sugar Co.; Socony Mobil	*Sugar Azucar 56* (5), 33 (1961); *Chem. Week 86* (6), 59 (February 6, 1960)
Combustion of S to SO_2	Celleco	*Brit. Chem. Eng. 8,* No. 6, 414 (1963)
SO_2 removal from gases	R. J. Best and J. G. Yates, Exxon-Esso, Abingdon, England; Mitsubishi, Japan; Westvaco Corp., N.Y.	*I. & E. C. Proc. Des. Dev., 16*, No. 3, 347-352 (1977); *Ind. Res.*, p. 23, May 1971; *Chem. Week*, p. 55, 4/15/67; *C. & E. N.*, p. 85, (2/15/71)
Sulfur and NH_3 recovery	Chemetics Intn'l., Can. Inds., Ltd.	*Chem. Week*, p. 68, (11/22/72)
Sulfur recovery from coke-oven gas	Appleby-Frodingham Steel, England	*Chem. Eng. 67* (9), 109 (May 2, 1960); *J. Inst. Fuel 42*, 319 (July 1958); *Chem. Eng. 65* (21), 74 (Oct. 20, 1958)
Terephthalonitrile	Lummus (Bloomfield, N.J.)	*Chem. Week*, p. 27, (4/11/73)

Table 10.1. *(Continued)*.

APPLICATION	ORGANIZATION	REFERENCE
Tetrafluorohydrazine from NF_3 and carbon	Stauffer Chemical Co.	*Chem. Eng. 68* (2), 124 (Jan. 23, 1961); *Chem. Eng. News*, p. 85 (Sept. 19, 1960)
Titanium dioxide production	LaPorte Titanium Ltd. (London)	*Chem. Week 88* (24), 74 (June 17, 1961); *Brit. Chem. Engrg., 16*, No. 1, 17, (Jan., 1971)
Titanium Tetrachloride	Fabriques De Produits Chimiques, Belgium	B. P. No. 1, 184, 199; *Brit. Chem. Engrg., 15*, No. 6, 735 (June, 1970)
Chlorinating rutile to $TiCl_4$ at 1800°F	Chemical and Metallurgical Research, Inc.	*Chem. Wk.*, p. 139, (9/15/62)
Tungsten from H_2 reduction of WF_6	Battelle Mem. Inst.; Allied Chem. Co.	*Nuclear Applic. 1*, 567 (1965)
H_2 reduction of Nb, Ta & W halides	E. I. duPont de Nemours	U.S. Pat. 3,020,148
Uranium from fluid-bed roasting of lignite	International Resources Corp., Custer, S. Dakota	*Chem. Eng. 67* (19), 113 (May 2, 1960)
Uranium from spent metal by halogenation	Brookhaven National Laboratory	*Chem. Eng. Progr. 56* (3), 96 (1960)
Uranium oxides and fluorides	General Chemical Div., Allied Chemical Corp.	*Chem. Eng. 67* (14), 70 (July 11, 1960); Chem. Week *87* (2), 41 (July 9, 1960)
Fluorination of U to UF_6	General Electric Co. Union Carbide Corp.	*Business Week*, p. 76, 11/16/74; Preprint 42D, 60th annual A. I. Ch. E. Mtg. N.Y., (11/26–30/67)
Direct conversion of UF_6 to UO_2 and UO_2 to UF_6	Argonne Nat'l. Lab., E. I. duPont de Nemours	*Nuc. Sci. & Eng. 20*, 259 (1964); I & E. C. Proc. Des. & Dev. *4*, No. 3, 338 (1965)
Flourination of UF_4 to UF_6 at 2000°F	Union Carbide Nuclear Co., Paducah, Ky.	*C. & E. N.* p. 42 8/20/62
Uranium dioxide by reduction of UO_2	Union Carbide Corp., Nuclear Div.	*Chem. Eng. News*, p. 40, (Mar. 23, 1959); *Chem. Eng. Progr. 56* (3), 4 (1960); *Chem. Eng. 66* (2), 140 (Mar. 23, 1959)
UO_3 from denitration of uranyl nitrate	Mallinckrodt Chemical Works	*Chem. Eng. 67* (6), 80 (March 21, 1960)
UO_2 from UF_6	Royal Inst. of Tech., Stockholm	*Nuclear Technology*, *18*, 177–184, (May, 1973)
UC and UN from UO_2	AERE, Harwell, England	*Chem. Eng.* (11/11/63), p. 134
UC from U plus hydrocarbons	Argonne Nat'l. Lab.; Ill. Inst. of Tech.	C. E. P. Symp. Ser. No. 67, Vol. 62, p. 76 (1966)

Table 10.1. *(Continued)*.

APPLICATION	ORGANIZATION	REFERENCE
Vinyl acetate production	K. Kawamichi, et. al.	Jap. Pat. 1863 (April 30, 1953)
Vinyl chloride production	M. G. Geiger, Jr. Pechiney-Saint-Gobain	Ph.D. thesis, Purdue Univ. 1953–1954; *C. & E. N.*, p. 39, (5/11/70)
Oxychlorination to vinyl chloride	B. F. Goodrich; Ethyl Corp.	*Chem. Week*, pp. 93–107 8/22/64; *Chem. Engrg.*, p. 105, 10/18/71; *C. E. P., 61*, No. 1, 21–26, (Jan. 1965)
Vibratory fluid bed cooler	Rexnord, Inc. Alfa-Lavalthermal, Inc.	*Chemical Processing*, p. 47, (mid-Nov. 1978); *Chem. Engrg.* p. 113, (4/10/78)
Vulcanization of rubber	Rubber and Plastics Research Assoc. of Gt. Britain	*C. & E. N.* p. 41, (8/13/62); *Chem. Eng.* p. 60, (8/20/62)
Hy-Flo process for treating ind. and municipal waste water	Ecolotrol-Dorr Oliver	*Chem. Week*, p. 40, 9/1/76; *Chem. Engrg.* p. 51, (2/13/78)
Fluid bed distillation of wood	Ga. Inst. of Tech.	*I. & E. C. Proc. Des. & Dev. 2*, No. 2, 148 (1963)
Wood distillation	M. S. Dimitri, et. al.	*Chem. Eng. 55* (12), 124 (1948)
Wood distillation	L. W. Morgan, et. al.	*Chem. Eng. Progr. 49* (2), 98 (1953)
Wood pulping	Univ. of Florida	*Chem. Week 86* (1), 34 (Jan. 2, 1960)
Regenerating spent NH_3 base pulping liquor	Copeland Systems, Inc.	U.S. 3,927,174
Zinc ore roasting for contact acid	T. T. Anderson and R. Bolduc	*Chem. Eng. Progr. 49* (10), 527 (1953)
Zinc production	P. W. Garbo	U.S. Pat. 2,475,607 (July 12, 1949)
Zinc ore roasting	Societé des Mines et Fonderies de la Vieille-Montagne, Belgium	*Chem. Week 87* (13), 66 (Sept. 30, 1961)
Recovery of zinc chloride	Conoco Coal Dev. Co.	*Chem. Week*, p. 37, 38, (12/21/77); I. & E. C. *Proc. Des. Dev., 8*, No. 4, 552–558 (1969)

a combination of the following three factors (listed in the order of their importance):

1. Turbulent agitation within the fluidized mass, which breaks and disperses any hot or cold spots throughout the bed before they grow to significant size. It should not be inferred from this statement that the temperature of every solid particle in a given fluidized-catalyst bed is the same. The catalytic activity will differ somewhat from particle to particle, and those with greater activity will accelerate the reaction in their neighborhood to a greater extent. As a consequence, their temperature will be different from that of the surrounding particles of lower activity. However, the departure of the individual particle temperature

Table 10.2. Some Commercial Fluid Bed Installations in the Mining and Metallurgical Industries.

ROASTING PYRITE OR PYRRHOTITE FOR SULFURIC ACID MANUFACTURE

	LOCATION	NO. UNITS	TONS/ DAY/ FEED	SIZE UNIT FT.	TONS H_2SO_4 /DAY	REMARKS
1) West Rand	S. Africa	1	22	14	35	
2) West Rand	S. Africa	1	22	14	35	
3) Daggafontein	S. Africa	3	235	20	250	
4) Western Keefs	S. Africa	3	235	20	250	
5) Bethlehem Steel	USA	3	180	18	250	Sulfating cobalt
6) Anaconda	USA	4	600	18	450	Low grade sulfur
7) Maria Cristina	P.I.	1	120	18	150	
8) Dowa Okayama	Japan	1	130	20	130	Sulfating Cu
9) Randfoutein	S. Africa	3	235	20	250	
10) Albuffoe	Holland	1	50	12	67	Gas recycle Flotation conc.
11) Rumianca Pieve V.	Italy	1	70	20	90	Flotation conc. Gas recycle.
12) Rumianca Pieve V.	Italy	1	70	10	90	Massive pyrite Gas recycle.
13) Rumianca Asenza	Italy	1	70	3	90	Massive pyrite Gas recycle.
14. Virginia mines	S. Africa	3	300	26	350	
15) Stilfontein	S. Africa	1	55	18	65	
16) Stilfontein	S. Africa	2	135	20	150	
17) Sumitomo	Japan	2	240	16	300	
18) Rico Argentine	USA	1	115	20	150	
19) Electrochemica Surden	Italy	1	10	5	13	Massive & Fle. Conc.
20) Chemische Werke Albert	Germany	1	35	7	50	
21) Soc. Intercon- zorsiale Romanagnola	Italy	1	100	10	125	Massive & Flot. Conc.
22) Soc. Monteca- tini Spinetta M.	Italy	2	155	9	200	Gas recycle.
23) Shin Nippon	Japan	1	100	22	125	
24) Dowa Okayama	Japan	1	125	20	110	
25) Hlyvooruitricht	S. Africa	2	130	20	150	
26) Lorado	Canada	1	60	14	50	
27) Atlas Fertilizer	P.I.	1	105	18	120	
28) St. Gebain St. Fons	France	2	235	10	300	Gas recycle.
29) Kennecott	USA	1	130	22	100	Calcine for sponge iron.
30) Anaconda	USA	1	175	22	150	Dry feed.
31) Chile Explora- tion	Chile	1	90	22	100	(10,000′ Elev.)
32) Tokal Ryuan	Japan	2	176	22	150	
33) Felli Nutti	Italy	1	25	6		H_2SO_4 and liquid SO_2
34) Buffelsfontein	S. Africa	1	85	24	100	
35) Harmony Gold Mine	S. Africa	2	100	18	120	
36) Buffelsfontein Est.	S. Africa	2	130	20	150	

Table 10.2. *(Continued).*

ROASTING PYRITE OR PYRRHOTITE FOR SULFURIC ACID MANUFACTURE

	LOCATION	NO. UNITS	TONS/ DAY/ FEED	SIZE UNIT FT.	TONS H_2SO_4 /DAY	REMARKS
37) British Titan Prod.	England	3	340 TPD $FeSO_4$ 1 8 H_2O	16	180	Decomposition of ferrous sulphate monohydrate for iron & sulfuric acid recovery.
38) Calvo Sutelo	Spain	2	160	10	200	
39) Stauffer Chemical	USA	1	120	10	150	
40) Transvaal Gold	S. Africa	1	15	8	20	Also Au & Cu
41) Abonos Sevdia	Spain	2	150	9	200	Massive pyrite Coils in bed.
42) Kaohsiung	Taiwan	1	250	20	300	Coils in bed.
43) Repesa	Spain	2	360	10′ & 18′	500	Massive pyrite 2 stage roaster.
44) Albatros	Holland	2	250	22	350	Coils in bed. Flot. Conc.
45) Sa. Montecatins Follonica	Italy	2	1,000	18.5	1,000	Massive pyrite 36% S–cooling coils in bed.
46) Kimoshima Chemical	Japan	1	94 MT/D	14	Chamber Acid	Pyrite
47) Kowa Seiko	Japan	1	440	26		Pyrite (Vanahara)
48) Kowa Seiko	Japan	2		28	600	Pyrite (Hanaoka)
49) Consolidated Mining & Smelting Co.	Canada	1	350	26		
50) Nehanga	S. Africa	1	500	22		Low grade Cu concentrate.
51) Hartebeest-fontein	S. Africa	1	12.9 ST/D	10		Pyrite Roaster
52) Kosovaka Mitrovica	Yugoslavia	2	370 MT/D	18		Pyrrhotite Flot. Conc.
53) Cinkarna Celje	Yugoslavia	1	140 MT/D	20		Flot. Conc.
54) Eseo	Philippines	2	540 T/D Pyrite +25 T/D of H_2S	30	750 T/D	Pyrite Roaster

ZINC–H_2SO_4 FOR ACID

	LOCATION	NO. UNITS	TONS/ DAY FEED	SIZE UNIT-FT.	TONS H_2SO_4
1) Mitsubishi (Talhei)	Japan	1	55	14	55
2) National Zinc	USA	2	500	18	200
3) Anaconda	USA	1	155	22	155
4) Alcan	Canada	1	150	22	150
5) Dowa Kosaka	Japan	1	20	10	20
6) Nippon Soda	Japan	1	55	14	55
7) Mitsui Mining & Smelt	Japan	1	140	24	140
8) Cinkarna Celje	Yugoslavia	2	110	15	

Table 10.2. (*Continued*).

ZINC–H_2SO_4 FOR ACID

	LOCATION	NO. UNITS	TONS/ DAY FEED	SIZE UNIT- FT.	TONS H_2SO_4
9) Mitsubishi Metal, Akita	Japan	1	140 MT/D	24	120
10) Mitsubishi Mining, Hosokura	Japan	1	96 T/D	18	70
11) Zorba Sabac	Yugoslavia	1	50 MT/D	11	
12) Toho Zinc	Japan	1	80	18	70

ROASTING OF MISCELLANEOUS SULFIDES FOR METALLURGICAL PURPOSES

	LOCATION	UNITS	TONS/ DAY/ FEED	SIZE UNIT FT.	MATERIAL TREATED
1) Cochenour Williams	Canada	1	20	6	Arsenic, removal for gold.
2) Golden Cycle	USA	1	65	14	Telurium removal for gold.
3) Campbell Red Lake	Canada	1	65	14	Arsenic removal for gold.
4) New Dickenson	Canada	1	20	6	Arsenic removal for gold.
5) Giant Yellow-knife	Canada	1	55	14	Arsenic removal for gold.
6) Negus Mines	Canada	1	30	10	Arsenic removal for gold.
7) Dowa Mining Co.	Japan	1	85	20	Copper-Zinc sulfating.
8) Falcon Mines	S. Rhodesia	1	50	16	Arsenic removal for Au.
9) St. John del Rey	Brazil	1	55	14	Arsenic removal for Au.
10) Campbell Red Lake	Canada	1	45	12	Arsenic removal for Au.
11) Kilembe	Uganda	1	70	16	Dead roast copper.
12) Union Minlere	Congo	1	85	14	Sulfate roast copper & cobalt.
13) Anaconda	USA	1	110	10	S removal for sponge iron.
14) Kerr-Addison	Canada	1	115	22	S removal for Au.
15) Macalder-Nyanza	Kenya	1	75	20	Cu-Zn sulfate roast.
16) Anaconda	USA	1	175	22	Calcines for sponge iron.
17) Kerr-Addison	Canada	1	115	22	S removal for Au.
18) Ndola Copper	N. Rhodesia	1	45	17	Cu Cobalt Sulfating.
19) Mitsubishi Metal	Japan	1	50	14	Dead roast Cu-Zn conc.
20) La Luz	Nicaragua	1	50	16	Cu dead roast.
21) Sherritt Gordon	Canada	1	30	3	Ni-Cu concentrate.
22) Giant Yellow-knife	Canada	2	150	14′ & 16′	Arsenic removal for gold.
23) Ndola Copper Refinery	N. Rhodesia	1	65	20	Cu-Co sulfating.
24) Rhokana Corp.	N. Rhodesia	1	130	20	Sulfate roast copper and Co.
25) Union Miniere	Congo	1	150	16	Cu-Co sulfating.
26) SGM d'Hoboken	Belgium	1	45	18	Co sulfating.
27) Union Miniere	Congo	1	150	16	Cu-Co sulfating.
28) International Nickel Co. (Thompson)	Canada	3	900	15	Partial roasting sulfide conc.
29) International Nickel Copper Cliff	Canada	3	500	12	Nickel sulfide roasts.
30) Tennessee Copper Co.	USA	1	300	12	Partial roasting of sulfide.
31) Getchell Mines	USA (Nevada)	1	1800	16	Gold ore.

Table 10.2. *(Continued)*.

ROASTING OF MISCELLANEOUS SULFIDES FOR METALLURGICAL PURPOSES

	LOCATION	UNITS	TONS/ DAY/ FEED	SIZE UNIT FT.	MATERIAL TREATED
32) Associated Lead Mfgs.	England (New Castle)	1	4–11	5	Oxidizing roasts of residue mattes for Sn & Cu recovery.
33) Johnson Matthey & Co.	England (Brimsdown)	3	3–9	4′–1½″	Oxidizing roast of residue matter for Ni and other recoveries.
34) Johson & Sons Smelting Works Limited	England	1	280–840+/Hr.	4′–1½″	Oxidation of minerals semiprecious mineral recovery.
35) Phelps-Dodge	USA	1	66 T/Hr.	22	Cu Conc. Partial Roast.

DRYING AND SIZING

	LOCATION	NO. UNITS	TONS/ HR.	SIZE UNIT FT.	MATERIAL
1) Nelco-Canaan	USA	1	50	6	Dolomite Dryer.
2) Nelco-Adams	USA	1	110	8	Limestone Dryer.
3) National Gypsum	USA	1	30	9	Dolomite Dryer.
4) Marquette Cement	USA	1	40	4	Cement Rock Dryer.
5) Columbia-Southern	USA	1	50	5	Limestone Dryer.
6) Peerless Cement	USA	1	35	8	Blast Furnace Slag.
7) Wyandotte Chemcial	USA	1	32	8	Blast Furnace Slag.
8) Lone Star Cement	USA	1	40	6	Oyster Shell.
9) Universal Atlas Cement (N.Y.)	USA	1	40	8	Blast Furnace Slag.
10) Umgababa Minerals	S. Africa	1	30	12	Ilmenite Concentrate.
11) Monsanto Chemical	USA	1	530 lbs.	5	Chlorinated Hydrate Dryer.
12) National Lime & Stone	USA	1	150	12	Limestone Dryer.
13) Universal Atlas (Ala.)	USA	1	40	8	Blast Furnace Slag.
14) J. G. Stein Co.	Scotland	2	10	6	Clay Dryer & Iron Carbonate to magnetite converter.
15) Associated Cement	India	1	30	7	Slag Dryer.
16) Guardite Company	USA	1	15	3	Sand Dryer.
17) Victor Chemical	USA	1	1.14	5	Chemicals.
18) Detergent Co.	USA	1	10	14	Detergent.
19) Detergent Co.	USA	1	10	14	Detergent.
20) Iron Ore Co.	Canada	1	10	3	Iron ore dryer.
21) Quebec Cartier	Canada	2	500	12	Iron ore dryer.
22) Dorr-Oliver-Long	Canada	1	15	3	Sand Dryer.
23) Victor Chemical	USA	1	2.75	9½	Chemicals.
24) Victor Chemical	USA	1	0.75	9½	Chemicals.
25) Huron Portland	USA	1	500	11	Limestone dryer.
26) Chemical Lime Inc.	USA	1	167	6	Limestone dryer.
27) Anglo Alpha Cement	S. Africa	1	20	4′–4″	Slag dryer (suction system) (Conversion of competitive dryer.)
28) Wabash Mines Ltd.	Canada	3	370	14	Iron Ore dryer.
29) IMC	USA	1	323 LT/Hr.	14	Phosphate Rock Con.

Table 10.2. (*Continued*).

DRYING AND SIZING

	LOCATION	NO. UNITS	TONS/ HR.	SIZE UNIT FT.	MATERIAL
30) Victorville	USA	1	150	5	Limestone dryer.
31) Glidden	USA	1	20	5′–8″	Ilmenite dryer.
32) Chevrolet Motors	USA	1	20	4′–9″	Foundry sand.
33) Victor Chemical	USA	1	6	15	Chemicals.
34) African Metals	S. Africa	2			Phosphate conc. dryer.
35) Montana Phosphate	USA	1	130 T/D	9	Phosphate rock.
36) U.S. Reduction	USA	1	4,000*f*/Hr.	3	Alumina Compounds.

CALCINING SYSTEMS

	LOCATION	NO. UNITS	SIZE FT.	TONS/DAY	MATERIAL
1) New England Lime Co.	USA	1	12	100 (CaO)	Limestone (5-compt.)
2) Wright Construction	USA	1	7.12. 14.11	360	Chrome Ore.
3) City of Lansing, Mich.	USA	1	6	30	Lime sludge.
4) New England Lime Co.	USA	1	12	100	Limestone.
5) J. G. Stein Co.	Scotland	1	9	175	Fire clay (4 compt.)
6) Central Farmers	USA	1	14	500	Phosphate rock (3 compt.)
7) Corn Products	USA	1	7	25	Carbon reactivator (3 compt.)
8) Cleveland-Cliffs	USA	1	3	24	Iron ore pilot plant.
9) San Francisco Chemical	USA	2	14	1000	Phosphate rock (3 compt.)
10) Anglo Lautaro	Chile	1	7	180	Caliche (2 compt.)
11) Caroline Tuff-Lite Co.	USA	1	8	960	Stone preheater (3 compt.)
12) J. G. Stein Co.	Scotland	1	9	175	Fire clay (4 compt.)
13) J. R. Simplot Co.	USA	1	15	1000	Phosphate rock calciner.
14) Chemical Lime Co.	USA	1	12	200 (CuO)	Limestone calciner.
15) Djebel Onk	Algeria	3	23	3000	3 compartment with aftercooler. Phosphate rock.
16) Soc. Montecatini Follonica	Italy	2	10	750	3 compartment magnetic roast.
17) Bay State Abrasives	USA	1	3	24	SiC grit.
18) W. S. Moore	USA	1	5	120 LT/D	Pilot Plant–iron ore.
19) GAFSA	Tunisia	1	4	50 LT/D	Phosphate Rock.
20) Djebel Onk	Algeria	1	4	50 LT/D	Phosphate Rock.
21) OCP	Morocco	1	4	50 LT/D	Phosphate Rock.
22) International Minerals	USA	1	4	25 LT/D	Phosphate Rock.
23) Smith Douglas	USA	1	4	25 LT/D	Phosphate Rock.
24) S. D. Warren Co.	USA	1	10	70 T/D	Lime Mud Reburning.

Table 10.2. (*Continued*).

CALCINING SYSTEMS

	LOCATION	NO. UNITS	SIZE FT.	TONS/DAY	MATERIAL
25) Kimberly-Clark	USA	1	9	50	1 compartment– paper mill Lime Sludge.
26) Billiton	Holland	1	3		Pilot (2 compt.) Tin Volenbering.
27) Anchor Minerals	USA	1	12	220 T/D	Limestone Calciner.
28) J. R. Simplot	USA	1	15	1000 T/24 hrs.	Phosphate Rock.
29) Chas. Pfizer (NELCO)	USA	1	12′	200 ST/D	Limestone.
30) Phelps-Dodge	USA	1	13′	220 T/D	Lime Calciner.
31) Ford Motor Co.	USA	1	9′	15 T/Hr.	Foundry Sand
32) Ford Motor Co.	Mexico	1	9′-6″	2 T/Hr.	Foundry Sand.
33) Gainesville	USA	1	$2\frac{1}{2}$″	4 T/D	Lime Mud reburning.

from the mean value for the bed will be much less than in a fixed-bed converter because of the turbulent mixing, the high heat-transfer rates and the high bed heat capacity.

2. High heat capacity of the bed relative to the gas within it. This factor stabilizes the temperature of the bed, permitting it to absorb relatively large heat surges with only small temperature change. For example, a bed of ordinary sand, fluidized with air at a solids concentration of about 70 lb/ft^3, would contain only about 0.05 lb of air/ft^3, corresponding to a mass ratio of 1,400 to 1.

3. High heat-transfer rates, which are possible because of the large amount of transfer surface per unit volume of the fluidized bed. This permits rapid leveling of any temperature surges either from the incoming gas or from reaction within the bed. Although the heat-transfer coefficients are not unusually high, the amount of surface per unit volume is very large; for example, the surface area of a bed of ordinary sand would be in the range of 1,000 to 5,000 ft^2/ft^3 of bed.

The remarkable uniformity of temperature in a well-fluidized bed has been noted in many references. Temperature traverses in large fluidized-catalyst beds indicate that the point-to-point variation is less than 10°C when the feed-gas temperature is not greatly different from that of the bed and particularly if the inlet gas is carefully distributed.

Continuity of Operation. The ability to handle the fluidized solid like a liquid permits the technique to be easily adapted to many continuous operations, thereby obtaining the advantages of lower labor requirements, precise and automatic control of process variables, and uniformity of product quality.

Heat Transfer. The fluidized-solid technique is a convenient method for transferring heat, either alone or in conjunction with other operations, such as catalysis, gas-solid reactions, and transport of solids and fluids incidental to these operations.

The advantages of the fluidized-solid technique for heat transfer are as follows:

1. The possibility of combining heat transfer with other operations.
2. For large heat-transfer units, the equipment volume can be smaller than with conventional heat exchangers, because so much transfer surface is available in the fluidized bed and there is a high rate of heat transfer between the bed and an external heat-transfer surface.
3. Corrosion resistance and extreme-tempera-

Table 10.3. Fluidized Bed Coal Combustion Developments among Major U.S. Interests.

1) Argonne Nat'l. Lab. (Argonne, Ill.)	PFBC	a) designing a 24,000–94,000 lb/hr steam boiler as a CTIU to fire 6 to 18 tons of coal/ day expected completion 1981; Stearns-Roger doing the actual design; Bed vel. ≈ 7 ft/sec, press, 3–12 atm., size 3′ × 3′ to 4′ × 4′
		b) operated a 6″ diameter PFBC at up to 10 atm. pressure
2) American Electric Power Co. (N.Y.)	PFBC	a) sponsored feasibility study of combined cycle plant (105 MW_e steam turbine & 65–70 MW_e gas turbine) with Stal-Laval and Woodall–Duckham; sponsoring pilot scale tests at CRE in Leatherhead
3) Babcock Contractors (subsidiary of Babcock & Wilcox Ltd.)	AFBC	a) installing a 60,000 lb/hr boiler in an Ohio State mental hospital (at a cost of $4,300,000)
4) Babcock & Wilcox (Alliance, Ohio)	AFBC	a) operating a 20,000 lb/hr steam 6′ × 6′ boiler under EPRI contract firing 22 tons/ day of coal
		b) operates a 39″ × 39″ unit under EPRI contracts to study sorbent utilization firing 500 lbs/hr of coal with 8 ft/sec fluidizing velocity
		c) operates a 1′ × 1′ unit also studying sorbent and fuel characteristics
5) Babcock & Wilcox (North Canton, Ohio)	AFBC	a) has designed a product line of FBC boilers producing steam in the 50,000 to 300,000 lb/hr range
6) Battelle, (Columbus, Ohio)	AFBC	a) has operated a 6″ dia. MS-FBC bench-scale unit with a combustor superficial velocity of 30 ft/sec.
		b) is designing (with Foster-Wheeler and A. G. McKee) a 40,000 lb/hr steam boiler prototype to serve Battelle's facility wherein the combustor will be 6′–0″ i.d. and the external heat exchanged 5′ × 10′
7) Burns & Roe Inc. (Oradell, N.J.)	AFBC	a) designing a 200–300 megawatt utility power plant under a DOE $1,300,000 study contract
		b) designed a 570 MW_e under a DOE contract
	PFBC	c) designed a 583 MW_e FBC combined cycle plant in association with Babcock & Wilcox and Pratt & Whiting div. of United Technologies, Inc.; unit will have 5 beds 71′ × 24′
8) Cleaver Brooks Div. of Aqua-Chem. Inc. (Milwaukee, Wisconsin)	AFBC	a) a mfr. of packaged boilers supplied the Alexandria Va. test facility of Pope, Evans & Robbins
9) Combustion Engineering (Windsor Locks, Conn.)	AFBC	a) designing a 50,000 lb/hr steam generator at the Great Lakes Naval Training Center scheduled to be in operation before the end of 1980
		b) operating a 3′ × 3′ test unit delivering 2300 lbs/hr steam; also have a Plexiglas cold flow model of this unit
		c) made a design study for a retrofit FBC unit for Con. Edison's 500 MW_e Arthur Kill #3

Table 10.3. *(Continued).*

		unit on Staten Island under EPRI and NYSERDA sponsorship
10) Combustion Power Corp. (Menlo Park, Calif.)	PFBC	a) operated a 40 ft^2 bed (~7′ i.d.) at 4 atm. press. burning Illinois No. 6 coal (4% Sulfur)
	APBC	b) operated a 2.2 ft. square combustor principally for corrosion testing
11) Curtiss-Wright Corp. (Woodridge, N.J.) with Stone & Webster (Boston) and Dorr-Oliver (Stamford, Conn.)	PFBC	a) constructing a combustor capable of producing 130,000 lbs/hr of steam from combustion of 109 tons/day of coal b) operating a bench scale unit $\frac{1}{4}$ the diameter of the above 13 MW_e pilot plant to provide design data
12) The Ducon Co. (Mineola, N.Y.)	AFBC or PFBC	a) has patented a modular combustor incorporating conceptual improvements in mechanical design, control, and operation
13) Energy Products of Idaho, a subsidiary of Energy, Inc. (Idaho Falls)	AFBC	a) has sold a dozen FB incinerators (Trade named Fluid Flame) in sizes producing 10,000 to 30,000 lbs/hr steam burning wood wastes, corn cobs, fruit pits etc.
14) Energy Resources Co. (Cambridge, Mass.)	AFBC	a) has operated a 2′ × 2′ FBC unit at 4 to 16 ft/sec testing 16 different coals b) is building a 6′ × 6′ unit for further test work & to establish credibility c) offering to design and build industrial size AFBC boilers producing up to 500,000 lbs/hr of steam
15) Exxon Res. & Engr'g. Co., (Florham Park, N.J.)	AFBC	a) operates a Plexiglas cold flow model 12″ × 90″ in cross section and 12′ high at superficial air velocities up to 15 ft/sec using horizontal simulated tubes 2″ to 6″ in diam. at various C–C spacings
16) Exxon Res. & Engr'g. Co., (Linden, N.J.)	PFBC	a) operates a 12″ i.d. miniplant burning 380 lbs of coal/hr with a max. operating press of 10 atm. at superficial velocities up to 10 ft/sec b) operates a 4.5″ i.d. bench scale unit burning 30 lbs coal/hr
17) Fluidyne Engrg. Corp., (Minneapolis)	AFBC	a) constructing two 5 × 11 foot beds (capable of producing up to 28,000,000 BTU/hr) at the Owatonna Tool Co. in Minnesota to supply high temp. air for paint-curing ovens, plating lines, and space heat b) are offering 15,000,000 BTU/hr modules on the open market c) have 3 test units in operation for demonstration and customer's systems studies
18) Fluidyne Engr'g. Co., The City of Wilkes-Barre, and the Shamokin Area Industrial Corp.	AFBC	a) have submitted proposals to DOE for construction of units to burn 910,000,000 yds of culm or anthracite wastes (equivalent to 250,000,000 tons of coal) scattered in the northeastern parts of Pennsylvania
19) Foster Wheeler Corp. (Livingston, N.J.)	AFBC	a) constructing a 100,000 lb steam/hr unit at Georgetown Univ.; this is a 2 bed unit with over bed feed of coal and limestone to operate a 8 ft/sec fluidizing velocity burning 125 tons/day of coal

Table 10.3. *(Continued).*

		b) operated a 6′ × 6″ cold model to study scale up correlations regarding tube spacing, solids distribution, etc. c) operates two test facilities in Livingston, one is 1.75′ × 1.75′ and the other a 36 ft^2 bed. d) offering commercial warrantees on FBC boilers up to 600,000 lbs/hr of steam e) operated an 18″ diameter unit burning 100 lbs/hr coal; this unit now at METC f) designed a 70,000 lb/hr steam AFBC boiler for Ford Motor Co.
20) The General Electric Co. (Schenectady, N.Y.)	AFBC & PFBC	a) operate a 1′ × 1′ and a 2′ × 2′ test bed to explore effects on in-bed exchanger tubes
21) General Electric and TVA	AFBC & PFBC	a) completed conceptual design and cost comparison study of a 750–925 MW_e FBC utility power plant
22) Grand Forks Energy Research Center (Grand Forks, North Dakota)	AFBC	a) this DOE Laboratory has been testing various coals and lignites for bed sulfur retention in a 6″ diameter combustor
23) International Boiler Works, subsidiary of Comb. Equip. Assoc., Inc. (East Stroudburg, Pa.)	AFBC	a) manufacturing an FBC from an Energy Resources Co. design
24) Johnston Boiler Co. (Ferrysburg, Mich.)	AFBC	a) operates a 10,000 lb/hr FBC steam generator at its Michigan offices b) offering 2500 to 50,000 lb steam/hr units (under license from Britains Fluidfire Ltd. through CSL) c) constructing a 23,500,000 BTU/hr boiler for the Ohio Center Convention Complex in Columbus, based on a design developed by Britain's CSL.
25) John Zink Co., (Tulsa, Ohio)	AFBC	a) is reported to be fabricating an in-house FBC
26) Morgantown Energy Technology Center (Morgantown, W.Va.)	AFBC	a) designing a 60,000 lb/hr steam capacity unit to burn 10 to 90 tons of coal/day as a Component Test & Integration Unit with two beds each 6′ × 6′ stacked vertically surmounted by a 3′ × 6′ carbon burnup cell. The beds will operate at superficial velocities of 7 to 15 ft/sec through beds 2 to 8 feet deep; this CTIU will be sited on the Medical Center Campus of W.Va. Univ. & is scheduled for operation end of 1979 b) have an 18″ diam. unit for test purposes
27) NASA-Lewis Res. Center (Cleveland, Ohio)	PFBC	a) operates a bench scale PFBC for technical support studies; burns 10–80 lbs/hr of coal at bed press of 25–200 psia in an 8.9″ to 20″ conical bed
28) Oak Ridge National Lab (Tenn.)	AFBC	a) carried out conceptual design of a unit geared to large apartment complexes b) operated a 4′ × 4″ cold flow model to observe bubble flow past tube arrays c) has developed a concept for an atmospheric fluidized bed coal combustor for Cogenera-

Table 10.3. *(Continued).*

		tion (ccc) to produce 300 to 500 kW of electricity plus some 2,500,000 BTU/hr of useful heat
29) Oregon State University (Corvalis, Oregon)	AFBC	a) operate a 3′ × 3′ × 35′ high cold flow model to study tube spacing effect, mining, etc.
30) Pope, Evans and Robbins (N.Y.)	AFBC	a) operating a 0.5 MW_e (5000 lb/hr Steam) pilot plant at Alexandria, Va., $1\frac{1}{2}'$ × 6′ in cross section b) operating a 30 MW_e plant at Rivesville comprising 3 modules each 10′ × 12′ rated at 30 MW_e plus a smaller "carbon burnup cell"; this unit produces 300,000 lbs steam/hr burning 330 tons of coal/day c) has licensed Mitsubishi and IHI of Japan to build a 300,000 lb/hr AFBC boiler to burn "coke breeze" waste
31) Process Equipment Modelling and Mfg. Co., (Nelsonville, N.Y.)	AFBC	a) designed and operated a Plexiglas cold model of a 6″ combustor and 3″ × 14″ external heat exchanger for Battelle's MS-FBC development
32) Stone & Webster Engr'g. Co. (Boston)	AFBC	a) designing a 500–600 megawatt utility power plant under a DOE $1,350,000 study contract
33) The Trane Co. (La Crosse, Wisconsin)	AFBC	a) has under development a proprietary design of FBC packaged boilers
34) Univ. of West Va. (Morgantown)	AFBC	a) operates a 2′ × 2′ cold model for mixing studies and has under construction a 2′ × 2′ FBC boiler
35) Wormser Engr'g. Inc. (Lynn, Mass.)	AFBC	a) operating a demonstration unit in a Lynn, Mass. factory where it is used for space heating b) expects to offer a line of FBC boilers in the 2 to 10 MW_{Th} range

ture resistance can be easily obtained by using ceramic materials for vessels and for granular solids.

4. The transfer can be effected in two stages with the fluidized solid acting as the heat reservoir to carry the heat from one fluid to the other. The stages may be physically close together or far apart.
5. The transfer may be effected with extreme rapidity because of the large surface available; this is important when undesirable reactions would occur at intermediate temperatures.
6. Similarly, a liquid may be heated, vaporized, and dispersed in a small fraction of a second by direct contact with hot fluidized solids.

Catalysis. Fluidized-solid technique is particularly adaptable to the contacting of free-flowing, nonsticky, granular solids with gases. It may therefore be applied in catalytic gas reactions in which solid catalysts are used. The technique has been most widely applied to the catalytic cracking of petroleum because of the unique combination of advantages that are inherent in fluidized-solid processing:

1. Control of reaction temperature.
2. Maintenance of uniform catalyst activity (as well as continuous catalyst regeneration)
3. Continuous removal of solid byproducts
4. Supply of heat to endothermic reaction
5. Simple equipment with few moving parts

6. Continuous operation with automatic control

Most catalysts gradually lose their activity during use because of poisoning or coating of the active surface with byproducts. Replacement or regeneration is therefore eventually required. As the activity decreases, operating conditions in conventional catalyst contactors must be altered in order to maintain the operating rate. Use of a higher temperature is one expedient, but this may increase the cost of the product. Lowering the operating rate is another expedient, but this also results in increased costs through higher investment because the plant must be large enough to manufacture at an average rate sufficient to compensate for the low-rate period.

In contrast, the fluidized-solid technique makes possible the maintenance of a definite level of catalyst activity because partially spent catalyst can be continuously withdrawn and fresh catalyst added. The level of activity is determined by the proportion between the rate of loss of catalyst activity and the rate of withdrawal of the catalyst. A stable catalyst will require a lower catalyst withdrawal rate than would one which loses activity rapidly; correspondingly, a higher level of activity would require a larger withdrawal rate. The activity in any actual bed must be a compromise based on economic considerations, taking into account the following factors:

1. Relationship between yields and catalyst activity
2. Value of incremental yield
3. Rate of catalyst degradation
4. Costs of handling and regenerating the catalyst
5. Catalyst losses during handling and regenerating
6. Cost of catalyst

Gas–Solid Reactions. Advantages of the fluidized-solid technique for carrying out gas–solid reactions are the isothermal reaction bed, the easily varied time of contact, the effective contact (as compared with rotary kilns or tray-type reactors), the simple methods of handling solids (no moving parts) and transferring heat, and the ease of continuous, automatic operation.

Disadvantages

Solids that do not flow freely or tend to agglomerate cannot be processed in a fluidized-solid reactor; rotary kilns and tray-type reactors are not thus limited. As the reaction proceeds, fine solid particles may be formed which will become entrained in the gas leaving the fluidized bed; recovery means must usually be included in the design. The pressure drop in the gas system of a fluidized-solid boiling-bed-type reactor is larger than in kilns or tray-type reactors because the gas supports and fluidizes the solid; this pressure drop may sometimes be a serious objection because of the larger compressors required.

Occasionally it is desirable to obtain a temperature gradient in a catalytic process; for example, a higher temperature may be desired at the upper part of a reactor in order to effect a cleanup of residual reactant. In the simple boiling-bed fluidized-solid reactor this is not possible; however, by proper design, including baffles or other staging or zoning means plus internal heat exchangers or by the use of series reactors, temperature gradients may be secured.

The pressure drop through a fluidized bed may be large as compared with that through an ordinary heat exchanger. If the fluidized bed is used solely for heat transfer, this high pressure drop might be detrimental unless circumvented by use of shallow beds.

The pressure drops must be balanced throughout the system in such a manner that gases will not flow to undesirable parts of the system; this may necessitate providing a gas purge at several points.

The observation that gas bubbles rising through the fluidized solids contain but little of the solids indicates that the efficiency of contact in a boiling bed may be much less (in terms of the availability of the active surface of the solids) than in a fixed-bed reactor; this disadvantage is at least partially overcome by the fact that the fluidized catalyst can be of much smaller particle size (and therefore much greater surface) than that used in fixed-bed reactors.

Since 1960, a great deal of experimental and theoretical work has been carried out in studying this bubble phenomenon in fluidized solids. When finally resolved it should be possible to predict the degree of reaction in any such gas-solids contactor and it is hoped, to indicate how bubbles might be controlled and contact maximized. To date, these studies have succeeded in rationalizing the scaleup of pilot plant results to commercial reactor designs. Without experimental reaction rates and conversions from an operating fluidized bed pilot plant, it is still a nebulous procedure to predict what might occur in a commercial scale plant; fixed bed reaction kinetics are not as yet, with any degree of certainty, transferable to a bubbling bed.

An appreciation of the operating limits and operating characteristics of a fluidized bed of particles is likely best gleaned from an outline of the steps involved in the design procedure.

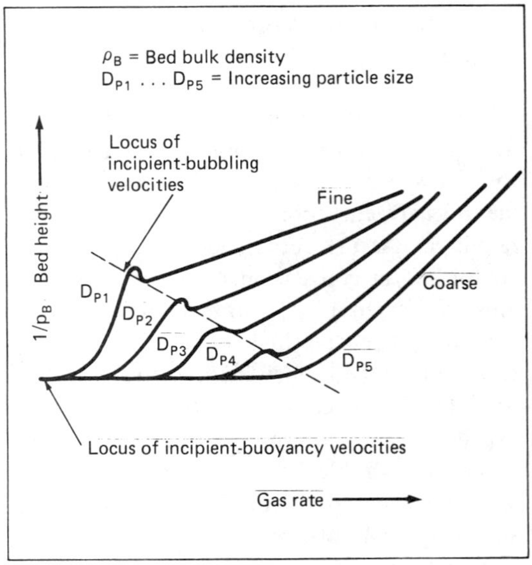

Figure 10.1. Particle size affects incipient-bouyancy and incipient-bubbling velocities.

10.3. OPERATING CHARACTERISTICS AND DESIGN PROCEDURES

Choice of Operating Gas Velocity. Fluidization occurs in a bed of particulates when an upward flow of fluid through the interstices of the bed attains a frictional resistance equal to the weight of the bed. At this point, an infinitesimal increase in the fluid rate will lift or support the particles. Hence, the particles are envisioned as barely touching, or as "floating" on a film of fluid.

In order to avoid ambiguities, this condition would better be considered as incipient buoyancy, since at such a fluid rate the particles are still so close as to have essentially no mobility—whereas the usual desire in fluidization is to create bed homogeneity. Such homogeneity can only be achieved by violent mixing. This is brought about by increasing the fluid velocity to the point of blowing "bubbles" or voids into the bed, which mix it as they rise. The increased fluid velocity at which bubbles first form is referred to as the incipient-bubbling velocity.

Fine powders can exist over a wide range of bulk densities and therefore exhibit substantial differences between incipient buoyancy and incipient bubbling, as illustrated qualitatively in Figure 10.1.

There is no way to predict precisely a powder's range of bulk densities. However, this is a relatively simple physical measurement. As an example, FCC (fluidized catalytic cracking) catalysts with particle-size distributions from about 10 to 200 micrometers in diameter exhibit incipient-buoyancy velocities in ambient air on the order of 0.01 to 0.03 ft/s, and incipient-bubbling velocities of about 0.1 to 0.3 ft/s. Many catalysts used in other processes consist of silica-alumina carriers (such as are used in FCC catalysts) impregnated with desired catalytic agents. Hence, these velocities are broadly representative. From a practical point, the incipient-bubbling velocity is the more significant one in reactor design.

The terms *particulate* and *aggregative* were coined[32] in the 1940s to differentiate between bubbling beds (aggregative) and nonbubbling beds (particulate). In general, liquid fluidized beds were nonbubbling, whereas gas fluidized beds bubbled. It is presently recognized that bubbling is related to fluid and particle properties in a manner permitting the prediction of a system's maximum attainable bubble size,[27] which if negligible leads to the observation of so-called particulate fluidization. Rather than employ the terms *aggregative* and *particulate*, it is more correct to refer to the maximum stable-bubble size for a particular system.

The "optimum" operating fluid velocity is not likely to have narrow limits. It generally represents physical compromises with entrainment, attrition, pressure drop, and economics. The lower limit is the bed's incipient bubbling velocity, and the upper limit usually approaches the terminal or free-fall velocity of the largest particle in the bed. There exists a nearly identical analogy to the optimum physical dimensions of a distillation or absorption tower.

The composition and physical properties of the fluidizing gas are fixed by the process conditions. If the density and size distribution of the particles to be fluidized are not also predetermined, then they must be obtained either from an incorporation of the cost of various degrees of grinding, or, in the case of a high-conversion gas-catalytic reaction, be taken as the size that optimally[46] leads to an incipient bubbling velocity between 0.1 and 0.3 ft/s under reaction conditions. Most frequently the optimum superficial fluidizing velocity will be based on pilot plant experience.

Calculation of Incipient Bubbling and Terminal Velocities

The superficial upward fluid velocity through a bed of particles that will initiate incipient bubbling is definable as the minimum flow that in passing through the interstices of the bed at its loosest bulk density suffers a particle-to-fluid frictional resistance just equalling the weight of the bed. Any additional fluid will therefore lift the bed (which must be unrestrained on its surface) and pass upwards in the form of a so-called bubble. Mechanistically it simply pushes the bed upward at its point of entry through the supporting grid port and thereby creates a void or hole into which the

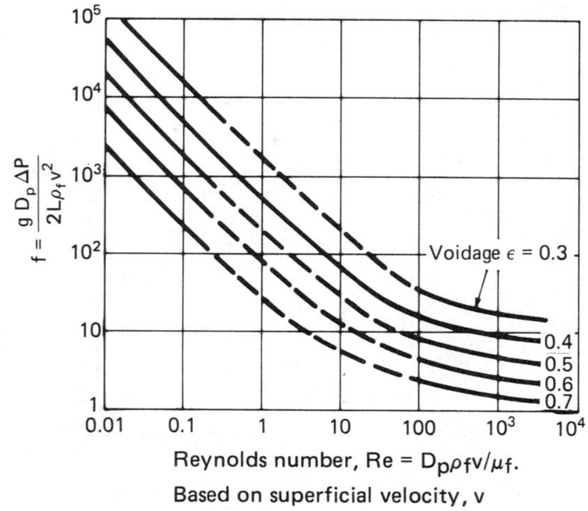

Figure 10.2. Friction factor for pressure drop through packed beeds of uniform particles.

bed solids can slide and thereby displace[39] the bubble-form void in the upward direction. The calculation of the incipient bubbling velocity therefore constitutes the substitution of the bulk density for the pressure drop in the conventional correlations for flow through packed beds and the determination of the attendant velocity. Correlations such as those of Ergun,[10] or of others[2,5,21] as suggested in Figure 10.2, are far more reliable than many published equations[11,18,29] for direct calculation of incipient fluidization velocities, since they empirically account for the voidage or loosest bulk density, which can have a very significant effect, as evidenced by the spread between the curves in Figure 10.2.

The superficial velocity at which the largest particle in the bed is able to be blown out of the reactor is frequently the practical or near-optimum operating velocity. This is simply calculable from the conventional drag coefficient versus Reynold's number curves[17] for the free-fall, or terminal, velocity of equivalent single spheres, as represented in Figure 10.3.

It is obvious from Figures 10.2 and 10.3 that the drag coefficient and friction factor differ solely in the constants $\frac{4}{3}$ and $\frac{1}{2}$ and are therefore interchangeable. Also, since both velocity and particle diameter appear in the abscissas as well as the ordinates, the prediction of

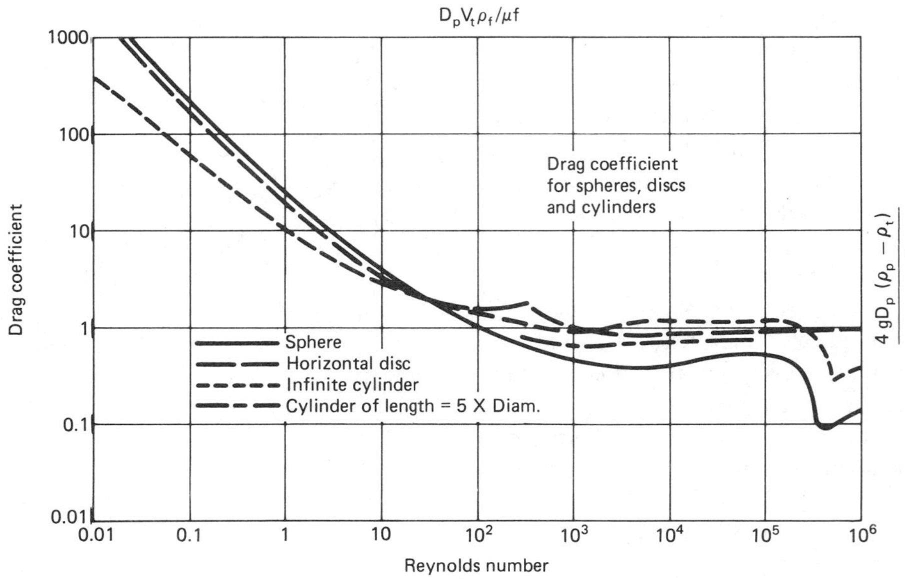

Figure 10.3. Drag coefficient for spheres, discs, and cylinders.

incipient bubbling velocity or of terminal velocity involves a trial and error procedure. This can be circumvented[35] by converting friction factor to drag coefficient ($C_D = (\frac{8}{3})f$) and then plotting $(C_D R_e^2)^{1/3}$ versus $(R_e/C_D)^{1/3}$, as shown in Figure 10.4, where now the denominators of the abscissa and the ordinate represent composites of the physico-chemical properties of the fluidizing medium and the solids (which composites are constants for any given system), so that Figure 10.4 is effectively a generalized correlation of velocity versus particle diameter at any known or desired bed density or void fraction. The incipient bubbling velocity is found from the value of the ordinate corresponding to an abscissa determined by the geometric weight mean particle diameter of the bed and the bed's void fraction at its loosest bulk density. At a value of the abscissa representing the largest particle in the bed, the ordinate determined from the $\epsilon = 1.0$ curve, yields the velocity that would blow the largest particle out of the bed. This is not necessarily an upper limit to bed operation since the particles entrained at any operating velocity must be separated from the exit gases and returned to the bed in order to maintain the bed. This particle recovery system usually consists of one or more cyclones[1] in series whose diplegs are submerged in the bed to discharge as near the grid level as feasible. The greater the superficial operating velocity, the greater the entrainment rate and the greater the need to design efficient cyclone recovery systems.

Predicting the Particle Entrainment Rate

The entrainment of particles from the surface of a bubbling bed is directly analogous to the entrainment of liquid droplets from a boiling or bubbling pool of liquid as occurs on a distillation tray. Since solid particles cannot coalesce in the manner of liquid droplets, the entrainment above a fluidized bed of solids, though declining with increasing height, inevitably reaches a constant rate representing all the bed particles with a free-fall velocity less than the operating superficial velocity, which were ejected from the bed surface. The force or mechanism of ejection is a function of the size and frequency of the bubbles erupting on the bed surface. Since the bursting bubbles represent intermittent locally higher-than-superficial velocity profiles, which dissipate with distance or so-called particle disengaging height above the bed, it is of interest to be able to predict at

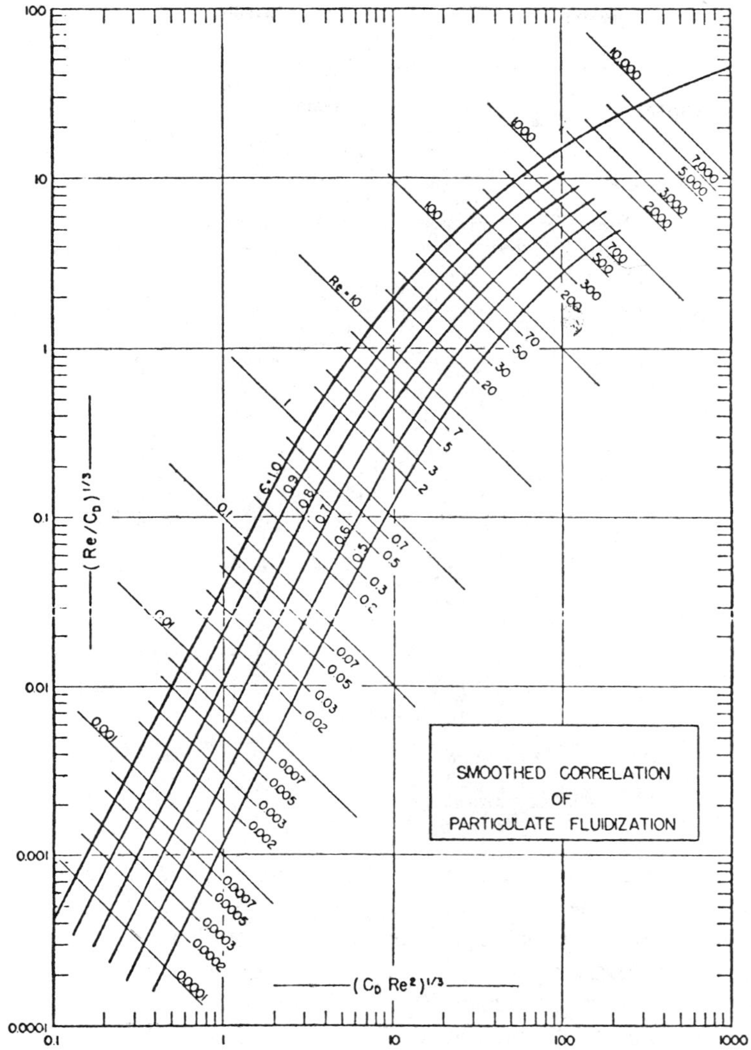

$$(\mathrm{Re}/C_D)^{1/3} = \left(\frac{D_P v_\epsilon \rho_f}{\mu_f} \bigg/ \frac{4gD_P(\rho_P - \rho_f)}{3\rho_f v_\epsilon^2}\right)^{1/3} = v_\epsilon \bigg/ \left(\frac{4g\mu_f(\rho_P - \rho_f)}{3\rho_f^2}\right)^{1/3}$$

$$(C_D\,\mathrm{Re}^2)^{1/3} = \left[\frac{4gD_P(\rho_P - \rho_f)}{3\rho_f v_\epsilon^2}\left(\frac{D_P v_\epsilon \rho_f}{\mu_f}\right)^2\right]^{1/3} = D_P \bigg/ \left(\frac{3\mu_f^2}{4g\rho_f(\rho_P - \rho_f)}\right)^{1/3}$$

Figure 10.4. Smoothed correlation of particulate fluidization.

what height above the bed the velocity profile will eventually be stabilized to the superficial velocity, with an attendant constant entrainment rate thereafter. This so-called Transport Disengaging Height[36] has been empirically correlated as shown in Figure 10.5. Increasing the reactor height to more than the TDH serves no purpose in decreasing the entrainment rate. The abscissa of Figure 10.5 is related to the frequency of surface bubble eruptions and the parameter to the physical size of the surface bubbles. As discussed subsequently, the size

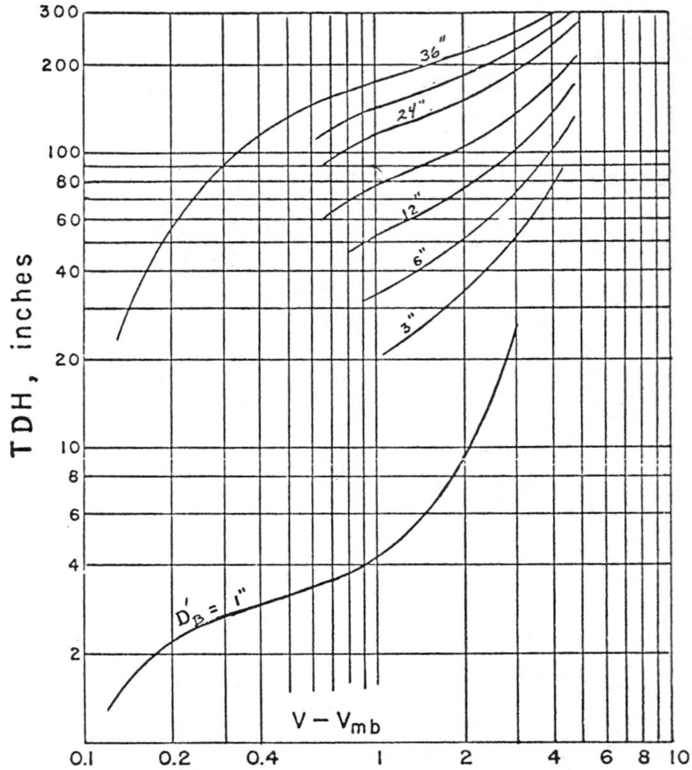

Figure 10.5. Empirical correlation of transport disengaging height.

of the surface bubbles is calculable from the grid characteristics, the bubble merger rate, and the maximum stable bubble size for the particle-fluid system under consideration.

The particle entrainment rate, at and above the TDH, can be determined from the empirical correlation[36] given in Figure 10.6. The detailed design of the cyclone recovery system can best be found in Chapter 11 of the A.P.I. Emissions Control Manual.[1]

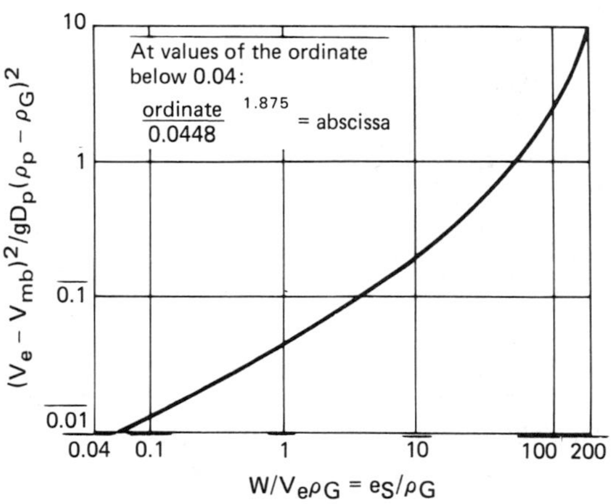

Figure 10.6. Maximum dilute phase entrainment in vertical gas-solids upflow.

Grid Design and Initial Bubble Size

Bubbles form at the grid ports where fluidizing gas enters the bed. They form simply because the velocity at the interface of the bed just above the hole represents a gas input rate in excess of what can pass through the interstices with a frictional resistance less than the bed weight, and hence the layers of solids above the holes are pushed aside until they represent a void through whose porous surface the gas can enter at the incipient fluidization velocity.[41] If the void attempts to grow larger, the interface velocity becomes insufficient to hold back the walls of the void and hence they cave in from the sides[39] cutting off the void and presenting a new interface to the incoming gas.

This sequence is illustrated in Figure 10.7. The depth of penetration of the grid gas jets has been correlated empirically,[31,34,40] as shown in Figure 10.8, and the diameter of the initial bubble resulting from a detached void has been observed, within experimental error, to be about half the penetration depth. Since the grid is the source of the bubbles, its design is relatively critical. The holes should be as small as is reasonable (considering cost, plate strength, and possible pluggage) and of such total area that the pressure drop of the fluidizing gas passing through the holes is sufficient to assure gas distribution[14,20,26,43] to all the holes. As in the case of a perforated plate distillation tray, bed weepage flow down through the holes is to be avoided. It has been demonstrated that the pressure drop through a flat plate grid necessary to assure that all the holes are bubbling must be at least 30% of the pressure drop through the bed atop the grid. This criterion establishes the grid hole velocity and, in conjunction with the lowest anticipated fluidizing gas volume, determines the total hole area. The number of holes is then dependent on the designer's choice of the hole diameter which simultaneously also determines the initial bubble size. In passing up through the bed these bubbles inevitably merge when they meet and hence their fluid mechanics must be understood in order to relate to the gas-solids contact occurring within the bed and to the size of the bubbles bursting on the bed surface.

Fluid Mechanics of Bubble Flow

Bubbles or "gas voids" rise in a fluidized bed by being displaced with an inflow of solids from their perimeter.[3,39] Since free-flowing and/or incipiently fluidized bulk solids have shallow

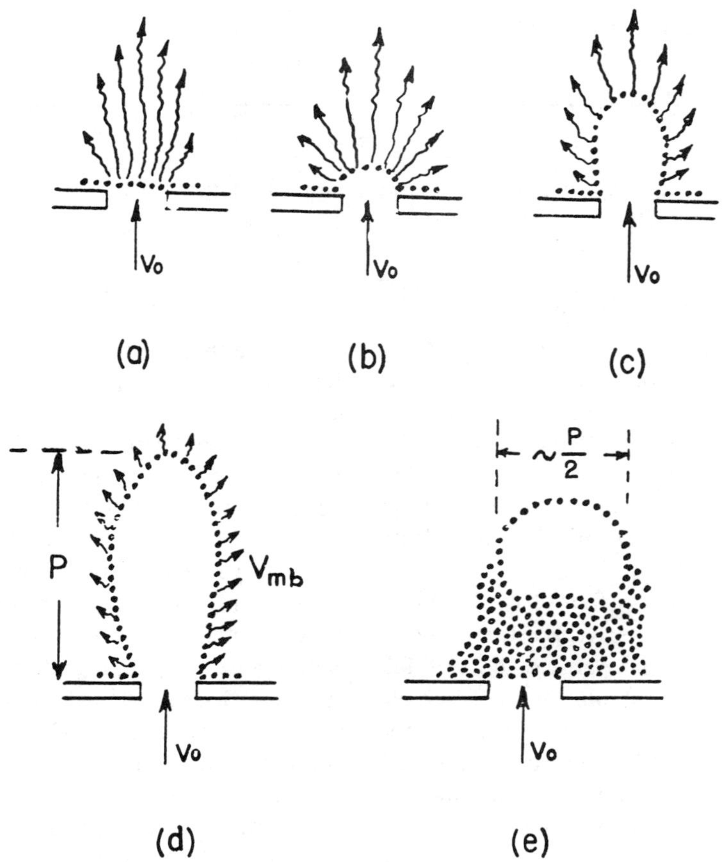

Figure 10.7. Bubble formation from bed-penetrating gas jets at the grid points.

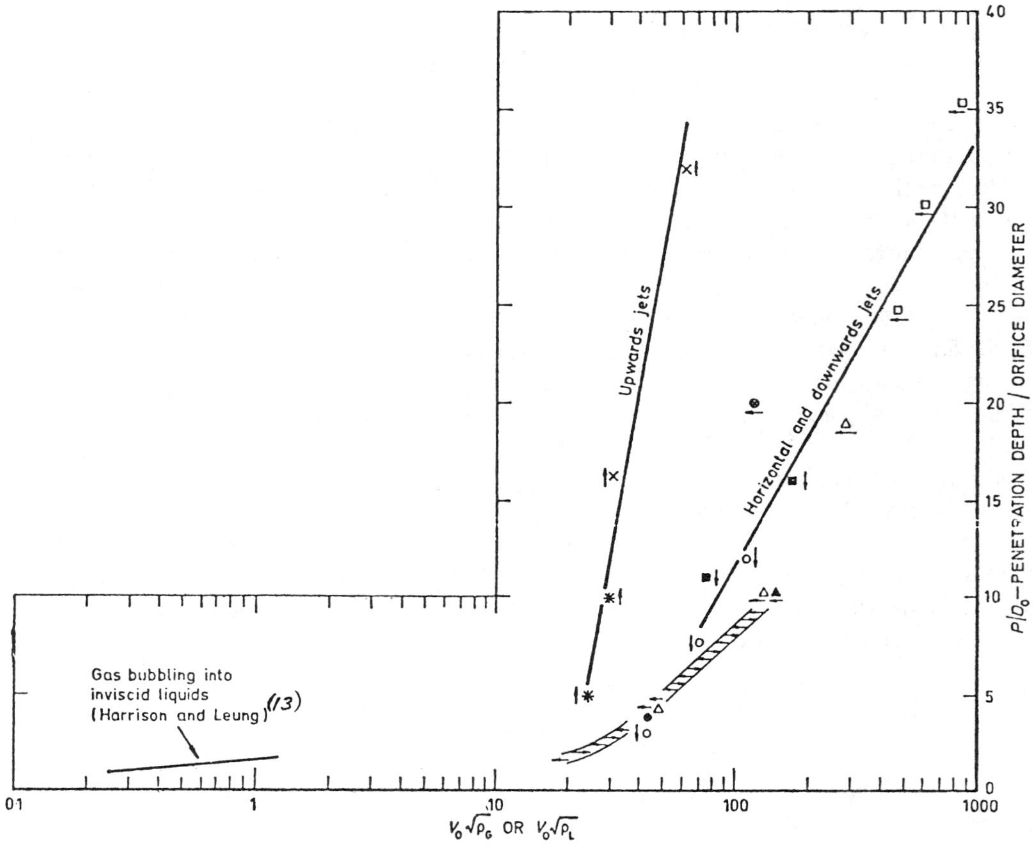

Figure 10.8. Jet penetration into fluid-particle media.

angles of repose, their walls cannot stand at 90° and hence the solids slide down the bubble's walls into its bottom where all the peripheral streams collide to form a so-called wake as illustrated in Figure 10.9. Observations of this downflow of solids in a "shell" around the bubble have shown it to occupy an annular thickness of $\frac{1}{4}$ of the bubble diameter so that the overall diameter within which a bubble can rise "freely," as it would in a bed of infinite diameter, can be defined as $1.5D_B$.

Since the peripheral surface of the bubble is simply a layer of particles, it is at first difficult to understand why the particles do not fall from its roof and annihilate the bubble. Danckwert's[7] simple bed support experiments, illustrated in Figure 10.10, provide the physical demonstration and Rowe and Henwood's[24] experiments the classical approach.

In Figure 10.10a the air rate is raised to the point of incipient fluidization and in Figure 10.10b through 10.10f this same gas rate is passed through the bed in the opposite direction. Note that in position (d) the solids do not slide to their angle or repose but instead are held at 90° and that on reaching (f) the bed is held up without solids falling from what is now its lower side or conversely the upper surface of a bubble in a fluidized bed. When the surface of a bed is traversed by an incipiently fluidizing flow the particles cannot separate from each other. This not only explains the bubble's surface stability but also the integrity of the walls of a bed-penetrating jet, as in Figure 10.7.

Rowe and Henwood carried out classical drag measurements which revealed that the drag on a downstream particle is reduced because of the presence of an adjacent upstream particle. This simply means that a particle cannot fall from the roof of a bubble because if it did, then it

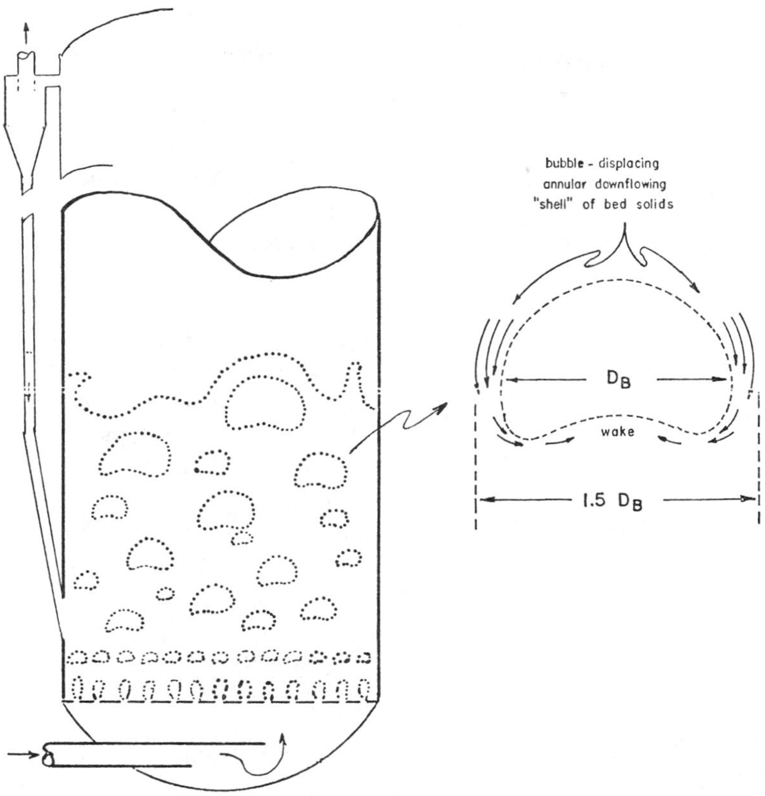

Figure 10.9. Bubble rise via displacement by inflow of a surrounding downflowing shell of bed solids.

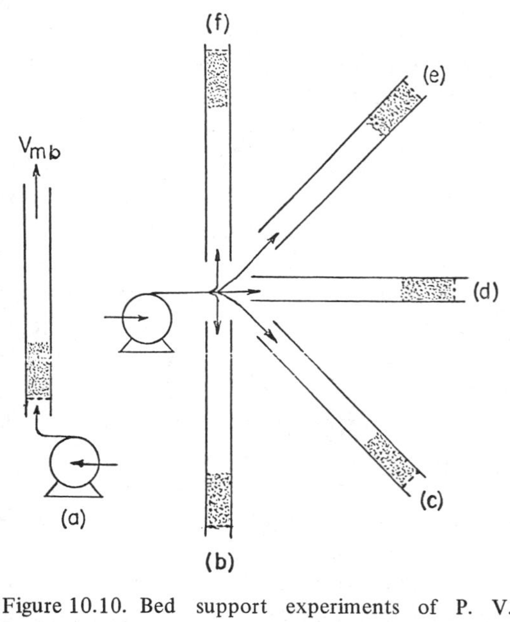

Figure 10.10. Bed support experiments of P. V. Danckwerts.

would immediately be followed by the particle above it, and that by the particle still further above, etc., so that the entire mass or bed above the bubble would have to collapse as a unit. For this to occur, the excess gas could not be passed through the bed unless the bed were physically held down or restrained at its upper surface.

The velocity at which bubbles rise in a gas fluidized bed has been measured photographically by several investigators. The results are in excellent agreement with what would be predicted for gas bubbles in liquids from the drag coefficient versus Reynolds number correlations of such investigators as Van Krevelen and Hoftijzer[33] illustrated in Figure 10.11. Over the range of Reynolds numbers corresponding to reasonable size bubbles the drag coefficient is essentially a constant so that simple substitution shows that if gas density is small relative to the bed density:

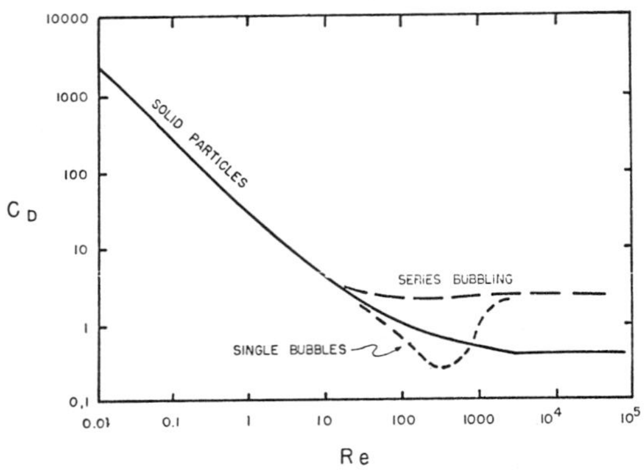

Figure 10.11. Rate of rise of gas bubbles in liquids.

$$C_D = \frac{4gD_B(\rho_B - \rho_G)}{3\rho_B V_B^2}$$

$$\therefore V_B = \sqrt{\frac{4gD_B(\rho_B - \rho_G)}{3C_D\rho_B}}$$

or

$$V_B = 4.01\sqrt{D_B}$$

This has been corroborated in experiments with freely bubbling beds.

Tarmy and Matsen[19] have shown that in slugging beds the full width of the downflowing solids shell (Figure 10.9) is restricted and the velocity of bubble rise then approximately $\frac{1}{2}$ that in a freely bubbling bed. In most instances this becomes a matter for consideration only in the design or scaleup of small pilot or bench scale fluidized bed processes.

Rate of Bubble Growth by Merger

That bubbles must grow by merger as they rise through the bed is obvious from the large and less frequent surface eruptions relative to a much higher frequency of small voids initiated from a usual multitude of grid ports. Growth by simple gas expansion resulting from the pressure reduction between bottom and top of a fluidized bed is generally relatively insignificant.

From the solids inflow model of Figure 10.9 it is obvious that a bed must be exceptionally homogeneous to expect the shell of downflowing solids around a bubble to be flowing at an equal rate in every plane. Any bed nonuniformity can cause a shift in the bubble shape or position. Merely the prior passage of another bubble could alter local densities or distributions so as to make bed solids in one local area more readily flowable in a given direction than the bed solids in an adjacent area. The solids inflow model therefore obviates a simple mechanism of bubble merger. If two bubbles get close enough that their shells of downflowing solids begin to interact, the touching shells will represent a local downflowing stream of solids faced with more than one path to the nearest void. The stream could be squeezed to the point of being insufficient to satisfy both bubbles and thereby drain off, leaving no wall between the voids and hence the appearance of a single somewhat larger bubble whose volume is the sum of the volumes of the two merged bubbles.

It is therefore readily acceptable that the idealized bubbling of Figure 10.12a will lead to a situation as in (b) where two bubbles of unit initial volume can merge into bubbles of twice this volume. Since larger bubbles rise more rapidly, these double volume bubbles will catch up and merge with other unit volume bubbles to yield bubbles of trice the initial bubble volume. These newer bubbles will rise even more rapidly and catch up with bubbles of 1 or 2 times the volume of the initial bubble resulting in bubbles of at most 5 times the volume of the initial bubble. The bubble of five-fold volume can now catch up with bubbles of 1, 2, or 3 times the volume of the initial bubble, resulting in bubbles of at most 8 times the volume of the initial bubble as illustrated in Figures 10.12c and 12d. Carrying on this process of overtaking bubbles results in a sequence of maximum multiples of the initial bubble volume in which each multiple is the sum of the two previous bubbles. This sequence, illustrated in Figure 10.13, is the well-known "Fibonacci" series.[42]

Since the levels at which the maxima exist represent the summation of the diameters of their forebearers and since their diameters are

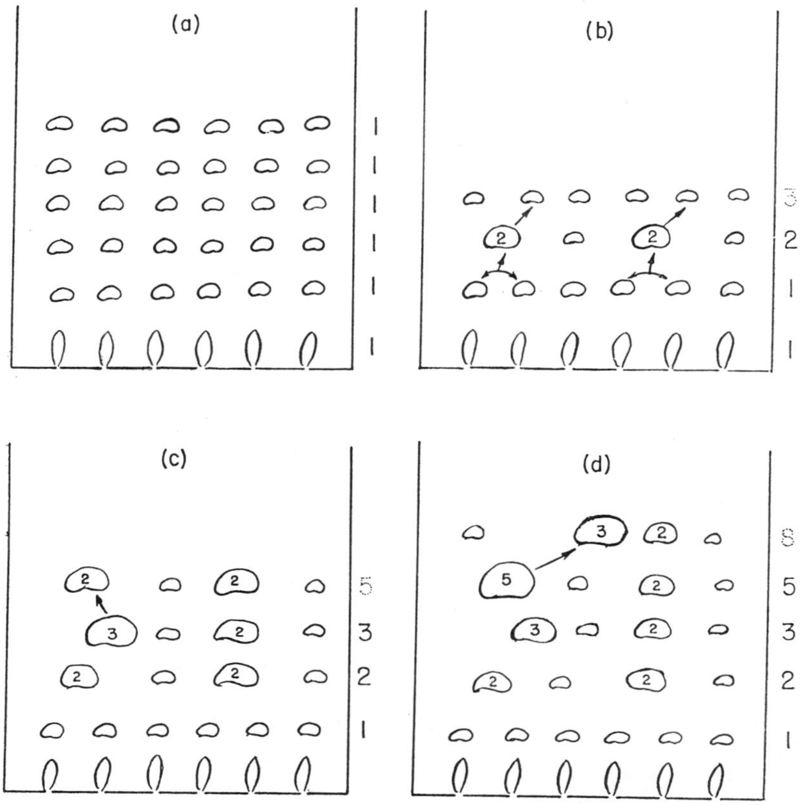

Figure 10.12. The "catch-up" mechanism of bubble growth.

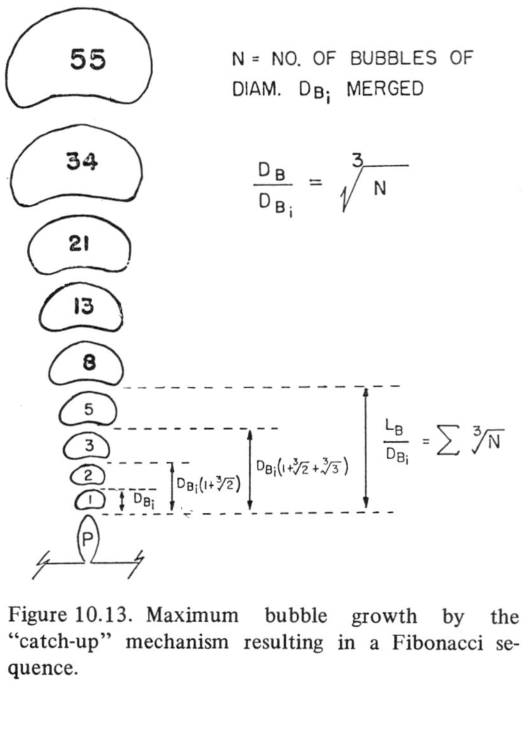

Figure 10.13. Maximum bubble growth by the "catch-up" mechanism resulting in a Fibonacci sequence.

proportional to the cube root of their volumes, it follows that the ratio of merged bubble diameter to initial bubble diameter is equal to the cube root of the number of initial bubbles consumed in the merger, and also that the level at which the merged bubbles exist relative to the height (or diameter) of the initial bubble is equal to the summation of the cube root of the number of initial bubbles consumed in the merged bubble.[41] For the case of the maximum size of merged bubble this is illustrated analytically in Figure 10.13 and shown graphically in Figure 10.14.

That the mechanism of Figures 10.13 and 10.14 appears in good agreement with experimental observations is illustrated in Figure 10.15 where the empirical bubble growth relationships proposed by Chavarie and Grace,[6] Werther,[30] and Rowe[23] are superimposed on the curve representing the Fibonacci series. In using Figure 10.15 to determine the maximum

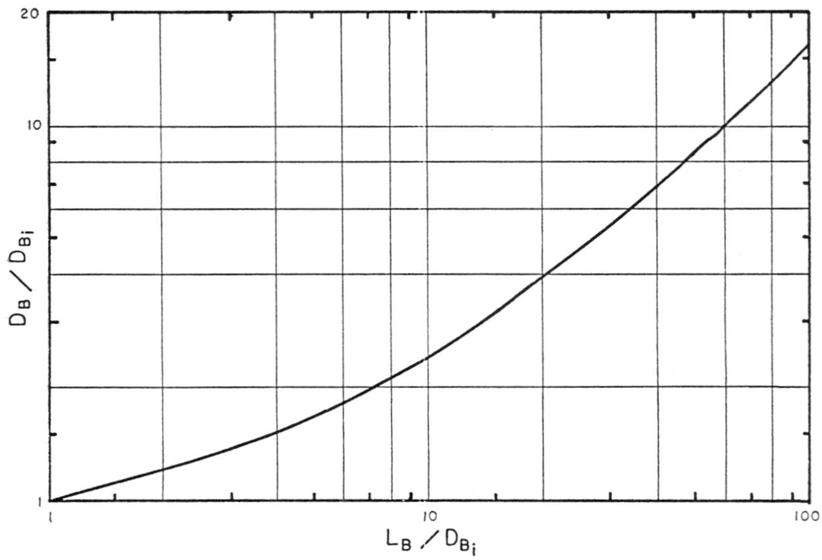

Figure 10.14. Bubble growth by merger represented by the Fibonacci sequence.

bubble diameter D_B at any bed level L_B above the grid, it is necessary to determine the initial bubble diameter D_{Bi}, which could exist at the grid level as a result of individual $(P/2)$ or merged jets. Figure 10.15 must also not be extrapolated beyond the maximum attainable stable bubble size.

Maximum Stable Bubble Size

Danckwert's bed support experiments (Figure 10.10) and those of Rowe and Henwood based on particle drag force measurements demonstrated that a bed interface (and hence a bubble) should be fundamentally stable against collapse

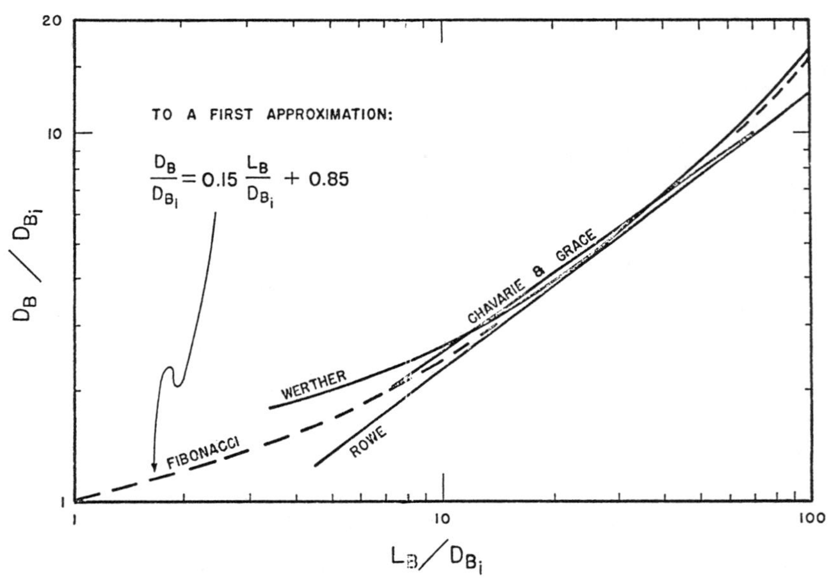

Figure 10.15. Comparison between empirical bubble growth correlations and the "catch-up" mechanism represented by the Fibonacci sequence.

as long as it is traversed by a superficial velocity equal to its incipient fluidization rate. Since the inflowing solids shell volume usually far exceeds the incipient fluidization rate, there would appear to be no limit to the attainable bubble size, or dome, apt to collapse. Presumably, if the dome cannot collapse amid free flowing bed solids then as the bubble grows it could only be limited by particles leaving the shell and being entrained into the bubble void. Such entrainment, or particle pickup, would be most likely to occur from the bubble walls as the result of the relative velocity between gas and surface particles at the interface.[27] Since against the downward velocity of bulk solids the bubble fluid (whether gas or liquid) rises at approximately an equal velocity, the relative flow of fluid past the particles at the bubble wall is twice the shell or bubble velocity. Equating twice the bubble velocity to the particle pickup velocity allows calculation of the minimum bubble size necessary to stir up the solids interface and thus thwart bubble appearance or growth. Since pickup velocity is approximately twice saltation,[15,37] this is equivalent to equating bubble velocity to saltation velocity. This procedure has given results in reasonable agreement with a broad range of observations reported to date.[41] For example 80 micrometer particles of sand fluidized with air could sustain a maximum bubble diameter of the order of 24 inches, whereas when fluidized with water the maximum bubble size would be indiscernible. Sand particles 600 micrometers in diameter when fluidized with water would permit a maximum stable bubble size of only $\frac{1}{4}$ inch, and 3,000 micrometer lead particles a water bubble of 7 inches.

Gas-Solids Contact

From grid design, operating superficial velocity, and fluid particle properties, it is possible to calculate the initial bubble size at the grid, the maximum stable size, and the bed depth over which the bubbles may grow from their initial to their stable diameter. Once having reached their maximum stable diameter, any further unlikely mergers would also lead to collapse, so that bubble diameter may be considered constant once having reached the stable size. Since the bubbles represent a flow superimposed on the superficial incipient bubbling velocity passing up through the bed, they are in effect being continuously purged as they rise. Since their local size, velocity, and residence times are calculable from grid to bed surface, it is also possible to calculate the degree to which they are purged before bursting at the surface and hence to make certain that no bubble gas bypasses contact with the bed solids. It may be assumed that the minimum bed depth required to avoid any feed gas breakthrough (e.g., 100% bubble purging) represents the minimum bed depth required for the desired reaction.[44] This is represented graphically in Figure 10.16 for a freely bubbling bed.

In choosing a pilot plant such as to entirely avoid scaleup considerations, superficial velocity and grid details must be identical to those anticipated in the commercial reactor, in addition, the diameter of the shell of downflowing bed solids surrounding the rising bubbles represents the minimum pilot reactor diameter necessary to simulate free bubbling and avoid approach to slugging. In addition to simulating free bubbling hydrodynamically, it may be argued that gas permeation from bubble into a surrounding bed should also be equalled. This only becomes significant or controlling with coarse and easily permeated beds having a high incipient fluidization velocity. The gas permeation or "cloud" diameter[25] is calculable from the depth of gas flow at incipient fluidization velocity over the time interval required for the bubble to rise a distance of one bubble diameter. Since the bubble rises at a velocity equal to 4 times the square root of its diameter it follows that:

$$\frac{\text{Thickness of gas penetrated "cloud"}}{\text{Thickness of downflowing solids "shell"}} = \frac{V_{mb}}{\sqrt{D_B}}$$

or since

$$\text{"Shell" O.D.} = 1.5D_B$$
$$\text{"Cloud" O.D.} = D_B + 0.5\sqrt{D_B}\, V_{mb}$$

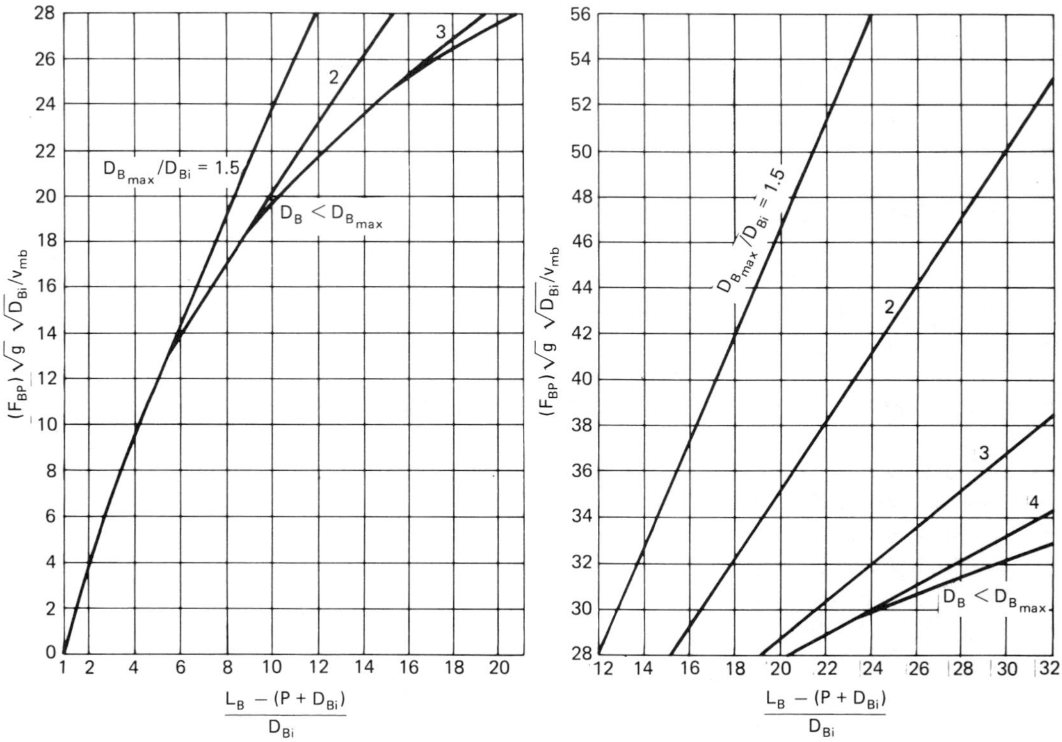

		Suggested units	
		English	Metric
F_{BP}	= Volume fraction of bubble gas purged from bubble	ft^3/ft^3	m^3/m^3
D_{Bi}	= Initial bubble diameter	ft	m
v_{mb}	= Minimum superficial bed bubbling velocity	ft/sec	m/sec
L_B	= Bed height above grid	ft	m
P	= Gas jet penetration at grid port	ft	m
$D_{B_{max}}$	= Maximum stable bubble diameter	ft	m
g	= Gravitational acceleration	32.2 ft/sec^2	9.71 m/sec^2

Note: Ordinate, abscissa, and parameter are dimensionless

Figure 10.16. Degree of bubble gas purging during its rise through a fluidized bed.

In applying free shell or cloud criteria in scaleup or scaledown, the relationship between bubble diameter and bed depth is obtainable from Figure 10.15 with the limitation of the system's maximum stable bubble size.

An unquestionably conservative approach to a minimal risk pilot plant reactor free of scaleup considerations would suggest it equal the larger of either "cloud" or "shell" diameter surrounding the system's maximum stable bubble.

Solids Mixing and Heat Transfer

Since rising gas bubbles are replaced with bed solids it is evident that the superficial gas velocity minus the incipient bubbling velocity also approximates the volumetric bulk solids movement across any unit bed cross section per unit time. This amounts to a relatively substantial mass movement, and hence it is not surprising that a fluidized bed exhibits reasonably uniform particle size distribution and bed

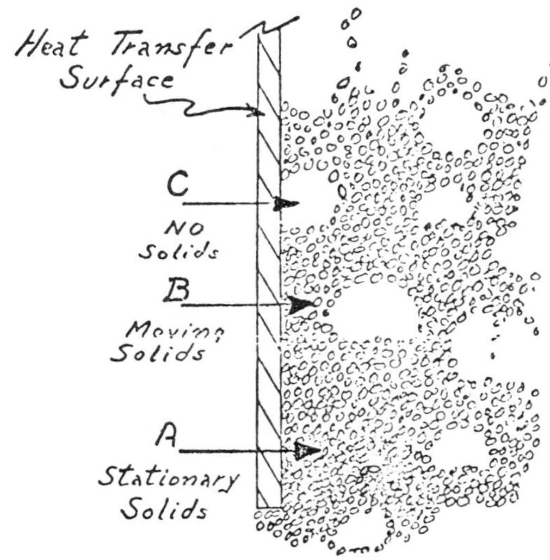

Figure 10.17. Comparison between overall bed coefficients and local flowing dense phase coefficients.

temperature throughout its volume. Reasonable quantitative estimates of such local solids mixing rates[16,25,45] are of major importance principally in determining allowable solids feed rates, since to avoid accumulation at any feed pipe location the bed mixing rate must be able to remove the feed material as rapidly as it enters. Such solids mixing rates have been reasonably well correlated by Talmor and Benenati.[28]

The substantial heat capacity of the bed solids relative to the gas inventory represents an enormous "flywheel," which coupled with the high solids mixing rate, leads to a rather uniform temperature throughout the bed. The heat transfer between gas and solids is nearly instantaneous, primarily because of the high particle surface area per unit of bed volume. However, the transfer of heat between the bulk bed and the vessel walls, or any other heat transfer surface, represents a composite of mechanisms such that the average may range from 10 to 100 BTU/hr $\times$ sq ft $\times$ °F depending on particle size, bubble size, fluid properties, and superficial fluidizing velocity. The homogeneity and relative uniformity of bed temperature make fluidized beds an attractive vehicle in which to conduct exothermic as well as endothermic reactions controlled by immersed boiler tubes, exchangers, platecoils, or other heat transfer surfaces when the bed walls do not offer sufficient area for cooling or heating via a fluid circulated through a surrounding jacket.

The transfer of heat from the bed to an immersed or wall surface depends instantaneously on whether the surface is bathed in stationary solids, in moving solids, or in a bubble void[38] as illustrated at points A, B, and C in Figure 10.17. Overall coefficients averaging these local mechanisms are summarized in Figure 10.18. Conductive transfer to or from stationary solids as at A in Figure 10.17, as well as transfer between bubble gas and metal surface as at C, are nearly negligible relative to the rate occurring when the surface is "wiped" by the shell of solids flowing down around a bubble passing within a distance of a quarter of its diameter from the transfer surface, as at B in Figure 10.17.

Though advantageously exploitable in industrial installations only under relatively rare circumstances, a number of experimental investigations[4,8,9,12,22] of the heat transfer coefficients under conditions such as at B in Figure 10.17 have also been correlated in terms

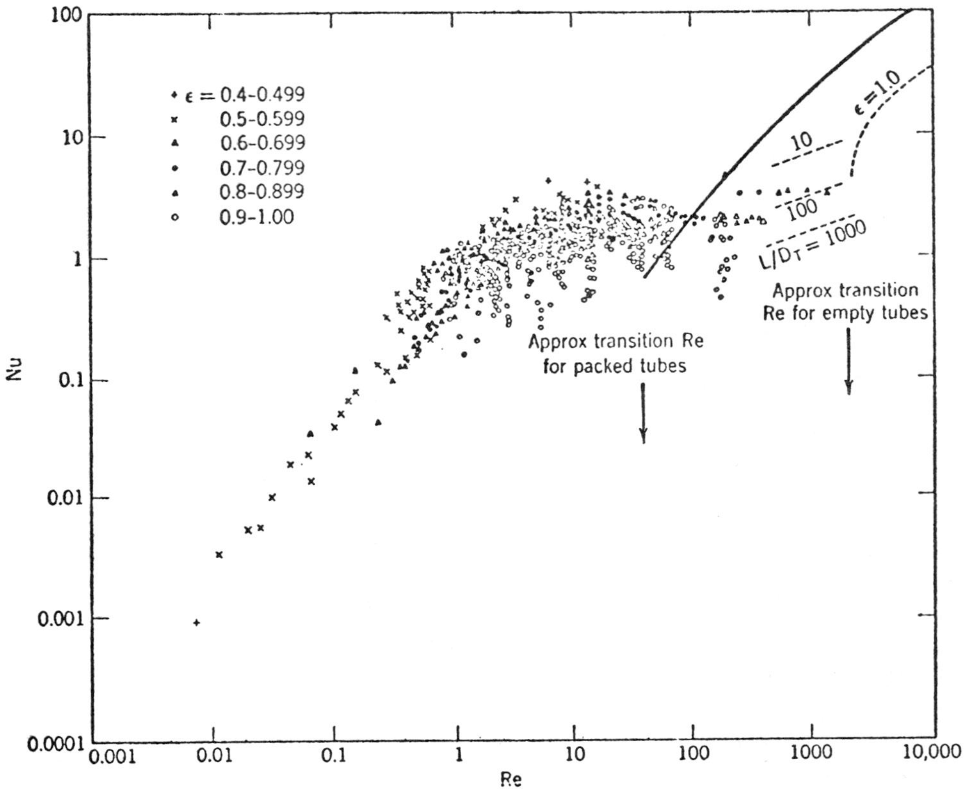

Figure 10.18. Heat transfer between fluid beds and tube walls (Correlation of overall coefficients).

Type of data	Value of *Re*	Value of *Nu*
fixed- and fluid-bed	$D_P v \rho_f / \mu_f$	$h D_P / K_f$
empty tube pipe flow	$D_T v_\epsilon \rho_f / \mu_f$	$h D_T / K_f$

Curves

(–) data for flow of fluid through fixed beds (M. Leva, *Ind. Eng. Chem.* **39**, 857–862 (1947); see also D. A. Plautz and H. F. Johnstone, *A.I.Ch.E. J.* **1**, 193–200 (1955)).

(---) correlation for flow of fluids through pipe (W. H. McAdams, *Heat Transmission*, 2nd ed., McGraw-Hill Book Co., Inc., New York, 1942).

Sources of Data

R. N. Bartholomew, Ph.D. Thesis, Univ. of Michigan, 1950; *Chem. Eng. Progr. Symp. Ser.* **48** (4), 3–10 (1952).
L. H. Collins, M.S. Thesis, Massachusetts Institute of Technology (MIT), Cambridge, Mass., 1946.
W. M. Dow, Ph.D. Thesis, Illinois Institute of Technology, 1949; *Chem. Eng. Progr.* **47**, 637–648 (1951).
H. Fischer and E. F. Dillon, B.S. Thesis, MIT, 1947.
W. Lazor and S. A. Murray, M.S. Thesis, MIT, 1947.
M. Leva, M. Weintraub, and M. Grummer, *Chem. Eng. Progr.* **45**, 563–572 (1949).
H. S. Mickley and C. A. Trilling, *Ind. Eng. Chem.* **41**, 1135–1147 (1949).
W. H. Millick and A. S. Humphrey, M. S. Thesis, MIT, 1948.
R. V. Trense, Ph.D. Thesis, Northwestern University, 1954.
R. W. Urie, M.S. Thesis, MIT 1948.

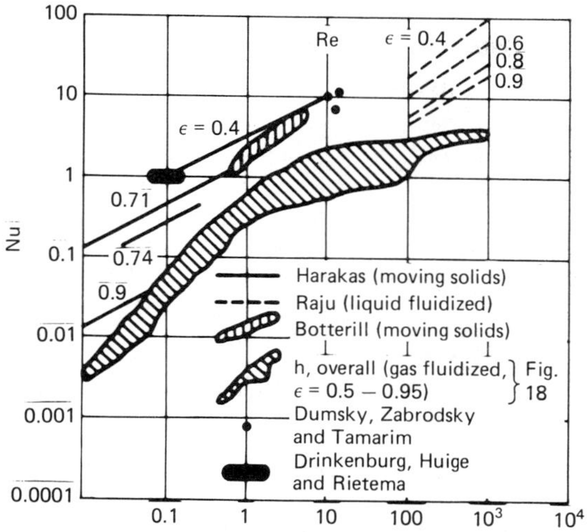

Figure 10.19. Comparison between overall bed coefficients and local flowing dense phase coefficients.

of Nusselt and Reynolds numbers, as shown in Figure 10.19. The cross hatched area in Figure 10.19 encompasses the overall heat transfer data of Figure 10.18.

LIST OF SYMBOLS

C_D	Drag coefficient, dimensionless (see Figs. 10.3 and 10.4)
D_B	Bubble diameter
D_B'	Bubble diameter bursting at bed surface
D_{B_i}	Initial bubble diameter at grid level
D_0	Grid hole diameter
$D_{B\,max}$	Maximum stable, or attainable, bubble diameter
D_p	Particle diameter
e_s	Maximum entrainable lbs of solids/ft.^{3} of gas
F_{BP}	Volume fraction of bubble gas purged
f	Friction factor (see Fig. 10.2)
g	Gravitational constant
h	Heat transfer coefficient
K_f	Thermal conductivity of fluidizing medium
L_B	Bed depth
N_u	Nusselt number, dimensionless (see Figs. 10.18 and 10.19)
P	Jet penetration depth
$\Delta P/L$	Pressure drop per unit length, lbs/sq. ft. × ft.
R_e	Reynolds number, dimensionless
TDH	Transport Disengaging Height
v	Superficial fluidizing medium velocity
v_ϵ	Superficial fluidizing medium velocity when bed voidage is ϵ
V_B	Bubble rise velocity
V_e	Effective superficial velocity governing rate of entrainment
V_{mb}	Superficial velocity at point of incipient bubbling
V_0	Fluidizing medium velocity through grid hole
V_t	Particle terminal or free fall velocity
W	Weight rate of solids entrained, lbs/sec × sq. ft. of vessel area
ϵ	Fractional void volume in bed
ρ_B	Bed or bulk density
ρ_f	Fluid or medium density
ρ_G	Gas density
ρ_L	Liquid density
ρ_p	Apparent particle density
μ_f	Viscosity of fluidizing medium

REFERENCES

1. Amer. Petrol. Inst., "Cyclone Separators," in *Emissions Control Manual*, Pub. no. 931, (May, 1975).
2. B. A. Bakameteff and N. V. Feodoroff, *J. Appl. Mechanics 4*:A97 (1937).
3. J. S. M. Botterill, J. S. George, and H. Besford, *Chem. Engr. Progr. Symp. Ser. 62*:7 (1966).
4. J. S. M. Botterill and J. R. Williams, *Trans. Instn. Chem. Engrs.* (London) *41*:217 (1963); idem, *Fluid Bed Heat Transfer* Academic Press, New York (1975).
5. P. C. Carman, *Trans. Instn. Chem. Engrs.* (London) *15*:150 (1937); idem, *J. Soc. Chem. Ind.* (London), *57*, 225 (1938).
6. C. Chavarie, J. R. Grace, *Ind. Eng. Chem. Fundamentals*, pp. 75–78 (May, 1975).
7. P. V. Danckwerts, Symp. on Fluidization held at AERE, Harwell, England (Oct. 5, 1959).
8. A. A. H. Drinkenburg, N. J. J. Huige, and K. Rietema, *Proc. Third Internat'l. Heat Transfer Conf.*, vol. IV, pp. 271–279, A.I.Ch.E., N.Y. (1966).
9. V. D. Dunsky, S. S. Zabrodsky, A. I. Tamarin,

Proc. Third Internat'l. Heat Transfer Conf., vol. IV, pp. 293–297, A.I.Ch.E., N.Y. (1966).

10. S. Ergun, *Chem. Engr. Progr. 48*(89) (1952); idem, *Ind. Eng. Chem. 41*:1179 (1949).
11. J. F. Frantz, *Chem. Engr. Progr. Symp. Ser. 62*(62):21–31 (1966).
12. N. K. Harakas and K. O. Beatty, *Chem. Engr. Progr. Symp. Ser. 59*(41):122 (1963).
13. D. Harrison and L. S. Leung, *Trans. Instn. Chem. Engrs.* (London) *39*:409 (1961).
14. J. W. Higby, *Chemie. Ingr. Tech. 36*:228 (1964).
15. P. J. Jones and L. S. Leung, *Ind. Eng. Chem. Proc. Des. Dev.*, *17*(4):571–575 (1978).
16. S. Katz and F. A. Zenz, *Petrol. Refiner 33*(5): 203–204 (1954).
17. C. E. Lapple and C. B. Shepherd, *Ind. Eng. Chem. 32*:605 (1940).
18. M. Leva, Shirai Takashi, and C. Y. Wen, *Genie Chim. 75*(2):33–42 (1956).
19. J. M. Matsen and B. L. Tarmy, *Chem. Engr. Progr. Symp. Ser. 66*(101):1–7 (1970); idem, *Chem. Eng. Sci. 24*:1743–1754 (1969).
20. Nat'l. Petrol. Refr's. Assoc., *Proceedings of the Question and Answer Session on Refining Technology*, p. 86 (1970).
21. A. O. Oman and K. M. Watson, *Natl. Petrol. News 36*:R795 (1944).
22. R. Raju, Ph.D. thesis, Andhra Univ., Waltair, India (1959).
23. P. N. Rowe, International Fluidization Conf., Asilomar, Cal. (June 15–20, 1975).
24. P. N. Rowe and G. A. Henwood, *Trans. Instn. Chem. Engrs.* (London), *39*, 43 (1961).
25. P. N. Rowe, B. A. Partridge and E. Lyall, At. Energy Res. Estab. (Gr. Brit.), Repts. R-3777, R-3846, "Particle Movement Caused by Bubbles in a Fluidised Bed." (Oct. 1961); R-4108, "Gas Flow through Bubbles in a Fluidised Bed," (Jan. 1963); R-4543, "Cloud Formation around Bubbles in Gas Fluidised Beds," (Feb. 1964); *Chem. Eng. Sci. 18*:973 (1964); *Chem. Eng. Progr. 60*:75 (March 1964); *Fluidisation*, Society of Chemical Industry, London (1964).
26. P. N. Rowe and F. A. Zenz, *The School Sci. Rev.* (Pub. by the Instn. of Chem. Engrs., London), *53*(182):94–102 (Sept., 1971).
27. A. M. Squires, Paper delivered at the 54th Annual A.I.Ch.E. Meeting, New York (Dec. 6, 1961); *Chem. Engr. Progr. Symp. Ser.*, *58*(38):57 (1962).
28. E. Talmor and R. F. Benenati, *A.I.Ch.E. Jnl. 9*(4): 536–540 (1963).
29. C. Y. Wen and Y. H. Yu, *Chem. Engr. Progr. Symp. Ser. 62*(62):100–111 (1966).
30. J. Werther, International Fluidization Conference, Asilomar, Cal. (June 15–20, 1975); idem, *Fluidization Technology*, ed. D. L. Keairns, vol. 1, Hemisphere Pub. Co., Wash. pp. 215–235 (1976).
31. J. Werther, *Fluidization*, ed. J. F. Davidson and D. L. Keairns, Cambridge Univ. Press, pp. 7–12 (1978).
32. R. H. Wilhelm and M. Kwauk, *Chem. Engr. Progr. 44*:201 (1948).
33. D. W. Van Krewelen and P. J. Hoftijzer, *Chem. Engr. Progr. 44*:529 (1948).
34. W. C. Yang and D. L. Keairns, *Fluidization*, ed. by J. F. Davidson and D. L. Keairns, Cambridge Univ. Press, pp. 208–214 (1978).
35. F. A. Zenz, *Petroleum Refiner*, *36*(8):147–155 (1957).
36. F. A. Zenz and N. A. Weil, *A.I.Ch.E.Jnl. 4*:472 (1958); idem, *Hydrocarbon Processing*, pp. 119–124 (April, 1974).
37. F. A. Zenz, *Ind. Eng. Chem.*, *Fundamentals 3*(1):65–76 (1964).
38. F. A. Zenz, *Proc. of the Third International Heat Transfer Conf.*, vol. VI, Pub. by A.I.Ch.E., pp. 311–313 (1966).
39. F. A. Zenz, *Hydrocarbon Processing 46*(4):171–175 (April, 1967).
40. F. A. Zenz, *Instn. of Chem. Engrs.* (London), *Symp. Ser.*, no. 30, pp. 136–139 (1968).
41. F. A. Zenz, *Chem. Engrg.*, pp. 81–91, (Dec. 19, 1977).
42. F. A. Zenz, *The Fibonacci Quarterly 16*(2):171–183 (April, 1978).
43. F. A. Zenz and D. F. Othmer, *Fluidization and Fluid-Particle Systems*, Reinhold Pub. Co., N.Y., p. 171 (1960).
44. Ibid., chapter 8.
45. Ibid., chapter 9.
46. F. A. Zenz and D. F. Othmer, *Fluidization and Fluid Particle-Systems*, annotated 1966 edition, p. 281,orig. pub. by Reinhold Pub. Corp., N.Y. (1960).

11
Spouting of Particulate Solids

N. Epstein and J. R. Grace
University of British Columbia
Vancouver, British Columbia

11.1. INTRODUCTION

The spouted bed technique has become established in recent years as an alternative to fluidization for handling particulate solids which are too coarse and uniform in size for good fluidization. Although the areas of application of spouted beds overlap with fluidized beds, the flow mechanisms in the two cases are very different. Agitation of particles in a spouted bed is caused by a steady axial jet and is regular and cyclic, as distinct from the more random and complex bubble-induced particle flow patterns in most fluidized beds.

Figure 11.1 illustrates a spouted bed schematically and photographically. Fluid, usually a gas, is injected vertically through a centrally located small opening at the base of a conical, cylindrical or conical-cylindrical (as in Fig. 11.1) vessel containing relatively coarse particulate solids (e.g. $d \gtrsim 1$ mm). If the fluid injection rate is high enough, the resulting jet causes a stream of particles to rise rapidly through a hollowed central core, or *spout*, within the bed of solids. These particles, after rising to a height above the surface of the surrounding packed bed, or *annulus*, rain back as a *fountain* onto the annulus, where they slowly move downward and, to some extent, inward as a loosely packed bed. Fluid from the spout leaks into the annulus and precolates through the moving packed solids there. These solids are re-entrained into the spout over the entire bed height. The overall system thereby becomes a composite of a centrally located dilute-phase cocurrent-upward transport region surrounded by a dense-phase moving packed bed with countercurrent percolation of fluid. A systematic cyclic pattern of solids movement is thus established, with effective contact between fluid and solids, and with unique hydrodynamics.[1]

The spouted bed regime, which occurs over a limited range of fluid velocity, is bracketed by fixed packed bed (i.e., static bed) operation at the lower velocities and by bubbling or slugging fluidized bed operation at the higher. For a given combination of fluid, solids and vessel configuration, the transitions between regimes can best be represented quantitatively by plots of bed depth versus fluid velocity. An example of such a plot, or "phase diagram," is given in Figure 11.2. The demarcation line obtained by decreasing the fluid velocity until the spout collapses to give a static bed in its random loose-packed condition represents the *minimum spouting velocity*, U_{ms}, at various bed depths. The horizontal transition line separating spouting and bubbling represents the *maximum*

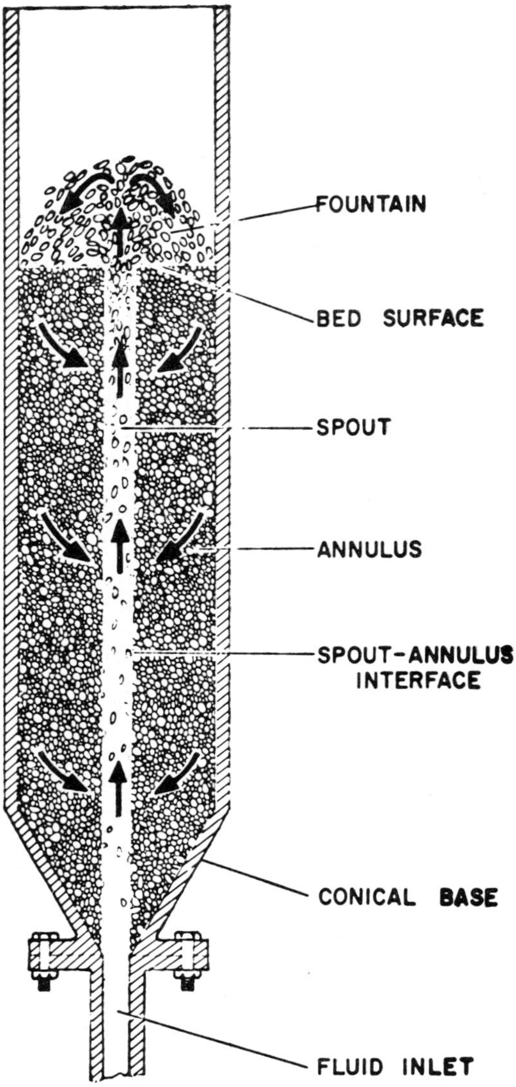

Figure 11.1a. Schematic diagram of a spouted bed. Arrows indicate direction of solids movement.

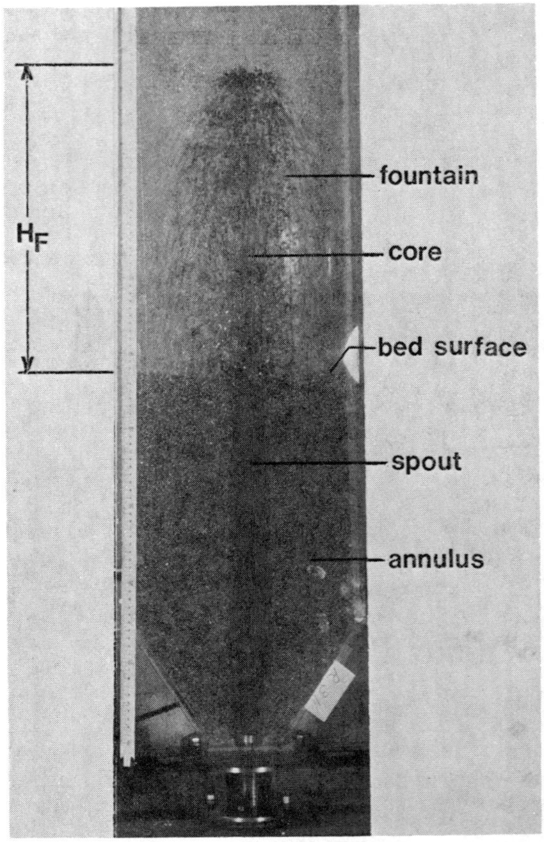

Figure 11.1b. Photograph of air-spouted wheat bed in half-cylindrical column.

spoutable bed depth, H_m, for the given system. Above some critical value of the inlet nozzle to column diameter ratio, D_i/D, there is no spouting regime. Instead, the bed changes directly from the fixed to the aggregatively fluidized state with increasing fluid velocity. This critical value, which can be approximated[2] by $(U_{mf}/U_t)^{1/2}$ or $\epsilon_{mf}^{n/2}$, decreases from 0.35 for gas spouting of coarse spheres (for which $\epsilon_{mf} \simeq 0.42$, $Re_T > 500$, $n = 2.39$) to 0.1 for finer particles[1] due to the accompanying increase of

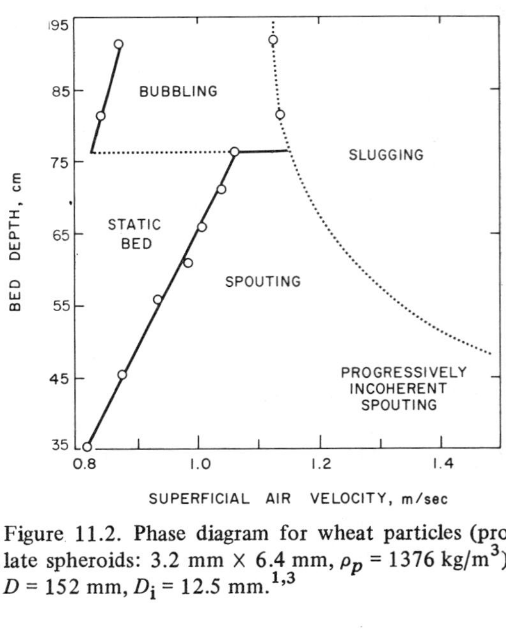

Figure 11.2. Phase diagram for wheat particles (prolate spheroids: 3.2 mm × 6.4 mm, $\rho_p = 1376$ kg/m^3), $D = 152$ mm, $D_i = 12.5$ mm.[1,3]

n, and can be expected to increase with decreasing particle sphericity due to the accompanying increase of ϵ_{mf}. The included angle of the conical base is a less critical parameter and need only exceed about 40° for stable spouting of most solid materials.[1]

11.2. MINIMUM SPOUTING VELOCITY

For cylindrical vessels up to about 0.6 m in diameter, with or without a conical base, the Mathur-Gishler[3] equation continues to be the most reliable predictor (within ±10%) of the minimum spouting velocity for a wide variety of solid materials, bed dimensions, nozzle diameters, and fluids ranging from air to water. The correlation is:

$$U_{ms} = \left(\frac{d}{D}\right)\left(\frac{D_i}{D}\right)^{1/3} \sqrt{\frac{2gH(\rho_p - \rho)}{\rho}} \tag{11.1}$$

where the particle diameter, d, is taken as the arithmetic average of bracketing screen apertures for closely sized near-spherical particles and as the volume-surface mean diameter for mixed sizes, using the equivolume sphere diameter d_p for nonspherical particles. An exception is the case of particles such as prolate spheroids that align themselves vertically in the spout, for which prediction by Equation (11.1) is best when d is taken as the horizontally projected diameter (i.e., the smaller of the two principal dimensions). For bed diameters exceeding 0.6 m, Equation (11.1), which can be rationalized qualitatively by jet-to-particle momentum transfer considerations, increasingly underestimates U_{ms}; there is no adequate equation available for such larger bed diameters.

The value of U_{ms} at the maximum spoutable bed depth for a given solid material in a given vessel is termed U_m, the *maximum value of the minimum spouting velocity*. In general, U_m is closely related to the minimum fluidization velocity, U_{mf}, for the given material, i.e.,

$$\frac{U_m}{U_{mf}} = b = 1.0 \text{ to } 1.5 \tag{11.2}$$

The value of b depends partly on the solid material spouted and partly on the spouting vessel geometry, its value decreasing towards unity as vessel diameter is increased for a fixed ratio of D/D_i, and increasing towards 1.5 as D_i is increased for a fixed value of D.[1]

11.3. MAXIMUM SPOUTABLE BED DEPTH

For irregularly shaped particles at the maximum spoutable bed depth, Equation (11.1) becomes

$$U_m = \left(\frac{d_p}{D}\right)\left(\frac{D_i}{D}\right)^{1/3} \sqrt{\frac{2gH_m(\rho_p - \rho)}{\rho}} \tag{11.1a}$$

For flow through a packed bed at the condition of minimum fluidization, the pressure gradient may be obtained from the Ergun[4] equation and is balanced by the buoyed weight of the bed per unit volume. Hence,

$$\left(\frac{-dP}{dz}\right)_{mf} = (\rho_p - \rho)(1 - \epsilon_{mf})g = \frac{150\mu(1 - \epsilon_{mf})^2 U_{mf}}{(\phi d_p)^2 \epsilon_{mf}^3} + \frac{1.75\rho(1 - \epsilon_{mf})U_{mf}^2}{\phi d_p \epsilon_{mf}^3} \tag{11.3}$$

With substitution of the empirical approximation of Wen and Yu,[5] i.e., $1/\phi\epsilon_{mf}^3 = 14$ and $(1 - \epsilon_{mf})/\phi^2\epsilon_{mf}^3 = 11$, Equation (11.3) can be solved to yield

$$Re_{mf} = \frac{d_p U_{mf} \rho}{\mu} = 33.7(\sqrt{1 + 35.9 \times 10^{-6} Ar} - 1) \tag{11.4}$$

where

$$Ar = d_p^3(\rho_p - \rho)g\rho/\mu^2 \tag{11.5}$$

When Equations (11.1a), (11.2), and (11.4) are combined to eliminate U_m and U_{mf}, the result is

$$H_m = \frac{D^2}{d_p}\left(\frac{D}{D_i}\right)^{2/3} \frac{568b^2}{Ar}(\sqrt{1 + 35.9 \times 10^{-6} Ar} - 1)^2 \quad (11.6)$$

McNab and Bridgwater[6] found that Equation (11.6) gave the best fit to existing experimental data for H_m in gas-spouted beds with $b = 1.11$, although there was considerable scatter. More empirical equations for H_m are available in References 1 and 2.

If Equation (11.6) is differentiated with respect to Ar, after substituting for d_p from Equation (11.5), and $dH_m/d(Ar)$ is set equal to zero, it is found that there is a critical value, $Ar = 223{,}000$, or

$$(d_p)_{crit} = 60.6\left(\frac{\mu^2}{g(\rho_p - \rho)\rho}\right)^{1/3} \quad (11.7)$$

below which H_m increases with d_p and above which H_m decreases as d_p increases. For gas spouting, the resulting value of $(d_p)_{crit}$ is typically in the range 1.0–1.5 mm, a result that agrees with experimental observation.[1,2,6] Substitution of $Ar = 223{,}000$ into Equation (11.4) gives a corresponding critical $Re_{mf} = 67$, which is also in close agreement with experiment for gas spouting.[1] It is noteworthy that these critical values are independent of vessel geometry, though H_m itself, including its maximum value at $Ar = 223{,}000$, varies as $(D^4/D_i)^{2/3}$ according to Equation (11.6).

A value of b in Equation (11.2) close to unity is consistent with the most frequently assumed mechanism for the termination of spouting, namely, fluidization of the upper solids layer in the annulus by the increasing annular fluid flow at the higher bed levels. The small excess of b over unity is attributable to the persistence of a higher superficial velocity in the spout than in the annulus even at $z = H = H_m$. Other postulated termination mechanisms are the onset of slugging (or "choking") in the spout due to the particle flow exceeding its conveying capacity, and the growth of surface instability waves at the spout-annulus interface.[1] Littman et al.[2] observed the spout-slugging termination mechanism in gas spouting of relatively small particles for which $Ar < 223{,}000$, and termination by fluidization at the top of the annulus using larger particles for which $Ar > 223{,}000$. It would seem then that the critical diameter given by Equation (11.7) may also represent the transition between these two termination mechanisms for gas spouting.

Equation (11.7) appears to be inapplicable to liquid spouting. Liquid spouted beds are characterized by a decrease of H_m as d_p increases for all values of d_p,[7] by the onset of fluidization in the annulus at $z = H = H_m$, and by persistence of spouting to a depth of H_m even when $H > H_m$, the spouted bed of height H_m then being capped by a particulately fluidized bed of height $(H - H_m)$.[8]

11.4. FLOW DISTRIBUTION OF FLUID

For a bed of height H_m, Mamuro and Hattori[9] considered a simplified force balance over a differential element dz of the annulus. Based on the assumption that Darcy's Law applies to the vertical component of flow through the annulus and on the boundary condition that the annular solids are incipiently fluidized at $z = H_m$, they derived the following expression for the superficial fluid velocity, U_a, in the annulus at height z in a cylindrical column:

$$\frac{U_a}{U_{mf}} = 1 - \left(1 - \frac{z}{H_m}\right)^3 \quad (11.8)$$

Grbavčić et al.,[8] amongst others[10], have shown that for given vessel geometry, given spouting fluid and given solids material, U_a at any level, z, is independent of total bed depth, H. Hence, Equation (11.8) should apply for $H \leqslant H_m$, and this is borne out by experiment[8,10] especially when $U \sim U_{ms}$. There is evidence that Equation (11.8) works well even where the annulus Reynolds number, $\rho d_p U_a/\mu$, is one or two orders of magnitude greater than the upper limit for Darcy's Law. This insensitivity of fluid flow distribution to deviation from Darcy's Law has been explained theoretically.[10,11]

If the spout diameter at any bed level, labeled D_s, is known, continuity at that level yields

$$Q_s + Q_a = U\frac{\pi}{4}D^2 \tag{11.9a}$$

or

$$U_{sz}D_s^2 + U_a(D^2 - D_s^2) = UD^2 \tag{11.9b}$$

Hence, the fraction of the total fluid flow that passes through the spout at any level, for a given superficial spouting velocity, U, is simply $U_{sz}D_s^2/UD^2$. At minimum spouting, the superficial velocity, $U = U_{ms}$, can be estimated from Equation (11.1); operating velocities for gas spouting are typically 10–50% above U_{ms}. As a first approximation, the additional gas flow above that required for minimum spouting may be assumed to pass through the spout, while the gas flow through the annulus is constant once $U \geqslant U_{ms}$. Recent data (Figure 11.3) indicate that increasing the spouting velocity above U_{ms} actually results in some *decrease* in the net gas flow through the annulus.[10] This effect is believed to be caused by the enhanced tendency toward downward gas entrainment in the annulus resulting from increased solids circulation (i.e., increased solids downflow in the annulus). The same effect is probably responsible for decreasing U_a at $z = H = H_m$ from U_{mf}, as given by Equation (11.8), to about $0.9U_{mf}$ at $U = 1.1U_{ms}$.[10] Equation (11.8) must then be modified by substituting U_{aH_m} ($\sim 0.9U_{mf}$) for U_{mf}.

Typical gas streamlines in the annulus are shown in Figure 11.4. There is some evidence[12] that, below the outermost streamline shown, i.e., immediately adjacent to the gas inlet, the gas reverses itself and flows downward and radially inward from the annulus to the spout,

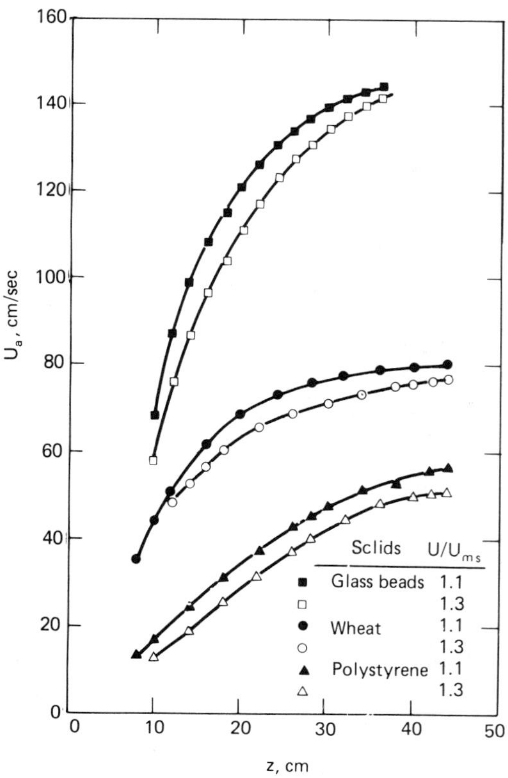

Figure 11.3. Effect of spouting air velocity on upward air velocity in the annulus, D = 152 mm, D_i = 19.0 mm, $\alpha = 60°$.[10]

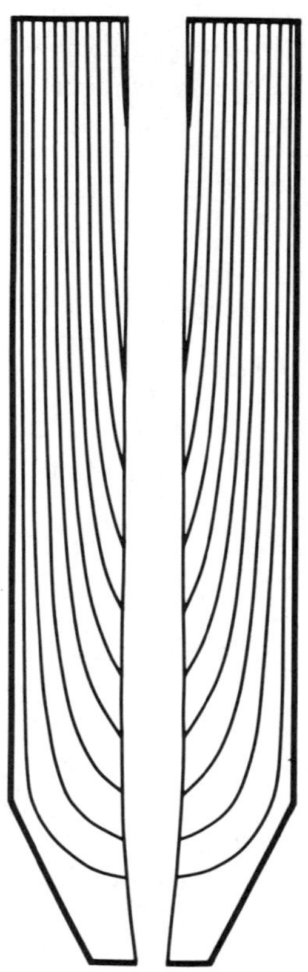

Figure 11.4. Calculated streamlines for a 0.24 m diam × 0.72 m deep bed of polystyrene pellets (d_p = 2.93 mm, D_i/D = 0.12, U/U_{ms} = 1.1), in substantial agreement with experimental observation.[14,18]

especially at $U > 1.1U_{ms}$.[13] This gas recirculation, which could be caused by the accelerated downward solids motion in the cone and/or the venturi effect of the inlet nozzle, does not, however, appear to affect the applicability of Equation (11.8) further up the bed. The fluid in each streamtube is in dispersed plug flow in the streamwise direction,[14] as in a moving packed bed, while that in the spout is essentially in plug flow. However, because of the large difference in flow velocities between spout and annulus, the residence time distribution of gas in the bed as a whole is intermediate between plug flow and perfect mixing.[1]

11.5. PRESSURE DROP

Figure 11.5 shows typical plots of $(-\Delta P)$ vs. U for the transition from flow through a fixed bed to a spouted bed. Before arriving at the fully spouting condition, the bed passes through a peak pressure drop, $-\Delta P_M$, associated with the energy required by the jet to disrupt the packing. As a first approximation, $-\Delta P_M$, which is bed-history-dependent, can be assumed to be equal to the buoyed weight per unit area, of the initial packed bed, i.e.,

$$-\Delta P_M \sim H(\rho_p - \rho)(1 - \epsilon) \quad (11.10)$$

Once the fully spouting condition is reached, the pressure drop stabilizes itself to a value, $-\Delta P_s$, which is essentially independent of superficial velocity.

The longitudinal pressure gradient in the annulus of a fully spouted bed at any level is given by

$$-dP/dz = K_1 U_a + K_2 U_a^2 \quad (11.11)$$

where K_1 and K_2 can be estimated as the coefficients of U_{mf} and U_{mf}^2, respectively, in Equation (11.3), and U_a may be evaluated from Equation (11.8). The pressure drop, $-\Delta P_f/H$, per unit length across a fluidized bed is

$$-\Delta P_f/H = (-dP/dz)_{mf} = K_1 U_{mf} + K_2 U_{mf}^2 \quad (11.3a)$$

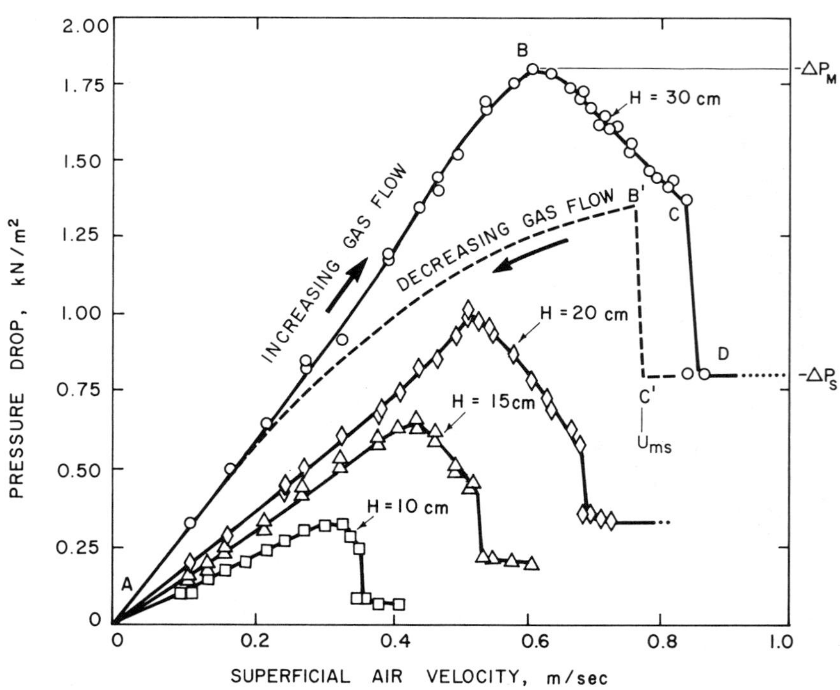

Figure 11.5. Typical pressure drop vs flow rate curves for the onset of wheat spouting, d_p = 3.6 mm, D = 152 mm, D_i = 12.7 mm, α = 60°.[1,15]

Combining Equations (11.11) and (11.3a), we obtain

$$\frac{-dP/dz}{-\Delta P_f/H} = \frac{2(\beta - 2)y + 3y^2}{2\beta - 1} \qquad (11.12)$$

where $\beta = 2 + (3K_1/2K_2 U_{mf})$ and $y = U_a/U_{mf}$. Substitution for y in Equation (11.12) from Equation (11.8) and integration between the limits (z, P) and (H, P_H) yields the following equation for the longitudinal pressure distribution in the annulus:

$$\frac{P - P_H}{-\Delta P_f} = \frac{1}{h(2\beta - 1)} \cdot [2(\beta - 2)\{1.5(h^2 - x^2) - (h^3 - x^3) + 0.25(h^4 - x^4)\} + 3\{3(h^3 - x^3) - 4.5(h^4 - x^4) + 3(h^5 - x^5) - (h^6 - x^6) + 0.143(h^7 - x^7)\}] \qquad (11.13)$$

where $h = H/H_m$ and $x = z/H_m$. The total pressure drop, $-\Delta P_s = P_0 - P_H$, across the spouted bed is obtained by putting $x = 0$ in Equation (11.13), i.e.,

$$\frac{-\Delta P_s}{-\Delta P_f} = \frac{2 - (4/\beta)}{2 - (1/\beta)}(1.5h - h^2 + 0.25h^3) + \frac{3}{2\beta - 1}(3h^2 - 4.5h^3 + 3h^4 - h^5 + 0.143h^6) \qquad (11.14)$$

Note that the final term of Equation (11.14) disappears when Darcy's Law prevails in the annulus ($\beta \to \infty$) while the first term on the right hand side drops out for the opposite extreme of inviscid flow ($\beta \to 2$). For a bed of maximum spoutable depth (i.e., $h = 1$), Equation (11.14) shows that $(\Delta P_s/\Delta P_f)_{max} = 0.75$ and 0.643 for the Darcy and inviscid regimes, respectively.[10]

A simpler empirical relation for vertical pressure distribution was proposed by Lefroy and Davidson:[16]

$$\frac{P - P_H}{-\Delta P_s} = \cos\left(\frac{\pi z}{2H}\right) \qquad (11.15)$$

Differentiating Equation (11.15) with respect to z yields

$$\frac{dP}{dz} = \frac{\Delta P_s}{2H} \sin\left(\frac{\pi z}{2H}\right) \qquad (11.16)$$

If incipient fluidization of the annulus is assumed at $z = H = H_m$,

$$\frac{\Delta P_f}{H_m} = \left(\frac{dP}{dz}\right)_{mf} = \frac{(\Delta P_s)_{max}\pi}{2H_m} \qquad (11.17)$$

i.e., $(\Delta P_s/\Delta P_f)_{max} = 2/\pi = 0.637$, in excellent agreement with the inviscid value above, but below the more realistic Darcy value. In view of the fact that all the above equations neglect radial pressure gradients, which in the vicinity of the fluid inlet may be sufficient to raise the total pressure drop by more than 25%, especially for shallow beds, the more conservative Equations (11.13) and (11.14) are preferred over Equation (11.15) and its corollaries.[10]

11.6. PARTICLE MOTION

The gas streamlines shown in Figure 11.4, if reversed in direction, also represent with little change the streamlines for the downward and inward flow of solids in the annulus. Because of the progressive cross-flow of solids from the annulus into the spout, downward particle velocities (v_w) measured at the transparent wall of a spouting vessel indicate values which decrease with decreasing level in the column. As shown in Figure 11.6, the mass flow of solids increases approximately linearly with height in the cylindrical part of the column, at least for $z \geqslant D$.[18] The slope of the plot at any level,

$$dW/dz = \rho_p(1 - \epsilon_a)d(v_w A_a)/dz \qquad (11.18)$$

is a measure of the cross-flow rate at that level. Here $\epsilon_a (\simeq \epsilon_{mf})$ represents the constant annulus voidage.

Individual particles in the spout are accelerated by the surrounding fluid from a vertical velocity

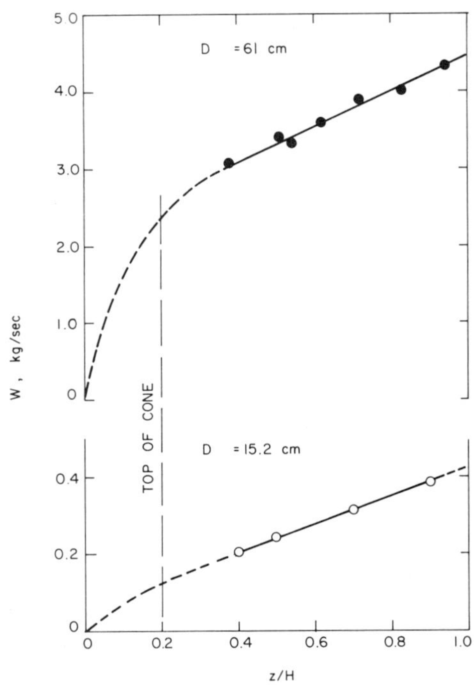

Figure 11.6. Solids flow in annulus, air-spouted wheat beds, $D/D_i = 6$, $H/D = 3$, $\alpha = 60°$, $U/U_{ms} = 1.1$.[1,3,17] $W = \rho_p(1 - \epsilon_a)A_a v_w$.

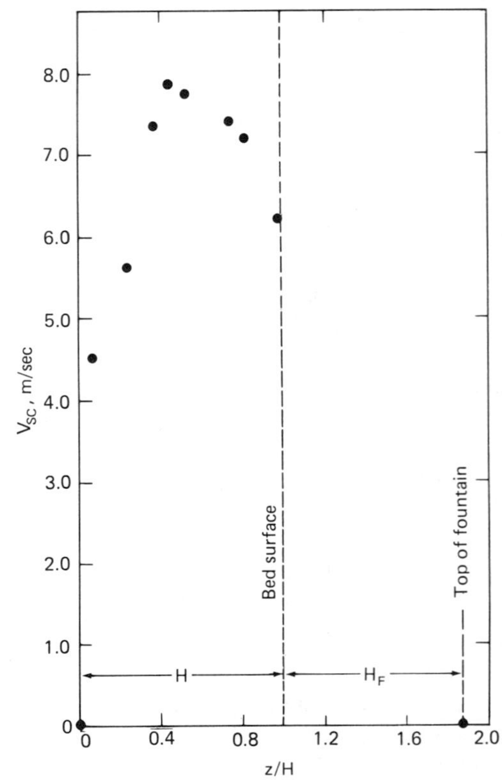

Figure 11.7. Experimental values of particle velocities along spout axis for 0.61 m diam × 1.22 m deep bed of 3.2 mm × 6.4 mm wheat particles; $\rho_p = 1376$ kg/m^3, $D_i = 102$ mm, $\overline{D}_s = 81$ mm, $U = 0.68$ m/s.[1,17]

of essentially zero, when they first enter, to some maximum value, after which the particles decelerate to achieve zero velocity again at the top of the fountain. The particles which enter the spout from the annulus rapidly become indistinguishable from the particles already in the spout at the same level. An experimental longitudinal profile of particle velocity, v_{sc}, along the spout axis appears in Figure 11.7. The point at the top of the fountain where $v_{sc} = 0$ is also shown.

At any horizontal level within the spout, the variation of upward particle velocity, v_s, with radial distance, r, from the spout axis may be represented by

$$\frac{v_s}{v_{sc}} = 1 - \left(\frac{r}{r_s}\right)^m \tag{11.19}$$

with $1.3 \leqslant m \leqslant 2.2$.[18] For $m = 2$ (parabolic profile), the radial-average particle velocity, $\overline{v}_s$, is simply $v_{sc}/2$, providing that the spout voidage, ϵ_s, is radially uniform.

A force balance on the spout particles over a differential height dz, using the relation

$$dt = \frac{dz}{v_z} \tag{11.20}$$

to convert from a Lagrangian to an Eulerian basis, yields

$$2v_s\frac{dv_s}{dz} - \frac{v_s^2}{1-\epsilon_s}\frac{d\epsilon_s}{dz} + \frac{v_s^2}{D_s^2}\frac{d(D_s^2)}{dz} = \frac{3\rho(u_s - v_s)|u_s - v_s|d^2C_D}{4d_p^3\rho_p} - \frac{(\rho_p - \rho)g}{\rho_p} \tag{11.21}$$

The drag coefficient C_D can be evaluated[20] as

$$C_D = C_{DT}/\epsilon_s^{2(n-1)} \tag{11.22}$$

where for gas spouting of spheres the terminal drag coefficient, C_{DT}, usually assumes the Newton's law value of 0.44 and the corresponding Richardson-Zaki[19] index $n = 2.39$. The upward interstitial fluid velocity in the spout, u_s, at any level is related to the corresponding superficial velocity, U_{sz}, by

$$u_s = U_{sz}/\epsilon_s \qquad (11.23)$$

Equations (11.21-11.23) and (11.9), in conjunction with experimental data for U_a or a relationship for U_a, such as Equation (11.8), may be solved[18] to yield vertical profiles of v_s, ϵ_s, and u_s, provided an auxiliary relationship involving at least one of these dependent variables is available. The boundary conditions at $z = 0$ are $v_s = 0$, $\epsilon_s = 1.0$ and $u_s = UD^2/D_i^2$.

If downward particle velocities at the vessel wall have been measured, then the auxiliary relationship which can be used is simply solids continuity at any level, i.e.,

$$W = \rho_p A_s(1 - \epsilon_s)v_s = \rho_p A_a(1 - \epsilon_a)v_w \qquad (11.24)$$

or its differential form, Equation (11.18). The solution using Equation (11.24) or (11.18) has been referred to as Model I.[18] Alternatively, an energy balance over a differential height of spout may be used, and this has been referred to as Model II.[18] Lim and Mathur[18] used a coefficient of unity instead of 2 in the first term of Equation (11.21), and their results for v_s by both models compared to experimental data for wheat spouting are shown in Figure 11.8. Model I gives good agreement with experiment over the entire bed height, but depends on measurements of W or dW/dz, for which no generalized correlations exist in the literature. Model II, which in theory requires no experimental input except D_s as a function of z, becomes unstable at $z/H < 0.2$, and reasonable values can only be obtained for $z/H > 0.2$, by starting with the experimentally measured values of v_s and ϵ_s at $z/H = 0.2$. Further work on modelling cross-flow of particles from the annulus into the spout, on the variation of D_s with z, and especially on the

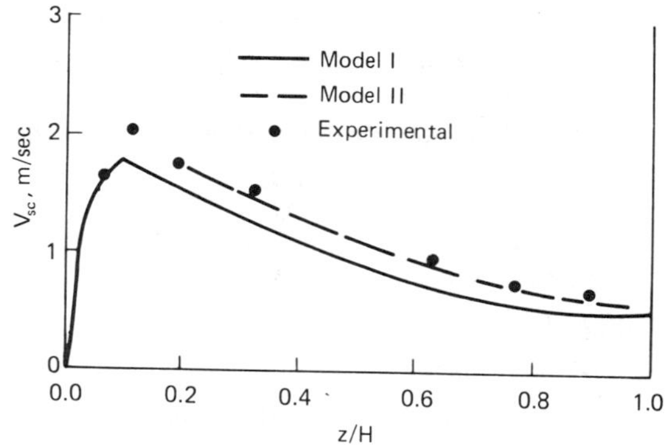

Figure 11.8. Radial-average particle velocity profile for air-spouting of 2.82 mm × 5.14 mm wheat particles, ρ_p = 1240 kg/m^3, D = 152 mm, D_i = 19 mm, $H/D = 3$, $U/U_{ms} = 1.1$.[18]

hydrodynamics of the fluid entrance region is required before the theoretical description of particle movement in the spout can be considered complete.

A similar analysis, but without the necessity of an auxiliary equation, has been applied[20] for particle motion in the fountain core (see Fig. 11.1b). For this case it is assumed that there is no cross-flow of solids and that the interstitial gas velocity is approximated by U/ϵ_s. The boundary conditions at $z = H$ are taken as $\epsilon_s = \epsilon_{sH}$ and $v_s = v_{sH} = (v_{sc})_H \epsilon_{sH}^{0.93}$, the last relationship having been obtained empirically.[20] Equation (11.21), the left-hand side of which reduces for the present conditions to $v_s dv_s/dz$, can then be solved in conjunction with Equations (11.22) and (11.23) for u_s, ϵ_s, and v_s, putting $U_{sz} = U$ and incorporating the additional requirements:

$$v_s(1 - \epsilon_s) = v_{sH}(1 - \epsilon_{sH}) \qquad (11.25a)$$

when

$$v_s > v_{sH}(1 - \epsilon_{sH})/(1 - \epsilon_a)$$

and

$$\epsilon_s = \epsilon_a \qquad (11.25b)$$

when

$$v_s \leqslant v_{sH}(1 - \epsilon_{sH})/(1 - \epsilon_a)$$

The height of the fountain, H_F, can also be predicted with little error by ignoring drag in the simplified Equation (11.21). Integration of this equation with the upper boundary condition, $z = H + H_F$, $v_s = 0$, gives

$$H_F = \frac{v_{sH}^2 \rho_p}{2g(\rho_p - \rho)} \tag{11.26}$$

where empirically v_{sH} is taken[20] as $(v_{sc})_H \cdot \epsilon_{sH}^{0.73}$; the decrease of the index on ϵ_{sH} from 0.93 to 0.73 arises from the neglect of drag.

Because of solids cross-flow from the annulus into the spout over the entire annulus height and because of the showering effect of the fountain, a spouted bed is a good solids mixer when a single species of solids is used. For most practical purposes, assuming that the solids feed and discharge ports are located to preclude any obvious short-circuiting, that the cone angle is sufficiently small to prevent any dead solids zones at the base, and that the mean residence time of the solids exceeds some minimum value of order minutes, perfect mixing of the species is a good approximation for a continuously fed spouted bed. This is illustrated in Figure 11.9.

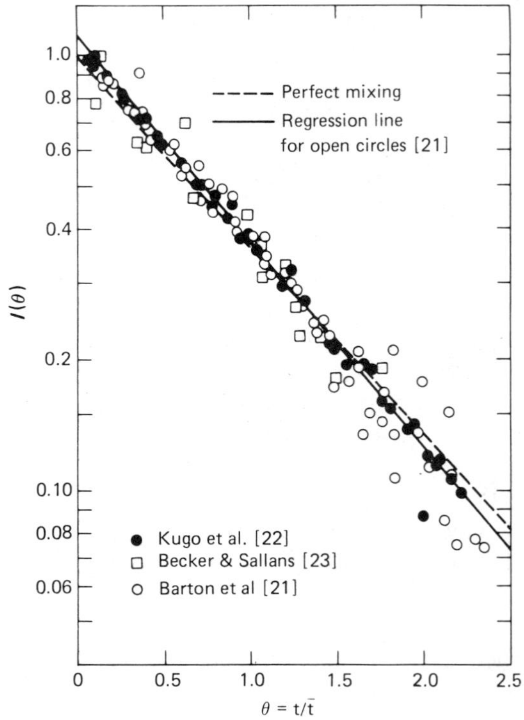

Figure 11.9. Solids mixing data correlated as internal age distribution function vs dimensionless time.[1]

When more than one species of solid material is used, e.g., particles of different size and/or density, considerable segregation occurs, especially at $U \sim U_{ms}$. The heavier and/or larger particles concentrate in the upper inside part of the annulus, and the concentration of these particles in the bed becomes significantly higher than in the steady state feed and discharge.[24] The primary cause of this segregation is the lower radial velocity imparted to the heavier particles when particle-particle collisions occur in the fountain region.[25] Segregation may therefore be largely countered if deflecting baffles are placed in the fountain region, or if U is increased so that the fountain particles strike the outer wall and bounce back towards the centre of the bed surface.

11.7. VOIDAGE DISTRIBUTION

The annulus of a spouted bed is a loose-packed bed of solids with its voidage, ϵ_a, very nearly the same as ϵ_{mf}. This voidage has a minimum of about 0.42 for closely sized smooth spheres; higher values are found for nonspherical particles.[10] Some negative deviations from ϵ_{mf} occur in certain regions of the annulus as a result of local variations in fluid percolation and solids flow rates,[26] but for most purposes these deviations can be ignored.

The voidage variation in the spout is roughly linear with height, as exemplified by the experimental data which appear in Figure 11.10. These data are for the same spouted bed as represented by Figure 11.8. The predictions of Lim and Mathur[18] for both particle circulation models described above are also plotted on this figure. As in the case of particle velocity, there is reasonable agreement with the experimental data—over the entire range of z/H for Model I, and starting at $z/H = 0.2$ for Model II.[18] Above the bed surface, the fountain analysis summarized above shows a continuous decrease in voidage from ϵ_{sH} to ϵ_a in the core of the fountain.[20]

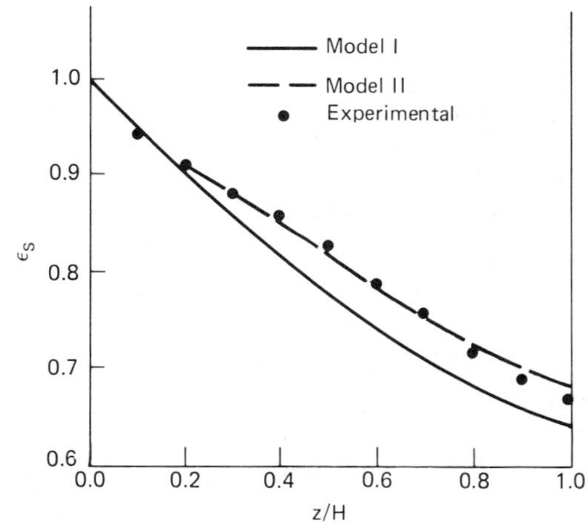

Figure 11.10. Spout voidage profile for system of Figure 11.8.[18]

11.8. SPOUT DIAMETER

The diameter of the spout is an important parameter for determining the flow distribution between spout and annulus via Equation (11.9), profiles of v_s and ϵ_s via Models I and II above, and predictions of fluid-particle heat transfer, mass transfer and chemical reaction via the two-zone model discussed below. Longitudinal variations of spout diameter which have been observed in conical-cylindrical columns are illustrated in Figure 11.11. Shape (a), the most common, tends to give way to (b) as column diameter increases, to (c) as particle size decreases, and to (d) for large inlet diameters.[1,27]

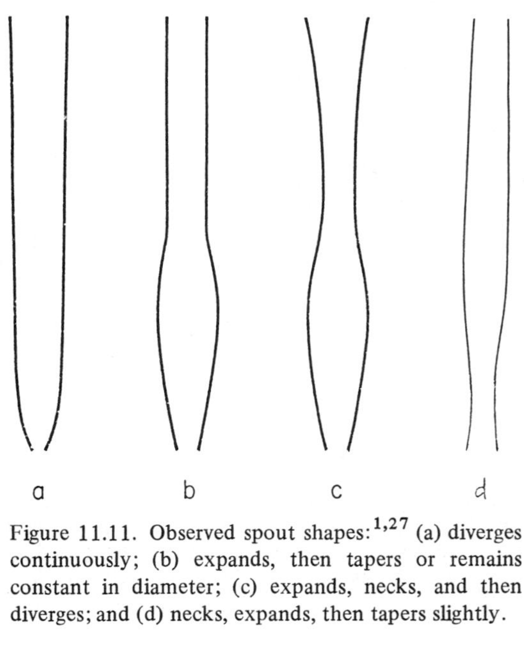

Figure 11.11. Observed spout shapes:[1,27] (a) diverges continuously; (b) expands, then tapers or remains constant in diameter; (c) expands, necks, and then diverges; and (d) necks, expands, then tapers slightly.

While the variation of D_s with z has not been quantified, the longitudinal average value of spout diameter, $\overline{D}_s$, has been correlated empirically by the dimensional correlation

$$\overline{D}_s = 2.00G^{0.49}D^{0.68}/\rho_b{}^{0.41} \pm 5.6\% \quad (11.27)$$

over a wide range of experimental data,[28] with $\overline{D}_s$ and D in m, $G(=\rho U)$ in kg/m²s and ρ_b [$= \rho_p(1 - \epsilon_{mf})$] in kg/m³. A more restrictive equation, which applies only to beds at their maximum spoutable depth, but which has the virtue of being dimensionally consistent, is[2]

$$\frac{\overline{D}_s}{D_i} = \left[\frac{2.10 \exp(-0.0180/A) + 1.00}{3.10}\right] \cdot \left[0.862 + 0.219\frac{D}{D_i} - 0.00530\left(\frac{D}{D_i}\right)^2\right] \quad (11.28)$$

where

$$A = \left(\frac{\rho}{\rho_p - \rho}\right)\left(\frac{U_T U_{mf}}{gD_i}\right) \quad (11.29)$$

Attempts to apply principles of solids flow mechanics to the determination of D_s have so far achieved only qualitative success.[1]

11.9. HEAT TRANSFER

Transfer of heat between the fluid and the solid particles in a spouted bed is most accurately described by means of the two-region model discussed below. A more conservative approach, based on the use of a fluid-particle heat transfer coefficient for a loose-packed bed, i.e., for annulus conditions, is presented in Reference 1. In the annulus, unlike the spout, thermal equilibrium between fluid and particles is achieved even in a shallow bed. For the relatively large particles used in spouted (as opposed to fluidized) beds, intraparticle heat transfer must be considered.[1]

Transfer between the bed and the wall is characterized by the development of a thermal

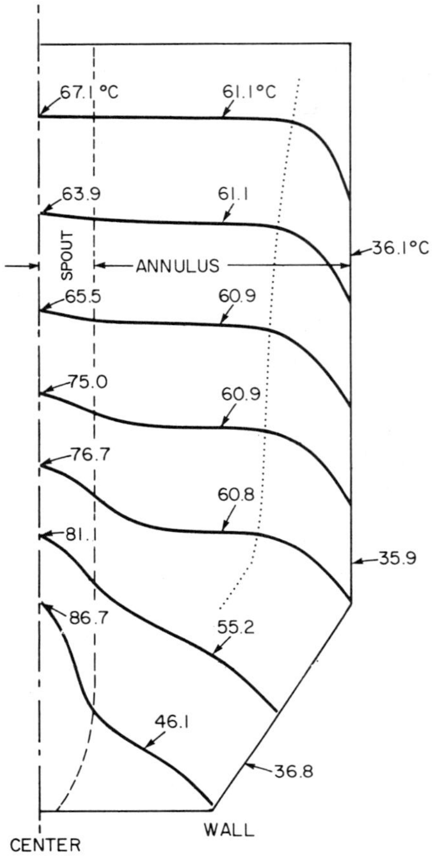

Figure 11.12. Local gas temperature profiles for a wall-cooled spouted-bed,[29] with thermal boundary layer profile (dotted line) added.[1]

boundary layer in the annulus, as exemplified by Figure 11.12 for gas spouting. For liquid spouting this boundary layer extends all the way to the spout. Over the range of conditions for which wall-to-bed heat transfer in gas-spouted beds has been studied,[1] the bed-to-wall heat transfer coefficient, h_w, can be predicted from the empirical equation of Malek and Lu:[29]

$$\frac{h_w d_p}{k_g} = 0.54 \left(\frac{d_p}{H}\right)^{0.17} \left(\frac{d_p{}^3 \rho_g{}^2 g}{\mu^2}\right)^{0.52} \left(\frac{\rho_b c_{pp}}{\rho_g c_{pg}}\right)^{0.45} \left(\frac{\rho_g}{\rho_b}\right)^{0.08} \quad (11.30)$$

An alternative theoretical approach, based on a two-dimensional penetration model,[1] results in:

$$h_w = 1.129\ [v_w \rho_b c_{pp} k_b/(H - z)]^{1/2} \quad (11.31)$$

where the heat transfer surface extends over a length $(H - z)$. The mean coefficient given by Equation (11.31) is twice the local coefficient at the level z. Equation (11.31) tends to over-predict h_w somewhat, due to the higher voidage at the wall than in the bulk of the annulus.[1]

A heating or cooling element submerged in the bed is a more efficient heater or cooler than a jacket around the column wall. Typical radial profiles of the immersed heat transfer coefficient, h_s, for a vertically aligned cylindrical heater are shown in Figure 11.13. It is seen that h_s reaches a maximum at the spout-annulus interface and increases with d_p.

Typical vertical profiles of h_s in the spout for a horizontally-aligned cylindrical heater are shown in Figures 11.14. It is seen that h_s decreases with increasing z, sharply near the bottom of the bed and then gradually further up, and that it increases with an increase in spouting velocity. Profiles of h_s within and around the fountain have also been reported for a spherical probe.[1,31] Generally speaking, values of h_s obtained in the annulus are similar to those for submerged objects in moving packed beds; values at the bottom of the spout

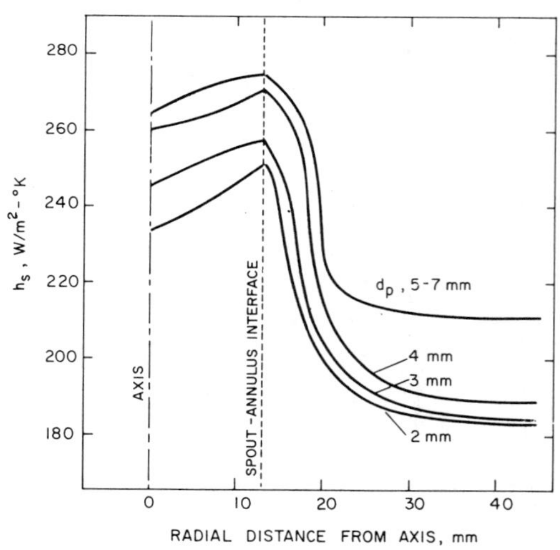

Figure 11.13. Radial profiles of submerged object-to-bed head transfer coefficient in upper part of bed measured with a vertically aligned cylindrical heater, 4 mm diam × 35 mm long:[30] air-spouted silica gel, d_p = 2 mm, D = 94 mm, D_i = 15 mm, H = 100 mm.[1]

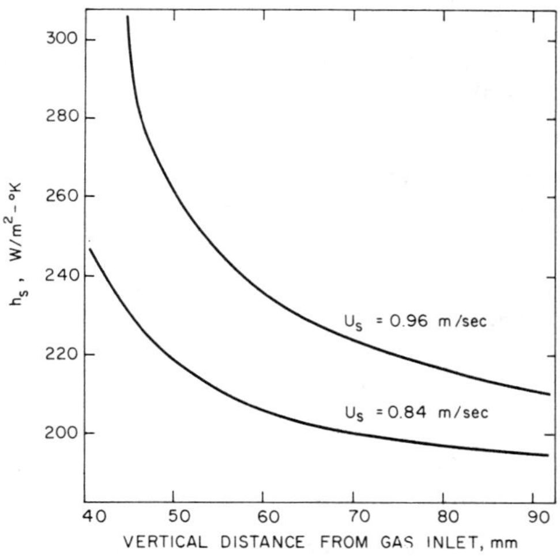

Figure 11.14. Vertical profiles of submerged object-to-bed heat transfer coefficient in the spout measured with a horizontally aligned cylindrical heater, 10 mm diam × 17 mm long:[30] air-spouted silica gel, d_p = 2 mm, D = 94 mm, D_i = 15 mm, H = 100 mm.[1]

are like those for the pure fluid flowing past the submerged object at comparable velocities, while coefficients higher in the spout and in the fountain are similar to those for a submerged object in a dense-phase fluidized bed.

11.10. MASS TRANSFER

As in the case of fluid-particle heat transfer, mass transfer between the fluid and the surface of the particles is best treated by the two-region model described below. Again, a more conservative approach for fluid-particle mass transfer under conditions of external control, e.g., constant rate drying, can be based on the use of a mass transfer coefficient for the loose-packed bed conditions which prevail in the annulus.[1]

However, drying of such materials as agricultural products and fertilizer granules, for which the spouted bed has proved to be most effective, is often carried out over ranges of moisture content which are well within the falling rate period. Moisture diffusion within the particles then controls the overall drying process. For such internal mass transfer control, Becker and Sallans[23] have derived equations which are particularly applicable to wheat drying, but which can be applied to other materials if the moisture diffusivity, $\mathfrak{D}$, within the given material is known as a function of the mean particle temperature, e.g., via a relationship of the Arrhenius form. The equations assume well-mixed solids which at any instant are isothermal at annulus conditions and which undergo negligible change of particle volume as drying progresses. For batchwise spouted bed drying,

$$\frac{\overline{m} - m_s}{m_0 - m_s} = \frac{6}{\pi^2} \sum_{p=1}^{\infty} \frac{1}{p^2} \exp\left[-\frac{p^2\pi^2\chi^2}{9}\right] \qquad (11.32)$$

where $\chi = (6/\phi d_p)(\mathfrak{D}t)^{1/2}$. Equation (11.32) relates uniform mean moisture content of the particles, $\overline{m}$, to their uniform drying time, t. For steady continuous spouted bed drying, the residence time, t, of individual particles tends to differ from the mean residence time $\overline{t}$. With the assumption of perfect solids mixing, discussed above, Equation (11.32) must be modified to

$$\frac{\overline{\overline{m}} - m_s}{m_i - m_s} = \int_0^{\infty} \frac{6}{\pi^2} \sum_{p=1}^{\infty} \frac{1}{p^2} \exp \left[-\frac{t}{\overline{t}}\left(\frac{p^2\pi^2\overline{\chi}^2}{9} + 1\right)\right] d\left(\frac{t}{\overline{t}}\right) \qquad (11.33)$$

where $\overline{\chi} = (6/\phi d_p)(\mathfrak{D}\overline{t})^{1/2}$. Equation (11.33), which relates the average moisture content of the product particles to $\overline{t}$, has been verified experimentally for wheat drying in all but the shallowest of beds.[1] Other shortcut or empirical methods are summarized in Reference 1.

11.11. CHEMICAL REACTION: TWO-REGION MODELS

Spouted beds share some of the principal advantages of fluidized beds as chemical reactors—solids mobility, relatively uniform temperature and, to some extent, favorable bed-to-surface heat transfer. Shared disadvantages between spouted and fluidized bed reactors are bypassing of gas, backmixing of solids, particle entrainment and attrition. Spouted beds give more reproducible flow patterns and have fewer flow regimes than fluidized beds, but their ranges of application in terms of mean particle size and vessel diameter are much more limited.

Bypassing in spouted beds is caused by fluid elements in the central spout traveling more quickly and with a much higher voidage than those in the annulus. For a catalytic gas-phase reaction, it is essential to distinguish between the two regions, since reaction will be much more favourable in the annulus, where gas elements are in intimate contact with the solids, than in the spout. Similar considerations apply when spouted beds are used for heat transfer between fluid and particles or for an analogous mass transfer process, e.g., adsorption of a component from a gas.

The earliest and simplest representation of a spouted bed for these purposes is in terms of a *one-dimensional model*[32] in which radial gradients within each region are ignored. For a first order reaction in an isothermal spouted bed reactor, steady state mass flow balances for an element of height dz of each region then result in the following equations:[33]

Spout:

$$Q_s \frac{dC_s}{dz} + k_{sa}\pi D_s(C_s - C_a) + K_r \frac{\pi}{4} D_s^2 (1 - \epsilon_s) C_s = 0 \qquad (11.34)$$

Annulus:

$$Q_a \frac{dC_a}{dz} + \frac{dQ_a}{dz}(C_a - C_s) + k_{sa}\pi D_s (C_a - C_s) + K_r (1 - \epsilon_a) A_a C_a = 0 \qquad (11.35)$$

The first and last terms in each of these equations are due to convection and chemical reaction respectively. Plug flow of fluid is assumed to prevail in each of the regions, and the reaction rate in each region is assumed to be controlled by chemical kinetics. The middle terms arise from interphase mass transfer, the second term in Equation (11.35) being due to net outflow from the spout into the annulus, as discussed above; the terms involving k_{sa} account for any additional transfer. The flow rates Q_a and Q_s through the two phases and the derivative dQ_a/dz can be obtained as functions of height from Equations (11.8) and (11.9). The spout diameter, D_s, can be calculated from Equation (11.27) or (11.28) or measured in a half-column, while A_a can be obtained from the geometry of the column. The rate constant, K_r, should be determined separately under isothermal conditions in a reactor whose hydrodynamics are well understood, e.g., in a packed bed or spinning basket reactor. For non-first order kinetic rate expressions, the final terms in Equations (11.34) and (11.35) must be replaced by the appropriate rate expressions. There is no reliable method of estimating k_{sa}, but values are typically less than 0.1 m/s and $k_{sa} \simeq 0$ appears to be a reasonable assumption when $d > 1$ mm.[33]

The boundary condition required for solution of Equations (11.34) and (11.35) is $C_s = C_{in}$ at $z = 0$. The equations can be integrated numerically from $z = 0$ to $z = H$. The exit concentration is then evaluated from the overall mass balance:

$$U \frac{\pi}{4} D^2 C_{\text{exit}} = [Q_s C_s + Q_a C_a]_H \qquad (11.36)$$

For the case where a component is removed from a gas stream in a spouted bed, e.g., for collection of aerosol particles on the bed particles,[34] the same one-dimensional model can be applied, but with the reaction terms in Equations (11.34) and (11.35) replaced by the respective adsorption rate per unit volume in that phase. For the annulus, the adsorption rate can be based on correlations for mass transfer between particles and fluid in packed beds. For the spout, mass transfer between the spouting fluid and particles can be estimated from the high-voidage correlation of Rowe and Claxton.[35] These equations should also be used when reaction rates within the individual regions are mass-transfer controlled.

Analogous equations can be developed for heat transfer in the case where a hot gas enters a bed of cold particles or a cold gas jet enters hot particles. Let us assume constant properties and spherical particles and neglect any interphase transfer, aside from that associated with crossflow of gas. In view of the rapid mixing of solids in spouted beds and the fact that the volumetric heat capacity of the solids, $\rho_p c_{pp}$, is much larger than that of the gas, $\rho_g c_{pg}$, we may, as a first approximation, treat the particles at any instant as being of uniform temperature, T_p. Then energy balances for gas in each of the regions yield the equations:

Spout:

$$Q_s \rho_g c_{pg} \frac{dT_{gs}}{dz} = (h_{pg})_s \frac{3\pi}{2d_p} D_s^2 \times (1 - \epsilon_s)(T_p - T_{gs}) \qquad (11.37)$$

Annulus:

$$\rho_g c_{pg} \left[Q_a \frac{dT_{ga}}{dz} + \frac{dQ_a}{dz}(T_{ga} - T_{gs}) \right] = (h_{pg})_a \frac{6A_a}{d_p} (1 - \epsilon_a)(T_p - T_{ga}) \qquad (11.38)$$

where the particle-gas heat transfer coefficients, $(h_{pg})_s$ and $(h_{pg})_a$, for the spout and the annulus may again be obtained from correlations for dilute suspensions and packed beds, respectively. Gas entering the annulus will equilibrate with the solids temperature within a small distance. To obtain the change of particle temperature for a batch process or the steady state particle temperature for a continuous process where solids are fed at a different temperature from the gas, a heat balance is also required for the solids. Equations (11.37) and (11.38) can be solved numerically with the boundary conditions $T_{gs} = T_{ga} = T_{gi}$ at $z = 0$. The outlet gas temperature is obtained from an energy balance, i.e.,

$$U \frac{\pi}{4} D^2 (T_g)_{\text{exit}} = [Q_s T_{gs} + Q_a T_{ga}]_H \qquad (11.39)$$

The one-dimensional model has been extended to spout-fluid beds (see section 11.13 below) by Hadžismajlović et al.[36] These workers also allowed for variation of ϵ_s with z rather than adopting an average value.

An alternative model, the *streamtube model*, has been applied[33] to the case of a first order gas phase reaction in an isothermal spouted bed. The model was first used[14] to describe gas residence time distributions in spouted beds. Whereas the one-dimensional model implicitly assumes perfect radial mixing of gas elements in the annulus, the streamtube model is based on a physical picture, shown in Figure 11.15, in which the gas entering the annulus fans outward and upward in a finite number of streamtubes. The coordinates of the streamlines bounding each of these streamtubes are calculated on the assumption that the vertical component of gas velocity is radially uniform at each section in the annulus. Streamwise dispersion is ignored (i.e., plug flow is assumed) in each of the streamtubes. Any interphase mass transfer, aside from the bulk flow obtained from Equations (11.8) and (11.9), is

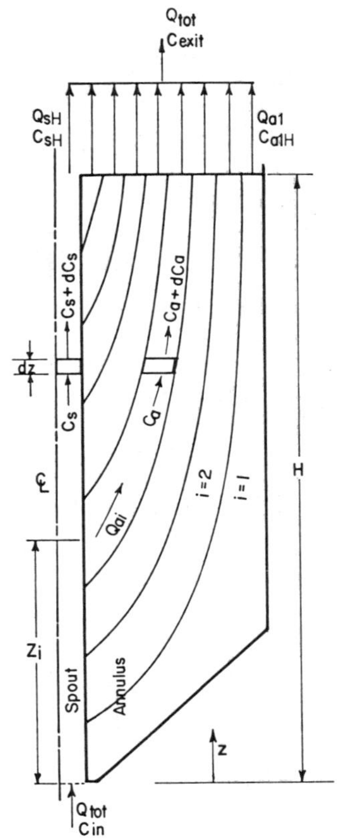

Figure 11.15. Vertical section through spouted bed showing flow distribution assumed in the streamtube model.

ignored in this model. Plug flow of gas is again assumed in the spout phase.

With these assumptions, a mass balance in the spout phase gives Equation (11.34) with k_{sa} = 0. The concentration of gas leaving the top of the ith streamtube is given[33] by

$$C_{aiH} = C_s(z_i)\exp\{-K_r(1 - \epsilon_a)\tau_i/\epsilon_a\} \tag{11.40}$$

where z_i is the midpoint entry height (i.e., the mean height of intersection of the bounding streamtubes with the spout-annulus interface) and τ_i is the mean residence time of gas within the streamtube. The exit concentration is again obtained by performing an overall mass balance at the top of the reactor:

$$U\frac{\pi}{4}D^2C_{\text{exit}} = \left[Q_sC_s + \sum_{i=1}^{N} Q_{ai}C_{ai}\right]_H \tag{11.41}$$

Piccinini et al.[33] found that N = 20, i.e., 20 streamtubes, gave a good compromise between accuracy and computational effort when applied to the streamtube model.

Experimental reaction data obtained in a spouted bed of catalyst pellets using the ozone decomposition reaction have been compared with both the one-dimensional model and the streamtube model.[33] Predicted concentration profiles from the latter model for one of the experimental runs are shown in Figure 11.16. Agreement between the experimental conversions and the predictions of both models was excellent. Further experimental results including in-bed concentration profile measurements have been obtained[40] to discriminate between the two models. While the one-dimensional model is simpler, and may be adequate for many cases, especially for beds of small diameter, the streamtube model gives a more

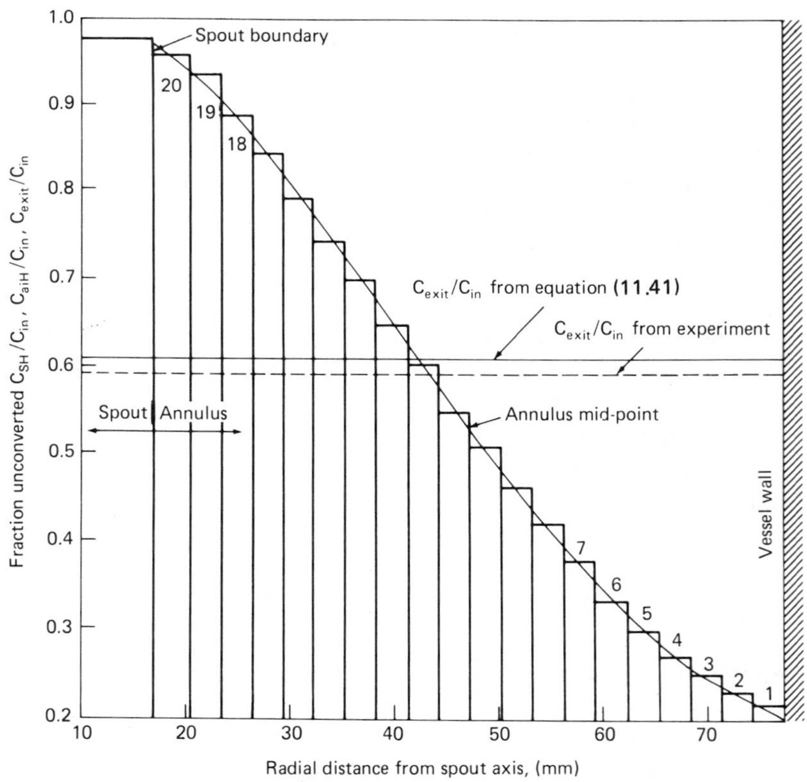

Figure 11.16. Predicted radial concentration profile at the bed surface for the streamtube model: d = 1.48 mm, H = 0.41 m, D = 0.155 m, U = 1.07 m/s = 1.1 U_{ms}, K_r = 4.2 s^{-1}, ρ_p = 2330 kg/m^3, ϵ_a = 0.48, ϵ_s = 0.85. Predicted and experimental overall exit concentrations are also shown.

accurate representation of the actual flow patterns and concentration profiles.

Neither of the models presented above makes allowance for the additional contacting between gas and solids that occurs in the fountain region above the bed surface. A procedure for estimating upper and lower limits on the additional reaction which occurs in the fountain has been suggested by Grace and Mathur.[20] In most cases the additional reaction will be small enough to be neglected.

The one-dimensional and streamtube models as described above can be used to predict the performance of gas-phase solid-catalyzed chemical reactions. In principle, they can also be applied to the case of gas-solid heterogeneous reactions, in much the same way that two-phase reactor models for fluidized beds have been extended to the case of heterogeneous reactions.[37,38] Foong et al.[39] used the one-dimensional model to describe conversion in a spouted bed coal gasifier. However, since the kinetics of the reaction were unknown, the reaction was treated like a gas phase reaction to yield an effective rate constant, and this value was then used to predict the influence of bed height, bed composition and column diameter. It is noteworthy that conversion is predicted to increase with increasing reactor diameter, in contrast to the case of fluidized beds where conversion invariably decreases as a reactor is scaled up. The improved performance with increasing D arises because the spout occupies a smaller fraction of the cross-sectional area of the spouted bed as the reactor is scaled up. The same trend has been predicted for spouted bed reactors previously,[1,32] but has recently been contradicted experimentally over a limited range, $D = 0.15$ to 0.22 m.[40]

Several complications arise in applying the reactor models to the more general case of gas–solid heterogeneous reactions. The kinetic rate expression must account for the way in which particles react, e.g., by assuming a shrinking core, surface reaction or homogeneous reaction throughout the particles.[37] The physical properties (size, density and shape) of the particles may change during their residence in the bed as a result of reaction, attrition or agglomeration. Solids residence time distributions (commonly approximated by perfect mixing) must be considered, since the extent of reaction of each particle depends on its residence time in the reactor. Population balances may be required to account for different sizes of particles as reaction, attrition, and entrainment proceed. Since heterogeneous reactions are often highly exothermic or endothermic, energy balances may also be required. Finally, the extent of reaction of gaseous components must be linked to that of the solids by means of the stoichiometry of the reactions. Application of spouted bed reactors has not yet advanced to the point where models complex enough to cope with these complications have been developed.

11.12. APPLICATIONS

Orginally developed for wheat drying,[41] Figure 11.17, gas-spouted beds have since been applied to a wide variety of operations[1] involving coarse (e.g., 1-5 mm) solid particles. These operations rely on one or more of the following features of the technique:

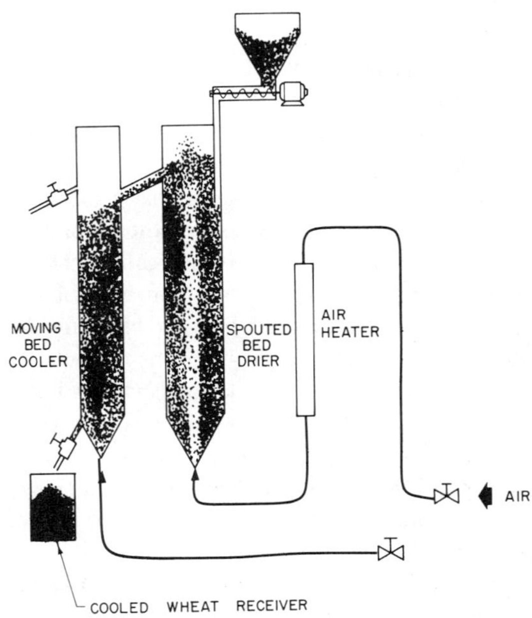

Figure 11.17. Original pilot wheat drier at National Research Council of Canada.[41,1]

1. Good solids mixing coupled with satisfactory gas-particle contact, i.e., the ability to accomplish for coarse solids what a fluidized bed does for fine solids.
2. Higher gas velocities and correspondingly lower gas residence times than for conventional fluidized beds of fine solids.
3. Systematic cyclic movement of solids, compared with the more random particle movement in other systems such as fluidized beds or rotary drums.
4. Solids deagglomeration and attrition caused by the high velocity interparticle collisions in the spout.
5. The absence of a distributor plate, in contrast to the case of a fixed or fluidized bed.

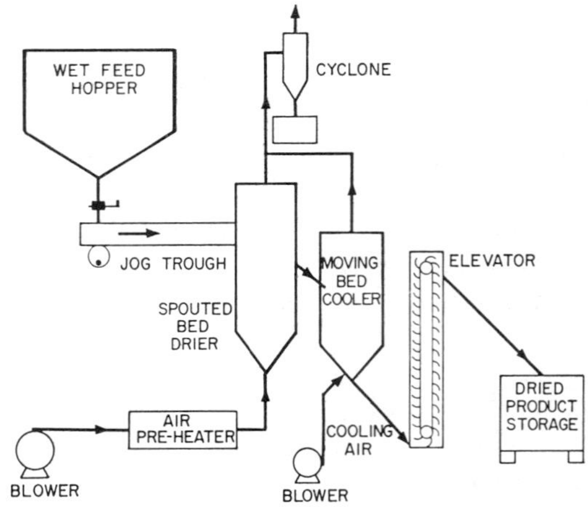

Figure 11.18. Layout of industrial drier for agricultural products.[42,1]

Good solids mixing, along with *effective gas-particle contact*, is the basis for spouted bed drying of noncaking granular solids.[1] The method is particularly suitable for heat-sensitive materials such as agricultural products or polymer granules, since the rapid agitation of the solids permits the use of higher temperature gas than in nonagitated driers, without the risk of thermal damage to the particles. Commercial driers of 0.6 m diameter with a bed depth of about 2 m are capable of safely drying up to 2 Mg/h of peas through an 8% moisture range, dry basis, using about 3 Mg/h of air at temperatures up to 557 K.[42] The layout of such an industrial unit for drying peas, lentils and flax is shown in Fig. 11.18.

Sensible heating or cooling of coarse solids in spouted beds also makes use of the favorable gas–solid contacting, but the good solids mixing is more important in heating than in cooling. In the use of a spouted bed for blending of solids, the intimate gas-particle contact is incidental, and only the good solids mixing is of importance. Multistage spouted bed preheating of coal feed to coke ovens has been successfully piloted, while commercial-scale rectangular (4.9 m $\times$ 1.8 m) two-stage multiple-spout fertilizer coolers with capacities up to 30 Mg/h and thermal efficiencies exceeding 85% have been developed by Fisons Ltd.; single-spout circular units of equal size have been operated by I.C.I. Fibres Ltd. for blending polyester polymer chips in batches exceeding 57 m^3.[1]

The relatively *high gas velocities* and correspondingly low gas residence times associated with spouting of coarse particles are the basis for the bench-scale development at Hokkaido University of a dual-spouted reactor-regenerator combination for the thermal cracking of petroleum feedstocks.[1] A similar combination has been developed by the same investigators for the catalytic desulfurization of residual fuel oil, using steam at 923 K plus the fuel oil as the spouting fluid in the reactor, and air in the regenerator.[43] High gas throughput per unit cross-section also makes the use of a spouted bed attractive for gas cleaning purposes, especially since high efficiencies at minimum spouting velocities have been measured for the bench-scale collection of liquid and electrified-solid aerosols from a gas in spouted beds of inert solids,[34] as well as for the chemical reduction of dilute SO_2 gas by a spouted bed of activated charcoal.[44] However, for both these gas cleaning processes, operation at velocities above minimum spouting sharply reduces the respective efficiencies, undoubtedly as a result of excessive gas bypassing through the spout, in addition to the lowering of gas residence time. Increased gas capacities can be effected without the loss of efficiency by the use of larger column cross-sections with multiple spouts, discussed below.

The highly systematic *cyclic movement of solids* in a spouted bed has proved to be a key advantage in such processes as granulation and particle coating. In granulation, a melt or solution is atomized into a bed containing seed granules spouted by hot gas. These granules build up by a mechanism of layer-by-layer growth as they cycle in the bed, and yield a final product which is well-rounded and uniform in structure.[45] Continuous operation requires that oversize product be crushed and recycled to the spouted bed along with undersize product,[46] as illustrated in Figure 11.19. Spouted granulation of fertilizers,[47,48] pigments and dyes[1] has been carried out on an industrial scale, and a number of 16 Mg/h sulphur granulators ("Perlomatic" system) have been built by PEC Engineering of France. In these units the spouting occurs in the 60° conical base, which expands up to and is topped by a 2 m diameter cylindrical disengaging section.[49] The units have encountered operating difficulties.

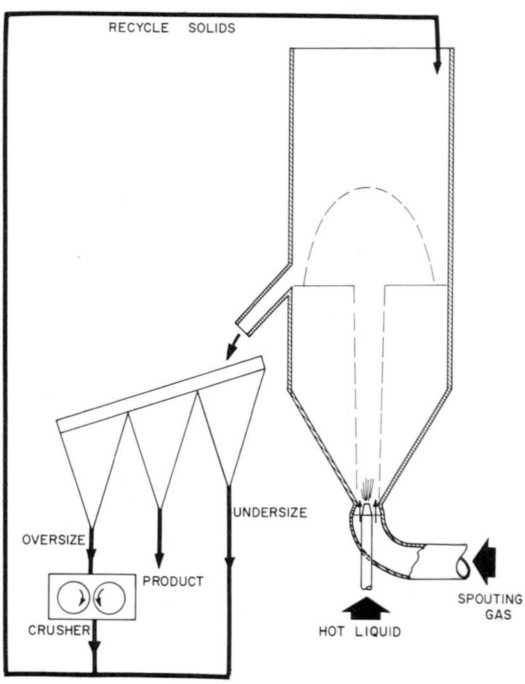

Figure 11.19. Spouted-bed granulation system, after Berquin.[46,1]

The coating of pharmaceutical tablets in a spouted bed is a well-established commercial operation, a typical batch being 100 kg.[1] A newer thermal coating development, still on a pilot scale but with promising larger scale economics, is the spouted bed coating of urea granules with sulphur to produce a slow release fertilizer.[50] The thermo-chemical coating of pyrolytic carbon and/or silicon carbide onto submillimeter nuclear fuel kernels of uranium oxide or carbide has been standardized in spouted bed furnaces of 75-125 mm internal diameter, with kernel loads of about 1 kg for each coating operation,[51] Fig. 11.20. Essentially the same technique is now being applied to the pyrolytic coating of prosthetic devices[52] by the Sorin Co. of Italy.

The *solids attrition* caused by the particle collisions in the spout is a liability for some spouted bed operations (e.g., granulation, tablet coating), but is an asset for several others. The most successful of these, developed at the Leningrad Institute of Technology, is the drying of suspensions and solutions by atomizing them into the lower region of a hot gas-spouted bed of inert particles.[53] The suspension or solution coats these particles and dries during the particle downward movement in the annulus. The fine product is broken away by interparticle collisions in the spout and collected from the overhead gas. Materials which lend themselves to this method of drying include organic dyes, dye intermediates, lacquers, salt and sugar solutions, and several chemical reagents. The more conventional spouted bed drying of granular materials with caking tendencies, for example ammonium nitrate, has also been industrially successful (where fluidized bed drying has failed), due to the breakdown of embryonic agglomerates in the high velocity spout.[1] Bench-scale spouted bed developments for which this property has been important include comminution of particulate solids, iron ore reduction, shale pyrolysis and coal carbonization (pyrolysis). The latter has been conducted at temperatures up to 925 K in Australia,[21] and up to 815 K in India[54] with coals of various caking tendencies.[1] Indian noncaking coals have been simi-

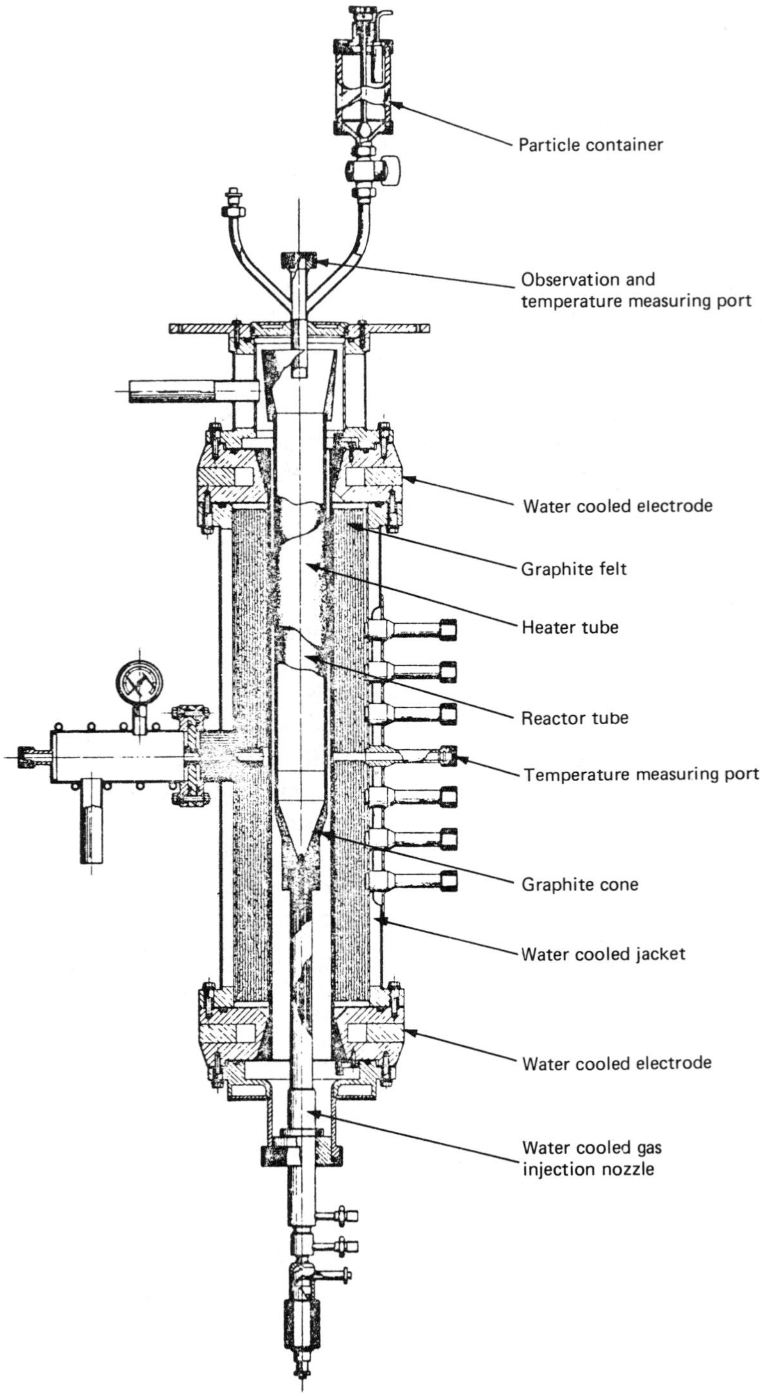

Figure 11.20. Spouted-bed pyrolytic coating furnace.[51,52]

larly gasified with air and steam.[55] A more impressive development is the gasification at atmospheric pressure and temperatures up to 1200 K of 0.8-3.6 mm highly caking coals from Western Canada in a 0.15 m diameter spouted bed containing a proportion of silica particles in the same size range.[39] This was achieved without the cumbersome and expensive procedures required for gasifying caking coals in a fluidized bed, and the process, shown schematically in Figure 11.21, has been successfully scaled up.[56]

The *absence of a distributor plate* in a spouted bed is a definite advantage in many of the above operations—especially in granulation and coating, in drying of solutions, suspensions and sticky solids, and in carbonization or gasification of caking coals. It is also an important consideration in a high temperature (1300-1800 K) industrial process for making granular activated carbon,[57] as well as in bench-scale spouted-bed kiln operations for calcination of limestone[58] and for production of cement clinker from decarbonated cement granules.[1] The decarbonation itself has been successfully accomplished in a system designated Kawasaki Spouted Bed and Vortex Chamber or KSV.[59] At least five cement plants with capacities of 350 Mg/day and one with 8500 Mg/day, using KSV calcining furnaces, were in satisfactory operation as of 1975.[60]

Over forty different patents on spouted beds were granted in various countries between 1957 and 1973.[60] Most of these were on applications of the "standard" spouted bed described above or minor variations thereof. Some, however, were on significant modifications to the standard geometry or mode of operation, and such modifications are discussed next.

11.13. MODIFIED SPOUTED BEDS

The following modifications of the standard spouted bed, most of which have been detailed elsewhere,[1] are worthy of note:

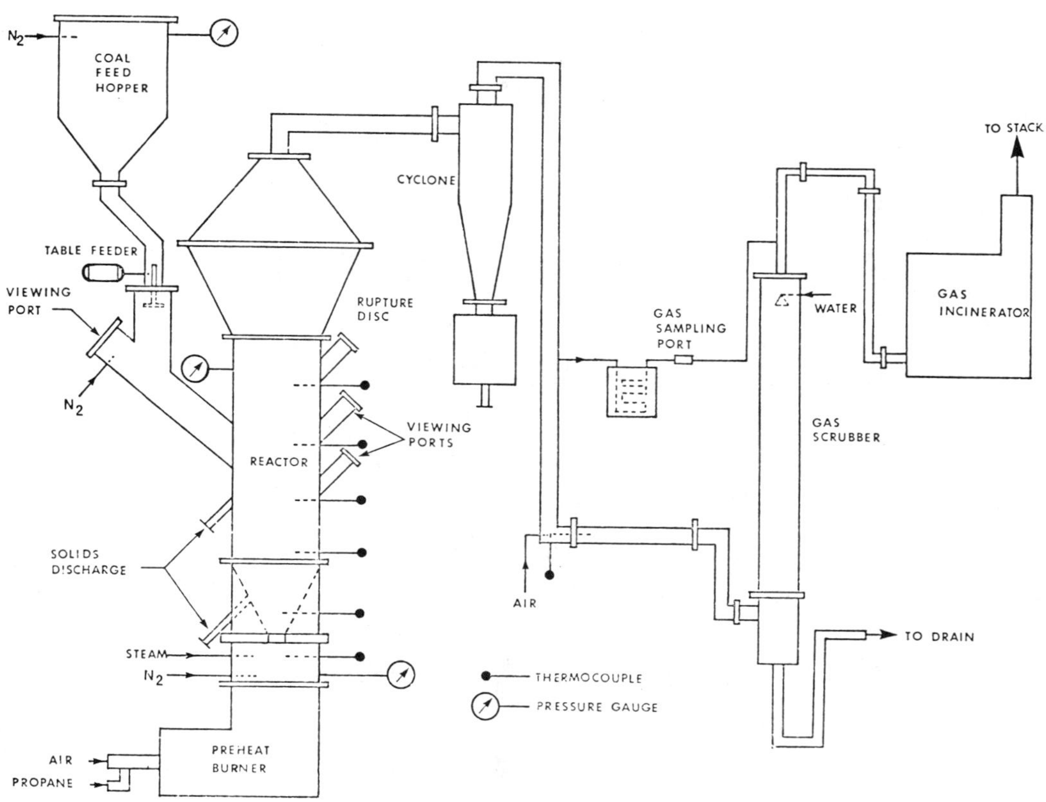

Figure 11.21. Schematic diagram of spouted-bed coal gasifier.[39]

Multiple Spouts

There is a practical limit to which vessel diameter can be increased using a simple fluid inlet. Since $H/D > 1$ for stable spouting, the large bed heights required for large column diameters would give rise to excessively high pressure drops. In addition, the long times spent by particles in the annulus of a large bed over the course of a single cycle, especially along the outer streamlines shown in Figure 11.4, could be a distinct disadvantage for certain processes, e.g., for particle drying where the excessive time in the hot region of the bed could cause thermal damage to the particles. The most obvious way of overcoming these difficulties is by using more than one fluid inlet nozzle in parallel, i.e., by multiple spouting.

Figure 11.22 shows a schematic of a multiple spouted bed with a flat base. Multiple cones have also been used.[1] Although the spouting fluid may originate from a single manifold, the flow to each inlet nozzle must be controlled separately. Even with such control, spouting stability problems arise when the inlet nozzles are too closely spaced, when the bed height is increased excessively (but still below H_m for a single nozzle), and when the ratio D_i/d is less than about 8.[61] Bed stability can be improved by using fluid inlet nozzles which project (e.g., about 3 mm) above the bed floor (as in Fig. 11.22), by cutting off lateral flow of fluid between spouting cells through the use of vertical partition walls,[1] and by the use of inverted-funnel spout deflectors above the fountains[1] to minimize interference between adjacent cells by spout wandering.

Foong et al.[61] have shown that, if one takes the diameter of each spouting cell as that of a circular cylinder having the same cross-sectional area, then both the minimum spouting velocity and the pressure drop across the spouted bed can, for a *stable* multispout bed, be predicted by the relationships applicable to a conventional single-spout bed. For handling equal inventories of solid, a multispout bed consumes

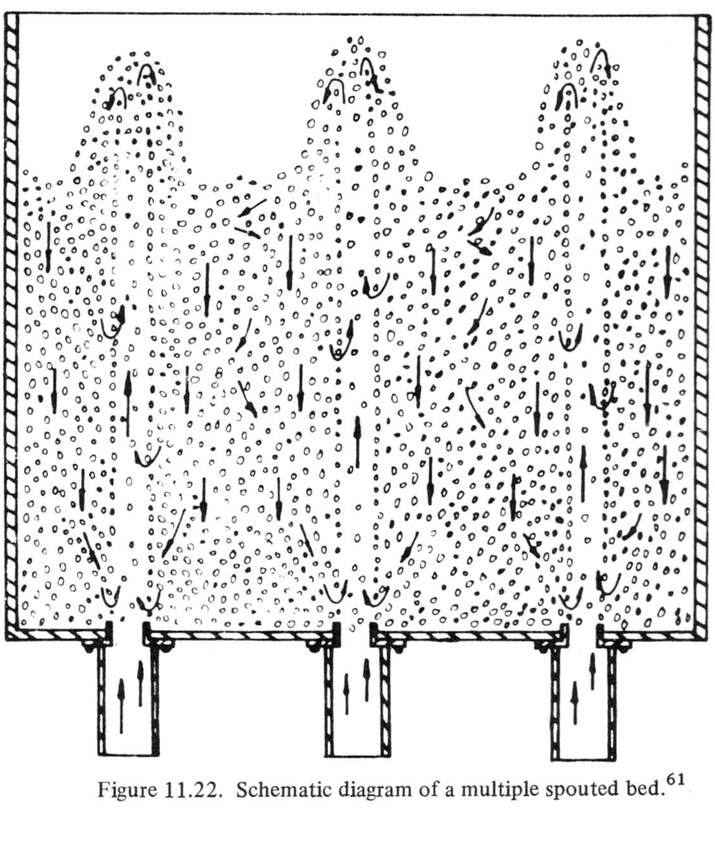

Figure 11.22. Schematic diagram of a multiple spouted bed.[61]

considerably more fluid than a single-spout unit, but results in correspondingly faster solids turnover.[1]

Draft Tube

Cross-flow of both fluid and solids between the spout and the annulus can be eliminated over most of a spouted bed's height by inserting in the spout region, starting at some distance (e.g., in excess of 10 d_p) above the fluid inlet nozzle, an open draft tube with walls that are impervious to both phases. The draft tube diameter is usually chosen to be similar to that of the spout which occurs without a draft tube, i.e., it is somewhat larger than the inlet nozzle diameter. The draft tube is aligned vertically with its axis collinear with the axis of the column. One result is that the bed can now function at depths greater than H_m. Other consequences are a large reduction in the fluid flow requirement for spouting, an even larger reduction in the solids circulation rate, and considerably reduced solids mixing.[1] These changes could be advantageous for granulation and particle coating, where plug flow of solids should increase the uniformity of the product. The method also reduces the size of solids that can be successfully spouted[62] and has been used in a so-called recirculating fluidized bed, Figure 11.23, which is an example of a spout-fluid bed, discussed below. The characteristics of a draft tube spouted bed grain drier have recently been detailed.[64]

If the draft tube is permeable, e.g., made of metallic screen, it can allow flow of the fluid but remain impervious to the solids. Such a screen is well-suited to applications in which it is desirable that all particles spend the maximum possible time in the annulus without any curtailment of annular fluid flow.[1]

Top-Sealed Vessel

By sealing the top of the spouting vessel and providing an alternative fluid outlet either at the bottom of the bed[1] or part way below the bed surface (see Fig. 11.24), the spout fluid is forced to travel downward through the annulus.

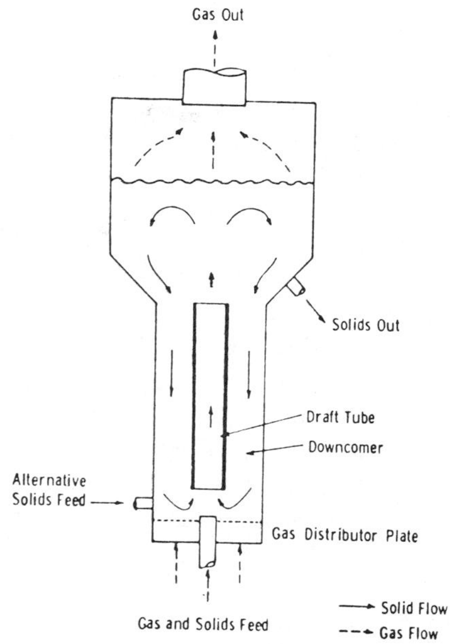

Figure 11.23. Schematic diagram of a "recirculating fluidized bed,"[63] i.e., a spout-fluid bed with a draft tube.

This results in a narrower range of fluid residence time distribution than otherwise. The residence time distribution can be further narrowed if, in addition, an inner draft tube is used.[65] The combination of side outlet and draft tube appears to give high gas conversions and great versatility in operable particle sizes.[62]

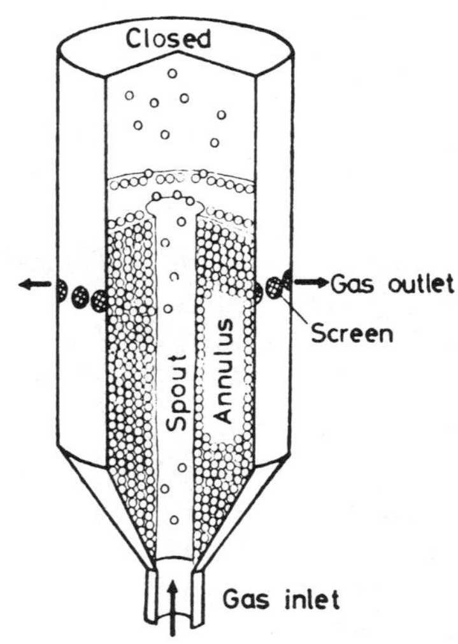

Figure 11.24. Schematic diagram of a top-sealed spouted bed with side-wall gas outlet.[65]

Spout-Fluid Bed

If, in addition to supplying spouting fluid through a centrally located inlet nozzle, sufficient extra fluid is also supplied through either a flat (Fig. 11.23) or conical (Fig. 11.25) distributor to fluidize the annular solids, the result is a "spouted-fluidized" or "spout-fluid" bed. The total fluid flow required for spout-fluidizing a given bed exceeds that for either spouting it or fluidizing it,[1] but the extra fluid may be well spent, since such a bed displays more vigorous solids agitation in the annulus than a standard spouted bed and a greater deagglomeration tendency than a conventional fluidized bed.

If the additional fluid fed to the annulus is insufficient to fluidize the annular solids, the total fluid flow required to maintain the bed in the spouted condition is greater than that for spouting without the auxiliary fluid, but less than that required for fluidizing the same solids.[66,67] This version of a "spout-fluid" bed has been applied to liquid contacting of an ion exchange resin,[68] as well as to gas spouting with a draft tube. In the latter case, aeration of the annulus tends to counteract the reduction in solids circulation rate caused by the presence of the draft tube.[69]

Three-Phase Spouting

In countercurrent gas-liquid spouting, low density solid spheres are spouted by an upward current of gas and irrigated by a downward flow of liquid.[70] In its performance and applications (e.g., gas absorption, dust removal) this type of operation is comparable to that of a turbulent bed (or "mobile bed" or "fluidized packing") contactor, that is, to fluidization of low density spheres by an upward continuous-phase gas flow counter to a downward trickle flow of liquid.[71] The three-phase spouted bed operation is characterized by a higher pressure drop than the three-phase fluidized bed, while the latter is characterized by a greater tendency to slugging and bed nonuniformity.[71] The disadvantages of both can be overcome by using a countercurrent gas-liquid spout-fluid bed, in which a portion of the gas is introduced via a centrally located nozzle and the rest through the surrounding gas distributor.[72]

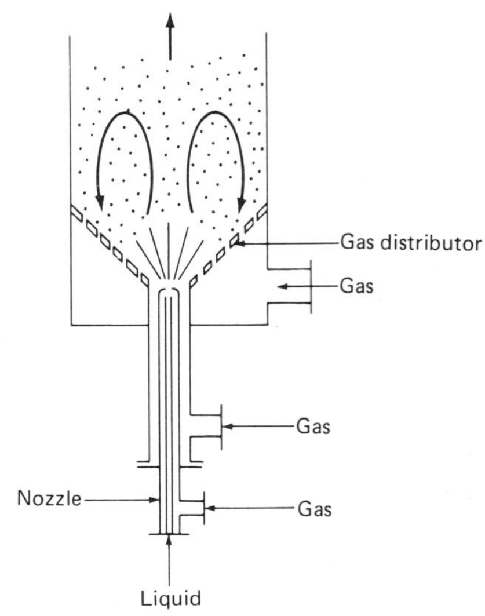

Figure 11.25. Conceptual diagram of a three-phase spout-fluid bed with cocurrent flow of gas and liquid.[73]

A cocurrent gas-liquid spout-fluid bed is illustrated in Figure 11.25. Gas is used to atomize the liquid feedstock through the inside of the entry nozzle and is additionally introduced around the periphery of this nozzle as well as through the conical gas distributor. Candidates for this type of vigorous gas–liquid–solid contacting include cracking of heavy hydrocarbons, gasification of residual oils, sintering of inorganic particles, granulation, particle coating and the production of adiponitrile from adipic acid using a B_2O_3 catalyst.[73]

Other Modifications

Spouted beds can be countercurrently staged,[1,60] directly vibrated[74] or subjected to flow pulsations.[1] The advantages of these modifications must in each case be weighed against corresponding increased costs. Fluid, instead of entering via a centrally located nozzle, can be introduced through concentric rings, vertical slots or tangential slots.[1,60] In a "vortex spouted bed" both the fluid stream and the solid particles are subjected to a circular motion, thus bringing about more intensive heat and mass transfer between the phases.[60]

11.14. PRACTICAL CONSIDERATIONS

We present here a number of practical suggestions, based on experience obtained on spouted beds, to help designers of spouted bed processes and equipment.

Particle Properties

Spouted beds operate best with dry, closely sized, rounded particles having a surface-to-volume mean diameter in the range 1-5 mm. Even when these conditions are met, and especially when they are not, it is best to test the spouting behavior in a small scale column. If the solids tend to be sticky or cohesive or if a liquid phase is to be present in addition to gas and solids, the possibility of operating a spout-fluid bed should be considered.

Test Column

Tests of spouting behavior for a given material should be carried out in a transparent column, not less than about 0.1 m in diameter. The column should be equipped with a conical base of included angle $\simeq 60°$, constructed in such a way that the inlet orifice diameter can be varied, but no larger than D/3. Among the features which can be investigated in the test column are:

- ease with which the material undergoes spouting
- tendency of the material to undergo attrition
- minimum spouting velocity, maximum spoutable bed depth and the agreement of these measured values with the principal correlations
- other hydrodynamic features such as the downward particle velocity at the wall and the fountain height.

Additional features can be observed and measured if a half-column (semicylindrical vessel) is employed:

- spout shape and diameter and the agreement of the latter with the correlations discussed herein
- tendency for solids segregation to occur
- dead zones, if any, within the column.

Fluid Inlet

A straight vertical approach section of 10-12 pipe diameters should generally precede the inlet orifice. This approach section may be of the same diameter as the orifice or of larger diameter, narrowing gradually to the entry diameter. A bundle of straightening tubes is sometimes fixed inside this approach section. Sometimes abrupt orifice plate is placed at the entry, which leads to increased spouting stability, but at the expense of an increased pressure drop. A coarse screen or a special inlet valve can be used to prevent dumping of solids when the bed is shut down.[1] Unless special startup measures[1] are taken to avoid the peak pressure drop, $-\Delta P_M$ (see Fig. 11.5), the blower or compressor feeding the bed must be sized to provide this pressure drop, in addition to the drop across the screen, entry section and upstream pipes, valves, and fittings. If the bed is to operate as a spout-fluid bed, the additional fluid should enter through orifices or nozzles on the conical lower section[73] or in the flat annular base if there is no lower cone. In either case the flow of auxiliary fluid should be controlled separately from the main spouting flow to allow the ratio of auxiliary to spouting fluid to be varied.

Solids Feeding

The simplest way to feed solids is to deliver them via gravity from a hopper to the bed surface. Agglomerating solids should be pneumatically conveyed into the column by the spouting gas. A third means of adding solids is from the side, using the suction created by the fluid jet entering the bed.[1]

Solids Discharge and Entrainment

Solid material can be discharged from a spouted bed, like liquid from an orifice in the side of a container, using the difference between the local pressure inside the vessel and that on the outside. Hence, solids efflux will be more rapid the lower the discharge port. For orifice-to-particle diameter ratios of about 30 or more, the discharge coefficient is expected to be about 0.5, as for solids discharging from fluidized beds.[75] The exit pipe should slope down at an angle of $\simeq 45°$, and it should be on the opposite side from any overhead solids feeder to prevent short-circuiting. For segregating solids, the position of the discharge port strongly affects the steady state bed composition,[76] and this should be taken into consideration.

For friable solids, columns with little freeboard space, or materials having a significant fines content, entrained solids leaving the column should be captured by means of one or more cyclones which may be followed by filters or other particle collection devices. Solids captured in cyclones may be returned to the annulus region by means of a dipleg which enters obliquely through the wall of the spouted bed vessel.

Baffles

Concave axisymmetric fountain deflectors (e.g., an inverted funnel) are sometimes used[1] to restrain the fountain, prevent flow-over of solids during startup, and induce greater symmetry and less wandering of the spout. For segregating solids a convex axisymmetric shape (e.g., a cone with its apex at the lowermost point) positioned near the top of the fountain can help to prevent segregation[25] by deflecting the heavier particles to the outside of the vessel. Either of these types of baffles in the fountain may, however, promote attrition.

ACKNOWLEDGMENT

The continuing financial support of the Natural Sciences and Engineering Research Council of Canada is gratefully acknowledged.

LIST OF SYMBOLS

A — $[\rho/(\rho_p - \rho)]\ [U_T U_{mf}/gD_i]$
A_a — Cross-sectional area of annulus at any level
A_s — Cross-sectional area of spout at any level
Ar — Archimedes No. $= gd_p^3\ (\rho_p - \rho)\rho/\mu^2$
b — U_m/U_{mf}
C_a — Species gas-phase concentration in the annulus
C_s — Species gas-phase concentration in the spout
C_D — Drag coefficient for particle in fluid stream $= F/(\pi d^2/4)\ (\rho/2)\ (u_s - v_s)^2$
C_{DT} — Drag coefficient under terminal settling conditions
C_{exit} — Exit species gas-phase concentration
C_{in} — Inlet species gas-phase concentration
c_{pg} — Specific heat capacity of gas
c_{pp} — Specific heat capacity of solid particles
$\mathfrak{D}$ — Internal moisture diffusivity of particle
D — Column diameter
D_i — Fluid inlet diameter
D_s — Spout diameter at any level
$\bar{D}_s$ — Longitudinal average spout diameter
d — Particle diameter; horizontally projected particle diameter
d_p — Diameter of sphere with same volume as particle
F — Drag force on particle
G — Superficial mass velocity of spouting fluid $= \rho U$
g — Acceleration of gravity
H — Height of loose-packed static bed; height of annulus, measured from fluid inlet orifice
H_F — Fountain height measured from bed surface
H_m — Maximum spoutable bed height
h — H/H_m
$(h_{pg})_a$ — Gas-to-particle heat transfer coefficient in the annulus
$(h_{pg})_s$ — Gas-to-particle heat transfer coefficient in the spout
h_s — Heat transfer coefficient between submerged object and bed

h_w Heat transfer coefficient between wall and bed, surface-mean value
$I(\theta)$ Internal age distribution function
i Integer
K_1 Viscous coefficient in Ergun equation $= 150\mu (1-\epsilon)^2/\phi^2 d_p{}^2 \epsilon^3$
K_2 Inertial coefficient in Ergun equation $= 1.75\rho (1-\epsilon)/\phi d_p \epsilon^3$
K_r First-order reaction rate constant referred to volume of solids
k_b Effective thermal conductivity of loose-packed bed
k_g Thermal conductivity of gas
k_{sa} Spout-annulus interphase mass transfer coefficient
m Index in Equation (11.19)
$\bar{m}$ Final moisture content of batch solids, dry basis
$\bar{\bar{m}}$ Moisture content of continuous solids product, dry basis
m_i Moisture content of continuous solids feed, dry basis
m_0 Initial moisture content of batch solids, dry basis
m_s Surface moisture content of particles, dry basis
N Total number of streamtubes
n Richardson-Zaki[19] index
P Fluid pressure
$-\Delta P_f$ Pressure drop for fluidized bed of height H
$-\Delta P_M$ Peak pressure drop
$-\Delta P_s$ Spouted bed pressure drop
p Dummy variable
Q_a Flow rate of fluid through annulus $= U_a A_a$
Q_{ai} Flow rate of fluid through ith streamtube in the annulus
Q_s Flow rate of fluid through spout $= U_{sz} A_s$
Q_{tot} Total fluid flow rate through spouted bed
Re Particle Reynolds number $= d_p U \rho/\mu$
Re_T Terminal particle Reynolds number $= d_p U_T \rho/\mu$
r Radial distance from spout axis
r_s Spout radius $= D_s/2$
T_{ga} Local temperature of gas in the annulus
T_{gs} Local temperature of gas in the spout
T_p Temperature of particles
t Time
$\bar{t}$ Mean residence time of solids
U Superficial fluid velocity
U_a Superficial fluid velocity in annulus at any level
U_m U_{ms} at maximum spoutable bed height
U_{ms} Minimum superficial fluid velocity for spouting
U_{sz} Superficial upward fluid velocity in spout or fountain core at any level
U_T Terminal settling velocity of isolated particle in spouting fluid
u_s Interstitial upward fluid velocity in spout or fountain core at any level
v_s Local upward particle velocity in spout or fountain core at any level
$\bar{v}_s$ Average upward particle velocity in spout or fountain core at any level
v_{sc} Upward particle velocity on spout axis at any level
v_w Downward particle velocity at column wall
W Mass downflow rate of solids in annulus at any level = mass upflow rate of solids in spout at same level
x z/H_m
y U_a/U_{mf} or U_a/U_{aH_m}
z Height coordinate measured from fluid inlet orifice
z_i Mean entry level of ith streamtube
α Included angle of cone
β $2 + (3K_1/2K_2 U_{mf})$
ϵ Voidage = 1 - volumetric fraction of solids
ϵ_a Voidage in annulus
ϵ_s Voidage in spout or fountain core at any level
θ Dimensionless time $= t/\bar{t}$
μ Fluid viscosity
ρ Fluid density
ρ_b Bulk density of loose-packed solids $= \rho_p(1-\epsilon_{mf})$
ρ_g Gas density
ρ_p Density of solid particles

τ_i	Mean residence time of gas in the ith streamtube
ϕ	Sphericity = surface area of equivolume sphere/surface area of particle
χ	$(6/\phi d_p)(\mathfrak{D}t)^{1/2}$
$\overline{\chi}$	$(6/\phi d_p)(\mathfrak{D}\overline{t})^{1/2}$

Subscripts

a	annulus
crit	critical
H	at $z = H$
H_m	at $z = H_m$
i	ith streamtube
max	at max spoutable bed depth
mf	at minimum fluidization
0	at $z = 0$
s	spout

REFERENCES

1. Kishan B. Mathur and Norman Epstein, *Spouted Beds*, Academic Press, New York (1974).
2. H. Littman, M. H. Morgan III, D. V. Vucović, F. K. Zdanski, and Ž. B. Grbavčić, "Prediction of the Maximum Spoutable Height and the Average Spout to Inlet Tube Diameter Ratio in Spouted Beds of Spherical Particles." *Can. J. Chem. Eng. 57*:684–687 (1979).
3. K. B. Mathur, and P. E. Gishler, "A Technique for Contacting Gases with Coarse Solid Particles." *A. I. Ch. E. J. 1*:157–164 (1955).
4. S. Ergun, "Fluid Flow through Packed Columns." *Chem. Eng. Progr. 48*(2):89–94 (1952).
5. C. Y. Wen, and Y. H. Yu, "A Generalized Method for Predicting the Minimum Fluidization Velocity." *A. I. Ch. E. J. 12*:610–612 (1966).
6. G. S. McNab and J. Bridgwater, "Spouted Beds–Estimation of Spouting Pressure Drop and the Particle Size for Deepest Bed." *Proc. European Congress on Particle Technology*, Nuremberg (1977).
7. H. Littman, M. H. Morgan III, D. V. Vucović, F. K. Zdanski, and Ž. B. Grbavčić, "A Theory for Predicting the Maximum Spoutable Height in a Spouted Bed." *Can. J. Chem. Eng. 55*:497–501 (1977).
8. Ž. B. Grbavčić, D. V. Vuković, F. K. Zdanski, and H. Littman, "Fluid Flow Pattern, Minimum Spouting Velocity and Pressure Drop in Spouted Beds." *Can. J. Chem. Eng. 54*:33–42 (1976).
9. T. Mamuro and H. Hattori, "Flow Pattern of Fluid in Spouted Beds." *J. Chem. Eng. Japan 1*:1–5 (1968).
10. N. Epstein, C. J. Lim, and K. B. Mathur, "Data and Models for Flow Distribution and Pressure Drop in Spouted Beds." *Can. J. Chem. Eng. 56*:436–447 (1978).
11. N. Epstein and S. Levine, "Non-Darcy Flow and Pressure Distribution in a Spouted Bed." *Fluidization. Proceedings of the Second Engineering Foundation Conference*, ed. J. F. Davidson and D. L. Keairns, Cambridge Univ. Press, pp. 98–103 (1978).
12. D. Van Velzen, H. J. Flamm and H. Langenkamp, "Gas Flow Patterns in Spouted Beds." *Can. J. Chem. Eng. 52*:145–149 (1974).
13. C. J. Lim and K. B. Mathur, "Residence Time Distribution of Gas in Spouted Beds." *Can. J. Chem. Eng. 52*:150–155 (1974).
14. C. J. Lim and K. B. Mathur, "A Flow Model for Gas Movement in Spouted Beds." *A. I. Ch. E. J. 32*:674–680 (1976).
15. L. A. Madonna, R. F. Lama, and W. L. Brisson, "Solids-Air Jets." *Brit. Chem. Eng. 6*:524–528 (1961).
16. G. A. Lefroy and J. F. Davidson, "The Mechanics of Spouted Beds." *Trans. Instn. Chem. Engrs. 47*:T120–T128 (1969).
17. B. Thorley, J. B. Saunby, K. B. Mathur, and G. L. Osberg, "An Analysis of Air and Solid Flow in a Spouted Wheat Bed." *Can. J. Chem. Eng. Eng. 37*:184–192 (1959).
18. C. J. Lim and K. B. Mathur, "Modeling of Particle Movement in Spouted Beds." in *Fluidization. Proceedings of the Second Engineering Foundation Conference*, ed. J. F. Davidson and D. L. Keairns, Cambridge Univ. Press, pp. 104–109 (1978).
19. J. F. Richardson and W. N. Zaki, "Sedimentation and Fluidisation: Part I." *Trans. Instn. Chem. Engrs. 32*:37–53 (1954).
20. J. R. Grace and K. B. Mathur, "Height and Structure of the Fountain Region above Spouted Beds." *Can. J. Chem. Eng. 56*:533–537 (1978).
21. R. K. Barton, G. R. Rigby, and J. S. Ratcliffe, "The Use of a Spouted Bed for the Low Temperature Carbonization of Coal." *Mech. Chem. Eng. Trans. 4*:105–112 (1968).
22. M. Kugo, N. Watanabe, O. Uemaki, and T. Shibata, "Drying of Wheat by Spouting Bed." *Bull. Hokkaido Univ. Sapporo Japan 39*:95–120 (1965).
23. H. A. Becker and H. R. Sallans, "On the Continuous, Moisture Diffusion-controlled Drying of Solid Particles in a Well-Mixed, Isothermal Bed." *Chem. Eng. Sci. 13*:97–112 (1961).
24. N. Piccinini, A. Bernhard, P. Campagna, and F. Vallana, "Segregation Phenomenon in Spouted Beds." *Can. J. Chem. Eng. 55*:122–125 (1977).
25. E. Kutluoglu, J. R. Grace, K. W. Murchie and P. H. Cavanagh, "Particle Segregation in Spouted Beds," Can. J. Chem. Eng. 61:308–316 (1983).

26. Y. Eljas, "Contribution to the Study of Spouted Beds with Particular Emphasis on Porosity." Paper No. 58d, *68th Annual A. I. Ch. E. Meeting*, Los Angeles (November, 1975).
27. C. J. Lim, "Gas Residence Time Distribution and Related Flow Patterns in Spouted Beds." Ph.D. thesis, Univ. of British Columbia (1975).
28. G. S. McNab, "Prediction of Spout Diameter." *Brit. Chem. Eng. & Proc. Tech. 17*:532 (1972).
29. M. A. Malek and B. C. Y. Lu, "Heat Transfer in Spouted Beds." *Can. J. Chem. Eng. 42*:14–20 (1964).
30. S. S. Zabrodsky and V. D. Mikhailik, "The Heat Exchange of the Spouting Bed with a Submerged Heating Surface." Collected papers on "Intensification of Transfer of Heat and Mass in Drying and Thermal Processes," *Nauka i Tekhnika BSSR*, Minsk, pp. 130–137 (1967).
31. Y. G. Klimenko, V. G. Karpenko, and M. I. Rabinovich, "Heat Exchange Between the Spouting Bed and the Surface of a Spherical Probe Element. *Heat Physics and Heat Technology*, no. 15, Ukr. SSR Acad. of Sci., Kiev, pp. 81–84 (1969).
32. K. B. Mathur and C. J. Lim, "Vapor Phase Chemical Reaction in Spouted Beds: A Theoretical Model," *Chem. Eng. Sci. 29*:789–797 (1974).
33. N. Piccinini, J. R. Grace, and K. B. Mathur, "Vapor Phase Chemical Reaction in Spouted Beds: Verification of Theory." *Chem. Eng. Sci. 34*:1257–1263 (1979).
34. M. Balasubramanian, A. Meisen, and K. B. Mathur, "Spouted Bed Collection of Solid Aerosols in the Presence of Electrical Effects." *Can. J. Chem. Eng. 56*:297–303 (1978).
35. Rowe, P. N. and K. T. Claxton, "Heat and Mass Transfer from a Single Sphere to Fluid Flowing through an Array." *Trans. Instn. Chem. Engrs. 43*:321–331 (1965).
36. Dž. E. Hadžismajlović, D. V. Vuković, F. K. Zdanski, Z. B. Grbavčić, and H. Littman, "A Theoretical Model of the First Order, Isothermic Catalytic Reaction in Spouted and Spout-fluid Beds." Paper C4.5, *6th CHISA Conference, Prague* (1978).
37. D. Kunii and O. Levenspiel, *Fluidization Engineering*. Wiley, New York (1969).
38. J. R. Grace, "Fluid Beds as Chemical Reactors." in *Fluid Bed Technology*, ed. D. Geldart, Wiley, New York (1981).
39. S.-K. Foong, C. J. Lim, and A. P. Watkinson, "Coal Gasification in a Spouted Bed." *Can. J. Chem. Eng. 58*:84–91 (1980).
40. G. Rovero, N. Piccinini, J. R. Grace, N. Epstein and C. M. H. Brereton, "Gas-phase Solid Catalysed Chemical Reaction in Spouted Beds." Chem. Eng. Sci., 38:557–566 (1983).
41. K. B. Mathur and P. E. Gishler, "A Study of the Application of the Spouted Bed Technique to Wheat Drying." *J. Appl. Chem. 5*:624–636 (1955).
42. W. S. Peterson, "Spouted Bed Drier." *Can. J. Chem. Eng. 40*:226–230 (1962).
43. O. Uemaki, M. Fujikawa, and M. Kugo, "Cracking of Residual Oil by Use of a Dual Spouted Bed Reactor with Three Chambers." *Sekiyu Gakkai Shi 20*(5):410–415 (1977).
44. S.-K. Foong, R. K. Barton, and J. S. Ratcliffe, "Reduction of Sulphur Dioxide with Carbon in a Spouted Bed." *Chem. Eng. in Australia* Instn. Engrs. Aust., *CE1* (1/2):1–8 (1976).
45. T. Robinson and B. Waldie, "Dependency of Growth on Granule Size in a Spouted Bed Granulator." *Trans. Instn. Chem. Engrs. 57*:121–127 (1979).
46. Y. F. Berquin, "Method and Apparatus for Granulating Melted Solid and Hardenable Fluid Products." U.S. Patent No. 3,231,413 (1966).
47. O. Uemaki and K. B. Mathur, "Granulation of Ammonium Sulfate Fertilizer in a Spouted Bed." *Ind. Eng. Chem. Proc. Des. Dev. 15*:504–508 (1976).
48. Y. F. Berquin, "Prospects for Full-Scale Development of Spouting Beds in Fertilizer Granulation." Paper 23, *First International Conference on Fertilizers*, London (November, 1977).
49. A. Meisen, private communication (July, 1978).
50. A. Meisen and K. B. Mathur, "Production of Sulphur-coated Urea by the Spouted Bed Process." Paper XIV, *Proc. Brit. Sulphur Corp.*-Part 1 (December, 1978).
51. N. Piccinini, "Coated Nuclear Fuel Particles." *Adv. Nuc. Sci. Technol. 8*:255–341 (1975).
52. E. H. Voice, "Coatings of Pyrocarbon and Silicon Carbide by Chemical Vapour Deposition." *The Chemical Engineer*, pp. 785–792 (December, 1974).
53. P. G. Romankov and N. B. Rashkovskaya, *Drying in a Suspended State*, 2nd ed (in Russian), Chem. Publ. House, Leningrad Branch (1968).
54. T. B. Ray and S. Sarkar, "Kinetics of Coal Pyrolysis in Spouted-Bed." *Indian Chemical Engineer 18*(2):11–19 (April–June, 1976).
55. A. N. Ingle and S. Sarkar, "Gasification of Coal in Spouted Bed." *Indian J. Technol. 14*:515–516 (1976).
56. J. Foong, G. Cheng and A. P. Watkinson "Spouted Bed Gasification of Western Canadian Coals." Can. J. Chem. Eng. *59*, 625–630 (1981).
57. C. Dumitrescu and D. Ionescu, "Contributions to the Spouted-Bed Studies as an Aspect of Fluidization." Paper E5.10, *4th CHISA Conference*, Prague (1972).
58. A. V. Golubkovich, "Study of the Limestone Calcination Regimes in Spouted-bed Kilns." *Khimicheskoe i Neftyanoe Machinostroenie*, no. 4, pp. 19–21 (April, 1976).

59. N. R. Iammartino, "Cement's Changing Scene." *Chem. Eng.* *81*(13):102–104 (June 24, 1974).
60. D. V. Vuković, F. K. Zdanski, and H. Littman, "Present Status of the Theory and Application of Spouted Bed Technique." Paper D2.20, *5th CHISA Conference*, Prague (1975).
61. S.-K. Foong, R. K. Barton, and J. S. Ratcliffe, "Characteristics of Multiple Spouted Beds." *Mech. & Chem. Eng. Trans.* Instn. Engrs. Aust., *MCII* (1,2):7–12 (1975).
62. H. Hattori and K. Takeda, "Side-outlet Spouted Bed with Inner Draft-tube for Small-sized Solid Particles." *J. Chem. Eng. Japan* *11*(2):125–129 (1978).
63. W. C. Yang and D. L. Keairns, "Design of Recirculating Fluidized Beds for Commercial Applications." *A. I. Ch. E. Symp. Series* no. 176, *74*: 218–228 (1978).
64. G. K. Khoe and J. van Brakel, "A Draft Tube Spouted Bed as Small Scale Grain Dryer." *I. Chem. E. Symp. Series* no. 59, Paper presented at International Conference on Solids Separation Processes, Dublin (April, 1980).
65. H. Hattori and K. Takeda, "Modified Spouted Beds–with the Gas Outlet Located in the Side Wall Surrounding the Annular Dense Bed." *J. Fac. Text. Sci. & Technol.*, Shinshu Univ., no. 70, ser. B. Engineering no. 12:1–13 (1976).
66. H. Littman, D. V. Vuković, F. K. Zdanski, and Z. B. Grbavčić, "Basic Relations for the Liquid Phase Spout-fluid Bed at the Minimum Spout Fluid Flowrate." *Fluidization Technology* ed. D. L. Keairns, *1*:373–386, Hemisphere (1976).
67. C. Dumitrescu, "The Hydrodynamical Aspects of a Spouted Bed Modified by the Introduction of an Additional Flow." *Revista de Chimie*, Roumania, *28*(8):746–754 (1977).
68. Dž. E. Hadžismajlović, D. V. Vuković, and Z. B. Grbavčić, "Mass Transfer in Liquid Spout-fluid Beds of Ion Exchange Resin." Paper D3.9, *5th CHISA Conference*, Prague (1975).
69. Ž. B. Grbavčić, F. K. Zdanski, D. V. Vuković, and Dž. E. Hadžismajlović, "Pressure Drop and Flow-rate Characteristics of a Spouted Bed with a Tubular Insert." Paper C3.7, *6th CHISA Conference*, Prague (1978).
70. D. V. Vuković, F. K. Zdanski, G. V. Vunjak, and Z. B. Grbavčić, "Pressure Drop, Bed Expansion and Liquid Holdup in a Three Phase Spouted Bed Contactor." *Can. J. Chem. Eng.* *52*:180–184 (1974).
71. G. Vunjak-Novaković, D. V. Vuković, F. K. Zdanski, and H. Littman, "Comparative Hydrodynamical Characteristics Relevant for Mass Transfer in Three-Phase Fluidized and Spouted Bed Contactors." Paper C2.7, *6th CHISA Conference*, Prague (1978).
72. D. V. Vuković and G. V. Vunjak-Novaković, "The Three-Phase Spout-fluid Bed–A Novel Gas-Liquid Contacting System." Paper C3.11, *6th CHISA Conference*, Prague (1978).
73. H. Kono, "A New Concept for Three Phase Fluidized Beds." *Hydrocarbon Processing* pp. 123–129 (January, 1980).
74. G. Rátkai, "Solid Mixing in Vibro-spouted Beds." Paper C3.5, *6th CHISA Conference*, Prague 1978.
75. L. Massimilla, "Flow Properties of the Fluidized Dense Phase," in *Fluidization*, ed. J. F. Davidson and D. Harrison, Academic Press, London (1971).
76. N. Piccinini, "Particle Segregation in Continuously Operating Spouted Bed," in *Fluidization*, ed. J. R. Grace and J. M. Matsen, Plenum, New York, pp. 279–286, (1980).

12
Transport of Solids in Pipelines

C. A. Shook
University of Saskatchewan
Saskatoon, Saskatchewan

12.1. CONSERVATION LAWS

Design equations involve volumetric flowrates and quantities derived from them. To relate these equations to known physical laws, additional variables must be defined. Although most of the applications involve steady flow, the set of definitions begins with instantaneous velocities.

If $v_f(x, y, z, t)$ and $v_s(x, y, z, t)$ are instantaneous fluid and particle velocities and if $c(x, y, z, t)$ denotes the instantaneous probability of finding the solid phase at a point in space, the time average velocities are

$$\langle v_f \rangle = \int_0^{t_1} v_f(1 - c)dt \bigg/ \int_0^{t_1} (1 - c)dt \tag{1}$$

$$\langle v_s \rangle = \int_0^{t_1} v_s c\, dt \bigg/ \int_0^{t_1} c\, dt \tag{2}$$

where

$$\langle c \rangle = \frac{1}{t_1} \int_0^{t_1} c\, dt \tag{3}$$

The period t_1 of the averaging process is usually considered to be short compared to the time scale of flow changes so that, for practical purposes, the averaging braces $\langle \ \rangle$ can be omitted in all the expressions that follow.

The mean velocity and concentration V and C_v of engineering interest are thus given as statements of the law of conservation of matter for the two phases

$$V = \frac{1}{A} \int_A [cv_s + (1 - c)v_f]\, dA \tag{4}$$

$$C_v = \frac{1}{AV} \int_A cv_s\, dA \tag{5}$$

Where changes occur fairly rapidly, or when the compressibility of the phases must be taken into account the continuity equations are useful[1]

$$\frac{\partial}{\partial t}\rho_f(1-c)+\nabla\cdot\rho_f(1-c)v_f=0 \quad (6)$$

$$\frac{\partial}{\partial t}\rho_s c+\nabla\cdot\rho_s c\, v_s=0 \quad (7)$$

For one-dimensional flows (in the z direction), if the variables are interpreted as spatial average values over the cross-section of the pipe, these can be integrated to yield

$$A\frac{\partial}{\partial t}\rho_f(1-c)+\frac{\partial}{\partial z}A\rho_f(1-c)v_f=0 \quad (6a)$$

$$A\frac{\partial}{\partial t}\rho_s c+\frac{\partial}{\partial z}A\rho_s c v_s=0 \quad (7a)$$

Momentum equations for two-phase mixtures have been explained by Jackson[2] and Wallis.[1] For pipe flows, the one-dimensional forms are most convenient. Thus, for either phase the inertial effect accompanying acceleration must balance the combined effect of pressure, body, interfacial drag and wall drag forces. Written as forces per unit volume of each phase these are[1]

$$\rho_f\left[\frac{\partial v_f}{\partial t}+\frac{v_f\partial v_f}{\partial z}\right]=-\frac{\partial P}{\partial z}-\rho_f g\frac{\partial h}{\partial z}+(\mathbf{f}_f)_s+\mathbf{f}_{fw} \quad (8)$$

$$\rho_s\left[\frac{\partial v_s}{\partial t}+\frac{v_s\partial v_s}{\partial z}\right]=-\frac{\partial P}{\partial z}-\rho_s g\frac{\partial h}{\partial z}+(\mathbf{f}_s)_f+\mathbf{f}_{sw} \quad (9)$$

if the only body force is due to gravity. The elevation above a horizontal datum is illustrated in Figure 12.1 and $\partial h/\partial z$ can be replaced by $\sin\theta$.

The interfacial drag forces $(\mathbf{f}_f)_s$ and $(\mathbf{f}_s)_f$ must cancel when the volumes of the phases are taken into account:

$$c(\mathbf{f}_s)_f+(1-c)(\mathbf{f}_f)_s=0 \quad (10)$$

If it is necessary to evaluate the interfacial force, Rowe's generalization of the Richardson-Zaki empirical relationship is useful for spherical particles:[1]

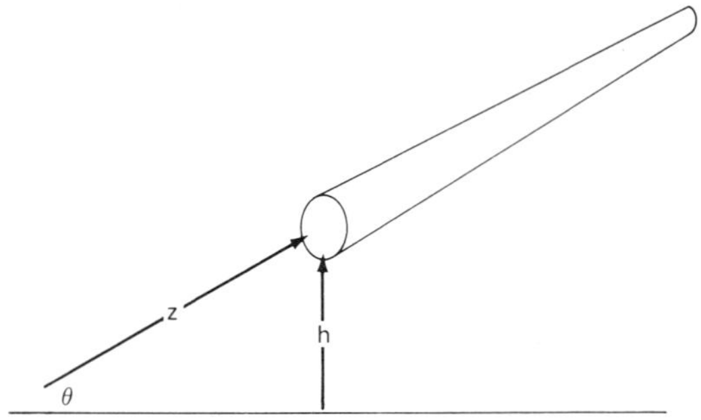

Figure 12.1. Oblique view of pipeline showing h and θ.

$$(\mathbf{f}_s)_f=3C_{Ds}\rho_f(v_f-v_s)|v_f-v_s|/4d(1-c)^{1.7} \quad (11)$$

where

$$C_{Ds}=\frac{24}{Re_s}\left[1+0.15Re_s^{0.687}\right]$$

for $Re_s<1000$
and $C_{Ds}=0.44$
for $Re_s>1000$
and $Re_s=d\rho_f|v_f-v_s|(1-c)/\mu_f$

These equations reduce to the conventional definitions for dilute systems as $c\to 0$.

The wall friction terms present the greatest difficulty to the designer. Thus, for steady horizontal pipe flow, combining Equations (8) and (9) under the constraint of (10) gives

$$-\frac{dP}{dz}=i\rho_f g=-[c\mathbf{f}_{sw}+(1-c)\mathbf{f}_{fw}] \quad (12)$$

where the negative sign on the right hand side results from the fact that $\mathbf{f}_{fw}$ and $\mathbf{f}_{sw}$ are negative because they oppose motion. Equation (12) illustrates the difficulty of slurry pipe flow: it is valid at all points in a horizontal pipe but what is required is a relationship involving V and C_V.

For a single phase flow, $\mathbf{f}_{fw}$ is written in terms of a wall shear stress or as the summation of viscous and Reynolds stress effects as

$$\mathbf{f}_{fw}=-\mu\nabla^2 v_f+\rho\left[\frac{1}{r}\frac{\partial}{\partial r}(r\overline{v'_{fr}v'_{fz}})+\frac{\partial}{\partial z}(\overline{v'^2_{fz}})\right] \quad (13)$$

for axisymmetric flow in the z direction. Integration over the cross-section allows this effect to be written in terms of the wall shear stress (neglecting minor terms) as $-4\tau_{wf}/D$.

In approximate calculations for reasonably homogeneous horizontal flows, one can use the estimate:

$$i\rho_f g = \frac{4\tau_w}{D} \simeq 4\frac{\tau_{wf}}{D}\left[\frac{c\rho_s + (1-c)\rho_f}{\rho_f}\right] \tag{14}$$

Equation (14) can be derived by neglecting the viscous stress contribution in Equation (13), writing an analogous equation for the solids, and assuming $v_{fz} = v_{sz} = v_z$. v_z is the instantaneous velocity for a single phase flow at the same V value.

Example 1

Estimate the in-situ concentration c in a pipe of 20 cm I.D. for a mixture flowing upward at 2 m/s if the particles are a narrowly sized gravel of diameter 4 mm. The specific gravity of the particles (relative to the fluid, water) is 2.65. C_V is 0.10 and the fluid is water at 20°C.

Solution

Since the particles are large and angular, single particle drag coefficients must be taken from experimental results. Figure 12.2 shows the drag coefficient for spheres as a function of Reynolds number. The relationship is that given by C_{Ds} in Equation (11) in the case where $c \to 0$. Figure 12.3 gives a correction factor for non-spherical particles.[3] The experimental values in Figure 12.3 were obtained in settling experiments using sand particles. Using Figures 12.2 and 12.3 C_{Ds} is estimated as 1 for the first iteration. The flow is vertical so that the variation of v_s, v_f, and c with position in the pipe is neglected. Combining equations (8), (9), (10), and (11) for steady flow gives

$$\frac{(\mathbf{f}_s)_f}{(1-c)} = \frac{3C_{Ds}\rho_f(v_f - v_s)|v_f - v_s|}{4d(1-c)^{2.7}} = (\rho_s - \rho_f)g + \mathbf{f}_{fw} - \mathbf{f}_{sw}$$

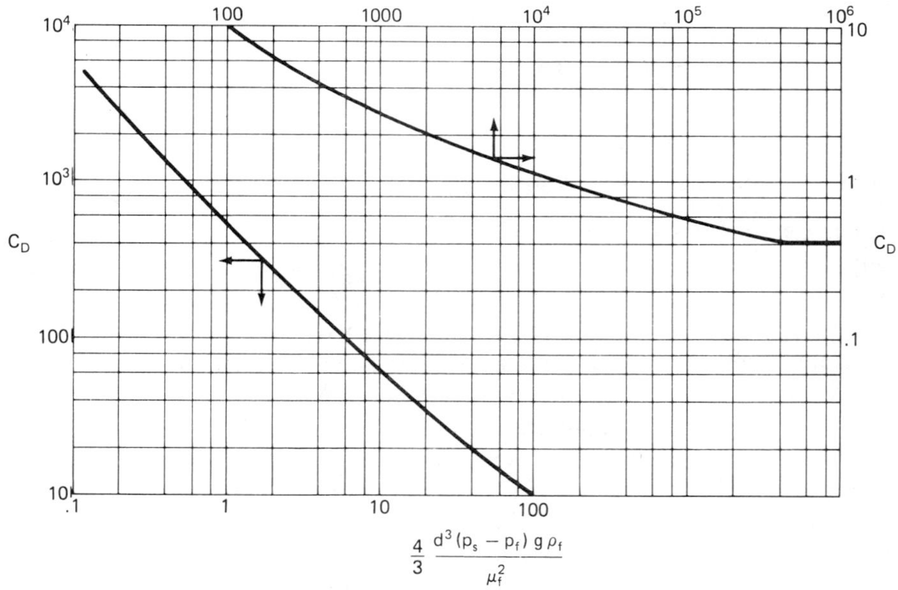

Figure 12.2. Drag coefficients for single spheres as a function of fluid and particle properties.

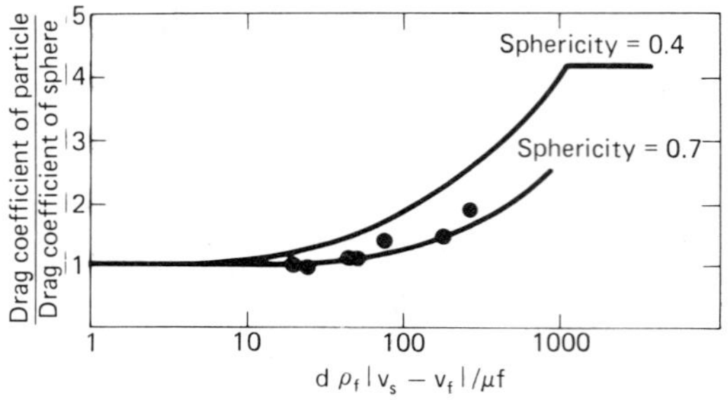

Figure 12.3. Effect of particle shape upon drag coefficient. The solid lines are taken from the relationships in Reference 3.

$\mathbf{f}_{fw}$ and $\mathbf{f}_{sw}$ will be estimated on the assumption that the Reynolds stress contributions (shown in Equation (13)) are dominant. Thus $\mathbf{f}_{fw}$ is taken as $-4\tau_{wf}/D$ or $-2fV^2\rho_f/D$. $\mathbf{f}_{sw}$ is taken as (ρ_s/ρ_f) times this value.

If $f = 0.004$, using $c \simeq 0.10$ in the first iteration, the solution of Equation (13) gives

$$v_f - v_s = 0.26 \text{ m/s}.$$

Using Equations (4) and (5), we can solve for v_s, v_f and a better value of c. The result is

$$c = 0.11$$

A second iteration confirms this value. We note in the calculation that the influence of the wall friction is minimal and the slip is essentially the same as in a stationary ($v_s = 0$) fluidized mixture of this concentration.

12.2. HOMOGENEOUS SLURRIES

In a truly homogeneous mixture the particles are sufficiently fine so that no settling tendency is apparent in an unstirred sample. The necessary resistance to gravitational segregation can be due to structural linkages between particles or to Brownian motion. Equation (11) indicates that as d becomes very small, the interfacial force per unit volume becomes very high even at low relative velocities $(v_f - v_s)$. For these slurries v_f and v_s are nearly identical so that Equations (8), (9), and (10) yield the homogeneous momentum equation

$$\rho\left[\frac{\partial v}{\partial t} + v\frac{\partial v}{\partial z}\right] = -\frac{\partial P}{\partial z} - \rho g \sin\theta + \mathbf{f}_w \quad (15)$$

with

$$\mathbf{f}_w = -[c\mathbf{f}_{sw} + (1-c)\mathbf{f}_{fw}] \quad (12a)$$

and

$$\rho = \rho_s c + \rho_f(1-c)$$

or in the integrated form, for steady flow

$$\int\frac{dP}{\rho} + \Delta\frac{V^2}{2\alpha} + \Delta(gh) + \frac{4\tau_w L}{\rho D} = 0 \quad (15a)$$

where α is the kinetic energy correction factor = $\int_A (v/V)dA / \int_A (v/V)^3 dA$.

Equation (15) usually requires the simplification

$$\mathbf{f}_w = -4\tau_w/D \quad (16)$$

rather than (12a) since it is impractical to separate $\mathbf{f}_{sw}$ and $\mathbf{f}_{fw}$ in most situations.

Particle deposition is not a significant problem with these slurries. However, they often display a high resistance to shear or high values of apparent viscosity μ_{app}

$$\mu_{app} = \frac{\tau}{\dot{\gamma}} \quad (17)$$

and are frequently non-Newtonian. Because of these high apparent viscosities, laminar flow is possible in full-scale pipelines and viscometric characterization is usually essential. These types of fluid have been thoroughly studied experimentally and theoretically. Standard reference books and texts are very useful[4,5,6,7] and there is a substantial body of well organized information available. The present section is intended to summarize this knowledge.

Tube viscometei testing of slurries is usually desirable for homogeneous mixtures. For a time-independent fluid whose rheological equation of state is of the form $\dot{\gamma} = f(\tau)$ we use the simplification of fully developed pipe flow in which

$$\dot{\gamma} = -\frac{dv_z}{dr}$$

Integration by parts, using the steady pipe flow shear stress decay law,

$$\tau/\tau_w = 2r/D$$

and the zero slip condition $v_z = 0$ at $r = D/2$ yields:[4]

$$\frac{8V}{D} = \frac{4}{\tau_w^3}\int_0^{\tau_w} \tau^2 f(\tau)d\tau \qquad (18)$$

In terms of

$$n' = \frac{d \ln \tau_w}{d \ln (8V/D)} \qquad (19)$$

Equation (18) yields, on differentiation with respect to τ_w

$$\left(-\frac{dv_z}{dr}\right)_w = \frac{3n' + 1}{4n'}\left[\frac{8V}{D}\right] = \dot{\gamma}_w \qquad (20)$$

Equation (18) shows that for a given fluid, $8V/D$ is a function of τ_w. The relationship is more conveniently stated in terms of K' defined by

$$\left(\frac{\tau_w}{K'}\right)^{1/n'} = \frac{8V}{D} \qquad (21)$$

so that we have two coefficients to characterize a fluid at a given wall shear stress τ_w. n' (dimensionless) indicates the degree of non-Newtonianism since it is unity for Newtonian fluids at all values of τ_w (as long as the flow remains laminar). For a fluid which obeys the power law model

$$\dot{\gamma} = \left[\frac{\tau}{K}\right]^{1/n} \qquad (22)$$

for all values of $\tau < \tau_w$, the value of n' obtained from the differentiation of experimental data according to Equation (19) will be equal to n. However, unlike n and K in Equation (22), use of n′ and K′ is not restricted to situations where they are constant. Their definitions show that they can be used as long as the τ_w value is that at which they were evaluated.

K' (S.I. units $kg \cdot m^{-1} s^{n'-2}$) is a measure of the consistency of the slurry, being related to the apparent viscosity at the wall shear stress by

$$K' = \mu_{appw} \frac{3n' + 1}{4n'}\left[\frac{8V}{D}\right]^{1-n'} \qquad (23)$$

Note that K' reduces to the Newtonian viscosity μ if $n' = \text{constant} = 1$.

The fact that $8V/D$ is a unique function of τ_w for a given fluid means that experiments conducted with a model pipeline can be scaled up to a full size pipeline (at the same value of τ_w) with complete confidence as long as wall slip is negligible and the flow remains laminar.

In practice it is necessary to conduct experiments with more than one viscometer tube diameter as a test for (apparent) wall slip. If slip does occur, the boundary condition is $v_z = v_w$ at $r = D/2$ so that in place of Equation (18) we have

$$\frac{8V}{D} = \frac{8v_w}{D} + \frac{4}{\tau_w^3}\int_0^{\tau_w} \tau^2 f(\tau)d\tau \qquad (24)$$

with $v_w = v_w(\tau_w)$ in general.[9] Figure 12.4 shows this sort of behavior. The wall slip prevents the plots of τ_w vs. 8V/D from coinciding.

It should be emphasized that v_w is an apparent slip as opposed to an actual wall slip. It is often caused by particle orientation during flow to leave a depleted region of low viscosity near the pipe wall. The effect is pronounced with fibrous particles such as wood pulp, or with equal density particles, and can also occur with mineral slurries.

Since the equations of state refer to laminar flow, care must be taken to evaluate the transition Reynolds number for all viscometric tests

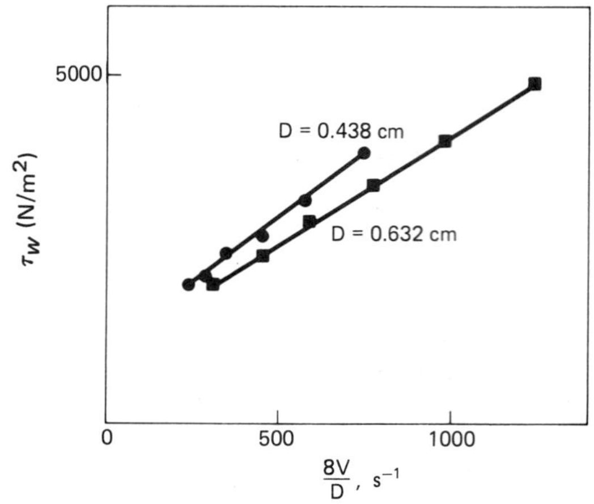

Figure 12.4. Typical rheograms for slurry flows showing displacement due to wall slip.

so that data taken in the turbulent region can be excluded. Figure 12.5 illustrates this. The transition from laminar to turbulent flow occurs at values of $DV\rho/\mu_e$ in the vicinity of 2100[4] where

$$\mu_e = \tau_w/(8V/D)$$

Alternative criteria, in terms of n′ and K′ are given by Skelland.[4]

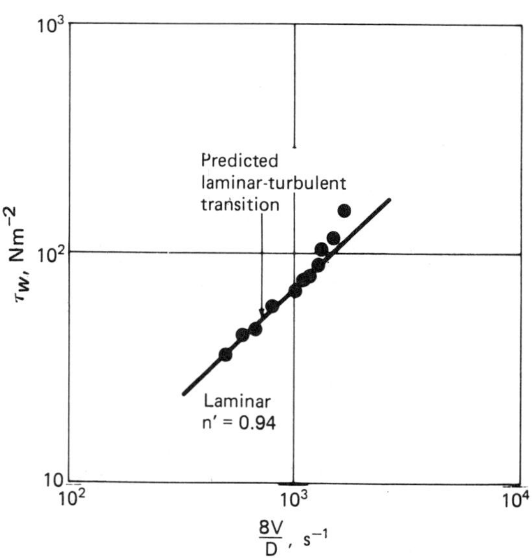

Figure 12.5. Rheogram showing laminar and lower turbulent flow regimes.

Using the values of n' and K' obtained in laminar flow tests, the method of Dodge and Metzner[7] can be employed to predict pressure drops in turbulent flow. If pipeline flow data are available for turbulent flow, a scaleup method such as Bowen's equation can be used.

This is suitable for turbulent flow with wall slip and some heterogeneity and will be described later. Figure 12.6 presents values of the Fanning friction factor $f = 2\tau_w/\rho V^2$ calculated from the Dodge and Metzner semitheoretical equation[7]

$$\frac{1}{\sqrt{f}} = \frac{4}{(n')^{0.75}} \log_{10} Re_g f^{1-n'/2} - \frac{0.4}{(n')^{1.2}} \tag{25}$$

where

$$Re_g = \frac{D^{n'} V^{2-n'} \rho}{K' 8^{n'-1}} \tag{26}$$

Note that Re_g has been defined so that in laminar flow, Equation (21) is equivalent to

$$f = \frac{16}{Re_g}$$

and the usual relationships are obtained for Newtonian fluids.

In addition to this method of dealing with experimental data, equations derived from rheological models are useful in laminar slurry flow. Thus using the expression for a Bingham fluid

$$\dot{\gamma} = \frac{\tau - \tau_y}{\eta_P} \tag{27}$$

Equation (18) yields, for zero wall slip:

$$\frac{8V}{D} = \frac{\tau_w}{\eta_P}\left[1 - \frac{4}{3}\left(\frac{\tau_y}{\tau_w}\right) + \frac{1}{3}\left(\frac{\tau_y}{\tau_w}\right)^4\right] \tag{28}$$

Numerous other cases are discussed in the literature. Figure 12.7 illustrates tube viscometer results for Newtonian, Power Law and Bingham fluids.[4] If the yield stress τ_y is low it

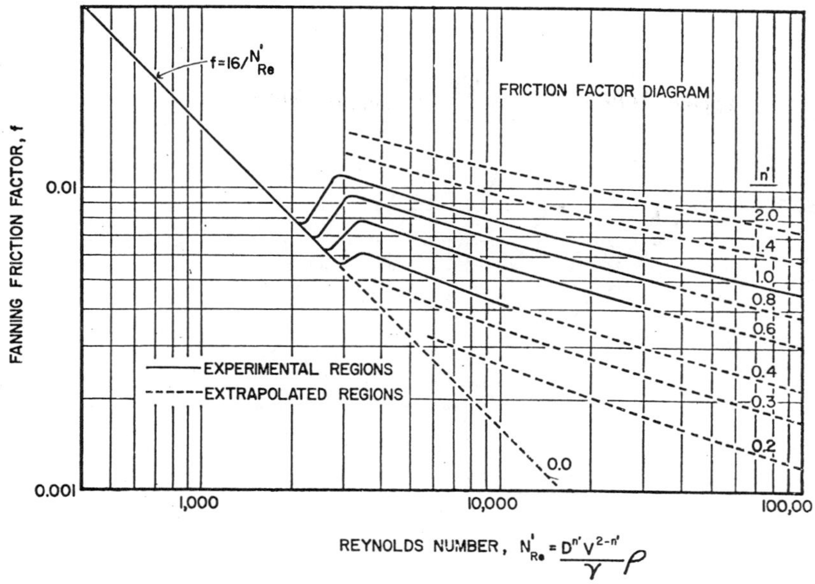

Figure 12.6. Friction factor as a function of generalized Reynolds number. Reproduced with permission from Reference 7.

may be difficult to distinguish between the latter two models unless the range of $8V/D$ is wide.

In addition to use of n' and K', for Bingham fluids Figure 12.8 can be used with values of the Reynolds number

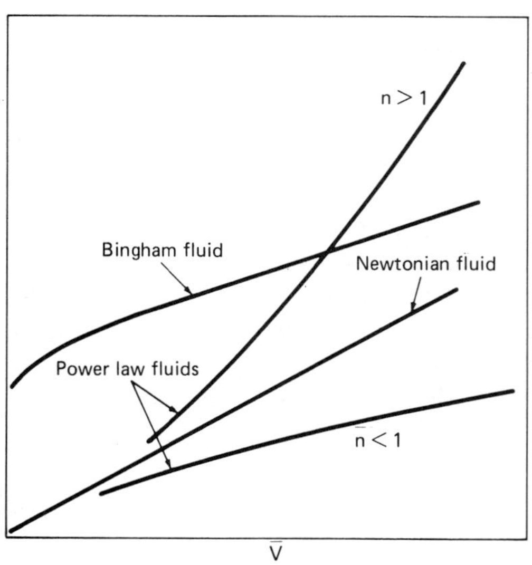

Figure 12.7. Headloss as a function of viscosity for the most common types of homogeneous flows in the laminar region.

$$Re = \frac{DV\rho}{\eta_P}$$

and the Hedstrom number

$$He = \frac{\rho D^2 \tau_y}{\eta_P^2} \qquad (29)$$

to predict Fanning friction factors in laminar and turbulent flow.[8] The turbulent boundary curve in this diagram is the Newtonian one. This could overestimate the friction factor since, according to Equation (19) or Figure 12.6, n' will be less than 1. Figure 12.6 shows reduced turbulent friction factors in this region.

Equations derived from rheological models must be used if the fluid is characterized by Couette or cone and plate viscometry. (Cone and plate viscometry is often impossible with slurries because of jamming effects). Wall slip can also occur with Couette viscometers although grooved walls can reduce this tendency. Test conditions must be chosen to avoid turbulence or Taylor instability. Limiting conditions for Newtonian fluids are, roughly:[5]

Outer cylinder rotates, inner stationary

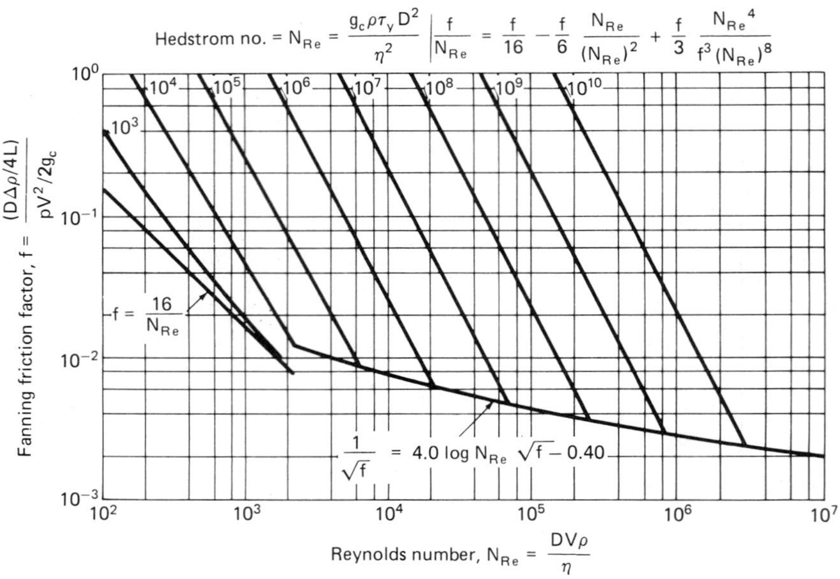

Figure 12.8. Friction factor as a function of Reynolds number and Hedstrom number for Bingham fluids. Reproduced with permission from Reference 8.

$$\frac{R_0^2 \omega \rho}{\mu} \leqslant 50{,}000 \tag{30}$$

Inner cylinder rotates, outer stationary

$$\frac{\omega R_0 R_i \rho}{\mu} \leqslant 41.3/(1 - R_i/R_0)^{3/2} \tag{31}$$

Tube and Couette viscometer data are often complementary since low shear rates can be more easily produced with the latter instruments. This is illustrated in Figure 12.9 for a limestone slurry. Couette viscometry is thoroughly discussed in standard references.[4,9] However, in addition to the limitations of Equations (30) and (31), end effects and gap width effects must be taken into account in data reduction.

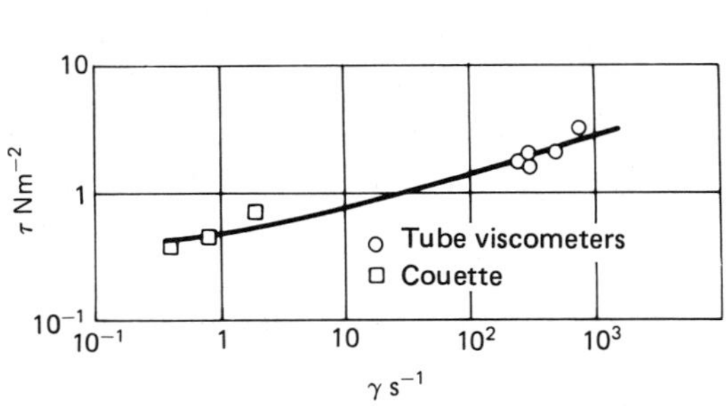

Figure 12.9. Tube and Couette flow viscometer results for a limestone slurry.

Transient flows have been studied thoroughly in the homogeneous or pseudohomogeneous regimes. In this case the mathematical techniques devised for single phase flows can be used. The momentum equation (14) is combined with equation (15) and solved simultaneously with the homogeneous conservation of mass equation:

$$\frac{\partial \rho}{\partial t} + \frac{\partial}{\partial z}(\rho v) = 0 \tag{32}$$

using the method of characteristics. Good agreement between predicted and observed pressure variations can be achieved, as shown in Figure 12.10. f in this case is the Fanning friction factor $2\tau_w/\rho V^2$.

For time-dependent fluids the equation of state is

$$\dot{\gamma} = f(\tau, t) \tag{33}$$

so that time appears as a parameter in the equations relating τ_w and V. Characterizing such fluids is correspondingly more difficult and

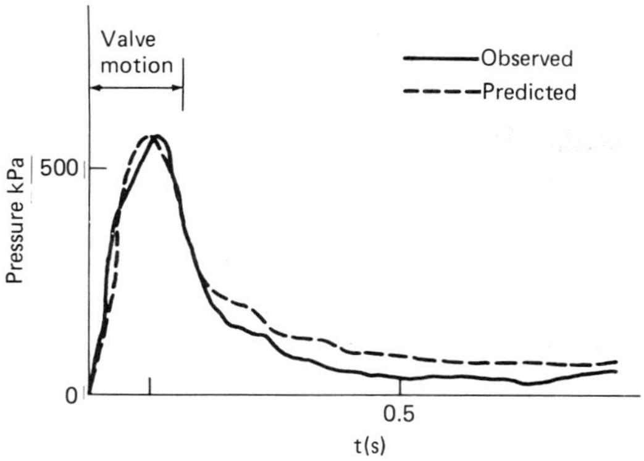

Figure 12.10. Predicted and observed pressure changes following partial valve closure for a slurry in turbulent (homogeneous) flow.

expensive. However, there is a growing body of theoretical and experimental knowledge concerning these slurries.[10] For pipeline flows, the complications of time dependence are confined to transient periods after startup and shutdown. In the time scale of most continuous process operations these periods are usually short. However if a structure is established on standing it may be necessary to avoid leaving the fluid in the pipeline at shutdown. Slurries of fine lignite particles in water exhibit this sort of behavior.

For long pipelines, the flow transient effects can interact with fluid time-dependence to produce more complicated pressure changes[11] during the transient period. After inertial effects diminish, the method of Ritter and Batycky[10] can be used to predict pressure drops.

Example 2

The slurry of Figure 12.9 is to flow under gravity through a pipe of I.D. 0.0525 m with an equivalent length of 89 m. The difference in elevation is 7.17 m and the slurry density is 1590 kg · m^{-3}. Estimate the velocity.

Solution

For steady pipe flow, Equation (15) reduces to

$$0 = -\frac{d}{dz}[P + \rho g h] - \frac{4\tau_w}{D}$$

Integrating over the pipe length L (89 m) gives

$$\tau_w = \frac{D\rho g\,\Delta h}{4L} = \frac{0.0525(1590)(9.8)(7.17)}{4(89)} = 16.5\text{Pa}$$

This high shear stress implies that it is the high τ portion of the diagram which is of greatest interest. Replotting the tube viscometer data suggests a Bingham fluid with

$$\tau_y = 1.2\text{Pa}$$

$$\eta_P = 0.0024\text{Pa}\cdot\text{s}$$

so

$$He = \frac{\rho D^2 \tau_y}{\eta_P^2} = 913000$$

Iteration is necessary to find V. However D is fairly high and η_P is low so turbulent flow is suspected. Try $f = 0.005$ so

$$V = \left(\frac{2\tau_w}{f\rho}\right)^{1/2} = 2.0\ m/s$$

checking:

$$Re = \frac{.0525(2.0)(1590)}{.0024} = 69500$$

confirming the choice of f.

12.3. PSEUDOHOMOGENEOUS FLOW

Truly homogeneous flows occur for slurries which show no settling tendency when stationary. The particles in such slurries are very fine, with high specific surfaces. Frequently, a settling test reveals a low but finite settling tendency. Under the influence of forces accompanying flow the particles may be distributed sufficiently uniformly to allow the use of the homogeneous flow equations to predict pressure drop. Such flows are called pseudohomogeneous and the symbolic relationship to other flow regimes is shown in Figure 12.11.

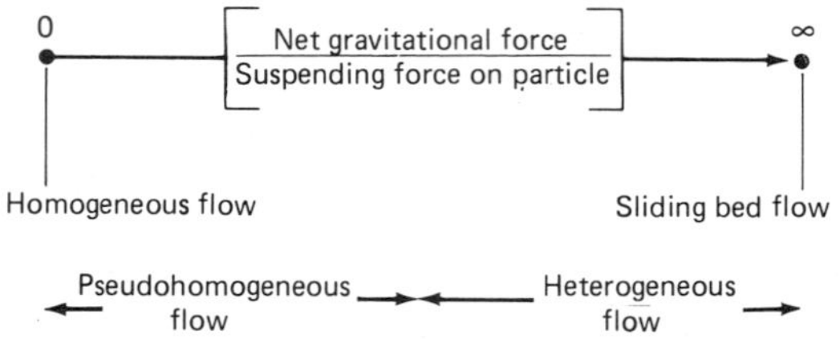

Figure 12.11. Symbolic relationship between flow regimes.

Figure 12.12 compares the hydraulic gradient-velocity relationships for the various types of slurry flow.

The pseudohomogeneous approximation is of course confined to a part of the range of possible velocities since at low velocities, stationary deposits can form. The flow forces which oppose the settling tendency in horizontal pipe flow can be those of turbulence or of particle-particle interaction. These latter forces have been investigated by Bagnold[12,13] and are present in laminar flow. Thus turbulence is not necessary to suspend particles. Figure 12.13 gives evidence of this for a suspension of sand particles in a viscous fluid. Between 1 and 2 m/s the slurry behaves almost as a Newtonian fluid; the viscosity is close to the values shown in Figure 12.14 obtained in Couette flow

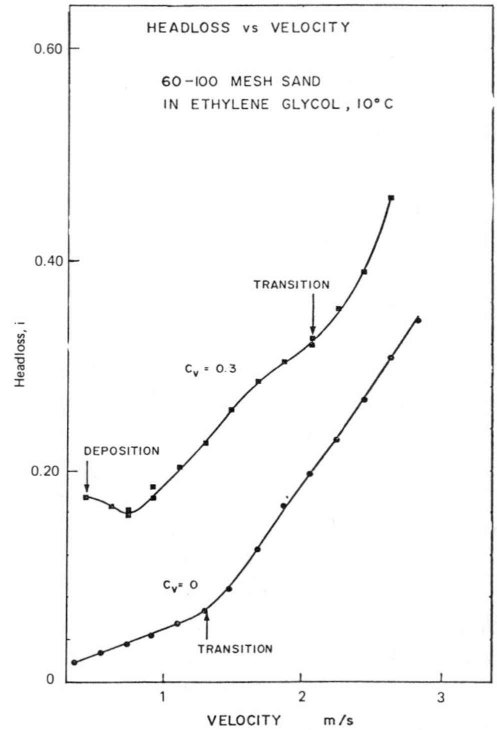

Figure 12.13. Headloss as a function of velocity for a slurry in the laminar and transitional flow regions.

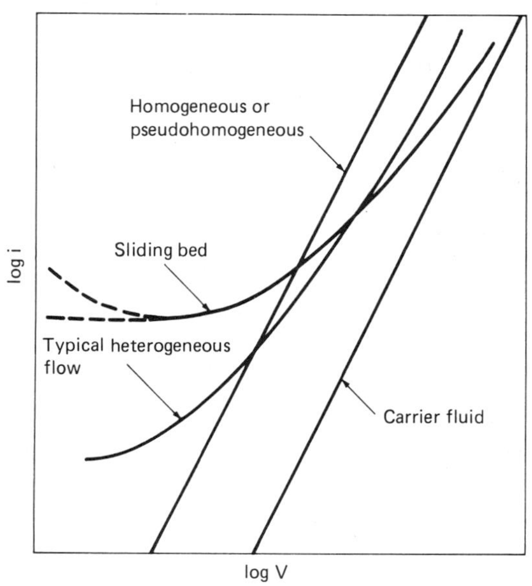

Figure 12.12. Headloss-velocity relationships for various types of turbulent slurry flow.

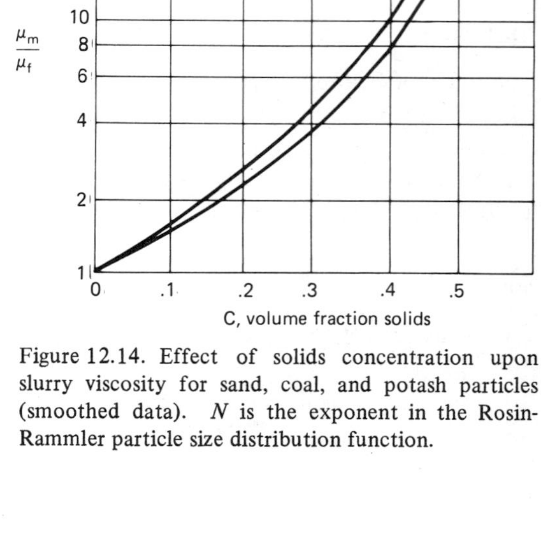

Figure 12.14. Effect of solids concentration upon slurry viscosity for sand, coal, and potash particles (smoothed data). N is the exponent in the Rosin-Rammler particle size distribution function.

viscometry using sand and coal particles. The results for the moderately broad particle size distributions were obtained for N equal to approximately 1 in a Rosin-Rammler distribution.

Since the settling tendency of the particles in combination with the flow forces determines particle distributions, slurry flows cannot be characterized adequately on the basis of settling velocity alone.

For many slurries which give pseudohomogeneous flows at high concentrations and velocities, viscometric characterization is possible. However minor quantities of coarse particles can cause difficulties in many cases so that approximate relationships such as Figure 12.14 may have to be used to estimate viscosities. Particle migration effects can also cause apparent slip with these slurries so that the appropriate viscosity can approach that of the carrier fluid. In this case Equation (14) can be used to determine $-4\tau_w/D$.

The local slip velocity $(v_f - v_s)$ will be negligible for horizontal pseudohomogeneous flows. Similarly the difference between C_v and the mean spatial concentration will be negligible.

12.4. SLIDING BED FLOW

Where the settling tendency of the particles is very high, horizontal pipe flow occurs with an obvious concentration variation. The mechanism of particle motion in the lower layer often resembles sliding although some rotation can occur. A simplified model is obtained[14, 15] by assuming the flow to be comprised of two layers as shown in Figure 12.15. In the lower layer the spatial concentration (time average) is c_2. The upper layer is particle-free. Thus for steady horizontal pipe flow the momentum equation for the upper layer is

$$0 = -\frac{dP}{dz} - \frac{\tau_1 S_1 + \tau_{12} S_{12}}{A_1} \tag{34}$$

where the wall force per unit volume has been written in terms of the mean interfacial stress τ_{12} and the wall stress τ_1.

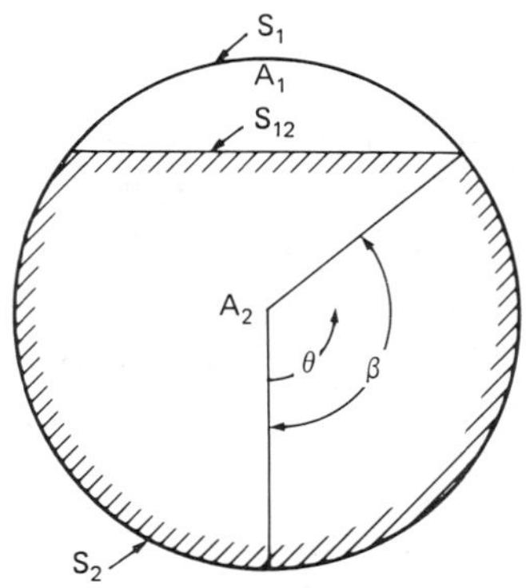

Figure 12.15. Cross section of pipe showing parameters used in sliding-bed model.

For the lower layer, momentum equations can be written for both particles and fluid

$$0 = -\frac{dP}{dz} + \frac{\tau_{12} S_{12}(1 - c_2) - \tau_{2f} S_2}{A_2(1 - c_2)} + (\mathbf{f}_f)_s \tag{35}$$

$$0 = -\frac{dP}{dz} + \frac{\tau_{12} S_{12} c_2 - \tau_{2s} S_2}{A_2 c_2} + (\mathbf{f}_s)_f \tag{36}$$

In these equations it is assumed that the stress τ_{12} is distributed between the particles and the fluid in proportion to the spatial concentration. Using the constraint of Equation (10)

$$c_2(\mathbf{f}_s)_f + (1 - c_2)(\mathbf{f}_f)_s = 0$$

the interfacial drag terms are eliminated to give the equation for the lower layer

$$0 = -\frac{dP}{dz} + \frac{\tau_{12} S_{12} - \tau_{2f} S_2 - \tau_{2s} S_2}{A_2} \tag{37}$$

Where the axial compressive stress between the particles is negligible, as in an unblocked pipe, the interparticle stress obeys a hydrostatic law. In this case the mean wall stress can be evaluated in terms of the angle β, which specifies the depth of the lower layer, and a coefficient of wall friction η_s.

$$\tau_{2s} S_2 = \frac{\tau_{12} S_2 c_2}{\tan\phi}\eta_s + 0.5(\rho_s - \rho_f) g c_2 D^2(\sin\beta - \beta\cos\beta)\eta_s \tag{38}$$

The first term on the right hand side results from the transmission of stress through the flowing particle mass to the pipe wall. The second term arises from particle immersed weight and the hydrostatic stress condition.

Stress τ_1 can be related to the velocity v_1 of the upper layer by a friction factor

$$\tau_1 = f_1 v_1^2 \rho_1 / 2$$

and Wilson[16] has suggested a similar type of relationship for τ_{12}:

$$\tau_{12} = f_{12} \frac{(v_1 - v_2)|v_1 - v_2|\rho_1}{2} \quad (39)$$

where

$$v_2 = v_{2s} c_2 + v_{2f}(1 - c_2) \quad (40)$$

with f_{12} evaluated from

$$f_{12} = 2\left[\frac{1}{4 \log_{10} D/d_{50} + 3.36}\right]^2 \quad (41)$$

τ_{f2} is related to the fluid velocity near the wall of the lower layer. Using the mean fluid velocity v_{2f}, one can write

$$\tau_{f2} = f_2 v_{2f}^2 \rho_f / 2 \quad (42)$$

Limited experimental evidence suggests that f_2 may be a function of particle diameter so that a range of f_2 values have to be used in design studies. To complete the model one uses the material balance relationships

$$AV = A_1 v_1 + A_2 [v_{2f}(1 - c_2) + v_{2s} c_2] \quad (43)$$

$$C_V A V = v_{2s} A_2 c_2 \quad (44)$$

and, since c_2 is near the limiting value, a packed bed relationship such as the Ergun equation may be more appropriate than Equation (11). The Ergun equation is

$$(f_s)_f = \frac{C_{fb} |v_{2f} - v_{2s}| (v_{2f} - v_{2s}) \rho_f}{d} \quad (45)$$

where

$$C_{fb} = \frac{150 \mu_f c_2}{d |v_{2f} - v_{2s}| \rho_f (1 - c_2)} + 1.75 \quad (46)$$

Although a solution of the set of equations that comprise the model can be fairly easily obtained, an approximate solution is of interest. If $v_{2s} \to v_{2f} \to v_2$, one obtains

$$-\frac{dP}{dz} = \frac{1}{A}\left[\frac{f_1 v_1^2 \rho_1 S_1}{2} + \frac{f_2 v_2^2 \rho_1 S_2}{2} + \frac{\tau_{12} S_2 \eta_s}{\tan\phi} c_2 + 0.5 c_2 (\rho_s - \rho_f) g D^2 \cdot (\sin\beta - \beta\cos\beta)\eta_s\right] \quad (47)$$

Since $v_1 \simeq V \simeq v_2$ this is approximately

$$-\frac{dP}{dz} = \frac{2fV^2\rho_1}{D} + 2c_2(\rho_s - \rho_f) \cdot g \frac{(\sin\beta - \beta\cos\beta)}{\pi}\eta_s \quad (48)$$

This equation is similar in form to empirical equations for the flow of mixtures of coarse particles.

12.5. PREDICTIONS FROM THE MODELS

There are numerous well-documented cases of pseudohomogeneous slurry flows. Fine coal slurries, with size distributions of the type which have been suggested for long distance transport, approach the pseudohomogeneous case at high concentrations and velocities.[17] Similar results were observed for iron ore concentrate slurries.[18] However at velocities near the deposition condition it is usual for the pseudohomogeneous assumption to underpredict pressure drops.

Pipe roughness effects have not been systematically studied for slurry flows so that the only course open to the designer is to use the pseudohomogeneous assumption. For vertical or inclined flows the potential energy contributions in equation 15A is often dominant. In this case even if $v_s \neq v_f$, the wall friction term can be estimated from the pseudohomogeneous flow equation assuming $c_2 = C_V$. This is what was done in Example 1.

For sliding bed flows, the quantities to be specified are c_2, η_s, $\tan\phi$ and the friction factors f_1 and f_2. $\tan\phi$ and η_s can be obtained from simple shearing tests.[14] f_1 can be estimated from an *f-Re* correlation using the hydraulic radius of the upper region. The need to consider the range of f_2 values was mentioned previously. For flows at high solids concentrations such as those from lock-hopper feeding devices, c_2 has been found to be approximately $(C_{max} - 0.05)$ where C_{max} is the concentration of a pack-settled deposit. At lower solids concentrations (C_V values), c_2 varies considerably with position and the appropriate mean value is much lower.[19] Thus, at low C_V values this model tends to overpredict pressure drops. However it can be useful for very large particles or where a conservative estimate is acceptable.

Example 3

For the upward-flowing slurry of Example 1, calculate the pressure drop per unit length of pipe.

Solution

Since the flow is steady and the pipe size is constant, no velocity changes occur. We can use Equations (15) and (16) with $\sin\theta = 1$. The pseudohomogeneous flow assumption will be used with the viscosity taken as that of the carrier fluid.

$$-\frac{dP}{dz} = \rho g + \frac{4\tau_w}{D}$$

but $\rho = \rho_s c + (1 - c)\rho_f$

$$\frac{4\tau_w}{D} \simeq \frac{4\tau_{wf}}{D}\left[\frac{c\rho_s + (1 - c)\rho_f}{\rho_f}\right]$$

by Equation (14)

$$\tau_{wf} = f\rho_f V^2/2$$

Therefore,

$$-\frac{dP}{dZ} = \rho_f\left[1 + c\left(\frac{\rho_s}{\rho_f} - 1\right)\right]\left[g + \frac{2fV^2}{D}\right]$$

$$= 1000[1 + 0.11(1.65)] \cdot \left[9.8 + \frac{2(0.004)(2)^2}{0.2}\right]$$

$$= 11770 \text{ Pa/m}$$

We observe that only 1.6% of the total pressure drop results from wall friction. The error resulting from use of C_V rather than c would also be small.

Example 4

The gravel particles of the previous problem are to be transported by pipeline using a lock-hopper feeding system. If necessary water is to be injected to bring C_V to a value of 0.42. c_2, the sliding bed concentration is 0.59, the coefficient of wall friction η_s is 0.40 and the angle of internal friction ϕ is 30°. Calculate the pressure gradient in a pipe of I.D. 0.0787 m if $V = 1\ m/s$.

Solution

For convenience in calculation we use v_2 defined by equation 40 and a mean concentration C_b defined by

$$AC_vV = A_2C_bv_2 \tag{49}$$

Calculations begin with the selection of an approximate value of C_b (say 0.55) and an assumed value of β (say 1.4 rad). A_1, A_2, S_1, S_{12}, and S_2 are then evaluated, and equations 49, 40, and 43 are solved simultaneously for v_1 and v_2. With v_1 and v_2 known, Equations (34) and (37) can be used to compute values of $-dP/dz$. These do not agree, primarily because of the choice of β. (Actually i values are more useful, here since they will range between 0 and 1). A new value of β is selected by comparing the two calculated values. This is not difficult since Equation (34) is much more sensitive to β than Equation (37). Thus, if i from Equation

(34) is higher than that from (37), the value of β should be reduced. A root-finding method such as Regula-Falsi speeds convergence in this step.

Once the selected β value gives agreement in the i values, the simplification involved in using C_b is relaxed. To do this the slip is calculated by eliminating $-dP/dz$ between Equations (35) and (36):

$$(f_s)_f - (f_f)_s = \frac{\tau_{2s}S_2}{A_2 c_2} - \frac{\tau_{2f}S_2}{A_2\,(1 - c_2)} = \frac{(f_s)_f}{(1 - c_2)}$$

$$= \frac{C_{fb}\rho_f U|U|}{(1 - c_2)d}$$

where

$$U = v_{f2} - v_{s2}$$

and C_{fb} is given by Equation (46).

From an iterative method U is calculated. Since v_2 is known, v_{2f} and v_{2s} are obtained by solving Equation (40) with the value of U. This gives

$$v_{2s} = v_2 - (1 - c_2)U$$

$$v_{2f} = v_2 + c_2 U$$

and, computing a new value of C_b from the mass balance

$$C_b = c_2\,[v_2 - U(1 - c_2)]\,/v_2$$

with the new value of C_b the process is repeated to convergence. Figure 12.16 shows the flow diagram for the calculation. Using $f_1 = f_2 = 0.046\,[(DV\rho f)/\mu_f]^{-0.2}$, we obtain

First iteration:

$C_b = 0.55, \beta = 1.4$ radians
$v_1 = 0.39$ m/s; $v_2 = 1.94$ m/s
i (Equation (34)) = 0.091; i (Equation (37)) = 0.508

At convergence (agreement within 2%).

$C_b = 0.562, \beta = 2.33$ radians

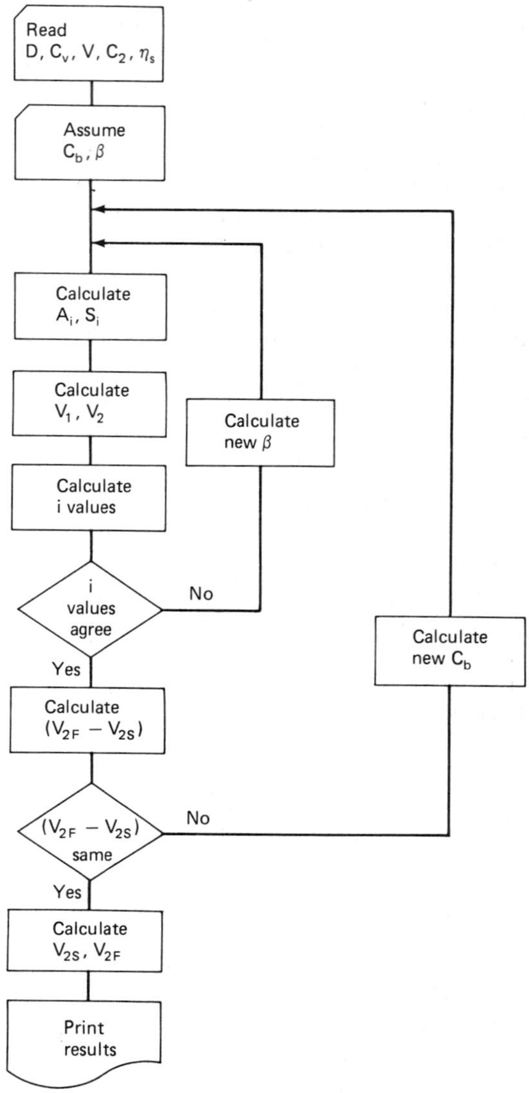

Figure 12.16. Flow diagram for calculations using sliding-bed model.

$v_1 = 2.546$ m/s $v_2 = 0.829$ m/s, $U = 0.094$ m/s
i (Equation (34)) = 0.698; i (Equation (37)) = 0.709

12.6. EMPIRICAL RELATIONSHIPS FOR FRICTIONAL LOSSES

Although many examples of pseudohomogeneous or sliding bed flow can be found, most slurry flows occupy an intermediate category.

For such slurries deposition occurs at low velocities and a density variation occurs in horizontal flow. Such flows could be described as heterogeneous, or at lower velocities, as flows with saltation because of the arched trajectory of the particles.

In the transitional region between the two extreme forms of behavior, the Durand-Condolios equation has been used for design for many years

$$\Phi = 81\left[\frac{gD(s-1)}{V^2\sqrt{C_D}}\right]^{3/2} \tag{50}$$

In this Equation C_D is the drag coefficient for a single particle settling at its terminal velocity W in the carrier fluid:

$$C_D = \frac{4gd(s-1)}{3W^2}$$

If the particle size distribution is broad, for the mixture one should use

$$\frac{1}{\sqrt{C_{D_m}}} = \sum_{j=1}^{N} \frac{w_j}{\sqrt{C_{D_j}}} \tag{51}$$

with N ranging between 5 and 10

Equation (50) is useful as an approximation but it does not reduce to either of the models for the limiting cases. For flows of the sliding bed type, the Newitt equation[20] is often used:

$$i = i_f + 0.8C_v(s-1) \tag{52}$$

or in the notation of Equation (50),

$$\Phi = \frac{0.8\mathrm{gD(s-1)}}{2\mathrm{fV}^2} \tag{52a}$$

Equations (52) and (48) are related but the empirical Equation (52) is actually more useful since in practice c_2 may not be known and is likely to vary.

A major deficiency of 50 is the lack of a viscosity-dependent term. The procedure of Wasp et al.[21,22] introduces a particle size dependency similar in principle to Equation (51). In addition, a viscosity correction term is suggested but exact details of the latter are not given.

The hydraulic gradient is written[21,22] as the sum of pseudohomogeneous (i_{hom}) and transitional or heterogeneous (i_{het}) contributions:

$$i = i_{hom} + i_{het} \tag{53}$$

Logically, the slurry viscosity correction should be applied to the calculation of i_{hom}. This would involve an expression of the form:

$$i_{hom} = i_f\,[1 + c_{v\,hom}\,(s-1)]\left[\frac{\nu_{hom}}{\nu_f}\right]^{0.2}$$

The Durand-Condolios equation is used to predict i_{het} in this model:

$$i_{het} = i_{hom}\,c_{v\,het}\,\phi_{het} \tag{54}$$

and

$$\Phi_{het} = \frac{81gD(s - s_{hom})}{V^2}\Sigma\frac{1}{\sqrt{C_{Dj}}}\frac{c_{vj\,het}}{c_{v\,het}} \tag{55}$$

with

$$s_{hom} = 1 + c_{v\,hom}\,(s-1) \tag{56}$$

$$c_{vj} = w_jC_v = c_{vj\,het} + c_{vj\,hom} \tag{57}$$

$$c_{vj\,hom} = c_{vj}\exp[-1.8W/\beta\kappa u_*] \tag{58}$$

$$c_{v\,het} = \Sigma c_{vj\,het}$$
$$c_{v\,hom} = \Sigma c_{vj\,hom} \tag{59}$$

An interative process is required but the solution converges rapidly. Equation (58) is the most doubtful step in the process and the method is probably best used for the near-homogeneous state. This condition is approached by the slurries of fine particles for which the method seems to have been developed.

In the transitional or intermediate region, the

method of dividing the solids into pseudo-homogeneous ('suspended load') and sliding bed ('bed load' or 'contact load') fractions has long been the practice in river engineering. Wilson[16] has proposed a procedure which has a theoretical basis. He relates the distribution of particles between the two modes of transport to a velocity V_{tj} for each size increment

$$V_{tj} = 0.85 \frac{W_j}{\sqrt{f_0}} \exp\,[45 d_j/D] \tag{60}$$

where f_0 is the fluid fanning friction factor at V. Then for each increment the distribution is calculated from the expression

$$\frac{C_{cj}}{C_{vj}} = \left[\frac{V_{tj}}{V}\right]^m \tag{61}$$

with m taken as 2. C_{cj} is conceptually similar to $C_{vj\,\text{het}}$. Wilson has constructed nomographs to aid in the prediction of pressure drop using a combination of the sliding bed model and a form of pseudohomogeneous model.[24] A simpler, but related expression is that of Vocadlo and Charles.[25] Still another simple procedure would be to use the Newitt equation together with Wilson's expressions:[60,61]

$$i = i_f[1 + (C_v - C_c)(s - 1)]\left(\frac{\nu_m}{\nu_f}\right)^{0.2} + 0.8C_c(s - 1) \tag{62}$$

with

$$C_c = \Sigma\, C_{cj}$$

Other empirical equations are available in the literature.[27]

Example 5

It is desired to estimate the pressure drop at V = 1.5 m/s of a slurry of coal particles (ρ_s = 1340 kg · m^{-3}) in water at $C_v = 0.51$. The viscosity of the slurry is 0.025 Pa·s and the pipe I.D. is 0.315 m. The size distribution is broken into five increments containing equal masses of particles and the mean sizes and settling velocities are given in tabular form below:

j	d_jm	W_j m/s	C_{Dj}	V_{tj}	C_{cj}
1	1.5×10^{-5}	4.2×10^{-5}	3.84×10^4	0	0
2	1.25×10^{-4}	2.55×10^{-3}	85	.040	0
3	2.8×10^{-4}	1.04×10^{-2}	11.6	.168	.0013
4	4.5×10^{-4}	1.78×10^{-2}	6.32	.295	.0040
5	8×10^{-4}	3.25×10^{-2}	3.36	.566	.0145

The observed value of f_0 is 0.003 so that i_f is 4.4×10^{-3}.

Using the Durand and Condolios equation with Equation (51) gives

$$\frac{1}{\sqrt{C_{Dm}}} = 0.270$$

$$\Phi = 81\,\frac{9.8\,(0.35)\,(0.34)\,(0.27)^{3/2}}{(1.5)^2} = 3.62$$

$$i = 0.0044[1 + 3.62\,(0.51)] = 0.0125$$

This is considerably larger than the observed value[17] of 0.0105 although the viscosity correaction was neglected in this calculation.

The sensitivity of predictive equations to the contact load component can be illustrated by using Equation (62). The remaining quantities in the table are calculated from Equations (60) and (61).

Thus, C_c = 0.019 and $(\nu/\nu_f)^{0.2}$ = 1.844

$$\begin{aligned} i &= 0.0044[1 + 0.491\,(0.34)]\,(1.844) \\ &\quad + 0.8\,(0.19)\,(0.34) = 0.00946 \\ &\quad + 0.00517 = 0.01463 \end{aligned}$$

It can be seen that although C_c is very low, its contribution to i is still about five times as high as it should be.

12.7. SCALEUP OF EXPERIMENTAL RESULTS

This refers to the problem of predicting the behavior of a slurry in a large pipeline from data gathered in small scale tests. Although the problem may be simple in principle, minor fluctuations in slurry compositions can easily occur. Kenchington[26] has shown the dramatic increase in the degree of uncertainty which arises from fairly minor data inadequacies. Carleton and Cheng[28] consider that some form of testing will always be required for major projects involving large pipes. Similarly, Example 5 illustrates the sensitivity of i for large pipes to the contribution from the contact load portion of the flow.

Any of the models or empirical equations previously discussed can be used to fit experimental data for purposes of prediction. In addition, Bowen's formula

$$D^x \tau_w = KV^y$$

can be used as long as neither the laminar-turbulent transition or deposition occurs to alter the relationship between τ_w, D, and V for a given slurry. A systematic procedure for scaleup is given below.

1. Evaluate Φ and $gD\ (s - 1)/V^2\sqrt{C_{Dm}}$ for each data point.
2. Plot the data with D as a parameter for given values of C_v. Minor variations in concentration, corresponding to viscosity changes of less than ±15%, should not affect the results unduly.

If the flow is pseudohomogeneous Φ will be constant and the effective value of ν_m/ν_f can be obtained from the results.

Figure 12.17 shows results obtained with iron ore concentrate at $C_v = 0.30$. $(\nu_m/\nu_f)^{0.2}$ is approximately 1.3 compared to a value of 1.2 suggested from Figure 12.14.

Figure 12.18 shows results for a narrow size distribution sand. In this case the limiting value of Φ at high values of $gD\ (s - 1)/V^2\sqrt{C_{Dm}}$ approaches $(s - 1)\ (\nu_m/\nu_f)^{0.2}$, indicating pseudohomogeneous flow in this region.

12.8. OPERATING VELOCITIES–HORIZONTAL FLOW

For vertical flows, limiting velocities can be obtained by determining the incipient fluidizing condition using Equations (8)-(11). For horizontal (or inclined) flows, unless the slurry is truly homogeneous a stationary layer of solids forms at the bottom of the pipe at low velocities. This is the usual lower limit of operation because the solids flow often shows concentration fluctuations and blockage is possible if the velocity is further reduced. Most economic analyses indicate that the optimum operating condition occurs near the incipient deposition condition. This is due in part to

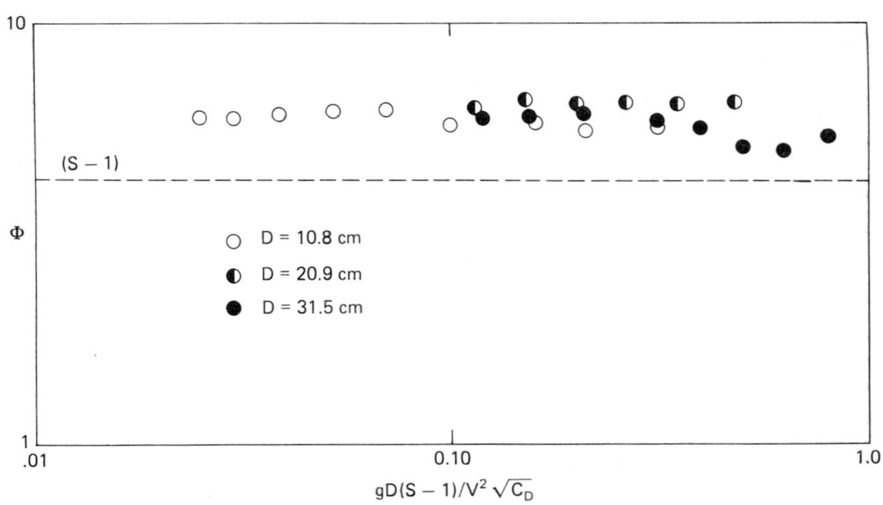

Figure 12.17. Experimental pressure drop data for a slurry of a fine iron ore concentrate at $C_v = 0.30$. The data were taken from Reference 18.

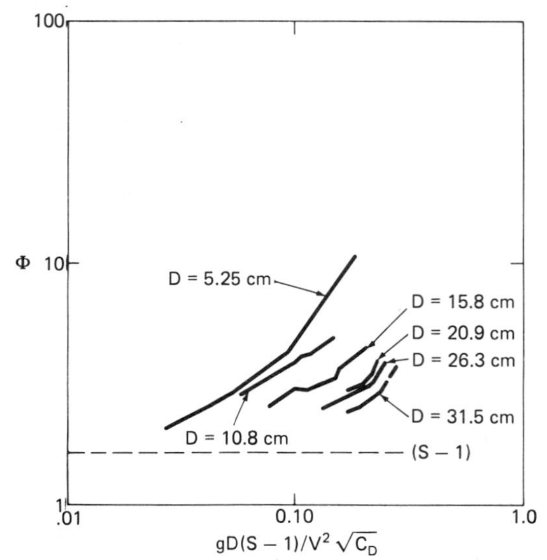

Figure 12.18. Experimental pressure drop data for a narrow PSD sand (d = 0.2 mm) at C_v 0.30–0.35. The data were taken from Reference 30.

the fact that, with small pipes at least, a minimum pressure drop often occurs just above the deposition velocity (the Durand-Condolios equation 50 predicts such a minimum). However, this minimum is much less well defined for large pipes and may not occur at all. Nevertheless, low operating velocities are desirable when the particles are harder than the pipe wall and erosion can occur. Thus for large projects, both wear rates and deposition velocities must be known for selection of operating velocities.

In any case, some information concerning deposition for a given slurry is invaluable. Even limited testing can reveal whether the deposition velocity increases or decreases with solids concentration. Empirical correlations often show an increase in deposition velocity V_c with solids concentration but the reverse is true at high concentrations for some slurries. The difference is important for processes subject to concentration fluctuations.

At least 60 predictive equations have been proposed. The best known was the Durand expression defining the factor F_L

$$V_c = F_L\sqrt{2gD(s - 1)} \qquad (63)$$

and values of F_L for sand slurries of $C_v \leq 0.2$ were quoted. A recent generalization is that of Wilson and Judge who suggest[23] that F_L varies as d/DC_D. Figure 12.19 shows data from References 17, 18, and 30 plotted in this way. The scatter is due in part to the effect of C_v and lack of precision in the determination of V_c values. Since there is not a satisfactory theoretical basis for the correlation it should be used with caution in selecting operating velocities. Wherever possible, tests should be

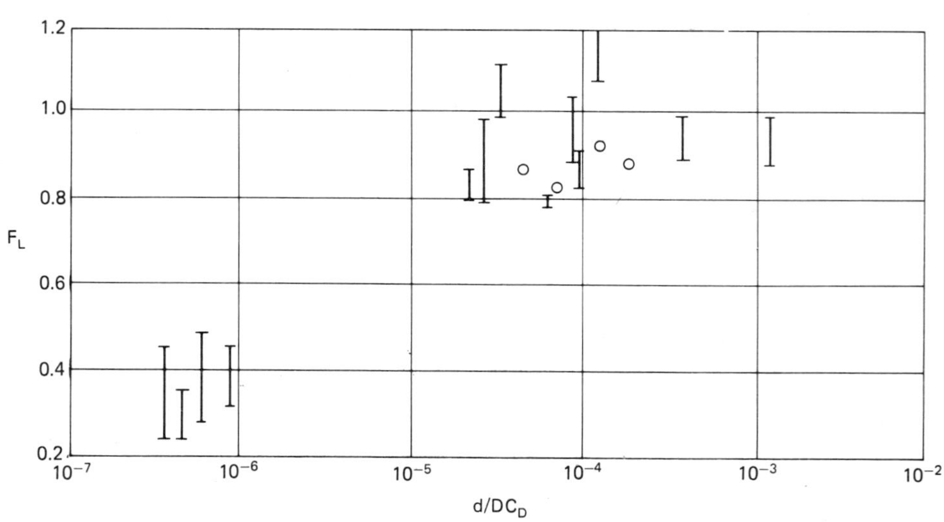

Figure 12.19. F_L as a function of d/DC_D.

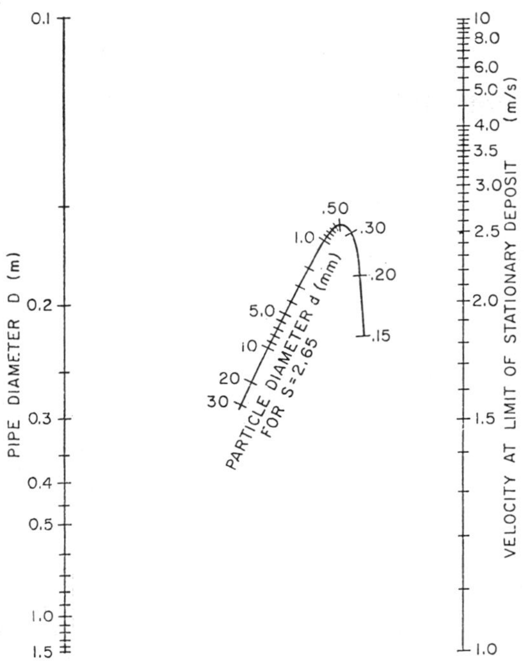

Figure 12.20. Nomogram for determining velocity for deposit-free flow for sand–water slurries. Reproduced with permission from Reference 24.

conducted using pipes of the sizes under consideration. Readers interested in other correlations should consult References 27, 28, and 29. Figure 12.20 may also be used for sand.

For most purposes, scaleup using the square root law ($V_c \propto \sqrt{D}$) will avoid deposition in larger pipes, at the cost of somewhat higher wear rates and off-optimum energy consumptions. Experimental data which can be of some use are given in References 17, 18, 30, and 31.

12.9. METERING

If the electrical conductivity of the slurry is sufficient, magnetic flux flowmeters are extremely useful for velocity measurement. Over wide ranges of velocity and solids concentration particles do not affect the calibration. If placed horizontally, some deviation is theoretically possible for sliding bed flows but unless a deposit is present in the flowmeter this effect is small. Zero shift effects attributable to the presence of gases have been observed in some flotation cell product slurries.

Venturi meters are also very useful and of course can be used with slurries of low conductivity. Even in horizontal flows, with highly stratified slurries, coefficients near unity are observed. Although in extreme cases coefficients greater than unity are predicted and observed, these effects will be negligible in the majority of cases.[32] The major problem with the venturi meter is that a reliable value of the density of the mixture (or c) is required. For practical purposes the pressure drop over a section of vertical pipe can often be used, together with the pseudohomogeneous flow assumption, to provide this information. Example 3 illustrates that the pressure gradient is primarily dependent on c and C_V is usually a reasonable approximation to c.

If a vertical U-bend is used, as in Figure 12.21 the effect of wall friction can be eliminated since, assuming c to be the same in both legs:

$$\rho = \rho_f[1 + c(s - 1)] = \frac{(P_1 - P_2) - (P_3 - P_4)}{2gH} \tag{64}$$

Of course c will be slightly larger in the upward flowing section than in the downflow portion.

Concentration measurement is more difficult, even in laboratory conditions. Gamma ray absorption can provide reliable values of concentration in two component systems. Where compositions of the liquid and/or solid phases vary, interpretation of these measurements becomes difficult. Direct measurements such as pipe weighing can be used but these are prob-

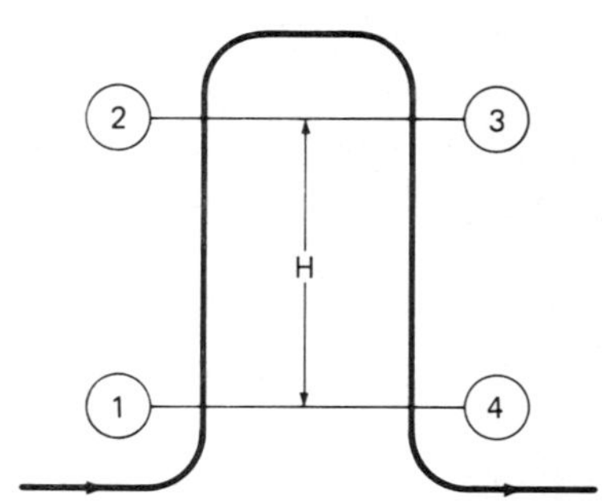

Figure 12.21. Vertical pipe loop for use in determining mixture density.

ably best restricted to laboratory studies where frequent recalibration is possible.

Sample removal is reliable if the flow is pseudohomogeneous or homogeneous. With increasing stratification or segregation, it becomes more difficult to obtain a representative sample for size distribution or C_v determination.

12.10. PUMP SELECTION

In addition to the head and capacity requirements particle size distribution affects pump selection. Piston and plunger pumps are useful for generating high pressures for slurries of fine particles such as drilling muds and fine cuttings. Typical values of mechanical and volumetric efficiency[33] are of the order of 90%. This high efficiency makes these pumps attractive for long distance transportation. Larger particles would be expected to lower the volumetric efficiency. The upper limit of particle size is of the order of 1 mm for coal slurries of high solids concentration.

Wear rates in valves and liners have been related to the properties of the slurry as expressed in an "abrasivity" test.[35] The results of this test are expressed in the form of a Miller Index which indicates the relative abrasion tendency on a scale extending from 0 for a non-abrasive slurry to 1000 for 200 mesh Corundum. Since the hardness of valves and liners is relatively invariant it is the slurry properties which are of importance.

Centrifugal pumps are commonly used for short distance slurry transport. As in the case of positive displacement pumps, the manufacturer's advice should be sought wherever possible to achieve optimum performance. Wiedenroth,[36] in studies conducted with sand slurries, found the head (expressed in meters of slurry) to be decreased by the presence of solids by an amount dependent upon the product

$$C_v d^{-1/3} N_P{}^{-2.46}$$

The reduction was significant even for particles of d_{50} = 0.2 mm. This indicates that at the high velocities and accelerations which occur during passage through the pump the pseudohomogeneous flow approximation fails, even for these fine particles. Efficiency was affected in a rather similar manner and Hunt and Faddick[34] show reduced efficiencies for slurries of simulated wood chips of density nearly equal to that of water. Further results are given in Reference 40.

Wear rates are also important in selecting centrifugal pumps. Rubber linings are suggested[37] for slurries of fine abrasive particles and hard metal pumps for coarse or angular particles. Internal flow patterns determine wear points and wear rates will be lowest when pumps are operated at best efficiency. This is because particle trajectories match the shape of solid surfaces. At these low impingement angles, wear rates are minimized.

Particle breakage can be significant in centrifugal pumps, especially for friable materials. If it is necessary to characterize the rate of change of the particle size distribution, the rate equation for rod or ball milling can be used:[38]

$$\frac{dw_k}{dt} = \sum_{j=1}^{k-1} b_{kj} w_j S_j - S_k w_k \tag{65}$$

where w is the weight fraction of material in a given size interval (say between successive screens in a $\sqrt{2}$ series). b_{kj} represents the contribution to the kth interval which arises from the jth interval: it is the fraction of the material originally in the jth interval which falls into the kth interval after breakage. The S values are rate constants which presumably depend on particle size. The b values indicate the type of breakage process. If abrasion tending to "round" the particles is dominant, breakage products are found in the next lowest size and in the very finest portion. Fracture tends to produce particles of intermediate size. In a study using lignite particles, rate constants appeared to vary as the tangential impeller velocity to the 2.5 power, as in an earlier study using coal.[39] Increasing clearances between impeller and casing reduced the breakage rates of the largest particles.

With high concentrations of coarse particles

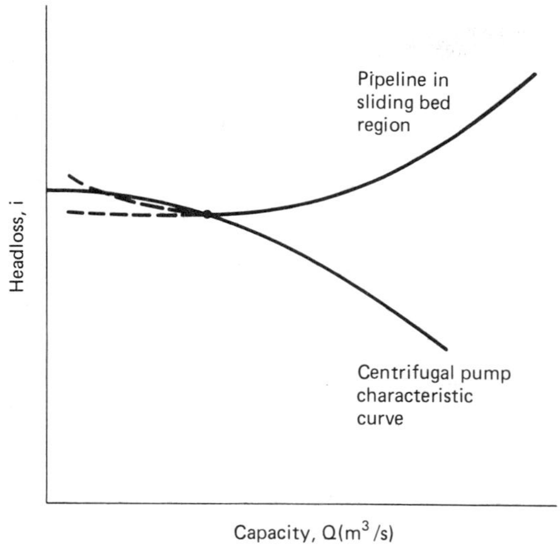

Figure 12.22. Characteristic curves for a centrifugal pump and a pipeline in sliding-bed flow.

the pressure drop–capacity curves for pipelines and centrifugal pumps are not satisfactory for stable flow. Figure 12.22 illustrates this for a slurry which obeys Newitt's equation. The characteristic curves intersect obliquely and, where a minimum is observed, more than one operating state can occur. If the pump size or maximum speed are insufficient, blockage can occur. If the concentration of the slurry in the pump fluctuates with time, or if cavitation or entrainment reduce the mixture density in the pump, blockage is even more likely because the pressure rise across the pump depends upon the mixture density.

In this case lock hopper operation or use of an indirect feeding system may be preferable. A positive displacement pump can be used as the source of high pressure with solids drawn into the system through a feeder. Various types of feeder exist, using either the jet-pump type of inductor[33] or rotary devices.[41,42]

12.11. EROSION AND CORROSION

Careful design, materials selection and fabrication can minimize the difficulties arising from these sources. It is known[43] that the angle of impingement of particles with surfaces dramatically affects the rate of erosion. Thus, at high impingement angles, as in pumps, and at bends, erosion rates are likely to be high unless resistant materials or longer radii of curvature are used. Similarly, misalignments can produce rapid erosion or scour at flanges or welds.

For longer pipelines, wear in the pipe itself is important since it determines the allowance of wall thickness for corrosion and erosion and contributes directly to the capital cost of the project. Amonton's Law of frictional resistance can be used to show that for abrasive wear the rate of erosion per contact varies directly as the normal force, is independent of the velocity and depends markedly on the relative hardness of the surfaces in contact. Since the number of contacts per unit time with a point on the eroded surface increases linearly with the velocity, the rate of erosion per unit time for sliding contact should vary directly with the velocity. However higher values than 1 are reported for m_1 in the erosion rate equation[44,45]

$$E = K_k V^{m_1} d^{m_2} D^{m_3} \tag{66}$$

so that presumably the kinetic energy of the particles is important in pipe wall erosion. Some of this effect can arise through the normal force. If the particles follow the fluid motion exactly (the pseudohomogeneous assumption) then on theoretical grounds the normal force at the wall should be proportional to

$$\frac{K_2 V^2}{D} + g \cos \beta$$

where K_2 is a weak function of the Reynolds number of the flow. θ is the angle shown in Figure 12.15. Observations of wear around the pipe periphery often confirm the angular dependence to be of this type. However the flow disturbances produced by valves or fittings often persist considerable distances downstream. Since these frequently contain a tangential velocity, normal forces at the pipe wall can be modified and the location of wear points can shift. Impingement angles will also vary with position on the pipe wall and with velocity.

This complicates the wear process and makes prediction of the effect difficult. Similarly, although the normal force effect implies that m_3 should approach −1, values between −1 and −4 have been quoted.[45]

For abrasive wear, the wear rate increases with particle size up to some limiting value determined, in part, by the clogging action of fine debris on the abraded surface.[46] Pipeline wear rates for silica sand slurries of $d < 128\ \mu$m were determined by Karabelas[44] for brass (Rockwell HRB = 70.8) and D = .0356 m with $K_1 D^{m_3}$ = 6.1, m_2 = 2.15 and m_1 = 3.27 in Equation (66). E was given in mm/year in this case. Karabelas suggests that these experiments were probably conducted below the limiting particle diameter.

Abrasive wear studies have shown the wear rate, for any particular abrading medium, to vary inversely with the hardness of the abraded surface. The rate shows a further marked decline when the hardness of the surface exceeds that of the abrading medium.[46] Further studies to confirm these effects for pipe wall materials would be of value. Kieselev[47] shows data indicating a linear decrease of erosion rate with wall hardness.

The principles of electrochemical corrosion and procedures to be taken to minimize corrosion costs are well known. The wiping action of the slurry, and perhaps more rapid transfer of oxygen to the surface, indicate that corrosion rates should be higher in the presence of particles. Ghandi[48] discusses the use of inhibitors or oxygen removal by other methods as cost-reducing techniques. In view of the variability of corrosion rates it is usually necessary to test a particular liquid for its corrosion tendency and this is equally important in slurries.

LIST OF SYMBOLS

A	cross-sectional area (m^2)
b	selection parameter in particle grinding rate equation
c	local solids concentration (time average value), volume fraction
c_2	solids concentration in lower layer, volume fraction
C_b	mean concentration in lower layer of sliding bed model, volume fraction
C_{Ds}	drag coefficient in multiparticle mixture
C_D	drag coefficient for single particle settling at terminal velocity = $(4gd(s-1)/3W^2)$
C_{fb}	friction coefficient in Ergun equation
C_v	discharge concentration, volume fraction
d	particle diameter (m)
d_{50}	mean particle diameter—diameter at 50% cumulative percentage retained (m)
D	pipe diameter (m)
E	erosion rate per unit time (mm s^{-1})
f	fanning friction factor (m)
$(\mathbf{f}_f)_s$	interfacial drag force of solids on fluid ($N \cdot m^{-3}$)
$(\mathbf{f}_s)_f$	interfacial drag force of fluid on solids ($N \cdot m^{-3}$)
$\mathbf{f}_{sw}$	drag force exerted by walls on solids ($N \cdot m^{-3}$)
$\mathbf{f}_{fw}$	drag force exerted by walls on fluid ($N \cdot m^{-3}$)
F_L	coefficient in equation for limit-deposit velocity
g	gravitational acceleration ($m \cdot s^{-2}$)
h	elevation above horizontal datum (m)
i	hydraulic gradient $(\Delta P/\rho_f g L)$
He	Hedstrom number $\rho D^2 \tau_y/\eta_P$
K	constant in power law model ($kg \cdot m^{-1}\ s^{n-2}$)
K′	coefficient defined by Equation (21) ($kg \cdot m^{-1}\ s^{n'-2}$)
L	length of pipe (m)
n	constant in power law model
n'	coefficient defined by Equation (19)
N_P	power number
P	pressure (Pa)
r	radial coordinate (m)
R	radius of cylinder in Couette viscometer (m)
Re	Reynolds number
Re_g	generalized Reynolds number $(D^{n'} \cdot V^{2-n'}\rho/K'8^{n'-1})$
s	density ratio (ρ_s/ρ_f)
S	perimeter (m)
S	selection rate constant (s^{-1})
t	time (s)
u_*	friction velocity $(\tau_w/\rho)^{1/2}$ ($m \cdot s^{-1}$)

U relative velocity or slip velocity ($v_f - v_s$) ($m \cdot s^{-1}$)
v' turbulent velocity fluctuation ($m \cdot s^{-1}$)
v velocity (time-average value) ($m \cdot s^{-1}$)
V mean velocity ($m \cdot s^{-1}$)
V_c limit-deposit velocity ($m \cdot s^{-1}$)
w fraction of total solids in a given size interval
W terminal falling velocity of a single particle ($m \cdot s^{-1}$)
z axial coordinate (m)
α kinetic energy correction factor
β constant modifying von Karman constant, typically 1
β angular coordinate defining lower layer (rad)
$\dot{\gamma}$ rate of shear strain (s^{-1})
κ von Karman constant
θ angle (rad)
η_P coefficient of rigidity ($Pa \cdot s$)
η_s coefficient of sliding friction
μ viscosity ($Pa \cdot s$)
ν kinematic viscosity ($m \cdot s^{-1}$)
ρ density ($kg \cdot m^{-3}$)
τ shear stress (Pa)
τ_y yield stress (Bingham fluid) (Pa)
ϕ angle of internal friction for solids (rad)
Φ dimensionless excess headloss $(i - i_f)/(C_v i_f)$
ω angular velocity ($rad \cdot s^{-1}$)

SUBSCRIPTS

f denotes fluid
m mixture of fluid and solids
s solids
w wall value
1 upper layer
12 interface
2 lower layer

GLOSSARY

Deposit Velocity (*limit-deposit velocity*, *critical velocity*)—the velocity below which a stationary deposit of particles is formed in the pipe.

Saltation—the arched trajectory of a large particle between successive contacts with the pipe bottom or with a particle layer. Also, a flow regime within the *heterogeneous* flow region (Figure 12.11) characterized by low frequency temporal variations of particle concentration. This regime occurs at velocities just above the deposit velocity condition.

Nonsettling Slurry—a slurry for which the continuum assumption is justified at all times. Such slurries produce *homogeneous* flows.

Settling Slurry—any other slurry. Such slurries may display *pseudohomogeneous*, *sliding bed* or any other form of *heterogeneous* flow.

REFERENCES

1. G. B. Wallis, *One-Dimensional Two Phase Flow*, McGraw-Hill Inc., New York (1969).
2. R. Jackson, Chapter 3 in J. F. Davidson and D. Harrison (eds.), *Fluidization*, Academic Press, London (1971).
3. H. A. Becker, "The Effects of Shape and Reynolds Number on Drag in the Motion of a Freely Oriented Body in an Infinite Fluid," *Can. J. Chem. Eng. 37*:85–91 (1959).
4. A. H. P. Skelland, *Non-Newtonian Flow and Heat Transfer*, Wiley and Sons Inc., New York (1967).
5. R. B. Bird, W. E. Stewart, and E. N. Lightfoot, *Transport Phenomena*, Wiley and Sons, Inc., New York (1960).
6. R. H. Perry and C. H. Chilton, (eds.), *Perry's Chemical Engineers' Handbook*, 5th edition, McGraw-Hill Inc., New York (1973).
7. D. W. Dodge and A. B. Metzner, "Turbulent Flow of Non-Newtonian Systems," *A.I.Ch.E. Journal 5*:189–204 (1959).
8. D. G. Thomas, "Heat and Momentum Transport Characteristics of Non-Newtonian Aqueous Thorium Oxide Suspensions," *A.I.Ch.E. Journal 6*:631–639 (1960).
9. W. L. Wilkinson, *Non-Newtonian Fluids and Heat Transfer*, Pergamon Press, New York (1960).
10. G. W. Govier and K. Aziz, *The Flow of Complex Mixtures in Pipes*, Van Nostrand Reinhold Co., New York (1972).
11. J. O. Liebe and C. A. Shook, "Transient Laminar Flow of some Non-Newtonian Fluids," Institute of Petroleum paper IP 76-003 (1976).
12. R. A. Bagnold, "Experiments on a Gravity-Free Dispersion of Large Solid Particles in a Newtonian Fluid under Shear," *Proc. Roy. Soc. A 225*:49–63 (1954).
13. R. A. Bagnold, "The Flow of Cohesionless Grains in Fluids," *Phil. Trans. Ray. Soc. A 249*:235–297 (1956).
14. K. C. Wilson, M. Streat, and R. A. Bantin, "Slip-

Model Correlation of Dense Two-phase Flow," paper B1, Hydrotransport 2, Second Int. Conf. on Hydraulic Transport of Solids in Pipes, BHRA Fluid Engineering, Cranfield, Bedford, U.K. (1970).

15. K. C. Wilson, "Slip Point of Beds in Solid-Liquid Pipeline Flow," *Proc. A.S.C.E. 96*:HY1 (Jan., 1970).
16. K. C. Wilson, "A Unified Physically Based Analysis of Solid-Liquid Pipeline Flow," paper A1, Hydrotransport 4, Fourth Int. Conf. on Hydraulic Transport of Solids in Pipes, BHRA Fluid Engineering, Cranfield, Bedford, U.K. (1976).
17. W. Schriek et al., "Experimental Studies of the Hydraulic Transport of Coal," Report E73-17, Saskatchewan Research Council, Saskatoon, Canada (1973).
18. W. Schriek et al., "Experimental Studies of the Transport of Iron Ore," Report E73-12, Saskatchewan Research Council, Saskatoon, Canada (1973).
19. A. J. Carleton et al., "Hydraulic Transport of Large Particles Using Conventional and High Concentration Conveying," paper D2, Hydrotransport 5, Fifth Int. Conf. on Hydraulic Transport of Solids in Pipes, BHRA Fluid Engineering, Cranfield, Bedford, U.K. (1978).
20. D. M. Newitt, J. F. Richardson, M. Abbott, and R. B. Turtle, "Hydraulic Conveying of Solids in Horizontal Pipes," *Trans. Inst. Chem. Engrs. 33*: 93–113 (1955).
21. E. J. Wasp et al., "Hetero-homogeneous Solids/Liquid Flow in the Turbulent Regime," paper 21 in I. Zandi (ed.), *Advances in Solid/Liquid Flow and Its Application*, Pergamon Press Inc., New York (1971).
22. E. J. Wasp et al., "Deposition Velocities, Transition Velocities and Spatial Distribution of Solids in Slurry Pipelines," paper H4. Hydrotransport 1, First Int. Conf. on the Hydraulic Transport of Solids in Pipes, BHRA Fluid Engineering, Cranfield, Bedford, U.K. (1970).
23. K. C. Wilson and D. G. Judge, "New Techniques for the Scale-up of Pilot Plant Results to Coal Slurry Pipelines," Inter. Symp. on Freight Pipelines, Washington, D.C. (Dec., 1976).
24. K. C. Wilson and D. G. Judge, "Analytically Based Homographic Charts for Sand-Water Flow," paper A1, Hydrotransport 5 (1978).
25. J. J. Vocadlo and M. E. Charles, "Prediction of Pressure Gradient for the Horizontal Flow of Slurries," paper C1, Hydrotransport 2 (1972).
26. J. M. Kenchington, "The Design of Large Pipelines for Cement Slurries," paper C4, Hydrotransport 2 (1972).
27. R. M. Turian and T-F. Yuan, "Flow of Slurries in Pipelines," *A.I.Ch.E. Journal 23*:232–243 (1977).
28. A. J. Carleton and D. C-H. Cheng, "Design Velocities for Hydraulic Conveying of Settling Suspensions," paper E5, Hydrotransport 3 (1974).
29. R. Durand and E. Condolios, "Transport hydraulique et decantation des materiau solids," *Compte. Rendu des Deuxièmes Journées de l'Hydraulique*, Grenoble (1952).
30. W. Schriek et al., "Experimental Studies of the Transport of Two Different Sands in Water in 2, 4, 6, 8, 10 and 12 inch Pipelines," Report E73-21, Saskatchewan Research Council, Saskatoon, Canada (1973).
31. W. Schriek et al., "Experimental Studies of the Hydraulic Transport of Potash," Report E73-16, Saskatchewan Research Council, Saskatoon, Canada (1973).
32. C. A. Shook and J. Masliyah, "Flow of a Slurry through a Venturi Meter," *Can. J. Chem. Eng. 52*:228–33 (1974).
33. D. M. Frazier, "New Solids Pumping Techniques and Analysis," paper 21 in I. Zandi (ed.), *Advances in Solid/Liquid Flow.*
34. W. A. Hunt and R. R. Faddick, "The Effects of Solids on Centrifugal Pump Characteristics," paper 20 in I. Zandi (ed.).
35. R. J. Prudhomme et al., "Reciprocating Pumps for Long Distance Slurry Pipelines," paper E4, Hydrotransport 1 (1970).
36. W. Wiedenroth, "The Influence of Sand and Gravel on the Characteristics of Centrifugal Pumps: Some Aspects of Wear in Hydraulic Transport Installations," paper E1, Hydrotransport 1 (1970).
37. G. Wilson, "The Design Aspects of Centrifugal Pumps for Abrasive Slurries," paper H2, Hydrotransport 2 (1972).
38. C. A. Shook, D. B. Haas, W. H. W. Husband, and M. Small, "Breakage Rates of Lignite Particles during Hydraulic Transport," *Can. J. Chem. Eng. 56*:448–454 (1978).
39. A. P. Pipilen, M. Weintraub, and A. A. Orning, "Hydraulic Transport of Coal," U.S.B.M. Rept. of Inv. 6743 (1966).
40. I. Cave, "Effects of Suspended Solids on the Performance of Centrifugal Pumps," paper H3, Hydrotransport 4 (1976).
41. E. D. Funk, M. D. Barrett, and D. W. Hunter, "Pilot Experiences with Run of Mine Coal Injection and Pipelining," paper F3, Hydrotransport 5 (1978).
42. M. Burnett et al., "A Coarse Coal Injector for Hydraulic Haulage," paper F4, Hydrotransport 5 (1978).
43. J. G. A. Bitter, "Study of Erosion Phenomena," Wear: 5–21 and 169–90 (1963).
44. A. J. Karabelas, "An Experimental Study of Pipe Erosion by Turbulent Slurry Flow," paper E2, Hydrotransport 5 (1978).

45. T. Kawashima et al., "Wear of Pipe for Hydraulic Transport of Solids," paper E3, Hydrotransport 5 (1978).
46. E. Rabinowicz, *Friction and Wear of Materials*, Wiley and Sons Inc., New York (1965).
47. G. I. Kieselev, "Wear Resistance of Metals Subjected to an Abrasive Solid Medium," *Russ. Eng. J. 51*(12) (1971).
48. R. L. Ghandi, B. L. Ricks, and T. C. Aude, "Control of Corrosion-Erosion in Slurry Pipelines," *Proc. 1st. Int. Symp. on Dredging Technology*, Canterbury, England, G5-53-64 (1975).

13
Size Reduction of Solids: Crushing and Grinding Equipment

L. G. Austin

The Pennsylvania State University
University Park, Pennsylvania

13.1. INTRODUCTION

The unit operation of the size reduction of solids by crushers and mills is a very important industrial operation involving many aspects of powder technology. It is estimated that mechanical size reduction of rocks, ores, coals, cement, plastics, grains, etc. involves at least a billion tons of material per year in the United States alone. The operation ranges in scale, for a single device, from a few kilograms per hour for speciality products to hundreds of tons per hour for metallurgical extractive purposes. In this chapter, the fundamental aspects will be emphasized rather than mechanical or process engineering aspects, to form a background for intelligent decision-making in the choice and analysis of size reduction systems.

In many operations, a material must be reduced from lumps of up to a meter in size to a fine powder, sometimes a powder essentially all less than 100 μm in size. It is clear that size reduction over many orders of magnitude in size cannot be efficiently achieved in a single machine and a sequence of different types of machine is used, each machine designed for efficient operation on a particular feed size. Machines for breakage of large lumps are called crushers and machines for smaller sizes are called mills, with a range of overlap where either a fine crusher or a coarse mill can be used. The operation of crushing does not normally give problems because the energy consumption and capital cost per ton per hour is not high. The principal requirements for crushers is a mechanical requirement—they must be very robust because of the high stress required to crush a large lump. On the other hand, fine grinding consumes a great deal of energy and may lead to high abrasive wear, so the major scientific and technical problems are concerned with fine grinding and most current research is on these problems.

Before discussing the various types of size reduction units in detail it is invaluable to have a clear idea of the fundamental physical laws involved in size reduction. These involve the areas of fracture mechanics, particle-fluid dynamics, agglomerative forces (dry and wet) and powder flow. The last four topics are covered elsewhere in this book and will only be mentioned here as they arise. Fracture mechanics will be discussed in some detail. Since the objective of size reduction is to obtain a suitable product size, the accurate measurement of powder size distributions is a basic feature of the process, and this is also of course covered in detail elsewhere. However, the prediction of size distributions and how they change with mill operation will be dealt with in depth.

13.2. A BRIEF REVIEW OF FRACTURE MECHANICS

Stress, Strain and Energy

In order to produce size reduction the lumps of solid must be fractured, and they must be stressed in order to produce fracture. Quantitative theoretical analysis is only possible for relatively simple states of stress, but the concepts which emerge are qualitatively useful for the complex stressing conditions of industrial crushers and mills. Materials are divided into two broad categories, elastic and ductile, with the corresponding failure under stress termed brittle or nonbrittle fracture, respectively. Consider a simple tensile stress, as illustrated in Figure 13.1. Stress is defined as $\sigma = F/A$, and Figure 13.2 shows the characteristics of elastic and ductile materials. An elastic material can be stressed, producing elongation, and

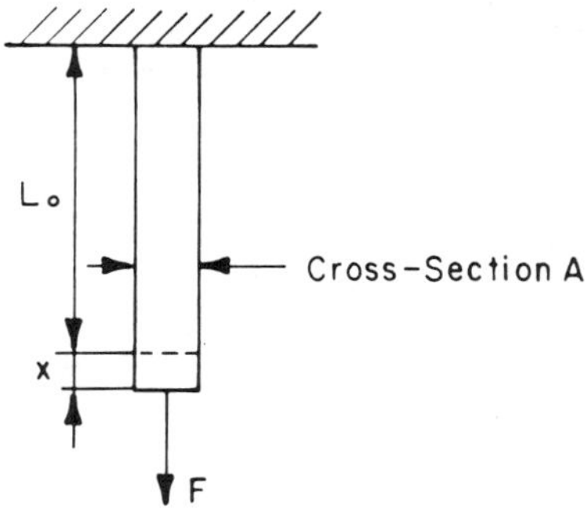

Figure 13.1. Simple tensile stress.

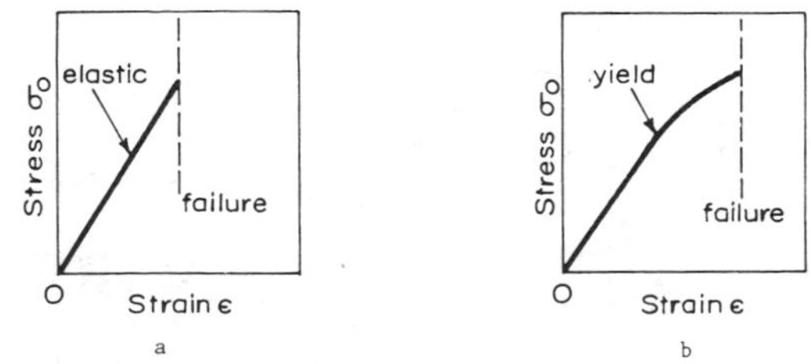

Figure 13.2. Illustration of stress-strain curves for simple tensile tests; σ_0 = force/original cross-section; strain $\epsilon = x/L_0$.

the material returns to its original shape when the stress is removed. However, if the solid is stretched too far, catastrophic failure occurs and the solid fractures at a stress termed the tensile strength. Ductile materials undergo a partially irreversible stretching before failure occurs.

Elastic materials fail at small strain so $\sigma \approx \sigma_0$, and the strain-stress relation up to where failure occurs is the empirical Hooke's law

$$\sigma = Y\epsilon = Y\frac{x}{L_0} \tag{1}$$

where Y is Young's modulus, ϵ is strain. For a perfect crystal Y depends on the orientation of the stress, but most brittle solids are polycrystalline with a random arrangement of crystallites, so Y is an effective isotropic elastic constant. The work done on the solid to go from zero external stress to a stressed state by slowly increasing F up to a final stress of σ is $\int_{L_0}^{L} F dx$ and using Hooke's law

$$\text{Work per unit volume} = Y\epsilon^2/2 = \sigma^2/2Y \tag{2}$$

This reversible strain energy is stored in the solid. If the solid is immediately loaded to σ the work done is $\sigma A \epsilon L$, which is σ^2/Y per unit volume. Half of this is strain energy and the other half will accelerate the solid and cause it to oscillate until frictional damping converts the kinetic energy to heat. Similarly, if a solid suddenly expands x at a constant σ, the work done per unit volume is σ^2/Y and again only half is reversible strain energy.

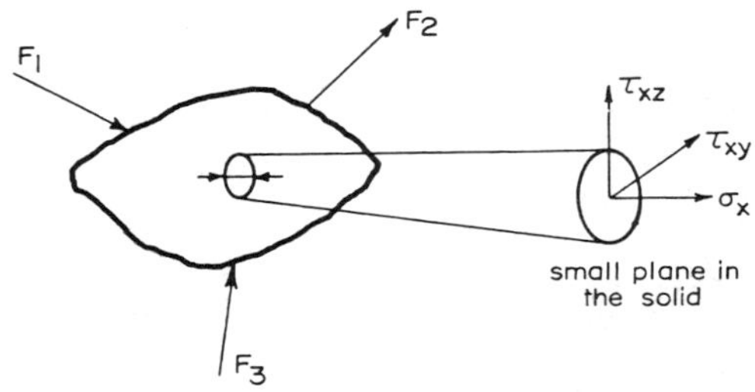

Figure 13.3. Illustration of stress through a point in a stressed solid at equilibrium.

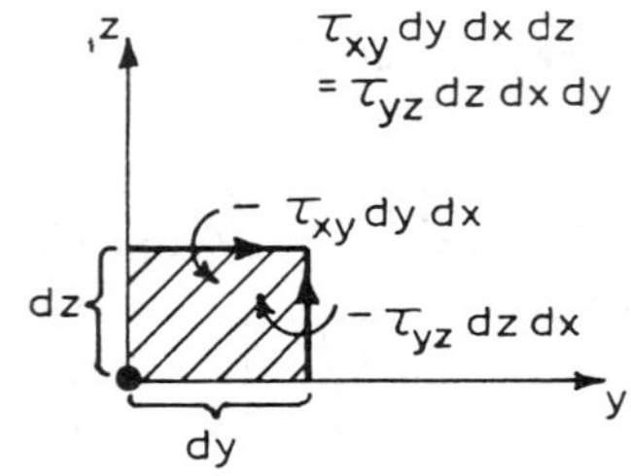

Figure 13.5. Moments about a point in the *zy* plane: Material outside square acts on material inside.

More generally, consider a stressed solid at equilibrium. At a differential plane at any point in the solid there is no net force (since there is no movement of one part of the solid with respect to another), as illustrated in Figure 13.3, and the force of material A acting on material B must equal the force of B acting on A. The force per unit area of A acting on B is called stress, and equals B on A, so stress is a force transmission through the solid. The stress at the point can be resolved into two components, the *normal* component perpendicular to the plane and the *shear* component in the plane. The normal stress tends to pull A away from B (tension) or force A into B (compression), whereas the shear stress tends to make A slip sideways with respect to B. From a molecular point of view, a solid consists of an array of atoms, molecules or ions at rest (although vibrating) with respect to one another, so that the attractive and repulsive forces between them are exactly balanced. Viewing these forces as acting like springs, Figure 13.4 illustrates the three stress states. Obviously, uneven compression or tension across a solid must produce shear stress.

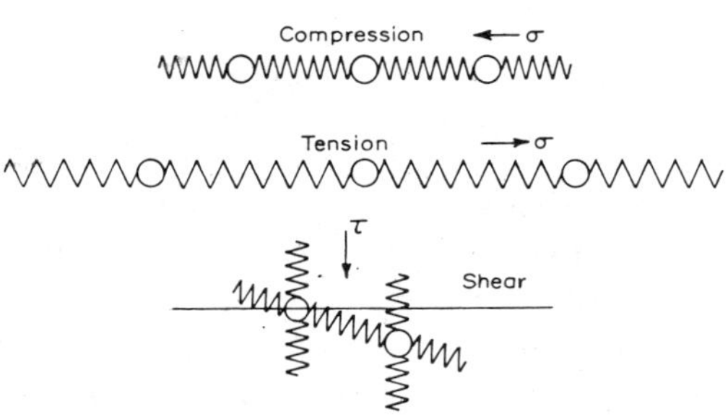

Figure 13.4. Illustration of states of stress on a molecular basis.

Drawing an arbitrary set of axes through the point which define x, y, z directions (see Fig. 13.3), the shear stress can be resolved into the components τ_{xy}, τ_{xz}. The sign convention is that material at $-x$ is dragging material at $+x$ in the y direction with a force per unit area of τ_{xy}, when the sign convention for normal stress is positive for compression, negative for tension. Taking moments about a point it is readily shown that $\tau_{xy} = \tau_{yx}$, $\tau_{yz} = \tau_{zy}$, $\tau_{zx} = \tau_{xz}$ (see Fig. 13.5).

Directions of Normal and Shear Stress

In order to describe the process of fracture it is necessary to know the normal and shear stresses and their directions in the solid. The relations between stress and direction can be readily developed for a planar solid (two-dimensional) as follows. Consider an arbitrary direction defined by α in Figure 13.6a, and imagine the shaded differential element of solid at equilibrium to be acted on by forces from the out-

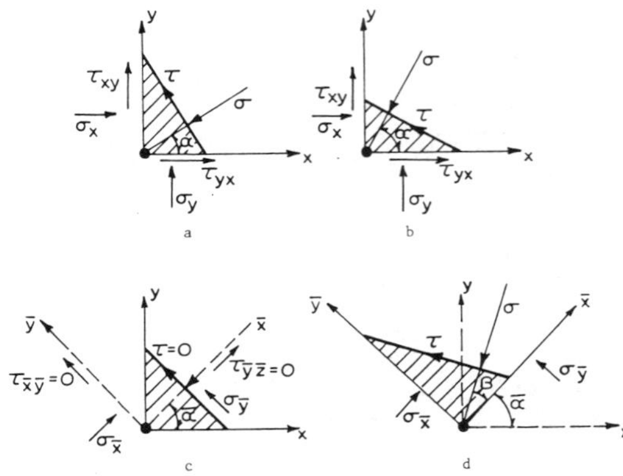

Figure 13.6. Equilibrium stress conditions in a planar element.

side material, as shown. Because the element is a differential element the forces are uniform over the small lengths of side and represent the forces at a point in the solid. The relative lengths along x, y, and the hypotenuse are $\cos\alpha$: $\sin\alpha$: 1, and since $\tau_{xy} = \tau_{yx}$ a force balance gives

$$\sigma = \sigma_x \cos^2\alpha + \sigma_y \sin^2\alpha + 2\tau \cos\alpha \sin\alpha \quad (3)$$

$$\tau = -\left(\frac{\sigma_x - \sigma_y}{2}\right) \sin 2\alpha + \tau_{xy} \cos 2\alpha \quad (4)$$

The force balance will apply for any other α, see Figure 13.6b, with a different σ and τ, of course, but the same $\sigma_x, \sigma_y, \tau_{xy}$.

From Equation (4), a particular value of α, that is, $\bar{\alpha}$, can be obtained which makes $\tau = 0$,

$$\tau_{xy}/(\sigma_x - \sigma_y) = \tfrac{1}{2} \tan 2\bar{\alpha} \quad (5)$$

New $\bar{x}, \bar{y}$ axes are defined along this direction, as shown in Figure 13.6c, then $\tau_{\bar{x}\bar{y}} = \tau_{\bar{y}\bar{x}} = 0$, and Equations (3) and (4) become

$$\sigma = \sigma_{\bar{x}} \cos^2\beta + \sigma_{\bar{y}} \sin^2\beta \quad (6)$$

$$\tau = -\left(\frac{\sigma_{\bar{x}} - \sigma_{\bar{y}}}{2}\right) \sin 2\beta \quad (7)$$

where β is now a general direction variable (angle) measured from the new axes and σ, τ are the stresses at angle β (at angle $\alpha = \beta + \bar{\alpha}$, see Figure 13.6d). These axes are called the directions of principal stress and $\sigma_{\bar{x}}$, $\sigma_{\bar{y}}$ are the *principal* stresses.

Eliminating β between Equations (6) and (7) gives the equation of a circle, so the relation between τ and σ at any angle β can be represented by the *Mohr* circle as shown in Figure 13.7. The maximum shear stress occurs in a direction of $\beta = 45°$ (= 135°) and

$$\tau_{max} = |\sigma_{\bar{y}} - \sigma_{\bar{x}}|/2 = \sqrt{\left(\frac{\sigma_x - \sigma_y}{2}\right)^2 + \tau_{xy}^2} \quad (8)$$

$$\sigma \text{ at } \tau_{max} = (\sigma_{\bar{x}} + \sigma_{\bar{y}})/2 \quad (9)$$

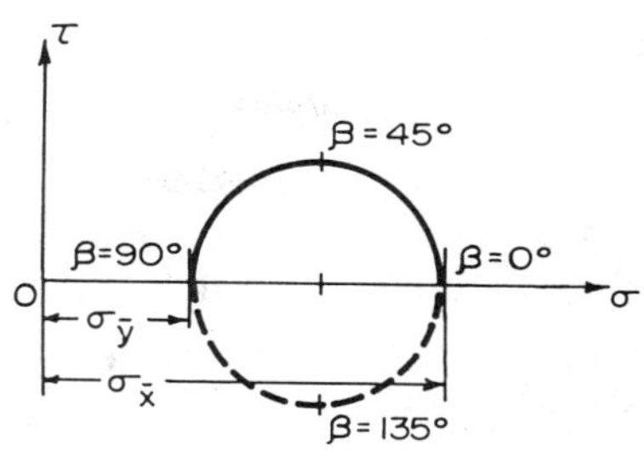

Figure 13.7. A Mohr stress circle for a planar system.

The maximum normal stress is clearly the larger of the principal stress values. Also, it is readily shown that the principal stresses are related to the normal stresses in the original coordinates by

$$\sigma_{\bar{x}}, \sigma_{\bar{y}} = \frac{\sigma_x + \sigma_y}{2} \pm \tau_{max} \quad (10)$$

Thus, knowing $\sigma_x, \sigma_y, \tau_{xy}$ at any point in the solid the direction and magnitudes of the maximum shear stress, tensile stress and compressive stress are readily calculated.

A similar treatment[1] in three dimensions, considering the six stress components, leads to Mohr circles for the three planes of principal stress as illustrated in Figure 13.8, where $\sigma_3, \sigma_2, \sigma_1$, are principal stresses ranked in order of magnitude. It is concluded that the maximum tensile stress has the magnitude and direction of the largest negative value of the three principal stresses and that the maximum shear stress occurs at 45° between the σ_1, σ_3 directions, with a magnitude given by Equation (8).

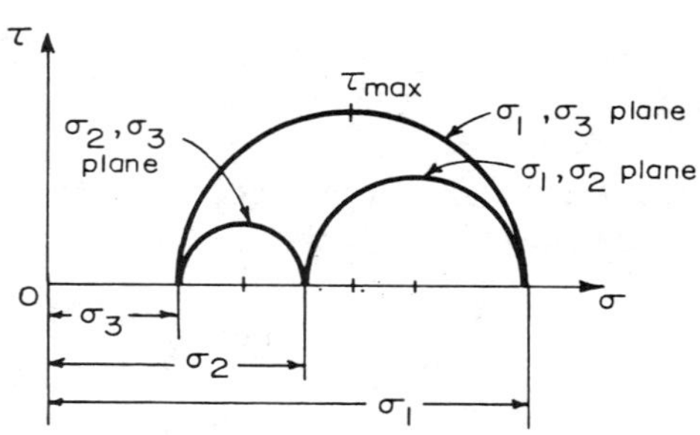

Figure 13.8. Mohr principal stress circles for a three-dimensional solid.

Differential Stress-Strain Equations

The second step is to find the values of $\sigma_x, \sigma_y, \tau_{xy}$ at all points in a solid, since these can be converted to maximum stresses and directions. For planar stress a differential force balance on a rectangular differential element at position x,y in the solid gives

$$0 = \frac{\partial \sigma_x}{\partial x} + \frac{\partial \tau_{yx}}{\partial y} + X \qquad (11)$$

$$0 = \frac{\partial \sigma_y}{\partial y} + \frac{\partial \tau_{xy}}{\partial x} + Y \qquad (12)$$

where X,Y are the body forces in the x and y direction at the point.

The differential strains at point x,y are defined by $\epsilon_x = \partial u/\partial x$, $\epsilon_y = \partial v/\partial y$ for the linear strains, where u is the change in x dimension from the non-stressed state at point x,y, v is change in y dimension. The differential planar shear strain γ_{xy} is illustrated in Figure 13.9 and is defined by γ_{xy} = angular deformation $\theta_1 + \theta_2$. Clearly, $\theta_1 = (\partial u/\partial y)\, dy/dy$ and $\gamma_{xy} = \gamma_{yx} = \partial u/\partial y + \partial v/\partial x$. The empirical physical laws relating stress and strain are Hooke's law, $\epsilon_x = \sigma_x/Y$, and the fact that a strain in the x direction causes a proportional dimensional change in the y direction (stretching in x gives a contraction in y, compression an expansion). Thus ϵ_y due to $\epsilon_x = -\nu\epsilon_x$, where ν is Poisson's ratio (≈ 0.25). For small elastic planar deformations the total strains are

$$\epsilon_x = \frac{\sigma_x}{Y} + \left(-\nu\frac{\sigma_y}{Y}\right) \qquad (13a)$$

$$\epsilon_y = \frac{\sigma_y}{Y} + \left(-\nu\frac{\sigma_x}{Y}\right) \qquad (13b)$$

Defining a modulus of rigidity $G = Y/2(1 + \nu)$, it can be shown from Hooke's law that

$$\gamma_{xy} = \tau_{xy}/G = \tau_{xy}\,\frac{2(1+\nu)}{Y} \qquad (14)$$

Using the definitions of strain

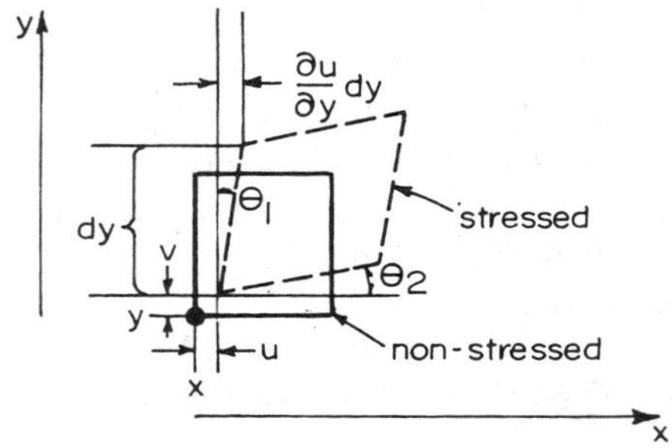

Figure 13.9. Illustration of differential strains at a point x,y in a planar solid.

$$\frac{\partial^2 \epsilon_x}{\partial y^2} + \frac{\partial^2 \epsilon_y}{\partial x^2} = \frac{\partial^2 \gamma_{xy}}{\partial x \partial y}$$

and from Equations (13) and (14)

$$\frac{\partial^2 \sigma_s}{\partial y^2} - \nu\frac{\partial^2 \sigma_y}{\partial y^2} + \frac{\partial^2 \sigma_y}{\partial x^2} - \nu\frac{\partial^2 \sigma_x}{\partial x^2}$$

$$= \frac{2(1+\nu)}{Y} = \frac{\partial^2 \tau_{xy}}{\partial x \partial y} \qquad (15)$$

If the body forces are known Equations (11), (12), and (15) are three simultaneous differential equations in the unknowns $\sigma_x, \sigma_y, \tau_{xy}$. They are solved using the stress and/or strain *boundary conditions*, that is, the stress-strains imposed on the solid from external action. For negligible X,Y the solution procedure is to define the Airy stress function $F(x,y)$ such that $\sigma_x = \partial^2F/\partial y^2$ and $\sigma_y = \partial^2F/\partial x^2$, for then $\tau_{xy} = -\partial^2F/\partial x\partial y$ and from Equations (13) and (15) $(\partial^4F/\partial y^4) + (\partial^4F/\partial x^4) + 2(\partial^4F/\partial x^2\partial y^2) = 0.$ Solving this equation with the transformed boundary conditions gives $F(x,y)$, and $\sigma_x, \sigma_y, \tau_{xy}$ follow by double differentiation. Equivalent but more complex equations exist for three dimensions.[1] The strain energy above the nonstressed state is calculated from

$$\frac{1}{2}\iiint_V (\sigma_x\epsilon_x + \sigma_y\epsilon_y + \sigma_z\epsilon_z + \tau_{xy}\gamma_{xy} + \tau_{xz}\gamma_{xz} + \tau_{yz}\gamma_{yz})\, dx\, dy\, dz \qquad (16)$$

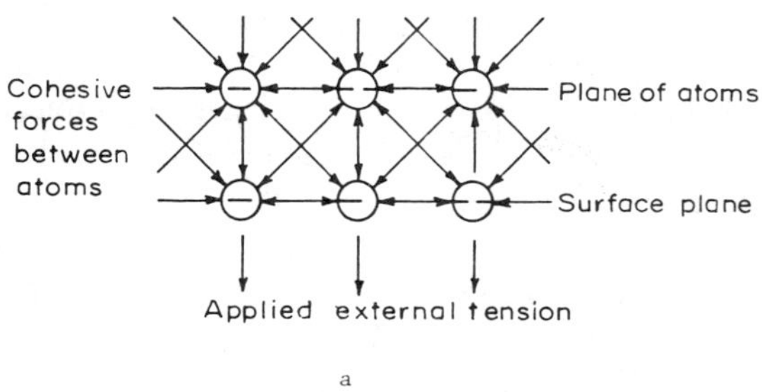

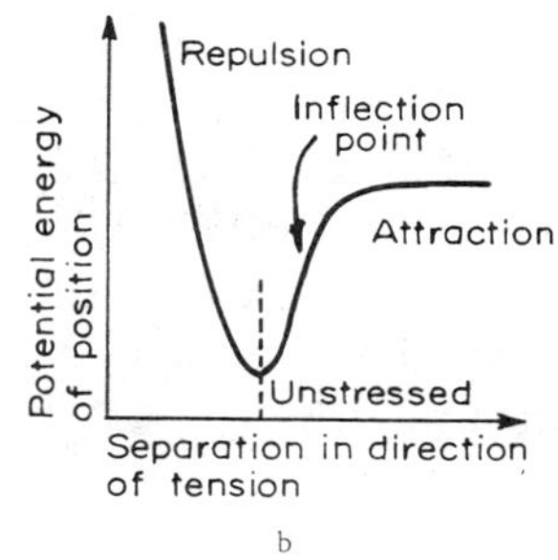

Figure 13.10. Illustration of forces between molecules in a solid: *a*, cohesive forces; *b*, energy of position.

Ideal Strength, Stress Concentration and the Griffith Crack Theory

The concept of *ideal strength* can be illustrated by considering an ideal solid made up of planes of molecules, subjected to simple one-dimensional tension. The tension stretches the bonds between the molecules, as illustrated in Figure 13.10, where the arrows indicate intermolecular attractive-repulsive forces. In the stretched state, any molecule still has a balance of forces on it but, as Figure 13.10b shows, the movement away from the nonstressed equilibrium against attractive forces requires addition of energy (integral of force x distance) and the solid reaches a new equilibrium at a higher energy state (stored strain energy). The maximum attractive force which the solid can exert on the surface layer is the inflection point of the potential energy curve since force = d(energy)/d(separation distance), and an external tension which exceeds this maximum causes an unbalance of forces and acceleration of one plane of molecules away from another. The solid would catastrophically disintegrate at all planes in the solid. Assuming Hooke's law to apply up to the inflection point, the strain energy per unit volume of solid is, from equation 2, $\sigma^2/2Y$. The area produced per unit volume is $2N$ where N is the number of planes per unit length; N equals $1/d$ where d is the interplanar spacing. Thus,

$$\sigma_{\text{ideal failure}} \approx \sqrt{\frac{4Y\gamma}{d}} \tag{17}$$

where γ is the surface energy defined as the work to create a unit area of surface from the unstressed solid. Equation (17) must underestimate the ideal strength since Hooke's law underestimates the force required to reach the inflection point. Since γ is known for simple solids, it is readily shown that the tensile force for real fracture is orders of magnitude *less* than ideal.

The concept of stress concentration or stress intensity factor can be illustrated by considering a planar solid containing a small hole, under a uniform externally applied tensile stress of S in the x direction and zero in the y direction. Without the hole, the solution is intuitively obvious as $\sigma_x = S, \sigma_y = 0, \tau_{xy} = 0$ for all values of x and y. With a small hole of radius a present (see Fig. 13.11), the added boundary condition is $\sigma_r(a, \theta) = 0$, $\tau_{r\theta}(a, \theta) = 0$ since there is no external stress inside the hole, and solution[2] is

$$\sigma_x(\text{r}, \theta) = \frac{S}{2}\left(1 + \frac{a^2}{r^2}\right) - \frac{S}{2}\left(1 + \frac{3a^4}{r^4}\right)\cos 2\theta \tag{18}$$

which gives a maximum stress of $3S$ in the x

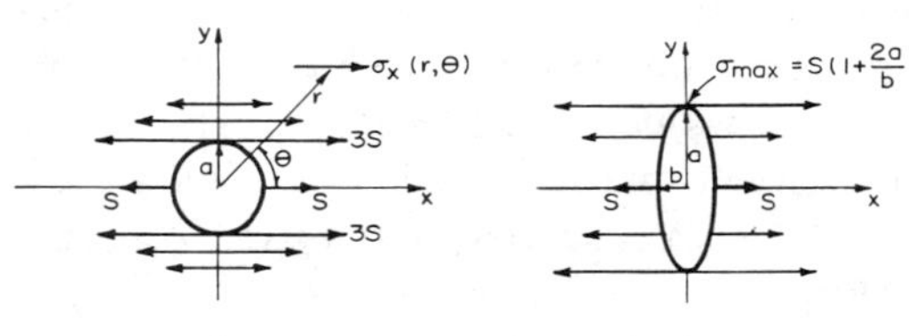

Figure 13.11. Illustration of stress concentration in a plane due to a circular or elliptical hole; s = applied tensile stress.

direction at $\theta = 90°$ and $270°$. Since a crack will open up under tension it is reasonable to expect that the solid will fail by cracks starting at the top and bottom of the hole and progressing in the $\pm y$ direction. The solution for a small elliptical hole is more complex but gives[1] a maximum stress of

$$\sigma_{max}/S = 1 + \frac{2a}{b} \quad (19)$$

where a is the ellipse axis in the y direction, b in the x direction. For an elliptical hole with its long axis perpendicular to the stress direction a is greater than b, and stress concentration can be very high if $a >> b$.

Griffith[2,3] argued that real solids contain many minute *flaws* corresponding to the three-dimensional equivalent of the elliptical holes discussed above and that these points of weakness initiate cracks at stress levels much below ideal. He made four basic assumptions: (i) that stress concentration occurs at the tip of the flaw (ii) that the solid is stressed to where the intermolecular bonds at the tip are stretched to breaking point (iii) that the stress state is reproduced at the tip for an infinitesinal expansion of the flaw (iv) that energy for expanding the flaw as a propagating crack is available because the solid cannot immediately relax from its externally applied stressed or strained state. The solution of the stress-strain equations for a long ellipse gives the extra strain energy due to the presence of the ellipse as $\Delta z \pi c^2 \sigma^2/Y$ where c is the long half-axis, that is, half the crack length, and Δz is the crack width. Thus, $dw_1/dc = \Delta z 2\pi c \sigma^2/Y$. A sudden, irreversible, change from c to $c + dc$ at the instant of fracture is like a loaded solid suddenly expanding dc at constant load, so that the work done is twice the (reversible) strain energy, $dw_3/dc = 2\Delta z 2\pi c\sigma^2/Y$. The energy necessary to break bonds is $4\gamma c \Delta z$ for a crack of half-length c, so $dw_2/dc = 4\gamma \Delta z$. Using the principle of virtual work, $dw_3 = dw_1 + dw_2$ at crack initiation and the critical tensile strength is

$$T_0 = -\sigma_c = \sqrt{\frac{2\gamma Y}{\pi c}} \quad (20)$$

Comparing Equation (20) with Equation (17), values of d are no more than a few angstroms, so a flaw with a half length of hundreds of angstroms can give orders of magnitude reduction in tensile strength T_0 from the ideal strength. As the crack progresses after initiation, $dw_3/dc > (dw_1/dc) + (dw_2/dc)$ and extra energy is available to accelerate the crack tip. The system is unstable and the crack rapidly expands, accelerating to high velocities. The strength is lower than ideal because the bulk stress does not have to be sufficient to break all the bonding forces at once, since only the bonds around the crack tip are breaking at any instance of time. In addition, Equation (20) is valid for a single flaw whereas the presence of many flaws close together will give further reductions in strength.

Obviously, pure compressive stress does not cause the flaw to open and will not cause crack propagation, so tensile stress is necessary for brittle failure. It might be thought that tensile stress will not exist under conditions of simple one-dimensional compression. However, a more detailed analysis considering all possible orientations of the flaws shows that tensile stresses are produced at the tip of an ellipse at a suitable orientation even under conditions of bulk compression. The result for a planar system with bulk normal stresses σ_1 and σ_2 and flaws of a size which would give a tensile strength of T_0 under one-dimensional tension (with the crack axis perpendicular to the stress) is shown in Figure 13.12. The compressive strength under one-dimensional compression is $8T_0$, that is, compressive strengths of brittle materials are about an order of magnitude higher than tensile strengths.

Under combined stress conditions the crack will propagate in a direction perpendicular to the local tensile stress conditions and may run into a region of compression that prevents further crack growth. Also, solutions of the stress-strain equations for simple compression of discs, cylinders in the "Brazilian" radial mode of testing, and spheres, show that tensile stresses are present, with maximum values along the loaded axis. Even for cubes and cylinders loaded along the axis, friction between the

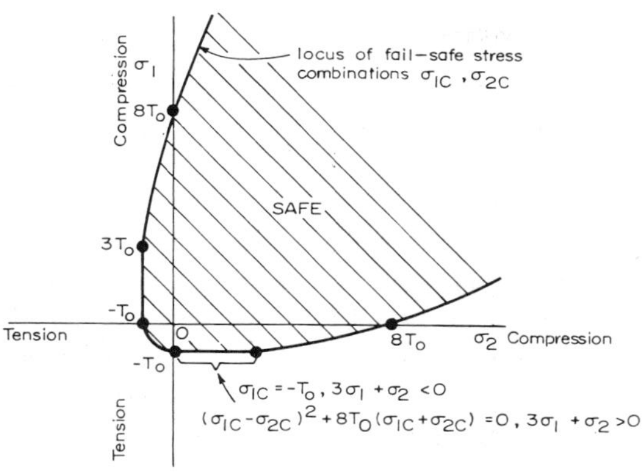

Figure 13.12. Illustration of effect of combined stress on failure from Griffith flaws with simple tensile strength of T_0: Equations are equations of locus.

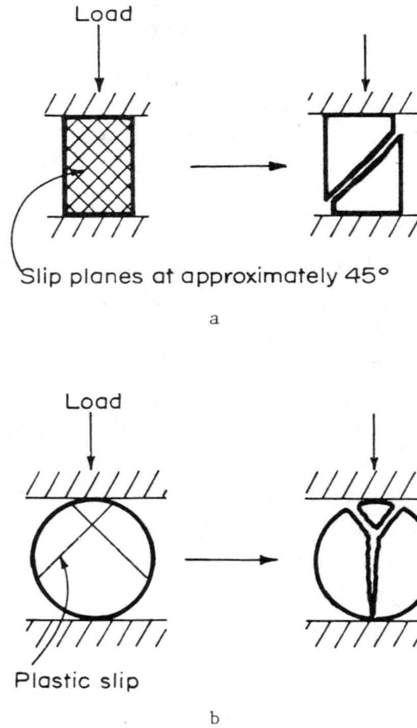

Figure 13.13. Illustration of failure by shear: Slip leads to brittle fracture.

loading platten and the sample leads to non-uniform compressive stress and regions of tensile stress. Thus compressive loading of irregularly shaped lumps or particles will certainly produce local regions of tensile stress and, hence, brittle fracture.

Ductile materials, on the other hand, undergo plastic deformation due to sliding of planes of solid over one another, with the fundamental mechanism being that of movement of dislocations under stress gradients. In this type of movement, the bonding forces between planes are not broken all at once, but only enough bonds are broken to allow the dislocation to move to the next position, the bonds reform behind the dislocation, and so on, thus leading to slip of one plane over another by series of low energy steps. We have already seen that the maximum shear force occurs at 45° to the direction of principal stress, so plasticity and failure by shear will appear as illustrated in Figure 13.13. The slip process appears as the region of yielding in Figure 13.1, and is quite unlike the unstable initiation of brittle failure. Slip may initiate from a suitably oriented flaw which gives stress concentration, but there is no opening of a crack comparable to that under tensile stress.

However, other factors come into play once plastic yield has commenced. The plastic slip may cause part of the solid to act as a wedge, thus creating tensile forces which then propagate brittle fracture, as illustrated in Figure 13.13. Also, the movement of dislocations can pile up dislocations at a grain boundary, thus leading to a small hole which can nucleate a Griffith crack. Highly ductile materials under simple one-dimensional tensile loading will neck down giving increased stress at the neck and, eventually, complete slip failure with possible cleavage along crystallographic planes of weakness.

The Griffith treatment is extended[4] to allow for plasticity by including a term, dw_4, for the energy required for plastic deformation caused by the moving stress field around the crack tip. Then the initiation condition is $dw_3 - dw_1 = dw_1 \geq dw_2 + dw_4$. The value of dw_4 depends on the size and density of dislocations in the solid and dominates over bond energy dw_2 for ductile materials. Thus, ductile materials are stronger than purely brittle materials. Once fracture commences, however, the term for plastic energy may decrease because the crack moves at high velocity in relation to the time scale for movement of the dislocations which give plasticity.

Some polymeric materials have the ability to deform to high strain without fracture, e.g., rubber, and the description of their fracture can be considered as a separate problem. They are difficult to break by compression and induced tension because it is difficult to apply enough strain to make the solid reach a highly stressed state, and because dw_4 is large. Rubber materials have this property because of the shape and flexibility of the long molecules, which can coil and uncoil, bend and straighten. A high degree of cross-linking bonds will reduce the flexibility, so these materials are weakly cross-linked, which means they are weak to shear stress. Thus, the best conditions for failure are tensile strain which straightens out the tangled and coiled molecules into a parallel array like a crystal, with superimposed shear which breaks the few cross-linking bonds.

Qualitative Applications of Fracture Theory: Grinding Energy

Rocks, ores, and coals being broken in size reduction machines will normally undergo brittle fracture via preexisting Griffith flaws. The strength or grindability of these materials will correlate only roughly with the hardness or chemical bond strength, because the number, size and orientations of the flaws are additional variables. The materials are stronger in compression than tension. In order to calculate the strength of a lump or particle being subjected to stress, from a priori theory of fracture mechanics, it would be necessary to: (a) solve the stress-strain equations for the geometry and conditions of applied stress; (b) convert the results to the *local* magnitude and direction of the principal stresses at all points in the solid; (c) consider the density (number per unit volume), distribution of sizes, and orientation (possibly random) of flaws in the solid; (d) determine the places where local *tensile* stress can activate the flaws to the point of fracture initiation, with failure commencing at the weakest location. Such a calculation is clearly impossibly complicated for most real conditions in a mill, and can only be attempted for idealized solids and simple stress conditions (see Eq. (23) for an example).

In addition, most grinding machines have some degree of impactive stress that propagates stress waves through the solid, activating flaws to tensile fracture in the process. The size distribution of the suite of fragments produced on fracture is as important as the fracture itself (see later), and there exists no known theory for its prediction. Theory predicts, and experiment confirms, that a fracture propagating under local tensile stress rapidly reaches high velocity (unless it reaches a zone of local compressive stress), of the order of the velocity of sound in the solid. This leads to a stress wave that propagates from the crack tip and this stress wave in turn initiates more fracture at flaws in the path of the crack. This leads to bifurcation of the crack, with bifurcation of each of the new arms, and so on, to give a "tree" of cracks through the solid (see Fig. 13.14). The energy associated with the rapidly moving stress wave is normally sufficient to pass the crack through grain boundaries and through regions of bulk compressive stress.

Ductile materials fail by initial shear, and it is again necessary to find the magnitude and direction of shear at all points through the solid. The Mohr-Coulomb criterion is that failure occurs when shear stress reaches the yield point given by

$$\tau_{y.p.} = \tau_0 + \mu\sigma \tag{21}$$

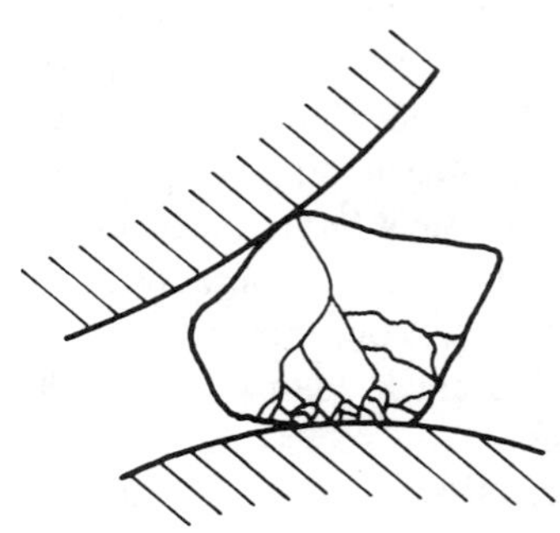

Figure 13.14. Tree of cracks in brittle failure.

where τ_0 is the yield shear stress under conditions of zero tensile or compressive stress perpendicular to the shear stress plane and μ is called the *coefficient of internal friction.* Equation (21) states that a high compressive stress perpendicular to the shear plane will tend to prevent slip thus requiring a higher τ to give yield, and vice versa. The slip surface is now along the direction of $\tau\text{-}\mu\sigma \geq \tau_0$. The value of μ is normally small so that the tensile strength is fairly close to the compressive strength, and slip surfaces tend to lie fairly close to 45° to the principal stress directions. From Equations (6) and (7) it is readily shown that

$$\mu = \frac{C_0 - T_0}{2\sqrt{C_0 T_0}} \tag{22}$$

where C_0, T_0 are the magnitudes of simple one-dimensional compressive and tensile stresses required to give yield. It will be remembered that the maximum shear stress for principal stresses of $\sigma_{\bar{x}}$, $\sigma_{\bar{y}}$ in two dimensions is $|\sigma_{\bar{y}} - \sigma_{\bar{x}}|/2$, so slip is aided by a combination of compressive and tensile stresses.

A comparison between the failure of brittle and ductile materials shows the following major features.

1. Pure brittle failure is almost independent of temperature, but as temperature increases to where dislocations are more mobile the failure may change to slip, and, hence, lower strengths. Pure ductile failure gives decrease of strength with increase of temperature due to greater mobility of dislocations. For brittle failure with a significant plastic energy term, strength increases with temperature due to the increase of the plastic zone around the tip, then decreases as failure changes to slip.
2. For failure from Griffith cracks, a smaller particle has a smaller probability of containing a large flaw and will be relatively stronger. Put another way, as brittle materials break the remaining fragments are stronger because the larger flaws have broken out. On the other hand, failure by yield is not very size sensitive because the dislocations are very small compared to lumps or particle sizes.
3. The rate of stress application is more important with ductile materials than with purely brittle materials, because a high rate of stress application may give brittle failure whereas the same stress reached by slow steps would give time for ductile behavior.
4. Ductile materials demonstrate work hardening, that is, initial deformation produces movement and pile-up of dislocations and further deformation is more difficult. They also demonstrate stress fatigue, again due to the gradual accumulation of dislocations on repeated cycles of stress.
5. Loading of brittle materials with uniform triaxial compressive stress, hydrostatically for example, leads to greatly increased strength by reducing local tensile forces and preventing cracks from opening.

In the case of tough, rubbery materials, the best stress application for size reduction is the scissors type of action, that is, a cutting action. This has three main features: (1) a large component of shear stress, (2) a high strain and stress caused by two forces applied in opposite directions by the blades (or stator and rotor), (3) the creation of a surface flaw by the very high local stress of a sharp blade penetrating the material. These features are illustrated in Figure 13.15. For rubbery polymers with a substantial degree of cross linking, which gives high shear strength, cooling the material to a low temperature can convert it to a brittle

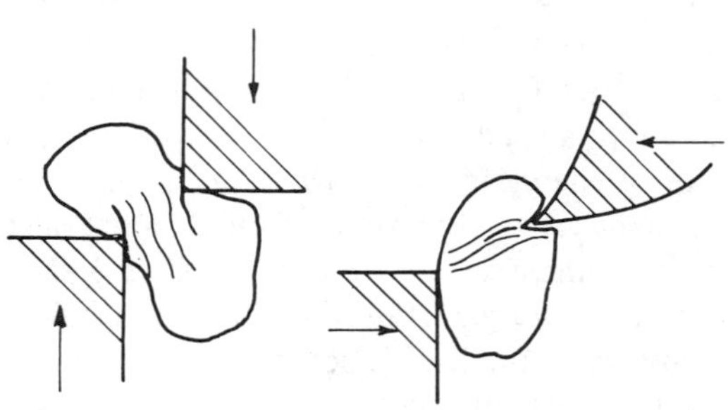

Figure 13.15. Illustration of shear-cutting actions.

material, which can then be broken like other brittle materials. The action of the cooling is to reduce the flexibility (ability to rotate and bend) of the bonds joining the groups making up the polymer chains; it is normally necessary to cool to very low temperatures, using liquid nitrogen (77°K).

There has been a great deal of misconception in the grinding literature concerning grinding energy. The previous discussions show that a strong solid must be raised to a higher state of stress for fracture to proceed, especially from applied compressive forces. Once the fracture has initiated, only a fraction of the *local* stored strain energy around the propagating cracks is used to break bonds (the γ term). The fragments of solid are removed from external stress when the solid disintegrates, and the rest of the strain energy stored in the solid is converted to heat and sound. Experiments on mills show that the fraction of the electric power input to the mill which is used directly to break bonding forces is very small ($<1\%$), usually less than the errors involved in the measurement of the energy balance. Rittingers law,[5] that the "energy of size reduction is proportional to the new surface produced," has no correct theoretical base.

In order to make size reduction more energy efficient it is necessary to: (1) match the machine to the particles being broken, so that mill energy is efficiently transferred to stressing the particles; (2) get nonuniform stress conditions in the particles, because nonuniform stress generates local tensile stress to activate flaws to the point where fracture can initiate; (3) generate the right type of stress to match the failure characteristics of the material. The specific energy consumption per unit of area produced, e.g., joules/m^2, can be used as a comparative guide to efficiency, because a higher value is certainly an index of more size reduction per unit of energy input. It will not necessarily be constant for a given machine and material since it may increase or decrease with a greater degree of size reduction. On the other hand, in many cases, the production of extra fine material is undesirable, and then the specific surface area of the product is obviously not a good guide to mill efficiency, because the specific surface area is contributed largely by the extra fine sizes.

Property Changes and Reactions

It is known that prolonged treatment of materials with repeated stress, by batch ball milling for long periods, for example, can cause massive changes in the properties of the materials. Rose[6] showed that quartz underwent phase change from one form to another during ball milling and the topic has been reviewed recently,[7] giving many examples. It has been suggested that shear stress will cause nucleation and growth of one phase from crystallites of another in a particle. In ball milling, tough organic polymers can undergo a delay period in which they hardly break at all, followed by breakage. Presumably the pounding by the balls makes the material weaker by causing some degree of crystallization (molecular alignment). It is known that repeated light taps on a friable coal create or extend cracks in the coals, so that it eventually fails. Coal is a brittle polymer with planes of weakness due to the geological process of laying down the material, but presumably other materials could show the same effect.

Benjamin[8] has discussed the formation of solid solutions of ductile metals by prolonged ball milling of a mixture of powders of the components, and the similar creation of a fine dispersion of a brittle material in a ductile matrix. The mechanism appears to be cold-welding of clean surfaces produced by fracture or flattening, so that size reduction and size growth occur simultaneously. In this case, the mill action must be such as to force particles together as well as fracture particles. It is known that organometallic compounds can be formed by ball milling chromium and nickel in organic liquids, accompanied by rearrangement of the organic molecule to other organic molecules with some H_2, CH_4, and CO_2 evolution. Similarly, reactions such as $Cr(s) + 3TiCl_4(l) \rightarrow CrCl_3(s) + 3TiCl_3(l)$ occur in anhydrous liquids. Again, the cause is undoubtedly the high reactivity of freshly fractured clean surfaces.

Abrasion

Abrasion is a special type of fracture—the tearing out of small pieces of material from the surfaces of the components used to apply stress to the material being fractured. It is obvious, of course, that these grinding components must be strong enough to stress the material being comminuted without bulk fracture themselves, but this is no guarantee that their surface will necessarily be abrasion resistant. The fracture mechanics of abrasion is not well developed theoretically, but it certainly involves high local surface stresses due to asperities in the rock and in the grinding surface, plus local surface microflaw structure, ductility, friction and possibly high local temperatures. High rates of surface stressing caused by high relative speed between stressing and stressed agents undoubtedly assist abrasive fracture.

The chemical environment at the surface can play a significant role, by two mechanisms. The first and most obvious is that an environment that attacks the grinding surfaces will cause surface flaws and weakness and accelerate abrasion. This effect is well recognized in wet grinding. Second, there is some evidence[9,10] that an environment can change the bond strength and ductility of material close to the surface, by strong chemical adsorption onto the surface. Such an effect will not change the bulk strength (unless conditions are such that fracture commences at surface flaws) but it can change the abrasive comminution.

The terms "hard" and "soft" are often indiscriminately used to characterize both the bulk strength or resistance to comminution of a material and its ability to penetrate or wear another material. It would be better to use the terms "strong" and "weak" for bulk comminution properties, and reserve "hard" or "soft" for the characteristics measured by one of the usual hardness tests such as the Rockwell or Vickers tests. For example, coal can be considered to be a weak rock, but certain coals

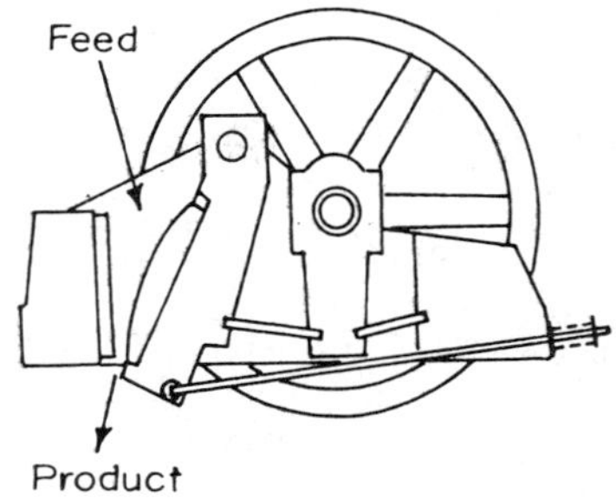

Figure 13.16. One type of jaw crusher.

are abrasive due to inclusions of quartz. Again, a tough plastic such as Teflon is not hard or abrasive but it can be very difficult to comminute.

13.3. SIZE REDUCTION MACHINES

There are many different types of machines for size reduction, and almost every method of breaking lumps which one might think of has been incorporated into a crusher or grinder. Figures 13.16, 13.17, and 13.18 show common types of industrial crushers, which are available in a wide range of sizes. The reciprocating action of the movable jaw in a jaw crusher strains solid to the point of fracture, as does the nonsymmetric movement of the rotating mantle in a cone or gyratory crusher. The size reduction ratio, defined approximately as the largest feed size divided by the largest product size, is of the order of 10 and is varied by the

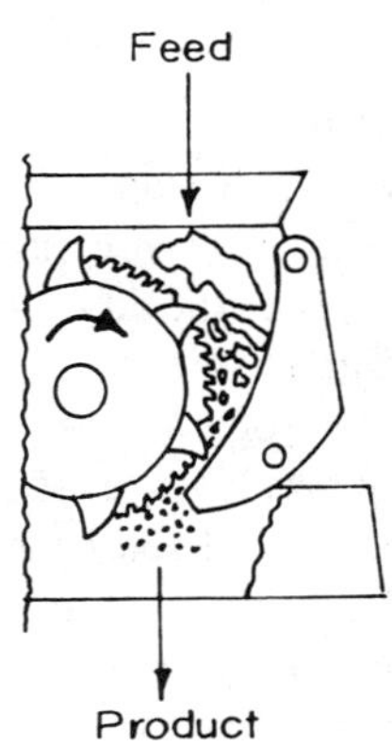

Figure 13.17. Toothed single-roll crusher.

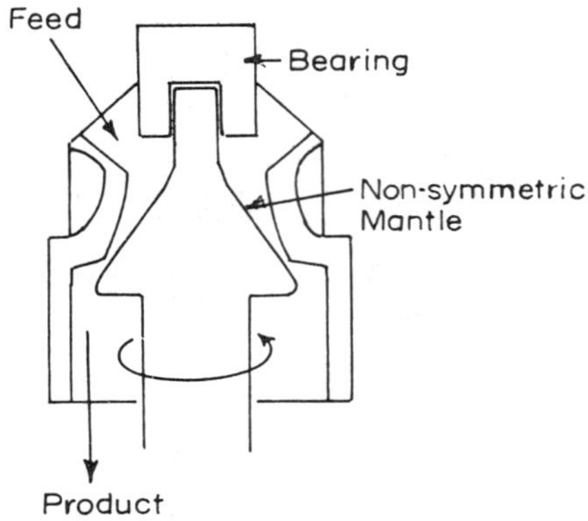

Figure 13.18. Gyratory crusher.

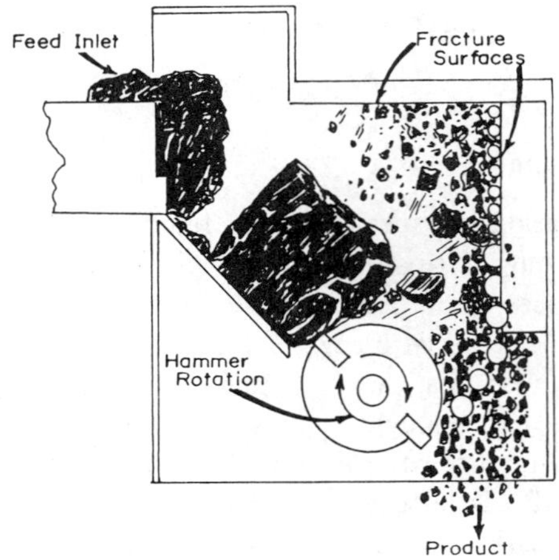

Figure 13.19. Low-speed hammer crusher (Jeffrey Coal buster).

adjustable gap setting. Feed or product fragments less than the gap setting pass rapidly out of the breakage zone and cannot be overground, so these devices can be referred to as *once-through* machines, although there may be several consecutive stages of breakage of larger lumps.

The machines are applying nonuniform compressive stress and the mill power must be sufficient to compress all the large pieces of rock to fracture point when the crusher is full of large lumps. The stronger the rock, the larger the power required. Solution of the equilibrium stress-strain equations for ideal diametral loading of spheres of brittle material gives the relation between force, diameter and maximum tensile stress as

$$\sigma_x = \sigma_y = -\frac{P}{\pi r^2}\frac{21}{(28+20\nu)} \propto -\frac{P}{r^2} \qquad (33)$$

where P is the force, r the sphere radius, ν Poissons' ratio, and tensile stress is negative. Thus, the force to give a comparable tensile stress for fracture is roughly proportional to r^2. The crushing surfaces of jaw and roll crushers are ribbed or toothed (a) to help prevent slippage of the rock as it is compressed thus ensuring additional shear stress and (b) to give higher local stress at the surface of the material, thus activating or even producing local flaws. The turning movement of the gyratory crusher also introduces shear stress.

Figure 13.19 illustrates a low-speed hammer crusher. Material is broken by direct impact of the hammers, by being thrown against the case or breaker bars, and by compression and shear when nipped between the hammers and the case. This type of crusher is best suited for nonabrasive materials, although it is sometimes used for fairly abrasive rock because of its low capital cost; the user must then resurface the hammers at frequent intervals. Figure 13.20 shows a Cage-Pactor crusher, in which solid is crushed as it passes between rows of rotating bars. Again, this is best suited for relatively weak non-abrasive material such as coal.

Figures 13.21, 13.22, and 13.23 show machines suitable for intermediate size reduction. Figure 13.21 shows a medium speed hammer mill, which is especially suitable for sticky or tough materials which can not be efficiently broken in rolls or rod mills, because of its ability to shear in addition to compress. The smooth roll crusher is widely used for size reduction in the laboratory, but it also has industrial application in preparing material with a top size of, say, 12 mesh (1.70 mm) and a minimum of fines (see Section 4). Because of abrasive wear the rolls have to be resurfaced at frequent intervals. The rod mill acts somewhat like a multiple set of rolls—as the cylinder rotates, the bed of rods is carried up until it lies at an angle to the horizontal. It is then unstable and rods start to roll down the bed

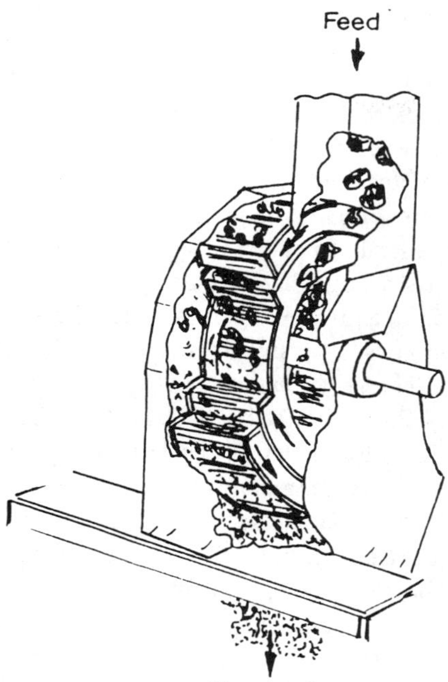

Figure 13.20. Cage mill (T. G. Grundlach Machine Company Cage-pactor).

surface, re-enter the bed, and get carried up again. The rods rolling over one another act like sets of rolls, stressing particles in a similar manner. The power to the mill is used to lift rods against gravity, the resulting potential energy of position is converted to kinetic

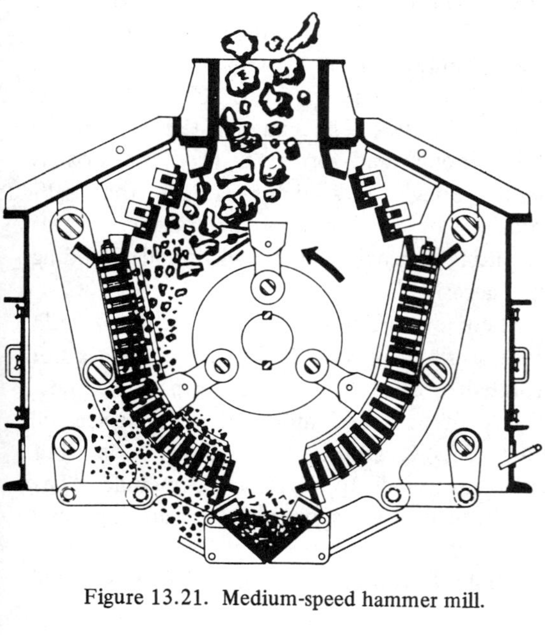

Figure 13.21. Medium-speed hammer mill.

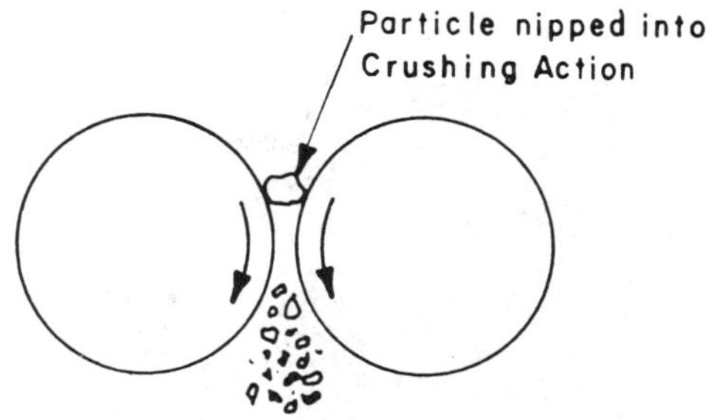

Figure 13.22. Smooth roll crusher.

energy as the rods fall, which in turn is converted to strain energy and, finally, to heat and sound.

However, there are two major differences between smooth roll crushers and rod mills. First, the rod mill is a *retention* device because fine fragments have to pass along the mill to overflow at the exit and can be rebroken again and again, so the mill is full of powder being acted upon. In this type of device, the *residence time distribution* (see below) of material in the mill is of great importance, and excess fines are produced. Second, there is obviously no controlled gap setting or controlled power to the turning rod, so it is not always possible to break large strong lumps, which can then leave in the overflow. The force available for fracture is increased by making the steel rods heavier (larger diameter) and the mill diameter larger, but this is limited by excessive damage

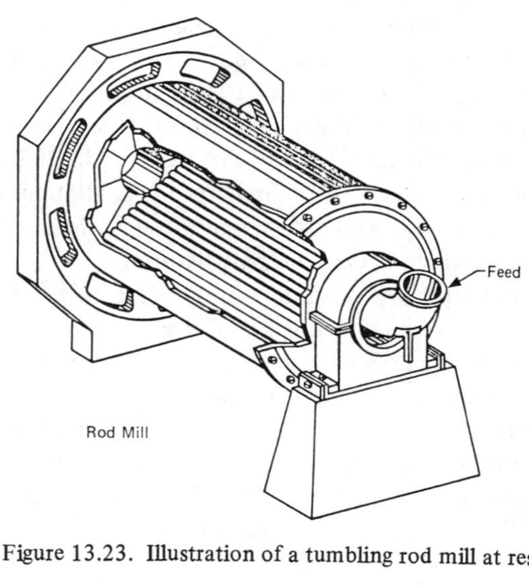

Figure 13.23. Illustration of a tumbling rod mill at rest.

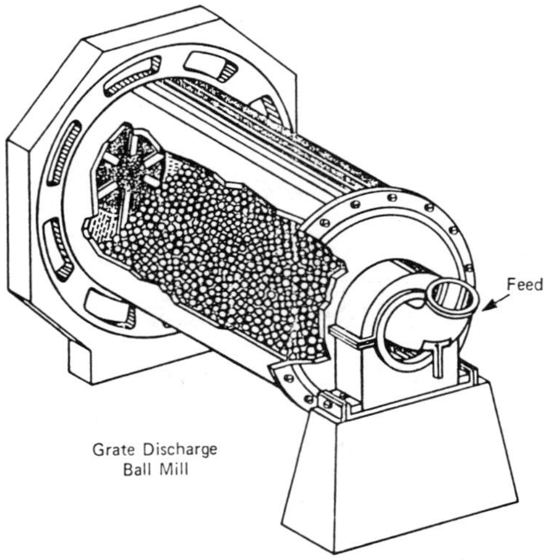

Figure 13.24. Illustration of a tumbling ball mill at rest.

to the mill lining by the falling rods. Thus, the feed to a rod mill is normally less than about 25 mm in top size, depending on material strength. It is normally used for wet grinding and abrasive wear on the rods means that worn down rods must be removed and replaced with fresh rods at suitable intervals.

Figure 13.24 shows the tumbling ball mill, which is very widely used for dry and wet grinding to relatively fine sizes. The principle is identical to that of the rod mill, but the maximum force available to break large strong lumps is even less, so the feed to the mill is rarely larger than 10 mm for strong rock. Because of its great industrial importance this type of mill has been widely investigated, and is discussed in detail below. Abrasive wear is easily handled by topping up the charge with fresh balls at frequent intervals and it is not even necessary to stop the mill to add the balls. The mill shown has an overflow discharge for continuous wet grinding, while discharge through slots or grates which retain the balls is often used for continuous dry grinding. For grinding coal, the mill is swept with hot air to dry the coal and the fine coal removed with the exit air stream.

Autogenous tumbling mills are similar in principle to the tumbling ball mill, but use the material being broken as the breakage media. There are four major types. The first is essentially identical in construction to a ball mill, but the feed consists of two streams, a narrow size range of lumps of rock (e.g., 75 mm by 150 mm) and the normal fine crushed feed. The large rocks wear to round pebbles (hence, the name pebble mills) on tumbling and then act like steel balls on the rest of the feed. The feed rate of large rock is adjusted to keep a suitable load of pebbles in the mill. The second type has a large diameter-to-length ratio (typically 2 : 1) and takes a natural crushed feed containing rock typically up to 200 to 300 mm, with discharge through slots of typically 20 mm width. Since the feed rate has to be in balance with the rate at which the large lumps break themselves to less than 20 mm by their own tumbling action, it is not possible to vary the product size distribution over a wide range. In fact, the third type, semiautogenous mills, are identical but add some charge of large (4 in.) steel balls (typically a few percent of the mill volume) to increase output capacity. The Scandinavian countries use a variant of the second type with a smaller diameter-to-length ratio (typically 0.5 to 1), which behave more like pebble mills.

Autogenous mills have lower capacity for a given mill volume than a ball mill and, hence, higher capital cost per unit of output, but they do not have the continuing cost of replacement steel balls. The fourth type of autogenous mill, the rotary breaker, is specific for coarse size reduction of coal. It has the added feature that the cylinder case is lightweight and contains many holes (typically 50 to 300 mm), so that material broken less than the desired top size falls through and forms the product. Coal is light enough and friable enough that self-breakage by tumbling gives high output without requiring a heavy shell to withstand pounding and abrasion.

There are two other variants of the ball mill. In the vibrating ball mill the cylinder is not rotated to cause tumbling but is packed almost full with balls and mounted on an eccentric that jerks it around the cylinder axis, thus causing the balls to vibrate in the cylinder. The mechanical stresses on the drive are high and the mill is not conveniently scaled to high continuous capacity. A small ball mill fitted into a shaking mechanism is similar in principle and very useful for preparing laboratory samples of fine powders. The planetary or centrifugal mill[11,12] contains two or more rotating cylinders partially filled with balls, mounted at the periphery (parallel to the axis) of a bigger cylinder or frame that is also rotated. The respective speeds of rotation are set by gears to use the centrifugal force of the outer rotation to throw the balls across their cylinders as they rotate, thus replacing gravitational fall with much higher centrifugal force and also greatly increasing the number of balls moved per unit volume per unit time. A recently developed mill[12] accomplishes the same purpose with a single horizontal mill shell mounted on an eccentric (with counter-balance weights), with the radius of gyration chosen to produce the effect of a centrifugal field moving around the mill with each gyration. This gives a high-force tumbling action of the ball charge but avoids the massive hammering of the charge against the case produced by the vibrating ball mill. The capacity per unit volume of the mill is very high and it is suitable for underground treatment of ores in mining tunnels, thus saving mill house construction costs. Abrasive wear is high and the mill is designed for rapid replacement of a removable lining in the mill.

Figure 13.25 gives an example of the class of mills known as spindle mills or roller mills. The rotating table throws material through the roll-race and the pulverized material passes over the rim and is swept up by an air stream flowing through the annulus between the rim and the case. The stream passes to a classifier which returns oversize to the table, so that the rollers are acting upon a fairly thick bed of material. These mills are widely used for coal grinding, again with hot air drying and conveying, and for raw material grinding for cement manufacture. They give a product size distribution of about 80% $<75\ \mu m$, from feeds with a top size of about 25 to 75 mm. The rollers are hydraulically loaded to give very high compressive forces and the mills work well only for brittle feeds of a natural size distribution. If the feed consists entirely of large lumps, the mill runs roughly and gives high abrasion of

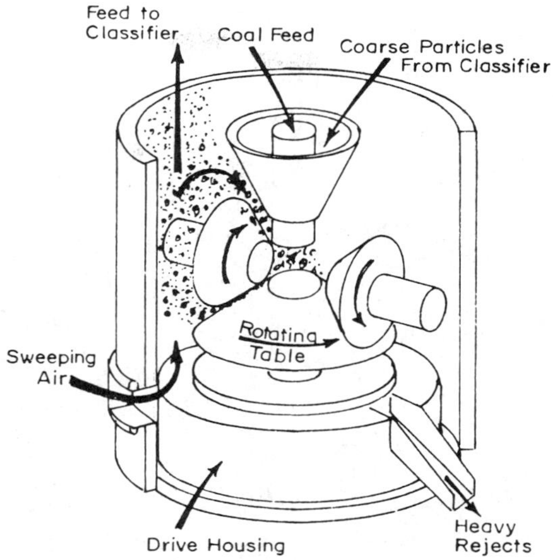

Figure 13.25. Illustration of roller mill.

the rollers and bed plates. If the feed contains only finer material, mill operation is unstable with gouging, chatter and slip of the rollers over the bed of powder. However, with a proper size range of feed they are efficient mills with a lower energy consumption per ton of product than other mill types.

High-speed hammer mills are also used for relatively fine dry grinding of nonabrasive materials like coal, with much of the grinding action being by shear between the hammer tips and the case. Figure 13.26 illustrates the disc mill, which also consists of surfaces rotating at high speed with respect to one another, but with the gap between the discs readily adjustable during operation. The force application is mainly by shear although particles will be compressed somewhat as they move into the narrower portions of the gap.

Figure 13.27 shows a sand mill or Attritor, which consists of paddles turning in a bed of water and sand or small steel balls. The large number of grinding particles give many breakage actions per unit time but the breakage action is mild, and the mill is used for comminution or deagglomeration of small relatively weak particles or agglomerates, such as dye-stuffs or crystals that can be removed with the water. A similar principle is used in the high energy ball

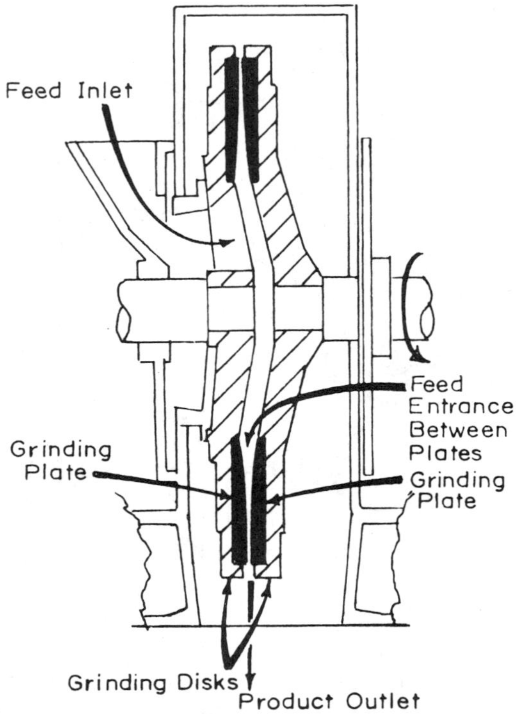

Figure 13.26. Disc mill.

mill,[8] with larger balls and high paddle speeds. These are used in relatively small scale for preparing mechanical alloys by dry grinding of metals. Shear mills consist of liquid flowing in a narrow annulus between a rotating drum and a stationary cylinder, with breakage caused by the high fluid shear forces across the annulus. Again, they are suitable only for small weak particles or agglomerates.

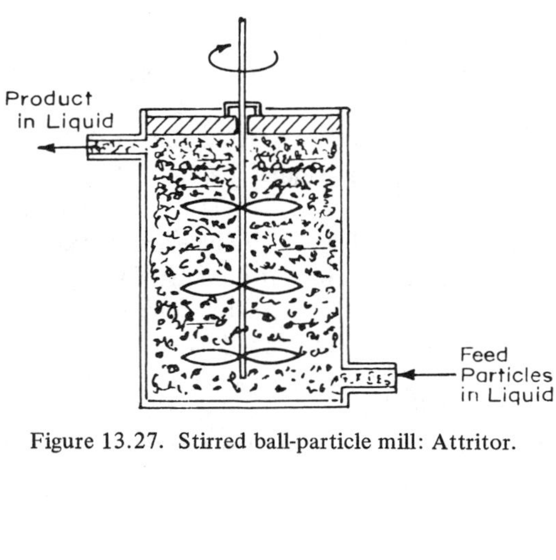

Figure 13.27. Stirred ball-particle mill: Attritor.

Figures 13.28 and 13.29 show types of fluid energy mill, in which small particles are suspended in high velocity streams of air or steam obtained by expansion from pressures of five to ten atmospheres. In the device illustrated in Figure 13.28, the tangential entry of high velocity fluid creates a doughnut of swirling particles and fluid in the grinding chamber, which retains coarser particles by centrifugal action. The microturbulence of the gas stream causes high-speed impact of particle-on-particle, and the centrifugal size classification allows only fine sizes to leave the breakage zones. In Figure 13.29, the opposed jets cause high-speed collision of the particles, and a size classifier and fan system in the device returns larger sizes into the jet stream. The mills are designed to give fluid boundary layers on the containing surfaces, to reduce particle impact on the surfaces and the consequent abrasion. The specific energy consumption calculated from the energy

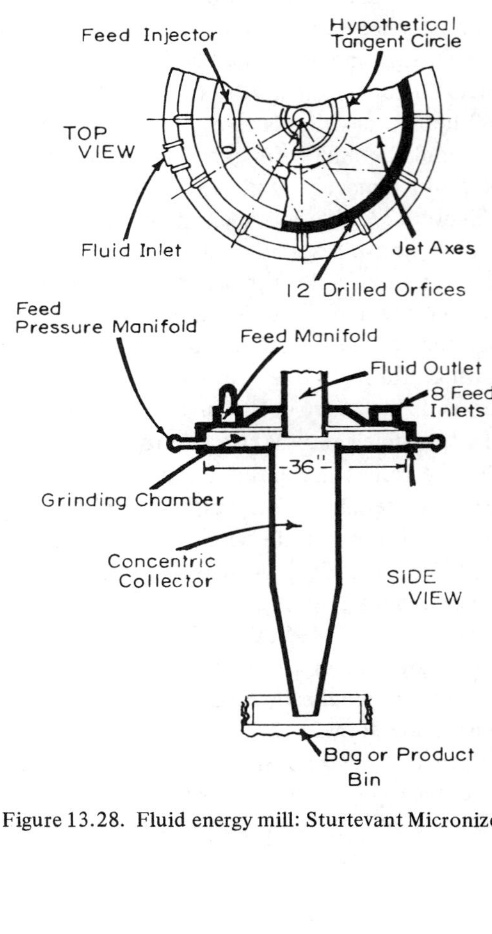

Figure 13.28. Fluid energy mill: Sturtevant Micronizer.

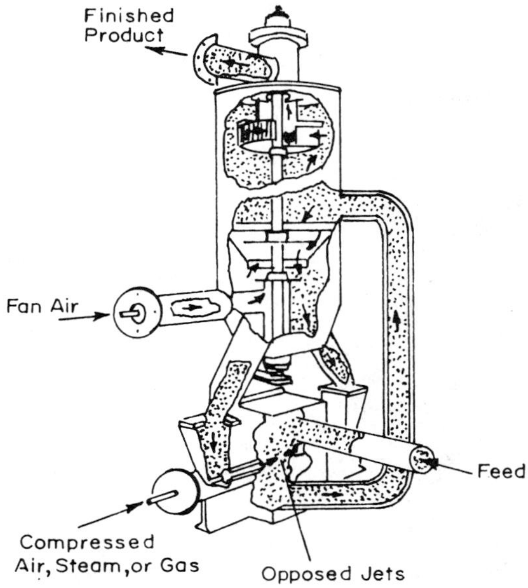

Figure 13.29. Fluid energy mill: Majac Jet Pulverizer.

required for air compression or steam-raising is high compared to mechanical grinders, but the mills are capable of producing very fine material (e.g., minus 5 μm) and are used primarily for speciality grinding of high-value materials or where cheap waste steam is available.

Figure 13.30 illustrates a whole class of mills designed specifically for size reduction of tough but non-abrasive materials such as PVC, Teflon, wood, etc. They rely on the cutting action, like a scissor action, between rotating and static sharp edges with narrow clearance. The efficiency of this type of mill is highly dependent on maintaining sharp cutting edges.

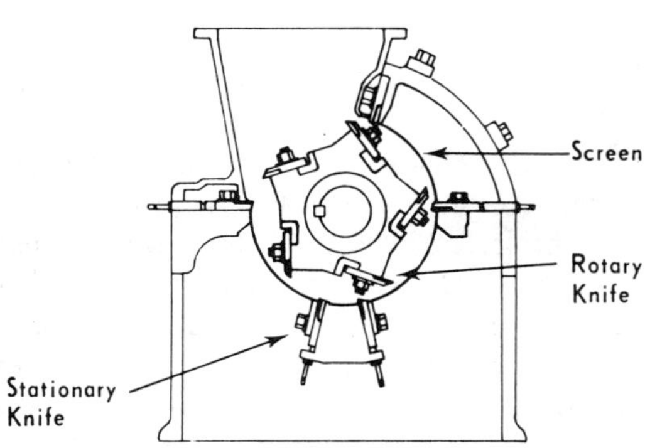

Figure 13.30. Illustration of rotary knife cutter mill.

13.4. THE ANALYSIS OF SIZE REDUCTION PROCESSES

General Concepts

It is clear from the previous section that the multiplicity of mill types and breakage actions make it virtually impossible to formulate a general theory of the unit operation of size reduction. In most cases good mill design has evolved by trial-and-error starting from common-sense applications of the concepts of fracture. However, for devices which reduce large tonnages of material, using substantial electrical energy, there is considerable impetus for accurate process design rules and for techniques for optimization of the system. As in other unit operations, it is invaluable to construct mathematical models of the operation to aid in its understanding and optimization. In the last decades, considerable advances have been made in this respect using concepts very similar to those of chemical reactor theory.[13,14] The mill is considered equivalent to a reactor which accepts feed components (the set of feed sizes) and converts them to products (the set of product sizes), and a size-breakage rate balance is performed on the reactor.

The rate at which a material breaks in a mill depends on its particle size as well as its strength characteristics. Normally, for any given mechanical action there will be particle sizes that are too big for efficient breakage because the action is not powerful enough, and particles that are too small for efficient breakage because the statistics of applying the action are not favorable. It is also apparent that the specific energy (kWh/ton) used for size reduction increases for breakage of finer and finer sizes because (a) it becomes more and more difficult to apply stress efficiently to millions of tiny particles and (b) the basic strength of brittle particles increases because large flaws (which can be stress-activated to fracture at low stress) become broken out as grinding proceeds to finer sizes. It is necessary, then, to analyze the breakage of each size range. It has been found convenient to use a $\sqrt{2}$ screen sequence to define the size ranges (e.g., "size" is defined as

16 × 20 mesh, 20 × 30 mesh, etc.) because material in one of these size intervals appears to behave like a uniform material, to a sufficient approximation. Since a geometric progression never reaches zero it is necessary to define a "sink" interval containing all material less than the smallest size measured: Thus, a feed size range can be split into n intervals, numbered 1 for the top size interval to n for the sink interval.

Using this basis, the size distribution from breaking a given "size" in one pass through the device is called the *daughter fragment distribution*, and is conveniently represented in the cumulative form "$D_{i,j}$ = weight fraction less than size x_i from breakage of larger size j," where x_i is the top size of interval i. Obviously, $D_{j,j} = 1$ and $1 - D_{j+1,j}$ is the fraction of size j which remains of size j after passing; the fraction of size j transferred to size i is $d_{i,j} = D_{i,j} - D_{i+1,j}$. The set of numbers $d_{i,j}$ is called the *transfer* matrix. For a once-through machine like a roll crusher these values can be determined experimentally by crushing each size independently.

For a retention machine like a ball mill, it is extremely valuable to define a *primary daughter fragment distribution*, $B_{i,j}$, which is the mean set of daughter fragments produced from one breakage action, with the products then mixed back into the mill contents to wait to be selected for a second breakage, and so on. It has been found that the form of the primary B values is similar to those of Figure 13.31 for many brittle materials and machine types: it is not difficult to see that this form is compatible with the tree of cracks illustrated in Figure 13.14. The slope γ of the finer end of the B plot is characteristic of the material and appears to be the same for all breaking sizes. In many cases, the B values are size normalizable, that is, the curves of Figure 13.31 fall on top of one another for different breaking sizes. Thus, the "weight fraction less than a given fraction of the breaking size" is constant and $B_{i,j} = B_{i+1,j+1} = B_{i+2,j+2}$, etc.

For retention grinders, the concept of *specific rate of breakage* S_i is applicable. Consider a mass W of powder in the mill, of which a

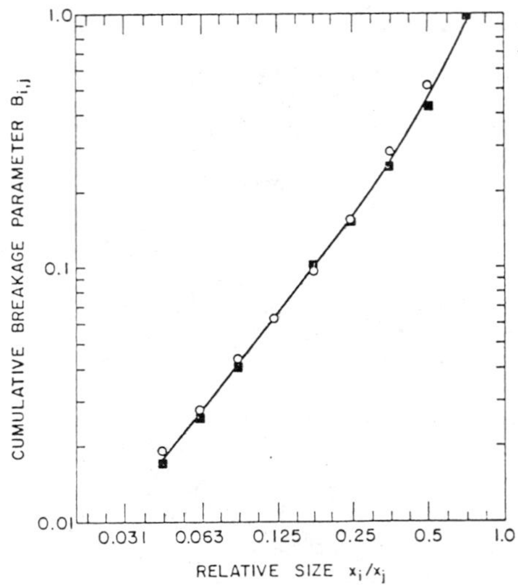

Figure 13.31. Typical cumulative primary daughter fragment distribution: ball milling of 20 × 30 mesh quartz: ■ dry, ○ wet.

weight fraction w_j is of size j. The specific rate of breakage S is defined by:

Rate of breakage of size j to smaller sizes

$$= S_j w_j W \quad (24)$$

It has units of time^{-1} and is comparable to a first-order rate constant in chemical kinetics. A batch grinding test on a feed of size j is comparable to an homogenous first-order "rate-of-reaction" experiment, and if S is constant Equation (24) goes to

$$d(w_j W)/dt = -S_j w_j W$$

and

$$w_j(t) = w_j(0) \exp(-S_j t) \quad (25)$$

Figure 13.32 shows a typical result. This first-order relation is observed so frequently that it can be called "normal" breakage, whereas nonfirst order kinetics indicate some "abnormal" feature. Methods of estimating S and B values from experimental tests have been described,[15,16,17] but they are primarily useful for laboratory or pilot-scale test data, and it is at present frequently necessary to infer values

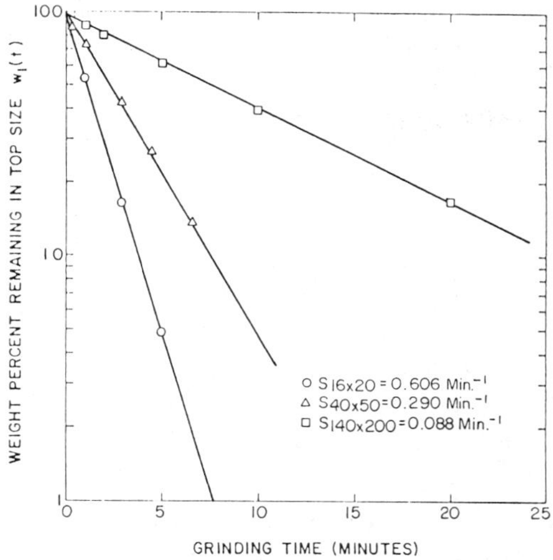

Figure 13.32. Typical first-order plot of batch grinding data, various sizes of cement clinker.

for large devices by extrapolation from smaller scale results.

Mill Models

The function of a mill model is to describe the product size distribution obtained from a given feed size distribution. The model can then be used to assist in the analysis of the influence of design and operating variables on mill performance. For example, consider a simple once-through device such as the smooth roll crusher (see Fig. 13.22). It can be assumed as a first approximation that each size of the feed lumps is stressed and fractured independently of the other sizes, as it reaches the region in the rolls where it is nipped and compressed. For a given gap setting, x_g say, breakage of each size will produce a mean set of daughter fragments denoted by $D_{i,j}$. Then, for a feed consisting of weight fraction f_1 of size 1, f_2 of size 2, etc., the fraction of product in size i is

$$\left.\begin{array}{l} p_i = d_{i,1}f_1 + d_{i,2}f_2 + \cdots + d_{i,i}f_i \\ \text{or} \\ P_i = D_{i,1}f_1 + D_{i,2}f_2 + \cdots + D_{i,i}f_i, \; n \geqslant i \geqslant 1 \end{array}\right\} \tag{26}$$

where P_i is the cumulative fraction of material less than x_i in the crusher product. The logical analysis of how the D values vary with conditions is given in Section 4 below as an example of the analysis of once-through devices.

Retention grinders such as the tumbling ball mill are very important industrially, and the mill model applicable to these is developed as follows. First, consider the simplest system of batch operation, with the powder getting finer and finer for longer and longer grinding times. Using the concept of primary daughter fragment distribution joined with Equation (24) the "net rate of production size i material is the sum of its production from all larger sizes minus its rate of breakage," or

$$W\frac{dw_i(t)}{dt} = -S_i w_i(t)W + W\sum_{\substack{j=1 \\ i>1}}^{i-1} b_{i,j}S_j w_j(t), \; n \geqslant i \geqslant j \geqslant 1 \tag{27}$$

where W is the mass of powder in the mill and $b_{i,j}$ is the *primary* daughter fragment distribution in the interval form, $b_{i,j} = B_{i,j} - B_{i+1,j}$. This set of n equations is known as the batch grinding equation. If $b_{i,j}$ and S_i do not vary with time, it has the solution[18, 19]

$$w_i(t) = \sum_{j=1}^{i} d_{i,j}w_j(0), \; n \geqslant i \geqslant j \geqslant 1 \tag{27a}$$

where the set of transfer numbers $d_{i,j}$ is computed from the algorithms

$$d_{i,j} = \begin{cases} e^{-S_i t}, & i = j \\ 0, & i < j \\ \sum_{k=j}^{i-1} a_{i,k}a_{j,k}(e^{-S_k t} - e^{-S_i t}), & i > j \end{cases}$$

$$a_{i,j} = \begin{cases} -\sum_{k=j}^{i-1} a_{i,k}a_{j,k}, & i < j \\ 1, & i = j \\ \frac{1}{(S_i - S_j)}\sum_{k=j}^{i-1} S_k b_{i,k} a_{k,j}, & i > j \end{cases}$$

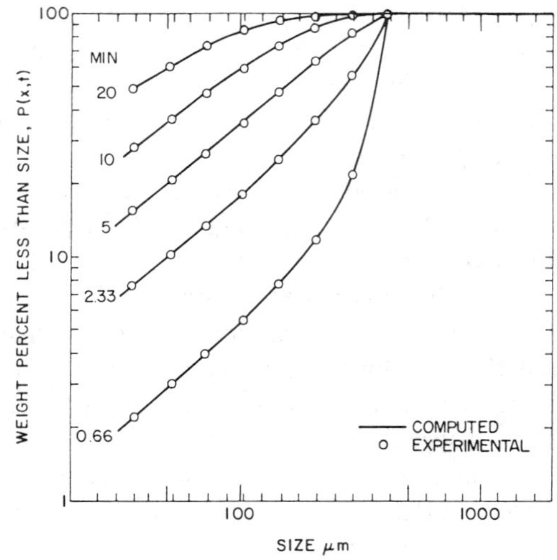

Figure 13.33. Comparison of computed to experimental size distributions for batch grinding.

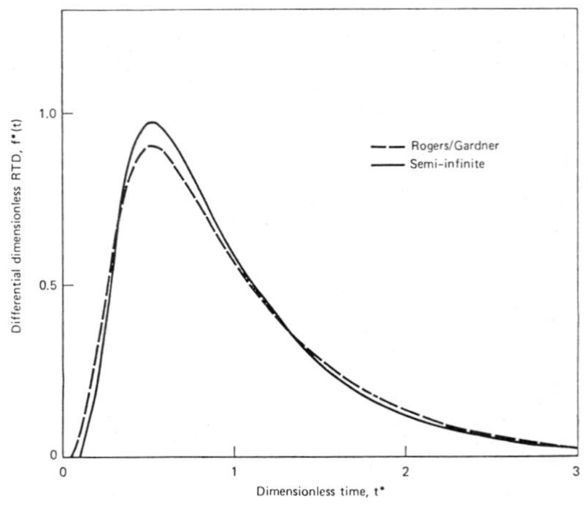

Figure 13.34. Residence time distribution for a 4.57 m diameter × 9.2 m long wet overflow discharge ball mill.

The equations are normally computed using a digital computer, starting with $i = 1$, knowing the feed size distribution $w_i(0)$. Figure 13.33 shows the computed solution compared to the smoothed experimental points for grinding of a narrow feed size, using experimentally determined values for S and B.

Second, consider a retention grinding machine where the powder flows uniformly, is ground, and is then fine enough to exit through an overflow or grate without preferential retention of larger sizes. If the flow through the mill was plug flow, Equation (27) would still apply with a grind time τ of $\tau = W/F$, F being the mass flow rate through the mill. However, retention mills will have a *residence time distribution* (RTD) defined by $\phi(t)dt$ = weight fraction of feed in at time 0 which leaves between time t and $t + dt$. Figure 13.34 gives an example determined by using a pulse of radiotraced powder in the mill feed and counting at the mill exit.[20] Then the steady product size distribution will be made up of material ground for all times over the RTD range, in a weighted sum[18]

$$p_i = \int_0^\infty w_i(t)\phi(t)\,dt \tag{28}$$

where $w_i(t)$ is the solution of Equation (27) for the mill feed.

For a fully mixed mill the mass-rate balance is "the rate of flow of size i out = rate of flow of size i in plus rate of production of size i by breakage of all larger sizes minus rate of breakage of size i." Thus,

$$Fp_i = Ff_i - S_i w_i W + W \sum_{\substack{j=1 \\ i>1}}^{i-1} b_{i,j} S_j w_j$$

However, $p_j = w_j$ for a fully mixed system with no size classification at the mill exit. Using $\tau = W/F$

$$p_i = \frac{f_i + \tau \sum_{\substack{j=1 \\ i>1}}^{i-1} b_{i,j} S_j p_j}{1 + S_i \tau}, \quad n \geqslant i \geqslant j \geqslant 1 \tag{29}$$

This set of equations is readily sequentially computed starting at $i = 1$.

The variable used in the computations is the mean residence time τ, and any model can be computed for a range of τ values. Since $\tau = W/F$, the value of τ which gives the desired product size also specifies the mass W necessary to get a desired production rate F. Then the mill size needed to contain W is calculated. Of course, it is also necessary to have equations which give mill power, in order to determine the specific energy of grinding.

An important general conclusion can be reached by considering Equations (27) or (29) applied to a comparison of two milling systems operating on the same feed. Suppose that the B values are the same between the two systems, but that the S values are different by a constant factor, $S_i' = kS_i$. Using Equation (29) as an example, applied to both mills,

$$p_i = \left(f_i + \tau \sum_{j=1}^{i-1} b_{i,j} S_j p_j\right) \Big/ (1 + S_i \tau) \quad \text{(Mill 1)}$$

$$p_i' = \left(f_i + \tau' \sum_{j=1}^{i-1} b_{i,j} S_j' p_j'\right) \Big/ (1 + S_i' \tau) \quad \text{(Mill 2)}$$

Substituting for S_i' in the second equation

$$p_i' = \left(f_i + k\tau' \sum_{j=1}^{i-1} b_{i,j} S_j p_j'\right) \Big/ (1 + k\tau S_i) \quad \text{(Mill 2)}$$

Obviously, $p_i' = p_i$ when $k\tau' = \tau$, that is, an identical set of size distributions is produced in mill 2 as in mill 1 *but with residence times decreased by the factor k.* If S values are doubled ($k = 2$), the required residence time is halved. Thus, there can be *similitude* between a small mill and a large mill, with only a difference in time scale. The same result is obtained for batch or plug flow grinding, and for Equation (28) providing the RTD is normalizable with respect to τ, that is, $f(t/\tau)$ is the same from one mill to another.

The use of these models is illustrated below, especially with respect to ball milling. Experimental measurement of the variation of the values of S_j with mill conditions is the most explicit and logical means for describing mill operation and mill efficiency.

It is useful to have an approximate mill model which is simple enough for quick hand calculations. The results of Figure 13.33 show that Bond's "law"[21] applies to a reasonable approximation,

$$E = m_p t/W = E_I\left(\sqrt{\frac{100\ \mu\text{m}}{x_{80P}}} - \sqrt{\frac{100\ \mu\text{m}}{x_{80F}}}\right) \quad (30)$$

where m_p is the net mill power, x_{80P} is the size in micrometers at which 80% passes that size in the product, x_{80F} is the 80%-passing size of the feed, and the energy index E_I is determined from the data. E is the specific energy of grinding (kWh/ton) required to go from a specified feed of x_{80F} to a desired product of x_{80P}. This empirical equation enables rapid estimation of the grinding time or specific energy to go from any feed to any product, *assuming* that E_I is a constant. It does not give any information on the size *distribution* of the product nor does it take into account the size *distribution* of the feed. As might be expected, E_I is not closely constant from one mill to another, or for different mill conditions. As used in practice, E_I is determined for a given material from an experiment under standard conditions[22] using an empirical correlating equation which converts it to the value expected for an 8 feet diameter wet overflow ball mill operated in closed circuit. E_I is then known as the Bond Work Index W_I, which has the physical meaning of the hypothetical kWh/ton necessary to go from a very large feed to 80% passing 100 μm, in the 8 feet diameter mill. Empirical correction factors based on prior experience are used to allow for different conditions and mill diameter.[21]

Mill Circuits

In industrial practice mills are frequently used in closed circuit, where the mill product is passed through a size classifier that gives two exit streams, a coarser stream returned to the mill feed and a finer stream, which is the final product. The operation of the classifier is best described by the set of classifier numbers s_i, defined as the weight fraction of size i presented to the classifier that is sent to the coarse stream. These are readily calculated from experimentally measured size distributions of the three streams. Figure 13.35 gives a typical example. The smaller the value of d_{50}, the bigger the overall fraction of the classifier feed which is directed

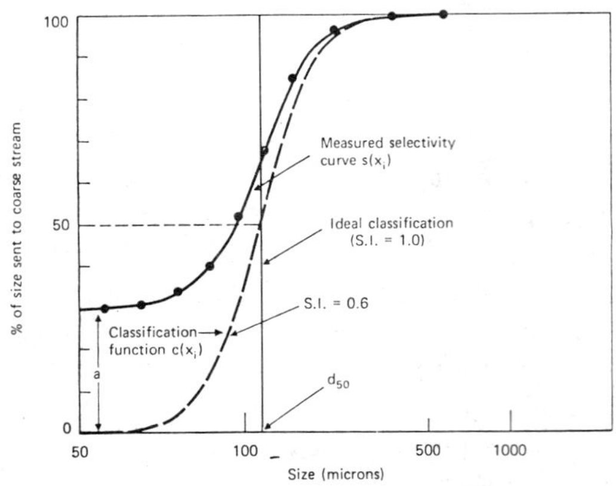

Figure 13.35. Illustration of selectivity values of a size classifier: a is an apparent by-pass.

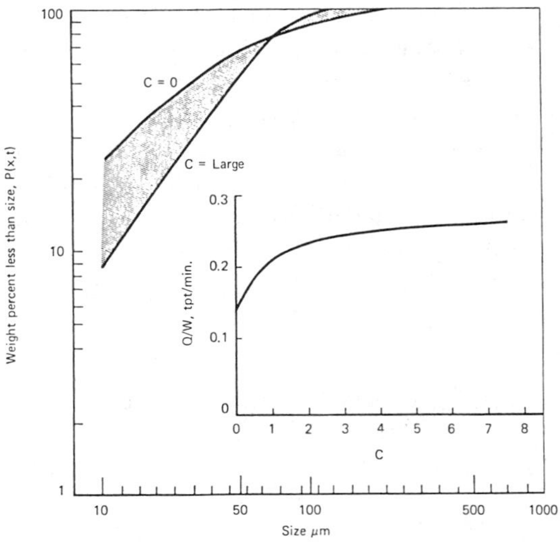

Figure 13.37. Permitted band of size distributions passing through a desired point, with varying circulating load.

into the coarser stream. The relation between the circuit feed and product and the mill feed and product are shown in Figure 13.36: defining the circulation ratio by $C = T/Q$ then $f_i(1 + C) = g_i + (1 + C)s_ip_i$ and $q_i = (1 + C)(1 - s_i)p_i$. These are used in conjunction with the appropriate mill model to predict the circuit product size distribution from a mill circuit simulation.[19]

Figure 13.37 shows one interesting result from a simulation of a tumbling ball mill. If a mill circuit is designed to produce a size distribution passing through a control point (ψ% passing size x^*) from a given mill, then this specification can be met by a suitable feed rate at open circuit, or by a suitable feed rate through a classifier with set s_i values, or by a different feed rate with the classifier adjusted to give lower d_{50} values (and, hence, more recycle and a larger C value). It is seen that there is a *permitted band* of size distributions through the control point, from $C = 0$ to $C = \infty$. Austin and Perez[23] have shown that the limiting (steepest) size distribution obtained at high circulating load depends *only* on the primary daughter fragment distribution. Thus, it is a material characteristic and it is not possible for a customer to specify a steeper distribution. The higher circulating load also gives a higher circuit output rate Q tph. The physical reason for these effects is that a high flow rate through the mill, $F = (1 + C)Q$, brings fine material rapidly to the classifier and removes it before it is overground. Thus, the mill contents contain on the average less fines and more coarser material, and coarser material breaks faster than fine material. *The general reason for closed*

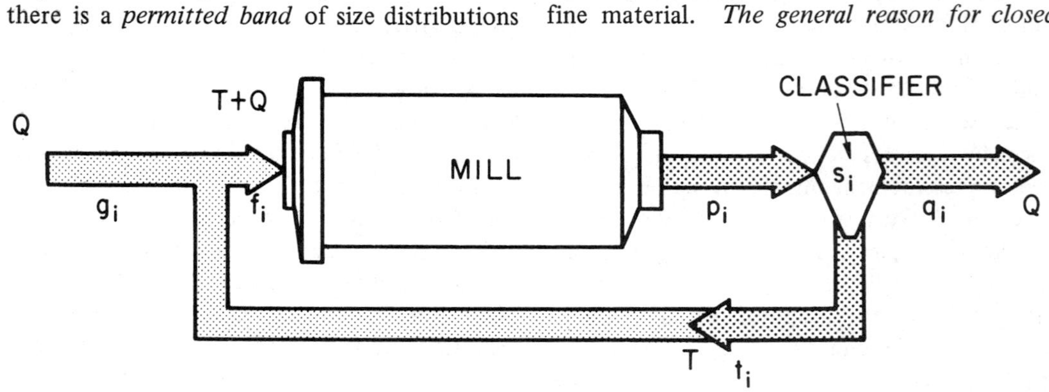

Figure 13.36. Normal closed circuit.

circuit operation is to remove particles that are already fine enough to prevent energy being wasted on grinding them even finer.

The return of fine material back to the mill feed, due to the apparent bypass of the classifier as shown in Figure 13.35, decreases efficiency by leading to overgrinding. In principle, this can be compensated by higher circulation, but in practice (a) it may not be possible to pass enough mass through the mill to approach this limit without *overfilling* the mill leading to poor breakage action and (b) increased mass flow through a classifier may also increase the bypass fraction, thus defeating the action. For these reasons it is advantageous for a classifier to approach as closely as possible the ideal classification shown in Figure 13.35. On the other hand, simulations show that if a circuit is run to produce essentially the same size distribution including fines (see Fig. 13.38), it does not make much difference whether it is produced using an efficient or inefficient classifier, a mill near plug flow or near a fully mixed condition, etc. *The function of efficient classification is to reduce the proportion of fine material by avoiding overgrinding of fines.*

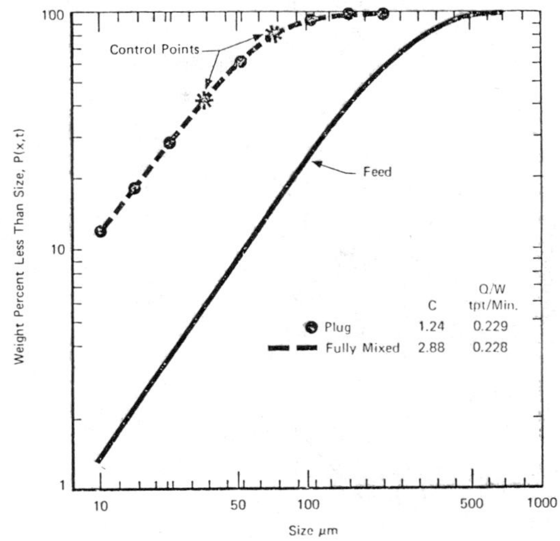

Figure 13.38. Comparison of capacities for two-point fitted size distributions produced by mills with plug flow versus fully mixed residence time distributions.

Nonfirst Order Grinding and Slowing of Grinding Rate

It can be reasoned from fracture mechanics and the difficulty of efficiently stressing unit mass of very small particles that the specific rates of breakage are smaller for small particles than for larger. This has been confirmed for every type of mill investigated to date. However, there is an additional effect of small size in retention devices such as ball mills and roller mills. As fine material builds up in the bed of powder, the breakage of *all* sizes slows down. This appears to be partly due to coating of the grinding surfaces but principally due to a *cushioning* action. In dry grinding, it is argued[24] that the agglomerative forces between fine particles impart a general rubbery nature to the bed which can absorb impact without giving high stress to particles directly under the stressing surfaces. This can be likened to trying to grind particles suspended in a sponge; the energy of a falling ball or passing roller is spread over a large elastic mass instead of being concentrated on a small mass of solid. In addition, air trapped in such a bed cannot rapidly flow out of the bed in the path of the stressing surface because of the high drag forces, so it moves away carrying particles with it, much like a liquid parting to let a solid ball fall through.

Figure 13.39 shows a Weibull plot for the dry ball milling of cement clinker. In the Weibull plot, the weight percentage less than a certain size is plotted versus grinding time on Rosin-Rammler coordinate paper.[25] Austin et al.[17] have shown that the solution of the batch grinding equation for a narrow feed size range will give almost straight lines on the Weibull plot, as shown in the figure, for constant S_i values. However, when the slowing-down process starts, the experimental line curves over as indicated. Because all values of S_i change by the same factor, the batch grinding model of Equation (27) still applies but the values of time used in the computation with constant S_i values must be equivalent first-order times θ instead of real time t.

A slowing-down factor K can be defined by $K = \theta/t$, which then also represents the ratio of the first-order value S_i to the mean value of S_i

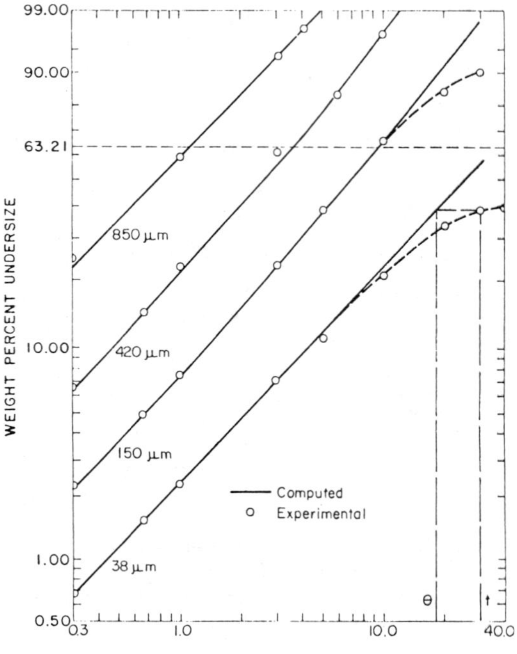

Figure 13.39. Weibull plot of % less than various mesh sizes with batch grinding time: feed 16 × 20 mesh cement clinker ground in a batch ball mill.

from time 0 to t. Figure 13.40 shows values of K for four different materials, plotted against the fraction of fine material less than 10 μm in size. It is apparent that different materials develop the slowing-down process at different amounts of fines. The magnitude of the effect

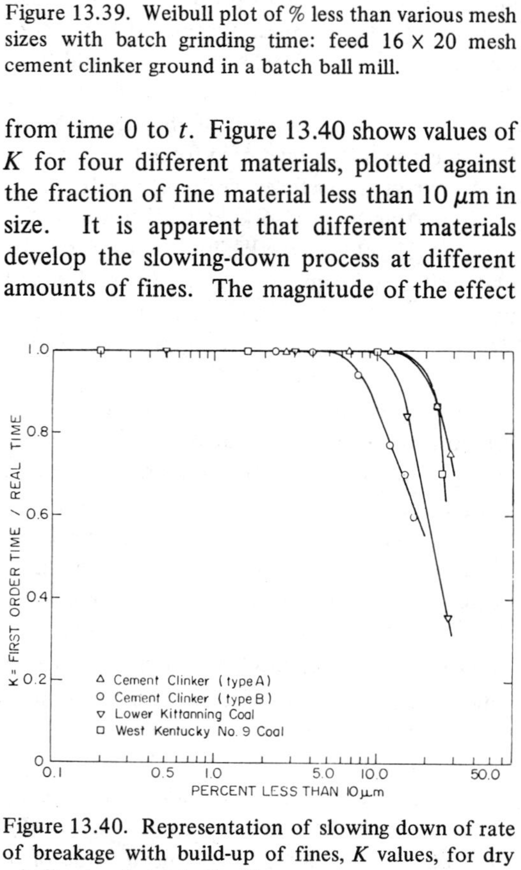

Figure 13.40. Representation of slowing down of rate of breakage with build-up of fines, K values, for dry grinding in a batch ball mill.

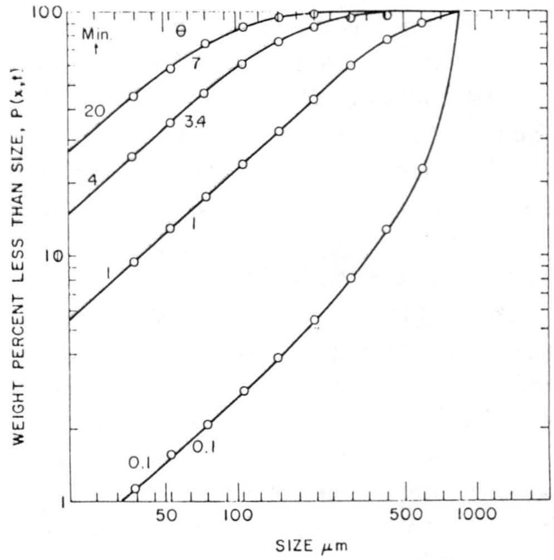

Figure 13.41. Comparison between computed and experimental size distributions of 20 × 30 mesh Lower Kittanning coal ground for different times in the ball mill. θ = first-order time, t = real time.

can be seen from Figure 13.41, where it takes 20 minutes to reach a size distribution which would have been obtained in 7 minutes if grinding had stayed first-order. Figure 13.40 suggests that normal size reduction will essentially stop as fines reach a certain level. Since the rate of energy input to the mill is almost constant, the slowing-down process produces greatly reduced grinding efficiency and leads to high specific grinding energy. A similar phenomenon is observed with wet grinding of slurries at high solid content.

Analysis of Smooth Roll Crushers

The unit operation of crushing does not usually give problems for brittle materials such as rocks and ores, so it has not yet received the amount of theoretical analysis given to fine grinding. However, the general concepts of the process engineering analysis of crushers can be illustrated by using smooth roll crushers as an example. There are five facets to the analysis: (1) the correct feed size; (2) the maximum force required; (3) the capacity in tons per hour; (4) the product size distribution; (5) the maximum power required.

Gaudin[26] gives the relation between angle of

nip Θ, coefficient of friction η, particle size x, gap x_g, and roll diameter d as

$$\left.\begin{aligned} \tan(\Theta/2) &\leqslant \eta \\ \cos(\Theta/2) &= \frac{d + x_g}{d + x} \end{aligned}\right\} \qquad (31)$$

see Figure 13.42. If x is too big it will not be nipped since Θ will then give $\tan(\Theta/2) > \eta$. Austin et al.[27] have pointed out that there is little published information on the effective values of η between the crushed materials and the rolls as a function of material, roll speed, surface roughness, etc. When the gap is small compared to roll diameter, Equation (31) gives

$$\frac{x}{d} \leqslant (1 + \eta^2)^{1/2} - 1 \qquad (31a)$$

For a value of $\eta = 0.5$ this states that the maximum lump size should be less than $\frac{1}{10}$ of the roller diameter.

The maximum force tending to separate the rollers can be estimated by assuming the worst possible case, that is, simultaneous compression

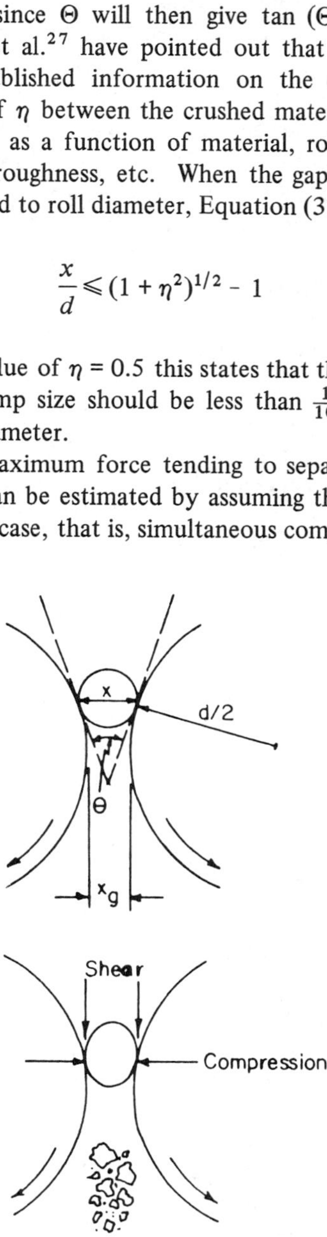

Figure 13.42. Illustration of nip angle in rolls and crushing-shear forces.

to failure of the maximum lumps of size x_m at all places along the rolls. Assuming that the lumps are small compared to the roll diameter, unit length of rolls contains $1/x_m$ large lumps each requiring the force P^* given by Equation (23). Thus,

$$\text{Maximum force} = x_m \sigma_c L/4 \qquad (32)$$

where σ_c is the compressive strength defined by $\sigma_c = P^*/(x_m/2)^2$, and L is the length of the rolls.

Gaudin[26] has described the concept of capacity calculation. Assuming that the feed is in sizes less than x_m, the maximum capacity per unit length of rolls is given by the ribbon of solid which can be pulled through the rolls

$$Q_{\max} = F(1 - \epsilon)\rho x_g u \qquad (33)$$

where $(1 - \epsilon)$ is the volume fraction of solids, ρ their true density, and u the circumferential velocity of the rolls. The ribbon of solid is actually a ribbon of packed powder with a porosity of ϵ, which is expected to be about 0.4. If the rolls are operating at different speeds it seems reasonable to use an arithmetic mean of the roll speeds. F is an empirical factor determined from test experience; ideally F would be 1 and in practice it varies from 0.2 to 0.9 for coals[28] depending on the strength and the roll diameter, with larger rolls and weaker material giving higher values.

A method of predicting the product size distribution obtained from any feed has been proposed by Austin et al.[27] They showed that the values of the transfer numbers $d_{i,j}$ depend on the ratio of the particle size x to the gap size x_g, providing the roll diameter is much bigger than x and x_g. Thus, values of $d_{i,j}$ measured experimentally for one gap setting can be converted to any other gap setting. Then Equation (26) is used to calculate the product size distribution. However, this method requires the experimental measurement of the total $d_{i,j}$ matrix, which is very time consuming.

To reduce the description of the $d_{i,j}$ matrix to a few parameters, they assumed that the breakage processes could be described as shown in Figure 13.43. First, feed particles of size j

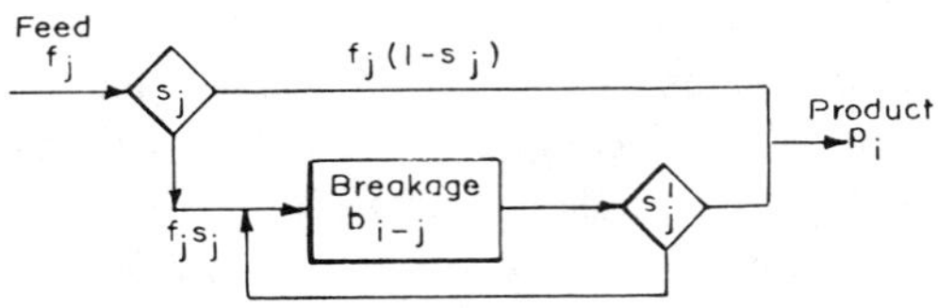

Figure 13.43. Equivalent circuit for a once-through roll crusher with multiple fracture actions.

have a probability s_j of breaking and a probability $1 - s_j$ of falling through the rolls without fracture: clearly, the fraction of larger sizes which fall through the roll gap without breakage is zero, $s_j = 1$, and sizes much smaller than the gap do not break, $s_j = 0$. Experimental values are shown in Figure 13.44 and were found to fit the empirical relation

$$1 - s_i = \frac{1}{1 + \left(\dfrac{x_i/x_g}{d_{50}/x_g}\right)^{6.6}} \tag{34}$$

where d_{50}/x_g is characteristic of the material. Second, it is assumed that all sizes break into a normalized primary daughter fragment distribution b_{i-j}, where b_1 is the weight fraction of breakage of one size which appears in the next lower size, b_2 appears in the size below that, etc. Third, it is assumed that a fragment of size i produced by fracture in the rolls has in turn a probability s'_i of being rebroken or $1 - s'_i$ of passing through the roll gap. Since this material results from fracture it is already in a favorable

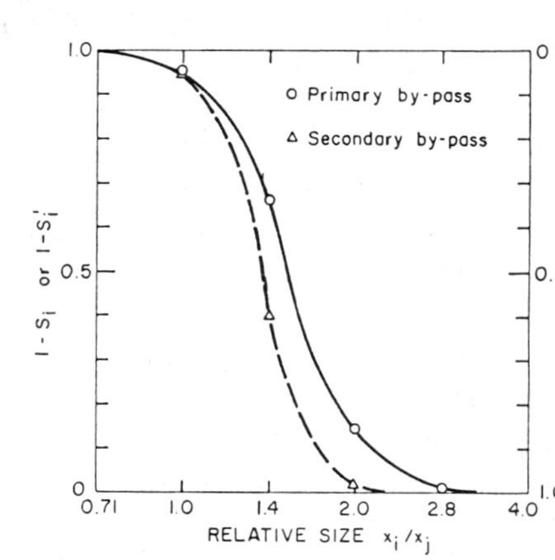

Figure 13.44. Measured primary bypass and estimated secondary bypass for Lower Freeport coal.

position to be nipped, so it is expected that s'_i will be greater than or equal to s_i.

Considering the repeated fracture of size 1 material, $1 - s_1$ falls through the rolls to size 1 product, and s_1 breaks. The material resulting from the breakage follows two routes—material that passes the gap to give product and material retained to fracture again. Let $a_{i,j} = b_{i,j}(1 - s'_i)$, which has the physical meaning "when size j breaks $a_{i,j}$ is the fraction sent to product size i." Let $c_{i,j} = b_{i,j}s'_i$, which is "when size j breaks $c_{i,j}$ is the fraction sent to size i for a further breakage." Then, the broken quantity s_1 distributes itself as $a_{2,1}$ in size 2, $a_{3,1} + c_{2,1}a_{3,2}$ in size 3, $a_{4,1} + c_{2,1}a_{4,2} + c_{3,1}a_{4,3} + c_{2,1}c_{3,2}a_{4,3}$ in size 4, and so on; $a_{4,1}$ is the product from breakage of size 1 to size 4, $c_{2,1}a_{4,2}$ is 1 breaking to 2 breaking to 4, $c_{3,1}a_{4,3}$ is 1 breaking to 3 breaking to 4, $c_{2,1}c_{3,2}a_{4,3}$ is 1 breaking to 2 breaking to 2 breaking to 4, and so on. Thus,

$$\begin{aligned} d_{1,1} &= 1 - s_1 \\ d_{2,1} &= s_1a_{2,1} \\ d_{3,1} &= s_1(a_{3,1} + c_{2,1}a_{3,2}) \\ d_{4,1} &= s_1(a_{4,1} + c_{2,1}a_{4,2} + c_{3,1}a_{4,3} \\ &\quad + c_{2,1}c_{3,2}a_{4,2}) \end{aligned} \tag{35}$$

etc., until c values become zero.

The equation is readily converted to $d_{i,j}$ by replacing 1 with j and 2 with $j + 1$, etc. Then the total size distribution from a feed of f_i is obtained from Equation (26).

Austin et al.[27,29] treated the above problem somewhat differently by developing the mass balance equations for the equivalent circuit of Figure 13.43 as if s_i and s'_i were due to external classifiers. Using the symbolism of this figure, the relation of p'_i of f'_i is $p'_i = \Sigma_{j=1}f'_jb_{i-j}$; $p'_1 = 0$ because all of f_1s_1 breaks and there is no production of size 1 from larger sizes. The effective steady-state feed f'_j is related to f_j and p'_j by $(1 + C)f'_j = f_js_j + (1 + C)p'_js'_j$ where C is an equivalent internal recycle for further breakage. It is convenient to use the dummy variable $p^*_j = (1 + C)p'_j$. Then $p^*_1 = 0$, and

$$\left.\begin{aligned} p_i^* &= \sum_{\substack{j=1\\ i>1}}^{i-1} b_{i-j}(p_j^* s_j' + f_j s_j),\ n \geqslant i > 1 \\ p_i &= f_i(1 - s_i) + p_i^*(1 - s_i') \end{aligned}\right\} \quad (36)$$

Given b_{i-j}, s_j, s_j', f_j this equation is sequentially computed starting with size 1 to get the values of p_i^* and p_i. For feed of size 1 only, $p_{i,1} = d_{i,1}$, $f_j = 0$ except for $j = 1$, and Equation (36) rearranges to

$$b_{i-1} = \frac{\dfrac{p_i}{1 - s_i'} - \displaystyle\sum_{\substack{j=2\\ i>2}}^{i-1} (b_{i-j}) \frac{p_j s_j'}{1 - s_j'}}{s_1}, \qquad n \geqslant i > 1 \quad (36a)$$

Thus b_{i-1} values can be sequentially computed from known $p_{j,1}$, s_j, s_j', starting with $b_1 = p_1/(1 - s_1')$.

They found that the values of b_{i-j} in the cumulative form $B_{i-j} = \Sigma_{k=n}^{i} b_{k-j}$ fitted the empirical function (see Fig. 13.31)

$$B_{i-j} = \Phi\left(\frac{x_{i-1}}{x_j}\right)^{\gamma} + (1 - \Phi)\left(\frac{x_{i-1}}{x_j}\right)^{\beta} \quad (37)$$

where Φ, γ, β are characteristic parameters for the material, as shown in Table 13.1.

By a trial-and-error matching of computed size distributions with experimental values they determined that s_i' could be estimated from s_i values by

$$s_i' = \begin{cases} s_{i-1}, & i < i_g - 1 \\ (s_{i_g-1} + s_{i_g-2})/2, & i = i_g - 1 \\ s_i, & i \geqslant i_g \end{cases} \quad (38)$$

where i_g is the interval number corresponding to the gap setting. Thus, a simulation model is constructed for smooth roll crushers, which has the experimentally determined material characteristics of Φ, γ, β and d_{50}/x_g. Figure 13.45 shows a typical match of computed versus experimental results.

They also simulated the effect of passing the product through a screen and recycling above-size material to the roll feed. The minimum production of fines was obtained when the gap and the screen were of the same size (see Fig. 13.46) even though the circulating load was relatively small. Larger gap settings and the associated higher circulating load produced very little change in the final product. This is because fine material is not acted on by the crusher as it passes through, so a high circulating load is no advantage.

The process of fracturing unit volume of feed to less than the gap setting requires stressing the original broken volume (s_1) of particles of size 1, stressing again the fraction of this volume which undergoes a second fracture, stressing again the fragments of these fragments which undergo a third fracture, and so on. The total stressed volume is readily calculated as s_1 plus the sum of all c terms, that is, $s_1(1 + c_{2,1} + c_{3,1} + c_{4,1} + \cdots + c_{2,1}c_{3,2} + c_{2,1}c_{4,2} + \cdots + c_{3,1}c_{4,3} + \cdots + c_{2,1}c_{3,2}c_{4,3} + \cdots)$. If it is

Table 13.1. Characteristic Breakage Parameters Determined from Smooth Roll Crusher Tests.[29]

MATERIAL	Φ	γ	β	d_{50}/x_g
Rhyolite	0.29	0.83	3.6	1.45
Diabase	0.40	0.84	4.0	1.40
Coals:				
Shamokin anthracite, Pa.	0.30	1.05	5.0	1.70
Illinois #6	0.36	0.81	3.0	1.66
Ohio #9	0.33	0.95	4.2	1.93
Western Kentucky #9	0.47	1.05	4.0	1.81
Belle Ayre, Wyoming	0.49	1.17	4.0	1.70
Pittsburgh E. Seam, Pa.	0.32	0.81	3.0	1.66
Upper Freeport, Pa.	0.39	0.96	4.0	1.56
Lower Freeport, Pa.	0.50	1.05	4.5	1.54

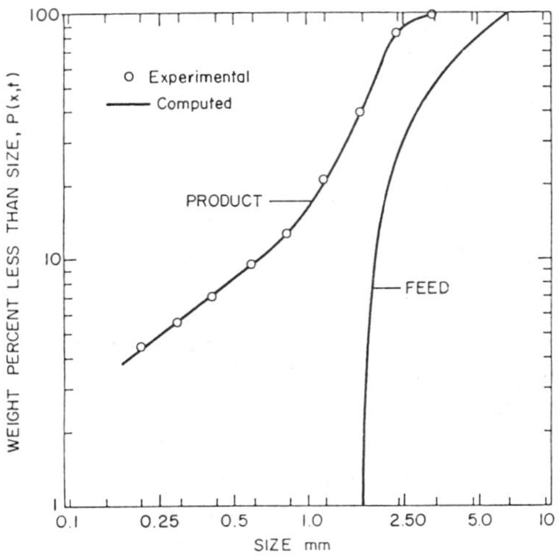

Figure 13.45. Crusher product size distribution of 3 X 12 mesh.

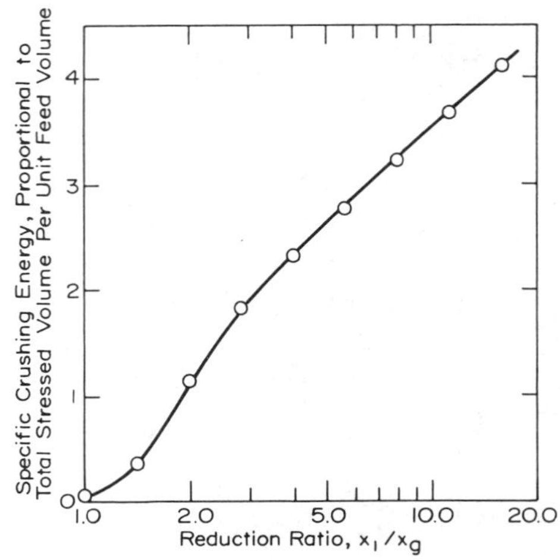

Figure 13.47. The total crushed volume per unit feed volume for roll crushing of a coal (Upper Freeport) through a smooth roll crusher, as a function of the particle size to gap size ratio.

assumed that the strain energy per unit stressed volume required to produce fracture is a constant, which is known[30] as Kicks' "Law," the total stressed volume is proportional to the ideal specific energy required to grind size 1 to less than the gap setting. Defining a reduction ratio by x_1/x_g, Figure 13.47 shows the relation of specific energy to reduction ratio. In practice, it is usually found that a larger reduction ratio requires a bigger

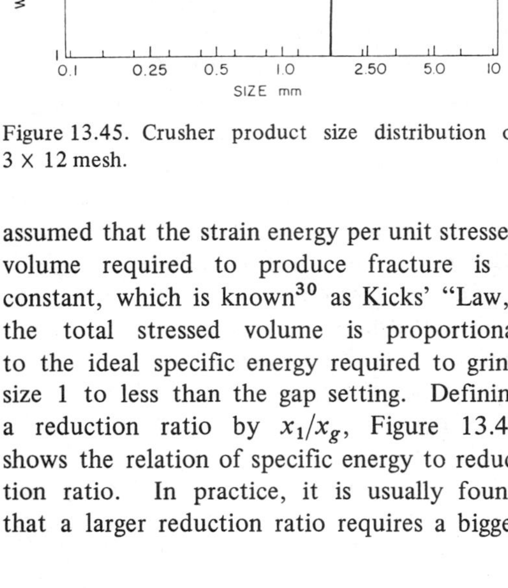

Figure 13.46. Simulated circuit product size distribution for 3 X 30 mesh Illinois #6 coal as a function of gap setting at ideal screening of 12 mesh.

increase of specific energy than that predicted by Figure 13.47 because smaller lumps become relatively stronger (require higher stress to cause breakage).

The maximum capacity of other crushers could be analyzed in a similar fashion. For example, a jaw crusher acts on a maximum solid volume rate of solid given by $V(1 - \epsilon)r$, where V is the volume of the crusher filled with solid and r is the strokes per unit time. However, not all of this material passes the crusher in one stroke, and the "repeated breakage and passing" analysis is required to find the mean volume passed per stroke. Similarly, in a gyratory crusher the maximum volume acted on per minute is $V(1 - \epsilon)$ (r.p.m.) since the narrowest cone-wall gap moves round once per revolution. Again, the cycles of repeated breakage must be analyzed.

The Analysis of Tumbling Ball Milling

Influence of Mill Conditions. The tumbling ball mill is the most widely used device for fine grinding of brittle materials on an industrial scale. Because of its simplicity it is mechanically reliable, which is very important in continuous process streams, and it is available in sizes ranging from small laboratory mills to industrial mills of 5 meters diameter by 10 meters long, or even larger. It is a retention device, where a bed of powder is acted upon by the tumbling balls and the mean residence time of solid in the bed is typically a few minutes to 30 minutes depending on the desired degree of size reduction. It has certain disadvantages. First, the mill power is almost independent of the level of filling by the powder, so a mill operated at lower than design capacity is inefficient because (a) if the powder level is held at a normal level, a low solid feed rate gives a long residence time ($\tau = W/F$) and the energy is used to grind finer than necessary or (b) if the level is dropped to keep τ constant, the energy is used to tumble balls on balls without enough powder between, also giving excess ball wear. Second, the cost of replacing steel balls as they wear is substantial, and the steel or rubber lining of the mill has also to be replaced every two or three years. Third, the "slowing-down" process discussed in Section 4 comes into play for very fine dry grinding or fine wet grinding of viscous pulps. Thus the grinding can become inefficient and consume high energy.

The major fraction of the direct power required to turn the mill (excluding motor and drive losses) is used in the act of raising the balls. On the other hand, the more balls raised per unit time, the higher the rates of breakage of powder in the mill because the tumbling of the raised balls gives the breakage action. Thus, the variation of power input with mill conditions is likely to be a direct index of the best breakage conditions. Figure 13.48 shows typical power variation with rotational speed and with ball filling. The critical rotational speed is defined as the speed where balls on the case would start to centrifuge and is readily shown to be

$$\text{critical speed} = \frac{76.6}{\sqrt{D - d}, \text{ in feet}} = \frac{42.2}{\sqrt{D - d}, \text{ in meters}} \text{ r.p.m.} \quad (39)$$

where D is mill diameter, d ball diameter. Thus

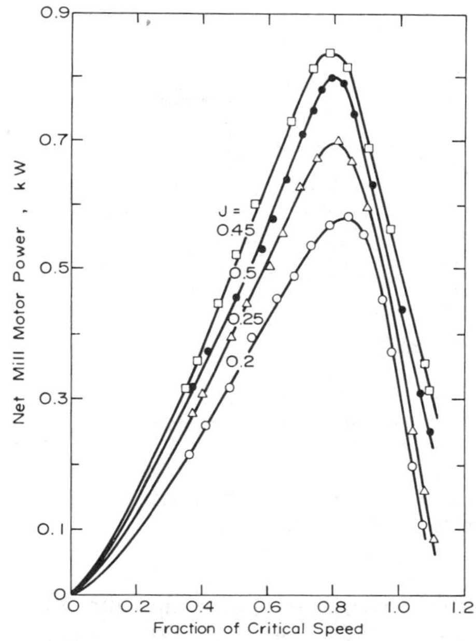

Figure 13.48. Variation of mill motor power with critical speed as a function of ball loading: 2 feet diameter laboratory mill.

Figure 13.48 shows that the power passes through a maximum at about 80% of critical speed. This varies somewhat with mill diameter and ball load because the force of a heavy ball charge acting on the case tends to prevent slip between the balls and the case (thus aiding the raising of balls) for larger mills. For large mills with steel balls the rotational speed is usually in the range 65% to 75% of critical speed. The figure also shows that the maximum power is obtained at about 45% filling of the mill volume by the ball bed at rest (calculated assuming the ball bed has a porosity of 0.4), $J = (M/\rho_b V)(1.0/0.6)$, where V is mill volume, M mass of balls, ρ_b true density of ball material. Continuous overflow mills, however, are normally run at $J < 0.4$, to prevent balls blocking the overflow.

The specific rates of breakage can be determined in laboratory or pilot-scale mills by batch tests with controlled powder and ball filling, and controlled pulp density if wet. Figure 13.49 shows a typical result for the variation of S_i with particle size. The rates of breakage are low for very large sizes because

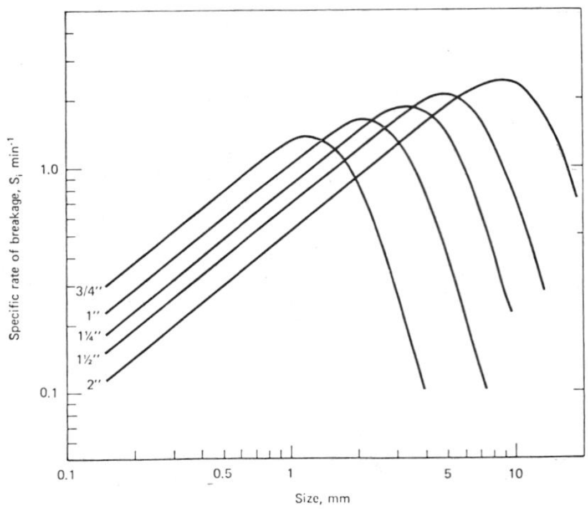

Figure 13.49. Predicted variation of S values with ball diameter for dry grinding of quartz.

(a) the particles are so big that the force required to break them is only achieved by relatively few of the tumbles and (b) the particles are too big to be nipped by a glancing blow (see Equation (31)). Small sizes also break slowly because (a) their basic strength is higher due to removal of large flaws and (b) the mass of particles captured in a ball-ball collision becomes smaller and smaller as particle size decreases with respect to ball size. It is difficult to put these common-sense concepts into precise quantitative form for such a complex force-strain assemblage as the variety of ball-particle-ball interactions in a real mill. However, the results of Figure 13.49 can be fitted by the empirical equation

$$S_i = a x_i^{\alpha} Q_i \tag{40}$$

where

$$Q_i = \frac{1}{1 + \left(\frac{x_i}{\mu}\right)^{\Lambda}} \tag{41}$$

where a, α, μ, and λ are descriptive parameters. The material to the left hand of the maximum in S has first-order breakage for $\sqrt{2}$ size intervals, and S_i varies as the α power of particle size (taken as the top size of the interval). The factor Q_i is 1 for small x_i and tends to zero as x_i increases. The maximum rate of breakage occurs at a particle size x_m related to μ and Λ by

$$x_m = \mu\left(\frac{\Lambda - \alpha}{\alpha}\right)^{-1/\Lambda}, \quad \Lambda > \alpha > 0 \tag{42}$$

The values of a, μ, and Λ vary with mill conditions but the value of α appears to be quite characteristic for a given material, provided the mill is not operated at very low or very high powder fillings.[31] Table 13.2 shows typical values of α for various materials. Similarly, the set of primary daughter fragments $B_{i,j}$ appear to be constant for normal mill conditions for a range of ball sizes, and not to vary with mill diameter. By the reasoning given in Section 4 this means that the size distributions obtained in a large mill can be identical to these in a small mill, when the feed sizes are to the left of x_m and, hence, $Q_i \approx 1$, but the time scale of the breakage depends on a. The B values fit the relation of Equation (37) and typical parameters are given in Table 13.2.

Figure 13.49 also shows the effect of ball diameter. Smaller balls break small particles more effectively because there are many more ball-ball collisions per unit mill volume per unit time for small balls, and the impact force is big enough to break small particles. On the other hand, the maximum particles size which can be broken effectively by a given ball size, x_m, increases for larger ball diameter because the required force of a properly nipped impact is more important than the total number of impacts. A simple empirical fit gives $a \propto 1/d$ and $x_m \propto d^2$. The hardness of the ball does not seem to affect the result,[32] providing the surface is not soft like rubber of course. For dry grinding, the value of a is directly proportional to the ball density.[33]

The variation of the value of a with ball and powder filling[31] is shown in Figure 13.50 in the form of relative values of $S_i f_c$ versus f_c for different J values: f_c is the fraction of the mill filled with powder, based on a formal bulk density of powder given by $\rho_p' = \rho_p (1 - \epsilon)$ where ρ_p is the true density of the material and ϵ is taken as 0.4. Thus $S_i f_c \rho_p' V$ has the physical meaning of the rate of breakage of size i, kg/min say, if the mill volume V was filled to a fraction f_c with material all of size i. The results can be explained qualitatively as follows. Considering a fixed ball charge J, the rate of breakage is clearly zero if there is no powder in the mill ($f_c = 0$) and the energy of the falling balls is used in ball-ball impact. As f_c is increased, some of the falling balls give ball-powder-ball impact, and the rates of breakage increase almost proportionally to the amount of powder. However, as more powder is added it eventually fills every ball-ball contact and the rate of breakage reaches a flat maximum. Further powder adds to a reservoir of powder not acted upon, and the cushioning action of excess powder eventually reduces the rates of breakage.

As expected, increasing the value of J increases the maximum breakage by providing proportionally more ball-powder-ball impact regions. However, as Figure 13.48 shows, overfilling the mill reduces the tumbling actions, and the rates of breakage drop proportionally: the maximum power and the maximum breakage are obtained at about 45% ball filling. The best powder level is related to the ball load, and it is conventional to express f_c in terms of the fraction of the ball bed porosity filled with powder, $U = f_c/0.4J$. Best results are for $U =$ 0.6 to 1.0 for this data, but it is normal to operate at higher levels to reduce ball wear, $U \approx 1$. The results of Figure 13.50 can be empirically expressed as

$$a \propto \frac{1}{1 + 6.6J^{2.3}} e^{-1.3U} \qquad 0.3 < U < 2, \quad 0 \leqslant J \leqslant 0.6 \tag{43}$$

which enables values of S_i measured for one set of conditions to be converted to other filling conditions. However, the values of the numerical constants in this equation will probably be somewhat different for large mills than for the small test mill used to derive the values quoted in Equation (43). The same relation appears to apply to the large particle sizes to the right of x_m. This region of breakage has not yet been adequately studied: experience is that breakage is nonfirst order in this region for strong materials. The B values have lower values of γ (relatively more fines) compared to the normal B values, for strong rock and ores, but sometimes the same for soft materials like coal

The other major variable is, of course, the mill diameter. It is found that S_i values increase with mill diameter to the about the 0.5 power, $S_i \propto D^{N_1}$, $N_1 \approx 0.4$ to 0.6. This can be explained as follows. It can be assumed that there is geometrical similarity in the tumbling action in two mills run at the same fraction of critical speed, that is, one revolution produces the same turnover of balls. Then, the number of balls tumbled per unit time is proportional to mill volume ($\pi D^2 L/4$) times r.p.m., but from Equation (39) the r.p.m. is proportional to $1/\sqrt{D}$. For a mill of diameter D each ball tumbles a mean distance proportional to D in going from its raised to its descended state. The number of impacts it makes as it tumbles down the bed surface is proportional to this distance. Thus, total breakage actions per unit time $\propto (D^2 L)(1/\sqrt{D})(D)$. The mass of powder acted upon is proportional to mill volume, so the breakage actions per unit mass per unit time is proportional to $(D^2L)(1/\sqrt{D})(D)/(D^2 L) \propto \sqrt{D}$, and S_i is proportional to $\sqrt{D}$. As expected, the position of S_{max} also increases with mill diameter, with an empirical fit of $x_m \propto D^{N_2}$, N_2 being about 0.1 to 0.2.

Finally, it is reasonable to assume that the sum effect of a ball loading J made up of a weight fraction m_1 of ball size d_1, m_2 of ball size d_2, etc., is given by the weighted sum

$$\bar{S}_i = \sum_k m_k S_{i,k}(d_k) \tag{44}$$

This means that a mixture of ball sizes can be more efficient than a single size of ball, because the larger balls will break larger particles efficiently and vice versa. The choice of a "perfect" mix of balls depends on the material, the feed size distribution and the desired product size distribution.

The effect of grinding additives in dry ball milling is undoubtedly to reduce the slowing-down process. The widest application in this area is the use of organic vapors in the grind-

Table 13.2. Breakage Parameters for Various Materials in a Laboratory Tumbling Ball Mill (One Inch Steel Balls).

COALS

BREAKAGE CHARACTERISTIC	PARAMETERS	SHAMOKIN ANTHRACITE	WESTERN KENTUCKY NO. 9	BELLE AYRE SO. WYOMING COAL	OHIO NO. 9	LOWER FREEPORT
Rate of Breakage	α	0.75	0.8	0.8	0.8	1.1
$S_{40 \times 50}$, min^{-1}	dry	0.5	0.8	0.9	0.7	1.0
	wet	0.5	1.0	1.1	1.0	1.9
	μ(mm)	1.8	3.6	3.7	2.5	3.4
	Λ	2.9	2.9	2.9	2.9	2.9
B Values, small sizes	γ	1.0	0.9	0.9	1.0	0.8
	β	3.1	2.8	2.8	3.5	2.3
	$\Phi_{40 \times 50}$	0.4	0.4	0.4	0.5	0.5
	δ	0.0	0.0	0.0	0.0	0.0
Experimental condition	Weight of powder, *g*	120	120	120	120	120
	Density (g/cm^3)	1.45	1.4	1.3	1.6	1.6
	Fractional ball filling, *J*	0.20	0.20	0.20	0.20	0.20
	Operational speed, ϕ_c	0.72	0.72	0.72	0.72	0.72
	Hardgrove grindability index	35	55	58	65	88

CEMENTS

BREAKAGE CHARACTERISTIC	PARAMETERS	CLINKER 1	CLINKER 2	CLINKER 3	CLINKER 4	SLAG CEMENT
Rate of breakage	α	0.9	0.95	1.1	0.95	1.6
$S_{40 \times 50}$, min^{-1}	dry	0.3	0.3	0.4	0.3	0.3
	wet	N.D.	N.D.	N.D.	N.D.	N.D.
	μ(mm)	1.45	1.7	1.75	2.0	1.5
	Λ	2.5	4.05	3.35	3.6	4.2

B Values, small sizes	γ	0.75	0.9	0.85	0.8	1.25
	β	4.0	4.0	4.0	3.3	4.25
	$\Phi_{40 \times 50}$	0.4	0.6	0.5	0.3	0.3
	δ	0.23	0.20	0.25	0.22	0.0
Experimental conditions	Weight of powder, *g*	300	300	300	300	300
	Density (g/cm^3)	3.2	0.20	3.2	3.2	3.2
	Fractional ball filling, *J*	0.20	0.20	0.20	0.20	0.20
	Operational speed, ϕ_c	0.80	0.80	0.80	0.80	0.80

ORES

BREAKAGE CHARACTERISTIC	PARAMETERS	PHOSPHATE ORE	QUARTZ	CU-MO ORE	LIMESTONE
Rate of breakage $S_{40 \times 50}$, min^{-1}	α	0.7	0.8	0.95	0.9
	dry	0.1	0.2	0.3	0.4
	wet	0.2	0.5	0.5	N.D.
	μ(mm)	8.6	1.9	1.4	1.4
	Λ	5.1	3.7	2.7	2.0
B Values, small sizes	γ	0.8	1.3	0.7	0.7
	β	4.0	5.8	4.3	3.2
	$\Phi_{40 \times 50}$	0.6	0.6	0.4	0.4
	δ	0.22	0.0	0.0	0.34
Experimental conditions	Weight of powder *g*	1345	328	343	338
	Density (g/cm^3)	3.0	2.7	2.7	2.7
	Fractional ball filling, *J*	0.37	0.20	0.20	0.20
	Operational speed, ϕ_c	$0.72\phi_c$	$0.72\phi_c$	$0.75\phi_c$	$0.75\phi_c$

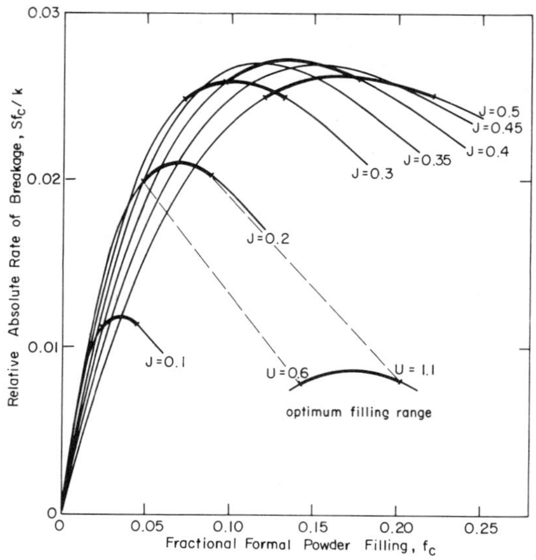

Figure 13.50. Variation of relative absolute rate of breakage with powder and ball filling for dry grinding.

ing of cement clinker to the fine powder of finished cement. It has long been known that such additives only affect the result in the production of fine cements with specific surface areas (measured by the Blaine test) of about 3000 cm^2/g or more. Locher and von Seebach[34] showed conclusively that the effect involved the chemisorption of the vapors on the powder at the relatively high temperatures in a cement mill, and that the chemisorbed vapors changed the agglomerative forces between the particles. Austin et al.[24] have shown that the slowing-down process described in Section 4 starts to come into effect during grinding of cement clinker when the Blaine area exceeds 2000 cm^2/g, for the cements they tested. It is also known that the organic grinding additives enable the finished cement to be packed more easily in bags, again demonstrating the change in the rheological properties of the powder due to changes in agglomerative forces.

On the other hand, the effect of water in wet grinding is quite different. If the slurry density is chosen to give a fluid slurry it is found[35] that the primary daughter fragment distributions and the shape of the value of S_i versus x *are almost* the same as for dry grinding of the same material at the same powder filling. However, the values of S_i are increased by a constant factor for all sizes, so that wet grinding produces the same size distributions but requires shorter grinding times and gives higher mill capacity. The factor varies from 1.1 to 1.8 for different materials. Thus water is a true grinding aid. For high slurry densities and fine grinding the process can again become non-first order as the pulp becomes viscous. Klimpel and Austin[36] have recently shown that polyelectrolyte dispersants increase the fluidity of these dense slurries and increase mill capacity. In this case, the effect is very comparable to that of a dry grinding aid. Whether a chemical will act as a wet grinding aid depends on its dispersant properties for the solid in water under mill conditions, on the slurry density and the degree of fineness of the grind, and on the rheological properties of the particular slurry.

Continuous Ball Milling. The analysis performed in the previous section has been entirely concerned with effects on the rates of breakage. However, in an operating industrial ball mill where material is flowing through the mill there are a number of additional factors involved. First, in principle it is necessary to have a physically based mass transfer law to describe how material moves through the rotating mill. Second, the stability and control of the system are of importance. Third, the effect of recirculation is an additional factor.

There appears to be no applicable mass transfer treatment available for tumbling ball mills. However, for overflow mills the mill has to fill to the overflow in order to pass any appreciable mass at all. Figure 13.50 shows that the variation in breakage rates over a narrow range of f_c values is small providing the mill is not underfilled (which can not happen with an overflow mill). Thus, although a higher mill flow rate will lead to a bigger value of W and, hence, τ, S_i will decrease correspondingly because $S_iW \approx$ constant in the flat maximum regions in Figure 13.50. Thus, the mill can be modeled or designed for a reasonable value of hold-up, W_0 say, and an apparent τ of W_0/F used in calculations with S_i values corresponding to W_0, since the real τ of W/F would be automatically compensated by a change in S_i. For grate discharge mills, the discharge area is empirically designed to give hold-up in the correct range of f_c values for the design flowrate. If the mill is then run at low flow rates, the filling may be so low as to put the mill in the inefficient region of low f_c in Figure 13.50. Recent work on residence time distributions in mills[37] suggests that mills operated at high flow rates are often overfilled, with $U \propto \sqrt{\text{feed rate}}$. In this case, chemical additives to reduce viscosity can increase output and reduce specific grinding energy by reducing the overfilling, moving the conditions closer to the optimum levels shown in Figure 13.50.

A true mass transfer law would enable the calculation of the mill residence time distributions under different condition. However, experimental measurements of RTD's on mills[20] can be used to construct an approximate model for predicting RTD. The model is based on the assumption that the backward and forward mixing process in the tumbling mill is the same for all particle sizes and can be considered analogous to a diffusional process described by the well-known convective-diffusion equation,

$$\frac{\partial c(x,t)}{\partial t} = \mathfrak{D} \frac{\partial^2 c(x,t)}{\partial x^2} - u \frac{\partial c(x,t)}{\partial x} \tag{45}$$

where $c(x,t)$ is the concentration of a tracer in the mill contents at position x along the mill at time t after admission of a short pulse of the tracer, u is the mean velocity of flow given by $u = L/\tau$, and $\mathfrak{D}$ is an effective mixing factor or diffusion coefficient. Mori et al.[38] have given a solution of this equation for the semi-infinite boundary conditions of "no flow of tracer through $x = 0$ (the feed end) at $t > 0$ and no influence of the wall at the mill discharge,"

$$f(t^*) = \frac{1}{2\sqrt{\pi \mathfrak{D}^* t^{*3}}} \exp\left[-(1 - t^*)^2/4\mathfrak{D}^* t^*\right] \tag{46}$$

where t^* is the reduced time t/τ, $f(t^*)\,dt^*$ is the fraction out in time t^* to $t^* + dt^*$ and $\mathfrak{D}^*$

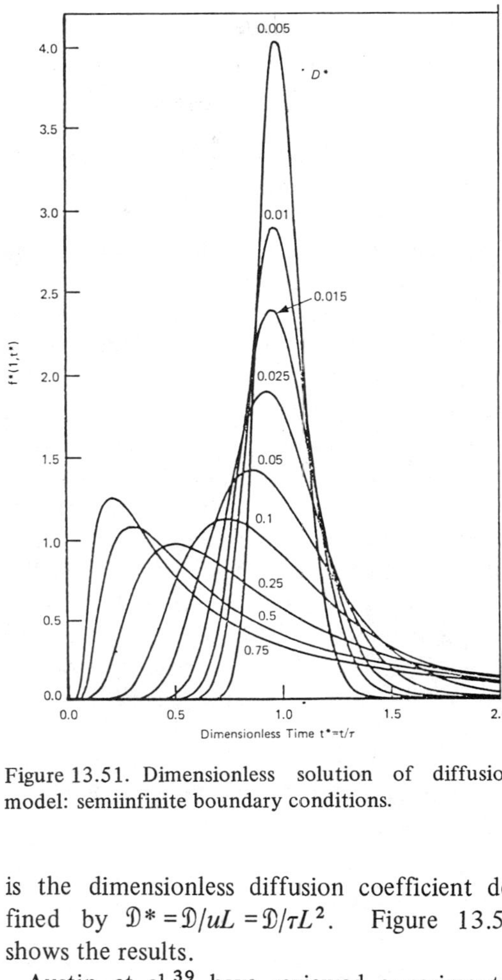

Figure 13.51. Dimensionless solution of diffusion model: semiinfinite boundary conditions.

is the dimensionless diffusion coefficient defined by $\mathfrak{D}^* = \mathfrak{D}/uL = \mathfrak{D}/\tau L^2$. Figure 13.51 shows the results.

Austin et al.[39] have reviewed experimental results of tracer tests on mills and concluded that they can be approximately described by this solution, with $\mathfrak{D}^*$ approximately 0.25. Fortunately, the size distribution produced by a closed circuit mill model is not very sensitive to the exact value of $\mathfrak{D}^*$. In addition, tests on wet mills show that traced water moves through the mills somewhat faster than traced solid, which suggests that there is a slight effect of particle size, since very small particles would move with the water.

A more important deficiency is the lack of equations to predict the maximum mass flow rate which can be passed by various tumbling mills. It is known that decreasing the size of balls causes increased hold-up W for a given flow rate and vice versa, but the influence of the resistance to flow of the ball bed compared to that of the discharge mechanism is not available in quantitative form.

The stability of this type of mill is good providing the powder reaching the discharge end is finer than any grate opening present, so that the material is not retained. If larger material is retained, the feed rate to the mill has a maximum stable value given by the condition where the rate of feed of large material equals the rate of breakage of the material to sizes less than the grate size. Similarly, when the circuit is closed through a classifier which returns all material greater than some cut of size x_c to the mill feed, then the same concept applies. The maximum rate can be calculated as follows. The maximum rate corresponds to material larger than x_c being recycled to the mill feed at infinite circulating load, so that all material less than x_c is rapidly removed from the circuit, leaving only large sizes in the mill. The high circulating load means that the mass flow rate through the mill is high and τ is low, tending to give a fully mixed mill. The classifier and mill behave like a fully mixed mill which has a screen of x_c in its exit.[23] Thus Equation (29) can be used in the form

$$0 = g_i - \gamma_i + \sum_{\substack{j=1 \\ i>1}}^{i-1} b_{ij}\gamma_j, \; i_c > i \geq j \geq 1$$

where g_i is the fraction of circuit feed of size i, $p_i = 0$ for all sizes existing in the mill because they are too large to leave the mill, and γ_i is a convenient grouping defined by $\gamma_i = S_i w_i \tau$. Rearranging the equation

$$\gamma_i = g_i + \sum_{\substack{j=1 \\ i>1}}^{i-1} b_{ij}\gamma_j \tag{47}$$

from which γ_i can be sequentially calculated from $i = 1$, $i = 2$, etc. down to $i = i_c - 1$. The values depend only on g_i and b_{ij}. Then, $w_i = \gamma_i/S_i\tau$ and since

$$\sum_{1}^{i_c-1} w_i = 1,$$

$$\tau = \sum_{1}^{i_c-1} (\gamma_i/S_i)$$

and

$$(Q/W)_{\max} = 1 \Big/ \sum_{1}^{i_c-1} (\gamma_i/S_i) \tag{48}$$

Any attempt to run the mill-classifier system to produce at a rate close to this maximum will produce unstable operation, because any slight decrease in the specific rates of breakage S_i would mean that material is being added faster than it is breaking, which would fill the mill more and lead to smaller values of WS_i on the right hand side of Figure 13.50, etc.

A number of conclusions can be reached by the simulation analysis of ball milling circuits. The permitted band of size distributions through a control point has already been mentioned, and that closing a circuit prevents overgrinding of fines. For open circuit grinding at the same capacity, a mill with a small $\mathfrak{D}^*$ (closer to plug flow, long L/D) gives a steeper size distribution and less fines than a larger $\mathfrak{D}^*$ For closed circuit, however, the difference is much smaller because of the recirculation, so it is not necessary to know the mill RTD with great accuracy nor is the L/D ratio crucial. If a one-point control specification is used, plug flow and closed circuit gives about 15% higher capacity than fully mixed and closed circuit, with any real system lying within this range. For industrial wet overflow mills, for example, the values of $\mathfrak{D}^*$ are almost independent of flow rate and L/D in the practical operating region, so that mill capacity is not significantly influenced by the RTD except via the mean residence time that material stays in the mill.

Similarly, Figures 13.52 and 13.53 illustrates other important features: A high bypass fraction reduces output and a less sharp classification also reduces output, for a product size distribution passing through a specified control point. The use of a classifier cut size (d_{50}) which is too large compared to the desired product size also reduced output very substantially.

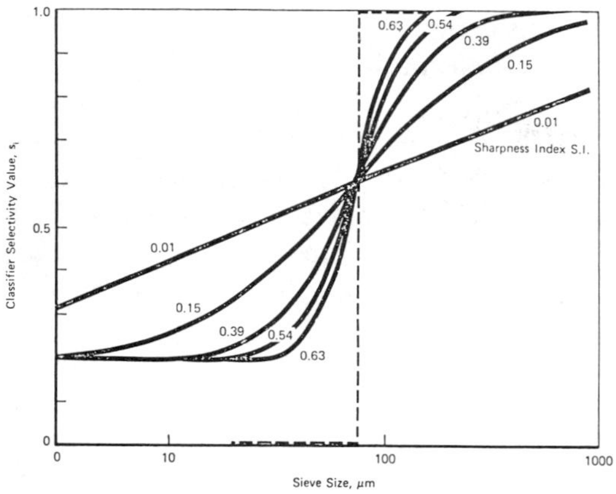

Figure 13.52. The variation of classifier selectivity values as the classifier parameter SI is varied: d_{50} and bypass held constant at 75 μm; a = 0.2.

Figure 13.54 illustrates the effect of choosing the correct make-up ball size when the feed contains relatively large material. This simulation uses the mixture of balls obtained by a natural wear process[39] and make-up addition of the top ball size d_{max},

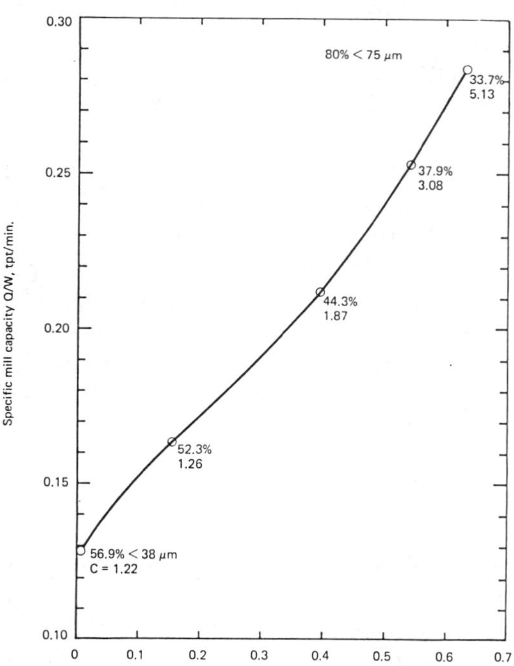

Figure 13.53a. Effect of steepness of Tromp curve, SI, on circuit output for size distributions passing through control point 80% < 75 μm: by-pass a = 0.2, d_{50} = 75 μm.

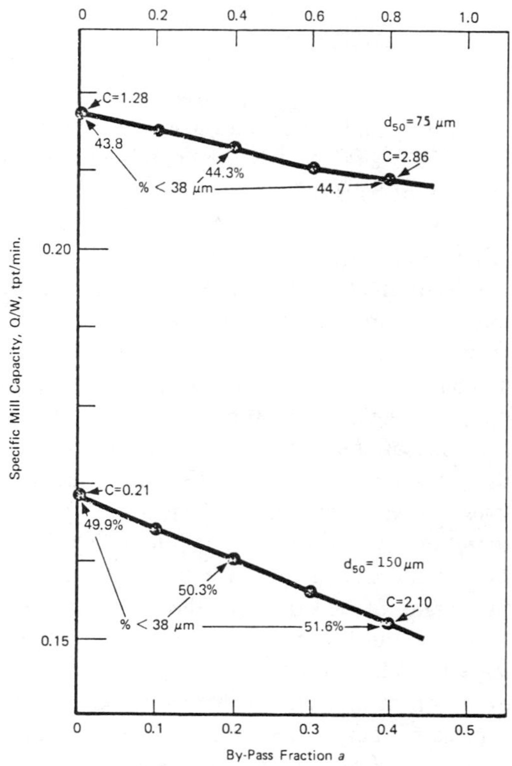

Figure 13.53b. Effect of classifier bypass on circuit output for size distributions passing through control point 80% < 75 μm, for two d_{50} values (SI = 0.39).

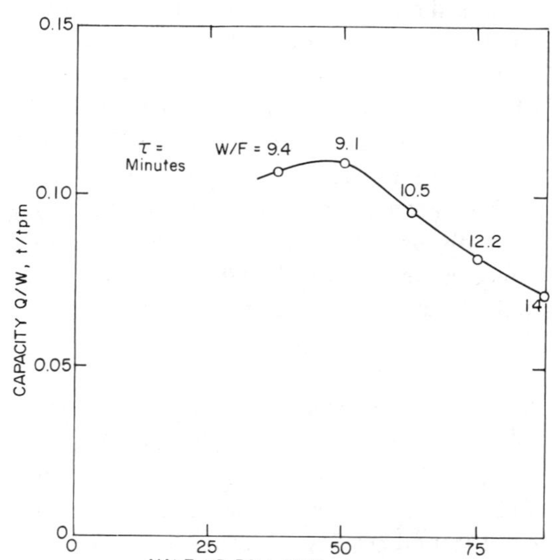

Figure 13.54. Illustration of effect of ball size on mill capacity for a feed of 0 × 20 mm.

$$M(d) = \frac{d^4 - d_{\min}^4}{d_{\max}^4 - d_{\min}^4} \qquad (49)$$

where $d_{\min}$ is the minimum size of ball retained in the mill and $M(d)$ is the cumulative weight fraction of balls of diameter less than or equal to d. Because of the fourth power relation the value used for $d_{\min}$ does not much influence the result and $\frac{5}{8}$ inch was used here. Although the final product size distributions are not changing much, the size distributions out of the mill show the cause of the low capacity at small $d_{\max}$ as being due to the low specific rates of breakage of the large feed sizes to the right of the maximum in S (see Fig. 13.49). These large sizes persist in the mill and are recycled until there are enough of them in the mill to bring their total rate of breakage into balance with their rate of addition via the make-up feed. Thus the mill is spending much of its action on large particles with low breakage rates, to the exclusion of breaking smaller sizes. Mill output decreases again for too large balls because they are not as efficient for breaking small sizes.

The use of grinding additives to reduce the slowing-down effect caused by the accumulation of fines has already been mentioned. The other common method of compensating for the slowing-down effect in dry tumbling ball mills is the use of air-swept mills, in which an air stream selectively sweeps out the finer particles, thus preventing the cushioning effect of the accumulation of fines. Austin et al.[40,41] have presented preliminary models which incorporate this effect. The power used for the air flow can be appreciable, and air-sweeping is primarily used for coals where the cushioning effect is very marked and it is in any case desired to produce a coal–air mixture for combustion in boiler furnaces or kilns.

Scale-up of Ball Mills. The exact performance of a circuit containing a large mill compared to one containing a smaller mill is complex to predict because residence time distributions and classification actions will generally be different, and the level and mixture of ball sizes can be different. However, if the simplifying assumption is made that the circuits are identical in all respects except for the rates of breakage in the two mills then a very simple result is obtained. The specific rates of breakage are proportional to $\sqrt{D}$. The mass of powder being acted upon is proportional to mill volume, that is, proportional to $\pi D^2 L/4$ where L is mill length. Thus to go from the *same feed* to the *same product* would give mill capacities in the ratio of $(\sqrt{D})(D^2L)$ or

$$Q \propto D^{2.5}L, \qquad \text{tph} \qquad (50)$$

Similarly, the net power drawn by the mill under comparable conditions is proportional to mass of balls raised per revolution × revolutions per minute × mean height balls are raised. Assuming the mean height is proportional to D, the r.p.m. is proportional to $1/\sqrt{D}$ and the mass of balls is proportional to D^2L,

$$\text{mill power} \propto D^{2.5}L, \qquad \text{kW} \qquad (51)$$

Comparing Equations (50) and (51) shows that the comparable specific grinding energy in kWh/t can be the same between different sizes of mill.

Bond[21] and Rowlands[22] state that actual plant data show that mill power per ton of grinding media is proportional to $D^{0.3}$ for mills larger than 8 feet in diameter so that mill power for a given ball filling is $\propto D^{2.3}L$ not $D^{2.5}L$; further, $Q \propto D^{2.5}L$ for $8 \leqslant D \leqslant 12.5$ feet. In this case the comparable specific grinding energy is somewhat less for the larger mill diameter. For $D \geqslant 12.5$, they give $Q \propto D^{2.3}L$ so that no further advantage of larger mill diameter is obtained with respect to specific grinding energy.

Analysis of Roller Mills

The Basic Relations of Load and Speed to Breakage Rates. Figure 13.25 shows an example of this type of mill, in which heavily loaded rollers or balls roll over material on a rotating table. The principle of operation is, essentially, that of a very large version of a pestle and mortar. The most detailed analysis of this type of mill which is avail-

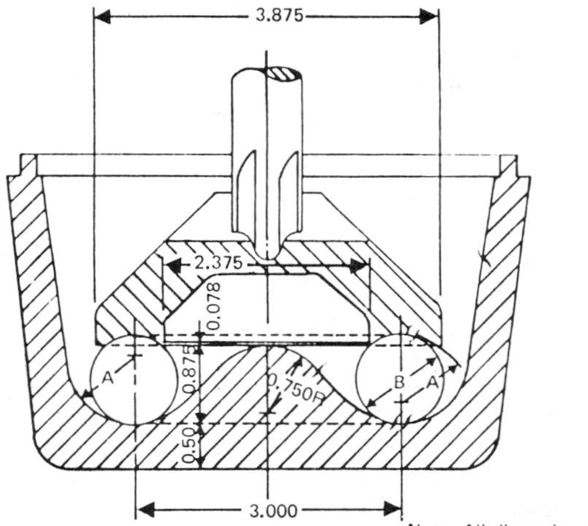

Figure 13.55. Grinding elements of the Hardgrove Machine.

able in the general literature is the work of Austin et al.[42-44] on ball-race mills, and this work is summarized here to show the general methodology of analysis.

It is convenient to perform initial detailed studies on small mills because of the difficulty of measuring S_i values in larger mills. The small test mill used was the standard Hardgrove mill,[45] shown in Figure 13.55, modified to enable the measurement of driving torque and, hence, power. The specific rates of breakage of various feed sizes under various conditions were determined by batch grinding tests, using quartz and various coals. It was found that the initial grinding of relatively coarse sizes of quartz gave first order breakage, while smaller feed sizes eventually demonstrated a slowing-down of rate as fines accumulated. Coals, on the other hand, gave only short regions of first-order grinding and soon exhibited the slowing-down effect. Using the initial specific rates of breakage the following conclusions were drawn.

1. Without powder in the race the torque (Υ) was small and did not depend on the weight loading on the balls (M), showing that the force was required primarily to overcome bearing friction.
2. With powder in the mill, the movement of balls over the powder caused friction and greater torque was required. To a reasonable approximation the torque was independent of rotational speed (ω). Since net mill power (m_p) is related to torque and speed by $m_p = \Upsilon\omega$, this means that net mill power $\propto$ rotational speed.
3. Torque was linearly proportional to crushing load M, up to the maximum load tested (50 kg).
4. Specific rates of breakage were linearly proportional to M and ω, hence, proportional to mill power.
5. Decreasing the number of balls gave a higher crushing load per ball, but the torque, power and specific rates of breakage were unchanged, showing that they were proportional to crushing load per ball × number of balls, that is, proportional to M.
6. Low powder levels in the race gave low torque, but beyond a certain filling level the torque was almost constant. The corresponding absolute rates of breakage, S_iW, were essentially constant (independent of W) in this region.
7. When the nonfirst order slowing-down of rate occurred the mill power fell somewhat, but the rates of grinding decreased much more, showing that transfer of mill energy to produce fracture was very inefficient in this region.
8. The primary daughter fragment distributions were normalized for quartz and the coals tested, with the same form as Equation (37), and they did not vary with mill conditions.
9. The specific rates of breakage varied with particle size in the same way as equation 40, $S_i = ax_i^\alpha Q_i$, passing through a maximum at about 5 mm.

From the above conclusion it is possible to relate torque and crushing load by an equivalent friction law

$$\Upsilon = \eta MD/2 \tag{52}$$

where D is the diameter of the center of the

race (3 in.). The apparent coefficient of friction η increased with powder load, lying between 0.085 and 0.10 for race filling levels from 0.075 to 0.5 fraction by volume of the empty space in the ball-race ring.

Pilot-Scale Studies. Some batch grind tests were performed on a Babcock and Wilcox Type E 1.7 mill, containing 10 balls of 5 in. diameter loaded at $M = 8400$ lb/h rotating in a race of 17 inches diameter. For these tests the mill was modified to run at low speed, and vertical walls were added to prevent crushed coal from being thrown out of the race. It was again found that breakage of larger sizes was first order and initial breakage of smaller sizes was first order. It was concluded that the values of the primary daughter fragment distribution were the same as these determined in the small mill, as was the value of α. Of course the absolute rates of breakage ($S_i W$) were much larger in the bigger mill.

Continuous tests were also run with the mill in its normal operating conditions, $\omega = 100$ r.p.m., with air-sweeping through a classifier built in to the top of the mill (for example, see Fig. 13.26). The function of the air-sweeping and classification actions is to remove fine material from the bed, so that the inefficient slowing-down of rate due to cushioning by fines is not present. Estimates were made of the internal classification action of larger particles falling back onto the race and of the built-on classifier action. The specific rates of breakage, B values and internal classifier actions were put into a mathematical simulation model to predict the size distributions produced by the system for various mass and air flow rates.

For most mill studies the above procedure would be sufficient to produce reasonable agreement between the model and the experimental results obtained in steady state, continuous, closed-circuit operation. However, for this type of mill there is an additional complication, as indicated in Figure 13.56, which shows that the power drawn by the mill depends on the feed rate of raw coal to the mill. The laboratory tests indicated that the absolute rates of breakage tend to vary in the same way as power, so

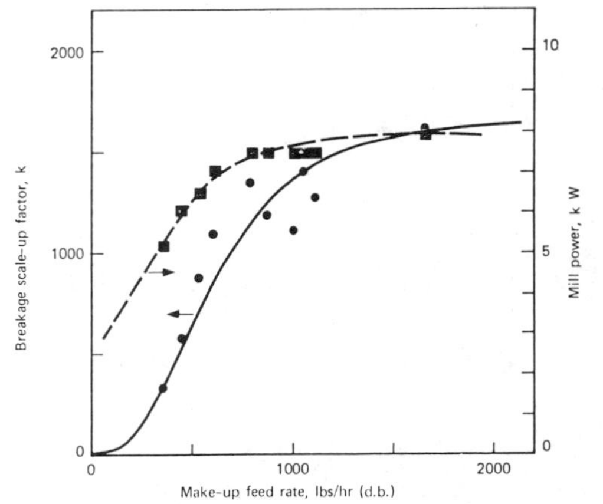

Figure 13.56. Scale-up factor and mill power as a function of coal feed rate: Babcock E1.7 mill (Western Kentucky Coal).

a low power draw by the mill implies low rates of breakage. There are two factors involved. First, the laboratory tests showed that torque and power decreased for low filling levels in the race, and in the continuous system it is apparent that a low feed rate will give a low filling level. Second, the discussion of the mode of operation of this type of mill based on industrial experience (Section 3) stated that the mill works well only when the race is maintained with a suitable mixture of particle sizes. If the material in the race is too fine the balls (or rollers) gouge the powder rather than roll over it with proper crushing action.

Consequently, the simulation model was adjusted to include a multiplying factor k to scale the rates of breakage measured in the laboratory mill to those in the big mill, where k varied depending on the make-up feed rate. Figure 13.55 shows the values of k deduced by trial-and-error matching of simulated to actual results. As expected, k varied in the same general way as net mill power. It is concluded that higher feed rates to the mill give better bed conditions in the race, lead to more power drawn by the mill, and give corresponding higher rates of breakage in the mill. Of course, when the best bed conditions are obtained any further increase in feed rate will decrease the mean residence time and give a coarser grind.

Figure 13.57 shows a typical result of grinding almost identical feeds, at almost the same

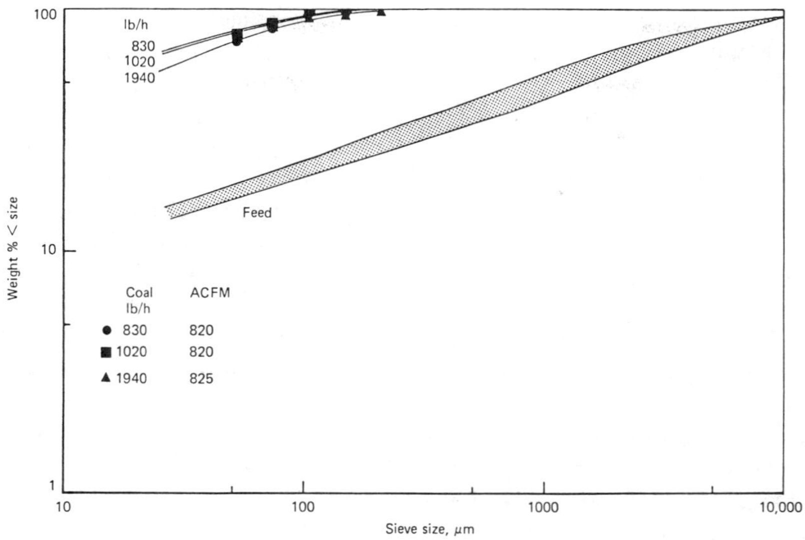

Figure 13.57. Results of grinding Lower Freeport coal (HGI = 88) in E1.7 mill with 50% vane setting on classifier.

air-sweeping rate, for different mass flows. Because the breakage action is not good at low feed rates the product size distribution is actually coarser at 360 lb/h than at 875 lb/h; the product again becomes coarser at the higher feed rate of 1660 lbs/h due to lower mean residence time. Thus this type of mill has two features which distinguish it from the tumbling ball mill. First, it is possible to design a ball mill to give a wide range of mean residence times at the same power draw, producing coarser or finer grinds. The vertical type of mill, however, is less flexible in this respect because it automatically compensates for changes in residence time. Second, if a given ball mill is run for some reason at lower than design capacity it will use the same energy, give a finer grind and use more specific energy per ton. Under the same circumstances, however, the roller mill will pull less power and produce almost the same size distribution with the same specific energy.

Industrial Scale Mills. A simple analysis of the scale-up from small mills to larger mills assuming geometrical similarity is as follows. For a race L wide containing a depth of powder δ, the mean time the feed powder spends passing through the race is L/V_D where V_D is radial powder velocity at the race. The fraction of the race area covered by rolling balls per unit time is $n\omega$ for n balls. The centrifugal action of the rotating table will produce a velocity of powder at the race of $V_D = \omega D/2$. Thus, the number of times the material is rolled over as it passes through the race is $(L/V_D)(n\omega) = 2nL/D$, and since $L \propto D$, the number is constant irrespective of mill size. The severity of the crushing action on each particle as the line contact of ball-on-race rolls over the particles is proportional to $M/nL\delta$. The rate of material passing is $\pi D\delta V_D = \pi D^2\delta\omega/2$. Multiplying the material acted on per unit time X severity of action X number of actions per pass gives "total rate of breakage $\propto MD\omega$." As far as M and ω are concerned, this simple analysis predicts the results found experimentally in Section 4.

Now, the manufacturers information[46] show that the crushing load M is scaled by

$$M = 1700D^2 \quad \text{lbs} \tag{53}$$

The rotational speed is scaled by

$$\omega = 109/\sqrt{D} \quad \text{r.p.m.} \tag{54}$$

Inserting into the total rate of breakage expression,

$$Q \propto D^{2.5} \tag{55}$$

This "law" is supported by the manufacturer's capacity versus mill size information for the Babcock and Wilcox E-type mill series.

Similarly, mill power is expected to be

$$\text{net mill power} = (1.356 \times 10^{-3})\eta M\pi D\omega, \text{ kW} \quad (56)$$

$$\propto D^{2.5} \quad (56a)$$

For M in lb, D in feet, ω in r.p.m. The data supplied for the E-type series gives net mill power = $1.23D^{2.5}$ kW, thus $\eta = 0.093$. This compared favorably with the range of η values found in the laboratory tests. Again, the analysis predicts the comparative specific grinding energy to be approximately constant from one size of mill to another.

The $D^{2.5}$ relation of Equations (55) and (56) is supported by information supplied[47] by the Polysius Corporation for their vertical roller mill. However, the major supplier of roller mills in the U.S.A., the Combustion Engineering Company, uses a law of $Q \propto D^{3.2}$ and $m_p \propto D^{3.0}$ for its bowl-roller mills.[48] The reason for the different scaling law between the different types of roller mill is not apparent.

13.5. FUTURE WORK

It must be apparent to the reader that much work remains to be done to raise the technical understanding of the unit operation of size reduction to that of other (perhaps fundamentally simpler) unit operations such as heat transfer, distillation, absorption, etc. The mechanical stressing conditions inside mills are complex, and the fracture and disintegration of natural materials is a complex phenomenon. It must be emphasized that for size reduction we are concerned not only with the conditions at which fracture occurs but also the size distribution of the set of fragments resulting from the fracture.

The conversion of electrical energy via mechanical action to surface energy of fracture is thermodynamically very inefficient. However, based on the industrial requirements on cost, throughput, wear and reliability of operation it is difficult to see how to improve existing devices substantially or how to invent new ones with much greater efficiency.

The methodology of characterizing a size reduction operation by examining the specific rates of breakage and the primary daughter fragment distributions has proved very informative. Again, however, there are no precise descriptions of why the values of S_i and $B_{i,j}$ vary in the way observed. The variations are often sensible from simple physical reasoning, but the quantitative relations involved are still essentially empirical.

The choice of a certain crusher-mill combination for a given job is made intuitively at present; the choice is not the logical result of a precise set of rules or calculations. This can only be improved by in-depth characterization of machines and materials.

The mechanisms of the slowing-down of rate which is observed as fines accumulate remain to be investigated in detail, and this branch of investigation will undoubtedly involve the nature of the cohesive interaction between particles, dry and in dense slurries, and the effect of grinding additives on these forces.

The better utilization of many ores and fuels in the future may involve requirements of mechanical reduction to ultrafine sizes. This represents a branch of investigation which has hardly been touched and which poses many problems in theory, experimental technique and engineering design.

REFERENCES

1. A. Nadai, *Theory of Flow and Fracture of Solids*, McGraw Hill, Inc., New York, p. 89 (1950). See also *Developments in Fracture Mechanics*, vol. 1, ed. G. G. Shell, Applied Science Publishers, (London) 1979.
2. A. A. Griffith, "Phenomena of Rupture and Flow in Solids," *Phil. Trans. Roy. Soc.* (London) 221A, 163 (1920).
3. A. A. Griffith, "The Theory of Rupture," *Proc. First Int. Cong. for Applied Mechanics*, Delft (1924).
4. G. R. Irwin, *Fracture Dynamics*: *Fracturing of Metals*, Am. Soc. Met., 1948; Orowan, E., "Fracture and Strength of Solids," *Reports of Progress in Physics*, Physical Society (London) *12*:185 (1949).

5. R. P. von, Rittinger, *Lehrbuch der Aufbereitungskunde*, Ernst v. Korn., Berlin (1857).
6. H. E. Rose, private communications (1964).
7. I. J. Lin and S. Nadir, "Review of the Phase Transformations and Synthesis of Inorganic Solids by Mechanical Treatment," *Mat. Sc. Eng. 39*:193–209 (1979).
8. J. S. Benjamin, "Mechanical Alloying," *Scientific American 234*:41–48 (May 1976).
9. N. H. Macmillan, "Chemisorption Induced Variations in the Plasticity and Fracture of Non-metals," in *Surface Effects in Crystal Plasticity*, Nordhoff, Leyden, p. 629 (1977).
10. A. R. C. Westwood and J. J. Mills, "Application of Chemo-mecahnical Effects to Fracture-dependent Industrial Processes," ibid., p. 835.
11. A. A. Bradley, P. S. Lloyd, D. A. White, and P. W. Willows, "High-speed Centrifugal Milling and Its Potential in the Milling Industry," *South African Mechanical Engineer 22*:129–134 (1972).
12. A. L. Hinde and F. B. Verardi, Studies on Design of Centrifugal Mill Grinding Circuits," *Proc. 3rd IFAC Symposium, Automation in Mining, Mineral and Metal Processing*, Montreal (Canada), pp. 283–294 (Aug., 1980). See also: L. P. Kitschen and P. J. Lloyd, "The Centrifugal Mill: Experience with a New Grinding System and Its Applications," *Proc. 14th IMPC*, Toronto, 1982.
13. L. G. Austin, "A Review Introduction to the Description of Grinding as a Rate Process," *Powder Technology 5*:1–17 (1971/72).
14. L. G. Austin, R. R. Klimpel, and P. T. Luckie, *The Process Engineering of Size Reduction: Ball Milling*, AIME, N.Y., in press.
15. L. G. Austin and V. K. Bhatia, "Experimental Methods for Grinding Studies in Laboratory Mills," *Powder Technology 5*:261–266 (April, 1972).
16. L. G. Austin and P. T. Luckie, "Methods for Determination of Breakage Distribution Parameters," *Powder Technology 5*:215–222 (March, 1972).
17. L. G. Austin and P. T. Luckie, "Estimation of Non-normalized Breakage Distribution Parameters from Batch Grinding," *Powder Technology 5*:267–277 (April, 1972).
18. K. Reid, "A Solution to the Batch Grinding Equation," *Chem. Eng. Sc. 20*:953 (1965).
19. P. T. Luckie and L. G. Austin, "A Review Introduction to the Solution of the Grinding Equations by Digital Computation," *Minerals Science and Engineering 4*:24–51 (April, 1972).
20. R. S. C. Rogers and R. P. Gardner, "Use of a Finite-stage Transport Concept for Analyzing Residence Time Distributions of Continuous Processes," *J.A.I.Ch.E. 25*:229 (1979).
21. F. C. Bond, "Crushing and Grinding Calculations," *Brit. Chem. Eng. 6*:378 (1965).
22. C. A. Rowland, Jr. and M. M. Kjos, "Rod and Ball Mills" in *Mineral Processing Plant Design*, ed. A. L. Mular and R. B. Bhappu, AIME, N.Y., 239–278 (1978).
23. L. G. Austin and J. W. Perez, "A Note on Limiting Size Distributions from Closed Circuit Mills," *Powder Technology 16*:291–293 (1977).
24. L. G. Austin and P. Bagga, "An Analysis of Fine Dry Grinding in Ball Mills," *Powder Technology, 28*:83–90 (1981).
25. T. Allen, *Particle Size Measurement*, 2d ed., Holstead Press, John Wiley, N.Y., p. 98 (1975).
26. A. M. Gaudin, *Principles of Mineral Dressing*, McGraw-Hill, N.Y., pp. 41–43 (1939).
27. L. G. Austin, D. R. Van Orden, and J. W. Perez, "A Preliminary Analysis of Smooth Roll Crushers," *Int. Journal of Mineral Processing 6*:321–336 (1980).
28. L. G. Austin and J. D. McClung, "Size Reduction of Coal," in AIME Handbook, *Coal Preparation*, Harvey Mudd Series, ed. J. Leonard, p. 7.14 (1979).
29. L. G. Austin, K. Shoji, D. R. Van Orden, B. McWilliams, and J. W. Perez, "Breakage Parameters of Some Materials in Smooth Roll Crushers," *Powder Technology, 28*:245–251 (1981).
30. F. Kick, *Dinger Polytech. J. 247*:1 (1883); *250*: 141 (1883).
31. K. Shoji, S. Lohrasb, and L. G. Austin, "The Variation of Breakage Parameters with Ball and Powder Loading in Dry Ball Milling," *Powder Technology 25*:109–114 (1979).
32. H. E. Rose and J. Sullivan, *Rod. Ball and Tube Mills*, Chemical Pub. Co., N.Y. (1958).
33. L. G. Austin, "Understanding Ball Mill Sizing," *Ind. Eng. Chem. Proc. Des. Devel. 12*:121–129 (February, 1973).
34. F. Locher and H. M. von Seebach, "Influence of Adsorption on Industrial Grinding," *I.E.C. Proc. Des. and Devel. 11*:190 (1972).
35. L. G. Austin, P. Bagga, and M. Celik, "Breakage Properties of Some Materials in a Laboratory Ball Mill," *Powder Technology, 28*:235–241 (1981).
36. R. R. Klimpel and L. G. Austin, "Chemical Additives for Wet Grinding of Minerals," *Powder Technology, 31*:239–253 (1982).
37. L. G. Austin, R. R. Klimpel, P. T. Luckie, and R. S. C. Rogers, "Chapter 20: Simulation of Grinding Circuits for Design," *Proc. Sym. on Milling*, Ed. A. L. Mular and G. V. Jergenson AIME, N.Y. (1982).
38. Y. Mori, G. Jimbo, and M. Yamazaki, "On the Residence Time Distribution and Mixing Characteristics of Powders in Open-circuit Ball Mill," *Kagaku Kogaku 28*:204 (1968).
39. L. G. Austin, R. R. Klimpel, and P. T. Luckie, "Residence Time Distributions," *Process Engineering of Size Reduction: Ball Milling*, AIME, N.Y., in press.

40. L. G. Austin, N. P. Weymont, and O. Knobloch, "The Simulation of Air-swept Cement Mills: Part I, The Simulation Model," *Proc. 5th European Symposium on Size Reduction*, Amsterdam (1980).
41. L. G. Austin, K. Shoji, "A Simulation Model for an Air-swept Ball Mill," *Powder Technology*, in press.
42. L. G. Austin, J. Shah, J. Wang, E. Gallagher, and P. T. Luckie, "An Analysis of Ball-and-Race Milling: Part I, The Hardgrove Mill," *Powder Technology*, *29*:263–275 (1981).
43. L. G. Austin, P. T. Luckie, and K. Shoji, "An Analysis of Ball-and-Race Milling: Part II, The Babcock E-1.7 Mill," *Powder Technology*, *33*: 113–125 (1982).
44. L. G. Austin, P. T. Luckie, and K. Shoji, "An Analysis of Ball-and-Race Milling: Part III, Scale-up to Industrial Mills," *Powder Technology*, *33*: 127–134 (1982).
45. ASTM Designation; D409-51, *19*:72 (1970).
46. Babcock and Wilcox (U.K.), Crawley, Surrey, private communication (1979).
47. Polysius Corporation, Beckum, W. Germany and Atlanta, Ga., private communication (1979).
48. Combustion Engineering, Inc., Raymond Division, Chicago, Ill., private communication (1979).

14
Sedimentation

Keith J. Scott
Council for Scientific and Industrial Research (CSIR)
Pretoria, South Africa

14.1. INTRODUCTION

Gravity sedimentation is a widely used method of separating solid–liquid mixtures and includes diverse applications such as clarification of sugar juices, thickening of milled gold ore pulps, flotation of suspended sewage solids and countercurrent washing of soluble metal values from acid-leached suspensions. These operations are performed typically in relatively large single-compartment tanks such as shown in Figs. 14.1, 14.2 and 14.3. The discussion in this chapter will concentrate on sedimentation in *liquids*.

Suspensions of solids normally sediment naturally, as the only driving force required is gravitational attraction which operates when there is a difference in density between solid and liquid. Given time a suspension separates into a clear liquid layer above, supernatant, and a zone of solids below, which remains "saturated" with liquid. Such *batch* sedimentation can be carried out on a large scale in tanks, termed collectors,[1] ponds, or lagoons.

To achieve *continuous* operation, it is necessary merely to supply a steady stream of fresh unsettled suspension, the *feed*, to the center (or end) of the sedimentation vessel and this results in the continuous automatic displacement of the lighter liquid phase, termed the *overflow*. Solids removal is normally achieved by continuous raking of the thickened sediment towards the center (or opposite end) of the tank, from where it is pumped out as the *underflow* stream. Once the effective solids concentration of the underflow exceeds a volume fraction of about 0.4 (65% solids by mass in the case of quartz) it rapidly becomes too stiff to pump. An appreciable fraction of the liquid therefore leaves the sedimentation tank together with the solids.

The relative simplicity of both the process and the mechanical equipment involved makes gravity sedimentation the least costly of the available solid-liquid separation techniques.[2,3] This is because of its capability of treating high water flow rates with relatively little hardware[1] while usually achieving a high degree of clarity in the overflow. Some solids which are difficult to filter can be separated effectively by sedimentation.

Other solid-liquid separation techniques need to be considered however, as an alternative, or in addition to, gravity sedimentation if:

1. the solids stream must have a lower moisture content than 30 to 40% by mass;

2. the loss to the underflow of 10 to 15% of the liquid in the feed is not acceptable;
3. the cost of the required floor space is excessive or space not available; and
4. the process must be carried out under pressure. Table 14.1 provides a *rough* guide as to the advantages of various particle-liquid separation methods.

Combinations of techniques may also be used to advantage, such as a *vacuum filter* immediately following a *thickener* to dewater, or wash the thickener underflow, or, *clarification* followed by *drying* of the solids sludge produced. Selection of optimum solid-liquid separation strategies is, however, outside the scope of this chapter.

One relevant "combination" discussed later under countercurrent decantation (CCD) on p. 670 is the use of a *series* of two to four additional thickeners to recover a valuable soluble component which would otherwise be lost in the underflow liquid of the *first* thickener. A series of wash liquors are mixed, each with an already thickened underflow stream containing liquid of *higher* concentra-

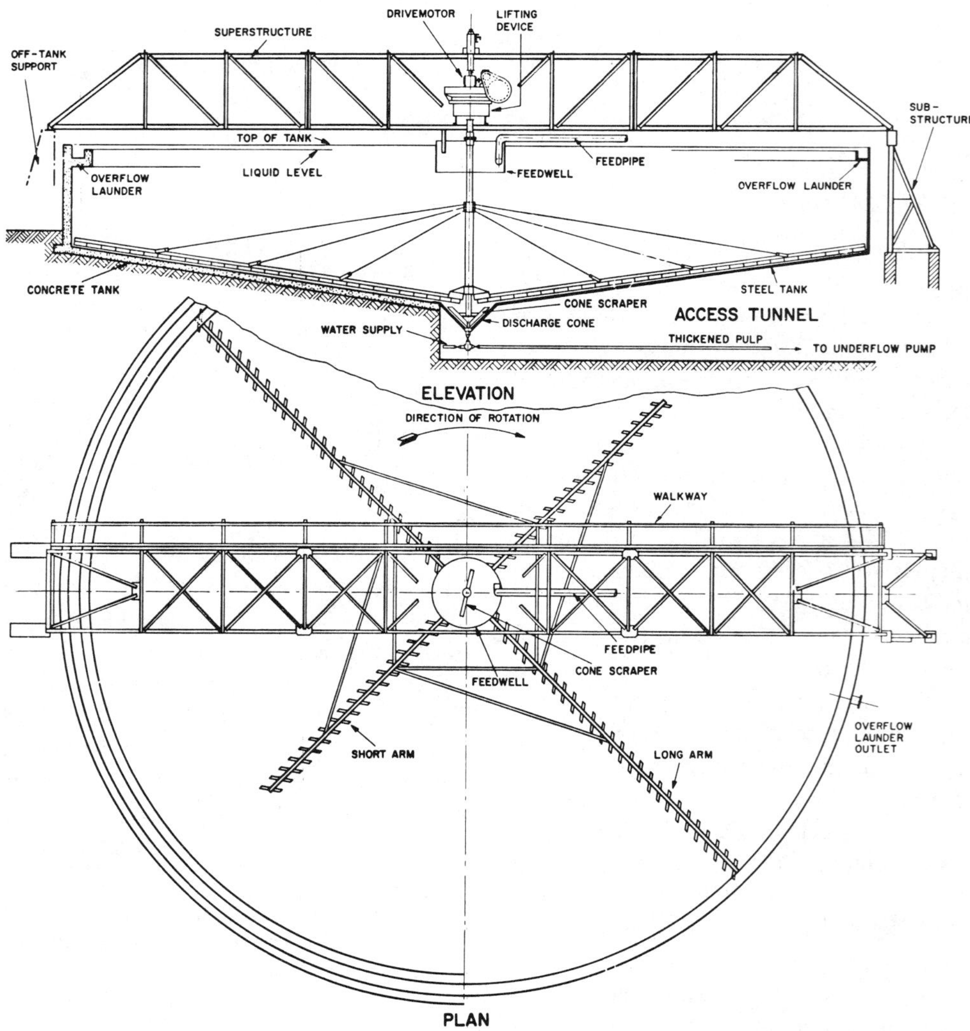

Figure 14.1. Cross-sectional view and plan of a thickener. The tank may be constructed of steel or concrete. The rake lifts vertically automatically if it encounters an unusual resistance. Excess lift results in a "trip-out."

Figure 14.2. View of an operating clarifier. The larger tank, the secondary clarifier, represents one of the final stages of producing clear drinking water from purified sewage in the Stander Water Reclamation Plant, a pilot plant at Daspoort, Pretoria. (Courtesy of National Institute for Water Research, CSIR)

Figure 14.3. View of an empty thickener at a Transvaal gold mine. Such tanks handle up to 15,000 ton/day of solids and 4 to 5 times as much water.

Table 14.1. A Qualitative Comparative Guide to the Particular Advantages of Various Solid-Liquid Separation Techniques.

SOLID-LIQUID SEPARATION TECHNIQUE	RATIO OF THROUGH-PUT TO FLOOR AREA	CLARITY OF LIQUID	MOISTURE CONTENT OF SOLIDS STREAM	EASE OF WASHING SOLIDS	OVERALL COST-CAPITAL AND OPERATING
Cyclones	Very high	Very poor	High	Requires 2-4 repeat operations	Medium
Drying	High	–	Extremely low	–	High
Filtration	High	Good	Low	Easy	High[a]
Screening	High	Very poor	Medium to high	Easy	Medium
Sedimentation					
–Centrifugal	Medium to high	Good	Low to medium	Easy	High
–Gravity	Low	Excellent	High	As for cyclones	Low[a]

[a]The operating cost of vacuum filters in the S.A. gold mining industry (~ 75 million tons/a) is 6-8 times the cost of thickening.[4]

tion of the valuable component. On resedimentation, about 80-85% of this component is therefore recovered in the overflow, and it in turn, becomes the wash liquor for the next higher stage. Such repeated operations, however, raise the overall cost of thickening to a level where sedimentation may no longer be economically competitive with other techniques.

In selecting a separation process it is essential therefore, to consider wider aspects than just the pros and cons of *single* techniques. The literature gives some guide lines for selecting equipment[2,5-7] but this should be supplemented by discussions with specialized consultants and with suppliers, who, because of their experience, should be able to help in formulating likely solutions for a given problem.

Objectives in Gravity Sedimentation

In sedimentation we distinguish two primary functions; *clarification*, in which absence of solids in the liquid overflow is the essential requirement and a relatively high proportion of liquid in the underflow can be tolerated and *thickening*, in which the minimum quantity of liquid in the underflow is the main objective and the presence of up to a few per cent of suspended solids in the overflow (often harmlessly recirculated) is of secondary concern. The distinction is therefore in the *end result* rather than the *process* as in thickening, the solids concentration in the feed stream is *increased* by sedimentation while in clarification the solids are *removed*, but also by the same process–sedimentation.

Each of the two functions can be maximized and controlled separately. The turbidity of a clarification tank overflow is related to slowly settling fine solids which may be flocculated to form larger faster settling units. The control of overflow clarity is therefore effected by the selection of flocculant, its dosage and by control of the volumetric feed rate.

Underflow density of a thickener depends on the height of pulp in the tank, the degree of flocculation in this pulp (flocculated material tends to incorporate more liquid than dispersed particles; p. 622) and on the volumetric feed rate. Because flocculation and feed rate affect the overflow clarity *and* the underflow density it is seldom possible to maximize both functions simultaneously.[8]

Applications of Gravity Sedimentation

Sedimentation processes are extensively used throughout the world in the mining industry, water treatment and sewage purification works. These large tonnage operations are often carried out in remote locations or nonurban areas where land is available and relatively inexpensive and hence the use of large tanks is not a serious disadvantage.

Sedimentation *is*, however, also practiced, although on a smaller scale, in semi-urban or industrial areas, in a variety of processes. Increasing attention is being paid to developing higher capacity units i.e., those of reduced area per unit throughput (p. 662). This interest in minimizing the main disadvantage of sedimentation, rather than selecting alternative types of equipment, is an indication of the desirability of the *positive* features of gravity sedimentation as a means of separating solid–liquid mixtures.

14.2. THEORY OF SEDIMENTATION

This heading covers the fundamental aspects of the sedimentation of particles, whether single in an infinite liquid or together with many other particles in a suspension of finite extent. As much of the published verification of this theory was carried out in laboratory measuring cylinders this section is also the theory of batch settling tests. How the settling behavior of suspensions observed in batch tests is used in the *design* of sedimentation equipment is covered separately (p. 634), while the nomenclature used, is given at the end of the chapter.

Sedimentation of a Sphere in an Infinite Fluid

The settling of a *single sphere*, in an *unbounded* fluid represents the simplest case of solid–liquid sedimentation. This ideal condition is seldom encountered in practice,[1] however, the complicating factors which arise in real situations, dealt with more fully in subsequent sections, are:

1. nonspherical and irregularly shaped particles,
2. wall effects,
3. the simultaneous presence of a large *number* of closely sized particles,
4. the result of a *range* of particle sizes, and finally
5. the presence of mutual particle attraction in which individual particles lose their identity and are grouped into agglomerates (flocs) by chemico-physical forces.

When a *single spherical particle* is suspended at rest in a liquid it experiences two opposing forces B_F and G_F as shown in Figure 14.4. Provided the density of the solid and liquid are not equal, there will be an unbalanced force, the difference between G_F, the downward gravitational attractive force, and B_F, the Archimedes' upthrust. This force, $(G_F - B_F)$, equal to $V(\rho_S - \rho_L)g$, causes the particle to accelerate (downwards if positive) and attain a velocity relative to the liquid. Skin-friction, i.e., the resistance offered by a fluid to the motion of a solid, then results in the development of a drag force, F_D, which opposes the motion and increases with increasing particle velocity. The drag reduces the acceleration and finally, the value of the drag force becomes equal to the original driving force $(G_F - B_F)$ and as there are then no further unopposed forces acting on the particle, it continues to travel at a constant rate called its terminal or free-settling velocity, u. We may then write

$$F_D = V(\rho_S - \rho_L)g \qquad (1)$$

This equation evaluates the magnitude of the

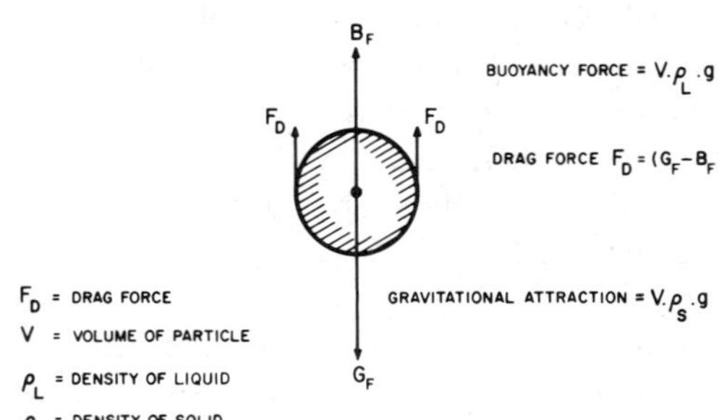

Figure 14.4. Forces acting on a spherical particle in a liquid.

drag force for any size particle but does not relate it to its unknown settling velocity u.

This relationship has been formulated for a sphere in an infinite fluid[9] for slow flows but its general solution depends upon the type and magnitude of flow around the particle as characterized by the dimensionless entity known as the Reynolds number.

Fluid Flow Around a Particle and the Reynolds Number. When the particle velocity is low, drag is due largely to the viscosity of the liquid and this flow is called viscous, laminar or streamlined. At higher velocities, the fluid streamlines do not continue completely around the particle but break up into vortices with the result that turbulent eddies and inertial forces also contribute to the drag. This finally develops into fully turbulent flow.

The criterion for distinguishing between flow conditions is the dimensionless *particle* Reynolds number

$$Re = \frac{d \cdot u \cdot \rho_L}{\mu} \tag{2}$$

where

d = particle diameter
u = relative velocity between particle and liquid
ρ_L = liquid density
μ = liquid viscosity

For particles,[12b] *streamlined* flow occurs below $Re \sim 0.2$, *turbulent* flow above $Re \sim 2 \times 10^5$, while in the intermediate region, in which inertial forces become increasingly significant, the flow is called *transitional*. These limiting values for the particle Re are orders of magnitude lower than for flow in pipes in which the fluid streamlines are constrained by the boundary walls.

Streamlined Flow. The analytical solution for the magnitude of the drag on a single *sphere*, sedimenting under *laminar* flow conditions in an unbounded liquid is given by Stokes[9] as

$$F_D = 3\pi\mu d u_\phi \tag{3}$$

where u_ϕ = terminal velocity of the sphere in an infinite fluid in streamline flow.

Even for this simplified condition, however, Equation (3) is only a close *approximation* and, for greater accuracy, additional terms have been found to be necessary. Proudman and Pearson[10] for example, advocate the equation

$$F_D = 3\pi\mu d u_\phi \left[1 + \frac{3}{16}Re + \frac{9}{160}(Re)^2 \ln\left\{\frac{Re}{2}\right\} + \cdots\right] \tag{4}$$

It has become common practice[11] to express the forces exerted on moving bodies by the fluid in terms of a dimensionless drag coefficient C_D, obtained by dividing the drag force by $\rho_L u^2/2$ and by the area of the body projected onto the plane normal to u. For a *sphere*, this area is $\pi d^2/4$, hence the coefficient is

$$C_D = \frac{F_D}{\frac{\rho_L u^2}{2} \times \frac{\pi d^2}{4}} \tag{5}$$

which, together with Equations (2) and (4) and putting $u = u_\phi$, becomes

$$C_D = \frac{24}{Re} + \frac{9}{2} + \frac{27}{20} Re \ln\left\{\frac{Re}{2}\right\} + \cdots \tag{6}$$

The first term, $24/Re$, is the equivalent of Equation (3). The *practical* significance of the two additional terms can be tested by considering the largest Reynolds number likely to be encountered in real situations. Thus the maximum overflow rate of operating sedimentation equipment quoted by Perry and Chilton[12a] is $u \sim 0.8$ mm/s; the 95% upper limiting diameter in their particulate slurries is estimated as $d \sim 0.2$ mm, and, if we accept the most common sedimentation liquid as being water at 20°C, $\rho_L = 1 \times 10^3$ kg/m^3, $\mu = 1 \times 10^{-3}$ kg/ms, and therefore Re should lie largely in the range 0 to 0.16.

The accurate value of C_D for $Re = 0.16$ according to Equation (6) is 153.95, while using the first term only, $C_D = 150.0$, representing a difference of 2.6%. This maximum error is quite acceptable in settler design as variations in ambient temperature of only 1 to 2°C can, by changing the liquid viscosity, result in greater variations in sedimentation rate. The second and third terms in Equation (6) may therefore be safely neglected and the equation simplified to give

$$C_D = \frac{24}{Re} \tag{7}$$

which is an alternative expression of Equation (3). As the particle is a *sphere* V in Equation (1) is $\pi d^3/6$ and from Equation (1) and (3) we can now write

$$u_\phi = \frac{d^2(\rho_S - \rho_L)g}{18\mu} \tag{8}$$

This equation provides a means of calculating the terminal settling velocity of a single sphere of diameter d in an infinite fluid, in streamlined flow, as determined by the physical properties of the sphere and fluid, i.e., their densities and the fluid viscosity. Alternatively, it permits the estimation of the diameter of a particle by observing its settling velocity under these prescribed conditions. For a spherical quartz particle in water at 20°C, the maximum permissible diameter is $d \sim 85\ \mu m$ if the error in u_ϕ using Equation (8) is not to exceed 5%.

Transitional and Turbulent Flow. Grit chambers, a special type of sedimentation basin used in sewage treatment, are designed for removal of coarse sands larger than 200 μm. Such large particles settle in transitional or turbulent flow but for the determination of their terminal velocities, no alternative simple expression similar to Equation (8) exists. To calculate their velocity, $u_{\phi t}$, the particle Reynolds number must first be known, but as this cannot be known until the value of $u_{\phi t}$ is determined, a tedious trial-and-error solution is one means of arriving at its value.[13]

The following more direct and accurate algorithm has, however, been found useful in avoiding both this repetitive procedure and the inaccurate graphical or cumbersome interpolation of the $Re - C_D Re^2$ values given in Perry and Chilton.[12b] The algorithm includes also the laminar flow region discussed previously.

Algorithm for Calculating $u_{\phi t}$ from d

1. Calculate the entity $J = C_D Re^2$ which does not contain $u_{\phi t}$:

$$J = C_D Re^2 = \frac{4}{3} \cdot \frac{(\rho_S - \rho_L)\rho_L d^3}{\mu^2} \cdot g$$

$$\text{Note: } J = \frac{4\,Ga}{3}$$

2. From J calculate the required values of a and b according to the data given in Table 14.2 using the appropriate range of J values.
3. Calculate

Table 14.2. Values of a and b for Calculating Re from $J = C_D Re^2$ in any Flow Regime.

FLOW REGIME		$J = C_D Re^2$	$\sim Re$	$\sim d$	a	b
Streamlined (Stokes' law)		0–10	0–0.4	1–75 μm	24	1
Transition region	I	10–10^3	0.4–20	75–350 μm	$33.7\,J^{-0.19}$	$1.05\,J^{-0.05}$
	II	10^3–2×10^5	20–600	0.35–2 mm	$63.6\,J^{-0.29}$	$1.01\,J^{-0.046}$
	III	2×10^5–1×10^8	600–1.5×10^4	2–15 mm	$91.7\,J^{-0.30}$	$0.96\,J^{-0.039}$
Newton's law		1×10^8–1.7×10^{10}	1.5×10^4–2×10^5	1.5–10 cm	0.67	0.5

$$J = C_D Re^2 = \frac{4}{3} \cdot \frac{(\rho_S - \rho_L)\rho_L d^3}{\mu^2} \cdot g = \frac{4}{3} Ga \qquad Re = \frac{J^b}{a} \qquad u_{\phi t} = \frac{Re \cdot \mu}{d \cdot \rho_L}$$

$$Re = \frac{J^b}{a}$$

4. Finally, determine

$$u_{\phi t} = \frac{Re \cdot \mu}{d \cdot \rho_L}$$

The values calculated in this way are accurate to within 5 to 6%. A subroutine for the trial-and-error solution mentioned earlier[13] is less accurate; e.g., at $Re \sim 70$ the error is 18%.

The algorithm covers a wide range of sphere diameters from 1 μm up to 10 cm and embraces the streamlined, transitional and Newton's flow regimes. When, however, the flow is known *beforehand* to be laminar ($Re < 0.2$), Equation (8) provides exactly the same answer in fewer steps.

A theoretically derived[199] equation:

$$Re = 20.52[(1 + 0.0921J^{0.5})^{0.5} - 1]^2$$

gives results which are within 7% of the experimental values for Re up to 7000, i.e., an upper limiting diameter of 7 mm.

A free-settling equation, valid not only for all particle sizes, but covering a wide range of naturally occurring *shapes* (see following section) is presented by Swanson;[14] his calculated values are within 20% of the measured values in most cases.

Nonspherical Particles

Spherical particles rarely occur in solid–liquid separation practice, some more common shapes being angular and flaky mineral particles derived from crushing or weathering, amorphous fluffy precipitates, randomly formed agglomerates and regular polyhedral or needle-shaped crystals.

The drag on a nonspherical particle depends upon its shape and its orientation with respect to the direction of motion.[12b] If a body possesses spherical isotropy (such as a cube or octahedron) and is placed initially in any orientation in a liquid and allowed to fall without initial spin, it will not rotate to a different position but fall vertically in its original orientation.

Most real particles are not symmetrical however and they experience not only drag forces parallel to the stream velocity, but lateral (lift) forces at right angles to the stream. This may cause drift to one side during settling, rotation to a position of maximum resistance, steady rotation or even a wobbling motion.

Even when such nontranslational motions are neglected, the calculation of the terminal settling velocity of *nonspherical* particles using an equation similar to Equation (8), i.e., $u_{ns} = f(\rho_S, \rho_L, \mu, d)$, requires first that the shape is known or can be determined, secondly that a representative "diameter" d can be assigned to this shape and thirdly, that a drag equation be available similar to Equation (3) for a sphere.[15] The problem is thus complicated, and certainly no analytical solution exists for the *irregular* shaped particles met with in practice.

However, empirical methods for dealing with shape are available. These are presented briefly with reference to some of the effects of increasing departure from spherical form shown in Table 14.3.

This table compares particles of various shapes on the common basis of having the same volume, (arbitrarily 1 mm^3), i.e., of each possessing an equal gravitational attractive force. Column 3 shows that as the shape departs increasingly from spherical, the surface area increases from 4.84 for a sphere to 8.35 for a cylindrical needle of the same volume, i.e., an increase of 73%. The skin friction and hence drag for nonspherical particles will thus be greater and their terminal settling velocities correspondingly lower than for a sphere of the same volume.

Skin friction is related to surface area. If, in the drag Equation (3), d is taken as the diameter of a sphere having the same *surface area* as a given shaped particle,[15] from Equation (1) and (3) it can be shown that

$$\frac{u_{ns}}{u_{\phi t}} = \frac{d}{d_A} = K \tag{9}$$

where

Table 14.3. Various "Diameters" of Particles of Equal Volume (1 mm³) but Differing in Their Shape.

SHAPE	DESCRIPTION	CALCULATED SURFACE AREA (mm^2)	DIAMETER OF A SPHERE (mm): HAVING THE SAME SURFACE AREA AS PARTICLE d_A	SHOWING THE SAME PROJECTED AREA AS PARTICLE (SIZING BY MICROSCOPE) d_M	PASSING THE SAME MINIMUM SQUARE APERTURE (SCREEN ANALYSIS) d_{scr}	RATIO d_M/d_{scr}
Sphere	–	4.84	1.24 = d	1.24	1.24	1.00
Icosahedron	20 equilateral triangles	5.15	1.28	1.32	1.23	1.07
Dodecahedron	12 pentagons	5.31	1.30	1.35	1.27	1.06
Cube-octahedron	6 octagons and 8 equilateral triangles	5.69	1.35	1.14	–	–
Octahedron	8 equilateral triangles	5.72	1.35	1.36	1.24	1.10
Hexahedron (cube)	6 squares	6.00	1.38	1.13	1.00	1.13
Tetrahedron	4 equilateral triangles	7.20	1.51	1.51	1.44	1.05
Plate (5:5:1)	–	8.19	1.61	1.93	1.45	1.33
Needle (10:1)	Right circular cylinder	8.35	1.63	1.80	0.50	3.60

$u_{\phi t}$ = settling rate of sphere of diameter d having the same *volume* as the nonspherical particle
d_A = diameter of sphere having same *surface area* as the nonspherical particle
K = shape factor

This treatment is not rigorous because Equation (3) applies strictly to spheres. However, the values of K calculated as above using the data in Table 14.3, compare reasonably well with the experimentally determined values[16] as shown in Table 14.4.

The agreement while not exact, serves to illustrate the principle of using shape correction factors, based upon more than one characteristic diameter of a particle, to estimate its settling rate. The particle must, however, be of *known* shape.

The more usual situation is that neither the shape *nor* the size is known, i.e., d and d_A *both* need to be determined experimentally.

Table 14.4. Comparison of Calculated and Observed Shape Factors *K*

SHAPE	SHAPE FACTOR K: CALCULATED, d/d_A	OBSERVED EXPERIMENTALLY[16]
Sphere	1.0	1.0
Cube-octahedron	0.92	0.96–0.98
Octahedron	0.92	0.93–0.95
Cube	0.90	0.92–0.94
Tetrahedron	0.82	0.82–0.86

Two commonly used methods of measuring particle size are microscopy and sieve analysis. For spheres both techniques give identical results but for nonspherical particles, the two methods, because they measure different properties, give *different* characteristic diameters for the *same* particle.

Thus, in microscopy the diameter of a particle, d_M, is the diameter of a circle with the same *projected* area as the particle which normally lies "flat," while in sieving the particle has a probability of being presented to the apertures in the most favourable position for passing, d_{scr}. Thus, for an elongated or platy particle, the two *larger* dimensions are recorded by means of the microscope while for sieving the two *smaller*.

Columns 5 to 7 in Table 14.3 compare the mean diameters of various shaped particles as measured by these two methods. It can be seen that d_M, is as expected, consistently larger than d_{scr}, except for a sphere where they are equal. The ratio d_M/d_{scr} therefore, also provides a basis for formulating shape correction factors but it is less consistent than the ratio d/d_A. The latter can, however, be used only if d and the shape are known.

Much work has been done on characterizing the shapes of microscopic images *alone*,[16-18] but the weakness in applying such results to sedimentation[19] is that only *two* dimensions are considered. Thus, a cube and a square plate both of side d settle at quite different rates but are indistinguishable under the normal microscope while a right circular cylinder of equal height and diameter may appear spherical to one observer and cubic to another depending on its position on the microscope slide. Different shape factors would therefore be applied to the identical particle!

Both the experimental procedures for measuring shape factors as well as their interpretation are therefore involved but the main purpose is merely to apply a *correction factor* to Stokes' law so as to be able to *calculate* the settling velocity of any nonspherical particle. The necessity for this however, is completely eliminated if for the measurement of d, we use an *alternative*, very commonly used method of particle sizing–the sedimentation method.

This technique supplies information on the settling velocity, u_{ns}, of *any* particle of unknown shape which when inserted into Equation (8) gives the exact equivalent Stokes' diameter, d_{ns}. This is the diameter of a sphere of settling velocity identical to that of the nonspherical particle. It combines both the "true diameter" of the particle and its shape correction factor into a single term. For the transitional and turbulent flow regimes such direct observation is the *only* means of determining the settling velocities of nonspherical particles.[15] Investigations of settling of such particles have therefore been mainly experimental.[16,20,21] Most particles encountered in industrial practice are far from spherical to such an extent that not only the overall shape plays a role but also the microsurface topography.

The experimental observations of Richards[21] on the terminal settling velocities of various sized quartz particles u_Q, obtained by screening, may be used to determine the shape factor for this irregular but commonly occurring material. Richards, using the screen apertures as the characteristic size d, calculated the expected velocities, u_R (assuming spheres) of the various sized particles using Rittinger's law[21] for *turbulent flow* and found that "for grains smaller than 1.55 mm in diameter, the value of f [his shape factor $= u_Q/u_R$] decreases in a most extraordinary degree." Selected values are shown in Table 14.5.

As the smaller particles settled within the *streamlined* regime, Rittinger's law does not apply in this range and his predicted settling velocities were therefore recalculated using the algorithm in Section 2 with viscosity and density values as given by Richards.[21] The results are plotted in Figure 14.5.

Table 14.5. Selected Shape Factors *f* as Calculated by Richards[21] for Various Sized Quartz Particles *d*.

d (mm)	f
11.93	0.6325
1.55	0.5634
0.121	0.1626
0.0126	0.00709

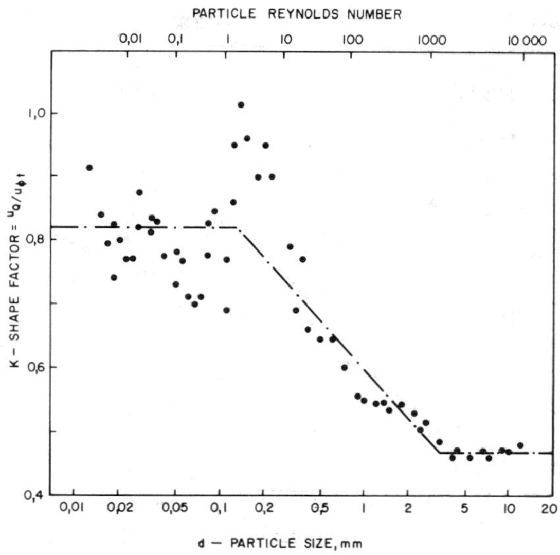

Figure 14.5. Variation of shape factor of quartz with particle size and *Re*.

It can be seen that the shape factor is now more consistent, varying over a 2, rather than the 90 fold range as before and that the value depends upon the size *d*. Relating the shape factor to the corresponding Reynolds number—at $Re < 1$, the data are scattered but average out at about 0.8; for $Re > 2000$ the shape factor is steady at 0.47, while in the transition region the factor shows a steady decrease with increasing *Re*.

The shape factor for a given nonspherical particle is therefore *not even a constant* for that particle but dependent also upon the prevailing conditions.

Accepting that for quartz particles in streamlined flow $u_Q = 0.8 \times u_{\phi t}$, then from Equation (9) $d/d_A = 0.8$ which from Table 14.3 gives a values of $d_A = 1.55$. Thus quartz particles apparently experience a drag approaching that of the extreme nonspherical shapes shown in Table 14.3 for which $d_A \sim 1.6$. From observation, quartz particles resemble neither needles nor plates and the additional effective drag area must therefore arise from their surface roughness.

Measurements of settling velocity of nonspherical particles are therefore simpler and of more use than *prediction* of these velocities from independent size and shape determinations.

The Effect of Container Walls

When a particle sediments in a closed column rather than in an infinite liquid, it displaces its volume of liquid from a lower to higher level and the wall interferes with the ideal liquid flow pattern. This results in an additional drag on the particle and a reduction in the free settling velocity $u_{\phi t}$ by a factor W, which decreases with increasing ratio d/D where D is the diameter of the container. The retardation effect may be expressed as follows

$$u_w = u_{\phi t} \times W \tag{10}$$

and many expressions for W have been proposed.[1]

According to Francis[22]

$$W = \left[\frac{1 - 0.475\,(d/D)}{1 - (d/D)}\right]^{-4}$$

for streamlined flow, while Garside and Al-Dibouni[23] give a simpler equation $W = [1 + 2.35\,(d/D)]^{-1}$; applicable for Reynolds numbers between 3 and 1200.

For a particle of $d = 200\ \mu m$, settling in a one-liter graduate cylinder ($D \sim 60$ mm) or in a 25 m diameter tank, the calculated settling velocities, u_w, are 99.3 and 99.998% respectively of its velocity in an unbounded fluid.

Although wall effects in full scale equipment may therefore be safely ignored, design of such equipment is often based upon settling rates observed in small laboratory glassware, for which corrections *may* therefore sometimes be required. In practice this correction is, however, usually neglected as the consequent error is both small and conservative.

Settling of a Particle in the Presence of Identical Neighbors

Many correlations have been presented to describe the effect of higher solids concentrations on the settling rate of similarly sized uniformly dispersed spherical particles.[23] Two effects have been observed. One is that some particles may loosely associate into a group, separated from each other by several diameters, and act as an entity descending at a higher rate than that corresponding to the expected terminal velocity of the individual particles. Such "clusters" are often transient and their occurrence has been observed[24-26] predominantly at low particle concentrations. Other particles in the same suspension remain single and may even show negative settling rates when being carried upward by the return flow from the rapidly descending clusters. Tory and Pickard[27] present a stochastic model that accounts for these wide variations in settling rate. They noted that in spite of variations of settling velocity between particles, the mean settling velocity as shown by their *overall* rate of descent remained remarkably constant.

The second effect is that as concentration increases *each* particle is subjected to increased drag due to the higher volume of return flow fluid displaced by the sedimenting particles. Alternatively, the ideal fluid flow around each particle is disturbed by the presence of its neighbors. This mutual retarding effect may therefore be regarded as a special form of the wall effect in which the container diameter is replaced by some characteristic mean interparticle distance. As this is much lower than D, the drag effect of the walls is relatively small.

Hindered Settling. With closely sized particles uniformly distributed in a settling cylinder, a visible interface between suspension and liquid forms at commencement of settling even at low concentrations because of the constant mean descent rate of each particle. In effect this interface is the one that exists between suspension and air before commencement of sedimentation. It is usually hazy, however, because particles are never *exactly* identical. At higher concentrations this interface becomes increasingly well defined and sharp, forming even for particles with a considerable range of sizes (see later). As such suspensions separate when dilute, because of the wide variation of settling velocities present, the formation of a distinct interface at higher concentrations indicates that interference between particles is such that particles of all sizes descend jointly, i.e. they are in hindered settling.

The settling velocity of the interface, u, was noted by Richardson and Zaki[28] to be related to the velocity of a single particle u_0, and the concentration c, by

$$u = u_0 (1 - c)^n \tag{11}$$

where

u = mean settling rate of particles (particle-supernatant interface) in a container in the presence of many others

u_0 = terminal velocity of a *single* representative particle, i.e., u when $c = 0$, under otherwise similar conditions. It is a constant for a given solid-liquid system and equivalent to u_w in Equation (10)

c = volume fraction of particles (dimensionless) $= C/\rho_S$

C = mass concentration of particles, e.g., kg of solid per m^3

$n = f[d/D, Re]$

u/u_0 = the hindrance factor $= (1 - c)^n$

For streamlined flow, which has been shown to be the most common in thickening practice and hindered settling, n becomes independent of Re and was determined experimentally[28] to be

$$n = 4.65 + 19.5\, d/D \quad (11a)$$

and hence a *second* constant for a given system. An exact value of n in the range 4.65 to 5 is, however, seldom critical and, as shown later, a value of $n = 4.7$ is found to give a satisfactory correlation for a large number of real suspensions even where d is unknown, i.e., an exact value of n cannot be calculated from Equation (11a). A similar compromise value of $n = 4.7$ (between $n = 4.65$ and 4.78) is arrived at also by Watanabe.[29]

Although Equation (11) was derived empirically, various authors[15,30,31] have shown its general validity on theoretical grounds. It is therefore soundly based both theoretically and experimentally and more importantly the hinderance factor term is a simple one[15] which permits of its modification to deal with sedimentation of irregular particles and particle aggregates dealt with later. Alternative hindered settling equations are much more complicated.[23] In some the concentration term appears in various forms up to five times in the one equation rather than once as in Equation (11).

The Mean Particle Diameter in a Suspension of Many Particles. As soon as there are two or more particles present in a suspension there are bound to be variations in particle size. This variation is often greater than expected even for carefully classified materials. Thus, *triple* screening of glass beads to yield a desired product between 61 and 74 μm, yielded 10% of beads coarser than this range, 33% finer and an overall spread[32] of 5-83 μm. Even with a natural product such as "nearly uniform spherical tapioca particles,"[33] the upper 10% of the distribution was ~ 15% coarser than the lower 10%. This difference represents variations in u_0 and therefore u, of over 30%.

To calculate the settling rate of suspension at various concentrations using Equation (11) it is necessary to know the value of u_0. For larger particles this may be done experimentally by timing the fall of a single particle in the fluid.[33] The value may however be calculated, with less possibility of bias because of the larger number of particles involved, by using Stokes' law and a mean particle size $\bar{d}$ based on the size distribution as measured by sedimentation.

It has been shown elsewhere[34] that the correct mean particle diameter to use is a composite diameter derived from the basic Equation (1), (5), and (7) in terms of a mean volume diameter d_V, a mean projected area diameter d_S and a mean length dimension d_L. For a single particle these simplify to a d^2 term as in Equation (8) but for a suspension of many particles differing in size, $d_V \neq d_S \neq d_L$.

Keeping these diameters separate therefore the appropriate diameter in hindered settling is

$$\bar{d} = \left[\frac{d_V^3 \cdot d_L}{d_S^2}\right]^{1/2} \quad (12)$$

where

d_V^3 = total volume of particles/their total number ($= N$)
d_S^2 = total projected area of particles/N
d_L = total length (diameter) of particles/N

As $N = \Sigma\, n_d$

where

n_d = number of particles in each size class d,
= mass of particles in size class d/mass of single particle
= Mass $(x)/\pi d^3 \cdot \rho_S/6$

it follows that:

$$\bar{d} = \left[\frac{\Sigma \dfrac{x}{d^2}}{\Sigma \dfrac{x}{d} \cdot \Sigma \dfrac{x}{d^3}}\right]^{1/2} \quad (12a)$$

where

x = mass fraction of particles observed in each small size interval of mean size d as determined by sedimentation.

Experimental Verification of the Richardson and Zaki Equation. As $(1 - c)$ is a fraction and n a positive number, Equation (11) indicates a decrease in particle settling rate with increasing

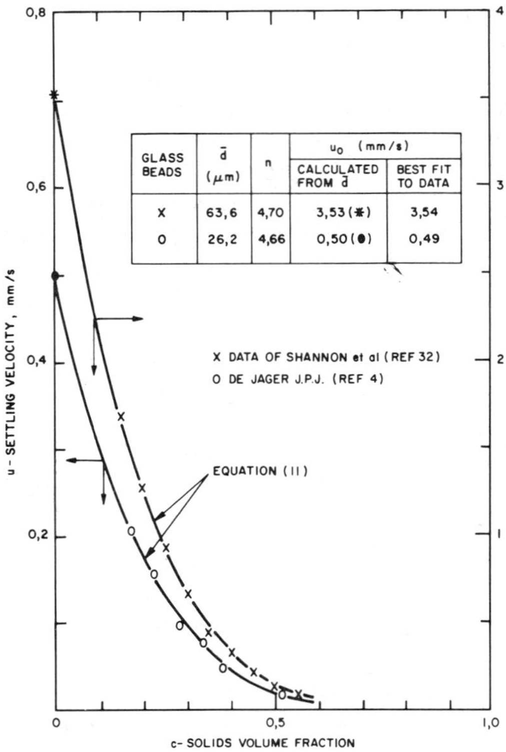

Figure 14.6. Settling velocity of glass bead suspensions as a function of solids concentration.

volume fraction of solids. Figure 14.6 compares the *experimentally* observed values of u for two suspensions of glass spheres with *predicted* values based upon Equation (11). The value of u_0 in this equation was calculated from Equation (8) with $d = \bar{d}$ as found from Equation (12) and n calculated from Equation (11a). The agreement can be seen to be good. Thus even for an uncharacterized suspension for which $\bar{d}$ and hence the constants u_0 and n are *not* known, it should be possible to estimate them from Equation (11) by means of a plot of ln u vs ln $(1 - c)$ and using the intercept and slope of the best fitting straight line. This is one basis for using hindered settling rates, u, to determine the *mean* particle size of a suspension of particles.

Two independent values of u_0 estimated in this way are compared in Figure 14.6 and the excellent agreement found between them has been shown[34] to apply also to a variety of solids and liquids ($d = 13$ to $1740\ \mu m$; $\rho_L = 890 - 1070\ kg/m^3$ and $\mu = 0.001 - 0.007$ kg/ms). The reliability of the Richardson and Zaki equation, together with their value of n, is therefore well established for relatively closely sized spherical particles up to the maximum attainable free-settling concentrations, $c \sim 0.55$.

Suspensions Consisting of a Range of Particle Sizes, Shapes, and Densities.

Suspensions consisting of uniform particles are rarely met with in practice. Real slurries contain a range of particle sizes and often different materials with various shapes and densities. Although such particles tend to segregate when dilute, at normal thickener feed concentration, mutual retardation of the particles in a batch test causes hindered or "zone-settling"[35] with a uniform particle settling velocity regardless of size. It is therefore less reliable to calculate u_0 from Equations (8) and (12a) and, for reasons that are discussed in the next section, use is made of an alternative form of Equation (11)

$$u^{1/n} = u_0^{1/n} - u_0^{1/n} \cdot c \qquad (11b)$$

where n may be taken as equal to 4.7 i.e., $1/n = 0.213$.

A plot of $u^{0.213}$ vs c should therefore give a straight line from which either the slope or the intercept can now be used to determine u_0. Although, because of differing $\bar{d}$ values, the values of u_0 vary between different suspensions, the hindrance factor of any suspension, $(1 - c)$,$^{4.7}$ should be constant at a given concentration. The *relative* velocity, u/u_0 should therefore be the same for each suspension at the same concentration and plots of $(u/u_0)^{0.213}$ vs c should yield a single straight line of slope $= -1$ because from Equation (11b)

$$\left[\frac{u}{u_0}\right]^{0.213} = (1 - c) \qquad (11c)$$

The data of Figure 14.6 are replotted on this basis in Figure 14.7 and they can be seen to fall closely on this theoretical line with intercepts at + 1 and a negative slope of 45°. In addition,

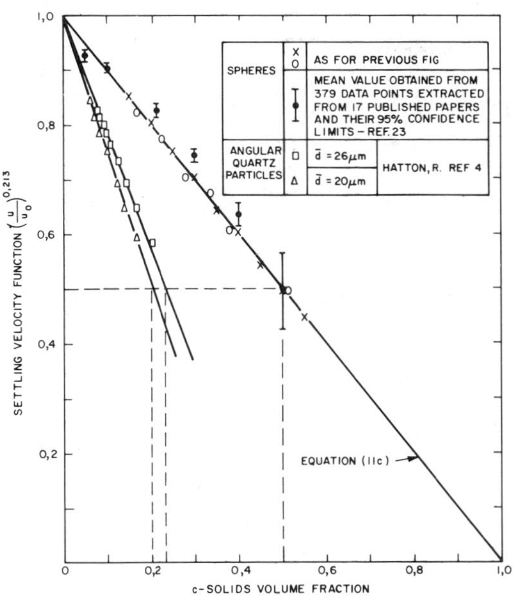

Figure 14.7. Settling velocity of spherical and angular particles plotted according to Equation (11c).

summarized data on spheres from an extensive survey of the literature[23] are included and the agreement with Equation (11c) is within 6%. Bearing in mind that no allowance was made in this work for fairly wide variations in d/D and its effect upon n and u_0, this agreement is reasonable.

When, however, the sedimentation rate of *angular* quartz particles rather than spheres are compared on the same basis, Figure 14.7 shows that the settling rate now decreases more rapidly with increasing concentration than predicted by Equation (11c). The spheres reach a hindrance factor u/u_0 = 0.0385 at c = 0.5 while this same retardation is experienced by quartz particles at a concentration as low as 0.2. As the retardation in settling velocity of a particle in hindered settling is due to the interference offered to its ideal return fluid flow pattern by the presence of its neighbors, it must be concluded from *hydraulic similarity* considerations that quartz particles present a greater effective blockage to the return flow than can be expected from their volume. A unit volume of quartz must in fact have the retardation effect of 0.5/0.2 = 2.5 volumes of an equivalent sphere. The most plausible inference is that such angular particles carry with them attached water due to their roughness[14,15] and this stagnant water behaves as if the volume of the particle were effectively increased. The net solids concentration is therefore *greater* than the volume of dry solids present.

By assuming the effective solids fraction for any degree of retardation to be similar to the volume fraction of spheres at the same retardation, we can calculate the proportion of fixed water. Thus, if a quartz suspension at c = 0.2, settles according to Equation (11c) as if it experienced the same drag as a suspension of spheres at c = 0.5, its effective total solids volume must in fact be equal to 0.5 with the extra volume being made up of stagnant water which moves with the particle. The w_i term of Steinour[36] was introduced to deal with this same phenomenon in suspensions of angular emery particles.

Because the stagnant water behaves as if solid, the lines for quartz in Figure 14.7 remain straight but because of the unknown quantity of water, they are of unknown slope. If we call the slope k, Equation (11c) can be rewritten

$$\left[\frac{u}{u_0}\right]^{0.213} = (1 - kc)$$

or

$$u = u_0\,(1 - kc)^{4.7} \tag{13}$$

Comparing with Equation (11) it can be seen that the original concentration term c is replaced by kc, which now represents the *effective* solids volume fraction. For spheres, where there is no stagnant liquid, k = 1, while for the two quartz suspensions $k \sim 2{,}5$ with the finer sample carrying relatively more water, i.e., having a slightly higher value of k.

Correlation of the settling data for quartz according to Equation (11) i.e., with k = 1 and *n variable*, gives values of n ranging from 11.8 to 14.6 depending upon c. The value of n is therefore not only much higher than expected from theory,[30] (n lies between 1 and 8) but more seriously is not a constant for a given system. Many such correlations have been attempted,[37-42] with values of the exponent as high as 466.7, but as shown by Capes,[43,44] these are reduced to expected levels if due allowance is made for the fixed water (reduced porosity) associated with particle aggregation or irregular shape.

Aggregated Suspensions

Natural aggregation is frequently present in particle suspensions,[1,45] especially at higher concentrations such as in thickener feed pulps, where the mutual proximity of the particles causes them to adhere and settle together as clumps rather than as single particles. This increase in "particle" size results in faster settling, and in thickening the effect is often exploited. In clarification, the solids concentration in the liquid is much lower, and natural aggregation is largely absent. Flocculation is therefore nearly always required. It can be brought about by reducing the mutually repellent charges on the particles by means of electrolytes (coagulation) or by bridging particles by the simultaneous adsorption of polymers pretreatment, p. 657).

In all cases, aggregates are produced, each consisting of a large number of varying size primary solid particles, associated together into a single relatively large sedimentation unit or floc. Such a floc includes not only this loosely held solids structure but also the interstitial stagnant water.[15] Flocs have a density lower than the solid particles, due to this water, but have a greatly increased diameter so that their settling rates are several orders of magnitude higher than the original individual particles. No distinction is made here between the terms flocs or aggregates nor their method of production as only the *sedimentation behavior* of the final aggregates is of concern at this stage.

Types of Settling Behavior in Aggregated Suspensions. Previous sections deal with discrete *individual* particles as the primary sedimentation units. Resuspension of these unaggregated pulps in a batch test to prepare a uniform suspension does not alter the size, shape or settling characteristics of these units from test to test. In flocculated suspensions, however, the sedimentation units (aggregates) are freshly formed only *after* agitation ceases. The shape of the resultant sedimentation curve (height of interface H, versus time, t) depends upon solids concentration,[46,47] i.e., on the number of primary particles present and their mutual proximity when agitation is stopped and the shearing force is removed.

When the suspension is *dilute* the aggregates (flocs) are formed independent of each other; they are widely spaced in the intervening liquid and descend through it as individual entities. After agitation ceases ($t = 0$) the floc formation time is fast[48] compared to the time over which sedimentation is observed in a batch test. For instance, in the presence of a coagulant, the silky appearance of dispersed micaceous clays noted during stirring, disappears within seconds after agitation is stopped. Dilute suspensions of flocs therefore show a constant interface descent rate from zero time (curve A_1 in Figure 14.8).

At *intermediate concentrations* the flocs have the same composition i.e., contain the same voidage fraction as those in the dilute range. This is evident from the density of the rising sediment zone formed at the base of a suspension of intermediate concentration[50] and the viscosity of these suspensions.[51] At the start of a batch settling test, i.e., after the cessation of agitation the suspension appears to gel[52] and thereafter one observes an "induction" period t_1 (curve B_1, Fig. 14.8) during which the relatively low initial sedimentation rate u_i increased with time either gradually,[46] or in discrete steps[47,53] and subsequently reaches a higher constant rate, u. The maximum steady value is accepted as the true settling rate in a static batch test.[45,54-56] Two phases in this acceleration process are shown in Figures 14.9A–C.

This rate is higher than would be expected from extrapolation of sedimentation data in the dilute range (Fig. 14.10), indicating that the mode of sedimentation is now different.[60] Higher settling rates can be attained only if the resistance to relative movement of flocs and

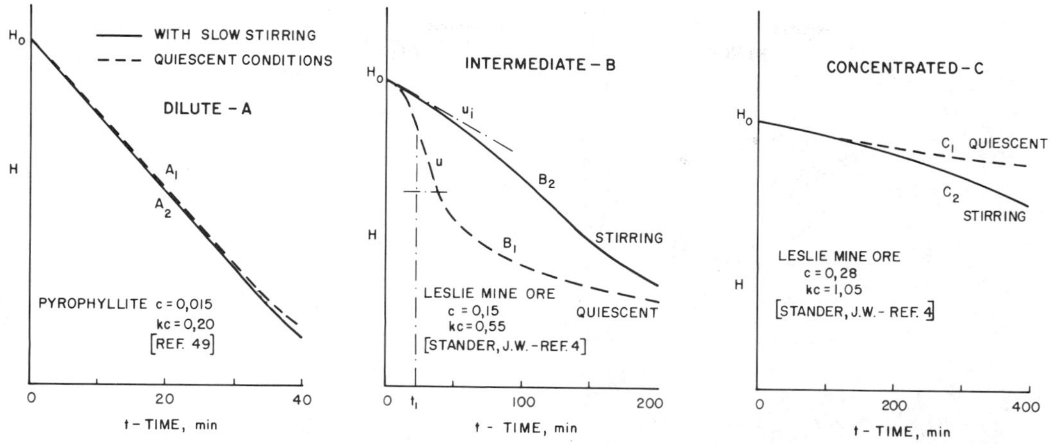

Figure 14.8. Mode of settling in aggregated pulps depends upon solids concentration and presence of mild agitation.

liquid is *less* than in a randomly dispersed suspension of flocs. This requires an anisotropic structure in the suspension with liquid channels of low flow resistance in an upward direction. The gradual formation of this structure after cessation of agitation causes the induction period. A similar argument was used to explain the accelerated settling of intermediate concentration suspensions in the presence of particles of density lower than,[61] equal to,[62] or greater than[63] the density of the fluid. Such particles rapidly destablize a uniformly setting suspen-

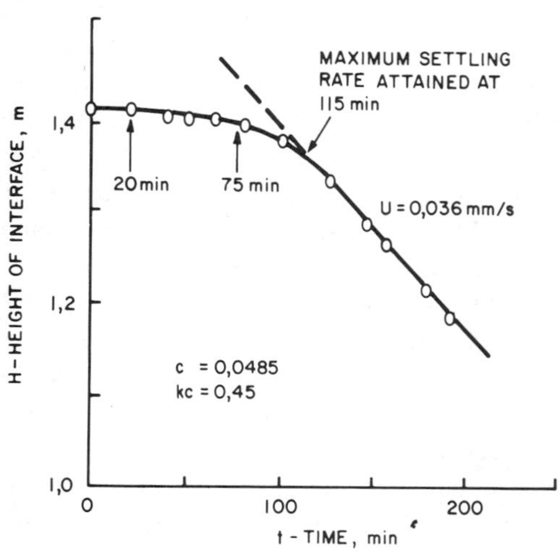

Figure 14.9A. Intermediate settling in a desanded mine pulp[49] showing the height of the pulp interface when photographs in Figures 14.9B and 14.9C were taken.

Figure 14.9B. Commencement of the break-up of the initially gelled mass; t = 20 minutes.

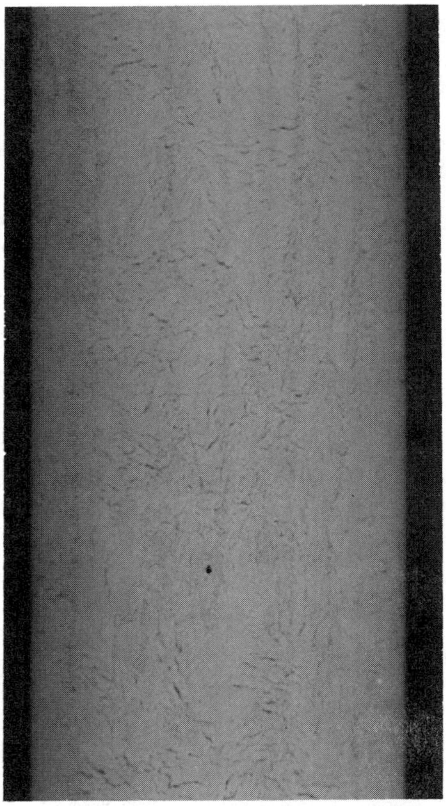

Figure 14.9C. Agglomerated structure beginning to appear; t = 75 minutes.

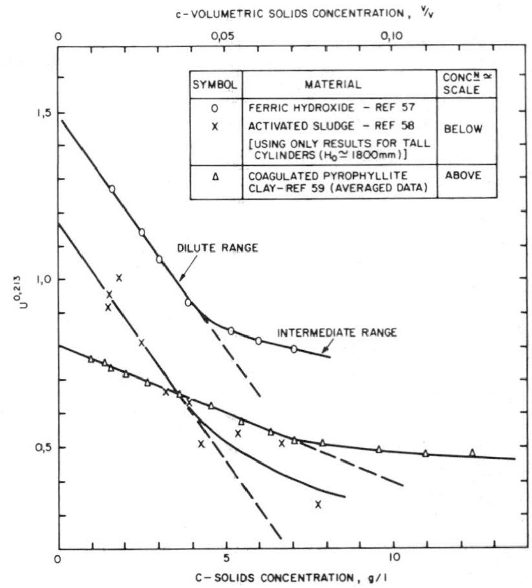

Figure 14.10. Settling "velocities" at low concentrations according to Equation (14) of three widely different types of suspension.

sion and set up a new stable flow pattern in the form of vertical streaming. As streaming was known to be present even without addition of foreign particles, it was concluded[61] than even in normal pulps, uniform settling is inherently *unstable* and can be upset by the presence of air bubbles.[52] The stability of the vertical streaming flow regime that subsequently develops is evident from the constant settling rate attained.

If instead of quiescent sedimentation conditions, mild slow stirring throughout the column is introduced,[49,64] the formation of the channel structure is prevented and a lower interface settling rate is maintained as expected (curve B_2 in Fig. 14.8).

At intermediate concentrations the flocs must be closer together than in dilute pulps and hence there is considerably less water surrounding each floc. Their probability of touching or bridging by particle growth is therefore much higher and this three dimensional interaction between solids is likely to be involved in the formation of the channel structure with flow channels being developed between the flocs.[46] It is therefore not surprising that the induction period increases with both concentration[200] and pulp height (Whitmore, R. L.).[49]

In the intermediate concentration range the maximum steady settling rates u decrease with concentration, but to a lesser extent than for pulps in the dilute range. (Fig. 14.10). The decrease is not unexpected because at higher floc concentrations less voidage between them is available for channel flow and fewer and somewhat narrower channels may be formed.

The "free-settling" of Coe and Clevenger,[54] i.e., the attaining of a steady descent rate "after flocculation is complete," is therefore not free settling in the sense of their being freely falling flocs present.

When individual flocs, formed in *dilute* suspensions, settle to the base of the container, they produce there a wide range of higher concentrations, including intermediate concentrations, merely by descending upon one another and packing closely together. The absence of any bridging and also the lack of time and height necessary to promote channel formation results however in these concentrations settling at a rate significantly *slower*[49] than B_1 in Figure 14.8. This gives rise to the discrepancies[65] between the settling rates of these higher concentrations when deduced from their rate of rise in dilute pulps (Talmage and Fitch) and as observed in initially homogeneous suspensions (Coe and Clevenger)—p. 638.

Concentrated suspensions may be defined as those in which curve B_1 with an induction period, Figure 14.8, is replaced by curve C_1. The suspension does not attain any degree of "mobility" but subsides at a sluggish and ever decreasing rate. There is usually less total water available in the suspension than is present proportionately within the flocs in the dilute regime and when agitation is stopped the *particles* are closer together than in a floc and are able to form a three-dimensional structure like a packed bed. The lower layers can be further compacted by the solids pressure from above i.e., the particle-particle links (probably maintained by friction) are sheared, the particles collapse inwardly towards each other and consequently liquid is expressed from these layers. This liquid moves upwards through the bed and because of the tight packing, mainly between the primary *particles.*[66] Some channelling may occur and often a few very large channel openings (volcanoes) are observed at the interface but only after considerable time has elapsed since the onset of sedimentation.[67] Concentrated suspensions are further discussed in p. 672.

Settling curves shown in Figure 14.8 A_1 to C_1 were obtained[68] also for sedimentation of coal particles in *oil* i.e., a nonaqueous system indicating their general nature.

Slow Stirring in Aggregated Suspensions. The effect of very mild stirring (0.1–2 rpm) depends upon the concentration regime present. In *dilute* pulps the formation and the subsequent sedimentation of the flocs is neither aided nor hindered and settling rates are therefore little affected (curves A_1 and A_2, Fig. 14.8). In *intermediate* pulps, horizontal shear hinders the formation of short-circuit flow channels and materially decreases the maximum settling rate attained (curve B_2 rather than B_1).

In *concentrated* pulps, mild mechanical disturbance *promotes* the shearing of the particle-particle links. Under quiescent conditions the three-dimensional structure which forms after cessation of agitation tends to resist collapse because of friction at the points of particle contact and the support from the base and the walls. The mass of solid above may not be sufficient to overcome the strength of the particle structure and hence slow shearing of the suspension results in bonds being broken, a realignment of the particles, release of water, and hence promotion of the subsidence of the interface, curve C_2, Figure 14.8.

Dilute Pulps. The water in the voids between the particles of a single aggregate (sedimentation unit) cannot contribute to the free fluid flow around flocs but it contributes, as in the case of angular particles, to the effective volumetric solids concentration. The possibility of a *permeable* floc is discussed later. The volume occupied by the flocs, defined as the total volume of the envelopes surrounding each, is therefore greater than the volume fraction of actual solids present, c (frequently measured by oven-drying a sample of pulp). The diameter, density and concentration of the *flocs*, should therefore replace these corresponding terms for spheres in Equations (8) and (11).

Because the effective volumetric concentration of the aggregates is relatively high, together with the improved size-homogeneity caused by the mutual aggregation of fine and coarse particles, a clear interface between suspension and supernatant may form at values of c as low[46,63] as 0.002.

Observation of the interface settling rate u of such suspensions over a range of concentrations c can be expected to be correlated according to Equation (13), which may be rewritten

$$u^{0.213} = u_F^{0.213}(1 - kc) \qquad (14)$$

where

u_F = terminal settling velocity of *one* representative *floc*
kc = volume fraction of *flocs*
k = volume of floc per unit volume of contained solid

Where, as in the case of hydroxide precipitates or organic suspensions, the volume fraction of the actual solids is not readily determined, k can be equally well expressed as k' = volume of floc per unit *mass* of contained solid and the concentration C = the *mass* of solids per unit volume of suspensions. As $k' = k/\rho_S$ and $C = c\rho_S$, $k'C$ becomes identical to kc i.e., it is not necessary to know ρ_S the rather ill defined density of such suspended solids. In general any units for C may be employed and the unit for k' is then merely the *inverse.*

As for dispersed particles in hindered settling, a plot of the experimental values $u^{0.213}$ and c, should give a straight line of intercept $I = u_F^{0.213}$, and of slope $I \cdot k$ over the concentration range pertaining to dilute pulps. This has been found to apply to a large variety of suspensions and plots for a typical metal hydroxide precipitate, clay-mineral and activated sludge are shown in Figure 14.10.

The values of u_F and k' derived from the straight line sections of these plots are shown in Table 14.6. Included are the corresponding floc diameters, d_F, calculated from Equation (8) modified as follows:

$$u_F = \frac{d_F^2(\rho_F - \rho_L)}{18\mu} \cdot g \qquad (15)$$

where ρ_F = floc density which, from a materials balance, is given by

$$(\rho_F - \rho_L) = \frac{(\rho_S - \rho_L)}{k} \qquad (15a)$$

Similar results were published by Michaels and Bolger[46] for platey kaolin slurries which they purposely flocculated in *different* ways to produce a variety of particle structures, e.g., face-to-face and edge-to-face. The k' value for the predictably voluminous card-house structure of kaolin was found, as expected, to be greater than for the more compact card-pack structure, in fact by a factor of ~2 (Table 14.6). This provides strong confirmation that the values of k' (or k) derived from batch sedimentation tests represents a real physical property of the flocs in suspension.

It has been shown that for similar materials the value of k increases with decreasing particle size of the primary particles.[51] (Or if k is assumed equal to unity, then the compensating value of n must increase with decreasing particle size.[39]) The increase in k is related to the specific surface area of the particles in such a way that the water in the floc voids tends to be proportional to the total surface area. This is the reason why increased grinding always decreases pulp settling rates even when the pulps are coagulated. For the same mass of solid in the floc, the surface and hence volume of water is increased, and this not only decreases floc density but increases the effective solids concentration.

Table 14.6. Properties of Single Representative Flocs of a Variety of Materials.

MATERIAL	SETTLING VELOCITY u_F (mm/s)	VOLUME OF FLOC PER g DRY SOLID k' (ml/g)	FLOC DIAMETER d_F (mm)
Ferric hydroxide precipitate $-Fe(OH)_3$	6.80	94.3	1.27
Activated sludge	2.43	127.0	1.06
Pyrophyllite	0.38	1.9	0.046
Kaolin[46] with card-house structure	1.00	25.2	0.26
Kaolin[46] with card-pack structure	0.28	13.6	0.099

A similar settling pattern is demonstrated by each of the suspensions shown in Figure 14.10 despite the variation in their nature and composition. There is the expected linear plot at low concentrations followed by settling rates which deviate *positively* at higher values.

The data of Figure 14.10 are replotted, replacing the abscissa c by the *floc* volume concentration kc, and normalizing the interface velocity with respect to u_F, in Figure 14.11. It is now the equivalent of Equation (11c) with (assumed) spherical flocs in the place of solid spheres and should therefore appear as the 45° line in Figure 14.7.

To demonstrate the general applicability of this plot, other aggregated suspensions have been included and, in addition, Figure 14.12 shows results for metallurgical pulps from ten mines, published as early as 1916 by Coe and Clevenger.[54] E is a limiting value described later.

For *all* pulps in the dilute range, where the flocs behave as separate entities there is a straight line of slope = −1 as expected, up to a characteristic value of kc, followed by increasing positive deviation from Equation (14). This marks the onset of channel flow and an increased permeability of the suspension as a whole. If however a change in flow regime is ignored then the observed increase in settling velocity must be attributed to a more permeable floc.[74] It has been shown however that there is no increase in settling floc porosity compared to dilute pulps[50] and that flocs have to be extremely porous to show significant permeability.[15,75] For a floc of porosity as high as 95% and a d_F/d ratio[49] of 40, the internal flow is 0.2% and therefore negligible compared to the assumptions made in the calculation.[76] Epstein and Neale[77] show that a swarm of permeable spheres will in fact always settle more *slowly* than a swarm of impermeable spheres of the same size and density. Increased settling rates must therefore arise from a more permeable *bed.*

For a specific pulp the point of departure from the straight line indicates the upper limiting concentration of its dilute range. Similar

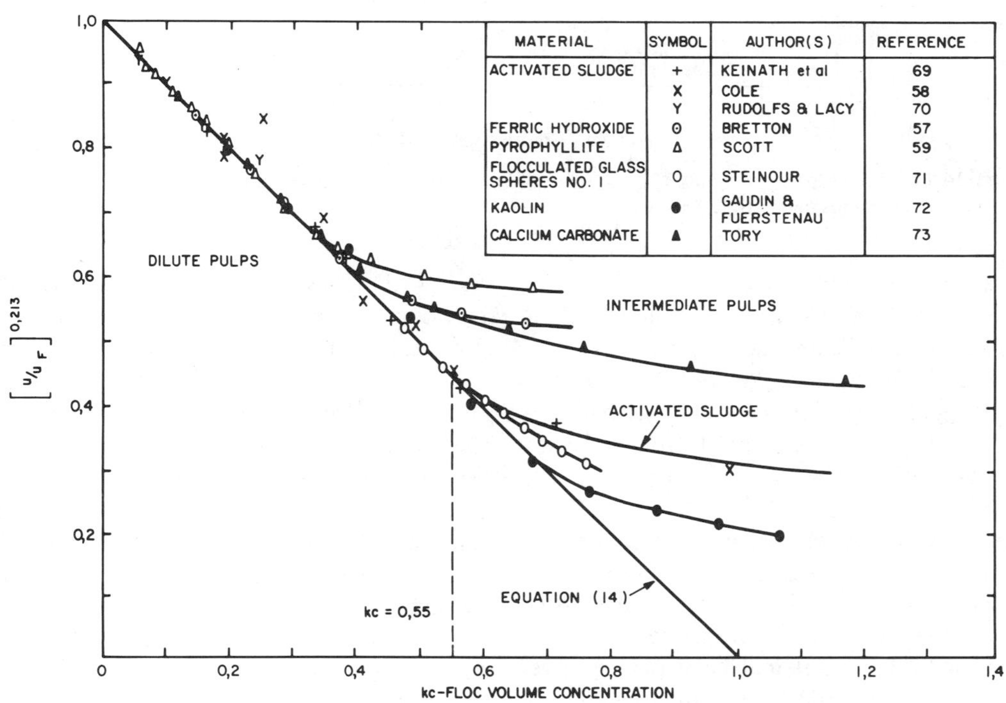

Figure 14.11. Dimensionless maximum interface sedimentation rates; data of Figure 14.10 and other sources plotted on "normalized" coordinates.

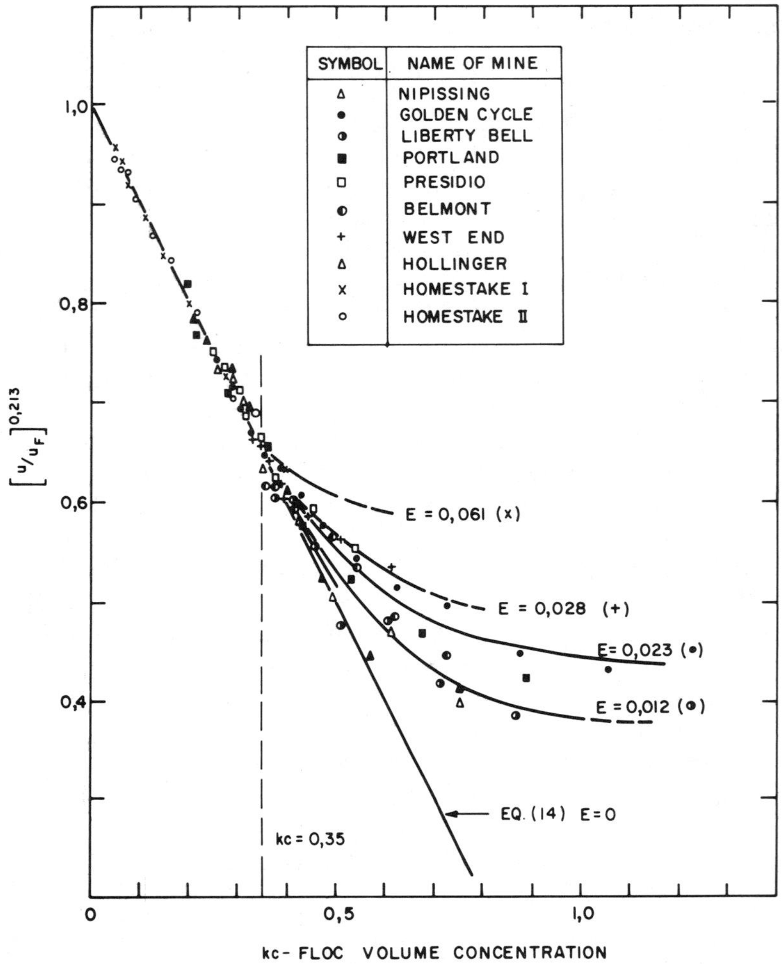

Figure 14.12. Dimensionless maximum interface sedimentation rates of coagulated pulps; data from Coe and Clevenger[54] on various metallurgical pulps.

materials, e.g., activated sludges (Fig. 14.11), milled ores (Fig. 14.12) or red mud[60] show similar limiting concentrations, i.e., *kc* at the point of departure from Equation (14) does not differ significantly for the same material even though they may be derived from different sources.

Settling Behavior at Intermediate Concentrations. The pulp interface at concentrations higher than for the dilute range settles *faster* than expected from a suspension of flocs assumed to consist of individual rigid spheres. This onset of intermediate settling behavior is observed at *kc* values from about 0.35 onward (Fig. 14.11). This value indicates that the formation of the channelled structure can take place at volumetric floc concentrations well below that describing the closest random packing of rigid spheres,[32] where $kc = 0.64$. The onset of the intermediate regime is, however, not observed at values higher than 0.7 although flocs are neither rigid nor strictly spherical. The onset would be expected to be related to the size and shape of the flocs and the interparticle and the interfloc forces, which all play a role in the formation of a structure in the flocculated suspension.

Figures 14.11 and 14.12 demonstrate that in the intermediate concentration range the deviation from Equation (14) increases with concentration. In terms of subsidence velocity the deviation is: $u - u_F(1 - kc)^{4.7}$ which normalized with respect to u_F becomes: $u/u_F - (1 - kc)^{4.7}$. Multiplication with kc gives the deviation in terms of solids flux (see p. 634) or

$$\Delta_{FLUX} = \left[\frac{u}{u_F} - (1 - kc)^{4.7}\right] \cdot kc \qquad (16)$$

Plots of Δ_{FLUX} vs kc are shown in Figure 14.13.

The curves can be seen to be exponential in form and asymptotic to a final value E. The flux deviations (from Equation (14)) are essentially zero for dilute pulps as these follow Equation (14) up to the point of deviation, but they then rise sharply at this concentration (corresponding to the onset of channeling at intermediate concentrations) and reach a plateau near a floc volume concentration of unity. The shape of the similar curves suggests the following *empirical* model:

$$\Delta_{FLUX} = E\left[1 - e^{-Bk(c - c_D)}\right] \qquad (17)$$

where

E = maximum deviation
c_D = concentration at which deviation commences, i.e., upper limit of dilute regime ($k \cdot c_D$ = floc volume concentration of limit)
B = "rate of rise" of the exponential curve.

Tabulated values of the three empirical parameters for the four different materials are included in Figure 14.13. Poorly cohesive flocs, typified by the glass spheres, do not permit channel formation until a relatively high concentration $kc_D = 0.55$ is reached and the maximum deviation $E = 0.0025$ is correspondingly small.

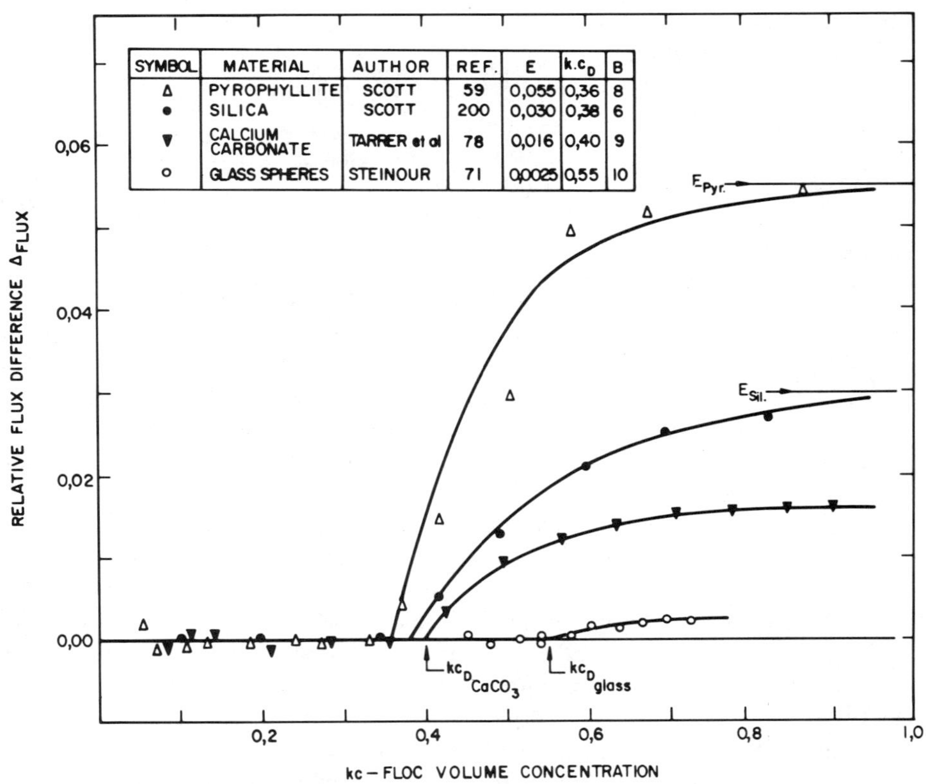

Figure 14.13. Deviations from Equation (14) expressed as a solids flux vs kc.

Where the nature of the suspension is such that a greater interaction between flocs is likely, the onset of the deviation occurs at a *lower* concentration, i.e., kc_D is smaller, and the final magnitude of the deviation E is larger.

E, the final value of Δ_{FLUX}, is reached only at $kc = \infty$, but by assuming that at $kc = 1$, Δ_{FLUX} lies between 95 and 99% of E (the experimental data are not sufficiently precise to determine this exactly) then B from Equation (17) can be shown to lie in the range 5-10. This parameter is thus not critical and adoption of a fixed mean value for all suspensions has the effect merely of altering the final estimated value of E by an insignificant but compensating amount. Also, as shown earlier, the value of kc_D tends to be the same for any one type of material. Equation (17) may therefore be reduced to an equation with one *variable* E and two empirical *constants* B and c_D which, for any given pulp, must be determined experimentally.

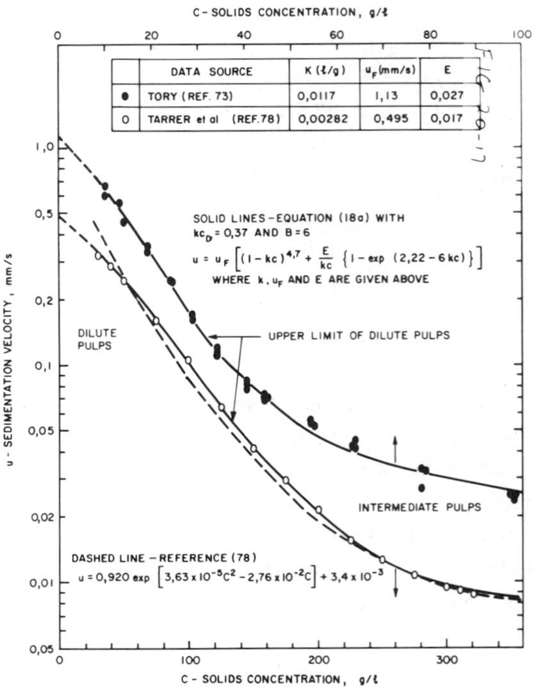

Figure 14.14. Settling rate of two calcium carbonate suspensions as described by Equation (18a).

Combining Equations (16) and (17) one obtains the following relation between u and c for the *intermediate* concentration range

$$u = u_F \left[(1 - kc)^{4.7} + \frac{E}{kc} \{ 1 - e^{-Bk(c - c_D)} \} \right] \tag{18}$$

where parameters u_F, k, E, and c_D have identifiable physical meanings and B is a scaling parameter.

Equation (18) can be converted to include *dilute* pulps when written in the form

$$u = u_F \left[(1 - kc)^{4.7} + \frac{E}{kc} \{ 1 - e^{-(\lambda + |\lambda|)/2} \} \right] \tag{18a}$$

where $\lambda = Bk(c - c_D)$. For $c < c_D$, i.e., for dilute pulps, the Δ_{FLUX} term becomes zero and Equation (14) is followed.

Equation (18a) therefore correlates experimental batch settling velocities u over concentrations c ranging from the most dilute pulp showing an interface right up to the point of compression, i.e., *all* "free settling" pulps. Two examples are shown in Figure 14.14 for calcium carbonate for this range and a third in Figure 14.15 for an oxygen-activated sludge where the data extend *beyond* the point of compression. All use common values of $B = 6$ and $kc_D = 0.37$. It can be seen that the model is able to give a fit as good as or better than the authors[78, 79] and provides a realistic value for u at $c = 0$. The data all support the ln u vs c curve being *convex* upward in the dilute range as given by Equation (18a) and concave in the intermediate range. Furthermore, they indicate a reduced value for E in the compression range (Fig. 14.15) to about 0.6 to account for the slower settling in this region. Empirical models that do not take into account the different modes of sedimentation in the dilute, intermediate and compression ranges, respectively, and in which the parameters have no physical meaning,[45] can be misleading. One example,[78] given in Figure 14.14 when extrapolated into the compression range, gives values of u which *increase* with increasing concentration above $C = 380$ g/liter. At their thickener underflow concentration of 766 g/liter, u actually *exceeds* the value of u_0 by ~20%.

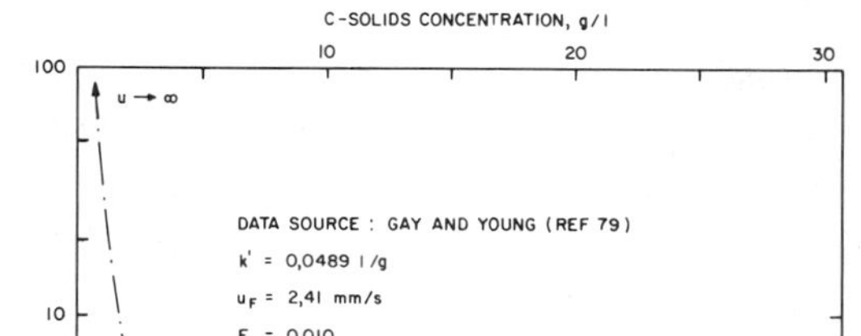

Figure 14.15. Settling rate of oxygen-activated sewage sludge as described by Equation (18a).

Stirring in Intermediate Pulps. As shown in Figure 14.8 slow stirring has the effect of eliminating the induction period and reducing the maximum settling rate. The data in Table 14.7 indicate that the settling rate under these conditions is reasonably close to the predictions of Equation (14). The transient value of u_i is more difficult to measure and therefore more variable than $\hat{u}$ and u_S.

Suspensions of Organic Materials. The free settling behavior of organic pulps (e.g., activated sludge) appears superficially to differ from that of mineral pulps in that the settling rate u depends not only upon solids concentration and the presence or absence of stirring but also markedly upon the dimensions of the batch cylinder, both its height H_0, and diameter, D, being important.[80]

This is clearly demonstrated by means of the sedimentation data of Cole.[58] Figure 14.16 and Table 14.8 show results of a contact stabilization sludge tested at three slightly increasing solids concentrations but in markedly different sized vessels.

Sludge number 6-10 (Fig. 14.16) behaved exactly as if in compression (curve C_1, Fig. 14.8) showing an ever-decreasing settling rate with time. The same material, although at a 25% higher concentration but now in a larger container, showed, after 40 minutes, a small (33%) but significant increase in settling rate. In a still larger container the same pulp, although at a slightly higher concentration, now behaved clearly as an intermediate pulp (curve B_1, Fig. 14.8) more than doubling its initial settling rate after a 110 minute induction time. Thus in spite of an increase in concentration of 30%, the sludge settled 50% *faster* in the larger column.

Similar results where D and H_0 are not *simultaneously* increased, are given in Table 14.9.

The individual variables H_0 and to a lesser extent, D, therefore each affect settling behavior. Increasing H_0 can convert a slowly subsiding pulp in compression into a faster

Table 14.7. Settling Rates of Dilute and Intermediate Pyrophyllite Pulp Under Quiescent and Slowly Stirred (2 rpm) Conditions.

CONCENTRATION			SETTLING RATE (mm/s)			
c	kc	REGIME	$\hat{u}$ EQUATION (14)	u_S STIRRING	u QUIESCENT	u_i
0.040	0.18	Dilute	0.17	0.15	0.14	~ 0.15
0.056	0.25	Dilute	0.11	0.14	0.15	~ 0.15
0.069	0.31	Dilute	0.077	0.070	0.067	~ 0.07
0.078	0.35	Intermediate	0.057	0.065	0.14	0.066
0.116	0.52	Intermediate	0.014	0.011	0.078	0.030

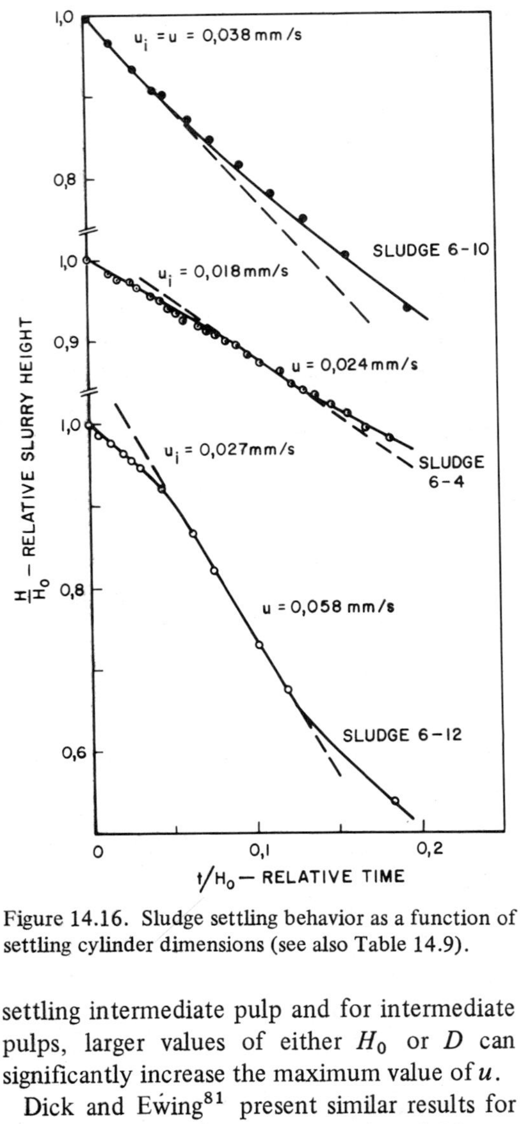

Figure 14.16. Sludge settling behavior as a function of settling cylinder dimensions (see also Table 14.9).

settling intermediate pulp and for intermediate pulps, larger values of either H_0 or D can significantly increase the maximum value of u.

Dick and Ewing[81] present similar results for three intermediate pulps in a tube of 90 mm diameter. Vesiland[82] considers that for estimating realistic values for u for settler design, D should be 200 mm and H_0 not less than 920 mm. Above these limits there is little further increase in settling rate and these maximum values are therefore considered to be close to the actual sedimentation values in larger full scale settling tanks.[81]

Small settling tubes therefore *retard* settling and fail to give the correct value of u for a given pulp. In such tubes intermediate concentration pulps (kc = 0.35 - 1.0) behave as if they were in compression and have been shown[83] to subside sluggishly at rates *below* that predicted by Equation (14) Figure 14.11 whereas in the correct size cylinders[69] they settle faster than Equation (14), as do mineral pulps. As container dimensions are not significant for *dilute* organic pulps, it is thought that in intermediate pulps in which a three-dimensional network develops after cessation of agitation, the dimensions become important because of the relatively lower driving force present compared to denser mineral pulps or to taller settling cylinders.

The magnitude of the solids driving force causing sedimentation is the unbuoyed solids pressure of the suspension, which is $C/\rho_S\,(\rho_S - \rho_L)gH$ Pa. Solids concentration, solids density and height therefore all play a role.

In calcium carbonate slurries, for example,[73] $\rho_S = 2680\ \text{kg/m}^3$ and as the onset of intermediate behavior appears at $C = 32.5\ \text{kg/m}^3$ the driving force is 200 Pa/m. For activated sludge[58] $C = 6\ \text{kg/m}^3$, $\rho_S \sim 2000\ \text{kg/m}^3$ and therefore the solids driving force is only 30 Pa/m. For the same force to operate in each, the test cylinders for organic sludges should be

Table 14.8. Effect of Vessel Dimensions on Maximum Settling Rate (Values Rounded Off).

SLUDGE NO. (REF. 58)	SOLIDS CONC. (g/l)	H_0 (mm)	D (mm)	SEDIMENTATION RATE u_i (mm/s) × 10^2	SEDIMENTATION RATE u (mm/s) × 10^2	INDUCTION TIME (t_1) MIN
6-10	4.1	260	50	3.8	–	–
6-4	5.1	420	75	1.8	2.4	40
6-12	5.3	1800	100	2.7	5.8	110

Table 14.9. Effect of H_0 and D on Maximum Settling Rate.

SLUDGE NO. (REF. 58)	SOLIDS CONC. (g/l)	VARIABLE	H_0 (mm)	D (mm)	SEDIMENTATION RATE u_i (mm/s) × 10^2	u (mm/s) × 10^2
4-4	3.1	H_0	350	60	3.3	–
4-7	4.3		1840	50	2.7	4.8
5-5	1.4	H_0	670	140	10.8	55.6
5-8	1.5		1820	50	4.2	75.0
4-7	4.3	D	1840	50	2.7	4.8
6-12	5.3		1790	100	2.7	5.8

6-7 times taller than for mineral pulps. This is close to the ratio of 2 m columns used by Cole[58] and the height of normal one litre graduates commonly used for sedimentation tests, 0.33 m i.e., 6 : 1.

In small cylinders the support of the base and wall-to-wall bridging effects therefore become significant compared to the gravitational force available to cause subsidence collapse in organic pulps. An organic pulp can therefore go into compression at a floc solids concentration kc below 1.0. Typical compression behavior, with channels formed from shear failure planes, was observed[83] in a pulp for which kc was only 0.58. As for compression pulps in general (Fig. 14.8c) slow stirring results in shear of the network causing the solids structure to collapse and the pulp to attain appreciably higher settling rates.[83]

There is therefore little fundamental difference between the behavior of organic sludges and mineral pulps. The onset of network structure typical of intermediate pulps occurs at relatively low solids concentrations (because k tends to be high) but, because of the subsequent absence of an adequate driving force, especially in normal (short) test cylinders and due to the lower solids density, organic sludges settle more sluggishly and can behave as concentrated pulps in true compression. A higher mass of solids and slow stirring both promote subsidence as they do with more concentrated *mineral* pulps when in compression.

It has been noted[84] that in large diameter cylinders, i.e., where the wall bridging effects are *not* important, slow stirring in organic pulps does not enhance settling and at higher stirring speeds, settling is in fact retarded. The apparent contradiction that stirring in organic pulps at intermediate concentration *enhances* settling whereas for mineral pulp it *retards* it (Fig. 14.8-B_2) is therefore resolved when wall effects are eliminated. Under similar conditions, stirring in organic sludges therefore, follows the behavior shown by mineral pulps.

Sedimentation Equations Based Upon Permeability and Other Principles. Several alternative approaches to the development of a sedimentation equation have been made.[23,85] A common one is to regard the pulp as consisting of a bed of particles at fixed positions and relate the subsidence rate of the bed to its permeability to liquid flow. Other approaches include consideration of momentum-balance relationships;[86] regarding the particles and fluid of a suspension as two superimposed continua in the framework of the continuum theory of mixtures[87] and the application of nonlinear Föcker–Planck equations[88] to Darcy's law. They all depend upon Kynch's assumption[94] in effect, and as none of them overcomes the difficulties of nonideal behavior in real pulps, they suffer from the same major defects as the simpler models.

Summary

The presented theory of sedimentation covers the settling behavior of suspensions from *single* spheres in an infinite fluid to batch-wise *hindered* settling of concentrated flocculated pulps. The hindered settling described includes suspensions with discrete primary particles of diverse size, shape and density, amorphous chemical precipitates and waste organic matter.

In hindered settling of *discrete* particles, the settling rate is largely a function of solids concentration. For cohesive particles and flocs, settling rates still depend largely upon concentration but container size, presence of stirring and whether a concentration is initially homogeneous or formed by sedimentation of flocs from a more dilute pulp, all have a significant effect. This is of obvious importance when comparing settling rates in a large continuously operated sedimentation tank and in small quiescent batch settling cylinders.

The settling rate at various concentrations measured in batch tests can be described by Equation (18a). The five parameters required u_F, k, B, c_D, and E (where B and kc_D are usually close to 6 and 0.4, respectively) can be obtained only by means of batch tests so the equation is descriptive and cannot be used for predicting the settling rate of an *unknown* pulp. For modeling continuous thickener performance by means of computer programs,[58,64,78] however, where the same pulp is used over an extended time, Equation (18a) provides a more meaningful expression for the required functional dependence of u on c than is given by completely empirical fits to a limited range of laboratory batch data.[64,78,89]

14.3. THICKENING

The primary function of a thickener is to remove the maximum quantity of liquid from a dilute feed slurry thereby producing a pulp of the highest possible solids concentration. The process is shown schematically in Figure 14.17 and the equipment used in Figures 14.1 and 14.3. In thickener technology, a slurry, sludge, pulp, or mud all describe a suspension of solid particles in a liquid.

Thickening is a process of reducing the mean interparticle distance. Therefore, sometime during their passage through the thickener, feed particles have to traverse the range of interparticle distances from the one representing the most dilute pulp present up to the most concentrated. All intermediate concentrations will therefore exist even if only as transients. The maximum solids throughput of a thickener is governed by that concentration layer which has the lowest solids mass descent rate and *the essential problem in selecting the correct diameter thickener for a given duty is therefore to identify this limiting layer, establish its concentration and determine its solids handling capacity or flux, using small scale equipment.*

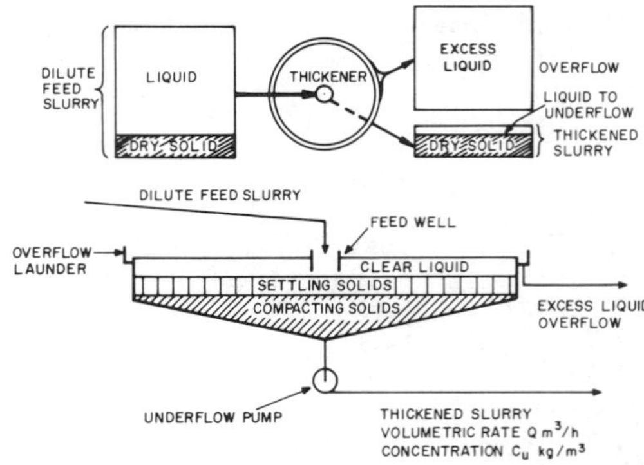

Figure 14.17. Schematic view of continuous thickening process.

Solids Flux in a Thickener

As the name implies the solids flux is the mass flow rate of solids per unit area and is given by the product of concentration and velocity. If the solids in a thickener descend at a velocity U (m/h) relative to the bulk of the suspension at a cross section where the solids concentration is C(kg/m^3) the solids *settling flux* is UC kg/m^{2}h.

Figure 14.18 shows a plot of several experimentally determined values of U as a function of C for a typical mineral slurry and also the flux UC as a function of C. (Methods of measuring U are described later.) It can be appreciated that although for dilute pulps the settling rate is high, the actual quantity of solids carried down, i.e., the solids flux is low and tends to zero as $C \rightarrow 0$. The solids flux is also reduced at high concentrations as U

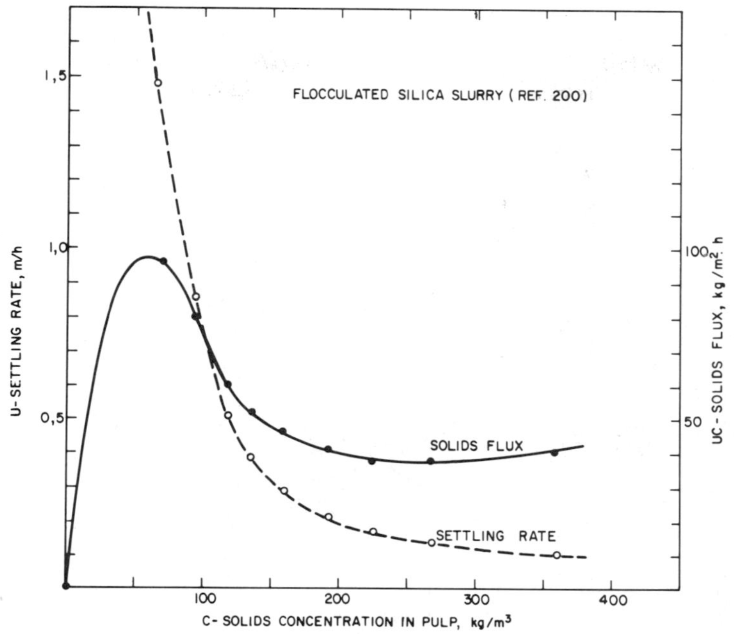

Figure 14.18. Settling rate and solids flux as a function of solids concentration.

decreases. At some intermediate concentration the settling flux therefore passes through a maximum.

The settling flux is taken *relative to the pulp*. In a continuous thickener the pulp as a whole is moving downward due to continuous volumetric draw-off at the base. If the underflow pumping rate is Q m^3/h and the area of the cylindrical section of the thickener is A m^2 then the bulk pulp velocity is Q/A m/h and the solids flux due to underflow pumping alone, called the *underflow flux*, is $(Q/A)C$. The *total flux* G_T, the solids flux *relative to the walls*, is the sum of the settling flux and the underflow flux, i.e.,

$$G_T = UC + (Q/A)C = (U + Q/A)C \quad (19)$$

The Hassett Construction. The graphical representation of the total flux G_T is due to Hassett[67] (Fig. 14.19). Disregarding the very dilute concentrations which normally apply only in clarification, the total flux curve can be seen to show a minimum value at M. From the typical values of feed, C_F, and underflow discharge concentrations, C_U, it is clear that, during thickening, feed pulps have to pass through the concentration of this minimum flux zone and its solids transmitting capability therefore controls the maximum steady state solids throughput of the thickener. The concentration of the minimum flux zone is (by definition) C_{crit} and its total flux (the minimum value of G_T) is

$$G_{min} = (U_{crit} + Q/A)C_{crit} \quad (20)$$

where U_{crit} = settling velocity at C_{crit}.

The concentration of the underflow pulp under *steady state* conditions must meet the materials balance: rate at which solids pass through the thickener equals the rate at which solids are discharged in the underflow pulp. i.e.,

$$G_{min} \times A = Q \times C_U$$

or

$$G_{min} = \frac{Q}{A} C_U \quad (21)$$

The underflow flux line $(Q/A)C$ reaches the

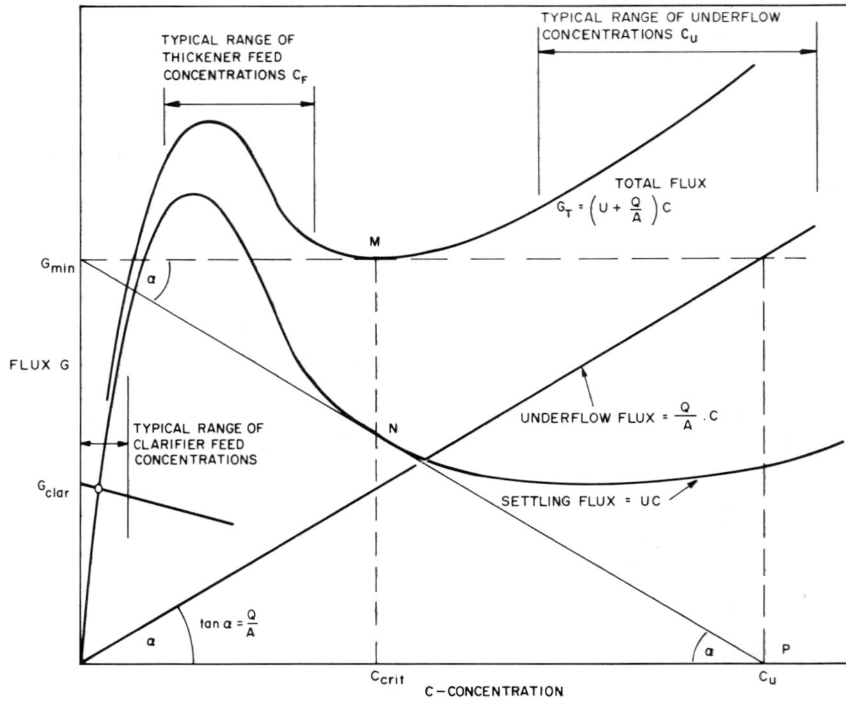

Figure 14.19. Underflow and settling fluxes may be summed graphically to give the total flux.

value G_{min} at the concentration $C = C_U$ (Fig. 14.19).

Determination of G_{min} according to Hassett's construction requires the drawing of a total flux curve for *each* value selected for the pumping rate Q and gives C_U only after determining G_{min}. Although it has the advantage of clearly illustrating the minimum value of G_T, it is cumbersome as C_U is normally the primary thickening objective.

The Yoshioka Construction. If in Figure 14.19, a line is drawn through point P (C_U on the abscissa) at an angle of which the tangent is $-Q/A$, congruency considerations dictate that this line intercepts the ordinate at the value of G_{min}. The equation for this line is

$$G = G_{min} - \frac{QC}{A}$$

which at C_{crit} attains the value $G_{min} - QC_{crit}/A$. At this concentration the equation for the *settling* flux, which is $UC = G_T - QC/A$ (eq. (19)) also attains the value $G_{min} - QC_{crit}/A$. The line drawn from G_{min} to P therefore coincides with the settling flux curve at C_{crit} and forms a tangent to this curve at N.

Yoshioka et al.[90] proposed therefore to determine G_{min} more directly than via the total flux curve by starting from C_U and drawing a tangent to the *UC* curve. Thus G_{min} is obtained from the intercept with the ordinate and the corresponding pumping rate Q, found from the slope of the tangent line.

From Figure 14.19 it is now apparent that if say a lower value of C_U is selected as an objective a second tangent to the settling flux curve can immediately be drawn. It will be steeper, corresponding to an increase in both G_{min} (maximum throughput) and Q/A and a decrease in C_{crit}. The reverse of course applies for an *increase* in C_U.

The Coe and Clevenger Formula. Coe and Clevenger[54] were the first authors to establish a sound method for the sizing of thickeners. They derived a total flux

$$G_C = \frac{U \cdot \rho_L}{F - D'}$$

where

U = settling velocity of suspension of consistency F
F = ratio by mass of fluid to solids in the pulp tested
D' = ratio of fluid to solids in the discharge required

An earlier but similar formula by Mishler[91] erroneously equates C_{crit} with C_F.

The Coe and Clevenger method involves determining the initial steady state batch settling rates U of test pulps of consistency F ranging from that of the feed to the thickest "free-settling" pulp, inserting corresponding values of U and F in the above equation, calculating G_C for each and noting the minimum value.

Their choice of symbol F was unfortunate as it has occasionally been misinterpreted[92,93] as referring only to the *feed* concentration; it probably referred to *f*ree-settling and their D' to *d*ischarge.

It can be shown that $(F - D') = \rho_L(C_U - C)/C\,C_U$ and therefore

$$G_C = \frac{UC\,C_U}{(C_U - C)}$$

or

$$\frac{G_C}{C_U} = \frac{UC}{(C_U - C)} \tag{22}$$

Equation (22) can be interpreted graphically using the settling flux curve in Figure 14.20 on which a line is drawn from P through L to G, where P is the selected underflow concentration as before and L is the flux at concentration C. From the similar triangles GOP and LCP, OG/OP = LC/CP. As G is the intercept on the flux ordinate it follows that $G/C_U = UC/(C_U - C)$ and therefore G is the total flux G_C of Coe and Clevenger.

As C is increased from the position shown, UC decreases and G_C first decreases and then increases as L moves along the curve LNR, i.e., it passes through a minimum value in a manner similar to that for the total flux curve of Figure

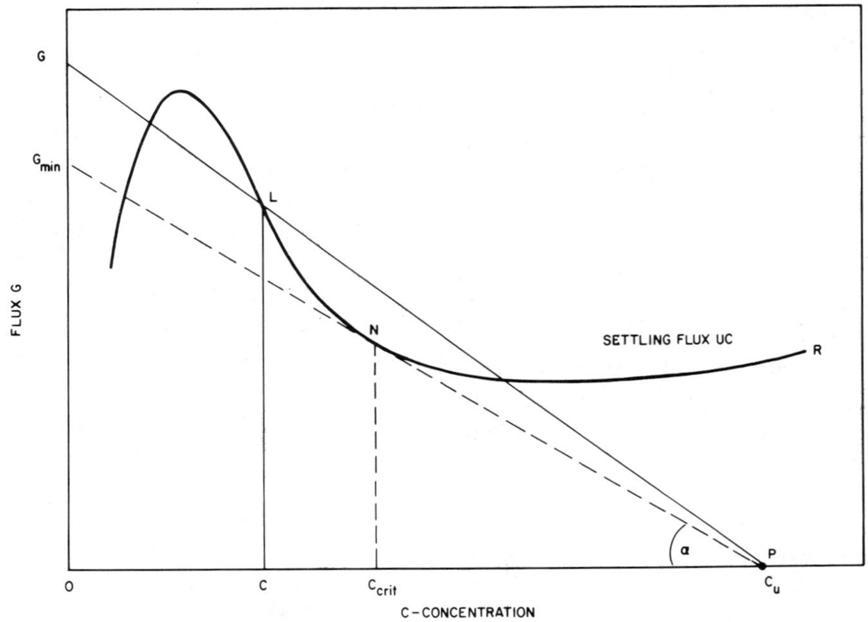

Figure 14.20. The total flux G as calculated by the Coe and Clevenger Equation (17).

14.19. In general the values differ, however, as the first assumes constant Q while Coe and Clevenger's formula is based upon a fixed C_U. Thus as $C \rightarrow C_U$, $G_C \rightarrow \infty$; that is Q can be made infinite (the thickener acting like a pipe) while G_T remains finite, limited by the *fixed* value of Q.

When G_C is at its minimum value as required by Coe and Clevenger then from Equation (22)

$$\frac{G_{min}}{C_U} = \frac{U_{crit}\, C_{crit}}{(C_U - C_{crit})} \tag{22a}$$

$$= \tan \alpha = \frac{Q}{A}$$

as given in Figure 14.19. The minimum value of G_C (G_{min}) therefore coincides with the minimum value of G_T at C_{crit} as at this point, the independently selected values of C_U and Q/A of each, mutually satisfy the condition

$$\frac{G_{min}}{C_U} = \frac{Q}{A}$$

The numerical and graphical methods of Hassett, Yoshioka, and Coe and Clevenger when applied to a given set of UC data therefore provide identical values for G_{min}. This critical flux, as determined from observations of the settling rates of uniform suspensions at various concentrations in batch experiments, will match this flux in a continuous thickener *only if the sedimentation conditions are the same in both.*

The Talmage and Fitch Construction. As only *one* concentration limits thickener throughput for a selected underflow concentration (or underflow pumping rate) it is an advantage if this value could be arrived at *directly*, thereby eliminating the need to measure the settling rates of many irrelevant concentration values. Talmage and Fitch[65] developed a procedure for obtaining the required minimum flux from a *single* batch settling test on the proposed thickener feed pulp by observing not only the initial steady settling rate but also the complete *H–t* settling curve.

The Kynch Model. Talmage and Fitch base their approach on the theoretical work of Kynch[94] on *batch* sedimentation but extend his concepts to *continuous* thickening.* Kynch showed that "a considerable amount can be learned by the single main assumption that at any point in a dispersion the velocity of fall of a particle depends only on the local concentration of particles." This defines an *ideal* pulp, i.e., one for which there is a *unique* value of U for a given C, and therefore a unique settling flux curve. Such pulps *must*, ipso facto, obey the predictions of Kynch. If has, however, been implicitly accepted that this *assumption* of ideality, used to develop the theory, actually applies to each and every suspension, some going so far as to state[80] that it "was proved mathematically . . . by Kynch."

In an ideal suspension, the particles nearest the base of the container are soon stopped after commencement of settling and immediately form a zone of *maximum* concentration. More dilute particles entering from above, add to this zone and cause a steady rise of its upper interface. The maximum concentration zone is, however, not the only one capable of forming such an interface. Depending upon the initial concentration, zones intermediate between initial and maximum values can also form, the prominent one being that with the highest upward propagating velocity. A fast rising interface results from a certain rate of solids entering from above and an inability to retransmit these solids downward. Such interfaces and zones below them are therefore closely related to the limiting concentration zone C_{crit}.

Figure 14.21 illustrates the conditions for such a rising discontinuity. The settling flux *into* the rising interface is $(U_1 + V_{int})C_1$ and, because solids cannot accumulate in an interface, this must be equal to the solids flux *out* $(U_2 + V_{int})C_2$. Solving for V_{int} gives

---

*The writer considers this extension of Kynch's mathematics to be a milestone in thickening theory, despite Fitch's statement[95] that they merely add "an inch to Kynch's mile." It is also inaccurate to refer to it as Kynch's method[81,96] as Kynch did not treat continuous thickening as such but confined himself to settling behavior in *batch* tests.

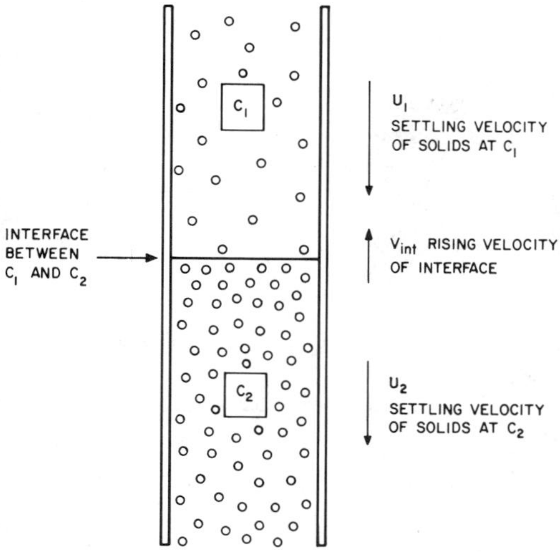

Figure 14.21. Rate of rise of a discontinuity between concentrations C_1 and C_2.

$$V_{\text{int}} = \frac{U_1 C_1 - U_2 C_2}{C2 \quad C_1}$$

$$= - \frac{\Delta G}{\Delta C}$$

On a flux plot (Fig. 14.22) this represents the tangent of the angle β drawn between the corresponding flux values at concentrations C_1 and C_2. The larger this angle therefore, the greater the rate of rise of the interface between C_1 and the higher concentration. It can hence be seen from Figure 14.22 that starting with an upper zone of concentration C_1, it is a zone of concentration C_3, angle δ, which will have the highest upward propagation rate. There will be a step change in concentration therefore from C_1 to C_3, and no concentrations between C_1 and C_3, including C_2, will appear.

If the upper zone has a concentration greater than the inflection point in the flux curve shown in Figure 14.22, no interface or step change in concentration can develop. The fastest rising concentration, because of the decreasing slope of the curve, is then the infinitesimally greater concentration $(C + dC)$. A concentration *gradient* therefore develops and the rate of rise of any concentration, C, in this region of the flux curve, can, by converting the above equation to differential form, be shown to be

$$V_c = - \frac{dG}{dC}$$

which is the slope of the flux curve at C.

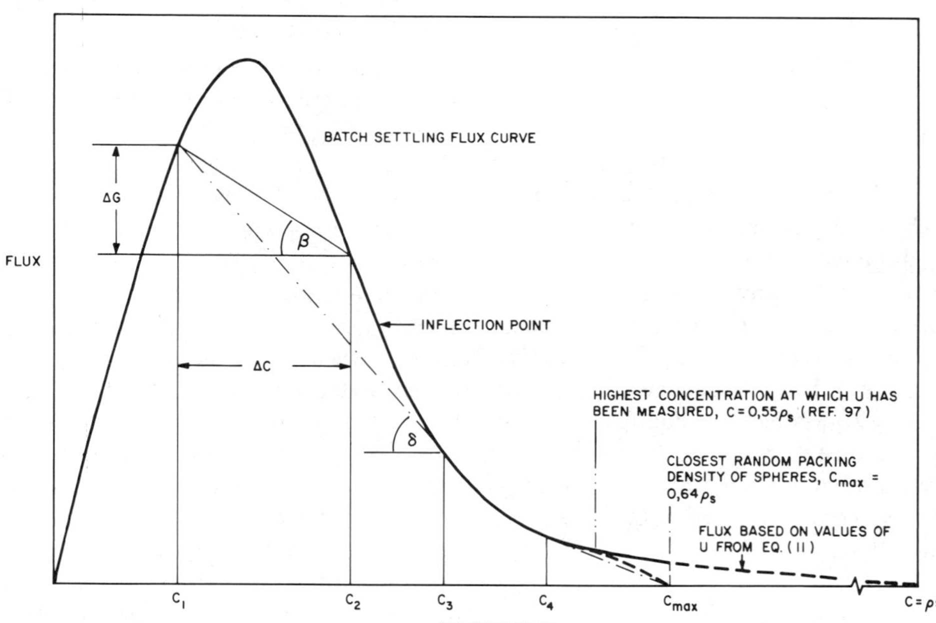

Figure 14.22. Flux plot showing the rate of rise of concentration discontinuities.

Application to Continuous Thickening. Reconsider now Figure 14.19 for which a value C_U has been selected. The values of C_{crit}, Q/A and G_{min} are found from a tangent drawn from C_U to the flux curve at angle α. We can *now* see that the rate of rise of C_{crit} in a batch test is tan α which has been shown to be Q/A or G_{min}/C_U.

Talmage and Fitch use a single batch test of initial concentration $C_0 = C_F$, to obtain the same information by allowing settling to proceed until the higher concentrations which form at the base at zero time reach the interface. Then in Figure 14.23 we can draw a line from the origin at angle α which therefore represents the rise of concentration layer C_{crit}. At time t_2 this concentration reaches the pulp interface which is now at height H_2 and therefore $V_{crit} = H_2/t_2 = Q/A$.

If, at the intersection of H_2 and t_2, a tangent is drawn to intersect H at H_1, then $(H_1 - H_2)/t_2$ is the rate of descent of the interface at t_2 and therefore equal to U_{crit}. Now

$$G_{min} = \left[U_{crit} + \frac{Q}{A}\right] C_{crit} \text{ (Eq. (20))}$$

$$= \left[\frac{(H_1 - H_2)}{t_2} + \frac{H_2}{t_2}\right] C_{crit}$$

$$= \frac{H_1 C_{crit}}{t_2}.$$

In the absence of a known value for α, which determines H_2 and t_2 by its intersection with the batch settling curve, some other means of locating this point is required.

The mass of solids present in the batch test is $C_0 H_0 A_b$ and at time t_2 this mass has completely passed through layer C_{crit} at a relative velocity $(U_{crit} + V_{crit})$. The solids flux is thus $(U_{crit} + V_{crit})C_{crit}$ which on multiplication by area A_b and time t_2 *also* gives the solids mass. This mass is of course constant, so by equating the two expressions, we get $H_1 C_{crit} = H_0 C_0$, and therefore

$$G_{min} = \frac{H_0 C_0}{t_2}.$$

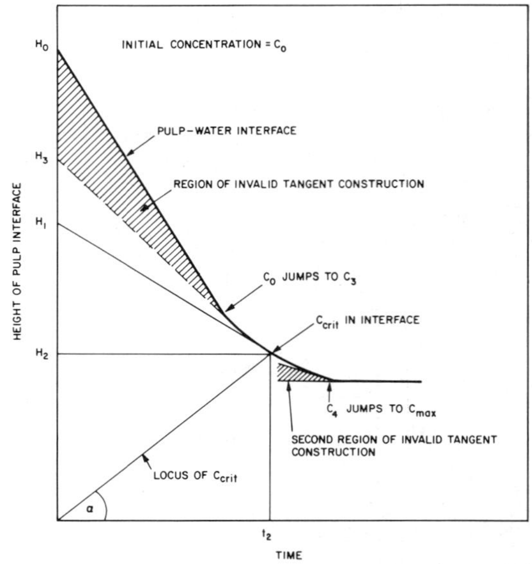

Figure 14.23. Sedimentation curve showing how discontinuities are related to the flux curve and regions where the tangent construction is invalid.

From Equation (21) $G_{min} = QC_U/A$, which is $(H_2/t_2)C_U$ and it therefore follows that $H_2 C_U$ also equals $H_0 C_0$.

Starting with a uniform pulp at intended thickener feed concentration $C_0 = C_F$ and at initial height H_0, then from a selected underflow concentration C_U, it is possible immediately to calculate H_2 and therefore the observed value of t_2, which is the time at which the pulp interface descends below H_2. Then, as shown, $G_{min} = H_0 C_0/t_2$.

This is essentially the procedure described by Talmage and Fitch. They called H_2 the "underflow line" or H_U while t_2 is their t_U.

It is clear from Fig. 14.23 that in general, from some point H on the ordinate, the slope of the tangent to the settling curve gives the rate of descent, U, of the concentration that is present at the interface at that time, and that this concentration, C, is given by $C = H_0 C_0/H$. Thus a *number* of pairs of $U - C$ values can be derived and hence the settling flux curve of Fig. 14.19 constructed from a *single* batch settling test. It must be borne in mind, however that theoretically not *all* values of C can appear in the interface. Thus, if the initial uniform concentration were equal to C_1 in Figure 14.22, then the concentration of the rising interface would be C_3 and therefore no tangents between H_0 and H_3 ($=H_0C_1/C_3$) in Figure 14.23 are valid. If drawn they merely represent the straight line joining the points U_1C_1 and U_3C_3 on the flux curve of Fig. 14.22 rather than the true UC curve as shown.

Rate of Rise of Maximum Concentration. Between C_4 and C_{max} no concentration rises faster than C_{max}. The shape of the flux curve, in this very concentrated region where batch settling behavior has not yet been measured,[97] can be regarded either as possessing a *discontinuity* at the concentration C_{max} (at which spheres settle into a packed bed) or as *continuous*, in which case a smooth curve is drawn from the highest experimental value of UC to $UC = 0$ at C_{max}. Although neither form of curve affects the rate of rise of C_{max} being equal to the rate of rise of C_4, the discontinuous curve is the more acceptable. From Equation (11) the settling velocity is proportional to the hindrance factor $(1 - C/\rho_S)^{4.7}$ which has a positive value right up to $C = \rho_S$ (100% solids), and the predicted flux therefore follows the upper dashed line in Figure 14.22. There is thus a *positive flux value associated with* C_{max} and no reason to suppose that it is not attainable in a concentration just below C_{max}. The solids flux for rigid spheres is therefore considered to follow the curve continuously up to the predicted value at C_{max} and then fall abruptly to zero[1] as settling ceases in the fully packed bed, i.e., the overall flux curve is *discontinuous.*

No concentrations between C_4 and C_{max} should therefore appear in a batch test as they rise too slowly, and a second discontinuity is therefore expected as the free falling particles at concentration C_4 abruptly increase their concentration and decrease their settling rate (flux) to zero as they "land" on the packed bed at concentration C_{max}.

This is an important feature of the batch settling curve and its existance becomes most evident when the rising packed zone C_{max} reaches the pulp-water interface. At this point the "fluidized" appearance of the interface gives place to a sharper ragged solid look. It was termed the "critical settling point" (CSP) by Coe and Clevenger[54] who define it as the "point [at which] the flocs at the surface just rest upon each other, but compression has not yet commenced in the surface layer. It is therefore obvious that any elimination of liquid from [this build-up] zone . . . cannot be accomplished by free settling but must be effected by compression of flocs."

Compression occurs when the roughly spherical flocs in a dilute suspension finally rest upon one another and transmit solids pressure to those below. These then deform to eliminate *interfloc* water, (which in the case of rigid spheres amounts of 36% of the packed volume) and later may compress to release *intrafloc* water by collapse of relatively open particle-particle structure in the flocs. In consequence C_{max} is (except for an ideal pulp) only a *transient* "maximum" concentration occurring only at the upper interface of the packed bed and readily exceeded when further settling solids cause compression from above.

The consequence of compression is that the interface continues to subside after the CSP but because the descent rate is lower than for free settling pulps the discontinuity in the $H - t$ curve is still usually observable. "Compression" effects are quite general and even observed with *rigid* glass spheres[98] for which small vibrations of the column cause the bed to assume a closer packing, i.e., the height of the originally packed bed can decrease further after complete settling.

Highly flocculated systems represent the other extreme where a small mass of overlying solids is sufficient to cause such a high degree of compression that the CSP may be completely masked. To overcome this, similar graphical procedures are advocated for both biological sludges[99] and polymer flocculated mineral pulps[96] (apparently independently of each other) to locate the CSP. A tangent is drawn from the *initial*, and a second from the *final* stage of the settling curve and from their intersection, either their bisection[99] or a normal to the curve[96] defines the CSP.

Two other techniques, described by Fitch,[55] are to plot ln H vs ln t or Roberts plot of ln

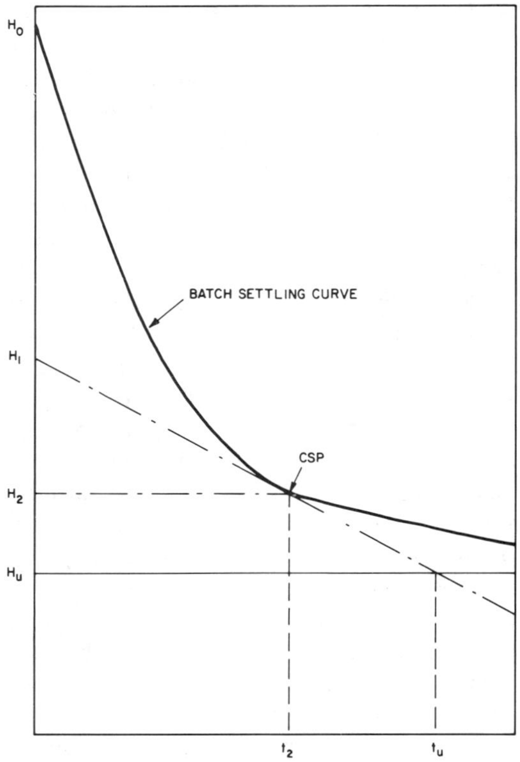

Figure 14.24. The Talmage and Fitch construction to obtain G_{min} for normal pulps.

$(H - H_\infty)$ vs t. Both accentuate the discontinuity of the normal H vs t plot at the CSP. H_∞, the height of settled solids at infinite time, must however be found by trial and error.

A further consequence of solids compression is that in full scale thickeners, with their greater relative depth, an even higher "maximum" concentration can be reached than is achieved in a batch settling test. The value of $H_U = C_0H_0/C_U$ (Fig. 14.24) then falls *below* the CSP meaning that the required thickener underflow concentration can be higher than the highest concentration observed in a batch test. The question then arises as to whether the subsidence solid fluxes existing in these unobservable *compression* zones will become rate controlling rather than the free-settling zones and how high a value of C_U can be safely selected. The answer, based on operating experience[65] with mineral pulps since 1916 is, that if C_U is maintained *within the constraints that the thickener sediment must remain sufficiently fluid to be pumped and also not overload the rakes*, then within normal thickener depths of 3 to 5 m, the height of the compression zone in the thickener can (in general) readily and automatically adjust so that each compression zone is able to transmit the imposed solids throughput G_{min} and finally attain the desired discharge concentration C_U. Too low a flux at any level in the compaction zone automatically results in a solids build-up which then causes an increase in flux at that level until eventually the bottleneck is eliminated.

The maximum thickener throughput therefore remains limited by *free settling concentrations* such as C_4 in Figure 14.22. Such concentrations are *not* subject to the effects of compression, while the flux curve for the higher concentrations up to C_U is "adjustable."[100] Tangents from C_U may therefore be drawn through or above these compression flux values[59] but not above the free settling portion of the batch flux curve. If a thickener were fed in accordance with such an operating line, the corresponding free settling C_{crit} zone would rise *steadily* to the overflow;[50, 59] while compression zones rise *only* until they reach a new steady state level. This is illustrated in Figure 14.25. For $C_U > C_{max}$ the line NMP of maximum allowable slope α is governed by the flux

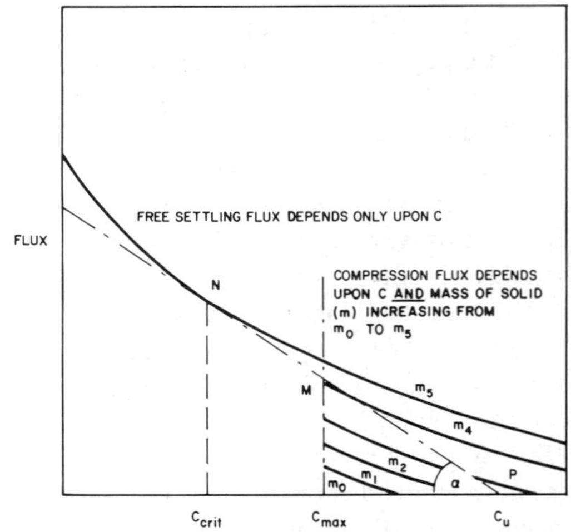

Figure 14.25. For compressive pulps, the Yoshioka construction line NC_U remains governed by the free settling flux curve but may "cross" the compression flux curves.

at N. At steady state and full load the thickener solids depth profile[59] shows C_{crit} near the top and lower down a compression (mud) zone of gradually increasing concentration. At each level in this zone, there will be just sufficient overlying solids m, to produce the required flux in that concentration as governed by the contour lines of m values in the region $C > C_{max}$.

In a batch test m is low, compression fluxes fall well below line MP for a thickener, and the rate of fall of the interface after the CSP is therefore not realistic. It is for this reason that Talmage and Fitch do not extend H_U in Figure 14.24 to intersect the pulp subsidence curve. Although a logical procedure for ideal pulps, and advocated by both Robins and Moncrieff, it has been shown[200] to lead to very poor thickener design data. The consequence of accepting line NMP (Fig. 14.25) as a legitimate tangent for thickener operation is an extension of tangent H_1 (Fig. 14.24) to cut H_U at t_U.

As before (Eq. 22a))

$$G_{min} = \frac{U_{crit} C_{crit} C_U}{(C_U - C_{crit})}$$

where

$$U_{crit} = \frac{(H_1 - H_2)}{t_2}$$

which, because of the construction

$$= \frac{(H_1 - H_U)}{t_U}$$

$$C_{crit} = \frac{H_0 C_0}{H_1}$$

and

$$C_U = \frac{H_0 C_0}{H_U}$$

from which it follows that

$$G_{TF} = \frac{H_0 C_0}{t_U} \tag{23}$$

as given by Talmage and Fitch. None of the previous equations involving a Q/A term can be applied here as at t_U in a batch test *not* all the solids have passed below H_U.

A similar, but empirical construction credited[101] to Oltmann, is to draw a straight line through the points H_0 (at $t = 0$) and the CSP, cutting H_U at some value of t, of necessity less than t_U. The required flux is then given by $G_{OLT} = H_0 C_0 / t$ from which it follows that $G_{OLT} > G_{TF}$.

Other empirical but confidential constructions have been developed by specific industries, for example, one involving the pulp height H after 1 and 24 hours settling. They rarely apply to other materials however and are seldom more precise than the conventional methods, which if necessary, can be multiplied by a correction factor for specific suspensions.

Comparison of Results of Batch Settling Methods

It has been shown that the graphical methods of Hassett and of Yoshioka et al. give estimates of G_{min} equivalent to the algebraic method of Coe and Clevenger. (A minor criticism of the latter is that the *true* C_{crit} and therefore G_{min} will usually fall between any two of the concentrations selected for the settling tests and hence the *apparent* minimum will be slightly in error. This may be overcome by plotting G_C vs C thus converting it to a continuous graphical method similar to the first two.) Flux curves derived from observed values of U (curve B_1, Fig. 14.8) of pulps in a batch settling test at various initially uniform concentrations C, *under quiescent conditions*, will be called the C and C method. Settling rates of the series of higher concentrations which form at the base of a batch test when a dilute pulp settles *under quiescent conditions* is the T and F method. Measuring $U - C$ values from tangents to the $H - t$ curve and constructing a flux plot, rather than determining t_U, does not constitute a different method.

The results of the C and C and T and F methods for a particular pulp can be directly compared without reference to full scale thickener operation. Talmage and Fitch[65] and Hassett[102] both show that the T and F technique gives consistently *lower* settling rates (except of course at C_0) and that these values may be as low as 30 to 50% of those of the C and C method.[45] This was confirmed by Scott[49] who showed further that the effect was confined to intermediate pulps for which, during the induction period in the C and C batch test, U increases with time and attains a settling rate significantly *greater* (because of channeling) than predicted for a concentrated suspension of individual flocs (Fig. 14.10). When such flocs, formed in the more dilute suspension of the T and F test however, settle to produce the same higher concentrations, there is neither sufficient height nor the necessary time for channel formation and in addition, the quantity of water displaced from below is already considerably reduced compared to a batch test of initially uniform concentration. *The T and F settling rates therefore fall closer to the values predicted by Equation (14) than the C and C values which follow Equation (18a).* As these may differ appreciably, it can be seen that the underlying assumption in thickener design that real pulps have a unique flux curve independent of method of measurement, *is at best, only a very rough approximation.* Therefore C and C and T and F design methods cannot *both* be correct. The Oltmann method which, as shown, gives values consistently higher than T and F, has been found[101] to give values corresponding closely with the C and C values.

A direct alternative to both these main methods is to carry them out in the presence of slow stirring (mild shear). Stirring in the C and C test has been done by Yoshioka for mineral pulps,[90] and by Dick and Javaheri for activated sludge.[103] A similar modification of the T and F method for activated sludge was made by Edde and Eckenfelder.[80]

A further modification is described by Hassett[102] and Fleming[104] to overcome some of the causes of the observed discrepancies in sedimentation batch tests, i.e.,

1. Variable floc structure
2. The effects of upward propagation of concentrations in compression, and
3. The difficulty of achieving an initial uniform concentration without residual disturbing circulatory currents.

Hassett and Fleming proposed that the pulp be held in suspension as a fluidized bed. Hassett employed fluidization merely to achieve uniform *starting* conditions and at $t = 0$, stopped the flow and measured U in the normal way (C and C). Fleming however, measured U during fluidization by timing the volumetric overflow rate, which could be done very accurately, and adjusted this rate from time to time to obtain a range of steady concentration-settling rate values.

A comparison of results for ferric hydroxide measured by the C and C method[105] and the fluidization technique[104] is shown in Figure 14.26. It can be seen that there is little to choose between them and therefore whatever suspected disadvantage may be inherent in the C and C technique affecting the results, has not been eliminated in the upflow procedure. Serious observed disadvantages of the fluidization procedure are

1. Particle classification in unflocculated pulps,[102]
2. Negative concentration gradients,[104] and
3. Floc break-up near the base of the column due to shear with gradual but irreversible loss of the coarser particles from the flocs[4] even with the use of a fluidized sand bed at the base to dissipate entry effects[104] as suggested by Fleming. The fluidization method is therefore not further considered.

Comparison of Batch Settling Tests with Thickener Performance

As pointed out by Pearse,[1] consideration of the construction and mode of operation of commercial thickeners, so different to that of a standard batch settling test, leads one to *expect* discrepancies between the two. The main features of the thickener are its continuous action and internal movement,[93] greater pulp depth and its rake action. However, as has been shown, there is more than one type of batch test available; static or with slow stirring, the

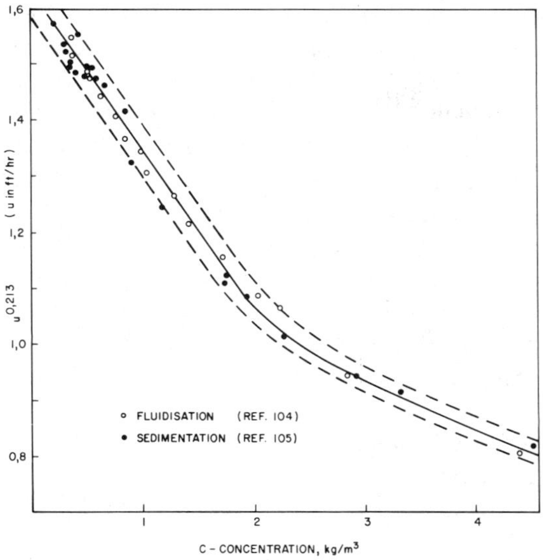

Figure 14.26. Comparison of batch settling and continuous fluidization relative settling velocities for ferric hydroxide pulps.

cylinder height can be increased, higher pulp concentrations can be made initially uniform or allowed to build up from the base, and some of the factors affecting the reliability of such batch methods in thickener design are as follows.

1. The formation of higher concentrations in a thickener, including C_{crit}, by descent of the solids in the *feed* pulp, is similar to the formation of this rising zone in the T and F test (C_{crit} does not of course rise in a thickener by virtue of the continuous pulp discharge). C_{crit} in the C and C test (although recognized only later) is produced and settled as a *uniform* pulp of height H_0.
2. The movement present in a thickener, caused by feed pulp entry and rotation of the rakes, (although not fully affecting the entire tank volume) could result in *decreased* settling rates for mineral pulps (Fig. 14.8, curve B_2) but *increased* rates for organic sludges that benefit from mild stirring when tested by the C and C method (p. 631).

3. Organic pulps show decreased settling rates in shorter test cylinders and the actual settling rates in a deeper thickener may therefore be underestimated even by the tallest batch test. This applies[84] to free settling C_{crit} and therefore G_{min} but for mineral pulps that are less affected by pulp height (except when in compression) this will not affect G_{min}.

The significance of these several effects can be quantitatively evaluated, however, only by experiment.

One way of checking the reliability of a batch test is to compare its settling rates at various concentrations with those measured directly in a thickener (operating at steady state) at corresponding concentrations.[45,49,64]

The mean discharge volume of a thickener operating at a "steady" rate and its mean concentration C_U over an extended time is measured, samples of pulp are taken at various depths and their concentration determined, and, from the known area A at that depth (A decreases in the conical bottom) the solids settling rate may then be calculated. The total throughput is QC_U kg/h and therefore the *flux* at any level is QC_U/A. This flux is made up of UC plus QC/A, so $U = Q/A\,[(C_U/C) - 1]$. From the various values of C found in the thickener profile, an *independent* flux curve can therefore be drawn and compared with that obtained from batch tests on the thickener feed pulp.

As alternative means of comparing the reliability of batch tests is to use them to predict G_{min} *directly* and compare it with the observed value in a fully loaded thickener. There is however, surprisingly little published data on this topic in spite of the fact that every thickener is designed from *some* estimated G_{min} and its subsequent actual performance (or nonperformance) is fairly readily ascertainable. It is nevertheless difficult to find representative results in the literature for a range of materials with which to compare each of the different batch tests. The available data also indicate that instead of G_{min} it is generally more convenient to compare total thickener area for a given throughput which is, of course, essentially equivalent. The ratio of "thickener area actually used" to "area predicted by a batch test" when *greater* than unity, indicates that the batch test *overestimates* U and therefore

G_{min}, i.e., the settling rates in the operating thickener are *lower* than predicted.

The view has been expressed[106] that *no* discrepancies found from experimental comparison of the batch test methods can be used as a reflection on the underlying *theory.* This theory is based on the behavior of an *ideal* pulp and if for any given pulp more than one value of U is observed at a particular concentration, it can merely be used as a measure of the degree of nonideality of that *pulp.*[107] In view of the large number of nonideal pulps requiring thickening and the absence of a theory specific to them, the present design procedure for real pulps is therefore based in effect, on the hypothesis that *settling rates in a batch test are equal to settling rates in a continuous thickener at the same concentration.* The required area is calculated from batch results using the ideal theory as presented and then multiplied by a factor to account for the nonideal behavior of the pulp. Coe and Clevenger's assumption is only as limited as this and it certainly need not be as stringent[1] as the assumption of Kynch.

Thickener Results. The results of a large number of comparisons of thickener areas expressed as a ratio R, of that actually required to that predicted, is given in Table 14.10 and summarized in Table 14.11.

The maximum and minimum values of the ratio for a given test shown in Table 14.10 give some indication of the *precision* of the data. This is not entirely due to variability in batch tests as a greater source of error is thought to lie in estimating the maximum solids throughput[113] which is not simple or straightforward. For example, over- or underfeeding into a large solids inventory can continue for many hours without the fact becoming evident and thus be erroneously taken as the actual throughput. As can be expected, the variation of the ratio is significantly greater for full scale plant work than for pilot plant runs, *even when the same investigator was involved.* The estimated standard deviation for individual small scale tests is 12% and for plant runs up to 30%.

From Table 14.11 it can be seen that except for a suspension of dispersed glass spheres (Class 3), the normal Coe and Clevenger procedure overestimates the thickener throughput, the average ratio for 86 runs on a variety of inorganic materials being 1.29 (Class 1). This means that if the C and C batch test predicted that for a given throughput, a thickener area of 100 square units would be sufficient, in actual practice 129 square units would be found to be necessary. This is in good agreement with contingency or scale-up factors for batch tests reported in the literature:

1. 1.2–1.5 Moore;[2]
2. "at least 20% should be added . . .";[114]
3. ". . . a 25% factor for safety is added . . .";[115] and
4. 1.10–1.25 for possible feed variations and 1.10–1.50 for feed turbulence with the product of the two being used for calculating the required area.[1]

Eklund is exceptional in that he found[107] a *higher* settling rate in a continuous thickener than in his batch test. His H_0 was 160 mm and this was undoubtedly too shallow as evidenced by his not reaching a constant settling rate for concentrations near C_{crit}. His values for C and C are therefore too low.

A scale-up factor of 1.29 (Class 1) implies that U_{crit} in the static batch test is some 30% higher than in a continuous thickener. It is probable that the enhanced settling rates that are observed following channel formation during the induction period in a batch test (Fig. 14.8B_1) cannot be attained in the continuous thickener because of the pulp movement. It is significant therefore that with slow stirring in a C and C batch test for inorganic pulps, the maximum settling rate is considerably reduced and it then predicts thickener areas which on the average are within 4% of the actual values (Class 4). A similar improvement is observed for activated sludge, an error of 40% in the static test (Class 2) being reduced to 11% with slow stirring (Class 5).

For this material, for which settling rates are increased by increased pulp height, it could, however, be expected that in continuous thickening an area ratio of *less* than unity would be obtained. The continuous tests reported, however, were done in a column less than 1.5 m while the batch tests in a tube "in excess of 1 m"[111] and the pulp height factor is therefore absent. Furthermore, a preflocculation step (passing air bubbles through the pulp) was introduced as it enabled "the high [induction] velocities to be obtained relatively quickly in the batch test . . ."[111] which therefore, appear to have reached their full potential so that they *exceed* somewhat more than usual, the settling rates in a continuous thickener.

In the T and F batch test, channeling, or other abnormal settling, is evidently absent in the region of the critical concentration as the predicted thickener areas for inorganic materials are with 5-6% of that actually required (Class 6), and stirring makes little or no difference (Class 8). For activated sludge, however, the improved "aggregation"[83] or pulp mobility brought about by stirring, results in settling rates which are four to five times *faster* in the batch test than in an operating thickener. These results are of course not general but depend on the size settling cylinders actually used in the tests.

For predicting thickener areas by means of batch tests for real slurries it is therefore necessary to ensure that the flocculated conditions of the slurry and the settling conditions in the test are as close as possible to conditions in practice.

All in all, the closest agreement between theory and practice is shown by a suspension of monosized spheres (99%) which is to be expected as it comes closest to Kynch's ideal pulp. For mineral suspensions, *slow stirring* in the C and C test and the *normal* T and F test, agree within 96% and 94% respectively of actual thickener performance. For organic pulps this agreement is not quite so good (111% and 85%, respectively) but in such pulps the *height* of the batch test is important and as this was (inevitably) not kept constant for the various tests reported, additional error is to be expected.

Of these two *batch tests*, the normal Talmage and Fitch method is preferred as it is:

1. simpler to carry out,
2. it gives more consistent results for a variety of different materials, and
3. it can be used for polymer-flocculated or other pulps in which the basic floc structure is affected by repeated resuspension as in the C and C method.

In addition, the "correct" stirring speed in the latter can be selected only if the thickener performance is known a priori,[49] which of course eliminates the need for any batch tests at all.

Two Examples of Predicting Thickener Areas from Batch Tests. *A mineral pulp.* A thickener is required to dewater a dilute pulp of 15% quartzite in water (by mass) to a relative density of 1.70 at a rate of 35 tons of dry solid per hour, what diameter is required? The water contains lime, $Ca(OH)_2$, at a pH of ~ 11 and this causes coagulation of the finer particles.

A *representative* sample of the feed pulp was placed in a one metre tall transparent cylinder of ~ 65 mm diameter. The pulp was gently agitated by means of a perforated plunger (alternatively, repeated inversions could have been used) and the cylinder then placed in a vertical position and the timer started. Readings were taken of the descending pulp-water interface height at suitable intervals and plotted as settling proceeded. The initial height was H_0 = 896 mm and Figure 14.27 shows the results of the settling test taken past the critical settling point, CSP, which is shown on the enlarged plot of Figure 14.28. This completes the experimental procedure for the Talmage and Fitch method.

The dry solids density is 2700 kg/m^3, therefore C_0, the thickener feed concentration, is

$$\frac{15 \text{ kg}}{\left(\dfrac{15}{2700} + \dfrac{(100-15)}{1000}\right) \text{m}^3} = 165.6 \text{ kg/m}^3$$

and C_U, the underflow concentration, is

$$\text{relative density} = \frac{(\text{mass of suspension})/\text{m}^3}{(\text{mass of water})/\text{m}^3}$$

Table 14.10. Comparison of Thickener Area Actually Required to Area Predicted by Batch Settling Tests.

TYPE OF MATERIAL	SPECIFIC PULP	AUTHOR(S)	REF.	DIAMETER OF CONTINUOUS THICKENER (m)	-R (RATIO OF ACTUAL THICKENER AREA/PREDICTED AREA)					
					COE AND CLEVENGER			TALMAGE AND FITCH		
					NO. RUNS	MIN MAX	MEAN	NO. RUNS	MIN MAX	MEAN
Minerals and metallurgical ores	Gold ore	Carman and Steyn	108	Sector of 7.3 m thickener [area equal to a 1.3 m circ. thickr.]	5	0.74 2.43	1.52	5	0.46 1.22	0.82
	Metallurgical pulps	Coe and Clevenger	54	Full scale operating plant units	10	0.70 1.87	1.12	–		
	Caustic mud	Couche and Goldney	96	1.52	–			10	0.84 1.11	0.98
	Acid-leached uranium ore							1	–	0.98
	Ground calcium carbonate			0.51				5	0.81 1.00	0.93
	Pyrophyllite	Scott	59	1.83	5^a	1.33 1.72	1.49	5	0.88 0.91	0.89
	Fine silica		200		13	1.02 1.28	1.20	6	0.87 1.05	0.97
	Milled gold ores-ex Durban-Roodepoort Deep		4		6	1.37 1.82	1.54	7	0.84 1.03	0.98
		Scott and Alderton	109	Full scale plant (15 m diameter)	30	0.73 2.54	1.62	–		
	ex Hartbees-fontein			do (23 m diameter)	8	1.04 2.33	1.66			
	ex West Driefontein	Sarosi	4		4	1.33 1.79	1.49	7	0.86 1.12	0.98
				1.83	3	1.19 1.25	1.22			

	Raw cement slurry	Yoshioka et al (slow stirring in bx tests)	90	0.1 and 0.2	4	0.78 1.17	0.94	4	0.88 1.20	1.02
Chemical products and precipitates	Precipitated calcium carbonate	Comings	110, 50	0.20	1	–	1.16	–		
		Hassett	67, 50	0.97			1.14			
	Magnesium carbonate	Yoshioka et al (slow stirring in bx tests)	90	0.1 and 0.2	12	0.73 1.19	0.99	12	0.67 1.17	0.91
	Barium sulphate				3	0.54 0.83	0.68	3	0.63 1.03	0.83
	Lithopone				1	–	0.82	1	–	1.01
	Stannous slime						1.39			0.93
Organic matter	Activated sludge	Hibberd and Jones	111	Full scale plant	40	1.13 2.20	1.40	No stirring in bx tests. Pulps preflocculated to achieve max. settling rate		
		Dick and Javaheri	103	0.15 and 0.19	35	0.94 1.37	1.11	Stirring in both bx and continuous tests		
		Dick and Ewing	81	Full scale plant	*Estimated* for quiescent 0.30 m deep bx test			3	0.35 0.80	0.63
		Edde and Eckenfelder	80		Stirring in bx test			10	1.05 7.50	4.20
	Sugar beet pulp	Talmage and Fitch	65					4	0.87 1.18	1.02
Ideal suspension	Kallodoc spheres	Oliver	112	0.10 × 0.015	6	0.86 1.08	0.99	–		

[a] Not previously published.

Table 14.11. Summarized Thickening Data from Table 14.10.

METHOD	MANNER	MATERIAL	NO. OF RUNS	MEAN R		CLASS NUMBER
Coe and Clevenger	Normal	Calcium carbonate	2	1.15	1.29	1
		Gold ores	56	1.51		
		Metallurgical pulps	10	1.12		
		Pyrophyllite	5	1.49		
		Silica	13	1.20		
		Activated sludge	40	1.40	1.40	2
		Monosized spheres	6	0.99	0.99	3
	Stirred	Chemical precipitates	17	0.97	0.96	4
		Raw cement	4	0.94		
		Activated sludge	35	1.11	1.11	5
Talmage and Fitch	Normal	Gold ores	19	0.93	0.95	6
		Metallurgical pulps	16	0.96		
		Pyrophyllite	5	0.89		
		Silica	6	0.97		
		Sugar beet pulp	4	1.02	0.85	7
		Activated sludge	3	0.63		
	Stirred	Chemical precipitates	17	0.92	0.94	8
		Raw cement	4	1.02		
		Activated sludge	10	4.2	4.2	9

$$= [(\text{mass of solid/m}^3 + \text{mass of water/m}^3)/1000]$$

i.e.,

$$1.7 = \frac{C_U + \left(1 - \frac{C_U}{2700}\right)1000}{1000}$$

and therefore

$$C_U = 1112 \text{ kg/m}^3$$

The underflow line H_U $(= C_0H_0/C_U)$ is therefore 0.133 m and the intersection of this horizontal line and the tangent drawn just *before* the critical settling point, H_cB, occurs at t_U = 112 min, i.e., 1.87 h. From Equation (23), $G_{min} = C_0H_0/t_U$, which immediately gives G_{TF} = 79.5 kg/m²h and therefore for 35 t/h, the required area is (35 × 1000)/79.5 = 440 m².

To compare this result with the normal Coe and Clevenger method, supernatant was decanted down to ~ 200 mm and the water height H noted. The residual liquid was then mixed with the settled solids below to form a homogeneous pulp of relatively high concentration C and its settling rate measured (Test No. 1 Table 14.12). These tests were terminated as soon as the maximum steady state settling rate had been obtained, and the concentration C was calculated from $C = C_0H_0/H$. Some of the supernatant was then returned and after resuspension of the solids the settling test repeated exactly as before. This was continued stepwise until the original concentration was restored. A check on its settling rate was then done to ensure that repeated reagitation had not altered the settling characteristics of the pulp. The results are shown in Table 14.12 for which the values of G_C were found from Equation (22) i.e., $G_C = UCC_U/(C_U - C)$.

The minimum flux value is 147.7 kg/m²h

Table 14.12. Coe and Clevenger Flux Values for Quartzite Pulp.

SETTLING TEST NO.	H (m)	C (kg/m^3)	U (m/h)	G_{C_2} (kg/m^{2}h)
1	0.205	723.8	0.088	182.5
2	0.260	570.7	0.126	147.7[a]
3	0.366	405.4	0.234	149.3
4	0.428	346.7	0.304	153.1
5	0.600	247.3	0.597	189.9
6	0.896 (=H_0)	165.6 (=C_0)	0.908	176.7

[a]Mininum value of G_C.

for a concentration of 570.7 kg/m^3. For a throughput of 35 tons/h, the required area according to the C and C method is therefore 237 m^2.

These settling tests were in fact done on a sample of pulp taken from an operating thickener while accepting its maximum steady state rate of feed, 35 t/h. The area of this thickener (D = 75 feet) was 410 m^2.

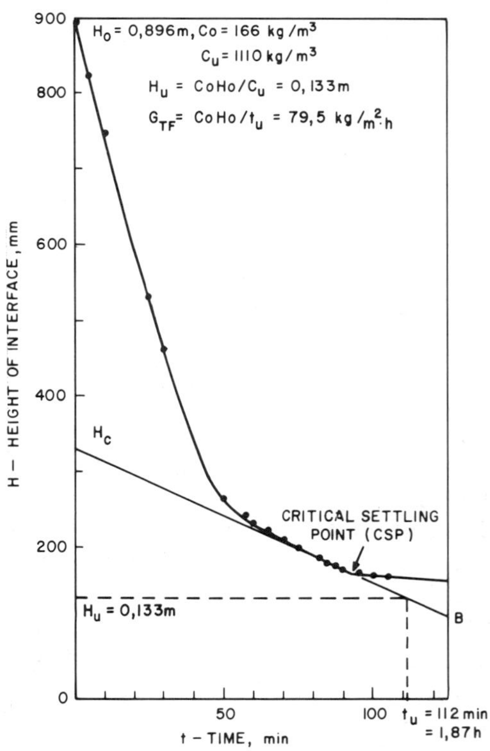

Figure 14.27. Talmage and Fitch construction for a mineral pulp.

The C and C test therefore underestimated the area by 42% and the T and F test overestimated it by 7% in general conformity with the average values given in Table 14.11. As postulated, this difference in predicted area can be shown to be largely due to the different settling rates for the same concentration. In Figure 14.27, C_{crit} is 449.6 kg/m^3 (C_0H_0/H_c) and its settling rate is 0.106 m/h $(H_c - H_U)/t_U$. From Table 14.12, U for this concentration (found by interpolation) is 0.193 m/h or 1.82 times higher, a factor virtually identical to the inverse ratio of the predicted areas, 1.86. C_{crit} for C and C however, is 570.7 kg/m^3 and U_{crit} 0.126 m/h the former being close to C_{crit} observed in the solids profile of the operating thickener.[116] The C and C method therefore predicts the correct value for C_{crit} but not its settling rate while the T and F method closely estimates its settling rate but via an incorrect concentration. This, together with its heavy leaning on the Kynch assumption of ideality and yet closely estimating thickener areas for nonideal pulps indicates the T and F method to be possibly fortuitously self compensating. But it does work.

Attempts to eliminate the Kynch assumptions as to the rate of rise of concentrations in the T and F method and measure rise and subsidence rates by means of direct *sampling*,[64] led to the unexpected result that actual settling rates then exceeded not only the normal T and F values but also those of C and C.

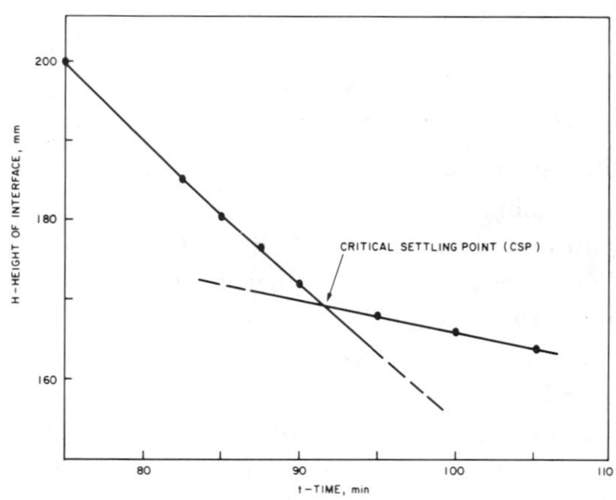

Figure 14.28. Enlarged view of sedimentation curve near the critical settling point.

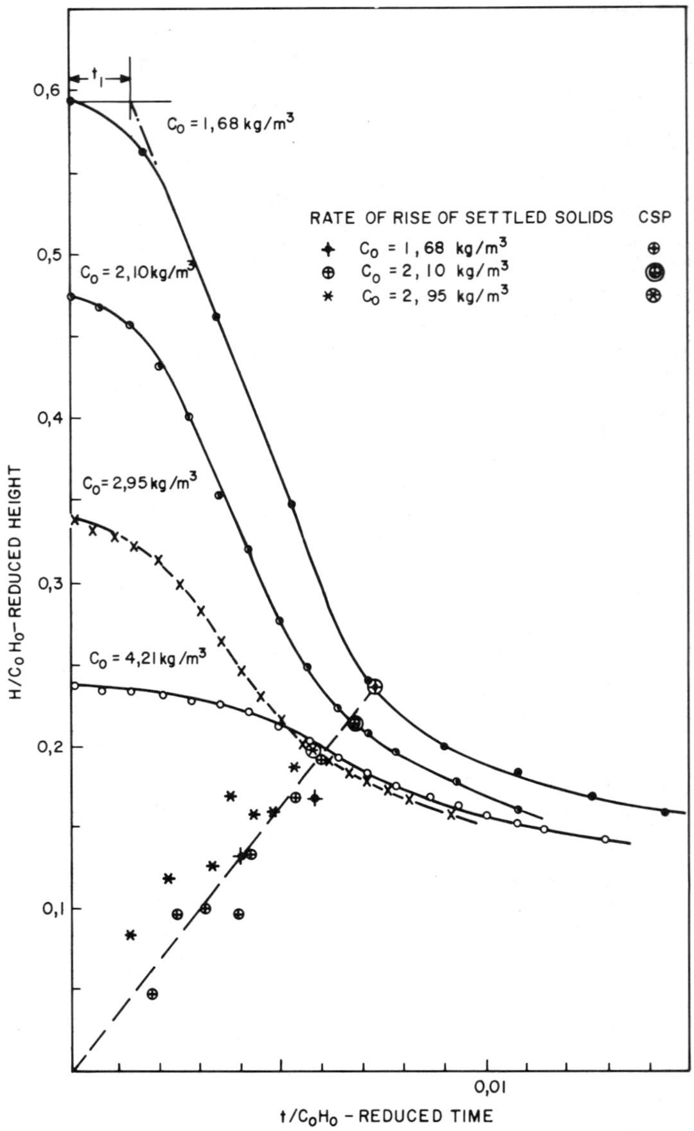

Figure 14.29 Reduced sedimentation curves for an activated sludge at four concentrations.

Activated Sludge. As a second example, what thickening area is required for final settling tanks to treat 75 × 10^6 liters/day of mixed liquor with an MLSS concentration of 4200 mg/l (4.2 kg/m^3) and give a return sludge (underflow) suspended solids concentration of 7.96 kg/m^3?

A Talmage and Fitch settling test was done on the "mixed liquid feed pulp" and on more dilute "feeds" made by taking 400, 500, and 700 ml of the normal feed pulp, and making each up to a liter (H_0 = 0.33 m) with clear settler overflow liquor.

The results are shown in Figure 14.29, using reduced (normalized) coordinates.[67] Each of the pulps tested showed intermediate behavior with a marked induction period, t_1, which increased with concentration. The value of t_1 for each was found by extending the linear portion of the H-t curve, used to determine U in the C and C test, backwards to intersect $H = H_0$ as shown for the most dilute pulp.

Settling is assumed to start only at t_1 and this value is finally subtracted from t_U.

In none of the *H-t* curves is there a distinct "kink", marking the CSP as observed for the mineral pulp in Figures 14.27 and 14.28. Besides observing the descending pulp-water interface, the demarcation between "already settled" and "still settling" solids was therefore carefully noted. The distinction was clear in the diluted pulps but not in the one at normal feed concentration. The rise of the height of settled solids right up to the pulp interface is included in Figure 14.29. Although the data are scattered, the points on the *H-t* curves intercepted by the rising zones fall remarkably close to a line drawn from the origin. These points define the true critical settling point (CSP) in each case. It is essential to know the location of the CSP for the T and F method and, wherever possible, *this* method of location should be used. The data can now be used to compare the four other methods advocated when the direct method fails. These methods are:

1. the log-log method,[55]
2. the Roberts' plot.[55] (A similar construction including the settling rate term of Gaudin et al.[201] is presented by Barnea[117]),
3. the bisection of the angle formed from lines drawn from the dilute and compression phases,[99] and
4. a normal drawn from the intersection of these two lines to the *H-t* curve.[96]

Figure 14.30 illustrates the first two methods, and shows the distinct change in slope in the plots which serves to locate the CSP.

Figure 14.31 shows an expanded portion of the *H-t* curve near the CSP to illustrate methods 3 and 4. This was the only pulp of the four tested where the bisection method and the normal to the *H-t* curve did not give identical results. Even here the difference is negligible. The bisection of angle LMN is an uncomplicated construction while the accuracy of a normal from M to the curve depends upon the symmetry of the curve near M. The simpler bisection method is therefore preferred.

The results of the five methods for locating the CSP are included in Figure 14.31. It can be seen that for this pulp they all agree in placing the CSP between 5.5 and 6 minutes, i.e., data points B and C on the curve. As free

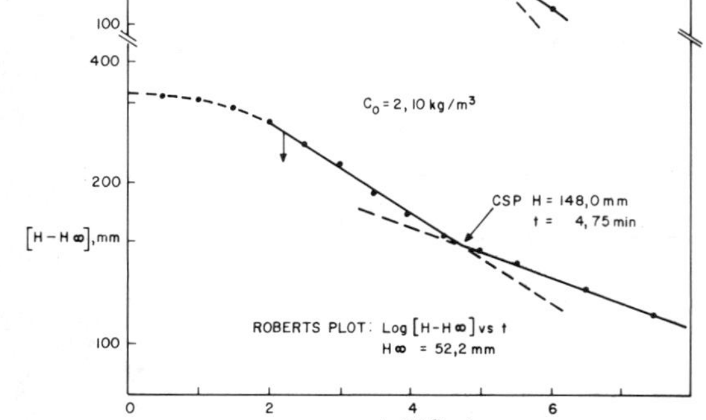

Figure 14.30. Constructions that serve to enhance the critical settling point. log *H* vs log *t* above, and the Roberts' plot below.

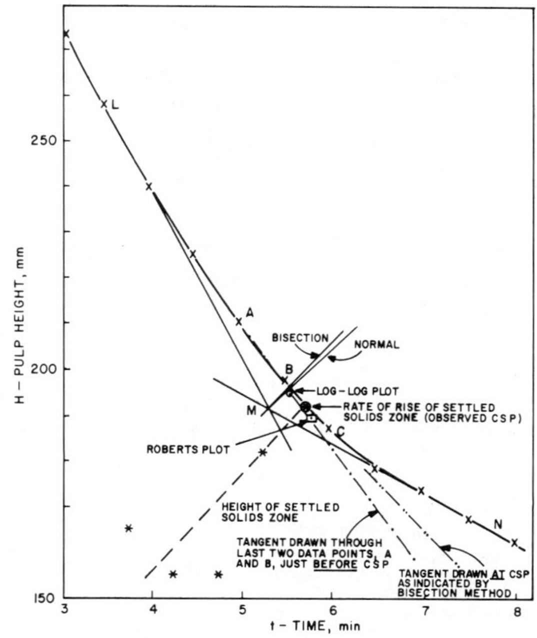

Figure 14.31. Enlarged view of sedimentation curve to compare various methods of locating the compression point.

settling concentrations govern throughput, the correct tangent is the one *before* the CSP with the most reliable values being drawn through the experimental points A and B.

A tangent drawn *at* the CSP gives a significantly larger value of t_U and is therefore conservative.

The bisection method generally indicated the CSP to be the highest on the settling curve (usually by one data point), the Roberts' plot the lowest, while the log-log plot gave the closest agreeement with the true value. This plot avoids both the lengthy graphical construction and the trial-and-error selection of H_∞, and is the recommended method if the rate of rise of the settled solids cannot be directly observed. The predicted thickener areas for the different techniques are shown in Table 14.13.

It can be seen that dilution of the feed, while facilitating *direct* observation of the CSP, did not have a more significant effect on the result than did different methods and that the log-log and Roberts' plots correctly located the true CSP and gave a ratio R = "actual area to predicted area" of 0.77 to 0.95. The two more empirical construction methods varied only slightly giving similar ratios of 0.73 to 0.97. The most reliable value, R = 0.77 based on log-log plot of pulp at feed concentration, may be compared with the values 0.80, 0.73 and 0.35 estimated by Dick and Ewing[81] for three different plants, A-C, respectively. In the present case, based on feed and discharge concentrations, the pulp falls between those of plants A and B.

Segregation of the Feed Pulp. Under circumstances where feed pulp entry results in a breakdown of the flocs, the finer size fraction may reflocculate preferentially[118] giving rise to a pulp with a settling rate lower than the same concentration when tested in the batch test (p. 643). Such floc-bed zones tend to form *above* the level of actual feed entry[119] and therefore limit solids throughput by restricting the permissible volume of *overflow* (and therefore feed) liquid.[109] This effect was observed for natural gold ores but not for its individual mineral components[59,200] nor for their mixtures, e.g., quartz and pyrophyllite.[4] The floc-bed phenomenon does not appear to have been observed outside the South African gold ore thickening practice[120] although the effects of size segregation is reported for metal concentrates in Russia.[121] The discrepancies are however so large ("the calculated thickener area may be dozens of times smaller than the actual value") that coagulation of the fines does not appear to have occurred.

Estimating Thickener Depth

The development of a design procedure for estimating the depth necessary to give the

Table 14.13. Predicted Minimum Areas and Ratio to Actually Required Area[a] for Different Feed Concentrations.

METHOD OF OBTAINING CSP AND t_U	PREDICTED MINIMUM AREA (m^2) AND RATIO R^a — FEED CONCENTRATION C_0: 1.68(kg/m³)		2.10(kg/m³)		2.95(kg/m³)		4.21(kg/m³)	
	m^2	R	m^2	R	m^2	R	m^2	R
Direct observation of rising pulp zone	1870	0.86	1700	0.95	1740	0.93	–	–
Log-log plot	do		do		do		2090	0.77
Roberts' plot	do		do		do		do	
Bisection of angles	2140	0.75	1790	0.90	1660	0.97	2220	0.73
Normal to curve	do		do		do		do	

[a]Actual area for this duty was 4035 m^2 with an overdesign factor of 2.5. The minimum required area is therefore 1610 m^2. R = predicted area/1610.

required pulp compression and to achieve a given underflow density has not received the same attention as that for estimating thickener area. This is unlikely to have been the case if compression-zone depth were as important as area in practice.

Any serious build-up of pulp in a thickener during normal operation should be regarded as due to running at too low a discharge rate which is equivalent to expecting an unrealistically high underflow density. Thus in Comings'[110] runs, at a constant feed rate of ~2.65 g/min, an underflow concentration C_U of 400 kg/m^3 was obtained when the discharge rate Q was 6.8 ml/min. Reduction of Q to 6.2 ml/min caused C_U to increase to 430 kg/m^3 a difference of only 7% but to achieve this, the depth of compression pulp almost doubled (X 1.8) and completely filled the thickener. The value of C_U = 430 kg/m^3 is therefore, in these circumstances, unrealistically high (the maximum concentration in a batch test was 330 kg/m^3) and if it *is* essential to dewater to this extent, filtration for example, rather than gravity thickening should be considered. In gravity sedimentation there is thus a practical upper limit to the concentration which can be reached. In continuous thickeners, as this limit is approached, sever things may happen:

1. the pulp may become too thick to be raked
2. it may be too thick to pump, and
3. a deep compression zone may build up extending to the overflow level and solids commence to slime.

In metallurgical pulps, 1 and 2 certainly limit the maximum attainable concentration and true compression height is negligible.[116] The nuisance of 3 in operation can be overcome by running at a slightly higher discharge rate but it would obviously be desirable if knowledge of the height-underflow density relationship were available at the design stage.

Adorján[42] tackles the problem by considering the compression strength of the pulp and its specific resistance (permeability) as unique functions of solids concentration. This implies ideal behavior in compression. These parameters are measured in a novel pressure-permeability cell, the experimental results expressed in the form of analytical functions, i.e., empirical equations, and inserted into the equilibrium of forces equation:

gravitational force = compressive strength
+ resistance to viscous flow of liquid.

From this he obtains an expression for dH/dc containing the variables c and c_U, the operating constants G and solid and liquid densities, and six empirical constants. The calculated results confirm that it is only as c_U approaches the maximum attainable value (in this case 93%) that the pulp height reaches a significant level 1.9 m. At 82% of the maximum, the level was only 0.5 m.

The disadvantages of the method include its neglect of the effects of channelling and absence of raking during compaction. It presents little advance on Fitch's[56] earlier proposal where, instead of *calculating* dH/dc from sophisticated experimental measurements, the pulp is allowed to settle in a deep tube for several days and sampled at different depths to determine c as a function of H; dH/dc is then obtained directly. The experimental procedure is simpler and may be carried out in the presence of slow raking,[55] an important feature in attaining maximum underflow concentrations.[110]

An even simpler procedure,[55] as it does not involve sampling the pulp, only noting the pulp height at various times, is essentially at follows. Note the time and height of the pulp as the critical settling point and then say after 24 or 48 hours, H'_U. The maximum $C_u = H_0C_0/H'_U$ and the height of compression pulp necessary to attain it is

$$H = \frac{G_{min} \times 24}{c_{AV}} \text{ m}$$

where

c_{AV} = average pulp concentration during compression[119]

When applied to Comings'[110] data, a height of 0.91 m is predicted while actual values range from 0.2 to 1.2 m indicating that the equation gives the correct order of magnitude. It is inconsistent however in that for any material, only *one* value of c_{AV} is obtained from the batch test, and hence when operating at a steady G as did Comings, but with different underflow rates (c_U), only *one* pulp height is predicted whereas such operation is possible *only* if the height varies.

The formula is however normally used merely to *prevent* a thickener being of too small an area and therefore of having too deep a compression zone. To overcome this, G and therefore H is reduced accordingly by increasing the diameter so that H is less than 1 m. This is known as the "three foot rule".[55] This is a safe procedure as even if the calculations are 300% out, the compression can still be contained in a normal size thickener.

The most realistic compression model presently available is that due to Kos and Adrian.[122] Based on measurements on actual thickeners they concluded that the compressive strength of the pulp was at least approximately a function of concentration but that its permeability (equivalent to pulp settling rate) was a function of both concentration and dynamic pressure gradient. In spite of this advance on the simpler models, it is nevertheless still considered more suitable for theoretical work than design[123] and pilot plant studies are recommended for sizing the thickening zone in the absence of actual full scale operating data. Scale-up factors for depth[1] are given as 1.75.

14.4. CLARIFICATION

As implied in the name, the purpose of clarification is to remove turbidity or suspended solids from a murky liquid and render it crystal clear. It is used in a wide variety of industries, applied to raw materials, intermediates and products and, increasingly in recent years, to liquid wastes.

The treatment of raw water supplies, clarification of solutions in the sugar, metal and inorganic chemical industries, removal of fine catalyst particles from petroleum intermediates, polishing of beer in racking tanks after addition of finings, and the disposal of waste water from paper, ceramic, and coal processing industries, are but a few examples of the use of sedimentation for the commercial scale clarification of liquids.

Sedimentation is, however, not the only means of achieving this,[5,6,124] and for a fuller coverage of clarification, the chapters dealing with filtration methods should also be consulted.

Clarification by gravity sedimentation is carried out in *circular tanks* similar to that shown in Figure 14.2, but of lighter construction than those shown in Figures 14.1 and 14.3, and also in rectangular tanks.[125] In the treatment of potable water, long *rectangular basins* are considered to be hydraulically more stable, with less short circuiting between feed and overflow points especially in larger plants.[126] In flocculated sewage treatment in the Toronto area,[127] a long, rectangular horizontal-flow settling tank was stated to be "much better" than a circular tank even when based on the same overflow rate and detention time.

In the food industry, the relatively prolonged residence time in normal gravity settling tanks sometimes leads to fermentation and deterioration. The processing time and the liquid inventory may both be reduced, however, by the use of *centrifugal* clarifiers. In these the gravitational force, g, is increased by two to four orders of magnitude with a corresponding increase in the settling velocity of the solids. Examples are the treatment of olive oil to avoid rancidity and the separation of yeast cells from beer to cause a rapid termination of their growth not possible in normal gravity sedimentation.[12c] Detailed discussion of centrifugal sedimentation[5] is, however, beyond the scope of this chapter.

Clarification of gold-bearing solutions in Southern Africa is traditionally done by precoat *filtration*. A recent full scale test[128] has shown, however, that a prior gravity *sedimenta-*

tion step can reduce the overall cost of clarification to 60% that of filtration alone. This specific example confirms that the relative lower cost of gravity sedimentation in general, indicated in Table 14.1, applies also to clarification.

Comparison of Clarifiers and Thickeners

The concentration of solids in the feed to clarifiers ranges from 0.1 to 10 kg/m^3 which is between 20 and 1000 times more *dilute* than for thickeners. As the main purpose of clarification is the *complete* removal of this solid matter from the liquid, the density at which these solids (usually waste) are rejected from the clarifier is of reduced importance. The solids are usually finer than in thickener feeds and require flocculation for efficient settling. In a fully loaded *thickener*, the deep layers of sedimenting and compacting solids, are a prominent feature of its depth-concentration profile whereas in clarifiers the major volume of the tank is occupied by quiescent liquid. As a consequence, the incoming feed, which may differ both in density and temperature from the contents of the tank, can readily upset the ideal flow pattern. The mode of feed entry and overflow removal is therefore more critical than in thickeners[126,129] and model studies should be used to investigate the hydraulic effects of novel designs[130] including the effects of baffles, weirs and distributor plates.[131]

In thickening, the solids settle by hindered settling with an interface between suspension and liquid so that size segregation of fines is minimal. Because of the lower solids concentration in clarifiers the flocs descend independently with the larger particles or flocs reaching the sludge level faster than the slower settling fine material. If it were not for this *range* of settling rates, the normal thickener flux curve (Fig. 14.18) could be used in design, with the maximum feed rate of solids, G_{clar}, being governed by the *feed* concentration as this represents the maximum solids flux available.

To apply thickener theory however, a different flux curve would be required for each species of particle size and particle density present but as maximum *solids* throughput is not of primary importance, a quite different approach is used in clarifier design.

More details of the conditions necessary for the separation of various sizes and densities of particles in a mixed suspension is given by Masliyah.[203]

Pretreatment for Sedimentation

Effective clarification often depends upon flocculating agents for success. Even when sedimentation is feasible without such aid,[1] pretreatment can result in both a reduction of the size of tanks required and an increased clarity of the product. Pretreatment also applies to *thickener* feed when throughputs have to be increased but this is considered to be a more expensive remedy than installing additional tanks for the long term.[108,132]

Flocculants are additives which cause suspended solids to agglomerate into flocs which act like single large particles and therefore settle more rapidly than their smaller components. Floc formation is brought about by coagulation, by capture in an hydrous precipitate or by the formation of polymer bridges between particles.

Coagulation occurs when the mutual electrostatic repulsive forces between particles are sufficiently reduced, by the addition of ions of opposite sign, to permit the London-Van Der Waals attractive forces to cause aggregation of the particles. This requires either a *p*H change or the addition of preferably polyvalent ions or a combination of these actions. Lime and alum are common coagulation additives.

At a suitable *p*H alum addition will lead to the formation of a hydrous aluminum hydroxide precipitate in which particles may be captured.

Polymer flocculation, whether by natural or synthetic neutral polymers or polyelectrolytes, may be considered[133] to take place in two stages:

1. adsorption of the polymer on to a particle

surface, attributed to hydrogen bonding or ion adsorption, and
2. flocculation of the particles either as a direct result of the London–Van Der Waals attractive forces or due to physical polymer bridges formed between the particles.

These bridges may be formed by the two ends of one polymer molecule being attracted to two different particles,[134] or by the "loops" of polymer chains on one particle being attracted to the loops of another.[133] Polymer flocculation is extremely sensitive to the molecular weight of the polymer used.[135] Because of the different mechanisms, coagulation is a reversible process while flocculation is not.

The *combined* use, first of an electrolyte to reduce repulsive charge, followed by a reduced quantity of the relatively more expensive polymer, often leads to a less costly pretreatment process than the use of either alone. There are of course restrictions in selecting flocculants for potable water and foodstuffs. Less efficient but edible natural products such as starches and gums find a useful application here.

The quantity of flocculant normally required to cause efficient flocculation is only a small fraction of that which can be adsorbed on the relatively large solid surface available, and polymer flocculant molecules are usually quickly and completely removed from solution. Contamination of the clarified liquid with residual free flocculant is therefore usually absent but not impossible and could lead to problems at another point in the circuit[1] or in the application of the product. A bigger problem at the flocculant addition stage is to ensure that the limited quantity of flocculant is equally distributed between all the particles in the suspension.

Although the principles of flocculation are reasonably well understood, selection and application of the best flocculant for a particular suspension is still an art.[1,5,136,137] Thus, negatively charged solids may be flocculated by cationic flocculants as expected but better results are obtained with an *anionic* polymer after addition of a divalent cation.[133] Determination of the best conditions and selecting the best product from a range of similar type flocculants is therefore usually based upon results of laboratory batch tests on the liquid to be clarified.[138] Such tests, if properly carried out, can indicate not only the chemicals to add, the required amounts and the order of their addition, but also the degree of stirring, method of application, and the wait period required either before the next addition or the commencement of sedimentation, i.e., the point in time at which the flocculated pulp should be admitted to the clarifier.

A typical sequence is:

1. Addition of lime solution to the feed under conditions of rapid mixing to ensure good dispersion and adsorption on to the solids
2. Addition of polymer solution, as dilute as possible and at more than one point of application, also under conditions of rapid mixing to ensure good distribution in the liquid, and finally
3. Slower mixing or gentle shear of the clarifier feed so that the treated particles, now ready to adhere to each other, grow into flocs and flocs into larger aggregates to cause rapid sedimentation and to incorporate the finer particles which otherwise cause residual turbidity after settling.

It must however be borne in mine that in designing flocculation equipment that the *detailed mechanisms* of polymer flocculation are generally not as well understood as the theory, and the process cannot therefore be applied without some empiricism and resort to pilot scale experiments, and even then the results cannot be scaled up with certainty.[139]

Floc Strength. Excess shear of the formed flocs should be avoided especially those produced by irreversible polymer bridging. In one difficult case, dropping from the end of the launder into the feedwell was sufficient to rupture the flocs. Removal of the launder from the feedwell and lowering its end a few centimeters below the water surface of the thickener provided a smoother entry for these fragile flocs and the solids throughput was increased 2 to 3 times as a result. Smith and Kitchener[140] present three techniques whereby the strength of particle adhesion in flocculating media may be measured. They found, as a quick check, that an increased floc size was a good indication of an increased strength of adhesion.

The processes of coagulation, flocculation and then settling of the flocs can be combined in a single vessel[141] on a large scale. It is claimed that well-formed flocs, produced in a central (feedwell) zone, are smoothly transferred to the clarification zone with minimal floc disruption. This leads to a simpler process and reduced reagent consumption.

In repeated settling of the same flocculated solids such as occurs in CCD (p. 670) floc shear in the transfer pumps is unavoidable. Although this can be minimized,[142] it is normally necessary to add a further quantity of flocculant at each stage, half or one third of the original dose being fairly typical. The actual quantity required can be adjusted by observing the performance of the thickeners involved.

Final Settled Volume. It may appear at first to be paradoxical, but the *greater the attraction* between particles, brought about by pretreatment, the further they stay apart in the final sediment. This is related to the stability of an open structure of adhering particles compared to the same structure when the particles are dispersed. The latter, as they descend individually from suspension, cannot remain in contact with the first two or three particles they encounter in the build-up zone, because of the repulsive charge, and continue to descend by sliding or rolling until they reach a point of maximum stability where they fill the lowest remaining gap in the structure. Dispersed sediments are therefore compact while flocculated sediments are voluminous[51] and more readily break down on shear.

Theory and Design

Clarifiers are sized on the basis of the permitted (or desired) upper limit of solids in the overflow and the settling rate of these solids. If the size distribution of the solids in the feed is known, it can be converted to a settling velocity distribution according to Equation (8). A typical plot is shown in Figure 14.32.

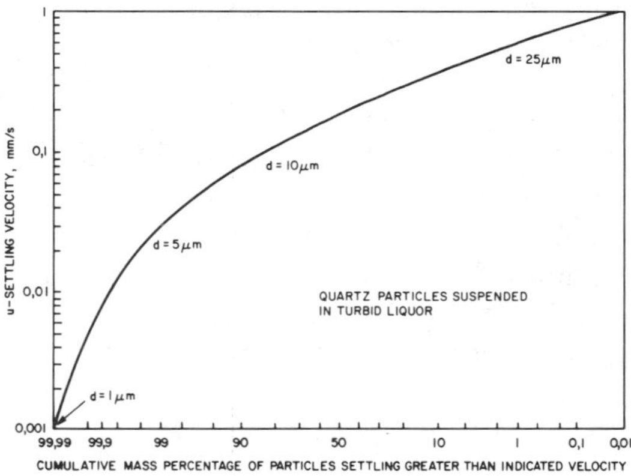

Figure 14.32. Settling velocity distribution of suspended quartz particles.

As the feed enters the quiescent conditions of the clarifier the particles with settling velocities greater than the upflow rate settle to the bottom while the rest are carried over. The *proportion* of feed particles appearing in the overflow can be determined from Figure 14.32 for any given overflow rate and their actual *concentration* then depends upon the total quantity of solids in the feed. The settler area is governed by the maximum allowable overflow rate.

When the "particles" to be settled are flocs or agglomerates of the originally dispersed fine solids, their size characterization is then not simple. The essential parameter, *settling velocity distribution*, may, however, be directly measured in the laboratory. This is illustrated in the examples below.

Two Examples of Estimating Clarifier Areas. (1) *A Liquor Containing Dispersed Solids.* A pregnant gold solution, obtained by rotary vacuum filtration of cyanide leached ore, contains 550 g/m^3 of 1–30 μm quartz fines and requires clarification to a limiting concentration of 20 g/m^3, what size sedimentation tank is required for a flow of 600 m^3/h?

The solids to be removed are (550–20) g/m^3 or 96.4% of the mass of incoming solids. As only solids which settle faster than the upward velocity of the overflow can be collected, the maximum overflow rate must be low enough to collect this percentage of the incoming solids. From Fig. 14-32, based on a size distribution of this material, 96.4% of the solids have settling velocities exceeding 0.052 mm/s. The maximum upflow rate, U, is therefore $0.052 \times \frac{3600}{1000} = 0.187$ m/h.

As throughput $Q_F = U \times$ area of clarifier (A)

$$600 \text{ m}^3/\text{h} = 0.187 \text{ m/h} \times A \text{ m}^2$$

$$\therefore A = 3200 \text{ m}^2$$

and diameter of circular tank = 64 m.

(2) *Clarification of a Liquor after Flocculation.* Assuming that the 64 m diameter tank is too large for the site, what steps can be taken to reduce it? Flocculant tests in the laboratory indicated good settling behavior of the solids in the feed after addition of 5 g/m^3 of ferric chloride coagulant followed by 0.2 g/m^3 of a polyacrylamide flocculant.

For sizing the clarifier a 10 liter sample of the pretreated turbid gold solution was then gently added to a transparent settling tube ~65 mm diameter and 2 to 3 m deep. Provision was made for periodically sampling the liquid at known depths either by lowering a syphon tube or through suitably spaced side ports.[2, 143]

At time zero, immediately the tube was filled, the contents were sampled at the top, middle and near the bottom to determine the *original* (feed) suspended solids concentration and check on even solids distribution. When the liquid started to clear, the time was noted and samples taken at all levels from top to bottom in that order, and analyzed for suspended solids. This could be *any* component such as BOD (biological oxygen demand) which it is necessary to remove. The sampling was repeated after four or five similar periods until all samples indicated a suspended solids concentration below the desired limit. Results of a typical test are shown in Table 14.14.

Sampling at H m below the surface after t hours static settling, is exactly equivalent to sampling the overflow liquid from a continuous clarifier operating at an upflow velocity $Q_F/A = H/t$ m/h. It can be seen from the results, as is to be expected, that the suspended solids (SS) at any level decreases with time and at any time increases with depth. These data provide the maximum permissible upflow velocity for any desired suspended solid in the overflow. By interpolating the SS data after 5 minutes settling, it may be estimated that for a clarified product containing no more than 20 g/m^3 of suspended solids (as before), the upflow must not exceed 4.8 m/h. For a volumetric throughput $Q_F = 600$ m^3/h this means a tank of diameter = 12.6 m.

Table 14.14. Suspended Solids Values Found at Various Depths at Different Times.

DEPTH FROM SURFACE (m)	TIME OF SAMPLING (MINUTES)										
	0	5		10		15		20		25	
	SS	U	SS	U	SS	U	SS	U	SS	U	SS
0.25		3	10	1.5	3	1	2	0.75	1	0.6	2
		(4.8)	(20)								
0.5	550	6	30	3	7	2	3	1.5	2	1.2	Nil
1		12	140	6	23	4	7	3	3	2.4	3
1.5	560	18	380	9	65	6	15	4.5	5	3.6	4
						(6.5)	(20)				
2		24	520	12	140	8	35	6	9	4.8	5
2.5	540	30	550	15	250	10	65	7.5	15	6	7

SS = suspended solids in g/m^3
U = the equivalent upflow velocity of sampling at a given depth, H at time t; $U = H/t$ m/h

If the exercise is repeated with the *15* min results, however, the diameter is found unexpectedly to decrease to 10.8 m. This could not occur unless there were a change in the nature of the settling solids with time. For the various times shown in Table 14.14 it can be seen, however, that the SS value for a *fixed* upflow rate say $U = 6$, decreases steadily with time from 30 at 5 min to 7 g/m^3 at 25 min. Both upflow rate *and* detention time are therefore important in clarifier design. (Note: This aspect of detention time is quite different from the idea that it may be required to achieve maximum sludge thickening.[144]) Particles or flocs continue to grow during settling either as a result of faster settling units overtaking and coalescing with slower ones or due to velocity gradients in the fluid.[145] The exploitation of particle capture and floc growth by contacting feed solids with *matured* flocs is presented on p. 664.

The detention time t_D in a cylindrical tank of diameter D, is $=\pi D^2 H/4Q_F$ while the upflow rate is $4Q_F/\pi D^2$. Therefore the dimensions of the tank are

$$D = \left[\frac{4Q_F}{\pi U}\right]^{1/2}$$

and

$$H = t_D \cdot U$$

In the example $Q_F = 600$ m^3/h and if we select a detention time

$$t_D = 5 \text{ min}, U \text{ must be } 4.8 \text{ m/h}$$

$$\therefore D = 12.6 \text{ m and}$$

$$H = 0.40 \text{ m}$$

If a longer detention time is selected, say t_D = 25 min, then U becomes 10.8 m/h (by extrapolation)

$$\therefore D = 8.4 \text{ m}$$

and

$$H = 4.5 \text{ m}$$

Table 14.15. Values of $t_D \times U$ calculated from Table 14.14.

TIME (MIN)	U FOR 20 g/m^3 SUSPENDED SOLIDS	$t_D \times U$ (m)
5	4.8	0.40
10	5.6	0.93
15	6.6	1.65
20	8.3	2.77
25	10.8	4.50

which represents a narrower tank but much deeper. This apparent wide choice of possibilities is however limited by considering the standard sizes available as settling tanks of this diameter normally come with a standard depth of ~2.5 m, and therefore $U \times t_D = 2.5$ m. The product $U \times t_D$ for various times taken from Table 14.14 are shown in Table 14.15. From the final column the required value of 2.5 m occurs at about 19 min, therefore

$$t_D = 19/60 = 0.315 \text{ h}$$

$$U = 2.5/0.315 = 7.9 \text{ m/h}$$

and

$$D = 9.8 \text{ m}$$

In actual practice a converted 10 m diameter pachuca tank was used[128] for this duty, in this case probably by reversing the procedure and adding the necessary quantity of flocculant to suit the size tank available. As for thickeners, the calculated value of D is rounded up to the next largest standard diameter tank manufactured and this has the double advantage for further lowering the upflow velocity *and* increasing the detention time; always useful for contingencies.

14.5. NONCONVENTIONAL SEDIMENTATION PROCESSES AND EQUIPMENT

Normal thickeners and clarifiers are large relatively empty tanks or basins which provide virtually quiescent conditions and sufficient surface area and volume for optimum settlement. This simple process has changed little since the turn of the century, and, based upon the percentage tonnage currently handled, appears to be as well established as ever.

However, situations do arise where the conventional design is not ideal for special conditions. These include sedimentation of solids whose density is only marginally greater than the fluid; the clarification of high-temperature or very volatile liquids which must in consequence be kept under pressure during treatment, or when new or additional sedimentation equipment must be installed in existing premises, and floor space has become severely limited. In some cases the cost of creating the additional *space* may be many times greater than the equipment itself. In underground mining for instance, 10 to 20 m^3 of virgin rock has to be blasted out for every m^2 of settling area[146] required. Such situations have led to the introduction of a number of innovations some of which have reached the commercial exploitation stage.

Their common purpose is to increase solids throughput per unit of superficial area. This can be done either through an improvement in actual particle sedimentation rates whereby more solids are settled per square metre of actual *settling* area or due to a multiple design where, for the same projected *floor* area, a larger net settling area is provided. Mechanical innovations such as improved designs of feedwell and rake are also presented.

Lamella Settlers

According to Hukki et al.,[147] the use of inclined baffles for settling fine solids dates back to 1882 when a French patent was granted to Gaillet and Huet. The principle has since then "been rediscovered many times in many countries."[147] Figure 14.33 shows the angle and general arrangement of the baffles in the early Hukki design, the mode of settlement of the solids and the method of calculating the floor area required. The ratio of floor area to settlement area is

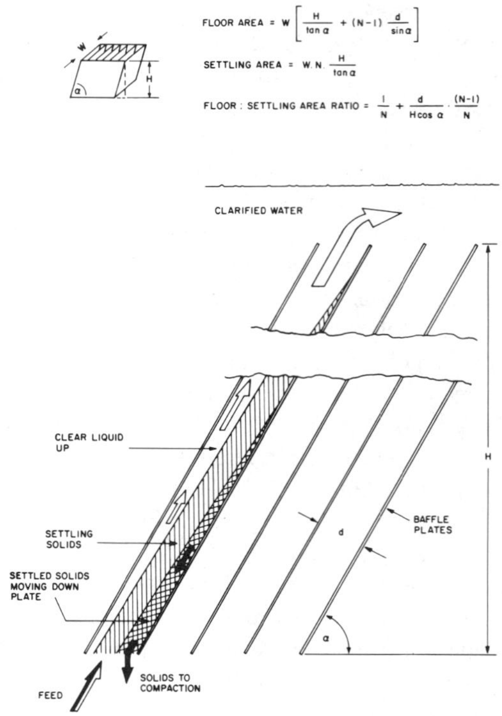

Figure 14.33. Principles of the lamella settler.

$$R' = \frac{1}{N} + \frac{d}{H\cos\alpha}\left[\frac{(N-1)}{N}\right]$$

and so to *minimize* R', the number of plates should be large, they should be deep, closely spaced and at as shallow an angle α is possible. The minimum value for α is prescribed by the flow properties of the sludge.

For plates of $H = 1.5$ m, at the usually accepted angle of $\alpha = 60°$ and spaced $d = 30$ mm apart, a 10 plate tank has an actual settling area about 7 times greater than a conventional settler in the same space while with 20 plates, this increases to 11. In the limit, as $N \to \infty$, the advantage becomes equal to $H/(d/\cos\alpha)$, which in this case is 25. As the denominator $d/\cos\alpha$ represents the *vertical* distance between adjacent plates, the maximum advantage is, as would be expected, the number of plates which overlap in the vertical direction.

This theoretical figure is rarely achieved in practice, however, for, besides the obvious end effect, factors contributing to the lower efficiencies in the earlier designs are:

1. Manifolding problems of feeding a large number of plates
2. Entrainment of solids in the rapidly rising liquid stream due to internal mixing,[148] giving rise to an increased overflow turbidity
3. Entrainment of solids *after* discharge from the plates by the ascending feed stream
4. Accumulation of solids on the plates, and
5. The fact that the solids have finally to be collected and compacted in a *conventional* hopper, i.e., one with no area advantage.

These shortcomings, which are not always serious, have been recognized by various manufacturers and the following modified designs are available commercially:

1. *Corrugated* plates to avoid a continuous curtain of descending solids, and guard gutters to separate the flow of these solids from the incoming feed
2. Feeding the plates from the *side* to avoid feeding and discharging solids at the same point
3. Feeding from the *top* with overflow return pipes extending to clear water zones near the base
4. Stacking *sets* of inclined plates, one above the other, each inclined in opposite directions to present a vertically zigzag profile for which settling solids should present the minimum disturbance to the clarified liquid
5. Continuous raking or low frequency vibrators to assist both removal of solids from the plates and promote their compaction, and
6. Use of *flexible* textile materials or rubber for the "plates" to permit periodic dislodgement of solids or cleaning.

The lamella settler has been successfully used in many fields; in coal preparation plants,[149, 150] removal of fine mill scale from hot rolling mill wastes for reuse of the water[151] and also for separating metal hydroxides, fly ash, nickel, and catalyst fines, cement dust and for clarification of phosphoric acid, lime kiln scrubber water and paint booth water curtain.[151]

The tubular settler[152-154] and the rotating spiral thickener[155] operate on the same general principle of providing increased settling area per unit of floor area. In the latter, a "Swiss roll" provides a number of flow channels between curved walls through which the slurry flows and concentrates while the unit is slowly rotated.

Willis[154] reviews the *practical* design factors for tubular settlers including important points such as sludge collection, how to specify overflow rate, and presents various shapes of tube which may be used. Much of it applies also to conventional plate settlers but tubes are considered to overcome the hydraulic instability of "wide horizontal plates."

An alternative view[148, 153] that lamella settlers offer the advantage of rapid sedimentation or additional clarifier capacity *because* of the decreased vertical fall height can lead to confusion between a reduced throughput time and a real increase in total solids throughput. Only the former applies here as solids throughput is governed by available area and not settling height.

The multiple compartment or tray thickener, although not a novel design[12d] is related to the lamella concept by virtue of the increased number of settling planes above a common base. The main differences of the tray thickener is its separate feeding, raking (although from a common shaft) and underflow pumping facilities for each compartment; these may number from 2 to 6.

Besides the economy of floor space, such units also reduce heat losses in the case of hot solutions. The design is no longer much in favour, however, the trend being towards the simpler single-compartment thickener. In thickening red mud on an industrial scale, a five-compartment thickener was in fact converted to one of three by removing two intercompartmental partitions.[156] The unit throughput actually *increased* by a factor of 2 to 3 in the simpler design and the overall output was increased by 30 to 50 per cent. The complications involved in uniform *parallel* feeding to the separate trays and achieving simultaneously high underflow densities falls away, however, if thickening is done in *series.* This is the situation called for in continuous countercurrent decantation[157] for which tray thickeners may be readily adapted.

Upflow Solids Contact

In contrast to the lamella settler which increases sedimentation "efficiency" by providing a multiplicity of settling planes, the upflow principle operates by improving the actual sedimentation characteristics of the feed. It is confined to flocculated suspensions or metal hydroxide precipitates in which the freshly formed loosely knit, voluminous flocs settle slowly due to their smaller initial mean size, their greater fragility or their decreased density differential compared to compact agglomerates.

Such flocs may, however, mature with time resulting in larger, stable, denser, fast settling units, and when these aged flocs contact freshly formed material the settling characteristics of the latter are promoted.[131] The upflow principle exploits this phenomenon by adding the freshly flocculated feed, not into a normal feedwell *above* a body of clear liquid, but *below* the pulp interface of a bed of aged flocs. These flocs are thereby kept in a state of fluidization and feed rates of up to 35 times higher than for conventional units are claimed. This can only be, if the matured flocs have a settling rate 35 times faster than freshly formed flocs. This represents a special case of the benefits of detention time (see p. 661) except that in this case the mass of the *solid* rather than the volume of the liquid is used to calculate detention.

Relatively long residence times are possible because of the high concentration of the solids in the fluidized zone. Escape by short circuiting of the slower settling fresh flocs with the water from this zone is reduced by providing a zone sufficiently deep and uniform. There must therefore be some "filtering" action whereby the new flocs or their fragments collide with and are held by older flocs until they too mature and trap fresh material. The need to maintain a stock of fluidized solids means however a relatively long start-up period (while running at low efficiency) and the inability to recover quickly following flow reductions or shut-downs.

The original application appears to have been for water clarification[158,159] and a simplified design is shown in Figure 14.34. More recently it has been applied to conventional thickener feeds such as 20% uranium slurries[160] or zinc mine tailings[161] in units shown schematically in Figure 14.35.

A useful comparison of solids contact clarifiers, lamella separators and inclined tube settlers is given by Mace and Laks[162] including cost data.

Deep Cone Thickeners

Faced with the need to improve the landscape and reduce river pollution, collieries are turning to a new geometry[163] of settling vessel to overcome their effluent thickening problems.[164] The space required by the deep cone is much less than for settling lagoons or conventional cylindrical thickeners[165] and the performance much higher.[166] The considerably reduced settling area is a direct consequence however of a relatively high[164] consumption of polymer flocculant ~ 200 g/t. (The average consumption for conventional thickeners in the South African gold mining industry is in the range 1 to 3 g/t while for upflow solids contact settlers it is 10 to 50 g/t.) Flocculant dosage at this

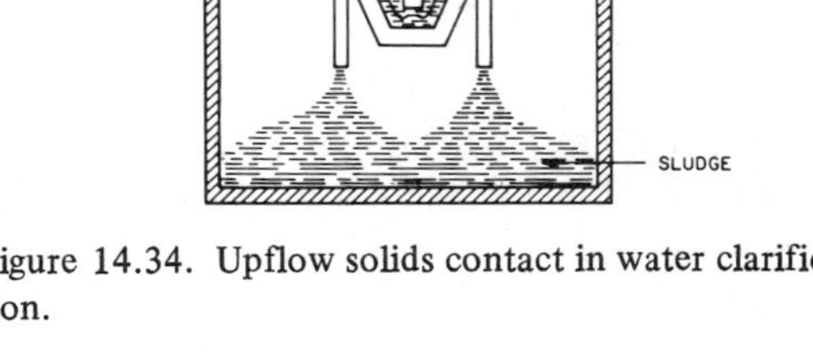

Figure 14.34. Upflow solids contact in water clarification.

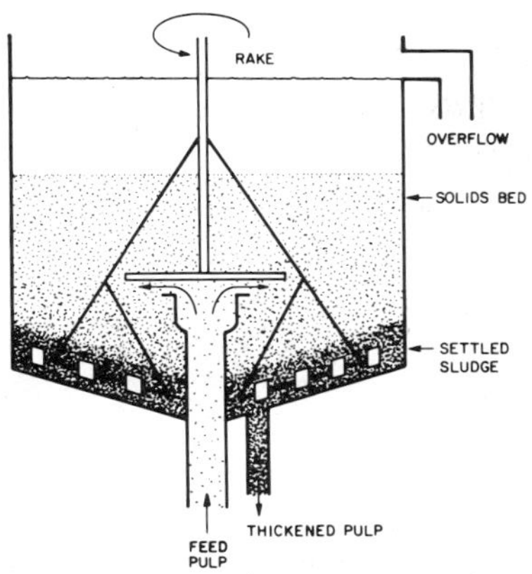

Figure 14.35. Upflow solids contact in thickening.

rate in the gold mining industry would increase *total* thickening costs 15 to 20 fold. Thus although the capital costs are reduced, running costs are high and flocculant solution make-up procedures because of their frequency pose a special problem.[164] Overflow clarity is not particularly good but the water is suitable for reuse in the process. Because of compression depth, underflow concentrations are high enough to permit direct discharge onto a conveyor belt.[165] To achieve these results the cone must be kept full of solids and stirred by a specially designed stirrer.

Manual control is inadequate to maintain the required conditions because the time constant is less than an hour[167] and automatic control is therefore warranted (see p. 670). The opening of the discharge valve is controlled by the density (pressure) at the base of the cone; variations in feed volume and concentration serve to regulate flocculant dosage while a turbidimeter monitors the quality of the clarified water.

Flotation

Although froth flotation is used on a wide scale for separating *between* solids, particularly the flotation of mineral particles from gangue, in this context it refers specifically to the removal of *all* solids suspended in a liquid to achieve complete clarification.

Where the density difference between solid and liquid is slight, as for example, the suspended organic solids in sugar juice and in sewage, sedimentation rates are low and sludge volumes usually high. Attachment of gas bubbles to the particles or flocs reduces their density still further to such an extent that the density *difference* between "particles" and liquid is reversed and increased and therefore also the driving force causing separation.

Due to the reversal in actual densities, the flocs now *rise* but at a rate more rapid than they descended previously and collect as a froth at the surface. They form a solids zone partially above the liquid surface so that the solids can drain to produce a "sludge" of considerably reduced liquid to solids ratio.[168]

The bubbles are normally generated by releasing the pressure of an air-saturated liquid stream[169] or by electrolysis.[170]

Besides application in sugar juice clarification and effluent treatment, the possibility of using flotation for treating fine hematite particles has been investigated.[171]

Further aspects of the process and detailed *design* of air flotation systems will be found elsewhere.[172]

Feedwell Design

Unconventional feedwell designs that have been proposed, aim to improve the hydraulics of feed entry or the settling characteristics of the solids.

Based on studies in a transparent model segment of a tank, Cross[173] observed that a higher rate of internal circulation produced a turbulent floc bed and a more compact, faster settling floc. This circulation was achieved by using a well of increased diameter and depth, containing rectangular slots below the water level. The higher density of the feed pulp results in a lower liquid level inside the well compared to the thickener body and hence a flow of water and floc solids *into* the slots. For a 23 m diameter thickener the feedwell proposed had a diameter 3 times and a depth 5 times the normal dimensions.

Tracer studies (Beyleveld, L. J.),[4] using irradiated feed solids, indicated that the benefit of the design probably lay in the opportunity for slower settling segregated flocs near the surface to be reincorporated into the feed pulp.

Although later studies on a larger model[108] indicated no improvement on solids throughput compared to a conventional feedwell, full scale plant tests carried out over a week (Alderton, J. L.)[4] on two adjacent 23 m diameter thickeners showed that for the same volumetric feed rate, the thickener fitted with the Cross feedwell operated at a significantly higher underflow density. The net increase in throughput was estimated as being between 10 and 12% based on a *similar* underflow density.

Tests of feedwell designs performed on *model* thickeners should be interpreted with caution and if possible repeated on full scale plant before making claims.[4] The dynamic effects are accentuated on models and whether good[173] or bad[202] generally decrease on scale-up (see p. 668).

In clarifiers such internal circulation (caused by density currents) is to be avoided because it leads to short circuiting and higher local upflow velocities near the overflow. Various ways of overcoming this defect are discussed and a new type of feedwell presented by Fitch and Lutz.[174] Within the normal cylindrical well are two adjacent, inwardly open, submerged channels between which the feed is split and introduced to each tangentially but in opposing directions. When these streams meet in "open water" near the center of the well their velocities are instantaneously dissipated as turbulence and residual velocity streams thereby eliminated.

Rake Design

In common with other materials throughout the world, the gold ores mined on the East Rand in South Africa tend to be especially "sticky."[175] The clayey solids accumulate[176] on the arms of conventional rakes and often cause disastrous shock overloads when they eventually fall off in massive chunks. It is then also extremely difficult at the next half turn to rake the resultant "islands" to the discharge point.

The Klopper Dragrake[177] was developed to overcome this type of problem. It consists of a tubular steel arm to which are welded the raking blades. This arm is pivoted at the base in the center and suspended by means of cables or chains from a drag arm rotating above the liquid. Solids build up on the arm is minimal and it has the ability to swing and lift automatically over any unusual loads. The great usefulness of this device is confirmed by its being used elsewhere[178] and by being taken up by international thickening manufacturers for the general treatment of thixotropic slurries and tacky muds.

A conventionally designed rake which lifts on a raisable platform to permit *planned* increases in solids storage rather than unexpected loads, is described on p. 676.

14.6. NONSTEADY STATE THICKENING AND THICKENER CONTROL

Although in the long term the total quantity of solids discharged from a continuous thickener must match the quantity fed, over shorter periods of hours or days, true steady state operation rarely prevails. The imbalances are absorbed by the stock of solids, which in a 23 m diameter continuously operating thickener, may vary from 200 to 900 tons,[116] representing 7 to 30 hours of total feed solids. Wide fluctuation in feed rates below and above the designed throughput are therefore permissible but there are obviously limits to excess overloads of long duration.

If the solids feed rate to the thickener is increased above G_{min}, the solids, because of their higher flux at concentrations below C_{crit}, will arrive at the level of the critical zone faster than they can be transmitted further. The result is an unsteady state in which the height of this zone will increase, i.e., the zone, with a distinct upper interface will rise in the thickener until it reaches the overflow level at which point the excess solids will leave the thickener with the overflow water. This condition, called *sliming*, is essentially different to the presence of nonsettling fine particles sometimes found in the overflow of a thickener or clarifier.

In contrast to the usual assumptions made in thickener design (p. 634) *non*steady state operation must be accepted as the norm. A regular *weekly* cycle during which gold ore thickeners receive 117% of their mean load during six days and discharge the excess on the seventh day (when mills are shut down, but not the less readily interrupted cyanidation operation) has operated in South Africa from the earliest days of the mining industry. The thickeners therefore provide a continuous supply of thickened pulp for seven days while being fed only for six.

Reclamation of a dump by hosing during periods of daylight and storing the excess solids in a thickener so as to ensure a continuous supply of pulp to a processing plant is a further example of unsteady state thickening, here with a *24 h cycle*.[114] The need to distinguish between solids capacity and solids throughput in this situation becomes clear.

Design of Thickeners that Accept Fluctuating Feed Loads

Unlike clarifiers, the efficiency of a thickener is seldom limited by the volumetric rate of feed and an increase in feed rate does not result in an *immediate* increase in solids carry over. The relatively high solids flux at feed concentration (Fig. 14.19) permits solids to settle even at high overloads relative to the available area and the excess solids then accumulate as a "mud." The mass of these solids depends on their mean concentration, i.e., the critical concentration which "backs up" on overload; this may be determined from the flux plot, Figure 14.36, as shown in the following example.

Thickener Design for a Cyclic Overload. A continuous thickener is required to accept a dilute pulp containing a total of 750 tons of solid each day during an eight-hour period, and deliver a concentrated pulp, $C_U = 1200$ kg/m^3 at a steady rate over 24 hours; what size thickener is required?

To determine the dimensions, a representative sample of the dilute feed pulp was taken and the slurry interface settling rate U observed at various concentrations C (see p. 651). The solids flux $G = UC$, was plotted against C as

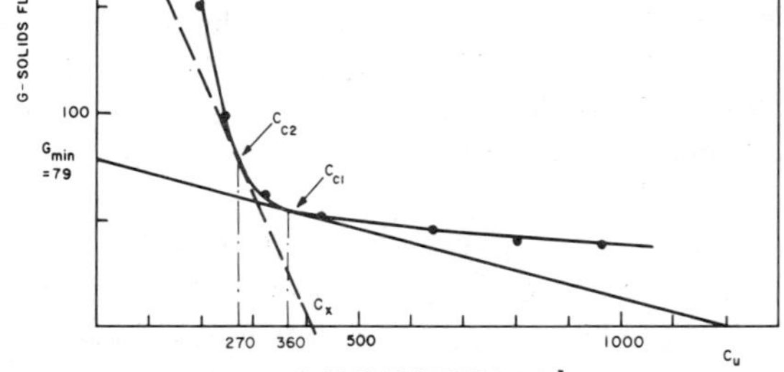

Figure 14.36. Fitch construction for temporary thickener overloads.

shown in Figure 14.36. From the Yoshioka construction, with $C_U = 1200$ kg/m^3, the critical flux G_{min} for steady state operation was found to be 79 kg/m^{2}h occurring at $C_{C1} = 360$ kg/m^3.

The minimum area A required for a steady 750 t/day is therefore

$$\frac{750 \times 1000}{A \times 24} = G_{min} = 79 \text{ kg/m}^2\text{h}$$

i.e.,

$$A = 396 \text{ m}^2$$

$$\therefore \text{diameter} = 22.4 \text{ m}$$

$$\equiv 73.6 \text{ Imp feet}$$

The next standard size manufactured would be one of 75 feet diameter (22.86 m) and this would be provisionally accepted (scale-up factors are discussed in the next section). Depending upon the supplier this thickener could have a side water depth of 2.44 m, a central cone depth of 1.68 m, a volume of 1230 m^3 and an area of 410 m^2.

What occurs when such a thickener is now presented with 750 tons of solid in *eight* hours, i.e., 3 times the load? The imposed flux now increases to

$$G_{OL} = \frac{750 \times 1000}{410 \times 8} = 229 \text{ kg/m}^2\text{h}$$

The operating line, drawn now from the calculated value of G_{OL} and not C_U is included in Figure 14.36. The required underflow for steady state operation would be C_X. The critical concentration zone can be seen to have shifted from $C_{C1} = 360$ to $C_{C2} = 270$ kg/m^3. The concentration of the rising zone is therefore C_{C2} but, as the solids flux between C_{C1} and C_{C2} is *greater* than required by either operating line, they settle faster than being fed and concentrate up to C_{C1} while moving down the thickener. The mean concentration of the stock in the thickener lies therefore *between* C_{C1} and C_{C2}.

The height of thickener required to store the excess solid can be computed from the rate of rise of C_{C2}, call it U_C.

Thus Fitch[114] has shown (Eq. (23)) that $U_C = G_{OL}/C_X$ in a *batch* test or a thickener with *zero* discharge (collector). There is however a downward velocity $Q/A = G_{min}/C_U$ (Eq. (21)) and hence the *nett* rise is

$$U_C = \frac{G_{OL}}{C_X} - \frac{G_{min}}{C_U}$$

which in this case is

$$U_C = 0.493 \text{ m/h}$$

Over 8 hours, neglecting the cone in which stock is normally maintained to prevent water short circuiting to the underflow (coning), this would rise 3.94 m. As a check, based on the stock in the thickener having an *estimated* concentration of $(C_{C1} + C_{C2})/2$, i.e., a mean of $C = 315$ kg/m^3, the required excess stock of $750 \times 16/24 = 500$ tons would require a volume V of $500 \times 1000/315 = 1587$ m^3. With an area A_1 of 410 m^2 this is equivalent to a height of 3.87 m, very close to the accurate value.

The options therefore are a 22.86 m diameter thickener with a side water depth *increased* to 4 m or a *larger diameter* standard thickener (say 85 ft) having a *normal* depth of 3.05 m, an area $A_2 = 527$ m^2 and the required storage volume of V m^3. The suppliers consider that normally the latter is more economical than building deeper walls. Choosing a larger area decreases G_{min} and a recalculation is required for a more *precise* value for the required height. This will however always be less than the height calculated from V/A_2 as C_{C1} and the mean concentration of stored pulp *increases*.

Dynamic Response of Overloaded Thickeners. Observations of the response of a stirred continuous model thickener of 150 mm diameter to overload conditions[179] indicated that the time needed for the solids to reach the overflow level was *less* than as predicted above. As the concentration of the rising zone was not lower than expected, the shorter storage period must be due to a reduced solids flux compared to the batch tests and the usual scale-up factors used for mean throughput should therefore apply.

A more serious failing, observed with model thickeners (Stander, J. W.[4]) was also noted by George.[89] The model behaved ideally with coagulated pyrophyllite pulp at steady state but on overloading the thickener when at a low solids inventory by means of a step increase in volumetric feed rate, the expected rising critical concentration with a definite upper boundary failed to appear. Instead the entire volume of the thickener below the feedwell increased gradually and uniformly in solids concentration, until the feed concentration was reached. Thereafter an interface rose above the feed level and sliming commenced as soon as it reached the overflow level. The storage period was only 0.5 to 0.7 times that predicted, and the stock of solids, being appreciably more dilute, was therefore lower than expected. Pearse's comment[1] that feeding arrangements are especially important in small thickeners due to the disturbance caused, appears relevant. Positive disturbance was also noted in a 300 mm diameter model[202] when a central well was replaced by a peripheral feed tube.

Until the behavior of full scale units is more fully understood, the feed concentration to storage thickeners should be maintained as high as possible and C_0 rather than C_{crit} used for a more conservative estimate of stock.

Thickener Control

Unlike other solid–liquid separation techniques (Table 14.1), gravity sedimentation does not require a strict control to be kept on the concentration and nature of the solids in the *feed* and, as shown previously, a thickener can in fact absorb up to three times its steady state volumetric feed rate for periods as long as eight hours. This useful buffer capacity[178] is often taken advantage of in helping to smooth out the inevitable fluctuations which occur in continuously operated plants. Besides feed rate these variations include changes in the *type* of raw material (e.g., the ore milled) giving rise to changes in the nature of solids to be settled, particle fineness and shape. Ambient temperatures also fluctuate, daily and seasonally. In addition, any dilute pulps arising from occasional clean-up or desludging operations, excess pulp arising from an adjacent thickener "trip-out" or material which cannot be accepted by the next stage in the process are all returned to the feed launder of the available thickeners.

Control of the volumetric feed rate is therefore rarely practiced and thickeners have to accept "all they are given."

Objectives of Thickener Control. *Control* objectives are, simply, the maintainance of normal thickener objectives in the face of fluctuating conditions. The objectives are:

1. To maintain a high underflow density
2. Prevent sliming
3. Prevent rake failure

In the case of clarifiers the primary objective is to maintain the clarity of the overflow in spite of fluctuating conditions in the feed,[162] which includes nature and concentration of the solids as well as its volumetric rate.

Manual Control. Because of their long response times, manual control of thickeners is normally adequate. The following describes typical practice in relation to the objectives.

Slime Level. Once every two hours the slime level (clear water depth) may be determined by successively lowering to increasing depths a long pole with a plate at the end and raising it quickly to see if it produces a cloud of solids. If a single thickener in a group has a high slime level, the overall feed distribution is adjusted by altering the positions of the feed gates, but if all the thickeners are so affected the flocculant dosage is increased. Two instruments for observing the complete solids profile without sampling were developed, one based on a series of density floats[4] and the other completely continuous based on an optical backscatter (Stander, J. W.[4]).

Underflow Density. At the same time during the operators round, a sample of the underflow pulp is taken, its density determined by observing its mass in a conical one-litre metal container. If too *low*, the pumping rate is *decreased* to a new setting based on the experience of the operator. As shown, in Equation (21), the constant long term average steady state throughput is $G_{min} \times A = QC_U$ t/h, so that a low C_U is caused by too high a value of Q and is therefore corrected by a reduction in pumping rate.

Rakes. The operator also observes the stress on the rakes either by reading a rake torque meter or noting the degree of lift in the case of balanced or automatic lifting devices. Correction of overload is achieved by reduction of feed rate, increasing Q or in severe cases, both.

Automation of these control actions on a continuous basis appears simple enough but very poor results with density control were obtained in initial trials where the controller took repetitive and finally disastrous corrective action when the thickener was, from the start, itself correcting the out-of-control situation. This can be understood from Equation (21) above. If at a pumping rate Q close to the required long term average value C_U *increases* due to plant fluctuations, then the total solids discharged will be higher than the average throughput $G_{min} \times A$. If this quantity of solids were pumped out at the *desired* value of C_U, then the pumping rate is already too high, i.e., *no* change in Q should by itself cause a *fall* in C_U. Such a density excursion is therefore both a symptom and its cure.

The result of operating for 24 hours at a fixed pumping rate without *any* control or adjustment is shown in Figure 14.37. It can be seen that the variation in relative density did not exceed 1.0% of its mean value over the entire period. Failure to recognize that a thickener underflow density tends to be self-adjusting therefore leads to over-compensation in the pumping rate either by inexperienced operators or by poorly designed automatic controllers. The fixed pumping rate selected for the test coincided with the average solids throughput. Adjustments must of course be made where this is not the case or where feed rates vary.

Automatic Control. Automatic control rather than manual operation is justified if:

1. the response of a fully stocked thickener to variable feed conditions is too fast for adequate manual control
2. manual labor is expensive or unreliable, or
3. the consequences of an out-of-control situation are unacceptable.

Such a situation arises in continuous countercurrent decantation (CCD) and Figure 14.38 shows a partial circuit of four thickeners with No. 4 being the last in line.

The CCD principle is exemplified in No. 3. In this thickener the values remaining in the liquor in the underflow of No. 2 is "washed" by mixing this slurry with the overflow liquor from No. 4. This forms the feed to No. 3. After thickening, i.e., decantation, about 85% of the combined soluble material originally

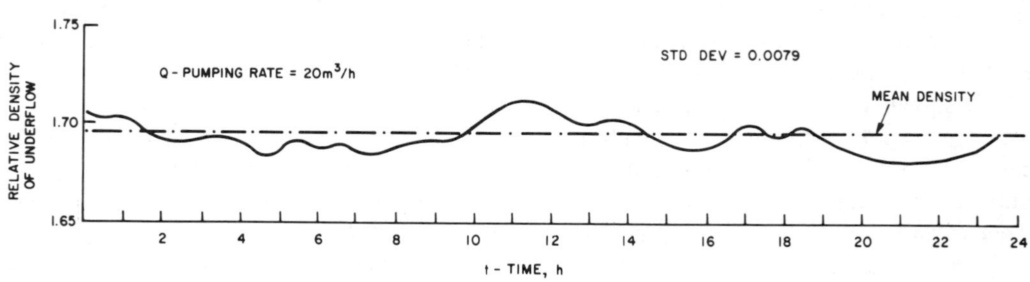

Figure 14.37. Variation of underflow density with time at a fixed pumping rate.

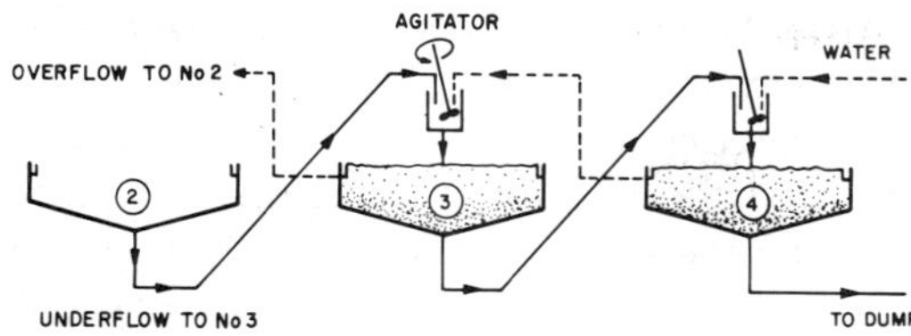

Figure 14.38. A section of a continuous counter-current decantation (CCD) process.

present in the underflow pulp of No. 2 and the overflow of No. 4, appears as clear liquor in the overflow of No. 3. The balance remains in the underflow and is washed in the next thickener and so on. In this case No. 4 is the last thickener in line and uses *water* for washing while its underflow solids together with its diluted remaining liquid[180] is discarded.

A group of such thickeners operates in series rather than parallel and therefore an imbalance in one affects them all, and reliable means of control assume greater significance.

A further situation calling for automatic control is when there is a high monetary value in reducing the amount of liquid in the underflow pulp so that tighter control of underflow density is desired. In a typical gold plant this liquid is pumped into a cyanide circuit of *fixed* capacity so that a corresponding volume of liquor has to be discarded. As this liquor contains lime, cyanide and traces of unprecipitated gold, reducing the volume of this dilution water, i.e., increasing underflow density, has a definite cost incentive.

Control of Underflow Density. Manual control of a test thickener at a mine[4] was replaced by installing an automatic controller[116] which took appropriate action based on the density signal from a gamma-ray attenuation gauge on the underflow pipe. Alternative means of measuring the density (and flow rates) of slurries are mentioned by Liptak.[181] A "wait period" was introduced to allow the thickener to respond before the previous corrective action was unnecessarily repeated.[176] The results were directly comparable with manual control of neighboring thickeners as the test thickener continued to be sampled manually as before but only for plant record purposes, not for control.

The results are shown in Figure 14.39 and indicate that automatic control reduced the density variations about 5 times. The low values at WE are due to the normal weekly operation of pumping out the stored thickener solids on Sunday (see p. 667) but the drop at LO is due solely to poor manual control as are the values of HI. As high density excursions are risky due to their liability of causing underflow pipe blockage or rake trip-outs, it is necessary in manual control to operate at a lower *mean* density as shown, to reduce the

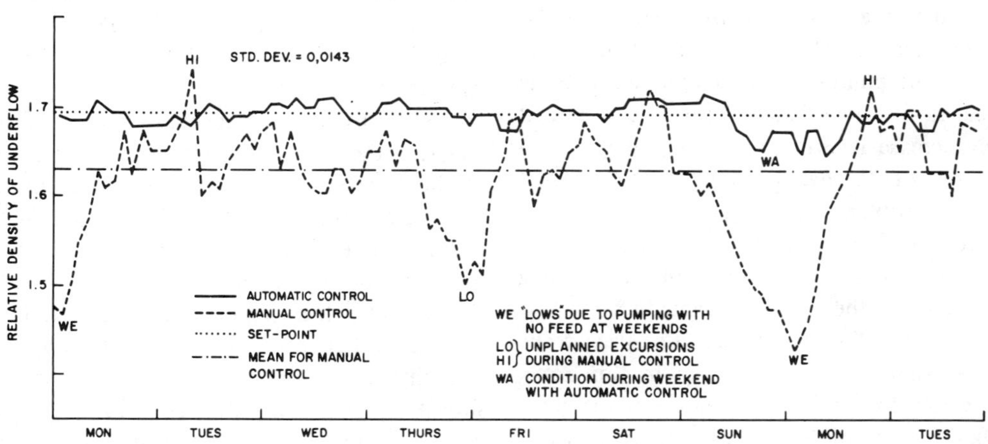

Figure 14.39. Comparison of the variation of underflow density, as measured by normal shift operator, of two adjacent thickeners, one controlled normally and the other automatically.

probability of their reaching these levels. In contrast, the steadier automatic controller could be safely set at a higher mean density, which in this case was equivalent to *20% less underflow water*, and still maintain the density peaks below the danger level. Furthermore, at the weekend, the low at WA is not nearly so severe as at WE with manual operation.

instead of a control based upon the *conditions* which might lead to attaining maximum density and ignoring the value obtained, it is more practical to use a *direct* signal which indicates whether the solids are reaching, or have already reached, this value. The maximum attainable density or concentration $c = C_U/\rho_S$, may be defined, however, in different ways.

	c	Rel. density
In normal plant practice at a particular mine the values were	0.38–0.41	1.65–1.70
Rakes began to lift at	0.42	1.71
Pump began to strain at	0.43	1.73
Quiescent 24 h batch settling of underflow pulp	0.46	1.78
Batch settling with raking at base (24 h)	0.51	1.87
Batch settling with a picket fence rake (24 h)	0.52	1.88
Batch settling with a picket fence followed by 12 h quiescent settling	0.54	1.92

A similar system is described by Okunev et al.[8] except that the pumping rate $Q\,m^3/h$ for any measured underflow concentration was adjusted not by a normal controller, but according to an empirical equation based upon plant experience $C_U(kg/m^3) = 322 - 0.63(Q - 60)$, which anticipates the change in C_U brought about by a change in Q.

Maximum Underflow Density. A weakness in underflow density control arises when the compressibility of the solids varies from day to day.[182] If the particle size becomes finer, clay content increases or the flocculant dosage varies, the maximum attainable density may fall *below* the set point of the controller. This would result in the pumping rate being reduced to an unrealistically low value with a consequent large decrease in throughput, or, steady operation consistently *below* the attainable density. This real problem was recognized by Cross, and to cater for such variations in attainable density, attempts were made[116] to control the pumping rate based on the stock of solids in the thickener, i.e., the "mud" height. Because of the lack of significant compression in this zone, however,[56] the underflow density commenced to fall within minutes after increasing pump speed and remained at an abnormally *low* value long before the controller had brought a high mud level down to its set value. Therefore

It is clear that for this mineral pulp, exceeding the normal concentration by 30 to 40% is physically possible by sedimentation and raking alone, but that an artificial barrier exists by virtue of the degree of pulp fluidity required for solids transport. This is confirmed by the fact that, following a rake "trip-out," the solids are able to settle and compact to such a high density that the thickener has to be dug out manually.

Figure 14.40 shows the result of plotting both the "maximum" density (24 h slow raking with a picket fence) and the normal *operating* plant density for a number of mines against the specific surface area, S_W, of the suspended solids (using the permeability technique). It can be seen that the maximum density varies from 1.48 to 2.00 and decreases with increasing fineness, the linear correlation being significant at the 1% level. The mines, however, operate well below this limit and 90% lie within the narrower range 1.50 to 1.67 regardless of their maximum *potential* density.

Marsden[183] has shown that for similar pulps at densities between 1.57 and 1.70 the apparent viscosity rapidly exceeds a value 100 times that shown at feed concentration. The operating density is thus governed by maximum pumpable *viscosity* rather than the maximum attainable density. As in *dispersed* systems, such viscosities are not reached until $c = 0.52$, some of the

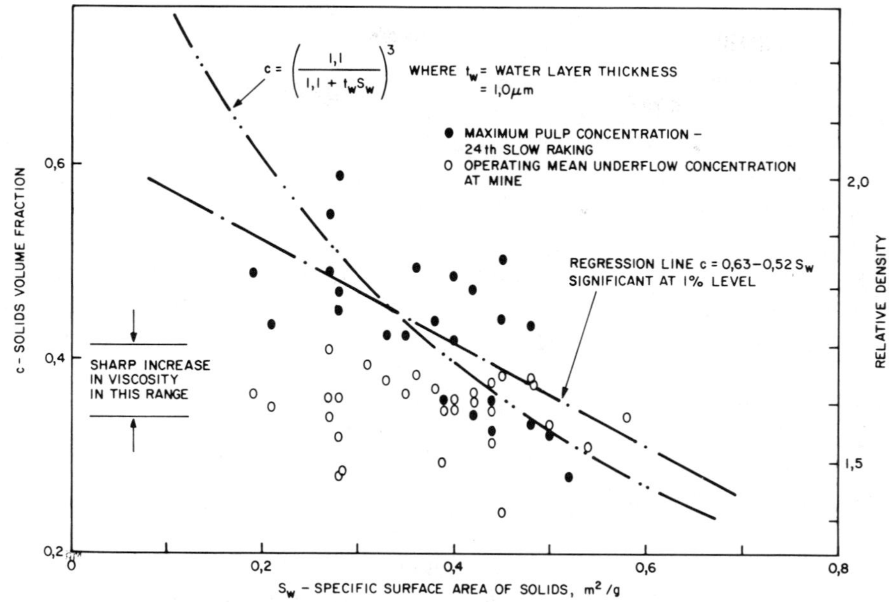

Figure 14.40. Variation of maximum pulp concentration and thickener underflow density with specific surface area of the dried pulp solids.

stagnant water originally present in the flocs must contribute to the "solids" concentration.[184] In the absence of a more robust method of extracting thickened pulp, some of the removable water must therefore be allowed to remain in the pulp discharge. A density of 1.7 is equivalent to a water-solids ratio by mass of 0.53; for the highest density shown in Figure 14.40 it is 0.26 while for the closest random packing of spheres[32] it is 0.21. The remaining water is therefore more than 100% of the minimum quantity.

Figure 14.40 shows that a model for maximum concentration which assumes that the volume of residual water is proportional to the surface area (representing what would be a surface film of thickness $t_W = 1$ μm) gives an adequate correlation with the data.

The situation where "maximum" operating density is governed by viscosity rather than actually attainable density is unlikely to apply to organic pulps or chemical precipitates. Their water–solids ratio at the maximum concentration of 10 kg/m³ is about 100 : 1 and the solids, while resisting further collapse by gravity compression, are readily sheared in a pump.

Rake and Pump Load. Although pulp viscosity places a constraint on the maximum density, it also provides a suitable direct signal[161] for automatic control at this limit. The degree of lift of the rakes,[162] a measure of the viscosity of the underflow pulp or strain on the pump could be used for this purpose.

A patent[185] issued to a Canadian firm is partially based upon this principle. A fluid-operated motor was used for rotating the rake assembly which, when a load was encountered, tended to slow down thereby causing a rise in pressure in the drive fluid. This increased pressure was used not only to lift the rake assembly clear of the obstruction, but also to increase the rate of solids removal via the speed of the discharge pump.

A sound method, based on the volume of excess water remaining in the pulp,[176] may be amenable to automation. A pulp sample is allowed to settle, and after 2 hours the volume of supernatant should (for this particular pulp) be ~100 ml per liter. Measuring the subsidence *rate* after say 15 minutes could speed up the test.

Control of Slime Level. When the feed rate of solids remains above the maximum throughput of a thickener for a long enough period the excess solids report to the overflow. It is normally desirable to prevent this, and the level of the water–slime interface may be monitored automatically by means of an optical turbidity meter or preferably an ultrasonic detector.[4] The former was found to be subject to liming of the windows, premature failure of the light bulb, variable operation between night and day depending upon the light level, and rapid fluctuations due to "wisps" of transient turbidity, none of which affected the ultrasonic instrument.

When a high slime level triggers the detector, the signal can be used to ring an alarm, increase the flocculant dosage rate or divert feed to another thickener. An alternative, increasing the underflow pumping rate, is only used in an emergency.

Pulp level control is essential in upflow solids thickeners[161] (p. 664) as it is important to maintain a steady depth of floc bed to prevent "breakthrough."

Economic Use of Flocculant. Polymer flocculants are relatively expensive reagents and their injudicious use can readily increase thickening costs.[108,186] Based on data[132] in South Africa in 1965, thickener running costs were 0.64 c/t solid, capital costs 0.74 c/t while addition of a polyelectrolyte at a rate of 1 g/ton (to increase throughput by ~10%) increased the running cost by 0.26 c/t or 40%. In the long term it clearly pays to extend the thickener plant rather than use polymer flocculants continuously.[108]

If however, flocculants are used only in winter or times of overload or to increase overflow clarity,[1] the situation is different but means of reducing its usage to a minimum would still be beneficial. This is especially so where the practice is to add flocculant at a steady rate to suit the worst conditions.

Hamza[187] discusses a graphical method for obtaining the optimum value of flocculant dosage by minimizing the total cost of thickening. It is based upon the slope of the plot of thickening cost vs reagent cost curves but runs into difficulties if the $H - t$ plots at various polymer dosages "cross over each other" especially at high dosages, 50–100 g/t.

To prevent overdosing of expensive flocculating reagents in coal-preparation plants, which is not only wasteful but can cause other problems in the process, a flocculation-control system was developed.[188] The *flocculated* feed pulp is sampled automatically and the settling rate measured by timing the descent of the pulp–water interface between two adjustable photoselective transmitter-receivers. A controller then compares the elapsed time with the desired settling time and adjusts the flocculant addition rate accordingly.

Automatic Control of Clarifiers

Proposals for automating the operation of clarifiers are made by Reuter.[189] Flocculant solution addition rate is regulated by the mass of solids entering in the feed pulp which is obtained by means of a combined signal from a magnetic flow meter and a gamma-ray density gauge. The underflow pumping rate *could* be controlled by means of the same signal but, because of the solids residence time, it is replaced by a signal derived from monitoring the height of the sludge bed. For this three methods are proposed, ultrasonic echo through the clear water, gamma-ray attenuation and continuously running sampling tubes from four depths, the turbidity of each of which affects the rate of discharge pumping.

Flocculant addition rate governed by overflow or tank turbidity should be a superior method of control in some cases. This would allow a more economic use of flocculant in cases in which there is a variable proportion of incoming feed particles which do not require flocculation for rapid settling or where the settling rate of the feed changes with time.

A device, based on this philosophy, was developed by Johnson[190] for flocculant addition control to thickeners which could also be applied to clarifiers in some cases. A small scale thickener with underflow and overflow rates set at a fixed ratio by metering pumps is fed at the sum of these rates from the feed launder together with just sufficient flocculant solution to maintain a constant level of the settled pulp interface. This is measured by means of a differential pressure range cell that responds to the pulp column density. The quantity of flocculant required in the model is then used as a signal to control the flocculant flow to the full plant stream.

Addition is based upon actual performance rather than on mass of solids or volume of liquid to be treated.

Optimization of Sedimentation Tanks

Optimization of settler operation, as opposed to *control at fixed preset conditions*, is a promising approach if the necessary measuring instruments and controllers are available. Cross' strategy in aiming at the *maximum attainable* density under varying plant conditions,[182] rather than running the thickener at a controlled (fixed) density, is pioneering contribution in this respect.

Addition of flocculant based upon the *settling rate* of the feed[182,189] rather the total quantity of solids fed is along the same lines, as also the use of rapid particle size analyzers, as a more sensitive measure of process performance, to complement the traditionally used turbidity measurements for control of water treatment plants.[191] Such optimization strategies can lead to substantial economies with little increase in the capital outlay than is required for normal automatic countrol.

14.7. ENGINEERING DESIGN AND CONSTRUCTION

Besides improvements in both the sedimentation process and the type of vessel in which it is carried out, discussed in Section 5, there have been less dramatic but nevertheless significant improvements in the engineering aspects such as methods of construction and underflow pump and rake drive mechanisms. In this section only a selection of the published *improvements*[192] are discussed. The normal procedures and designs are rightly the know-how and expertise of the equipment suppliers.

Construction of Earth Bottom Thickeners

Earth bottom thickeners are generally considered to be the lowest cost solution for thickener bottom construction presently available in the mining industry;[193] quoted diameters range from 38 to 132 m. The essential steps comprise testing and selection of soil materials, grading of the site, excavation of the thickener access tunnel, center foundation and ringwall footing and then forming and pouring the concrete. The walkway, thickener mechanism and rake arms are installed next and after completion of the wall, the thickener is flooded with waterproofing agent and after treatment for a specified time is pumped out and ready for service.

A novel means of constructing concrete walls for three 126 m diameter thickeners at Sishen,[194] was to slide them horizontally. The task was completed in 10 days instead of 2 months.

Rake and Drive Mechanisms

In rectangular tanks, reciprocating rakes are used, supported by a beam or truss structure spanning the width of the tank and traveling along its entire length.

In cylindrical vessels, the diameter determines the arrangement of the mechanical system. From 2 to 30 m diameter, a truss or beam-supported mechanism is usual. The feed pipe or launder and the walkway to the center of the tank are also supported by a braced truss or beams that span the tank (see Figs. 14.1 and 14.3).

Such a structure becomes increasingly costly above 30 m however and rake mechanisms are then supported by a concrete or steel center-pier. A structure is built from the side, but only to the center of the tank, to provide access to the drive and to support the feed pipe. Center-pier supported mechanisms are used in tanks ranging from as little as 10 to as large as 120 m.

When thickener rakes must span diameters ranging from 75 to 180 m, caisson-type units become more economical. The solids are pumped up through pipes in a hollow center shell, replacing the conventional tunnel which can be costly[195] in poor soil conditions.

The caissons, 6 m diameter and 12 m high, contain the underflow pumps, the pumping lines and the controls for the thickener drive. Rakes for pier or bridge-supported mechanisms are usually powered by a central drive, but traction drives supported at the center of the tank and driven at the wall by the motor-driven carriage running on a rail or pad are becoming increasingly common. The advantage of the center drive is a positive drive and hence absence of wheel skidding in the presence of ice, dust or rain, and the ease of self-elevation.

The drive heads and bearings must be able to withstand eccentric or unbalanced loads and designing the mechanism for proper torque must be based on possible mud *overloads*, not on average conditions. Shut down of a thickener due to rake trip-out involves not only throwing additional load on the remaining thickeners of a group, or worse, a plant shut down but the accumulated solid has then to be dug out manually.

In dealing with corrosive materials the normal structural steel used for rakes is either rubber-lined, painted, or replaced with alloys such as stainless steel. Tubular structural sections have advantages in these circumstances because of its minimum surface area and ease of painting and cleaning.

To increase the underflow density, stilt-type rakes[176] or vertical pickets on the rake arms may be installed. These pass slowly through the thickening zone and are thought thereby to promote release of trapped water. Details of raking mechanisms are presented in useful descriptive overviews of thickeners and clarifiers by Dahlstrom and Cornell[142] and also Raynor and Porter.[157]

A special rake design allowing large solids inventories was successfully used for storage of a flotation concentrate.[196]

The depth was 4.9 m, instead of the "normal" 2.4 m, and as the solids filled up during one shift of storage, the rakes were lifted by raising a movable platform on which the main drive gear was mounted. During the next two shifts of solids withdrawal, programmed automatic controls then relowered the drive platform and rakes. A reasonably smooth operation was achieved during a three year period. During the raising of the rakes, including the cone-scraper which keeps the underflow outlet clear, a false-bottom of solids evidently formed with a "rat-hole" in the center through which the solids raked at the top were conveyed to the discharge pump.[176]

Underflow Pumping

Various locations and designs of pumps for pumping thickener slurry from sedimentation vessels may be used.[197] For larger tanks, the conventional method of piping the raked solids from the center through an access tunnel to a pump on the *periphery* of the tank is becoming increasingly costly and may be prohibitive with certain soils.

Alternative lay-outs include either a pump room at the base of the center column where there is positive flooded suction but maintenance is risky and difficult, or bridge-mounting where there is no submerged valving, and pumps are completely accessible, but special priming arrangements are required.[197]

Pumps are traditionally rubber diaphragm types with variable stroke and speeds mounted at an elevation just above the overflow level. Rubber-lined centrifugal pumps although not recommended for accurate control[176] have proved satisfactory in the gold-mining and other industries[178] and free discharge into a common launder via a variable size spigot valve has also been practiced. The latter has the disadvantage, however, that the thickener cannot then be readily "recirculated." This is necessary when, for some reason, the next stage in the plant cannot accept thickened pulp. To prevent *over*thickening, the pulp is pumped from the thickener as usual and then gravitated back to the feedwell.

A suction type unit,[141] in which solids are removed by syphon tubes rather than standard rakes and a pump, is used in the final sedimentation tanks in activated sludge treatment plants. The unit withdraws sludge continuously across the entire tank bottom through a number of suction ducts mounted on the rotating rake arms. Radial V-plates (squeegees) funnel the settled sludge to each syphon opening. The maximum underflow concentration is low,

~1%, but this is no disadvantage as recycle of *liquor* as well as solids is usually required in this process.

Plant Layout

The arrangement of sedimentation tanks when more than one is used is normally dictated by the available ground space, the shape of the tanks, whether gravity feed is preferred, the benefits or otherwise of proximity to a central building and the need to keep different feeds separate.

In thickening milled gold ore for example it has been preferred until recently to have a larger number of "small" thickeners and feed four or five in a "leg" in parallel from a common feed launder. Two such legs may treat an ore fraction with a lower grade of gold while a third leg treats the high grade pulp or one containing a sufficiently high uranium content for later treatment. The legs run parallel to form a simple grid and allows cross-overs when the volumetric feed proportions change. Thickeners may share a pump house in which their individual pumps are located and the underflow pulps are then recombined into a common launder. The siting of the plant is selected for gravity flow of feed, underflow and overflow streams, if possible.

Sometimes it may be advantageous to exploit special features of a site. At a magnesium-from-seawater plant in the North of England, for instance, use is made of large natural depressions in the sand dunes to locate the thickeners, rather than some fixed geometry. Care must be taken to investigate the geological formation under a thickener site and not to build unknowingly over *underground* depressions or sinkholes.[108]

In water purification and sewage treatment rectangular tanks are commonly used. This shape permits close stacking, and hence space saving, common sidewalls and raking drives.

To ensure uniformity of operation throughout a battery of sedimentation vessels, the method of distributing the feed requires careful consideration. An open splitter box which permits size segregation prior to settling is not suitable. Experience suggests that a circular tank with offtakes symmetrically arranged achieves a uniform size consistently in each branch stream.[164] For equal distribution of flow volumes, the main feed is admitted via a bell mouth overflow into a common distribution chamber.[198] This breaks the velocity advantage which the feed pipe may afford any tank by running in that particular direction.

LIST OF SYMBOLS

A	Area of horizontal cross-section of thickener or clarifier; maximum value in cylindrical section; decreases to zero in the cone.
A_b	Area of batch test cylinder.
a	Divisor for calculation of *Re* defined in Table 14.2.
B	"Rate of rise" of Δ_{FLUX} with c.
B_F	Buoyancy force or Archimedes' upthrust $= V \cdot \rho_L \cdot g$.
b	Exponent for calculation of *Re* defined in Table 14.2.
C	Mass concentration of particles i.e., mass of oven dry solids per unit volume of suspension.
C_{AV}	Average pulp concentration during compression, i.e., between CSP and arbitrary period of 24 or 48 hours.
C_{crit}	Concentration of the minimum flux zone.
C_D	Dimensionless drag coefficient defined in Equation (5).
C_F	Concentration of thickener feed pulp.
C_{max}	Maximum concentration attained by random close packing of spheres i.e., in absence of compression.
C_0	Initial uniform concentration in a batch test. In Talmage and Fitch[65] test C_0 must be made equal to C_F.
C_U	Concentration of thickener underflow pulp.
C_X	Required value of C_U corresponding to "overload" condition G_{OL}, to avoid sliming in the long term.
C and C	*UC* values measured in batch tests with uniform initial concentrations C between C_F and C_U without stirring (Coe and Clevenger[54]).

CCD Continuous countercurrent decantation.
CSP Critical settling point.
c Volume fraction of particles in a suspension $= C/\rho_S$.
c_D Concentration at which deviation from Equation (14) commences, upper limit of dilute regime.
D Diameter of settling vessel.
D' Mass of liquid per unit mass of solid in thickener underflow pulp.
d (1) Particle diameter, (2) Distance between plates in lamella settler (normal to plates).
$\bar{d}$ Mean sedimentation particle diameter of a suspension of multisized particles (Eqs. 12 and 12a).
d_A Diameter of a sphere having same *surface area* as a nonspherical particle.
d_F Mean floc diameter.
d_L Mean length diameter = total length of particle/their total number.
d_M Diameter of circle with same projected area as image of particle in microscope.
d_{ns} Diameter of sphere with same terminal settling velocity as nonspherical particle.
d_S Mean projected area diameter = (total projected area of particles/their total number)$^{1/2}$.
d_{scr} Diameter of sphere just passing same square aperture as particle.
d_V Mean volume diameter = (total volume of particles/their total number)$^{1/3}$
E Asymptote value of Δ_{FLUX}.
F Mass of liquid per unit mass of solid in free-settling pulps tested for sedimentation rates.
F_D Drag force on particle moving relative to liquid.
f Shape factor of Richards[21] $= u_Q/u_R$.
G Solids flux.
Ga Galilei number $= (\rho_S - \rho_L)\rho_L d^3 g/\mu^2$.
G_C Total flux due to Coe and Clevenger.[54]
G_{clar} Solids flux in a clarifier.
G_F Gravitational force of attraction $= V \cdot \rho_S \cdot g$.
G_{min} Minimum value of total flux G_T.
G_{OL} Solids flux during overload.
G_{OLT} Minimum flux due to Oltmann,[101] $G_{OLT} \nless G_{TF}$ and $\sim G_{C(min)}$.
G_T Total flux due to Hassett,[67] Equation (19).
G_{TF} Minimum flux due to Talmage and Fitch.[65]
g Acceleration due to gravity.
H Height (1) to suspension-supernatant interface in a batch test, (2) vertical height of lamella settler.
H_0 Height of pulp in batch test at zero time.
H_U Underflow line $= H_0 C_0/C_U$.
H_∞ Height of pulp at infinite time; used in Roberts plot and found by trial-and-error.
I Intercept value of a plot of $u^{0.213}$ vs c.
J $C_D Re^2 = 4Ga/3$. Property of a particle-liquid system independent of u.
K Shape factor $= d/d_A$.
k Effective volume of a sedimentation entity (single rough particle, cluster or floc) per unit volume of contained solids, i.e., ratio volume of particles plus fixed water to volume of particles.
k' Effective volume of a floc per unit *mass* of contained solids.
$kc = k'C$ Effective volume fraction of flocs (solids) in suspension.
MLSS Mixed liquor suspended solids.
m Mass of solids causing compression.
N (1) Total number of particles in a sample $= \Sigma n_d$; (2) Number of plates in a lamella settler.
n Exponent in the Richardson and Zaki[28] equation $= 4.65 + 19.5\, d/D$ in streamlined flow, ~ 4.7 for most pulps in gravity sedimentation practice.
n_d Number of particles in each size class d.
Q Volumetric underflow discharge rate from a continuous thickener.
Q_F Volumetric feed rate.
R Ratio of thickener area for a given

	duty actually required in practice to area predicted by batch tests.
R'	Ratio of floor area required to locate lamella settler to vertically projected settling area of all its plates.
Re	Reynolds number of a particle = $d.u.\rho_L/\mu$.
SS	Suspended solids.
S_W	Specific surface area of dry powder in m^2/g measured by air permeability of packed bed.
T and F	UC values estimated from the position and rate of descent of the pulp-water interface in a batch test initially at concentration C_F (Talmage and Fitch[65]).
t	Time.
t_D	Detention time in a clarifier.
t_U	Time at which H_U intersects extension of tangent to batch settling curve drawn before CSP.
t_W	Equivalent thickness of water around each particle to account for voidage water to solids ratio in packed bed.
t_1	Period during which initial settling rate u_i increases to a maximum steady value u; the induction period.
t_2	Time for C_{crit} to reach descending pulp interface (Fig. 14.23).
U	Settling velocity of pulp-water interface.
U_C	Rate of rise of upper level of stock of pulp at overload.
U_{crit}	Settling velocity of minimum flux zone, C_{crit}.
U_1	Settling velocity of concentration C_1.
U_2	Settling velocity of concentration C_2.
u	Settling velocity of a particle, a concentration zone, or a suspension-supernatant interface relative to the container.
$\hat{u}$	Settling velocity predicted according to Equation (14) i.e., extrapolated for dilute pulps.
u_F	u_0 for a suspension of flocs.
u_i	Initial settling rate of an intermediate pulp.
u_{ns}	Terminal settling velocity of a non-spherical particle.
u_Q	Observed settling velocities of individual quartz particles.
u_R	Calculated settling velocities of quartz particles of diameter d_{scr} using Rittinger's law.[21]
u_S	Settling velocity under conditions of slow stirring.
u_W	Settling velocity in presence of container walls as opposed to settling in an infinite liquid.
u_ϕ	Terminal settling velocity of a sphere in an unbounded liquid in *streamlined* flow.
$u_{\phi t}$	As for u_ϕ but in *transitional* or *turbulent* flow.
u_0	Settling velocity of a single particle, representative of a suspension i.e., suspension settling velocity at $c = 0$, equivalent to u_W.
V	(1) Volume of particle (2) Volume of thickener.
V_C	Slope of settling flux curve at concentration C.
V_{crit}	Rate of rise of C_{crit} in a batch test.
V_{int}	Rate of rise of interface between C_1 and C_2.
W	(1) Retardation factor due to presence of container walls Equation (10); (2) Width of lamella settler.
w_i	Term to account for stagnant water that accompanies particle during sedimentation (Steinour[36]).
x	Mass fraction of particles in each size interval of mean size d determined by sedimentation.
α	(1) Slope of "underflow flux line," $\tan \alpha = Q/A$. Equals rate of rise of C_{crit} in batch settling test; (2) Angle of plates in a lamella thickener to the horizontal.
β	for concentrations C_1 and C_2, $\tan \beta = -\Delta G/\Delta C$.
ΔC	Concentration difference across an interface.
Δ_{FLUX}	Deviation of actual settling velocity of intermediate concentration pulp from velocity extrapolated for dilute pulps, Equation (14), expressed as a normalized solids flux.

ΔG	Flux difference across an interface.
δ	Maximum slope of $-\Delta G/\Delta C$ starting from concentration C_1.
λ	$Bk(c - c_D)$ Equation (18a).
ρ_F	Floc density. Mass of solid plus water in floc per unit volume of floc.
ρ_L	Liquid density.
ρ_S	Particle or solids density.
μ	Liquid viscosity.

TERMS

Concentrated Pulp: A pulp in compression; settling rates increase with cylinder height and presence of slow stirring or raking and H-t curve concave upwards.

Critical Settling Point (CSP): Point during batch sedimentation at which the last settling flocs just arrive at the rising sediment zone and compression has not commenced at the pulp surface.

Dilute Pulp: A pulp showing a steady settling rate from $t = 0$.

Draw Off: Underflow pulp discharge from a continuous thickener, Q.

Floc: (1) An agglomerate of discrete particles, bound together by van der Waals forces or polymer links, acting as a single entity in sedimentation, comprising both the solid material and the voidage water between the particles; (2) A metal hydroxide precipitate.

Free-Settling: According to Coe and Clevenger,[54] a pulp which shows a constant settling rate either at zero time or after an induction period, t_1. Includes dilute and intermediate pulps in present nomenclature, i.e., free-settling dilute flocs and also channel-flow in 3-D floc structure.

Intermediate Pulp: A pulp which tends to set or gel on cessation of agitation, shows a gradually increasing settling rate during an induction period t_1, and thereafter reaches a maximum steady value. This settling behavior is a function of container dimensions, presence or absence of stirring, degree of aggregation, as well as particle concentration. A *dilute* pulp in a large cylinder may become *intermediate* or *concentrated* in a smaller one.

Rat Hole: Opening or channel in sedimented mud in a thickener through which thinner pulp from above may flow to the underflow pipe.

Relative Density: Absolute density of pulp divided by density of water in same units.

Scale-up Factor: A factor to cover unknown contingencies when extrapolating thickener areas estimated from batch tests to full-scale plant.

Sink Hole: Underground cavity produced, for example, in a dolomite layer through action of percolating CO_2-bearing water. When the roof is weakened by moisture or mechanical vibration it may collapse catastrophically.

Solids Flux: Mass of solids descending below any horizontal plane of unit area in unit time.

Trip-out: Occurs when thickener rakes, after lifting to their maximum extent to avoid excess load, actuate an automatic switch and cease to rotate.

Underflow Line: (1) Underflow flux line = QC/A; (2) Horizontal line drawn at $H = H_U$ on Talmage and Fitch construction, Figure 14.24.

REFERENCES

1. M. J. Pearse, *Gravity Thickening Theories: A Review.* Warren Spring Laboratory, Stevenage, England, No. LR 261 (MP) (1977).
2. R. W. Moore, Jr., *Liquid-Solids Separation.* Presented at a symposium organized by Hunslet Taylor-Eimco Sales, Johannesburg (6 August 1971).
3. H. W. Campbell, R. J. Rush, and R. Tew, *Sludge Dewatering Design Manual.* Res. Rep. No. 72. Res. Program Abatement Munic. Pollut. Provi. Canada-Ontario Agreement Great Lakes Water Quality (1978).
4. K. J. Scott, Previously unpublished work of co-

worker (when mentioned) and the author. CERG-CSIR Internal Reports (1965–1970).

5. F. M. Tiller, (ed). "How to select solid-liquid separation equipment." *Chem. Eng. 81*:(9) 116–36 (April 29), (10) 98–104 (May 13, 1974).
6. B. Fitch, "Choosing a separation technique." *Chem. Eng. Progress 70*:(12) 33–7 (1974).
7. D. B. Purchas, "Solid/liquid separation equipment: A preliminary experimental selection programme." *The Chem. Eng.* No. 328:47–9 (January, 1978).
8. B. S. Okunev, I. I. Morosova, Yu. A. Tisunov, and N. F. Kotov, "Automatic control of radial thickener operation." *Coke Chem. USSR*:(12) 49–52 (1977).
9. G. G. Stokes, "On the effect of the internal friction of fluids on the motion of pendulums." *Trans. Cambridge Phil. Soc. 9*(II):8–106 (1851).
10. I. Proudman and J. R. A. Pearson, "Expansions at small Reynolds numbers for the flow past a sphere and a circular cylinder." *J. Fluid Mech. 2*:237–62 (1957).
11. G. K. Batchelor, *An Introduction to Fluid Dynamics*. Cambridge University Press, London, p. 233, (1967).
12. R. H. Perry and C. H. Chilton, *Chemical Engineers' Handbook*, Fifth Ed. McGraw-Hill Kogakusha, Tokyo, 1973. 12a p. 9–55, 12b p. 5–61, 12c p. 1011 [Third Ed.–(1950)] 12d p. 19–46, (1973).
13. M. B. Sonnen, "Subroutine for settling velocities of spheres." *Amer. Soc. Civil Engrs. J. Hydraulics Div. 103*:(HY9) 1097–101 (1977).
14. V. F. Swanson, "A free-settling equation valid for all particle sizes." Conference: *Proc. Tech. Program*: *Int. Powder Bulk Solids Handl. Pr*, pp. 82–9 (Published 1978). (Available Ind. Sci. Conference Man., Inc., Chicago, Ill.)
15. P. G. W. Hawksley, "Survey of the relative motion of particles and fluids." *Brit. J. Appl. Phys. 5*:S1–S5 (1954).
16. E. S. Pettyjohn and E. B. Christiansen, "Effect of particle shape on free-settling rates of isometric particles." *Chem. Eng. Prog. 44*(2):157–72 (1948).
17. A. I. Medalia, "Dynamic shape factors of particles." *Powder Technol. 4*:117–38 (1970/71).
18. J. Tsubaki and G. Jimbo, "A proposed new characterization of particle shape and its application." *Powder Technol. 22*:161–78 (1979).
19. S.-T. Fong, J. K. Beddow, and A. F. Vetter, "A refined method of particle shape representation." *Powder Technol. 22*:17–21 (1979).
20. J. M. Zeigler and B. Gill, *Tables and Graphs for the Settling Velocity of Quartz in Water, Above the Range of Stokes' Law*. Woods Hole Oceanographic Institution, Reference No. 59–36 (July, 1959).
21. R. H. Richards and C. E. Locke, *Textbook of Ore Dressing*, Third Ed. New York: McGraw-Hill Company, Inc., Table 30, p. 129, (1940).
22. A. W. Francis, "Wall effect in falling ball method for viscosity." *Physics 4*:403–6 (1933).
23. J. Garside and M. R. Al-Dibouni, "Velocity-voidage relationships for fluidization and sedimentation in solid-liquid systems." *Ind. Eng. Chem., Process Des. Dev. 16*:206–14 (1977).
24. B. H. Kaye and R. P. Boardman, "Cluster formation in dilute suspensions." *Proc. Symp. Interaction between Fluids and Particles, Inst. Chem. Engr.* A17–21 (1962).
25. R. Johne, "Einfluss der Konzentration einer monodispersen Suspension auf die Sinkgeschwindigkeit ihrer Teilchen." *Chemie-Ing.-Techn. 38*:428–30 (1966).
26. B. Koglin, "Dynamic equilibrium of settling velocity distribution in dilute suspensions of spherical and irregularly shaped particles." *Proc. Conf. Particle Technology*, IIT Res. Inst. Chicago, pp. 266–71, August 21–24 (1973).
27. E. M. Tory and D. K. Pickard, "A three-parameter Markov model for sedimentation." *Can. J. Chem. Eng. 55*:655–65 (1977).
28. J. F. Richardson and W. N. Zaki, "Sedimentation and fluidization: Part 1." *Trans. Instn Chem. Engrs 32*:35–53 (1954).
29. H. Watanabe, "Voidage function in particulate fluid systems." *Powder Technol. 19*:217–25 (1978).
30. A. D. Maude and R. L. Whitmore, "A generalized theory of sedimentation." *Brit. J. Appl. Phys. 9*:477–82 (1958).
31. W. C. Thacker and J. W. Lavelle, "Stability of settling of suspended sediments." *Phys. Fluids 21*:291–2 (1978).
32. P. T. Shannon, E. Stroupe, and E. M. Tory, "Batch and continuous thickening. Basic theory. Solids flux for rigid spheres." *Ind. Eng. Chem. Fundam. 2*:203–11 (1963).
33. H. H. Steinour, "Rate of sedimentation. Nonflocculated suspensions of uniform spheres." *Ind. Eng. Chem. 36*:618–24 (1944).
34. K. J. Scott and W. G. B. Mandersloot, "The mean particle size in hindered settling of multisized particles." *Powder Technol. 24*:99–101 (1979).
35. B. Fitch, "Sedimentation process fundamentals." *Trans. AIME 223*:129–37 (1962).
36. H. H. Steinour, "Rate of sedimentation: Suspensions of uniform-size angular particles." *Ind. Eng. Chem. 36*:840–7 (1944).
37. J. P. Mogan, R. W. Taylor, and F. L. Booth, "The value of the exponent *n* in the Richardson and Zaki equation, for fine solids fluidized with gases under pressure." *Powder Technol. 4*:286–9 (1970/71).

38. D. K. Vohra, "Sedimentation in a viscous suspending medium." *Inst. Eng. (India) Jnl Chem. Eng. 57*:97-100 (1977).
39. A. K. Korol'kov, M. N. Kell, and A. A. Vasil'eva, "Dependance of the rate of the hindered settling of grains on their specific surface.", *Obogashch. Rud 15*:24-6 (1970). (Russ.) [*Chem. Abstr. 74*:89097 k (1971).]
40. L. Davies, D. Dollimore, and J. H. Sharp, "Sedimentation of suspensions: Implications of theories of hindered settling." *Powder Technol. 13*:123-32 (1976); *17*:147-52 (1977).
41. G. B. Wallis, "A simplified one-dimensional representation of two-component vertical flow and its application to batch sedimentation." In *Proc. Symp. Interaction Between Fluids and Particles*. London: Instn Chem. Engrs 9-16 (1963).
42. L. A. Adorján, "Determination of thickener dimensions from sediment compression and permeability test results." *Trans. Inst. Min. Metall. Sect. C, 85*:C157-63 (1976).
43. C. E. Capes, "Particle agglomeration and the value of the exponent *n* in the Richardson-Zaki equation." *Powder Technol. 10*:303-6 (1974).
44. A. E. Fouda and C. E. Capes, "Hydrodynamic particle volume and fluidized bed expansion." *Can. J. Chem. Eng. 55*:386-91 (1977).
45. L.-G. Eklund and Å. Jernqvist, "Experimental study of the dynamics of a vertical continuous thickener - I." *Chem. Eng. Sci. 30*:597-605 (1975).
46. A. S. Michaels and J. C. Bolger, "Settling rates and sediment volumes of flocculated kaolin suspensions." *Ind. Eng. Chem. Fundam. 1*:24-33 (1962).
47. E. K. Obiakor and R. L. Whitmore, "Settling phenomena in flocculated suspensions." *Rheologica Acta 6*:353-9 (1967).
48. P. G. Cooper, J. G. Rayner, and S. K. Nicol, "Flow equation for coagulated suspensions." *J. Chem. Soc. Faraday Trans.* I, *74*:785-94 (1978).
49. K. J. Scott, "Theory of thickening: Factors affecting settling rate of solids in flocculated pulps." *Trans. Instn. Min. Metall. (Sect. C: Mineral Process. Extr. Metall.) 77*:C85-97 (1968); *78*:C116-9, 244-5 (1969).
50. K. J. Scott, "Thickening of calcium carbonate slurries. Comparison of data with results for rigid spheres." *Ind. Eng. Chem. Fundam.* 7:484-90 (1968).
51. K. J. Scott, "The water content of flocs." *J. S. Afr. Inst. Min. Metall. 65*:357-67 (1965).
52. C. C. Harris, P. Somasundaran, and R. R. Jensen, "Sedimentation of compressible materials: Analysis of batch sedimentation curve." *Powder Technol. 11*:75-84 (1975).
53. P. A. Vesilind, "Evaluation of activated sludge thickening theories." *Jnl San. Eng. Div., Proc. ASCE 94*:SA1, 185-91 (1968).
54. H. S. Coe and G. H. Clevenger, "Methods for determining the capacities of slime-settling tanks." *Trans. AIME. 55*:356-84 (1916).
55. B. Fitch, "Batch tests predict thickener performance." *Chem. Eng. 78*:(19) 83-8 (August 23, 1971).
56. B. Fitch, "Current theory and thickener design." *Ind. Eng. Chem. 58*(10):18-28 (1966).
57. R. H. Bretton, *The design of continuous thickeners*. Ph.D. thesis, Yale University (August, 1949).
58. J. A. Cole, *A model for activated sludge thickening*. Ph.D. thesis, University of Wisconsin (January, 1970).
59. K. J. Scott, "Continuous thickening of flocculated suspensions." *Ind. Eng. Chem. Fundam. 9*:422-7 (1970).
60. G. Sarmiento and P. H. T. Uhlherr, "The effect of temperature on the sedimentation parameters of flocculated suspensions." *Powder Technol. 22*:139-42 (1979)
61. R. H. Weiland and R. R. McPherson, "Accelerated settling by addition of buoyant particles." *Ind. Eng. Chem. Fundam. 18*:45-9 (1979).
62. R. L. Whitmore, "The sedimentation of suspensions of spheres." *Brit. J. Appl. Phys. 6*:239-45 (1955).
63. P. Somasundaran, E. L. Smith, Jr., and C. C. Harris, "Effect of coarser particles on the settling characteristics of phosphatic slimes." *Proc. Conf. Particle Technology*, IIT Res. Inst, Chicago, pp. 145-50, August 21-24 (1973).
64. J. B. McVaugh, Jr., *Mathematical and experimental investigation of nonsteady state thickening of an ideal slurry*. M.Sc. thesis, Delaware University (May, 1975).
65. W. P. Talmage and E. B. Fitch, "Determining thickener unit areas." *Ind. Eng. Chem. 47*:38-41 (1955).
66. K. J. Scott, "Mathematical models of mechanism of thickening." *Ind. Eng. Chem. Fundam. 5*:109-13 (1966).
67. N. J. Hassett, "Mechanism of thickening and thickener design." *Trans. Instn Min Metall. 74*:627-56 (1964/65).
68. D. J. Slagle, Y. T. Shah, G. E. Klinzing, and J. G. Walters, "Settling of coal in coal-oil slurries." *Ind. Eng. Chem. Process Des. Dev. 17*:500-4 (1978).
69. T. M. Keinath, M. D. Ryckman, C. H. Dana, and D. A. Hofer, *A unified approach to the design and operation of the activated sludge system*. Presented at the 21st Annual Industrial Wastes Conference, Purdue University (4 May 1976).

70. W. Rudolfs and I. O. Lacy, "Settling and compacting of activated sludge." *Sewage Works Jnl. 6*:647–75 (1934).
71. H. H. Steinour, "Rate of sedimentation: Concentrated flocculated suspensions of powders." *Ind. Eng. Chem. 36*:901-7 (1944).
72. A. M. Gaudin and M. C. Fuerstenau, *On the Mechanism of Thickening*. International Mineral Processing Congress, Institution of Mining and Metallurgy, London, pp. 115–27 (1960).
73. E. M. Tory, *Batch and continuous thickening of slurries*. Ph.D. thesis, Purdue University (June, 1961).
74. K. Matsumoto, A. Suganuma, and D. Kunii, "Effect of permeability on settling velocity of an actual floc." *Chem. Eng. Sci. 33*:1554–6 (1978).
75. K. Matsumoto and A. Suganuma, "Settling velocity of a permeable model floc." *Chem. Eng. Sci. 32*:445–7 (1977).
76. D. N. Sutherland and C. T. Tan, "Sedimentation of a porous sphere." *Chem. Eng. Sci. 25*:1948–50 (1970).
77. N. Epstein and G. Neale, "On the sedimentation of a swarm of permeable spheres." *Chem. Eng. Sci. 29*:1841–2 (1974).
78. A. R. Tarrer, H. C. Lim, L. B. Koppel, and C. P. L. Grady, Jr., "A model for continuous thickening." *Ind. Eng. Chem., Process Des. Dev. 13*:341–6 (1974).
79. D. W. Gay and K. W. Young, *The Use of High-Purity Oxygen in the Activated Sludge Process*., vol. 1, edited by J. R. McWhirter, CRC, West Palm Beach, Fla., Chap. 7, 195–225 (1978).
80. H. J. Edde and W. W. Eckenfelder, Jr. "Theoretical concept of gravity sludge thickening; scaling-up laboratory units to prototype design." *Jnl. WPCF 40*:1486–98 (1968).
81. R. I. Dick and B. B. Ewing, "Evaluation of activated sludge thickening theories." *Jnl. San. Eng. Div., Proc. ASCE 93*:SA4, 9–29 (1967).
82. P. A. Vesilind, "The design of thickeners from batch settling tests." *Water Sewage Works 115*:418–9 (1968).
83. P. A. Vesilind, "Quiescent batch thickening of activated sludge in small cylinders." *Vatten 69*:456–62 (1969).
84. P. A. Vesilind and R. I. Dick, "Initial depth as a variable in activated sludge settling tests." *EWT Jnl. 9*:263–7 (1969).
85. E. Barnea and J. Mizrahi, "A generalized approach to the fluid dynamics of particulate systems. Part I. General correlation for fluidization and sedimentation in solid multiparticle systems." *Chem. Eng. J. 5*:171–89 (1973).
86. D. C. Dixon, "Momentum-balance aspects of free-settling theory. I. Batch thickening." *Separation Sci. 12*:171-91 (1977).
87. A. Bedford and C. D. Hill, "A mixture theory formulation for particulate sedimentation." *AIChE Jnl. 22*:938–40 (1976).
88. J. R. Blake, P. M. Colombera, and J. H. Knight, "A one-dimensional model of sedimentation using Darcy's law." *Separation Sci. Technol. 14*:291-304 (1979).
89. D. B. George, *Dynamics of continuous thickening*. Ph.D. thesis, Clemson University, December (1976).
90. N. Yoshioka, Y. Hotta, S. Tanaka, S. Naito, and S. Tsugami, "Continuous thickening of homogeneous flocculated slurries." *Chem. Eng. (Japan) 21*:66–74 (1957).
91. R. T. Mishler, "Settling slimes at the Tigre Mill." *Engng Min. J. 94*:643–6 (1912).
92. F. Yeates, "Discussion and contributions: Maximum solids handling capacity of continuous thickeners." *Trans. Instn Min. Metall. (Sect. C: Mineral Process. Extr. Metall.) 76*:C136–7 (1967).
93. A. G. Moncrieff, "Theory of thickener design based on batch sedimentation tests." *Trans. Instn Min. Metall. 73*:729–59 (1963/64); *74*:153–63, 579–80 (1964/65).
94. G. J. Kynch, "A theory of sedimentation." *Trans. Faraday Soc. 48*:166–76 (1952).
95. B. Fitch, "Current theory and thickener design (1975) - Part 1. Introduction; zone settling theories." *Filtration and Separation 12*:355–9 (1975).
96. R. A. Couche and L. H. Goldney, "The design of continuous thickeners for flocculated materials." *Proc. Austral. Instn Min. Metall*, no. 191:117–39 (1959).
97. P. T. Shannon and E. M. Tory, "Batch and continuous thickening. Prediction of batch settling behavior from initial rate data with results for rigid spheres." (Correspondence) *Ind. Eng. Chem. Fundam. 4*:367–8 (1965).
98. P. T. Shannon, R. D. Dehaas, E. P. Stroupe, and E. M. Tory, "Batch and continuous thickening." *Ind. Eng. Chem. Fundam. 3*:250–60 (1964).
99. W. W. Eckenfelder, Jr. and N. Melbinger, "Settling and compaction characteristics of biological sludges." *Sewage and Ind. Wastes 29*:1114–22 (1957).
100. P. T. Shannon and E. M. Tory, "The analysis of continuous thickening." *Trans. ASME 235*:376–82 (1966).
101. B. Fitch, "Current theory and thickener design (1975)–Part 3. Design procedures." *Filtration and Separation 12*:636–8 (1975).
102. N. J. Hassett, "Design and operation of continuous thickeners." *Ind. Chem. 34*:169–72; 489–94 (1958).
103. R. I. Dick and A. R. Javaheri, "Continuous

thickening of non-ideal suspensions." UILU-WRC-72-0045 *WRC Research Report No. 45* (1972).
104. W. M. Fleming, *Alternative to quiescent batch settling tests for defining settling flux curve.* Ph.D. thesis, Clemson University (December, 1976).
105. K. Tracy, *Mathematical modelling of unsteady-state thickening of compressible slurries.* Ph.D. thesis, Clemson University (December, 1973).
106. P. T. Shannon and E. M. Tory, "Settling of slurries." *Ind. Eng. Chem. 57*:18-25 (1965).
107. L.-G. Eklund, "The working conditions in a continuous pilot thickener at optimal load." *Chem. Eng. Sci. 31*:881-91 (1976).
108. E. H. D. Carman and D. P. Steyn, "Some observations on thickening." *8th Commonwealth Min. Metall. Congress Australia 6*:443-54 (1965).
109. K. J. Scott and J. L. Alderton, "Maximum solids handling capacity of continuous thickeners." *Trans. Instn Min. Metall. (Sect. C: Mineral Process. Extr. Metall.) 75*:C201-10 (1966).
110. E. W. Comings, "Thickening calcium carbonate slurries." *Ind. Eng. Chem. 32*:663-7 (1940).
111. R. L. Hibberd and W. F. Jones, "The design and operation of activated-sludge final settling tanks." *Wat. Pollut. Control 73*:14-32 (1974).
112. D. R. Oliver, "Continuous vertical and inclined settling of model suspensions." *Can. J. Chem. Engng 42*:268-72 (1964).
113. N. J. Hassett, "Thickening in theory and practice." *Min. Sci. Engng 1*(1):24-39 (1969).
114. B. Fitch, "The design of intermittently fed thickeners." *Ind. Eng. Chem. 60*:8-14 (July, 1968).
115. F. A. Seeton, "The cyanide process for gold and silver ores." Denver Equipment Company Bulletin No. M3-B16.
116. K. J. Scott, "Solids-settling zones observed on continuous thickeners and their bearing on automatic control." *Proc. Conf. Particle Technology* IIT Res. Inst. Chicago, pp. 179-85, (August 21-24, 1973).
117. E. Barnea, "New plot enhances value of batch-thickening tests." *Chem. Eng. 84*(18):75-8 (August 29, 1977).
118. K. J. Scott and J. L. Alderton, "Maximum solids handling capacity of continuous thickeners." *Trans. Instn Min. Metall. (Sect. C: Mineral Process. Extr. Metall.) 76*:C220-4 (1976).
119. F. C. Bond, "Settling and thickening." *Canadian Inst. Min. Metall. Trans. 49*:546-56 (1946).
120. Anon. "CSIR Tests behavior of feed in thickeners." *S. Afr. Min. Eng. Jnl.*" pp. 2803-4 (November 10, 1967).
121. L. G. Goryainov, O. K. Shcherbakov, M. I. Nasonova, and V. M. Korus, "Calculation of the thickener area." *Sov. Jnl. Non-Ferrous Metals 13*(7):85-7 (1972).
122. P. Kos and D. D. Adrian, "Gravity thickening of water treatment plant sludges." Paper presented at 94th Annual Conference AWWA, Boston, Mass. (June, 1974).
123. P. Kos, "Fundamentals of gravity thickening." *Chem. Eng. Prog. 73*(11):99-105 (1977).
124. Anon. "Electrostatic separation of solids from liquids." *Filtration and Separation 14*:140-4 (March/April, 1977).
125. Anon. "Rectangular versus circular settling tanks." *Amer. City 88*(10):98-99 (1973).
126. J. D. Walker, "Sedimentation maximizes clarity, minimizes turbidity in potable water treatment." *Water Sewage Works.* Reference number *1978*: R136-150 (11 pp.) (1978).
127. J. H. Tay, "Study of settling characteristics of physical-chemical flocs in sedimentation tanks." *Diss. Abstr. 38*:1921 B (1978).
128. L. E. Kun, R. O. Oelofsen, and E. J. J. van Vuuren, "Hopper clarification of gold pregnant solution at Vaal Reefs South." *J. S. Afr. Inst. Min. Metall. 79*:201-6 (1979).
129. C. A. Lee, "Increasing settling tank efficiency." *Plant Engineering* 126-7, April 19, (1973).
130. G. E. Hubbell, "Hydraulic characteristics of various circular settling tanks." *Am. Water Works, Assn. J. 30*:335-53 (1938).
131. Anon. "Sludge settling tank more than halves retention time." *Chem. Processing 13*(5):4-7 (May, 1967).
132. C. G. Bruckmann, *Economic Aspects of Thickener Operations.* Presented at Symposium S 19 on Thickener Design and Operation organized by the NCRL-CSIR, Pretoria (28 June 1966).
133. G. H. Matheson and J. M. W. MacKenzie, "Flocculation and thickening coal-washery refuse pulps." *Coal Age 67*:94-100 (December, 1962).
134. D. G. Hall, "The role of bridging in colloid flocculation." *Colloid and Polymer Sci. 252*: 241-3 (1974).
135. W. E. Walles, "Role of flocculant molecular weight in the coagulation of suspensions." *Jnl. Colloid Interface Sci. 27*:797-803 (1968).
136. Committee Report, "State of the Art of Coagulation. Mechanism and Stoichiometry." *Am. Water Works Ass. J. 63*:99-108 (1971).
137. S. L. Daniels, "A survey of flocculating agents—process descriptions and design considerations." *AIChE Symposium Series* No. 136, *70*:266-81 (1973).
138. F. M. Tiller, J. Wilensky, and P. J. Farrell, "Pretreatment of slurries." *Chem. Eng. 81*(9):123-6 (April 29, 1974).
139. R. J. Akers, *Flocculation.* Publ. by I. Chem. E. Services for Inst. Chem. Eng., 15 Belgrave Square, London, England SWIX 8PT (1975).

140. D. K. W. Smith and J. A. Kitchener, "The strength of aggregates formed in flocculation." *Chem. Eng. Sci. 33*:1631-6 (1978).
141. R. N. Kovalcik, "Single waste-treatment vessel both flocculates and clarifies." *Chem. Eng. 85*(14):117-20 (June 19, 1978).
142. D. A. Dahlstrom and C. F. Cornell, "Thickening and clarification." *Chem. Eng.* Deskbook Issue *78*(4):63-9 (February 15, 1971).
143. R. A. Conway and V. H. Edwards, "How to design sedimentation systems from laboratory data." *Chem. Eng. 68*(19):167-70 (September 18, 1961).
144. H. C. Bramer and R. D. Hoak, "Design criteria for sedimentation basins." *I and EC Process Des. Develop. 1*:185-9 (1962).
145. E. B. Fitch, "The significance of detention in sedimentation." *Sewage and Industrial Wastes 29*:1123-33 (1957).
146. W. H. Mitchell, "The preparation and treatment of mine water prior to pumping and notes on maintenance of pumping plant." *Instn. Certificated Engrs.* (S. Afr.) J. *26*:86-99 (1953).
147. R. T. Hukki, G. Diehl, and P. Vanninen, "Principles of construction and operation of the channel and syphon thickener." *7th Int. Mineral Processing Congress*, vol. 1, New York: Gordon and Breach, pp. 115-23 (1965).
148. R. F. Probstein and R. E. Hicks, "Lamella settlers: A new operating mode for high performance." *Ind. Water Eng. 15*:6-8 (1978).
149. R. L. Cook, "Compact lamella thickeners in coal preparation plants." *Pap. Symp. Coal Prep.* (Washington, DC) 2nd, at NCA/BCR (Natl Coal Assoc./Bitum. Coal Res.) *Coal Conf. and Expo.* 3, Louisville, Ky. (October 19-21, 1976). (Available from NCA.)
150. R. L. Cook and J. J. Childress, "Performance of lamella thickeners in coal preparation plants." *Min. Eng.* (New York) *30*:566-71 (1978).
151. L. C. Meitzler and G. H. Weyermuller, "Compactness of thickener permits installation in limited space." *Chem. Processing 37*(5):18 (1974).
152. F. C. McMichael, "Sedimentation in inclined tubes and its application for the design of high rate sedimentation devices." *J. Hydraulic Research 10*:59-75 (1972).
153. D. Davis, S. Rogel, and K. Robe, "Increases clarifier capacity 85% at 1/20th cost of new unit." *Chem. Process* (U.S.) *36*:10 (January, 1973).
154. R. M. Willis, "Tubular settlers–a technical review." *J. Am. Water Works Assoc. 70*:331-5 (1978).
155. Axel Johnson Institute, "A new separation technique." *Trade Brochure* Box 13, Nynäshamn, Sweden.
156. A. S. Panov and Z. V. Ermolenko, "Increasing the productivity of multichamber red mud thickeners" (in Russian). *Tsvet met* no. 3:37-9 (1977).
157. R. C. Raynor and E. F. Porter, "Thickeners and clarifiers." *Chem. Eng. 73*(13):198-202 (June 20, 1966).
158. J. W. de Villiers, "An investigation into the design of underground settlers." *J. S. Afr. Inst. Min. Metall. 61*:501-21 (1961).
159. A. W. Bond, "Upflow solids contact basin." *Jnl., San. Eng. Div., Proc. ASCE* (SA6) 73-99 (1961).
160. Private communication from T. M. Stielau, Delkor Technik, Randburg. S.A.
161. Annon, "Revolutionary thickener design tackles heavy flow of zinc mine tailings." *Engineering and Mining J. 179*:78-9 (April, 1978).
162. G. R. Mace and R. Laks, "Developments in gravity sedimentation." *Chem. Eng. Prog. 74*(7): 77-83 (1978).
163. J. L. Chandler, "Design of deep thickeners." *Inst. Chem. Eng. Symp. Ser.* No. 45, Des. Dev. Relat. to Saf., Costs and Solids Syst., Sel. Pap. from the Des. Congr. 1976 Symp., Univ. of Aston, Birmingham, Engl., September 9-10, 1976. Publ. by Inst. Chem. Eng., Rugby, Engl., 1977 p G.1.1-G.1.11. [*Engng Index 16*:item 006214 (1978)].
164. J. Abbott, C. C. Dell, B. Denison, D. Knott, and N. W. Hill, Coal preparation plant effluent disposal by means of deep cone thickeners. Int. Coal Prep. Cong., 6th, Paris, March, 1973, Paper 20 E, 18 pp. [*Mine & Quarry 2*:37-50 (8 pp) (October, 1973).]
165. N. P. Chironis, "From UK: deep-cone thickeners." *Coal Age 82*:126-9 (October, 1977).
166. Anon. "Thickener makes slurries easier to bulk handle." *Bulk 3*:107 and 109 (January/February, 1977).
167. L. M. Maiskaya and I. N. Evstigneev, "The thickener with cake consolidation as an object for automatic control." *Coke Chem. USSR 1*:56-60 (1976).
168. J. R. Bratby, "Aspects of sludge thickening by dissolved-air flotation." *Water Pollut. Control (London) 77*:421-32 (1978).
169. M. T. Turner, "Use of dissolved air flotation for the thickening of waste activated sludge." *Effluent Water Treat. J. 15*:243-51 (7 p.) (May, 1975).
170. E. R. Ramirez, "Dewatering skimmings and sludges with a lectro-thic unit." Water and Wastewater Equip. Manuf. Assoc. Pollut. Control Conf., 3rd Annu: Ind. Solutions '75: *Air-Water-Noise Solid Waste, Proc.*, pp. 467-78. Publ. by ANN, April 1-4 1975.
171. S. Takeda and I. Matsuoka, "Fundamental study

on dissolved air pressure flotation of artificial fine hematite particles." (In Japanese) *J. Min. Metall. Inst. Jpn. 94*:335–40 (1978).

172. V. Gulas and R. Lindsey, "Factors affecting the design of dissolved air flotation systems." *J. Water Pollut. Control Fed. 50*:1835–40 (1978).
173. H. E. Cross, "A new approach to the design and operation of thickeners." *J. S. Afr. Inst. Min. Metall. 63*:271–89 (1963).
174. E. B. Fitch and W. A. Lutz, "Feedwells for density stabilization." *J. Water Pollut. Control Fed. 32*:147–56 (1960).
175. Anon. "Improved thickener rake." *S. Afr. Min. Eng. J.*, p. 415 (August 23, 1968).
176. F. R. Weber, "How to select the right thickener." *Coal Min. Process. 14*(5):98–100, 104, 116 (1977).
177. V. S. Dillon, "Special features of the Kinross Mines, Ltd., reduction plant." *Proc. 9th Commonwealth Mining Metall. Congress, London, Mineral Process. Extr. Metall. 3*:485–508 (1969).
178. R. P. Plasket and D. A. Ireland, "Ancillary smelter operations and sulphuric acid manufacture at Impala Platinum Limited." *Jnl. S. Afr. Inst. Min. Metall.* 1–10 (August, 1976).
179. M. V. Maljian and J. A. Howell, "Dynamic response of a continuous thickener to overloading and underloading." *Trans. I. Chem. E. 56*: 56–61 (1978).
180. H. R. Moyer, III and R. F. Brindisi, "Start-up of the world's largest countercurrent decantation thickeners." *Min. Congr. J. 62*:26–31 (1976).
181. B. G. Liptak, "Instruments to measure and control slurries and viscous materials." *Chem. Eng. 74*(4):151–8 (February 13, 1967).
182. H. E. Cross, "Automation of thickeners used for dewatering milled gold ore pulps." *Coal 14*(7): 38–44 (September, 1966).
183. D. D. Marsden, "The effect of *p*H value, temperature and density on the kinematic viscosity of some South African gold mine slurries." *J. S. Afr. Inst. Min. Metall.* 391–8 (1962).
184. Ir. R. Rutgers, "Relative viscosity and concentration." *Rheologica Acta 2*:305–48 (1962).
185. Sherritt Gordon Mines Limited, U.K. Patent 1 359 019 (September 16, 1971).
186. D. C. McLean, "Chemical aspects of thickening and clarification." *Soc. Min. Eng.* of *AIME* Preprint No. 68-B-62. New York (February 25–29, 1968).
187. H. A. Hamza, "Least-cost flocculation of clay minerals by polyelectrolytes." *Trans. Instn. Min. Metall.* (Sect C: Mineral Process. Extr. Metall.), *87*:C212–14 (1978).
188. D. B. Neal, "Using flocculant economically." *Mine Quarry 6*(7/8):63–4 (1977).
189. J. Reuter, "Possibilities of automating the operation of clarifying thickeners in processing plants by using organic-synthetic flocculants" (in German) *Aufbereit.-Tech.*, no. 4:187–93 (1974).
190. W. Johnson, "Atlas Minerals saves money with automatic thickener control." *Eng. Min. J. 167*:92–3 (October, 1966).
191. M. C. Kavanaugh, C. H. Tate, A. Trussell, R. R. Trussell, and G. P. Treweek, "Use of particle-size measurements for water treatment plant process selection and control." *Am. Chem. Soc. Div. Environ. Chem. Prepr. 18*(1); Prepr. for 175th ACS National Meeting, Anaheim, California, March 12–17, 1978. Publ. by ACS, New York, N.Y. pp. 246–52, (1978).
192. Anon. "Screens, cyclones, and thickeners." *Eng. Min Jnl. 177*:232–234 (June, 1976).
193. E. S. Hsia and F. W. Reinmiller, "How to design and construct earth bottom thickeners." *Min. Eng. (N.Y.) 29*:36–39 (August, 1977).
194. Anon. "Horizontal sliding for thickeners." Saldanha-Sishen Progress Report. Supplement to *Construction in S.A.* and Coal, Gold and Base Minerals of Southern Africa *22*:25–27 (August, 1974).
195. Anon. "Kidd Creek uses caisson tailings thickener of advanced design." *Eng. and Min. Jnl. 177*:32 (September, 1976).
196. M. V. Lowry, "Concentrate storage in a platform-lift thickener." *Soc. Min. Engrs 18*:57–59 (1966).
197. Anon. "Underflow pumping from thickeners." *Chem. Process. Eng. 52*:3 (January, 1971).
198. Stanbridge (ed.) *Unit Processes*: *Primary Sedimentation.* Manuals of British Practice in Water Pollution Control. Inst. WPC, Maidstone, Kent (1973).
199. F. Concha and E. R. Almendra, "Settling velocities of particulate systems. 1. Settling velocities of individual spherical particles." *Int. J. Mineral Process. 5*:349–67 (1979).
200. K. J. Scott, "Experimental study of continuous thickening of a flocculated silica slurry." *Ind. Eng. Chem. Fundam. 7*:582–95 (1968).
201. A. M. Gaudin, M. C. Fuerstenau, and S. R. Mitchell, "Effect of pulp depth and initial pulp density in batch thickening." *AIME Trans. 214*: 613–616 (1959).
202. L.-G. Ecklund, "Influence of feed conditions on continuous thickening." *Chem. Eng. Sci. 34*: 1063–1066 (1979).
203. J. H. Masliyah, "Hindered settling in a multi-species particle system." *Chem. Eng. Sci. 34*: 1166–1168 (1979).

15
Filtration of Solids from Liquid Streams

Derek B. Purchas
Moreton-in-Marsh, Gloucestershire, England

15.1 INTRODUCTION

Filtration is the mechanism whereby solid particles suspended in a liquid are separated from the liquid as a consequence of the passage of that liquid through a porous medium, on or in which the solid particles are retained.

This general definition covers an extremely wide variety of industrial equipment, the performance of which is dependent upon many different parameters. The latter include not only the driving force (gravity, vacuum, pressure or centrifugal), but also the nature of the filter medium (e.g., cloth, paper, sintered metal, sand, etc.,) and also the type of pretreatment to which the solid/liquid suspension may have been subjected before filtration (e.g., temperature changes, adjustment of *p*H, addition of flocculants, etc.).

The characteristics of the solid/liquid suspension to be processed (e.g., concentration and settling characteristics of the solids, volatility and flamability of the liquid, toxicity, etc.) influence profoundly the suitability of a given type of filter for a specific duty. Other important criteria are the mode of operation (intermittent or continuous), the ability to wash and/or dry the solids before discharge, and the degree of manual or automatic operation.

The practical choice of filter for a specific application is generally based on a combination of the suitability of the basic type, the determination of the necessary size, and an assessment of the likely capital and operating costs. The selection of one or more suitable types is comparatively straight-forward, although it may be complicated by the immense choice available. Estimating the size is often much more difficult, since it generally requires knowledge of filtration rate data, prediction of which from known fundamental properties (e.g., density, viscosity, particle size) is rarely if ever possible; in the absence of data from a comparable operating plant, laboratory bench tests are used, even though the subsequent scale-up may be relatively inaccurate. As a consequence, accurate cost projections present even more difficulty.

An understanding of the fundamental mechanisms of filtration can be of great value, both qualitatively and especially quantitatively. The basic differences between surface and depth filtration are readily grasped, even by those with no mathematical aptitude, thereby providing an understanding of the practical differences between using on the one hand a precoat filter and on the other a sand filter for the same clarification duty.

The quantitative use of filtration theory, however, is potentially a much more complex matter as the 444 pages of the book *The Scientific Basis of Filtration*[1] amply illustrate; fortunately, great practical benefits can be gained by use of some of the simpler less rigorous mathematical relationships, such as the cake filtration equations developed by Ruth.

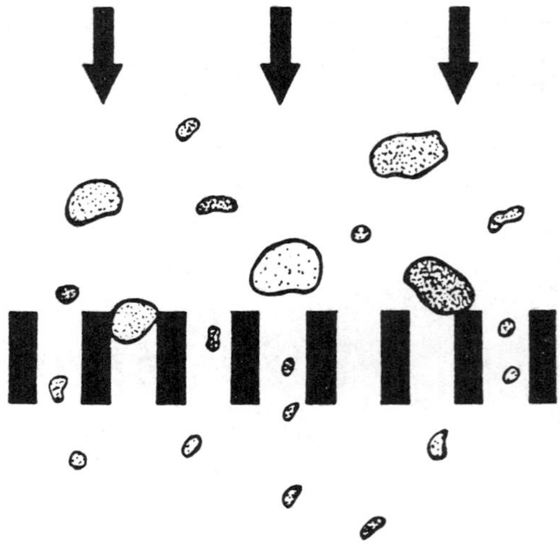

Figure 15.1. Surface straining mechanisms.

15.2. FILTRATION MECHANISMS

Whereas in practice it is common for two or more filtration mechanisms to occur either simultaneously or in very rapid succession, it is convenient to distinguish between four basic types, which are surface straining, depth straining, depth filtration and cake filtration.

Surface Straining involves the deposition of solid particles on the outer (up-stream) surface of the medium. As shown in Figure 15.1, it will occur provided the particles are larger than the pores of the medium; smaller particles pass through the medium and remain in the filtrate. This is the operating mode of a plain weave mesh made from a monofilament of plastic or metal wire.

Depth Straining resembles surface straining but recognizes that with certain media, the cross-sectional area of a pore may be less within the depth of the medium than at the surface. With such a pore, a particle may be small enough to escape being trapped on the surface but is then caught by the same mechanism of straining at some point within the depth of the medium, as illustrated in Figure 15.2. This is the operating mode of felts and of cellulose ester membranes.

Depth Filtration is the mechanism whereby, even though the particles are very much smaller than the pores of the medium, they are nonetheless trapped in the depth of the medium as the liquid flows through it. More strictly, depth filtration is a combination of mechanisms rather than a single mechanism, the overall

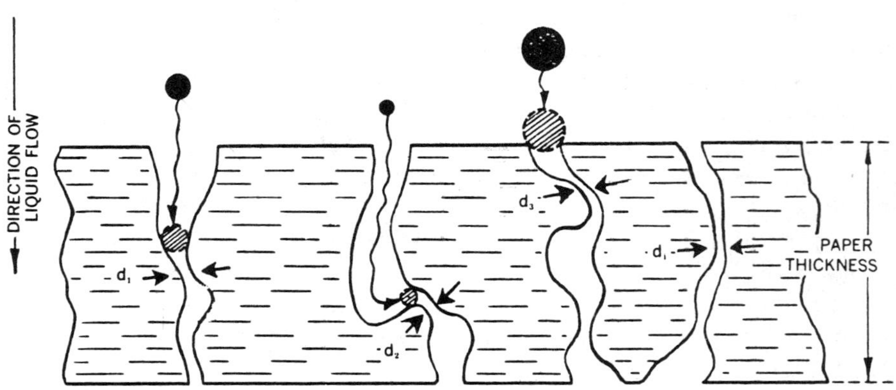

Figure 15.2. Depth straining mechanism.

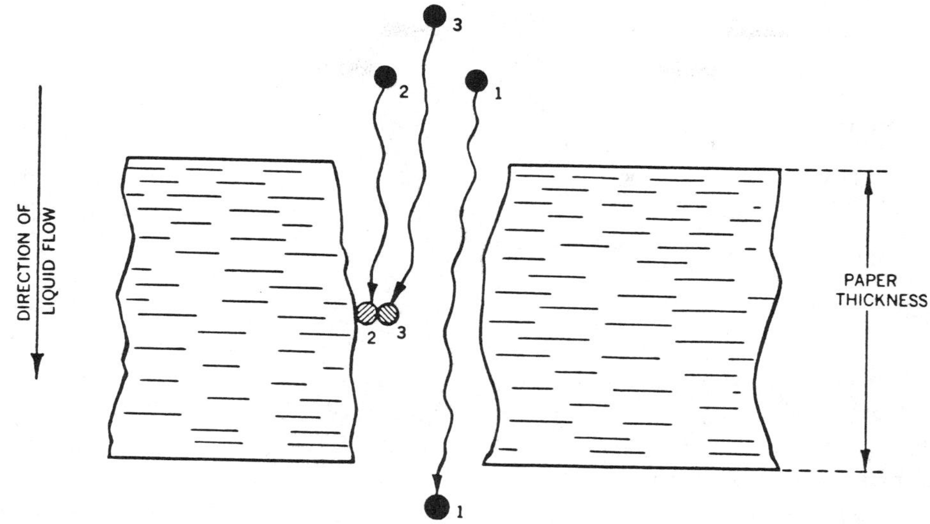

Figure 15.3. Depth filtration mechanism.

result being that the solid particles adhere either to the internal surfaces of the medium or to each other, as in Figure 15.3. This is the operating mode of filter sheets (e.g., cellulose/asbestos) and also of deep bed filters (e.g., sand).

Cake Filtration describes the accumulation of a thick layer of particles on the surface of the medium, as illustrated in Figure 15.4. If the solid particles are larger than the pores, then cake filtration is preceded by a period of surface and/or depth straining. Frequently, however, the particles are substantially smaller than the pores; cake filtration is then preceded by a bridging process, as in Figure 15.4. Cake filtration is the general mode of operation whenever solids are being recovered (i.e., as opposed to clarification).

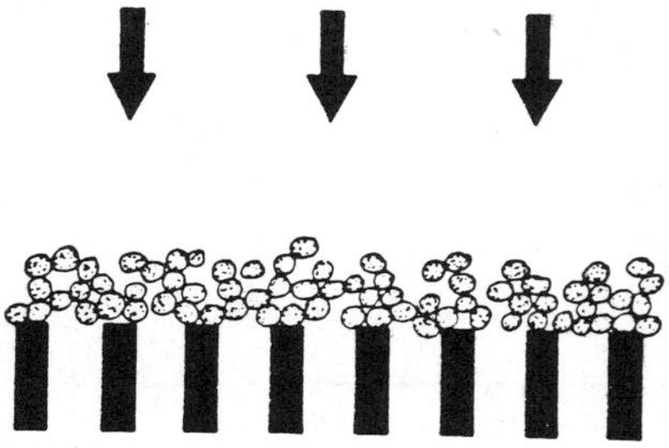

Figure 15.4. Cake filtration mechanism.

Bridging, as a preliminary to cake filtration, cannot occur if the concentration of solids in the suspension is too low, or if the particles are too small in comparison with the size of the pores in the medium. Whereas the relationship between these criteria has been only partly explored, a valuable contribution has been made by Rushton,[2] who investigated various different cloth weaves and slurries of five different organic solids; results from one series are reproduced in Figure 15.5.

There is some evidence that the critical concentration is not that of the mean particle diameter of the solids, but the concentration of the largest particles; thus, if the particles have a narrow size distribution, a higher concentra-

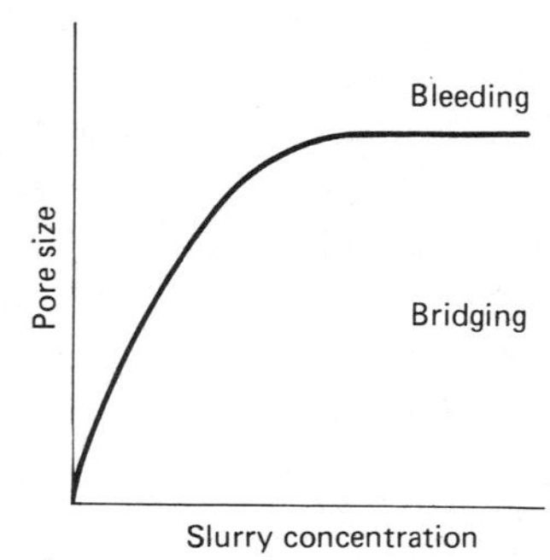

Figure 15.5. Pore bridging as a function of slurry concentration.

Table 15.1. Filtration Mechanisms and Practical Applications.

MECHANISM	APPLICATION	
	SOLIDS RECOVERY	CLARIFICATION
1. Surface straining	(X)	X
2. Depth straining	(X)	X
3. Depth filtration		X
4. Cake filtration	X	(X)

tion is required for bridging to occur. Rushton reports that cloths generally bridge at concentrations in the range 0.01-2%.

The relevance of these four filtration mechanisms to applications involving solids recovery on the one hand and clarification on the other hand is summarized in Table 15.1; the symbol X denotes a major mechanism, while (X) denotes a minor mechanism.

15.3. FILTRATION THEORY

Space precludes doing more than touching very cursorily upon the great body of theoretical knowledge and speculation which continues to be evolved describing in mathematical terms the fundamental mechanism of filtration. Inevitably but unfortunately, as fundamental knowledge has advanced, it has grown increasingly complex, thereby becoming progressively more the domain of the academic specialist than of the industrial practitioner, even if the latter is himself academically well qualified. Nonetheless, there are a few of the simpler basic relationships which merit restating since they are valuable, even if only approximate, practical tools.

Common to all filtration mechanisms, is the flow of liquid through porous medium; as stated in the opening paragraph of this chapter, the distinctive feature of filtration is that solid particles suspended in the liquid are retained on the surface or in the depth of the medium. The process needs a driving force (e.g., pressure) to overcome the resistance both of the medium itself and of the solid particles after they have been deposited. With a clean liquid containing no particles, the resistance of the system does not change with time, so that with a constant driving force the flow rate remains constant; by contrast, the presence of particles in the liquid results in filtration, in the progressive increase in resistance of the system, and in a decline in flow rate at constant driving force. These two cases are summarized graphically in Figure 15.6.

The flow of clean liquid through a porous medium can be described by the modified Darcy equation:

$$\frac{Q}{A} = \frac{P\phi}{L\mu} = \frac{P}{R_m \mu} \tag{1}$$

where P is the pressure drop, L is the thickness, Q is the volumetric rate of flow, μ is the viscosity of the fluid, A is the area of the filter, R_m is the resistance of the medium, and ϕ is its permeability.

This relationship assumes that the flow regime is streamlined and not turbulent. While

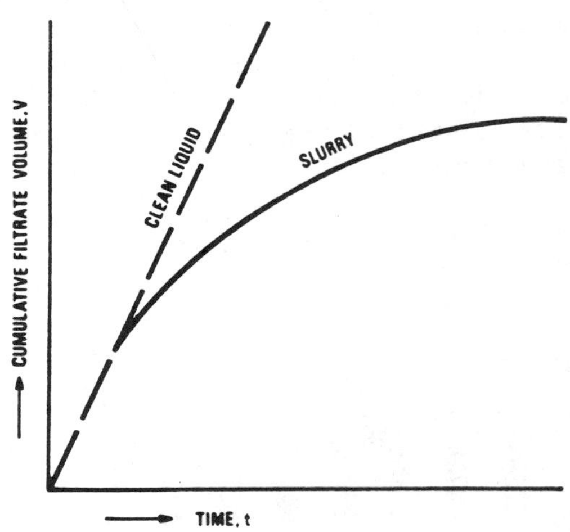

Figure 15.6. Cumulative filtrate volume versus time for a, clean liquid; b, slurry.

this assumption is almost always true so far as the filter medium is concerned, it may in practice be obscured by turbulent flow in the inlet and outlet ducts and piping which form part of the filter itself. As a consequence, whereas doubling the pressure drop across the medium alone would double the flow rate, the practical effect of doubling the pressure drop across a filter may be substantially less advantageous.

Where filtration is occurring, Equation (1) must be modified to include the resistance of the deposited solids. This resistance increases progressively with time at a rate which depends upon the type of filtration mechanism, typically being much less with depth filtration than with cake filtration. Partly because of this, and also because of its dominance in the processing industry, a consideration of filtration theory is frequently limited to the cake mechanism; a further factor is no doubt the ease with which the elementary cake theory can be understood and put to practical use.

The general cake filtration equation may be written in the form:

$$\frac{Q}{A} = \frac{1}{A} \cdot \frac{dV}{dt} = \frac{P}{\mu\left(\frac{\alpha w V}{A} + R_m\right)} \quad (2)$$

where V is the cumulative volume of filtrate collected in time t, α is the specific resistance to filtration per unit weight of the dry solids deposited as a cake, and w is the weight of dry cake per unit volume of filtrate. This equation states the instantaneous conditions at any moment during filtration and can be integrated over a finite time interval for either constant pressure or constant rate filtration.

For Constant Pressure Filtration, the only variables are flow rate and total filtrate. The above equation may therefore be simplified to give:

$$\frac{1}{Q} = \frac{dt}{dV} = V\left(\frac{\mu\alpha w}{A^2 P}\right) + \left(\frac{\mu R_m}{AP}\right) \quad (3)$$

which may be written in the abbreviated form:

$$\frac{1}{Q} = \frac{dt}{dV} = aV + b \quad (4)$$

in which a and b are constants. Simple manipulation of this equation makes it possible to calculate the total volume of filtrate when the rate of filtration has fallen to a final level, Q_f, after a time t.

Rearrangement of Equation (4) permits V to be evaluated from Equation (5), while integration of Equation (4) gives t from Equation (6):

$$V = \frac{1 - bQ_f}{aQ_f} \quad (5)$$

$$t = (a/2)V^2 + bV \quad (6)$$

For Constant Rate Filtration, the variables are pressure and the total volume filtered, so that Equation (3) may conveniently be arranged in the form:

$$P = \frac{\mu\alpha w Q_c}{A^2} \cdot V + \frac{\mu R Q_c}{A} \quad (7)$$

This gives directly the final pressure required to permit collection of a total filtrate volume V at constant rate, Q_c. The time required, t, is given by the separate simple Equation (8):

$$t = \frac{V}{Q_c} \quad (8)$$

Continuous Filters, such as rotary vacuum drum filters, are in reality a specialized case of batch constant pressure filtration, utilizing a very short repetitive cycle. The throughput per cycle is therefore described by Equation (6); this may be adapted to give the more useful throughput per unit time, since $t = F \cdot t_c$ where t_c is the time of a complete cycle and F is the fraction of the cycle devoted to cake formation.

With continuous filtration (and often also with batch filtration), it is customary to assume that the resistance of the filter medium is negligible by comparison with that of the filter cake. The second term in the right hand side of Equations (3), (4), and (6) therefore

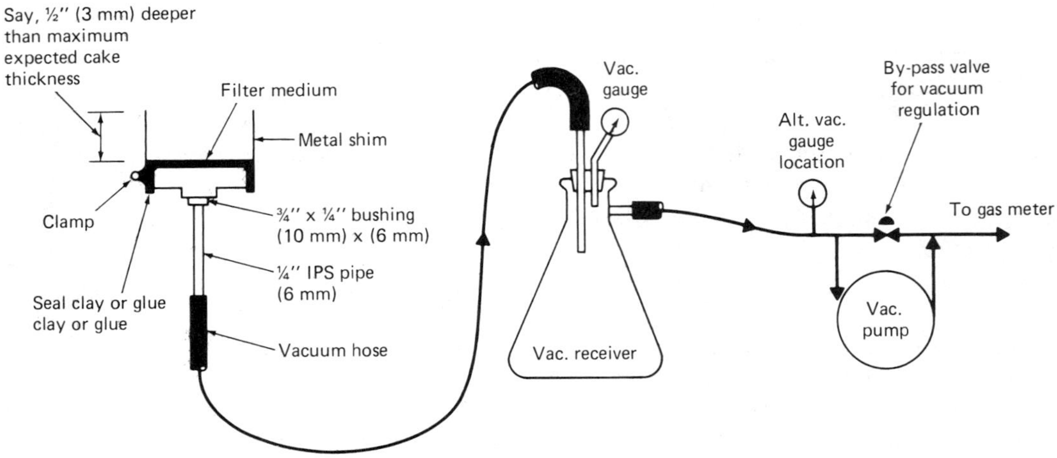

Figure 15.7. Typical filter leaf test set-up.

disappears. Hence, the throughput of a continuous filter may be expressed as the rate of flow of filtrate per unit area, Y, where

$$Y = \frac{V}{At_c} = \left[\frac{2PF}{\mu\alpha w t_c}\right]^{0.5} \tag{9}$$

This may be converted to give the output of dry solids per unit time, Z by multiplying by w:

$$Z = \frac{Vw}{At_c} = \left[\frac{2wPF}{\mu\alpha t_c}\right]^{0.5} \tag{10}$$

15.4. SIZING

Filters are generally sized on the basis of data obtained from tests on small or bench scale filters. Valuable initial information can frequently be obtained using a laboratory suction-operated Buchner funnel, although it is preferable to employ a special test leaf (Figure 15.7), because of the very poor drainage characteristics of the normal Buchner funnel.

The simple filtration theories summarized in Section 3 can be very beneficial in interpreting the resultant data. On the basis of Equation (3), a plot of the cumulative filtrate volume, V, against the inverse of the instantaneous filtration rate, dt/dV, provides a convenient method to evaluate both the specific filtration resistance of the cake, α, and also the resistance of the filter medium, R_m. As shown in Figure 15.8, the slope of the line gives the constant a in Equation (4), while the intercept gives b; from these α and R_m may then be calculated.

The filter area thus estimated is generally multiplied by a scale-up factor of at least 1.25, to allow for experimental error. The procedure is open to considerable error, because of the various assumptions made in the development of the equations summarized in Section 3. Quite apart from the explicit assumption in relation to Equations (9) and (10) that R_m is

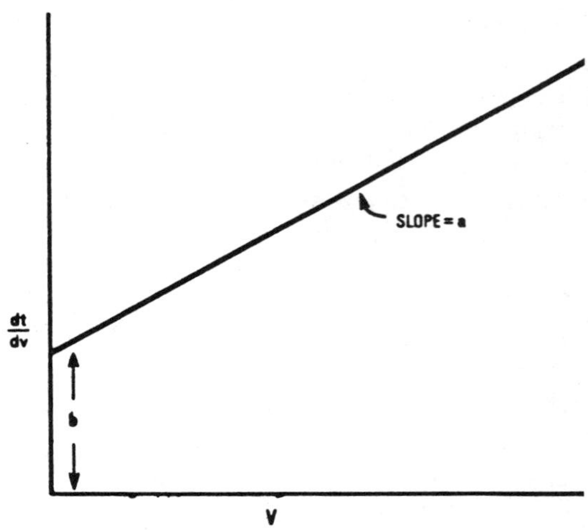

Figure 15.8. Constant pressure filtration.

negligible, there is also the implicit assumption that the cake resistance, α, has a constant value for a given material. In reality, α is known to vary with parameters such as pressure, slurry concentration and rate of filtration, so that it is not constant even through the depth of a given cake.

Other implicit assumptions are that the quantity of liquid retained within the cake is negligible, and that liquid enters and leaves the cake at exactly the same rate. To overcome the errors implicit in these and other assumptions, fuller and more complex mathematical models have been developed.[1]

Even greater error frequently arises from major differences between the samples of slurry tested and the actual slurry encountered by the operating equipment. The difficulty of obtaining a representative sample of slurry is a notorious difficulty but is frequently overlooked; major errors may also occur even where representative slurry is available, due to changes in the properties of the sample during storage and subsequent handling before filtration tests are carried out.

An extensive and detailed description of the procedures for the performance of filtration tests, and for the interpretation of the resultant data, is available in Reference 3.

Table 15.2. Generalized Summary of Filter Media on Basis of Rigidity.

MAIN TYPE	SUBDIVISIONS
1. Solid fabrications	(a) Bar screens (b) Wire-wound tubes (c) Stacks of rings
2. Rigid porous media	(a) Ceramics and stoneware (b) Plastics (c) Sintered metals
3. Cartridges	(a) Sheet fabrications (b) Yarn wound (c) Bonded beds
4. Metal sheets	(a) Perforated (b) Woven wire
5. Plastic sheets	(a) Woven monofilaments (b) Porous sheets (c) Membranes
6. Woven fabrics	(a) Monofilaments (b) Multifilaments (c) Staple fiber yarns
7. Nonwoven media	(a) Filter sheets (b) Felts and needle felts (c) Paper media (d) Bonded media
8. Loose media	(a) Fibers (b) Powders

15.5. FILTER MEDIA

Description of Available Media

Table 15.2 presents a convenient and highly condensed summary of the immense variety of filter media available for liquid filtration. In this table, media have been arranged in the order of decreasing rigidity, primarily as an arbitrary device to impose a degree of order on this confusing field; in practice, rigidity is an important criterion controlling the suitability of a given medium for use with a specific type of filter.

Each type of filter medium is discussed below, following the sequence of Table 15.2. Space limitations preclude an extensive discussion; for fuller information the reader is referred to References 4 and 5.

Solid Fabrications. Bar screen comprise assemblies of parallel 1-3 mm wide bars or wires, the cross section of which is typically wedge shaped; aperture widths vary upwards from about 50 microns (corresponding to a free flow area of about 4%). By contrast, Figure 15.9 illustrates a typical filter element constructed by winding a wire of circular cross section around a cylindrical former, the latter having machined into its surface a fine thread to provide the wire with a precise seating; the pitch of the thread controls the final aperture which is generally of the order of 200 micrometers. Other tubular constructions include an adaptation of the welded wedge wire screen with apertures as small as 20 micrometers, and also stacks of specially shaped discs or rings, such as in Figure 15.10.

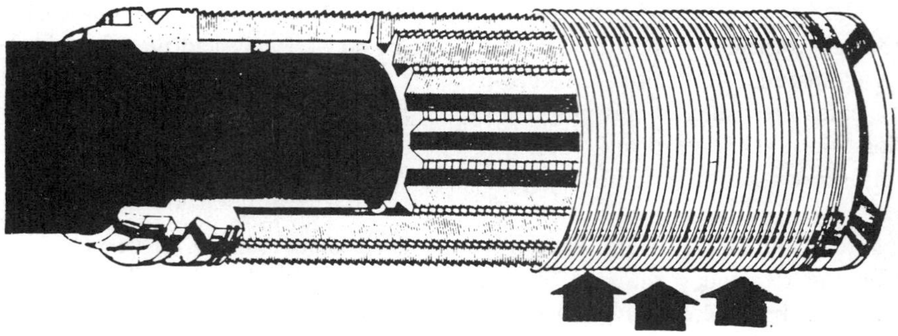

Figure 15.9. Wire-wound filter element.

Porous Ceramics are valuable filter media because of their high chemical and thermal stability, but their use is limited because they are generally available only in the form of thick tiles or tubes; grades are available with average pore sizes ranging from approximately 500 down to 10 micrometers, the corresponding porosities being typically between 45 and 35%.

Porous Plastics, while more limited than porous ceramics in their chemical and thermal properties, are more adaptable mechanically and are available in the form of tubes or sheets, typically with a thickness of the range 0.75–5 mm; pore sizes tend to be relatively coarse, generally extending upward from about 40 micrometers.

Sintered Metal Media are available in four main types, depending upon the form of the original metal: (a) powders, (b) fibers, (c) woven meshes, (d) composites of several layers, such as two or more grades of woven mesh, or powder plus a mesh. This inevitably leads to major structural differences, which are reflected in

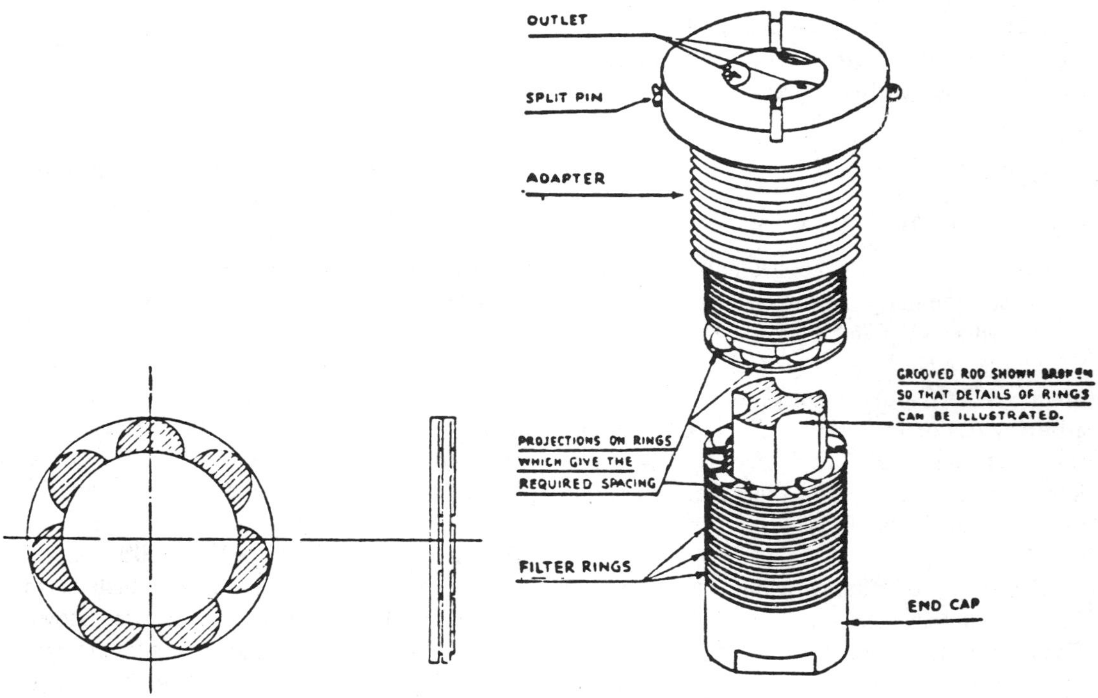

Figure 15.10. Element of Meta-filter.

properties such as porosity and pore size distribution, and which also influence the sizes and shapes of components which may be produced. Stainless steel is the most common metal used but many others are also available in sintered form. A great advantage of sintered metals is that in general they may be cut, welded, shaped and generally fabricated by standard metal working techniques, while their mechanical properties are similar or identical to those of the virgin metal. A wide range of grades is available, with pore sizes from about 0.5 to 200 micrometers; the porosities of sintered powder media are generally in the range 20%–50%, this extending up to 90% for sintered fiber media.

Cartridges. Many cartridges are constructed from one or more layers of sheet type media (e.g., paper, membranes), the filtration characteristics of which potentially determine the ultimate performance of the cartridge; whether this potential is reached in reality depends upon the care and skill with which the cartridge is constructed, as well as on the quality of the housing in which it is inserted for use, deficiences in either permitting solid particles to leak into the filtrate.

Important categories of cartridge are also often made either by winding a yarn around a perforated tube to form a deep matrix as in Figure 15.11, or by bonding onto a similar tubular former a thick layer of fibers impregnated with suitable resin. Each constructional technique can produce cartridges of various grades, typically with nominal ratings from 10 to about 100 microns; these numbers should not be taken literally, as Section 4 makes clear.

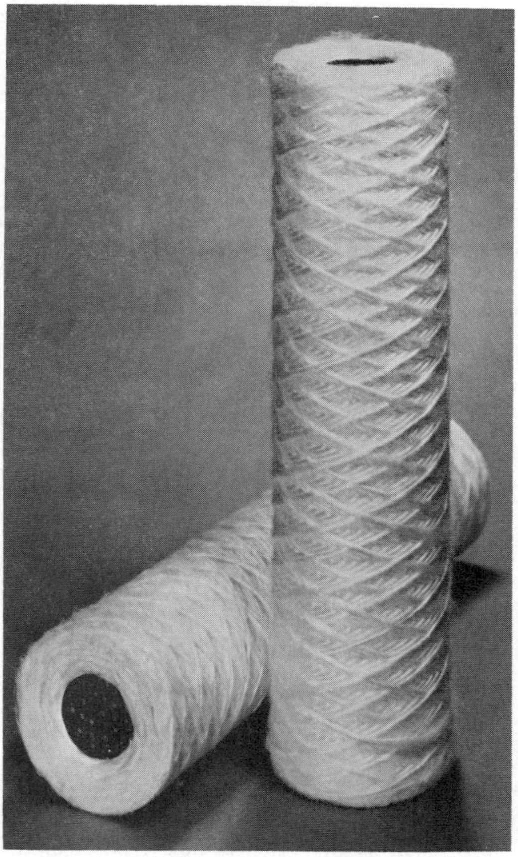

Figure 15.11. Yarn-wound cartridge.

Perforated Metal Sheets made by traditional metal working processes are generally too coarse to be of significant interest as filter media, excepting for use in roughing strainers; for example, the finest grade in the range of one major manufacturer has holes of 660 micrometers. A major advance has resulted from modern electro-forming and photo-etching techniques, whereby it is possible to produce regular and uniform holes as small as 5 micrometers (the tolerance being ±2 micrometers on holes up to 500 micrometers); unfortunately the porosities are generally very low, such as less than 2% with 5 micron holes, 3.5% with 40 micron holes, increasing to 22.5% with 100 micron holes.

Woven Metals have been widely used for many years and are available in a large variety of weaves made from many different metals. In making a wire cloth, a balance has to be struck between using a very fine wire to achieve a minimum mesh size, and a larger wire, which is required if the cloth is to have acceptable mechanical properties, such as strength and wear resistance.

The popular answer is one of the so-called dutch or hollander weaves, in which a greater number of fine wires is used in the weft compared with fewer heavier wires in the warp. A

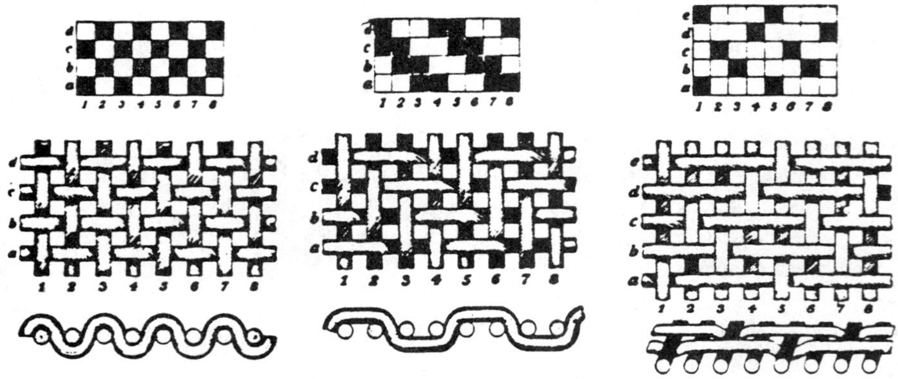

(i) PLAIN (ii) TWILL (iii) SATIN

Figure 15.12. Basic weaves.

simple over-and-under weave then gives plain dutch, while twilled dutch is a denser cloth achieved by the weft wires crossing over two warp wires at a time, in a staggered arrangement which gives the typical diagonal pattern of twill; these two common weaves are illustrated schematically in Figure 15.12. Numerous variations on this basic theme are used so as to control certain characteristics of the finished cloth, such as the fineness of the pores or the smoothness of one or both surfaces. Grades are available in pore sizes down to about 5 micrometers.

Woven Monofilament Plastics. Whereas the heavier grades of woven monofilament plastic media are as rigid as the woven wire products discussed in the preceding section, the finest grades are more like woven textiles in their flexibility and fragility. Plain weave is virtually standard throughout this very wide range, extending from 3-4 mm pores with media woven from 1 mm diameter filaments, down to pores and filaments of only a few micrometers in size. The widest choice is generally available in polyamide (nylon), with polyester grades less numerous, and both polypropylene and teflon available in only a limited number of grades. Because of the different extrusion and weaving characteristics of the various plastic materials and of metal, corresponding grades often differ in terms of both aperture size and percentage open area; in general, plastic monofilament meshes have more open area than metal meshes of equal aperture.

Membranes were originally developed for laboratory use in the middle of the nineteenth century and were manufactured from the esters of cellulose such as the nitrate. Until comparatively recently, the fragility of such membranes inhibited their use industrially. During the last decade this situation has been revolutionized as a result of the development of a new generation of far more robust membranes. This has been achieved in four different ways, either alone or in combination: (1) the use of polymers such as PVC and PTFE in place of cellulose esters; (2) the formation of the membrane around a woven monofilament web (typically of nylon); (3) laminating the membrane to a substrate such as paper or a nonwoven material, either during formation or as a subsequent separate operation; (4) the development of new manufacturing techniques, such as irradiation and leaching (for the manufacture of "Nuclepore") in place of the traditional solvent evaporation system.

Membranes are traditionally available as discs of various diameters, the largest generally being 293 mm. Rectangular sheets of approximately 250 × 200 mm are also commonly available. A few of the stronger recently developed membranes can also be bought in the form of rolls, one example being a 0.5 m wide roll containing some 400 m of membrane.

A very wide range of different grades of membranes is available, with pore sizes extending down to a few Angstroms in diameter. The finest membranes are used for reverse osmosis, whereby soluble ionic materials such as sodium chloride can be removed from solution; somewhat coarser membranes permit the separation of organic materials of very high molecular weight by ultra filtration. At an ill-defined point, but in the region of a pore diameter of approximately 0.05 micrometers, ultra filtration yields place to particulate filtration with a range of membranes extending to a coarsest grade with a pore size of about 12 micrometers. The thickness, porosity and hence resistance to flow of membranes varies considerably according to their nature and to the type of manufacturing process used; Table 15.3 summarizes the properties of one well-known commercial range together with an example of their typical applications.

Woven Fabrics. This is by far the largest group of filter media as a consequence of three major constructional variables, namely, the nature of the fiber, the form of the yarn, and the weave and finish of the cloth.

The fiber may be a natural material, such as wool or cotton, or one of the increasing number of synthetics, such as nylon; confusion is added by the common use of trade names instead of generic chemical names, nylon strictly being polyamide with various competitive equivalents such as Enkalon and Amilan. It is the generic nature of a fiber that determines its suitability for an application, both chemically and physically, from data such as that summarized in Table 15.4.

There are three basic forms of yarn; monofilament, multifilament, and staple. Monofilament yarns comprise single continuous fibers, the cross-sectional shape and dimensions of which are determined by the extrusion head through which the original molten polymer is extruded. Multifilament yarns comprise groups of several single filaments that are twisted together to form a yarn; the number of filaments per yarn, their diameter and shape, and the degree of twisting are all important variables. Staple yarns are made from very short lengths of fibers, either from natural fibers such as wool or by cutting up synthetic monofilaments; the number of twists per unit length in the spinning process influences the density and diameter of the yarn. The characteristics of the yarn profoundly influence the filtration properties of the finished cloth; this is summarized in Table 15.5 for various performance parameters, in decreasing order of preference.[6]

Three basic weave patterns are of importance in the manufacture of filter cloths: plain, twill, and satin (or sateen), as illustrated in Figure 15.12; various other common weaves, such as basket and duck, are essentially variations on this basic trio. The effects of the weave pattern on the filtration characteristics of the finished cloth are summarized in Table 15.6, again in order of decreasing preference under various operation parameters.[6]

The woven fabric may be subject to a considerable variety of finishing processes, of which the three most important are calendering, napping, and heat treatment. Calendering uses high pressure hot rollers, which cause the cloth to shrink and also give to its surface a smooth polish, which may improve cake release. Napping is the use of a fine steel comb, which produces a soft fuzz on the face of the cloth (one or both sides); the effect can be to improve the ability of the cloth to retain fine particles and also to increase the dirt holding capacity. Heat treatment is sometimes applied to synthetic fibers to stabilize the fabrics and make them more suitable for use at elevated temperatures.

Felts are made from random individual fibers that are held together in sheet form either by mechanical entanglement or by some form of adhesive. Almost any fibrous material can be made into a felt, provided the fibers are not so brittle as to disintegrate during the manufacturing process; a virtually limitless range of felts can thereby be formed, of varying densities, porosities, permeabilities, and pore sizes, by

Table 15.3. Properties of Millipore Membranes.

FILTER TYPE	MEAN PORE SIZE (μm)	FLOW RATE[a] WATER[b]	FLOW RATE[a] AIR[c]	POROSITY %	REPRESENTATIVE APPLICATIONS
MF (mixed esters of cellulose)					
SC	8.0	850	55	74	General prefiltration/clarification especially viscous liquids.
SM	5.0	540	35	84	Exfoliative cytology, gravimetric analysis of industrial hydraulic fluids.
SS	3.0	400	20	83	General microfiltration, especially oils and other viscous liquids.
RA	1.2	300	15	82	Stabilizing finished beer, ultracleaning of water and solvents, serum prefiltration.
AA	0.80	212	11	82	General air sampling, gravimetric analysis of fuels, stabilizing wines and juices.
(black)		212	11	82	Fluorescence and microscopic study of airborne particles, yeast and mold analyses of sugar.
DA	0.65	150	10	81	Microbiological analysis of whole milk, stabilizing wines.
HA	0.45	52	4	79	Microbiological analysis of liquids, FDA sterility testing, hydrocarbon fluid analysis, ultracleaning of water.
(black)		52	4	79	Fluorescent antibody studies, yeast and mold analysis of beverages, analysis of silica in liquids.
PH	0.30	40	3.7	77	Analysis of radioactive particulates, final prefiltration of serum, photomicroscopy.
GS	0.22	21	2.5	75	Sterilizing filtration, gravimetric analysis of used lube oils.
VC	0.10	2	0.49	74	Preparation of optically clear liquids (i.e., for light scattering), virus filtration.
VM	0.05	1	0.31	72	Virus filtration, ultrafiltration of proteins.
VS	0.025	0.2	0.22	70	Ultrafiltration of separation of high mol. wt. proteins, virus filtration.
Duralon (nylon)					
NC	14.0	1020	130	68	Prefiltration/clarification of ketone and ester solutions, especially photo resists.
NS	7.0	610	50	65	Microfiltration of high mol. wt. alcohol solutions, reagents, fine chemicals.
NR	1.0	270	13	63	Ultracleaning and analysis of ketones, ester-based hydraulic fluids, photo resists.
Solvinert (solvent resistant)					
UR	1.5	260	9	79	Ultracleaning of missile propellants/oxidizers, precision plating solutions.
UH	0.5	42	3.5	73	Analysis of missile propellants/oxidizers, alcohols, plating solutions.
UG	0.25	8	1.5	67	Sterilizing filtration of alcohol solutions and other "exotic" pharmaceuticals.
Celotate (cellulose acetate)					
EA	1.0	250	11	74	Final filtration of distilled beverages, ultracleaning of rinse alcohol.
EH	0.5	67	5	73	Nitrogen-free analysis, contamination analysis of selected solvents.

Table 15.3. (*Continued*).

FILTER TYPE	MEAN PORE SIZE (μm)	FLOW RATE[a] WATER[b]	AIR[c]	POROSITY %	REPRESENTATIVE APPLICATIONS
EG	0.2	21	2.5	71	Sterilizing filtration of alcohol-containing solutions.
Mitex (Teflon)					
LC	10.0	170	9	68	Clarification of concentration of acids, alkalies, photo resists, other reactive and corrosive liquids.
LS	5.0	70	6	60	Analysis of gross particulate matter in above liquids and cryogenics.
Polyvic (polyvinyl chloride)					
BS	2.0	312	10	79	Ultracleaning of moderately strong acids and alkalies; heat-sealed porous devices.
BD	0.6	45	4	73	Particle analysis in alcohols, mod. acids and alkalies; construction of diffusion chambers.
Microweb (nylon-reinforced MF)					
WS	3.0	155	7.1	45	General clarification.
WH	0.45	55	4.5	43	Microfiltration of test stand hydraulic fluids.

[a]Depending upon filter type, actual flow rates may vary by ±5 to ±33% from the average values given here.
[b]Water flow rates are $ml.cm^{-2}\ min^{-1}$ of filtration area, at 25°C with a pressure differential of 70 cm of mercury (13.5 psi). Flow rates for Solvinert and Polyvic filters are based on methyl alcohol instead of water.
[c]Air flow rates are $l.cm^{-2}\ min^{-1}$ of filtration area, at 25°C with a pressure differential of 70 cm of mercury (13.5 psi).

appropriate combinations of fiber dimensions and consolidation of the original loose mat.

For filtration applications, adhesive bonding is generally only used where the fibers are too brittle to be adequately entangled mechanically, as with glass; the disadvantages of using an adhesive are that it inevitably reduces the permeability of the felt and also introduces an additional material with its own chemical and thermal limitations. Needle felts are therefore far more common; the name arises from the manufacturing process, which involves passing a very open web of fibers under a multiplicity of reciprocating barbed needles that tangle the fibers both around each other and also around an inner coarse monofilament mesh that gives the finished felt much of its strength.

Paper is in a real sense a special felt, the fibers being traditionally cellulose but more recently glass and other inorganic materials. Because of their structure, cellulose fibers are able to bond strongly to each other without the need of an adhesive; nonetheless, papers for liquid filtration are often strengthened by being impregnated with a suitable resin.

With glass and other inorganic fibers the addition of a bonding agent is essential. In liquid filtration, cellulose papers are widely used on both small and large scale; whereas glass papers are important in the laboratory, they find little use industrially except in certain specialized cartridges, as a consequence of their mechanical fragility.

Bonded media are frequently superficially physically similar to paper, and sometimes are made by the same type of wet-laying process. The earlier grades used for liquid filtration (e.g., the clarification of machine tool coolants) were made from viscose fibers bonded together with cellulose; more typical of current technology are polyester fibers bonded with acrylonitrile resin, although many other variations occur. An important recent development is the range of spunbonded media, made by the

Table 15.4. Physical Properties of Natural and Synthetic Fibers.

							SUITABILITY IN CONTACT WITH		
FIBER	MAX SAFE TEMP. °F	SPECIFIC GRAVITY	ABSORBENCY OF WATER %	WET BREAKING TENACITY gm/den	ELONGATION AT BREAKING %	RESISTANCE TO WEAR	ACIDITY	ALKALI	COMMON SOLVENT
Acetate	210	1.30	9–14	0.8–1.2	30–50	Poor	No	No	No
Acrylic	275–300	1.14–1.17	3–5	1.8–3	25–70	Good	Yes	No	Yes
Cotton	200	1.55	16–22	3.3–6.4	5–10	Fair	No	Yes	Yes
Fluorocarbon	400	2.3	Nil	1–2	13–27	Fair	Yes	Yes	Yes
Glass	550–600	2.50–2.55	up to 0.3	3–6	2–5	Poor	No	No	Yes
Aramid (Nomex)	400–450	1.38		4.1	14	Excellent	No	Yes	Yes
Polyamide (Nylon)	225–250	1.14	6.5–8.3	3–8	30–70	Excellent	No	Yes	Yes
Polyester	300	1.38	0.04–0.08	3–8	10–50	Excellent	Yes	No	Yes
Polyethylene									
low density	150–165	0.92	0.01	1–3	20–80	Good	Yes	Yes	No
high density	200–230	0.92	0.01	3.5–7	10–45	Good	Yes	Yes	No
Polypropylene	250	0.91	0.01–0.1	4–8	15–35	Good	Yes	Yes	No
PVC	150–160	1.38	2	1–3	11–18	Fair	Yes	Yes	No
Saran	160–180	1.7	0.1–1.0	1.2–2.3	15–30	Fair	Yes	Yes	Yes
Wool	180–200	1.3	16–18	0.76–1.6	25–35	Fair	Yes	No	Yes

Table 15.5. Effect of Type of Yarn on Filtration Performance of Cloth.

ORDER OF PREFERENCE	MAX. FILTRATE CLARITY	MIN. RESISTANCE TO FLOW	MIN. MOISTURE IN CAKE	EASIEST CAKE DISCHARGE	MAX. CLOTH LIFE	LEAST TENDENCY TO BLIND
Best	Staple	Monofil	Monofil	Monofil	Staple	Monofil
	Multifil	Multifil	Multifil	Multifil	Multifil	Multifil
Worst	Monofil	Staple	Staple	Staple	Monofil	Staple

simultaneous extrusion of a multiplicity of fine filaments that are then bonded together, either by partially melting some of the filaments or by the addition of an adhesive.

While a very wide range of different grades of bonded media is available, a characteristic of them is that in general they are of very open structure compared with paper, and therefore are of limited retentive power until a cake has formed. On the other hand, their advantages include high strength and very low resistance to flow.

Filter Sheets are also closely related to paper, being manufactured from cellulose fibers by a similar wet-laying process; in appearance they are similar to very thick paper, their thickness being typically 2-6 mm. For many years, all commercially available filter sheets were derived from a mixture of the fibers of cellulose and of chrysotile asbestos, in varying proportions. To avoid the risk to health from the traditional asbestos, alternative grades have been developed during the last few years, using other raw materials such as diatomaceous earth and synthetic inorganic fibers. Ultimately this program has led to the evolution of filter sheets comprising only cellulose fibers, the properties of which have been modified by chemical treatment so as to induce on the cellulose fibers surface charges similar to those that occur naturally on chrysotile asbestos fibers; it is the presence of these positive surface charges that gives filter sheets their unique depth filtration characteristics and permits the finer grades to be used for sterilizing filtration applications.

Filter Powders, such as diatomaceous earth and expanded perlite, may be used either as filter media or as filter aids. As filter media, they are used to form a thin layer or precoat on another filter medium (e.g., woven wire), so that the main filtration takes place through this precoat layer, on top of which the dirt or other solid particles accumulate during filtration; periodically the precoat plus accumulated solids are discarded and replaced with a fresh layer of filter powder. As filter aids, small quantities of filter powders are dispersed in the liquid to be filtered, so that the particles of filter powder and of the solids in the original liquid deposit simultaneously during filtration; the purpose is to encourage the formation of a comparatively porous filter cake with low resistance to flow.

The two major types of filter powder, diatomite and expanded perlite, both occur as natural minerals, which are processed by operations such as crushing, grinding, calcining and classifying, thereby yielding a range of different grades of product. Alternative materials of

Table 15.6. Effect of Weave Pattern on the Filtration Performance of Cloth.

ORDER OF PREFERENCE	MAX. FILTRATE CLARITY	MIN. RESISTANCE TO FLOW	MIN. MOISTURE IN CAKE	EASIEST CAKE DISCHARGE	MAX. CLOTH LIFE	LEAST TENDENCY TO BLIND
Best	Plain	Satin	Satin	Satin	Twill	Satin
	Twill	Twill	Twill	Twill	Plain	Twill
Worst	Satin	Plain	Plain	Plain	Satin	Plain

Table 15.7. Comparative Filter Powder Characteristics.

PRODUCT	COMPOSITION	NO. OF GRADES AVAILABLE	RELATIVE FLOW RATE	PRIME ADVANTAGES AND APPLICATIONS	WET CAKE DENSITY gm/cc	RETENTION ON COARSE SCREENS	SOLUBILITY	
							IN ALKALIS (ROOM TEMP)	IN ACIDS (ROOM TEMP)
Diacalite[a] Diatomite Filter-aids	Silica	11	1–20	General filtration for highest clarity	.25–.35	Good	Slight in dilute alkalis	Slight in dilute acids
Diacalite[a] Perlite Filter-aids	Glassy Silicate	7	1–13	Outstanding on rotary vacuum Reduced dosage on both pressure and vacuum filters	.15–.30	Good	Slight in dilute alkalis	Slight in dilute acids
Solka-Floc[a]	Cellulose	1	7–15	Excellent for pre-coating coarse screens Highest purity for absorption of oil from condensate, and removal of iron from caustic	.14–.32	Excellent	Slight in dilute or strong alkalis Excellent resistance	None in dilute acids Excellent resistance
Gellfilt[b]	Polyurethane	2		Inert to most liquids up to 200°C.	.08	Good	Insoluble	Insoluble

[a]Great Lakes Carbon Corporation.
[b]Herbert Smith & Co. (Grinding) Limited.

Table 15.8. Equipment-orientated Properties of Filter Media.

1. Rigidity
2. Strength
3. Creep/stretch resistance
4. Stability of edges
5. Resistance to abrasion
6. Stability to vibration
7. Dimensions of available supplies
8. Ability to be fabricated
9. Sealing/gasketing function

Table 15.9. General Application-orientated Properties of Filter Media.

1. Chemical stability
2. Thermal stability
3. Biological stability
4. Dynamic stability
5. Adsorptive characteristics
6. Wettability
7. Health/safety hazard characteristics

both natural and synthetic origin are also available, notable examples being cellulose fibers and carbon. A useful summary of the major types of filter powder is given in Table 15.7, which shows also the general range of their application and properties; while these data mostly relate specifically to the products of one manufacturer, they are indicative of the materials generally available.

Sand is the traditional material for use in deep bed filters, although many other materials have been used such as crushed quartz, crushed marble, coke, charcoal, anthracite, and garnet sand.

Selecting the Right Filter Medium

With so wide a choice of filter media, selecting the right one for a specific application can be a source of great difficulty. Account must be taken of many different properties which for convenience may be divided into three broad categories: (1) equipment-orientated properties, (2) general application-orientated properties, (3) filtration-specific properties; the individual properties in each category are summarized in Tables 15.8, 15.9, and 15.10. Of these three groups of properties, the filtration-specific one listed in Table 15.10 presents the greatest difficulty, because of a simultaneous lack of data combined with a frequent ambiguity of such data as is available, as the following brief discussion of each of the four questions posed in the table illustrates. The final choice of the optimum filter medium for a specific application should obviously also take account of cost factors. This is a complex topic, which to receive the attention which it merits, requires a comprehensive study of the total annual costs of each alternative scheme under consideration; on this basis, it by no means follows that the cheapest medium per unit area is the appropriate choice.

Minimum Particle Retained. Only with the simplest type of filter medium, such as perforated sheet, is it possible to define the pore dimensions by direct measurement. With most media an indirect method is used, the most common being the bubble point test and the second choice generally being filtration using particles, such as glass beads, of known dimensions.

The bubble point test[7] is so-called since a sample of the medium is immersed in suitable liquid and is then pressurized with compressed air, to permit observation of the pressure at which bubbles of air just begin to appear; the corresponding diameter of an ideal circular

Table 15.10. Filtration-specific Properties of Filter Media.

1. What is the minimum size particle it will retain?
2. What is its efficiency against particles of a defined size?
3. What is the resistance to flow of the clean medium?
4. How does the resistance to flow alter with time?

pore can hence be calculated and is generally known as the "bubble point diameter."

Pore measurement by filtration using particles of known dimensions (especially if the particles are spherical, as with glass beads) is obviously likely to give a more realistic result than the bubble point test. Unfortunately, the observed results may be heavily influenced by numerous experimental factors, such as the precise nature of the particles, the concentration of the test suspension, and the velocity of filtration, making it extremely difficult to compare data from different sources.

Retention Efficiency. Figure 15.13 illustrates the common fact that two different types of filter medium (in this example felt and wire gauze) can each be 100% efficient in retaining particles down to a certain minimum size (in this example 35 microns) but very different in their efficiencies against particles of smaller size. Such complete data on filtration efficiency is rarely available, since it requires special facilities and considerable skilled manpower; moreover, it is important to note that such efficiency data are valid only for the test conditions and could well alter dramatically if, for example, different solid particles were used.

Clean Medium Flow Resistance. Since flow through the filter medium is almost always laminar, it should be possible to characterize the resistance to flow of a medium by application of the classical fluid flow relationships such as Equation (1), derived from Darcy's equation. On this basis, the "Darcy" is sometimes used as a unit of permeability, a filter medium having a permeability of 1 Darcy if it will permit a flow rate of 1 l./s, through an area of 1 cm^2 and a thickness 1 cm, under a pressure differential of 1 atmosphere, the viscosity of the liquid being 1 cp. In practice, the Darcy is only rarely used, a confusing multitude of other units being preferred, with very little standardization. Probably the most common unit is the Frazier scale, expressed as cfm. air/ft^2 at 0.5 in.w.g.

Change of Resistance With Time. The significance of how the resistance to flow alters with time depends upon the application, there being two possibilities. First, for clarification applications, which generally utilize depth filtration, it is desirable for the medium to have a high solids-holding capacity and a correspondingly long operating life before the filter needs to be serviced; inevitably, solids are deposited in the depth of the medium, with a progressive increase in the resistance to flow. Appropriate small-scale tests enable this solids-holding capacity of a medium to be measured.

The second possibility is totally different. For solids recovery applications, utilizing cake filtration, the ideal is for there to be no penetration of solids into the medium; the progressive penetration, which in practice frequently occurs, results in partial or total blinding of the medium. Unfortunately, the speed with which a medium will blind during cake filtration can generally only be determined by operating experience.

15.6. PRETREATMENT

The ease with which particles can be separated from a solid/liquid suspension is often profoundly influenced by the conditions to which the suspension is exposed prior to the filtration or other separation operation. This statement applies equally to both deliberate and accidental events, so that the critical characteristics of the suspension at the moment of separation are in reality dependent upon the total history of the suspension up to that moment. The validity of this statement is apparent from the wide variety of techniques which are used to pretreat suspensions, with the intention of improving their separation characteristics, as summarized in Table 15.11. Because of the particular importance of the various types of chemical pretreatment, Table 15.12 reproduces a summary of some typical performance data from a variety of industries.[8]

15.7. DESCRIPTION OF FILTRATION EQUIPMENT

The immense variety of filtration equipment available has evolved (and is continuing to

Table 15.11. Summary of Pretreatment Techniques.

TYPE	DESCRIPTION	COMMENTS
Orthokinetic flocculation	Shear flow induced, e.g., by gentle agitation or baffles	Promotes contact between solid particles, whereby they adhere to each other.
Chemical dosing:		
1. *p*H adjustment	Addition of acid or alkali	Can induce coagulation by direct adhesion of solid particles to each other.
2. Inorganic ions	Typically ferrous sulphate and aluminium sulphate	Induces coagulation by neutralizing the electrical charges on suspended particles. Polyvalent ions preferred.
3. Floc precipitates	Typically ferrous or aluminium hydroxides	Function by mechanically trapping particles as well as by charge-neutralization.
4. Organic compounds	Natural products, such as starch, or synthetic polyelectrolytes	Function primarily by linking particles into flocs, but also may involve charge-neutralization.
Heating:		
1. Temperature adjustment	Raise temperature of liquid being filtered	Reduces viscosity, thereby speeding filtration or settling and reducing moisture in filter cake.
2. Sludge treatment	Maintain at elevated temperature to permit coagulation	Coagulation of silver chloride in the standard laboratory analysis for chloride in the standard laboratory analysis for chloride is typical. The Porteous process for sewage sludge treatment involves pressure cooking at 190–200°C for about 40 minutes.
Freeze/thaw process	The solid/liquid mixture is first frozen solid, held thus for a limited time, then thawed	Induces coagulation. Devised originally for sewage sludge but more successful with inorganic sludges (e.g., in water treatment).
Ultrasonics	Exposure to ultrasonic radiation	Can induce coagulation, but wrong energy level has opposite effect.
Ionized radiation	Exposure to x-rays, etc	Known to induce coagulation with sewage sludge and bentonite clays though at present generally not economic.
Prethickening	Use of, for example, gravity settling prior to filter press, or add further solids	Low solids concentrations result in closely-packed cakes which impede or make impossible cake-type filtration; the same material will often filter easily if the concentration is raised to 3–4% or more.

evolve) because of the great variations that occur in the three interacting areas of suspension characteristics, process needs, and mechanical possibilities. The following brief summary of this extensive subject has been structured in terms of the first two factors since it is generally these, rather than mechanical considerations, which are the important criteria in choosing a filter for a specific application. The pattern follows closely that of Reference 4 to which the reader is referred for more detailed information.

The field is divided first by distinguishing between three basic types of solid/liquid suspension:

1. Rapid settling suspensions, where the particles are coarse and dense, but where their concentration is generally unimportant. This corresponds broadly with particles greater than say 50 μm, which, with sand in water, corresponds to a settling rate of about 12 cm/min. It may be noted that 50 μm is the smallest particle visible to the naked eye.

Table 15.12. Examples of Use of Pretreatment Chemicals.

TRADE NAME	COMPOSITION	TYPE OR MECHANISM	TYPICAL APPLICATION	NORMAL RANGE OF *p*H EFFECTIVENESS	NORMAL EFFECTIVE CONCENTRATION	APPROX. PRICE PER lb.	MANUFACTURER
Alum	$Al_2(SO_4)_3 \cdot XH_2O$	Electrolytic and coagulation	Water treatment	5 to 10	15 ppm	2¢	Inorganic chemical mfgrs.
Ferric sulfate	$Fe_3(SO_4) \cdot XH_2O$	Electrolytic coagulation	Water treatment and chemical processing	any	5 to 100 ppm	2¢	Inorganic chemical mfgrs.
Sodium CMC	Sodium carboxy-methylcellulose	Coagulation and bridging	Mineral processing	3 to 9	0.03 to 0.5 lb/ton	50¢	Hercules, Du Pont
Kelgin W	Algins	Coagulation and bridging	Water treatment	4 to 11	Up to 5 ppm	$1.50	Kelco Co.
Separan	Acrylamide polymer	Bridging	Chemical processing	2 to 10	0.2 to 10 ppm	$1.00 to $2.00	Dow Chemical Co.
Fibrefloc	Animal glue	Electrolytic	Waste treatment	1 to 9	5 to 30 ppm	18¢	Armour & Co.
Corn starch	Corn starch	Bridging	Mineral processing	2 to 10	10 lb/ton	7¢	–
Polynox	Polyethylene oxide	Bridging	Chemical processing	2 to 10	1 to 50 ppm	$2.00	Union Carbide
Silica sol	Activated silica sol	Electrolytic coagulation	Waste treatment	4 to 6	1 to 20 ppm	1.5¢ as sodium silicate	Inorganic chemical mfgrs.
Sodium aluminate	Sodium aluminate	Coagulation	Water treatment	3 to 12	2 to 10 ppm	10¢	National Aluminate
Guar gum	Guar gum	Bridging	Mineral proc.	2 to 12	0.02 to 0.3 lb/ton	35¢	General Mills
Sulphuric acid	H_2SO_4	Electrolytic	Waste treatment	1 to 5	Highly variable	1¢	Inorganic chemical mfgrs.

2. Medium and slow settling suspensions, where the particles are smaller and where their concentration is between about 1 and 50% by weight. This concentration is low enough for the slurry to flow readily as a liquid, but is high enough to enable a cake of say 2–3 mm or more in thickness to be formed within a few minutes on a vacuum filter. Such slurries are very common and include, for example, many chemical precipitates, starch, and sewage sludge.
3. Very dilute suspensions requiring clarification, in which the particles are generally very small and have little if any tendency to settle. This covers an extremely wide field, such as water purification, the polishing of beer and wines, and both lubricating and hydraulic oil filters.

A second division is then made in terms of two common process needs, the first depending on whether or not it is required to wash the filter cake before it is discharged, and the second being whether this discharge should be batchwise or continuous.

Further subdivisions are made as appropriate within the text on the basis of mechanical factors such as the driving force (gravity, vacuum, pressure, or centrifugal) and the equipment geometry.

Cake Filtration: Rapid Settling Suspensions

(i) No Cake Washing Required. *Batch Discharge of Cake.* While simple gravity operated strainers are sometimes used for small quantities of coarse material, vacuum or pressure operation is much more common. The laboratory vacuum operated Buchner funnel has its industrial equivalent in the simple Nutsch filter, the cake being dug or scooped out manually; an elaboration of this is the tipping pan vacuum filter, which is inverted for cake discharge. A more sophisticated variation, which can operate under vacuum or pressure, is the Rosenmund (Fig. 15.14) incorporating mechanical facilities to smooth the cake (overcoming cracking), reslurry the cake or discharge the solids; full automation is possible.

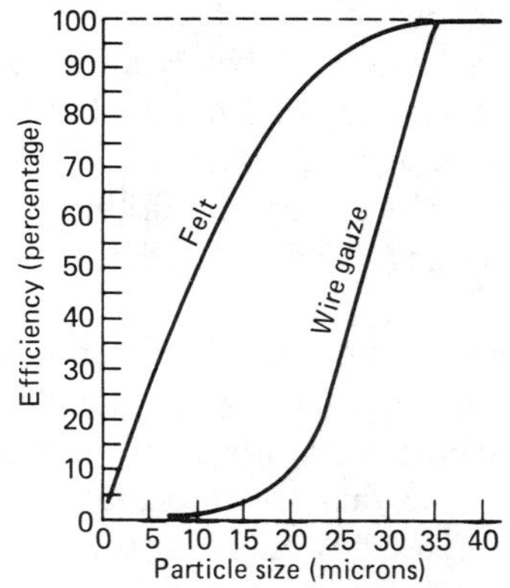

Figure 15.13. Two media (felt and woven wire) with the same cut-off point.

While many filters operate under pressure, this often occurs solely because they are incorporated in a pumping system to convey liquid, a typical example being the common Y-type in-line strainer, the pressure drop across which is normally restricted to say 0.1–0.2 bar. To avoid interrupting the flow, special duplex forms of strainer are frequently used, permitting one to be cleaned while the other is in operation. An alternative, especially for larger installations, is some form of mechanized or automated cleaning system, typically involving back flushing a portion of the filtering element either intermittently or continuously.

The batch discharge filtering centrifuge is widely used, both for simple dewatering and also where washing is required, because of its simplicity and versatility. The perforated metal

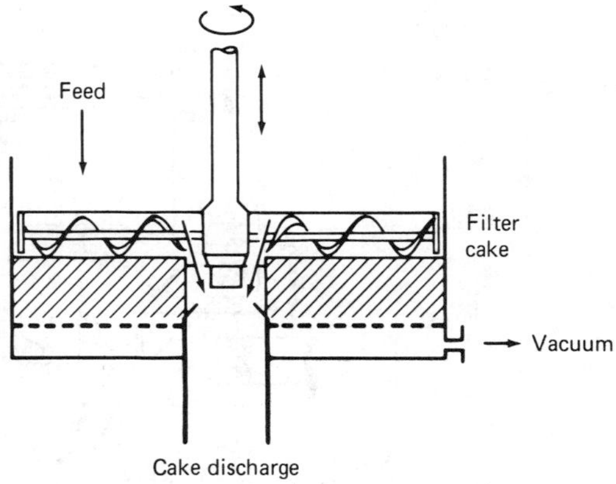

Figure 15.14. The Rosenmund filter.

basket or bowl is generally lined with a wire mesh or a cloth. Its axis is most commonly either vertical or horizontal, but in some models is set at an angle; if vertical, it may be either underdriven or overdriven, the latter generally being used only for free filtering materials.

The simplest type of centrifuge relies upon manual discharge, such as by literally digging out the solids; to facilitate this, the basket is underdriven so that the top is fully accessible. Figure 15.15 illustrates a typical modern machine in which the whole assembly of casing, basket, and drive is suspended by links from three supporting posts or pillars, so that in operation it can find its own center of rotation.

Cake discharge is generally mechanized and very frequently is automatic. Two basic types of device are used, the plough and the peeler, both of which dislodge the cake as the basket continues to rotate; a plough is a long blade or scraper that extends the full height of the basket, whereas a peeler is a very narrow blade which moves back and forth across the face of the cake while it simultaneously advances toward the screen surface. A plough is thus mechanically simpler but has the disadvantage that it tends to glaze the surface of the thin heel of cake left in the basket at the end of discharge; the more complex peeler also leaves a heel but does not glaze its surface. For plough discharge, the basket speed is generally reduced to 40-45 rpm, although sometimes ploughing at full speed is carried out with horizontal axis centrifuges, as is normal with peeler discharge.

The constructional details and dimensions of centrifuges vary considerably depending upon the manufacturer. Basket diameters generally lie in the range 0.3-2 m; depth is usually roughly half the diameter. The speed of rotation is about 700-3000 rpm, although probably 2000 rpm is a more common maximum; corresponding centrifugal forces are typically 600-1300 g.

The capacity of the centrifuge is a function both of its dimensions and its operating cycle. Cycle times are generally 5-30 minutes, but certain fully automated designs in which all operations are carried out at full speed utilize cycles as short as 30 seconds. Additional sophistication can include inert gas blanketing, so that flamable materials may safely be handled without the risk of explosion; an excellent source of advice on this specialized topic is available in Reference 9.

Continuous Discharge of Cake. Backwashing is frequently used to remove the solids both from

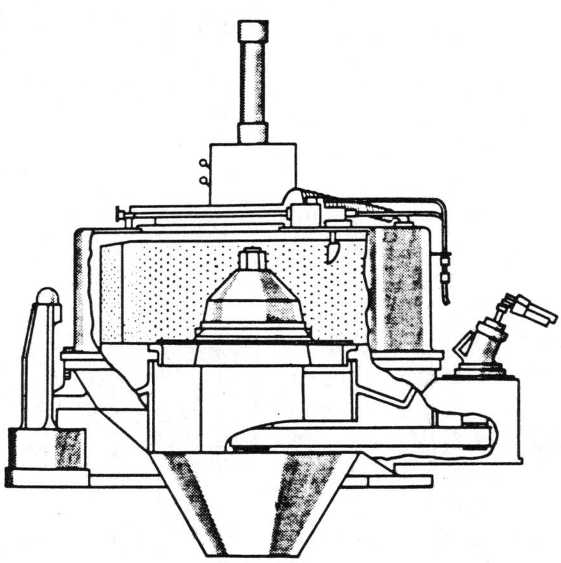

Figure 15.15. Broadbent type 46A batch centrifuge.

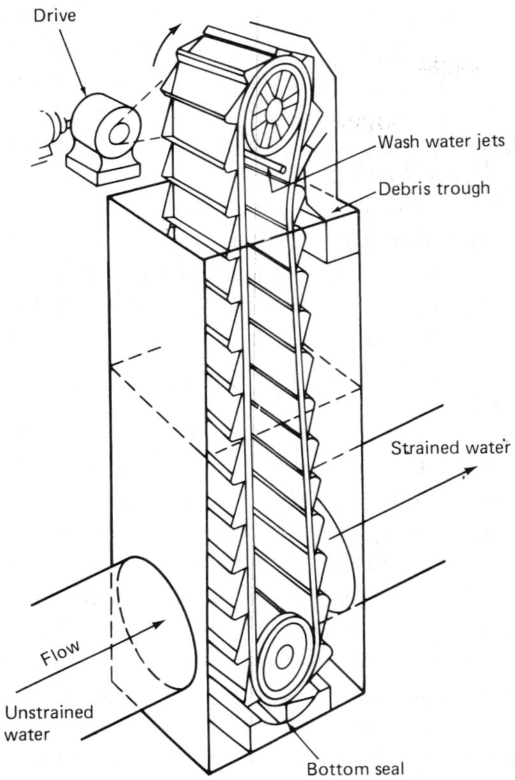

Figure 15.16. Traveling band water screen.

in-line pressure strainers and also from the various types of gravity screen and strainer used in water treatment. One example of the former incorporates a number of individual tubular strainer elements mounted horizontally in a circular pattern, so that each can be rotated in turn to be opposite the backwashing point.

An example of backwashing strainer widely used in the water purification industry is the traveling screen or band type illustrated in Figure 15.16. Similar operating principles are used in the rotary disc and rotary drum strainers, which also function under gravity. Various grades of filter media are used, according to the duty, extending down to about 20 μm monofilament mesh; with a build up of slimes on the screen this can result in the removal of particles as small as 3 μm.

Backflush discharge of solids has the disadvantages of consuming liquid and of resulting in the solids emerging in the form of a slurry. An example of a device that avoids these disadvantages is the Callow horizontal traveling belt screen used in the mineral preparation industry; in essence this is no more than a wire mesh belt traveling at 15–30 m/min with the dewatered solids discharging as the belt reverses over the end roller. Another form of horizontal belt is the flat-bed filter widely used for the clarification of machine tool coolants, illustrated in Figure 15.17; this uses a roll of disposable nonwoven medium, and may operate either under gravity or using the suction of the filtrate pump.

Vibrating screens also enable the solids to be recovered in a relatively dry state. Heavy duty

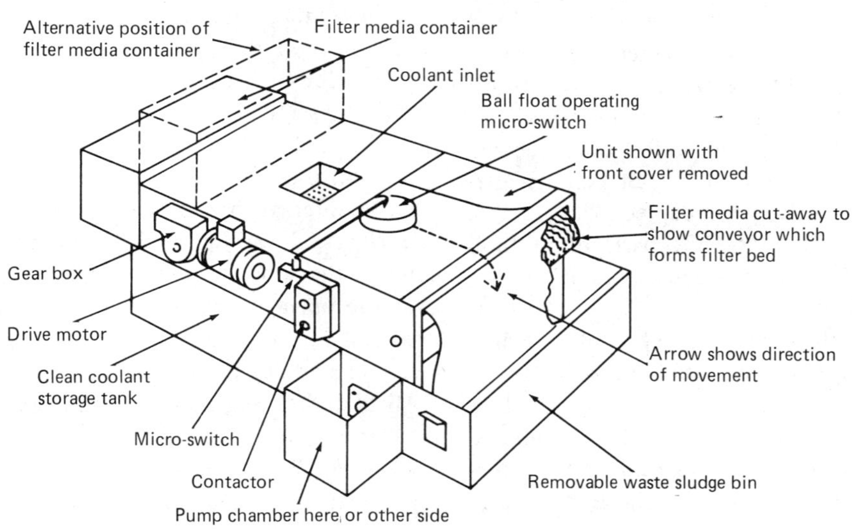

Figure 15.17. Flat-bed filter (M.E.L. Equipment Ltd.)

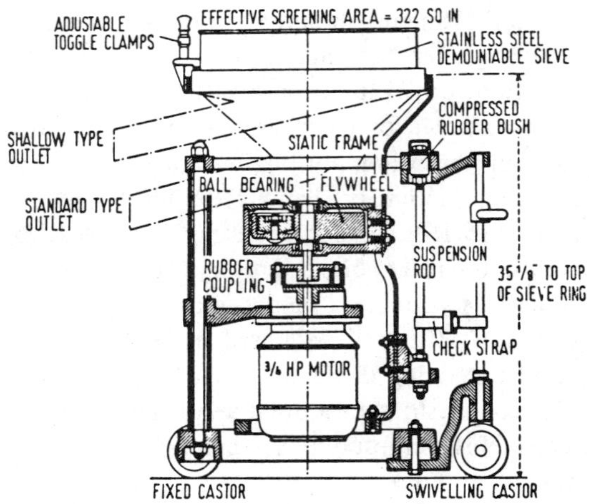

Figure 15.18. Russell gyratory screen.

screens are generally rectangular in shape and inclined at an angle between about 15° and 45°, 20-30° being most common; these screens, which are widely used in the mining industry, may be vibrated mechanically, electromagnetically or pneumatically. For lighter duties, the gyratory type screen (Fig. 15.18) permits the use of screens down to 350 mesh, although the limit of about 200 mesh is more normal; this type of vibrating screen is not generally used for fast settling slurries but is extremely useful to handle difficult materials such as paint and fruit pulp.

Various forms of continuous vacuum filters are well suited to handling fast settling slurries, the major types being horizontal belt and pan filters; since these filters are generally used in conjunction with cake washing, discussion of them is deferred until the following section. A further possibility is the top-fed rotary vacuum drum filter (Fig. 15.19), which permits little washing but is sometimes adapted to incorporate a hot air drying stage; the basic drum filter from which it is derived, is discussed below.

Filtering centrifuges permit the solids to be recovered with a minimum of residual liquid. While any filtering centrifuge can achieve this result, the only types that it is proposed to describe at this point are those that are limited to a simple dewatering operation; the other types which also permit the solids to be washed are described in the following section. While several distinct forms of dewatering centrifuge exist, a feature that they have in common is that the basket is conical; their differences arise from the different methods used to discharge the dewatered solids.

The simplest is the self-discharging centrifugal filter illustrated in Figure 15.24 from which the solids discharge solely under the influence of centrifugal force provided the angle of the cone has been correctly chosen to suit the specific solids being handled; typically, the angle of slope of the wall is in the range 25-35°. Successful operation also depends upon the liquid being moderately viscous and having some lubricating properties, such as in sugar refining and corn fiber separation, where this type of centrifuge has found wide use.

The need for a particular bowl angle for each material is avoided in the Krauss-Maffei "tumbler" centrifuge, so named because the horizontal axis of the bowl is made to gyrate as the bowl spins, thereby imparting a tumbling motion to the bowl. As a result, the actual angle of inclination of each part of the wall is constantly changing as it rotates, so that at some point in each rotation it will coincide with the optimum for whatever material is being handled.

Other modes of discharge are utilized in the vibrating centrifuge and the scroll type. In the former, the conical basket is mounted with its axis horizontal, while the angle of the cone is varied according to the material in the range 13-18°; vibration, caused by rotating eccentric weights, results in the discharge of the cake. Both vertical and horizontal forms of scroll type centrifuges exist, their common feature being an internal concentric scroll, which rotates at a slightly lower speed than the basket, and conveys the solids so that they are discharged over the open mouth of the cone; unlike the previously described machines, this form permits a very limited degree of washing.

All of these are genuinely continuous centrifugal filters, with the cake being continuously discharged. While the short cycle batch centrifuge obviously does not conform with this description, in practice the form with a very short cycle (e.g., 30 seconds) may provide a convenient and acceptable compromise.

(ii) Including Cake Washing. *Batch Discharge of Cake.* Solid material collected in a simple gravity strainer can obviously be washed before being discharged. If washing is required, however, it is more usual to use the vacuum Nutsch or tipping pan filter (see page 707) so as to minimize the residual quantity of mother liquor in the cake before washing commences. Pressurized versions of Nutsch filters (e.g., the Rosenmund), which operate under a gas pressure, are similarly well suited to washing.

Any of the batch centrifuges described above may conveniently be used for washing the cake prior to discharge.

Continuous Discharge of Cake. While wash water is sometimes applied to the gravity screen described above, it is important to note that generally the objective is to remove the fines from the cake, so that they pass into the filtrate; the operation is thus strictly classification. Washing to remove mother liquor adhering to the cake is likely to be very inefficient, since the very coarse particles typically handled on this type of screen result in the formation of a very open cake.

The top-fed vacuum drum filter (Fig. 15.19) permits only a very limited wash due to the geometry of the filter surface. By contrast, those forms of continuous vacuum filter in which the filtering surface is totally horizontal are ideally suited to washing application, especially if the fast settling nature of the solids also demands a horizontal surface. Numerous different forms are available, divided broadly between those which rotate (e.g., the rotary table filter) and those incorporating linear travel (e.g., the belt filter); with all, the piping and ancillary arrangements are broadly similar to those for the disc filter, as shown in Figure 15.20.

In its simplest form (Fig. 15.21), the rotary table filter is constructed so that the top face presents a continuous surface which can be

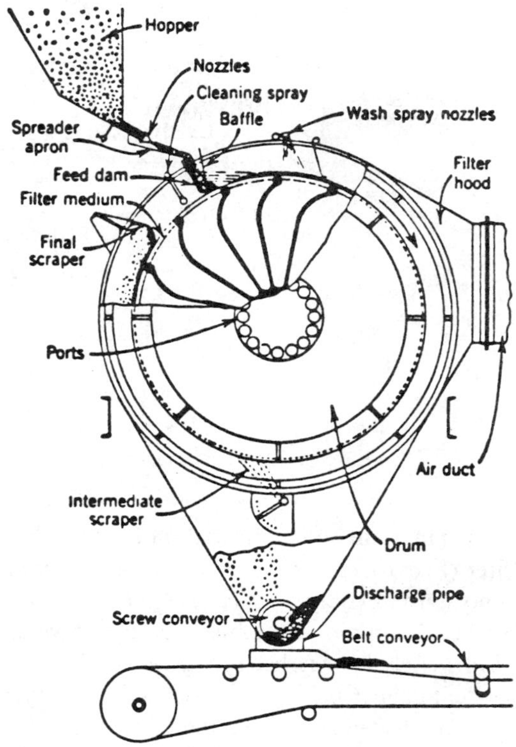

Figure 15.19. Top-fed vacuum drum filter, with hot air drying. (Swenson Evaporator Co.)

covered with a suitable cloth. The underside is more complex and comprises a number of identical segmental compartments, each with its own drainage pipe connecting to a central multiport valve about which the table rotates, each compartment thereby being submitted to a sequence of operations; in this the table filter is very similar to the rotary drum filter (see below, Fig. 15.41). Slurry is fed onto the filter through a weir-type distributor extending radially across the width of the surface; the cake thus formed is then subjected to the desired series of dewatering and washing operations and is finally discharged by means of a scroll-type elevator. The limitations of this simple form of cake discharge device, which requires a fairly thick cake to function but which also leaves behind a significant heel of undischarged cake, have provoked the development of numerous alternative forms of filter, including the tipping pan filter discussed below.

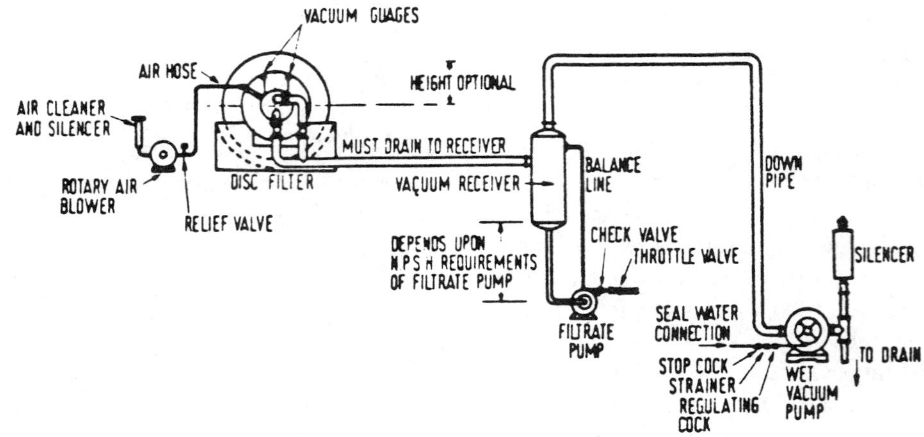

Figure 15.20. Typical arrangement for rotary vacuum disc filter (Esmil-Envirotech).

An important modern variation is the Ucego filter (Fig. 15.22), in which the discharge scroll is no longer required to elevate the cake over the vertical rim of the table since the previous fixed metal rim has been replaced by an independent rubber belt; at the discharge zone, this belt is deflected away from the table by a series of rollers, thereby providing an opening through which the cake can be readily discharged. Whereas the traditional table filter is restricted in size to a maximum of approximately 20 m^2, the range of the Ucego starts at the latter figure and extends up to 150 m^2 or more.

The tipping pan filter, as the name implies, solves the discharge problem by dividing the circular table into a number of independent pans, each of which can be inverted when it reaches the discharge zone (Fig. 15.23). While the pan is inverted, vacuum suction is interrupted, so that the cake simply falls out; sometimes a blow back with compressed air is used both to assist the discharge of the cake and also to keep the cloth slightly distended while it is washed with water sprays. Filters of this design are available with areas in the range approximately 2-150 m^2.

Although numerous types of traveling belt filter are available, the basic principles of them all are typified by Figure 15.34, in which an endless belt passes over a series of separate vacuum chambers between two rotating drums; slurry is fed on at one end of the machine, forming cake which is finally discharged as the belt reverses its travel at the far end. The details of the discharge technique vary considerably, partly to suit the properties of different materials, the

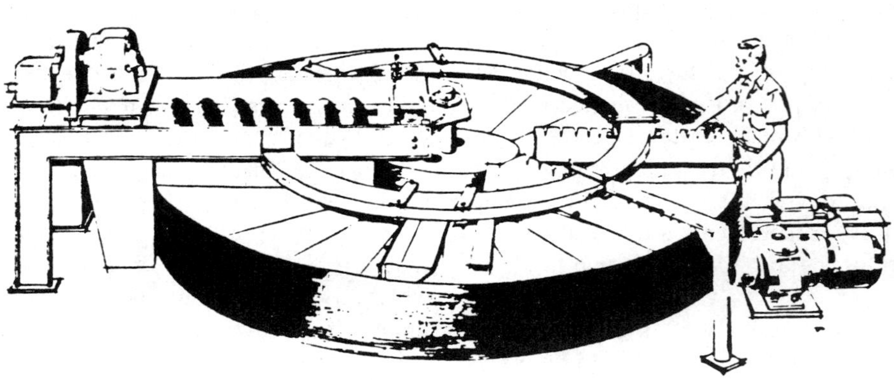

Figure 15.21. Table filter. (Ametek Inc.)

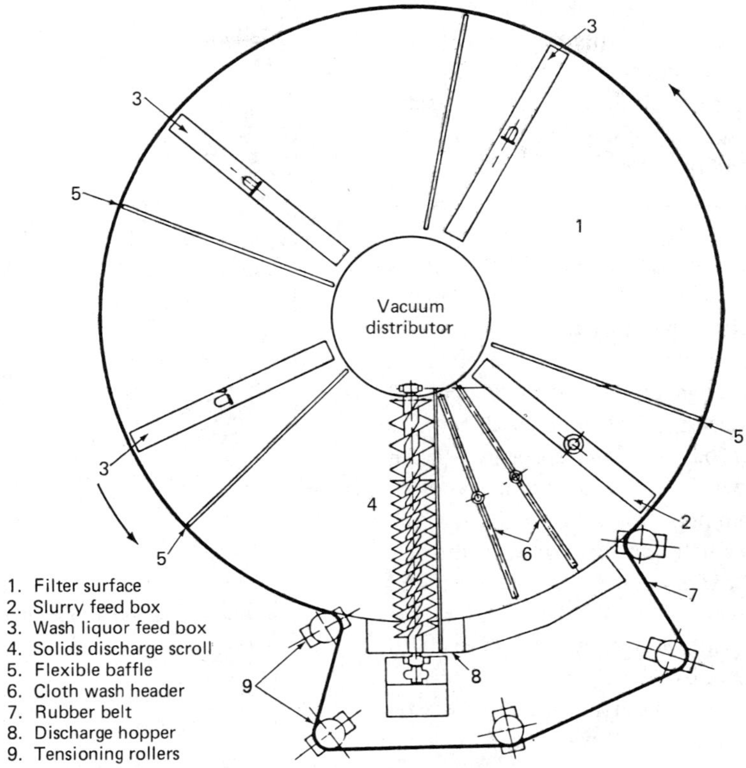

Figure 15.22. Ucego table filter.

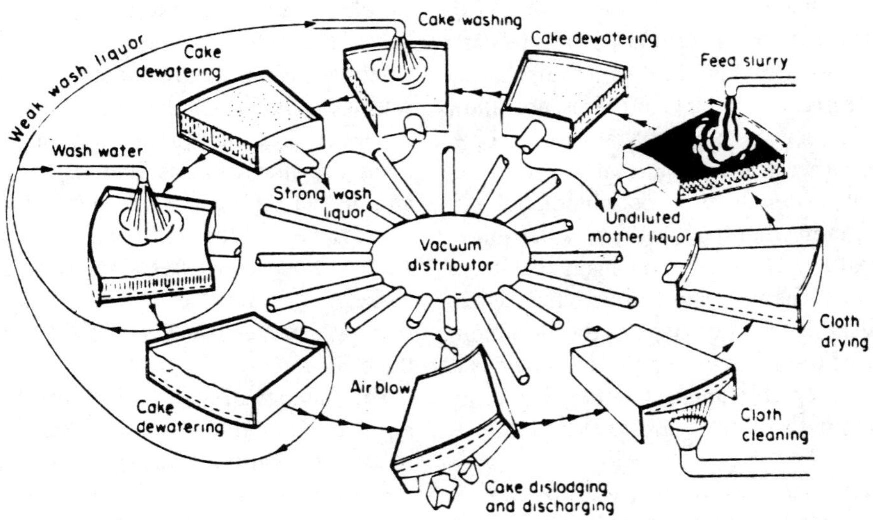

Figure 15.23. The Preyon tipping pan filter. (Bird Machine Co.)

simplest depending on the cake to break away from the cloth and fall off by gravity; if necessary, this may be aided by devices such as a scraper knife, reverse air blow, etc. Incomplete discharge can be overcome if necessary by washing the belt on one or both sides during the return run, so as to reduce the risk of blinding.

The majority of belt filters incorporate a heavy duty flexible rubber belt that supports and conveys the separate belt of filter cloth. This makes it possible to maintain a constant vacuum even though the belt is moving continuously; the success of this depends upon a sliding contact between the specially shaped portion of the belt and a continuous suction duct. Less common are forms of filter which omit the rubber belt and drive directly the belt of filter cloth, examples being the Pannevis and the Adpec filters. Both of these require the vacuum to be interrupted cyclically, so as to permit relative movement between the filter cloth belt and the supporting metal or plastic surfaces. Belt filters are available with areas ranging from about 0.5 m^2 up to 200 m^2, although the forms that exclude a rubber belt are restricted to a considerably lower maximum.

Centrifugal filters comprise the other category of machine for use with fast settling materials where washing is required. Of those already described, the scroll discharge centrifuge (with a scroll incorporated in the basic machine shown in Fig. 15.24) permits only a limited degree of washing but does not allow the wash liquor and main filtrate to be segregated; nonetheless it is important in the manufacture of materials such as ammonium sulphate. A more thorough wash, plus the ability to segregate wash liquor and main filtrate, is possible with the very short cycle batch centrifuge; while this does not give strictly continuous discharge of the solids, its 30–60 second cycle is often an acceptable approximation, especially since the length of the washing period can be freely selected.

With the pusher centrifuge, the discharge is effectively totally continuous as a consequence

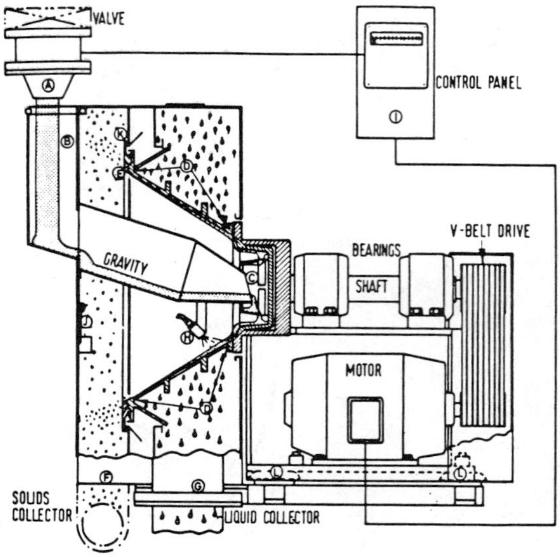

A. automatic feed valve
B. feed pipe
C. accelerating cup
D. conical basket
E. deflecting ring to arrest discharging solids
F. solids outlet
G. liquid outlet
H. wash nozzles
I. control panel
J. sampling door
K. seal ring
L. rubber mountings

Figure 15.24. Self-discharging centrifuge. (Allis-Chalmers Co.)

of a piston which reciprocates at the rate of 30–70 strokes per minute; attached to the piston is a pusher plate at the base of the basket (see Fig. 15.25), the repeated small movements of which cause the cake to move progressively across the screen and out of the mouth of the rotating basket. Unfortunately the degree of washing possible is limited by the time of residence of solids in the washing zone, this being dependent on the dimensions of the bowl and on the throughput; more effective washing is possible in the multistage machine (Fig. 15.26) which comprises a series of concentric baskets, alternate ones of which reciprocate, so that each basket acts as a pusher for the cake in the next outer one.

The number of stages which can be incorporated in one pusher centrifuge varies according to the application, but may be as high as 6 or even 8. The success of this design depends upon two factors, one being the rapid drainage

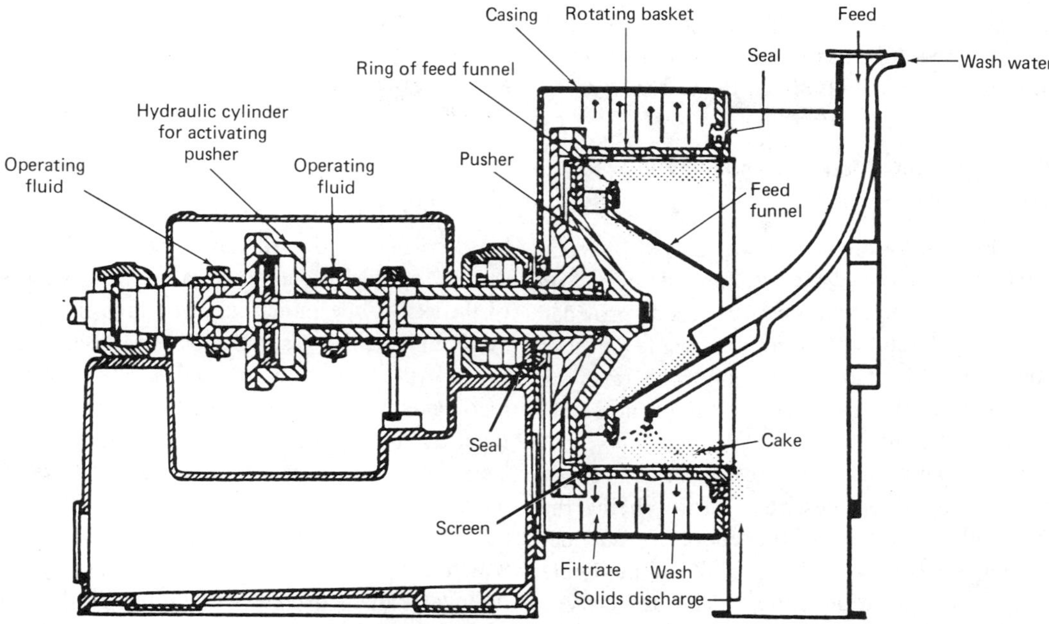

Figure 15.25. Pusher discharge centrifuge.

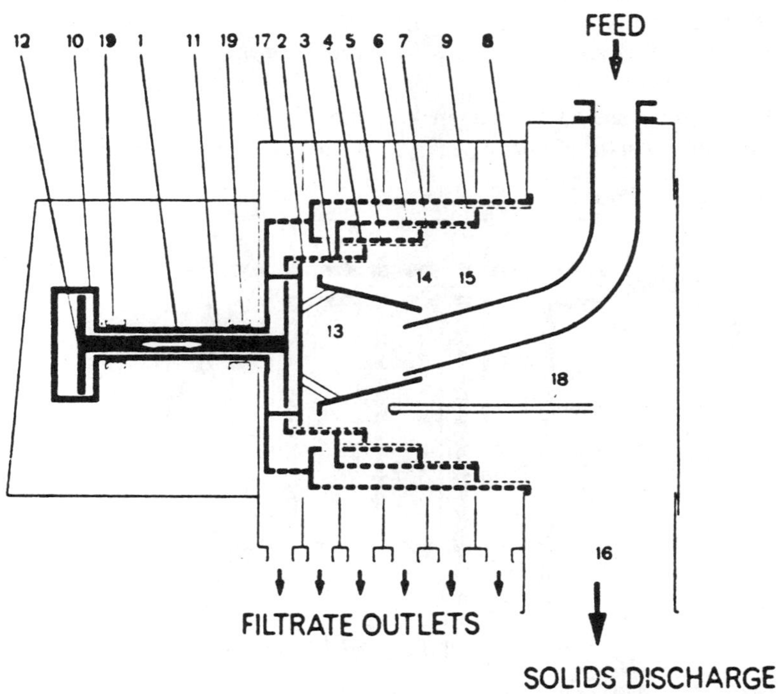

Figure 15.26. Multistage pusher centrifuge. (Escher-Wyss).

of the liquor, and the other being that the bed of deposited solids must be sufficiently strong to transmit the impulse from the piston.

Cake Filtration: Moderate and Slow Settling Suspensions

(i) No Cake Washing Required. *Batch Discharge of Cake.* Simple gravity strainers described on page 707, can sometimes be used, provided they are fitted with a suitable grade of filter medium. Inevitably, very low flow rates are achieved, so that the method is applicable only to small scale operations.

The vacuum (and pressure) Nutsch and single tipping pan filters described earlier, being the industrial equivalents of the laboratory Buchner funnel, are widely used for small and moderate scale batch filtration duties. The basket type centrifuge (Fig. 15.15), which can be fitted with filter cloths of many different grades, is another popular and versatile device.

In general, however, this field of application is dominated by pressure filters, notably filter presses and pressure vessel filters. Before considering these, however, mention should first be made of the small in-line type, the counterpart of the in-line pressure strainer discussed on page 707. A combination of these with the correct grade of filter medium can make them suitable for slurries in the present category, provided their small cake-holding capacity is acceptable, which can only be true for small and specialized applications.

(a) Filter Presses

The filter press has rightly been termed the 'workhorse' of the filtration world, since it can be applied to a remarkably wide variety of duties. While numerous variations exist, the basic unit is very simple: it comprises a sequence of hollow perforated plates (usually rectangular) alternating with frames, held together by pressing, as the name implies. Generally, each face of each plate is covered with a filter cloth, so that the assembly is a number of cloth-covered chambers, into each of which slurry may be pumped, separated by the hollow plates covered with filter cloth; the filtrate flows through the filter cloth and into the hollow plates before being discharged via a suitable drain point (Fig. 15.27). As this occurs, a cake of solids builds up more or less evenly on the cloth, which forms each side of each chamber; filtration can proceed until the chamber is filled with solids, at which point operation must be interrupted so that the press can be opened and emptied.

The filter illustrated in Figure 15.27 is commonly called the 'flush plate and frame press', to distinguish it from the 'recessed plate press'.

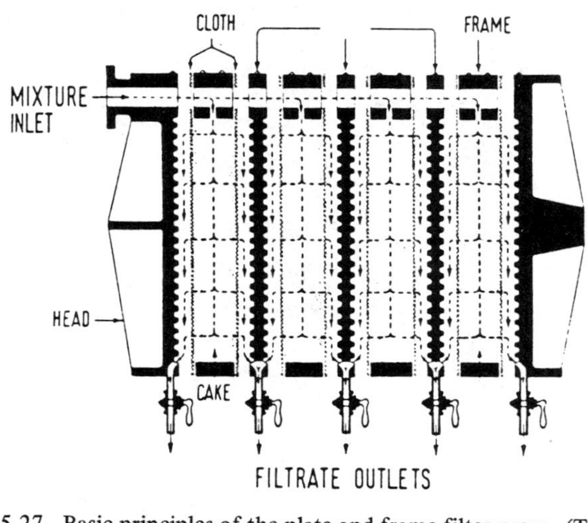

Figure 15.27. Basic principles of the plate-and-frame filter press. (T. Shriver & Co. Inc.)

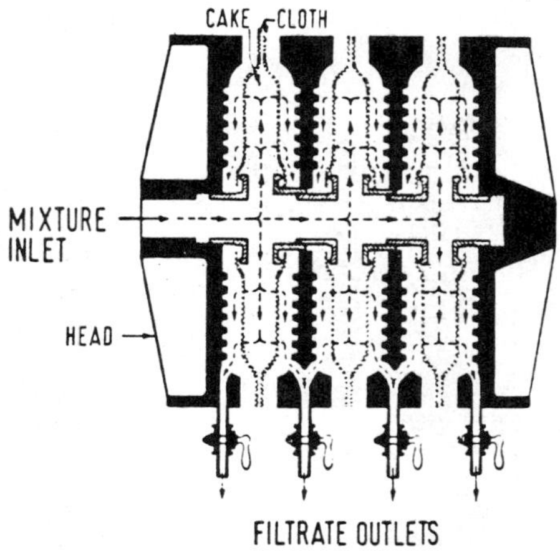

Figure 15.28. Recessed-plate filter press (T. Shriver & Co. Inc.)

In the latter, the chamber in which the cake accumulates is formed by constructing each plate with a protruding rim on each face; by this means the use of a frame is unnecessary, since a chamber is formed merely by placing two plates in contact, as in Figure 15.28. Nevertheless, frames are sometimes used in conjunction with recessed plates so as to increase the depth of the chamber from the normal maximum of about 32 mm (i.e., a 16 mm recess on each plate). By contrast, by using frames the depth of chamber may be selected to suit the particular duty, typical figures being 25-40 mm. Where washing is not required, recessed plate presses are generally preferred.

Commonly, the filter cloth is draped over a plate so that it hangs equally down each side, holes being cut into it to correspond with the various inlet and outlet ports that are an essential part of the plates and frames; when the latter are clamped together, these ports line up so as to form continuous channels. Each cloth is large enough to protrude beyond the edges of the plates and is therefore nipped between adjacent plates or plates and frames when they are clamped together; thus, whereas the main part of the cloth performs its natural function as a filter medium, the outer perimeter is required to act as a gasket. Only in rare (and costly) instances are these two functions separated, with the cloth caulked securely into position and with provision of a separate gasket. This dual duty imposes great restrictions on the choice of filter cloths; the likelihood of any leakage similarly restricts the potential use of filter presses.

The individual plates are made in a wide variety of dimensions, especially with the recessed design, ranging upwards from about 20 X 20 cm up to 1 X 1.5 m and even larger. The number of chambers per press may also vary widely, typically up to about 20, though occasionally as many as 200. As a consequence, a filter of specific area or volumetric cake holding capacity may comprise either a few large plates or a larger number of smaller ones, the appropriate choice generally being a combination of optimizing the cost and factors such as operational convenience.

A major operational factor with the traditional filter press is the heavy work load it imposes on the operators each time the press needs to be emptied. Each plate and each frame must be manually shifted through a sufficient distance to permit the filter cake to fall out; often the operator may have to assist this discharge, using a scraper, with the risk that he damages the cloth. After discharge is complete, the plates and frames must all be moved back into position and the whole assembly clamped tightly together with a device such as a capstan wheel or a ratchet gear.

Mechanization or automation of part or all of this sequence of operations are common features with modern filter presses, especially with larger recessed plates which are too heavy to be moved manually. While many different and sophisticated designs have been evolved, and are continuing to be developed, it is probably true to say that, with one possible exception, there is still no genuinely fully automatic version of the basic filter press. The reason is simply the fundamental difficulty of achieving one hundred percent success with the discharge of every cake from every chamber each time the press is emptied; in practice, it is generally necessary to have an operator available during discharge so that he can intervene whenever the occasional cake fails to discharge completely.

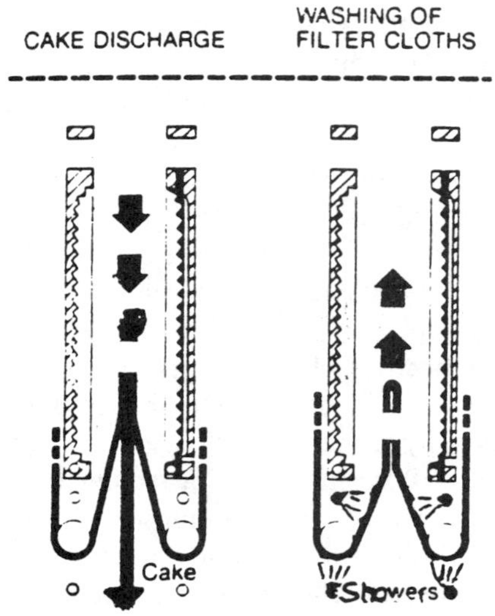

Figure 15.29. "Lasta" fully automatic recessed-plate filter press. (Ishigaki Mechanical Industry Co.)

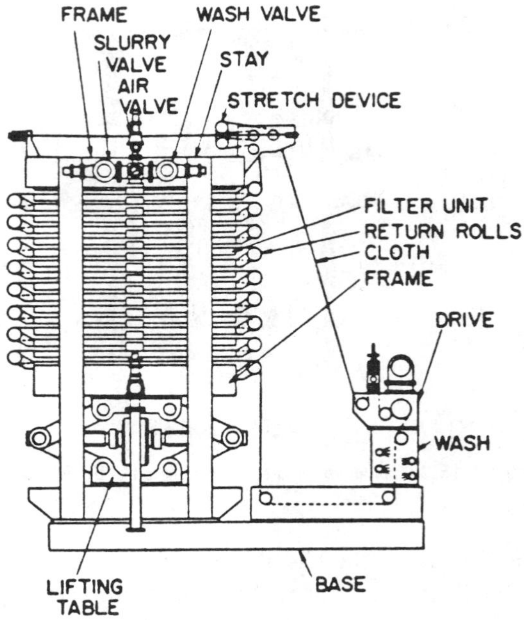

Figure 15.30. Hoesch tower press.

The possible exception to this generalization is the recently developed Japanese 'Lasta' press in which discharge is achieved by physically moving each filter cloth around a roller, as shown schematically in Figure 15.29; the cloth moves downwards to discharge the cake and then reverses back into the chamber ready for the next filtration cycle after being washed on both sides by spray nozzles. This technique, adopted from rotary vacuum filters, imposes a uniquely positive discharge mechanism on the traditional filter press.

The Hoesch tower press also moves the filter cloth to give positive discharge but radically departs from the traditional filter press by effectively standing it on end (Fig. 15.30).

Other radical features it incorporates include the use of one continuous filter cloth in the form of a belt, and the fact that filtration in each chamber is through one face only. This filter was one of the earliest successfully to incorporate a compression diaphragm in each chamber, so that the cake formed by filtration can be hydraulically squeezed to reduce its residual moisture content before it is discharged.

Compression diaphragms (or membranes) are now a relatively common feature in filter presses, including in the 'Lasta' press described above. As shown schematically in Figure 15.31, they give an extra degree of flexibility to the traditional filter press. With the latter, to minimize the moisture content of the final cake, it is necessary to continue filtering until the entire chamber has been filled with cake; typically this involves a 'compression period,' during which the feed pump slowly forces more material into the chamber after it has been nominally filled with a relatively wet cake by filtration. With the membrane plate, filtration may be terminated at any desired point and the residual heel of unfiltered slurry in contact with the cake in the chamber can be eliminated by inflating the diaphragm; moreover, the cake is thereby compressed uniformly in a way not possible by traditional operation. The effect is the production of dryer cakes and also the ability to operate with thin cakes, with a corresponding increase in the rate of output per unit area.

Filter presses generally operate at pressures up to about 5 or 6 bar, occasionally extending up to approximately 10 bar. The unusually high pressure of 100 bar is used in the unique "tube press," which was originally developed by English Clays for use in the production of

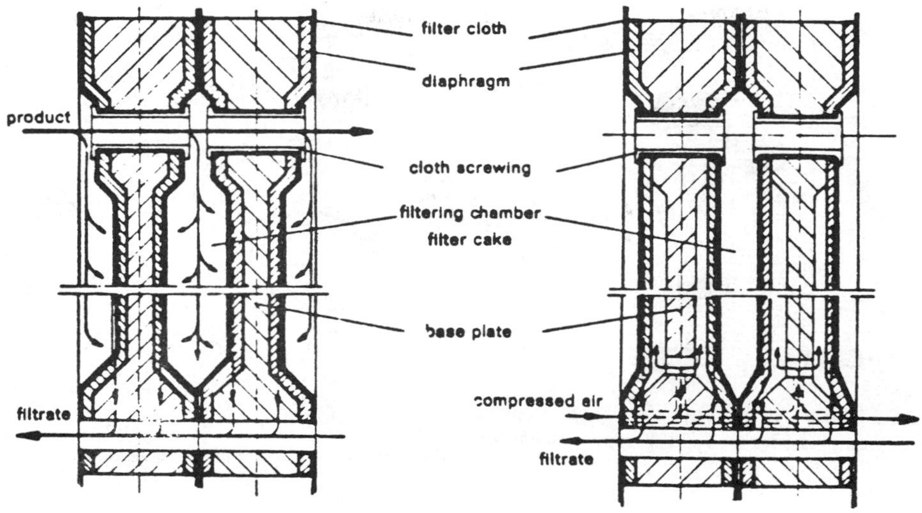

Figure 15.31. Principles of the membrane plate filter press (Eberhard Hoesch and Söhne).

their china clay products. As the name implies, the 'tube press' comprises multiples of individual tubular filters, each about 25 cm diameter and 2 m high; inside this tubular shell is a tubular filter element covered with cloth, surrounded by an inflatable rubber sleeve. The space between this rubber sleeve and the filter element is rapidly filled with a batch of slurry which is then dewatered by inflating the rubber sleeve, the very high pressure then giving an exceptionally dry cake that is discharged through the bottom of the tube. The filter is designed to operate on a short cycle totaling three or four minutes, and is therefore fully automated, several identical units generally being actuated by one set of ancillaries.

While any of the above forms of filter press can be used for dewatering where washing is not required, the absence of the need to wash generally favours the recessed plate rather than the plate and frame filter press, and is likely to make the costly Hoesch tower press uneconomic.

(b) Pressure Vessel Filters

A pressure vessel filter is almost always cylindrical, but may be arranged with its axis either vertical or horizontal; filter elements inside such a vessel may be either flat leaves or tubes, and may be assembled with their surfaces either horizontal or vertical. Tubular elements, typically 2-8 cm diameter and up to 2 meters long, are almost always housed in vertical vessels; total filtering areas extend generally up to 20 m^2 but occasionally to 100 m^2. Vertical vessels are also used to house vertically mounted leaf type elements, up to an area of about 60 m^2. Very large areas, up to about 300 m^2, are generally obtained with vertical leaves in a horizontal vessel, these being arranged either parallel to or across the axis of the vessel.

Around these basic parameters, a wide variety of filters is available commercially. A significant point of variation is the method used to effect cake discharge, depending partly upon whether the cake is required in a dry form or whether it is acceptable as a slurry. Slurry discharge is the easiest and is generally effected by various combinations of spray nozzles, reverse flow, scrapers and vibration, to dislodge the cake from the elements. Figure 15.32 represents a typical method used with tubular element filters.

For dry discharge, a vertical cylindrical vessel needs a wide easily opened access door or valve at the base, through which the cake can fall. With horizontal vessels, the usual choice is between moving back the filter shell to give

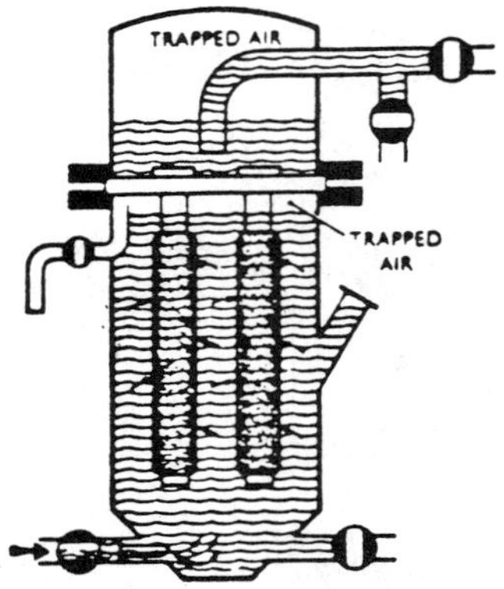

Figure 15.32. The Stella filter uses two trapped pockets of compressed air to aid slurry discharge of cake.

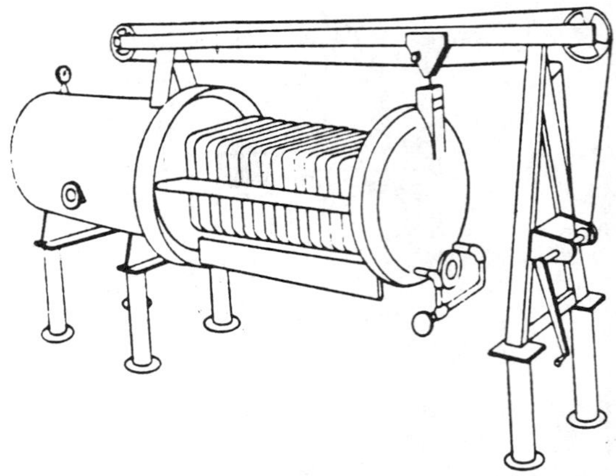

Figure 15.33. Typical dry cake discharge, requiring the filter to be opened at one end.

access to the elements, and withdrawing from the shell the elements either individually or as a battery (Fig. 15.33). Occasionally as with the Sweetland filter, the filter shell is split along its length so that half of it can be swung open.

Continuous Discharge of Cake. Continuous vacuum filters are the main category used in this type of application. A very simple version is the flat bed filter illustrated earlier in Figure 15.17; this uses disposable nonwoven filter medium and operates either by gravity or under a low suction generated by the filtrate pump. Also potentially of use is the horizontal-belt vacuum filter (Figure 15.34), although this is not often used in the absence of the need to wash the cake. Vacuum drum filters, which are suited to this type of slurry, are also generally used in combination with washing; they are therefore considered fully below.

The rotary vacuum disc filter is the major type used when washing is not required. It comprises one or more circular discs (Figure 15.35) mounted on a common horizontal central shaft which is driven so that the discs rotate. Each disc is constructed from a number of segmental elements which can be detached and individually covered with filter cloth, generally in the form of a bag. The central shaft houses a series of pipes into which the individual segments are connected; the pipes lead into a rotary valve connected to the vacuum suction system (Fig. 15.20).

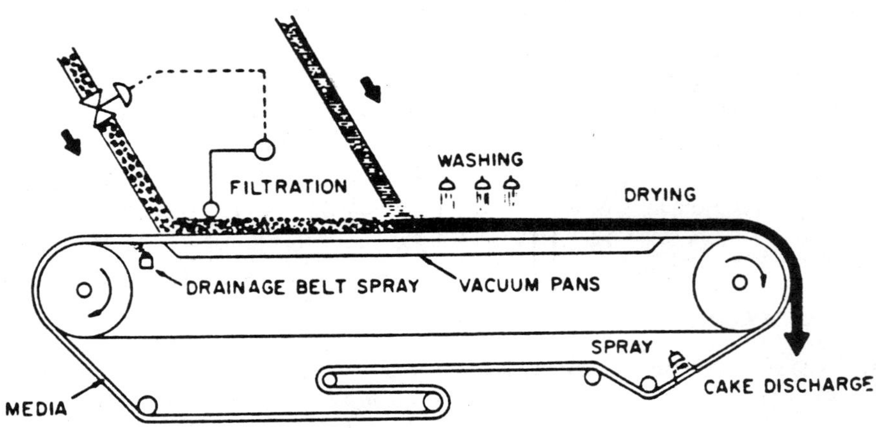

Figure 15.34. Horizontal vacuum belt filter.

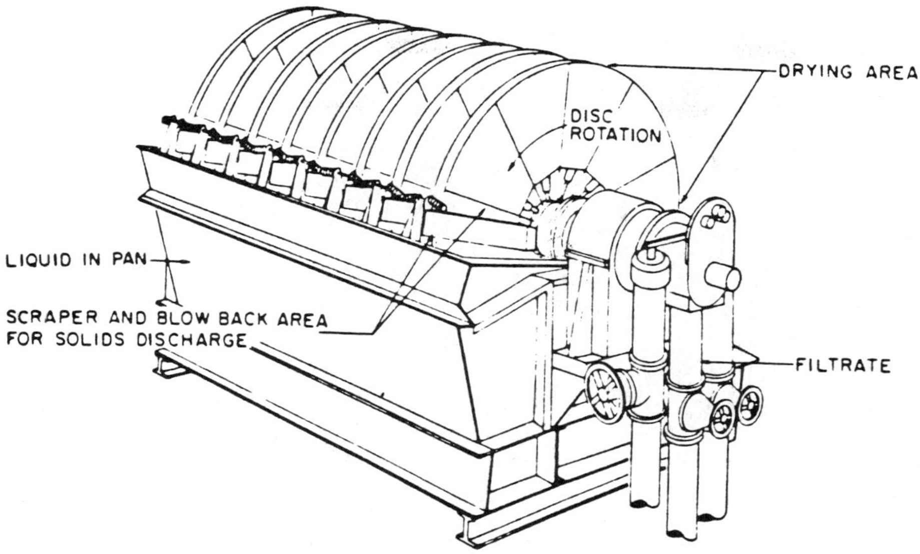

Figure 15.35. Rotary vacuum disc filter.

As the discs rotate, each segment in turn is submerged for a period in the tank of slurry which forms an integral part of the filter. A layer of cake is formed as the segment passes through the slurry; continued suction after the cake emerges from the slurry dewaters and partially dries the cake by drawing air through it. Finally the cake is discharged from the disc, generally by a combination of a reverse blow of compressed air plus a scraper knife; an alternative is a discharge roll to which the cake sticks preferentially, being then removed from this by a scraper.

So as to avoid the mechanical complication of stuffing boxes, the slurry level is generally low enough to avoid submerging the drive shaft. As a consequence, the portion of the disc surface available for cake formation is limited to 20%–45% of the total, the actual figure increasing with the diameter of the disc. Discs are made in diameters from about 0.5 up to 5 m, with up to twelve discs assembled on one shaft, thus giving effective areas ranging from about 1–300 m^2. The speed of rotation is typically 2–5 mpr.

Although not standard on all disc filters, a paddle type agitator is frequently included in the spaces between adjacent discs, since many slurries handled on disc filters have a tendency to settle. For effective discharge, traditional disc filters require a minimum cake thickness of about 10 mm; considerably thinner cakes can be handled using certain forms of modern filter.

Pressurized versions of the vacuum disc filter are achieved by enclosing the total filter inside a vessel that can be pressurized with compressed air or gas. The pressure differential across the filter is generally little if any higher than with a vacuum filter, since higher pressures greatly inflate the construction costs; the advantage of using a positive rather than a negative pressure is that it is then possible to filter highly volatile liquids such as very hot water. In the sugar beet industry, where this filter is frequently used, it is common practice to wash the cake using spray nozzles, although in general washing on a disc filter is uncommon.

The Artisan filter (Fig. 15.36) is an example of the new family of 'delayed cake' filters which have been developed during the last few years. The term 'delayed cake' arises from the incorporation in this type of filter of a high speed mechanical rotor that inhibits the formation of the normal cake, the solids instead discharging continuously as a concentrated slurry. The Artisan filter may be used in combination

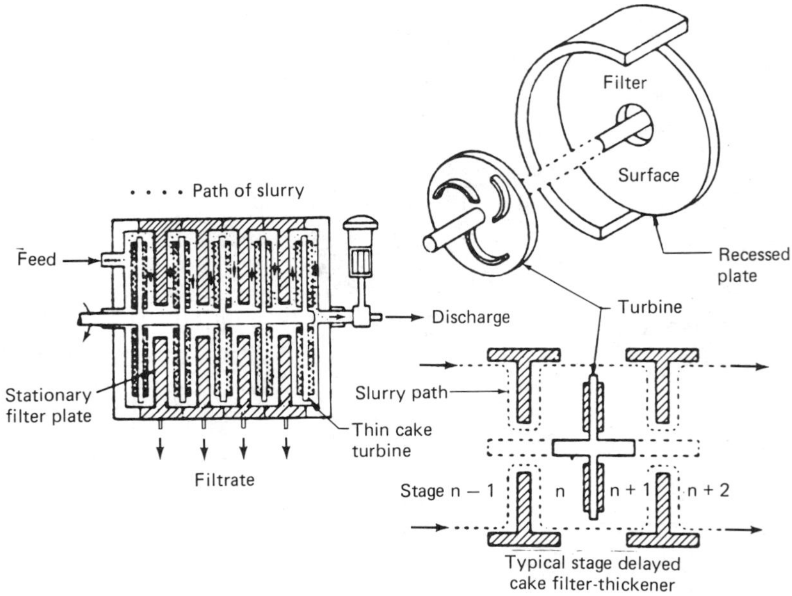

Figure 15.36. The Artisan "delayed cake" filter.

with washing, although its main advantages are its ability to handle very fine material (e.g., dyestuffs) and to give unusually low residual moisture content.[10] It typically operates at a pressure of approximately 20 bar, generated by the slurry feed pump, and is available in sizes with areas from 1–20 m^2.

Compression belt filters constitute a separate important group, of which many versions have been developed during recent years, following the success achieved with the early designs in the dewatering of materials such as sewage sludges. As shown in Figure 15.37, the basic principle is of a pair of converging continuous belts of filter cloth; the sludge introduced between them is subjected to compressive and shear forces, being squeezed between the belts as they pass round a succession of rollers. More recent designs incorporate a more complex path for the belts, so as to increase the shear forces. The attractive simplicity of a belt press is reflected in the low capital cost; on the other hand its successful operation generally depends upon the feed slurry being dosed with substantial quantities of polyelectrolyte, which can impose a significant operating cost.

(ii) Including Cake Washing. *Batch Discharge of Cake.* For small duties, the various forms of vacuum Nutsch and single tipping pan filters

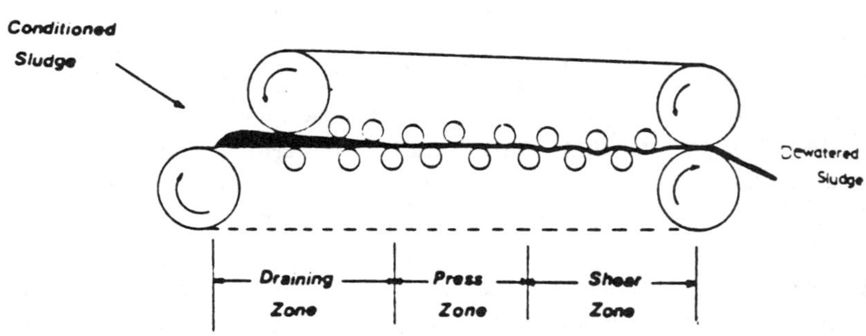

Figure 15.37. Principles of the compression belt filter. (William Jones [Chemical] Engineers, Ltd.)

described above are able to handle a very wide variety of materials. Vacuum filtration is occasionally used also for large scale duties where washing is particularly important but where the fineness of the solids makes it a slow and difficult operation, examples being the manufacture of titanium dioxide and the processing of sugar beet. Thus, the Moore filter comprises a group of parallel vertical filter leaves suspended from an overhead frame in such a way that they can be immersed in sequence in a series of fixed open-topped tanks; an alternative is for the filter leaves to be fixed and the tanks to be moved.

A pressurized version of the Moore filter is the 'Universal' filter developed by the French manufacturer Filtres Gaudfrin for the sugar beet industry; the vertical filter leaves are housed inside a tall cylindrical vessel, twice the height of the leaves, the lower half of the vessel being divided into several segmental compartments into which the leaves are immersed in sequence as they rotate about a central drive shaft.

The simpler forms of vertical leaf pressure vessel filters described above are often used where washing is required. In practice, the switch from unfiltered slurry to wash water frequently presents operating difficulties, especially since it may involve a disturbance to the pressure and thereby cause one or more of the cakes to slip, leading to bypassing, which prevents effective washing. A horizontal filtering surface is therefore preferable, especially if only the upward facing surface on each element is used for filtration.

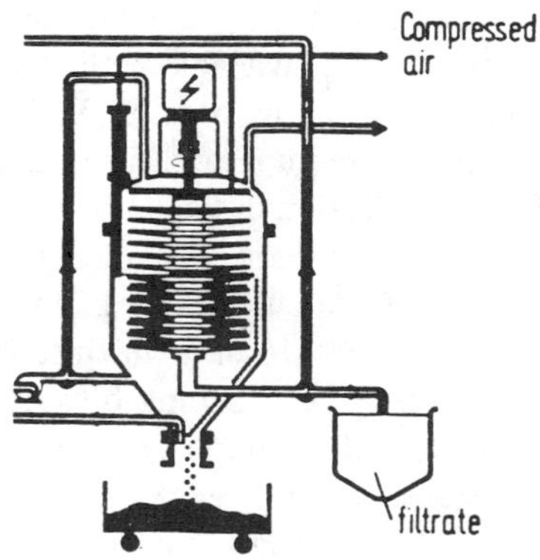

Figure 15.39. The "Funda" filter, with dry cake discharge.

The pressure Nutsch filter is the simplest form of horizontal leaf filter; its disadvantages are that, with only a single filtering element, it is of limited size and generally requires manual discharge (excepting with the Rosenmund filter).

Multiple-element horizontal-leaf pressure filters take three basic forms, typified by Figures 15.38, 15.39, and 15.40. The first (Fig. 15.38), known sometimes both as the "plate filter" and the "enclosed filter press," comprises a stack of circular filter elements which can be pre-

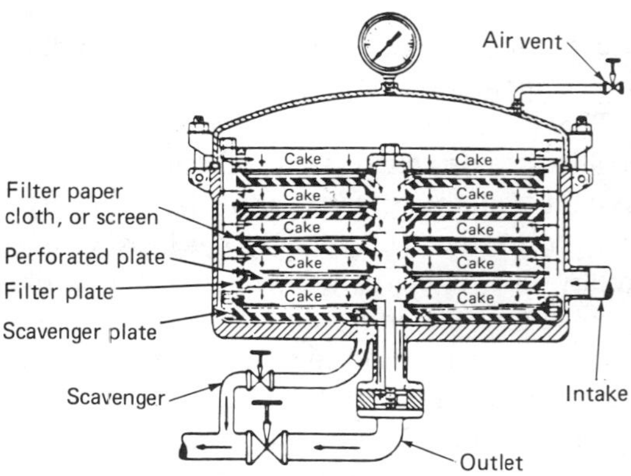

Figure 15.38. The "plate filter" or "enclosed filter press." (Sparkler Mfg. Co.)

Figure 15.40. Manually discharged horizontal leaf filter. (T. Shriver & Co.)

assembled and inserted in a vessel and then removed for cleaning; available with areas from about 0.3 to 25 m^2, this type of filter is used both to collect cakes up to 5 cm in thickness and also for clarification duties.

The centrifical discharge filter (Fig. 15.39) comprises a set of circular filter plates assembled on a central hollow shaft which serves two very different purposes. During filtration, this shaft acts as the filtrate outlet; to discharge the cake, the plates are rotated by a drive connected to the hollow shaft, so that the cake is thrown off by centrifugal action. Depending upon the application, it is thus possible to discharge solid cake from the base of the filter housing, while in other instances the plates are simultaneously sluiced with a suitable liquid so that the cake emerges as a slurry. Filters of this type are available in sizes up to about 25 m^2 and may be fully automated.

The horizontal filter in Figure 15.40 enables each filter element or tray to be removed in turn manually; it is available in sizes from about 1-25 m^2. Mechanized discharge is achieved with the Sparkler HRC, the stack of filter plates being rotated from their normal horizontal position through 90° so that they are vertical for cake discharge; the cake then falls by gravity into the bottom of the filter which is hopper shaped and includes a screw conveyor.

The various batch centrifuges described above can also be applied to slurries in this category, where washing is required, provided the centrifuge basket is lined with a suitable grade of filter medium.

Continuous Discharge of Cake. The versatility of continuous vacuum filters makes various types of them well suited to duties which require washing. Of greatest value, since this can permit almost unlimited washing including six or more counter-current stages, is the horizontal belt filter (Fig. 15.34); this has been discussed above. The horizontal table and tipping pan filters discussed in the same section also permit extensive washing although they are generally not quite so versatile as belt filters; moreover, they are not generally suitable for slow filtering slurries. At the extremity is the disc filter (Fig. 15.35), the vertical filtering surface of which generally permits little if any washing.

The vacuum drum filter, in terms of washing capability, occupies a position midway between the extremes of the belt and disc filters, since the upward facing part of its face can be thoroughly wetted by a well designed spray bar system. This statement applies to the most common form, in which the rotating filtering drum is partially immersed in a tank of slurry, during passage through which a cake is formed; as it emerges from the slurry, continued suction partly dries the cake and also allows it to be washed before it reaches the discharge zones.

Figure 15.41, which shows the typical construction of a drum filter, incorporates the use of string discharge; the cake is formed on top of the series of endless strings which, at the discharge zone, leave the drum and pass over a small diameter external roller, taking the cake with them; as the strings pass round this roller, the cake falls off leaving the strings free to pass through a comb device to maintain their alignment as they return to the drum. More common than string discharge is a reverse blast of compressed air which lifts the cake so that it is caught by a scraper, ideally without the latter actually coming into contact with the filter cloth.

The filter cloth is typically fixed to the surface of the drum by means of a series of longitudinal caulking grooves, which effectively divide the surface into a number of segmental panels; frequently the cloth is finally locked in place by a wire spirally wound at a pitch of 0.5-8 cm. The grade of cloth varies according to the application.

The great majority of drum filters are of annular construction with an internal piping system connecting each segmental panel separately to the rotary valve, much as with a rotary disc filter. Much less common is the form where the entire drum is under vacuum. The rotary valve leads to a receiver and vacuum pump system similar to that used with a disc filter (see Fig. 15.20); the main filtrate and wash liquor can be collected separately if desired.

The portion of the surface of a drum which is

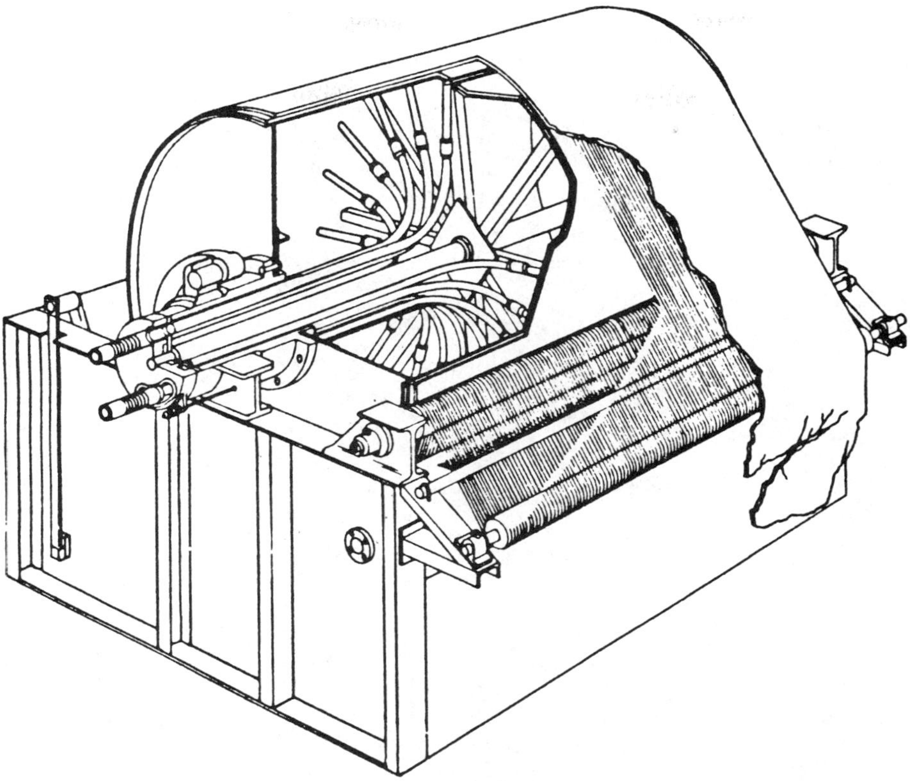

Figure 15.41. Vacuum drum filter, with string discharge.

submerged in the slurry tank may vary widely, generally between a minimum of 15% and a maximum of 70%; 35%–40% is the limit if the complication of submerged trunions is to be avoided. A limiting factor is the number of individual cells into which the annular vacuum space is divided, this depending partly on the size of the drum. The diameter is typically 1-5 m with the length usually not more than twice the diameter, giving areas generally in the range 1-100 m^2; the speed of rotation is typically 1-10 mpr.

A popular adaptation of the basic drum filter is the belt-drum filter. The filter cloth is not directly attached to the drum but is instead in the form of an endless belt which discharges the cake by passing over external rollers, in a manner similar to the string discharge mechanism illustrated in Figure 15.41. Its particular advantage is its ability to handle very thin cakes (e.g., 2 mm) which cannot be discharged by other methods. Furthermore, it gives the opportunity to wash the filter cloth, either continuously or intermittently, on both faces, to combat blinding. Where cake washing is required, however, the belt-drum filter is generally less effective than the standard drum filter because the washing zone is more restricted.

Pressurized versions of the vacuum drum filter are available, the constructional principle being essentially to enclose the whole machine inside a vessel, which is pressurized by air or gas, thus replacing the function of the usual vacuum pump. Even with the operating pressure restricted to 5-10 bar, the bulk and cost of this type of filter impose severe limitations on its application. More common is the corresponding pressure disc filter, notably in the sugar beet industry.

The BHS-Fest is a German rotary pressure filter which has proved useful for specialized applications, such as dyestuffs manufacture. Pressure is provided by pumping slurry directly

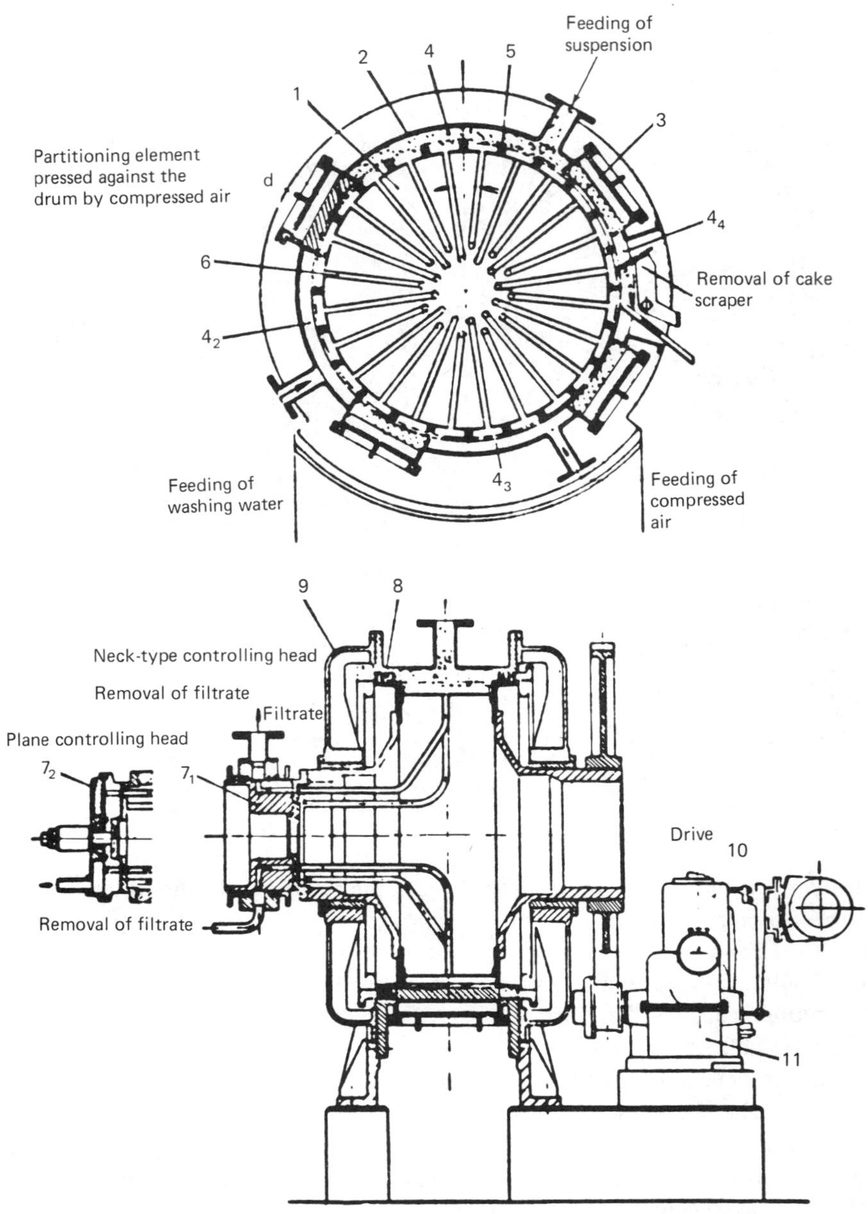

1. Filter drum
2. Pressure-tight housing
3. Partitioning elements
4. Filter chambers
5. Dividing ledges
6. Drain pipes
7. Control head
8. Stuffing box
9. End plate
10. Variable gear
11. Reduction gear

Figure 15.42. The BHS-Fest pressure drum filter.

into one of the annular chambers (Fig. 15.42) in which filtration occurs; as the drum rotates, the cake is transferred to the washing zone, fed by a separate pump. An alternative pump-fed pressure filter suited to washing is the delayed cake type illustrated in Figure 15.36 and discussed above. Wash water can be injected if desired into one or more of the filtration chambers, using either simple or countercurrent washing.

Clarification

Where clarification is the objective, the selection of the filter medium for a specific application is particularly critical since it has a profound effect on the subsequent choice of equipment in which to use it. In practice, this choice is almost always limited to one of the forms of filter described in earlier sections of this chapter. Since the variety of filter media suitable for clarification is relatively limited, it is convenient to consider each type of filter medium in turn and to indicate briefly the types of filter with which it may be used.

Figure 15.43 provides a useful background for this discussion since it summarizes the approximate limit of the clarifying power of the different types of media discussed in Section 5. In this figure, the limiting clarifying powder is expressed as the approximate cut-off point in micrometers, assuming simple surface filtration or depth filtration.

Paper. Cellulose paper is frequently used in plate and frame filter presses (Fig. 15.27), in combination with a filter cloth, in the 'plate filter' (Fig. 15.38) and in the simple Nutsch suction filter. It is also the basis of many cartridges, frequently with a pleated construction to maximize the available surface area; cartridges are used singly or in multiples up to large numbers in pressure housing. Less common is the form of pressure vessel filter with tubular elements, into each of which a disposable sheet of rolled up paper can be inserted, filtration being outward from inside each tube. The fragile nature of glass paper generally inhibits its use industrially, excepting in the form of cartridges.

Sheets. Sheets are most often used in special forms of filter press, a distinctive feature of which is the high quality finish (e.g., polished stainless steel). Another common form is the "enclosed filter press," which closely resembles

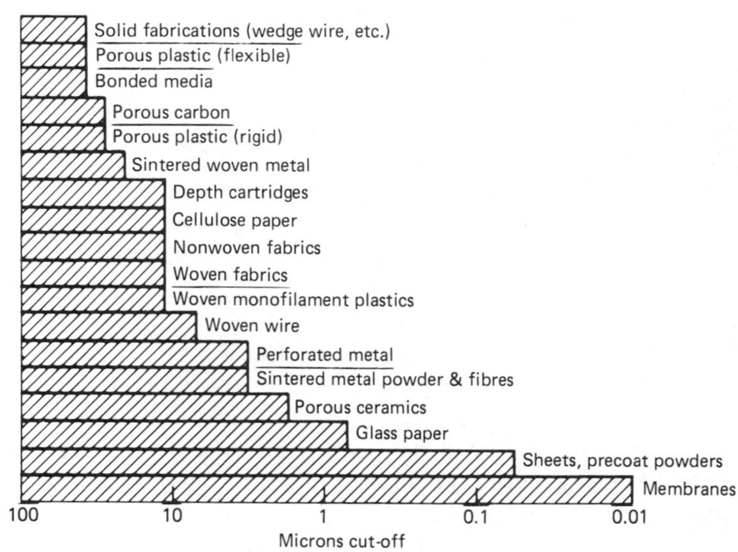

Figure 15.43. Approximate clarifying limits of filter media.

the pressure plate filter illustrated in Figure 15.38. Cartridges are also available but are less common.

Membranes. The fragile nature of traditional cellulose ester membranes limits their industrial application to very special forms of pressure housing able to accommodate flat circular discs of membrane; up to 20 discs, providing a filter area of 1 m^2 or more, can be accommodated in a housing somewhat similar to Figure 15.38.

The combination of cellulose membranes with paper, as a laminate, provides a stronger material which can be used in ordinary filter presses or in the plate type filter shown in Figure 15.38.

Cellulose membranes are also incorporated in some cartridges, although cartridge construction is better suited to the tougher membranes made from other polymeric materials, since these may safely be pleated. These other membranes are also strong enough to be used if desired in filter presses (Fig. 15.27) and plate filters (Fig. 15.38).

Nonwoven Fabrics. Cartridges are frequently made from various forms of bonded and unbonded felts as well as from deep beds of bonded fibers or powders, for use in suitable pressure housings.

Precoat Powders. Precoat powders are widely used with many types of pressure vessel filter, with both tubular and leaf type elements. Generally vertically orientated elements are preferred (Fig. 15.32 and 15.33), since this facilitates discharge of the waste cake as a slurry; if the cake is required in dry form, the centrifugal discharge type of filter as shown in Figure 15.39 is frequently used. Filter presses are also fairly often used for precoat filtration.

The vacuum precoat drum filter is an adaptation of the basic rotary vacuum drum filter (Fig. 15.41). Its distinguishing feature is that filtration takes place through a thick layer of diatomite or expanded perlite, which is deposited on the drum during a preliminary coating operation, and which is progressively removed by a slowly advancing knife which shaves off the surface layer complete with deposited dirt, thereby ensuring that a fresh surface is continuously available. The frequency with which the 6–12 cm bed must be replaced depends on the application but is typically 1–3 days. This form of filter is particularly useful where clarification involves the removal of slimes which blind the surface of the filter medium.

As with the basic rotary vacuum drum filter, pressurized forms of the precoat drum filter are available.

Granular Beds. The sand filter is the classical form of granular bed filter and is of major importance for large scale clarification filtration, such as in water purification. Traditional designs incorporate about 1 m of graded sand and gravel, with downward flow during filtration and upward flow (often augmented with compressed air) intermittently to remove deposited solids. Such beds are housed either in open rectangular concrete tanks, with gravity filtration, or in pump fed metal or plastic vessels. Important developments during recent years include variations in flow pattern, multilayer beds formed from several media of different densities (e.g., sand, anthracite, garnet sand), and moving bed designs which permit fully continuous operation.

LIST OF SYMBOLS

a Constant in Equation (4)
A Filter area
b Constant in Equation (4)
F Fraction of cycle devoted to cake formation
L Thickness
P Differential pressure
Q Volumetric flow rate
Q_c Constant volumetric flow rate
Q_f Final volumetric flow rate
α Specific resistance to filtration per unit weight of dry cake
R_m Resistance of medium to flow

t_c Cycle time
V Volume of filtrate
W Weight of dry cake per unit volume of filtrate
Y Filtrate rate per unit area of continuous filter
Z Output rate of dry cake per unit area of continuous filter
μ Viscosity of liquid
ϕ Permeability of medium.

REFERENCES

1. K. J. Ives (ed.), *The Scientific Basis of Filtration*, Noordhoff International Publishing, Leyden (1975).
2. Albert Rushton and A. Rushton, "Size and Concentration Effects in Filter Cloth Pore Bridging," *Filtration and Separation* 9(3):274–278 (1972).
3. Derek B. Purchas (ed.), *Solid/Liquid Separation Equipment Scale-up*, Uplands Press Ltd., Croydon, England (1977).
4. Derek B. Purchas, *Solid/Liquid Separation Technology*, Uplands Press Ltd., Grayden, England, (1981).
5. Derek B. Purchas, *State of the Art Report: The Selection and Performance of Filter Media for Liquids*, Separation Processes Service, Harwell (1980).
6. Sperry Ehlers, "The Selection of Filter Fabrics Re-examined," *Ind. Eng. Chem.* *53*:552–563 (1961).
7. "Standard Method of Test for Pore Size Characteristics of Membrane Filters for Use with Aerospace Fluids." ASTM F 316-70: American National Standard Z11 270 (1970).
8. H. F. Porter, J. E. Flood, and F. W. Rennie, *Chem. Eng.*, p. 142 (June 20, 1966).
9. *User Guide for the Safe Operation of Centrifuges*, The Institution of Chemical Engineers, London (1978).
10. Alex Bagdasarian and Frank M. Tiller, "Operational Features of Staged High-pressure Thin-cake Filters," *Filtration and Separation* *15*(6):594–598 (1978).

16
Cyclones

David Leith
Harvard School of Public Health
Boston, Massachusetts

16.1. INTRODUCTION

A cyclone is a device without moving parts which spins a gas stream to remove entrained particles by centrifugal force. Cyclones are simple and inexpensive to make, relatively economical to operate and are adaptable to a wide range of operating conditions. Although they are inefficient for collecting particles smaller than about five micrometers in diameter, cyclones have found applications throughout industry for nearly a century. They are probably the dust collectors most frequently used.

Cyclones have been used since the 1880s for the removal of dust from gases.[1] By the turn of the century, they were used to collect sawdust and shavings in woodworking shops. Ten years later, cyclones began to control the dust from cement kilns and shortly thereafter, they were first used to remove fly ash from flue gas.[2] The first published efforts to predict cyclone performance did not appear until about 1930. Extensive studies of the gas flow pattern in cyclones made during the 1940s led to the development of many models for predicting cyclone pressure drop and dust collection efficiency; efforts at modeling cyclone performance have continued to the present. Although our knowledge of what goes on inside a cyclone has increased over the years, the basic cyclone design shown in German patent 39,219 (1885) looks a good deal like a cyclone which might be used today.

16.2. PERFORMANCE CHARACTERISTICS

Types

Over the years, many different types of cyclones have been built. However, the "reverse-flow," or cone-under-cylinder design shown in Figure 16.1 is the type used most often for industrial gas cleaning. In this design, aerosol enters the cyclone at the cylinder top, where the shape of the entry causes the gas to spin. Tangential, scroll, and swirl vane entries have been used as shown in Figure 16.2; tangential entries are most common. After entering the cyclone, the gas forms a vortex with a high tangential velocity which gives particles entrained in the gas a high centrifugal force, throwing them to the cyclone wall for collection. Below the bottom of the gas exit duct, the spinning gas gradually migrates inward, to a "central core" along the cyclone axis, and from here up, out the gas exit duct. Collected dust descends the cyclone walls to the dust

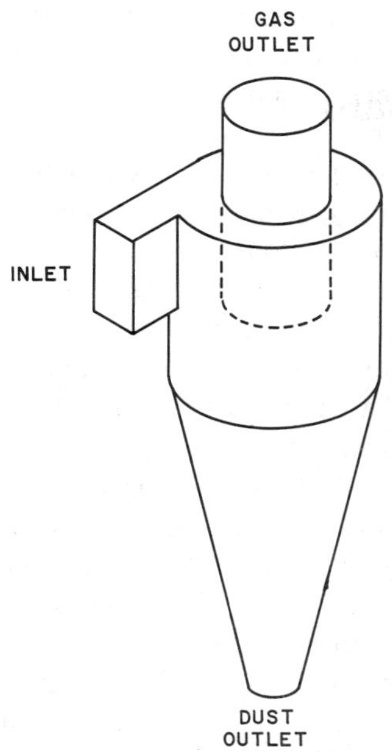

Figure 16.1. Reverse-flow cyclone.

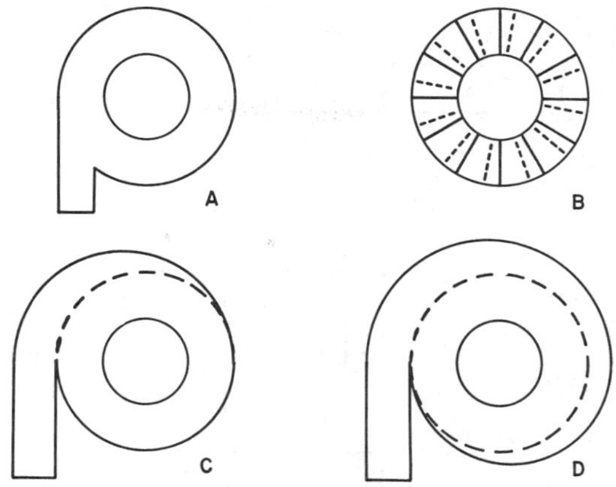

Figure 16.2. Cyclone entries: A, tangential; B, swirl vane; C, half scroll; D, full scroll.

outlet at the bottom of the cone, primarily due to the downward component of the gas velocity at the cyclone wall rather than due to gravity.

Figure 16.3 shows a "straight-through" cyclone. Here, dusty gas enters at one end while cleaned gas and separated dust exit separately at the opposite end. Again, the entry shape causes the inlet gas to spin. Swirl vane entries are used most often on straight-through cyclones. Within the cyclone, centrifugal force pushes particles to the wall. Cleaned gas leaves from a central exit duct while separated particles flow out with a small amount of purge gas through the annular dust discharge. The remainder of this chapter will be devoted to the reverse-flow cyclone, as it is the type used by far the most often for industrial dust control. Particle collection theory for straight-through cyclones[3] is not as well developed as that for conventional reverse-flow cyclones.

Standard Designs

Figure 16.4 shows the eight dimensions of reverse-flow cyclone. It is convenient to express cyclone dimensions as multiples of the diameter, D. The "dimension ratios," a/D, b/D, De/D, S/D, h/D, H/D, and B/D, allow comparing the shape of two or more cyclones which might differ in size. Several sets of dimension ratios, or "standard designs" are given in Table 16.1. The data from Stern et al. are average values found in a survey of 23 cyclones.

A comparison of the designs in Table 16.1 reveals that cyclone shape varies with recom-

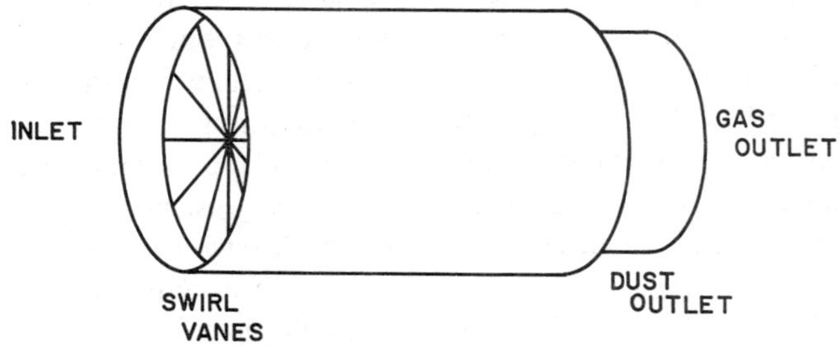

Figure 16.3. Straight-through cyclone.

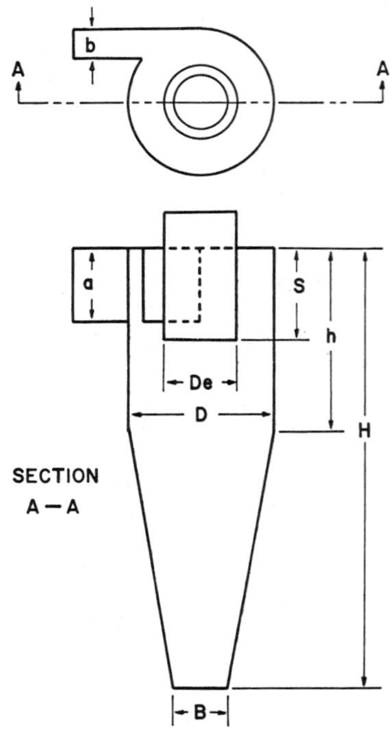

Figure 16.4. Dimensions of reverse-flow cyclone.

mended duty. A high efficiency cyclone has a smaller inlet area (a/D and b/D) and exit area (De/D) than does a high throughput cyclone. Gas outlet length (S/D) is less in the high efficiency designs, probably because inlet height (a/D) is less. Outlet length should be greater than inlet height in order to be sure that a stable vortex is formed within the cyclone body. The high efficiency and general purpose standard designs have tangential gas entries, while the high throughput designs have scroll entries. High efficiency can be traded against high throughput for cyclones operating at the same pressure drop.

Because cyclone design changes with recommended duty, no single optimum cyclone design exists which will work best for all possible applications. Design of a cyclone appropriate for a particular task involves compromises among number of cyclones, throughput per cyclone, pressure drop, and efficiency.

Application Areas

Cyclones generally collect particles larger than five to ten micrometers in diameter with reasonable efficiency. For smaller particles, efficiency declines rapidly. For particles larger than several hundred micrometers, a settling chamber can often be used with lower first cost and operating cost than would be necessary for a cyclone system. Cyclone pressure drop is similar to that found in other particle collection devices, except for high energy scrubbers which can require much higher pressure drops. Values for cyclone pressure drop range from about 0.1 to 2 kPa (0.5 to 8 inches of water).

A properly designed cyclone can process effectively dusts in very high concentrations, and in practice loadings of over 2000 g/m^3 (1000 gr/ft^3) have been accommodated.[4] Cyclones have the fortunate ability simultaneously to increase efficiency[9-12] and decrease pressure drop[13-15] with an increase in dust loading. This may come about due to the increased number of particles which move radially outward through the cyclone vortex when dust loading increases. This movement might hinder the formation of the vortex and thereby decrease pressure drop. Despite vortex suppression, efficiency might increase due to the increased opportunity for larger particles to strike and collect smaller particles while they move toward the cyclone wall.

Cyclones are available in many sizes, and can be made from materials able to withstand extreme operating conditions. They are commercially available in sizes to process 50 to 50,000 m^3/hr. While smaller diameter units generally are more efficient,[4,16-17] a manifold must be used to connect many small cyclones together to process a large gas flow. Refractory lined cyclones have been operated at temperatures of 1000°C, while other units have run at pressures of several hundred atmospheres.[18] However, special materials of construction chosen to allow operation under extreme conditions may not always have good resistance to erosion of the cyclone walls by collected dust. Sticky, hygroscopic dusts may not discharge readily through the dust outlet[8] and these dusts may be better suited to collection in a scrubber. Small diameter cyclones housed together and operating in parallel, sometimes called multiclones, are frequently connected to the same

Table 16.1. Standard Designs for Reverse-Flow Cyclones.

SOURCE (REF.)	RECOMMENDED DUTY	D	a/D	b/D	De/D	S/D	h/D	H/D	B/D	C	ΔH	Q/D^2 (m/hr)
Stairmand (4)	High efficiency	1	0.5	0.2	0.5	0.5	1.5	4.0	0.375	55.1	5.4	5500
Swift (5)	High efficiency	1	0.44	0.21	0.4	0.5	1.4	3.9	0.4	64.6	9.2	4940
Lapple (6, 7)	General purpose	1	0.5	0.25	0.5	0.625	2.0	4.0	0.25	50.4	8.0	6860
Swift (5)	General purpose	1	0.5	0.25	0.5	0.6	1.75	3.75	0.4	47.7	7.6	6680
Stern et al. (8)	Consensus	1	0.45	0.2	0.5	0.625	0.75	2.0			no data	
Stairmand (4)	High throughput[b]	1	0.75	0.375	0.75	0.875	1.5	4.0	0.375	na[a]	7.2	16500
Swift (5)	High throughput[b]	1	0.8	0.35	0.75	0.85	1.7	3.7	0.4	na[a]	7.0	12500

[a]Not applicable.
[b]Scroll type gas entry used.

dust bin without valves on the dust discharge of each unit. Unequal inlet pressure distribution across the inlet and exhaust manifolds may cause gas to flow out the exit duct of some cyclones, through the dust bin, then up and out through the dust exit and gas exit of other units. This flow pattern will adversely affect the performance of the cyclone system. The performance of multiclones is almost never as good as that of each small cyclone operating individually. However, multiclone performance should be better than that of a single large diameter cyclone operating at the same pressure drop and handling the same gas flowrate as the manifolded design. The small diameter cyclone, manifolded design does offer the advantage of compact installation.

Industrial processes use cyclones for unloading material from process gas streams, and for controlling particulate emissions to the atmosphere. The dry product collected in a cyclone can often by recycled to the plant for further processing. Among the processes using cyclones are coal driers, grain elevators, grain driers and milling operations, sawmills and wood-working shops, asphalt plant rotary rock driers, and detergent manufacturing processes. Teams of cyclones operating in series are used under high temperature and pressure conditions to collect catalyst dust from catalytic cracking units at oil refineries and to collect fly ash generated from coal combustion in a pressurized fluidized bed.

16.3. PERFORMANCE MODELING

Flow Pattern

To understand pressure loss and particle collection in a cyclone, it is important to understand the cyclone gas flow pattern. Cyclone flow pattern has been studied in some detail; the overall trend of gas motion has been generally confirmed by all workers. However, no generalized model of the flow pattern is available which will allow the prediction of all velocity components at any point in the cyclone. Each experimenter has concentrated on at most several cyclones of similar design, none of which is particularly similar to the units studied by others. While each set of results is internally consistent, ambiguities arise when comparing the trends reported in several studies. The most complete review of the gas flow patterns in cyclones is probably that by Jackson.[1]

Tangential Gas Velocity. After entering the cyclone, the gas stream forms a confined vortex such that the tangential gas velocity, v_t, is related to the distance, r, from the cyclone axis by Equation (1).

$$v_t r^n = \text{constant} \tag{1}$$

The vortex exponent, n, is +1 for an ideal fluid, 0 for velocity which is constant regardless of radial position, and −1 for rotation as a solid body. While theoretical descriptions of cyclone flow patterns[17,19,20] employ values of n from +1 to −1, data[14,21–23] indicate that the usual range is from 0.5 to 0.9, with most values around 0.5. Tangential gas velocity, therefore, increases from a minimum at the cyclone wall to higher values approaching the cyclone axis. Tangential velocities may be lower than the gas inlet velocity at the cyclone wall, but can exceed the inlet velocity by several times at some distance from the wall. Alexander[24] has given an empirical expression for dependence of vortex exponent on cyclone diameter in meters, D_m, at temperature of 283° K.

$$n = 0.67\, D_m^{0.14} \tag{2}$$

He gives another expression to allow adjustment of the vortex exponent for gas temperature variations.

$$\frac{1 - n_1}{1 - n_2} = \left(\frac{T_1}{T_2}\right)^{0.3} \tag{3}$$

Figure 16.5 shows the relationship between vortex exponent, n, cyclone diameter, D, and gas temperature as given by Equations (2) and (3).

If Equation (1) applied throughout the cyclone, tangential gas velocity would increase

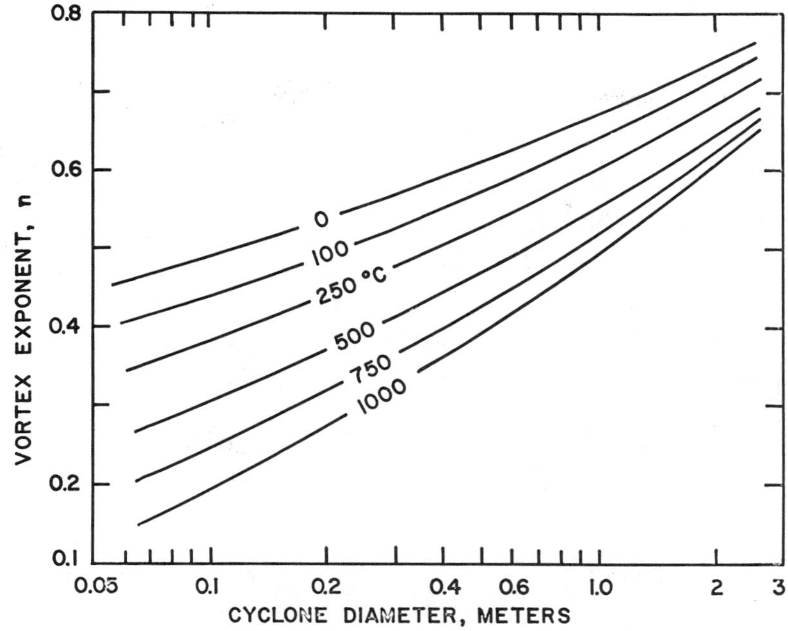

Figure 16.5. Vortex exponent n as function of cyclone diameter and gas temperature, according to Equations (2) and (3).

with decreasing radius to infinity at the cyclone axis. Actually, within the portion of the cyclone above the bottom of the gas outlet duct, tangential velocity is limited by the gas outlet wall. At the cyclone body wall, and at the gas exit wall, the tangential velocity falls off rapidly. If the gas exit is large relative to body diameter, wall effects may hinder the formation of the vortex to the point that tangential velocity becomes almost constant rather than increasing with decreasing radius as is normally the case.[14]

Below the gas exit, tangential velocity increases with decreasing radius, up to a maximum at a point defined as the radius of the cyclone central core. Within this core, tangential gas velocity decreases with decreasing radius, and falls ultimately to a value near zero at the cyclone axis. Within the core, Equation (1) still describes the tangential gas velocity but with a revised value for the vortex exponent ranging[23] from about -0.5 to less than -2. The diameter of the central core ranges from about 0.5 to 1.0 times the gas outlet diameter.[4,25] The core diameter seems to decrease from a value close to that of the gas exit diameter directly below the gas exit to a value of about a third of the gas exit diameter near the bottom of the cyclone.[1] Figure 16.6 presents ter Linden's measurements[22] of tangential velocity at different vertical positions in a cyclone with scroll entry.

Vertical and Radial Gas Velocity. In general, the gas within the cyclone flows downward near the cyclone wall and upward near the cyclone axis; these vertical velocities, both downward and upward, are much less than tangential gas velocities. The radial position at which vertical velocity changes from down to up is relatively closer to the cyclone wall at the top of the unit than at the base of the cone. At all vertical locations, the velocity changeover point appears to be outside the central core. However, once within the core the upward gas velocity increases substantially. Figure 16.7 shows measurements of vertical velocity for ter Linden's cyclone.[22]

The radial component of gas velocity has not been measured as extensively as have the tan-

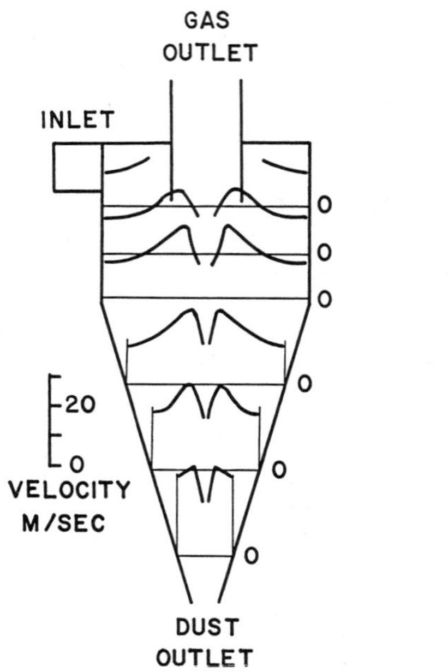

Figure 16.6. Tangential gas velocities in reverse-flow cyclone.

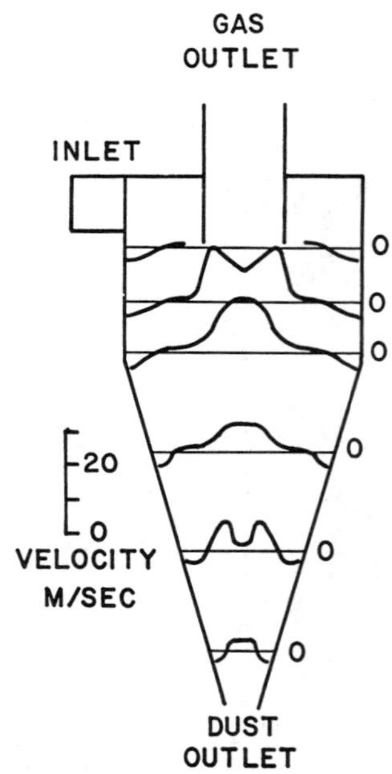

Figure 16.7. Vertical gas velocities in reverse-flow cyclone.

gential and vertical components. Data show that inward radial velocity is low, constant with radial position, and approximately equal at all vertical positions within the cyclone below the gas outlet duct. However, these data are not self-consistent as radial velocity must be greater near the central core from conservation of mass principles. Radial gas velocity is the most difficult velocity component to measure experimentally. Still, knowledge of this component is essential for determining particle collection efficiency through the "static particle" approach described below. Lack of data on this point has led to unproven speculation on the variability of the radial velocity which is used to explain the inadequacies of this efficiency theory. ter Linden's measurements[22] of radial velocity are given in Figure 16.8.

Pressure Distribution. The total pressure at any point in a cyclone is the sum of the static pressure and velocity pressure at that point. Total pressure slowly decreases from a maximum value at the cyclone wall to a minimum value near the cyclone axis. With the high tangential gas velocities present in a cyclone, velocity pressures can be so high that static pressure becomes negative relative to the atmosphere. The static pressure within the central core can be negative, even when the cyclone is installed on the discharge side of a fan. The zone of negative static pressure can extend from the core through the dust outlet and if a suitable valve is not used at the dust outlet, into the dust collection bin. If no valve or a leaky valve is fitted and the dust collection bin is not airtight, dusty air from the bin will be drawn into the central core, up and out of the cyclone. For this reason the cyclone dust hopper should always be airtight. ter Linden's measurements of static and total pressure[22] in a cyclone are shown in Figure 16.9.

Overall Gas Flow Pattern. As gas enters the cyclone, it forms a vortex in the annulus above the gas outlet duct. Below the gas outlet, spinning gas gradually migrates into the central core. Near the cyclone walls, gas flows downward, whereas gas closer to and within the central core flows upward toward the gas outlet duct.

As the gas spirals down through the cyclone,

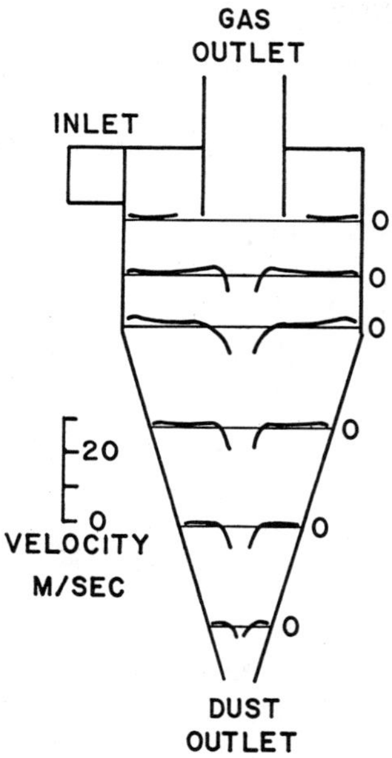

Figure 16.8. Radial gas velocities in reverse-flow cyclone.

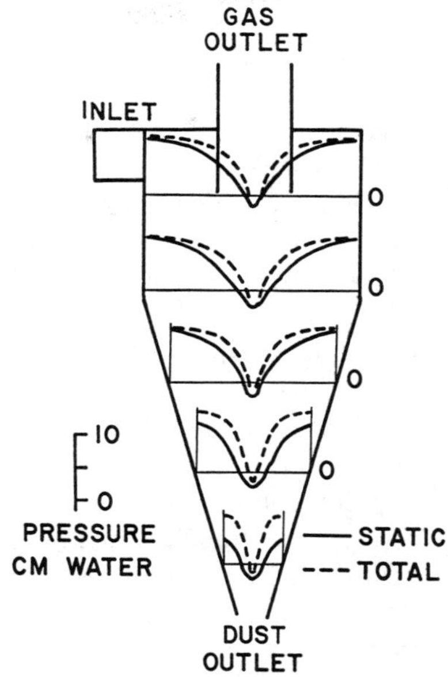

Figure 16.9. Static and total pressures in reverse-flow cyclone.

more and more of it flows into the central core and out the outlet duct. At some point, all the gas will have flowed into the central core. Alexander[22] called this point the cyclone "natural length," that distance below the bottom of the exit duct where the vortex turns. He stated that the natural length, Z, depends only upon the proportions of the cyclone inlet and outlet, and is independent of the gas inlet velocity.

$$Z = 2.3\, De \left(\frac{D^2}{ab} \right)^{1/3} \tag{4}$$

Presumably, that portion of the cyclone below the natural length will not have value for collecting dust because it will not contain a vortex. Conversely, a cyclone whose natural length potentially extends below the actual base of the cyclone will not develop its full separating potential, as the gas will be forced into the central core before it would have entered naturally.

First[23] showed that after the gas has enters the cyclone and makes one full revolution, it is not entirely displaced downward by gas entering subsequently. Some older gas is forced toward the cyclone axis in an inward spiral, a phenomenon First calls "lapping." As the newer gas squeezes the older gas toward the cyclone axis, the tangential velocity of the older gas increases through conservation of momentum.

Pressure Drop

Factors[14] that contribute to cyclone pressure drop, static pressure differential across the cyclone, include:

1. Gas expansion as it enters the cyclone,
2. Formation of the vortex,
3. Wall friction, and
4. Regain of the rotational kinetic energy as pressure energy.

The first three factors are probably the most important. Controversy exists over the importance of wall friction on pressure drop, as Iinoya[15] has shown that sand glued onto the cyclone walls, increasing wall roughness, actually decreases the pressure drop. If this is correct, then energy consumption due to vortex formation plays a greater role in pressure drop than does wall friction. First[23] also found that wall friction makes an insignificant contribution to overall pressure drop.

Devices such as an inlet vane, an extension of the inner wall of the tangential gas entry within the cyclone body up to a position close to the gas exit duct, and a cross baffle in the gas outlet duct will lower pressure drop. However, these devices probably suppress vortex formation,[1,8,26] and so decrease efficiency as well as pressure drop. Because a cyclone is a device for vortex generation, it is not logical to put attachments within it which inhibit vortex formation. Cyclones can be designed for low pressure drop without resorting to internal attachments which may impair efficiency.

Many investigators have developed expressions to predict pressure drop; some are empirical, some theoretical, and most a mixture of both. Despite the complexity of some pressure drop relationships, no single expression has been developed that will give a reliable estimate of pressure drop for all cyclones operating under all conditions.

Cyclone pressure loss is expressed most conveniently as a number on inlet velocity heads, ΔH. Velocity heads can be converted to loss in pressure units, ΔP, by Equation (5).

$$\Delta P = \Delta H(\tfrac{1}{2}\,\rho_G v_i^2) \tag{5}$$

The number of inlet velocity heads, ΔH, will be constant for any cyclone design although the pressure loss, ΔP, varies with different operating conditions. Pressure drop for a cyclone can best be established by determining ΔH experimentally for a particular cyclone design. The static pressure loss, ΔP, for geometrically similar cyclones can then be found from Equation (5) for different operating conditions. Values of ΔH are listed in Table 16.1 for the standard design cyclones listed there.

Many analytical expressions for determining ΔH from cyclone geometry have been presented in the literature. Several are listed in Table 16.2. One review[27] found that the Barth, Stairmand, and Shepherd and Lapple equations work better than those by Alexander and First. The Barth and Stairmand approaches are complex and require knowledge of all cyclone dimensions. The Shepherd and Lapple approach, Equation (6), is simpler to use, and while it does not include all cyclone dimensions it nevertheless gives results about as good as those produced by the more complex calculation methods.

$$\Delta H = 16\,\frac{ab}{De^2} \tag{6}$$

Values of cyclone pressure drop calculated from theory may give results in error by 50 percent or more. There is currently no alternative to experimental testing when cyclone pressure drop must be known accurately.

Efficiency

Collection efficiency, η, is defined as that fraction of particles of a certain size which are collected by the cyclone. Experience in dealing with cyclones has shown that collection efficiency increases with:

1. Increasing particle diameter and density,
2. Increasing gas inlet velocity,
3. Decreasing cyclone diameter,
4. Increasing cyclone length,
5. Drawing some of the gas from the cyclone through the dust exit duct, and
6. Wetting the cyclone walls.

A plot of collection efficiency against particle diameter is called a fractional efficiency curve or grade efficiency curve. A typical fractional efficiency curve for a cyclone is shown in Figure 16.10. Fractional efficiency rises rapidly at first, then flattens out and approaches unity for vary large particles.

Particles are separated from the gas stream in a cyclone by spinning to the cyclone wall through centrifugal force. Figure 16.11 shows the forces acting on a particle rotating with

Table 16.2. Equations for Predicting Pressure Loss as Number of Inlet Velocity Heads, ΔH.

SOURCE	PRESSURE LOSS EQUATION	
Shepherd and Lapple[28]	$\Delta H = 16 \frac{ab}{De^2}$	(6)
First[23]	$\Delta H = \frac{ab}{De^2}\left(\frac{24}{\left(\frac{h\,(H-h)}{D^2}\right)^{1/3}}\right)$	(7)
Alexander[24]	$\Delta H = 4.62 \frac{ab}{DDe}\left(\left(\left(\frac{D}{De}\right)^{2n} - 1\right)\left(\frac{1-n}{n}\right) + f\left(\frac{D}{De}\right)^{2n}\right)$	(8)
	$n = 0.67\, D_{\text{m}}^{0.14}$ at 283° K	(2)
	$\frac{1-n_1}{1-n_2} = \left(\frac{T_1}{T_2}\right)^{0.3}$	(3)
	$f = 0.8\left(\frac{1}{n(1-n)}\left(\frac{4-2^{2n}}{3}\right) - \left(\frac{1-n}{n}\right)\right) + 0.2\left((2^{2n}-1)\left(\frac{1-n}{n}\right) + 1.5(2^{2n})\right)$	(9)
Stairmand[29]	$\Delta H = 1 + 2\,\phi^2\left(\frac{2\,(D-b)}{De} - 1\right) + 2\left(\frac{4ab}{\pi De^2}\right)^2$	(10)
	$\phi = \frac{-\left(\frac{De}{2\,(D-b)}\right)^{1/2} + \left(\frac{De}{2\,(D-b)} + \frac{4\,G^* A}{ab}\right)^{1/2}}{\frac{2\,G^*\,A}{ab}}$	(11)
	$A = \frac{\pi}{4}(D^2 - De^2) + \pi Dh + \pi\, DeS + \frac{\pi}{2}(D+B)\left((H-h)^2 + \left(\frac{D-B}{2}\right)^2\right)^{1/2}$	(12)
	$G^* = 0.005$	(13)
Barth[25]	$\Delta H = \left(\frac{4ab\theta}{\pi De^2}\right)^2 \epsilon$	(14)
	$\epsilon = \frac{De}{D}\left(\frac{1}{(1 - 2\theta(H-S)(\lambda/De))^2} - 1\right) + 4.4\,\theta^{-2/3} + 1$	(15)
	$\theta = \frac{\pi De(D-b)}{4ab\alpha^* + 2\,(H-S)(D-b)\pi\lambda}$	(16)
	$\alpha^* = 1 - \frac{1.2\,b}{D}$	(17)
	$\lambda = 0.02$	(18)

tangential velocity u_t at radial position r. The particle moves radially outward with velocity u_r. The tangential velocity of the gas and that of the particle will be assumed equal, $v_t = u_t$. This is probably a reasonable assumption for small particles, for which efficiency is most difficult to determine.

The centrifugal force, F_c, acting on the particle is given in Equation (19):

$$F_c = \frac{\pi d^3 \rho_p v_t^2}{6r} \qquad (19)$$

The drag force, F_d, acting on the particle as it

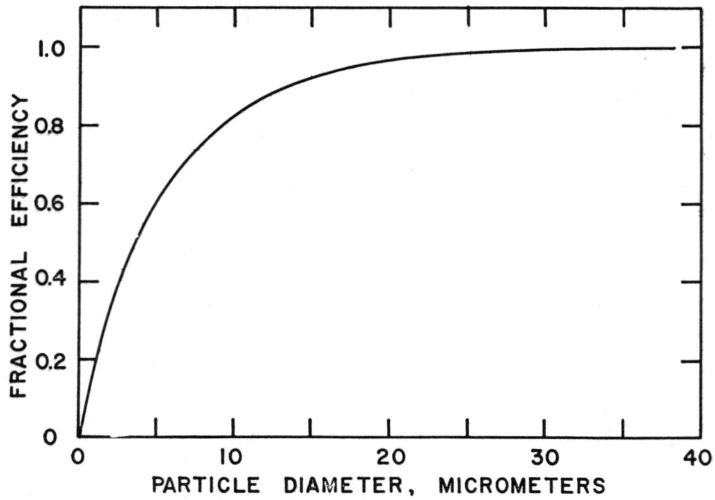

Figure 16.10. Typical cyclone fractional efficiency curve.

moves rapidly outward can be given by Stokes' law; for larger particles with higher radial velocities Stokes' law becomes a progressively poorer approximation.

$$F_d = 3\pi\mu d(u_r - v_r) \tag{20}$$

Equation (1), which describes gas tangential velocity as a function of radial position, gives tangential velocity at position r as a function of cyclone wall radius, r_w, and the tangential velocity at the wall, v_{tw}.

$$v_t r^n = \text{constant} = v_{tw} r_w^n \tag{1}$$

Strictly speaking, the gas tangential velocity at the cyclone wall must be zero. However, the boundary layer at the cyclone wall is thin, as can be seen in Figure 16.6, and Equation (1) describes the tangential velocity near the wall with little error. The sum of the centrifugal and drag forces acting on the particle will equal its mass times its acceleration.

$$\left[\frac{\pi d^3 \rho_p v_t^2}{6r} - 3\pi\mu d(u_r - v_r)\right] = \frac{\pi d^3}{6}\rho_p \frac{d^2 r}{dt^2} \tag{21}$$

Simplifying, and making the substitutions $u_r = dr/dt$ and $v_t = v_{tw} r_w^n / r^n$ yields Equation (22).

$$\frac{d^2 r}{dt^2} + \frac{18\mu}{d^2 \rho_p}\frac{dr}{dt} - \left(\frac{v_{tw}^2 r_w^{2n}}{r^{(2n+1)}}\right) + \frac{18\mu v_r}{d^2 \rho_p} = 0 \tag{22}$$

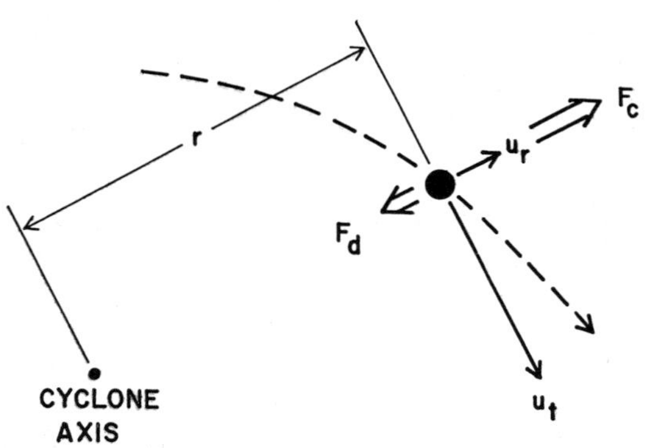

Figure 16.11. Forces on a particle in a cyclone.

Equation (22) describes radial particle motion within a vortex and underlies many approaches used to calculate cyclone collection efficiency. Unfortunately, Equation (22) has not been solved analytically. Approximate solutions can be found by postulating various flow conditions within the cyclone allowing deletion of some terms in the equation. All these approximations are open to criticism. The relative importance of each term will change with each cycone design and particle diameter. It is unlikely that any approximation will yield good results for all applications.

The theoretical efficiency of a cyclone can be characterized in terms of a "critical particle" diameter, d_{100}. The critical particle is that which, according to theory, is collected with 100% efficiency. Since collection efficiency increases gradually with increasing particle diameter and only approaches 100% as a limit, the critical particle is not observed experimentally. Calculations of critical particle diameter will locate the approximate position of the "knee" in the fractional efficiency curve. Two approaches for calculating critical particle diameter have been developed: the "static particle" approach and the "timed flight" approach.

Critical Diameter: The Static Particle Approach. Gas at the edge of the central core will have maximum tangential velocity, v_{max}. Gas flowing radially inward to the cyclone central core will flow past the core edge with an average inward radial velocity given by Equation (23).

$$v_r = \frac{-Q}{2\pi r_{core}\,(H - S)} \tag{23}$$

For particles of the critical diameter, the inward drag force caused by the inrushing gas will just balance the outward centrifugal force caused by their rapid rotation about the cyclone axis. These "static particles" will theoretically remain suspended at the edge of the central core. Larger particles should spin out to the cyclone wall and become collected, and smaller particles should flow past the static particles into the central core and out the cyclone. As they are stationary, the critical particles will have no radial acceleration or velocity ($d^2r/dt^2 = dr/dt = 0$). From Equation (1), $v_{tw}^2 r_w^{2n} = v_{max}^2 r_{core}{}^{2n}$, which when substituted into Equation (22) yields the critical particle diameter.

$$d_{100} = \left(\frac{9Q\mu}{\pi(H - S)\rho_p v_{max}^2}\right)^{1/2} \tag{25}$$

According to static particle theory, "fractional degree of dust separation in the case of a (critically sized) particle suddenly increases from 0 to 100%."[25] The shape of experimental fractional efficiency curves, which show a gradual increase in efficiency with increasing particle diameter, is explained as being due to variations in the gas inward drift velocity, v_r, along the cyclone axis.

Critical Diameter: The Timed Flight Approach. The timed flight approach is another way to calculate critical diameter, and involves a different set of assumptions for solving Equation (22). Let r_i be the innermost radial position at which particles enter the cyclone. Particles entering at this point must cross the distance from r_i to the cyclone wall to be collected, and if a particle is not in the cyclone long enough to travel this distance it will escape collection. In the timed flight approach, the particle's radial acceleration and the gas radial velocity are arbitrarily set equal to zero and neglected ($d^2r/dt^2 = v_r = 0$). The gas velocity in the entrance duct is assumed equal to the velocity at the cyclone wall, $v_i = v_{tw}$. These assumptions allow solution of Equation (22) for particle radial position as a function of particle residence time within the cyclone, t. The particle with critical diameter will travel from its initial position at r_i and just reach the cyclone wall in time t.

$$d_{100} = \left[\frac{9\mu D^2\left(1 - \dfrac{2r_i}{D}\right)^{2n+2}}{4(n+1)v_i^2\rho_p t}\right]^{1/2} \tag{26}$$

Investigators have made assumptions about the initial particle radial position, r_i, and the value for vortex exponent, n. Residence time, t, is sometimes defined in terms of an empirical "number of turns," N, which the gas stream makes within the cyclone. The value for N reportedly varies from 0.3 to 10, with a mean value of about 5.[30] Table 16.3 gives several sets of assumptions for r_i, n, and t along with the resultant equations for critical particle diameter. Table 16.3 also lists Lapple's Equation (6) for cut particle diameter, d_{50}, the particle size collected with 50% efficiency. Particles of this size theoretically enter the cyclone at the midpoint of the gas entry.

Table 16.3. Equations Derived from Equation (21) for Predicting Collection Characteristics.

$$\frac{d^2r}{dt^2} + \frac{18\mu}{d^2\rho_p}\frac{dr}{dt} - \left(\frac{v_{tw}^2 r_w^{2n}}{r^{(2n+1)}} + \frac{18\mu v_r}{d^2\rho_p}\right) = 0 \qquad (22)$$

STATIC PARTICLE APPROACH	$\frac{d^2r}{dt^2}$	$\frac{dr}{dt}$	ASSUMPTIONS r_{core}	v_{max}	RESULTANT EQUATION	
Stairmand (3)	0	0	$De/4$	$\frac{Q\phi}{ab}\left(\frac{2\left(D-\frac{b}{2}\right)}{De}\right)^{1/2}$	$d_{100} = \left(\frac{9Q\mu}{2\pi(H-S)\rho_p v_{max}^2}\right)^{1/2}$	(24)
Barth (25)	0	0	$De/2$	$\frac{4Q\theta}{\pi De^2}$	$d_{100} = \left(\frac{9Q\mu}{\pi(H-S)\rho_p v_{max}^2}\right)^{1/2}$	(25)

TIMED FLIGHT APPROACH	$\frac{d^2r}{dt^2}$	v_r	ASSUMPTIONS r_i	n	t	RESULTANT EQUATION	
Rosin et al. (20)	0	0	$\frac{D}{2} - b$	0	$\frac{\pi DN}{v_i}$	$d_{100} = \left(\frac{9\mu b}{\pi\rho_p v_i N}\left(1-\frac{b}{D}\right)\right)^{1/2}$	(27)
Lapple and Shepherd (32)	0	0	$\frac{D-De}{2}$	0	$\frac{\pi DN}{v_i}\left(1-\frac{De}{2D}\right)$	$d_{100} = \left(\frac{9\mu De}{2\pi\rho_p v_i N}\right)^{1/2}$	(28)
Davies (33)	0	0	$\frac{De}{2}$	1	$\frac{H}{v_i}$	$d_{100} = \left(\frac{9\mu D^2\left(1-\left(\frac{De}{D}\right)^4\right)}{8\rho_p v_i H}\right)^{1/2}$	(29)
Lapple (6)	0	0	$\frac{D-b}{2}$	0	$\frac{\pi DN}{v_i}\left(1-\frac{De}{2D}\right)$	$d_{50} = \left(\frac{9\mu b}{2\pi\rho_p v_i N}\right)^{1/2}$	(30)
Leith and Licht (31)	0	0	0	n depends on geometry and throughput		$\eta = 1 - \exp\left(-2(C\psi)^{1/(2n+2)}\right)$	(31)

The Fractional Efficiency Curve. Critical particle diameter is useful only as a rough estimator for cyclone efficiency. For more precise work, as when estimating overall cyclone efficiency on a dust with a range of particle sizes, the entire fractional efficiency curve is necessary. Lapple[6] and Barth[25] have developed generalized plots of efficiency versus a dimensionless particle parameter. Lapple's parameter is defined as particle diameter over the cut diameter calculated from Equation (30). This plot is given in Figure 16.12 and is valid for cyclones of the Lapple design listed in Table 16.1. No figures for cyclones of other design are available.

Leith and Licht[31] combined an approximate solution to Equation (22) with the assumption that uncollected dust is remixed within the cyclone gas stream due to gas stream turbulence. The assumptions they made for solving Equation (22) are also listed in Table 16.3. The resultant equation predicts the fractional efficiency curve.

$$\eta = 1 - \exp\left(-2(C\psi)^{1/(2n+2)}\right) \qquad (31)$$

Here, the vortex exponent, n, can be calculated from Equations (2) and (3), or found from Figure 16.5. The influence of particle and gas properties are combined into ψ, a dimensionless inertia parameter or Stokes' number.

$$\psi = \frac{d^2\rho_p v_i(n+1)}{18\mu D} \qquad (32)$$

The effect of cyclone geometry is consolidated in C, a dimensionless geometry parameter. The geometry parameter depends only on the cyclone dimension ratios, and is independent of size.

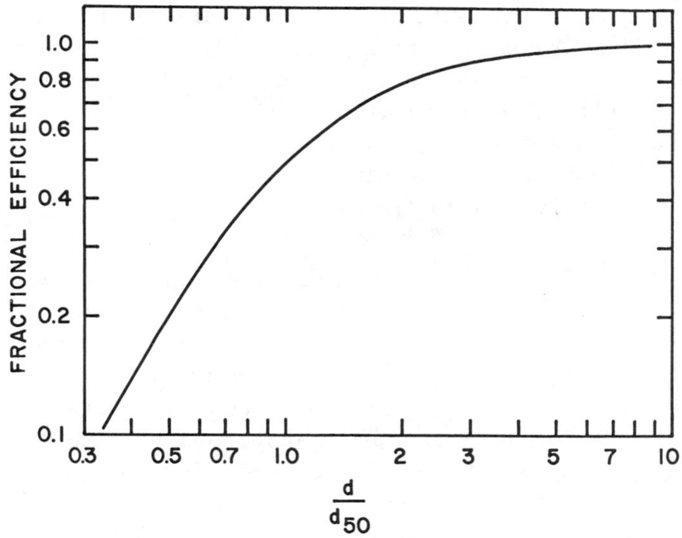

Figure 16.12. Fractional efficiency vs d/d_{50} for Lapple design cyclone.

$$C = \frac{\pi D^2}{ab}\left(2\left(1 - \left(\frac{De}{D}\right)^2\right)\left(\frac{S}{D} - \frac{a}{2D}\right) + \frac{1}{3}\left(\frac{S + Z - h}{D}\right)\left(1 + \frac{d_c}{D} + \left(\frac{d_c}{D}\right)^2\right) + \frac{h}{D} - \left(\frac{De}{D}\right)^2 \frac{Z}{D} - \frac{S}{D}\right) \tag{33}$$

Natural length, Z, is found from Equation (4), while d_c, the diameter of the cone at the natural length, can be determined from Equation (34).

$$d_c = D - (D - B)\left(\frac{S + Z - h}{H - h}\right) \tag{34}$$

Equation (31) implies that a cyclone with a high value of geometry parameter, C, should have a higher efficiency than a unit with a low value of C for particles of all sizes and for all operating conditions. The efficiency capabilities of alternative cyclone designs can be evaluated by comparing their values of C in the same way that pressure drop requirements are evaluated by comparing values of ΔH. Values of C are listed in Table 16.1 for the standard design cyclones given there. Equation (31) was tested against experimental data from the literature[31] and was found to predict the data reasonably well. It was probably the best approach for calculating efficiency currently available, short of experimental work. Equation (31) is useful for determining the efficiency of industrially sized cyclones, a few hundred millimeters in diameter. Smaller cyclones may not generate a gas stream sufficiently turbulent to justify the assumptions made in deriving this expression, and for these small cyclones alternative collection efficiency expressions may be more appropriate.[35,36]

Although theoretical calculations of critical particle diameter and fractional efficiency are useful, they, like theoretical pressure drop calculations, may predict performance substantially in error from that experienced in the field. All efficiency theories discussed have been for tangential entry, reverse-flow cyclones as shown in Figure 16.1. Their applicability is unknown to other cyclone designs, such as those with either scroll or swirl vane entries, or to straight-through cyclones of the type shown in Figure 16.2.

The best way to determine cyclone fractional efficiency characteristics is to test the cyclone in the laboratory. Once an experimental fractional efficiency curve has been developed for a cyclone operating under known conditions, the fractional efficiency curve can be determined for a cyclone of the same design under different operating conditions by adjusting the efficiency curve of the test cyclone. According to Equation (31), two cyclones of the same design (same value of C) will have the same efficiency, η, when their values of inertia parameter, ψ, are the same. If the experimental cyclone has known efficiency on particles of size d_1, a similar cyclone will have the same efficiency on particles of size d_2, where:

$$d_2 = d_1 \left(\frac{Q_1}{Q_2} \frac{\rho_{p1}}{\rho_{p2}} \frac{\mu_2}{\mu_1} \frac{D_2}{D_1} \right)^{1/2} \quad (35)$$

This analysis assumes that the diameters of the two cyclones are close enough that the value of the vortex exponent, n, does not change appreciably. The fractional efficiency curve for the similar cyclone can be constructed from the curve for the tested cyclone by picking a series of coordinates from the experimentally derived efficiency curve and calculating the analogous coordinates for the similar cyclone from Equation (35).

The accuracy of this procedure decreases as each of the ratios in Equation (35), Q_1/Q_2, etc. departs more and more from unity. The procedure is especially suspect when predicting the performance of cyclones with much greater diameter and throughput than the test model. Also, when adapting results based on an experimental dust to a different dust, particle shape may change as well as density. Nevertheless, a fractional efficiency curve calculated using this procedure is strongly preferred over one determined strictly from theory.

Other Variables Affecting Performance

Although cyclone performance theories express the effect of many variables on cyclone performance, several variables known to influence pressure drop and efficiency are not considered.

Increasing inlet dust concentration, c_i, simultaneously increases collection efficiency and decreases pressure drop. Briggs[13] quantified the influence of dust loading on pressure drop.

$$\Delta P_{\text{dusty}} = \frac{\Delta P_{\text{clean}}}{1 + 0.0086(c_i)^{1/2}} \quad (35)$$

Here, c_i has the dimensions of grams per cubic meter. The effect on efficiency of changing inlet loading from c_{i1} to c_{i2} can be found[11] from:

$$\frac{100 - \eta_1}{100 - \eta_2} = \left(\frac{c_{i2}}{c_{i1}} \right)^{0.182} \quad (37)$$

Presumably the values of efficiency and concentration in Equation (37) are for polydisperse dusts and the relationship applies to overall dust concentration and efficiency rather than to values for any one particle size.

If the tangential velocity of the gas near the cyclone wall is too high, saltation will occur; particles will bounce along the cyclone wall and not be separated effectively from the gas stream. Kalen and Zenz[37] have examined this phenomenon, and its implications for cyclone design are discussed by Koch and Licht.[38] An empirical equation, (37), which gives the cyclone inlet velocity above which saltation occurs, v_{is}, is:

$$v_{is} = 2400 \frac{\mu \rho_p}{\rho_f^2} D^{0.2} \frac{(b/D)^{1.2}}{1 - b/D} \quad (38)$$

SI units (m, kg, s) must be used in this equation. Cyclone efficiency increases with inlet velocity up to about 1.25 v_{is}; further increases in inlet velocity cause a decrease in efficiency as saltation and reentrainment of collected dust become more important.

Stairmand[4] showed that the overall efficiency of a well designed cyclone increases from its normal value of 92% to an increased value of 93.6% when about ten per cent of the gas flow is drawn through the dust outlet. A similar "base purge" increased the efficiency of a poorer cyclone design from 89.1% to 92.2%. Stairmand believes this efficiency increase is due to a reduced reentrainment of separated dust in the dust outlet region. The disadvantages of this practice are that it requires the use of otherwise unnecessary auxiliary fans and ducts to draw off the purge, and that if the purge is recycled to the cyclone inlet, the cyclone must be sized to handle the purge air as well as the process air. In practice, base purge is seldom used.

Table 16.4. Overall Collection Efficiency Calculation Using Numerical Integration of Equation (39).

(1) SIZE RANGE (MICROMETERS)	(2) MEAN SIZE (MICROMETERS)	(3) FRACTION IN RANGE	(4) EFFICIENCY ON MEAN SIZE	(5) (3) × (4)
0–2	1	0.10	0.35	0.035
2–5	3.5	0.10	0.60	0.060
5–10	7.5	0.10	0.77	0.077
10–20	15	0.15	0.89	0.134
20–30	25	0.10	0.95	0.095
30–40	35	0.10	0.97	0.097
40–60	50	0.15	0.99	0.149
60–76	68	0.10	1.00	0.100
76–104	90	0.07	1.00	0.070
104–150	127	0.03	1.00	0.030
		1.00		0.847

Stairmand[4] also reported that efficiency increases from a normal value of 92 to 93.7% for the well designed cyclone and 89.1 to 93.2% for the poorer design when these cyclones operate with wetted walls. The wetted walls may reduce reentrainment of collected dust throughout the cyclone. Disadvantages of this practice are that water piping is required and that the collected dust is in a slurry.

Overall Efficiency on Polydisperse Dusts

Industrial dusts contain particles of many sizes. To calculate the overall cyclone collection efficiency, $\eta_{overall}$, on such a dust one must multiply efficiency for each particle size by the fraction of particles in the dust which are of that size. The sum of these products is the overall fractional efficiency for the cyclone. Table 16.4 illustrates this process for the cyclone whose fractional efficiency curve is shown in Figure 16.10. The dust size distribution is plotted in Figure 16.13. Equation (38) is the formal mathematical statement of this process.

$$\eta_{overall} = \int_0^1 \eta \, dG \qquad (39)$$

Here, η is the efficiency on particles of a certain size, and dG is the fraction of all particles of that size in the dust. The overall efficiency for this cyclone on this dust is found to be about 85%.

16.4. CYCLONE DESIGN

Necessary Design Information

Before attempting to design a cyclone or cyclone system it is important to consider cyclone limitations, to be sure that an alternative control device might not work better. Cyclones

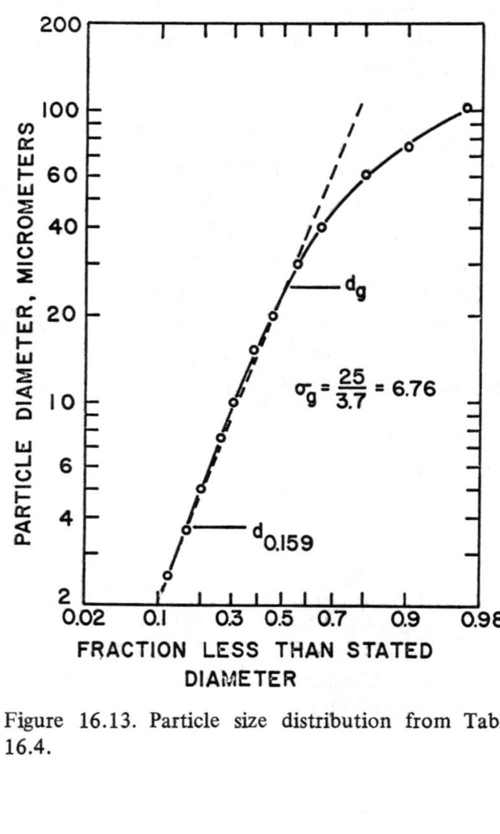

Figure 16.13. Particle size distribution from Table 16.4.

are unsuited for collection of particles less than about five micrometers in diameter, as efficiency falls off rapidly for particles smaller than this. Other types of collectors such as fabric filters, electrostatic precipitators and some kinds of scrubbers will be able to collect these small particles more efficiently. If the inlet dust loading is high and the desired outlet concentration is low, it may be necessary to use a higher efficiency collector such as a fabric filter either instead of, or in conjunction with (usually after) a cyclone system. Sticky or hygroscopic dusts may stick to the cyclone walls, and not discharge into the collection hopper. For dusts of this type, a scrubber may be a better collector choice than a cyclone.

To design a cyclone or any collection device, the inlet dust concentration and size distribution must be known. Although preliminary estimates of expected dust properties are available from the literature,[39-41] this information should always be obtained by stack sampling when possible. Of course, when designing control equipment for a plant that has yet not been constructed, stack testing is impossible and in this case the design will have to be based on data obtained from similar plants in conjunction with the design plans for the process to be controlled.

Design criteria such as gas flow rate, temperature, and particle density–material density, not apparent or bulk density–special conditions of corrosivity, particle abrasiveness and fluctuations in gas flow should be noted. These data requirements are summarized in Table 16.5. All the data necessary for the design of a cyclone system can be obtained from a stack test performed on the gas stream to be cleaned.

Table 16.5. Data Necessary for Cyclone Design.

Particle size distribution
Inlet dust loading (g/m^3)
Particle density (kg/m^3)
Gas flowrate (m^3/hr)
Gas temperature (°C)
Special conditions of corrosivity, abrasiveness, fluctuations in gas flow, etc.

Cyclone Specification

Usually cyclones are not custom designed. Rather an accepted standard design is selected, such as one listed in Table 16.1 or a manufacturer's proprietary design. Cyclone diameter can be determined from gas flowrate Q, using the value for Q/D^2 tabulated for each standard design given in Table 16.1. Once diameter is known, the remaining seven dimensions can be determined from the dimension ratios of the standard design selected. For volumetric gas flows larger than about 20,000 m^3/hr it is often better to use several smaller cyclones in parallel rather than one large cyclone. This is because collection efficiency decreases with increasing cyclone diameter and also because of possible problems with space or headroom requirements for very large cyclones.

A fractional efficiency curve for the selected design can be determined by one of the methods discussed above. The value of cyclone geometry parameter, C, for use with Equation (31) is listed in Table 16.1 for each of the standard designs listed there, except for the high throughput designs with scroll gas entries for which Equation (31) cannot be used. The overall collection efficiency for the selected cyclone design, inlet dust size distribution and concentration to be processed, and outlet dust concentration desired can then be determined from the methods described previously.

A cyclone can be custom designed to perform a specific dust collection job.[27] This approach may give a cyclone with greater collection efficiency, smaller in size or with lower pressure drop than would be found for a standard design. This design procedure requires trial and error calculations. First, choose an approximate diameter for the cyclone, from Equation (40):

$$D_{\text{meters}} = \frac{(Q_{\text{meters}^3/\text{hr}})^{1/2}}{80} \tag{40}$$

The flow can be divided, and the system designed on the basis of a number of cyclones operating in parallel.

Next, pick as a design basis a value for pressure drop, ΔP. This is an arbitrary, preliminary choice; a final choice will be made later as discussed below. A value for cyclone pressure drop factor, P, can now be found from:

$$P = \frac{D^4 \, \Delta P}{8 \rho_G Q^2} \tag{41}$$

Figures 16.14–16.16 contain plots of geometry factor, C, against pressure drop factor, P, for cyclones with overall height, $H = 4D$, $5D$, and $6D$ respectively. These plots were drawn using the technique developed by Leith and Mehta.[27] As expected, the taller cyclones give higher C values, hence higher efficiency, for a fixed value of pressure drop factor, P. These figures also contain plots of the cyclone dimension ratios, so that identification of P fixes the entire cyclone design. Once geometry factor, C, is known, the fractional efficiency curve for the cyclone can be found using Equation (31), and the overall efficiency for collection of a polydisperse dust can be found as discussed previously. If the overall efficiency determined is not high enough, a higher value for pressure drop, ΔP, can be selected and the process repeated until a suitable cyclone design has been found. Of course, in some cases no cyclone will provide adequate collection efficiency; in this case teams of cyclones in series or alternative control devices should be evaluated.

The customized cyclone design is now complete. However, it is possible to go back and repeat the process for cyclone diameters about 20% larger or smaller than the one determined from Equation (40). The largest diameter cyclone will operate with the lowest pressure drop, while the smallest cyclone will have the highest pressure drop. The designer can now exercise economic judgement and compromise between the greater size and initial cost against lower pressure drop and operating cost. The decision will have to be made on a case by case basis.

Design Example

The design procedures described in the previous section can be illustrated by the following design

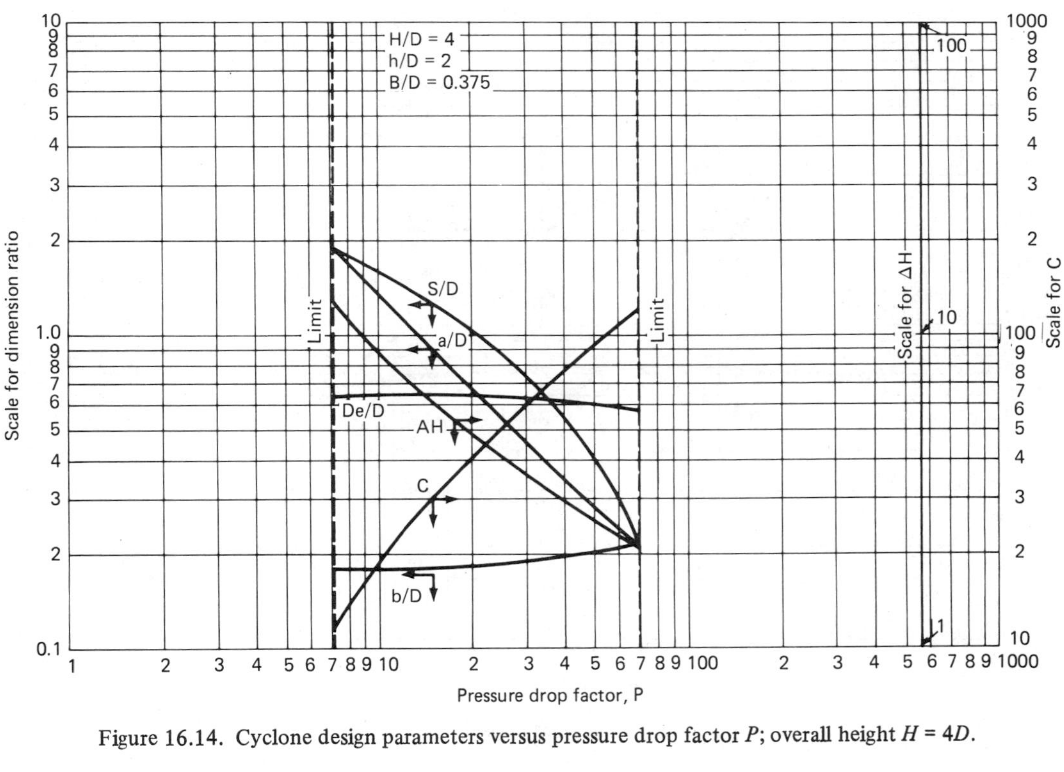

Figure 16.14. Cyclone design parameters versus pressure drop factor P; overall height $H = 4D$.

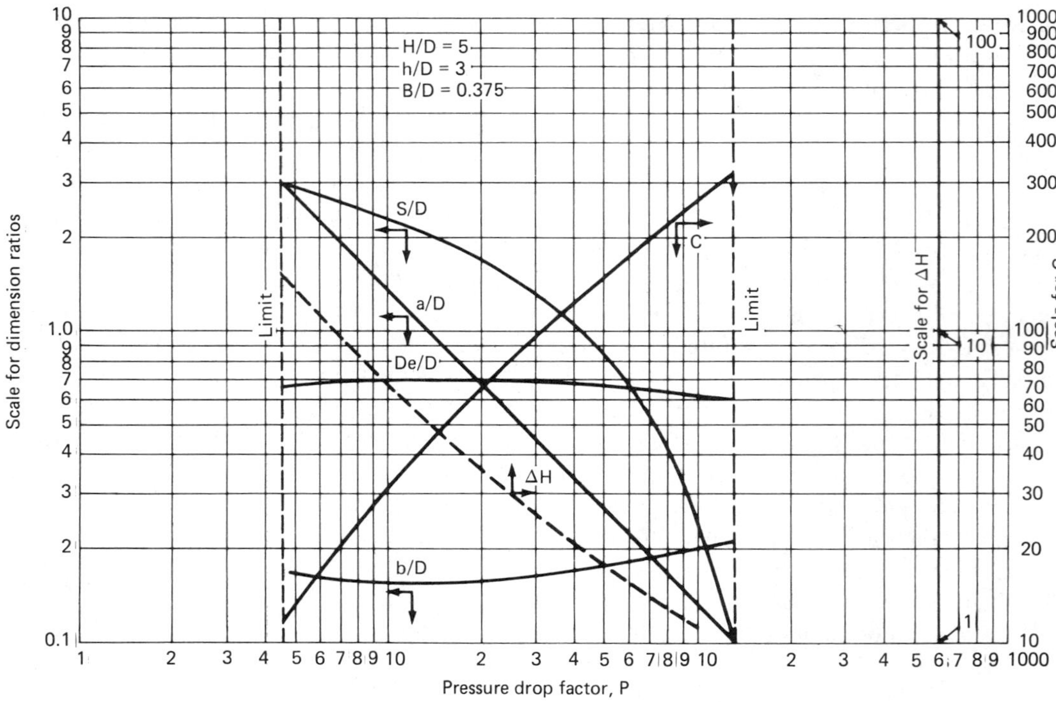

Figure 16.15. Cyclone design parameters versus pressure drop factor P; overall height $H = 5D$.

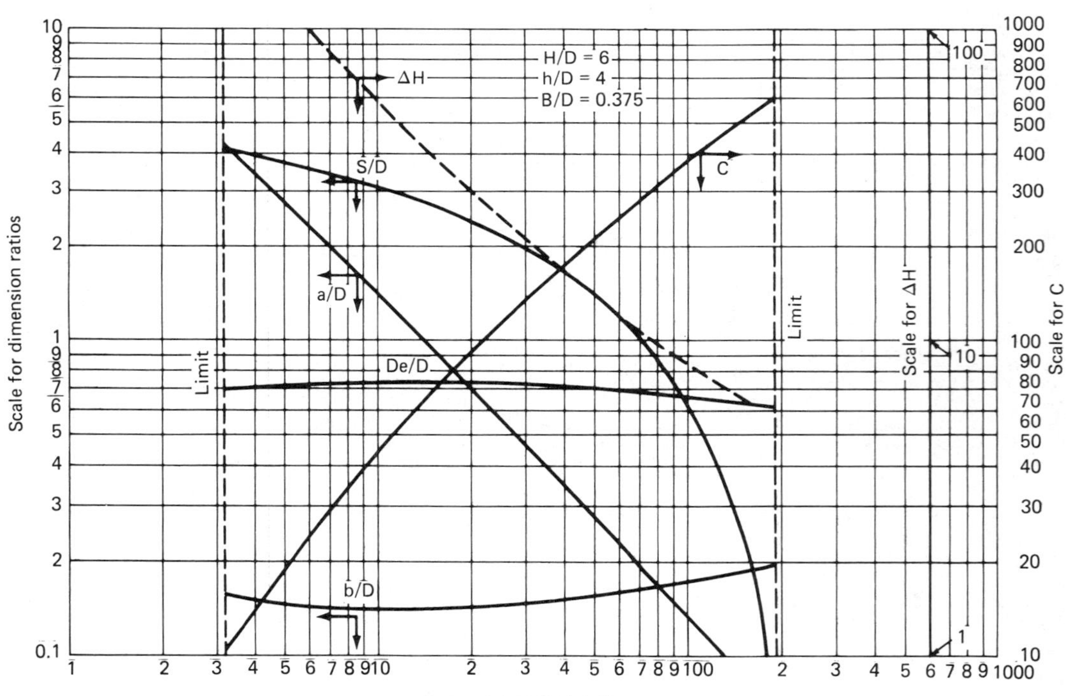

Figure 16.16. Cyclone design parameters versus pressure drop factor P; overall height $H = 6D$.

example. Suppose 40,000 m^3/hr of air containing rock dust comes from a rotary drier at 100°C. It is desired to collect as much dust as possible in a cyclone system for recycle to the process. The effluent from the cyclone system will go to a scrubber for final control before release to the atmosphere. An inlet loading of 10 g/m^3 or less to the scrubber is the design objective. A stack test finds that the dust loading from the drier is 50 g/m^3, and that the dust has the size distribution shown in Figure 16.13. The density of the rock is 2500 kg/m^3.

Example Using Standard Design. The Lapple general purpose design cyclone might be used. Since the air flow rate is appreciable, it will probably be best to operate several, say three cyclones in parallel, rather than building a single, large cyclone. The diameter of each cyclone can be found from the appropriate entry in Table 16.1.

$$D = (40{,}000/3)^{1/2}/(6860)^{1/2} = 1.39 \text{ m} \qquad (42)$$

The other dimensions can be calculated from the dimension ratios for the chosen design listed in Table 16.1 and are given in Table 16.6. Inlet gas velocity will be found from Equation (43).

$$(40{,}000 \text{ m}^3/\text{hr}/3)(1 \text{ hr}/3600 \text{ s})/((0.70 \text{ m})(0.35 \text{ m})) = 15.2 \text{ m/s} \qquad (43)$$

From Table 16.1, the pressure drop should be 8.0 inlet velocity heads. This can be converted to pressure loss in Pascals by Equation (5).

$$\Delta P = 8(\tfrac{1}{2}(0.944 \text{ kg/m}^3)(15.2 \text{ m/s})^2) = 872 \text{ Pa or 89 mm of water column} \qquad (44)$$

The expected efficiency and outlet dust concentration can be calculated using the method outlined previously. The vortex exponent, n, can be found from Equations (2) and (3).

$$n_{283°\text{K}} = 0.67(1.39)^{0.14} = 0.70 \qquad (45)$$

$$n_{373°\text{K}} = 1 - (1 - 0.70)(373°\text{K}/283°\text{K})^{0.3} = 0.68 \qquad (46)$$

Table 16.6. Cyclone Specifications for Design Example.

DIMENSION	LAPPLE STANDARD DESIGN (m)	CUSTOMIZED DESIGN, H/D = 5: REGULAR (m)	LARGE DIA. (m)	SMALL DIA. (m)
D	1.39	1.44	1.73	1.15
a	0.70	0.94	0.54	1.83
b	0.35	0.23	0.29	0.17
De	0.70	0.99	1.14	0.81
S	0.87	2.35	1.66	2.81
h	2.78	4.32	5.19	3.45
H	5.56	7.20	8.65	5.75
B	0.35	0.54	0.65	0.43
Pressure drop, Pa:	872	500	500	500
mm of water:	89	51	51	51
Inlet velocity m/s	15.2	17.2	23.5	11.7
Maximum inlet velocity for no saltation. m/s	48	25	29	22
Pressure drop parameter, P	–	20.8	43.1	8.52
Geometry parameter, C	50.4	67	130	26
Overall fractional efficiency:	0.85	0.86	0.90	0.81

Next the value of Ψ for particles of diameter d can be calculated from Equation (32).

$$\Psi = \frac{(d, \mathrm{m})^2(2500\ \mathrm{kg/m^3})(15.2\ \mathrm{m/s})(1 + 0.68)}{(18)(2.1 \times 10^{-5}\ \mathrm{kg/m \cdot s})(1.39\ \mathrm{m})} = 1.22 \times 10^8 (d, \mathrm{m})^2 \quad (47)$$

Table 16.1 shows that geometry factor, C, is 50.4 for the Lapple standard design. Cyclone efficiency can now found using Equation (31) for particles of any size, d.

$$\eta = 1 - \exp\left(-2((50.4)(1.22 \times 10^8 (d, \mathrm{m})^2))^{1/(2(0.68 + 2)}\right) \quad (48)$$

The relationship between particle diameter and collection efficiency for the Lapple cyclone given by Equation (48) is plotted in Figure 16.10, already discussed. The overall efficiency for this cyclone on particles with the distribution given in Figure 16.13 is determined through the calculations shown previously in Table 16.4; overall fractional efficiency was found to be about 0.85.

Outlet dust concentration will be:

$$c_o = c_i(1 - \eta_{\mathrm{overall}}) = 50\ \mathrm{g/m^3}(1 - 0.85) = 7.50\ \mathrm{g/m^3} \quad (48)$$

The cyclone system should meet the design objective of 10 $\mathrm{g/m^3}$ outlet dust concentration. Had the calculations shown that the design objective would not be met, the designer could try a cyclone design with larger geometry parameter, C, or break the gas stream up and use many smaller diameter cyclones rather than a few larger ones.

Example Using Customized Design. As in the standard design procedure, the system will consist of three cyclones operating in parallel. The diameter of each can be found from Equation (40).

$$D_{\mathrm{meters}} = (40{,}000/3)^{1/2}/80 = 1.44\ \mathrm{m} \quad (49)$$

Next, choose a value for operating pressure drop, say $\Delta P = 500$ Pa (51 mm water column), a value somewhat less than that found for the Lapple standard design cyclone. Equation (41) can now be used to find the appropriate value for pressure drop factor, P.

$$P = \frac{D^4\,\Delta P}{8\rho_G Q^2} = \frac{(1.44\ \mathrm{m})^4(500\ \mathrm{kg/s^2 \cdot m})}{(8)(0.944\ \mathrm{kg/m^3})(3.70\ \mathrm{m^3/s})^2} = 20.8 \quad (50)$$

Figure 16.15, for a cyclone with overall height $H/D = 5$, shows that for this P value, a cyclone with geometry factor $C = 67$ can be made. Because the C value for this cyclone is higher than that for the Lapple cyclone, overall efficiency will be higher as well. Dimension ratios taken from the same figure can be used to determine the dimensions for this cyclone which are also listed in Table 16.6. The equation to describe the relationship between particle diameter and efficiency that is analogous to Equation (48) for the Lapple cyclone is:

$$\eta = 1 - \exp\left(-2((67)(1.33 \times 10^8 (d, \mathrm{m})^2))^{1/(2(0.68) + 1)}\right) \quad (51)$$

The overall efficiency for this cyclone by using Equation (51) to determine the efficiencies tabulated in column (4) of Table 16.4. The overall efficiency for this cyclone determined from this procedure is 0.863; the corresponding outlet concentration is 6.9 $\mathrm{g/m^3}$, a little better than the Lapple standard design. A similar procedure for diameters twenty per cent larger and smaller than that found from Equation (49) and the same pressure drop will give the alternate cyclone designs also listed in Table 16.6. Any of the four cyclone designs listed in Table 16.6 should achieve the design objective. This table also lists the maximum inlet velocities below which saltation occurs, as given in Equation (38). In each case, the specified inlet velocity is below that for saltation so that reentrainment of separated dust should not be a problem.

Other Design Considerations

After the shape and size of a cyclone or cyclone system have been decided, it is important to consider the additional design criteria which will assure long, trouble-free operation of the system.

When small cyclones, less than 300 mm or so in diameter are used, wall erosion may pose a serious problem[8] if the dust is abrasive. Larger dust particles strike the cyclone wall more forcefully and have more effect than smaller particles. Abrasion can be especially troublesome around welded seams, and occurs whether the seams are horizontal or vertical. The seam itself may not be as susceptible as the cyclone wall around the seam, which may have been softened through annealing during the welding process.[8] To minimize the effect of wall erosion, several steps can be taken. Oftentimes, commercially available small diameter cyclones are cast rather than fabricated from sheet metal. Casting eliminates the problem with erosion around weldments, and may provide a thicker wall. Replaceable wear plates are sometimes installed on the cyclone wall opposite to the tangential gas inlet. When installing a wear plate, it is important that the plate be fitted to maintain a smooth interior wall. Failure to maintain a smooth wall will hasten the erosion of the wear plate or the wall around the edges of the plate, and may also adversely affect cyclone efficiency. Wear plates and entire cyclone interior walls have been rubber coated to reduce erosion.

It is essential that air not be allowed to leak into the cyclone through the dust outlet. Leakage at this point can keep the cyclone from discharging dust to the dust bin, and if sufficiently severe, can lower collection efficiency to zero.[1] Leakage through the dust exit can even occur with the cyclone on the pressure side of a fan due to the low static pressure in the cyclone central core, although the problem is more pronounced if the cyclone is on the suction side. When possible, cyclones are mounted on the upstream or suction side of a fan to minimize wear of the fan impeller from the dust in an uncleaned gas stream.

If the cyclone operates on the effluent from a batch process, the unit is often directly connected to a dust bin below the dust exit without an intervening valve. When using this arrangement, it is essential that the dust bin be emptied before it fills and blocks the cyclone dust exit. In this case, the gas flowing to the cyclone must be diverted before the dust bin can be emptied. The dust bin must be airtight to prevent leakage from the bin entering the cyclone dust exit.

A better solution to dust exit sealing is through use of a valve between the exit and dust bin. The valve must allow for the continuous discharge of collected dust, but not permit backflow of air. A valved dust discharge arrangement is essential for a continuous process as it allows the collection bin to be emptied at any time. Rotary valves are often used[8] when the negative pressure at the dust exit is less than about 1 kPa (100 mm of water column).

Cyclone gas inlet velocities are frequently of the order of 15 m/s while duct velocities are usually lower than this. To minimize pressure drop through the cyclone system it is important to provide a good transition between the inlet ductwork and the cyclone inlet.

Attempts have been made to regain some of the rotational energy in the outlet gas stream by modifying the shape of the gas outlet. A thorough review of these devices is provided by Stern, Caplan, and Bush.[8] Reverse scrolls mounted above the gas outlet duct, and curved or straight vanes within the gas outlet duct have been used. These devices usually provide a reduction in pressure drop in the ten per cent range, but despite careful design, collection efficiency may be adversely affected. Pressure recovery devices are not generally used.

Buildup of collected dust on the cyclone walls can be a problem, especially where soft, small diameter, hygroscopic particles are collected. When buildup occurs, it can sometimes be scoured out by feeding some large diameter, hard particles as an abrasive. Wall deposition of hygroscopic dusts is aggravated by condensation of moisture from the gas stream on the cyclone walls, when the cyclone is mounted outdoors in winter. If the problem occurs only on startup, preheating the cyclone by either warming the inlet gas stream or running gas through without dust may help. Cyclones whose inlet walls are smooth and which operate at inlet velocities in excess of 15 m/s will be less prone to wall buildup.

LIST OF SYMBOLS

a	Gas entry height
A	Inside surface area of cyclone
b	Gas entry width
B	Dust outlet diameter
c_i	Inlet dust concentration
c_o	Outlet dust concentration
C	Cyclone geometry parameter
d	Particle diameter
d_c	Diameter of cyclone cone at the natural length
d_{50}	Cut particle diameter, theoretically collected with 50% efficiency
d_{100}	Critical particle diameter, theoretically collected with 100% efficiency
D	Cyclone cylinder diameter
De	Gas outlet diameter
f	Factor in Equation (9)
F_c	Centrifugal force acting on particle
F_d	Drag force acting on particle
g	Acceleration of gravity, 9.81 m/s^2
G	Dust cumulative size distribution
G^*	Friction factor, 0.005
h	Cyclone cyliner height
H	Cyclone overall height
n	Vortex exponent
N	Number of turns gas makes within cyclone
Q	Volumetric gas flowrate
r	Radial distance from cyclone axis
r_{core}	Radial distance from cyclone axis to edge of central core
r_i	Radial distance from cyclone axis to innermost particle at entry
r_w	Radial distance from cyclone axis to cyclone wall, $D/2$
S	Gas outlet height
t	Time
T	Absolute temperature, °K
u	Particle velocity
u_r	Radial component of particle velocity, dr/dt
u_t	Tangential component of particle velocity
v	Gas velocity
v_i	Gas inlet velocity, Q/ab
v_{is}	Gas inlet velocity above which saltation occurs
v_{max}	Maximum gas tangential velocity
v_r	Radial component of gas velocity
v_t	Tangential component of gas velocity
v_{tw}	Gas tangential velocity at cyclone outer wall
Z	Natural length
α^*	Factor in Equation (17)
ϵ	Loss factor
η	Fractional collection efficiency of particles of one size
$\eta_{overall}$	Overall collection efficiency for polydisperse dust
ΔH	Pressure drop expressed as number of inlet velocity heads
ΔP	Pressure drop expressed as static pressure head
θ	Ratio of maximum tangential gas velocity to gas velocity in gas outlet, see Equation (16)
λ	Friction factor, 0.02
μ	Gas viscosity
ρ_G	Gas density
ρ_p	Particle density
ϕ	Ratio of maximum tangential gas velocity to velocity within gas entry
ψ	Inertia parameter

REFERENCES

1. R. Jackson, *Mechanical Equipment for Removing Grit and Dust from Gases*, Cheney and Sons, Banbury, England (1963).
2. W. Barth, *Staub 21*:382 (1961).
3. J. I. T. Stenhouse and M. Trow, in *Proceedings of Second World Filtration Congress*, 1 Katharine St., Croydon CR9 1LB, England (1979).
4. C. J. Stairmand, *Trans. Inst. Chem. Eng. 29*:356 (1951).
5. P. Swift, *Steam and Heating Engineer 38*:453 (1969).
6. C. Lapple, *Chem. Eng.* (*New York*) *58*:144 (1951).
7. R. H. Perry and C. H. Chilton, *Chemical Engineer's Handbook*, 5th ed., McGraw Hill, New York (1973).
8. A. C. Stern, K. J. Caplan, and P. D. Bush, *Cyclone Dust Collectors*, American Petroleum Institute, New York (1956).

9. H. J. van Ebbenhorst Tengbergen, *De Ingenieur*. 77th Year of Publication, W 1 (1965).
10. H. J. van Ebbenhorst Tengbergen, *Staub 25*:44 (1965).
11. W. A. Baxter, In *Source Control by Centrifugal Force and Gravity*. K. J. Caplan, In *Air Pollution*, 2d ed. A. C. Stern (ed.), vol. 3., Academic, New York (1968).
12. L. C. Whiton, *Chem. Met. Eng. 39*:150 (1932).
13. L. W. Briggs, *Trans. Amer. Inst. Chem. Eng. 42*:511 (1946).
14. C. B. Shepherd and C. E. Lapple, *Ind. Eng. Chem. 31*:972 (1939).
15. K. Iinoya, *Mem. Fac. Eng. Nagoya Univ. 5* (Sept., 1953).
16. E. Anderson, *Chem. Met. Eng. 40*:525 (1933).
17. M. A. Lissman, *Chem. Met. Eng. 37*:630 (1930).
18. C. E. Lapple, *Amer. Ind. Hyg. Ass. Quart. 11*:40 (1950).
19. M. Seillan, *Chal. Ind. 10*:233 (1929).
20. P. Rosin, E. Rammler, and W. Intelmann, *V. D. I. (Ver. Deut. Ing.) Z. 76*:433 (1932).
21. F. Procket, *Glasers Ann. 107*:43 (1930).
22. A. J. ter Linden, *Proc. Inst. Mech. Engrs. (London) 160*:233 (1949).
23. M. W. First, Sc.D. thesis. Harvard University, Boston (1950).
24. R. McK. Alexander, *Proc. Australas. Inst. Mining Met. N. S. 152–153*:203 (1949).
25. W. Barth, *Brennst.-Waerme-Kraft 8*:1 (1956).
26. K. J. Caplan, In *Air Pollution*, 3d ed., A. C. Stern, (ed.), vol. 4, Academic, New York (1977).
27. D. Leith and D. Mehta, *Atmos. Environ. 7*:527 (1973).
28. C. B. Shepherd and C. E. Lapple, *Ind. Eng. Chem. 32*:1246 (1940).
29. C. J. Stairmand, *Engineering (London) 168*:409 (1949).
30. S. K. Friedlander, L. Silverman, P. Drinker, and M. W. First, *Handbook on Air Cleaning*. U. S. A. E. C., AECD-3361, NYO-1572, Washington (1952).
31. D. Leith and W. Licht, *A.I.Ch.E. Symposium Scr. 68*:196 (1972).
32. C. E. Lapple and C. B. Shepherd, *Ind. Eng. Chem. 32*:605 (1940).
33. C. N. Davies, *Proc. Inst. Mech. Engrs. (London) 10*:185 (1952).
34. N. A. Fuchs, *The Mechanics of Aerosols*, Pergamon, New York (1964).
35. T. Chan and M. Lippmann, *Environ. Sci. Technol. 11*:377 (1977).
36. W. Licht, T. Chan, and M. Lippman, *Environ. Sci. Technol. 11*:1021 (1977).
37. B. Kalen and F. A. Zenz, *A. I. Ch. E. Symposium Ser. 70*(137):388 (1974).
38. W. Koch and W. Licht, *Chem. Engin. 84*:80 (Nov. 4, 1977).
39. Midwest Research Institute. *Handbook of Emissions, Effluents and Control Practices for Stationary Particulate Pollution Sources*. NAPCA contract CPA 22-69-104, NTIS Publication No. PB 203-522, Springfield, VA (1970).
40. U.S. Environmental Protection Agency. *Compilation of Air Pollution Emission Factors*, 2d, ed., Publication No. AP 42 (April, 1973).
41. J. A. Danielson, *Air Pollution Engineering Manual*, 2d ed., EPA. Publication No. AP 40 (May 1973).

17
Electrostatic Precipitator Application and Concepts

Jacob Katz
Precipitator Technology, Inc.
Munhall, Pennsylvania

17.1. INTRODUCTION

General Comments

The electrostatic precipitator uses electrical forces to capture either liquid or solid particulate matter from a flue gas system. We tend to classify the precipitator as a high-efficiency collector, comparable to the fabric bag-house or high-pressure venturi scrubber. As such, collection efficiencies of 99.5% plus are within a design range for most applications. A prime characteristic separating the various methods for high efficiency collection is that the precipitator concentrates its primary energy forces on the particle, rather than on the carrier gas stream. However, the gas stream or process characteristics will generally determine whether the particle will be easily collected or prove difficult to contain by electrical forces.

An interesting facet of precipitation is that even after many years of application, simple fundamental knowledge has eluded personnel involved with this equipment. That is why we continually have to face the same field problems and why performance of the precipitators often varies greatly from its design criteria. This lack of understanding cannot be overcome with more theoretical coverage, but rather there is a need for practical concepts and field information to be clearly identified and distributed.

For that reason, this presentation attempts to provide a brief description for some of the key areas of precipitation without regard for detailed theory. It actually consists of excerpts from the book *The Art of Electrostatic Precipitation* written for the practitioner. The bibliography at the end of this chapter also includes sources of literature that can be used to upgrade the theoretical knowledge of the precipitator.

Performance Comments

Each industrial process presents its own specific problems or application factors when precipitators are considered. Subtle changes in the process, of raw materials and equipment can often produce a wide band of performance characteristics in a specific precipitator. These subtle changes probably produce the greatest number of installations which fail to meet expected collection efficiencies because of reduced electrical power input. A secondary contribution to subperformance levels is a combination of design short comings, including among others, gas distribution, gas sneakage, and insufficient sectionalization. A third group of factors that keeps the collector from performing consistently on a satisfactory basis is the balance of reliability, either in the evacuation of material from hoppers or in the failure of precipitator components.

Applications

The use of precipitators has been applied in all the basic as well as some exotic industries over the years. Collection of particulate matter in a dry-type precipitator with flue gas temperatures between 250 to 700°F has been the most popular application. However, specific process characteristics will usually determine the design and type of precipitator utilized. There are process situations where the effectiveness of the electrical collector is questionable because the material is difficult to handle. Some typical industrial processes that have successfully employed the precipitator include:

PROCESSES	PRINCIPLE MATERIAL COLLECTED
Utility	Fly ash (SiO_2, Al_2O_3, Fe_2O_3)
Industrial boiler houses	Fly ash
Oxygen steelmaking furnaces	Iron oxide (Fe_3O_4)
Cement kilns	Calcium oxide, silicon oxide
Pulip and paper	Sodium sulfate

A large number of precipitators have also cleaned process gases from blast furnaces, sinter plants, open hearths, coke oven, gypsum plants, catalytic cracking, smelters, sulfuric acid, and phosphoric acid plants. Other processes that use precipitators include electric-arc furnaces, scarfing machines, incinerators, and the carbon black industry. Other precipitators have also recovered valuable metals in special situations.

Precipitator Arrangements

A convenient and logical method of comparing precipitators is the total collecting surface area and the amount of mechanical and electrical sectionalization. The basic terminology used to describe sectionalization follows and an illustration is shown on Figure 17.1.

Precipitator. A single precipitator is an arrangement of collecting surfaces and discharge electrodes contained within an independent housing.

Bus Section. The smallest portion of the precipitator which can be independently deenergized.

High Voltage Power Supply. The power supply unit to produce the high voltage required for precipitation, consisting of a transformer-rectifier combination and assorted controls. Numerous bus sections can be energized by one power supply.

Field. A field of a precipitator is an arrangement of bus sections in the direction of gas flow that is energized by one or more power supplies situated laterally across the gas flow.

Collecting Surfaces. The individually ground components which make up the collecting system and which collectively provide the total area of the precipitator for the deposition of particulate.

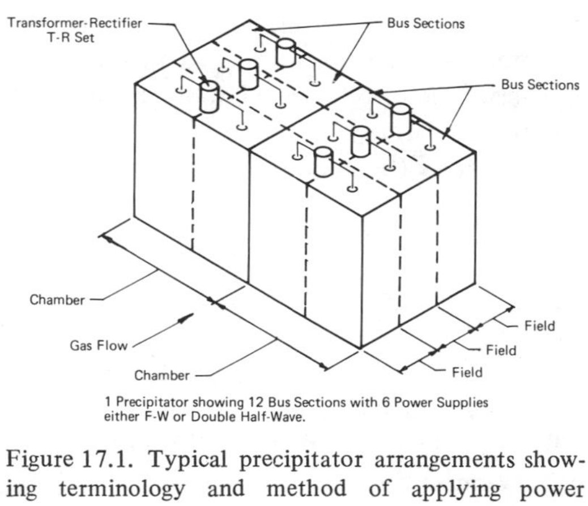

Figure 17.1. Typical precipitator arrangements showing terminology and method of applying power supplies.

Collecting Surface Area. The total flat projected area of collecting surface exposed to the electrostatic field (effective length × effective height × number of sides).

Effective Height. Total height of collecting surface measured from top to bottom.

Effective Width. Total number of gas passages multiplied by the center to center spacing of the collecting surfaces. (Disregard shape of collection surface.)

Effective Cross-Sectional Area. Effective width times effective height.

Gas Passage. Formed by two adjacent rows of collecting surfaces.

Discharge Electrode. The component which is installed in the high voltage system to provide the function of ionizing the gas and creating the electric field.

Collecting Surface Rapper. A device for imparting vibration or shock to the collecting surface to dislodge the deposited particulate.

Basic Concepts of Precipitation

Probably the best way to gain an insight into the process of precipitation is to study a relationship generally known as the Deutsch-Anderson equation. This equation and adaptations of it are well covered in several books.[1,2] It describes the factors involved in the collection efficiency of the precipitator as shown in its simplest form:

Collection Efficiency

$$N = 1 - e^{-(A/V)W}$$

where

A = effective collecting electrode area of the precipitator, sq. m
V = gas flow rate through the precipitator, acm/s
e = base of natural logarithm = 2.178
W = migration velocity, m/s

This equation has been used extensively in the above form in past years. Unfortunately, while the relationship is scientifically valid, there are a number of operating parameters that can cause the exponent to be in error by as much as a factor of two or more. It is well to remember that the basic D-A equation can be used as an indicator or tool, but has limitations more often than not unless equated with some practical and empirical considerations by the designer. Values used can either be in the English or metric systems.

The exponent term W, known as the migration velocity, actually represents the speed of movement of the particle toward the collector surface under the influence of an electrical field. While we would consider it more an indicator than actual velocity, it does have a finite value that can be used for comparison purposes. This migration velocity is comprised of:

$$W = \frac{aE_0E_\rho}{2\pi\theta}$$

where

a = particle radius, microns
E_0 = strength of field in which particles are charged, statvolts per cm (represented by the peak voltage)
E_ρ = strength of field in which particles are collected, statvolts per cm (normally the field close to the collecting plates)
θ = viscosity of frictional resistance coefficient of the gas.

High levels of voltage and useful corona power in the precipitator, all other conditions being equal, are synonymous with high collection efficiencies. Figure 17.2 shows a typical performance curve of the effect on efficiency by changes in the peak voltage of a precipitator. This simple curve can only represent one situation because each precipitator will have its own characteristic curve based on many factors. The important point to remember is that small changes can produce substantial changes in power, and hence in the efficiency of the collector. This is especially true at the lower

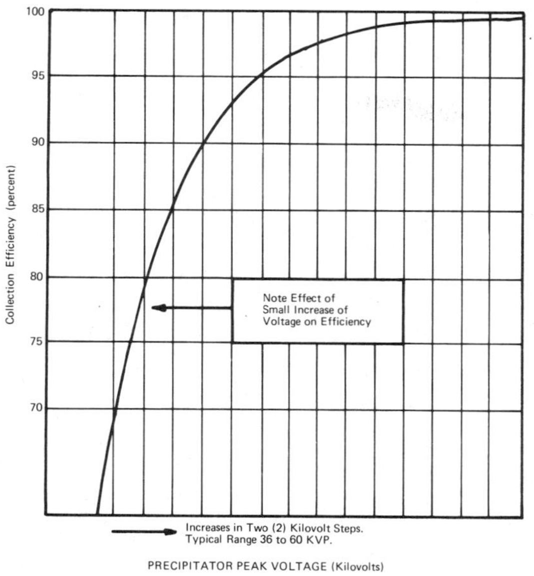

Figure 17.2. A typical electrostatic precipitator peak voltage versus dust collection efficiency curve shows how efficiency increases with voltage.

levels of power input. It is therefore important to understand the factors that effect the electrical characteristics of the precipitator.

Main Factors Effecting Electrical Characteristics

Optimum power input to the precipitator varies among processes and even changes on a minute to minute basis for certain applications. There are seven basic factors that directly effect the electrical characteristics. These are:

1. Design of power supply
2. Physical design of precipitator
3. Design of electrode system
4. Characteristics of gas stream
5. Effect of process changes
6. Characteristics of particulate
7. Maintenance factors

The power supply must be matched correctly for the precipitator section or service expected, or several difficulties can arise:

1. The impedance of the power supply, including a ballast resistance or reactor in the primary winding of the transformer, may not be sufficient to dampen the severity of electrical breakdowns in the precipitator. This condition is especially likely if the power supply rating is much larger than the actual operating level.
2. If the physical size of the precipitator is too large compared to the size of the power supply, then lower than desirable precipitator voltages may exist because the current rating of the supply becomes the limiting factor.
3. The gas and particulate matter conditions can drastically alter the voltage-current relationship and produce lower voltage fields than expected because a small power supply becomes current limited.

The average precipitator can be sensitive to process changes in the following ways:

1. Changes to gas temperature (effect on density)
2. Changes in gas pressure (effect on density)
3. Changes in gas flow rate
4. Changes in gaseous composition
5. Changes in particulate chemical characteristics
6. Changes in particulate concentration or loading
7. Changes in the size distribution of the particulate
8. Changes in the electrical conducting characteristics of the particulate

It is difficult to separate the effect of one process change on another. If the rate of process change is rapid, the readings can change almost instantaneously. On the other hand, rapid changes of temperature may not be seen readily on the meters because of the heat sink effect of the precipitator. Some changes in the process will cause large variations of voltage-current readings, while others will cause subtle effects.

The size distribution of the particulate matter can have a bearing on electrical readings. For example, iron oxide fume from a basic oxygen vessel contains a predominance of submicron particles that will react like a space-charge in a vacuum tube. This can actually impede the flow of precipitator current and thereby elevate the voltage potential across the space. This condition can become serious enough to completely nulify the precipitator process depending on the electrode geometry and the concentration level of the submicron particles.

17.2. FACTORS AND EFFECTS

Gas Flow Rate

While the true measurement of the gas flow rate commands an important place in specifications and performance tests, in practice, it is often less critical than other factors. This comment is especially true with the larger designs that exist today. Even with the 1.8 m/s (6 ft/s) or more velocity designs, it is often the quality of gas that weighs most importantly.

Whether or not the relationship of higher gas flow rate to reduced efficiency becomes critical is dependent in large part on the characteristics of the particle. Certainly, large porous particles such as combustible grit found in fly ash applications will be sensitive to increased velocities. On the other hand, fine-sized particulate matter that tends to agglomerate in the deposited layer of the collecting surface will resist easy reentrainment into the gas stream. With low levels of power input and low Aspect Ratios, high gas flow rates can often be observed in reduced performances of the precipitator.

Gas Flow Distribution

Gas distribution problems are of concern from the standpoint of velocity, temperature, and concentration of material as well as particle size. If one area of precipitation has become worse in recent years, it is in gas distribution. The trend toward larger collectors has meant greater difficulty in transferring the gas leaving the inlet nozzle to an acceptable pattern at the face of the precipitator. Granted optimum gas distribution is not as critical in the larger units with all fields serviceable, but the margin can be quickly lost with outages of equipment.

Probably one fallacy in gas distribution is placing too much emphasis on the results of model studies. The model cannot forsee the fallout of material during periods of reduced operation that will often distort the actual flow pattern.

Gas Temperature

The level of gas temperature in the precipitator opens up many areas of interest, especially the effect on the viscosity of the gas stream. But the major effects of temperature lie in the modification of the electrical characteristics and the reactions of the particles as they deposit on surfaces. The effect on metal corrosion by changes in flue gas temperature must also be considered.

Practically all of the particulate matter handled in precipitators will go through a wide spectrum of electrical characteristics for the temperature range of 200 to 750°F. Much of this has to do with condensation effects and surface leakage at the lower range and conductivity changes in the bulk material at the higher temperatures. The true effects at any given temperature will depend on the moisture level and chemical composition of the particles. Of greater interest would be whether the precipitator is operating in critical temperature zones for that particulate material. For example:

1. High sulfur coal for pulverized coal-fired precipitators would be critical in the 250 to 280°F zone.
2. Lower sulfur coal for this same precipitator might find its most critical zone between 310 to 360°F.
3. Cement precipitators might find its most critical range in the 350 to 400°F.

A variation in electrical readings may occur with as little as 10 to 15°F movements in the process gases. In some fly ash installations a 15°F change has meant a three-to-fourfold increase in emissions.

The ability to change flue gas temperatures from critical zones is as important to successful precipitator performance as any other design feature. As with variations in gas flow, short term variations in flue gas temperature should be controlled in order to minimize losses from the collector. In fact, it is usually better to operate at a less than optimum uniform temperature rather than experience variations. The heat sink effect of the internal structure will tend to mask effects of the temperature cycle if it is less than 10 minutes in duration.

Rappers and Reentrainment

Nowhere in the original Deutsch-Anderson equation is an allowance made for the losses that occur in transferring the collected material from electrode to hoppers. The interplay between the electrical forces holding the material on the collecting surface and the rapping device attempting to remove it provides a real challenge for effective precipitation. But this challenge does not merit the priority some people have placed on high rapping forces. This statement is valid as long as the rapper mechanism is sufficient to impart at least 10 to 25 Gs to the support structure holding a group of collector plates. With many process conditions, even a substandard rapper system will not effect performance adversely. But when the build-up on the collecting surface reaches over 1.9 cm ($\frac{3}{4}$ in.) it is prudent to assess whether the rapper system is sufficient. Great emphasis should be placed on the reliability of the rapper and control circuitry components. This has become more important as collection efficiency levels have increased.

The effect of rapping on precipitator performance is whether puffing losses are observed or measured since this can denote a significant reentrainment of material from the collector surfaces caused by the rapper operation. While the reentrainment puff is usually a mechanical occurrence, the operation of the rapping device can sometimes effect the electrical characteristics at the same time, thus aggravating the magnitude of the problem. The vibration of the high voltage frames could produce an electrical disturbance dependent on the structural integrity of the discharge electrode system.

Power Supply Characteristics

As precipitators have grown in size so have the power supplies grown in Kva ratings. This trend to larger transformer-rectifier capacities has injected some difficulties in stability, and yes, even in the performance of the precipitator if a gross mismatch occurs between the size of the power supply and the field to be energized.

It is well to understand a basic concept of precipitation that each field of any installation will only effectively absorb the amount of power that the existing gas, dust, and internal structure integrity allows. Therefore, the actual voltage-current requirements of a precipitator field may be drastically different than shown by the full load rating of the power supply.

Operation and Maintenance Factors

Many success stories of the high collection performance of precipitators are well documented. But to many users, a constant battle is waged to maintain these performance levels. A major reason for this situation lies in the basic design of components for the overall system that produce sensitivity for breakdowns. A concerted effort must be made by the user to understand all the inputs to the potential problems of maintenance.

Obviously, it is exceedingly difficult to predict where some of the maintenance troubles may occur, but there are eight key areas that can be emphasized.

1. Raw material and operation forecasts—original design
2. Design concepts
3. Construction phase
4. Initial check and training
5. Personnel assignment
6. Control of process
7. Record keeping
8. The actual maintenance program

Construction Phase

The best precipitator design can be adversely affected in the fabricating and erection phases. Just how the quality of welding is controlled, or the shaping of the component is finally accomplished in the shop could have a significant effect on the final operating characteristics of the precipitator.

How the material is handled on the job site

can be important to prevent a tendency to distort long electrode elements. Weather protection is of primary concern if long storage time is required.

There are advantages for the user to assign an inspector during the actual erection phase. Cross checks of the actual erection procedures are important.

Personnel Assignment

Just who to assign to the precipitator system should be given much thought. The value of the initial check-out and contacts with the manufacturer can be lost if the user representative is moved to another assignment.

A person who can be assigned long term to oversee the precipitator system and monitor the process as to how it affects the collector is probably the best investment a company can make. Recent years have shown the advantage of close supervision for the large precipitator installations.

17.3. RESISTIVITY

Introduction

Much emphasis has already been placed on the fact that effective precipitation coincides with the occurrence of optimum amounts of electrical power input in the corona process. While power input is sometimes limited by structure or individual component defects, the performance of limited power installations occurs under conditions of excessive-electrical resistivity of the collected material, usually expressed in ohm-centimeters.

All finely divided particles that are generated in the basic industrial processes have critical temperature zones which can effect the electrical operation of the precipitator. The chemical composition of the bulk of dust particles contain common constituents even if the make-up varies somewhat in weight fractions. Given similar gas conditions, it might be hard to differentiate electrically whether it was fly ash, cement dust, or iron oxide being handled in the collector. Of course, that is a simplistic statement because in practice there are an infinite number of process conditions where differences in raw material can alter the electrical characteristics of the particles.

Fortunately high moisture contents in the flue gas stream (such as found in wet process cement applications or other applications where spray water is used to cool the gases) will usually nullify the subtle chemical particle composition and provide ample power inputs. You can call water vapor a primary conditioning agent that will control resistivity problems if the quantity of water used is effectively matched to the gas temperature levels of the flue gas entering the collector.

When moisture levels in the flue gas are low—usually below 10% by volume—the chemical make-up of the particle becomes a dominant factor in controlling electrical characteristics.

The classification of these characteristics of the particle is simply related to its ability to conduct or resist the passage of electric current. This ability is not critical for the individual particle as it drifts in the gas stream, but becomes important after it deposits on the collecting surface.

One of several power input limits can occur:

1. The voltage limit of the T-R set can be reached before any other limitation.
2. Either the primary or secondary winding current limit could be reached.
3. Or spark-over can occur within the field limiting the available power from the T-R set.

These three limits should be well understood. First, the voltage limit is rarely observed on normal precipitator applications, but it can occur with an appreciable mismatch of the T-R set to the load requirement. That is, it can occur when a large capacity of supply is connected to a small surface area field and is combined with high concentrations of finer-sized particles.

Second, the current limit is observed more often. Adjusting the conduction angle of the secondary current to approximately 86% by the correct application of linear reactors will generally produce a match of the primary and secondary currents. As the conduction angle decreases from 86%, the primary current will trend toward higher readings relative to the secondary current reading. The rated secondary current will sometimes be achieved before the primary current limit if conduction angles rise past the 86% point when higher levels of impedance are applied in the primary circuit.

This brings us to the third limit, which is a sparkover between the discharge electrode and collecting surface. When this occurs, the power supply voltage must be reduced to keep the breakdowns within a reasonable level. This level could range from a nominal 150 sparks/min for the inlet field to the occasional spark for the outlet field.

However, the designer predicated his precipitator performance on a power parameter that now may not be attained because of the limitation imposed by spark-over. Basically, precipitator spark-over can occur by one of two mechanisms:

1. The impressed voltage is greater than the spacing and the physical contour or conditions between electrode surfaces will allow, regardless of the characteristic of the particulate matter. This condition if often observed with electrode misalignment where one or more of the discharge electrodes has moved too close to the collecting surface.

Another important factor that can cause premature breakdowns is the presence of severe discontinuities on the collector surface opposite the corona-emitting zones of the discharge electrode. This type of breakdown tends to provide a greater electrical disturbance compared to the spark-over caused by high resistivity.

2. Spark-over caused by high resistivity levels is the most common reason for low power inputs to the precipitator. The resistance of the layer of collected material on the collecting surface is the prime reason spark-over will occur. This layer will develop a voltage drop based on three factors: the resistivity value × the layer thickness × the current density. If the voltage drop is greater than the dust layer can withstand, then breakdown within the layer occurs.

This phenomenon is not unlike the breakdown of a capacitor. Reduction of the layer resistivity can be achieved by a number of methods including flue gas additives and process modifications. Sufficient reduction of the layer thickness is often difficult to obtain because as resistivity increases, so does the tenacity of the particles to stick together and adhere to the collecting surface.

The third component of the voltage drop is the current density. This means that the amount of corona current attempting to pass through the dust layer must be reduced if either the resistivity value or layer thickness increases. Otherwise, spark-over can occur. For example, 0.43 ma/sq m (40 ma/1000 sq ft) may occur with a material resistivity of 10^{10} ohm-cm. But if the actual resistivity was 10^{11} ohm-cm the current density might have to decrease to 0.27 ma/sq m (25 ma/1000 sq ft) to keep spark-over at a reasonable level. In other words, the higher the resistivity level, the lower the current density must be to keep the precipitation process functional. It is not uncommon to see current densities below 0.054 ma/sq m (5 ma/1000 sq ft) on certain fly ash applications.

Effects of Resistivity on Power Levels

What this means in a practical sense is that an infinite number of voltage and current readings can occur in the precipitator that will not in anyway match the name-plate data of the T-R sets. The important thing to remember is that higher resistivity conditions will decrease power inputs because of the spark-over limitation. Superimposed on a resistivity problem is the possible condition of the internal structure causing the spark-over to occur at a much lower power level. Once the dust resistivity reaches a critical point, its deposition on a sharp edge, or for that matter any kind of discontinuity on the collecting surface, will cause a localized electrical stress build-up point that will draw the spark. This is why uniformity of alignment and elimination of all internal irregularities becomes more important as the resistivity moves up from the moderate range.

It is difficult to specify various resistivity levels as denoting good or bad operation. The poor physical design of the precipitator components from a high-voltage standpoint can alter spark-over levels. It is advantageous to group resistivity into three basic zones; low, moderate, and high. The moderate range would generally encompass a resistivity from 10^9 to 10^{11} ohm-cm and is considered the best zone for effective precipitation. A finer grouping might show the following:

COMMENTS	RESISTIVITY RANGE
10^4 to 10^7 ohm-cm	Usually highly conductive material–hard to retain–low voltage fields present.
10^8 to 10^9 ohm-cm	Sensitive stage where lack of resistive characteristics can sometimes hurt–especially in fly ash cases.
10^{10} to 10^{11} ohm-cm	Appears to be the best range to shoot for–should show some spark-over in precipitator.
10^{12} to 10^{13} ohm-cm	Range usually associated with low sulfur coals–reduced power in all fields can exist.
Over 10^{13} ohm-cm	Not commonly observed in basic industries with normal moisture contents. Can produce severe electrical disturbances.

The description of spark-over can be defined as an electrical breakdown through an isolated gas path between the negative and positive electrodes. The case of the threshold resistive spark-over where the discharge streamers occur from the deposited dust layer is considered the start of back corona. This situation is not unlike that which occurs in atmospheric lightning when positive streamers from earth actually draw the localized stroke from the negatively charged clouds. But in the case of very high resistivity, a severe back-corona condition can occur characterized by greatly reduced voltage and high current densities without spark-over.

When low resistivity exists with low voltage conditions, it is difficult to achieve high collection because of dust reentrainment losses. Power consumption is high because of the high current flow through the dust layer caused by a practically nil electrical resistance. Internal inspections usually show collecting surfaces devoid of any buildup. During this low resistivity, it will be difficult achieving the guaranteed efficiency at even half the design gas velocity.

Moderate resistivity will allow dust particles to bond together in the dust layer by forming a charged dipole relationship not unlike those found in a magnet. The opposite polarities provide good adhesiveness at the tangent contact points of adjacent particles and even aid in holding these particles together as they become dislodged by the rapping procedure. Reentrainment losses are at a minimum level.

When the resistivity increases into the high range, the bonding can become severe and layer dislodgment by rapping is difficult. However, at some point the reentrainment portion of the total precipitator losses can be higher than in the moderate range and are caused by the reduced power levels. A sizeable part of the precipitator dust loss in high-resistivity cases can also occur during high spark-over conditions because each spark blows out a small volume of dust from the layer.

The Deutsch equation showed that the migration velocity contained the product of two precipitator voltage gradients without regard to the corona current component. The magnitude of the current flow will partially depend on the resistivity conditions for any given voltage. For a fly ash example showing 280 volts on a primary voltmeter, a current may indicate 750 ma with low resistivity, 400 ma in the moderate range, or 150 ma in a higher resistivity range. In this example, the 400 ma condition would probably provide the better collection performance on a higher velocity precipitator. It is always advantageous to work toward higher voltage gradients and take whatever corona current results. The only exceptions would be current suppression caused by discharge electrode build-up or excessive space charge caused by high concentration levels of fine-sized particles.

One important concept is that each process will produce a particulate matter whose resistivity will usually decrease rapidly on the low temperature side of a peak, while decreasing at a lesser rate on the high temperature side. Figure 17.3 shows a typical plot of resistivity

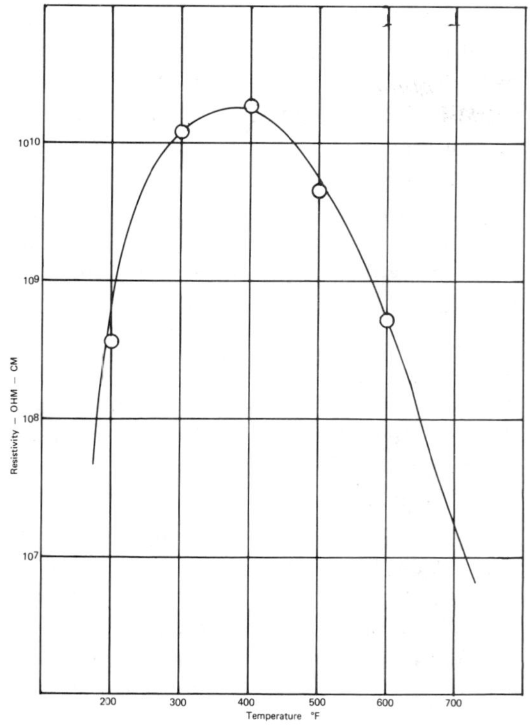

Figure 17.3. Resistivity curve for dust at inlet to cement precipitator: Lab measurement at 4,000 volts and 25% H_2O.

obtained by laboratory analysis for dust entering a cement precipitator. A typical fly ash from an eastern bituminous coal source with 2 to 3% sulfur might show a similar pattern at approximately 6% moisture by volume.

17.4. OPERATION AND MAINTENANCE

This section intends to provide many practical items and advice for personnel who continually work with the precipitator. All precipitators have common trouble areas, yet, all precipitator designs will perform well given certain operating conditions. However, it is the recurring trouble areas that nullify good performance and often cause costly production losses because of the outages required to correct these difficulties.

Internal Inspection

The internal inspection of the precipitator is an important operating and maintenance tool that can provide many benefits if done thoroughly. This is where a person who is trained to do the job on a periodic basis can build-up a knowledge of observations and data that will help define causes of difficulties and even catch areas of impending trouble. The internal condition of the small precipitator with only one field may require this careful inspection more than the multifield unit. Internal defects in any one field of the bigger collectors can often be detected electrically during operation, while it might be difficult in a small unit to ascertain whether a change in the voltage-current characteristics is due to the process or to an internal defect.

Alignment of Electrode Systems

The net effect on the electrical readings by the characteristics of the gas and particulate matter will often be contingent on the proper spacing or alignment between the electrodes. Meter readings may indicate a resistivity problem, while close spacing or even a specific electrode design may be causing the spark-sensitive precipitator. There is no substitute for careful measurements and inspections.

The degree of misalignment allowable in a precipitator is dependent on a number of factors. In highly resistive conditions, deviations of 0.63 cm ($\frac{1}{4}$ in.) in some discharge electrodes from the center of a passage could provide sensitive electrical breakdowns at slightly reduced voltages. Other installations can stand electrode misalignment of 2.54 cm (1 in.) and more, and the effect on performance cannot be easily observed. Nevertheless, significant misalignment between the high voltage and collecting electrodes, whether with weighted wire or rigid frame, should be cause for alarm.

Particulate Removal from Hoppers

Difficulties with the evacuation of material from hoppers can provide a continuous headache for plant personnel. The large-sized precipitator can have a good potential performance nullified by hopper troubles. This does not have to occur. Unfortunately, in the quest to produce high-performance units in order to meet stringent regulations, the emphasis on good hopper design was less than the concerns with SCA or power input. Hopper difficulties are often fought for years because key modifications are not applied, especially to correct the underlying causes. This lack of commitment to eliminate hopper troubles has produced a sad reputation for precipitators. The problems are many, the solutions often hard to come by, but do not hesitate to expend time and money to obtain a satisfactory performance of the hopper system.

Process characteristics, the type of material being handled, and moisture and gas temperature levels are important inputs in understanding the reasons for hopper problems. Without a reliable evacuation system from the hopper flange to the storage area, the burden on the hopper components may be too much. Based on many field experiences, here are several main factors.

Insufficient Heat and Insulation

Lack of proper insulation and heat input, especially at the bottom apex of the hopper has probably accounted for most of the troubles experienced. Regardless of the gas and dust characteristics, the ability to keep the wall surface temperature of the lower hopper no less than 250°F is most important.

Air Inleakage into Hoppers

Entry of outside air into any part of the hopper system is considered poor practice. Aside from the effect on performance, excessive air can cool down the inside wall surfaces, or condense moisture in some of the high water-vapor installations. Unfortunately, most of the screw conveyor installations coupled with process precipitators are conducive to this condition.

Level Detectors

While it is always better to place time and money in the prevention of hopper difficulties, the detection of build-ups by some method is desirable in most applications. These devices can utilize gamma radiation, sound, capacitance, pressure differential, temperature, or even paddle-wheel methodology for the detection of excessive build-up. Any method that does not require components within the hopper appears the most desirable.

Several comments: Use detectors as maintenance tools rather than to identify full hoppers. For example, if an automatic batch cycle allows a maximum 0.91 m (3 ft) of build-up in the inlet hoppers, it is well to locate the detectors no more than 1.52 m (5 ft) above the apex flange. The object is to alert the operator before a major hopper fill-up exists, yet minimize frequent detector alarms. A rule of thumb would allow the lapsed time from normal dust height to alarm level to equal the same length of time it takes the dust to rise from the apex to the normal height. Remember that the pyramid design allows for a greater volume of material to accumulate in each foot of hopper height. One problem arises in the uneven build-up that occurs by the slope and corner effect of the hoppers. The detector should recognize this effect and be placed a little lower at the center of the wall or a little higher near the corners.

Outage Clean-Down of Electrodes

If good shut-down procedures are followed on most installations, the degree of build-up on electrode surfaces will usually require no further cleaning during the outage. That does not mean that 0.63 to 0.95 cm ($\frac{1}{4}$ to $\frac{3}{8}$ in.) mounds of deposit will not exist, but the build-up will be spotty with most of the surface holding less than 0.31 to 0.63 cm ($\frac{1}{8}$ to $\frac{1}{4}$ in.) thick mounds. There are exceptions, especially caused with high resistivity materials or other operating characteristics.

Whether any manual cleaning is implemented during the outage depends on several factors. If it is an annual outage with certain planned work on the electrode system, then a water wash of the unit might be considered. It is not recommended to use this type of cleaning unless it is necessary to perform major work on the system. Depending on the time of year and thoroughness of the washing, some rusting and corrosion pockets can be accelerated.

Important Trouble-Shooting Approaches

Because of the high-voltage danger, familiarity with all the safety aspects of the system cannot be overstressed. Even portions of equipment inside each control cabinet will be at a 480 volt potential, so care must be taken in any measurement procedure. The manufacturer's manual should be well studied for guidelines in the handling of certain control difficulties.

If any troubles occur initially with control circuit components, fuses, or any other low voltage trouble source, correct these problems posthaste, since the high voltage portion of the precipitator tends to supply enough potential difficulties of its own.

Normal Versus Abnormal Power Characteristics

An understanding of the electrical readings of the precipitator must be a starting point in coping with this collector. It was already stressed that the name-plate electrical values will not be observed on many fields of the precipitator, so the patterns of meter readings become an important tool of evaluation.

The key word is "uniformity" of patterns of each precipitator because gross power values used to compare one unit to another sometimes provide questionable evaluation results. It is recommended that comparison of the voltage to current flow value of each field be ascertained under normal conditions as well as process variations.

By now, you should well understand that the control panel readings are a reflection of everything that is occurring in the precipitator. The magnitude, as well as the trends of the readings will generally tell a story of normal precipitator performance as well as abnormal conditions. A sound knowledge of the effective voltage-current characteristics can allow one to judge emissions on reproducible process operations almost as well as actual stack tests.

The primary winding voltage and current readings should provide valid reflections on what is occurring in the secondary circuit of the high-voltage transformer. However, the presence of secondary as well as primary meters does provide added monitor capability. In all difficulties within the precipitator, the two voltmeters and the two ammeters will work in unison for specific characteristics. That is, when the primary voltage is low, the secondary voltage should also be low, while the ammeters could both be showing relatively high values (see Fig. 17.4).

Probably 80 to 90% of the problems that occur in precipitators will tend to reduce voltages and raise currents at the same time. These problems are usually associated with a multitude of difficulties with electrode failure, dust buildups in hoppers, and electrical leakage over insulator surfaces.

As each field of a multifield precipitator does its work, the reduction of suspended particulate matter in the flue gas will alter the voltage-current relationship from inlet to outlet. This phenomena is best observed in the moderate resistivity range. A change in the resistivity of the material in each field can alter the patterns, but the slope of the pattern is mostly space charge oriented. For example:

	PRI. VOLTS	SEC. MA	SPARKS/ MIN.
Inlet	360	400	50
Center	330	550	20
Outlet	300	750	occ.

In other words, for identically sized fields, we are generally looking for a stepped decrease in voltage and stepped increase in current from inlet to outlet. Some key concepts in meter observations include:

1. Patterns in voltage and current readings from inlet to outlet should form some type of recognizable pattern unless internal defects cloud the issue.
2. The high resistivity range can produce a relatively low flat voltage and current pattern, but only in the very high resistivity zones (10^{12} and above). Generally you should see an increasing pattern of current flow in the direction of gas flow.

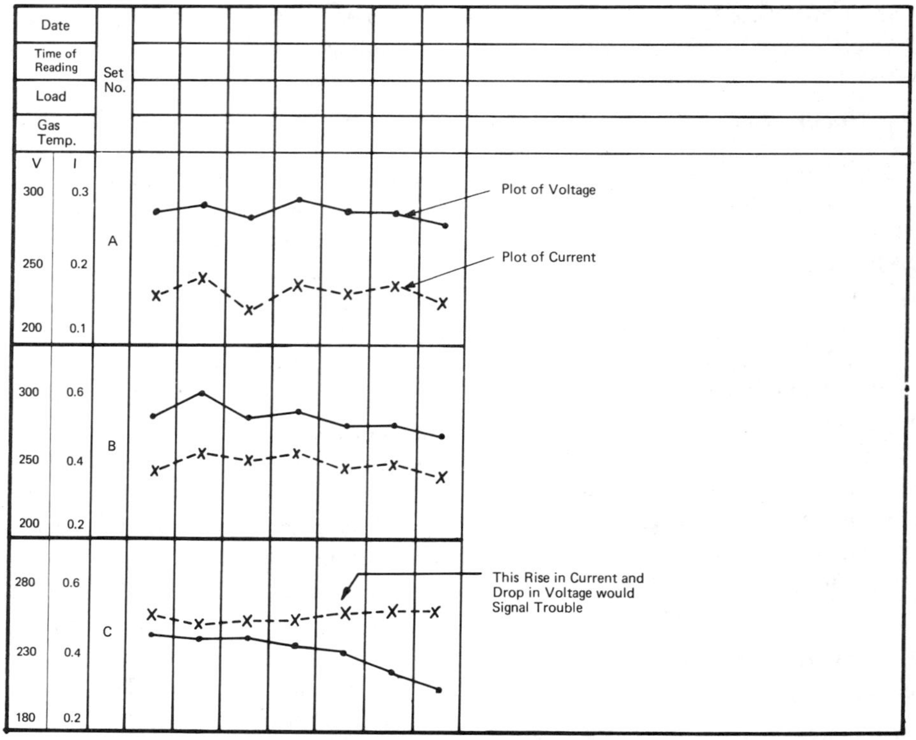

Figure 17.4. Suggestion for a daily plot of voltage and current to help detect start of troubles.

3. The moderate range of resistivity would show the greatest magnitude of change from inlet to outlet.
4. As the resistivity becomes low enough so that all sparkover ceases from this cause, then the voltage and current readings tend to flatten out again from inlet to outlet, but at a much higher power level.
5. It is only in the moderate to high resistivity ranges that internal electrode defects will show major distortions in the voltage-current patterns between fields and adjacent cells.
6. Effects of resistivity can completely nulify the effects of the space charge on the inlet to outlet patterns.

Reentrainment

Some additional comments in the area of reentrainment of material are warranted. The effects or resistivity on reentrainment has already been mentioned, but the subject is much more complex. How far the material moves out into the gas system, where it redeposits in relation to its original position, and has it changed its physical character are but some of the unknown factors in the reentrainment syndrome.

A common description of the dust layer sliding down the collecting surface does not usually occur. The shock or tremor imparted by the rapper appears to more often dislodge some percentage of material from its resting place. If the particles have had a chance to agglomerate, the adverse effects of reentrainment are minimized. It is when the particles bounce back into the gas stream in the same condition as they were collected that troubles begin to mount. This is where proper resistivity and the timing between raps can play an important part in this interesting phase of precipitation.

Some key concepts include:

1. Always use the internal inspection and other visual means to help ascertain the lowest

rapping intensity possible commensurate with other performance observations.

2. Always attempt to match the rapping to the dust characteristics or resistivity. For example, a low resistivity requires soft rapping, the moderate range requires a harder blow, and the high-resistivity zone means real trouble. Remember that hard rapping with high-resistivity materials usually exhibits limited success and changing the resistivity is usually a much better way to achieve a satisfactory performance.

3. Do not feel that the inlet field must be overrapped because it handles the bulk of material. Be aware that puffs out of the stack or other signs of reentrainment are not unique to the outlet fields. Actually, the material collected in the inlet field of some precipitators is often easier to dislodge, and excessive carryover adds to the reentrainment potential of the following fields.

4. Rapping loss is not usually uniform across the precipitator, even discounting the effects of resistivity gradients. When reentrainment losses are observed, investigations into possible problems with gas distribution is highly recommended.

5. When reentrainment losses are severe, lengthening the time between raps on the collecting surfaces in the direction of inlet to outlet is usually recommended. As a first adjustment it might be advisable to double the rap time on succeeding fields. For example, if the inlet field rapped every 5 minutes, then the second field would be rapped every 10 minutes, the third field every 20 minutes and so on.

6. The more the power characteristics are improved, the better the chance for reentrainment losses to diminish. Rappers should always operate across one field before the cycle moves on to the next field. As discussed in the text, excessive dust disturbances on the collected layer can lead to adverse electrical field activity in certain cases. Allowing the surface contour to smooth out slightly between raps can relieve localized stress points and reduce the spark-over potential. Usually 5 to 10 minutes time duration is needed to observe this phenomenon where it will occur.

Gas Distribution

Whether gas distribution is effecting the precipitator performance adversely can be related to many factors. Efforts are usually worthwhile in exploring improvements in the gas distribution pattern if one or more of the following conditions exist:

1. Aspect ratios of 1.0 or less.
2. Average calculated gas velocity of more than 1.8 m/s (6 ft/s) through the precipitator.
3. Reentrainment dust losses above 50% of the total ESP emission losses.
4. Low-resistivity characteristics are apparent with an absence of spark-over.
5. High-resistivity characteristics are apparent with less than an average of 0.11 ma/sq m (10 ma/1000 sq ft) of collecting surface.

Field observations have pointed out a number of concepts:

1. The gas flow vectors in a dynamic system will tend to keep going in the direction pointed, until striking another obstruction. This concept is fundamental to the understanding of why some installations have problems.

2. The velocity level of the gases entering an expansion plenum will determine the final patterns at the face of the precipitator. If there is a poor vector pattern at the entry of the nozzle, then higher flow rates will usually aggravate the distribution by the time the gases reach the precipitator.

3. The 40 to 50% open area diffuser plates will provide little correction of a poor gas pattern if the pressure drop across the plate is less than 1.27 cm (0.5 in.) H_2O. However, these plates will reduce the rolling action of the gas, and most of the kinetic energy will be transferred into smaller jets. Generally, below the 3.0 to 4.6 m/s (10 to 15 ft/s) range, only minimal benefits will accrue in the gas spreading effects of the low pressure drop diffuser.

4. With a 40 to 50% open area diffuser plate, which is commonly used, any gas vectors striking the plate at 45° or more from the perpendicular will have a sizeable fraction of that gas flow slide across the plate.

5. Any flue expansion with more than about a 8° slope will generally have some separation of gas from the surface. The common practice of 30 to 45° plenum expansions tend to present distribution problems for that reason.

6. Any process whose flue gases contain particles over 30 micrometers in diameter could get into distribution troubles by the settling of dust in the expansion plenum. This condition becomes worse during long periods of reduced process operation with its attendent low gas velocities.

17.5. GAS CONDITIONING

The preferred method to improve the performance of existing precipitators involves the use of higher power inputs. Poor resistivity levels can be overcome by the modification of the flue gas characteristics. The term *gas conditioning* normally refers to the various methods used for injection of chemical constituents into the flue gas stream, primarily to help alter resistivity levels in the precipitator. This term should include any method, whether or not it is inherent in the process or supplied from an external source.

Concepts

The process should be first explored to determine if inherent changes in operation equipment can modify resistivity levels. Some of these techniques include:

1. Substitute, blend, or prepare some of the bad actors in the raw materials or fuel in a manner conducive to precipitation. This may require some modification of the handling equipment.
2. More efforts on the maintenance of moisture levels in the flue gas is important. Just the elimination of inleakage air will have this net effect.
3. Awareness of the temperature effect on resistivity must be uppermost for any process change. Even the elimination of high to low gas temperature zones may help moderate a poor performance to one that is acceptable.

Probably the use of additional moisture in the flue gas by way of water sprays or steam injection can be considered a primary method of gas conditioning. Water addition to many process gas streams is often part of the operation. Water forms part of the raw material in some cases, or in others, is primarily used to control gas temperature levels at the discharge of the process. Fuel supplies another source of moisture. As previously covered, moisture contents over 20% by volume tends to nullify resistivity problems depending on the gas temperature range at the precipitator. The use of steam is much less utilized because of the cost factor, but is useful on a short-term basis where water may present condensation problems.

The use of chemical additives offers a secondary approach if the time factor or economics dictates that any modification of the process is not a satisfactory route. Fly ash collection has been the greatest area of implementation for this method in recent years. There are a number of companies and techniques available in the gas conditioning field and success has been achieved on difficult installations.

17.6. DESIGN AND PERFORMANCE CONCEPTS

The simple Deutsch equation is a valid way to understand how the various critical inputs can effect the performance of the precipitator. As mentioned earlier, the exponent can be low as much as a factor of two because of a number of problems that the designer did not forsee. Excessive reentrainment and poor gas distribution were two of the prime reasons for the disparity between the theoretical and actual results.

Recent designs have taken the migration exponent to another $\frac{1}{2}$ power or less to correct for previous problem areas and provide additional margin for the fine-sized particles existing in the latter fields of the precipitator. What this means is that a doubling of physical precipitator is indicated compared to what was considered a standard design of the early 1970s. Whether this is warranted is based on two things: whether or not good design concepts are applied and, second, the confidence of the user that he can exercise some control over the process gas and dust characteristics. In retrospect, the design of 8 to 12 years ago could meet their guarantees if conditions were optimized. Designs of 40 to 50% greater surface area over those of the past now appear quite reasonable if you weigh all the factors of today's environment. This still means carefully addressing the process characteristics and applying a commitment to proper operating and maintenance techniques.

Each precipitator field should be considered a separate collector unit, and for that matter, each gas passage of a field must perform well to attain the best bottom line of the overall system. For high collection efficiencies to be achieved, the inlet field must perform near design levels usually in the nominal 80% range. Theoretically, that means about 80% of the particulate matter would deposit in the front hopper. This is why the inlet field looms important in any upgrading program.

I would stress a few points:

1. Each succeeding field works on the residual of the preceeding field, but the potential collection efficiency tends to decrease in the direction of flow. Part of the reason is that collection values are harder to achieve as the magnitude of particles decrease.

2. Another reason is that the particles that are left in the gas stream in the latter half of the precipitator are tougher to collect since they usually consist of the finer sized segment. Unless current densities above 0.22 ma/sq m (20 ma/1000 sq ft) are observed in these latter fields, their collection efficiencies can deviate substantially from design.

REFERENCES

For Theoretical Background

1. H. J. White, *Industrial Electrostatic Precipitation*, Addison-Wesley, Reading, Mass. (1963).
2. M. Robinson, "Electrostatic Precipitation" in *Air Pollution Control I*, ed. W. Strauss, Wiley-Interscience, New York, N.Y. (1971).
3. J. D. Cobine, *Gaseous Conductors*, Dover Publications, New York, N.Y. (1958).

Operation and Maintenance

4. J. Katz, *The Art of Electrostatic Precipitation*, Precipitator Technology, Inc., Munhall, Pa. (2d printing) (1981).

For Manuals and Publications on ESP's

5. The National Technical Information Service, Springfield, Va.
6. The McIlvaine Company, Northbrook, Ill.
7. Air Pollution Control Association, Pittsburgh, Pa.
8. Industrial Gas Cleaning Institute, Alexandria, Va.

18
Granular Bed Filters

Frederick A. Zenz
Manhattan College, Riverdale, New York

18.1. INTRODUCTION

The suggestion that extraneous solids could be removed from gaseous streams by passage through a bed of particles is to this day regarded by many as novel, if not improbable, despite the fact that the average engineer is well aware of the technology of treating liquid streams by passage through beds of particles for purposes of removing undesirable elements via ion exchange and more simply for removing any suspended solids. The suggestion becomes more acceptable and credible by considering the analogy between the well accepted bag filter, which represents a tortuous path through the interstices in a layer, or bed, of interlaced cylinders (fibers), as opposed to the granular bed which represents a tortuous path through the interstices in a layer of interlaced spherical or angular shapes.

The novelty of the concept, in principle, vanishes rapidly when a search of the patent and technical literature reveals that shallow bed, granular, filtration devices in various forms have been proposed, explored, tested and even marketed over the period from at least 1896 to the present day.

18.2. PURPOSES AND APPLICATIONS

The principal interest in development of granular bed filters today rests in removal of the particulates from hot pressurized process gases which could then be utilized to drive turbines and other equipment for efficient recovery of otherwise lost energy. A secondary interest lies in insuring that exhaust gases meet EPA emission standards for particulate matter, and a third in the simple incentive for an efficient filter not subject to high maintenance costs.

Although no universally accepted standards have been specified or agreed upon for the allowable particulate loadings of gases driving turbines, it is generally agreed that the lifetime and performance efficiency of such turbines can justifiably be increased by reducing the particulate content of the feed gas stream via a reasonable cost filter. As an example, some typical particulate loading specifications are summarized in Table 18.1. Electrostatic precipitators, scrubbers and fabric filters have been successfully used for removing particulates from exhaust gases at moderate levels of temperature and pressure, but not under high temperature (e.g., 1000° to 2000°F) and pressure conditions. Lowering the temperature and pressure of the turbine inlet gas stream for the purpose of facilitating particle removal by using proven techniques would result in large losses of energy. The increased prospective utilization of combined cycle gas/steam turbine/electrical generating systems, coal gasification, fluidized bed combustion, and synthetic gaseous fuels

Table 18.1. Typical Turbine Specifications.[a]

SOURCE	PARTICULATE LOADING (MAXIMUM ALLOWABLE)
United Aircraft	0.8 lb/million SCF low BTU fuel gas (or approximately 12 ppm)
Westinghouse	0.03 % (in fuel oil) (or 300 ppm by weight)
	0.002 grains/SCF all smaller than 6 microns (for Fluidized Bed Coal Combustors)
General Electric (for aircraft-type turbines)	30 ppm by weight in fuel gas (10 micron maximum)
Brown Boveri	1–2 ppm by weight (in gases entering turbine)
DOE (estimates)	0.75 grains/SCF (or approx. 2.6 ppm by weight) in 0 to 2 micron range
	0.001 grains/SCF (or approx. 35 ppb by weight) in 2 to 6 micron range
Exxon (estimates)	45 to 1 mg/m^3 (0.02 to 0.0004 grains/SCF)

[a]NTIS-BP 266 231 Feb. 77.

has intensified the interest in granular bed filters. Unfortunately the effect on turbine blades appears to be a function of the physical properties of the impinging particles. In catalytic cracking of petroleum fractions, carry-over silica-alumina catalyst fines remaining in the exhaust gases after passing through three stages of cyclones have shown power recovery turbine blade erosion to a degree necessitating re-blading only at 5 year intervals.[1] The severity with coal combustion or gasification fines is reportedly far greater.

Other than the incentives of power recovery the increasingly stringent EPA emissions standards create definitive advantages for dry collection to avoid the handling of slurries from wet scrubbers. With cyclones not likely to meet the standards, with electrostatic precipitators representing high installation and maintenance costs, and with bag houses at 2 to 10 CFM/sq. ft. representing large installations requiring considerable space, the potential of a 40 to 100 CFM/sq. ft. high efficiency, low maintenance, granular bed filter is an inevitable attraction even in conventional electrical utility applications.

Other than power recovery and meeting emission standards there exist a host of industrial applications in which a "sand" medium would offer considerable process advantage over a fabric. Condensibles in a gas stream can cause severe bag failures particularly in systems carrying cement kiln and similar calcined effluents, which can solidify on the bags. Sand media can be effectively dried and even washed in situ, and light accretions removed by the grinding action of the media. Where collected particulates can represent a fire hazard as in the collection of carbon black, cellulosic solids, and similar materials, a noncombustible filter medium would have immeasurable advantage.

In the popular literature are found such terms as "gravel bed filters," "panel bed filters," "expandable bed filters," "moving bed filters," "sand filters," "loose-surface filters," "porous bed filters," "MB filters," and a host of others all of which pertain to versions of what are generically referred to as granular bed filters.

All potential commercializations of the concept rest on 4 basic interrelated factors:

1. Collection efficiency
2. Cleaning or regeneration capability
3. Capacity
4. Competitive cost

Their practicality lies simply in the operational details of their technology which might best be illustrated by a relatively chronological review of the principal attempts at commercial development over the past 30 years and the status of such work today.

18.3. POROUS SINTERED GRANULE BEDS

Particle filtration via porous membranes rests primarily on the formation of a filter cake, removable by a reverse flow of fluid when the resistance of the cake (or the pressure drop) exceeds any desired level. The porous membrane may take the form of a bed of particles or for that matter a mat of fibers such as is the case with cloth bag filters. The thinner the membrane the lower the overall resistance with or without a filter cake and hence by analogy to the near mono layer of fibers in a cloth collector one could construct a thin bed or sheet of granules by sintering a shallow layer of metallic granules in a high temperature furnace. Such sheets of sintered metal granules in various forms are sold commercially for filtration purposes. Their principal application is in liquid systems, such as the maintenance of dirt-free fuel lines in aircraft engines. They have had limited application in the recovery of carryover from fluid bed reactors and similar fine-particle processing, but are not broadly acceptable. Their commercialization stemmed from the work of Dr. David Pall whose interest in the early 1940s lay in the development of a gaseous diffusion barrier for isotope separation in connection with the Manhattan District project during World War II.

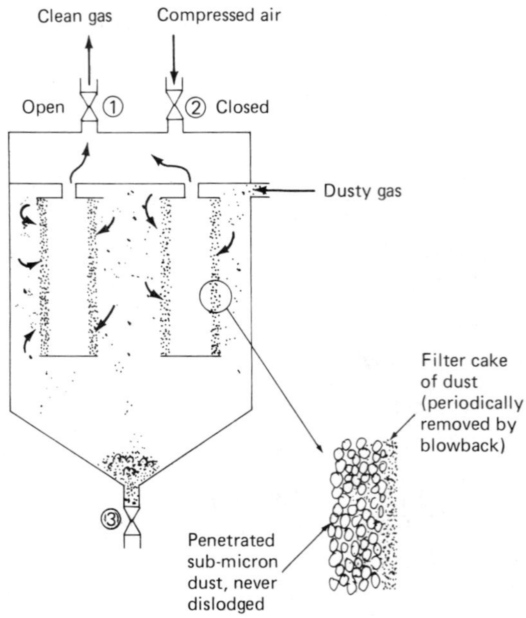

Figure 18.1. Porous sintered granule filter.

In the late 1940s the Pall Corporation (now Pall Trinity Corp.) supplied such porous metal in tubular form with the suggestion that their application would be analogous to a cloth bag filter in which the cloth is replaced by a rigid thin bed of sintered metal particles.[2] As illustrated in Figure 18.1, in normal operation valve 1 would be open, allowing dusty inlet gas to flow through the elements, depositing a filter cake on their outer upstream surfaces and leaving through the exhaust plenum. When the cake has grown to a thickness exhibiting high pressure drop, valve 1 is closed, valve 2 is opened, and a short, high-pressure pulse of air is admitted in reverse flow through the porous elements to dislodge the filter cake which falls to the bottom of the containing vessel, to be eventually withdrawn through valve 3. In continuous use, the valves operate on a timed cycle and the containing vessel is provided with a multiplicity of porous elements and separate plenums for localized reverse cleaning.

There are a number of inherent disadvantages to this form of filter. In order to obtain structural strength, the granules making up the porous element must be small in size to present sufficient bonding surfaces. This results in high-pressure drop or low gas capacity despite the only approximately $\frac{1}{8}$ inch wall thickness of the elements. Simultaneously, the thin-wall tubes are very subject to cracking as a result of repeated thermal shocks between normal high-temperature operation and relatively colder reverse-flow cleaning blasts. In addition, over extended operation, the submicron fines penetrate the interstices of the elements. These fines become trapped and are unable to be blown out with the reverse cleaning blast. Thus, these elements build up a residual pressure drop which further limits their capacity. They are not recommended where only very fine particles are to be collected; they operate most satisfactorily in handling streams containing sufficient coarse particles, in effect, to build up their own precoat. Such instances are rare, since where they occur the particle loading is usually so high that the cost of the elements is prohibitive even if they were never to fail by thermal shock.

18.4. CONTINUOUS MOVING-BED FILTERS

As early as 1924 Cramp,[3,4] proposed a "dust curtain" system for filtration of blast furnace effluent gases. The dust curtain was a vertical layer of the collected dust itself, held in place by an arrangement of grids and louvers packed under its own weight. Dusty gas passed through this curtain horizontally adding more dust to it. The excess dust gradually overflowed the louvers and fell to the bottom. The curtain was renewed periodically by dropping dust out of the bottom of the grids and adding more at the top. It is not clear whether this concept was ever put to industrial use.

In the early 1950s the late Morton Dorfan (Mechanical Industries, Inc.) introduced the Dorfan Impingo Filter,[5,6] which consisted basically of a downflowing bed of gravel retained between louvered walls. Dust-laden gas was filtered by blowing through the bed normal to the direction of gravel flow, and filtered dust, hopefully carried downward with the gravel, was trapped within its interstices, as illustrated in Figure 18.2.

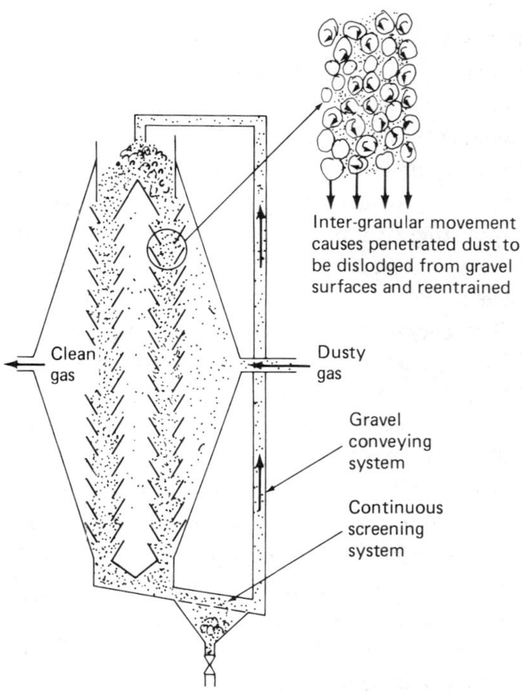

Figure 18.2. Continuous moving-bed filter.

Four units, each of 17,000 CFM air capacity and consisting of two cells each, were installed in a plant to collect asbestos rock dust from a stream of flue gas coming from a direct-fired dryer in which the rock was dried prior to milling. The dust was 100% finer than 100 mesh and 60% finer than 10 microns. The concentration entering the collector was approximately 6 grains/cu. ft. and that leaving about 0.2 grains/cu. ft.

Dorfan's filter circumvented the thermal shock failure of Pall's sintered beds by avoiding reverse-flow cleaning and instead circulating the bed granules, plus collected dust, through a vibrating screen. However, this consequently required continual replenishment of the bed and hence an enormous and costly gravel circulation system. In order to provide reasonable gas throughput capacity, the bed granules had to be relatively large so as not to be blown off the retaining louvers; in practice "gravel" of plus $\frac{1}{4}$ to minus $1\frac{1}{2}$ inch was recommended. The commensurately large size of the bed interstices resulted in lower collection efficiency, and required staging beds in series, as shown in Figure 18.2, to achieve reasonable overall collection efficiencies. Such staging only added to enormity of the required gravel circulation. Basically, micron-size particle collection efficiencies of the order of 99+% could never be attained with such a moving bed system since intergranule movement in a downflowing mass physically dislodges collected particles and makes them subject to reentrainment.

Simultaneous with Dorfan's development a similar device was reported in the Russian literature by Zhitkevich[7] for the removal of peat dust from air by passing it through a vertical moving bed of pieces of peat of 5 mm in diameter. The rate of downward movement of the bed was controlled by the speed of a screw conveyor removing the dusty peat at the base of the unit in much the same fashion as rotary valves (not shown in Figure 18.2) at the bottom of each louvered column in Dorfan's arrangement. It is not known to what degree Zhitkevich' device found industrial application.

In 1954 Egleson et al.[8] published information

on pilot tests of a downwardly flowing coke bed proposed as a possible means of filtering coal dust from a stream of synthesis gas produced in the gasification of pulverized coal. An 8.5 foot deep bed of 0.125 to 0.40 in. diameter coke particles moved vertically downward by gravity flow through a 12 in. i.d. column counter to an upwardly flowing stream of dusty air. At inlet loadings of 2 to 8 grains/cu. ft. of 200 mesh dust, collection efficiencies were reported as high as 99.9% with superficial air velocities of 33 CFM/sq. ft. giving pressure drops of only 0.3 in. of water per foot of bed depth. The dust laden coke leaving the bottom of the column was washed with water and continuously recirculated to the top of the column. Though achieving, by means of counterflow, a higher efficiency than the Dorfan or similar cross flow devices, it became obvious that the enormity of the granular media handling equipment would again make industrial application impractical.

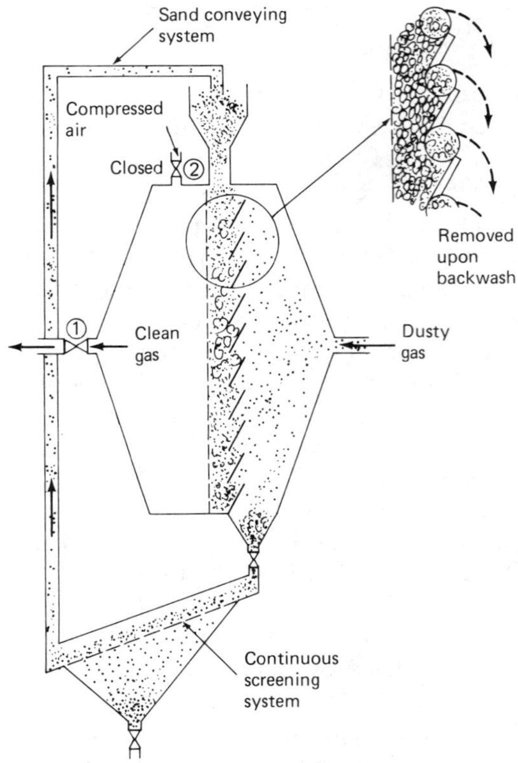

Figure 18.3. Intermittent moving-bed filter.

18.5. INTERMITTENT MOVING-BED FILTERS

In the late 1950s it occurred to A. M. Squires (presently associated with the Chemical Engineering Department at the Virginia Polytechnic Institute) that the Dorfan filter could be made far more efficient by using a finer medium and arresting the bed motion during a filtration period, after which an accumulated cake of filtered fines would be removed by moving just the dusty sand. The desirable arrangement illustrated in Figure 18.3 would circumvent the fouling action of unremovable penetrated fines (as experienced with Pall type filters) and hopefully lessen the burden of circulating the immense quantities of sand associated with continuously moving beds (as exemplified by Dorfan type units).

In the arrangement pictured in Figure 18.3 the dusty gas passes through a thin bed of stationary sand held between a panel of louvers and a fine mesh screen. The filtered fines build up a cake on the exposed bed surfaces and to some extent penetrate the interstices. When the resistance of the cake has reached an undesirable level, the clean gas outlet valve 1 is closed and a short pulse of compressed air is blasted in reverse flow through the sand bed by opening valve 2. In continuous use, the valves operate on a timed cycle and the containing vessel is provided with a multiplicity of panels exhausting to a partitioned plenum, permitting localized reverse flow cleaning of individual panels.

The backwash pulse is sufficient to physically lift the sand beds as a mass, with minimum interparticle movement, so that a surface layer of sand between each pair of louvers is physically ejected from the panel and falls to the bottom of the filter vessel along with the collected filter cake. The expelled sand is immediately replaced by downward movement of fresh sand from the overhead hoppers. This only intermittent downward movement, coupled with a fine size of "gravel" permits high-efficiency collection and avoids the buildup of resistance due to penetrating submicron fines

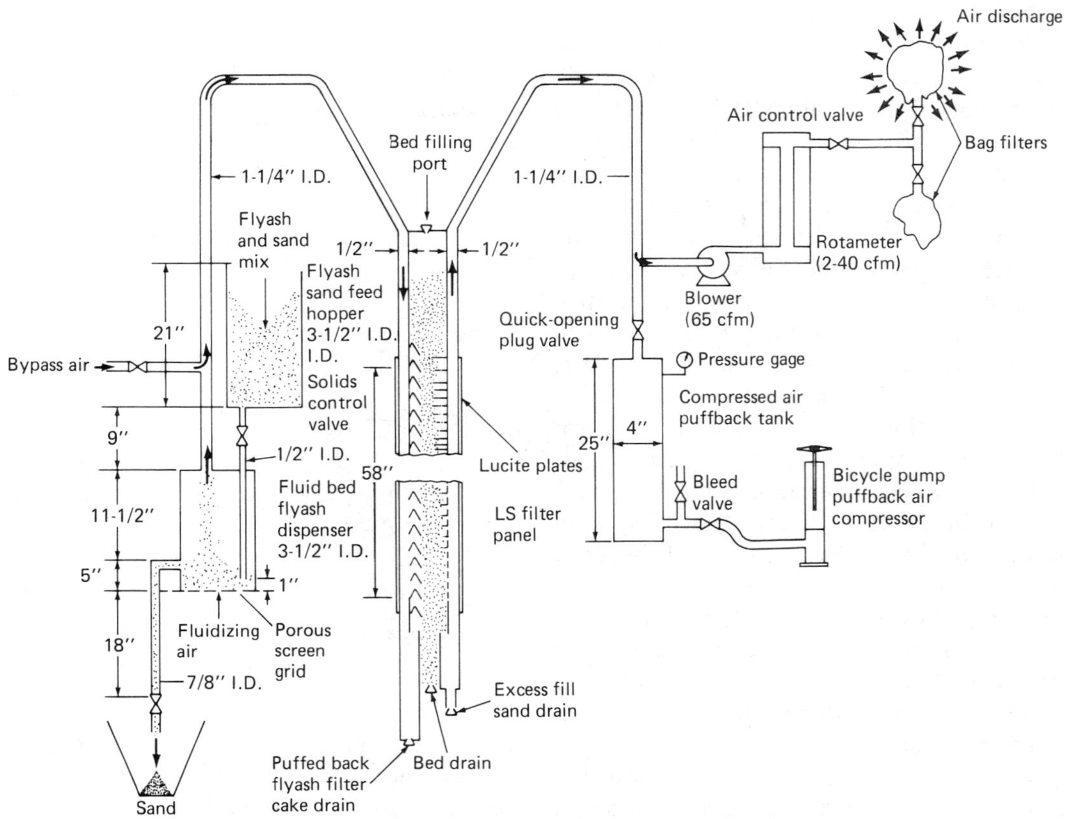

Figure 18.4. Flow diagram of test unit LS number 2.

since these are expelled along with the surface layer of their surrounding sand. Incomplete removal of the sand containing penetrated fines would lead to their eventual accumulation near the bottom of the panel, by gravity flow, after repeated cleaning cycles, causing a gas flow maldistribution and excessive pressure drop. In order to minimize penetration and hence minimize sand circulation, the gas throughput was limited to approximately 10 to 12 ACFM/sq. ft. of exposed bed surface.

Referred to as the "Loose-Surface" or LS filter, several models were built and tested in the period of 1959 to 1961 at the laboratories of F. A. Zenz then in Roslyn Harbor, L. I. Figure 18.4 illustrates the largest of these. Typical experimental results using hand-sieved bank sand as the filter medium, and bagged flyash from a Consolidated Edison power plant as the redispersed dust, are given in Table 18.2.

The simplicity of the laboratory unit of Figure 18.4, appeared sufficiently attractive to

Table 18.2. Typical Loose Surface, Panel Bed Filter, Test Results.

SAND MEDIUM PARTICLE SIZE	16 TO 30 MESH	30 TO 100 MESH
Flyash loading in feed gas, grains/cu. ft.	2 to 6	7 to 9
Face velocity, ACFM/sq. ft.	8 to 12	6
Initial (clean) bed pressure drop, H_2O	0.1	0.38
Maximum pressure drop (prior to blowback), H_2O	0.6	0.6
Flyash laydown between cleanings, oz.	14	2 to 3
Collection efficiency	99.7%	99.8%

warrant its purchase in 1960 by the Fuller Company (of Catasauqua, Pennsylvania, a subsidiary of General American Transportation). After filing their patent[9] in Squires' name and spending three years in an attempt to build equally successful test models in their own laboratories Fuller abandoned the commercial development of Squires' filter and currently maintain solely a nonexclusive interest in the original patent. Fuller's decision to halt their development was based on (1) an inability to restrict the amount of sand lost on each cleaning cycle to levels commensurate with economical recirculation costs, (2) the realization that the cost of steel and support structures was far greater than Squires' original estimates, and (3) that at 10 CFM/sq. ft. the capacity was not competitive with bag filters and hence a limited market would possibly exist only in a few very high temperature applications. At the time of its purchase by Fuller the LS filter was considered a potential competitor of electrostatic precipitators for power plant flyash collection.

In view of subsequent developments it is interesting to note that during the early period of laboratory development of the LS filter it was suggested that the circulation of sand could be entirely avoided by installing a retaining screen on the clean side of the panel thus trapping the sand in a number of superimposed beds which could be cleaned by a reverse fluidizing flow of gas to elutriate out the collected fines. Squires reasoned that this would not permit high filtration efficiency because the mixing action accompanying fluidization would result in backmixing of collected fines so that eventually the sand bed would be dusty throughout and therefore some amount of collected dust would find its way into the clean side of the filter. Exploration of this suggestion was therefore not carried further.

Upon joining the faculty of City College in 1966 Squires obtained substantial financial support from EPA and similar government agencies to essentially repeat the LS development with, in some instances, minor modifications in bed depth or louver configuration.[10,11,12]

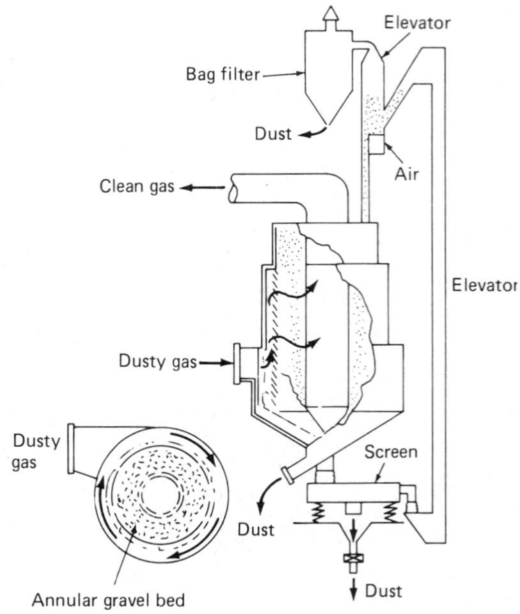

Figure 18.5. CPC annular bed filter.

Inspired by Squires' modification, the principles of the Dorfan filter was investigated anew at the U.S. Bureau of Mines.[13,14] In this instance the test apparatus, designed by Combustion Power Corporation (CPC) utilized downward moving beds of sand grains. The CPC design comprises essentially the panel form of the Dorfan filter rotated into a cylindrical arrangement and operated with intermittent media downflow. With the expectation that the CPC filter would live up to the claims of its developers on economically competitive grounds, the management of Weyerhauser Industries purchased all the outstanding stock of Combustion Power Corporation and installed such filters on a number of their own bark boilers. Unfortunately such complete installations were found to require the following component steps illustrated in Figure 18.5:

1. sieving of the collected fines from the bed media
2. conveyance of the media to a superimposed vessel
3. elutriation of residual fines from the media by fluidization

4. recovery of the elutriated fines in a "small" bag filter
5. redistribution of the granular media to the filter annulus.

A complete installation therefore takes on many undesirable and costly complexities quite apart from the otherwise simple principle. In addition to an inadequate overall collection efficiency (including the periods during media movement) a complete installation reaches levels of cost which encounter a severe sales resistance. Weyerhauser has reportedly been disenchanted with their own installations and further development is presently only minimally supported with U.S. Department of Energy (DOE) funds.

18.6. FLUIDIZED BED FILTERS

It should be obvious from the foregoing experiences with cross flow moving beds that high efficiency collection is not possible unless the downwardly moving "wall" is very thick (e.g., representing a deep bed). The rubbing action of intergranule movement and the continuously changing interstitial configurations allow entrapped dust particles to be blown by the gas stream to deeper and deeper penetrations of the filter bed. Therefore very deep beds are required to achieve only reasonable collection efficiencies. It would, therefore, be expected that gas fluidized beds would similarly yield relatively low collection efficiencies. This conclusion is generally borne out by several such reported investigations.[15,16,17]

Under U.S. Army Chemical Corps Contract DA-18-064-CML-2758, a mechanically (vibrated) fluidized bed of sand was investigated for its possible filtration potential. The point of this investigation was to determine whether the poorer efficiency of gas fluidized beds was solely attributable to intergranule movement (as in moving bed filters) or significantly affected by the passage of unfiltered gas in the bubbles rising through the bed. The vibratory fluidized bed was operated with gas downflow thus avoiding bubble formation and the accompanying bypassing of dusty gas. The results showed insignificant improvement over a bubbling fluidized bed thus leading again to the conclusion that high filtration efficiency is compatible only with stationary media.

It is, however, of interest to note that a shallow moving fluidized bed of raw bauxite feed has been used to filter the carryover of particulates from the off-gases of electrolytic cells used in the production of aluminum metal. Such a unit has been operated by Aluminum Co. of America following the description in their patent.[18] The filtered particulates are primarily pitch or asphalt employed as a binder in the cell anodes. Though they appear to be dry they are sufficiently tacky to adhere to most surfaces on contact. Kaiser Aluminum installed an electrostatic precipitator at their Chalmette, Louisiana plant which operated for only a short time until it became completely coated with a layer resembling the undercoating on an automobile. The sticky nature of the collected particulates might be the principal reason for the practical application of a fluid bed filter in this instance; particularly since the filter medium is itself the raw material continuously fed to the cells thus requiring no recirculation equipment and efficiently returning the asphalt fines to the cells as further fuel. The filtered off gases leaving the fluid bed are in any case subsequently passed through a bag house for cleanup of nonsticky fines to achieve the level of overall efficiency necessary to pass EPA standards. A filter of the Dorfan type was also proposed for such aluminum plant operations.[19]

18.7. GRANULAR BED FILTERS MECHANICALLY CLEANED

As opposed to the LS filter's attempt to compromise between the principles illustrated in Figures 18.1 and 18.2 (e.g., fixed bed filtration but minimal sand circulation) Max and Wolfgang Berz[20] in 1957 proposed an arrangement requiring no sand circulation. The cleaning would be carried out by flowing a gas in reverse direction through the bed while subjecting it to mechanical vibration of sufficient magnitude to cause the intergranule movement necessary for removal of entrapped fines. The principle of their proposed filter is illustrated in Figure 18.6.

In normal operation dusty feed gas would flow upward through a stationary fixed bed of sand held on a horizontal screen of a mesh size smaller than the bed granules. A filter cake might develop and adhere to the lower surface of the screen and some amount of fines penetration would occur within the interstices of the bed. Since they employed a granule size many times larger than the approximately 30 mesh sand investigated by Fuller, they experienced a greater degree of fines penetration (as opposed to cake buildup), required deeper beds, and operated at considerably higher throughput capacities per unit of bed surface area.

When resistance to flow reached a level requiring bed cleaning, the dusty feed gas was diverted to another unit and a reverse flow of gas at low velocity passed down through the bed, while it was simultaneously vibrated in a manner such as to impart an orbital rotation to the particle mass. Thus, the mechanically induced interparticle motion dislodged the penetrated fines so that they could be swept out of the bed by the downflowing breeze. In continuous use, the reverse flow and bed vibration were automated on a timed cycle and the containing vessel again provided with multiple units cleaned individually in sequence.

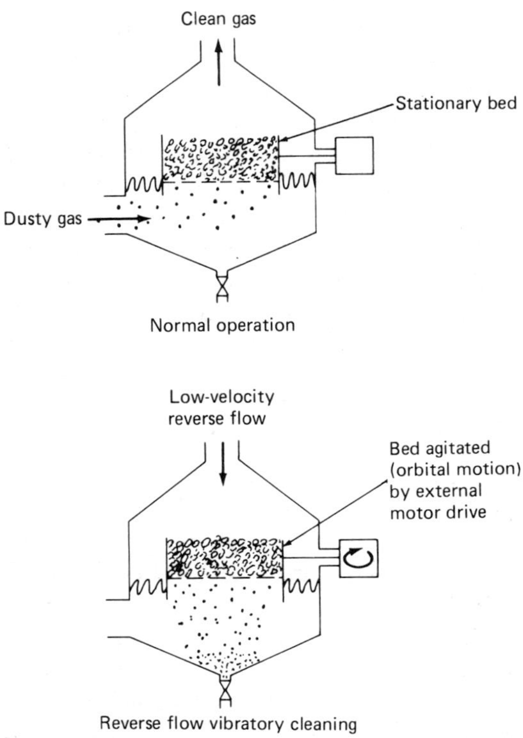

Figure 18.6. Mechanically cleaned granular bed filter.

The Berz' MB filter was sold commercially by Lurgi Apparatebau G.M.bH. of Frankfurt, Germany, until about the end of 1969 when it was abruptly withdrawn from the market. Its shortcoming lies in the strain imposed on the necessary flexing membranes associated with the vibrating technique. It is conceivable that at low temperatures where rubber membranes and spring-supported bed mounts are feasible, such a filter could operate with a reasonable life. However, at low temperatures it could never economically compete with the simplicity of bag houses. At high temperatures only metal bellows would suffice as the flexing membrane, and their life expectancy in the hot and dusty environment is impractically short.

Undaunted by their mechanical failure Berz devised an alternate arrangement now marketed through Gesellschaft fur Entstaubungsanlagen m.b.H. of Munich, Germany. In this version, illustrated in Figure 18.7, the gravel medium is agitated during the cleaning cycle by the stirring action of a rotating rake whose fingers are imbedded in the gravel medium. Though the long-term mechanical integrity of the raking system as well as its possible introduction of flow paths of lower efficiency might be objectively questioned, it appeared that this filter might receive reasonable acceptance in at least the cement industry. Users, however, were not as completely satisfied as might be implied in Berz' optimistic presentation to the 1972 IEEE cement industry technical conference held in May, 1972. The cleaning period requires several minutes and the module undergoing cleaning is isolated from the dusty gas by a valve system during the cleaning period. Carrier Corp. (Rexnord Division) installed such filters at over a dozen cement plants,[21] however rake

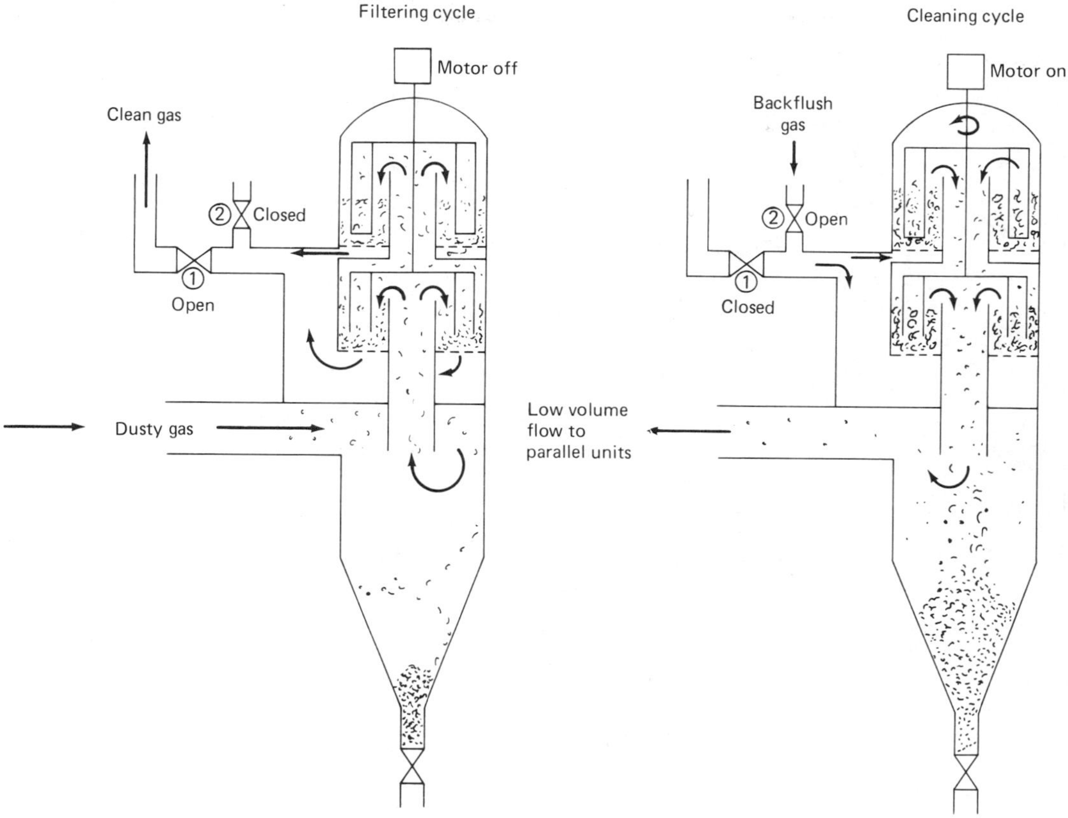

Figure 18.7. Alternate form of mechanically cleaned granular bed filter.

failures began, as anticipitated, within about a year. Present efforts are concentrating on possible incorporation of a means of periodically removing and replacing the bed medium.

18.8. GRANULAR BED FILTERS PNEUMATICALLY CLEANED

In 1970 the Ducon Co. introduced the "Expandable Bed" filter,[22,23,24] in which a number of superimposed beds of sand, trapped between retaining screens, would be cleaned by a reverse flow of fluidizing gas. Its principles are illustrated in Figure 18.8. In normal operation dusty gas passes through vertical arrays of parallel shallow granular beds held within metal-walled compartments sealed at top and bottom by perforated meshes finer in aperture than the size of the bed granules. The dusty gas enters the beds at relatively high velocities, in the range of 40 to 100 CFM/sq. ft. of bed surface area. When flow resistance reaches a level requiring the bed to be cleaned, a sufficient momentary reverse flow of compressed gas is admitted to fluidize the bed granules, or "expand" the bed, and thus by entrainment from these agitated beds expel the particles collected and agglomerated in the bed interstices. No loss of sand can occur since any granules which might reach the end of a compartment cannot escape through the perforated retaining mesh.

The fluidized expansion of the granule bed and the accompanying intergranule movement, which allows efficient cleaning, are directly analogous to the action in bag filters. Upon cyclically reversing the flow of gas, particulates trapped between the fibers are expelled as the cloth flexes or expands. This movement of the fibers in the expansion of a bag allows such

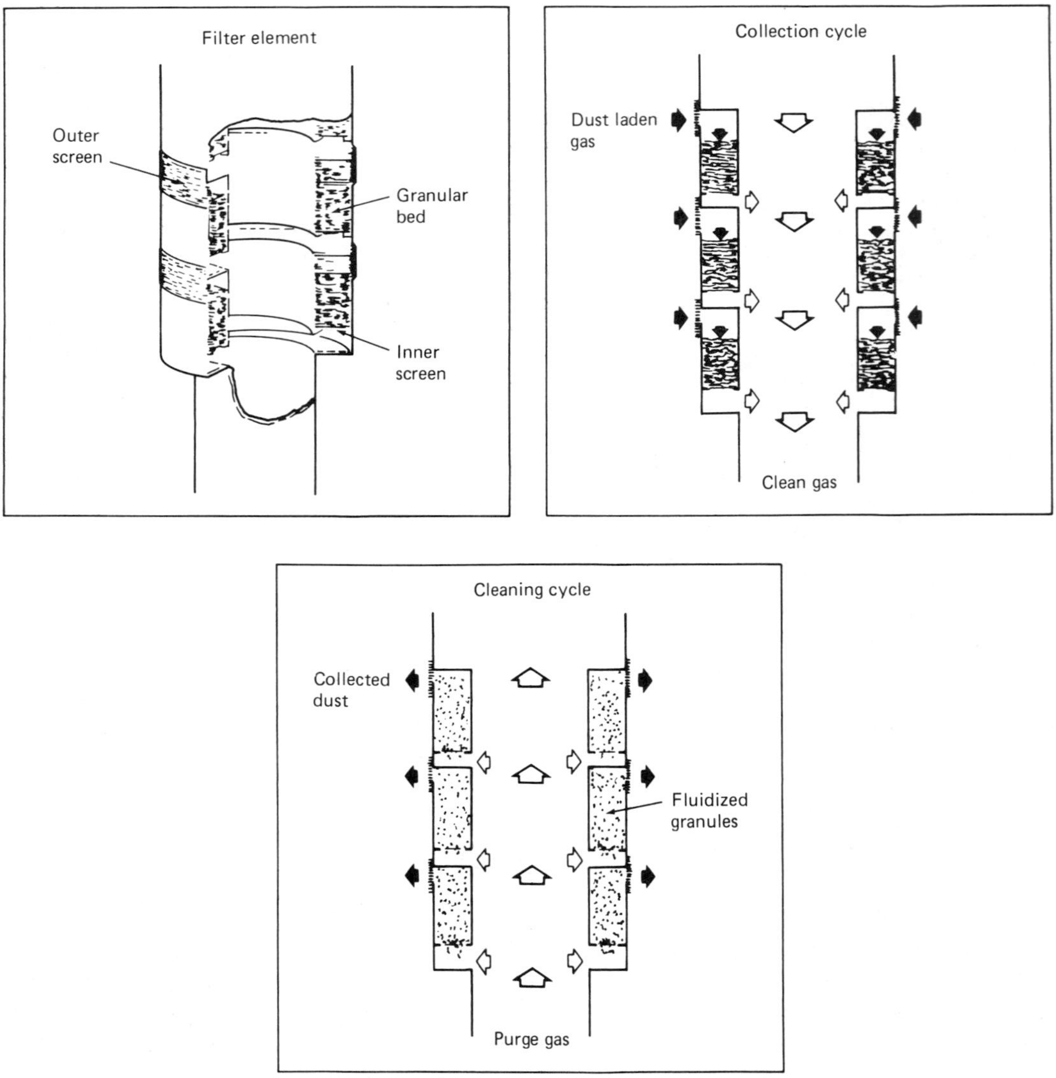

Figure 18.8. Pneumatically cleaned granular bed filter.

complete cleaning (depending upon the number of flexings during each cycle) that bags can always be returned to near their original pressure drop characteristics. By sufficient and proper duration of backwash flows, the trapped fines in the beds of Figure 18.8 can also be nearly completely expelled by the analogous intergranular bed flexing, bed expansion, or bed fluidization. Thus fluidization affords an economical nonmechanical cleaning mechanism compatible with high-temperature filter operation, requiring no costly recycle conveying of sand, no complex redistribution hoppers, no subsequent need to separate collected fines from filter element sand, and hence no attrition of bed granules which normally accompanies repeated solids handling operations.

In practice it was discovered that the intermingling of collected fines within the granular bed, which inevitably occurs as a result of the fluidized agitation during cleaning, resulted in an eventual buildup to an equilibrium fines content which then increased the collection efficiency, presumably simply because it presented a more tortuous interstitial path for the dusty gas flow. It was also found that the economic optimum operation lay in high gas capacity with minor sacrifice in efficiency. In this connection it is interesting to note that, in a paper dealing with the panel bed filter, presented at the Washington, D.C. A.I.Ch.E. meeting December 3, 1974, R. Pfeffer[12] reported also observing higher efficiencies when using dusty as opposed to fresh sand. However, it is also obvious that following a reverse flow cleaning there could exist a small amount of the collected dust adhering to "sand" media lying directly atop the bed supporting grid. This dust might well be blown off into the clean side of the filter when it is again placed in filtering operation and so contribute to a less than 100% filtration efficiency depending on cleaning frequency.

18.9. TECHNOLOGICAL STATUS OF SYSTEMS UNDER DEVELOPMENT AND UNDER COMMERCIALIZATION

The technological success and eventual commercialization, of any one or more of the many granular bed filter concepts, lies in

a. satisfactory dust collection efficiency
b. reproducible regeneration or filter media cleaning
c. competitive total installation cost

Unfortunately these three aspects are rather intimately interrelated and in all likelihood only eventual full-scale testing will narrow the field; this becomes partially evident in even a cursory review of the major commercial contenders.

The Fuller-Squires Panel Bed concept though abandoned by Fuller has undergone a series of studies related principally to improvements in louver design. There is no doubt that this concept affords excellent collection efficiency and that following each blowback the dusty gas faces a reproducibly cleaned filter bed. The questions that remain as yet incompletely answered are whether sieving the dust laden sand drawn from the filter will produce sufficient cleaning. Will some residual dust particles cling to the media and escape into the cleaned gas stream upon return to the louvers? Would an additional fluidized bed elutriation column provide sufficient cleaning; how would this fluidizing gas stream then be cleaned and what would such additional equipment add to the total cost? How many louvers can be uniformly cleansed upon each blast of blowback gas? What is the investment increment for the media circulation and distribution system and what degree of media attrition might be anticipated. No tests of the Fuller-Squires panels have ever been conducted with continuous media recirculation or in an industrial environment.

The Combustion Power Corporation Annular Bed as illustrated in Figure 18.5 has undergone full scale operation on bark boiler effluent giving results such as shown in Table 18.3 on the following page. It is unknown whether the reported media efficiencies include periods during which the media were in downward movement or only periods when the media were stationary. The media pressure drops may also refer to clean conditions prior to dust buildup. In any event the media efficiencies are substantially below the 99+% levels normally sought in granular bed filtration. Efficiencies in the 90% range could more effectively be attained with multiclone centrifugal separators. Reliable cost figures for a complete installation including all the elements in Figure 17.5 are unavailable.

C.P.C.'s principal development experience centered around a pilot unit installed at the Snoqualmie, Washington plant of Weyerhaeuser Lumber Co. to clean up effluent from a hog mill waste combustor. The pebble bed filter was in this instance a single downflowing sand annulus $8\frac{1}{2}$ ft. O.D. and about 6 ft. I.D. with an effective filtering height of 16 feet. The face velocity and effective inlet area are quoted by CPC on the basis of the mean diameter of the annulus and in this instance correspond to 100 to 163 ft./min. (or ACFM/sq. ft.) and to 365 ft.^{2}, respectively. The pebbles were angular in shape and ranged from $\frac{1}{8}$ to $\frac{1}{4}$ in. in average diameter. They move downward

Table 18.3. CPC Annular Bed Filter Test Data[a] (Power Boiler Effluent–Waste Wood and Fuel Oil Fired).

MEDIA SIZE, INCHES	MEDIA FACE VELOCITY, ft./min.	MEDIA PRESSURE DROP, H_2O	CYCLONE PRESSURE DROP, H_2O	LOADING, Gr/dry SCF			COLLECTION EFFICIENCY, %		
				CYCLONE INLET	MEDIA INLET	MEDIA OUTLET	CYCL.	MEDIA	TOTAL
0.125–0.250	125	6	1.2	2.768	0.875	0.075	68.4	91.4	97.3
0.125–0.250	150	9.3	2.0	1.486	0.609	0.080	59	86.9	94.6
0.065–0.130	150	11.8	1.4	2.542	0.80	0.07	68.5	91.3	97.3
0.065–0.130	125	9.7	1.0	4.719	0.618	0.026	86.9	95.7	99.4

[a]Weyerhaeuser New Bern N.C. Pulp Mill, January 1975

through the annulus in gravity flow at a bulk velocity of 2 to 4 feet/hour. The pebble inventory is of the order of 40 tons though only about 20 tons are in the region exposed to gas flow. At 35000 ACFM gas flow (~100 ft./min) with inlet loadings of 0.17 to 0.3 grains/dry cu. ft. and outlet loadings of 0.053 grains/dry cu. ft. the pressure drop across the downflowing pebble bed was of the order of 12 in. of water. The unit was operated at 300° to 350°F. Collection efficiencies in all tests ranged from 75% to 97%. The feed particulates showed at best about 27% smaller than 0.7 microns and 17% larger than 20 microns. The loss analysis showed 57% smaller than 0.7 microns and 11% larger than 20 microns.

Relative to public utility flue gas cleanup requirements the above performance figures fall far short of the necessary goals. Nevertheless CPC appears interested in promoting their filter for such application based on the premise that the Weyerhaeuser unit was not intended for high efficiency and that a finer size of pebble combined with removal of the coarser feed solids by tangential feed gas entry into the containing vessel shell might yield the necessary overall efficiency for utility purposes.

CPC's conceptual design for a 1000 MW electric utility burning 12,000 tons of coal/day with 3,300,000 ACFM of 400°F flue gas, would consist of 3 banks of scrubbers each having 8 cylindrical filter vessels and each vessel containing 4 annular elements. They estimated a pressure drop of 8 to 10 in. of water across these filters and in 1974 a turnkey installed cost including fans and motors of \$2.00 to \$2.50 per ACFM. This would involve the sieving and circulation of 225 to 450 tons of pebbles per hour or as much as 10 times the ash rate assuming the flyash loading to the filters is of the order of 0.2 grains/cu. ft. Higher inlet loadings would require proportionately higher pebble circulation rates. The conceptual design is based on recirculating the flyash-freed pebbles pneumatically to the top of the filters where the pebbles are collected via cyclones and the conveying gas vented through bag filters. The pebbles are again freed of the collected flyash via screens and elutriation; the latter would involve more bag house area but less screen wear. It was not clear whether the \$2 to \$2.50/ACFM cost estimate included the solids handling and/or separation equipment.

In terms of a test unit for installation in a public utility's flue gas system operating on a slip stream, CPC felt that a 100,000 ACFM unit would be the minimum size for obtaining reliable and scalable data.

The Canmet-Prasco Hitec System[25,26] represents a development very similar to the CPC design as illustrated in Figure 18.9. Initially, the packed-bed region is filled with appropriately sized granular material, as shown by the shaded portion of the drawing. The hot "dirty" gases enter at the location marked '1' and are directed by pipes to pass horizontally through the packed bed, which is held in place by louvres. As the dust is removed, the pores

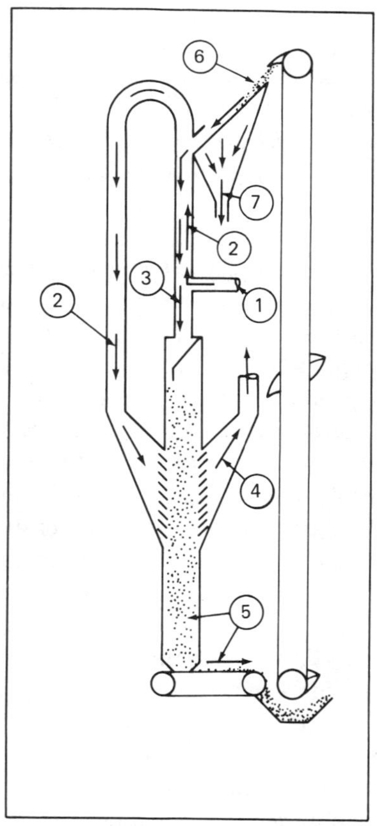

Figure 18.9. The CANMET-Prasco Ltd. Hitec System.

of the filter material begin to plug up, causing an increase of pressure. When this happens, some filter material drops from a chute onto a conveyor belt at '5,' and fresh, properly sized material moves down the column '3' from above the louvred section to replace it. The dirty filter material goes to the top of the unit in a bucket elevator and is dumped onto an inclined screen '6' where most of the dust '7' is sifted out for disposal. The remainder of the filter material is then fed back into the unit near the top of the vertical column into what is called an elutriation column. The gases to be cleaned enter at the bottom of this elutriation column and come in contact with the falling filter material. By controlling the speed of these gases, the finer particles of dust still in the filter material that were not removed by the screen are carried up with the gas and follow the path shown by the arrows marked '2,' leading into the packed bed. After cleaning, the gases are vented to the atmosphere at '4.' The granular filter material is too heavy to be carried up by the gas stream and thus falls down into the storage area above the louvers for reuse. In this manner the problem of the plugging of the filter material has been overcome.

Tests carried out in CANMET laboratories have shown the system to be capable of capturing dust particles down to 1 micron (0.00004 inch) in diameter at efficiencies of over 99.9%. Under appropriate operating conditions, the results obtained on the experimental unit showed the exhaust gases to contain less than 0.05 grain per cubic foot. A commercial-size unit capable of treating 7,500 cubic feet of "dirty" gas per minute has been constructed and installed in Winnipeg on a furnace used to melt cast iron. This unit is undergoing testing to determine its operational capabilities and whether it will meet the Manitoba Government regulations for pollution abatement devices.

The concurrent sizing of the filtration bed material by the gas being treated plus the baffle above the bed which results in a desired size gradation across the packed bed, are the novelties claimed to be essential to the achievement of the performance goals. Typical laboratory scale tests with ambient air are summarized in Table 18.4. Subsequent tests on an Ancast Industries Ltd. cupola in Winnipeg, Manitoba showed average collection efficiencies of 99.4% for particles larger than 25 microns, 97.3% for particles larger than 1 micron, and 50.3% for particles smaller than 1 micron. The resulting overall efficiency of 87.9% was insufficient to bring the effluent loading to below the Manitoba Department of the Environments' limit of 0.25 grains per standard cubic foot. Failure to reach the efficiencies anticipated from laboratory tests was attributed to poor cupola management which led to an unusual amount of submicron material in its effluent. License for fabrication and sale of the Hitec system has been granted to Prasco Ltd. of Winnipeg, Manitoba.

The Kawasaki louvred bed development is again patterned on the Dorfan moving bed principle. Little is known of this work[27] other than tests carried out at 85 to 295 CFM passing through a 10 in. wide × 40 in. high louvred panel holding 8 in. thick beds of downflowing silica sand either 1250 to 2500 microns or 2500 to 5000 microns in diameter. Figure 18.10 summarizes results obtained with an oil-fired utility flyash as the test dust.

The Ducon Fluidizing Reverse Flow filter described in principle in Figure 18.8 has undergone several stages of detailed development as earlier models displayed performance failures in piloted industrial environments for a variety of reasons. The modular design presently offered[28] is illustrated in Figure 18.11; its operating principle is still the same as portrayed in Figure 18.8, but its design details are now regarded with extreme care. For example:

1. bed depth has been optimized at 1.5 in. since shallower beds impose unrealistic fabrication and erection tolerances and deeper beds add unnecessary and undesirable pressure drop;
2. filter media sand particle size has been optimized at 250 to 600 microns since coarser materials require excessive fluidizing velocity (blowback gas consumption) and entail

Table 18.4. Laboratory Scale Filtration Data for the CANMET-HITEC System.

PACKED BED FILTER							
MATERIAL			DUST CHARACTERISTICS				
MEAN DIAM., MICRONS	RATE lb/min.	PRESSURE DROP, IN. OF WATER	TYPE[b]	INLET CONC., gr/SCF	OUTLET CONC., gr/SCF	FILTER COLLECTION EFFICIENCY, %	DUSTY AIR FLOW RATE, SCFM[c]
1100	0.97	2.5–7.5	MCD	135	0.050	99.96	10
1100	0.62	5.0	MCD	134	0.036	99.97	10
1100	0.44	7.0	MCD	134	0.054	99.96	10
1100	0.96	5.0	MCD	45.5	0.175	99.62	15
1100	0.98	5.0	MCD	87.8	0.196	99.78	15
1100	0.98	3.5–6.5	MCD	81.6	0.156	99.81	15
600	0.62	5.0	MCD	135	0.011	>99.99	10
600	0.36	5.0	MCD	132	0.023	99.98	10
600	0.96	5.0	MCD	82.4	0.056	99.93	15
600	1.21	5.0	MCD	85.1	0.196	99.77	15
600	0	1.4	GB	56.6	0	100	10
600	0	1.6	GB	65.8	0.14	99.79	10
600	0	1.9	GB	137	1.28	99.07	10
600	0	2.0	GB	150	3.02	97.98	10
600	0.50	2.2	GB	148	4.22	97.15	10
600	1.22	2.2	GB	118	1.18	99.00	10
600	1.74	2.2	GB	153	0.880	99.42	10
600	2.20	2.2	GB	130	0.420	99.67	10
250	1.44	5.0[a]	MCD	30.5	1.91	93.74	10
250	1.44	5.0[a]	MCD	31.4	1.59	94.94	10
250	1.68	5.0[a]	MCD	35.7	0.637	98.22	10
800	0.19	3.5	MCD	141	0.061	99.96	10
800	0.38	3.5	MCD.	144	0.082	99.94	10
800	0.76	3.5	MCD	153	0.109	99.93	10
900	0.28	3.5	MCD	144	0.010	>99.99	10
900	0.53	3.5	MCD	142	0.016	99.99	10
900	0.89	5.0	MCD	82.6	0.091	99.89	15
1500	0.04	3.5	MCD	38.0	0.023	99.94	10
1500	0.23	3.0	MCD	81.2	0.051	99.94	10
1500	0.44	3.0	MCD	118	0.068	99.94	10

[a]Pressure varied widely; figure reported is average pressure drop.
[b]MCD–mixed cupola dust (8% < 10μM, 10% < 30μM, 50% < 150μM, 100% < 1000μM) GB–glass beads (1 to 25μM)
[c]Filter face area: 6 × 6 in.
Filter bed depth: 3 in.

reduced collection efficiency whereas finer particles would entrain too readily upon fluidization and would require too fragile a bed support;

3. reverse flow fluidization at 120 CFM/sq. ft. for at most 6 to 8 seconds appears an optimum for cleaning to maintain an equilibrium low concentration fines inventory without excessive pressure drop; the degree of cleaning is calculable from fluidized bed entrainment correlations;
4. the bed supports are designed as screened perforated plates with a multiplicity of tiny holes to distribute the fluidizing gas bubbles over the entire bed in close proximity and with sufficient pressure drop to distribute equally to each bed in the stacked module;
5. the disengaging height above each bed is set

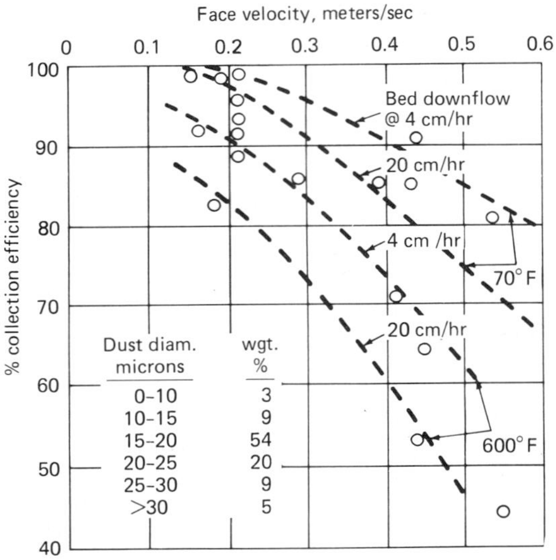

Figure 18.10. Reported filtration efficiencies of 8 in. deep beds of 1,250–5,000 micrometer diameter sands.

at no less than $10\frac{1}{2}$ in. in order to avoid filter media loss (without an inlet screen) compatible with the sand size, density, and reverse flow velocity.

No tests have as yet been carried out in an industrial environment with a unit bearing all of these design considerations and limited to face velocities of about 40 CFM/sq. ft. with a maximum 1 psi buildup of pressure drop before reverse flow cleaning. As in the case of all granular bed filters, laboratory scale tests show excellent filtration efficiencies as illustrated for example in Figure 18.12 from the results published by Westinghouse researchers.[29] The ultimate technical feasibility of this concept lies in the ability to achieve cleaning of the beds in a practical manner.

For purposes of illustration Figure 18.13 represents a filter composed of four modules each containing two filter elements. In the normal continuous mode of operation valves D-1, D-2, etc. would always be open so that the dusty gas would be distributed from the feed manifold equally into all elements of all modules continuously. To maintain any desired pressure difference between the feed and product manifolds, with minimal fluctuations with time as the filter elements accumulated dust, the elements would be blown back individually on a timed cycle by sequentially opening and closing valve B-1, then B-2, then B-3, etc. Such sequential blowback cleaning of individual elements follows conventional bag house practice. Consider for the moment the effect of such flow reversal on the other elements within the filter. During the time that valve B-1 is open no dusty gas can enter element E-1; therefore, elements E-2 through E-8 are subject to a dusty gas feed rate increased by a factor of $\frac{8}{7}$ or 14%. However, in addition to this rela-

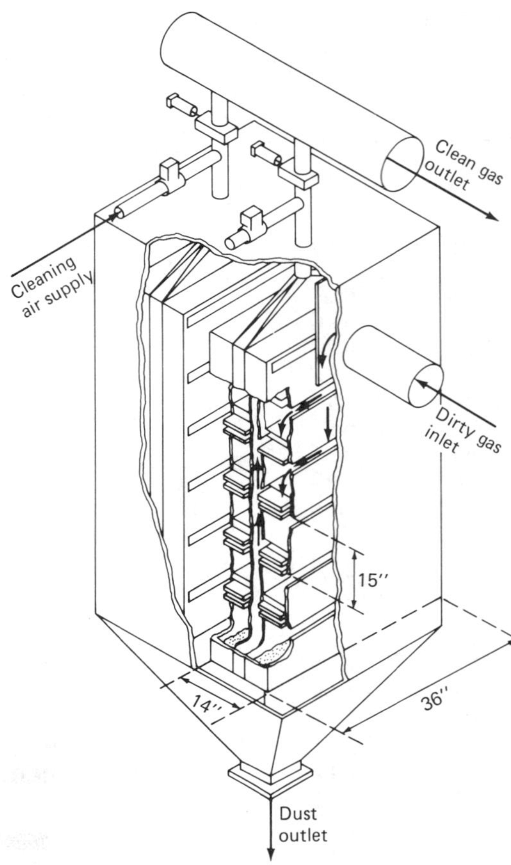

Figure 18.11. The Ducon fluidizing reverse flow granular bed filter. 900 ACFM nominal capacity of each 14 × 36 × 90 in. element consisting of 12 beds in 2 stacks of 6 each, at 50 ACFM/ft^2; 5 in. of water "clean" pressure drop with ambient air through $1\frac{1}{2}$ in. deep beds of 30 to 60 mesh sand at rated capacity.

Figure 18.12. Typical GBF Filtration Efficiencies: Gas Velocity 50 fpm ± 10%. Bed Material 16-20 mesh Ottawa sand. Bed Depth 4 in. (Ref. 29)

PARTICLE SIZE microns	INLET DUST, mg/m^3	OUTLET DUST, mg/m^3	COLLECTION EFFICIENCY, %
0-0.3	32	0.39	98.8
0.3-0.45	20	0.08	99.6
0.45-0.75	33	0.12	99.6
0.75-1.5	53	0.08	99.8
1.5-2.3	47	0.04	99.9
2.3-3.3	32	0.04	99.9
3.3-5.0	53	0.04	99.9
5.0-8.0	46	–	100
8.0	108	–	100
	513	–	100

RUN NO.	INLET LOADING, g/m^3	OVERALL EFFICIENCY, %
6-24	5.8	99.96 (99.92)
6-25	0.94	99.92 (99.8)
6-26	1.17	99.90 (99.7)

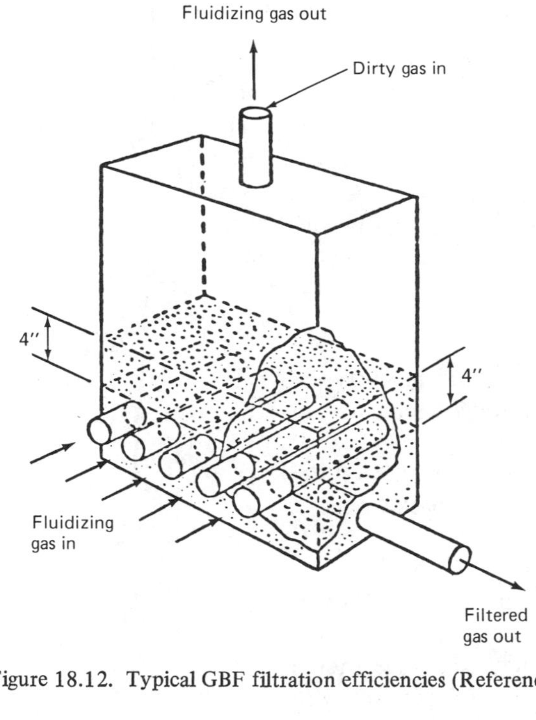

Figure 18.12. Typical GBF filtration efficiencies (Reference 29).

tively minor increase, these seven elements must also absorb the blowback gas from element E-1. Though dependent upon such factors as bed particle size, frequency of blowback, size and density of collected fines, etc., it is current practice to specify a blowback rate which creates a superficial fluidizing velocity of 2 f./s or 120 ACFM/sq. ft. of filter bed surface. If the elements were operating at a design face velocity of 50 ACFM/sq. ft. then obviously blowing back E-1 would increase the flow rate to the remaining seven elements from 50

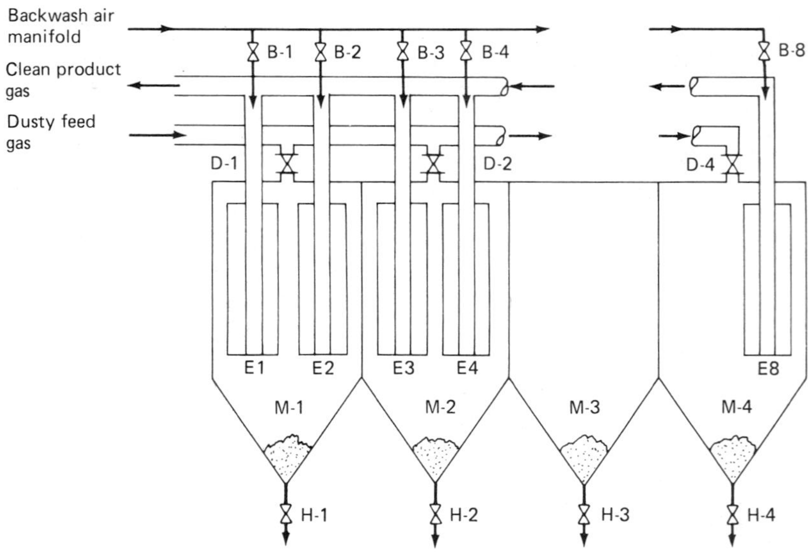

Figure 18.13. Illustrative arrangement of GBF modules.

ACFM/sq. ft. to:

$$50\left(\tfrac{8}{7}\right) + (120/7) = 74 \text{ ACFM/sq. ft.}$$

during the blowback period. This can be expressed analytically in the form:

$$\% OL = 100 \left[\frac{T_E + (12/5)}{T_E - 1} \right] - 1$$

where

$\% OL$ = % increase or overload in dusty gas face velocity through elements not being cleaned

T_E = total number of elements in filter and only one element backwashed at any moment.

Represented graphically in Figure 18.14 it becomes obvious that any such installation whether on a pilot or industrial scale should preferably be provided with no less than 10 to 20 elements no matter what their size. Lower gas capacity for pilot tests should be provided with elements containing a smaller number of stacked beds rather than a smaller total number of elements.

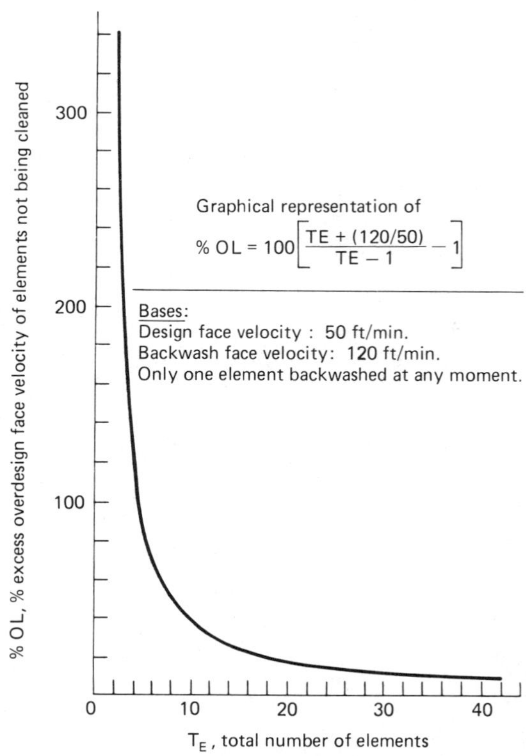

Figure 18.14. Graphical representation of % OL = 100 [TE + (120/50)/TE − 1] −1. Bases: design face velocity; 50 ft/min; backwash face velocity: 120 ft/min. Only one element backwashed at any one moment.

Note in relation to Figure 18.13 it is assumed the reverse blowback from E-1 would be equally distributed to all seven remaining elements; however, this is probably unrealistic since it requires an absolutely zero frictional resistance in valves B-1, B-2, etc. as well as in the entire dusty gas manifold, and in addition, ignores the inertia transients. In practice, element E-2 would be considerably more overloaded. Some reasonable limits must be placed on this overload since it can result in dust breakthrough by the force of the high velocity interstitial gas circumventing particle impact by heightened particle pickup. For example, 100 ACFM/sq. ft. through a bed of 40% voids amounts to an interstitial velocity a little over 4 ft./s which will exceed the saltation velocity of a variety of particulates depending on their properties. If the bed interstices were filled to some equilibrium content with collected fines, this effective interstitial velocity could easily reach 8 ft./s and cause reentrainment. The permissible overload cannot be specified without detailed knowledge of the system characteristics and properties.

The PEMMCO Restricted Circulation Filter illustrated in Figure 18.15 is one[30] of two concepts under development designed to minimize the filter media losses experienced in the Fuller-Squires type of unit or to permit cleaning[31] without any reverse flowing gas. Only the shallow layers facing the dusty gas in Figure 18.15 are blown off by the reverse flow cleaning gas; the gaps at the bases of the louvers prevent immediate media replacement and therefore excessive media blow off. A basic advantage of such granular bed filters, in which the medium is removed and replenished, lies in their adaptability to simultaneously act as gaseous adsorbers or chemical reactors as in the RESOX process for flue gas desulfurization and conversion to elemental sulfur.[32,33]

The Melcher Electrofluidized Bed Filter consisting typically of millimeter sized particles is stressed by an imposed electric field that effectively polarizes the particles. They then act as collection sites for previously charged fine particulates entrained in the dusty fluidizing feed

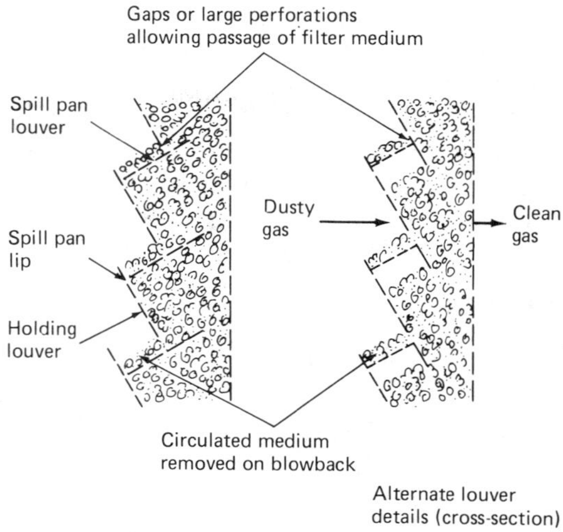

Figure 18.15. Design principle of restricted circulation granular bed filter-adsorber.

gases. The advantages claimed for the electrofluidized bed derive from the greatly reduced residence time for effective cleaning of the gas, realized by virtue of the bed's large collection surface area per unit volume, and the ease of handling the bed particles for removal of the collected fines.[35,36]

Collection efficiencies in excess of 98% have been reported[37] for submicron asphaltic particles in 10 cm deep beds of 2 mm sand. Superficial velocities ranged from 1.5 to 2 m/s with bed pressure drops typically about 12 cm of water. Electrical energization required less than 80 W/1000 CFM. Though these conditions are a vast improvement over even multistaged[38] nonelectrified fluidized beds they still fall short of such typical fixed bed results as illustrated in Figure 18.12 despite their substantially greater face velocity.

The application of an electrical potential gradient to a bed of ceramic beads for filtration of petroleum fluid catalytic cracking fractionator liquid bottoms has been reported[39] as achieving exit solids contents as low as 0.01% by weight. Offered by Gulf Science & Technology Co. it again suggests the possibility of application to fixed bed gas filters though this concept had been considered beyond the limits of economic feasibility.

It appears to be the general concensus that some form of granular bed filter will eventually emerge as a significant industrial tool very likely some day supplanting the electrostatic precipitator. The time will be determined by economic pressures, developmental details and further air quality regulations.

REFERENCES

1. A. P. Krueding, *Chem. Eng. Prog. 71*(10):56–61 (1975).
2. D. B. Pall, *Ind. Eng. Chem. 45*:1197 (1953).
3. G. B. Cramp, *Chem. and Met. Eng. 30*:400–401 (March 10, 1924).
4. G. B. Cramp, *Blast Furnace Steel Plant, 12*:101–103 (February, 1924).
5. *The Dorfan Impingo Filter*, bull. no. 4, Mechanical Industries, Inc., 541 Wood Street, Pittsburgh 22, Pa. (1954).
6. M. I. Dorfan, *Electric Furnace Steel Proc. 10*:41–60 (1952).
7. L. K. Zhitkevich, *Trudy Inst. Energ. Akad. Nauk. Belorus. SSR 1*:150–160 (1954).
8. G. C. Egleson, H. P. Simons, L. J. Kane, and A. E. Sands, *Ind. Eng. Chem. 46*:1157–1162 (1954).
9. A. M. Squires, U.S. Pat. No. 3,296,775; filed 10/16/62, issued January 10, 1967.
10. L. Paretsky, et al, *Jnl. A. P. C. A. 21*:204–209 (April, 1971).
11. A. M. Squires and R. Pfeffer, *Jnl. A. P. C. A. 20*:534–538 (August, 1970).
12. K. C. Lee, A. M. Squires, and R. Pfeffer, "Filtration of Fly Ash and Puffback in a Panel Bed Filter," paper No. 25a; 67th annual A. I. Ch. E. meeting, Washington, D.C. (December 3, 1974).
13. Bureau of Mines, Report of Investigations No. 7276 (July, 1969).
14. E. P. R. I. Report 243-1 (November, 1974).
15. H. P. Meisner and H. S. Mickley, *Ind. Eng. Chem. 41*:1238–1242 (1949).
16. J. P. Pilney and E. E. Erickson, *Jnl. A. P. C. A. 18*(10):64–685 (October, 1968).
17. C. H. Black and R. W. Boubel, *I. and E. C. Proc. Des. and Dev. 8*(4):573–578 (October, 1969).
18. L. L. Knapp and C. C. Cook, U.S. Pat. No. 3,503,184; filed 3/7/68, issued March 31, 1970.
19. A. F. Johnson, U.S. Patent No. 3,470,075; filed 2/6/67, reissued as Re 27,383, May 30, 1972.
20. M. Berz and W. Berz, U.S. Pat. No. 3,090,180; filed May 19, 1960; see *Staub, 24*(11):449–452 (1964).
21. Rexnord Corp., *Environmental Sci. and Tech. 8*(7):600–601 (July, 1974).
22. Bulletin No. F-9671, The Ducon Co., 147 E. Second Street, Mineola, L.I., N.Y. (1971).
23. U.S. Patent No. 3,410,055; filed 10/26/66; issued November 12, 1968.
24. B. Kalen, U.S. Patent No. 3,798,882; filed 9/28/71; issued March 26, 1974.
25. R. K. Buhr and E. Darke, CANMET Phys. Met. Res. Labs. Reports MRP/PMRL (CF) 76-4 (R) and 75-10 (FT).
26. R. K. Buhr and R. D. Warda, CANMET Report MRP/PMRL-75-2 (R) (January 14, 1975).
27. Private communication (1977) Kawasaki Corp.
28. U.S. Patent No. 4,067,704; filed 10/18/76; issued January 10, 1978.
29. D. F. Ciliberti, D. L. Keairns, and D. H. Archer, "Particulate Control for Pressurized Fluidized-Bed Combustion Processes." presented at the 5th International Conference on Fluidized Bed Combustion (December 13, 1977).
30. U.S. Patent No. 3,770,388; filed 5/24/71; issued November 6, 1973.
31. U.S. Patent No. 3,800,508; filed 10/26/70; issued April 2, 1974.
32. W. F. Bischoff, Jr., and P. Steiner, *Chem. Engrg.* pp. 74, 75 (January 6, 1975).
33. G. O. Layman, *Environ. Sci. and Tech. 9*(8):712–713 (August, 1975).
34. K. Zahedi and J. R. Melcher, *Jnl. A. P. C. A. 26*:345 (1976).
35. J. C. Alexander and J. R. Melcher, *I. and E. C. Fundamentals 16*(3):311–317 (1977).
36. *Popular Science*, p. 10 (August, 1975).
37. P. B. Zieve, K. Zahedi, J. R. Melcher, and J. F. Denton,; *Envir. Sci. and Tech. 12*(1):96–99 (January, 1978).
38. R. G. Patterson and M. L. Jackson, *A. I. Ch. E. Symposium Series*, no. 161, vol. 73, pp. 64–73 (1977).
39. Gulftronic Separator Systems, *Chem. Proc.*, p. 20, (mid-April, 1978).

BIBLIOGRAPHY

Including some references to liquid filtration, where theory or equipment arrangements are pertinent to granular bed filtration development.

"Backing Germany's Magnesium Bid" (Knapsack-Griesheim Flowing Coke Bed Filter), *Chemical Week*, p. 71–72 (June 17, 1961).

"Big Dryer," *Chem. & Eng. News*, p. 131 (October 23, 1961).

J. S. M. Botterill and E. Aynsley, "The Collection of Airborne Dusts Parts 1 and 2," *British Chem. Engrg.*,

12(10):1593-1598 (October, 1967); ibid., *12*(12): 1899-1903 (December, 1967).

"Braided-Wire Tubes Increase Filtering Efficiency," *Chemical Engrg.*, p. 116 (June 20, 1966).

British Chem. Engrg. 15(4):549 (April, 1970).

Chemical Engrg., pp. 90-91 (September 27, 1965).

Chemical Week, pp. 71-73 (June 17, 1961).

"Coal Filters for Waste Treatment," *Chem. Engrg.*, p. 122 (March 14, 1966).

C. N. Cochran, W. C. Sleppy, and W. B. Frank, "Chemistry of Evolution and Recovery of Fumes in Aluminum Smelting," Paper no. A70-22, Metallurgical Society of the AIME meeting held February 16-19, 1970.

J. T. Cookson, "Removal of Submicron Particles in Packed Beds," *Environmental Sci., & Tech. 4*(2): 128-134 (February, 1970).

E. D. Ermenc, *Chemical Engrg.*, pp. 87-94 (May 29, 1961).

N. Fuchs and A. Kirsch, "The Effect of Condensation of a Vapor on the Grains and of Evaporation from their Surfaces on the Deposition of Aerosols in Granular Beds," *Chem. Eng. Sci. 20*:181-185 (1965).

G. Funke, "Pollution and Nuisance Control Activities," Zement-Kalks-Gips, no. 5, pp. 209-219 (1968).

L. Goldman, "Is the Gravel Layer Suitable as an Air Filter," *Wasser, Luft und Betrieb, 6*(7):233-236 (May, 1962).

J. P. Herzig, D. M. Le Clerc, and P. Le Goff, "Flow of Suspensions through Porous Media-Application to Deep Filtration," *I. & E. C. 62*(5):8-35 (May, 1970).

S. Jackson and S. Calvert, "Entrained Particle Collection in Packed Beds," *A. I. Ch. E. Jnl. 12*(6):1075-1078 (November, 1966).

B. Kalen and F. A. Zenz, "Filtering Effluent from a Cat Cracker," *Chem. Engrg. Progress 49*:67-71 (June, 1973).

Y. V. Krasovitskii and V. A. Zhuzhikov, "Separation of Dust from a Gas Stream by Filtration at Constant Velocity," *Khim. Prom., 49*(2):49-52 (1963); Translated by E. K. Wilip as ANL-Trans-572 (February, 1968).

W. D. Lovett and F. T. Cuniff, "Air Pollution Control by Activated Carbon" (Moving Bed Adsorber Panel), *Chem. Engrg. Progress, 70*(5):43-47 (May, 1974).

A. Maroudas and P. Eisenklam, "Clarification of Suspensions, A Study of Particle Deposition in Granular Media, Part I–Some Observations on Particle Deposition," *Chem. Engrg. Sci. 20*:867-873 (1964).

A. Maroudas and P. Eisenklam, "Part II–A Theory of Clarification," *Chem. Engrg., Sci. 20*:875-888 (1965).

R. E. Pasceri and S. K. Freidlander, "The Efficiency of Fibrous Aerosol Filters: Deposition by Diffusion of Particles of Finite Diameter," *Can. Jnl. of Chem. Engrg. 38*(6):212-213 (December, 1960).

R. D. Rea, "Plume-Free Stacks Achieved in Sulfuric Acid Production," *Chemical Processing*, p. 13-14 (January, 1971).

"Sand Filter Saves Space," *Chemical Engrg.*, p. 112 (September 21, 1970).

E. W. Schmidt, J. A. Giesche, P. Gelfand, T. W. Lugar, and D. A. Furlong, "Filtration Theory for Granular Beds," *Jnl. A. P. C. A. 28*(2):143-146 (1978).

"Simultaneous Sulfur Dioxide and Fly Ash Removal," *Environmental Sci. & Tech. 5*(1):18-19 (January, 1971).

L. Spielman and S. L. Goren, "Model for Predicting Pressure Drop and Filtration Efficiency in Fibrous Media," *Environmental Sci. & Tech. 2*(4):270-287 (April, 1968).

L. A. Spielman and S. L. Goren, "Capture of Small Particles by London Forces from Low-Speed Liquid Flows," *Environmental Sci. & Tech. 4*(2):135-140 (February, 1970).

D. I. Tardos, N. Abuaf, and C. Gutfinger, "Dust Deposition in Granular Bed Filters: Theories and Experiments," *Jnl. A. P. C. A. 28*(4):354-363 (1978).

R. M. Werner and L. A. Clarenburg, "Aerosol Filters," *I. & E. C. Proc. Des. & Dev. 4*(3):288-299 (1965).

R. L. Zahradnik, J. Anyigbo, R. A. Steinberg, and H. L. Toor, "Simultaneous Removal of Fly Ash and Sulfur Dioxide from Gas Streams by a Shaft-Filter-Sorber," *Environmental Sci. & Tech. 4*:663 (1970).

F. A. Zenz and H. Krockta, "The Shallow Expandable Bed–A Versatile Processing Tool," *A. I. Ch. E. Sym. Ser. 67*(116):245-250 (1971).

F. A. Zenz and H. Krockta, "The Evolution of Granular Beds for Gas Filtration and Adsorption," *Brit. Chem. Eng. & Proc. Tech. 17*(3):224-227 (March, 1972).

LIST OF PATENTS

The following list of patents is a composite of the most pertinent references obtained by independent searchers as well as those cited by various patent office examiners. As such, it contains a relatively broad spectrum including some referring solely to liquid systems, but which nevertheless bear on related techniques or structures; it is not intended to be exhaustive.

U.S. Patents

557,177	668,441
631,143	746,255

746,261
758,345
890,625
989,665
1,212,199
1,608,678
1,702,311
1,731,223
1,782,823
1,895,601
1,988,798
1,995,293
2,046,816
2,145,049
2,287,983
2,303,717
2,384,932
2,423,013
2,430,420
2,441,724
2,448,493
2,493,218
2,493,356
2,559,784
2,564,837
2,604,187
2,608,474
2,639,000
2,641,364
2,780,310
2,780,363
2,804,168
2,867,497
2,951,551
3,016,147
3,090,180
3,090,744
3,166,382
3,296,775
3,311,466
3,410,055
3,410,056
3,470,075
3,594,991
3,708,981
3,770,388
3,798,882
3,800,508
3,854,902
3,892,551
3,912,466
4,000,995
4,004,350
4,004,897
4,006,533
4,012,210
4,017,278

French

26,145
57,131
354,140
373,181
412,042
474,615
992,225
1,062,580
1,140,347
1,146,213
1,368,615
1,411,474

Swiss

508,409

British

110,235
194,730
243,117
257,879
450,048
984,844
1,026,826

Swedish

91,075

Japanese

116,665

German

44,492
74,734
176,943
207,579
381,500
409,942
457,430
522,573
524,982
676,508
1,063,015

Australian

432,317
465,058
429,424
46579/68
51375/69
80089/75

Canadian

592,973
613,352
721,176

Italian

741,236
820,467

Spanish

346,278

So. African

67-5801

19
Wet Scrubber Particulate Collection

Douglas W. Cooper
Harvard School of Public Health
Boston, Massachusetts

19.1. INTRODUCTION

Emission Control Goals

Activities involving powders can result in the generation of airborne particulate material, aerosols, which may need to be controlled because of concern about health, because of laws and regulations, or because of an economic incentive for process material recovery. The principal alternatives for fine particle control are cyclones, filters, scrubbers, and electrostatic precipitators. Alternatives are generally compared with respect to effectiveness and cost.

Scrubbers are air pollution control devices that use liquid to collect particles or gases or both from a gas stream. Usually, the liquid used is water, occassionally with a surface-active agent added. The uses of scrubbers to remove gases is not discussed here.

Control System Options

The considerations for the selection among the types of air pollution control equipment are summarized in Figure 19.1. Scrubbers are used not only for air pollution control, however, but also to recover valuable materials, to cool gas streams and to add liquid or vapor to gas streams.

A succinct comparison of wet collectors (scrubbers) and dry collectors (cyclones, filters, electrostatic precipitators) was presented by Strauss[2] and is shown in Table 19.1. Economic comparisons will be given at the end of this chapter, but as a rough guideline: scrubbers have higher efficiencies and higher costs than cyclones, they can be made to have efficiencies comparable to those for filters and electrostatic precipitators but at higher operating costs and, generally, lower equipment costs. Further information on various air and gas cleaning devices is presented in Figure 19.2.

Figure 19.3 gives approximate collection efficiency as a function of particle size (fractional efficiency) for each of the major collector types. It is useful for qualitative comparisons only, as each of these devices has collection efficiency characteristics more complicated than the relationship shown.

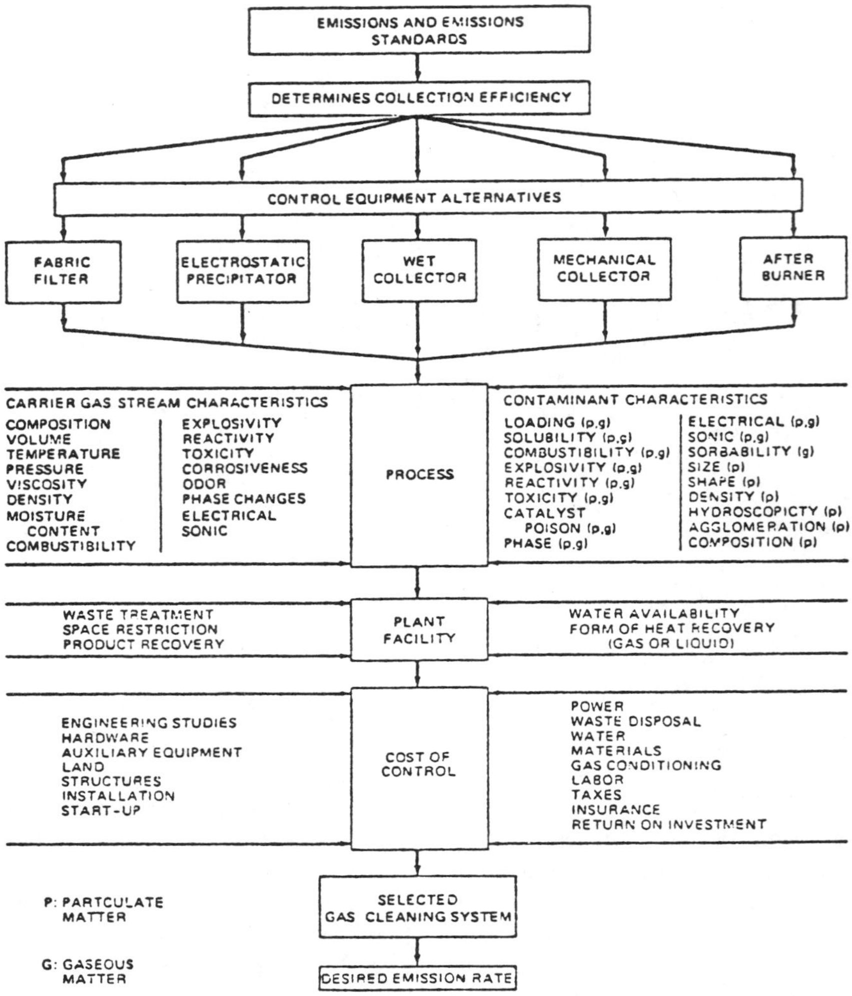

Figure 19.1. Process for selection of gas cleaning equipment.[1]

Types of Scrubbers

Scrubbers capture particles on droplets, liquid surfaces, or liquid-coated surfaces. The droplets may be formed independently of the gas flow, or they may be atomized by the flow. Scrubbers using preformed sprays include: spray towers, cyclone spray scrubbers, water-jet scrubbers, and mechanical scrubbers. Venturi and orifice scrubbers are usually designed to produce a spray by gas atomization of the scrubbing liquid. Impingement scrubbers and sieve plates involve flow into or through a volume of liquid. Some of these scrubber types are shown in Figure 19.4. Particles are captured primarily on liquid-coated surfaces in packed-bed scrubbers, fluidized-bed scrubbers, and fibrous bed scrubbers. Many of these different types use about the same amount of power to achieve the same degree of particle collection. The choice of scrubber type is therefore dictated by space constraints, the availability of certain kinds of power (e.g., waste heat) and equipment (such as pumps, fans, ducting, piping), and the aspects of the dust/scrubber combination which affect plugging, corrosion, and the handling of liquid and solid wastes.

In a *spray tower*, the particle-laden gas stream flows upward through a spray falling down-

Table 19.1. Advantages and Disadvantages of Wet and Dry Collectors.[2]

WET COLLECTORS	DRY COLLECTORS
Advantages	Advantages
1. Can collect gases and particles at the same time.	1. Recovery of dry material may give final product without further treatment.
2. Recovers soluble material, and the material can be pumped to another plant for further treatment.	2. Freedom from corrosion in most cases.
3. High-temperature gases cooled and washed.	3. Less storage capacity required for product.
4. Corrosive gases and mists can be recovered and neutralized.	4. Combustible filters may be used for radioactive wastes.
5. No fire or explosion hazard if suitable scrubbing liquor used (usually water).	5. Particles greater than 0.05 micrometers may be collected with long equipment life and high collection efficiency.
6. Plant generally small in size compared to dry collectors such as baghouses or electrostatic precipitators.	
Disadvantages	Disadvantages
1. Soluble materials must be recrystallized.	1. Hygroscopic materials may form solid cake and be difficult to shake off.
2. Insoluble materials require settling in filtration plant.	2. Maintenance of plant and disposal of dry dust may be dangerous to operatives.
3. Waste liquids require disposal, which may be difficult.	3. High temperatures may limit means of collection.
4. Mists and vapors may be entrained in effluent gas streams.	4. Limitation of use for corrosive mists for some plants (e.g., baghouses).
5. Washed air will be saturated with liquid vapor, have high humidity and low dew point.	5. Creation of secondary dust problem during disposal of dust.
6. Very small particles (submicrometer sizes) are difficult to wet, and so will pass through plant.	
7. Corrosion problems.	
8. Liquid may freeze in cold weather.	

ward. In a *spray chamber*, the gas flow is generally horizontal and the spray is often in a cross-current orientation. In a *cyclone spray scrubber*, a spray is introduced near the cyclone entrance. The relative motion of spray and particles induced by the cyclones aids collection of the particles and also of the drops containing the captured particulate material. *Mechanical scrubbers* use sprays and moving baffling (fans, etc.) to induce particle capture. A *venturi scrubber* accelerates the gas in a converging channel, introduces a spray near the throat section, then decelerates the gas through a very gradually tapered diverging section. *Orifice scrubbers* work in a similar fashion, except that the throat is an orifice, an abrupt change in duct cross-sectional area, of negligible length, followed by an abrupt expansion.

Impingement scrubbers direct the gas at the surface of a liquid, causing intimate liquid/particle mixing, using a variety of geometries. *Tray towers* (*sieve plates*) have a series of multiply perforated plates arranged vertically in such a way that water introduced at the top of the array flows downward from plate to plate, and the particle-laden gas is passed through the plate perforations and through the liquid in essentially cross-current flow at each stage; for the array, the flow is counter-current. *Packed* and *fluidized beds* may have sprays providing scrubbing liquid either cocurrently or in a counter-current fashion; a great variety of packing materials are employed; often these are designed to remove gases as well as particles from the flow, for which their relatively large surface-to-volume ratios and residence times are advantageous.

Pressure drop and collection efficiency equations will be given in this chapter for many of these scrubber types. The *Scrubber Handbook*[6] and manufacturer publications provide fuller descriptions.

Typically, a scrubber system will include not only the scrubber but also a demister for re-

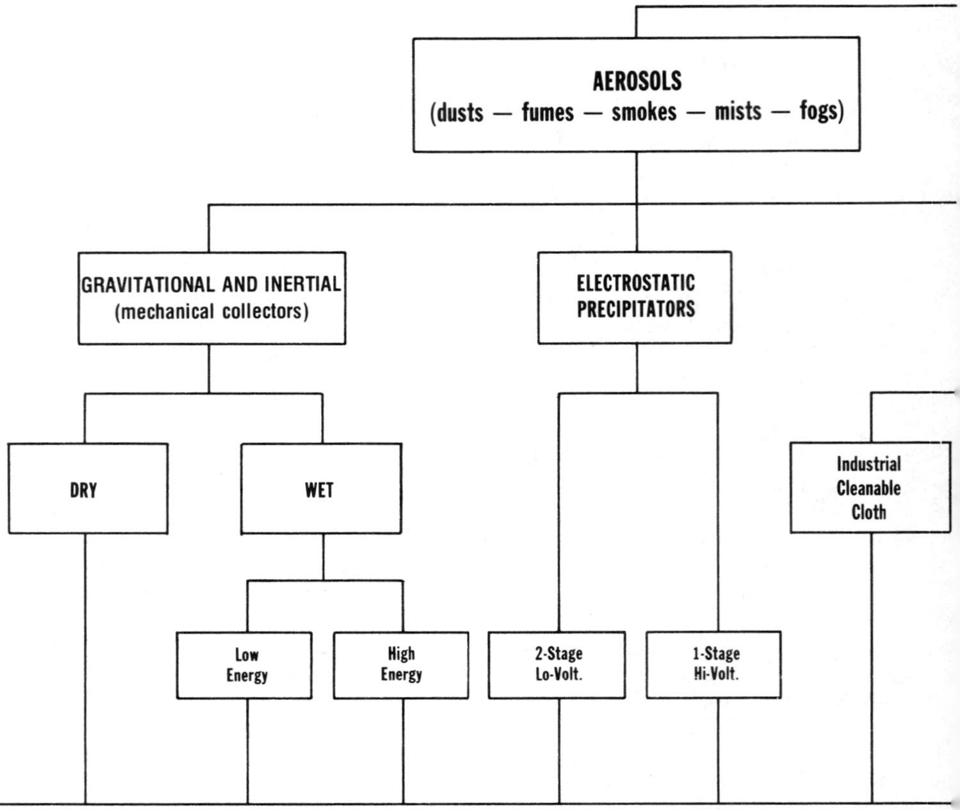

	Settling chambers. Inertial separators. Cyclones Dynamic (Fan & Sep.)	Wetted baffles Wetted cyclone Wetted dynamic Ejector	Venturi scrubber Fog tower	Furnace air cleaner	Plate and wire Tube and wire	
TYPES OF CONTAMINANTS	Crushing Grinding Machining	Same as dry	Chemical & Metallurgical fumes	atm. air oil mist	flyash tar H_2SO_4	all dry powders
LOADINGS	.1-100 g/m³	.1-100 g/m³	.1-100 g/m³	<0.1 g/m³	.1-2 g/m³	.1-20 g/m³
COLLECTION EFFICIENCY	high for >10 μm	high for >2 μm	high for >.25 μm	high for >.25 μm	high for >.1 μm	high for >.1 μm
ENERGY REQUIREMENTS (cm. w.g.)	5-15	5-15	50-200 (Vent. scrub.) 30atm nozzles (fog twr.)	0.25	0.6	7-15
INITIAL COST	Low	moderate	low	moderate	high	moderate
OPERATING COST	moderate	moderate	high	low	low	moderate
SERVICEABILITY DURABILITY	good (erosion)	good (corrosion)	good (corrosion)	poor	fair	

Figure 19.2. Characteristics of air and gas cleaning methods and equipment.[3]

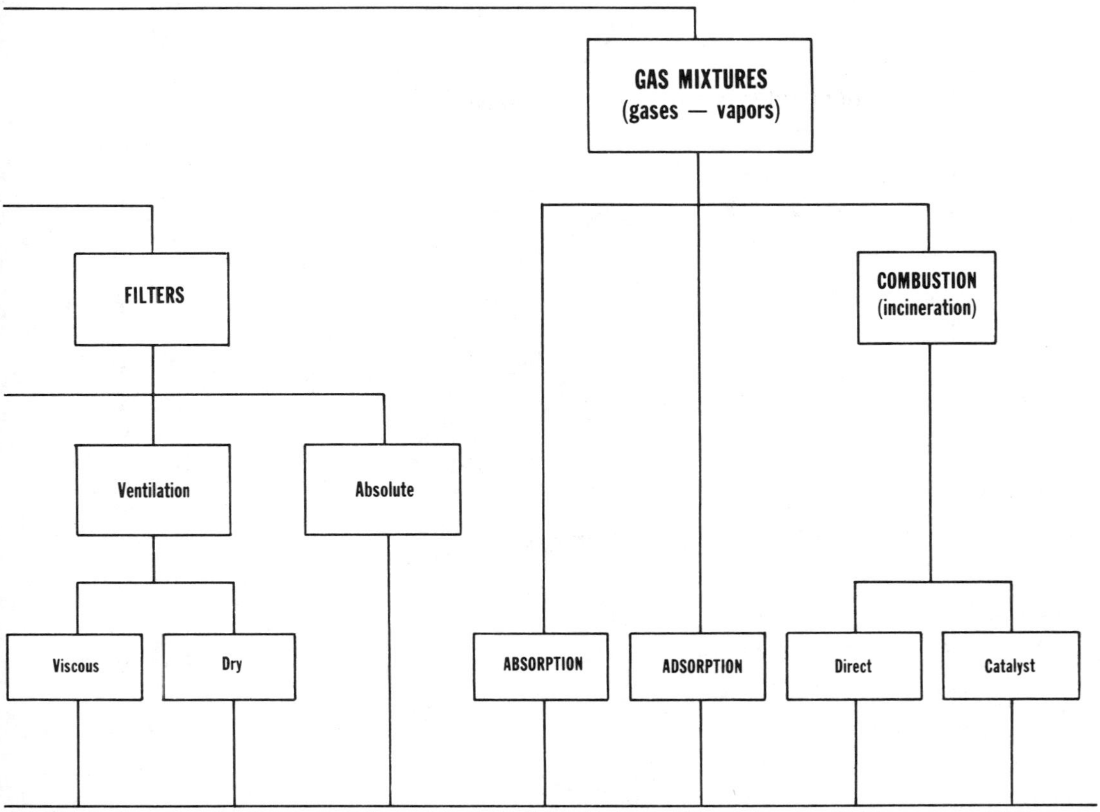

Cleanable Throwaway	Pleated filters Automatic renewable	Paper Deep bed	Packed tower	Carbon bed	Direct flame incinerator	Catalytic converter
atm. air	atm. air	pre-cleaned atm. air	inorganic gases(HC1, HF, SO_2, $C1_2$)	organic gases & vapors (odors)	organic gases & vapors (odors)	organic gases & vapors (odors)
<.01 g/m^3	<.01 g/m^3	<.001 g/m^3	ppm to %	ppb to %	ppb to %	ppb to %
high for >.1 μm	high for >.5 μm	high for all sizes		95-99+	90-99	90-99
<0.3	0.3-2	2-15	10-30	5-15	0.3	2-5
low	low	moderate	moderate	high	low	moderate
low	low	high	moderate	moderate	extremely high	high
good	poor	fair to poor	poor	fair	good	poor

Figure 19.2 (Continued)

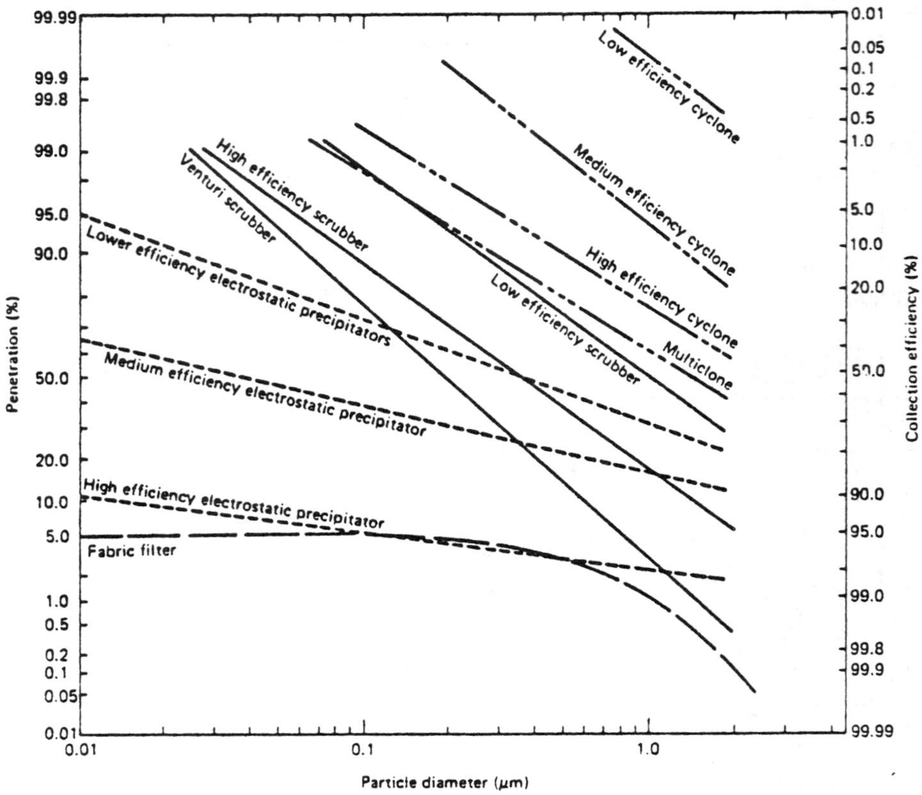

Figure 19.3. Extrapolated fractional efficiency of control equipment.[4]

moving the droplets separated in the scrubbing process, and a clarifier for concentrating the solids and removing them from the liquid effluent. For pollution control, both the demister and the clarifier are important. Demisters will be covered in this chapter. The removal of solids from liquid streams by filtration is covered in Chapter 15.

19.2. POWER CONSUMPTION

Introduction

The difference in gas pressure at the inlet and outlet of the scrubber, due to the resistance to gas flow of the scrubber, is the pressure drop, ΔP ($N/m^2 = Pa$), the energy consumption per unit volume of air. For scrubbers having appreciable collection efficiency for submicron particulates, the energy costs can outweigh all the other costs, so minimizing pressure drop, while maintaining adequate collection efficiency, is important.

Definition

Pressure represents potential energy per unit volume, and the product of the pressure drop and the volume rate of flow of the gas or liquid represents power consumption. The metric units for pressure drop are N/m^2 and for the product of pressure drop and volume flow rate is $(N/m^2)(m^3/s) = N\text{-}m/s$ or watts (W). Frequently, pressure differences are measured by manometers and are given as inches of water (in. WC or in. WG); 6.3 inches of water pressure drop is equivalent to one horsepower per 1000 ft^3/min; 1.0 in. WG is 249 N/m^2. The electrical power consumed by a fan moving the gas through this pressure drop will be the product of the pressure drop and the gas volume flow

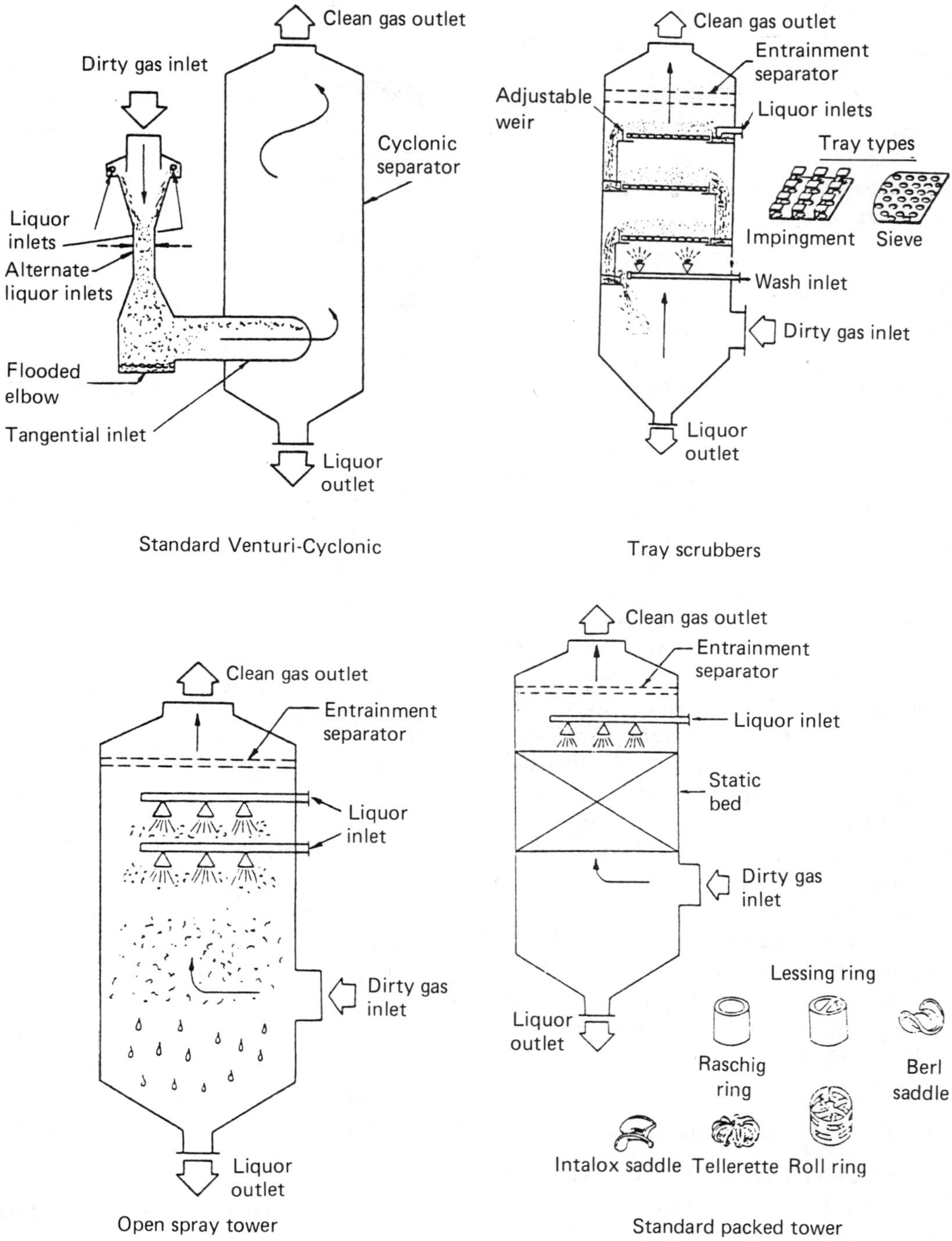

Figure 19.4. Examples of scrubbers.[5]

rate, divided by the fan/motor efficiency (typically about 0.6), $Q\Delta P/E_f$. The pressure drop across spray nozzles, the volume flow rate of spray, and the electrical energy consumed by the pump have the analogous relationship. As energy costs rise, pressure drops across fans and pumps become more important as a design consideration.

Contacting Power

For many years, some scrubber experts thought that the collection efficiency of *any* type of scrubber would be the same for a given aerosol if the power consumption were the same.[7] Although this is a useful rule of thumb, it is not strictly correct. Testing an orifice scrubber, a multiple orifice contactor, and a variety of spray type scrubbers, Semrau et al.[8] found, for test aerosols that were primarily submicrometer, that the various spray scrubbers gave the same collection efficiency for the same aerosol at given levels of power consumption, but the orifice scrubber did somewhat better. Calvert[9] also found that at conditions appropriate for scrubbing submicrometer particles efficiently, power consumption differences among scrubber types became appreciable. The difference becomes evident for high-energy scrubbing, which is precisely where it is most important. As total pressure drop is increased for a venturi or orifice scrubber, it becomes advantageous, in terms of collection efficiency, to divide the pressure drop equally between two or more scrubbers in series rather than concentrate the power consumption in a single-stage scrubber.[10, 11]

Other Types of Power Consumption

Besides the power which is used for maintaining the pressure drop across the scrubber and across spray nozzles (where used), power may be required in the following:

1. monitoring scrubber performance;
2. keeping scrubber elements above freezing temperatures;
3. filtering the scrubbing liquid;
4. drawing air to the scrubber and forcing it through a demister and to a stack;
5. heating scrubber outlet to decrease or prevent condensation in stack or plume;
6. electrostatic augmentation, if charged droplet scrubbing is used;
7. rotating a mechanical element within the scrubber to enhance droplet disintegration and particle capture;
8. generating steam for subsequent condensation to enhance scrubbing;
9. cooling the gas in association with condensation scrubbing; and
10. handling and disposing of the solid and liquid wastes.

Most of these aspects are discussed at greater length below.

19.3. COLLECTION EFFICIENCY

Introduction

Scrubbers are designed to achieve adequate efficiency (or acceptable penetration) at minimum cost, and for high-energy scrubbers ($\Delta P > 3 \times 10^3$ N/m^2 or 12 in. WG), this means at nearly the minimum power consumption.[12]

Collection Efficiency and Penetration

The total mass collection efficiency (often called "total efficiency") is the difference between the inlet mass flux ($\dot{M}_0$) and the outlet mass flux ($\dot{M}$), divided by the inlet mass flux:

$$\overline{E} = (\dot{M}_0 - \dot{M})/\dot{M}_0 \tag{19.1}$$

The total penetration is just $\overline{Pt} = \dot{M}/\dot{M}_0$; it is the fraction of the mass that penetrates the device. The efficiency for a particular particle size (or narrow size range) is called the "fractional efficiency" and is, for particles in size range i:

$$E_i = (\dot{M}_0 - \dot{M})_i/\dot{M}_{0i} \tag{19.2}$$

To compare gas–particle separation devices, one generally needs their fractional efficiencies over the particle size range of interest. To get total mass efficiency from fractional efficiencies, one multiplies the fractional efficiencies, size interval by size interval, with the fractions of aerosol mass in each size interval and sums the products, closely related to the numerical integration of Equation (19.4). For the scrubbers discussed below, equations will be given for determining collection efficiency as a function of particle aerodynamic diameter, because impaction is almost always the predominant collection mechanism for scrubbers, especially for particles larger than about one micrometer, and impaction is a function of the particle aerodynamic diameter. The particle aerodynamic diameter, d_{pa}, is the diameter of a solid particle having the density of water that would have the same terminal settling velocity due to gravity as does the particle in question. From Stokes's law (with the Cunningham correction), this means that $(1 + 2.5\lambda/d_p)\rho_p d_p^2 = (1 + 2.5\lambda/d_{pa})\rho_w d_{pa}^2$. λ is the mean free path of the gas molecules, 0.065 μm at standard temperature and pressure; ρ_p is the particle density, ρ_w is the density of water.

Figure 19.5 gives the aerodynamic diameters of spherical particles of the densities

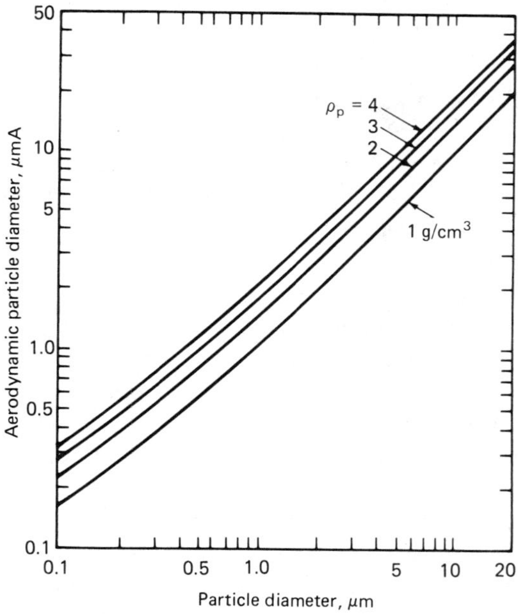

Figure 19.5. Relation between physical and aerodynamic diameter.[13]

indicated, as functions of particle physical (geometric) diameter.[13] For particles for which the Cunningham correction is negligible ($d_p >> 10\lambda$), the aerodynamic diameter is $d_{pa} = (\rho_p/\rho_w)^{1/2} d_p$.

Generally, rather than using efficiency, one works with penetration, Pt, one minus the fraction collected. The aerosol aerodynamic diameter mass size distribution $m(d_{pa})$ is defined so that $m(d_{pa})$ represents the fraction (of the total mass concentration of the aerosol) having aerodynamic diameters between d_{pa} and $d_{pa} + dd_{pa}$. Thus, the distribution is normalized to unity:

$$\int_0^\infty m(d_{pa})dd_{pa} = 1 \qquad (19.3)$$

and the fraction (by mass) of particles which will penetrate the scrubber is given by:

$$\overline{Pt} = \int_0^\infty Pt(d_{pa})m(d_{pa})dd_{pa} \qquad (19.4)$$

This is sometimes called the "integrated penetration" or the "total penetration." The product of the total penetration and the inlet mass concentration gives the outlet mass concentration, often the quantity of interest. The outlet particle size distribution becomes $Pt(d_{pa})m(d_{pa})/\overline{Pt}$.

Often it is convenient and sufficiently accurate to approximate the size distribution of an aerosol with a log-normal distribution. (This is the same as saying the logarithms of the particle diameters are distributed normally). The two parameters describing a log-normal distribution are its median (d_g), which for a log-normal distribution equals its geometric mean, and its geometric standard deviation (σ_g). Of the aerosol mass, 68% is due to particles having diameters between d_g/σ_g and $d_g\sigma_g$; 95% of the mass is due to particles of diameters between d_g/σ_g^2 and $d_g\sigma_g^2$. The log-normal distribution is used in venturi scrubber design algorithms: Calvert[9] presented several figures which are convenient to use for scrubber design for particles having log-normal size distributions, once the cut diameter (d_{pc}), the diameter for which $E(d_{pc}) = 0.50$, is determined. Others have used the log-normal assumption for closed-form evaluations of Equation (19.4) by approximating the fractional efficiency curve as a cumulative log-normal curve.[15, 17, 106]

Single Obstacle Efficiency

Formulas for scrubber collection efficiency require the single collector (obstacle) efficiency, η, which involves a number of physical mechanisms. It is defined for a single collector in an unbounded stream as:

$$\eta = \frac{\text{flow area cleaned}}{\text{collector cross-sectional area}} \tag{19.9}$$

The calculation of η depends in part upon the flow past the collector. Two flow models are commonly in use: viscous flow and potential flow. Viscous flow is an appropriate model when the obstacle Reynolds number is small; that is, when:

$$Re_c = \rho_G(U_G - U_c)D_c/\mu_G << 1 \tag{19.10}$$

Although the motion of the dust particles in the gas stream often meets this Reynolds number criterion, the flow around the collectors usually does not. (A flow of air at 0.1 m/s past a fiber or droplet 100 μm in diameter gives $Re_c \sim 1$). The model of potential flow is derived for $Re_c >> 1$, but even in this regime it is appropriate only up to where the flow separates and forms a wake which trails behind the obstacle.

The single collector efficiency can be calculated for various collection mechanisms separately and then combined as though the mechanisms acted independently.[18] It is more accurate and more difficult to solve for the particle trajectories in the appropriate flow field, including the collection forces and mechanisms.[19, 20, 21]

Collection Mechanisms

When a dust particle strikes the collection surface because of its inertia, the collection is said to be due to *impaction*. The impaction process can be characterized by the impaction parameter, ψ:

$$\psi = \frac{U_G C(d_{ae})\rho_W d_{ae}^2}{18\mu_G D_c} \tag{19.11}$$

where D_c is the collector diameter and μ_G is the gas viscosity.

The following expression approximates the single sphere efficiency for impaction:[6]

$$\eta_I = \psi^2/(\psi + 0.35)^2 \tag{19.12}$$

Impaction is usually the most important collection mechanism for scrubbers, for particles larger than 0.1 μm. Figure 19.6 gives the impaction efficiency for several collector geometries versus the impaction parameter.[22]

"Interception" occurs when a particle strikes a collector even though the particle center would not have. Incorporating it correctly with other collection mechanisms really means altering the boundary conditions for the problem. Define N_R as the ratio of (spherical) particle radius to (spherical or cylindrical) collector radius. The incremental efficiency due to "interception" (above that of impaction, if operative) is between $2N_R$ and $3N_R$ for potential flow around a spherical collector and between N_R and $2N_R$ for potential flow around a cylinder, for inertialless and highly massive particles, respectively.[23, 107]

Capture by *diffusion* occurs because of the Brownian motion of the particles. It becomes appreciable only as the Peclet number (the gas velocity times the collector diameter divided by the particle diffusivity) becomes much less than one, which is rarely the case for $d_p > 0.1$ μm.

Electrostatic forces have been employed to augment the collection efficiency of scrubbers. The case in which the particles and the collectors are charged (Coulombic interaction) typically produces a much greater effect than those cases in which either the collectors or the

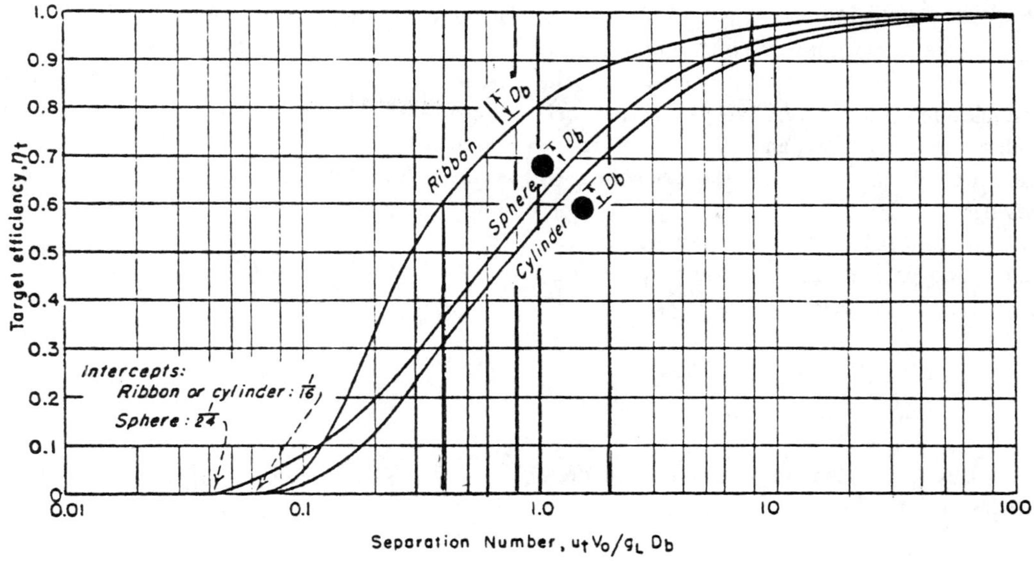

Figure 19.6. Target efficiency of spheres, cylinders, and ribbons.[22]

particles are charged, but not both (image force interactions). The "migration velocity" is the terminal velocity of a particle at the surface of the collector, due to the electrical forces. For electrostatic interaction to be important, the migration velocity should not be very much smaller than the product of the relative velocity and the collection efficiency due to all other mechanisms, $\eta(U_G - U_c)$. Figure 19.7 gives the migration velocities calculated by assuming that the particles were charged to saturation in a 10 kV/cm field (or they are uncharged) and the same field is produced by the collectors (or they are uncharged), for particles of the size indicated and a 100-μm spherical collector.[24] Electrostatic collection is intrinsically energy-efficient because the collection force can be applied directly to the particles, rather than indirectly to the particles through moving the gas.

The collection efficiency for a scrubber can be obtained from the collection efficiency (η) of its obstacles (droplets, beads, fibers, etc.) as follows. The number of particles collected in time, dt, as the particles flow (parallel to the x axis) through an infinitessimal volume, $dV = Adx$, is:

$$dN_p = -\eta n_p A_c (U_G - U_c) dt \qquad (19.13)$$

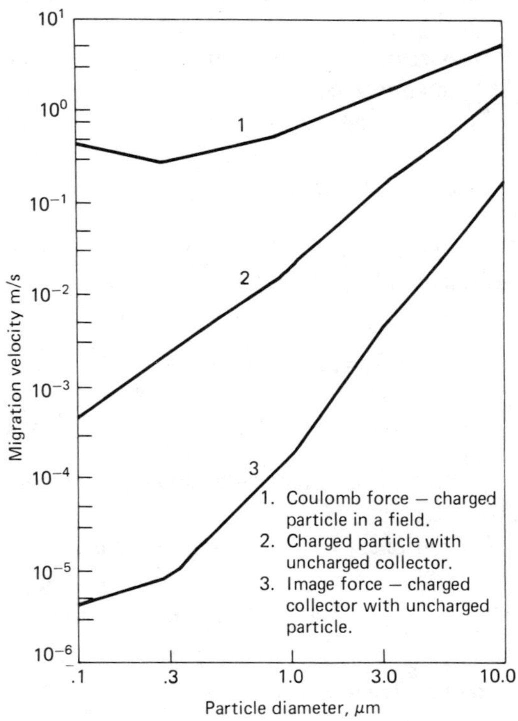

Figure 19.7. Theoretical collection migration velocities for three electrostatic mechanisms.[24]

where dN_p is the number of particles collected, n_p is the particle number concentration, A is the scrubber cross-sectional area, A_c is the obstacles (collectors) cross-sectional area, perpendicular to the flow, and U_G and U_c are the velocities of the gas and the collectors (if moving).

If the concentrations and velocities of collectors and particles are uniform perpendicular to the flow, then (using $dt = dx/U_G$):

$$\frac{dn_p}{n_p} = -\eta\left(\frac{U_G - U_c}{U_G}\right)\frac{A_c dx}{V} \quad (19.14)$$

From this point, the derivation can be done with various degrees of sophistication:

1. The calculation of the single droplet efficiency η can include some or all of the following mechanisms: impaction, interception, diffusion, electrostatic interactions, diffusiophoresis, thermophoresis.
2. The velocities U_G and U_c can be calculated in detail, including their dependence on position.
3. Any spatial variation of the collectors can be taken into account.
4. Various averages of collector areas, A_c, can be used, or a functional form for their distribution employed.

If the various quantities in the right-hand side of Equation (19.14) are uniform, then

$$Pt = \frac{n_p}{n_{po}} = \exp\left(-\eta\left[\frac{U_G - U_c}{U_G}\right]\frac{A_c L}{V}\right) \quad (19.15)$$

n_{po} = concentration at $x = 0$, and n_p is the concentration at $x = L$.

For stationary collectors, this becomes:

$$Pt = \exp\left(-\eta A_c L/V\right) \quad (19.16)$$

Predicting Total Efficiency

In sections to follow, fractional efficiency equations will be given for various scrubber types. In general, they come from assuming particle and collector concentrations to be uniform perpendicular to the mean flow and inertial impaction to be the dominant collection mechanism. Once the fractional efficiencies are known, the total efficiency is determined from:

$$\overline{E} = 1 - \overline{Pt} = \int_0^\infty E(d_{pa})m(d_{pa})dd_{pa} \quad (19.17)$$

or

$$\overline{E} = \sum_{i=1}^{m} E_i(d_{pa_i})(\dot{M}_{0i}/\dot{M}_0) \quad (19.18)$$

19.4. SCRUBBER SELECTION

Introduction

Generally, power consumption costs and other operating costs for scrubbers operating at high pressure drops (greater than about 2.5 kPa) are greater than or comparable to the annualized equipment costs. Further, the collection efficiency of one type of scrubber compared with another, on a given aerosol, will be about the same for a given pressure drop (thus a given energy consumption). Thus, the choice among scrubber types may depend upon factors other than collection efficiency and power consumption. Krockta and Lucas[25] presented a detailed list of the factors to be considered in selecting a scrubber for a particular application, a list prepared by a committee of the Air Pollution Control Association. Among these factors are: economic aspects, including capital expenditures, operating and maintenance costs; environmental factors, such as climate and the resources for power and waste treatment; engineering factors, including such particle characteristics as size distribution, concentration, solubility in scrubbing liquid, chemical reactivity, abrasiveness; gas characteristics, such as temperature, humidity, pressure, and chemical composition; and such scrubbing liquid characteristics as viscosity, density, surface tension, solids concentration.

Table 19.2. "Minimum" Particle Size for Various Types of Scrubbers.[a]

	PRESSURE DROP, IN. WATER	PRESSURE DROP, kPa	MIN. PARTICLE SIZE, μm[b]
Spray towers	0.5-1.5	0.12–0.38	10
Cyclone spray scrubbers	2-10	0.5-2.5	2-10
Impingment scrubbers	2-50	0.5-12	1-5
Packed-and fluidized-bed scrubbers	2-50	0.5-12	1-10
Orifice scrubbers	5-100	1.2-25	1
Venturi scrubbers	5-100	1.2-25	0.8
Fibrous-bed scrubbers	5-110	1.2-28	0.5

[a]Adapted from Reference 26.
[b]Smallest particle size for which the scrubber has at least 80% collection efficiency.

Although a scrubber type might be operable at pressure drops outside its conventional design range, the information in Table 19.2 is useful in suggesting what scrubbers are appropriate for achieving at least 80% efficiency on the particle sizes indicated.[26]

The following special considerations apply:[26]

1. Gas absorption: For collecting gases and vapors as well as particles, counter-current flow is to be preferred, with steps taken to maximize surface area and contact time, using, for example, a packed bed if plugging can be avoided.
2. Plugging: Fibrous beds and packed beds are susceptible to plugging, and generally one should use the more open types of scrubbers (venturi, orifice, preformed spray) for heavy aerosol and concentrations ($\gtrsim 10$ g/m^3); high recirculation rates may also lead to plugging of spray nozzles; it may be advantageous to have a low-energy scrubber or a cyclone upstream of a high-energy scrubber to help prevent plugging.
3. Reentrainment: Once the scrubbing liquid has captured the particles, the liquid must be retained; scrubbing droplets must be captured by an efficient demister; failure to do so for dyes and pigments, for example, can be serious.
4. Stack-condensate fallout: The condensation of scrubbing liquid within the exhaust stack can cause spray to be generated from the stack walls if stack velocities become too high.
5. Freezing: Cold-weather operation must include provisions for preventing freezing during operation and for preventing damage due to freezing when the scrubber is not in operation.

A limited survey of scrubber applications was carried out by Calvert et al.,[6] the results of which are shown in Table 19.3.[27] (It was noted that the number of sources surveyed was small.) The following comments were among those made:[27]

1. Packed bed and fibrous scrubbers were used in applications requiring the collection of gases, liquids and those particles (soluble or nonadhering) which tend not to plug.
2. Preformed (hydraulic) sprays were mainly used to capture gases.
3. Mechanically aided scrubbers were rarely found.

From this table, it appears that centrifugal scrubbers were preferred for the coarse dusts ($\gtrsim 10$ μm) from crushing operations, but for the fine dusts from smelting operations (much of it $\lesssim 1$ μm), gas-atomized scrubbers were dominant, and hydraulic ("preformed") spray scrubbers were of secondary importance.

Summary

For control of coarse aerosols, such as powders formed by disintegration of bulk material, low-energy, open-structure scrubbers such as

Table 19.3. Results of Survey of Scrubber Applications.[27]

PROCESS	SCRUBBER TYPE									
	PLATE[a]	MASSIVE PACKING	FIBER BED	PREFORMED SPRAY	GAS-ATOMIZED SPRAY	CENTRIFUGAL	BAFFLE	IMPINGEMENT	MECH. AIDED	MOVING BED
Calcining	6	2	–	13	21	–	–	43	–	–
	(1)[b]	(1)	(0)	(5)	(23)	(0)	(0)	(3)	(0)	(0)
Combustion	17	–	–	5	2	2	–	29	–	9
	(3)	(0)	(0)	(2)	(2)	(1)	(0)	(2)	(0)	(2)
Crushing	6	–	–	–	–	26	–	14	–	5
	(1)	(0)	(0)	(0)	(0)	(11)	(0)	(1)	(0)	(1)
Drying	39	–	–	10	18	70	100	–	25	64
	(7)	(0)	(0)	(4)	(19)	(30)	(1)	(0)	(1)	(14)
Gas removal	17	72	40	45	9	2	–	14	50	5
	(3)	(33)	(2)	(18)	(10)	(1)	(0)	(1)	(2)	(1)
Liquid-mist	0	24	60	7	–	–	–	–	–	–
recovery	(0)	(11)	(3)	(3)	(0)	(0)	(0)	(0)	(0)	(0)
Smelting	17	2	–	20	50	–	–	–	25	18
	(3)	(1)	(0)	(8)	(54)	(0)	(0)	(0)	(1)	(4)

[a]Read vertically. Example: 39% of all plate-type scrubbers are used to control discharges from drying processes.
[b]Numbers in parentheses refer to number of operators reporting information to the survey.

those with pre-formed sprays or impingement scrubbers, are often applicable with little risk of plugging. For fine aerosols, such as from powders made from condensation processes (including gas-phase reactions), high-energy scrubbers, such as venturi scrubbers, can be applied successfully, with attention to the prevention of plugging and reentrainment.

19.5. ATOMIZED SPRAY SCRUBBERS (VENTURI, ORIFICE, IMPINGEMENT)

Introduction

Directing a high velocity flow of gas across a liquid surface forms drops, which can then be used as collectors of particles in the gas stream. A variety of atomizing scrubbers work this way. Three different examples are shown in Figure 19.8. In atomizing scrubbers the air flow controls both the droplet size distribution

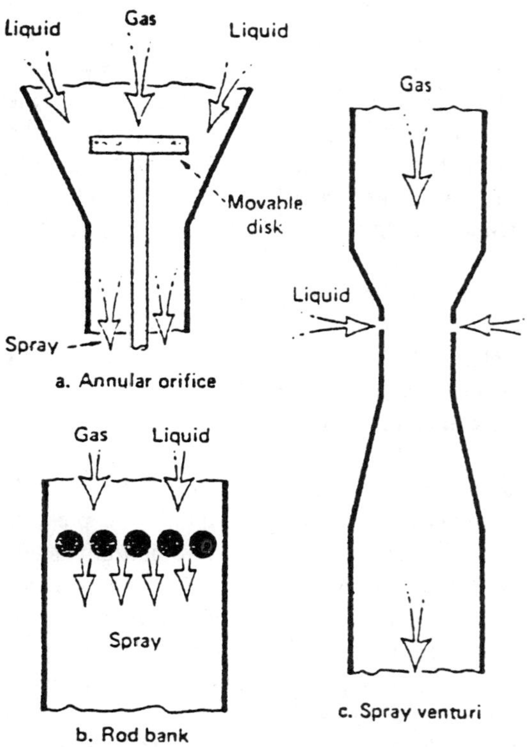

Figure 19.8. Three atomizing scrubber types.[27] Excerpted by special permission from *Chemical Engineering* (Aug. 29, 1977), copyright © 1977 by McGraw-Hill, Inc., New York, N.Y. 10020.

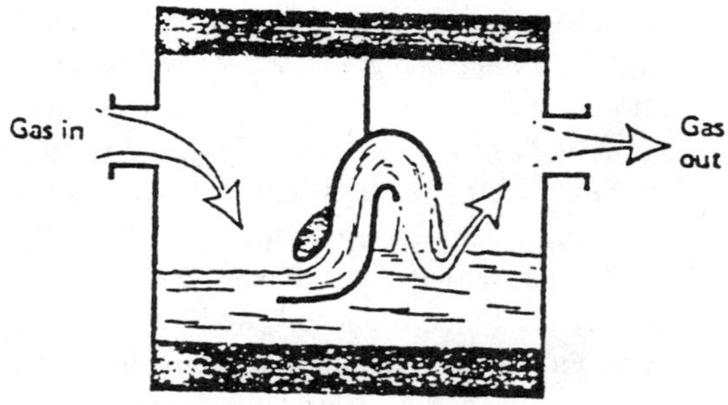

Figure 19.9. An impingement scrubber.[27] Excerpted by special permission from *Chemical Engineering* (Aug. 29, 1977), copyright © 1977 by McGraw-Hill, Inc., New York, N.Y. 10020.

and the ratio of droplet volume to gas volume, the liquid-to-gas flow ratio (Q_L/Q_G), but in hydraulic spray scrubbers the droplet size distribution can be changed independently of the liquid-to-gas ratio, and in various column-type scrubbers, the liquid flow rate can also be changed without affecting the collector size, which in turn affects the collector efficiency. A *venturi scrubber* has a converging section, a throat, and a diverging section. It accelerates the gas in the converging channel, introduces liquid (often as a spray) near the throat, where most of the particle collection occurs, then decelerates the gas and droplets in the diverging sections, generally quite gradually tapered. It is very widely used and receives special attention here. *Orifice scrubbers* use much the same principles. They are generally made from a single opening put in a place in the duct, with the plate being wetted by a flow of liquid which is then atomized at the plate edge. *Impingement scrubbers* direct a flow of gas at the surface of a liquid, using a variety of geometries, causing intimate mixing of liquid and particles due to atomization and turbulence. An example is shown in Figure 19.9.

Venturi Scrubber

Venturi scrubbers are quite popular, especially in applications (such as metallurgical emissions control) where efficiencies of 90% or more are required for particles of one micron diameter or smaller. Such applications may require pressure drops of 10 kPa or more. Venturis are relatively simple to build, using geometries whose cross sections are either circular or rectangular. Figure 19.8a shows an adjustable-throat venturi scrubber. Here liquid is introduced near the top of the converging section, to be atomized by the high-velocity gas at the throat. The diverging section is often followed by a flooded elbow, and the material not caught at the elbow is captured in a mist eliminator, such as a cyclone. Adjustable throats are needed where the gas volume flow is variable. For rectangular throats, the area can be changed by adjusting the throat width; for circular throats, usually a disk will be inserted to form an annular throat, which can be adjusted conveniently by moving the disk to various positions in the converging section.

Power Consumption. The pressure drop is the main contributor to power consumption. Without liquid flow, a venturi would have a pressure drop of about one-tenth the gas "velocity pressure," the latter being $0.5\rho_G U^2$. This is small in comparison with the energy consumed in accelerating the droplets to the gas velocity, some of which energy is regained in the expanding section. After reviewing several correlations for pressure drop, Yung et al.[28] recommended the following equation (a misprint has been corrected):

$$\Delta P = -\ 2\rho_L U_G^2 \left(\frac{Q_L}{Q_G}\right) [1 - X^2 + (X^4 - X^2)^{0.5}] \qquad (19.19a)$$

where Q_L and Q_G are the liquid and gas volume flow rates, U_G is the gas velocity in the throat, and X is the dimensionless throat length:

$$X = 1 + 3LC_{D1}\rho_G/16D_d\rho_L \qquad (19.19b)$$

in which L is the length of the throat and C_{D1} is the drag coefficient for the droplets at the throat:

$$C_D = 0.22 + (24/Re_T)(1 + 0.15Re_T{}^{0.6}) \qquad (19.20)$$

$$Re_T = \rho_G U_G D_T/\mu_G \qquad (19.21)$$

D_T is the throat diameter and D_d is the diameter which characterizes the droplets. For air and water, the expression for atomized droplet D_d of Nukiyama and Tanasawa becomes[29] (SI units):

$$D_d = 0.0050/U_G + 0.92(Q_L/Q_G)^{1.5} \qquad (19.22)$$

with D_d in m, U_G in m/s, and the flows in m^3/s. (See Table 19.4.) Equation (19.19a) for pressure drop compared well with data for liquid-to-gas ratios of 10^{-4} to 10^{-3}. It was assumed in deriving the equation that all the drops are accelerated in the venturi throat and that none of the momentum thus imparted is recovered as pressure gain when the drops decelerate in the diffuser, that there is no initial axial component of velocity for the droplets, that the flow is one dimensional, incompressible and adiabatic, that at any cross section the liquid fraction is small, and that the net pressure difference of wall friction minus pressure recovery in the diffuser is negligible. If the throat length is long enough to accelerate the droplets to the velocity of the gas, the term in brackets becomes 0.5 and Equation (19.19a) reduces to that presented earlier by Calvert:[30]

$$\Delta P = \beta\rho_L(Q_L/Q_G)U_G^2 \qquad (19.23)$$

Calvert's original value for β was 1.00,[30] but it has been found that β = 0.85 agrees better with experimental data.[31]

Collection Efficiency. Mathematical models developed to predict penetration and pressure drop for venturi scrubbers have limitations due to the assumptions that go into their derivations, and conclusions based on model results must be interpreted cautiously. However, good models can help in obtaining improvements in scrubber performance.

Models for venturi scrubber performance have been developed and reported by Johnstone et al.,[32] Calvert,[30,33] Boll,[34] Taheri and Shieh,[35] Goel and Hollands[36] and by Yung

Table 19.4. Droplet Sizes Predicted by Nukiyama-Tanasawa Equation for Various Gas Velocities for Air and Water.

U_G (m/s)	$0.0050/U_G$ (mm)	Q_L/Q_G (10^{-3})	$0.92(Q_L/Q_G)^{1.5}$ (mm)	D_d (mm)
1	5.0	1	0.029	5.0
		10	0.92	5.9
3	1.7	1	0.029	1.7
		10	0.92	2.6
10	0.50	1	0.029	0.53
		10	0.92	1.42
30	0.17	1	0.029	0.20
		10	0.92	1.09
100	0.050	1	0.029	0.079
		10	0.92	0.97
300	0.017	1	0.029	0.046
		10	0.92	0.94

et al.[28,29] Assumptions made in these derivations differ; they use various relationships between impaction parameter and single droplet collection efficiency, make a variety of assumptions regarding drop velocity at the time atomization occurs, and assume particle collection occurs in various parts of the venturi. All the models assume monodisperse droplets and complete liquid utilization, except that Taheri and Shieh[35] incorporated particle and droplet concentration distribution.

The model most frequently used for penetration and pressure drop in a venturi scrubber is probably the model presented by Calvert[30,33] and Calvert et al.[16] These equations are used in this section. This approach considers the same processes described in all venturi models, but its equations are more tractable. Calvert et al.[16] showed that agreement between the theoretical predictions and data is generally good, although this agreement is helped by an adjustable constant, f, in the equation for the penetration. Drop velocity at atomization is assumed to be "f" times the gas velocity in the venturi throat, where f is between zero and one. With proper selection of this constant, the theory and data can be made to agree. The utility of the Calvert penetration model largely depends on the extent to which f remains constant for all venturi scrubbers and for all aerosols. Values of f from 0.25 to 0.5 were reported by Calvert,[30,31] the larger values being appropriate for more hydrophilic particles and larger gas flow rates.

The Calvert[6,16] equation for penetration through a venturi is:

$$-\ln Pt = \frac{2Q_L\rho_L D_d U_G}{55 Q_G \mu_G K}\left[0.7 + fK - 1.4\ln\left(\frac{0.7+fK}{0.7}\right) - \frac{0.49}{0.7+fK}\right]. \tag{19.24}$$

The parameter K is just 2ψ (Equation (19.11)).

This equation can fairly readily be programmed on a computer or even a programmable calculator.[38] Using dimensional analysis, we identified a group which helps describe scrubber performance, the performance number, N_p[12]:

$$N_p = C\rho_p d_p^2 \Delta P/18\mu_G^2 = \tau\Delta P/\mu_G \tag{19.25}$$

where τ is the particle aerodynamic relaxation time.[23] Let the pressure drop be given by Equation (19.23); then this model predicts the following minimum penetration (Pt^*) at a given pressure drop:[39]

$$Pt^* = \exp[-0.124(f^2/\beta)N_p] \tag{19.26}$$

Figure 19.10 shows the number of transfer units

$$N_{tu} = -\ln Pt \tag{19.27}$$

versus performance number for experiments with small venturis and for the minimum penetration conditions, as calculated with Equation (19.26). Ideally, all liquid will be fully atomized to droplets immediately upon injection, and all droplets will accelerate to the full gas throat velocity. In this case, which represents full liquid utilization with no pressure regain

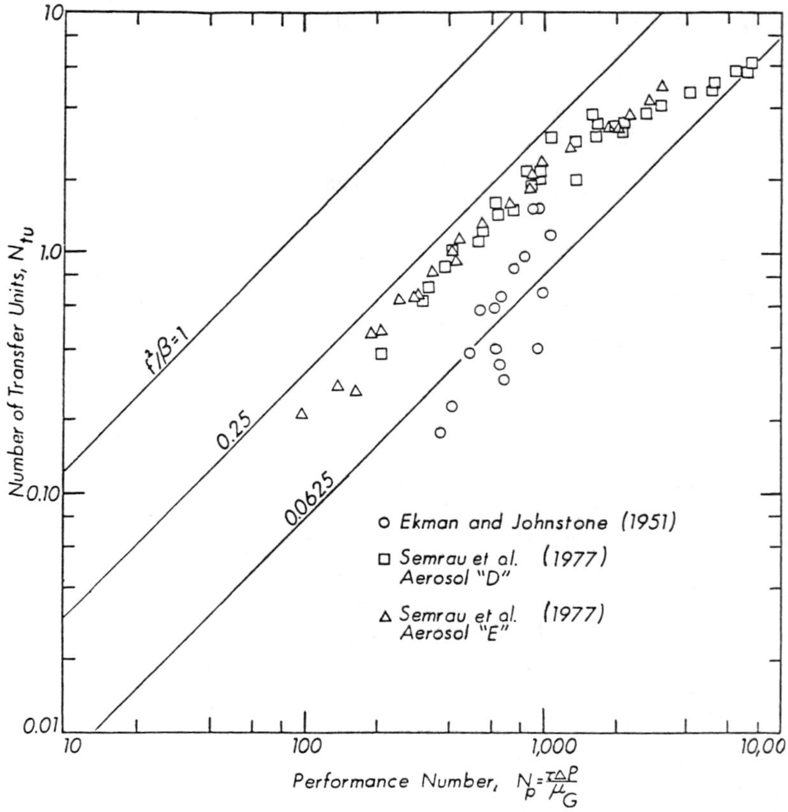

Figure 19.10. Transfer units versus performance number for atomizing scrubbers.[39]

due to droplet deceleration, the values of constants f and β are unity and f^2/β becomes unity as well. Smaller values of f^2/β represent less complete liquid utilization and little or no gain.

The dependence of f or β on variables under the control of the venturi designer has not yet been quantified. Calvert[31] suggested f and β may depend upon the size of the venturi, the method of liquid injection, or other factors. In practice, f and β are seldom if ever known with certainty; in designing a venturi they are estimated on the basis of past experience.

Two more venturi scrubber models have recently appeared in the literature.[111, 112] The results of the first indicated "dispersity of the droplet size distribution only slightly affects collection efficiency over the operating range normally encountered."[111] The other author concluded that polydispersity makes a difference and that "calculations based on the assumption that droplets are monodisperse result in an under-estimation of the efficiency."[112]

Optimization of Design. The factors to be decided upon in the design of the scrubber include: angles of convergence and divergence, throat cross-sectional area and length, liquid-to-gas ratio. The angles of convergence and divergence are thought not to be critical within the range of the conventional designs (20–25° and 5–7°). The cross-sectional area will be determined by the gas volume flow rate and the desired throat velocity. The throat length criterion has been proposed as:[29]

$$4/3 \leq C_{D1} L \rho_G / D_d \rho_L \leq 2 \qquad (19.28)$$

It represents a compromise between increased particle collection and increased frictional flow resistance as throat length is increased. The liquid-to-gas ratio (typically around 10^{-3}) affects both the pressure drop and the collection efficiency; in general, increased values of the ratio Q_L/Q_G improve collection efficiency at a given pressure drop but also increase the amount of water to be handled for recirculation and disposal.

For a fixed value of scrubber performance number, N_p, and predetermined value of f^2/β, Equation (19.27) can be shown to depend only upon the fK product.[39] The value of fK for which penetration is minimized is:

$$fK = 1.10 \tag{19.29}$$

according to the Calvert model, assuming $X >> 1$. Smaller values of throat length lead to larger values of fK being optimal. Reasons for an optimum value for fK can be discussed in terms of an optimum droplet diameter. A drop larger than the optimum will sweep through a larger volume of particle-laden gas and have a larger surface area. However, a larger droplet will also have a smaller single droplet collection efficiency, due to impaction, contribute more to pressure drop, and, for a given amount of liquid used, fewer such drops will be produced. The optimum droplet diameter reflects the best compromise among these factors. Although Semrau et al.[8] did not find, experimentally, an independent effect of liquid-to-gas ratio and pressure drop, Ekman and Johnstone[40] and Muir et al.[41] found that venturi efficiency improved at a given pressure drop when the liquid-to-gas ratio increased, thus when the droplet size increased.

For any selected value of pressure drop, for particles of specified d_{pa} and with f and β fixed, it is possible to predict the gas velocity and liquid-to-gas ratio that should produce drops of optimum size and allow operation at theoretically maximum efficiency, using Equations (19.25), (19.26), and (19.28). Three equations can be written, using as unknowns the gas velocity, liquid-to-gas ratio, and droplet diameter at optimum conditions. The solutions to these equations show the operating conditions necessary to produce theoretically optimum performance. The simultaneous solution of these equations allows determination of optimum gas velocity, U_G^*, and optimum liquid-to-gas ratio, $(Q_L/Q_G)^*$:[39]

$$U_G^* = \sqrt{\frac{0.005 + (2.5 \times 10^{-5} + 6.69 f\tau\,(\Delta P/\rho_L)^{3/2})^{1/2}}{3.65 f\tau}} \tag{19.30}$$

The diameter of the "optimum drops" that correspond to these U_G^* and $(Q_L/Q_G)^*$ conditions can be found from Equation (19.22). (U_G must be in units of m/s and $\Delta P/\rho_L$ in N·m/kg, however.)

Nukiyama and Tanasawa found Equation (19.22) empirically for the following conditions: 70 m/s $< U_G <$ 230 m/s, $8 \times 10^{-5} < Q_L/Q_G < 1 \times 10^{-3}$, corresponding to atomized drops 20 μm $< d_d <$ 100 μm in diameter.[39] The present analysis indicates that for particles larger than about 0.5 μm in diameter, for venturis with long throats, optimum sized atomized drops are larger than those for which Equation (19.22) can be used with confidence. Other relationships for the diameter of atomized drops might be used with greater confidence for larger atomized drop diameters.[34]

This analysis is for collection of monodisperse particles by monodisperse droplets. The extension of this work to polydisperse aerosols requires further investigation. As a first approximation, one might use the simple expression for penetration:[6,9,27]

$$Pt \doteq \exp(-A d_{pa}^B) \tag{19.31}$$

($B \doteq 2$) and use the curves presented in these references for the total mass (integrated) penetration for log-normal aerosols of various values of σ_g as functions of d_{pc}/d_{pg}. For this approximation, performance is optimized by determining the particle cut diameter needed to achieve the integrated penetration desired, then finding the optimal conditions as described above to give 50% efficiency for a particle of that cut diameter (See Fig. 19.11).

Design Example.[39] Consider the design of a venturi scrubber to collect particles 0.5 μm in diameter with 90% efficiency from 5 m^3/s of gas, using the minimum possible pressure drop. Assume $\rho_p = 2000\ kg/m^3$, $\rho_G \doteq 1.2\ kg/m^3$ and $\mu_G = 1.8 \times 10^{-5}\ kg/m \cdot s$, respectively. Further, follow Calvert's suggestion that often $f = 0.5$ and $\beta = 0.85$.[31]

First, determine the number of transfer units required:

$$N_{tu} = -\ln Pt = -\ln(0.10) = 2.30 \qquad (19.32)$$

Next, find the minimum scrubber performance number:

$$N_p = \frac{N_{tu}}{0.0124\, f^2/\beta} = \frac{2.30}{0.0124(0.5)^2/0.85} = 631 \qquad (19.33)$$

Particle relaxation time can be found from:

$$\tau = \frac{d_p^2 \rho_p C}{18\mu_G} = \frac{(5 \times 10^{-7})^2\,(2000)\,(1.33)}{18(1.8 \times 10^{-5})} = 2.1 \times 10^{-6}\ \text{s} \qquad (19.34)$$

The lowest pressure drop theoretically necessary under these conditions can now be obtained from Equation (19.25):

$$\Delta P = \frac{N_p \mu_G}{\tau} = \frac{(631)\,(1.8 \times 10^{-5})}{2.1 \times 10^{-6}} = 5410\ \text{Pa} \qquad (19.35)$$

The gas velocity in the venturi throat (assumed very long) required to generate optimum-sized drops at this pressure drop is given by Equation (19.30):

$$U_G = \sqrt{\frac{0.005 + \left(2.5 \times 10^{-5} + 6.69\,(0.5)\,(2.1 \times 10^{-6})\left(\frac{5410}{1000}\right)^{3/2}\right)^{1/2}}{3.65\,(0.5)\,(2.1 \times 10^{-6})}} = 64\ \text{m/s} \qquad (19.36)$$

The cross-sectional area of the venturi throat to provide this gas velocity is:

$$A = Q_G/U_G = (5\ m^3/s)/(64\ m/s) = 0.078\ m^2 \qquad (19.37)$$

A circular throat 0.32 m in diameter will serve.

The liquid flow rate required can now be found from Equation (19.31):

$$Q_L = \frac{Q_G\,\Delta P}{\rho_L \beta U_G^2} = \frac{(5)\,(5410)}{(1000)\,(0.85)\,(64)^2} = 7.8 \times 10^{-3}\ m^3/s \qquad (19.38)$$

With the diameter of the venturi throat and liquid flow rate fixed, the essential design of the venturi is complete.

Orifice Scrubbers

Orifice scrubbers are often made by inserting a plate with a hole or slit in it into a vertical run of ducting, irrigating the plate so that scrubbing droplets are formed at the aperture. They operate quite similarly to venturi scrubbers and are effectively venturis with zero throat length and 180° angles of convergence and divergence. The pressure drop is given by Equation (19.23) for venturi scrubbers. In their summary of pressure drop and efficiency equations useful for various scrubbers, Yung and Calvert[42] used the same Equations, (19.23) (with $\beta = 0.85$) and (19.24). Venturi and orifice scrubbers behave almost identically.[8,21]

A wetted butterfly valve was tested by Taheri et al.[43] and found to be an inexpensive variable orifice scrubber, convenient for use on variable gas flows. Pressure drop and collection efficiency data were given in their article, and although they did not compare the perfor-

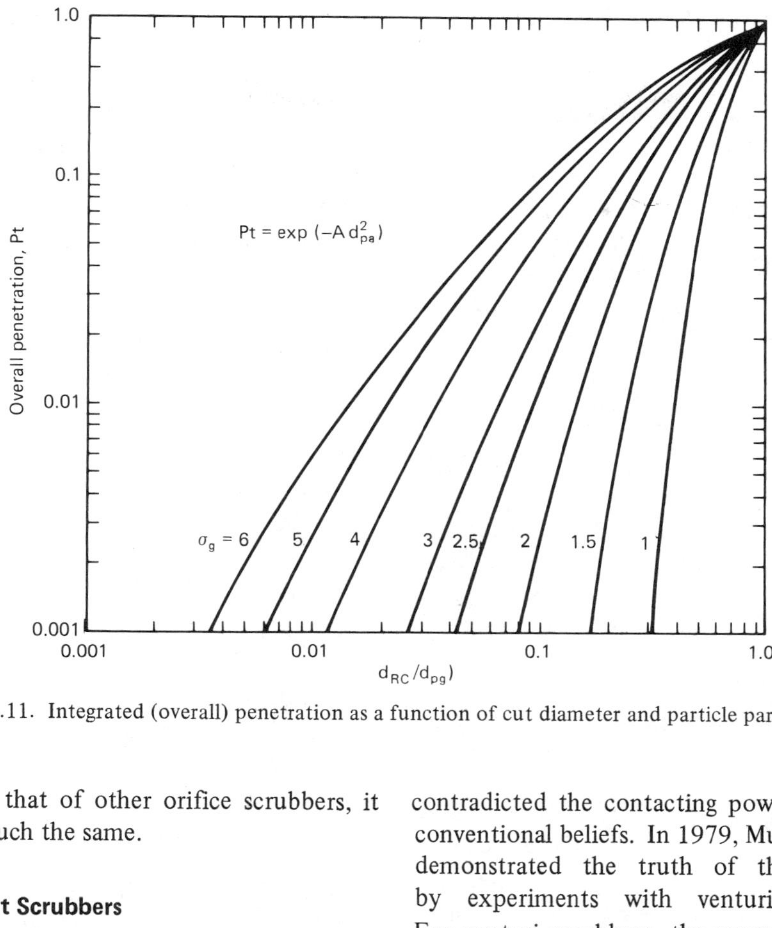

Figure 19.11. Integrated (overall) penetration as a function of cut diameter and particle parameters.[13]

mance with that of other orifice scrubbers, it should be much the same.

Impingement Scrubbers

There are a variety of designs for relatively low-energy impingement scrubbers, such as that shown in Figure 19.9. These gas-atomizing scrubbers have collection efficiencies similar to those of venturi or orifice scrubbers operating at the same pressure drop.[27,42] Equations (19.23) for pressure drop and (19.24) for efficiency apply, with $\beta = 1.0$.[42]

Multiple-Stage Scrubbing

In 1976, we proposed that for high-energy scrubbing there were situations in which multiple-stage devices, such as two venturis in series, would be more efficient for a given level of energy consumption than would a single-stage device, even without particle growth or other condensation effects.[10] This contradicted the contacting power theory and conventional beliefs. In 1979, Muir and Meheisi demonstrated the truth of this hypothesis by experiments with venturi scrubbers.[11] For venturi scrubbers, the power advantage of multiple-stage scrubbing becomes appreciable when the scrubber performance number, $N_p = \tau \Delta P/\mu_G$, becomes greater than or on the order of 10^3.[12] It is likely the same will hold true for other atomization scrubbers. Added improvement due to particle growth and water vapor flux forces make multiple-stage scrubbing potentially even more attractive.

Summary

Venturi and other gas-atomized scrubbers have similar pressure drop and efficiency characteristics. They can be used over a wide range of operating conditions, are simple, rugged, and resistant to plugging. The formation of fine spray means demisters are essential to their successful operation.

19.6. HYDRAULIC SPRAY SCRUBBERS

Introduction

The scrubbers discussed in this section all have preformed sprays, produced by nozzles and having their droplet size distributions determined by nozzle geometry and the properties of the liquid but not by the properties of the gas or its flow rate. The types of scrubbers covered here are: spray chambers, ejectors, and centrifugal (cyclone) spray scrubbers.

Spray Chambers

Spray chambers are conceptually quite simple. Gas and particles flow through a chamber with sprays directed co-current, cross-current, or counter-current to the flow, the latter being advantageous if gases are also to be removed in the scrubbing process. A demister is generally used as well.

Power Consumption. The power consumed in a spray chamber is the product of the gas volume flow rate and gas pressure drop (usually negligible) plus the power consumed by the nozzles, the sum of the products of their volume flow rates and pressure drops. Similarly, the demister consumes power due to the added gas flow resistance. The particle size for which collection efficiency becomes negligible is that for which the impaction parameter is substantially less than one; therefore, droplet size and velocity are important. Throughout the chamber, the droplet velocity relative to the gas will be within the range of the initial droplet velocity and the droplet terminal settling velocity, the larger limiting how small a particle can be captured. The initial velocity of the droplet as it leaves the nozzle will be, in the potential flow approximation:

$$U_d = (2\Delta P_L/\rho_L)^{1/2} \qquad (19.39)$$

Since the impaction parameter depends upon $d_{pa}^2 U_d$, the pressure on the nozzles would have to be increased by about $2^4 = 16$ to get the same single droplet efficiency on a particle of half the size. Actually, the situation is more complicated: the liquid and gas flow rates, the trajectory of the droplet and its stopping distance, and the geometry of the scrubber should go into such an analysis. It has been found that spray scrubbers are not quite as efficient in collecting particles as are venturi scrubbers for the same power consumption.[8] Preformed spray scrubbers use rather high liquid-to-gas ratios, 4 to 12 $\times$ 10^{-3}, and this high water usage is often coupled with problems of corrosion, erosion, and plugging of the spray nozzles.[27]

Collection Efficiency. The penetration of a cross-current spray chamber can be approximated by:[31]

$$Pt = \exp\left[-\eta_I 3Q_L h/2Q_G D_d\right] \qquad (19.40)$$

where η_I is the impaction collection efficiency (Equation (19.12)) and h is the dimension of the scrubber traversed by the drops. For D_d it is consistent with the derivation of this equation to use the ratio of the mean cubed diameter to the mean squared diameter, $\overline{D_d^3}/\overline{D_d^2}$, a ratio also known as the "Sauter mean diameter." For a counter-current spray operated vertically and having height z, Pt can be estimated from Equation (19.40) by replacing h with $zU_d/(U_d - U_G)$.[31] The choice of velocity at which to evaluate Equation (19.40) can be problematic, and we emphasize the equation is quite approximate.

Ejectors

The motion of droplets ejected from a nozzle spraying co-currently with the gas flow can be used to collect particles and to move the gas which bears them. This eliminates the need for fans in corrosive and erosive atmospheres.[44] Ideally, the momentum transferred from the nozzle would go entirely to the mixture of gas and droplets. Some units need $Q_L/Q_G \sim 10^{-2}$ to generate drafts of even 250 Pa, however. Harris[44] analyzed theoretically such a scrubber to predict collecting efficiencies for particles, vapors, and gases.

Where the pressure for the nozzles can be obtained from utilizing waste heat, such scrubbers may be economically advantageous.[45] One ejector was found to have collection efficiency quite similar to a venturi scrubber for the same power consumption.[46]

Centrifugal Scrubbers

By imparting a rotary motion to a gas, a cyclone can remove particles by a mechanism similar to impaction. Introducing a spray at the inlet of a cyclone can enhance particle collection by both capturing particles on the droplets and by preventing reentrainment of captured particles from the walls. The cyclone, or other centrifugal scrubber, may serve as its own demister and is resistant to plugging.

The pressure drop across the cyclone will be somewhat greater than what it would be without the spray. The penetration is approximately what would be predicted without the spray times $\exp\,[-3Q_L h\eta_I/2Q_G D_d]$, where h is the difference between the inner and outer radius of the cyclone.[42]

A centrifugal scrubber was tested[42] and found to have collection efficiency equal to that predicted for a venturi scrubber, Equation (19.24), with $f = 0.4$, which is within the range of performance found for venturis ($f = 0.25$ to 0.50).

19.7. WETTED PACKED BEDS AND FIBROUS MATS

Introduction

Wetted packed beds and fibrous mats can be advantageously used for the collection of mists, gases and vapors. They tend to plug, however, when used to capture insoluble particulate material, so they may not find much use in powder technology. The tendency to plug depends upon the particle size distribution, the gas and liquid flow rates, the particle concentrations in the gas and in the liquid, and the dimensions of the interstices in the bed, so that there are situations in which such scrubbers might be employed.

Packed Beds

Packed beds have been used for the separation of one gaseous constituent from another and to a lesser extent for the separation of particulates from gases. For a collector having packing with a mean surface-to-volume diameter D_{sv} (equal to six times the total solid volume of the packing material divided by total surface area) the Ergun equation holds for the pressure drop when operated dry:[48]

$$\Delta p = \frac{150\mu_G U_G L(1-e)^2}{D_{sv}^2 e^3} + \frac{1.75\rho_G U_G^2(1-e)L}{D_{sv} e^3} \qquad (19.41)$$

where e is the volume void fraction (dimensionless) and L is the length of the bed. This equation is just the sum of the Blake–Kozeny equation for laminar flow plus the Burke–Plummer equation for turbulent flow. For $Re = \rho_G U_G \cdot D_{sv}/\mu_G$ greater than 10^3, the first term is negligible.

The pressure drop for the dry bed will be less than that for the wet bed, but the calculations for predicting the latter are beyond our scope here, for which the reader should consult Reference 26.

Calvert[31] presented this equation for the penetration of a packed column for particles caught due to inertia:

$$Pt = \exp\,[-3.5\psi L/eD_c] \qquad (19.42)$$

where ψ is the impaction parameter (19.11) and D_c is the diameter of the collectors making up the bed. This relationship can also be written as:[31]

$$Pt = \exp\,[-Ad_{pa}^2] \qquad (19.43)$$

with

$$A = 0.69/d_{pc}^2 \qquad (19.44)$$

where d_{pc} is the cut diameter, $Pt(d_{pc}) = 0.5$:

$$d_{pc} = D_c(3.6e\mu_G/LU_G\rho_p)^{1/2} \qquad (19.45)$$

Figure 19.11 can be used to determine the overall mass penetration for an aerosol which has a log-normal distribution with a mass median = geometric mean diameter d_{pg} and a geometric standard deviation σ_g.

Fluidized Beds

A fluidized bed results when the upward flow of gas through packing which is unconstrained at its top becomes sufficient to support the weight of the bed. At this velocity, the packing material moves freely. At greater velocities the packing material may be carried off in the gas stream, but for wetted fluidized beds, unacceptable levels of liquid entrainment would likely occur before this velocity was reached.

As gas velocity is increased, pressure drop across such a bed increases as it would for any packed bed. The bed becomes fluidized when the pressure drop equals the weight per unit area of the bed and its associated liquid; further increases in velocity give much less added pressure drop increase per unit velocity increase.

Dry fluidized beds are receiving much attention for coal desulfurization, but their tendency to form channels and bubbles has limited their use in particle collection. Initial tests of a wetted fluidized bed were unusually promising. The results of subsequent tests have not shown that they have an energy advantage over most other scrubbers in collecting particulate matter.[27] Where mass-transfer as well as particulate collection is important, the wetted fluidized bed may be advantageous.

Wetted Fibrous Mats

Dry fiber mats are covered more extensively in the chapter on filtration. Wetted fibrous filter mats are attractive for scrubbing, in that the collection due to interception on fine fibers offers the hope of doing somewhat better in terms of efficiency versus pressure drop than do other scrubbers, which rely on impaction.[31,49]

The pressure drop across a fibrous filter can be estimated from the traditional Kozeny-Carman equation for pressure drop in laminar flow:[50]

$$\Delta P = k'S^2 \mu_G U_G L(1 - e)^2/e^3 \quad (19.46)$$

which for circular cylinders becomes:

$$\Delta P = 16k'\mu_G U_G L(1 - e)^2/D_c^2 e^3 \quad (19.47)$$

where k' is the Kozeny constant, equal to about 5 for porosities between 0.2 and 0.8. (The surface-to-volume ratio of the fibers is $S = 4/D$ for cylinders of diameter D.) For fibers oriented transverse to the flow, k' is 6.0, and for fibers parallel to the flow it is 3.1.[2]

Pressure drop depends strongly on porosity. In Table 19.5 $(1 - \epsilon)^2/\epsilon^3$ is given for $\epsilon = 0.2$, 0.3, . . . , 0.8. Over that range of porosities, the pressure drop changes a factor of 1000.

The Kozeny-Carman equation is for $\epsilon < 0.8$. Davies[51] cited his prior research with filter pads of different materials, having porosities from 0.7 to 0.994 as support for the equation:

$$\Delta P = 64\mu_G U_G L(1 - e)^{1.5}(1 + 56(1 - e)^3)/D_c^2 \quad (19.48)$$

Clearly, the wetted mat will have greater flow resistance than when dry, however.

The collection efficiency of a clean fibrous bed is approximately:[52]

$$E = 1 - \exp\,[-4(1 - e)L\eta_c'/\pi e D_c] \quad (19.49)$$

where η_c is the collection efficiency of a single fiber transverse to the flow, and η_c' is the collection efficiency of that fiber as part of a mat. If

Table 19.5. Values of $(1 - e)^2/e^3$ for e from 0.2 to 0.6.

POROSITY e	$(1 - e)^2/e^3$
0.2	80.
0.3	18.1
0.4	5.6
0.5	2.0
0.6	0.74
0.7	0.26
0.8	0.078

the collection is due to impaction, then:[53]

$$\eta_c' = \eta_c[1 + a(1 - e)] \qquad (19.50)$$

where a is about 5 to 20. For impaction, the collection efficiency of a single fiber is approximately (from fitting data presented by May and Clifford[54]):

$$\eta_c = \psi^2/(\psi^2 + 0.64) \qquad (19.51)$$

where ψ is the impaction parameter.

A wetted fiber filter was tested in the laboratory and modeled mathematically in a fashion quite similar to the analysis above.[49] It produced a somewhat higher efficiency than would be predicted for a venturi scrubber operating at the same pressure drop (about 7.5 kPa), which was attributed in part to the interception mechanism. (The fiber diameters were approximately 50 μm and the mat porosity was 0.97.) The model correctly predicted a sharp decline in penetration as particle aerodynamic diameter became greater than 0.5 μm.

19.8. TRAY TOWERS

Tray towers have one or more perforated plate trays which are irrigated with water and through which gas travels and is scrubbed. Often a series of such plates will be used, with the liquid introduced at the top of the scrubber to travel from tray to tray via "downcomers" or by trickling through the holes in the plate. (See Fig. 19.4.) If the holes have (submerged) baffles or targets connected to them at which the jet of gas and liquid are directed, one has an "impingement plate" scrubber; if there are holes but no impingement targets, one has a "perforated plate."

Sieve Plates

A "sieve plate" is a common type of plate scrubber adapted from gas-liquid contacting uses. At liquid or gas flow rates which are too high for the design, flooding will occur, marked by a sharp decrease in liquid througput and an increase in pressure drop. Avoiding this condition is one important goal of the design. If the gas flow becomes too low, liquid can seep through the perforations, decreasing contacting efficacy. Design equations are available to prevent either of these malfunctions.[26]

The pressure drop across the plates is due to the resistance to gas flow due to the geometrical arrangement itself, as when dry, and the added resistance of the flow through the scrubbing liquid. For each of the dry plates, the gas flow can be apportioned among the holes, and this equation for pressure drop used:[13]

$$\Delta P_{dry} = 1.14[0.4(1.25 - f_h) + (1 - f_h)^2](\rho_G U_G^2/2) \qquad (19.52)$$

in which f_h is the fraction of the plate area represented by the holes and U_G is the gas velocity through the holes. The other major contribution to pressure drop for each plate comes from the hydrostatic pressure represented by the height of the liquid on each plate (H_{weir}) as determined by the weir (often about 5 cm in height):

$$\Delta P_{weir} = \rho_L g H_{weir} \qquad (19.53)$$

The pressure drop across each plate is approximately $\Delta P_{dry} + \Delta P_{weir}$. More exact formulas and more design details are available elsewhere.[26]

For hydrophilic (wettable) aerosols, Taheri and Calvert found the following relationship for the penetration through a sieve plate scrubber as a function of particle size:[55]

$$Pt = \exp(-80 F_L^2 \psi) \qquad (19.54a)$$

for

$$0.38 \leqslant F_L \leqslant 0.65 \qquad (19.54b)$$

where F_L is the volume of clear liquid per volume of froth, m^3/m^3, and ψ is an impaction parameter, Equation (19.11), based on the hole diameter and the velocity of the gas through the hole. Taheri and Calvert found that hydrophobic aerosols were collected less effectively than hydrophilic, and that the addition of wetting agents lessened collection efficiency, by creating a less dense froth.[55]

Impingement Plates

The pressure drop for impingement plates can be estimated by the equations given above for perforated plates.

Impingement plate penetration is predicted to be:[31]

$$Pt = \exp(-0.693 d_{pa}^2 / d_{pc}^2) \quad (19.55)$$

with

$$d_{pc} = (1.37 \mu_G n_h D_h^3 / Q_G)^{1/2} \quad (19.56)$$

in which n_h is the number of holes per unit area and D_h is the hole diameter; the source of this design equation noted a lack of reliable experimental data to support it. Calvert also estimated the cut diameters of two-plate and three-plate systems as 88% and 83% of the one plate system.[31] Note that increasing the number of trays from one to three often will not greatly increase collection efficiency for particles, though it may for gases.[27]

Equation (19.56) can be used with Figure 19.11 to estimate total penetration for aerosols with log-normal distributions.

Foam Scrubbers

The formation of low density foam ($F_L << 1$) from a perforated plate has been the basis of several foam scrubber designs.[56,57,58] Unlike most scrubbers, impaction may not be the predominating mechanism. The longer residence times characteristic of such scrubbers and the small dimensions of the foam bubbles give sedimentation and diffusion more importance than usual, augmented by the interception effect. An important design problem is the breaking up of the foam and the capture of the fine particle-bearing droplets from the breaking up. One evaluation made in 1977 had this conclusion: "The operating cost of foam scrubbing with 99% surfactant recycle is an order of magnitude higher than that of the most expensive conventional method."[59] Further, residence times were $\gtrsim 10^1$ s, suggesting substantial construction costs for high-volume-flow operations.

19.9. CONDENSATION SCRUBBING

Theory and Experiment

Decades ago, Schauer[60] and Lapple and Kamack[61] found that the addition of steam to the gas to be scrubbed could bring about marked improvements in scrubber collection efficiency. Semrau[62] noted anomalously high collection efficiencies reported for scrubbers in which condensation occurred. (An extensive literature review of the work done before 1973 is available in the report by Calvert and coworkers.[63]) Several factors act:

Condensation of water vapor on spray scrubber droplets, caused by the droplets being at temperatures below the saturation temperature of the gas, can enhance particulate capture due to diffusiophoresis, the principal component of which is the net flow of water molecules toward the droplets; this is accompanied by a more subtle force due to the concentration gradient of the water molecules. Diffusiophoresis is accompanied by thermal forces tending to oppose it, however.

The diffusiophoresis "flux force" mechanism was discussed in detail by Waldmann and Schmitt. The existence of this mechanism is evident from the experimental results of Lapple and Kamack[61] and Semrau et al.[65] The latter, for example, noted a large difference in efficiency between wet scrubbers operating with hot versus cold water sprays. They suggested the differences could be caused by evaporation from the hot water droplets, which would produce a diffusiophoretic force away from the drop surface and therefore would result in reduced efficiency.

Sparks and Pilat[66] calculated particle collection efficiencies by droplets, assuming that (1) condensation, or (2) evaporation, or (3) neither, occurred. The collection mechanisms studied were inertial impaction and diffusiophoresis. Condensation was shown to enhance, while evaporation was shown to diminish, the collection by inertial impaction, the effects being more pronounced for the smaller particles.

Condensation of water vapor on droplets will also cause a temperature gradient. The latent heat of vaporization must be conducted away from the droplet. This may offset the effects of diffusiophoresis.[67]

Calculations were made for collection by droplets of 100, 500, and 1000 μm diameters in a spray tower.[68] The gas was assumed saturated and the droplets were taken to be cooler, warmer, or at the same temperature of the gas, so that particle growth due to condensation was not a factor. Single droplet collection efficiencies for the condensing case were quite insensitive to particle size, the mechanisms considered being impaction, diffusion, diffusiophoresis and thermophoresis. The condensation/evaporation effects were greater for the larger droplets due to the longer maintenance of the temperature gradients.

Whitmore[69] found that the fraction of particles collected due to the flow of water vapor to scrubber surfaces and droplets was approximately equal to the fraction of the gas which condensed.[70]

Condensation of water vapor on particles can lead to enhanced capture due to the increase in particle aerodynamic diameter. Soluble particles will become droplets at humidities greater than their "transition" humidities,[71] the humidities a solution made from the bulk material would produce in air in a closed vessel. (For NaCl, for example, this is 75% relative humidity.) A hydrophilic liquid such as sulfuric acid does not have a transition humidity; such a droplet changes size to be in equilibrium with any ambient humidity. Aerosols made of hygroscopic liquids and solids change their volumes approximately in proportion to $1/(1 - H)$, where H is the fractional humidity.[72,73] Hydrophobic particles will not grow until the gas is supersaturated, often to a multiple of the saturation vapor concentration; this condition is hard to create because soluble condensation nuclei, almost always present, will compete for the water vapor, making it hard to achieve supersaturation. Typically, about 75% of the condensing vapor goes to the cold surfaces of the scrubber and 25% of the particles.[70] Lancaster and Strauss[74] concluded that diffusiophoresis was less important than particle growth in conventional scrubbers in which steam is injected. Calvert et al.[63] and Calvert and Jhaveri[75] showed that condensation scrubber efficiency is insensitive to particle size. (Therefore, condensation scrubbing would be potentially competitive with high-energy scrubbers when high collection efficiencies for submicron particles are required.) They found also that condensation collection increases as the concentration of particles decreases.[63,75] The available moisture is shared by fewer particles, which thereby grow larger and are collected more easily than otherwise. Thermophoresis was shown by them to be of minor importance compared with diffusiophoresis and the effect of particle enlargement by condensation.[63,75]

In experiments using hydrophobic oil drop aerosols with diameters of roughly 2 μm, Jacko and Holcomb[77] determined that the penetration of a multiple-tray sieve plate scrubber decreased from 0.44 to 0.03 as steam injection was added; the steam injection ratio, the mass of water per mass of dry gas, was 0.43. Lowering the scrubber water temperature from 57°C to 15°C decreased the penetration from 0.11 to 0.05 at an injection ratio of 0.25.

Application

Humidification of a gas, by addition of steam for example, consumes energy. The use of condensation scrubbing, therefore, is most likely to be economically attractive in those applications where waste heat is available. A summary of condensation scrubbing was prepared by Calvert and Parker,[42] from which Table 19.6 is taken, showing the major industrial sources of particulate emissions for which this technique is attractive. Figure 19.12, from the same report, shows a conceptual design. An analysis

Table 19.6. Major Industrial Particulate Sources for Which Condensation Scrubbing is Attractive.[42]

INDUSTRY	SOURCE
Iron and steel	Sinter plants
	Coke manufacture
	Blast furnaces
	Steel furnaces
	Scarfing
Forest products	Wigwam burners
	Pulp mills
Lime	Rotary kilns
	Vertical kilns
Primary nonferrous	
Aluminum	Calcining
	Reduction cells
Copper	Roasting
	Reverberatory furnaces
	Converters
Zinc	Roasting
	Sintering
	Distillation
Lead	Sintering
	Blast furnaces
	Dross reverberatory furnaces
Asphalt	Paving materials
	Roofing materials
Ferroalloys	Blast furnaces
	Electric furnaces
Iron foundry	Furnaces
Secondary nonferrous metals	
Copper	Material preparation
	Smelting and refining
Aluminum	Sweating furnaces
	Refining furnaces
	Chlorine fluxing
Lead	Pot furnaces
	Blast furnaces
	Reverberatory furnaces
Zinc	Sweating furnaces
	Distillation furnaces

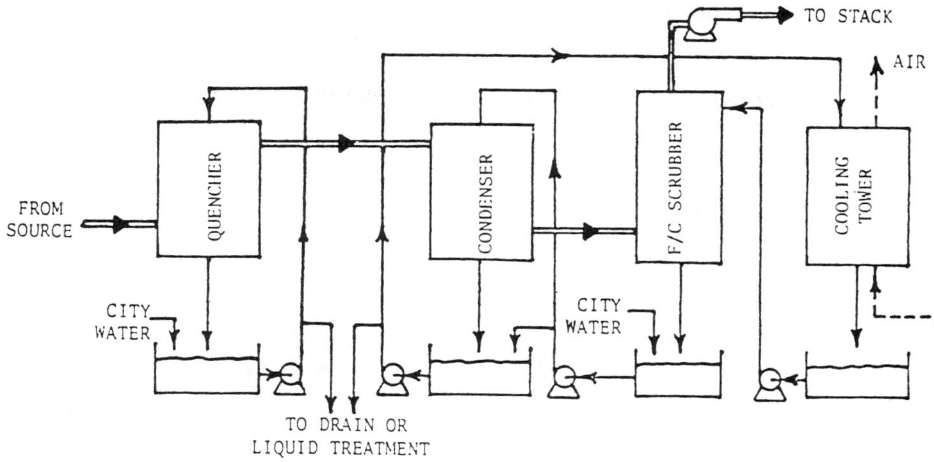

Figure 19.12. Conceptual design for a condensation scrubber.[42]

is presented in that report which concludes that condensation scrubbing would be economically superior (in both capital and operating costs) to a conventional high-energy scrubber for a grey iron cupola.

The feasibility of a "flux force/condensation" system was demonstrated in the control of emissions from a secondary metals recovery furnace, controlling a flow rate of 3.3 m^3/s maximum, using a quencher, a sieve plate column, and an entrainment separator: "The system was generally capable of 90 to 95% efficiency on particles with a mass median aerodynamic diameter of 0.75 μm."[76] The pressure drop was 7 kPa (27 in. WG) and it was estimated that a conventional high energy scrubber would have required 3 to 7 times as much pressure drop to achieve the same range of efficiencies.

The use of condensation scrubbing seams likely to increase. In some cases it would be a relatively low-cost modification to upgrade a scrubber already in operation.

19.10. ELECTROSTATIC AUGMENTATION

Introduction

Collecting particles by impaction requires accelerating the gas in which the particles are suspended to cause particle deposition due to particle inertia, an inherently inefficient approach, considering that particle mass concentrations are roughly one-thousandth or less of gas densities. Charging the scrubber surfaces and/or charging the particles produces electrostatic forces operating on the particles directly, not using the gas as an intermediary, and is inherently more energy efficient.

Theory

The two major types of electrostatic force which are significant in the collection of particles in scrubbers are the Coulomb force, which occurs when a charged particle is subjected to an electric field (such as from a charged droplet), and the induced charge ("dipole," "image") force, which is caused by the presence of an inhomogeneous field. Two other electrostatic forces can sometimes be significant: the image force due to the interaction between a charged particle and an uncharged collector and the mutual repulsion (or attraction) of the aerosol particles themselves.[23]

The Coulomb force (F_{Qq}) exerted on a particle of charge Q_p is:

$$F_{Qq} = Q_p E \qquad (19.57)$$

in which E is the electrical field created by the collector. The force due to induced polarization in the particle in an inhomogeneous field is:[23]

Table 19.7. Definitions of the Electrical Force Parameter (K) for Spherical Particles.[21,78,79]

	SPHERE	CYLINDER
Coulombic force (F_{Qq})	$\dfrac{CQ_cQ_p}{24\pi^2\epsilon_0 r_p R_c^2 \mu_G U_G}$	$\dfrac{C(Q_c/L)Q_p}{12\pi^2\epsilon_0 r_p R_c \mu_G U_G}$
Charged-collector image force (F_{Q0})	$\left(\dfrac{\epsilon_p - 1}{\epsilon_p + 2}\right)\dfrac{CQ_c^2 r_p^2}{12\pi^2\epsilon_0 R_c^5 \mu_G U_G}$	$\left(\dfrac{\epsilon_p - 1}{\epsilon_p + 2}\right)\dfrac{C(Q_c/L)^2 r_p^2}{6\pi^2\epsilon_0 R_c^3 \mu_G U_G}$
Charged-particle image force (F_{0q})	$\left(\dfrac{\epsilon_c - 1}{\epsilon_c + 2}\right)\dfrac{CQ_p^2}{24\pi^2\epsilon_0 r_p R_c^2 \mu_G U_G}$	$\left(\dfrac{\epsilon_c - 1}{\epsilon_c + 1}\right)\dfrac{CQ_p^2}{96\pi^2\epsilon_0 r_p R_c^2 \mu_G U_G}$

$$F_{q0} = \frac{1}{4\pi\epsilon_0}\chi_E V_p \operatorname{grad}(E^2) \quad (19.58)$$

in which

$$\chi_E = (3/8\pi)(\epsilon_p - 1)/(\epsilon_p + 2) \quad (19.59)$$

for spherical particles, where ϵ_p is the dielectric constant of the particle relative to the dielectric constant of a vacuum and V_p is the volume of the particle. The gradient of homogeneous electric fields is zero, so this force occurs only in inhomogeneous fields.

The collection efficiency (η) of an obstacle is defined as the area of the oncoming gas it cleans divided by the cross-sectional area it presents to the flow. Where both the particles and the collector are charged, the collection efficiency of a collector of any shape for particles assumed to have negligible inertia can be shown to be:[78,79]

$$\eta_E = 4K \quad (19.60)$$

in which K is the ratio of the particle terminal velocity, calculated for the force as evaluated at the surface of the collector, to the velocity of the free stream (U_G, the superficial or mean gas velocity). See Table 19.7 for definitions of K for spherical and cylindrical collectors.[79]

The parameter K can be used to sum up experimental and theoretical results concerning collection efficiencies for various conditions, as done in Table 19.8.[78,79] In Table 19.8 are listed the collector geometry, the force type, its radial dependence (the forms for F_{0q} are approximate), the range of K for which the efficiency expression is correct, and the efficiency, η_E. The expression "$O(K^{2/5})$" means that the efficiency is roughly $K^{2/5}$, with a correction factor of order unity which will be somewhat different depending upon the flow field. Calculation of K (Table 19.7) and the use of the

Table 19.8. Summary of Experimental, Theoretical Results for Collection Efficiencies for Electrostatic Interactions and Inertialess Particles.[21,78,79]

COLLECTOR	FORCE	R^n	K RANGE	EFFICIENCY
Sphere	F_{Qq}	R^{-2}	all	$4K$
	F_{Q0}	R^{-5}	$K \gg 1$	$O(K^{2/5})$
			$K \ll 1$	$4K$
	F_{0q}	$\sim R^{-5}$	$K \gg 1$	$O(K^{2/5})$
			$K \ll 1$	$O(K^{1/2})$
Cylinder	F_{Qq}	R^{-1}	all	πK
	F_{Q0}	R^{-3}	$K \gg 1$	$O(K^{1/3})$
			$K \ll 1$	πK
	F_{0q}	$\sim R^{-2}$	$K \gg 1$	$O(K^{1/2})$
			$K \ll 1$	$O(K^{1/2})$

material in Table 19.8 provide a simple method for estimating the electrostatic contribution to collection efficiency for these cases.

To increase the effect of electrostatics, one can charge the particles. The saturation charge due to charging a particle ($d_p \gtrsim 1$ μm) in the presence of an electric field of strength E_0 is:[80]

$$q_s = 3[\epsilon_p/(\epsilon_p + 2)] d_p^2 \pi \epsilon_0 E_0 \quad (19.61)$$

Particles much smaller than 1 μm can be charged by diffusion of ions at relatively high concentrations. (See the chapter on electrostatic precipitation.)

Applications

Three electrostatic scrubbers that have received attention are:

1. A scrubber developed at the University of Washington that uses particles charged to one polarity and droplets from spray nozzles kept at a high voltage of the opposite polarity.[81,82]
2. A scrubber developed by TRW, Inc. that uses the geometry of an electrostatic precipitator but replaces the corona-producing central wire with an array of electrohydrodynamic sprays to produce droplets that both transfer charge to the particles, so they will be captured by the grounded collector plates, and that capture the particles by impaction and electrostatic interactions.[83,84]
3. A scrubber developed by Air Pollution Systems, Inc. that uses a novel particle charging geometry[85] to produce high levels of charge on the particles, which are then collected by inertial and electrostatic forces in a venturi scrubber.

Table 19.9 (adapted from one presented in Reference 86) gives information determined by experiments done during development programs, so these results are not definitive. The power which a conventional venturi would use to produce 90% efficiency at 0.5 μm aerodynamic diameter is about 5kW/(m³/s), much greater than the power used in these devices.[86]

Compared with electrostatic precipitators of similar efficiency, high efficiency venturi scrubbers are typically smaller and less expensive in capital costs but use more power and have higher operating costs. Electrostatic scrubbers are likely to show capital and operating costs which are between those for scrubbers and those for electrostatic precipitators and should be judged by their annualized costs rather than by their power consumptions alone. (Cost comparison methodology is treated briefly at the end of this chapter). Corrosion problems and electrical isolation problems can also be significant.

19.11. DEMISTERS AND ENTRAINMENT SEPARATORS

Introduction

A scrubber uses liquid surfaces to rid the gas stream of particles. Inevitably, the scrubber produces droplets containing solid and dissolved material which must be captured before the gas

Table 19.9. Comparisons of Electrostatic Droplet Scrubbers Based on Developmental Units (Adapted from Reference 86).

ELECTROSTATIC SCRUBBER	POWER USED (W/(m³/s))	LIQUID-TO-GAS RATIO (10^{-3})	COLLECTION EFFICIENCY AT (AERODYNAMIC DIAMETER): 0.5 μm	1.0 μm
University of Washington	0.22	0.8 to 2.3	0.99	0.97
APS, Inc.	1.3	1.4	0.96	0.90
TRW, Inc.	0.5–0.7	0.12	0.90	0.85–0.95

is emitted to the atmosphere, either to meet emissions limits or to prevent damage to fans, ducting, etc. (Wetted surfaces produce droplets due to atomization or to the liquids' falling from the surfaces.) The droplet size produced will be a function of scrubber type, geometry, power consumption, and flow velocity; for example, for a packed bed cross-flow scrubber, Bell and Strauss[87] found the number mean droplet size to decrease from 400 to 100 μm as superficial velocity in the scrubber increased 50% (from 3 to 5 m/s). As liquid usage is increased, so is droplet entrainment; as scrubber energy input is increased, through increased pressure drop in the gas or increased spray nozzle pressure, the entrained liquid droplets can be expected to become smaller, their mass concentration greater. Droplets may contain captured particulate matter; even without captured matter, they dry to become fine solid particles due to dissolved minerals ("hardness") in the water. To prevent emission of material due to droplet reentrainment, the scrubber should be followed by a demister (also called an "entrainment separator" or a "mist eliminator").

The design goals for demisters were summed up by Bell and Strauss:

> In general, mist eliminators should have the following characteristics: low cost, ease of manufacture and installation, low pressure drop, and high efficiencies over a wide range of superficial gas velocities and mist loadings. The units should be self-draining and self-cleaning, with low operating and maintenance charges, able to operate for long periods without attention.[87]

Entrainment separator design can be improved by using guidelines recently published:[109, 110] For fibrous beds or packed beds, optimal efficiency at a fixed pressure drop (or minimum pressure drop at a fixed efficiency) can be obtained by choosing a collector element size and collector face velocity such that the impaction parameter is approximately one for the droplet size of interest.

Types

Droplets formed from atomization from scrubber surfaces have number mean diameters $\sim 10^2$–10^3 μm. Generally, the higher the gas velocity, the smaller the droplets. Droplets from hydraulic spray scrubbers will be similar to the spray. In cases where a mist forms from condensation of water vapor, the droplets will be $\lesssim 1$ μm; these are much more difficult to collect and are not discussed further here. (This condition should be avoided.) For the larger droplets, the collection mechanisms which come into use are: gravity settling, centrifugal collection, and impaction.[88]

Centrifugal collection or inertial impaction are really the same collection mechanism: the gas stream has its direction (and sometimes speed) changed, and the droplet's inertia gives it a velocity component toward the collector wall, perpendicular to the mean gas flow. Devices operating on this principle include cyclones, baffled chambers (using chevrons, corrugated sheets, etc.) and packed beds, with packing material of many geometries and a wide range of characteristic collector dimensions and volume void fractions.

Pressure Drop

The pressure drop across the mist eliminator can be identified as friction drag and form drag, proportional to velocity and velocity squared, respectively. The pressure drop as a function of superficial velocity (gas volume flow divided by scrubber cross-sectional area before any baffles or obstacles are introduced) will be of the form:

$$\Delta P = aU_G + bU_G^2 \tag{19.62}$$

As the gas flow Reynolds number in the scrubber increases the U_G^2 term will predominate. Packed bed pressure drop can be estimated using the Ergun Equation (19.23) and predictive correlations are available for cyclones.[89] The pressure drop required will be determined by the collection efficiency needed, so the relationship between pressure drop and efficiency is discussed next.

Collection Efficiency

"Primary" collection efficiency is the fraction of the droplets which are caught on collection surfaces. Total collection efficiency is the fraction of droplets that are retained by the mist eliminator. The difference is due to reentrainment of the captured droplets.

Inertial collection of a spray droplet is correlated with the droplet impaction parameter. Thus, demister collection efficiency can be expected to change as a function of droplet size. Calculation of the total efficiency requires integrating the collection efficiency as a function of droplet size over the droplet size distribution. The method Calvert[9] described for obtaining the total collection efficiency for droplets (or aerosols) assumed to be log-normally distributed in droplet size with known mass median diameter and geometric standard deviation has been presented above (See Fig. 19.11). Figure 19.13 shows the droplet cut diameter as a function of entrainment separator pressure drop for several types of separators (note: 1 cm WC = 98 Pa).[90] The power advantage of the wire mesh is apparent, although this may be offset by cleaning/plugging problems.

Porous fibrous structures or wire meshes are often used as mist eliminators. Strauss[91] presented Table 19.10 (from Griwatz et al.[92]). The eliminator "type" descriptions were:

1. 20-μm diameter Teflon (DuPont) fibers combined with 152-μm wire;
2. American Air Filter Type T bonded fiberglass;
3. Mine Safety Appliances bonded fiberglass;
4. 9-μm diameter fiberglass mixed and knitted with 127-μm wire (Mine Safety Appliances);
5. knitted wire mesh (Farr Type 68-44 MHZ).

Reentrainment of Droplets

Increasing the velocity increases the pressure drop and may lead to increased collection efficiency. Beyond some velocity, however, flooding of the separator or reentrainment of droplets from the separator surfaces will produce an increase in emissions, thus an apparent decrease in efficiency. Approximate values of superficial velocity at which entrainment begins are given below:[88]

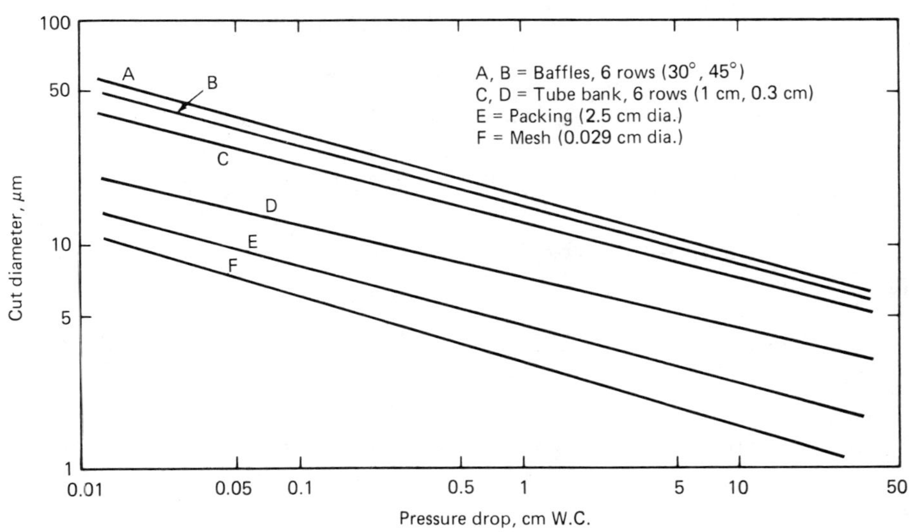

Figure 19.13. Entrainment separator performance cut diameters.[90]

Table 19.10. Operating Characteristics and Efficiencies for Fiber Mist Eliminators.[91,92]

TYPE (SEE TEXT)	1	2	3	4	5
Bed depth (mm)	67	61	125	125	100
Flow velocity (m/s)	2.0	1.44	2.0	2.0	2.26
Pressure drop (Pa)	322–555	200–300	250–475	250–425	65–87
Efficiency (%)					
at 100 μm	26	100	100	100	100
at 10 μm	36	100	100	100	90
at 0.6 μm	31	7	20	22	1
at 0.3 μm	7	5	4	–	0

SEPARATOR	GAS VELOCITY, m/s
Zigzag with upward gas flow and horizontal baffles	3.7–4.6
Zigzag with horizontal gas flow and vertical baffles	4.6–6.1
Cyclone (gas inlet velocity)	30.5–39.6
Knitted mesh with vertical gas flow	3.1–4.6
Knitted mesh with horizontal gas flow	4.6–7.0
Tube bank with vertical gas flow	3.7–4.9
Tube bank with horizontal gas flow	5.5–7.0

As Calvert concluded, "Liquid drainage is best when the gas flow is horizontal and collection surfaces are near vertical; also, with this configuration, reentrainment occurs at higher flow rates than for horizontal elements."[88]

Plugging

Captured solids and solute material that precipitates from solution can build up on scrubber and entrainment separator surfaces, leading to increased flow resistance. Recirculation of the scrubbing liquid will aggravate this condition. Some actions which may help reduce this problem are:

1. Reduce slurry concentrations.
2. Design collection elements to have nearly vertical surfaces.
3. Provide for washing of the collection surface.
4. Avoid drying of the surfaces; if scrubber is shut off, clean before reusing.
5. Design using geometries having larger rather than smaller minimum flow path dimensions (i.e., choose a chevron over a knitted mesh, other things being equal).

Summary

Mist eliminators ("demisters") are needed for almost all scrubbers. They capture droplets by inertial mechanisms, generally with superficial velocities (except for cyclones) less than 7 m/s, to prevent reentrainment. Horizontal gas flow is preferable to vertical flow, but plugging can be a problem for either, especially where dissolved solids are being used to scrub gases from the effluent stream. For more detailed information, see the work by Strauss.[91]

19.12. SUNDRY DESIGN CONSIDERATIONS

Introduction

Covered here are several factors which should be taken into consideration in design but that did not fit conveniently into other sections of this chapter.

Corrosion

Corrosion problems are specific to the particular source type under control.

Case histories of scrubber applications and problems in the metallurgical industry were presented by Steiner and Thompson:[93] Abrasion of mild steel piping used to carry slurry occurred in a venturi used to control gaseous and particulate emissions from a boiler; the problem was cured by using rubber-lined piping and valving. In a sinter plant application, corrosion of carbon steel in the liquid flow lines was a problem, perhaps due to inadequate *p*H control; no such problem occurred in the air flow passages, where type 304 stainless steel was used. In a third situation, control of an open-hearth furnace, nonstainless steel components were corroded severely; lime neutralization led to problems of scaling due to calcium sulfate, later mitigated by switching to caustic for neutralization.

Hoxie and Tuffnell summarized extensive tests in scrubbers used for flue gas desulfurization:[94] carbon steel and type 304L stainless steel were inadequate in the wet areas, and type 316L steel was occasionally attacked, specifically by certain combinations of *p*H and chloride concentration. They presented detailed information for more than a dozen steels.

Three options for corrosion protection were identified by Busch et al.:[95] liners, different materials of construction, thicker materials. They presented cost comparisons for various steels and a steel and rubber liner combination.

Further information on corrosion control may be obtained from the National Association of Corrosion Engineers, 2400 West Loop South, Houston, Texas 77027, U.S.A.

Wetting Agents

There is some belief among pollution control engineers that the addition of wetting agents to the scrubbing liquid can improve scrubber performance.[108] In certain cases, it is certainly possible that the droplet size distribution of the hydraulic or atomized spray will become somewhat better suited to scrubbing the aerosol, but it is as likely that the size distribution will become less suited. It seems preferable to change the nozzles, the flow rates, or the pressures than to introduce wetting agents to the liquid, which will represent added materials expense and perhaps added water pollution control costs.

In some cases, the scrubbing improvement noted in scrubbing wettable versus nonwettable particles may have been due to hygroscopic growth of the former. Experiments have shown that wettable and nonwettable particles are caught with equal efficiency by drops (at a given impaction parameter value), except in the rare instances where nonwetting particles coat the droplet to the degree that other particles strike them and are not retained.[96]

Scale-Up

Even some companies with extensive experience with scrubbers have made it a policy to use pilot scale scrubbers to help design the full-scale scrubber to be used in a particular application.[97] Even so, some assumptions must be made in scaling up the results of such tests.

Two sets of investigators[98,99] found improved performance in larger scrubbers at a given pressure drop, perhaps due to increased turbulence at the higher Reynolds numbers. On the other hand, Behie and Beeckmans[100] reviewed many previous investigations and concluded there were no appreciable effects due to scaling up a scrubber.

Water Pollution

As water quality standards and water pollution control requirements become more stringent, scrubber design must increasingly take water treatment into account, influencing water usage rates, recycling rates, construction materials, and additive selection (such as for *p*H control). This is well beyond our scope, however.

Mechanical Aids

The use of wetted fans or other blade-type mechanical methods for disintegrating droplets and capturing particles is not particularly attractive; collection efficiency is not quite as good for such scrubbers as for venturi scrubbers at the same power consumption and problems of corrosion, erosion, and vibration are inherent in such designs.[27]

19.13. COSTS

A great many factors contribute to the total cost of a scrubber. Figure 19.14 shows a generalized cost evaluation scheme.[1] The source and its operating characteristics will influence the choice of control type, its capacity, efficiency, construction materials, and thus the costs of control. Handling the collected materials is costly, though there may be salvage value. Note that the cost of the control hardware is only a part of the total cost, especially for high-energy scrubbers.

One approach for cost comparisons of various particulate control options is that described by Edmiston and Bunyard.[101] The goal is to develop a single cost parameter, here the *total annualized cost*, with which to compare different air pollution control devices. This is quite useful because, for example, electrostatic precipitators have relatively high initial costs and relatively low operating costs in comparison to scrubbers of similar collection efficiency.

The costs can be divided into three categories:[101]

1. *Capital investment cost.* This includes the control hardware cost, the cost of auxiliary equipment and the cost of installation, including initial studies.
2. *Maintenance and operating costs.* These are taken on a yearly basis, averaged over the life of the equipment.
3. *Capital charges.* These are what it costs to borrow the money equivalent to the capital investment, plus taxes and insurance.

To convert these various costs into a single number, the total annualized cost, one sums the annual capital investment depreciation, the operating and maintenance costs, and the capital charges.

The usual method of depreciation in such contexts is to assume straight-line depreciation: Estimate the life of the equipment, Edmisten and Bunyard[101] suggested 15 years, and figure the yearly depreciation as the capital investment divided by the life expectancy. Thus, the total annualized cost is given by the sum of capital investment divided by the lifetime plus yearly maintenance and operating costs and

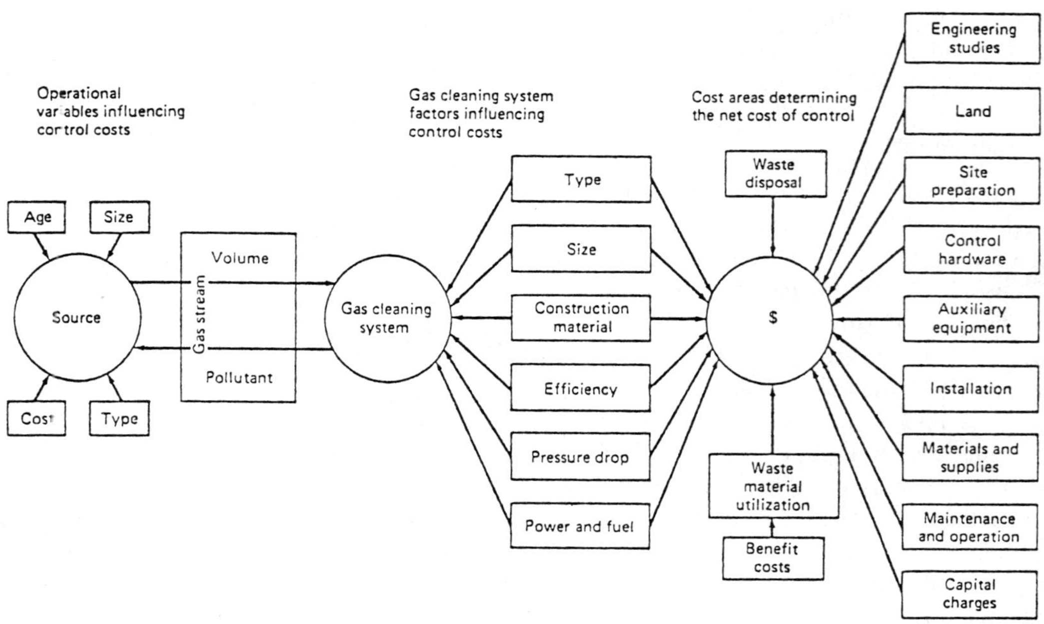

Figure 19.14. Diagram of a cost evaluation scheme for a pollutant control system.[1]

capital charges. Generally, maintenance and capital charges will be nearly proportional to the capital investment. Table 19.11 is based on a survey done as part of the preparation of the *Scrubber Handbook*[6] and allows one to make a rough estimate of the installed cost of a scrubber, based on the current Marshall and Stevens Index. The fixed capital investment is about three times the installed cost.[31] Table 19.12 shows the conditions that can affect the installed costs of control devices, factors that are reflected in the ranges attributed to costs in Table 19.11.

For more details on conventional control device costs, see the article by Edmisten and Bunyard[101] and the articles by Hanf and MacDonald[104] and by Fraser and Eaton[103] and Neveril et al.[104] who presented graphs and equations for estimating the prices for electrostatic precipitators, venturi scrubbers, fabric filters, incinerators, and absorbers, as well as the costs of auxiliary equipment, ductwork and dampers, and such other costs as operating, maintenance, and installation.

Although much of the necessary information on costs will have to be obtained from manufacturers for a specific application, one can readily estimate power costs. Power consumption figures are often given in terms of kW per m^3/s flow rate or horsepower per 1000 actual cubic feet per minute flow rate. (Note: 1 hp = 0.746 kW; 1000 acfm = 0.472 m^3/s). When the power is given as hydraulic power (pressure drop times volume flow rate), a pump/fan/motor efficiency factor must be used (as a divisor) to convert to actual electrical power; this efficiency factor is generally about 0.6, whether fans are moving gas or pumps are moving liquid. The power cost is given by the product of: volume rate of gas flow, power consumption per unit flow of gas, cost per unit of energy, and operating time.

Certain forms of power may be nearly free: the recovery of waste heat is free with regard to operating costs, although it will add to the capital investment and to the costs associated with capital investment. As with other costs, the power costs will vary considerably from situation to situation.

Specific circumstances will also greatly affect waste disposal costs. Scrubbers produce waste water that must be handled properly; wastewater treatment produces solid wastes that must be used or disposed. Generally the final phase is to convert the captured material into a solid for such uses as landfill or to recycle some or all of the captured material. Solid waste disposal cost can be broken down into costs of hauling and cost of disposal. Hauling costs are dependent on the type of equipment, length of hauls, type of route, and traffic encountered, and the number of employees necessary. Cost of disposal usually means the cost of a sanitary landfill. Components of sanitary landfill cost are cost of site, degree of compaction, and cost of developing such things as access roads, water supply, fences, landscaping, water runoff diversion facilities, etc.

Table 19.11. Reported Costs of Complete, Installed Scrubber Systems.[31]

	MEAN COST/acfm[a,b] (AT acfm LISTED)					
SCRUBBER TYPE	1000	10,000	50,000	100,000	HIGH/MEAN	LOW/MEAN
Venturi	\$14.00	\$5.50	\$3.00	\$2.20	3	$\frac{1}{3}$
Packed bed	\$14.00	\$3.00	\$0.80	–	3	$\frac{1}{3}$
Spray	\$50.00	\$5.00	\$1.00	\$0.70	2	$\frac{1}{2}$
Centrifugal	\$3.00	\$1.30	\$0.70	–	2	$\frac{1}{2}$
Impingement and entrainment	\$8.00	\$3.50	\$2.00	\$1.50	1.5	0.7
Mobile bed	–	\$3.00	\$2.00	–	1.5	0.7

[a]Costs are for Marshall and Stevens Index of about 280.
[b]acfm: Actual cubic feet per minute. (1000 acfm = 0.47 m^3/s.)

Table 19.12. Conditions Affecting Cost of Control Devices Installed.[105]

COST CATEGORY	LOW COST	HIGH COST
Equipment transportation	Minimum distance; simple loading and unloading procedures	Long distance; complex procedure for loading and unloading
Plant age	Hardware designed as an integral part of new plant	Hardware installed into confines of old plant requiring structural or process modification or alteration
Available space	Vacant area for location of control system	Little vacant space requires extensive steel support construction and site preparation
Corrosiveness of gas	Noncorrosive gas	Acidic emissions requiring high alloy accessory equipment using special handling and construction techniques
Complexity of start-up	Simple start-up, no extensive adjustment required	Requires extensive adjustments; testing; considerable downtime
Instrumentation	Little required	Complex instrumentation required to assure reliability of control or constant monitoring of gas stream
Guarantee on performance	None needed	Required to assure designed control efficiency
Degree of assembly	Control hardware shipped completely assembled	Control hardware to be assembled and erected in the field
Degree of engineering design	Autonomous "package" control system	Control system requiring extensive integration into process, insulation to correct temperature problem, noise abatement
Utilities	Electricity, water, waste disposal facilities readily available	Electrical and waste treatment facilities must be expanded, water supply must be developed or expanded
Collected waste material handling	No special treatment facilities or handling required	Special treatment facilities and/or handling required
Labor	Low wages in geographical area	Overtime and/or high wages in geographical area

REFERENCES

1. A. E. Vandergrift, L. J. Shannon, E. W. Lawless, P. G. Gorman, E. E. Sallee, and M. Reichel, "Particulate Systems Study," vol. III, *Handbook of Emission Properties.* APTD-0745 (NTIS PB 203 522), US EPA (1971).
2. W. Strauss, *Industrial Gas Cleaning*, Pergamon, New York (1966).
3. M. W. First, Harvard School of Public Health, Boston, Massachusetts (1979).
4. L. J. Shannon, P. G. Gorman, and M. Reichel, "Particulate Pollutant Systems Study," vol. II, *Fine Particle Emissions.* APTD-0744 (NTIS PB 203 522), US EPA (1971).
5. Courtesy of the Industrial Gas Cleaning Institute, Alexandria, Virginia.
6. S. Calvert, J. Goldschmid, D. Leith, and D. Mehta, *Scrubber Handbook*, US EPA, NTIS PB 213 016 (1972).
7. K. Semrau and C. L. Witham, *Wet Scrubber Liquid Utilization.* Stanford Research Institute, Menlo Park, Calif., EPA-650/2-74-108, US EPA (October, 1974).
8. K. T. Semrau, C. L. Witham, and W. W. Kerlin, *Energy Utilization by Wet Scrubbers.* EPA-600/2-77-234, US EPA (1977).
9. S. Calvert, "Engineering Design of Wet Scrubbers," *J. Air Pollut. Contr. Assn. 24*:929–934 (1974).
10. D. W. Cooper, "Theoretical Comparison of Efficiency and Power for Single-Stage and Multiple-Stage Particulate Scrubbing," *Atmos. Environ. 10*:1001–1004 (1976).

11. D. M. Muir and Y. Miheisi, "Comparison of the Performance of a Single- and Two-stage Variable-throat Venturi Scrubber," *Atmos. Environ. 13*: 1187–1196 (1979).
12. D. W. Cooper, "Optimizing Venturi Scrubber Performance Through Modeling," Presented at 2d Symposium on Transfer and Utilization of Particulate Control Technology, Denver, Particulate Control Technology, Denver, Colo., 23–27 July 1979, sponsored by US EPA (1979).
13. S. C. Yung and S. Calvert, *Particulate Control Highlights*: *Performance and Design Model for Scrubbers.* EPA-600/8-78-005b, US EPA (1978).
14. J. Aitchison and J. A. C. Brown, *The Log-Normal Distribution*, Cambridge University Press, Cambridge (1957).
15. G. Kubie, "A Note on the Treatment of Impactor Data for Some Aerosols," *J. Aerosol Sci. 2*:23–30 (1971).
16. S. Calvert, D. Lundgren, and D. S. Mehta, "Venturi Scrubber Performance," *J. Air Pollut. Contr. Assn. 22*:529–532 (1972).
17. J. L. Held and D. W. Cooper, "Theoretical Investigation of the Effects of Relative Humidity on Aerosol Respirable Fraction," *Atmos. Environ. 13*:1419–1425 (1979).
18. P. Knettig and J. M. Beeckmans, "Inertial Capture of Aerosol Particles by Swarms of Accelerating Spheres," *J. Aerosol Sci. 5*:225–233 (1974).
19. L. E. Sparks, "The Effect of Scrubber Operating and Design Parameters on the Collection of Particulate Air Pollutants," Ph.D. dissertation (Civil Engineering), University of Washington, Seattle (1971).
20. H. F. George and G. W. Poehlein, "Capture of Aerosol Particles by Spherical Collectors: Electrostatic, Inertial, Interception, and Viscous Effects," *Env. Sci. Technol. 8*:46–49 (1974).
21. K. A. Nielsen and J. C. Hill, "Collection of Inertialess Particles on Spheres With Electrical Forces," *Ind. Eng. Chem., Fundam.* 15:149–157 (1976). "Capture of Particles on Spheres by Inertial and Electrical Forces," *Ind. Eng. Chem., Fundam.* 15:157–163 (1976).
22. I. Langmuir and K. B. Blodgett, "A Mathematical Investigation of Water Droplet Trajectories," U.S. Army Air Forces Tech. Report. 5418, (NTIS: PB 27565) (February, 1946).
23. N. A. Fuchs, *Mechanics of Aerosols*, Pergamon, New York (1964).
24. D. W. Cooper, "Fine Particle Control by Electrostatic Augmentation of Existing Methods," Paper 75-02.1 presented at the 68th Annual Meeting of the Air Pollution Control Assoc., Boston, Mass. (1975).
25. H. Krockta and R. L. Lucas, "Information Required for the Selection and Performance Evaluation of Wet Scrubbers," *J. Air Pollut. Contr. Assn. 22*:459–462 (1972).
26. R. H. Perry and C. H. Chilton, *Chemical Engineers' Handbook*, 5th ed., McGraw-Hill, New York (1973).
27. S. Calvert, "How to Choose a Particulate Scrubber," *Chem. Eng.*, pp. 54–68 (August 29, 1977).
28. S. C. Yung, H. F. Barbarika, and S. Calvert, "Pressure Loss in Venturi Scrubbers," *J. Air Pollut. Contr. Assn. 27*:348–351 (1977).
29. S. Yung, S. Calvert, and H. F. Barbarika, "Venturi Scrubber Performance Model," EPA-600/2-77-172, US EPA (1977).
30. S. Calvert, "Source Control by Liquid Scrubbling," in A. C. Stern, ed., *Air Pollution*, Academic Press, New York (1968).
31. S. Calvert, "Scrubbing," in A. C. Stern, ed., *Air Pollution*, Academic Press, New York (1977).
32. H. F. Johnstone, R. B. Field, and M. C. Tassler, "Gas Absorption and Aerosol Collection in Venturi Scrubber," *Ind. Eng. Chem. 46*:1602–1608 (1954).
33. S. Calvert, "Venturi and Other Atomizing Scrubbers Efficiency and Pressure Drop," *A.I.Ch.E. Journal 16*:392–396 (1970).
34. R. H. Boll, "Particle Collection and Pressure Drop in Venturi Scrubbers," *Ind. Eng. Chem. Fund. 12*:40–50 (1973).
35. M. Taheri and C. M. Shieh, "Mathematical Modeling of Atomizing Scrubbers," *A.I.Ch.E. Journal 21*(1):153–157 (1975).
36. K. C. Goel and K. G. T. Hollands, "Optimum Design of Venturi Scrubbers," *Atmos. Environ. 11*:837–845 (1977).
37. S. Calvert, D. Lundgren, and D. S. Mehta, "Venturi Scrubber Performance," *J. Air Pollut. Control Assn. 22*:529–532 (1972).
38. L. E. Sparks, "SR-52 Programmable Calculator Programs for Venturi Scrubbers and Electrostatic Precipitators," EPA 600/7-78-026, Office of Research and Development, US EPA (March, 1978).
39. D. Leith and D. W. Cooper, "Venturi Scrubber Optimization," *Atmos. Environm. 14*:657–664 (1980).
40. F. O. Ekman and H. F. Johnstone, "Collection of Aerosols in a Venturi Scrubber," *Ind. Eng. Chem. 43*:1358–1370 (1951).
41. D. M. Muir, C. D. Grant, and Y. Miheisi, "Relationship between Collection Efficiency and Energy Consumption of Wet Dust Collectors," *Filtrn. and Sepn. 15*:332–340 (1978).
42. S. Calvert and R. Parker, "Particulate Control Highlights: Flux Force/Condensation Wet Scrubbing," EPA-600/8-78-005c, US EPA (June, 1978).

43. M. Taheri, S. A. Beg, and M. Beizie, "Gas Cleaning in a Wetted Butterfly Valve," *J. Air Pollut. Contr. Assn. 22*:794–798 (1972).
44. L. S. Harris, "Fume Scrubbing with the Ejector Venturi System," *Chem. Eng. Progress 62*:55–59 (1966).
45. H. E. Gardenier, "Submicron Particulate Scrubbing with a Two Phase Jet Scrubber." *J. Air Pollut. Contr. Assn. 24*:954–957 (1974).
46. D. W. Cooper and D. P. Anderson, "Dynactor Scrubber Evaluation," EPA-650/2-74-083a, US EPA (June, 1975).
47. S. Calvert, N. C. Jhaveri, and S. Yung, "Fine Particle Scrubber Performance Tests," EPA-650/2-74-093, US EPA (October, 1974).
48. R. B. Bird, W. E. Stewart, and E. N. Lightfoot, *Transport Phenomena*, Wiley & Sons, New York (1960).
49. J. D. Brady, D. W. Cooper, and M. T. Rei, "A Wet Collector of Fine Particles," *Chem. Eng. Prog. 73*(8):45–53 (1977).
50. J. Happel and H. Brenner, *Low Reynolds Number Hydrodynamics*, Prentice-Hall, Englewood Cliffs (1965).
51. C. N. Davies, *Air Filtration*, Academic Press, New York (1973).
52. K. Iinoya and C. Orr, Jr., "Filtration," in A. C. Stern, ed., *Air Pollution*, Academic Press, New York (1977).
53. C. Y. Chen, "Filtration of Aerosols by Fibrous Media." *Chem. Rev. 55*:595–623 (1955).
54. K. R. May and R. Clifford, "The Impaction of Aerosol Particles on Cylinders, Spheres, Ribbons, and Discs." *Ann. Occup. Hyg. 10*:83–95 (1967).
55. M. Taheri and S. Calvert, "Removal of Small Particles from Air by Foam in a Sieve-plate Column," *J. Air Pollut. Contr. Assn. 18*:240–245 (1968).
56. B. S. Javorsky, "Gas Cleaning with the Foam Scrubber," *Filtrn. and Sepn. 9*:173 (1972).
57. B. Javorsky, "Fume Control and Gas Cleaning with an Industrial Scale Foam Bed Scrubber," *Filtrn. and Sepn. 10*:21 (1973).
58. T. E. Ctvrtnicek, H. H. S. Yu, C. M. Moscowitz, and G. H. Ramsey, "Fine Particulate Control Using Foam Scrubbing," in T. Ariman, ed., *Novel Concepts and Advanced Technology in Particulate-Gas Separation*, University of Notre Dame, Notre Dame, Ind. (1978).
59. G. Ramsey, "Evaluation of Foam Scrubbing as a Method for Collecting Fine Particulate," EPA-600/2-77-197, US EPA (September, 1977).
60. P. J. Schauer, "Removal of Submicron Aerosol Particles from a Moving Gas Stream," *Ind. Eng. Chem. 43*(9):1532–1538 (July, 1951).
61. C. E. Lapple and H. J. Kamack, "Performance of Wet Dust Scrubbers," *Chem. Eng. Prog. 51*:110–121 (1955).
62. K. T. Semrau, "Dust Scrubber Design–A Critique on the State of the Art," *J. Air Pollut. Contr. Assn. 13*:587–594 (1963).
63. S. Calvert, J. Goldschmid, D. Leith, and N. C. Jhaveri, "Feasibility of Flux Force/Condensation Scrubbing for Fine Particulate Collection," APT, Inc., Riverside, Calif., EPA-650/5-73-076, US EPA (1973).
64. L. Waldmann and K. H. Schmitt, "Thermophoresis and Diffusiophoresis of Aerosols," in C. N. Davies, ed. *Aerosol Science*, Academic Press, New York (1966).
65. K. T. Semrau, C. W. Marynowski, K. E. Lunde, and C. E. Lapple, "Influence of Power Input on Efficiency of Dust Scrubber," *Ind. Eng. Chem. 50*:1615–1620 (1958).
66. L. E. Sparks and M. J. Pilat, "Effect of Diffusiophoresis on Particle Collection by Wet Scrubbers," *Atmos. Environ. 4*:651–660 (1970).
67. W. G. N. Slinn and J. M. Hales, "A Re-evaluation of the Role of Thermophoresis as a Mechanism of In- and Below-cloud Scavenging," *J. Atmos. Sci. 28*:1465–1471 (1971).
68. M. J. Pilat and A. Prem, "Effect of Diffusiophoresis and Thermophoresis on the Overall Particle Collection Efficiency of Spray Droplet Scrubbers," *J. Air Pollut. Contr. Assn. 27*:982–988 (1977).
69. P. J. Whitmore, "Diffusiophoretic Particle Collection Under Turbulent Conditions," Ph.D. thesis, University of British Columbia, Canada (1976).
70. S. Calvert and R. Parker, "Particulate Control Highlights: Fine Particle Scrubber Research," EPA-600/8-78-005a, US EPA (June, 1978).
71. C. Orr, Jr., F. K. Hurd, and W. J. Corbett, "Aerosol Size and Relative Humidity," *J. Coll. Sci. 13*:472–482 (1958).
72. M. Neiburger and M. G. Wurtele, "On the Nature and Size of Particles in Haze, Fog and Stratus of the Los Angeles Region," *Chem. Rev. 44*:321–335 (1949).
73. D. W. Cooper, D. W. Underhill, and M. J. Ellenbecker, "A Critique of the U.S. Standard for Industrial Exposure to Sodium Hydroxide Aerosols," *Am. Indus. Hyg. Assn. J. 40*:365–371 (1979).
74. B. W. Lancaster and W. Strauss, "A Study of Stream Injection into Wet Scrubbers," *Ind. Eng. Chem. Fund. 10*:362–369 (1971).
75. S. Calvert and N. C. Jhaveri, "Flux Force/Condensation Scrubbing," *J. Air Pollut. Contr. Assn. 24*:947–952 (1974).
76. S. Calvert, S. Gandhi, D. L. Harmon, and L. E. Sparks, "FF/C Scrubber Demonstration on a Secondary Metals Recovery Furnace," *J. Air Pollut. Contr. Assn. 27*:1076–1080 (1977).
77. R. B. Jacko and M. L. Holcomb, "A Parametric

Study of Flux Force/Condensation Scrubber for the Removal of Fine Hydrophobic Particles." Paper 78-17.2 presented at the 71st Annual Meeting of the Air Pollution Control Association, Houston, Tex. (June, 1978).

78. D. W. Cooper, "Approximate Equations for Predicting Electrostatic Particle Collection," in T. Ariman, ed., *Novel Concepts and Advanced Technology in Particulate-Gas Separation*, University of Notre Dame, Notre Dame, Ind. (1978).
79. K. A. Nielsen, "Written Discussion," in T. Ariman, ed., *Novel Concepts and Advanced Technology in Particulate-Gas Separation*, University of Notre Dame, Notre Dame, Ind. (1978).
80. S. Oglesby, Jr. and G. B. Nichols, "Electrostatic Precipitation," in A. C. Stern, ed., *Air Pollution*, Academic Press, New York (1977).
81. M. J. Pilat, S. A. Jaasund, and L. E. Sparks, "Collection of Aerosol Particles by Electrostatic Droplet Spray Scrubbers," *Env. Sci. Technol.* *4*:360-362 (1974).
82. M. J. Pilat, "Collection of Aerosol Particles by Electrostatic Droplet Spray Scrubbers," *J. Air Pollut. Contr. Assn. 25*:176-178 (1975).
83. C. W. Lear, W. F. Krieve, and E. Cohen, "Charged Droplet Scrubbing for Fine Particle Control," *J. Air Pollut. Contr. Assn. 25*:184-189 (1975).
84. S. Calvert, S. C. Yung, H. Barbarika, and R. G. Patterson, "Evaluation of Four Novel Fine Particulate Collection Devices," EPA-600/2-78-062, US EPA, March (1978).
85. M. T. Kearns, "High Intensity Ionization Applied to Venturi Scrubbing," *J. Air Pollut. Contr. Assn. 29*:383-385 (1979).
86. D. C. Drehmel, "Advanced Electrostatic Collection Concepts," *J. Air Pollut. Contr. Assn. 27*: 1090-1092 (1977).
87. C. G. Bell and W. Strauss, "Effectiveness of Vertical Mist Eliminators in a Cross Flow Scrubber," *J. Air Pollut. Contr. Assn. 23*:967-969 (1973).
88. S. Calvert, "Guidelines for Selecting Mist Eliminators," *Chem. Eng.*, 109-112 (February 27, 1978).
89. D. Leith and D. Mehta, "Cyclone Performance and Design," *Atmos. Environ.* 7:527-549 (1973).
90. S. Calvert and R. Parker, "Particulate Control Highlights: Fine Particle Scrubber Research," EPA-600/8-78-005a, US EPA (June, 1978).
91. W. Strauss, "Mist Eliminators," in A. C. Stern, ed., *Air Pollution*, Academic Press, New York (1977).
92. G. H. Griwatz, J. V. Friel, and J. L. Creehouse, Report 71-45, U.S. Atomic Energy Commission, Mine Safety Applicances Research Corp., Evans City, Pa. (1971).
93. B. A. Steiner and R. J. Thompson, "Wet Scrubbing Experience for Steel Mill Applications," *J. Air Pollut. Contr. Assn. 27*:1069-1075 (1977).
94. E. C. Hoxie and G. W. Tuffnell, "A Summary of INCO Corrosion Tests in Power Plant Flue Gas Scrubbing Processes," in *Resolving Corrosion Problems in Air Pollution Equipment.* National Association of Corrosion Engrs., Houston, Tex. (1976).
95. J. S. Busch, W. E. MacMath, and M. S. Lin, "Design and Cost of High Energy Scrubbers: 1. The Basic Scrubber," *Pollut. Engrg.*, pp. 28-32 (January, 1973).
96. L. D. Stulov, F. I. Murashkevich, and N. A. Fuchs, "The Efficiency of Collision of Solid Aerosol Particles with Water Surfaces," *J. Aerosol Sci. 9*:1-6 (1978).
97. R. W. McIlvaine, "When to Pilot and When to Use Theoretical Predictions of Required Venturi Pressure Drop." Paper 77-17.1 presented in the 70th Annual Meeting of the Air Pollution Control Association, Toronto, Canada (1977).
98. M. Taheri, S. A. Beg, and M. Beizie, "The Effect of Scale-up on the Performance of High Energy Scrubbers," *J. Air Pollut. Contr. Assn. 23*:963-966 (1973).
99. N. S. Balakreshnan and G. H. S. Cheng, "Scale-up Effect of Venturi Scrubber." Paper 78-17.3 presented at the 71st Annual Meeting of the Air Pollution Control Association, Houston, Tex., (June, 1978).
100. S. W. Behie and J. M. Beeckmans, "Effects of Water Injection Arrangement on the Performance of a Venturi Scrubber," *J. Air Pollut. Contr. Assn. 24*:943-945 (1974).
101. N. G. Edmisten and F. L. Bunyard, "A Systematic Procedure for Determining the Cost of Controlling Particulate Emissions from Industrial Sources," *J. Air Pollut. Contr. Assn. 20*:446-452 (1970).
102. E. M. Hanf and J. W. MacDonald, "Economic Evaluation of Wet Scrubbers," *Chem. Eng. Prog.* 7(3):48-52 (1975).
103. M. D. Fraser and D. R. Eaton, "Cost Models for Venturi Scrubber System." Presented at 68th Annual Meeting of the Air Pollution Control Association, Boston (1975).
104. R. B. Neveril, J. U. Price, and K. L. Engdahl, "Capital and Operating Costs of Selected Air Pollution Control Systems-I.-V." *J. Air Pollut. Contr. Assn. 28*:829-836, 963-968, 1069-1072, 1171-1174, 1253-1256 (1978).
105. A. C. Stern, H. C. Wohlers, R. W. Boubel, and W. P. Lowry, *Fundamentals of Air Pollution*, Academic Press, New York (1973).
106. D. W. Cooper, "On the Products of Lognormal and Cumulative Lognormal Particle Size Distributions," *J. Aerosol Sci. 13*:111-120 (1982).

107. K. W. Lee and J. A. Gieseke, "A Note on the Approximation of Interceptional Collection Efficiencies," *J. Aerosol Sci.*, *11*:335-341 (1980).
108. D. S. F. Atkinson and W. Strauss, "Droplet Size and Surface Tension in Venturi Scrubbers," *J. Air Pollut. Contr. Assn.* *28*:1114-1118 (1978).
109. D. W. Cooper, "Filter Beds: Energy-Efficient Packing Diameter," *J. Air Polut. Contr. Assn.*, *32*:205-208 (1982).
110. D. W. Cooper, "Optimizing Filter Fiber Diameter," *Atmos. Environ.*, *16*:1529-1533 (1982).
111. T. D. Placek and L. K. Peters, "Analysis of Particulate Removal in Venturi Scrubbers–Effect of Operating Variables on Performance," *AICHE J.*, *27*:984-993 (1981).
112. L. P. Bayvel, "The Effect of the Polydispersity of Drops on the Efficiency of a Venturi Scrubber," *TransICHemE*, *60*:31-34 (1982).

LIST OF SYMBOLS

a	Subscript: aerodynamic; coefficient in Equation (19.62), N-s/m^3
b	Coefficient in Equation (19.62) N-s^2/m^4
c	Subscript: collector, cut
d	Diameter, m; subscript: droplet
e	Void volume fraction
f	Subscript: fan; empirical parameter for venturi efficiency
g	Subscript: geometric mean; gravitational acceleration, 9.8 m/s^2
h	Length, m; subscript: hole
i	Index number; subscript: particle size interval
k	Kozeny constant
m	Particle size distribution by mass, m^{-1}
n	Number concentration, m^{-3}; number per area, m^{-2}
o	Subscript: initial, vacuum
p	Subscript: particle
q	Subscript: charge on particle
r	Radius, m
t	Time, s
w	Subscript: water
x	Coordinate axis, m
y	Coordinate axis, m
z	Coordinate axis, m
A	Area, m^2; coefficient in Equation 19.31, m^{-B}
B	Exponent in Equation 19.31
$C(d_p)$	Cunningham correction (approximately $1 + 2.5\lambda/d_p$)
C_D	Drag coefficient
D	Diameter, m
E	Efficiency
F	Volume fraction; force, N
G	Subscript: gas
H	Relative humidity (fraction)
I	Subscript: impaction
K	Ratio of particle terminal velocity to gas free stream velocity; an impaction parameter = 2ψ
L	Length, m; subscript: liquid
M	Mass, kg
$\dot{M}$	Mass flux, kg/s
N	Number, ratio
P	Pressure, N/m^2
Q	Volume flow rate, m^3/s; electrical charge, coul; subscript: charge on collector
R	Radius, m; Subscript: interception
S	Surface-to-volume ratio, m^{-1}
T	Temperature, °K; subscript: throat
U	Velocity, m/s
V	Volume, m^3
Pt	Penetration
Re	Reynolds number
Stk	Impaction parameter
β	Coefficient in pressure drop Equation 19.23
ϵ	Dielectric constant
η	Efficiency
λ	Mean free path, m
μ	Viscosity, N-s/m^2
ρ	Density, kg/m^3
σ	Standard deviation
ψ	Impaction parameter
Δ	Difference or change

Index